Canadian Edition

GENETICS

From Genes to Genomes

Leland H. Hartwell
Fred Hutchinson Cancer Research Center

Leroy Hood
The Institute for Systems Biology

Michael L. Goldberg
Cornell University

Ann E. Reynolds
Fred Hutchinson Cancer Research Center

Lee M. Silver
Princeton University

Jim Karagiannis
University of Western Ontario

Maria Papaconstantinou
University of Toronto

McGraw-Hill Ryerson

McGraw-Hill
Ryerson

Genetics: From Genes to Genomes
Canadian Edition

ISBN-13: 978-0-07-094669-9
ISBN-10: 0-07-094669-8

1 2 3 4 5 6 7 8 9 10 TCP 1 9 8 7 6 5 4

Printed and bound in Canada.

Care has been taken to trace ownership of copyright material contained in this text; however, the publisher will welcome any information that enables them to rectify any reference or credit for subsequent editions.

Director of Product Management: *Rhondda McNabb*
Group Product Manager: *Leanna MacLean*
Marketing Manager: *Jeremy Guimond*
Product Developer: *Kamilah Reid-Burrell*
Senior Product Team Associate/Product Team Associate: *Stephanie Giles and Amelia Chester*
Supervising Editor: *Jessica Barnoski*
Photo/Permissions Researcher: *Monika Schurmann*
Copy Editor: *Eileen Jung*
Proofreader: *May Look*
Plant Production Coordinator: *Scott Morrison*
Manufacturing Production Coordinator: *Lena Keating*
Cover Design: *Vince Satira*
Cover Image: *Chad Baker/Getty Images (RF)*
Interior Design: *Vince Satira*
Page Layout: *Laserwords Private Limited*
Printer: *Transcontinental Printing Group*

Library and Archives Canada Cataloguing in Publication

Hartwell, Leland, author

    Genetics : from genes to genomes / Leland H. Hartwell,
Leroy Hood, Michael L. Goldberg, Ann E. Reynolds, Lee M.
Silver, Jim Karagiannis, Maria Papaconstantinou.—Canadian edition.
Includes index.
ISBN 978-0-07-094669-9
    1. Genetics—Textbooks. I. Karagiannis, Jim, 1972-, author
II. Papaconstantinou, Maria, author III. Title.
QH430.H27 2014   576.5   C2013-906009-X

About the Authors

Dr. Leland Hartwell is President and Director of Seattle's Fred Hutchinson Cancer Research Center and Professor of Genome Sciences at the University of Washington.

Dr. Hartwell's primary research contributions were in identifying genes that control cell division in yeast including those necessary for the division process as well as those necessary for the fidelity of genome reproduction. Subsequently, many of these same genes have been found to control cell division in humans and often to be the site of alteration in cancer cells.

Dr. Hartwell is a member of the National Academy of Sciences and has received the Albert Lasker Basic Medical Research Award, the Gairdner Foundation International Award, the Genetics Society Medal, and the 2001 Nobel Prize in Physiology or Medicine.

Dr. Lee Hood received an M.D. from the Johns Hopkins Medical School and a Ph.D. in biochemistry from the California Institute of Technology. His research interests include immunology, cancer biology, development, and the development of biological instrumentation (e.g, the protein sequencer and the automated fluorescent DNA sequencer). His early research played a key role in unravelling the mysteries of antibody diversity. More recently he has pioneered systems approaches to biology and medicine.

Dr. Hood has taught molecular evolution, immunology, molecular biology, genomics, and biochemistry. In addition, he has co-authored textbooks in biochemistry, molecular biology, and immunology, as well as *The Code of Codes*—a monograph about the Human Genome Project. He was one of the first advocates for the Human Genome Project and directed one of the federal genome centres that sequenced the human genome. Dr. Hood is currently the president (and co-founder) of the cross-disciplinary Institute for Systems Biology in Seattle, Washington.

Dr. Hood has received a variety of awards, including the Albert Lasker Award for Medical Research (1987), the Distinguished Service Award from the National Association of Teachers (1998), and the Lemelson/MIT Award for Invention (2003). He is the 2002 recipient of the Kyoto Prize in Advanced Biotechnology—an award recognizing his pioneering work in developing the protein and DNA synthesizers and sequencers that provide the technical foundation of modern biology. He is deeply involved in K–12 science education. His hobbies include running, mountain climbing, and reading.

Dr. Michael Goldberg is a professor at Cornell University, where he teaches introductory genetics and human genetics. He was an undergraduate at Yale University and received his Ph.D. in biochemistry from Stanford University. Dr. Goldberg performed postdoctoral research at the Biozentrum of the University of Basel (Switzerland) and at Harvard University. He received an NIH Fogarty Senior International Fellowship for study at Imperial College (England) and fellowships from the Fondazione Cenci Bolognetti for sabbatical work at the University of Rome (Italy). His current research uses the tools of *Drosophila* genetics and the biochemical analysis of frog egg cell extracts to investigate the mechanisms that ensure proper cell cycle progression and chromosome segregation during mitosis and meiosis.

Dr. Ann Reynolds is an educator and author. She began teaching genetics and biology in 1990. Her research has included studies of gene regulation in *E. coli*, chromosome structure and DNA replication in yeast, and chloroplast gene expression in marine algae. She is a graduate of Mount Holyoke College and received her Ph.D. from Tufts University. Dr. Reynolds was a postdoctoral fellow in the Harvard University Department of Molecular Biology and Genome Sciences at the University of Washington. She was also an author and producer of the laser disc and CD-ROM *Genetics: Fundamentals to Frontiers*.

Dr. Lee M. Silver is a professor at Princeton University in the Department of Molecular Biology and the Woodrow Wilson School of Public and International Affairs. He has joint appointments in Princeton's Program in Science, Technology, and Environmental Policy, the Program in Law and Public Policy, and the Princeton Environmental Institute. He received a Bachelor's and Master's degree in physics from the University of Pennsylvania, a doctorate in biophysics from Harvard University, postdoctoral training in mammalian genetics at the Sloan-Kettering Cancer Center, and training in molecular biology at Cold Spring Harbor Laboratory. Silver was elected a lifetime Fellow of the American Association for the Advancement of Science and was a recipient of an unsolicited National Institutes of Health MERIT award for outstanding research in genetics. He has been elected to the governing boards of the Genetics Society of

America and the International Mammalian Genome Society. He is currently on the Board of Trustees of the American Council on Science and Health, the Advisory Board of The Reason Project, and the Scientific Advisory Board of the Institute of Systems Biology in Seattle.

Silver has published over 180 research articles in the fields of developmental genetics, molecular evolution, population genetics, behavioural genetics, and computer modelling. He is the lone author of three books: *Mouse Genetics: Concepts and Applications* (1995), *Remaking Eden* (1997), and *Challenging Nature* (2006). He has also published essays in *The New York Times*, *Washington Post*, *Time*, and *Newsweek International* and has appeared on numerous television and radio programs including the *Charlie Rose Show*, *20/20*, *60 Minutes*, *PBS*, *NBC* and *ABC News*, *Nightline*, *NPR*, and the *Steven Colbert Report*. Recently, Silver collaborated with the playwright Jeremy Kareken on the script of "Sweet, Sweet, Motherhood," which won first prize in the 2007 Two-headed Challenge from the Guthrie Theater, awarded to the best play written by a playwright and a non-theatre partner.

Dr. Jim Karagiannis is an associate professor at the University of Western Ontario where he teaches introductory genetics as well as senior/graduate level courses in molecular biology, genetics, and functional genomics. After receiving his Ph.D. from Queen's University in Kingston, he went on to conduct postdoctoral research at Temasek Life Sciences Laboratory in Singapore as a Millenium Foundation Fellow. His current research makes use of the model eukaryote *Schizosaccharomyces pombe* to explore the complex post-translational modifications that take place on the carboxy-terminal domain (CTD) of the largest subunit of RNA polymerase II. Through an empirical examination of the informational properties and regulatory potential of the CTD, Dr. Karagiannis hopes to decipher the "programming language" used by eukaryotes to control aspects of gene expression.

Dr. Maria Papaconstantinou is a lecturer in the Human Biology Program at the University of Toronto. She coordinates and teaches a variety of courses, including introductory genetics and other senior-level courses in human and molecular genetics. Dr. Papaconstantinou completed her undergraduate studies at the University of Toronto, received her doctorate in molecular genetics and biology from McMaster University, and performed postdoctoral research at McMaster University. Her research has included genetic and biochemical analyses of the function and signalling pathways of the *Mnn1* tumour suppressor gene in the fruit fly *Drosophila melanogaster*. Her current research interests lie in the examination of the impact and effectiveness of various pedagogical approaches on undergraduate learning outcomes.

Brief Contents

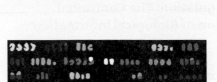

UNIT II NEXT-GENERATION GENETICS: ANALYZING BIOLOGICAL INFORMATION AT THE SYSTEMS LEVEL

Genome **The End of the Beginning**

Contents

Preface

A Note from the Authors

The science of genetics is less than 150 years old, but its accomplishments within that short time have been astonishing. In 1865, Gregor Mendel first described genes as abstract units of inheritance; his work was ignored and then "rediscovered" in 1900. During the years of 1910–1920, Thomas Hunt Morgan and his students provided experimental verification of the idea that genes reside within chromosomes. By 1944, Oswald Avery and his co-workers had established that genes are made of DNA. In 1953, James Watson and Francis Crick published their pathbreaking structure of DNA; less than 50 years later (in 2001), an international consortium of investigators deciphered the sequence of the three billion nucleotides in the human genome. Remarkably, a mere nine years after that, J. Craig Venter and colleagues synthesized, assembled, and transplanted the first complete human-made genome into a host bacterium—thus creating the first cell to be born, not of another cell, but of the efforts of the research team.

This, the Canadian edition of *Genetics: From Genes to Genomes* emphasizes not only the core concepts of genetics, but also the cutting-edge discoveries, modern tools, and analytical methods that have made the science of genetics the exciting, vibrant, and dynamic discipline that it is today. To facilitate the balanced delivery of both foundational genetics concepts as well as more recent advances, we have divided the text into two units. The first unit, entitled *Foundations of Genetics*, takes students on a journey from the Austrian monastery, where Mendel conducted his experiments on pea plants, to the beginnings of the Human Genome Project. The second unit, entitled *Next-Generation Genetics*, highlights recent advances in genomics, proteomics, systems biology, and the newly emerging discipline of synthetic biology. We expect this second unit to expand considerably in future editions to keep pace with the rapid advances taking place within the field.

In addition to foundational concepts, the text goes to great lengths to highlight the research contributions of Canadian geneticists. From Oswald Avery (who demonstrated that genes were made of DNA) to Murray Barr (discoverer of the sex chromatin body now called the Barr body) to modern day researchers like Jack Greenblatt (who was the first to purify a transcriptional anti-termination protein), and Charlie Boone (synthetic genetic array analysis), students will be made well aware of the significant and continuing contributions of Canadians to society's understanding of genetics.

Our Focus—An Integrated Approach

Genetics: From Genes to Genomes, Canadian Edition, represents a new approach to an undergraduate course in genetics. It is one that integrates genetics concepts (from Mendel to Venter) to create an up-to-date vantage point from which students can explore the molecular basis of life. The strength of this integrated approach is that students who complete the book will have a strong command of genetics as it is practiced today by both academic and corporate researchers. These scientists are rapidly changing our understanding of living organisms, including ourselves.

We integrate the following:

- **Formal genetics:** the rules by which genes are transmitted.
- **Molecular genetics:** the structure of DNA and how it directs the structure of proteins.
- **Genomics and proteomics:** recent technologies that allow a comprehensive analysis of the entire gene-set and its expression in an organism.
- **Human genetics:** how genes contribute to health and diseases, including cancer.
- **Molecular evolution:** the molecular mechanisms by which biological systems and whole organisms have evolved and diverged.
- **Systems biology:** the multidisciplinary, integrated study of life processes that may lead to new ways to analyze, detect, and treat disease.
- **Synthetic biology:** a new and exciting discipline that combines knowledge derived from the fields of molecular biology and genetics with the principles of engineering.

The Genetic Way of Thinking

Modern genetics is a molecular-level science, but an understanding of its origins and the discovery of its principles is an essential context. To encourage a genetic way of thinking, we begin by reviewing Mendel's principles and the chromosomal basis of inheritance. From the outset, however, we aim to integrate organism-level genetics with fundamental molecular mechanisms.

Chapter 1 presents the foundation of this integration by summarizing the main biological themes we explore. In Chapter 2, we tie Mendel's studies of pea-shape inheritance to the action of an enzyme that determines whether a pea is round or wrinkled. In the same chapter, we point

to the relatedness of the patterns of heredity in all organisms, and cover extensions to Mendel. Chapters 3 and 4 discuss the chromosome theory of inheritance, and the fundamentals of gene linkage and mapping. In Chapters 5–9, we focus on the physical characteristics of DNA, mutation, and on the manner in which DNA encodes, copies, and transmits biological information.

Chapters 10 and 11 discuss gene regulation in both prokaryotes and eukaryotes. These chapters clearly illustrate the importance of molecular-level interactions for the controlled expression of biological information. Chapters 12 and 13 cover population genetics, with a view of how molecular tools have provided information on species relatedness and on genome changes at the molecular level over time.

Beginning in Chapters 14 and 15, we move into the digital revolution in DNA analysis with a look at modern genetics techniques, including gene cloning, hybridization, and PCR. Next, Chapters 16–18 describe the power of genetic analysis and its role in elucidating (1) the mechanisms of cell cycle control, (2) the complex interactions of eukaryotic development, and (3) the biology of prokaryotes (as well as eukaryotic organelles).

Finally, in Chapters 19–24, we explore more recent advances in genetic research; these chapters include in-depth discussions of functional genomics, proteomics, systems biology (an integrated field utilizing input from several disciplines), and synthetic biology (a relatively new discipline with the potential to forever change how scientists view biological processes).

Throughout the Canadian Edition, we present the scientific reasoning of some of the ingenious researchers of the field—from Mendel, to Avery, and all the way to J. Craig Venter. We hope student readers will see that genetics is not simply a set of data and facts, but also a human endeavour that relies on contributions from exceptional individuals.

Student-friendly Features

We have taken great pains to help the student make the leap to a deeper understanding of genetics. Numerous features of this book were developed with that goal in mind.

- *One Voice* Genetics: From Genes to Genomes, Canadian Edition, has a friendly, engaging reading style that helps students master the concepts throughout this book. The writing style provides the student with the focus and continuity required to make the book successful in the classroom.

- *Visualizing Genetics* The highly specialized art program integrates photographs and line art in a manner that provides the most engaging visual presentation of genetics available. Our Feature Figure illustrations break down complex processes into step-by-step illustrations that lead to greater student understanding.

- *Accessibility* Our intention is to bring cutting-edge content to the student level. A number of more complex illustrations are revised and segmented to help the student follow the process. Legends have been streamlined to highlight only the most important ideas, and throughout the book, topics and examples have been chosen to focus on the most critical information.

- *Problem Solving* Developing strong problem-solving skills is vital for every genetics student. The authors have carefully created problem sets at the end of each chapter that allow students to improve upon their problem-solving ability.

- *Solved Problems* These problems cover topical material, with complete answers providing insight into the step-by-step process of problem solving.

- *Review Problems* These problems offer more than 600 questions involving a variety of levels of difficulty that develop excellent problem-solving skills. They are organized by chapter section and in order of increasing difficulty within each section for ease of use by instructors and students.

Detailed List of Changes

Chapter 1

- New opening figure integrates the concepts of information theory with genome sequencing.
- New **Fast Forward** box "Synthetic Genomics."
- New figure 1.2a illustrating complementarity.
- New figure 1.3 showing an automated next-generation DNA sequencer.
- New figures 1.9c and 1.9d compare wild-type and mutant fruit flies (defective in the *Pax6* gene).
- Figure 1.12 is updated to accurately reflect gene numbers in various model organisms.
- New figure 1.14 highlighting Ozzy Osbourne's genome sequence.
- New figure 1.15 "Canadian Coalition for Genetic Fairness."

Chapter 2

- Mendel's principles of heredity and extensions to Mendel's laws are combined into one chapter for a more concise presentation of the content.
- New introductory figure illustrates Mendel's principles and extensions to his laws.
- The scientific contributions of Canadians Lap-Chee Tsui and John Riordan are highlighted in the **Fast Forward** box "Genes Encode Proteins."
- A new **Tools of Genetics** box "Temperature-Sensitive Mutations" highlights the scientific contributions of Canadian researcher David Suzuki.

Chapter 3

- New introductory figure provides a visualization of chromosomal behaviour during mitosis in *Drosophila*.
- Chorionic villus sampling (CVS) and a new figure depicting this technique is added to the **Genetics and Society** box "Prenatal Genetic Diagnosis."
- New Figure 3.14 depicts an electron micrograph of a synaptonemal complex.
- New Figure 3.23 illustrates male pattern baldness in the British royal family as a sex-influenced trait.

Chapter 4

- New introductory figure depicts mapping analogies.
- Additional information on William Bateson's, Reginald Punnett's, and T.H. Morgan's observations is provided. New Figures 4.4 and 4.5 illustrate these findings, and provide the student with a more complete introduction to genetic linkage.
- New Figure 4.7 depicts a photomicrograph of chiasmata.
- New Table 4.3 lists human disease-related genes identified by linkage mapping.
- New Figure 4.16 portrays a saturated genetic map of the crop species *Brassica napus* from the University of Manitoba, highlighting Canadian research.
- New information in the **Fast Forward** box "Mapping the Cystic Fibrosis Gene" illustrates the scientific contributions of Canadian researchers Lap-Chee Tsui and John Riordan.

Chapter 5

- New opening figure illustrates the multifaceted nature of the DNA molecule.
- New figure 5.2 illustrating Friedrich Miescher and the first isolation of DNA.
- New figure 5.4 "Chromosomes from the Onion (*Allum cepa*) Visualized Using the Feulgen Reaction."
- New **Focus on Inquiry** box "Genes Are Made of DNA, not Protein."
- New **Focus on Critical Thinking** box "Information Theory."

Chapter 6

- New **Fast Forward** box "What Is a Gene?"
- New **Focus on Inquiry** box "The Discovery of the Barr Body."

Chapter 7

- New opening figure illustrates a modern interpretation of the central dogma.
- Description of the work of University of Western Ontario researcher, Yong Kang, is added to a **Genetics and Society** box.

Chapter 8

- New introductory figure illustrates the phenotypic consequences of mutations.
- Expanded description of germline and somatic mutations provides students with further clarification.
- New **Tools of Genetics** box "Site-directed Mutagenesis" portrays the scientific contributions of Canadian researcher and Nobel laureate Michael Smith.

Chapter 9

- New introductory figure illustrates the detection of large-scale chromosomal changes by karyotype analysis.
- Additional information on transposable elements (e.g., replicative transposition, conservative transposition, target-site duplication) provides the student with a more complete knowledge base.
- New **Focus on Genetics** box "Transposable Elements in Corn" highlights the scientific contributions of Nobel laureate Barbara McClintock.

Chapter 10

- New introductory figure illustrates binding of the *E. coli* Lac repressor protein (LacI) to different *lac* operon regulatory elements.
- New **Focus on Genetics** box "Transcriptional Regulation in Bacteriophage Lambda" portrays Canadian Jack Greenblatt's scientific contributions.

Chapter 11

- New introductory figure shows how epigenetic changes result in phenotypic differences.
- New **Focus on Genetics** box "The Environment and the Epigenome" depicts the epigenetic regulation of the *agouti* gene in mice and how the epigenome can be influenced by the environment.
- New **Genetics and Society** box "Epigenetics and Complex Disease" highlights Canadian researcher Art Petronis's scientific contributions.

Chapter 12

- New introductory figure illustrates the range of diversity in different organisms.
- The section on analyzing quantitative variation has been updated and expanded to include narrow-sense heritability.
- New **Focus on Genetics** box "Population and Quantitative Genetic Approaches in the Wild" portrays the scientific contributions of Canadian researcher David Coltman.

Chapter 13

- New opening figure illustrates the evolution of the human glucocorticoid receptor.
- New **Tools of Genetics** box "The Burgess Shale of Southeastern British Columbia."
- New **Focus on Inquiry** box "What Makes Us Human?"
- New figure 13.5 illustrates the role of conserved genetic networks in the evolution and development of the heart.

- New **Fast Forward** box "Network Motifs."

Chapter 14

- New opening figure shows the genetically modified Black Tetra.
- New **Fast Forward** box "Next-Generation Sequencing."
- New figure 14.18 "Site-directed Mutagenesis of Haemoglobin."

Chapter 15

- New opening figure of *Canadian Association of Genetic Counsellors* introduces the student to genetic counselling.
- New **Genetics and Society** box "Correcting a Miscarriage of Justice."
- New figure 15.14 describing the use of DNA fingerprinting to refute Ann Anderson's claim to be Anastasia Nilolaevna.
- New **Focus on Inquiry** box "Cloning of the Cystic Fibrosis (*CF*) Gene—25 Years Later."

Chapter 16

- New introductory figure shows the results of spectral karyotype (SKY) analysis of a cancer patient's tumour cells.
- New and updated information in Figure 16.1 and Table 16.1 highlight Canadian statistics.
- New **Focus on Genetics** box "Cell Signalling Mechanisms and Cancer" highlights Canadian researcher Tony Pawson's scientific contributions.

Chapter 17

- New introductory figure illustrates Adams–Oliver syndrome and its associated developmental phenotypes.
- Updated information is added to the **Genetics and Society** box "Stem Cells and Human Cloning" and the contributions of Canadian scientists James Till and Ernest McCulloch are highlighted.
- New **Focus on Genetics** box "Stem Cells and the Genetic Control of Embryonic Development" portrays the scientific contributions of Canadian researcher Janet Rossant.

Chapter 18

- New **Fast Forward** box "Defining the Minimal Gene-Set Required for Life."
- New **Focus on Inquiry** box "Understanding the Role of Mitochondria in Human Disease."

Chapters 19–24

• These chapters (forming the second unit "Next-Generation Genetics: Analyzing Biological Information at the Systems Level") represent the most significant departure from the most recent U.S. edition of the text. This unit is composed of predominantly novel material exploring recent technical and conceptual advances in genetic and genomic research. These chapters include in-depth discussions of genome sequencing and analysis, functional genomics, proteomics, systems biology, and synthetic biology. We expect this unit to expand considerably in future editions so as to keep pace with the rapid advances taking place within the field.

Acknowledgements

The creation of a project of this scope is never solely the work of the authors. We are grateful to our colleagues around the world who took the time to review the previous edition and make suggestions for improvement. Their willingness to share their expectations and expertise was a tremendous help to us.

Owen Rowland, *Carleton University*
T. Michael Stock, *Grant MacEwan University*
Ann Marie Davison, *Kwantlen Polytechnic University*
Steven Carr, *Memorial University of Newfoundland*
William Bendena, *Queen's University*
Ian D. Chin-Sang, *Queen's University*
Jeffrey Fillingham, *Ryerson University*
Scott F. Briscoe, *Simon Fraser University*
Joanna Freeland, *Trent University*
Michael Deyholos, *University of Alberta*
J. Thomas Beatty, *University of British Columbia*
Ray Lu, *University of Guelph*
Dana Schroeder, *University of Manitoba*
Jason Addison, *University of New Brunswick*
Fiona Rawle, *University of Toronto*
Anthony Percival-Smith, *University of Western Ontario*
Michael Crawford, *University of Windsor*

We would also like to thank the highly skilled publishing professionals at McGraw-Hill Ryerson who guided the development and production of the Canadian edition of ***Genetics: From Genes to Genomes***: Leanna MacLean for her sponsorship and support; Kamilah Reid-Burrell for her organizational skills and tireless work to tie up all loose ends; and Jessica Barnoski, Monika Schurmann, Eileen Jung, May Look, and the entire production team for their careful attention to detail and ability to move the schedule along.

Guided Tour

Integrating Genetic Concepts

Genetics: From Genes to Genomes, Canadian Edition, takes an integrated approach in its presentation of genetics, thereby giving students a strong command of genetics as it is practised today by academic and corporate researchers. Principles are related throughout the text in examples, essays, case histories, and Connections sections to make sure students fully understand the relationships between topics.

Chapter Outline ▶

Every chapter opens with a brief outline of the chapter's contents.

Chapter Outline
2.1 Background: The Historical Puzzle of Inheritance
2.2 Genetic Analysis According to Mendel
2.3 Mendelian Inheritance in Humans
2.4 Extensions to Mendel for Single-Gene Inheritance
2.5 Extensions to Mendel for Gene Interactions

Learning Objectives

1. Discuss the physical basis for genetic linkage.
2. Explain why Mendel's law of independent assortment does not apply to linked genes.
3. Examine how linkage between two genetic loci is determined.
4. Describe when during the cell cycle a crossover occurs, the mechanism, and its potential outcomes.
5. Demonstrate how genetic distance is estimated between two or three linked loci.
6. Construct a simple genetic map based upon data from testcrosses.

◀ NEW! Learning Objectives

This list provides you with an overview of what you are to know. Your instructor can assign activities through Connect™ to help you achieve these objectives.

Section Summary ▶

After several major headings within the chapter, the authors have provided a short summary to help the students focus on the critical items of that section.

F_2 phenotypic ratios of 9:3:3:1 or its derivatives indicate the combined action of two independently assorting genes. For heterogeneous traits caused by recessive alleles of two or more genes, a mating between affected individuals acts as a complementation test, revealing whether they carry mutations in the same gene or in different genes.

Fast Forward Essays ▶

This feature is one of the methods used to integrate the Mendelian principles presented early in the book with the molecular principles that will follow.

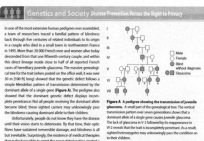
◀ Tools of Genetics Essays

Current readings explain various techniques and tools used by geneticists, including examples of applications in biology and medicine.

Genetics and Society Essays ▶

Dramatic essays explore the social and ethical issues created by the multiple applications of modern genetic research.

◀ Focus on Boxes

Focus on Inquiry, Focus on Critical Thinking, and Focus on Genetics essays place the spotlight on the groundbreaking research conducted by scientists from the nineteenth century to the present.

▼ Comprehensive Examples

Comprehensive Examples are extensive case histories or research synopses that, through text and art, summarize the main points in the preceding section or chapter and show how they relate to each other.

3.3 Meiosis: Cell Divisions That Halve Chromosome Number

During the many rounds of cell division within an embryo, most cells either grow and divide via the mitotic cell cycle just described, or they stop growing and become arrested in G_0. These mitotically dividing and G_0-arrested cells are the so-called somatic cells whose descendants continue to make up the vast majority of each organism's tissues throughout the lifetime of the individual. Early in the embryonic development of animals, however, a group of cells is set aside for a different fate. These are the germ cells: cells destined for a specialized role in the production of gametes. Germ cells arise later in plants, during floral development instead of during embryogenesis. The germ cells become incorporated in the reproductive organs—ovaries and testes in animals; ovaries and anthers in flowering plants—where they ultimately undergo meiosis, the special two-part cell division that produces gametes (eggs and sperm or pollen) containing half the number of chromosomes as other body cells.

The union of haploid gametes at fertilization yields diploid offspring that carry the combined genetic heritage of two parents. Sexual reproduction therefore requires the alternation of haploid and diploid generations. If gametes were diploid rather than haploid, the number of chromosomes would double in each successive generation such that in humans, for example, the children would have 92 chromosomes per cell, the grandchildren 184, and so on. Meiosis prevents this lethal, exponential accumulation of chromosomes.

During meiosis I, homologues pair, exchange parts, and then segregate

The events of meiosis I are unique among nuclear divisions (see **Figure 3.12**, meiosis I). The process begins with the replication of chromosomes, after which each one consists of two sister chromatids. A key to understanding meiosis I is the observation that the centromeres joining these chromatids remain intact throughout the entire division, rather than splitting as in mitosis.

As the division proceeds, homologous chromosomes align across the cellular equator to form a coupling that ensures proper chromosome segregation to separate nuclei. Moreover, during the time homologous chromosomes face each other across the equator, the maternal and paternal chromosomes of each homologous pair may exchange parts, creating new combinations of alleles at different genes along the chromosomes. Afterward, the two homologous chromosomes, each still consisting of two sister chromatids connected at a single, unsplit centromere, are pulled to opposite poles of the spindle. As a result, it is homologous chromosomes (rather than sister chromatids as in mitosis) that segregate into different daughter cells at the conclusion of the first meiotic division. With this overview in mind, let us take a closer look at the specific events of meiosis I, bearing in mind that we analyze a dynamic, flowing sequence of cellular events by breaking it down somewhat arbitrarily into the easily pictured, traditional phases.

▼ Connections

Each chapter closes with a Connections section that serves as a bridge between the topics in the just-completed chapter and those in the upcoming chapter or chapters.

Connections

Mendel answered the three basic questions about heredity as follows: To "What is inherited?" he replied, "alleles of genes." To "How is it inherited?" he responded, "according to the principles of segregation and independent assortment." And to "What is the role of chance in heredity?" he said, "for each individual, inheritance is determined by chance, but within a population, this chance operates in a context of strictly defined probabilities."

Within a decade of the 1900 rediscovery of Mendel's work, numerous breeding studies had shown that Mendel's laws hold true not only for seven pairs of antagonistic characteristics in peas, but for an enormous diversity of traits in a wide variety of sexually reproducing plant and animal species,

including four-o'clock flowers, beans, corn, wheat, fruit flies, chickens, mice, horses, and humans. Some of these same breeding studies, however, raised a challenge to the new genetics. For certain traits in certain species, the studies uncovered unanticipated phenotypic ratios, or the results included F_1 and F_2 progeny with novel phenotypes that resembled those of neither pure-breeding parent.

These phenomena could not be explained by Mendel's hypothesis that for each gene, two alternative alleles, one completely dominant, the other recessive, determine a single trait. We now know that most traits, including skin colour, eye colour, and height in humans, are determined by interactions between two or more genes. We also know that within

Visualizing Genetics

Full-colour illustrations and photographs bring the printed word to life. These visual reinforcements support and further clarify the topics discussed throughout the text.

FEATURE FIGURE 3.12

Meiosis: One Diploid Cell Produces Four Haploid Cells

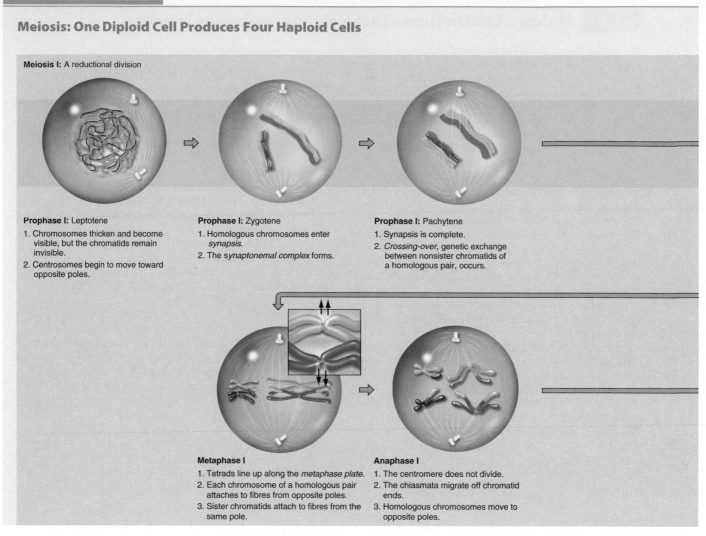

Meiosis I: A reductional division

Prophase I: Leptotene

1. Chromosomes thicken and become visible, but the chromatids remain invisible.
2. Centrosomes begin to move toward opposite poles.

Prophase I: Zygotene

1. Homologous chromosomes enter *synapsis*.
2. The *synaptonemal complex* forms.

Prophase I: Pachytene

1. Synapsis is complete.
2. *Crossing-over*, genetic exchange between nonsister chromatids of a homologous pair, occurs.

Metaphase I

1. Tetrads line up along the *metaphase plate*.
2. Each chromosome of a homologous pair attaches to fibres from opposite poles.
3. Sister chromatids attach to fibres from the same pole.

Anaphase I

1. The centromere does not divide.
2. The chiasmata migrate off chromatid ends.
3. Homologous chromosomes move to opposite poles.

▲ Feature Figures

Special multipage spreads integrate line art, photographs, and text to summarize important genetic concepts in detail.

▼ Process Figures

Step-by-step descriptions allow the student to walk through a compact summary of important details.

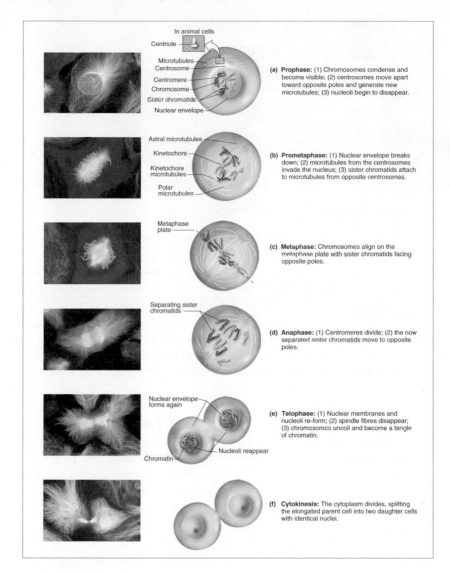

In animal cells

Centriole

Microtubules
Centrosome
Centromere
Chromosome
Sister chromatids
Nuclear envelope

(a) Prophase: (1) Chromosomes condense and become visible; (2) centrosomes move apart toward opposite poles and generate new microtubules; (3) nucleoli begin to disappear.

Astral microtubules
Kinetochore
Kinetochore microtubules
Polar microtubules

(b) Prometaphase: (1) Nuclear envelope breaks down; (2) microtubules from the centrosomes invade the nucleus; (3) sister chromatids attach to microtubules from opposite centrosomes.

Metaphase plate

(c) Metaphase: Chromosomes align on the metaphase plate with sister chromatids facing opposite poles.

Separating sister chromatids

(d) Anaphase: (1) Centromeres divide; (2) the now separated sister chromatids move to opposite poles.

Nuclear envelope forms again

Chromatin — Nucleoli reappear

(e) Telophase: (1) Nuclear membranes and nucleoli re-form; (2) spindle fibres disappear; (3) chromosomes uncoil and become a tangle of chromatin.

(f) Cytokinesis: The cytoplasm divides, splitting the elongated parent cell into two daughter cells with identical nuclei.

▼ Micrographs

Stunning micrographs bring the genetics world to life.

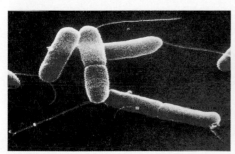
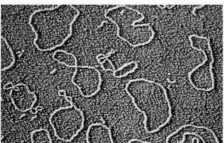

▼ Experiment and Technique Figures

Illustrations of performed experiments and genetic analysis techniques highlight how scientific concepts and processes are developed.

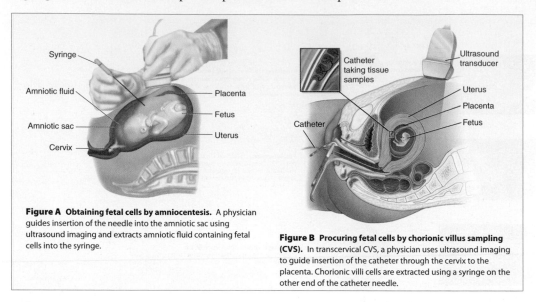

Figure A Obtaining fetal cells by amniocentesis. A physician guides insertion of the needle into the amniotic sac using ultrasound imaging and extracts amniotic fluid containing fetal cells into the syringe.

Figure B Procuring fetal cells by chorionic villus sampling (CVS). In transcervical CVS, a physician uses ultrasound imaging to guide insertion of the catheter through the cervix to the placenta. Chorionic villi cells are extracted using a syringe on the other end of the catheter needle.

Solving Genetics Problems

The best way for students to assess and increase their understanding of genetics is to practise through problems. At the end of each chapter, there are problem sets that assist students in evaluating their understanding of key concepts and allow them to apply what they have learned to real-life issues.

Solved Problems

Solved problems offer step-by-step guidance needed to understand the problem-solving process.

Review Problems

Problems are organized by chapter section and in order of increasing difficulty to help students develop strong problem-solving skills. The answers to select problems can be found in the back of this text.

Technology and Supplements

McGraw-Hill Connect

McGraw-Hill Connect™ is a web-based assignment and assessment platform that gives students the means to better connect with their coursework, with their instructors, and with the important concepts that they will need to know for success now and in the future.

With Connect, instructors can deliver assignments, quizzes and tests online. Instructors can edit existing questions and author entirely new problems. Track individual student performance—by question, assignment or in relation to the class overall—with detailed grade reports. Integrate grade reports easily with Learning Management Systems (LMS).

By choosing Connect, instructors are providing their students with a powerful tool for improving academic performance and truly mastering course material. Connect allows students to practice important skills at their own pace and on their own schedule. Importantly, students' assessment results and instructors' feedback are all saved online—so students can continually review their progress and plot their course to success.

Connect content was reviewed and developed by Sara Nolte and Devika Sharanya. Our Connect provides 24/7 online access to an eBook—an online edition of the text—to aid them in successfully completing their work, wherever and whenever they choose.

Key Features

Simple Assignment Management

With Connect, creating assignments is easier than ever, so you can spend more time teaching and less time managing.

- Create and deliver assignments easily with multiple choice quizzes and testbank material to assign online.
- Streamline lesson planning, student progress reporting, and assignment grading to make classroom management more efficient than ever.
- Go paperless with the eBook and online submission and grading of student assignments.

Smart Grading

When it comes to studying, time is precious. Connect helps students learn more efficiently by providing feedback and practice material when they need it, where they need it.

- Automatically score assignments, giving students immediate feedback on their work and side-by-side comparisons with correct answers.

- Access and review each response; manually change grades or leave comments for students to review.
- Reinforce classroom concepts with practice tests and instant quizzes.

Instructor Library

The Connect Instructor Library is your course creation hub. It provides all the critical resources you'll need to build your course, just how you want to teach it.

- Assign eBook readings and draw from a rich collection of textbook-specific assignments.
- Access instructor resources:
  - **Computerized Test Bank** Prepared by Michael Deyholos of the University of Alberta, the computerized test bank has been extensively revised and technically checked for accuracy. The computerized test bank contains a variety of questions, including true/false, multiple-choice, and short-answer questions requiring analysis and written answers. The computerized test bank is available through EZ Test Online—a flexible and easy-to-use electronic testing program that allows instructors to create tests from book-specific items. EZ Test accommodates a wide range of question types and allows instructors to add their own questions. Test items are also available in Word format (rich text format). For secure online testing, exams created in EZ Test can be exported to WebCT and Blackboard. EZ Test Online is supported at mhhe.com/eztest, where users can download a *Quick Start Guide*, access FAQs, or log a ticket for help with specific issues.
  - **Microsoft® PowerPoint® Lecture Slides** Prepared by Fiona Rawle of the University of Toronto, the PowerPoint slides draw on the highlights of each chapter and provide an opportunity for the instructor to emphasize the most relevant visuals in class discussions.
  - **Solution's Manual/Study Guide** Extensively revised by Drs. Papaconstantinou and Karagiannis, this manual presents the solutions to the end of chapter problems and questions along with the step-by-step logic of each solution. The manual also includes a synopsis, the objectives, and problem-solving tips for each chapter. Key figures and tables from the text are referenced throughout to guide student study.
- View assignments and resources created for past sections.
- Post your own resources for students to use.

eBook

Connect reinvents the textbook learning experience for the modern student. Every Connect subject area is seamlessly integrated with Connect eBooks, which are designed to keep students focused on the concepts key to their success.

- Provide students with a Connect eBook, allowing for anytime, anywhere access to the textbook.
- Merge media, animation and assessments with the text's narrative to engage students and improve learning and retention.
- Pinpoint and connect key concepts in a snap using the powerful eBook search engine.
- Manage notes, highlights and bookmarks in one place for simple, comprehensive review.

Presentation Centre

In addition to the images from your book, this online digital library contains photographs, artwork, animations, and other media from an array of McGraw-Hill textbooks that can be used to create customized lectures, visually enhanced tests and quizzes, compelling course websites, or attractive printed support materials.

SUPERIOR LEARNING SOLUTIONS AND SUPPORT

The McGraw-Hill Ryerson team is ready to help you assess and integrate any of our products, technology, and services into your course for optimal teaching and learning performance. Whether it is helping your students improve their grades, or putting your entire course online, the McGraw-Hill Ryerson team is here to help you do it. Contact your Learning Solutions Consultant today to learn how to maximize all of McGraw-Hill Ryerson's resources!

For more information on the latest technology and Learning Solutions offered by McGraw-Hill Ryerson and its partners, please visit us online: www.mcgrawhill.ca/he/solutions.

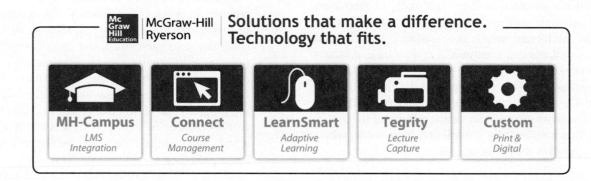

Genetics: The Study of Biological Information

Information can be stored in many ways: in the pattern of letters and words on the printed page, or in the sequence of nucleotides in a molecule of DNA. For example, in the bookcase *to the left* sits a complete copy of the entire sequence of the human genome. A total of 116 volumes, each containing approximately 1000 pages, are required to represent the sequence of nucleotide bases that make up our genome. The sequence in each book is printed in a font size so small that it is barely legible. In fact, it would take a human working 40 hours a week over 50 years to type the sequence manually. Luckily, a typical eukaryotic cell is much more efficient; it is able to replicate the genome in only eight hours!

Genetics is at its core the study of biological information. All living organisms—from single-celled bacteria and protozoa to multicellular plants and animals—utilize the information stored in molecules of DNA (deoxyribonucleic acid) to develop, grow, and reproduce. This information is precious, and as such, must be easily accessible to direct the function of an organism. At the same time it must be carefully packaged for safe-keeping and replicated with accuracy. Furthermore, it must be faithfully transmitted to the next generation; in this way, the physiological, physical, and behavioural traits that define an organism are passed on, from parent to offspring, as part of the process of heredity.

While the information is undoubtedly precious, it is not immutable. On the contrary, **mutations** (i.e., heritable changes in DNA sequence) can accumulate within populations of individuals to produce widespread variation. Moreover, the selection of individuals possessing DNA variants that make them particularly suited to survive and reproduce in a given environment can result in the **evolution** of specific traits or even new species. This process of "descent with modification" is responsible for the great variety, and the great beauty, of the living things that make Earth their home (**Figure 1.1**; see also wellcometreeoflife.org to further explore the diversity of living things in a fully interactive manner).

This book introduces you to the subject of genetics as it is currently understood in the second decade of the twenty-first century. In the broadest sense, it concerns the quest for understanding (1) how the "character" of living things is encoded, and (2) how this encoded information is passed on from parent to offspring. Several important themes recur throughout this presentation. First, that biological information is encoded and transmitted within DNA. Second, that the proteins responsible for an organism's many functions and characteristics are built from this DNA code. Third, that these protein elements are dynamic and interact to form intricate systems that govern life processes. Fourth, that genomes have a modular construction that has allowed for the rapid evolution of complexity. And finally, that this complexity—although daunting at times—is made more manageable by the fact that all living forms are closely related at the molecular level. In the remainder of this chapter, we introduce these themes. Make sure to keep them in mind as you delve deeper into the world of genetics.

Chapter Outline
1.1 DNA: The Molecule That Stores Digitized Biological Information
1.2 Proteins: The Molecules That Govern Life Processes
1.3 Complex Systems and Molecular Interactions
1.4 The Molecular Similarity of All Life-Forms
1.5 The Modular Construction of Genomes
1.6 Modern Genetic Techniques
1.7 Human Genetics

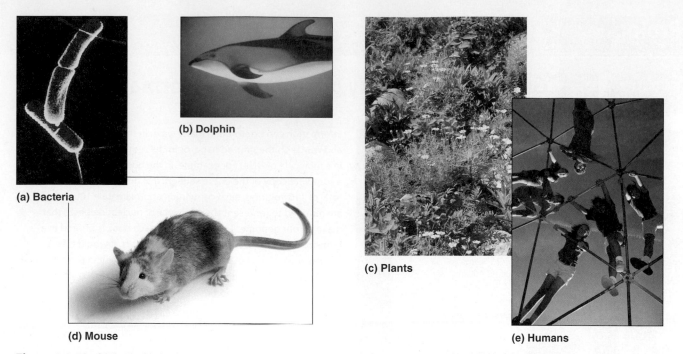

(a) Bacteria

(b) Dolphin

(c) Plants

(d) Mouse

(e) Humans

Figure 1.1 The biological information in DNA generates an enormous diversity of living organisms.

Learning Objectives

1. Explain how biological information is digitally encoded within living organisms.

2. Describe how one-dimensional biological information is transformed into a four-dimensional state.

3. Provide examples of the molecular similarity of living things.

4. Explain how modularity is related to the development of complexity.

5. Appraise the potential benefits and drawbacks of current genetic technology.

1.1 DNA: The Molecule That Stores Digitized Biological Information

The process of evolution has taken close to 4 billion years to generate the amazingly efficient mechanisms used for storing, replicating, expressing, and diversifying biological information. The DNA molecule stores biological information in units known as nucleotides. Amazingly, this simple alphabet, comprised of only four different characters—G, C, A, and T—has the capacity to specify which proteins an organism will make, which tissues they will be made in, and in what quantities they will be produced. The letters refer to the bases—guanine, cytosine, adenine, and thymine—that are components of the nucleotide building blocks of DNA. The DNA molecule itself exists as two linear strands that have intertwined to form a double helix carrying complementary G–C or A–T base pairs (**Figure 1.2a**). These complementary base pairs can bind together through hydrogen bonds. The molecular complementarity of double-stranded DNA is one of its most important properties and one of the keys to understanding how DNA functions (**Figure 1.2b**).

Although the DNA molecule is physically three-dimensional, its most important attribute—its information content—can be thought of as being both one-dimensional and digital. The information is considered one-dimensional because it is possible to encode the information by simply listing the sequence of bases in the correct order. This

characteristic can be formally defined by the fact that any nucleotide in the DNA molecule can be designated by providing only one coordinate (e.g., the 2346th nucleotide base in the sequence). In contrast, a two-dimensional object would require two coordinates to designate its position.

The information content of DNA is considered digital because each unit of information—one of the four letters of the DNA alphabet—is discrete (i.e., only four values are allowed, A, G, C, or T; this is in contrast to analog systems where an infinite number of values is possible). The take-home message here is simple. Because genetic information is one-dimensional and digital, it can be stored just as readily in the form of characters on a printed page, or in computer memory, as it can be in the DNA molecule itself. Indeed, the combined power of DNA sequencers (**Figure 1.3**), computers, and DNA synthesizers, makes it possible to interpret, store, replicate, and transmit genetic information electronically from one place to another anywhere on the planet. This information can then be used to synthesize an exact replica of a portion of the originally sequenced DNA molecule. Remarkably, DNA synthesis technology has progressed to the point where it is possible to synthetically construct entire bacterial genomes (see the **Fast Forward** box **"Synthetic Genomics"**).

In 2010, the J. Craig Venter Institute synthesized the first completely human-made bacterial genome. By transplanting this genome into a host cell, they produced the first synthetic self-replicating bacterium, *Mycoplasma mycoides* JCVI-syn1.0 (*upper right and lower left*). This remarkable work led by J. Craig Venter (*upper left*), Hamilton O. Smith (*lower right, standing to the left*), and Clyde A. Hutchison (*lower right, wearing glasses*) has provided the framework for the future creation of synthetic genomes in the laboratory. In theory, it is now possible to create a bacterial life-form bearing characteristics that have been rationally incorporated by the researcher. In practice, the technology is limited due to inadequacies in our fundamental understanding of biological systems. In the same way that knowing how to write does not guarantee that one will be able to compose a document of Shakespearean quality, neither does having the ability to synthesize genome-length DNA sequences guarantee the creation of a functional or useful organism. You will learn more about these techniques and their mind-boggling ramifications in Chapter 24 on "Synthetic Biology".

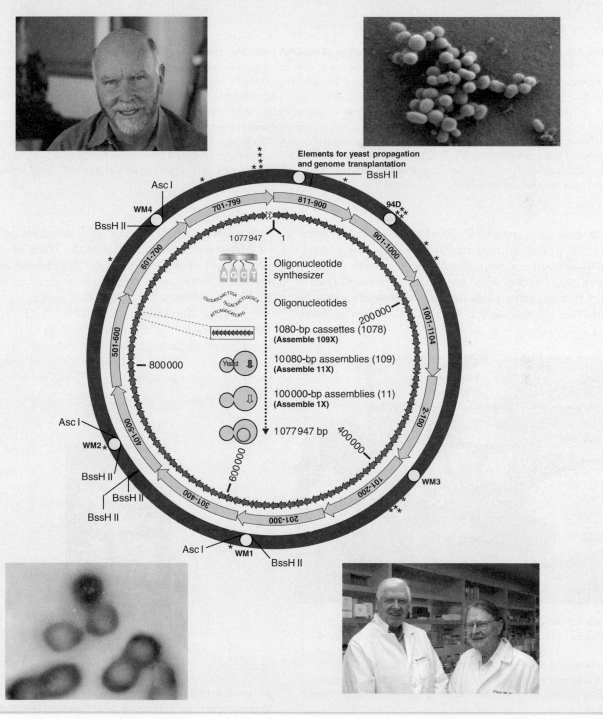

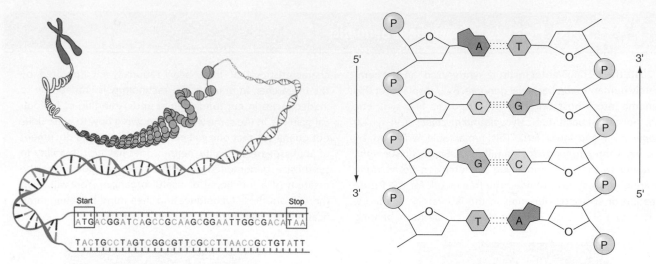

Figure 1.2 Complementary base-pairing is a key feature of the DNA molecule. (a) In eukaryotic organisms biological information is stored in the form of chromosomes located within the nucleus of each cell. If an individual chromosome could be carefully unwrapped from its proteinaceous packing material (see Chapter 6), one would see that it consisted of a single, linear DNA double helix (on the order of millions of base pairs in length). Each individual strand of the double helix is composed of nucleotide subunits (schematically represented by the letters A, T, G, and C) whose sequence encodes the information necessary to produce the proteins that govern cellular physiology. **(b)** In this close-up view of a short (four-base-pair) segment of a double helix, one can see that each nucleotide consists of a deoxyribose sugar (depicted here as a *white pentagon*), a phosphate (depicted as a *yellow circle*), and one of four nitrogenous bases—adenine, thymine, cytosine, or guanine (designated as *lavender* or *green* A's, T's, C's, or G's). The chemical structure of the bases enables A to associate tightly with T, and C to associate tightly with G, through hydrogen bonding. Thus the two strands are complementary to each other. The arrows labelled 5′ to 3′ show that the strands have opposite orientation.

A **gene** is a specific segment of DNA in a discrete region of a chromosome that serves as a unit of function by encoding a particular RNA (see Section 1.4) or protein (see Section 1.2). Just as the limited number of letters in a written alphabet places no restrictions on the stories one can tell, so too the limited number of letters in the DNA alphabet places no restrictions on the kinds of RNAs and proteins (and thus the kinds of organisms) genetic information can define. Within the cells of an organism, DNA molecules carrying genes are assembled into **chromosomes**: structures that package and manage the storage, duplication, expression, and evolution of DNA (**Figure 1.4**). The entire collection of chromosomes in each cell of an organism is its **genome**. Human cells, for example, contain 24 distinct kinds of chromosomes carrying approximately 3×10^9 base pairs and roughly 21 000 genes.

Figure 1.3 An automated "next generation" DNA sequencer. This state of the art sequencing instrument, located at the London Regional Genomics Centre in London, Ontario, can sequence up to 5 000 000 000 base pairs/day. Previous generation technology was capable of sequencing only 1 000 000 base pairs/day. The new instrument is also much more cost efficient on a per-base-pair basis.

Figure 1.4 One of 24 different types of human chromosomes. Each chromosome contains thousands of genes.

To appreciate the long journey from a finite amount of genetic information to the production of a human being, we must examine proteins, the molecules that determine how complex systems of cells, tissues, and organisms function.

> DNA, a macromolecular chain built using four different nucleotide subunits, is the repository of biological information. Genes are specific regions of DNA that encode particular RNAs or proteins.

1.2 Proteins: The Molecules That Govern Life Processes

Although no single characteristic distinguishes living organisms from inanimate matter, you would have little trouble deciding which entities in a given group of objects are alive. Over time, these living organisms, governed by the laws of physics and chemistry as well as a genetic program, would be able to reproduce themselves. Most of the organisms would also have an elaborate and complicated structure that would change over time—sometimes drastically, as when an insect larva metamorphoses into an adult. Yet another characteristic of life is the ability to move. Animals swim, fly, walk, or run, while plants grow toward or away from light. Still another characteristic is the capacity to adapt selectively to the environment. Finally, a key characteristic of living organisms is the ability to use sources of energy and matter to grow; that is, the ability to convert foreign material into their own body parts. The chemical and physical reactions that carry out these conversions are known as **metabolism**.

Most properties of living organisms ultimately arise from the class of molecules known as **proteins**—large polymers composed of hundreds to thousands of amino acid subunits strung together in long chains; each chain folds into a specific three-dimensional conformation dictated by the sequence of its amino acids (**Figure 1.5**). There are 20 different amino acids. The information in the DNA of genes dictates, via a genetic code, the order of amino acids in a protein molecule. An astonishing amount of diversity in three-dimensional structure can be generated in this way. This extraordinary diversity of protein structure is in turn reflected in the extraordinary diversity of protein function that is the basis of each organism's complex and adaptive behaviour. The structure and shape of the haemoglobin protein, for example, allow it to transport oxygen in the bloodstream and release it to the tissues. The proteins myosin and actin can slide together to allow muscle contraction. Chymotrypsin and elastase are enzymes that help break down other proteins. Most of the properties associated with life emerge from the constellation of protein molecules that an organism synthesizes according to instructions contained in its DNA.

> Proteins, macromolecules containing up to 20 different amino acids in a sequence encoded in DNA, are responsible for most biological functions.

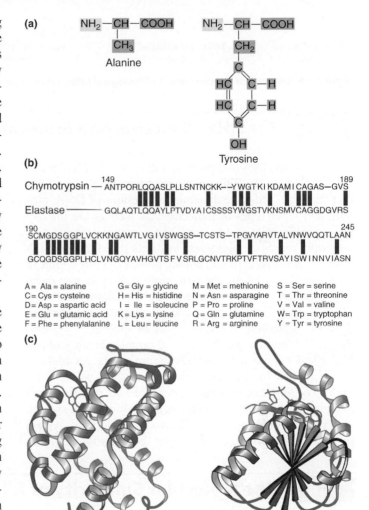

A = Ala = alanine
C = Cys = cysteine
D = Asp = aspartic acid
E = Glu = glutamic acid
F = Phe = phenylalanine

G = Gly = glycine
H = His = histidine
I = Ile = isoleucine
K = Lys = lysine
L = Leu = leucine

M = Met = methionine
N = Asn = asparagine
P = Pro = proline
Q = Gln = glutamine
R = Arg = arginine

S = Ser = serine
T = Thr = threonine
V = Val = valine
W = Trp = tryptophan
Y = Tyr = tyrosine

Figure 1.5 Proteins are polymers of amino acids that fold in three dimensions. The specific sequence of amino acids in a chain determines the precise three-dimensional shape of the protein. **(a)** Chemical formulas for two amino acids: alanine and tyrosine. All amino acids have a basic amino group (–NH) at one end and an acidic carboxyl group (–COOH) at the other. The specific side chain determines the amino acid's chemical properties. **(b)** A comparison of equivalent segments in the chains of two digestive proteins, chymotrypsin and elastase. The *red lines* connect sites in the two sequences that carry identical amino acids; the two chains differ at all the other sites shown. **(c)** Schematic drawings of the haemoglobin β chain (*green*) and lactate dehydrogenase (*purple*) show the different three-dimensional shapes determined by different amino acid sequences.

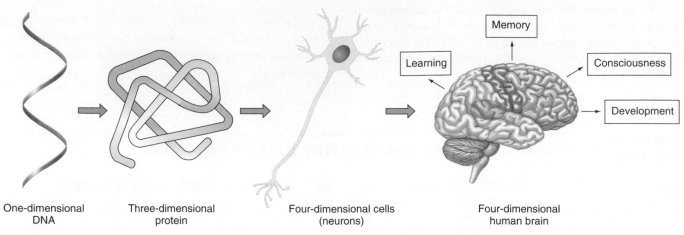

One-dimensional DNA

Three-dimensional protein

Four-dimensional cells (neurons)

Four-dimensional human brain

Figure 1.6 Diagram of the conversion of biological information from a one- to a three- and finally a four-dimensional state.

1.3 Complex Systems and Molecular Interactions

In addition to DNA and protein, a third level of biological information exists. This level encompasses the properties that "emerge" from the dynamic interactions of DNA, protein, and other types of molecules, as well as the interactions among cells and tissues. These complex interactive networks represent **biological systems** that function both within individual cells and among groups of cells within an organism. Here we use the term biological system to mean any complex network of interacting molecules or groups of cells that function in a coordinated manner through dynamic signalling. Several layers of biological systems exist. The human pancreas, for example, is an isolated biological system that operates within the larger biological system of the human body and mind. A whole community of animals, such as a colony of ants that functions in a highly coordinated manner, is also a biological system.

The information that defines any biological system is four-dimensional because it is constantly changing over the three dimensions of space and the one dimension of time. One of the most complex examples of this level of biological information is the human brain with its 10^{11} (100 000 000 000) neurons connected through perhaps 10^{18} (1 000 000 000 000 000 000) junctions known as synapses. From this enormous biological network, based ultimately on the information in DNA and protein, arise properties such as memory, consciousness, and the ability to learn (**Figure 1.6**). Understanding how these properties emerge from these systems is one of the main goals of modern genetic analysis and defines a relatively new field called "systems biology" (see Chapter 23).

A biological system is a network of interactions between molecules or groups of cells to accomplish coordinated function.

1.4 The Molecular Similarity of All Life-Forms

The evolution of biological information is a fascinating story spanning the 4 billion years of Earth's history. Many biologists think that RNA was the first information-processing molecule to appear. Very similar to DNA, RNA molecules are also composed of four subunits: the bases G, C, A, and U (for uracil, which replaces the T of DNA). Like DNA, RNA has the capacity to store, replicate, mutate, and express information; like proteins, RNA can fold in three dimensions to produce molecules capable of catalyzing the chemistry of life. RNA molecules, however, are intrinsically unstable. Thus, it is probable that the more stable DNA took over the linear information storage and replication functions of RNA, while proteins, with their far greater capacity for diversity, pre-empted the functions derived from RNA's three-dimensional folding. With this division of labour, RNA became an intermediary in converting the information in DNA into the sequence of amino acids in protein (**Figure 1.7a**). The separation that placed information storage in DNA and biological function in proteins was so successful that all organisms alive today descend from the first organisms that happened upon this molecular specialization. The flow of information from DNA to RNA to protein is often referred to as the "**central dogma**" of molecular biology (a term coined in 1957 by Francis Crick, the co-discoverer of the structure of DNA). As with all generalizations there are exceptions to the rule. In some cases a segment of DNA will code for an RNA that is not converted into a protein. Some examples include transfer RNAs, ribosomal RNAs, and microRNAs. We will learn more about these molecules and their biological roles in upcoming chapters.

The evidence for the common origin of all living forms is present in their DNA sequences. All living organisms

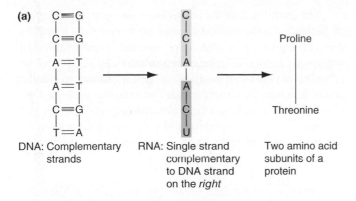

(a)

DNA: Complementary strands → RNA: Single strand complementary to DNA strand on the *right* → Two amino acid subunits of a protein

Proline

Threonine

(b)

Second letter

	U	C	A	G	
U	UUU UUC } Phe UUA UUG } Leu	UCU UCC UCA UCG } Ser	UAU UAC } Tyr UAA Stop UAG Stop	UGU UGC } Cys UGA Stop UGG Trp	U C A G
C	CUU CUC CUA CUG } Leu	CCU CCC CCA CCG } Pro	CAU CAC } His CAA CAG } Gln	CGU CGC CGA CGG } Arg	U C A G
A	AUU AUC } Ile AUA AUG Met	ACU ACC ACA ACG } Thr	AAU AAC } Asn AAA AAG } Lys	AGU AGC } Ser AGA AGG } Arg	U C A G
G	GUU GUC GUA GUG } Val	GCU GCC GCA GCG } Ala	GAU GAC } Asp GAA GAG } Glu	GGU GGC GGA GGG } Gly	U C A G

First letter (rows) / Third letter (columns)

Figure 1.7 RNA is an intermediary in the conversion of DNA information into protein via the genetic code. (a) The linear bases of DNA are copied through molecular complementarity into the linear bases of RNA. The bases of RNA are read three at a time (i.e., as triplets) to encode the amino acid subunits of proteins. (b) The genetic code dictionary specifies the relationship between RNA triplets and the amino acid subunits of proteins.

use essentially the same genetic code in which various triplet groupings of the four letters of the DNA and RNA alphabets encode the 20 letters of the amino acid alphabet (**Figure 1.7b**).

The relatedness of all living organisms is also evident from comparisons of genes with similar functions in very different organisms. For example, there is striking similarity between the genes for many proteins in bacteria, yeast, plants, worms, flies, mice, and humans (**Figure 1.8**). Moreover, it is often possible to place a gene from one organism into the genome of a very different organism and see it function normally in the new environment. Human genes that help regulate cell division, for example, can replace related genes in yeast and enable the yeast cells to function normally.

S. cerevisiae	GPNLHGIFGRHSGQVKGYSYTDANINKNVKW
A. thaliana	GPELHGLFGRKTGSVAGYSYTDANKQKGIEW
C. elegans	GPTLHGVIGRTSGTVSGFDYSAANKNKGVVW
D. melanogaster	GPNLHGLIGRKTGQAAGFAYTDANKAKGITW
M. musculus	GPNLHGLFGRKTGQAAGFSYTDANKNKGITW
H. sapiens	GPNLHGLFGRKTGQAPGYSYTAANKNKGIIW
	** *** . ** . * . * . ** * .. *

S. cerevisiae	DEDSMSEYLTNPKKYIPGTKMAFAGLKKEKDR
A. thaliana	KDDTLFEYLENPKKYIPGTKMAFGGLKKPKDR
C. elegans	TKETLFEYLLNPKKYIPGTKMVFAGLKKADER
D. melanogaster	NEDTLFEYLENPKKYIPGTKMIFAGLKKPNER
M. musculus	GEDTLMEYLENPKKYIPGTKMIFAGIKKKGER
H. sapiens	GEDTLMEYLENPKKYIPGTKMIFVGIKKKEER
	... ** *********** * . ** . *

* Indicates identical and . indicates similar

Figure 1.8 Comparisons of gene products in different species provide evidence for the relatedness of living organisms. This chart shows the amino acid sequence for equivalent portions of the cytochrome C protein in six species: *Saccharomyces cerevisiae* (yeast), *Arabidopsis thaliana* (a weedlike flowering plant), *Caenorhabditis elegans* (a nematode), *Drosophila melanogaster* (the fruit fly), *Mus musculus* (the house mouse), and *Homo sapiens* (humans). Consult **Figure 1.5** for the key to amino acid names.

One of the most striking examples of relatedness at this level of biological information was uncovered in studies of eye development. Both insects and vertebrates (including humans) have eyes, but they are of very different types (**Figures 1.9a** and **b**). Biologists had long assumed that the evolution of eyes occurred independently, and in many evolution textbooks, eyes are used as an example of **convergent evolution**, in which structurally

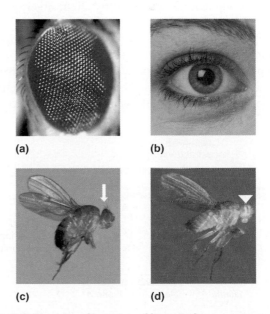

(a) **(b)**

(c) **(d)**

Figure 1.9 The eyes of insects and humans have a common ancestor. (a) A fly eye and (b) a human eye. (c) A fruit fly with normally developed eyes (*white arrow*). (d) A mutant fruit fly expressing a defective version of the *Pax6* gene; no eyes develop (*white arrowhead*).

unrelated but functionally analogous organs emerge in different species as a result of natural selection. Studies of a gene called *Pax6* have turned this view upside down.

Mutations in the *Pax6* gene lead to a failure of eye development in both people and fruit flies. Furthermore, molecular studies have suggested that *Pax6* might play a central role in the initiation of eye development in all vertebrates (**Figures 1.9c** to **d**). Remarkably, when the human *Pax6* gene is expressed in cells along the surface of the fruit fly body, it induces numerous little eyes to develop there. This result demonstrates that after 600 million years of divergent evolution, both vertebrates and insects still share the same main control switch for initiating eye development!

The usefulness of the relatedness and unity at all levels of biological information cannot be overstated. It means that, in many cases, the experimental manipulation of organisms known as **model organisms** can shed light on complex networks in humans. If genes similar to human genes function in simple model organisms such as fruit flies or bacteria, scientists can determine gene function and regulation in these experimentally manipulable organisms and bring these insights to an understanding of the human organism.

> Living organisms exhibit marked similarities at the molecular level; certain genes have been carried through the evolution of widely divergent species.

1.5   The Modular Construction of Genomes

We have seen that roughly 21 000 genes direct human growth and development. How did such complexity arise? Recent technical advances have enabled researchers to complete structural analyses of the entire genome of many organisms. The information obtained reveals that families of genes have arisen by duplication of a primordial gene. After duplication of the primordial gene, mutations and rearrangements (in either of the two copies) may cause the genes to diverge from each other (**Figure 1.10**). In both mice and humans, for example, five different haemoglobin β genes produce five different haemoglobin molecules at successive stages of development, with each protein functioning in a slightly different way to fulfill different needs for oxygen transport. The set of five haemoglobin genes arose from a single primordial gene by several duplications followed by slight divergences in structure.

Duplication followed by divergence underlies the evolution of new genes with new functions. This principle appears to have been built into the genome structure of all eukaryotic organisms. The protein-coding region of most genes is subdivided into as many as ten or more small

pieces (called exons), separated by DNA that does not code for protein (called introns) as shown in Figure 1.10. This modular construction facilitates the rearrangement of different modules from different genes to create new combinations during evolution. It is likely that this process of modular reassortment facilitated the rapid diversification of living forms about 570 million years ago.

The tremendous advantage of the duplication and divergence of existing pieces of genetic information is evident in the history of life's evolution (**Table 1.1**). **Prokaryotic** cells such as bacteria, which do not have a membrane-bounded nucleus, evolved about 3.7 billion years ago; **eukaryotic** cells such as algae, which have a membrane-bounded nucleus, emerged around 2 billion years ago; and multicellular eukaryotic organisms appeared 600–700 million years ago. Then, at about 570 million years ago, within the relatively short evolutionary time of roughly 20–50 million years known as the Cambrian Explosion, multicellular life-forms diverged into a bewildering array of organisms, including primitive vertebrates.

A fascinating question is, how could the multicellular forms achieve such enormous diversity in only 20–50 million years? The answer lies, in part, in the hierarchical organization of the information encoded in chromosomes. Exons are arranged into genes; genes duplicate and diverge to generate multigene families; and multigene families sometimes rapidly expand to gene superfamilies containing hundreds of related genes. In both mouse and human adults, for example, the immune system is encoded by a gene superfamily composed of hundreds of closely related but slightly divergent genes. With the emergence of each successively larger informational unit, evolution gains the ability to duplicate increasingly complex informational modules through single genetic events.

Probably even more important for the evolution of complexity is the rapid change of regulatory networks that specify how genes behave (i.e., when, where, and to what degree they are expressed) during development. For

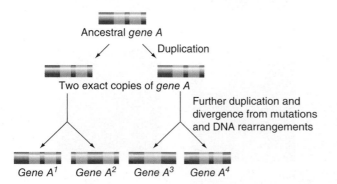

Figure 1.10 How genes arise by duplication and divergence. Duplications of ancestral *gene A* followed by mutations and DNA rearrangements have generated a family of related genes. The *dark blue* and *red bands* indicate the different exons of the genes while the *light blue bands* represent introns.

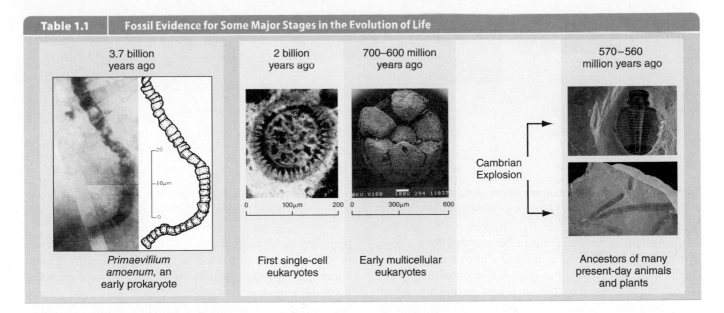

3.7 billion years ago	2 billion years ago	700–600 million years ago	570–560 million years ago
Primaevifilum amoenum, an early prokaryote	First single-cell eukaryotes	Early multicellular eukaryotes	Ancestors of many present-day animals and plants

Cambrian Explosion

example, the two-winged fly evolved from a four-winged ancestor not because of changes in gene-encoded structural proteins, but rather because of a rewiring of the regulatory network, which converted one pair of wings into two balancing organs known as halteres (**Figure 1.11**).

> The duplication of genes, and their subsequent divergence, has allowed for the evolution of new gene functions. In eukaryotes, separated exons composing a single gene allow potential rearrangements and rapid diversification.

1.6 Modern Genetic Techniques

The complexity of living systems has developed over 4 billion years from the continuous amplification and refinement of genetic information. The simplest bacterial cells contain about 1000 genes that interact in complex networks. Yeast cells, the simplest eukaryotic cells, contain about 6000 genes. Nematodes (roundworms) and fruit flies contain roughly 15 000–20 000 genes; humans have approximately 21 000 genes. The Human Genome Project, in addition to completing the sequencing of the entire human genome, has sequenced the genomes of *E. coli*, yeast, the nematode, the fruit fly, and the mouse (**Figure 1.12**). Each of these organisms has provided valuable insights into biology in general and human biology in particular.

With modern genetic techniques, researchers can dissect the complexity of a genome piece by piece. The logic used in genetic dissection is quite simple: inactivate a gene in a model organism and observe the consequences. For example, loss of a gene for visual pigment produces fruit flies with white eyes instead of eyes of the normal red colour. One can thus conclude that the protein product of this gene plays a key role in the development of eye pigmentation. From their study of model organisms, researchers are amassing a detailed picture of the complexity of living systems.

Even though the power of genetic techniques is astonishing, the complexity of biological systems is still difficult to comprehend. Knowing everything there is to know about each of the human genes and proteins would not reveal how a human results from this particular ensemble. For example, the human nervous system is a network of 10^{11} neurons with perhaps 10^{18} connections. The complexity

Figure 1.11 Two-winged and four-winged flies. Geneticists converted a contemporary normal two-winged fly to a four winged insect resembling the fly's evolutionary antecedent. They accomplished this by mutating a key element in the fly's regulatory network. Note the club-shaped halteres (*arrows*) behind the wings of the fly at the top.

Organism	E. coli	S. cerevisiae	C. elegans	D. melanogaster	Mus musculus
Genome size: (in megabases)	4.5 Mb	16 Mb	100 Mb	130 Mb	3000 Mb
Number of genes	3244	6201	20 470	15 016	23 786

Figure 1.12 Five model organisms whose genomes were sequenced as part of the Human Genome Project. The chart indicates genome size in millions of base pairs, or megabases (Mb). It also shows the approximate number of genes for each organism.

of the system is far too great to be encoded by a simple correspondence between genes and neurons or genes and connections. Moreover, the remarkable properties of the system, such as learning, memory, and personality, do not arise solely from the genes and proteins; network interactions and the environment also play a role. The goal of understanding higher-order processes that arise from interacting networks of genes, proteins, cells, and organs is one of the most challenging aspects of modern biology.

The new global tools of genomics—such as high-throughput DNA sequencers, genotypers, and large-scale DNA arrays (also called DNA chips)—have the capacity to analyze thousands of genes rapidly and accurately. These global tools are not specific to a particular system or organism; rather, they can be used to study the genes of all living things.

The DNA chip is a powerful example of a global genomic tool. Individual chips are subdivided into arrays of microscopic blocks that each contains a unique string of DNA units (**Figure 1.13a**). When a chip is exposed to a complex mixture of fluorescently labelled nucleic acid—such as DNA or RNA from any cell type or sample—the unique string in each microscopic block can bind to and detect a specific complementary sequence. This type of binding is known as **hybridization** (**Figure 1.13b**). A computer-driven microscope can then analyze the bound sequences of the hundreds of thousands of blocks on the chip, and special software can enter this information into a database (**Figure 1.13c**).

The potential of DNA chips is enormous for both research and clinical purposes. Already chips with over 400 000 different detectors can provide simultaneous information on the presence or absence of 400 000 discrete DNA or RNA sequences in a complex sample. And they can do it within hours. Here is one example. Now that the

sequence of all human genes is known, unique stretches of DNA representing each of the approximately 21 000 human genes can be placed on a chip and used to determine the complete set of genes copied into RNA in any human cell type at any stage of development or differentiation. Computer-driven comparisons can contrast the genes expressed in different cell types, for example, in neurons versus muscle cells, making it possible to determine which genes contribute to the construction of various cell types. Scientists have already created catalogues of the genes expressed in different cell types and have discovered that some genes, called "housekeeping genes", are expressed in nearly all cell types, whereas other genes are expressed only in certain specialized cells. This knowledge of the relation between particular genes and particular cell types is helping us understand how the cellular specialization necessary for the construction of all human organs arises.

In medicine, clinical researchers have used DNA chip technology to identify genes whose expression increases or decreases when tumour cells are treated with an experimental cancer drug (Figures 1.13b and c). Changes in the patterns of gene expression may provide clues to the mechanisms by which the drug might inhibit tumour growth. In a related but slightly different application of the same idea, researchers can assess the inherent differences between breast cancers that respond well to a particular drug therapy and those that do not (i.e., that recur despite treatment). Using microarray analysis of patients' tumours can predict with considerable accuracy whether a specific drug will be effective against their particular type of cancer.

> Modern techniques such as computerized processing and automated sequencing, DNA amplification, and hybridization have provided knowledge of genomes at the sequence level.

1.7 Human Genetics

In the mid-1990s, a majority of scientists who responded to a survey conducted by *Science* magazine rated genetics as the most important field of science in the coming decades. One reason is that the powerful tools of genetics open up the possibility of understanding biology, including

human biology, from the molecular level up to the level of the whole organism.

The Human Genome Project, by changing the way we view biology and genetics, has led to a significant paradigm change: the systems approach to biology and medicine.

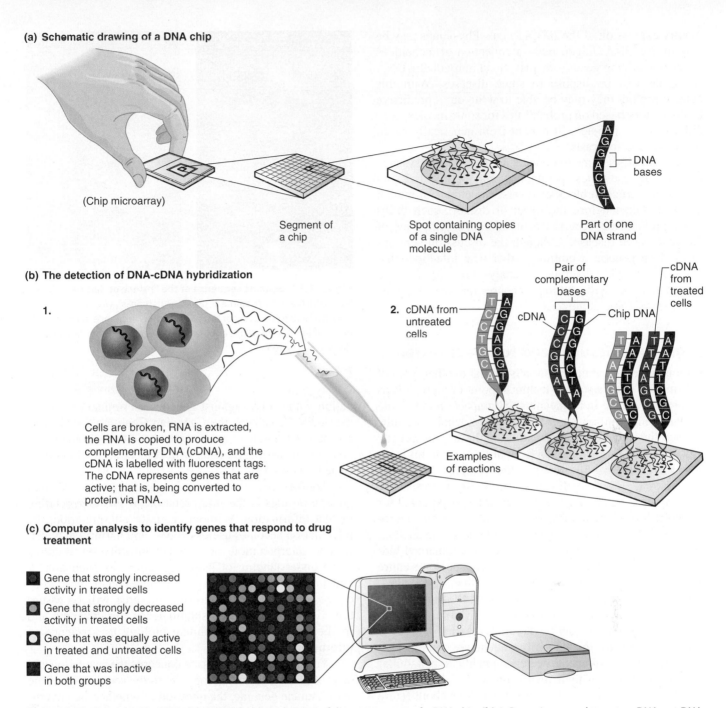

(a) Schematic drawing of a DNA chip

(Chip microarray)

Segment of a chip

Spot containing copies of a single DNA molecule

Part of one DNA strand

DNA bases

(b) The detection of DNA-cDNA hybridization

1.

Cells are broken, RNA is extracted, the RNA is copied to produce complementary DNA (cDNA), and the cDNA is labelled with fluorescent tags. The cDNA represents genes that are active; that is, being converted to protein via RNA.

Pair of complementary bases

cDNA from treated cells

2. cDNA from untreated cells

cDNA

Chip DNA

Examples of reactions

(c) Computer analysis to identify genes that respond to drug treatment

Gene that strongly increased activity in treated cells

Gene that strongly decreased activity in treated cells

Gene that was equally active in treated and untreated cells

Gene that was inactive in both groups

Figure 1.13 One use of a DNA chip. (a) Schematic drawing of the components of a DNA chip. **(b) 1.** Preparing complementary DNA, or cDNA, with a fluorescent tag from the RNA of a group of cells. **2.** The hybridization of chip DNA to fluorescent cDNA from untreated and drug-treated cells. **(c)** Computerized analysis of chip hybridizations makes it possible to compare gene activity in any two types of cells.

The systems approach seeks to study the relationships of all the elements in a biological system as it undergoes genetic perturbation or biological activation (see Chapter 23). This is a fundamental change from the study of complex systems one gene or protein at a time.

Molecular studies may lead to predictive and preventive medicine

Over the next 25 years, geneticists will identify hundreds of genes with variations that predispose people to many types of disease: cardiovascular, cancerous, immunological, mental,

and metabolic. Some mutations will always cause disease; others will only predispose to disease. For example, a change in a specific single DNA base (i.e., a change in one DNA unit) in the β-*globin* gene will nearly always cause sickle-cell anaemia, a painful, life-threatening condition that leads to severe anaemia. By contrast, a mutation in the *breast cancer 1* (*BRCA1*) gene increases the risk of breast cancer to between 40 percent and 80 percent in a woman carrying one copy of the mutation. This conditional state arises because the *BRCA1* gene interacts with environmental factors that affect the probability of activating the cancerous condition, and because various forms of other genes

modify expression of the *BRCA1* gene. Physicians may be able to use DNA diagnostics—a collection of techniques for characterizing genes—to analyze an individual's DNA for genes that predispose to some diseases. With this genetic profile, they may be able to write out a predictive health history based on probabilities for some medical conditions. Many people will benefit from genetically based diagnoses and forecasts.

As scientists come to understand the complex systems in which disease genes operate, they may be able to design therapeutic drugs to block or reverse the effects of mutant genes. If taken before the onset of disease, such drugs could prevent the occurrence or minimize symptoms of the gene-based disease. Although the discussion here has focused on genetic conditions rather than infectious diseases, it is possible that ongoing analyses of microbial and human genomes will lead to procedures for controlling the virulence of some pathogens.

Many social issues need to be addressed

Although biological information is similar to other types of information from a strictly technical point of view, it is as different as can be in its meaning and impact on individual human beings and on human society as a whole. The difference lies in the personal nature of the unique genetic profile carried by each person from birth. Within this basic level of biological information are complex life codes that provide greater or lower susceptibility or resistance to many diseases, as well as greater or lesser potential for the expression of many physiological, physical, and neurological attributes that distinguish people from one another. Until now, almost all this information has remained hidden away. But today, it is possible to read a person's entire genetic profile at a cost of only a few thousand dollars. While at present, it is only practical for the affluent (e.g., heavy metal and reality TV star, Ozzy Osborne) to obtain a personal genome sequence (**Figure 1.14**), it will soon be commonplace—as the cost of sequencing continues to decrease—for members of the general public to be familiar with the sequence of their own genomes. With this information comes the power to make predictions about future possibilities and risks.

As you will see in many of the **Genetics and Society** boxes throughout this book, society can use genetic information not only to help people but also to restrict their lives (e.g., by denying insurance or employment). Many believe that, just as society respects an individual's right to privacy in other realms, it should also respect the privacy of an individual's genetic profile and work against all types of discrimination. In the United States, United Kingdom, Germany, France, Italy, Japan, and Spain, federal governments have passed legislation prohibiting insurance companies and employers from discrimination on the basis of genetic tests. Surprisingly, Canada remains the only G8

Figure 1.14 Genome sequence of the "prince of darkness"?! In late 2010, the self-proclaimed "prince of darkness", Ozzy Osborne, obtained a personal copy of his own genome sequence in the hope of learning more about his ancestry, and the origin of his Parkinson's disease-like systems.

nation with no law against genetic discrimination. Fortunately for Canadians, the Canadian Coalition for Genetic Fairness (CCGF) has been established to lobby both federal and provincial governments with respect to this important issue (**Figure 1.15**).

Another issue raised by the potential for detailed genetic profiles is the interpretation or misinterpretation of that information. Without accurate interpretation, the information becomes useless at best and harmful at worst. Proper interpretation of genetic information requires some understanding of statistical concepts such as risk and probability. To help people understand these concepts, widespread education in this area will be essential. Children especially should learn the concepts and implications of modern human biology as a science of information.

Yet another pressing issue concerns the regulation and control of the new technology. With the sequencing of the entire human genome, the question of whether the government should establish guidelines for the use of genetic and genomic information, reflecting society's social and ethical values, remains in open debate.

To many people, the most frightening potential of the new genetics is the development of technology that can alter or add to the genes present within the **germ line** (reproductive cell precursors) of human embryos. This technology, referred to as "transgenic technology" in scientific discourse and "genetic engineering" in public discussions, has become routine in hundreds of laboratories working with various animals other than humans.

Some people caution that developing the power to alter our own genomes is a step we should not take, arguing that

Who has perfect genes?
Genetic discrimination affects us all!

Figure 1.15 Canadian Coalition for Genetic Fairness (CCGF). The CCFG is dedicated to preventing genetic discrimination against Canadians. Their mission is to "educate Canadians about genetic discrimination and to influence federal and provincial governments, and other relevant organizations, to create positive change." You can view their website at ccgf-cceg.ca.

if genetic information and technology are misused, as they certainly have been in the past, the consequences could be horrific. Attempts to use genetic information for social purposes were prevalent in the early twentieth century, leading to enforced sterilization of individuals thought to be inferior, to laws that prohibited interracial marriage, and to laws prohibiting immigration of certain ethnic groups. The scientific basis of these actions has been thoroughly discredited.

Others agree that the mistakes of the past must not be repeated, but warn that if the new technologies could help children and adults lead healthier, happier lives, then society should think carefully about whether the reasons for objecting outright to their use are valid. Most agree that the ongoing biological revolution will have a greater impact on human society than any technological revolution of the past and that education and public debate are the keys to preparing for the consequences of this revolution.

The focus on human genetics in this book looks forward into the new era of biology and genetic analysis. These new possibilities raise serious moral and ethical issues that will demand wisdom and humility. It is in the hope of educating young people for the moral and ethical challenges awaiting the next generation that this book is written.

> Advances in human genetics have great promise for the treatment or prevention of disease. Guidelines must be established, however, to prevent misuse of this knowledge.

Connections

Genetics, the study of biological information, is also the study of the DNA and RNA molecules that store, replicate, transmit, and evolve information for the construction of proteins. At the molecular level, all living things are closely related, and as a result, observations of model organisms as different as yeast and mice can provide insights into general biological principles as well as human biology.

Remarkably, more than 75 years before the discovery of DNA, Gregor Mendel, an Augustinian monk, delineated the basic laws of gene transmission with no knowledge of the molecular basis of heredity. He accomplished this by following simple traits, such as flower or seed colour, through several generations of the pea plant (*Pisum sativum*). We now know that his findings apply to all sexually reproducing organisms. Chapter 2 describes Mendel's studies and insights, which became the foundation of the field of genetics.

1. Biological information is digitally encoded in the DNA molecule. **[LO1–3, LO5]**

2. Biological function emerges primarily from protein molecules. **[LO1–4]**

3. Complex biological systems emerge from the functioning of regulatory networks that specify the behaviour of genes and proteins. **[LO1–2, LO4]**

4. All living forms are descended from a common ancestor and therefore are closely related at the molecular level. **[LO3–4]**

5. The modular construction of genomes has allowed rapid evolution of biological complexity. **[LO1, LO3–4]**

6. Modern genetic technology permits detailed analysis and dissection of biological complexity. **[LO1–2, LO5]**

7. Application of modern technology to human genetics shows great promise for prediction, prevention, and treatment of disease. **[LO5]**

connect For more information on the resources available from McGraw-Hill Ryerson, go to www.mcgrawhill.ca/he/solutions.

The white-flowered (*far left*) and purple-flowered (*middle*) garden pea plants that Mendel used as his experimental model allowed him to propose the basic laws of heredity. In garden pea plants, the gene controlling flower colour has two alleles. However, most genes have several alleles. For example, in the petunia plant (*far right*), multiple alleles can result in flowers with a variety of different colours, such as the three depicted here.

A quick glance at an extended family portrait is likely to reveal children who resemble one parent or the other or who look like a combination of the two (**Figure 2.1**). Some children, however, look unlike any of the assembled relatives and more like a great, great grandparent. What causes the similarities and differences of appearance and the skipping of generations?

The answers lie in our **genes**, the basic units of biological information, and in **heredity**, the way genes transmit physiological, anatomical, and behavioural traits from parents to offspring. Each of us starts out as a single fertilized egg cell that develops, by division and differentiation, into a mature adult made up of 10^{14} (a hundred trillion) specialized cells capable of carrying out all the body's functions and controlling our outward appearance. Genes, passed from one generation to the next, underlie the formation of every heritable trait. Such traits are as diverse as the presence of a cleft in your chin, the tendency to lose hair as you age, your hair, eye, and skin colour, and even your susceptibility to certain cancers. All such traits run in families in predictable patterns that impose some possibilities and exclude others.

Genetics, the science of heredity, pursues a precise explanation of the biological structures and mechanisms that determine inheritance. In some instances, the relationship between gene and trait is remarkably simple. A single change in a single gene, for example, results in sickle-cell anaemia, a disease in which the haemoglobin molecule found in red blood cells is defective. In other instances, the correlations between genes and traits are bewilderingly complex. An example is the genetic basis of facial features, in which many genes determine a large number of molecules that interact to generate the combination we recognize as a friend's face.

Gregor Mendel (1822–1884; **Figure 2.2**), a stocky, bespectacled Augustinian monk and expert plant breeder, discovered the basic principles of genetics in the mid-nineteenth century. He published his findings in 1866, just seven years after Darwin's *On the Origin of Species* appeared in print. Mendel lived and worked in Brünn, Austria (now Brno in the Czech Republic), where he examined the inheritance of clear-cut alternative traits in pea plants, such as purple versus white flowers or yellow versus green seeds. In so doing, he inferred genetic laws that allowed him to make verifiable predictions about which traits would appear, disappear, and then reappear, and in which generations.

Chapter Outline
2.1 Background: The Historical Puzzle of Inheritance
2.2 Genetic Analysis According to Mendel
2.3 Mendelian Inheritance in Humans
2.4 Extensions to Mendel for Single-Gene Inheritance
2.5 Extensions to Mendel for Gene Interactions

Figure 2.1 A family portrait. The extended family shown here includes members of four generations.

Figure 2.2 Gregor Mendel. Photographed around 1862, holding one of his experimental plants.

Figure 2.3 Like begets like and unlike. A Labrador retriever with her litter of pups.

Mendel's laws are based on the hypothesis that observable traits are determined by independent units of inheritance that are not visible to the naked eye. We now call these units *genes*. The concept of the gene continues to change as research deepens and refines our understanding. Today, a gene is recognized as a region of DNA that encodes a specific protein or a particular type of RNA. In the beginning, however, it was an abstraction—an imagined particle with no physical features, the function of which was to control a visible trait by an unknown mechanism.

We begin our study of genetics with a detailed look at what Mendel's laws are and how they were discovered. We will also discuss logical extensions to these laws and describe how Mendel's successors grounded the abstract concept of hereditary units (genes) in an actual biological molecule (DNA).

Four general themes emerge from our detailed discussion of Mendel's work. The first is that variation, as expressed in alternative forms of a trait, is widespread in nature. This genetic diversity provides the raw material for the continuously evolving variety of life we see around us. Second, observable variation is essential for following genes from one generation to the next. Third, variation is not distributed solely by chance; rather, it is inherited according to genetic laws that explain why like begets both like and unlike. Dogs beget other dogs—but hundreds of breeds of dogs are known. Even within a breed, such as Labrador retrievers, genetic variation exists: Two black dogs could have a litter of black, brown, and golden puppies (**Figure 2.3**). Mendel's insights help explain why this is so. Fourth, the laws Mendel discovered about heredity apply equally well to all sexually reproducing organisms, from protozoans to peas to people.

Learning Objectives

1. Explain how monohybrid crosses led Mendel to infer the law of segregation.

2. Distinguish between the terms *gene* and *allele* and contrast dominant alleles with recessive alleles.

3. Explain Mendel's law of segregation.

4. Differentiate between the terms *homozygous, heterozygous, genotype,* and *phenotype.*

5. Analyze Mendel's law of independent assortment, and explain how Mendel proposed the law from dihybrid crosses.

6. Compare and contrast complete dominance, incomplete dominance, and codominance relationships, and demonstrate how a dominance series can be established.

7. Explain the terms *wild-type allele, mutant allele, monomorphic,* and *polymorphic.*

8. Describe pleiotropy and how it arises.

9. Compare and contrast complementary gene action, recessive epistasis, and dominant epistasis.

10. Evaluate the significance of the complementation test as a tool for genetic analysis.

11. Distinguish between penetrance and expressivity.

12. Explain the inheritance of continuous traits.

2.1   Background: The Historical Puzzle of Inheritance

Several steps lead to an understanding of genetic phenomena: the careful observation over time of groups of organisms, such as human families, herds of cattle, or fields of corn or tomatoes; the rigorous analysis of systematically recorded information gleaned from these observations; and the development of a theoretical framework that can

explain the origin of these phenomena and their relationships. In the mid-nineteenth century, Gregor Mendel became the first person to combine data collection, analysis, and theory in a successful pursuit of the true basis of heredity. For many thousands of years before that, the only genetic practice was the selective breeding of domesticated plants and animals, with no guarantee of what a particular mating would produce.

Artificial selection was the first applied genetic technique

A rudimentary use of genetics was the driving force behind a key transition in human civilization, allowing hunters and gatherers to settle in villages and survive as shepherds and farmers. Even before recorded history, people practised applied genetics as they domesticated plants and animals for their own uses. As a result of this **artificial selection**—purposeful control over mating by choice of parents for the next generation—the domestic dog (*Canis lupus familiaris*), for example, slowly arose from ancestral wolves (*Canis lupus*). The oldest bones identified indisputably as dog (and not wolf) are that of a skull excavated from a 20 000-year-old Alaskan settlement. Many millennia of evolution guided by artificial selection have produced hundreds of modern breeds of dog. By 10 000 years ago, people had begun to use this same kind of genetic manipulation to develop economically valuable herds of reindeer, sheep, goats, pigs, and cattle that produced life-sustaining meat, milk, hides, and wools.

Farmers also carried out artificial selection of plants, storing seed from the hardiest and tastiest individuals for the next planting, eventually obtaining strains that grew better, produced more, and were easier to cultivate and harvest. In this way, scrawny weedlike plants gradually, with human guidance, turned into rice, wheat, barley, lentils, and dates in Asia; corn, squash, tomatoes, potatoes, and peppers in the Americas; and yams, peanuts, and gourds in Africa. Later, plant breeders recognized male and female organs in plants and carried out artificial pollination.

Desirable traits sometimes disappear and reappear

In 1822, the year of Mendel's birth, what people in Austria understood about the basic principles of heredity was not much different from what the people of ancient Assyria had understood. By the nineteenth century, plant and animal breeders had created many strains in which offspring often carried a prized parental trait. Using such strains, they could produce plants or animals with desired characteristics for food and fibre, but they could not always predict why a valued trait would sometimes disappear and then reappear in only some offspring. For example, selective breeding practices had resulted in valuable flocks of merino sheep producing large quantities of soft, fine wool, but at the 1837 annual meeting of the Moravian Sheep Breeders Society, one breeder's dilemma epitomized the state of the art. He possessed an outstanding

ram that would be priceless "if its advantages are inherited by its offspring," but "if they are not inherited, then it is worth no more than the cost of wool, meat, and skin." Which would it be? According to the meeting's recorded minutes, current breeding practices offered no definite answers. In his concluding remarks at this sheep-breeders meeting, the Abbot Cyril Napp pointed to a possible way out. He proposed that breeders could improve their ability to predict what traits would appear in the offspring by finding the answers to three basic questions: What is inherited? How is it inherited? What is the role of chance in heredity?

This is where matters stood in 1843 when 21-year-old Gregor Mendel entered the monastery in Brünn, presided over by the same Abbot Napp. Although Mendel was a monk trained in theology, he was not a rank amateur in science. The province of Moravia, in which Brünn was located, was a centre of learning and scientific activity. Mendel was able to acquire a copy of Darwin's *On the Origin of Species* shortly after it was translated into German in 1863. Abbot Napp, recognizing Mendel's intellectual abilities, sent him to the University of Vienna—all expenses paid—where he prescribed his own course of study. His choices were an unusual mix: physics, mathematics, chemistry, botany, paleontology, and plant physiology. Christian Doppler, discoverer of the Doppler effect, was one of his teachers. The cross-pollination of ideas from several disciplines would play a significant role in Mendel's discoveries. One year after he returned to Brünn, he began his series of seminal genetic experiments. **Figure 2.4** shows where Mendel worked and the microscope he used. In **Figure 2.4a**, red and white begonias have been planted in a grid to represent the inheritance patterns that Mendel obtained with his pea plants.

Mendel devised a new experimental approach

Before Mendel, many misconceptions clouded people's thinking about heredity. Two of the prevailing errors were particularly misleading. The first was that one parent contributes most to an offspring's inherited features; Nicolaas Hartsoeker, one of the earliest microscopists, contended in 1694 that it was the male, by way of a fully formed "homunculus" inside the sperm (**Figure 2.5**). Another deceptive notion was the concept of *blended inheritance*, the idea that parental traits become mixed and forever changed in the offspring, as when blue and yellow pigments merge to green on a painter's palette. The theory of blending may have grown out of a natural tendency for parents to see a combination of their own traits in their offspring. While blending could account for children who look like a combination of their parents, it could not explain obvious differences between biological brothers and sisters or the persistence of variation within extended families.

The experiments Mendel devised would lay these myths to rest by providing precise, verifiable answers to the three questions Abbot Napp had raised almost 15 years

(a)  **(b)**

Figure 2.4 Mendel's garden and microscope. **(a)** Gregor Mendel's garden was part of his monastery's property in Brünn. **(b)** Mendel used this microscope to examine plant reproductive organs and to pursue his interests in natural history.

earlier: What is inherited? How is it inherited? What is the role of chance in heredity? A key component of Mendel's breakthrough was the way he set up his experiments.

What did Mendel do differently from those who preceded him? First, he chose the garden pea (*Pisum sativum*)

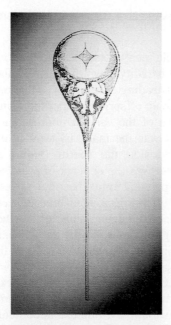

Figure 2.5 The homunculus: A misconception. Well into the nineteenth century, many prominent microscopists believed they saw a fully formed, miniature fetus crouched within the head of a sperm.

as his experimental organism (**Figures 2.6a** and **b**). Peas grew well in Brünn, and with male and female organs in the same flower, they were normally self-fertilizing. In **self-fertilization** (or *selfing*), both egg and pollen come from the same plant. The particular anatomy of pea flowers, however, makes it easy to prevent self-fertilization and instead to **cross-fertilize** (or *cross*) two individuals by brushing pollen from one plant onto a female organ of another plant, as illustrated in **Figure 2.6c**. Peas offered yet another advantage. For each successive generation, Mendel could obtain large numbers of individuals within a relatively short growing season. By comparison, if he had worked with sheep, each mating would have generated only a few offspring and the time between generations would have been several years.

Second, Mendel examined the inheritance of clear-cut alternative forms of particular traits—purple versus white flowers, yellow versus green peas. Using such "either-or" traits, he could distinguish and trace unambiguously the transmission of one or the other observed characteristic, because there were no intermediate forms. (The opposite of these so-called *discrete traits* are *continuous traits*, such as height and skin colour in humans. Continuous traits show many intermediate forms.)

Third, Mendel collected and perpetuated lines of peas that bred true. Matings within such **pure-breeding lines** produce offspring carrying specific parental traits that remain constant from generation to generation. Mendel observed his pure-breeding lines for up to eight

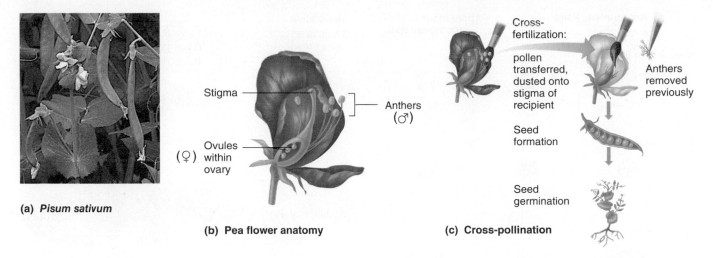

(a) *Pisum sativum*

(b) **Pea flower anatomy**

(c) **Cross-pollination**

Figure 2.6 Mendel's experimental organism: The garden pea. (a) Pea plants with white flowers. **(b)** Pollen is produced in the anthers. Mature pollen lands on the stigma, which is connected to the ovary (which becomes the pea pod). After landing, the pollen grows a tube that extends through the stigma to one of the ovules (immature seeds), allowing fertilization to take place. **(c)** To prevent self-fertilization, breeders remove the anthers from the female parents (here, the white flower) before the plant produces mature pollen. Pollen is then transferred with a paintbrush from the anthers of the male parent (here, the purple flower) to the stigma of the female parent. Each fertilized ovule becomes an individual pea (mature seed) that can grow into a new pea plant. All of the peas produced from one flower are encased in the same pea pod, but these peas form from different pollen grains and ovules.

generations. Plants with white flowers always produced offspring with white flowers; plants with purple flowers produced only offspring with purple flowers. Mendel called constant but mutually exclusive alternative traits, such as purple versus white flowers or yellow versus green seeds, "antagonistic pairs" and settled on seven such pairs for his study (**Figure 2.7**). In his experiments, he not only perpetuated pure-breeding stocks for each member of a pair, but he also cross-fertilized pairs of plants to produce **hybrids**, offspring of genetically dissimilar parents, for each pair of antagonistic traits. Figure 2.7 shows the appearance of the hybrids he studied.

> Gregor Mendel performed genetic crosses in a systematic way, using mathematics to analyze the data he obtained and to predict outcomes of other experiments.

2.2 Genetic Analysis According to Mendel

In early 1865 at the age of 43, Gregor Mendel presented a paper entitled "Experiments on Plant Hybrids" before the Natural Science Society of Brünn. Despite its modest heading, it was a scientific paper of uncommon clarity and simplicity that summarized a decade of original observations and experiments. In it Mendel describes in detail the transmission of visible characteristics in pea plants, defines unseen but logically deduced units (genes) that determine when and how often these visible traits appear, and analyzes the behaviour of genes in simple mathematical terms to reveal previously unsuspected principles of heredity.

Published the following year, the paper would eventually become the cornerstone of modern genetics. Its stated purpose was to see whether there is a "generally applicable law governing the formation and development of hybrids." Let us examine its insights.

Monohybrid crosses reveal units of inheritance and the law of segregation

Once Mendel had isolated pure-breeding lines for several sets of characteristics, he carried out a series of matings between individuals that differed in only one trait, such as seed colour or stem length. In each cross, one parent carries one form of the trait, and the other parent carries an alternative form of the same trait. **Figure 2.8** illustrates one such mating. Early in the spring of 1854, for example, Mendel planted pure-breeding green peas and pure-breeding yellow peas and allowed them to grow into the **parental (P) generation**. Later that spring when the plants had flowered, he dusted the female stigma of "green-pea" plant flowers with pollen from "yellow-pea" plants. He also performed the reciprocal cross, dusting "yellow-pea" plant stigmas with "green-pea" pollen. In the fall, when he collected and separately analyzed the progeny peas of these reciprocal crosses, he found that in both cases the peas were all yellow.

These yellow peas, progeny of the P generation, were the beginning of what we now call the **first filial (F_1)** generation. To learn whether the green trait had disappeared entirely or remained intact but hidden in these F_1 yellow peas, Mendel planted them to obtain mature F_1 plants that he allowed to self-fertilize. Such experiments involving hybrids for a single trait are often called

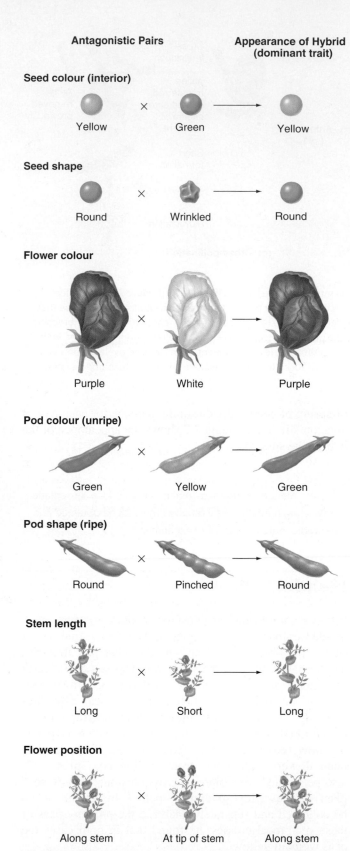

Antagonistic Pairs   **Appearance of Hybrid (dominant trait)**

Seed colour (interior)

Yellow × Green → Yellow

Seed shape

Round × Wrinkled → Round

Flower colour

Purple × White → Purple

Pod colour (unripe)

Green × Yellow → Green

Pod shape (ripe)

Round × Pinched → Round

Stem length

Long × Short → Long

Flower position

Along stem × At tip of stem → Along stem

Figure 2.7 The mating of parents with antagonistic traits produces hybrids. Note that each of the hybrids for the seven antagonistic traits studied by Mendel resembles only one of the parents. The parental trait that shows up in the hybrid is known as the "dominant" trait.

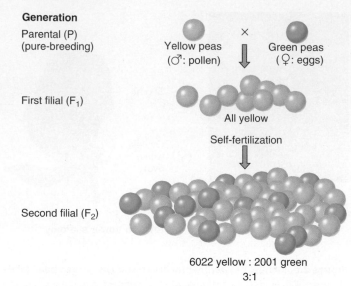

Generation

Parental (P) (pure-breeding)

Yellow peas (♂: pollen)   ×   Green peas (♀: eggs)

First filial (F₁)

All yellow

Self-fertilization

Second filial (F₂)

6022 yellow : 2001 green
3:1

Figure 2.8 Analyzing a monohybrid cross. Cross-pollination of pure-breeding parental plants produces F₁ hybrids, all of which resemble one of the parents. Self-pollination of F₁ plants gives rise to an F₂ generation with a 3:1 ratio of individuals resembling the two original parental types. (For simplicity, we do not show the plants that produce the peas or that grow from the planted peas.)

monohybrid crosses. He then harvested and counted the peas of the resulting **second filial (F₂)** generation, progeny of the F₁ generation. Among the progeny of one series of F₁ self-fertilizations, there were 6022 yellow and 2001 green F₂ peas, an almost perfect ratio of 3 yellow : 1 green. F₁ plants derived from the reciprocal of the original cross produced a similar ratio of yellow to green F₂ progeny.

Reappearance of the recessive trait

The presence of green peas in the F₂ generation was irrefutable evidence that blending had not occurred. If it had, the information necessary to make green peas would have been irretrievably lost in the F₁ hybrids. Instead, the information remained intact and was able to direct the formation of 2001 green peas actually harvested from the second filial generation. These green peas were indistinguishable from their green grandparents.

Mendel concluded that there must be two types of yellow peas: those that breed true like the yellow peas of the P generation, and those that can yield some green offspring like the yellow F₁ hybrids. This second type somehow contains latent information for green peas. He called the trait that appeared in all the F₁ hybrids—in this case, yellow seeds—**dominant** (see Figure 2.7) and the "antagonistic" green-pea trait that remained hidden in the F₁ hybrids but reappeared in the F₂ generation **recessive.** But how did he explain the 3:1 ratio of yellow to green F₂ peas?

Genes: Discrete units of inheritance

To account for his observations, Mendel proposed that for each trait, every plant carries two copies of a unit of inheritance, receiving one from its maternal parent and the other from the paternal parent. Today, we call these units of inheritance *genes*. Each unit determines the appearance of a specific characteristic. The pea plants in Mendel's collection had two copies of a gene for seed colour, two copies of another for seed shape, two copies of a third for stem length, and so forth.

Mendel further proposed that each gene comes in alternative forms, and combinations of these alternative forms determine the contrasting characteristics he was studying. Today we call the alternative forms of a single gene **alleles**. The gene for pea colour, for example, has yellow and green alleles; the gene for pea shape has round and wrinkled alleles. In Mendel's monohybrid crosses, one allele of each gene was **dominant** because it manifested itself regardless of the presence of the other allele, and the other allele was **recessive** because its effect was hidden or masked when the dominant allele was present. In the P generation, one parent carried two dominant alleles for the trait under consideration; the other parent, two recessive alleles. The F_1 generation hybrids carried one dominant and one recessive allele for the trait. Individuals having two different alleles for a single trait are **monohybrids**. (The **Fast Forward** box **"Genes Encode Proteins"** in this chapter describes the biochemical and molecular mechanisms by which different alleles determine different forms of a trait.)

The law of segregation

If a plant has two copies of every gene, how does it pass only one copy of each to its progeny? And how do the offspring then end up with two copies of these same genes, one from each parent? Mendel drew on his background in plant physiology and answered these questions in terms of the two biological mechanisms behind reproduction: gamete formation and the random union of gametes at fertilization. **Gametes** are the specialized cells—eggs within the ovules of the female parent and sperm cells within the pollen grains—that carry genes between generations. He imagined that during the formation of pollen and eggs, the two copies of each gene in the parent separate (or *segregate*) so that each gamete receives only one allele for each trait (**Figure 2.9a**). Thus, each egg and each pollen grain receives only one allele for pea colour (either yellow or green). At fertilization, the pollen with one or the other allele unites at random with an egg carrying one or the other allele, restoring the two copies of the gene for each trait in the fertilized egg, or **zygote** (**Figure 2.9b**). If the pollen carries yellow and the egg green, the result will be a hybrid yellow pea like the F_1 monohybrids that resulted when pure-breeding parents of opposite types mated. If the yellow-carrying pollen unites with a yellow-carrying egg, the result will be a yellow pea that grows into

(a) The two alleles for each trait separate during gamete formation.

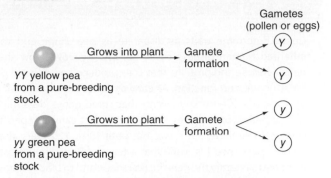

(b) Two gametes, one from each parent, unite at random at fertilization.

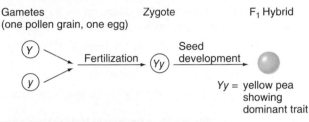

Y = yellow-determining allele of pea colour gene
y = green-determining allele of pea colour gene

Figure 2.9 The law of segregation. (a) The two identical alleles of pure-breeding plants separate (segregate) during gamete formation. As a result, each pollen grain or egg carries only one of each pair of parental alleles. **(b)** Cross-pollination and fertilization between pure-breeding parents with antagonistic traits result in F_1 hybrid zygotes with two different alleles. For the seed colour gene, a *Yy* hybrid zygote will develop into a yellow pea.

a pure-breeding plant like those of the P generation that produced only yellow peas. And finally, if pollen carrying the allele for green peas fertilizes a green-carrying egg, the progeny will be a pure-breeding green pea.

Mendel's **law of segregation** encapsulates this general principle of heredity: *The two alleles for each trait separate (segregate) during gamete formation, and then unite at random, one from each parent, at fertilization.* Throughout this book, the term **segregation** refers to such *equal segregation* in which one, and only one, allele of each gene goes to each gamete. Note that the law of segregation makes a clear distinction between organisms, whose cells have two copies of each gene, and gametes, which bear only a single copy of each gene.

The Punnett square

Figure 2.10 shows a simple way of visualizing the results of the segregation and random union of alleles during gamete formation and fertilization. Mendel invented a system of symbols that allowed him to analyze all his crosses in the same way. He designated dominant alleles with a capital

Genes determine traits as disparate as pea shape and the inherited human disease cystic fibrosis. We now know that genes encode the proteins that cells produce and depend on for structure and function. As early as 1940, investigators had uncovered evidence suggesting that some genes determine the formation of enzymes, the proteins that catalyze specific chemical reactions. But it was not until 1991, 126 years after Mendel published his work, that a team of British geneticists was able to identify the gene for pea shape and to pinpoint how the enzyme it specifies influences a seed's round or wrinkled contour. At about the same time, medical researchers in Canada and the United States identified the cystic fibrosis gene. They discovered how a mutant allele causes unusually sticky mucus secretion and a susceptibility to respiratory infections and digestive malfunction, once again, through the protein the gene determines.

The pea shape gene encodes an enzyme known as SBE1 (for <u>s</u>tarch-<u>b</u>ranching <u>e</u>nzyme 1), which catalyzes the conversion of amylose, an unbranched linear molecule of starch, to amylopectin, a starch molecule composed of several branching chains (**Figure A**). The dominant allele of the pea shape gene (denoted by a capital *R*) causes the formation of active SBE1 enzyme that functions normally. As a result, *RR* homozygotes produce a high proportion of branched starch molecules, which allow the peas to maintain a rounded shape. In contrast, the enzyme determined by the recessive allele of the pea shape gene (denoted by a lowercase *r*) is abnormal and does not function effectively. In homozygous recessive *rr* peas, sucrose builds up because less of it is converted into starch. The excess sucrose modifies osmotic pressure, causing water to enter the young seeds. As the seeds mature, they lose water, shrink, and wrinkle. The single dominant allele in *Rr* heterozygotes apparently produces enough of the normal enzyme to prevent wrinkling. In summary, a specific gene determines a specific enzyme whose activity affects pea shape.

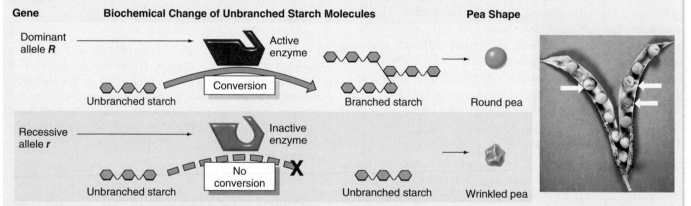

Figure A **Round and wrinkled peas: How one gene determines an enzyme that affects pea shape.** The *R* allele of the pea shape gene directs the synthesis of an enzyme that converts unbranched starch to branched starch, indirectly leading to round pea shape. The *r* allele of this gene determines an inactive form of the enzyme, leading to a buildup of linear, unbranched starch that ultimately causes seed wrinkling. The photograph (*at right*) shows two pea pods, each of which contains wrinkled (*arrows*) and round peas; the ratio of round to wrinkled in these two well-chosen pods is 9:3 (or 3:1).

The human disease of cystic fibrosis was first described in 1938, but doctors and scientists did not understand the biochemical mechanism that produced the serious respiratory and digestive malfunctions associated with the disease. As a result, treatments could do little more than relieve some of the symptoms, and most cystic fibrosis sufferers died before the age of 30.

In 1989, molecular geneticists Lap-Chee Tsui and John Riordan, working at the Hospital for Sick Children in Toronto, and Francis Collins, a researcher at the University of Michigan, Ann Arbor in the United States (see the **Fast Forward** box **"Mapping the Cystic Fibrosis Gene"** in Chapter 4), found that the normal allele of the cystic fibrosis gene determines a protein that forges a channel through the cell membrane (**Figure B**). This protein, called the <u>c</u>ystic <u>f</u>ibrosis <u>t</u>ransmembrane conductance <u>r</u>egulator (CFTR), controls the flow of chloride ions into and out of the cell. The normal allele of this gene

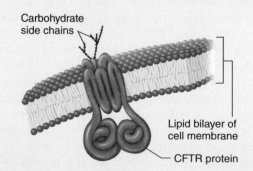

Figure B **The cystic fibrosis gene encodes a cell membrane protein.** A model of the normal CFTR protein that regulates the passage of chloride ions through the cell membrane. A small change in the gene that codes for CFTR results in an altered protein that prevents proper flow of chloride ions, leading to the varied symptoms of cystic fibrosis.

produces a CFTR protein that correctly regulates the back-and-forth exchange of ions, which, in turn, determines the cell's osmotic pressure and the flow of water through the cell membrane. In people with cystic fibrosis, however, the two recessive alleles produce only an abnormal form of the CFTR protein. The abnormal protein cannot be inserted into the cell membranes, so patients lack functional CFTR chloride channels. The cells thus retain water, and a thick, dehydrated mucus builds up outside the cells. In cells lining the airways and the ducts of secretory organs, such as the pancreas, this single biochemical defect produces clogging and blockages that result in respiratory and digestive malfunction.

Identification of the cystic fibrosis gene brought not only a protein-based explanation of the disease symptoms but also the promise of a cure. In the early 1990s, medical researchers placed the normal allele of the gene into respiratory tissue of mice with the disease. These mice could then produce a functional CFTR protein. Such encouraging results in these small mammals suggested that in the not-too-distant future, gene therapy might bestow relatively normal health on people suffering from this once life-threatening genetic disorder. Unfortunately, human trials of CFTR gene therapy have not yet achieved clear success.

A, *B*, or *C* and recessive ones with a lowercase *a*, *b*, or *c*. Modern geneticists have adopted this convention for naming genes in peas and many other organisms, but they often choose a symbol with some reference to the trait in question—a *Y* for yellow or an *R* for round. Throughout this book, we present gene symbols in italics, a common scientific writing practice (see the *Guidelines for Gene Nomenclature*, directly following Chapter 24). In Figure 2.10, we denote the dominant yellow allele by a capital *Y* and the recessive green allele by a lowercase *y*. The pure-breeding plants of the parental generation are either *YY* (yellow peas) or *yy* (green peas). The *YY* parent can produce only *Y* gametes, the *yy* parent only *y* gametes. You can see from the diagram why every cross between *YY* and *yy* produces exactly the same result—a *Yy* hybrid—no matter which parent (male or female) contributes which particular allele.

Next, to visualize what happens when the *Yy* hybrids self-fertilize, we set up a Punnett square (named after British mathematician Reginald Punnett, who introduced it in 1906; Figure 2.10). The square provides a simple and convenient method for tracking the kinds of gametes produced as well as all the possible combinations that might occur at fertilization. As the Punnett square shows, each hybrid produces two kinds of gametes, *Y* and *y*, in a ratio of 1:1. Thus, half the pollen and half the eggs carry *Y*, the other half *y*. At fertilization, 1/4 of the progeny will be *YY*, 1/4 *Yy*, 1/4 *yY*, and 1/4 *yy*. Since the gametic source of an allele (egg or pollen) for the traits Mendel studied had no influence on the allele's effect, *Yy* and *yY* are equivalent. This means that 1/2 of the progeny are yellow *Yy* hybrids, 1/4 *YY* true-breeding yellows, and 1/4 true-breeding *yy* greens. The diagram illustrates how the segregation of alleles during gamete formation and the random union of egg and pollen at fertilization can produce the 3:1 ratio of yellow to green that Mendel observed in the F$_2$ generation.

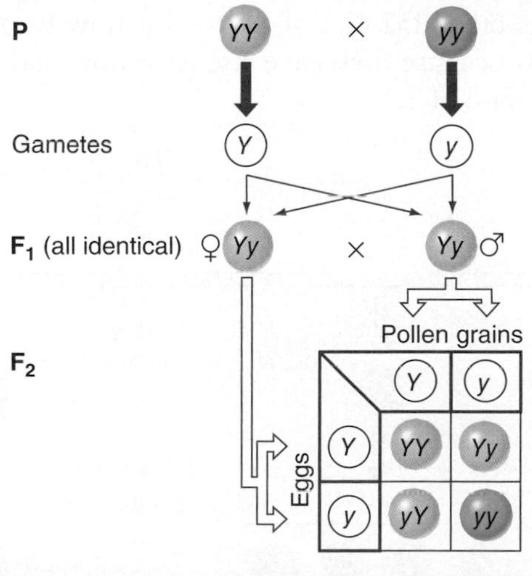

Mendel's law of segregation states that alleles of genes separate during gamete formation and then come together randomly at fertilization. The Punnett square is one tool for analyzing allele behaviour in a cross.

Mendel's results reflect basic rules of probability

Though you may not have realized it, the Punnett square illustrates two simple rules of probability—the product rule and the sum rule—that are central to the analysis of genetic crosses. These rules predict the likelihood that a particular combination of events will occur.

The product rule

The **product rule** states that the probability of two or more *independent events* occurring together is the *product* of the probabilities that each event will occur by itself. With independent events:

Probability of event 1 *and* event 2 =

Probability of event 1 × probability of event 2

Figure 2.10 The Punnett square: Visual summary of a cross. This Punnett square illustrates the combinations that can arise when an F$_1$ hybrid undergoes gamete formation and self-fertilization. The F$_2$ generation should have a 3:1 ratio of yellow to green peas.

Consecutive coin tosses are obviously independent events; a heads in one toss neither increases nor decreases the probability of a heads in the next toss. If you toss two coins at the same time, the results are also independent events. A heads for one coin neither increases nor decreases the probability of a heads for the other coin. Thus, the probability of a given combination is the product of their independent probabilities. For example, the probability that both coins will turn up heads is

$$1/2 \times 1/2 = 1/4$$

Similarly, the formation of egg and pollen are independent events; in a hybrid plant, the probability is 1/2 that a given gamete will carry Y and 1/2 that it will carry y. Because fertilization happens at random, the probability that a particular combination of maternal and paternal alleles will occur simultaneously in the same zygote is the product of the independent probabilities of these alleles being packaged in egg and sperm. Thus, to find the chance of a Y egg (formed as the result of one event) uniting with a Y sperm (the result of an independent event), you simply multiply $1/2 \times 1/2$ to get 1/4. This is the same fraction of YY progeny seen in the Punnett square of Figure 2.10, which demonstrates that the Punnett square is simply another way of depicting the product rule.

The sum rule

While we can describe the moment of random fertilization as the simultaneous occurrence of two independent events, we can also say that two different fertilization events are mutually exclusive. For instance, if Y combines with Y, it cannot also combine with y in the same zygote. A second rule of probability, the **sum rule**, states that the probability of either of two such *mutually exclusive events* occurring is the *sum* of their individual probabilities. With mutually exclusive events:

Probability of event 1 *or* event 2 =

Probability of event 1 + probability of event 2

To find the likelihood that an offspring of a Yy hybrid self-fertilization will be a hybrid like the parents, you add 1/4 (the probability of maternal Y uniting with paternal y) and 1/4 (the probability of the mutually exclusive event where paternal Y unites with maternal y) to get 1/2, again the same result as in the Punnett square. In another use of the sum rule, you could predict the ratio of yellow to green F_2 progeny. The fraction of F_2 peas that will be yellow is the sum of 1/4 (the event producing YY) plus 1/4 (the mutually exclusive event generating Yy) plus 1/4 (the third mutually exclusive event producing yY) to get 3/4. The remaining 1/4 of the F_2 progeny will be green. So the yellow-to-green ratio is 3/4:1/4, or more simply, 3:1.

In the analysis of a genetic cross, the product rule multiplies probabilities to predict the chance of a particular fertilization event. The sum rule adds probabilities to predict the proportion of progeny that share a particular trait such as pea colour.

Further crosses verify the law of segregation

Although Mendel's law of segregation explains the data from his pea crosses, he performed additional experiments to confirm its validity. In the rigorous check of his hypothesis illustrated in **Figure 2.11**, he allowed self-fertilization of all the plants in the F_2 generation and counted the types of F_3 progeny. Mendel found that the plants that developed from F_2 green peas all produced only F_3 green peas, and when the resulting F_3 plants self-fertilized, the next generation also produced green peas (not shown). This is what we (and Mendel) would expect of pure-breeding lines carrying two copies of the recessive allele. The yellow peas were a different story. When Mendel allowed 518 F_2 plants that developed from yellow peas to self-fertilize, he observed that 166, roughly 1/3 of the total, were pure-breeding yellow through several generations, but the other 352 (2/3 of the total yellow F_2 plants) were hybrids because they gave rise to yellow and green F_3 peas in a ratio of 3:1.

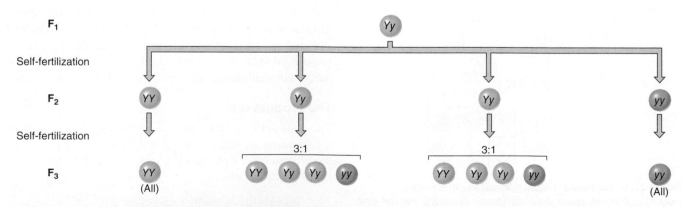

Figure 2.11 Yellow F_2 peas are of two types: pure breeding and hybrid. The distribution of a pair of contrasting alleles (Y and y) after two generations of self-fertilization. The homozygous individuals of each generation breed true, whereas the hybrids do not.

It took Mendel years to conduct such rigorous experiments on seven pairs of pea traits, but in the end, he was able to conclude that the segregation of dominant and recessive alleles during gamete formation and their random union at fertilization could indeed explain the 3:1 ratios he observed whenever he allowed hybrids to self-fertilize. His results, however, raised yet another question, one of some importance to future plant and animal breeders. Plants showing a dominant trait, such as yellow peas, can be either pure-breeding (*YY*) or hybrid (*Yy*). How can you distinguish one from the other? For self-fertilizing plants, the answer is to observe the appearance of the next generation. But how would you distinguish pure-breeding from hybrid individuals in species that do not self-fertilize?

Testcrosses: A way to establish genotype

Before describing Mendel's answer, we need to define a few more terms. An observable characteristic, such as yellow or green pea seeds, is a **phenotype**, while the actual pair of alleles present in an individual is its **genotype**. A *YY* or a *yy* genotype is called **homozygous**, because the two copies of the gene that determine the particular trait in question are the same. In contrast, a genotype with two different alleles for a trait is **heterozygous**; in other words, it is a hybrid for that trait (**Figure 2.12**). An individual with a homozygous genotype is a **homozygote**; one with a heterozygous genotype is a **heterozygote**. Note that the phenotype of a heterozygote (i.e., of a hybrid) defines which allele is dominant: Because *Yy* peas are yellow, the yellow allele *Y* is dominant to the *y* allele for green. If you know the genotype and the dominance relation of the alleles, you can accurately predict the phenotype. The reverse is not true, however, because some phenotypes can derive from more than one genotype. For example, the phenotype of yellow peas can result from either the *YY* or the *Yy* genotype.

With these distinctions in mind, we can look at the method Mendel devised for deciphering the unknown genotype (we will call it *Y–*) responsible for a dominant phenotype; the dash represents the unknown second allele,

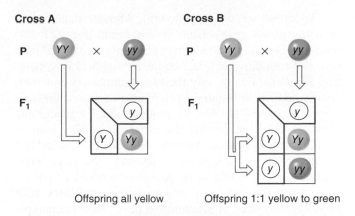

Cross A   Cross B

Offspring all yellow   Offspring 1:1 yellow to green

Figure 2.13 How a testcross reveals genotype. An individual of unknown genotype, but dominant phenotype, is crossed with a homozygous recessive. If the unknown genotype is homozygous, all progeny will exhibit the dominant phenotype (*cross A*). If the unknown genotype is heterozygous, half the progeny will exhibit the dominant trait and the other half the recessive trait (*cross B*).

either *Y* or *y*. This method, called the **testcross**, is a mating in which an individual showing the dominant phenotype, for instance, a *Y–* plant grown from a yellow pea, is crossed with an individual expressing the recessive phenotype, in this case a *yy* plant grown from a green pea. As the Punnett squares in **Figure 2.13** illustrate, if the dominant phenotype in question derives from a homozygous *YY* genotype, all the offspring of the testcross will show the dominant yellow phenotype. But if the dominant parent of unknown genotype is a heterozygous hybrid (*Yy*), 1/2 of the progeny are expected to be yellow peas, and the other half green. In this way, the testcross establishes the genotype behind a dominant phenotype, resolving any uncertainty.

As we mentioned earlier, Mendel deliberately simplified the problem of heredity, focusing on traits that come in only two forms. He was able to replicate his basic monohybrid findings with corn, beans, and four-o'clocks (plants with tubular, white or bright red flowers). As it turns out, his concept of the gene and his law of segregation can be generalized to almost all sexually reproducing organisms.

The results of a testcross, in which an individual showing the dominant phenotype is crossed with an individual showing the recessive phenotype, indicate whether the individual with the dominant phenotype is a homozygote or a heterozygote.

Dihybrid crosses reveal the law of independent assortment

Having determined from monohybrid crosses that genes are inherited according to the law of segregation, Mendel turned his attention to the simultaneous inheritance of two or more apparently unrelated traits in peas. He asked how two pairs of alleles would segregate in a **dihybrid** individual; that is, in a plant that is heterozygous for two genes at the same time.

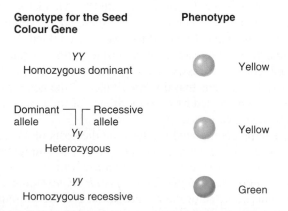

Genotype for the Seed Colour Gene	Phenotype
YY Homozygous dominant	Yellow
Dominant allele ⌐ ⌐ Recessive allele *Yy* Heterozygous	Yellow
yy Homozygous recessive	Green

Figure 2.12 Genotype versus phenotype in homozygotes and heterozygotes. The relationship between genotype and phenotype with a pair of contrasting alleles where one allele (*Y*) shows complete dominance over the other (*y*).

To construct such a dihybrid, Mendel mated true-breeding plants grown from yellow round peas (*YY RR*) with true-breeding plants grown from green wrinkled peas (*yy rr*). From this cross, he obtained a dihybrid F_1 generation (*Yy Rr*) showing only the two dominant phenotypes, yellow and round (**Figure 2.14**). He then allowed these F_1 dihybrids to self-fertilize to produce the F_2 generation. Mendel could not predict the outcome of this mating. Would all the F_2 progeny be **parental types** that looked like either the original yellow round parent or the green wrinkled parent? Or would some new combinations of phenotypes occur that were not seen in the parental lines, such as yellow wrinkled or green round peas? New phenotypic combinations like these are called **recombinant types**.

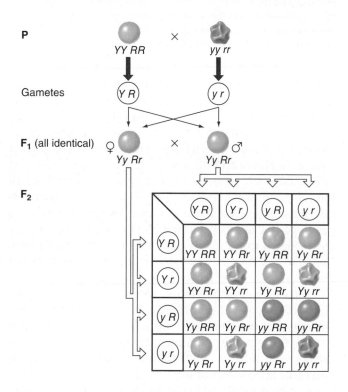

Type	Genotype	Phenotype		Number	Phenotypic Ratio
Parental	*Y– R–*		yellow round	315	9/16
Recombinant	*yy R–*		green round	108	3/16
Recombinant	*Y– rr*		yellow wrinkled	101	3/16
Parental	*yy rr*		green wrinkled	32	1/16

Ratio of yellow (dominant) to green (recessive)	= 12:4 or 3:1
Ratio of round (dominant) to wrinkled (recessive)	= 12:4 or 3:1

Figure 2.14 A dihybrid cross produces parental types and recombinant types. In this dihybrid cross, pure-breeding parents (P) produce a genetically uniform generation of F_1 dihybrids. Self-pollination or cross-pollination of the F_1 plants yields the characteristic F_2 phenotypic ratio of 9:3:3:1.

When Mendel counted the F_2 generation of one experiment, he found 315 yellow round, 101 yellow wrinkled, 108 green round, and 32 green wrinkled peas. There were, in fact, yellow wrinkled and green round recombinant phenotypes, providing evidence that some shuffling of the alleles of different genes had taken place.

The law of independent assortment

From the observed ratios, Mendel inferred the biological mechanism of that shuffling—the **independent assortment** of gene pairs during gamete formation. Because the genes for pea colour and for pea shape assort independently, the allele for pea shape in a *Y* carrying gamete could with equal likelihood be either *R* or *r*. Thus, the presence of a particular allele of one gene, say, the dominant *Y* for pea colour, provides no information whatsoever about the allele of the second gene. Each dihybrid of the F_1 generation can therefore make four kinds of gametes: *Y R*, *Y r*, *y R*, and *y r*. In a large number of gametes, the four kinds will appear in an almost perfect ratio of 1:1:1:1, or put another way, roughly 1/4 of the eggs and 1/4 of the pollen will contain each of the four possible combinations of alleles. That "the different kinds of germinal cells [eggs or pollen] of a hybrid are produced on the average in equal numbers" was yet another one of Mendel's incisive insights.

At fertilization then, in a mating of dihybrids, four different kinds of eggs can combine with any one of four different kinds of pollen, producing a total of 16 possible zygotes. Once again, a Punnett square is a convenient way to visualize the process. If you look at the square in Figure 2.14, you will see that some of the 16 potential allelic combinations are identical. In fact, there are only nine different genotypes—*YY RR*, *YY Rr*, *Yy RR*, *Yy Rr*, *yy RR*, *yy Rr*, *YY rr*, *Yy rr*, and *yy rr*—because the source of the alleles (egg or pollen) does not make any difference. If you look at the combinations of traits determined by the nine genotypes, you will see only four phenotypes—yellow round, yellow wrinkled, green round, and green wrinkled—in a ratio of 9:3:3:1. If, however, you look at just pea colour or just pea shape, you can see that each trait is inherited in the 3:1 ratio predicted by Mendel's law of segregation. In the Punnett square, there are 12 yellow for every 4 green and 12 round for every 4 wrinkled. In other words, the ratio of each dominant trait (yellow or round) to its antagonistic recessive trait (green or wrinkled) is 12:4, or 3:1. This means that the inheritance of the gene for pea colour is unaffected by the inheritance of the gene for pea shape, and vice versa.

The preceding analysis became the basis of Mendel's second general genetic principle, the **law of independent assortment**: *During gamete formation, different pairs of alleles segregate independently of each other* (**Figure 2.15**). The independence of their segregation and the subsequent random union of gametes at fertilization determine the phenotypes observed. Using the product rule for assessing the probability of independent events, you can see mathematically how the 9:3:3:1 phenotypic ratio observed in a

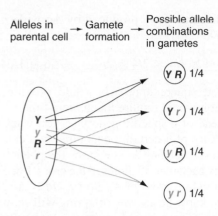

Figure 2.15 **The law of independent assortment.** In a dihybrid cross, each pair of alleles assorts independently during gamete formation. In the gametes, *Y* is equally likely to be found with *R* or *r* (i.e., *Y R* = *Y r*); the same is true for *y* (i.e., *y R* = *y r*). As a result, all four possible types of gametes (*Y R*, *Y r*, *y R*, and *y r*) are produced in equal frequency among a large population.

dihybrid cross derives from two separate 3:1 phenotypic ratios. If the two sets of alleles assort independently, the yellow-to-green ratio in the F_2 generation will be 3/4:1/4, and likewise the round-to-wrinkled ratio will be 3/4:1/4. To find the probability that two independent events such as yellow and round will occur simultaneously in the same plant, you multiply as follows:

Probability of yellow round = 3/4 × 3/4 = 9/16

Probability of yellow wrinkled = 3/4 × 1/4 = 3/16

Probability of green round = 1/4 × 3/4 = 3/16

Probability of green wrinkled = 1/4 × 1/4 = 1/16

Thus, in a population of F_2 plants, there will be a 9:3:3:1 phenotypic ratio of yellow round to yellow wrinkled to green round to green wrinkled.

Branched-line diagrams

A convenient way to keep track of the probabilities of each potential outcome in a genetic cross is to construct a **branched-line diagram** (**Figure 2.16**), which shows all the possibilities for each gene in a sequence of columns. In Figure 2.16, the first column shows the two possible pea colour phenotypes; and the second column demonstrates that each pea colour can occur with either of two pea shapes. Again, the 9:3:3:1 ratio of phenotypes is apparent.

Testcrosses with dihybrids

An understanding of dihybrid crosses has many applications. Suppose, for example, that you work for a wholesale nursery, and your assignment is to grow pure-breeding plants guaranteed to produce yellow round peas. How would you proceed? One answer would be to plant the peas produced from a dihybrid cross that have the

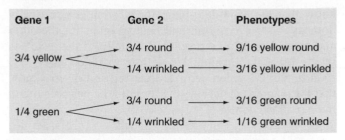

Figure 2.16 **Following crosses with branched-line diagrams.** A branched-line diagram, which uses a series of columns to track every gene in a cross, provides an organized overview of all possible outcomes. This branched-line diagram of a dihybrid cross generates the same phenotypic ratios as the Punnett square in **Figure 2.14**, showing that the two methods are equivalent.

desired yellow round phenotype. Only one out of nine of such progeny—those grown from peas with a *YY RR* genotype—will be appropriate for your uses. To find these plants, you could subject each yellow round candidate to a testcross for genotype with a green wrinkled (*yy rr*) plant, as illustrated in **Figure 2.17**. If the testcross yields all yellow round offspring (testcross A), you can sell your test plant, because you know it is homozygous for both pea colour and pea shape. If your testcross yields 1/2 yellow round and 1/2 yellow wrinkled (testcross B), or 1/2 yellow round and 1/2 green round (testcross C), you

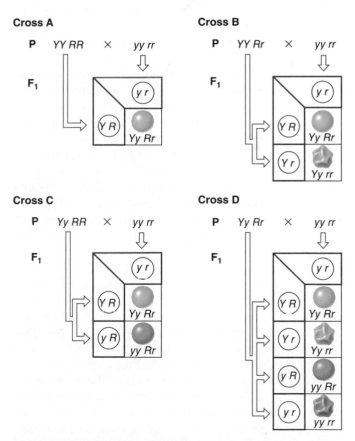

Figure 2.17 **Testcrosses on dihybrids.** Testcrosses involving two pairs of independently assorting alleles yield different, predictable results depending on the tested individual's genotype for the two genes in question.

know that the candidate plant in question is genetically homozygous for one trait and heterozygous for the other and must therefore be discarded. Finally, if the testcross yields 1/4 yellow round, 1/4 yellow wrinkled, 1/4 green round, and 1/4 green wrinkled (testcross D), you know that the plant is a heterozygote for both the pea colour and the pea shape genes.

> The law of independent assortment states that the alleles of genes for different traits segregate independently of each other during gamete formation.

Geneticists use Mendel's laws to calculate probabilities and make predictions

Mendel performed several sets of dihybrid crosses and also carried out **multihybrid crosses**—matings between the F_1 progeny of true-breeding parents that differed in three or more unrelated traits. In all of these experiments, he observed numbers and ratios very close to what he expected on the basis of his two general biological principles: the alleles of a gene segregate during the formation of egg or pollen, and the alleles of different genes assort independently of each other. Mendel's laws of inheritance, in conjunction with the mathematical rules of probability, provide geneticists with powerful tools for predicting and interpreting the results of genetic crosses. But as with all tools, they have their limitations. We examine here both the power and the limitations of Mendelian analysis.

First, the power: Using simple Mendelian analysis, it is possible to make accurate predictions about the offspring of extremely complex crosses. Suppose you want to predict the occurrence of one specific genotype in a cross involving several independently assorting genes. For example, if hybrids that are heterozygous for four traits (or genes) are allowed to self-fertilize—*Aa Bb Cc Dd × Aa Bb Cc Dd*—what proportion of their progeny will have the genotype *AA bb Cc Dd*? You could set up a Punnett square to answer the question. Because for each trait (or gene) there are two different alleles, the number of different eggs or sperm is found by raising 2 to the power of the number of differing traits (2^n, where n is the number of traits). By this calculation, each hybrid parent in this cross with four traits would make $2^4 = 16$ different kinds of gametes. The Punnett square depicting such a cross would thus contain 256 boxes (16 × 16). This may be fine if you live in a monastery with a bit of time on your hands, but not if you are taking a one-hour exam. It would be much simpler to analyze the problem by breaking down the multihybrid cross into four independently assorting monohybrid crosses. Remember that the genotypic ratios of each monohybrid cross are one homozygote for the dominant allele, to two heterozygotes, to one homozygote for the recessive allele = 1/4:2/4:1/4. Thus, you can find the probability of *AA bb Cc Dd* by multiplying the probability of each independent event:

AA (1/4 of the progeny produced by *Aa × Aa*); *bb* (1/4); *Cc* (2/4); *Dd* (2/4):

$$1/4 \times 1/4 \times 2/4 \times 2/4 = 4/256 = 1/64$$

The Punnett square approach would provide the same answer, but it would require much more time.

If instead of a specific genotype, you want to predict the probability of a certain phenotype, you can again use the product rule as long as you know the phenotypic ratios produced by each pair of alleles in the cross. For example, if in the multihybrid cross of *Aa Bb Cc Dd × Aa Bb Cc Dd*, you want to know how many offspring will show the dominant A trait (genotype *AA* or *Aa* = 1/4 + 2/4, or 3/4), the recessive b trait (genotype *bb* = 1/4), the dominant C trait (genotype *CC* or *Cc* = 3/4), and the dominant D trait (genotype *DD* or *Dd* = 3/4), you simply multiply

$$3/4 \times 1/4 \times 3/4 \times 3/4 = 27/256$$

In this way, the rules of probability make it possible to predict the outcome of very complex crosses.

You can see from these examples that particular problems in genetics are amenable to particular modes of analysis. As a rule of thumb, Punnett squares are excellent for visualizing simple crosses involving a few genes, but they become unwieldy in the dissection of more complicated matings. Direct calculations of probabilities, such as those in the two preceding problems, are useful when you want to know the chances of one or a few outcomes of complex crosses. If, however, you want to know all the outcomes of a multihybrid cross, a branched-line diagram is the best way to go as it will keep track of the possibilities in an organized fashion.

Now, the limitations of Mendelian analysis: Like Mendel, if you were to breed pea plants or corn or any other organism, you would most likely observe some deviation from the ratios you expected in each generation. What can account for such variation? One element is chance, as witnessed in the common coin toss experiment. With each throw, the probability of the coin coming up heads is equal to the likelihood it will come up tails. But if you toss a coin ten times, you may get 30 percent (3) heads and 70 percent (7) tails, or vice versa. If you toss it 100 times, you are more likely to get a result closer to the expected 50 percent heads and 50 percent tails. The larger the number of trials, the lower the probability that chance significantly skews the data. This is one reason Mendel worked with large numbers of pea plants. Mendel's laws, in fact, have great predictive power for populations of organisms, but they do not tell us what will happen in any one individual. With a garden full of self-fertilizing monohybrid pea plants, for example, you can expect that 3/4 of the F_2 progeny will show the dominant phenotype and 1/4 the recessive, but you cannot predict the phenotype of any particular F_2 plant. In Chapter 4, we discuss mathematical methods for assessing whether the chance variation observed in a sample of individuals within a population is compatible with a genetic hypothesis.

(a) Gregor Mendel

(b) Carl Correns

(c) Hugo de Vries

(d) Erich von Tschermak

Figure 2.18 The science of genetics begins with the rediscovery of Mendel. Working independently near the beginning of the twentieth century, Correns, de Vries, and von Tschermak each came to the same conclusions as those Mendel had summarized in his laws.

> Branched-line diagrams or direct calculations of probabilities are often more efficient methods than Punnett squares for the analysis of genetic crosses involving two or more genes.

Mendel's work was unappreciated before 1900

Mendel's insights into the workings of heredity were a breakthrough of monumental proportions. By counting and analyzing data from hundreds of pea plant crosses, he inferred the existence of genes—independent units that determine the observable patterns of inheritance for particular traits. His work explained the reappearance of "hidden" traits, disproved the idea of blended inheritance, and showed that mother and father make an equal genetic contribution to the next generation. The model of heredity that he formulated was so specific that he could test predictions based on it by observation and experiment.

With the exception of Abbot Napp, none of Mendel's contemporaries appreciated the importance of his research. Mendel did not teach at a prestigious university and was not well known outside Brünn. Even in Brünn, members of the Natural Science Society were disappointed when he presented "Experiments on Plant Hybrids" to them. They wanted to view and discuss intriguing mutants and lovely flowers, so they did not appreciate his numerical analyses. Mendel, it seems, was far ahead of his time. Sadly, despite written requests from Mendel that others try to replicate his studies, no one repeated his experiments. Several citations of his paper between 1866 and 1900 referred to his expertise as a plant breeder but made no mention of his laws. Moreover, at the time Mendel presented his work, no one had yet seen the structures within cells, the *chromosomes*, that actually carry the genes. That would happen only in the next few decades (as described in Chapter 3). If scientists had been able to see these structures, they might have more readily accepted Mendel's ideas, because the chromosomes are actual physical structures that behave exactly as Mendel predicted.

Mendel's work might have had an important influence on early debates about evolution if it had been more widely appreciated. Charles Darwin (1809–1882), who was unfamiliar with Mendel's work, was plagued in his later years by criticism that his explanations for the persistence of variation in organisms were insufficient. Darwin considered such variation a cornerstone of his theory of evolution, maintaining that natural selection would favour particular variants in a given population in a given environment. If the selected combinations of variant traits were passed on to subsequent generations, this transmission of variation would propel evolution. He could not, however, say how that transmission might occur. Had Darwin been aware of Mendel's ideas, he might not have been backed into such an uncomfortable corner.

For 34 years, Mendel's laws lay dormant—untested, unconfirmed, and unapplied. Then in 1900, 16 years after Mendel's death, Carl Correns, Hugo de Vries, and Erich von Tschermak independently rediscovered and acknowledged his work (**Figure 2.18**). The scientific community had finally caught up with Mendel. Within a decade, investigators had coined many of the modern terms we have been using: phenotype, genotype, homozygote, heterozygote, gene, and genetics, the label given to the twentieth-century science of heredity. Mendel's paper provided the new discipline's foundation. His principles and analytical techniques endure today, guiding geneticists and evolutionary biologists in their studies of genetic variation.

| 2.3 | **Mendelian Inheritance in Humans** |

Although many human traits clearly run in families, most do not show a simple Mendelian pattern of inheritance. Suppose, for example, that you have brown eyes, but both your parents' eyes appear to be blue. Because blue is normally considered recessive to brown, does this mean that you are adopted or that your father is not really your father? Not necessarily, because eye colour is influenced by more than one gene.

Like eye colour, most common and obvious human phenotypes arise from the interaction of many genes. In contrast, single-gene traits in people usually involve an abnormality that is disabling or life-threatening. Examples are the progressive mental deterioration and other neurological damage of Huntington disease and the clogged lungs and potential respiratory failure of cystic fibrosis. A defective allele of a single gene gives rise to Huntington disease; defective alleles of a different gene are responsible for cystic fibrosis. There were roughly 10 000 such single-gene traits known in humans in 2013, and the number continues to grow as new studies confirm the genetic basis of more traits. **Table 2.1** lists some of the most common single-gene traits in humans.

Pedigrees aid the study of hereditary traits in human families

Determining a genetic defect's pattern of transmission is not always an easy task because people make slippery genetic subjects. Their generation time is long, and the families they produce are relatively small, which makes statistical analysis difficult. They do not base their choice of mates on purely genetic considerations. There are thus no pure-breeding lines and no controlled matings. And there is rarely a true F_2 generation (like the one in which Mendel observed the 3:1 ratios from which he derived his rules) because brothers and sisters almost never mate.

Geneticists circumvent these difficulties by working with a large number of families or with several generations of a very large family. This allows them to study the large numbers of genetically related individuals needed to establish the inheritance patterns of specific traits. A family history, known as a **pedigree**, is an orderly diagram of a family's relevant genetic features, extending back to at least both sets of grandparents and preferably through as many more generations as possible. From systematic pedigree analysis in the light of Mendel's laws, geneticists can tell if a trait is determined by alternative alleles of a single gene and whether a single-gene trait is dominant or recessive. Because Mendel's principles are so simple and straightforward, a little logic can go a long way in explaining how traits are inherited in humans.

Figure 2.19 shows how to interpret a family pedigree diagram. Squares (□) represent males, circles (○) are females, and diamonds (◇) indicate that the sex is unspecified; family members affected by the trait in question are indicated by a filled-in symbol (e.g., ■). A single horizontal line connecting a male and a female (□—○) represents a mating, a double connecting line (□=○) designates a **consanguineous mating** (i.e., a mating between relatives), and a horizontal line above a series of symbols (○ □ ○) indicates the children of the same parents (a *sibship*) arranged and numbered from left to right in order of their birth. Roman numerals to the left or right of the diagram indicate the generations.

To reach a conclusion about the mode of inheritance of a family trait, human geneticists must use a pedigree that supplies sufficient information. For example, they could not determine whether the allele causing the disease depicted at the bottom of Figure 2.19 is dominant or recessive solely on the basis of the simple pedigree shown. The data are consistent with both possibilities. If the trait is dominant, then the father and the affected son are heterozygotes, while the mother and the unaffected son are homozygotes for the recessive normal allele. If instead the trait is recessive, then the father and affected son are homozygotes for the recessive disease-causing allele, while the mother and the unaffected son are heterozygotes.

Table 2.1	Some of the Most Common Single-Gene Traits in Humans	
Disease	**Effect**	**Incidence of Disease**
Caused by a Recessive Allele		
Thalassemia (chromosome 16 or 11)	Reduced amounts of haemoglobin; anaemia; bone and spleen enlargement	1/10 in parts of Italy
Sickle-cell anaemia (chromosome 11)	Abnormal haemoglobin; sickle-shaped red cells, anaemia, blocked circulation; increased resistance to malaria	1/625 African-Americans
Cystic fibrosis (chromosome 7)	Defective cell membrane protein; excessive mucus production; digestive and respiratory failure	1/2000 Caucasians
Tay-Sachs disease (chromosome 15)	Missing enzyme; buildup of fatty deposit in brain; buildup disrupts mental development	1/3000 Eastern European Jews
Phenylketonuria (PKU) (chromosome 12)	Missing enzyme; mental deficiency	1/10 000 Caucasians
Caused by a Dominant Allele		
Hypercholesterolaemia (chromosome 19)	Missing protein that removes cholesterol from the blood; heart attack by age 50	1/122 French Canadians
Huntington disease (chromosome 4)	Progressive mental and neurological damage; neurological disorders by ages 40–70	1/25 000 Caucasians

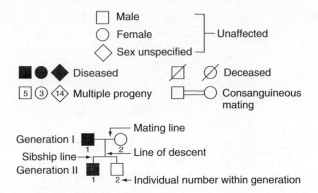

Figure 2.19 Symbols used in pedigree analysis. In the simple pedigree at the bottom, I-1 is the father, I-2 is the mother, and II-1 and II-2 are their sons. The father and the first son are both affected by the disease trait.

Several kinds of additional information could help resolve this uncertainty. Human geneticists would particularly want to know the frequency at which the trait in question is found in the population from which the family came. If the trait is rare in the population, then the allele giving rise to the trait should also be rare, and the most likely hypothesis would require that the fewest genetically unrelated people carry the allele. Only the father in Figure 2.19 would need to have a dominant disease-causing allele, but both parents would need to carry a recessive disease-causing allele (the father two copies and the mother one). However, even the information that the trait is rare does not allow us to draw the firm conclusion that it is inherited in a dominant fashion. The pedigree in the figure is so limited that we cannot be sure the two parents are themselves unrelated. As we discuss later in more detail, related parents might have both received a rare recessive allele from their common ancestor. This example illustrates why human geneticists try to collect family histories that cover several generations.

We now look at more extensive pedigrees for the dominant trait of Huntington disease and for the recessive condition of cystic fibrosis. The patterns by which these traits appear in the pedigrees provide important clues that can indicate modes of inheritance and allow geneticists to assign genotypes to family members.

A vertical pattern of inheritance indicates a rare dominant trait

Huntington disease is named for George Huntington, the physician who first described its course. This illness usually shows up in middle age and slowly destroys its victims both mentally and physically. Symptoms include intellectual deterioration, severe depression, and jerky, irregular movements, all caused by the progressive death of nerve cells. If one parent develops the symptoms, his or her children have a 50 percent probability of suffering from the disease, provided they live to adulthood. Because symptoms are not present at birth and manifest only later in life, Huntington disease is known as a **late-onset** genetic condition.

How would you proceed in assigning genotypes to the individuals in the Huntington disease pedigree depicted in **Figure 2.20**? First, you would need to find out if the disease-producing allele is dominant or recessive. Several clues suggest that Huntington disease is transmitted by a dominant allele of a single gene. Everyone who develops the disease has at least one parent who shows the trait, and in several generations, approximately half of the offspring are affected. The pattern of affected individuals is thus vertical: If you trace back through the ancestors of any affected individual, you would see at least one affected person in each generation, giving a continuous line of family members with the disease. When a disease is rare in the population as a whole, a vertical pattern is strong evidence that a dominant allele causes the trait; the alternative would require that many unrelated people carry a rare recessive allele. (A recessive trait that is extremely common might also show up in every generation; we examine this possibility in Problem 34 at the end of this chapter.)

In tracking a dominant allele through a pedigree, you can view every mating between an affected and an unaffected partner as analogous to a testcross. If some of the offspring do not have Huntington disease, you know the parent showing the trait is a heterozygote. You can check your genotype assignments against the answers in the caption to Figure 2.20.

No effective treatment yet exists for Huntington disease, and because of its late onset, there was until the 1980s no way for children of a parent with the disease to know before middle age—usually until well after their own childbearing years—whether they carried the Huntington disease allele (*HD*). Children of parents with the disease have a 50 percent probability of inheriting *HD* and, before they are diagnosed, a 25 percent probability of passing the defective allele on to one of their children. In the mid-1980s, with new knowledge of the gene, molecular geneticists developed a DNA test that determines whether an individual carries the *HD* allele. Because of the lack

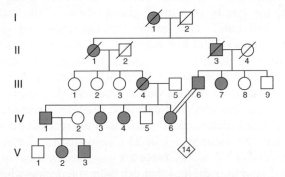

Figure 2.20 Huntington disease: A rare dominant trait. All individuals represented by filled-in symbols are heterozygotes (except I-1, who could have been homozygous for the dominant *HD* disease allele); all individuals represented by open symbols are homozygotes for the recessive *HD*⁺ normal allele. Among the 14 children of the consanguineous mating, DNA testing shows that some are *HD HD*, some are *HD HD*⁺, and some are *HD*⁺ *HD*⁺. The diamond designation masks personal details to protect confidentiality.

of effective treatment for the disease, some young adults whose parents died of Huntington disease prefer not to be tested so that they will not prematurely learn their own fate. However, other at-risk individuals employ the test for the *HD* allele to guide their decisions about having children. If someone whose parent had Huntington disease does not have *HD*, he or she has no chance of developing the disease or of transmitting it to offspring. If the test shows the presence of *HD*, the at-risk person and his or her partner might choose to conceive a child, obtain a prenatal diagnosis of the fetus, and then, depending on their beliefs, elect an abortion if the fetus is affected.

> If an individual is affected by a rare dominant trait, the trait should also affect at least one of that person's parents, one of that person's grandparents, and so on.

A horizontal pattern of inheritance indicates a rare recessive trait

Unlike Huntington disease, most confirmed single-gene traits in humans are recessive. This is because, with the exception of late-onset traits, deleterious dominant traits are unlikely to be transmitted to the next generation. For example, if people affected with Huntington disease died by the age of 10, the trait would disappear from the population. In contrast, individuals can carry one allele for a recessive trait without ever being affected by any symptoms. **Figure 2.21** shows three pedigrees for cystic fibrosis (CF), the most commonly inherited recessive disease among Caucasian children in Canada and the United States. A double dose of the recessive *CF* allele causes a fatal disorder in which the lungs, pancreas, and other organs become clogged with a thick, viscous mucus that can interfere with breathing and digestion. Approximately 1 in every 3500 Caucasian Canadians is born with cystic fibrosis, with the average age of survival in the early 30s.

There are two salient features of the CF pedigrees. First, the family pattern of people showing the trait is often horizontal: The parents, grandparents, and great-grandparents of children born with CF do not themselves manifest the disease, while several brothers and sisters in a single generation may. A horizontal pedigree pattern is a strong indication that the trait is recessive. The unaffected parents are heterozygous **carriers**: They bear a dominant normal allele that masks the effects of the recessive abnormal one. An estimated 1 in 25 Canadians are carriers of the recessive *CF* allele. **Table 2.2** summarizes some of the clues found in pedigrees that can help you decide whether a trait is caused by a dominant or a recessive allele.

The second salient feature of the CF pedigrees is that many of the couples who produce afflicted children are blood relatives; that is, their mating is consanguineous (as indicated by the double line). In Figure 2.21a, the consanguineous mating in generation V is between third cousins. Of course, children with cystic fibrosis can also have unrelated carrier parents, but because relatives share

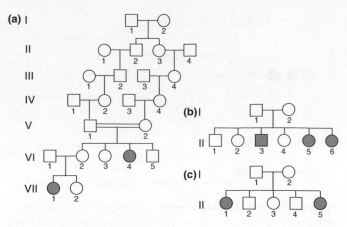

Figure 2.21 Cystic fibrosis: A recessive condition. In **(a)**, the two affected individuals (VI-4 and VII-1) are *CF CF*; that is, homozygotes for the recessive disease allele. Their unaffected parents must be carriers, so V-1, V-2, VI-1, and VI-2 must all be *CF CF⁺*. Individuals II-2, II-3, III-2, III-4, IV-2, and IV-4 are probably also carriers. We cannot determine which of the founders (I-1 or I-2) was a carrier, so we designate their genotypes as *CF⁺–*. Because the *CF* allele is relatively rare, it is likely that II-1, II-4, III-1, III-3, IV-1, and IV-3 are *CF⁺CF⁺* homozygotes. The genotype of the remaining unaffected people (VI-3, VI-5, and VII-2) is uncertain (*CF⁺–*). In **(b)** and **(c)**, these two families demonstrate horizontal patterns of inheritance. Without further information, the unaffected children in each pedigree must be regarded as having a *CF⁺–* genotype.

genes, their offspring have a much greater than average chance of receiving two copies of a rare allele. Whether or not they are related, carrier parents are both heterozygotes. Thus among their offspring, the proportion of unaffected to affected children is expected to be 3:1. To look at it another way, the chances are that one out of four children of two heterozygous carriers will be homozygous CF sufferers.

You can gauge your understanding of this inheritance pattern by assigning a genotype to each person in Figure 2.21 and then checking your answers against the caption. Note that for several individuals, such as the generation I individuals in part (a) of the figure, it is impossible

Table 2.2	How to Recognize Dominant and Recessive Traits in Pedigrees

Dominant Traits

1. Affected children always have at least one affected parent.

2. As a result, dominant traits show a *vertical pattern* of inheritance: The trait shows up in every generation.

3. Two affected parents can produce unaffected children, if both parents are heterozygotes.

Recessive Traits

1. Affected individuals can be the children of two unaffected carriers, particularly as a result of consanguineous matings.

2. All the children of two affected parents should be affected.

3. *Rare* recessive traits show a *horizontal pattern* of inheritance: The trait first appears among several members of one generation and is not seen in earlier generations.

4. Recessive traits may show a vertical pattern of inheritance if the trait is extremely common in the population.

to assign a full genotype. We know that one of these people must be the carrier who supplied the original *CF* allele, but we do not know if it was the male or the female. As with an ambiguous dominant phenotype in peas, the unknown second allele is indicated by a dash.

In Figure 2.21a, a mating between the unrelated carriers VI-1 and VI-2 produced a child with cystic fibrosis. How likely is such a marriage between unrelated carriers for a recessive genetic condition? The answer depends on the gene in question and the particular population into which a person is born. As Table 2.1 shows, the incidence of genetic diseases (and thus the frequency of their carriers) varies markedly among populations. Such variation reflects the distinct genetic histories of different groups. The area of genetics that analyzes differences among groups of individuals is called *population genetics*, a subject we cover in detail in Chapter 12. Notice that in Figure 2.21a, several unrelated, unaffected people, such

as II-1 and II-4, married into the family under consideration. Although it is highly probable that these individuals are homozygotes for the normal allele of the gene (CF^+ CF^+), there is a small chance (whose magnitude depends on the population) that any one of them could be a carrier of the disease.

Genetic researchers in Canada and the United States identified the cystic fibrosis gene in 1989, but they are still in the process of developing an effective gene therapy that would ameliorate the disease's debilitating symptoms (review the **Fast Forward** box **"Genes Encode Proteins"** in this chapter).

> If an individual is affected by a rare recessive trait, it is likely that none of that person's ancestors displayed the same trait. In many cases, the affected individual is the product of a consanguineous mating.

2.4 Extensions to Mendel for Single-Gene Inheritance

Unlike the pea traits that Mendel examined, most human characteristics do not fall neatly into just two opposing phenotypic categories. These complex traits, such as skin and hair colour, height, athletic ability, and many others, seem to defy Mendelian analysis. The same can be said of traits expressed by many of the world's food crops; their size, shape, succulence, and nutrient content vary over a wide range of values.

Lentils (*Lens culinaris*) provide a graphic illustration of this variation. Lentils, a type of legume, are grown in many parts of the world as a rich source of both protein and carbohydrate. The mature plants set fruit in the form of diminutive pods that contain two small seeds. These seeds can be ground into meal or used in soups, salads, and stews. Lentils come in an intriguing array of colours and patterns (**Figure 2.22**), and commercial growers always seek to produce combinations to suit the cuisines of different cultures. But crosses between pure-breeding lines of lentils result in some startling surprises. A cross between pure-breeding tan and pure-breeding grey parents, for

example, yields an all-brown F_1 generation. When these hybrids self-pollinate, the F_2 plants produce not only tan, grey, and brown lentils, but also green.

Beginning with the first decade of the twentieth century, geneticists subjected many kinds of plants and animals to controlled breeding tests, using Mendel's 3:1 phenotypic ratio as a guideline. If the traits under analysis behaved as predicted by Mendel's laws, then they were assumed to be determined by a single gene with alternative dominant and recessive alleles. Many traits, however, did not behave in this way. For some, no definitive dominance and recessiveness could be observed, or more than two alleles could be found in a particular cross. Other traits turned out to be controlled by two or more genes, as for example in the analysis of lentil seed coat colour in Figure 2.22, while others were **multifactorial** or **complex**; that is, determined by two or more genes and their interaction with the environment.

Because such traits arise from an intricate network of interactions, they do not necessarily generate straightforward Mendelian phenotypic ratios. Nonetheless, simple extensions of Mendel's hypotheses can clarify the relationship between genotype and phenotype, allowing explanation of the observed deviations without challenging Mendel's basic laws.

One general theme stands out from these breeding studies: To make sense of the enormous phenotypic variation of the living world, geneticists usually try to limit the number of variables under investigation at any one time. Mendel did this by using pure-breeding, inbred strains of peas that differed from each other by one or a few traits, so that the action of single genes could be detected. Similarly, twentieth-century geneticists used inbred populations of fruit flies, mice, and other experimental organisms to study specific traits. Of course, geneticists cannot approach

Figure 2.22 Some phenotypic variation poses a challenge to Mendelian analysis. Lentils show complex speckling patterns that are controlled by a gene that has more than two alleles.

people in this way. Human populations are typically far from inbred, and researchers cannot ethically perform breeding experiments on people. As a result, the genetic basis of much human variation remained a mystery. The advent of molecular biology in the 1970s provided new tools that geneticists now use to unravel the genetics of complex human traits as described later in Chapters 14–15 and 19–20.

Consistent exceptions to simple Mendelian ratios thus revealed unexpected patterns of single-gene inheritance. By distilling the significance of these patterns, William Bateson (an early interpreter and defender of Mendel, who coined the terms "genetics", "allelomorph" [later shortened to "allele"], "homozygote", and "heterozygote") and other early geneticists extended the scope of Mendelian analysis and obtained a deeper understanding of the relationship between genotype and phenotype. We now look at the major extensions to Mendelian analysis elucidated over the last century.

Dominance is not always complete

A consistent working definition of dominance and recessiveness depends on the F_1 hybrids that arise from a mating between two pure-breeding lines. If a hybrid is identical to one parent for the trait under consideration, the allele carried by that parent is deemed dominant to the allele carried by the parent whose trait is not expressed in the hybrid. If, for example, a mating between a pure-breeding white line and a pure-breeding blue line produces F_1 hybrids that are white, the white allele of the gene for colour is dominant to the blue allele. If the F_1 hybrids are blue, the blue allele is dominant to the white one (**Figure 2.23**).

Mendel described and relied on complete dominance in sorting out his ratios and laws, but it is not the only kind of dominance he observed. Figure 2.23 diagrams two situations in which neither allele of a gene is completely dominant. As the figure shows, crosses between true-breeding

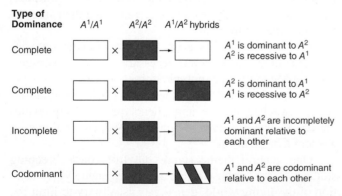

Type of Dominance	A^1/A^1		A^2/A^2	A^1/A^2 hybrids	
Complete	☐	×	■	→	☐	A^1 is dominant to A^2 A^2 is recessive to A^1
Complete	☐	×	■	→	■	A^2 is dominant to A^1 A^1 is recessive to A^2
Incomplete	☐	×	■	→	▨	A^1 and A^2 are incompletely dominant relative to each other
Codominant	☐	×	■	→	◨	A^1 and A^2 are codominant relative to each other

Figure 2.23 Different dominance relationships. The phenotype of the heterozygote defines the dominance relationship between two alleles of the same gene (here, A^1 and A^2). Dominance is complete when the hybrid resembles one of the two pure-breeding parents. Dominance is incomplete when the hybrid resembles neither parent; its novel phenotype is usually intermediate, but typically is more similar to one parent than to the other. Codominance occurs when the hybrid shows the traits from both pure-breeding parents.

strains can produce hybrids with phenotypes that differ from both parents. We now explain how these phenotypes arise.

Incomplete dominance: The F_1 hybrid resembles neither pure-breeding parent

A cross between pure late-blooming and pure early-blooming pea plants results in an F_1 generation that blooms in between the two extremes. This is just one of many examples of **incomplete dominance**, in which the hybrid does not resemble either pure-breeding parent. F_1 hybrids that differ from both parents often express a phenotype that is intermediate between those of the pure-breeding parents, falling somewhere along a kind of continuum, although it typically more closely resembles one parental phenotype than the other. Thus, with incomplete dominance, neither parental allele is dominant nor recessive to the other; both contribute to the F_1 phenotype. Mendel observed plants that bloomed midway between two extremes when he cultivated various types of pure-breeding peas for his hybridization studies, but he did not pursue the implications. Blooming time was not one of the seven characteristics he chose to analyze in detail, almost certainly because in peas, the time of bloom was not as clear-cut as seed shape or flower colour.

In many plant species, flower colour serves as a striking example of incomplete dominance. With the tubular flowers of four-o'clocks or the floret clusters of snapdragons, for instance, a cross between pure-breeding red-flowered parents and pure-breeding white yields hybrids with pink blossoms, as if a painter had mixed red and white pigments to get pink (**Figure 2.24a**). If allowed to self-pollinate, the F_1 pink-blooming plants produce F_2 progeny bearing red, pink, and white flowers in a ratio of 1:2:1 (**Figure 2.24b**). This is the familiar *genotypic* ratio of an ordinary single-gene F_1 self-cross. What is new is that because the heterozygotes look unlike either homozygote, the *phenotypic* ratios are an exact reflection of the genotypic ratios.

The modern biochemical explanation for this type of incomplete dominance is that each allele of the gene under analysis specifies an alternative form of a protein molecule with an enzymatic role in pigment production. If the "white" allele does not give rise to a functional enzyme, no pigment appears. Thus, in snapdragons and four-o'clocks, two "red" alleles per cell produce a double dose of a red-producing enzyme, which generates enough pigment to make the flowers look fully red. In the heterozygote, one copy of the "red" allele per cell results in only enough pigment to make the flowers look pink. In the homozygote for the "white" allele, where there is no functional enzyme and thus no red pigment, the flowers appear white.

Codominance: The F_1 hybrid exhibits traits of both parents

A cross between pure-breeding spotted lentils and pure-breeding dotted lentils produces heterozygotes that are both spotted and dotted (**Figure 2.25a**). These F_1 hybrids

(a) *Antirrhinum majus* (snapdragons)

(b) A Punnett square for incomplete dominance

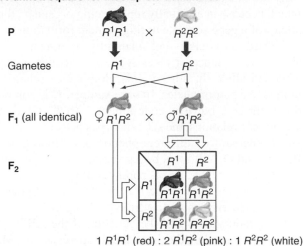

1 R^1R^1 (red) : 2 R^1R^2 (pink) : 1 R^2R^2 (white)

Figure 2.24 Pink flowers are the result of incomplete dominance. **(a)** Colour differences in these snapdragons reflect the activity of one pair of alleles. **(b)** The F_1 hybrids from a cross of pure-breeding red and white strains of snapdragons have pink blossoms. Flower colours in the F_2 appear in the ratio of 1 red : 2 pink : 1 white. This ratio signifies that the alleles of a single gene determine these three colours.

illustrate a second significant departure from complete dominance. They look like both parents, which means that neither the "spotted" nor the "dotted" allele is dominant or recessive to the other. Because both traits show up equally in the heterozygote's phenotype, the alleles are termed **codominant**. Self-pollination of the spotted/dotted F_1 generation generates F_2 progeny in the ratio of 1 spotted : 2 spotted/dotted : 1 dotted. The Mendelian 1:2:1 ratio among these F_2 progeny establishes that the spotted and dotted traits are determined by alternative alleles of a single gene. Once again, because the heterozygotes can be distinguished from both homozygotes, the phenotypic and genotypic ratios coincide.

In humans, some of the complex membrane-anchored molecules that distinguish different types of red blood cells exhibit codominance. For example, one gene (*I*) with alleles I^A and I^B controls the presence of a sugar polymer that protrudes from the red blood cell membrane. The alternative alleles each encode a slightly different form of an enzyme that causes production of a slightly different form of the complex sugar. In heterozygous individuals, the red blood cells carry both the I^A-determined and the I^B-determined sugars on their surface, whereas the cells of homozygous individuals display the products of either I^A or I^B alone (**Figure 2.25b**). As this example illustrates, when both alleles produce a functional gene product, they are usually codominant for phenotypes when analyzed at the molecular level.

(a) Codominant lentil coat patterns

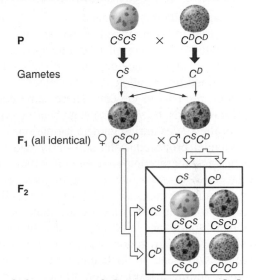

1 C^SC^S (spotted) : 2 C^SC^D (spotted/dotted) : 1 C^DC^D (dotted)

(b) Codominant blood group alleles

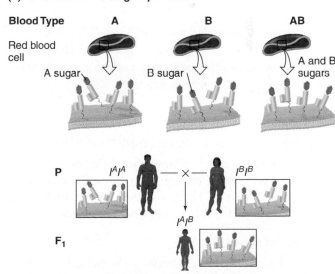

Figure 2.25 In codominance, F_1 hybrids display the traits of both parents. **(a)** A cross between pure-breeding spotted lentils and pure-breeding dotted lentils produces heterozygotes that are both spotted and dotted. Each genotype has its own corresponding phenotype, so the F_2 ratio is 1:2:1. **(b)** The I^A and I^B blood group alleles are codominant because the red blood cells of an I^AI^B heterozygote have both kinds of sugars at their surface.

Figure 2.23 summarizes the differences between complete dominance, incomplete dominance, and codominance for phenotypes reflected in colour variations. Determinations of dominance relationships depend on what phenotype appears in the F_1 generation. With complete dominance, F_1 progeny look like one of the true-breeding parents. Complete dominance results in a 3:1 ratio of phenotypes in the F_2. With incomplete dominance, hybrids resemble neither of the parents and thus display neither pure-breeding trait. With codominance, the phenotypes of both pure-breeding lines show up simultaneously in the F_1 hybrid. Both incomplete dominance and codominance yield 1:2:1 F_2 ratios.

Mendel's law of segregation still holds

The dominance relations of a gene's alleles do not affect the alleles' transmission. Whether two alternative alleles of a single gene show complete dominance, incomplete dominance, or codominance depends on the kinds of proteins determined by the alleles and the biochemical function of those proteins in the cell. These same phenotypic dominance relations, however, have no bearing on the segregation of the alleles during gamete formation.

As Mendel proposed, cells still carry two copies of each gene, and these copies—a pair of either similar or dissimilar alleles—segregate during gamete formation. Fertilization then restores two alleles to each cell without reference to whether the alleles are the same or different. Variations in dominance relations thus do not detract from Mendel's laws of segregation. Rather, they reflect differences in the way gene products control the production of phenotypes, adding a level of complexity to the tasks of interpreting the visible results of gene transmission and inferring genotype from phenotype.

In cases of incomplete dominance or codominance, mating of F_1 hybrids produces an F_2 generation with a 1:2:1 phenotypic ratio. The reason is that heterozygotes have a phenotype different from that of either homozygote.

A gene may have more than two alleles

Mendel analyzed "either-or" traits controlled by genes with two alternative alleles, but for many traits, there are more than two alternatives. Here, we look at two such traits: human ABO blood types and lentil seed coat patterns.

ABO blood types

If a person with blood type A mates with a person with blood type B, it is possible in some cases for the couple to have a child that is neither A nor B nor AB, but a fourth blood type called O. The reason for this? The gene for the ABO blood types has three alleles: I^A, I^B, and i (**Figure 2.26a**). Allele I^A gives rise to blood type A by specifying an enzyme that adds sugar A, and I^B results in blood type B by specifying an enzyme that adds sugar B; i does not produce a functional sugar-adding enzyme. Alleles I^A and I^B are both dominant to i, and blood type O is therefore a result of homozygosity for allele i.

Note in Figure 2.26a that the A phenotype can arise from two genotypes, $I^A I^A$ or $I^A i$. The same is true for the B blood type, which can be produced by $I^B I^B$ or $I^B i$. But a combination of the two alleles $I^A I^B$ generates blood type AB.

We can draw several conclusions from these observations. First, as already stated, a given gene may have more than two alleles, or **multiple alleles**; in our example, the series of alleles is denoted I^A, I^B, i.

Second, although the ABO blood group gene has three alleles, each person carries only two of the alternatives—$I^A I^A$, $I^B I^B$, $I^A I^B$, $I^A i$, $I^B i$, or ii. There are thus six possible ABO genotypes. Because each individual carries no more than two alleles for each gene, no matter how many alleles there are in a series, Mendel's law of segregation remains intact, because in a sexually reproducing organism, the two alleles of a gene separate during gamete formation.

Third, an allele is not inherently dominant or recessive; its dominance or recessiveness is always relative to a second allele. In other words, dominance relations are unique to a pair of alleles. In our example, I^A is completely dominant to i, but it is codominant with I^B. Given these dominance relations, the six genotypes possible with I^A, I^B, and i generate four different phenotypes: blood groups A, B, AB, and O. With this background, you can understand how a type A and a type B parent could produce a type O child: The parents must be $I^A i$ and $I^B i$ heterozygotes, and the child receives an i allele from each parent.

An understanding of the genetics of the ABO system has had profound medical and legal repercussions. Matching ABO blood types is a prerequisite of successful blood transfusions, because people make antibodies to foreign blood cell molecules. A person whose cells carry only A molecules, for example, produces anti-B antibodies; B people manufacture anti-A antibodies; AB individuals make neither type of antibody; and O individuals produce both anti-A and anti-B antibodies (**Figure 2.26b**). These antibodies cause coagulation of cells displaying the foreign molecules (**Figure 2.26c**). As a result, people with blood type O have historically been known as universal donors because their red blood cells carry no surface molecules that will stimulate an antibody attack. In contrast, people with blood type AB are considered universal recipients, because they make neither anti-A nor anti-B antibodies, which, if present, would target the surface molecules of incoming blood cells.

Information about ABO blood types can also be used as legal evidence in court, to exclude the possibility of paternity or criminal guilt. In a paternity suit, for example, if the mother is type A and her child is type B, logic dictates that the I^B allele must have come from the father, whose genotype may be $I^A I^B$, $I^B I^B$, or $I^B i$. In 1944, the actress Joan Barry (phenotype A) sued Charlie Chaplin (phenotype O) for support of a child (phenotype B) whom she claimed he fathered. The scientific evidence indicated that Chaplin could not have been the father, since he was apparently ii

(a)

Genotypes	Corresponding Phenotypes: Type(s) of Molecule on Cell
$I^A I^A$ $I^A i$	A
$I^B I^B$ $I^B i$	B
$I^A I^B$	AB
ii	O

(b)

Blood Type	Antibodies in Serum
A	Antibodies against B
B	Antibodies against A
AB	No antibodies against A or B
O	Antibodies against A and B

(c)

Blood Type of Recipient	Donor Blood Type (Red Cells)			
	A	B	AB	O
A	+	−	−	+
B	−	+	−	+
AB	+	+	+	+
O	−	−	−	+

Figure 2.26 ABO blood types are determined by three alleles of one gene. (a) Six genotypes produce the four blood group phenotypes. **(b)** Blood serum contains antibodies against foreign red blood cell molecules. **(c)** If a recipient's serum has antibodies against the sugars on a donor's red blood cells, the blood types of recipient and donor are incompatible and coagulation of red blood cells will occur during transfusions. In this table, a plus (+) indicates compatibility, and a minus (−) indicates incompatibility. Antibodies in the donor's blood usually do not cause problems because the amount of transfused antibody is small.

and did not carry an I^B allele. This evidence was admissible in court, but the jury was not convinced, and Chaplin had to pay. The injustice of the jury's decision later led to a change in law that allowed blood tests to be permissible as evidence. Today, the molecular genotyping of DNA (*DNA fingerprinting*, see Chapter 15) provides a powerful tool to help establish paternity, guilt, or innocence, but juries still often find it difficult to evaluate such evidence.

Lentil seed coat patterns

Lentils offer another example of multiple alleles. A gene for seed coat pattern has five alleles: spotted, dotted, clear (pattern absent), and two types of marbled. Reciprocal crosses between pairs of pure-breeding lines of all patterns (marbled-1 × marbled-2, marbled-1 × spotted, marbled-2 × spotted, and so forth) have clarified the dominance relations of all possible pairs of the alleles to reveal a **dominance series** in which alleles are listed in order from most dominant to most recessive. For example, crosses of marbled-1 with marbled-2, or of marbled-1 with spotted or dotted or clear, produce the marbled-1 phenotype in the F_1 generation and a ratio of three marbled-1 to one of any of the other phenotypes in the F_2. This indicates that

the marbled-1 allele is completely dominant to each of the other four alleles.

Analogous crosses with the remaining four phenotypes reveal the dominance series shown in **Figure 2.27**. Recall that dominance relations are meaningful only when comparing two alleles; an allele, such as marbled-2, can be recessive to a second allele (marbled-1) but dominant to a third and fourth (dotted and clear). The fact that all tested pairings of lentil seed coat pattern alleles yielded a 3:1 ratio in the F_2 generation (except for spotted × dotted, which yielded the 1:2:1 phenotypic ratio reflective of codominance) indicates that these lentil seed coat patterns are determined by different alleles of the same gene.

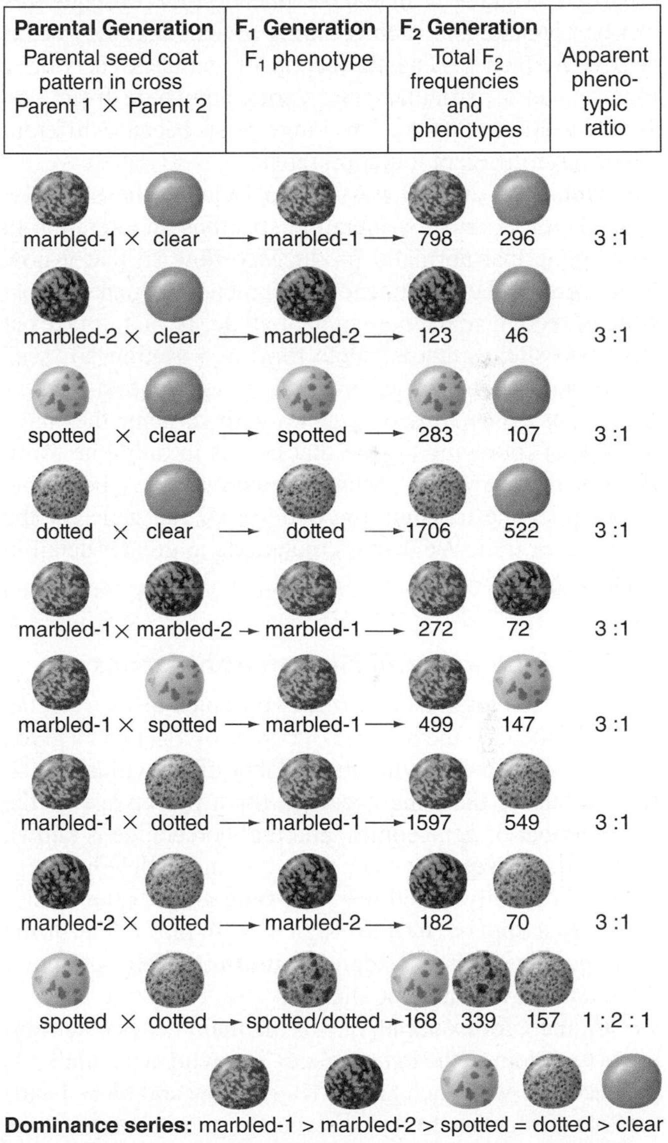

Parental Generation	F_1 Generation	F_2 Generation			Apparent phenotypic ratio
Parental seed coat pattern in cross Parent 1 × Parent 2	F_1 phenotype	Total F_2 frequencies and phenotypes			

marbled-1 × clear	→ marbled-1 →	798	296		3 :1
marbled-2 × clear	→ marbled-2 →	123	46		3 :1
spotted × clear	→ spotted →	283	107		3 :1
dotted × clear	→ dotted →	1706	522		3 :1
marbled-1 × marbled-2	→ marbled-1 →	272	72		3 :1
marbled-1 × spotted	→ marbled-1 →	499	147		3 :1
marbled-1 × dotted	→ marbled-1 →	1597	549		3 :1
marbled-2 × dotted	→ marbled-2 →	182	70		3 :1
spotted × dotted	→ spotted/dotted →	168	339	157	1 : 2 : 1

Dominance series: marbled-1 > marbled-2 > spotted = dotted > clear

Figure 2.27 How to establish the dominance relations between multiple alleles. Pure-breeding lentils with different seed coat patterns are crossed in pairs, and the F_1 progeny are self-fertilized to produce an F_2 generation. The 3:1 or 1:2:1 F_2 monohybrid ratios from all of these crosses indicate that different alleles of a single gene determine all the traits. The phenotypes of the F_1 hybrids establish the dominance relationships (*bottom*). Spotted and dotted alleles are codominant, but each is recessive to the marbled alleles and is dominant to clear.

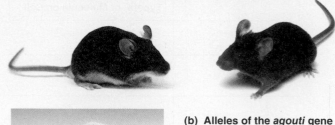

(a) *Mus musculus* (house mouse) coat colours

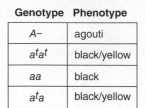

(b) Alleles of the *agouti* gene

Genotype	Phenotype
A–	agouti
a^{t}a^t	black/yellow
aa	black
a^{t}a	black/yellow

Within a population, a gene may have multiple alleles, but any one individual can have at most two of these alleles. Considered in pairs, the alleles can exhibit a variety of dominance relationships.

Mutations are the source of new alleles

How do the multiple alleles of an allelic series arise? The answer is that chance alterations of the genetic material, known as **mutations**, arise spontaneously in nature. Once they occur in gamete-producing cells, they are faithfully inherited. Mutations that have phenotypic consequences can be counted, and such counting reveals that they occur at low frequency. The frequency of gametes carrying a mutation in a particular gene varies anywhere from 1 in 10 000 to 1 in 1 000 000. This range exists because different genes have different mutation rates.

Mutations make it possible to follow gene transmission. If, for example, a mutation specifies an alteration in an enzyme that normally produces yellow so that it now makes green, the new phenotype (green) will make it possible to recognize the new mutant allele. In fact, it takes at least two alleles, that is, some form of variation, to "see" the transmission of a gene. Thus, in segregation studies, geneticists can analyze only genes with variants; they have no way of following a gene that comes in only one form. If all peas were yellow, Mendel would not have been able to decipher the transmission patterns of the gene for the seed colour trait. We discuss mutations in greater detail in Chapter 8.

Allele frequencies and monomorphic genes

Because each organism carries two copies of every gene, you can calculate the number of copies of a gene in a given population by multiplying the number of individuals by 2. Each allele of the gene accounts for a percentage of the total number of gene copies, and that percentage is known as the **allele frequency**. The most common allele in a population is usually called the **wild-type allele**, often designated by a superscript plus sign ($^+$). A rare allele in the same population is considered a **mutant allele**. (A mutation causes a new mutant allele.)

In mice, for example, one of the main genes determining coat colour is the *agouti* gene. The wild-type allele (*A*) produces fur with each hair having yellow and black bands that blend together from a distance to give the appearance of dark grey, or agouti. Researchers have identified in the laboratory 14 distinguishable mutant alleles for the *agouti* gene. One of these (*a^t*) is recessive to the wild type and gives rise to a black coat on the back and a yellow coat on the belly; another (*a*) is also recessive to *A* and produces a pure black coat (**Figure 2.28**). In nature, wild-type agoutis (*AA*) survive to reproduce, while very few black-backed or pure black mutants (*a^{t}a^t* or *aa*) do so because their dark coat makes it hard for them to evade the eyes of predators.

(c) Evidence for a dominance series

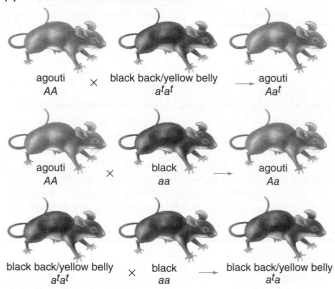

Dominance series: $A > a^t > a$

Figure 2.28 The mouse *agouti* gene: One wild-type allele, many mutant alleles. (a) Black-backed, yellow-bellied (*top left*); black (*top right*); and agouti (*bottom*) mice. **(b)** Genotypes and corresponding phenotypes for alleles of the *agouti* gene. **(c)** Crosses between pure-breeding lines reveal a dominance series. Interbreeding of the F$_1$ hybrids (*not shown*) yields 3:1 phenotypic ratios of F$_2$ progeny, indicating that *A*, *a^t*, and *a* are, in fact, alleles of one gene.

As a result, *A* is present at a frequency of much more than 99 percent and is thus the only wild-type allele in mice for the *agouti* gene. A gene with only one common, wild-type allele is **monomorphic**.

In contrast, some genes have more than one common allele, which makes them **polymorphic**. For example, in the ABO blood type system, all three alleles—I^A, I^B, and *i*—have appreciable frequencies in most human populations. Although all three of these alleles can be considered to be wild-type, geneticists instead usually refer to the high-frequency alleles of a polymorphic gene as *common variants*.

One gene may contribute to several characteristics

Mendel derived his laws from studies in which one gene determined one trait; but, always the careful observer, he himself noted possible departures. In listing the traits selected for his pea experiments, he remarked that specific seed coat colours are always associated with specific flower colours.

The phenomenon of a single gene determining a number of distinct and seemingly unrelated characteristics is known as **pleiotropy**. Because geneticists now know that each gene determines a specific protein and that each protein can have a cascade of effects on an organism, we can understand how pleiotropy arises. Among the aboriginal Maori people of New Zealand, for example, many of the men develop respiratory problems and are also sterile. Researchers have found that the fault lies with the recessive allele of a single gene. The gene's normal dominant allele specifies a protein necessary for the action of cilia and flagella, both of which are hair-like structures extending from the surfaces of some cells. In men who are homozygous for the recessive allele, cilia that normally clear the airways fail to work effectively, and flagella that normally propel sperm fail to do their job. Thus, one gene determines a protein that indirectly affects both respiratory function and reproduction. Because most proteins act in a variety of tissues and influence multiple biochemical processes, mutations in almost any gene may have pleiotropic effects.

Recessive lethal alleles

A significant variation of pleiotropy occurs in alleles that not only produce a visible phenotype but also affect viability. Mendel assumed that all genotypes are equally viable—that is, they have the same likelihood of survival. If this were not true and a large percentage of, say, homozygotes for a particular allele died before germination or birth, you would not be able to count them after birth, and this would alter the 1:2:1 genotypic ratios and the 3:1 phenotypic ratios predicted for the F$_2$ generation.

Consider the inheritance of coat colour in mice. As mentioned earlier, wild-type agouti (*AA*) animals have black and yellow striped hairs that appear dark grey to the eye. One of the 14 mutant alleles of the *agouti* gene gives rise to mice with a much lighter, almost yellow colour. When inbred *AA* mice are mated to yellow mice, one always observes a 1:1 ratio of the two coat colours among the offspring (**Figure 2.29a**). From this result, we

can draw three conclusions: (1) All yellow mice must carry the *agouti* allele even though they do not express it; (2) yellow is therefore dominant to agouti; and (3) all yellow mice are heterozygotes.

Note again that dominance and recessiveness are defined in the context of each pair of alleles. Even though, as previously mentioned, agouti (*A*) is dominant to the a^t and *a* mutations for black coat colour, it can still be recessive to the yellow coat colour allele. If we designate the allele for yellow as A^y, the yellow mice in the preceding cross are A^yA heterozygotes, and the agoutis, *AA* homozygotes. So far, there are no surprises. But a mating of yellow to yellow produces a skewed phenotypic ratio of two yellow mice to one agouti (**Figure 2.29b**). Among these progeny, matings between agouti mice show that the agoutis are all pure-breeding and therefore *AA* homozygotes as expected. There are, however, no pure-breeding yellow mice among the progeny. When the yellow mice are mated to each other, they produce 2/3 yellow and 1/3 agouti offspring, a ratio of 2:1, so they must therefore be heterozygotes (A^yA). In short, one can never obtain pure-breeding yellow mice.

(a) All yellow mice are heterozygotes.

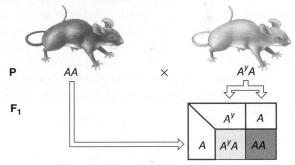

(b) Two copies of A^y cause lethality.

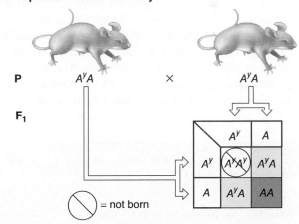

= not born

Figure 2.29 A^y: A recessive lethal allele that also produces a dominant coat colour phenotype. (a) A cross between inbred agouti mice and yellow mice yields a 1:1 ratio of yellow to agouti progeny. The yellow mice are therefore A^yA heterozygotes, and for the trait of coat colour, A^y (for yellow) is dominant to *A* (for agouti). **(b)** Yellow mice do not breed true. In a yellow × yellow cross, the 2:1 ratio of yellow to agouti progeny indicates that the A^y allele is a recessive lethal.

How can we explain this phenomenon? The Punnett square in Figure 2.29b suggests an answer. Two copies of the A^y allele prove fatal to the animal carrying them, whereas one copy of the allele produces a yellow coat. This means that the A^y allele affects two different traits: It is dominant to A in the determination of coat colour, but it is recessive to A in the production of lethality. An allele, such as A^y, that negatively affects the survival of a homozygote is known as a **recessive lethal allele**. Note that the same two alleles (A^y and A) can display different dominance relationships when looked at from the point of view of different phenotypes; we return later to this important point.

Because the A^y allele is dominant for yellow coat colour, it is easy to detect carriers of this particular recessive lethal allele in mice, but such is not the case for the vast majority of recessive lethal mutations that do not simultaneously show a visible dominant phenotype for some other trait. Lethal mutations can arise in many different genes, and as a result, most animals, including humans, carry some recessive lethal mutations. Such mutations usually remain "silent," except in rare cases of homozygosity, which in people are often caused by consanguineous matings (i.e., matings between close relatives). If a mutation produces an allele that prevents production of a crucial molecule, homozygous individuals would not make any of the vital molecule and would not survive. Heterozygotes, by contrast, with only one copy of the deleterious mutation and one wild-type allele, would be able to produce 50 percent of the wild-type amount of the normal molecule; this is usually sufficient to sustain normal cellular processes such that life goes on.

Delayed lethality

In the preceding discussion, we have described recessive alleles that result in the death of homozygotes prenatally, *in utero*. With some mutations, however, homozygotes may survive beyond birth and die later from the deleterious consequences of the genetic defect. An example is seen in human infants with Tay-Sachs disease. The seemingly normal newborns remain healthy for five to six months but then develop blindness, paralysis, mental retardation, and other symptoms of a deteriorating nervous system; the disease usually proves fatal by the age of six. Tay-Sachs disease results from the absence of an active lysosomal enzyme called hexosaminidase A, leading to the accumulation of a toxic waste product inside nerve cells. The approximate incidence of Tay-Sachs among live births is 1/35 000 worldwide, but it is 1/3000 among Jewish people of Eastern European descent. Reliable tests that detect carriers, in combination with genetic counselling and educational programs, have all but eliminated the disease in Canada and the United States.

Recessive alleles that cause prenatal or early childhood lethality can only be passed on to subsequent generations by heterozygous carriers, because affected homozygotes die before they can mate. However, for late-onset diseases causing death in adults, homozygous patients can pass on the lethal allele before they become debilitated. An example is provided by the degenerative disease Friedreich ataxia: Some homozygotes first display symptoms of ataxia (loss of muscle coordination) at age 30–35 and die about five years later from heart failure.

Dominant alleles causing late-onset lethality can also be transmitted to subsequent generations; Figure 2.20 illustrates this for the inheritance of Huntington disease. By contrast, if the lethality caused by a dominant allele occurs instead during fetal development or early childhood, the allele will not be passed on, so all dominant early-onset lethal mutant alleles must be new mutations.

Table 2.3 summarizes Mendel's basic assumptions about dominance, the number and viability of one gene's alleles, and the effects of each gene on phenotype, and then compares these assumptions with the extensions contributed by his twentieth-century successors. Through carefully controlled monohybrid crosses, these later geneticists analyzed the transmission patterns of the alleles of single genes, challenging and then confirming the law of segregation.

> A mutant allele can disrupt many biochemical processes; as a result, mutations often have pleiotropic effects that can include lethality at various times in an organism's life cycle.

Table 2.3	For Traits Determined by One Gene: Extensions to Mendel's Analysis Explain Alterations of the 3:1 Monohybrid Ratio		
What Mendel Described	**Extension**	**Extension's Effect on Heterozygous Phenotype**	**Extension's Effect on Ratios Resulting from an $F_1 \times F_1$ Cross**
Complete dominance	Incomplete dominance Codominance	Unlike either homozygote	Phenotypes coincide with genotypes in a ratio of 1:2:1
Two alleles	Multiple alleles	Multiplicity of phenotypes	A series of 3:1 ratios
All alleles are equally viable	Recessive lethal alleles	No effect	2:1 instead of 3:1
One gene determines one trait	Pleiotropy: One gene influences several traits	Several traits affected in different ways, depending on dominance relations	Different ratios, depending on dominance relations for each affected trait

A comprehensive example: Sickle-cell disease illustrates many extensions to Mendel's analysis

Sickle-cell disease is the result of a faulty haemoglobin molecule. Haemoglobin is composed of two types of polypeptide chains, alpha (α) globin and beta (β) globin, each specified by a different gene: $Hb\alpha$ for α globin and $Hb\beta$ for β globin. Normal red blood cells are packed full of millions upon millions of haemoglobin molecules, each of which picks up oxygen in the lungs and transports it to all the body's tissues.

Multiple alleles

The β-globin gene has a normal wild-type allele ($Hb\beta^A$) that gives rise to fully functional β globin, as well as close to 400 mutant alleles that have been identified so far. Some of these mutant alleles result in the production of haemoglobin that carries oxygen only inefficiently. Other mutant alleles prevent the production of β globin, causing a haemolytic (blood-destroying) disease called β-*thalassemia*. Here, we discuss the most common mutant allele of the β-globin gene, $Hb\beta^S$, which specifies an abnormal polypeptide that causes sickling of red blood cells (**Figure 2.30a**).

Pleiotropy

The $Hb\beta^S$ allele of the β-globin gene affects more than one trait (**Figure 2.30b**). Haemoglobin molecules in the red blood cells of homozygous $Hb\beta^S\ Hb\beta^S$ individuals undergo an aberrant transformation after releasing their oxygen. Instead of remaining soluble in the cytoplasm, they aggregate to form long fibres that deform the red blood cell from a normal biconcave disk to a sickle shape (see Figure 2.30a). The deformed cells clog the small blood vessels, reducing oxygen flow to the tissues and giving rise to muscle cramps, shortness of breath, and fatigue. The sickled cells are also very fragile and easily broken. Consumption of fragmented cells by phagocytic white blood cells leads to a low red blood cell count, a condition called anaemia.

On the positive side, $Hb\beta^S\ Hb\beta^S$ homozygotes are resistant to malaria, because the organism that causes the disease, *Plasmodium falciparum*, can multiply rapidly in normal red blood cells, but cannot do so in cells that sickle. Infection by *P. falciparum* causes sickle-shaped cells to break down before the malaria organism has a chance to multiply.

Recessive lethality

People who are homozygous for the recessive $Hb\beta^S$ allele often develop heart failure because of stress on the

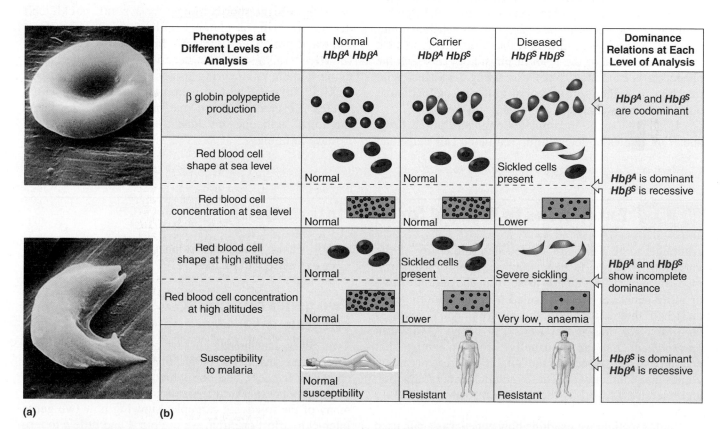

(a)   (b)

Figure 2.30 Pleiotropy of sickle-cell anaemia: Dominance relations vary with the phenotype under consideration. (a) A normal red blood cell (*top*) is easy to distinguish from the sickled cell in the scanning electron micrograph at the *bottom*. (b) Different levels of analysis identify various phenotypes. Dominance relationships between the $Hb\beta^S$ and $Hb\beta^A$ alleles of the $Hb\beta$ gene vary with the phenotype and sometimes even change with the environment.

circulatory system. Many sickle-cell sufferers die in childhood, adolescence, or early adulthood.

Different dominance relations

Comparisons of heterozygous carriers of the sickle-cell allele—individuals whose cells contain one $Hb\beta^A$ and one $Hb\beta^S$ allele—with homozygous $Hb\beta^A\,Hb\beta^A$ (normal) and homozygous $Hb\beta^S\,Hb\beta^S$ (diseased) individuals make it possible to distinguish different dominance relations for different phenotypic aspects of sickle-cell anaemia (Figure 2.30b).

At the molecular level—the production of β globin—both alleles are expressed such that $Hb\beta^A$ and $Hb\beta^S$ are *codominant.* At the cellular level, in their effect on red blood cell shape, the $Hb\beta^A$ and $Hb\beta^S$ alleles show *incomplete dominance.* Although under normal oxygen conditions, the great majority of a heterozygote's red blood cells have the normal biconcave shape, when oxygen levels drop, sickling occurs in some cells. All $Hb\beta^A\,Hb\beta^S$ cells, however, are resistant to malaria because like the $Hb\beta^S\,Hb\beta^S$ cells described previously, they break down before the malarial organism has a chance to reproduce. Thus, for the trait of resistance to malaria, the $Hb\beta^S$ allele is *dominant* to the $Hb\beta^A$ allele. But luckily for the heterozygote, for the phenotypes of anaemia or death, $Hb\beta^S$ is *recessive* to $Hb\beta^A$. A corollary of this observation is that in its effect on general health under normal environmental conditions and its effect on red blood cell count, the $Hb\beta^A$ allele is *dominant* to $Hb\beta^S$.

Thus, for the β-globin gene, as for other genes, dominance and recessiveness are not an inherent quality of alleles in isolation; rather, they are specific to each pair of alleles and to the level of physiology at which the phenotype is examined. When discussing dominance relationships, it is therefore essential to define the particular phenotype under analysis.

In the 1940s, the incomplete dominance of the $Hb\beta^A$ and $Hb\beta^S$ alleles in determining red blood cell shape had significant repercussions for certain soldiers who fought in World War II. Aboard transport planes flying troops across the Pacific, several heterozygous carriers suffered sickling crises similar to those usually seen in $Hb\beta^S\,Hb\beta^S$ homozygotes. The reason was that heterozygous red blood cells of a carrier produce both normal and abnormal haemoglobin molecules. At sea level, these molecules together deliver sufficient oxygen, although less than the normal amount, to the body's tissues, but with a decrease in the amount of oxygen available at the high-flying altitudes, the haemoglobin picks up less oxygen, the rate of red blood cell sickling increases, and symptoms of the disease occur.

The complicated dominance relationships between the $Hb\beta^S$ and $Hb\beta^A$ alleles also help explain the puzzling observation that the normally deleterious allele $Hb\beta^S$ is widespread in certain populations. In areas where malaria is endemic, heterozygotes are better able to survive and pass on their genes than are either type of homozygote. $Hb\beta^S\,Hb\beta^S$ individuals often die of sickle-cell disease, while those with the genotype $Hb\beta^A\,Hb\beta^A$ often die of malaria. Heterozygotes, however, are relatively immune to both conditions, so high frequencies of both alleles persist in tropical environments where malaria is found. We explore this phenomenon in more quantitative detail in Chapter 12 on population genetics.

New therapies have improved the medical condition of many $Hb\beta^S\,Hb\beta^S$ individuals, but these treatments have significant shortcomings; as a result, sickle-cell disease remains a major health problem. The **Fast Forward** box **"Gene Therapy for Sickle-Cell Disease in Mice"** in this chapter describes recent success in using genetic engineering to counteract red blood cell sickling in mice whose genomes carry human $Hb\beta^S$ alleles. Researchers hope that similar types of "gene therapies" will one day lead to a cure for sickle-cell disease in humans.

2.5   Extensions to Mendel for Gene Interactions

Although some traits are indeed determined by allelic variations of a single gene, the vast majority of common traits in all organisms arise from the action of two or more genes, or are controlled by two or more genes and their interaction with the environment (known as *multifactorial* or *complex* traits). In genetics, the term *environment* has an unusually broad meaning that encompasses all aspects of the outside world an organism comes into contact with. These include temperature, diet, and exercise as well as the uterine environment before birth.

In this section, we examine how geneticists again used breeding experiments and the guidelines of Mendelian ratios to analyze the network of interactions between two or more genes. Multifactorial inheritance is also discussed more extensively in Chapter 12.

Two genes can interact to determine one trait

Two genes can interact in several ways to determine a single trait, such as the colour of a flower, a seed coat, a chicken's feathers, or a dog's fur, and each type of interaction produces its own signature of phenotypic ratios. In many of the following examples showing how two genes interact to affect one trait, we use big *A* and little *a* to represent alternative alleles of the first gene and big *B* and little *b* for those of the second gene.

The most widespread inherited blood disorder in Canada and the United States is sickle-cell disease. It is caused, as you have seen, by homozygosity for the $Hb\beta^S$ allele of the gene that specifies the β-globin constituent of haemoglobin. Because heterozygotes for this allele are partially protected from malaria, $Hb\beta^S$ is fairly common in people of African, Indian, Mediterranean, and Middle Eastern descent; four out of five African-Canadians are carriers of the sickle-cell allele. Before the 1980s, most people with sickle-cell disease died during childhood. However, advances in medical care have improved the outlook for many of these patients so that about half of them now live beyond the age of 50.

The main therapies in use today include treatment with the drug hydroxyurea, which stimulates the production of other kinds of haemoglobin; and bone marrow transplantation, which replaces the patient's red-blood-cell-forming haematopoietic stem cells with those of a healthy donor. Unfortunately, these treatments are not ideal. Hydroxyurea has toxic side effects, and bone marrow transplantation can be carried out successfully only with a donor whose tissues are perfectly matched with the patient's. As a result, medical researchers are exploring an alternative: the possibility of developing gene therapy for sickle-cell disease in humans.

In 2001 a research team from Harvard Medical School announced the successful use of gene therapy to treat mice that had been genetically engineered to have sickling red blood cells. These transgenic mice (called SAD mice) express an allelic form of the human $Hb\beta$ gene, closely related to $Hb\beta^S$.

The research team began by removing bone marrow from the SAD mice and isolating the haematopoietic stem cells from the marrow. They next used genetic engineering to add an antisickling transgene to these stem cells. The transgene was a synthetically mutated allele of the human $Hb\beta$ gene; it encoded a special β-globin protein designed to prevent sickling in red blood cells that also contain $Hb\beta^S$. When the genetically modified stem cells were transplanted back into the SAD mice, healthy, nonsickling red blood cells were produced. The new genetically modified transgene thus counteracted the effects of the $Hb\beta^S$ allele and prevented sickling, as predicted.

For human gene therapy, adding a transgene to haematopoietic stem cells derived from the sickle-cell patient would in theory mean no threat of tissue rejection when these engineered stem cells are transplanted back into the patient. However, researchers must overcome several potential problems. First, the method is not guaranteed to work in humans because SAD mice do not exhibit all aspects of sickle-cell disease in humans. Another difficulty is how to make sure the therapeutic gene gets into enough target cells to make a difference. The Harvard group resolved this issue in mice by using a modified version of the HIV virus causing AIDS (Acquired Immune Deficiency Syndrome) to transport the genetically engineered antisickling transgene into the stem cells. It has not been proven that virus-treated cells will be safe when reintroduced into the human body. Finally, successful gene therapy of this type requires that all the haematopoietic stem cells without the transgene must be removed. The Harvard researchers did this by destroying the bone marrow in the SAD mice with large doses of X-rays before putting the transgene-containing stem cells back into the mice. However, such a treatment in humans would be extremely toxic. Despite these potential complications, the successful application of gene therapy to a mouse model for sickle-cell disease suggests an exciting pathway for future clinical research.

Novel phenotypes resulting from gene interactions

Earlier we described a mating of tan and grey lentils that produced a uniformly brown F_1 generation and then an F_2 generation containing lentils with brown, tan, grey, and green seed coats. An understanding of how this can happen emerges from experimental results demonstrating that the ratio of the four F_2 colours is 9 brown: 3 tan: 3 grey: 1 green (**Figure 2.31a**). Recall that this is the same ratio Mendel observed in his analysis of the F_2 generations from dihybrid crosses following two independently assorting genes. In Mendel's studies, each of the four classes consisted of plants that expressed a combination of two unrelated traits. With lentils, however, we are looking at a single trait—seed coat colour. The simplest explanation for the parallel ratios is that a combination of genotypes at two independently assorting genes interacts to produce the phenotype of seed coat colour in lentils.

Results obtained from self-crosses with the various types of F_2 lentil plants support this explanation. Self-crosses of F_2 green individuals show that they are pure-breeding, producing an F_3 generation that is entirely green. Tan individuals generate either all tan offspring, or a mixture of tan offspring and green offspring. Greys similarly produce either all grey, or grey and green. Self-crosses of brown F_2 individuals can have four possible outcomes: all brown, brown plus tan, brown plus grey, or all four colours (**Figure 2.31b**). The two-gene hypothesis explains why there is

- only one green genotype: pure-breeding $aa\ bb$, but
- two types of tans: pure-breeding $AA\ bb$ as well as tan- and green-producing $Aa\ bb$, and

(a) A dihybrid cross with lentil coat colours

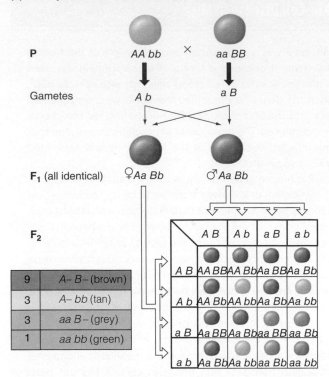

(b) Self-pollination of the F₂ to produce an F₃

Phenotypes of F₂ Individual	Observed F₃ Phenotypes	Expected Proportion of F₂ Population*
Green	Green	1/16
Tan	Tan	1/16
Tan	Tan, green	2/16
Grey	Grey, green	2/16
Grey	Grey	1/16
Brown	Brown	1/16
Brown	Brown, tan	2/16
Brown	Brown, grey	2/16
Brown	Brown, grey, tan, green	4/16

*This 1 : 1 : 2 : 2 : 1 : 1 : 2 : 2 : 4 F₂ genotypic ratio corresponds to a 9 brown : 3 tan : 3 grey : 1 green F₂ phenotypic ratio.

(c) Sorting out the dominance relations by select crosses

Seed Coat Colour of Parents	F₂ Phenotypes and Frequencies	Ratio
Tan × green	231 tan, 85 green	3 : 1
Grey × green	2586 grey, 867 green	3 : 1
Brown × grey	964 brown, 312 grey	3 : 1
Brown × tan	255 brown, 76 tan	3 : 1
Brown × green	57 brown, 18 grey, 13 tan, 4 green	9 : 3 : 3 : 1

Figure 2.31 How two genes interact to produce seed colours in lentils. (a) In a cross of pure-breeding tan and grey lentils, all the F₁ hybrids are brown, but four different phenotypes appear among the F₂ progeny. The 9:3:3:1 ratio of F₂ phenotypes suggests that seed coat colour is determined by two independently assorting genes. **(b)** Expected results of self-pollinating individual F₂ plants of the indicated phenotypes to produce an F₃ generation, if seed coat colour results from the interaction of two genes. The third column shows the proportion of the F₂ population that would be expected to produce the observed F₃ phenotypes. **(c)** Other two-generation crosses involving pure-breeding parental lines also support the two-gene hypothesis. (In this table, the F₁ hybrid generation has been omitted.)

- two types of greys: pure-breeding *aa BB* and grey- and green-producing *aa Bb*, yet
- four types of browns: true-breeding *AA BB*, brown- and tan-producing *AA Bb*, brown- and grey-producing *Aa BB*, and *Aa Bb* dihybrids that give rise to plants producing lentils of all four colours.

In short, for the two genes that determine seed coat colour, both dominant alleles must be present to yield brown (*A– B–*); the dominant allele of one gene produces tan (*A– bb*); the dominant allele of the other specifies grey (*aa B–*); and the complete absence of dominant alleles (i.e., the double recessive) yields green (*aa bb*). Thus, the four colour phenotypes arise from four **genotypic classes**, with each class defined in terms of the presence or absence of the dominant alleles of two genes: (1) both present (*A– B–*), (2) one present (*A– bb*), (3) the other present (*aa B–*), and (4) neither present (*aa bb*). Note that the *A–* notation means that the second allele of this gene can be either *A* or *a*, while *B–* denotes a second allele of either *B* or *b*. Note also that only with a two-gene system in which the dominance and recessiveness of alleles at both genes is complete can the nine different genotypes of the F₂ generation be categorized into the four genotypic classes described. With incomplete dominance or codominance, the F₂ genotypes could not be grouped together in this simple way, as they would give rise to more than four phenotypes.

Further crosses between plants carrying lentils of different colours confirmed the two-gene hypothesis (**Figure 2.31c**). Thus, the 9:3:3:1 phenotypic ratio of brown to tan to grey to green in an F₂ descended from pure-breeding tan and pure-breeding grey lentils tells us not only that two genes assorting independently interact to produce the seed coat colour, but also that each genotypic class (*A– B–*, *A– bb*, *aa B–*, and *aa bb*) determines a particular phenotype.

Complementary gene action

In some two-gene interactions, the four F₂ genotypic classes produce fewer than four observable phenotypes, because some of the phenotypes include two or more genotypic classes. For example, in the first decade of the twentieth century, William Bateson conducted a cross between two lines of pure-breeding white-flowered sweet peas (**Figure 2.32**). Quite unexpectedly, all of the F₁ progeny were purple. Self-pollination of these novel hybrids produced a ratio of 9 purple : 7 white in the F₂ generation. What was the explanation for this? Two genes work in tandem to produce purple sweet-pea flowers, and a dominant allele of both genes must be present to produce that colour.

A simple biochemical hypothesis for this type of **complementary gene action** is shown in **Figure 2.33**. Because it takes two enzymes catalyzing two separate biochemical reactions to change a colourless precursor into

(a) *Lathyrus odoratus* (sweet peas)

(b) A dihybrid cross involving complementary gene action

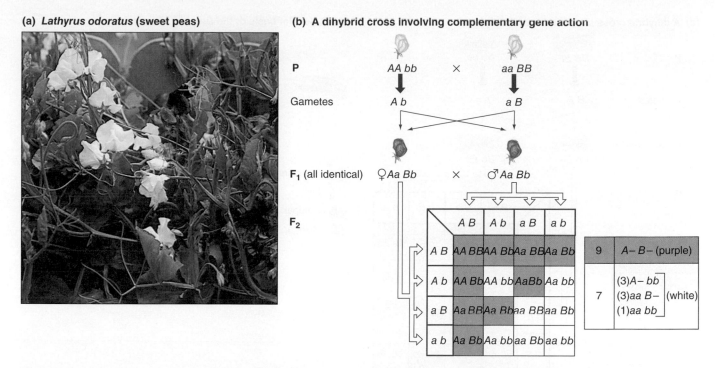

Figure 2.32 **Complementary gene action generates colour in sweet peas.** **(a)** White and purple sweet pea flowers. **(b)** The 9:7 ratio of purple to white F₂ plants indicates that at least one dominant allele for each gene is necessary for the development of purple colour.

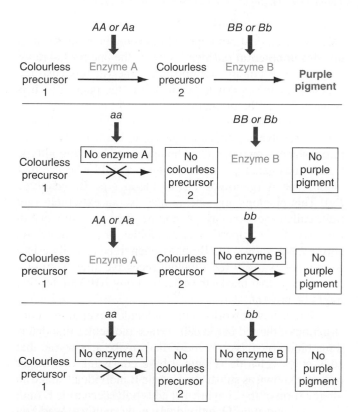

Figure 2.33 **A possible biochemical explanation for complementary gene action in the generation of sweet pea colour.** Enzymes specified by the dominant alleles of two genes are both necessary to produce pigment. The recessive alleles of both genes specify inactive enzymes. In *aa* homozygotes, no intermediate precursor 2 is created, so even if enzyme B is available, it cannot create purple pigment.

a colourful pigment, only the *A– B–* genotypic class, which produces active forms of both required enzymes, can generate coloured flowers. The other three genotypic classes (*A– bb*, *aa B–*, and *aa bb*) become grouped together with respect to phenotype because they do not specify functional forms of one or the other requisite enzyme and thus give rise to no colour, which is the same as white. It is easy to see how the "7" part of the 9:7 ratio encompasses the 3:3:1 of the 9:3:3:1 ratio of two genes in action. The 9:7 ratio is the phenotypic signature of this type of complementary gene interaction in which the dominant alleles of two genes acting together (*A– B–*) produce colour or some other trait, while the other three genotypic classes (*A– bb*, *aa B–*, and *aa bb*) do not (see Figure 2.32b).

Epistasis

In some gene interactions, the four Mendelian genotypic classes produce fewer than four observable phenotypes because one gene masks the phenotypic effects of another. An example is seen in the sleek, short-haired coat of Labrador retrievers, which can be black, chocolate brown, or golden yellow (Figure 2.3). Which colour shows up depends on the allelic combinations of two independently assorting coat colour genes (**Figure 2.34a**). The dominant *B* allele of the first gene determines black, while the recessive *bb* homozygote is brown. The *E* gene controls the deposition of pigment in the hair shaft of Labrador retrievers. The dominant *E* allele has no visible effect on black or brown coat colour since pigment is deposited in the hair shaft, but a double dose of the recessive allele (*ee*) hides the effect

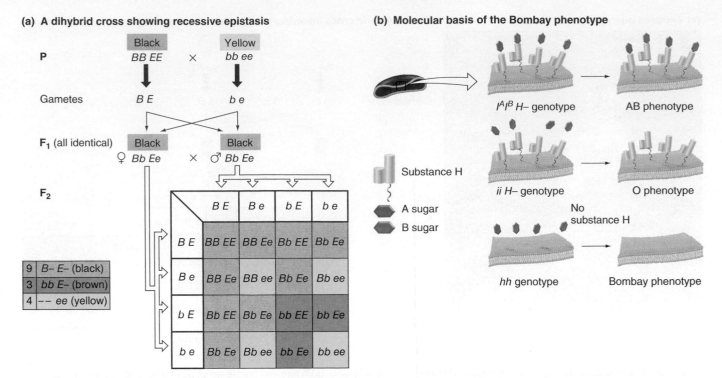

(a) A dihybrid cross showing recessive epistasis

P: Black *BB EE* × Yellow *bb ee*

Gametes: *B E*   *b e*

F₁ (all identical): Black ♀ *Bb Ee* × ♂ Black *Bb Ee*

F₂:

	B E	*B e*	*b E*	*b e*
B E	*BB EE*	*BB Ee*	*Bb EE*	*Bb Ee*
B e	*BB Ee*	*BB ee*	*Bb Ee*	*Bb ee*
b E	*Bb EE*	*Bb Ee*	*bb EE*	*bb Ee*
b e	*Bb Ee*	*Bb ee*	*bb Ee*	*bb ee*

9	*B– E–*	(black)
3	*bb E–*	(brown)
4	*– – ee*	(yellow)

(b) Molecular basis of the Bombay phenotype

I^A I^B H– genotype → AB phenotype

Substance H

A sugar

B sugar

ii H– genotype → O phenotype

No substance H

hh genotype → Bombay phenotype

Figure 2.34 Recessive epistasis: Coat colour in Labrador retrievers and a rare human blood type. (a) Golden Labrador retrievers are homozygous for the recessive *e* allele, which masks the effects of the *B* or *b* alleles of a second coat colour gene. In *E–* dogs, a *B–* genotype produces black and a *bb* genotype produces brown. **(b)** Homozygosity for the *h* Bombay allele is epistatic to the *I* gene determining ABO blood types. *hh* individuals fail to produce substance H, which is needed for the addition of A or B sugars at the surface of red blood cells.

of any combination of the *black* or *brown* alleles to yield gold. A gene interaction in which the effects of an allele at one gene hide the effects of alleles at another gene is known as **epistasis**; the allele that is doing the masking (in this case, the *e* allele of the *E* gene) is **epistatic** to the gene that is being masked (the *hypostatic gene*). In this example, where homozygosity for a recessive *e* allele of the second gene is required to hide the effects of another gene, the masking phenomenon is called **recessive epistasis** (because the allele causing the epistasis is recessive), and the recessive *ee* homozygote is considered epistatic to any allelic combination at the first gene.

Recessive Epistasis   Let us look at the phenomenon in greater detail. Crosses between pure-breeding black retrievers (*BB EE*) and one type of pure-breeding golden retriever (*bb ee*) create an F₁ generation of dihybrid black retrievers (*Bb Ee*). Crosses between these F₁ dihybrids produce an F₂ generation with nine black dogs (*B– E–*) for every three brown (*bb E–*) and four gold (*– – ee*) (Figure 2.34a). Note that there are only three phenotypic classes because the two genotypic classes without a dominant *E* allele—the three *B– ee* and the one *bb ee*—combine to produce golden phenotypes. The telltale ratio of recessive epistasis in the F₂ generation is thus 9:3:4, with the 4 representing a combination of 3 (*B– ee*) + 1 (*bb ee*). Although the *ee* genotype completely masks the influence of the other gene for coat colour, you can still tell by looking at a golden Labrador what its genotype is for the *black* or *brown* (*B* or *b*) gene by looking at the colour of its nose or lips since pigment is

still produced, just not deposited into the hair shaft. Thus, in a **developmental pathway**, such as the one for Labrador retriever coat colour, the epistatic gene (in this case the *E* gene) encodes a downstream step in the pathway—it is downstream of the pigment-producing *B* gene.

An understanding of recessive epistasis made it possible to resolve an intriguing puzzle in human genetics. In rare instances, two parents who appear to have blood type O, and thus genotype *ii*, can produce a child who is either blood type A (genotype *I^A i*) or blood type B (genotype *I^B i*). This phenomenon occurs because an extremely rare trait, called the Bombay phenotype after its discovery in Bombay, India, superficially resembles blood type O. As **Figure 2.34b** shows, the Bombay phenotype actually arises from homozygosity for a mutant recessive allele (*hh*) of a second gene that masks the effects of any ABO alleles that might be present.

Here is how it works at the molecular level. In the construction of the red blood cell surface molecules that determine blood type, type A individuals make an enzyme that adds polysaccharide A onto a base consisting of a sugar polymer known as substance H; type B individuals make an altered form of the enzyme that adds polysaccharide B onto the base; and type O individuals make neither A-adding nor B-adding enzymes and thus have an exposed substance H in the membranes of their red blood cells. All people of A, B, or O phenotype carry at least one dominant wild-type *H* allele for the second gene and thus produce some substance H. In contrast, the rare Bombay-phenotype individuals, with genotype *hh* for the second gene, do not make

substance H at all, so even if they make an enzyme that would add A or B to this polysaccharide base, they have nothing to add it onto; as a result, they appear to be type O. For this reason, homozygosity for the recessive *h* allele of the H-substance gene masks the effects of the ABO blood group gene, making the *hh* genotype epistatic to any combination of I^A, I^B, and *i* alleles. Thus, in a **biosynthetic** or **biochemical** pathway, such as the one for determination of blood type, the epistatic gene (in this case the *H* gene) codes for an upstream step in the pathway—it is located biochemically upstream of the *I* gene.

A person who carries I^A, I^B, or both I^A and I^B but is also an *hh* homozygote for the H-substance gene may appear to be type O, but he or she will be able to pass along an I^A or I^B allele in the sperm or egg. The offspring receiving, let us say, an I^A allele for the *ABO* gene and a recessive *h* allele for the H-substance gene from its mother plus an *i* allele and a dominant *H* allele from its father would have blood type A (genotype $I^A i$, *Hh*), even though neither of its parents is phenotype A or AB.

Dominant Epistasis Epistasis can also be caused by a dominant allele. In summer squash, two genes influence the colour of the fruit (**Figure 2.35a**). With one gene, the dominant allele (*A*–) determines yellow, while homozygotes for the recessive allele (*aa*) are green. A second gene's dominant allele (*B*–) produces white, while *bb* fruit

may be either yellow or green, depending on the genotype of the first gene. In the interaction between these two genes, the presence of *B* hides the effects of either *A*– or *aa*, producing white fruit, and *B*– is thus epistatic to any genotype of the *Aa* gene. The recessive *b* allele has no effect on fruit colour determined by the *Aa* gene. Epistasis in which the dominant allele of one gene hides the effects of another gene is called **dominant epistasis**. In a cross between white F_1 dihybrids (*Aa Bb*), the F_2 phenotypic ratio is 12 white : 3 yellow : 1 green (Figure 2.35a). The "12" includes two genotypic classes: 9 *A*– *B*– and 3 *aa B*–. Another way of looking at this same phenomenon is that dominant epistasis restores the 3:1 ratio for the dominant epistatic phenotype (12 white) versus all other phenotypes (4 green plus yellow).

A variation of this ratio is seen in the feather colour of certain chickens (**Figure 2.35b**). White leghorns have a doubly dominant *AA BB* genotype for feather colour; white wyandottes are homozygous recessive for both genes (*aa bb*). A cross between these two pure-breeding white strains produces an all-white dihybrid (*Aa Bb*) F_1 generation, but birds with colour in their feathers appear in the F_2, and the ratio of white to coloured is 13:3 (Figure 2.35b). We can explain this ratio by assuming a kind of dominant epistasis in which *B* is epistatic to *A*; the *A* allele (in the absence of *B*) produces colour; and the *a*, *B*, and *b* alleles produce no colour. The interaction is characterized by

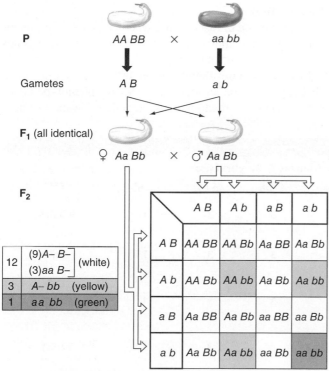

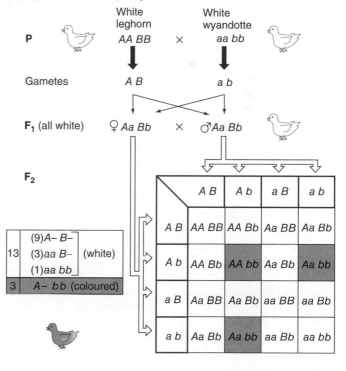

Figure 2.35 Dominant epistasis produces telltale phenotypic ratios of 12:3:1 or 13:3. (a) In summer squash, the dominant *B* allele causes white colour and is sufficient to mask the effects of any combination of *A* and *a* alleles. As a result, yellow (*A*–) or green (*aa*) colour is expressed only in *bb* individuals. **(b)** In the F_2 generation resulting from a dihybrid cross between white leghorn and white wyandotte chickens, the ratio of white birds to birds with colour is 13:3. This is because at least one copy of *A* and the absence of *B* is needed to produce colour.

a 13:3 ratio because the 9 *A– B–*, 3 *aa B–*, and 1 *aa bb* genotypic classes combine to produce only one phenotype: white. This type of epistatic interaction is also referred to as **dominant suppression**.

So far we have seen that when two independently assorting genes interact to determine a trait, the 9:3:3:1 ratio of the four Mendelian genotypic classes in the F_2 generation can produce a variety of phenotypic ratios, depending on the nature of the gene interactions. The result may be four, three, or two phenotypes, composed of different combinations of the four genotypic classes. **Table 2.4** summarizes some of the possibilities, correlating the phenotypic ratios with the genetic phenomena they reflect.

Heterogeneous traits and the complementation test

Approximately 50 different genes have mutant alleles that can cause deafness in humans. Many genes generate the developmental pathway that brings about hearing, and a loss of function in any part of the pathway, for instance, in one small bone of the middle ear, can result in deafness. In other words, it takes a dominant wild-type allele at each of these 50 or so genes to produce normal hearing. Thus, deafness is a **heterogeneous trait**: A mutation at any one of a number of genes can give rise to the same phenotype.

It is not always possible to determine which of many different genes has mutated in a person who expresses a heterogeneous mutant phenotype. In the case of deafness, for example, it is usually not possible to discover whether a particular nonhearing man and a particular nonhearing woman carry mutations at the same gene, unless they have children together. If they have only children who can hear, the parents most likely carry mutations at two different genes, and the children carry one normal, wild-type allele at both of those genes (**Figure 2.36a**). By contrast, if all of their children are deaf, it is likely that both parents are homozygous for a mutation in the same gene, and all of

(a) Complementation: mutations in two different genes

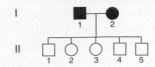

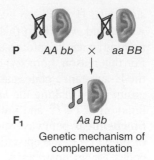

Genetic mechanism of complementation

(b) Noncomplementation: mutations in the same gene

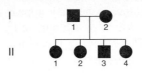

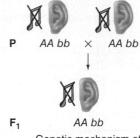

Genetic mechanism of noncomplementation

Figure 2.36 Genetic heterogeneity in humans: Mutations in many genes can cause deafness. (a) Two deaf parents can have hearing offspring. This situation is an example of genetic complementation; it occurs if the nonhearing parents are homozygous for recessive mutations in different genes. **(b)** Two deaf parents may produce all deaf children. In such cases, complementation does not occur because both parents carry mutations in the same gene.

their children are also homozygous for this same mutation (**Figure 2.36b**).

This method of discovering whether a particular phenotype arises from mutations in the same or separate genes is a naturally occurring version of an experimental genetic tool called the **complementation test**. Simply put, when what appears to be an identical *recessive* phenotype arises in two separate breeding lines, geneticists want to know

Table 2.4	Summary of Discussed Gene Interactions					
		F_2 **Genotypic Ratios from an** F_1 **Dihybrid Cross**				F_2 **Phenotypic Ratio**
Gene Interaction	**Example**	*A– B–*	*A– bb*	*aa B–*	*aa bb*	
None: Four distinct F_2 phenotypes	Lentil: seed coat colour (see Figure 2.31a)	9	3	3	1	**9:3:3:1**
Complementary: One dominant allele of each of two genes is necessary to produce phenotype	Sweet pea: flower colour (see Figure 2.32b)	9	3	3	1	**9:7**
Recessive epistasis: Homozygous recessive of one gene masks both alleles of another gene	Retriever coat colour (see Figure 2.34a)	9	3	3	1	**9:3:4**
Dominant epistasis I: Dominant allele of one gene hides effects of both alleles of another gene	Summer squash: colour (see Figure 2.35a)	9	3	3	1	**12:3:1**
Dominant epistasis II (also referred to as dominant suppression): Dominant allele of one gene hides effects of dominant allele of another gene	Chicken: feather colour (see Figure 2.35b)	9	3	3	1	**13:3**

whether mutations at the same gene are responsible for the phenotype in both lines. They answer this question by setting up a mating between affected individuals from the two lines. If offspring receiving the two mutations— one from each parent—express the wild-type phenotype, **complementation** has occurred. The observation of complementation means that the original mutations affected two different genes, and for both genes, the normal allele from one parent can provide what the mutant allele of the same gene from the other parent cannot. Figure 2.36a illustrates one example of this phenomenon in humans. By contrast, if offspring receiving two recessive mutant alleles—again, one from each parent—express the mutant phenotype, complementation does not occur because the two mutations independently alter the same gene—therefore, the two mutations are alleles of the same gene (Figure 2.36b). Thus, the occurrence of complementation reveals genetic heterogeneity. Note that complementation tests cannot be used if either of the mutations is dominant to the wild type. Chapter 8 includes an in-depth discussion of complementation tests and their uses.

To summarize, several variations on the theme of gene interactions and genetic heterogeneity can be identified:

1. Genes can interact to generate novel phenotypes.
2. The dominant alleles of two interacting genes can both be necessary for the production of a particular phenotype.
3. One gene's alleles can mask the effects of alleles at another gene.
4. Mutant alleles at one of two or more different genes can result in the same phenotype.

In examining each of these categories, for the sake of simplicity, we have looked at examples in which one allele of each gene in a pair showed complete dominance over the other. But for any type of gene interaction, the alleles of one or both genes may exhibit incomplete dominance or codominance, and these possibilities increase the potential for phenotypic diversity. For example, **Figure 2.37** shows how incomplete dominance at both genes in a dihybrid cross generates additional phenotypic variation.

Although the possibilities for variation are manifold, none of the observed departures from Mendelian phenotypic ratios contradicts Mendel's genetic laws of segregation and independent assortment. The alleles of each gene still segregate as he proposed. Interactions between the alleles of many genes simply make it harder to unravel the complex relation of genotype to phenotype.

> F$_2$ phenotypic ratios of 9:3:3:1 or its derivatives indicate the combined action of two independently assorting genes. For heterogeneous traits caused by recessive alleles of two or more genes, a mating between affected individuals acts as a complementation test, revealing whether they carry mutations in the same gene or in different genes.

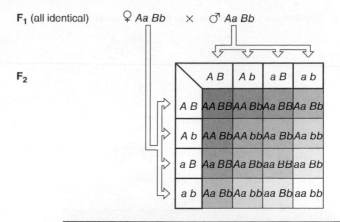

1	AA BB	purple shade 9
2	AA Bb	purple shade 8
2	Aa BB	purple shade 7
1	AA bb	purple shade 6
4	Aa Bb	purple shade 5
1	aa BB	purple shade 4
2	Aa bb	purple shade 3
2	aa Bb	purple shade 2
1	aa bb	purple shade 1 (white)

Figure 2.37 With incomplete dominance, the interaction of two genes can produce nine different phenotypes for a single trait. In this example, two genes produce purple pigments. Alleles A and a of the first gene exhibit incomplete dominance, as do alleles B and b of the second gene. The two alleles of each gene can generate three different phenotypes, so double heterozygotes can produce nine (3 × 3) different colours in a ratio of 1:2:2:1:4:1:2:2:1.

Breeding studies help decide how a trait is inherited

How do geneticists know whether a particular trait is caused by the alleles of one gene or by two genes interacting in one of a number of possible ways? Breeding tests can usually resolve the issue. Phenotypic ratios diagnostic of a particular mode of inheritance (e.g., the 9:7 or 13:3 ratios indicating that two genes are interacting) can provide the first clues and suggest hypotheses. Further breeding studies can then show which hypothesis is correct. We have seen, for example, that yellow coat colour in mice is determined by a dominant allele of the *agouti* gene, which also acts as a recessive lethal. We now look at two other mouse genes for coat colour. Because we have already designated alleles of the *agouti* gene as *Aa*, we use *Bb* and *Cc* to designate the alleles of these additional genes.

A mating of one strain of pure-breeding white albino mice with pure-breeding brown results in black hybrids; and a cross between the black F$_1$ hybrids produces 90 black, 30 brown, and 40 albino offspring. What is the genetic constitution of these phenotypes? We could assume that we are seeing the 9:3:4 ratio of recessive epistasis and hypothesize that two genes, one epistatic to the other, interact to produce the three mouse phenotypes (**Figure 2.38a**). But how

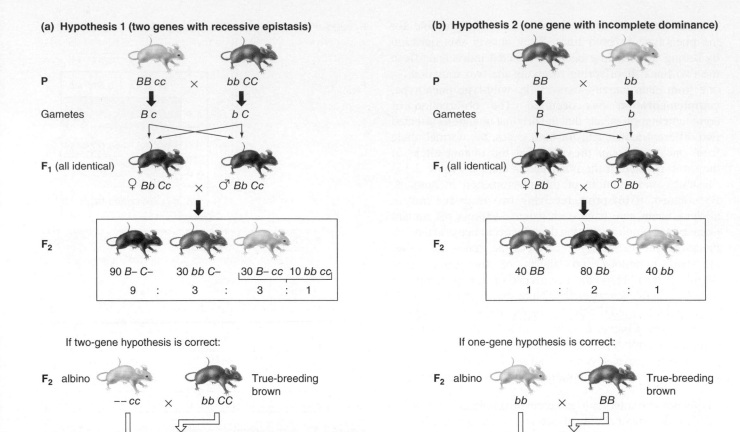

(a) Hypothesis 1 (two genes with recessive epistasis)

P   BB cc   ×   bb CC

Gametes   B c   b C

F₁ (all identical)

♀ Bb Cc   ×   ♂ Bb Cc

F₂

90 B– C–   30 bb C–   30 B– cc   10 bb cc
9   :   3   :   3   :   1

If two-gene hypothesis is correct:

F₂ albino   – – cc   ×   bb CC   True-breeding brown

	b C
B c	Bb Cc

or

	b C
B c	Bb Cc
b c	bb Cc

or

	b C
b c	bb Cc

(b) Hypothesis 2 (one gene with incomplete dominance)

P   BB   ×   bb

Gametes   B   b

F₁ (all identical)

♀ Bb   ×   ♂ Bb

F₂

40 BB   80 Bb   40 bb
1   :   2   :   1

If one-gene hypothesis is correct:

F₂ albino   bb   ×   BB   True-breeding brown

	B
b	Bb

Figure 2.38 Specific breeding tests can help decide between hypotheses. Either of two hypotheses could explain the results of a cross tracking coat colour in mice. **(a)** In one hypothesis, two genes interact with recessive epistasis to produce a 9:3:4 ratio. **(b)** In the other hypothesis, a single gene with incomplete dominance between the alleles generates the observed results. One way to decide between these models is to cross each of several albino F₂ mice with true-breeding brown mice. The two-gene model predicts several different outcomes depending on the – – cc albino's genotype at the *B* gene. The one-gene model predicts that all progeny of all the crosses will be black.

do we know if this hypothesis is correct? We might also explain the data—160 progeny in a ratio of 90:30:40—by the activity of one gene (**Figure 2.38b**). According to this one-gene hypothesis, albinos would be homozygotes for one allele (*bb*), brown mice would be homozygotes for a second allele (*BB*), and black mice would be heterozygotes (*Bb*) that have their own "intermediate" phenotype because *B* shows incomplete dominance over *b*. Under this system, a mating of black (*Bb*) to black (*Bb*) would be expected to produce 1 *BB* brown : 2 *Bb* black : 1 *bb* albino, or 40 brown : 80 black : 40 albino. Is it possible that the 30 brown, 90 black, and 40 albino mice actually counted were obtained from the inheritance of a single gene? Intuitively, the answer is yes: the ratios 40:80:40 and 30:90:40 do not seem that different. We know that if we flip a coin 100 times, it does not always come up 50 heads : 50 tails; sometimes it is 60:40 just by chance. So, how can we decide between the two-gene versus the one-gene model?

The answer is that we can use other types of crosses to verify or refute the hypotheses. For instance, if the one-gene hypothesis were correct, a mating of pure white F₂ albinos with pure-breeding brown mice similar to those

of the parental generation would produce all black heterozygotes (brown [*BB*] × albino [*bb*] = all black [*Bb*]) (Figure 2.38b). But if the two-gene hypothesis is correct, with recessive mutations at an albino gene (called *C*) epistatic to all expression from the *B* gene, different matings of pure-breeding brown (*bb CC*) with the F₂ albinos (– – *cc*) will give different results—all progeny are black; half are black and half brown; all are brown—depending on the albino's genotype at the *B* gene (see Figure 2.38a). In fact, when the experiment is actually performed, the diversity of results confirms the two-gene hypothesis. The comprehensive example presented at the end of this chapter outlines additional details of the interactions of the three mouse genes for coat colour.

With humans, pedigree analysis replaces breeding experiments

Breeding experiments cannot be applied to humans, for obvious ethical reasons. But a careful examination of as many family pedigrees as possible can help elucidate the genetic basis of a particular condition.

In a form of albinism known as ocular-cutaneous albinism (OCA), for example, people with the inherited condition have little or no pigment in their skin, hair, and eyes (**Figure 2.39a**). The horizontal inheritance pattern seen in **Figure 2.39b** suggests that OCA is determined by the recessive allele of one gene, with albino family members being homozygotes for that allele. But a 1952 paper on albinism reported a family in which two albino parents produced three normally pigmented children (**Figure 2.39c**). How would you explain this phenomenon?

The answer is that albinism is another example of heterogeneity: Mutant alleles at any one of several different genes can cause the condition. The reported mating

(a) Ocular-cutaneous albinism (OCA)

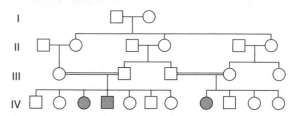

(b) OCA is recessive

(c) Complementation for albinism

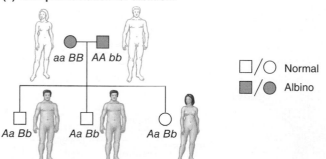

□/○ Normal
■/● Albino

Figure 2.39 Family pedigrees help unravel the genetic basis of ocular-cutaneous albinism (OCA). (a) An albino Nigerian girl and her sister celebrating the conclusion of the All Africa games. **(b)** A pedigree following the inheritance of OCA in an inbred family indicates that the trait is recessive. **(c)** A family in which two albino parents have nonalbino children demonstrates that homozygosity for a recessive allele of either of two genes can cause OCA.

was, in effect, an inadvertent complementation test, which showed that one parent was homozygous for an OCA-causing mutation in gene *A*, while the other parent was homozygous for an OCA-causing mutation in a different gene, *B* (compare with Figure 2.36).

The same genotype does not always produce the same phenotype

In our discussion of gene interactions so far, we have looked at examples in which a genotype reliably fashions a particular phenotype. But this is not always what happens. Sometimes a genotype is not expressed at all; that is, even though the genotype is present, the expected phenotype does not appear. Other times, the trait caused by a genotype is expressed to varying degrees or in a variety of ways in different individuals. Factors that alter the phenotypic expression of genotype include modifier genes, the environment (in the broadest sense, as defined earlier), and chance.

Penetrance and expressivity

Geneticists use the term **penetrance** to describe how many members of a population with a particular genotype show the expected phenotype. Penetrance can be *complete* (100 percent), as in the traits that Mendel studied, or *incomplete*, as in hereditary juvenile glaucoma (see the **Genetics and Society** box **"Disease Prevention Versus the Right to Privacy"** in this chapter).

Expressivity refers to the degree or intensity with which a particular genotype is expressed at the phenotypic level. Expressivity can be *variable*, as in retinoblastoma, the most malignant form of eye cancer, where one or both eyes may be affected, or *unvarying*, as in pea colour. Incomplete penetrance and variable expressivity can be the result of chance, but modifier genes and/or the environment can also cause such variations in the appearance of phenotype.

Modifier genes

Not all genes that influence the appearance of a trait contribute equally to the phenotype. Major genes have a large influence, while **modifier genes** have a more subtle, secondary effect. Modifier genes alter the phenotypes produced by the alleles of other genes. There is no formal distinction between major and modifier genes. Rather, there is a continuum between the two, and the cutoff is arbitrary.

Modifier genes influence the length of a mouse's tail. The mutant *T* allele of the tail-length gene causes a shortening of the normally long wild-type tail. But not all mice carrying the *T* mutation have the same tail length. A comparison of several inbred lines points to modifier genes as the cause of this variable expressivity. In one inbred line, mice carrying the *T* mutation have tails that are approximately 75 percent as long as normal tails; in another inbred line, the tails are 50 percent of normal length; and in a third line, the tails are only 10 percent as long as wild-type tails.

Because all members of each inbred line grow the same length tail, no matter what the environment (e.g., diet, cage temperature, or bedding), geneticists conclude it is genes and not the environment or chance that determines the length of a mutant mouse's tail. Different inbred lines most likely carry different alleles of the modifier genes that determine exactly how short the tail will be when the *T* mutation is present.

Environmental effects on phenotype

Temperature is one element of the environment that can have a visible effect on phenotype. For example, temperature influences the unique coat colour pattern of Siamese cats (**Figure 2.40**). These domestic felines are homozygous for one of the multiple alleles of a gene that encodes an enzyme catalyzing the production of the dark pigment melanin. The form of the enzyme generated by the variant "*Siamese*" allele does not function at the cat's normal body temperature. It becomes active only at the lower temperatures found in the cat's extremities, where it promotes the production of melanin, which darkens the animal's ears, nose, paws, and tail. The enzyme is thus *temperature sensitive*. Under the normal environmental conditions in temperate climates, the Siamese phenotype does not vary much in expressivity from one cat to another, but one can imagine the expression of a very different phenotype—no dark extremities—in equatorial deserts, where the ambient temperature is at or above normal body temperature.

Temperature can also affect survivability. In one type of experimentally bred fruit fly (*Drosophila melanogaster*), some individuals develop and multiply normally at temperatures between 18°C and 29°C; but if the thermometer climbs beyond that cutoff for a short time, they become reversibly paralyzed, and if the temperature remains high for more than a few hours, they die. These insects carry a temperature-sensitive (*ts*) allele of the *shibire* gene, which encodes a protein essential for nerve cell transmission. This type of allele is known as a **conditional lethal** because it is lethal only under certain conditions. The range of temperatures under which the insects remain viable are **permissive conditions**; the lethal temperatures above that are **restrictive conditions**. Thus, at one temperature, the allele gives rise to a phenotype that is indistinguishable from the wild type, while at another temperature, the same allele generates a mutant phenotype (in this case, lethality). Flies with the wild-type *shibire* allele are viable even at the higher temperatures. The fact that some mutations are lethal only under certain conditions clearly illustrates that the environment can affect the penetrance of a phenotype. These temperature-sensitive *shibire* alleles in *Drosophila* were isolated by Canadian researcher and environmentalist Dr. David Suzuki (see the **Tools of Genetics** box **"Temperature-Sensitive Mutations"** in this chapter).

Some types of environmental change may have a positive effect on an organism's survivability, as in the following example, where a straightforward application of medical science artificially reduces the penetrance of a mutant phenotype. Children born with the recessive trait known as phenylketonuria, or PKU, will develop a range of neurological problems, including convulsive seizures and mental retardation, unless they are put on a special diet. Homozygosity for the mutant PKU allele eliminates the activity of a gene encoding the enzyme phenylalanine hydroxylase. This enzyme normally converts the amino acid phenylalanine to the amino acid tyrosine. Absence of the enzyme causes a buildup of phenylalanine, and this buildup results in neurological problems. Today, a reliable blood test can detect the condition in newborns. Once a baby with PKU is identified, a protective diet that excludes

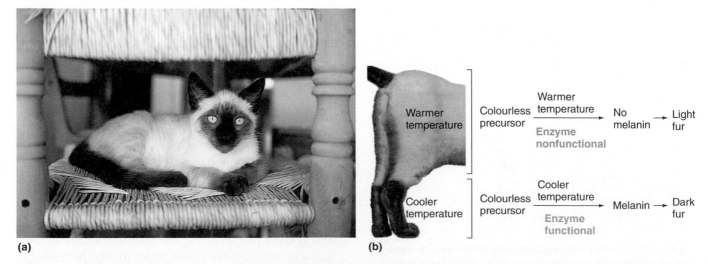

(a) (b)

Figure 2.40 In Siamese cats, temperature affects coat colour. (a) A Siamese cat. (b) Melanin is produced only in the cooler extremities. This is because Siamese cats are homozygous for a mutation that renders an enzyme involved in melanin synthesis temperature sensitive. The mutant enzyme is active at lower temperatures but inactive at higher temperatures.

phenylalanine is prescribed; the diet must also provide enough calories to prevent the infant's body from breaking down its own proteins, thereby releasing the damaging amino acid from within. Such dietary therapy—a simple change in the environment—now enables many PKU infants to develop into healthy adults.

Finally, two of the top killer diseases in Canada and the United States—cardiovascular disease and lung cancer—also illustrate how the environment can alter phenotype by influencing both expressivity and penetrance. People may inherit a propensity to heart disease, but the environmental factors of diet and exercise contribute to the occurrence (penetrance) and seriousness (expressivity) of their condition. Similarly, some people are born genetically prone to lung cancer, but whether or not they develop the disease (penetrance) is strongly determined by whether they choose to smoke.

Thus, various aspects of an organism's environment, including temperature, diet, and exercise, interact with its genotype to generate the functional phenotype, the ultimate combination of traits that determines what a plant or animal looks like and how it behaves.

By contributing to incomplete penetrance and variable expressivity, modifier genes, the environment, and chance give rise to phenotypic variation. Unlike dominant epistasis or recessive lethality, however, the probability of penetrance and the level of expressivity cannot be derived from the original Mendelian principles of segregation and independent assortment; they are determined empirically by observation and counting.

> Because modifier genes, the environment, and chance events can affect phenotypes, the relationship of a particular genotype and its corresponding phenotype is not always absolute: An allele's penetrance can be incomplete, and its expressivity can be variable.

Tools of Genetics   Temperature-Sensitive Mutations

Dr. David Suzuki (**Figure A**) is a Canadian icon who has been known internationally for many decades as an advocate of scientific and environmental issues. Although most individuals may know him best as an environmentalist, a broadcaster, and a world leader in sustainable ecology, Dr. Suzuki is also an award-winning scientist. Born in Vancouver, British Columbia, a strong interest in genetics early on in his academic career led him to study *Drosophila melanogaster*, the common fruit fly, as a model in his studies, which gained him worldwide recognition. In 1963, David Suzuki became a faculty member at the University of British Columbia (UBC) and was appointed Professor of Zoology in 1969. (He is now Professor Emeritus at UBC.) While at UBC, he conducted large-scale genetic screens and isolated numerous *Drosophila* mutations, although his main focus was the study of temperature-sensitive (*ts*) mutations, including *ts* alleles of the *shibire* and *paralytic* genes, which are conditional lethal alleles that affect neurological development. These *ts* alleles are useful as tools for investigations into the nature of genetic lesions. By manipulating the conditional and restrictive conditions, one could determine at what developmental stage(s) lethality occurs, as well as any phenotypically distinct (or pleiotropic) defects that may arise upon brief exposure to the restrictive temperature at various stages of development.

Over his prolific career Dr. Suzuki has authored numerous books and articles for a variety of audiences, and has hosted many television and radio shows on science and nature, including CBC's *The Nature of Things*. Co-founder and chair of the *David Suzuki Foundation*, an environmental nonprofit organization registered in Canada, he has received several awards and distinctions, including Officer of the Order of Canada, Companion of the Order of Canada, Fellow of the Royal Society of Canada, Fellow of the American Association for the Advancement of Science, the E.W.R. Steacie Memorial Fellowship, the United Nations Environmental Program Medal, UNESCO's Kalinga Prize for Science, the John Drainie Award for broadcasting excellence, and four Gemini Awards as best host of various Canadian television series. His popularization of many scientific and environmental issues, including climate change, pollution, sustainable ecology, genetically modified foods, and alternative energy sources, has helped to raise awareness about them around the globe.

Figure A David Suzuki

In one of the most extensive human pedigrees ever assembled, a team of researchers traced a familial pattern of blindness back through five centuries of related individuals to its origin in a couple who died in a small town in northwestern France in 1495. More than 30 000 French men and women alive today descended from that one fifteenth-century couple, and within this direct lineage reside close to half of all reported French cases of hereditary juvenile glaucoma. The massive genealogical tree for the trait (when posted on the office wall, it was over 30 m [100 ft] long) showed that the genetic defect follows a simple Mendelian pattern of transmission determined by the dominant allele of a single gene (**Figure A**). The pedigree also showed that the dominant genetic defect displays incomplete penetrance: Not all people receiving the dominant allele become blind; these sighted carriers may unknowingly pass the blindness-causing dominant allele to their children.

Unfortunately, people do not know they have the disease until their vision starts to deteriorate. By that time, their optic fibres have sustained irreversible damage, and blindness is all but inevitable. Surprisingly, the existence of medical therapies that make it possible to arrest the nerve deterioration created a quandary in the late 1980s. Because treatment, to be effective, has to begin before symptoms of impending blindness show up, information in the pedigree could have helped doctors pinpoint people who are at risk, even if neither of their parents is blind. The researchers who compiled the massive family history therefore wanted to give physicians the names of at-risk individuals living in their area, so that doctors could monitor certain patients and recommend treatment if needed. However, a long-standing French law protecting personal privacy forbids public circulation of the names in genetic pedigrees. The French government agency interpreting this law maintained that if the names in the glaucoma pedigree were made public, potential carriers of the disease might suffer discrimination in hiring or insurance.

France thus faced a serious ethical dilemma: On the one hand, giving out names could save perhaps thousands of people from blindness; on the other hand, laws designed to protect personal privacy precluded the dissemination of specific names. The solution adopted by the French government at the time was a massive educational program to alert the general public to the problem so that concerned families could seek medical advice. This approach addressed the legal issues but was only partially helpful in dealing with the medical problem, because many affected individuals escaped detection.

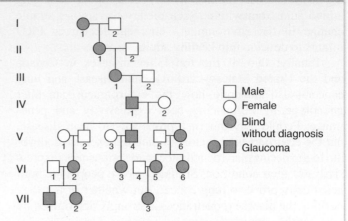

Figure A A pedigree showing the transmission of juvenile glaucoma. A small part of the genealogical tree: The vertical transmission pattern over seven generations shows that a dominant allele of a single gene causes juvenile glaucoma. The lack of glaucoma in V-2 followed by its reappearance in VI-2 reveals that the trait is incompletely penetrant. As a result, sighted heterozygotes may unknowingly pass the condition on to their children.

By 1997, molecular geneticists had identified the gene whose dominant mutant allele causes juvenile glaucoma. This gene specifies a protein called myocilin whose normal function in the eye is at present unknown. The mutant allele encodes a form of myocilin that folds incorrectly and then accumulates abnormally in the tiny canals through which eye fluid normally drains into the bloodstream. Misfolded myocilin blocks the outflow of excess vitreous humor, and the resulting increased pressure within the eye (glaucoma) eventually damages the optic nerve, leading to blindness.

Knowledge of the specific disease-causing mutations in the *myocilin* gene has more recently led to the development of diagnostic tests based on the direct analysis of genotype. (We describe methods for direct genotype analysis in Chapters 14 and 15.) Not only can these DNA-based tests identify individuals at risk, but they can also improve disease management. Detection of the mutant allele before the optic nerve is permanently damaged allows for timely treatment. If these tests become sufficiently inexpensive in the future, they could resolve France's ethical dilemma. Doctors could routinely administer the tests to all newborns and immediately identify nearly all potentially affected children; private information in a pedigree would thus not be needed.

Mendelian principles can also explain continuous variation

In Mendel's experiments, height in pea plants was determined by two segregating alleles of one gene (in the wild, it is determined by many genes, but in Mendel's inbred populations, the alleles of all but one of these genes were invariant). The phenotypes that resulted from these alternative alleles were clear-cut, either short or tall, and pea plant height was therefore known as a **discontinuous trait**. In contrast, because people do not produce inbred populations, height in humans is determined by segregating

alleles of many different genes whose interaction with each other and the environment produces continuous variation in the phenotype; height in humans is thus an example of a **continuous trait**. Within human populations, individual heights vary over a range of values that when charted on a graph produces a bell curve. In fact, many human traits, including height, weight, and skin colour, show continuous variation, rather than the clear-cut alternatives analyzed by Mendel. The more genes, the more phenotypic classes, and the more classes, the more the variation appears continuous.

As a hypothetical example, consider a series of genes (A, B, C, . . .) all affecting the height of pole beans. For each gene, there are two alleles, a "0" allele that contributes nothing to height and a "1" allele that increases the height of a plant by one unit. All alleles exhibit incomplete dominance relative to alternative alleles at the same gene. The phenotypes determined by all these genes are additive. What would be the result of a two-generation cross between pure-breeding plants carrying only 0 alleles at each height gene and pure-breeding plants carrying only 1 alleles at each height gene? If only one gene were responsible for height, and if environmental effects could be discounted, the F_2 population would be distributed among three classes: homozygous A^0A^0 plants with 0 height (they lie prostrate on the ground); heterozygous A^0A^1 plants with a height of 1; and homozygous A^1A^1 plants with a height of 2 (**Figure 2.41a**). This distribution of heights over three phenotypic classes does not make

a continuous curve. But for two genes, there will be five phenotypic classes in the F_2 generation (**Figure 2.41b**); for three genes, seven classes (**Figure 2.41c**); and for four genes, nine classes (not shown).

The distributions produced by three and four genes thus begin to approach continuous variation, and if we add a small contribution from environmental variation, a smoother curve will appear. After all, we would expect bean plants to grow better in good soil, with ample sunlight and water. The environmental component effectively converts the stepped bar graph to a continuous curve by producing some variation in expressivity within each genotypic class. Moreover, additional variation might arise from more than two alleles at some genes (**Figure 2.41d**), unequal contribution to the phenotype by the various genes involved (review Figure 2.37), interactions with modifier genes, and chance. Thus, from what we now know about the relation between genotype and phenotype, it is possible to see how just a handful of genes that behave according to known Mendelian principles can easily generate continuous variation.

Continuous traits (also called **quantitative**, **multifactorial**, or **complex** traits) vary over a range of values and can usually be measured: the length of a tobacco flower in millimetres, the amount of milk produced by a cow per day in litres, or the height of a person in metres. Continuous traits are **polygenic**—controlled by multiple genes—and show the additive effects of a large number of alleles, which creates an enormous potential for

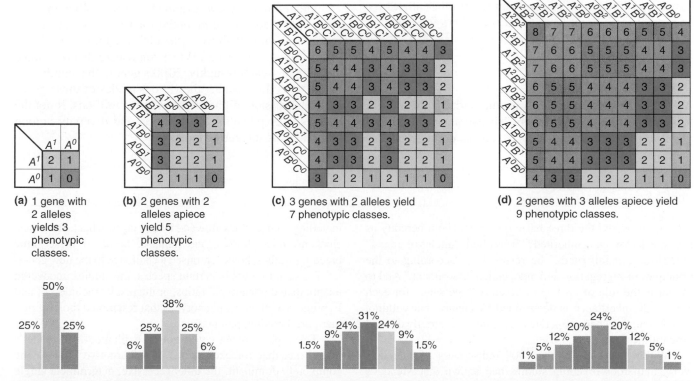

(a) 1 gene with 2 alleles yields 3 phenotypic classes.

(b) 2 genes with 2 alleles apiece yield 5 phenotypic classes.

(c) 3 genes with 2 alleles yield 7 phenotypic classes.

(d) 2 genes with 3 alleles apiece yield 9 phenotypic classes.

Figure 2.41 A Mendelian explanation of continuous variation. The more genes or alleles, the more possible phenotypic classes, and the greater the similarity to continuous variation. In these examples, several pairs of incompletely dominant alleles have additive effects. Percentages shown at the *bottom* denote frequencies of each genotype expressed as fractions of the total population.

variation within a population. Differences in the environments encountered by different individuals contribute even more variation. We discuss the analysis and distribution of multifactorial traits in Chapter 12 on population genetics.

> The action of a handful of genes, combined with environmental effects, can produce an enormous range of phenotypic variation for a particular trait.

A comprehensive example: Mouse coat colour is determined by multiple alleles of several genes

Most field mice are a dark grey (agouti), but mice bred for specific mutations in the laboratory can be grey, tan, yellow, brown, black, or various combinations thereof. Here we look at the alleles of three of the genes that make such variation possible. This review underscores how allelic interactions of just a handful of genes can produce an astonishing diversity of phenotypes.

Gene 1: Agouti or other colour patterns

The *agouti* gene determines the distribution of colour on each hair, and it has multiple alleles. The wild-type allele A specifies bands of yellow and black that give the agouti appearance; A^y gets rid of the black and thus produces solid yellow; a gets rid of the yellow and thus produces solid black; and a^t specifies black on the animal's back and yellow on the belly. The dominance series for this set of *agouti* gene alleles is $A^y > A > a^t > a$. However, although A^y is dominant to all other alleles for coat colour, it is recessive to all the others for lethality: A^yA^y homozygotes die before birth, while A^yA, A^ya^t, or A^ya heterozygotes survive.

Gene 2: Black or brown

A second gene specifies whether the dark colour of each hair is black or brown. This gene has two alleles: B is dominant and designates black; b is recessive and generates

brown. Because the A^y allele at the *agouti* gene completely eliminates the dark band of each hair, it acts in a dominant epistatic manner to the B gene. With all other *agouti* alleles, however, it is possible to distinguish the effects of the two different B alleles on phenotype. The A– B– genotype gives rise to the wild-type agouti having black with yellow hairs. The A– bb genotype generates a colour referred to as cinnamon (with hairs having stripes of brown and yellow); $aa\ bb$ is all brown; and $a^ta^t\ bb$ is brown on the animal's back and yellow on the belly. A cross between two F_1 hybrid animals of genotype $A^ya\ Bb$ would yield an F_2 generation with yellow ($A^ya\ –\ –$), black ($aa\ B$–), and brown ($aa\ bb$) animals in a ratio of 8:3:1. This ratio reflects the dominant epistasis of A^y and the loss of a class of four ($A^yA^y\ –\ –$) due to prenatal lethality.

Gene 3: Albino or pigmented

Like other mammals, mice have a third gene influencing coat colour. A recessive allele (c) abolishes the function of the enzyme that leads to the formation of the dark pigment melanin, making this allele epistatic to all other coat colour genes. As a result, cc homozygotes are pure white, while C– mice are agouti, black, brown, yellow, or yellow and black (or other colours and patterns), depending on what alleles they carry at the A and B genes, as well as at some 50 or so other genes known to play a role in determining the coat colour of mice. Adding to the complex colour potential are other alleles that geneticists have uncovered for the albino gene; these cause only a partial inactivation of the melanin-producing enzyme and thus have a partial epistatic effect on phenotype.

This comprehensive example of coat colour in mice gives some idea of the potential for variation from just a few genes, some with multiple alleles. Amazingly, this is just the tip of the iceberg. When you realize that both mice and humans carry roughly 20 000 genes, the number of interactions that connect the various alleles of these genes in the expression of phenotype is in the millions, if not the billions. The potential for variation and diversity among individuals is staggering indeed.

Connections

Mendel answered the three basic questions about heredity as follows: To "What is inherited?" he replied, "alleles of genes." To "How is it inherited?" he responded, "according to the principles of segregation and independent assortment." And to "What is the role of chance in heredity?" he said, "for each individual, inheritance is determined by chance, but within a population, this chance operates in a context of strictly defined probabilities."

Within a decade of the 1900 rediscovery of Mendel's work, numerous breeding studies had shown that Mendel's laws hold true not only for seven pairs of antagonistic characteristics in peas, but for an enormous diversity of traits in a wide variety of sexually reproducing plant and animal species,

including four-o'clock flowers, beans, corn, wheat, fruit flies, chickens, mice, horses, and humans. Some of these same breeding studies, however, raised a challenge to the new genetics. For certain traits in certain species, the studies uncovered unanticipated phenotypic ratios, or the results included F_1 and F_2 progeny with novel phenotypes that resembled those of neither pure-breeding parent.

These phenomena could not be explained by Mendel's hypothesis that for each gene, two alternative alleles, one completely dominant, the other recessive, determine a single trait. We now know that most traits, including skin colour, eye colour, and height in humans, are determined by interactions between two or more genes. We also know that within

a given population, more than two alleles may be present for some of those genes.

Simple embellishments provided explanations for apparent exceptions to Mendelian analysis. These embellishments included the ideas that dominance need not be complete; that one gene can have multiple alleles; that one gene can determine more than one trait; that several genes can contribute to the same trait; and that the expression of genes can be affected in a variety of ways by other genes, the environment, and chance. Each embellishment extends the range of Mendelian analysis and deepens our understanding of the genetic basis of variation. And no matter how broad the view, Mendel's basic conclusions, embodied in his first law of segregation, remain valid.

But what about Mendel's second law that genes assort independently? As it turns out, its application is not as universal as that of the law of segregation. Many genes do assort independently, but some do not; rather, they appear to be linked and transmitted together from generation to generation. An understanding of this fact emerged from studies that located Mendel's hereditary units, the genes, in specific cellular organelles, the chromosomes. In describing how researchers deduced that genes travel with chromosomes, Chapters 3 and 4 establish the physical basis of inheritance, including the segregation of alleles, and clarify why some genes assort independently while others do not.

Essential Concepts

1. Discrete units called genes control the appearance of inherited traits. [LO2]

2. Genes come in alternative forms called alleles that are responsible for the expression of different forms of a trait. [LO2]

3. Body cells of sexually reproducing organisms carry two copies of each gene. When the two copies of a gene are the same allele, the individual is homozygous for that gene. When the two copies of a gene are different alleles, the individual is heterozygous for that gene. [LO4]

4. The genotype is a description of the allelic combination of the two copies of a gene present in an individual. The phenotype is the observable form of the trait that the individual expresses. [LO4]

5. A cross between two parental lines (P) that are pure-breeding for alternative alleles of a gene will produce a first filial (F_1) generation of hybrids that are heterozygous. The phenotype expressed by these hybrids is determined by the dominant allele of the pair, and this phenotype is the same as that expressed by individuals homozygous for the dominant allele. The phenotype associated with the recessive allele will reappear only in the F_2 generation in individuals homozygous for this allele. In crosses between F_1 heterozygotes, the dominant and recessive phenotypes will appear in the F_2 generation in a ratio of 3:1. [LO1–2]

6. The two copies of each gene segregate during the formation of gametes. As a result, each egg and each sperm or pollen grain contains only one copy, and thus, only one allele, of each gene. Male and female gametes unite at random at fertilization. Mendel described this process as the law of segregation. [LO3]

7. The segregation of alleles of any one gene is independent of the segregation of the alleles of other genes. Mendel described this process as the law of independent assortment. According to this law, crosses between *Aa Bb* F_1 dihybrids will generate F_2 progeny with a phenotypic ratio of 9 (*A– B–*) : 3 (*A– bb*) : 3 (*aa B–*) : 1 (*aa bb*). [LO5]

8. The F_1 phenotype defines the dominance relationship between each pair of alleles. One allele is not always completely dominant or completely recessive to another. With incomplete dominance, the F_1 hybrid phenotype resembles neither parent. With codominance, the F_1 hybrid phenotype includes aspects derived from both parents. Many allele pairs are codominant at the level of protein production. [LO6]

9. A single gene may have any number of alleles, each of which can cause the appearance of different phenotypes. New alleles arise by mutation. Common alleles in a population are considered wild types; rare alleles are mutants. When two or more common alleles exist for a gene, the gene is polymorphic; a gene with only one wild-type allele is monomorphic. [LO7]

10. In pleiotropy, one gene contributes to multiple traits. For such a gene, the dominance relation between any two alleles can vary according to the particular phenotype under consideration. [LO8]

11. Two or more genes may interact in several ways to affect the production of a single trait. These interactions may be understood by observing characteristic deviations from traditional Mendelian phenotypic ratios (review Table 2.4). [LO9]

12. In epistasis, the action of an allele at one gene can hide traits normally caused by the expression of alleles at another gene. In complementary gene action, dominant alleles of two or more genes are required to generate a trait. In heterogeneity, mutant alleles at any one of two or more genes are sufficient to elicit a phenotype. The complementation test can reveal whether a particular phenotype seen in two individuals arises from mutations in the same or separate genes. [LO9–10]

13. In many cases, the route from genotype to phenotype can be modified by the environment, chance, or other genes. A phenotype shows incomplete penetrance when it is expressed in fewer than 100 percent of individuals

with the same genotype. A phenotype shows variable expressivity when it is expressed at a quantitatively different level among individuals with the same genotype. **[LO11]**

14. A continuous trait can have any value of expression between two extremes. Traits of this type are polygenic; that is, determined by the interactions of multiple genes. **[LO12]**

Solved Problems

Solving Genetics Problems

The best way to evaluate and increase your understanding of the material in the chapter is to apply your knowledge in solving genetics problems. Genetics word problems are like puzzles. Take them in slowly—do not be overwhelmed by the whole problem. Identify useful facts given in the problem, and use the facts to deduce additional information. Use genetic principles and logic to work toward the solutions. The more problems you do, the easier they become. In doing problems, you will not only solidify your understanding of genetic concepts, but you will also develop basic analytical skills that are applicable in many disciplines.

Solving genetics problems requires more than simply plugging numbers into formulas. Each problem is unique and requires thoughtful evaluation of the information given and the question being asked. The following are general guidelines you can follow in approaching these word problems:

a. Read through the problem once to get some sense of the concepts involved.

b. Go back through the problem, noting all the information supplied to you. For example, genotypes or phenotypes of offspring or parents may be given to you or implied in the problem. Represent the known information in a symbolic format—assign symbols for alleles; use these symbols to indicate genotypes; make a diagram of the crosses including genotypes and phenotypes given or implied. Be sure that you do not assign different letters of the alphabet to two alleles of the same gene, as this can cause confusion. Also, be careful to discriminate clearly between the upper- and lowercases of letters, such as $C(c)$ or $S(s)$.

c. Now, reassess the question and work toward the solution using the information given. Make sure you answer the question being asked!

d. When you finish the problem, check to see that the answer makes sense. You can often check solutions by working backwards; that is, see if you can reconstruct the data from your answer.

e. After you have completed a question and checked your answer, spend a minute to think about which major concepts were involved in the solution. This is a critical step for improving your understanding of genetics.

For each chapter, the logic involved in solving two or three types of problems is described in detail.

1. In cats, white patches are caused by the dominant allele P, while pp individuals are solid-coloured. Short hair is caused by a dominant allele S, while ss cats have long hair. A long-haired cat with patches whose mother was solid-coloured and short-haired mates with a short-haired, solid-coloured cat whose mother was long-haired and solid-coloured. What kinds of kittens can arise from this mating, and in what proportions?

Answer

The solution to this problem requires an understanding of dominance/recessiveness, gamete formation, and the independent assortment of alleles of two genes in a cross.

First make a representation of the known information:

Mothers:	solid, short-haired		solid, long-haired
Cross:	cat 1		cat 2
	patches, long-haired	×	solid, short-haired

What genotypes can you assign? Any cat showing a recessive phenotype must be homozygous for the recessive allele. Therefore, the long-haired cats are ss; solid-coloured cats are pp. Cat 1 is long-haired, so it must be homozygous for the recessive allele (ss). This cat has the dominant phenotype of patches and could be either PP or Pp, but because the mother was pp and could only contribute a p allele in her gametes, the cat must be Pp. Cat 1's full genotype is $Pp\ ss$. Similarly, cat 2 is solid-coloured, so it must be homozygous for the recessive allele (pp). Because this cat is short-haired, it could have either the SS or Ss genotype. Its mother was long-haired (ss) and could only contribute an s allele in her gamete, so cat 2 must be heterozygous Ss. The full genotype is $pp\ Ss$.

The cross is therefore between a $Pp\ ss$ (cat 1) and a $pp\ Ss$ (cat 2). To determine the types of kittens, first establish the types of gametes that can be produced by each cat and then set up a Punnett square to determine the genotypes of the offspring. Cat 1 ($Pp\ ss$) produces Ps and ps gametes in equal proportions. Cat 2 ($pp\ Ss$) produces pS and ps gametes in equal proportions. *Four types of kittens can result from this mating with equal probability: Pp Ss (patches, short-haired), Pp ss (patches, long-haired), pp Ss (solid, short-haired), and pp ss (solid, long-haired).*

		Cat 1	
		$P\ s$	$p\ s$
Cat 2	$p\ S$	$Pp\ Ss$	$pp\ Ss$
	$p\ s$	$Pp\ ss$	$pp\ ss$

You could also work through this problem using the product rule of probability instead of a Punnett square. The principles are the same: Gametes produced in equal amounts by either parent are combined at random.

Cat 1 gamete		Cat 2 gamete		Progeny
1/2 *P s*	×	1/2 *p S*	→	1/4 *Pp Ss* patches, short-haired
1/2 *P s*	×	1/2 *p s*	→	1/4 *Pp ss* patches, long-haired
1/2 *p s*	×	1/2 *p S*	→	1/4 *pp Ss* solid-coloured, short-haired
1/2 *p s*	×	1/2 *p s*	→	1/4 *pp ss* solid-coloured, long-haired

II. In tomatoes, red fruit is dominant to yellow fruit, and purple stems are dominant to green stems. The progeny from one mating consisted of 305 red fruit, purple stem plants; 328 red fruit, green stem plants; 110 yellow fruit, purple stem plants; and 97 yellow fruit, green stem plants. What were the genotypes of the parents in this cross?

Answer

This problem requires an understanding of independent assortment in a dihybrid cross as well as the ratios predicted from monohybrid crosses.

Designate the alleles:

R = red, *r* = yellow

P = purple stems, *p* = green stems

In genetics problems, the ratios of offspring can indicate the genotype of parents. You will usually need to total the number of progeny and approximate the ratio of offspring in each of the different classes. For this problem, in which the inheritance of two traits is given, consider each trait independently. For red fruit, there are 305 + 328 = 633 red-fruited plants out of a total of 840 plants. This value (633/840) is close to 3/4. About 1/4 of the plants have yellow fruit (110 + 97 = 207/840). From Mendel's work, you know that a 3:1 phenotypic ratio results from crosses between plants that are hybrid (heterozygous) for one gene. Therefore, the genotype for fruit colour of each parent must have been *Rr*.

For stem colour, 305 + 110 or 415/840 plants had purple stems. About half had purple stems, and the other half (328 + 97) had green stems. A 1:1 phenotypic ratio occurs when a heterozygote is mated to a homozygous recessive (as in a testcross). The parents' genotypes must have been *Pp* and *pp* for stem colour.

The complete genotype of the parent plants in this cross was Rr Pp × Rr pp.

III. Tay-Sachs is a recessive lethal disease in which there is neurological deterioration early in life. This disease is rare in the population overall but is found at relatively high frequency in Ashkenazi Jews from Central Europe. A woman whose maternal uncle had the disease is trying to determine the probability that she and her husband could have an affected child. Her father does not come from a high-risk population. Her husband's sister died of the disease at an early age.

a. Draw the pedigree of the individuals described. Include the genotypes where possible.

b. Determine the probability that the couple's first child will be affected.

Answer

This problem requires an understanding of dominance, recessiveness, and probability. Designate the alleles:

T = normal allele; *t* = Tay-Sachs allele

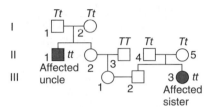

The genotypes of the two affected individuals, the woman's uncle (II-1) and the husband's sister (III-3) are *tt*. Because the uncle was affected, his parents must have been heterozygous. There was a 1/4 chance that these parents had a homozygous recessive (affected) child, a 2/4 chance that they had a heterozygous child (carrier), and a 1/4 chance they had a homozygous dominant (unaffected) child. However, you have been told that the woman's mother (II-2) is unaffected, so the mother could only have had a heterozygous or a homozygous dominant genotype. Consider the probability that these two genotypes will occur. If you were looking at a Punnett square, there would be only three combinations of alleles possible for the normal mother. Two of these are heterozygous combinations and one is homozygous dominant. There is a 2/3 chance (2 out of the 3 possible cases) that the mother was a carrier. The father was not from a high-risk population, so we can assume that he is homozygous dominant. There is a 2/3 chance that the wife's mother was heterozygous and, if so, a 1/2 chance that the wife inherited a recessive allele from her mother. Because both conditions are necessary for inheritance of a recessive allele, the individual probabilities are multiplied, and the probability that the wife (III-1) is heterozygous is 2/3 × 1/2.

The husband (III-2) has a sister who died from the disease; therefore, his parents must have been heterozygous. The probability that he is a carrier is 2/3 (using the same rationale as for II-2). The probability that the man and woman are both carriers is 2/3 × 1/2 × 2/3. Because there is a 1/4 probability that a particular child of two carriers will be affected, *the overall probability that the first child of this couple (III-1 and III-2) will be affected is 2/3 × 1/2 × 2/3 × 1/4 = 4/72, or 1/18.*

IV. Imagine you purchased an albino mouse (genotype *cc*) in a pet store. The *c* allele is epistatic to other coat colour genes. How would you go about determining the genotype of this mouse at the brown locus? (In pigmented mice, *BB* and *Bb* are black, *bb* is brown.)

Answer

This problem requires an understanding of gene interactions, specifically epistasis. You have been placed in the role of experimenter and need to design crosses that will answer the question. To determine the alleles of the *B* gene present, you need to eliminate the blocking action of the *cc* genotype. Because only the recessive *c* allele is epistatic, when a *C* allele is present, no epistasis will occur. To introduce a *C* allele during the mating, the test mouse you mate to your albino can have the genotype *CC* or *Cc*. (If the mouse is *Cc*, half of the progeny will be albino and will not contribute useful information, but the nonalbinos from this cross would be informative.) What alleles of the *B* gene should the test mouse carry? To make this decision, work through the expected results using each of the possible genotypes.

Test mouse genotype		Albino mouse	Expected progeny
BB	×	*BB*	all black
	×	*Bb*	all black
	×	*bb*	all black
Bb	×	*BB*	all black
	×	*Bb*	3/4 black, 1/4 brown
	×	*bb*	1/2 black, 1/2 brown
bb	×	*BB*	all black
	×	*Bb*	1/2 black, 1/2 brown
	×	*bb*	all brown

From these hypothetical crosses, you can see that a test mouse with either the *Bb* or *bb* genotype would yield distinct outcomes for each of the three possible albino mouse genotypes. However, a *bb* test mouse would be more useful and less ambiguous. First, it is easier to identify a mouse with the *bb* genotype because a brown mouse must have this homozygous recessive genotype. Second, the results are completely different for each of the three possible genotypes when you use the *bb* test mouse. (In contrast, a *Bb* test mouse would yield both black and brown progeny whether the albino mouse was *Bb* or *bb*; the only distinguishing feature is the ratio.) *To determine the full genotype of the albino mouse, you should cross it to a brown mouse (which could be CC bb or Cc bb).*

V. In a particular kind of wildflower, the wild-type flower colour is deep purple, and the plants are true-breeding. In one true-breeding mutant stock, the flowers have a reduced pigmentation, resulting in a lavender colour. In a different true-breeding mutant stock, the flowers have no pigmentation and are thus white. When a lavender-flowered plant from the first mutant stock was crossed to a white-flowered plant from the second mutant stock, all the F_1 plants had purple flowers. The F_1 plants were then allowed to self-fertilize to produce an F_2 generation. The 277 F_2 plants were 157 purple : 71 white : 49 lavender. Explain how flower colour is inherited. Is this trait controlled by the alleles of a single gene? What kinds of progeny would be produced if lavender F_2 plants were allowed to self-fertilize?

Answer

Are there any modes of single-gene inheritance compatible with the data? The observations that the F_1 plants look different from either of their parents and that the F_2 generation is composed of plants with three different phenotypes exclude complete dominance. The ratio of the three phenotypes in the F_2 plants has some resemblance to the 1:2:1 ratio expected from codominance or incomplete dominance, but the results would then imply that purple plants must be heterozygotes. This conflicts with the information provided that purple plants are true-breeding.

Consider now the possibility that two genes are involved. From a cross between plants heterozygous for two genes (*W* and *P*), the F_2 generation would contain a 9:3:3:1 ratio of the genotypes *W– P–*, *W– pp*, *ww P–*, and *ww pp* (where the dash indicates that the allele could be either a dominant or a recessive form). Are there any combinations of the 9:3:3:1 ratio that would be close to that seen in the F_2 generation in this example? The numbers seem close to a 9:4:3 ratio. What hypothesis would support combining two of the classes (3 + 1)? If *w* is epistatic to the *P* gene, then the *ww P–* and *ww pp* genotypic classes would have the same white phenotype. With this explanation, 1/3 of the F_2 lavender plants would be *WW pp*, and the remaining 2/3 would be *Ww pp*. Upon self-fertilization, *WW pp* plants would produce only lavender (*WW pp*) progeny, while *Ww pp* plants would produce a 3:1 ratio of lavender (*W– pp*) and white (*ww pp*) progeny.

VI. Huntington disease (HD) is a rare dominant condition in humans that results in a slow but inexorable deterioration of the nervous system. HD shows what might be called "age-dependent penetrance", which is to say that the probability that a person with the HD genotype will express the phenotype varies with age. Assume that 50 percent of those inheriting the *HD* allele will express the symptoms by age 40. Susan is a 35-year-old woman whose father has HD. She currently shows no symptoms. What is the probability that Susan will show symptoms in five years?

Answer

This problem involves probability and penetrance. Two conditions are necessary for Susan to show symptoms of the disease. There is a 1/2 (50 percent) chance that she inherited the mutant allele from her father and a 1/2 (50 percent) chance that she will express the phenotype by age 40. Because these are independent events, *the probability is the product of the individual probabilities, or 1/4.*

Vocabulary

1a. For each of the terms in the left column, choose the best matching phrase in the right column.

i. phenotype	**1.** having two identical alleles of a given gene		
ii. alleles	**2.** the allele expressed in the phenotype of the heterozygote		
iii. independent assortment	**3.** alternate forms of a gene		
iv. gametes	**4.** observable characteristic		
v. gene	**5.** a cross between individuals both heterozygous for two genes		
vi. segregation	**6.** alleles of one gene separate into gametes randomly with respect to alleles of other genes		
vii. heterozygote	**7.** reproductive cells containing only one copy of each gene		
viii. dominant	**8.** the allele that does not contribute to the phenotype of the heterozygote		
ix. F_1	**9.** the cross of an individual of ambiguous genotype with a homozygous recessive individual		
x. testcross	**10.** an individual with two different alleles of a gene		
xi. genotype	**11.** the heritable entity that determines a characteristic		
xii. recessive	**12.** the alleles an individual has		
xiii. dihybrid cross	**13.** the separation of the two alleles of a gene into different gametes		
xiv. homozygote	**14.** offspring of the P generation		

1b. For each of the terms in the left column, choose the best matching phrase in the right column.

i. epistasis	**1.** one gene affecting more than one phenotype
ii. modifier gene	**2.** the alleles of one gene mask the effects of alleles of another gene
iii. conditional lethal	**3.** both parental phenotypes are expressed in the F_1 hybrids
iv. permissive condition	**4.** a heritable change in a gene
v. reduced penetrance	**5.** genes whose alleles alter phenotypes produced by the action of other genes
vi. multifactorial trait	**6.** less than 100 percent of the individuals possessing a particular genotype express it in their phenotype
vii. incomplete dominance	**7.** environmental conditions that allow conditional lethals to live
viii. codominance	**8.** a trait produced by the interaction of alleles of at least two genes with the environment
ix. mutation	**9.** individuals with the same genotype have related phenotypes that vary in intensity
x. pleiotropy	**10.** a genotype that is lethal in some situations (e.g., high temperature) but viable in others
xi. variable expressivity	**11.** the heterozygote resembles neither homozygote

Section 2.1

2. During the millennia in which selective breeding was practiced, why did breeders fail to uncover the principle that traits are governed by discrete units of inheritance (i.e., by genes)?

3. Describe the characteristics of the garden pea that made it a good organism for Mendel's analysis of the basic principles of inheritance. Evaluate how easy or difficult it would be to make a similar study of inheritance in humans by considering the same attributes you described for the pea.

Section 2.2

4. An albino corn snake is crossed with a normal-coloured corn snake. The offspring are all normal-coloured. When these first generation progeny snakes are crossed among themselves, they produce 32 normal-coloured snakes and 10 albino snakes.

a. Which of these phenotypes is controlled by the dominant allele?

b. In these snakes, albino colour is determined by a recessive allele *a*, and normal pigmentation is determined by the *A* allele. A normal-coloured female snake is involved in a testcross. This cross produces 10 normal-coloured and 11 albino offspring. What are the genotypes of the parents and the offspring?

5. Two short-haired cats mate and produce six short-haired and two long-haired kittens. What does this information suggest about how hair length is inherited?

6. Piebald spotting is a condition found in humans in which there are patches of skin that lack pigmentation. The condition results from the inability of pigment-producing cells to migrate properly during development. Two adults with piebald spotting have one child

who has this trait and a second child with normal skin pigmentation.

    a. Is the piebald spotting trait dominant or recessive? What information led you to this answer?

    b. What are the genotypes of the parents?

7. As a *Drosophila* research geneticist, you keep stocks of flies of specific genotypes. You have a fly that has normal wings (dominant phenotype). Flies with short wings are homozygous for a recessive allele of the wing-length gene. You need to know if this fly with normal wings is pure-breeding or heterozygous for the wing-length trait. What cross would you do to determine the genotype, and what results would you expect for each possible genotype?

8. A mutant cucumber plant has flowers that fail to open when mature. Crosses can be done with this plant by manually opening and pollinating the flowers with pollen from another plant. When closed × open crosses were done, all the F_1 progeny were open. The F_2 plants were 145 open and 59 closed. A cross of closed × F_1 gave 81 open and 77 closed. How is the closed trait inherited? What evidence led you to your conclusion?

9. In a particular population of mice, certain individuals display a phenotype called "short tail", which is inherited as a dominant trait. Some individuals display a recessive trait called "dilute", which affects coat colour. Which of these traits would be easier to eliminate from the population by selective breeding? Why?

10. In humans, a dimple in the chin is a dominant characteristic.

    a. A man who does not have a chin dimple has children with a woman with a chin dimple whose mother lacked the dimple. What proportion of their children would be expected to have a chin dimple?

    b. A man with a chin dimple and a woman who lacks the dimple produce a child who lacks a dimple. What is the man's genotype?

    c. A man with a chin dimple and a nondimpled woman produce eight children, all having the chin dimple. Can you be certain of the man's genotype? Why or why not? What genotype is more likely, and why?

11. Among native Americans, two types of earwax (cerumen) are seen: dry and sticky. A geneticist studied the inheritance of this trait by observing the types of offspring produced by different kinds of matings. He observed the following numbers:

Parents	Number of mating pairs	Offspring	
		Sticky	Dry
sticky × sticky	10	32	6
sticky × dry	8	21	9
dry × dry	12	0	42

    a. How is earwax type inherited?

    b. Why are there no 3:1 or 1:1 ratios in the data shown in the chart?

12. Imagine you have just purchased a black stallion of unknown genotype. You mate him to a red mare, and she delivers twin foals, one red and one black. Can you tell from these results how colour is inherited, assuming that alternative alleles of a single gene are involved? What crosses could you do to work this out?

13. If you roll a die (singular of dice), what is the probability you will roll (a) a 6? (b) an even number? (c) a number divisible by 3? (d) If you roll a pair of dice, what is the probability that you will roll two 6s? (e) an even number on one and an odd number on the other? (f) matching numbers? (g) two numbers both over 4?

14. In a standard deck of playing cards, there are four suits (red suits = hearts and diamonds, black suits = spades and clubs). Each suit has thirteen cards: Ace (A), 2, 3, 4, 5, 6, 7, 8, 9, 10, and the face cards Jack (J), Queen (Q), and King (K). In a single draw, what is the probability that you will draw a face card? A red card? A red face card?

15. How many genotypically different eggs could be formed by women with the following genotypes?

    a. *Aa bb CC DD*

    b. *AA Bb Cc dd*

    c. *Aa Bb cc Dd*

    d. *Aa Bb Cc Dd*

16. What is the probability of producing a child that will phenotypically resemble either one of the two parents in the following four crosses? How many phenotypically different kinds of progeny could potentially result from each of the four crosses?

    a. *Aa Bb Cc Dd* × *aa bb cc dd*

    b. *aa bb cc dd* × *AA BB CC DD*

    c. *Aa Bb Cc Dd* × *Aa Bb Cc Dd*

    d. *aa bb cc dd* × *aa bb cc dd*

17. A mouse sperm of genotype *a B C D E* fertilizes an egg of genotype *a b c D e*. What are all the possibilities for the genotypes of (a) the zygote and (b) a sperm or egg of the baby mouse that develops from this fertilization?

18. Galactosaemia is a recessive human disease that is treatable by restricting lactose and glucose in the diet. Susan Smithers and her husband are both heterozygous for the galactosaemia gene.

    a. Susan is pregnant with twins. If she has fraternal (non-identical) twins, what is the probability both of the twins will be girls who have galactosaemia?

    b. If the twins are identical, what is the probability that both will be girls and have galactosaemia?

    For parts *c–g*, assume that none of the children is a twin.

c. If Susan and her husband have four children, what is the probability that none of the four will have galactosaemia?

d. If the couple has four children, what is the probability that at least one child will have galactosaemia?

e. If the couple has four children, what is the probability that the first two will have galactosaemia and the second two will not?

f. If the couple has three children, what is the probability that two of the children will have galactosaemia and one will not, regardless of order?

g. If the couple has four children with galactosaemia, what is the probability that their next child will have galactosaemia?

19. Albinism is a condition in which pigmentation is lacking. In humans, the result is white hair, nonpigmented skin, and pink eyes. The trait in humans is caused by a recessive allele. Two normal parents have an albino child. What are the parents' genotypes? What is the probability that the next child will be albino?

20. A cross between two pea plants, both of which grew from yellow round seeds, gave the following numbers of seeds: 156 yellow round and 54 yellow wrinkled. What are the genotypes of the parent plants? (Yellow and round are dominant traits.)

21. A third-grader decided to breed guinea pigs for her school science project. She went to a pet store and bought a male with smooth black fur and a female with rough white fur. She wanted to study the inheritance of those features and was sorry to see that the first litter of eight contained only rough black animals. To her disappointment, the second litter from those same parents contained seven rough black animals. Soon the first litter had begun to produce F_2 offspring, and they showed a variety of coat types. Before long, the child had 125 F_2 guinea pigs. Eight of them had smooth white coats, 25 had smooth black coats, 23 were rough and white, and 69 were rough and black.

a. How are the coat colour and texture characteristics inherited? What evidence supports your conclusions?

b. What phenotypes and proportions of offspring should the girl expect if she mates one of the smooth white F_2 females to an F_1 male?

22. The self-fertilization of an F_1 pea plant produced from a parent plant homozygous for yellow and wrinkled seeds and a parent homozygous for green and round seeds resulted in a pod containing seven F_2 peas. (Yellow and round are dominant.) What is the probability that all seven peas in the pod are yellow and round?

23. The achoo syndrome (sneezing in response to bright light) and trembling chin (triggered by anxiety) are both dominant traits in humans.

a. What is the probability that the first child of parents who are heterozygous for both the achoo gene and trembling chin will have achoo syndrome but lack the trembling chin?

b. What is the probability that the first child will not have achoo syndrome or trembling chin?

24. A pea plant from a pure-breeding strain that is tall, has green pods, and has purple flowers that are terminal (at tip of stem) is crossed to a plant from a pure-breeding strain that is dwarf, has yellow pods, and has white flowers that are axial (along stem). The F_1 plants are all tall and have purple axial flowers as well as green pods.

a. What phenotypes do you expect to see in the F_2?

b. What phenotypes and ratios would you predict in the progeny from crossing an F_1 plant to the dwarf parent?

25. The following chart shows the results of different matings between jimsonweed plants that had either purple or white flowers and spiny or smooth pods. Determine the dominant allele for the two traits and indicate the genotypes of the parents for each of the crosses.

Parents	Offspring			
	Purple Spiny	White Spiny	Purple Smooth	White Smooth
a. purple spiny × purple spiny	94	32	28	11
b. purple spiny × purple smooth	40	0	38	0
c. purple spiny × white spiny	34	30	0	0
d. purple spiny × white spiny	89	92	31	27
e. purple smooth × purple smooth	0	0	36	11
f. white spiny × white spiny	0	45	0	16

26. A pea plant heterozygous for plant height, pod shape, and flower colour was selfed. The progeny consisted of 272 tall, inflated pods, purple flowers; 92 tall, inflated, white flowers; 88 tall, flat pods, purple; 93 dwarf, inflated, purple; 35 tall, flat, white; 31 dwarf, inflated, white; 29 dwarf, flat, purple; and 11 dwarf, flat, white. Which alleles are dominant in this cross?

27. In the fruit fly *Drosophila melanogaster*, the following genes and mutations are known:

Wingsize: recessive allele for tiny wings *t*; dominant allele for normal wings *T*.

Eye shape: recessive allele for narrow eyes *n*; dominant allele for normal (oval) eyes *N*.

For each of the following crosses, give the genotypes of each of the parents.

Male		Female		
Wings	Eyes	Wings	Eyes	Offspring
a. tiny	oval	× tiny	oval	78 tiny wings, oval eyes
				24 tiny wings, narrow eyes
b. normal	narrow	× tiny	oval	45 normal wings, oval eyes
				40 normal wings, narrow eyes
				38 tiny wings, oval eyes
				44 tiny wings, narrow eyes
c. normal	narrow	× normal	oval	35 normal wings, oval eyes
				29 normal wings, narrow eyes
				10 tiny wings, oval eyes
				11 tiny wings, narrow eyes
d. normal	narrow	× normal	oval	62 normal wings, oval eyes
				19 tiny wings, oval eyes

28. Based on the information you discovered in Problem 27 above, answer the following:

a. A female fruit fly with genotype *Tt nn* is mated to a male of genotype *Tt Nn*. What is the probability that any one of their offspring will have normal phenotypes for both characters?

b. What phenotypes would you expect among the offspring of this cross? If you obtained 200 progeny, how many of each phenotypic class would you expect?

Section 2.3

29. For each of the following human pedigrees, indicate whether the inheritance pattern is recessive or dominant. What feature(s) of the pedigree did you use to determine the inheritance? Give the genotypes of affected individuals and of individuals who carry the disease allele.

(a)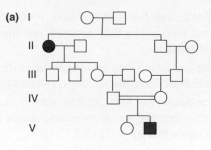

(b)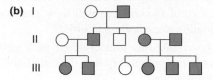

(c)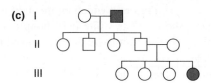

30. Consider the pedigree that follows for cutis laxa, a connective tissue disorder in which the skin hangs in loose folds.

a. Assuming complete penetrance and that the trait is rare, what is the apparent mode of inheritance?

b. What is the probability that individual II-2 is a carrier?

c. What is the probability that individual II-3 is a carrier?

d. What is the probability that individual III-1 is affected by the disease?

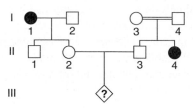

31. A young couple went to see a genetic counsellor because each had a sibling affected with cystic fibrosis. (Cystic fibrosis is a recessive disease, and neither member of the couple nor any of their four parents is affected.)

a. What is the probability that the female of this couple is a carrier?

b. What are the chances that their child will be affected with cystic fibrosis?

c. What is the probability that their child will be a carrier of the cystic fibrosis mutation?

32. Huntington disease is a rare fatal, degenerative neurological disease in which individuals start to show symptoms,

on average, in their 40s. It is caused by a dominant allele. Joe, a man in his 20s, just learned that his father has Huntington disease.

    a. What is the probability that Joe will also develop the disease?

    b. Joe and his new wife have been eager to start a family. What is the probability that their first child will eventually develop the disease?

33. Is the disease shown in the following pedigree dominant or recessive? Why? Based on this limited pedigree, do you think the disease allele is rare or common in the population? Why?

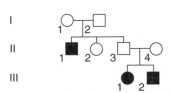

34. Figure 2.20 shows the inheritance of Huntington disease in a family from a small village near Lake Maracaibo in Venezuela. The village was founded by a small number of immigrants, and generations of their descendents have remained concentrated in this isolated location. The allele for Huntington disease has remained unusually prevalent there.

    a. Why could you not conclude definitively that the disease is the result of a dominant or a recessive allele solely by looking at this pedigree?

    b. Is there any information you could glean from the family's history that might imply the disease is due to a dominant rather than a recessive allele?

35. The common grandfather of two first cousins has hereditary haemochromatosis, a recessive condition causing an abnormal buildup of iron in the body. Neither of the cousins has the disease nor do any of their relatives.

    a. If the first cousins mated with each other and had a child, what is the chance that the child would have haemochromatosis? Assume that the unrelated, unaffected parents of the cousins are not carriers.

    b. How would your calculation change if you knew that 1 out of every 10 unaffected people in the population (including the unrelated parents of these cousins) was a carrier for haemochromatosis?

36. People with nail-patella syndrome have poorly developed or absent kneecaps and nails. Individuals with alkaptonuria have arthritis as well as urine that darkens when exposed to air. Both nail-patella syndrome and alkaptonuria are rare phenotypes. In the following pedigree, vertical red lines indicate individuals with nail-patella

syndrome, while horizontal green lines denote individuals with alkaptonuria.

    a. What are the most likely modes of inheritance of nail-patella syndrome and alkaptonuria? What genotypes can you ascribe to each of the individuals in the pedigree for both of these phenotypes?

    b. In a mating between IV-2 and IV-5, what is the chance that the child produced would have both nail-patella syndrome and alkaptonuria? Nail-patella syndrome alone? Alkaptonuria alone? Neither defect?

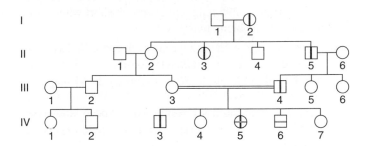

37. Midphalangeal hair (hair on top of the middle segment of the fingers) is a common phenotype caused by a dominant allele M. Homozygotes for the recessive allele (mm) lack hair on the middle segment of their fingers. Among 1000 families in which both parents had midphalangeal hair, 1853 children showed the trait while 209 children did not. Explain this result.

Section 2.4

38. In four-o'clocks, the allele for red flowers is incompletely dominant over the allele for white flowers, so heterozygotes have pink flowers. What ratios of flower colours would you expect among the offspring of the following crosses: (a) pink × pink, (b) white × pink, (c) red × red, (d) red × pink, (e) white × white, and (f) red × white? If you specifically wanted to produce pink flowers, which of these crosses would be most efficient?

39. A cross between two plants that both have yellow flowers produces 80 offspring plants, of which 38 have yellow flowers, 22 have red flowers, and 20 have white flowers. If one assumes that this variation in colour is due to inheritance at a single locus, what is the genotype associated with each flower colour, and how can you describe the inheritance of flower colour?

40. In the fruit fly *Drosophila melanogaster*, very dark (ebony) body colour is determined by the e allele. The e^+ allele produces the normal wild-type, honey-coloured body. In heterozygotes for the two alleles, a dark marking called the trident can be seen on the thorax, but otherwise the body is honey-coloured. The e^+ allele is thus considered to be incompletely dominant to the e allele.

a. When female e^+e^+ flies are crossed to male e^+e flies, what is the probability that progeny will have the dark trident marking?

b. Animals with the trident marking mate among themselves. Of 300 progeny, how many would be expected to have a trident, how many ebony bodies, and how many honey-coloured bodies?

41. A wild legume with white flowers and long pods is crossed to one with purple flowers and short pods. The F₁ offspring are allowed to self-fertilize, and the F₂ generation has 301 long purple, 99 short purple, 612 long pink, 195 short pink, 295 long white, and 98 short white. How are these traits being inherited?

42. In radishes, colour and shape are each controlled by a single locus with two incompletely dominant alleles. Colour may be red (*RR*), purple (*Rr*), or white (*rr*) and shape can be long (*LL*), oval (*Ll*), or round (*ll*). What phenotypic classes and proportions would you expect among the offspring of a cross between two plants heterozygous at both loci?

43. Familial hypercholesterolaemia (FH) is an inherited trait in humans that results in higher than normal serum cholesterol levels (measured in milligrams of cholesterol per decilitre of blood [mg/dL]). People with serum cholesterol levels that are roughly twice normal have a 25 times higher frequency of heart attacks than unaffected individuals. People with serum cholesterol levels three or more times higher than normal have severely blocked arteries and almost always die before they reach the age of 20. The pedigrees below show the occurrence of FH in four Japanese families.

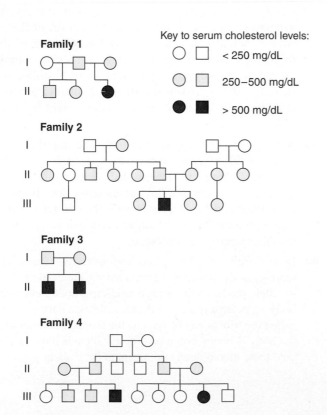

Key to serum cholesterol levels:
○ □ < 250 mg/dL
○ □ 250–500 mg/dL
● ■ > 500 mg/dL

Family 1

Family 2

Family 3

Family 4

a. What is the most likely mode of inheritance of FH based on this data? Are there any individuals in any of these pedigrees who do not fit your hypothesis?

b. Why do individuals in the same phenotypic class (unfilled, yellow, or red symbols) show such variation in their levels of serum cholesterol?

44. Describe briefly the following:

a. The genotype of a person who has sickle-cell anaemia.

b. The genotype of a person with a normal phenotype who has a child with sickle-cell anaemia.

c. The *total* number of different alleles of the β-globin gene that could be carried by five children with the same mother and father.

45. Assuming no involvement of the Bombay phenotype:

a. If a girl has blood type O, what could be the genotypes and corresponding phenotypes of her parents?

b. If a girl has blood type B and her mother has blood type A, what genotype(s) and corresponding phenotype(s) could the other parent have?

c. If a girl has blood type AB and her mother is also AB, what are the genotype(s) and corresponding phenotype(s) of any male who could *not* be the girl's father?

46. There are several genes in humans in addition to the ABO gene that give rise to recognizable antigens on the surface of red blood cells. The *MN* and *Rh* genes are two examples. The *Rh* locus can contain either a positive or negative allele, with positive being dominant to negative. *M* and *N* are codominant alleles of the *MN* gene. The following chart shows several mothers and their children. For each mother–child pair, choose the father of the child from among the males in the right column, assuming one child per male.

	Mother	Child	Males
a.	O M Rh pos	B MN Rh neg	O M Rh neg
b.	B MN Rh neg	O N Rh neg	A M Rh pos
c.	O M Rh pos	A M Rh neg	O MN Rh pos
d.	AB N Rh neg	B MN Rh neg	B MN Rh pos

47. Alleles of the gene that determines seed coat patterns in lentils can be organized in a dominance series: marbled > spotted = dotted (codominant alleles) > clear. A lentil plant homozygous for the marbled seed coat pattern allele was crossed to one homozygous for the spotted pattern allele. In another cross, a homozygous dotted lentil plant was crossed to one homozygous for clear. An F₁ plant from the first cross was then mated to an F₁ plant from the second cross.

a. What are the expected phenotypes of the F₁ plants from the two original parental crosses?

b. What phenotypes in what proportions are expected from this mating between the two F₁ types?

48. In clover plants, the pattern on the leaves is determined by a single gene with multiple alleles that are related in a dominance series. Seven different alleles of this gene are known; an allele that determines the absence of a pattern is recessive to the other six alleles, each of which produces a different pattern. All heterozygous combinations of alleles show complete dominance.

a. How many different kinds of leaf patterns (including the absence of a pattern) are possible in a population of clover plants in which all seven alleles are represented?

b. What is the largest number of different genotypes that could be associated with any one phenotype? Is there any phenotype that could be represented by only a single genotype?

c. In a particular field, you find that the large majority of clover plants lack a pattern on their leaves, even though you can identify a few plants representative of all possible pattern types. Explain this finding.

49. In a population of rabbits, you find three different coat colour phenotypes: chinchilla (C), himalaya (H), and albino (A). To understand the inheritance of coat colours, you cross individual rabbits with each other and note the results in the following table.

Cross number	Parental phenotypes	Phenotypes of progeny
1	H × H	3/4 H : 1/4 A
2	H × A	1/2 H : 1/2 A
3	C × C	3/4 C : 1/4 H
4	C × H	all C
5	C × C	3/4 C : 1/4 A
6	H × A	all H
7	C × A	1/2 C : 1/2 A
8	A × A	all A
9	C × H	1/2 C : 1/2 H
10	C × H	1/2 C : 1/4 H : 1/4 A

a. What can you conclude about the inheritance of coat colour in this population of rabbits?

b. Ascribe genotypes to the parents in each of the ten crosses.

c. What kinds of progeny would you expect, and in what proportions, if you crossed the chinchilla parents in crosses #9 and #10?

50. Spherocytosis is an inherited blood disease in which the erythrocytes (red blood cells) are spherical instead of biconcave. This condition is inherited in a dominant fashion, with Sph^- dominant to Sph^+. In people with spherocytosis, the spleen "reads" the spherical red blood cells as defective, and it removes them from the bloodstream, leading to anaemia. The spleen in different people removes the spherical erythrocytes with different efficiencies. Some people with spherical erythrocytes suffer severe anaemia and some mild anaemia, yet others have spleens that function so poorly there are no symptoms of anaemia at all. When 2400 people with the genotype $Sph^- Sph^+$ were examined, it was found that 2250 had anaemia of varying severity, but 150 had no symptoms.

a. Does this description of people with spherocytosis represent incomplete penetrance, variable expressivity, or both? Explain your answer. Can you derive any values from the numerical data to measure penetrance or expressivity?

b. Suggest a treatment for spherocytosis and describe how the incomplete penetrance and/or variable expressivity of the condition might affect this treatment.

51. Fruit flies with one allele for curly wings (Cy) and one allele for normal wings (Cy^+) have curly wings. When two curly-winged flies were crossed, 203 curly-winged and 98 normal-winged flies were obtained. In fact, all crosses between curly-winged flies produce nearly the same curly:normal ratio among the progeny.

a. What is the approximate phenotypic ratio in these offspring?

b. Suggest an explanation for these data.

c. If a curly-winged fly was mated to a normal-winged fly, how many flies of each type would you expect among 180 total offspring?

52. You have come into contact with two unrelated patients who express what you think is a rare phenotype—a dark spot on the bottom of the foot. According to a medical source, this phenotype is seen in 1 in every 100 000 people in the population. The two patients give their family histories to you, and you generate the pedigrees that follow.

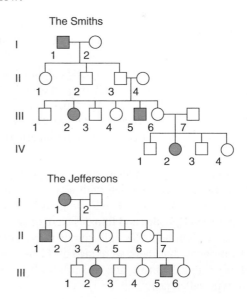

a. Given that this trait is rare, do you think the inheritance is dominant or recessive? Are there any special conditions that appear to apply to the inheritance?

b. Which nonexpressing members of these families must carry the mutant allele?

c. If this trait is instead quite common in the population, what alternative explanation would you propose for the inheritance?

d. Based on this new explanation (part *c*), which non-expressing members of these families must have the genotype normally causing the trait?

53. In a species of tropical fish, a colourful orange and black variety called montezuma occurs. When two montezumas are crossed, 2/3 of the progeny are montezuma and 1/3 are the wild-type, dark greyish green colour. Montezuma is a single-gene trait, and montezuma fish are never true-breeding.

a. Explain the inheritance pattern seen here and show how your explanation accounts for the phenotypic ratios given.

b. In this same species, the morphology of the dorsal fin is altered from normal to ruffled by homozygosity for a recessive allele designated *f*. What progeny would you expect to obtain, and in what proportions, from the cross of a montezuma fish homozygous for normal fins to a green, ruffled fish?

c. What phenotypic ratios of progeny would be expected from the crossing of two of the montezuma progeny from part *b*?

54. Polycystic kidney disease is a dominant trait that causes the growth of numerous cysts in the kidneys. The condition eventually leads to kidney failure. A child with polycystic kidney disease is born to a couple, neither of whom shows the disease. What possibilities might explain this outcome?

Section 2.5

55. A black mare was crossed to a chestnut stallion and produced a bay son and a bay daughter. The two offspring were mated to each other several times, and they produced offspring of four different coat colours: black, bay, chestnut, and liver. Crossing a liver grandson back to the black mare gave a black foal, and crossing a liver granddaughter back to the chestnut stallion gave a chestnut foal. Explain how coat colour is being inherited in these horses.

56. A rooster with a particular comb morphology called walnut was crossed to a hen with a type of comb morphology known as single. The F_1 progeny all had walnut combs. When F_1 males and females were crossed to each other, 93 walnut and 11 single combs were seen among the F_2 progeny, but there were also 29 birds with a new kind of comb called rose and 32 birds with another new comb type called pea.

a. Explain how comb morphology is inherited.

b. What progeny would result from crossing a homozygous rose-combed hen with a homozygous pea-combed rooster? What phenotypes and ratios would be seen in the F_2 progeny?

c. A particular walnut rooster was crossed to a pea hen, and the progeny consisted of 12 walnut, 11 pea, 3 rose, and 4 single chickens. What are the likely genotypes of the parents?

d. A different walnut rooster was crossed to a rose hen, and all the progeny were walnut. What are the possible genotypes of the parents?

57. You do a cross between two true-breeding strains of zucchini. One has green fruit and the other has yellow fruit. The F_1 plants are all green, but when these are crossed, the F_2 plants consist of 9 green : 7 yellow.

a. Explain this result. What were the genotypes of the two parental strains?

b. Indicate the phenotypes, with frequencies, of the progeny of a testcross of the F_1 plants.

58. Filled-in symbols in the pedigree that follows designate individuals suffering from deafness.

a. Study the pedigree and explain how deafness is being inherited.

b. What is the genotype of the individuals in generation V? Why are they not affected?

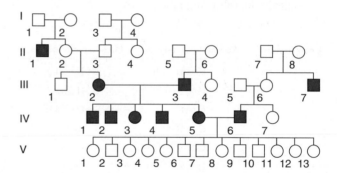

59. Explain the difference between epistasis and dominance. How many loci are involved in each case?

60. Two true-breeding white strains of the plant *Illegitimati noncarborundum* were mated, and the F_1 progeny were all white. When the F_1 plants were allowed to self-fertilize, 126 white-flowered and 33 purple-flowered F_2 plants grew.

a. How could you describe inheritance of flower colour? Describe how specific alleles influence each other and therefore affect phenotype.

b. A white F_2 plant is allowed to self-fertilize. Of the progeny, 3/4 are white-flowered, and 1/4 are purple-flowered. What is the genotype of the white F_2 plant?

c. A purple F_2 plant is allowed to self-fertilize. Of the progeny, 3/4 are purple-flowered, and 1/4 are white-flowered. What is the genotype of the purple F_2 plant?

d. Two white F_2 plants are crossed with each other. Of the progeny, 1/2 are purple-flowered, and 1/2 are white-flowered. What are the genotypes of the two white F_2 plants?

61. "Secretors" (genotypes *SS* and *Ss*) secrete their A and B blood group antigens into their saliva and other body fluids, while "nonsecretors" (*ss*) do not. What would be the apparent phenotypic blood group proportions among the offspring of an $I^A I^B$ *Ss* woman and an $I^A I^A$ *Ss* man if typing was done using saliva?

62. As you will learn in later chapters, duplication of genes is an important evolutionary mechanism. As a result, many cases are known in which a species has two or more nearly identical genes.

    a. Suppose there are two genes, *A* and *B*, that specify production of the same enzyme. An abnormal phenotype results only if an individual does not make any of that enzyme. What ratio of normal versus abnormal progeny would result from a mating between two parents of genotype *Aa Bb*, where *A* and *B* represent alleles that specify production of the enzyme, while *a* and *b* are alleles that do not?

    b. Suppose now that there are three genes specifying production of this enzyme, and again that a single functional allele is sufficient for a wild-type phenotype. What ratio of normal versus abnormal progeny would result from a mating between two triply heterozygous parents?

63. The following table shows the responses of blood samples from the individuals in the pedigree to anti-A and anti-B sera. A "+" in the anti-A row indicates that the red blood cells of that individual were clumped by anti-A serum and therefore the individual made A antigens, and a "−" indicates no clumping. The same notation is used to describe the test for the B antigens.

    a. Deduce the blood type of each individual from the data in the table.

    b. Assign genotypes for the blood groups as accurately as you can from these data, explaining the pattern of inheritance shown in the pedigree. Assume that all genetic relationships are as presented in the pedigree (i.e., there are no cases of false paternity).

	I-1	I-2	I-3	I-4	II-1	II-2	II-3	III-1	III-2
anti-A	+	+	−	+	−	−	+	+	−
anti-B	+	−	+	+	−	−	+	−	−

64. Normally, wild violets have yellow petals with dark brown markings and erect stems. Imagine you discover a plant with white petals, no markings, and prostrate stems. What experiment could you perform to determine whether the non-wild-type phenotypes are due to several different mutant genes or to the pleiotropic effects of

alleles at a single locus? Explain how your experiment would settle the question.

65. In mice, the A^y allele of the *agouti* gene is a recessive lethal allele, but it is dominant for yellow coat colour. What phenotypes and ratios of offspring would you expect from the cross of a mouse heterozygous at the agouti locus (genotype $A^y A$) and also at the albino locus (*Cc*) to an albino mouse (*cc*) heterozygous at the agouti locus ($A^y A$)?

66. Three different pure-breeding strains of corn that produce ears with white kernels were crossed to each other. In each case, the F_1 plants were all red, while both red and white kernels were observed in the F_2 generation in a 9:7 ratio. These results are tabulated here.

	F_1	F_2
white-1 × white-2	red	9 red : 7 white
white-1 × white-3	red	9 red : 7 white
white-2 × white-3	red	9 red : 7 white

    a. How many genes are involved in determining kernel colour in these three strains?

    b. Define your symbols and show the genotypes for the pure-breeding strains white-1, white-2, and white-3.

    c. Diagram the cross between white-1 and white-2, showing the genotypes and phenotypes of the F_1 and F_2 progeny. Explain the observed 9:7 ratio.

67. You picked up two mice (one female and one male) that had escaped from experimental cages in the animal facility. One mouse is yellow in colour, and the other is brown agouti. You know that this mouse colony has animals with different alleles at only three coat colour genes: the agouti or non-agouti or yellow alleles of the *A* gene, the black or brown allele of the *B* gene, and the albino or non-albino alleles of the *C* gene. However, you do not know which alleles of these genes are actually present in each of the animals that you have captured. To determine the genotypes, you breed them together. The first litter has only three pups. One is albino, one is brown (non-agouti), and the third is black agouti.

    a. What alleles of the *A*, *B*, and *C* genes are present in the two mice you caught?

    b. After raising several litters from these two parents, you have many offspring. How many different coat colour phenotypes (in total) do you expect to see expressed in the population of offspring? What are the phenotypes and corresponding genotypes?

68. A student whose hobby was fishing pulled a very unusual carp out of Cayuga Lake: It had no scales on its body. She decided to investigate whether this strange nude phenotype had a genetic basis. She therefore obtained some inbred carp that were pure-breeding for the wild-type scale phenotype (body covered with scales in a regular pattern) and crossed them with her nude fish. To her

surprise, the F_1 progeny consisted of wild-type fish and fish with a single linear row of scales on each side in a 1:1 ratio.

a. Can a single gene with two alleles account for this result? Why or why not?

b. To follow up on the first cross, the student allowed the linear fish from the F_1 generation to mate with each other. The progeny of this cross consisted of fish with four phenotypes: linear, wild type, nude, and scattered (the latter had a few scales scattered irregularly on the body). The ratio of these phenotypes was 6:3:2:1, respectively. How many genes appear to be involved in determining these phenotypes?

c. In parallel, the student allowed the phenotypically wild-type fish from the F_1 generation to mate with each other and observed, among their progeny, wild-type and scattered carp in a ratio of 3:1. How many genes with how many alleles appear to determine the difference between wild-type and scattered carp?

d. The student confirmed the conclusions of part *c* by crossing those scattered carp with her pure-breeding wild-type stock. Diagram the genotypes and phenotypes of the parental, F_1, and F_2 generations for this cross and indicate the ratios observed.

e. The student attempted to generate a true-breeding nude stock of fish by inbreeding. However, she found that this was impossible. Every time she crossed two nude fish, she found nude and scattered fish in the progeny, in a 2:1 ratio. (The scattered fish from these crosses bred true.) Diagram the phenotypes and genotypes of this gene in a nude × nude cross and explain the altered Mendelian ratio.

f. The student now felt she could explain all of her results. Diagram the genotypes in the linear × linear cross performed by the student (in part *b*). Show the genotypes of the four phenotypes observed among the progeny and explain the 6:3:2:1 ratio.

69. Three genes in fruit flies affect a particular trait, and one dominant allele of *each* gene is necessary to get a wild-type phenotype.

a. What phenotypic ratios would you predict among the progeny if you crossed triply heterozygous flies?

b. You cross a particular wild-type male in succession with three tester strains. In the cross with one tester strain (*AA bb cc*), only 1/4 of the progeny are wild type. In the crosses involving the other two tester strains (*aa BB cc* and *aa bb CC*), half of the progeny are wild type. What is the genotype of the wild-type male?

70. Figure 2.37 and Figure 2.41b both show traits that are determined by two genes, each of which has two incompletely dominant alleles. But in Figure 2.37 the gene interaction produces nine different phenotypes, while the situation depicted in Figure 2.41b shows only five

possible phenotypic classes. How can you explain this difference in the amount of phenotypic variation?

71. In foxgloves, there are three different petal phenotypes: white with red spots (WR), dark red (DR), and light red (LR). There are actually two different kinds of true-breeding WR strains (WR-1 and WR-2) that can be distinguished by two-generation intercrosses with true-breeding DR and LR strains:

Cross number	Parental	F_1	WR	LR	DR
				F_2	
1	WR-1 × LR	all WR	480	39	119
2	WR-1 × DR	all WR	99	0	32
3	DR × LR	all DR	0	43	132
4	WR-2 × LR	all WR	193	64	0
5	WR-2 × DR	all WR	286	24	74

a. What can you conclude about the inheritance of the petal phenotypes in foxgloves?

b. Ascribe genotypes to the four true-breeding parental strains (WR-1, WR-2, DR, and LR).

c. A WR plant from the F_2 generation of cross #1 is now crossed with an LR plant. Of 500 total progeny from this cross, there were 253 WR, 124 DR, and 123 LR plants. What are the genotypes of the parents in this WR × LR mating?

72. The garden flower *Salpiglossis sinuata* ("painted tongue") comes in many different colours. Several crosses are made between true-breeding parental strains to produce F_1 plants, which are in turn self-fertilized to produce F_2 progeny.

Parents	F_1 phenotypes	F_2 phenotypes
red × blue	all red	102 red, 33 blue
lavender × blue	all lavender	149 lavender, 51 blue
lavender × red	all bronze	84 bronze, 43 red, 41 lavender
red × yellow	all red	133 red, 58 yellow, 43 blue
yellow × blue	all lavender	183 lavender, 81 yellow, 59 blue

a. State a hypothesis explaining the inheritance of flower colour in painted tongues.

b. Assign genotypes to the parents, F_1 progeny, and F_2 progeny for all five crosses.

c. In a cross between true-breeding yellow and true-breeding lavender plants, all of the F_1 progeny are bronze. If you used these F_1 plants to produce an F_2 generation, what phenotypes in what ratios would you expect? Are there any genotypes that might produce a phenotype that you cannot predict from earlier experiments, and if so, how might this alter the phenotypic ratios among the F_2 progeny?

73. A married man and woman, both of whom are deaf, carry some recessive mutant alleles in three different "hearing genes": *d1* is recessive to *D1*, *d2* is recessive

to *D2*, and *d3* is recessive to *D3*. Homozygosity for a mutant allele at any one of these three genes causes deafness. In addition, homozygosity for any two of the three genes together in the same genome will cause prenatal lethality (and spontaneous abortion) with a penetrance of 25 percent. Furthermore, homozygosity for the mutant alleles of all three genes will cause prenatal lethality with a penetrance of 75 percent. If the genotypes of the mother and father are as indicated here, what is the likelihood that a live-born child will be deaf?

Mother: *D1 d1, D2 d2, d3 d3*

Father: *d1 d1, D2 d2, D3 d3*

74. In a culture of fruit flies, matings between any two flies with hairy wings (wings abnormally containing additional small hairs along their edges) always produce both hairy-winged and normal-winged flies in a 2:1 ratio. You now take hairy-winged flies from this culture and cross them with four types of normal-winged flies; the results for each cross are shown in the following table. Assuming that there are only two possible alleles of the hairy-winged gene (one for hairy wings and one for normal wings), what can you say about the genotypes of the four types of normal-winged flies?

	Progeny obtained from cross with hairy-winged flies	
Type of normal-winged flies	Fraction with normal wings	Fraction with hairy wings
1	1/2	1/2
2	1	0
3	3/4	1/4
4	2/3	1/3

The Chromosome Theory of Inheritance

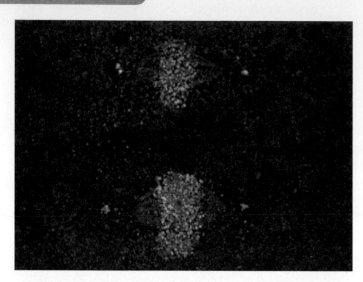

Mitosis in *Drosophila* requires katanin (*green*), a microtubule severing protein, that localizes to kinetochores of chromosomes (*blue*) during both metaphase (*top*) and anaphase (*bottom*). Spindle microtubules are shown in *red*.

In the spherical, membrane-bounded nuclei of plant and animal cells prepared for viewing under the microscope, chromosomes appear as brightly coloured, threadlike bodies. The nuclei of normal human cells carry 23 pairs of chromosomes for a total of 46. There are noticeable differences in size and shape among the 23 pairs, but within each pair, the two chromosomes appear to match exactly. (The only exceptions are the male's sex chromosomes, designated X and Y, which constitute an unmatched pair.)

Down syndrome was the first human genetic disorder attributable not to a gene mutation but to an abnormal number of chromosomes. Almost all children born with Down syndrome have 47 chromosomes in each cell nucleus because they carry three, instead of the normal pair, of a very small chromosome referred to as number 21. The aberrant genotype, known as trisomy 21, gives rise to an abnormal phenotype, including a wide skull that is flatter than normal at the back, an unusually large tongue, learning disabilities caused by the abnormal development of the hippocampus and other parts of the brain, and a propensity to respiratory infections as well as heart disorders, rapid aging, and leukaemia (**Figure 3.1**).

How can one extra copy of a chromosome that is itself of normal size and shape cause such wide-ranging phenotypic effects? The answer has two parts. First and foremost, chromosomes are the cellular structures responsible for transmitting genetic information. In this chapter, we describe how geneticists concluded that chromosomes are the carriers of genes, an idea that became known as the chromosome theory of inheritance. The second part of the answer is that proper development depends not just on what type of genetic material is present but also on how much of it there is. Thus the mechanisms governing gene transmission during cell division must vigilantly maintain each cell's chromosome number.

Proof that genes are located on chromosomes comes from both breeding experiments and the microscopic examination of cells. As you will see, the behaviour of chromosomes during one type of nuclear division called *meiosis* accounts for the segregation and independent assortment of genes proposed by Mendel. Meiosis figures prominently in the process by which most sexually reproducing organisms generate the gametes—eggs or sperm—that at fertilization unite to form the first cell of the next generation. This first cell is the fertilized egg, or *zygote*. The zygote then undergoes a second kind of nuclear division, known as *mitosis*, which continues to occur during the millions of cell divisions that propel development from a single cell to a complex multicellular organism. Mitosis provides each of the many cells in an individual with the same number and types of chromosomes.

Figure 3.1 Down syndrome: One extra chromosome 21 has widespread phenotypic consequences. Trisomy 21 usually causes changes in physical appearance as well as in the potential for learning. Many children with Down syndrome, such as the Grade 5 student at the centre of the photograph, are able to participate fully in regular activities.

of an organism. When the machinery does not function properly, errors in chromosome distribution can have severe repercussions on the individual's health and survival. Most cases of Down syndrome, for example, are the result of a failure of chromosome segregation during meiosis. The meiotic error gives rise to an egg or sperm carrying an extra chromosome 21, which if incorporated in the zygote at fertilization, is passed on via mitosis to every cell of the developing embryo. Trisomy—three copies of a chromosome instead of two—can occur with other chromosomes as well, but in nearly all of these cases, the condition is prenatally lethal and results in a miscarriage.

Two themes emerge in our discussion of meiosis and mitosis. First, direct microscopic observations of chromosomes during gamete formation led early twentieth-century investigators to recognize that *chromosome movements parallel the behaviour of Mendel's genes, so chromosomes are likely to carry the genetic material.* This chromosome theory of inheritance was proposed in 1902 and was confirmed in the following 15 years through elegant experiments performed mainly on the fruit fly *Drosophila melanogaster.* Second, the chromosome theory transformed the concept of a gene from an abstract particle to a physical reality—part of a chromosome that could be seen and manipulated.

The precise chromosome-parcelling mechanisms of meiosis and mitosis are crucial to the normal functioning

Learning Objectives

1. Examine the evidence for genes residing on chromosomes.

2. Explain how the union of maternal and paternal haploid gametes forms the diploid zygote.

3. Evaluate the reasoning that led scientists to suggest that chromosomes are related to heredity.

4. Compare and contrast the sequence of events and outcomes of mitosis and meiosis.

5. Discuss the genetic and phenotypic consequences of synaptonemal crossing-over.

6. Relate the chromosomal theory of inheritance to Gregor Mendel's principles of genetics.

7. Analyze sex-linked inheritance patterns.

3.1 Chromosomes: The Carriers of Genes

One of the first questions asked during pregnancy or at the birth of an infant—is it a boy or a girl?—acknowledges that male and female are mutually exclusive characteristics like the yellow versus green of Mendel's peas. In addition, among humans and most other sexually reproducing species, a roughly 1:1 ratio exists between the two genders. Both males and females produce cells specialized for reproduction—sperm or eggs—that serve as a physical link to the next generation. In bridging the gap between generations, these gametes must each contribute half of the genetic material for making a "normal," healthy son or daughter. Whatever part of the gamete carries this material, its structure and function must be able to account for the either-or aspect of sex determination as well as the generally observed 1:1 ratio of males to females. These two features of sex determination were among the earliest clues to the cellular basis of heredity.

Genes reside in the nucleus

The nature of the specific link between sex and reproduction remained a mystery until Anton van Leeuwenhoek, one of the earliest and most astute of microscopists, discovered in 1667 that semen contains spermatozoa (literally "sperm animals"). He imagined that these microscopic creatures might enter the egg and somehow achieve fertilization, but it was not possible to confirm this hypothesis for another 200 years. Then, during a 20-year period starting in 1854 (about the same time Gregor Mendel was beginning his pea experiments), microscopists studying fertilization in frogs and sea urchins observed the union of male and female gametes and recorded the details of the process in a series of drawings. These drawings, as well as later micrographs (photographs taken through a microscope), clearly show that egg and sperm nuclei are the only

elements contributed equally by maternal and paternal gametes. This observation implies that something in the nucleus contains the hereditary material. In humans, the nuclei of the gametes are less than 2 millionth of a metre in diameter. It is indeed remarkable that the genetic link between generations is packaged within such an exceedingly small space.

Genes reside on chromosomes

Further investigations, some dependent on technical innovations in microscopy, suggested that yet smaller, discrete structures within the nucleus are the repository of genetic information. In the 1880s, for example, a newly discovered combination of organic and inorganic dyes revealed the existence of the long, brightly staining, threadlike bodies within the nucleus that we call chromosomes (literally "coloured bodies"). It was now possible to follow the movement of chromosomes during different kinds of cell division.

In embryonic cells, the chromosomal threads split lengthwise in two just before cell division, and each of the two newly forming daughter cells receives one-half of every split thread. The kind of nuclear division followed by cell division that results in two daughter cells containing the same number and type of chromosomes as the original parent cell is called mitosis (from the Greek *mitos* meaning "thread" and *osis* meaning "formation" or "increase").

In the cells that give rise to male and female gametes, the chromosomes composing each pair become segregated, so that the resulting gametes receive only one chromosome from each chromosome pair. The kind of nuclear division that generates egg or sperm cells containing half the number of chromosomes found in other cells within the same organism is called meiosis (from the Greek word for "diminution").

Fertilization: The union of haploid gametes to produce diploid zygotes

In the first decade of the twentieth century, cytologists—scientists who use the microscope to study cell structure—showed that the chromosomes in a fertilized egg actually consist of two matching sets, one contributed by the maternal gamete, the other by the paternal gamete. The corresponding maternal and paternal chromosomes appear alike in size and shape, forming pairs (with one exception—the *sex chromosomes*—which we discuss in a later section).

Gametes and other cells that carry only a single set of chromosomes are called haploid (from the Greek word for "single"). Zygotes and other cells carrying two matching sets are diploid (from the Greek word for "double"). The number of chromosomes in a normal haploid cell is designated by the shorthand symbol n; the number of chromosomes in a normal diploid cell is then $2n$. **Figure 3.2** shows diploid cells as well as the haploid gametes that arise from them in *Drosophila*, where $2n = 8$ and $n = 4$. In humans, $2n = 46$; $n = 23$.

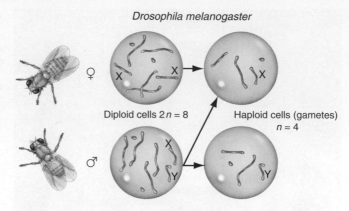

Figure 3.2 Diploid versus haploid: 2*n* versus *n*. Most body cells are diploid: They carry a maternal and paternal copy of each chromosome. Meiosis generates haploid gametes with only one copy of each chromosome. In *Drosophila*, diploid cells have eight chromosomes ($2n = 8$), while gametes have four chromosomes ($n = 4$). Note that the chromosomes in this diagram are pictured before their replication.

You can see how the halving of chromosome number during meiosis and gamete formation, followed by the union of two gametes' chromosomes at fertilization, normally allows a constant $2n$ number of chromosomes to be maintained from generation to generation in all individuals of a species. The chromosomes of every pair must segregate from each other during meiosis so that the haploid gametes will each have one complete set of chromosomes. After fertilization forms the zygote, the process of mitosis then ensures that all the cells of the developing individual have identical diploid chromosome sets.

> Microscopic studies suggested that the nuclei of egg and sperm contribute equally to the offspring by providing a single set of n chromosomes. The zygote formed by the union of haploid gametes is diploid ($2n$).

Species variations in the number and shape of chromosomes

Scientists analyze the chromosomal makeup of a cell when the chromosomes are most visible—at a specific moment in the cell cycle of growth and division, just before the nucleus divides. At this point, known as *metaphase* (described in detail later), individual chromosomes have duplicated and condensed from thin threads into compact rodlike structures. Each chromosome now consists of two identical halves, known as sister chromatids, attached to each other at a specific location called the centromere (**Figure 3.3**). In metacentric chromosomes, the centromere is more or less in the middle; in acrocentric chromosomes, the centromere is very close to one end. Modern high-resolution microscopy has failed to find any chromosomes in which the centromere is exactly at one end. As a result, the sister chromatids of all chromosomes actually have two "arms" separated by a centromere, even if one of the arms is very short.

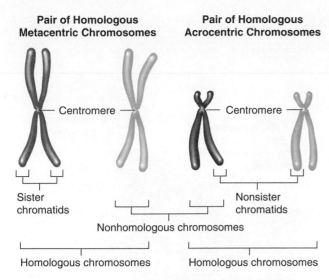

Pair of Homologous Metacentric Chromosomes

Pair of Homologous Acrocentric Chromosomes

Centromere — — Centromere

Sister chromatids

Nonsister chromatids

Nonhomologous chromosomes

Homologous chromosomes

Homologous chromosomes

Figure 3.3 Metaphase chromosomes can be classified by centromere position. Before cell division, each chromosome replicates into two sister chromatids connected at a centromere. In highly condensed metaphase chromosomes, the centromere can appear near the middle (a metacentric chromosome), very near an end (an acrocentric chromosome), or anywhere in between. In a diploid cell, one homologous chromosome in each pair is from the mother and the other from the father.

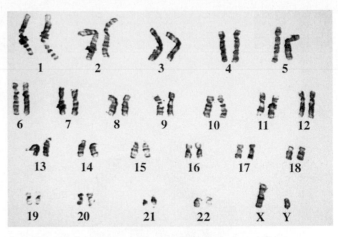

Figure 3.4 Karyotype of a human male. Photographs of metaphase human chromosomes are paired and arranged in order of decreasing size. In a normal human male karyotype, there are 22 pairs of autosomes, as well as an X and a Y ($2n = 46$). Homologous chromosomes share the same characteristic pattern of dark and light bands.

Cells in metaphase can be fixed and stained with one of several dyes that highlight the chromosomes and accentuate the centromeres. The dyes also produce characteristic banding patterns made up of lighter and darker regions. Chromosomes that match in size, shape, and banding are called homologous chromosomes, or homologues. One homologous chromosome of a pair is paternal in origin and the other maternally inherited. The two homologues of each pair contain the same set of genes, although for some of those genes, they may carry different alleles. The differences between alleles occur at the molecular level and do not show up under the microscope.

Figure 3.3 introduces a system of notation employed throughout this book, using colour to indicate degrees of relatedness between chromosomes. Thus, sister chromatids, which are identical duplicates, appear in the same shade of the same colour. Homologous chromosomes, which carry the same genes but may vary in the identity of particular alleles, are pictured in different shades (light or dark) of the same colour. *Nonhomologous chromosomes*, which carry completely unrelated sets of genetic information, appear in different colours.

To study the chromosomes of a single organism, geneticists arrange micrographs of the stained chromosomes in homologous pairs of decreasing size to produce a karyotype. Karyotype assembly can now be speeded and automated by computerized image analysis. **Figure 3.4** shows the karyotype of a human male, with 46 chromosomes arranged in 22 matching pairs of chromosomes and one nonmatching pair. The 44 chromosomes in matching pairs are known as autosomes. The two unmatched chromosomes in this male karyotype are called *sex chromosomes*, because they determine the sex of the individual. (We discuss sex chromosomes in more detail in subsequent sections.)

Modern methods of DNA analysis can reveal differences between the maternally and paternally derived chromosomes of a homologous pair, and can thus track the origin of an extra chromosome 21 that causes Down syndrome in individual patients. In 80 percent of cases, the third chromosome 21 comes from the egg; in 20 percent, from the sperm. The **Genetics and Society** box in this chapter describes how physicians use karyotype analysis and techniques called *amniocentesis* and *chorionic villus sampling* (CVS) to diagnose Down syndrome prenatally, roughly three months after a fetus is conceived.

Through thousands of karyotypes on normal individuals, cytologists have verified that the cells of each species carry a distinctive diploid number of chromosomes. Among three species of fruit flies, for example, *Drosophila melanogaster* carries 8 chromosomes in 4 pairs, *Drosophila obscura* carries 10 (5 pairs), and *Drosophila virilis* carries 12 (6 pairs). Mendel's peas contain 14 chromosomes (7 pairs) in each diploid cell, macaroni wheat has 28 (14 pairs), giant sequoia trees have 22 (11 pairs), goldfish have 94 (47 pairs), dogs have 78 (39 pairs), and people have 46 (23 pairs). Differences in the size, shape, and number of chromosomes reflect differences in the assembled genetic material that determines what each species looks like and how it functions. As these figures show, the number of chromosomes does not always correlate with the size or complexity of the organism.

> Karyotyping, the analysis of stained images of all the chromosomes in a cell, reveals that different species have different numbers and shapes of chromosomes.

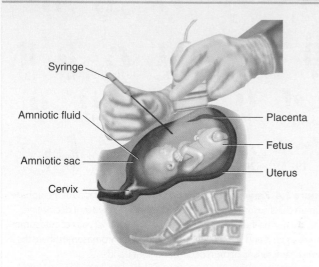

Figure A Obtaining fetal cells by amniocentesis. A physician guides insertion of the needle into the amniotic sac using ultrasound imaging and extracts amniotic fluid containing fetal cells into the syringe.

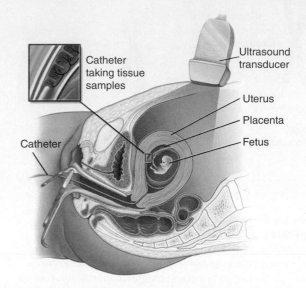

Figure B Procuring fetal cells by chorionic villus sampling (CVS). In transcervical CVS, a physician uses ultrasound imaging to guide insertion of the catheter through the cervix to the placenta. Chorionic villi cells are extracted using a syringe on the other end of the catheter needle.

With new technologies for observing chromosomes and the DNA in genes, modern geneticists can define an individual's genotype directly. They can use this information to predict aspects of the individual's phenotype, even before these traits manifest themselves. Doctors can even use this basic strategy to diagnose, before birth, whether or not a baby will be born with a genetic condition.

The first prerequisite for prenatal diagnosis is to obtain fetal cells whose DNA and chromosomes can be analyzed for genotype. The most frequently used methods for acquiring these cells are **amniocentesis** (**Figure A**) and **chorionic villus sampling** (**CVS**) (**Figure B**). To carry out amniocentesis, a doctor inserts a needle through a pregnant woman's abdominal wall into the amniotic sac in which the fetus is growing; this procedure is performed about 16 weeks after the woman's last menstrual period. By using ultrasound imaging to guide the location of the needle, the physician can minimize the chance of injuring the fetus. The doctor then withdraws some of the amniotic fluid, in which the fetus is suspended, back through the needle into a syringe. This fluid contains living cells called *amniocytes* that were shed by the fetus. When placed in a culture medium, these fetal cells undergo several rounds of mitosis and increase in number. Once enough fetal cells are available, clinicians look at the chromosomes and genes in those cells.

CVS may sometimes be preferred over amniocentesis because larger samples are collected and it may be performed earlier in the pregnancy, usually between 10 and 13 weeks after a pregnant woman's last menstrual period. However, the procedure has a slightly higher risk of miscarriage in comparison with amniocentesis. The chorion, the outermost extraembryonic membrane that forms the fetal part of the placenta,

develops fingerlike projections known as villi that extend from its external surface. The cells that make up the chorionic villi have the same genetic makeup as the fetus. CVS involves removing chorionic villi cells from the placenta at the point where it attaches to the uterine wall. Samples are usually collected transcervically by a physician, whereby ultrasound imaging is used to guide a thin catheter or tube through the cervix to the placenta and the chorionic villi cells are then gently suctioned into the catheter. Alternatively, a doctor can perform the procedure transabdominally, in which ultrasound imaging is used to guide a needle through the abdominal wall to the placenta. The needle draws a sample of tissue into the syringe and then is removed. The chorionic villus sample is placed into a nutrient medium and sent to the laboratory for chromosomal analysis and other specialized tests. In later chapters, we describe techniques that allow the direct examination of the DNA constituting particular disease genes.

Amniocentesis and CVS also allow the diagnosis of Down syndrome through the analysis of chromosomes by karyotyping. Because the risk of Down syndrome increases rapidly with the age of the mother, more than half the pregnant women in North America who are over the age of 35 currently undergo amniocentesis or CVS. Although the goal of this karyotyping is usually to learn whether the fetus is trisomic for chromosome 21, many other abnormalities in chromosome number or shape may show up when the karyotype is examined.

The availability of these techniques of prenatal diagnosis is intimately entwined with the personal and societal issue of abortion. The large majority of these procedures are performed

with the understanding that a fetus whose genotype indicates a genetic disorder, such as Down syndrome, will be aborted. Some prospective parents who are opposed to abortion still elect to undergo amniocentesis or CVS so that they can better prepare for an affected child.

The ethical and political aspects of the abortion debate influence many of the practical questions underlying prenatal diagnosis. For example, parents must decide which genetic conditions would be sufficiently severe that they would be willing to abort the fetus. They must also assess the risk that amniocentesis and CVS might harm the fetus. The normal risk of miscarriage at 16 weeks of gestation is about 2 to 3 percent; amniocentesis increases that risk by about 0.5 percent (about 1 in 200 procedures). For CVS, the risk is higher, at about 1 percent (1 in 100 procedures).

In current practice, the risks and costs of prenatal testing generally restrict amniocentesis and CVS to women over age 35 or to mothers whose fetuses are at high risk for a testable genetic condition because of family history. The personal and societal equations determining the frequency of prenatal testing may, however, need to be overhauled in the not-too-distant future because of technological advances that will simplify the procedures and thereby minimize the costs and risks. As one example, clinicians are now able to take advantage of new methods to purify the very small number of fetal cells that find their way into the mother's bloodstream during pregnancy. Collecting these cells from the mother's blood is much less invasive and expensive than amniocentesis and CVS and would pose no risk to the fetus, and their karyotype analysis is just as accurate. In fact, sometime in the near future, sequencing the whole genome of the fetus using cell-free fetal DNA present in the mother's blood may become more commonplace than amniocentesis and CVS. By comparing this free-floating fetal DNA with genome sequences obtained from the father and the mother, scientists have recently been able to almost accurately reconstruct the whole-genome sequence of a fetus.

Sex chromosomes

Walter S. Sutton, a young American graduate student at Columbia University in the first decade of the twentieth century, was one of the earliest cytologists to realize that particular chromosomes carry the information for determining sex. In one study, he obtained cells from the testes of the great lubber grasshopper (*Brachystola magna*) and followed them through the meiotic divisions that produce sperm. He observed that prior to meiosis, precursor cells within the testes of a great lubber grasshopper contain a total of 24 chromosomes. Of these, 22 are found in 11 matched pairs and are thus autosomes. The remaining two chromosomes are unmatched. He called the larger of these the X chromosome and the smaller the Y chromosome. After meiosis, the sperm produced within these testes are of two equally prevalent types: one-half have a set of 11 autosomes plus an X chromosome, while the other half have a set of 11 autosomes plus a Y chromosome. By comparison, all of the eggs produced by females of the species carry an 11-plus-X set of chromosomes like the set found in the first class of sperm. When a sperm with an X chromosome fertilizes an egg, an XX female grasshopper results; when a Y-containing sperm fuses with an egg, an XY male develops. Sutton concluded that the X and Y chromosomes determine sex.

Several researchers studying other organisms soon verified that in many sexually reproducing species, two distinct chromosomes—known as the sex chromosomes—provide the basis of sex determination. One sex carries two copies of the same chromosome (a matching pair), while the other sex has one of each type of sex chromosome (an unmatched pair). The cells of normal human females, for example, contain 23 pairs of chromosomes. The two chromosomes of each pair, including the sex-determining X chromosomes, appear to be identical in size and shape. In males, however, there is one unmatched pair of chromosomes: the larger of these is the X; the smaller, the Y

(Figure 3.4 and **Figure 3.5a**). Apart from this difference in sex chromosomes, the two sexes are not distinguishable at any other pair of chromosomes. Thus, geneticists can designate women as XX and men as XY and represent sexual reproduction as a simple cross between XX and XY.

If sex is an inherited trait determined by a pair of sex chromosomes that separate to different cells during gamete formation, then an XX × XY cross could account for both the mutual exclusion of genders and the near 1:1 ratio of males to females, which are hallmark features of sex determination (**Figure 3.5b**). And if chromosomes carry information defining the two contrasting sex phenotypes, we can easily infer that chromosomes also carry genetic information specifying other characteristics as well.

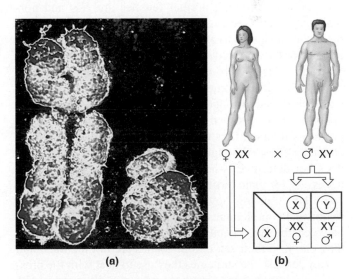

Figure 3.5 How the X and Y chromosomes determine sex in humans. (a) This colourized micrograph shows the human X chromosome on the *left* and the human Y on the *right*. **(b)** Children can receive only an X chromosome from their mother, but they can inherit either an X or a Y from their father.

Table 3.1 — Sex Determination in Fruit Flies and Humans

| | **Complement of Sex Chromosomes** | | | | | | |
	XXX	XX	XXY	XO	XY	XYY	OY
Drosophila	Dies	Normal female	Normal female	Sterile male	Normal male	Normal male	Dies
Humans	Nearly normal female	Normal female	Klinefelter male (sterile); tall, thin	Turner female (sterile); webbed neck	Normal male	Normal or nearly normal male	Dies

Humans can tolerate extra X chromosomes (e.g., XXX) better than can *Drosophila*. Complete absence of an X chromosome is lethal to both fruit flies and humans. Additional Y chromosomes have little effect in either species.

Species variations in sex determination

You have just seen that humans and other mammals have a pair of sex chromosomes that are identical in the XX female but different in the XY male. Several studies have shown that in humans, it is the presence or absence of the Y that actually makes the difference; that is, any person carrying a Y chromosome will look like a male. For example, rare humans with two X and one Y chromosomes (XXY) are males displaying certain abnormalities collectively called *Klinefelter syndrome*. Klinefelter males are typically tall, thin, and sterile, and they sometimes show mental retardation. That these individuals are males shows that two X chromosomes are insufficient for female development in the presence of a Y. In contrast, humans carrying an X and no second sex chromosome (XO) are females with *Turner syndrome*. Turner females are usually sterile, lack secondary sexual characteristics such as pubic hair, are of short stature, and have folds of skin between their necks and shoulders (webbed necks). Even though these individuals have only one X chromosome, they develop as females because they have no Y chromosome.

Other species show variations on this XX versus XY chromosomal strategy of sex determination. In fruit flies, for example, although normal females are XX and normal males XY (see Figure 3.2), it is ultimately the ratio of X chromosomes to autosomes (and not the presence or absence of the Y) that determines sex. In female *Drosophila*, the ratio is 1:1 (there are two X chromosomes and two copies of each autosome); in males, the ratio is 1:2 (there is one X chromosome but two copies of each autosome). Curiously, a rarely observed abnormal intermediate ratio of 2:3 produces intersex flies that display both male and female characteristics. Although the Y chromosome in *Drosophila* does not determine whether a fly looks like a male, it is necessary for male fertility; XO flies are thus sterile males. **Table 3.1** compares how humans and *Drosophila* respond to unusual complements of sex chromosomes. Differences between the two species arise in part because the genes they carry on their sex chromosomes are not identical and in part because the strategies they use to deal with the presence of additional sex chromosomes are not the same. The molecular mechanisms of sex determination in *Drosophila* are covered in detail in Chapter 11.

The XX = female / XY = male strategy of sex determination is by no means universal. In some species of moths, for example, the females are XX, but the males are XO. In *C. elegans* (one species of nematode), males are similarly XO, but XX individuals are not females; they are instead self-fertilizing hermaphrodites that produce both eggs and sperm. In birds and butterflies, males have the matching sex chromosomes, while females have an unmatched set; in such species, geneticists represent the sex chromosomes as ZZ in the male and ZW in the female. The gender having two different sex chromosomes is termed the **heterogametic sex** because it gives rise to two different types of gametes. These gametes would contain either X or Y in the case of male humans, and either Z or W in the case of female birds. Yet other variations include the complicated sex-determination mechanisms of bees and wasps, in which females are diploid and males haploid, and the systems of certain fish, in which sex is determined by changes in the environment, such as fluctuations in temperature. **Table 3.2** summarizes some of the astonishing variety in the ways that different species have solved the problem of assigning gender to individuals.

In spite of these many differences between species, early researchers concluded that chromosomes can carry the genetic information specifying sexual identity—and

Table 3.2 — Mechanisms of Sex Determination

	♀	♂
Humans and *Drosophila*	XX	XY
Moths and *C. elegans*	XX (hermaphrodites in *C. elegans*)	XO
Birds and Butterflies	ZW	ZZ
Bees and Wasps	Diploid	Haploid
Lizards and Alligators	Cool temperature	Warm temperature
Tortoises and Turtles	Warm temperature	Cool temperature
Anemone Fish	Older adults	Young adults

In the species highlighted in *purple*, sex is determined by sex chromosomes. The species highlighted in *green* have identical chromosomes in the two sexes, and sex is determined instead by environmental or other factors. Anemone fish (*bottom row*) undergo a sex change from male to female as they age.

probably many other characteristics as well. Sutton and other early adherents of the chromosome theory realized that the perpetuation of life itself therefore depends on the proper distribution of chromosomes during cell division. In the next sections, you will see that the behaviour of chromosomes during mitosis and meiosis is exactly that expected of cellular structures carrying genes.

In many species, the sex of an individual correlates with a particular pair of chromosomes termed the sex chromosomes. The segregation of the sex chromosomes during gamete formation and their random reunion at fertilization explains the 1:1 ratio of the two sexes.

3.2 Mitosis: Cell Division That Preserves Chromosome Number

The fertilized human egg is a single diploid cell that preserves its genetic identity unchanged through more than 100 generations of cells as it divides again and again to produce a full-term infant ready to be born. As the newborn infant develops into a toddler, a teenager, and an adult, yet more cell divisions fuel continued growth and maturation. Mitosis, the nuclear division that apportions chromosomes in equal fashion to two daughter cells, is the cellular mechanism that preserves genetic information through all these generations of cells. In this section, we take a close look at how the nuclear division of mitosis fits into the overall scheme of cell growth and division.

If you were to peer through a microscope and follow the history of one cell through time, you would see that for much of your observation, the chromosomes resemble a mass of extremely fine tangled string—called chromatin—surrounded by the nuclear envelope. Each convoluted thread of chromatin is composed mainly of DNA (which carries the genetic information) and protein (which serves as a scaffold for packaging and managing that information, as described in Chapter 7). You would also be able to distinguish one or two darker areas of chromatin called *nucleoli* (singular, nucleolus; literally "small nucleus"); nucleoli play a key role in the manufacture of ribosomes, organelles that function in protein synthesis. During the period between cell divisions, the chromatin-laden nucleus houses a great deal of invisible activity necessary for the growth and survival of the cell. One particularly important part of this activity is the accurate duplication of all the chromosomal material.

With continued vigilance, you would observe a dramatic change in the nuclear landscape during one very short period in the cell's life history: The chromatin condenses into discrete threads, and then each chromosome compacts even further into the twin rods clamped together at the centromere that can be identified in karyotype analysis (review Figure 3.3). Each rod in a duo is called a chromatid; as described earlier, it is an exact duplicate of the other sister chromatid to which it is connected. Continued observation would reveal the doubled chromosomes beginning to jostle around inside the cell, eventually lining up at the cell's midplane. At this point, the sister chromatids comprising each chromosome separate to opposite poles of the now elongating cell, where they become identical sets of chromosomes. Each of the two identical sets eventually ends up enclosed in a separate nucleus in a separate cell. The two cells, known as *daughter cells*, are thus genetically identical.

The repeating pattern of cell growth (an increase in size) followed by division (the splitting of one cell into two) is called the cell cycle (**Figure 3.6**). Only a small part of the cell cycle is spent in division (or **M phase**); the period between divisions is called interphase.

During interphase, cells grow and replicate their chromosomes

Interphase consists of three parts: gap 1 (G_1), synthesis (S), and gap 2 (G_2) (Figure 3.6). G_1 lasts from the birth of a new cell to the onset of chromosome replication; for the genetic material, it is a period when the chromosomes are neither duplicating nor dividing. During this time, the

(a) The cell cycle

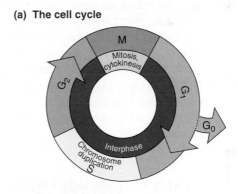

(b) Chromosomes replicate during S phase

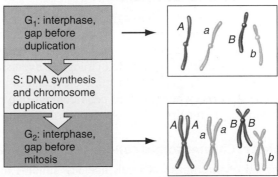

Figure 3.6 The cell cycle: An alternation between interphase and mitosis. (a) Chromosomes replicate to form sister chromatids during synthesis (S phase); the sister chromatids segregate to daughter cells during mitosis (M phase). The gaps between the S and M phases, during which most cell growth takes place, are called the G_1 and G_2 phases. In multicellular organisms, some terminally differentiated cells stop dividing and arrest in a "G_0" stage. **(b)** Interphase consists of the G_1, S, and G_2 phases together.

cell achieves most of its growth by using the information from its genes to make and assemble the materials it needs to function normally. G_1 varies in length more than any other phase of the cell cycle. In rapidly dividing cells of the human embryo, for example, G_1 is as short as a few hours. In contrast, mature brain cells become arrested in a resting form of G_1 known as G_0 and do not normally divide again during a person's lifetime.

Synthesis (S) is the time when the cell duplicates its genetic material by synthesizing DNA. During duplication, each chromosome doubles to produce identical sister chromatids that will become visible when the chromosomes condense at the beginning of mitosis. The two sister chromatids remain joined to each other at the centromere. (Note that this joined structure is considered a single chromosome as long as the connection between sister chromatids is maintained.) The replication of chromosomes during S phase is critical; the genetic material must be copied exactly so that both daughter cells receive identical sets of chromosomes.

Gap 2 (G_2) is the interval between chromosome duplication and the beginning of mitosis. During this time, the cell may grow (usually less than during G_1); it also synthesizes proteins that are essential to the subsequent steps of mitosis itself.

In addition, during interphase, an array of fine microtubules crucial for many interphase processes becomes visible outside the nucleus. The microtubules radiate out into the cytoplasm from a single organizing centre known as the centrosome, usually located near the nuclear envelope. In animal cells, the detectable core of each centrosome is a pair of small, darkly staining bodies called centrioles (**Figure 3.7a**); the microtubule-organizing centre of plants does not contain centrioles. During the S and G_2 stages of interphase, the centrosomes replicate, producing two centrosomes that remain in extremely close proximity.

During mitosis, sister chromatids separate and two daughter nuclei form

Although the rigorously choreographed events of nuclear and cellular division occur as a dynamic and continuous process, scientists traditionally analyze the process in separate stages marked by visible cytological events. The artist's sketches in **Figure 3.7** illustrate these stages in the nematode *Ascaris*, whose diploid cells contain only four chromosomes (two pairs of homologous chromosomes).

Prophase: Chromosomes condense (Figure 3.7a)

During all of interphase, the cell nucleus remains intact, and the chromosomes are indistinguishable aggregates of chromatin. At prophase (from the Greek *pro* meaning "before"), the gradual emergence, or condensation, of individual chromosomes from the undifferentiated mass of chromatin marks the beginning of mitosis. Each condensing chromosome has already been duplicated during interphase and thus consists of sister chromatids attached at the

centromere. At this stage in *Ascaris* cells, there are therefore four chromosomes with a total of eight chromatids.

The progressive appearance of an array of individual chromosomes is a truly impressive event: Interphase DNA molecules as long as 3–4 cm condense into discrete chromosomes whose length is measured in microns (millionths of a metre). This is equivalent to compacting a 200 m length of thin string (as long as two football fields) into a cylinder 8 mm long and 1 mm wide.

Another visible change in chromatin also takes place during prophase: The darkly staining nucleoli begin to break down and disappear. As a result, the manufacture of ribosomes ceases, providing one indication that general cellular metabolism shuts down so that the cell can focus its energy on chromosome movements and cellular division.

Several important processes that characterize prophase occur outside the nucleus in the cytoplasm. The centrosomes, which replicated during interphase, now move apart and become clearly distinguishable as two separate entities under the light microscope. At the same time, the interphase scaffolding of long, stable microtubules disappears and is replaced by a set of dynamic microtubules that rapidly grow from and shrink back toward their centrosomal organizing centres. The centrosomes continue to move apart, migrating around the nuclear envelope toward opposite ends of the nucleus, apparently propelled by forces exerted between interdigitated microtubules extending from both centrosomes.

Prometaphase: The spindle forms (Figure 3.7b)

Prometaphase ("before middle stage") begins with the breakdown of the nuclear envelope, which allows microtubules extending from the two centrosomes to invade the nucleus. Chromosomes attach to these microtubules through the kinetochore, a structure in the centromere region of each chromatid that is specialized for transport. Each kinetochore contains proteins that act as molecular motors, enabling the chromosome to slide along the microtubule. When the kinetochore of a chromatid originally contacts a microtubule at prometaphase, the kinetochore-based motor moves the entire chromosome toward the centrosome from which that microtubule radiates. Microtubules growing from the two centrosomes randomly capture chromosomes by the kinetochore of one of the two sister chromatids. As a result, it is sometimes possible to observe groups of chromosomes congregating in the vicinity of each centrosome. In this early part of prometaphase, for each chromosome, one chromatid's kinetochore is attached to a microtubule, but the sister chromatid's kinetochore remains unattached.

During prometaphase, three different types of microtubule fibres together form the mitotic spindle; all of these microtubules originate from the centrosomes, which function as the two "poles" of the spindle apparatus. Microtubules that extend between a centrosome and the kinetochore of a chromatid are called kinetochore microtubules, or

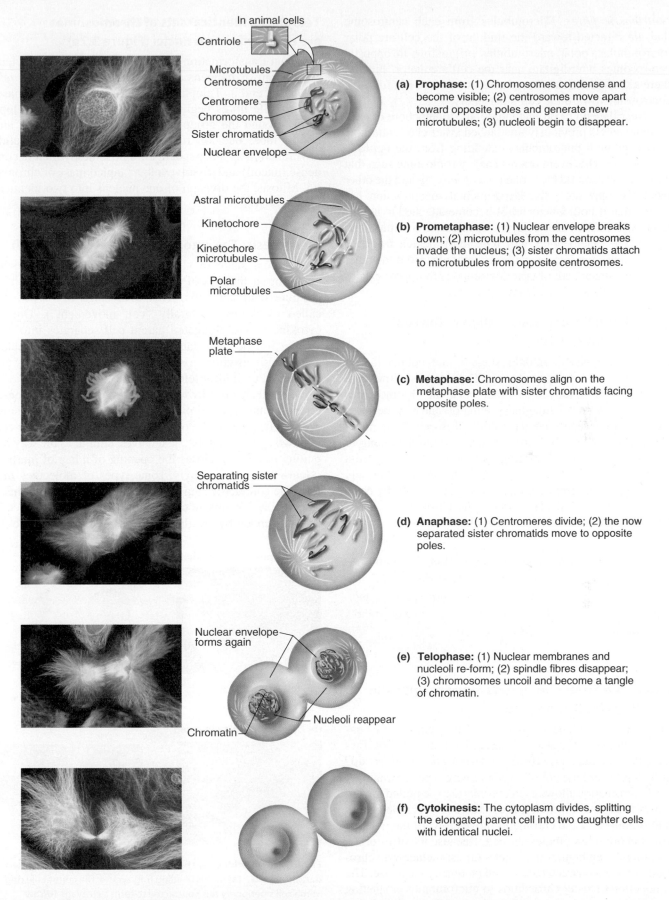

(a) Prophase: (1) Chromosomes condense and become visible; (2) centrosomes move apart toward opposite poles and generate new microtubules; (3) nucleoli begin to disappear.

In animal cells
Centriole
Microtubules
Centrosome
Centromere
Chromosome
Sister chromatids
Nuclear envelope

(b) Prometaphase: (1) Nuclear envelope breaks down; (2) microtubules from the centrosomes invade the nucleus; (3) sister chromatids attach to microtubules from opposite centrosomes.

Astral microtubules
Kinetochore
Kinetochore microtubules
Polar microtubules

(c) Metaphase: Chromosomes align on the metaphase plate with sister chromatids facing opposite poles.

Metaphase plate

(d) Anaphase: (1) Centromeres divide; (2) the now separated sister chromatids move to opposite poles.

Separating sister chromatids

(e) Telophase: (1) Nuclear membranes and nucleoli re-form; (2) spindle fibres disappear; (3) chromosomes uncoil and become a tangle of chromatin.

Nuclear envelope forms again
Nucleoli reappear
Chromatin

(f) Cytokinesis: The cytoplasm divides, splitting the elongated parent cell into two daughter cells with identical nuclei.

Figure 3.7 **Mitosis maintains the chromosome number of the parent cell nucleus in the two daughter nuclei.** In the photomicrographs of newt lung cells at the *left*, chromosomes are stained *blue* and microtubules appear either *green* or *yellow*.

centromeric fibres. Microtubules from each centrosome that are directed toward the middle of the cell are polar microtubules; polar microtubules originating in opposite centrosomes interdigitate near the cell's equator. Finally, there are short astral microtubules that extend out from the centrosome toward the cell's periphery.

Near the end of prometaphase, the kinetochore of each chromosome's previously unattached sister chromatid now associates with microtubules extending from the opposite centrosome. This event orients each chromosome such that one sister chromatid faces one pole of the cell, and the other faces the opposite pole. Experimental manipulation has shown that if both kinetochores become attached to microtubules from the same pole, the configuration is unstable; one of the kinetochores will repeatedly detach from the spindle until it associates with microtubules from the other pole. The attachment of sister chromatids to opposite spindle poles is the only stable arrangement.

Metaphase: Chromosomes align at the cell's equator (Figure 3.7c)

During metaphase ("middle stage"), the connection of sister chromatids to opposite spindle poles sets in motion a series of jostling movements that cause the chromosomes to move toward an imaginary equator halfway between the two poles. The imaginary midline is called the metaphase plate. When the chromosomes are aligned along it, the forces pulling and pushing them toward or away from each pole are in a balanced equilibrium. As a result, any movement away from the metaphase plate is rapidly compensated by tension that restores the chromosome to its position equidistant between the poles.

> The essence of mitosis is the arrangement of chromosomes at metaphase. The kinetochores of sister chromatids are connected to fibres from opposite spindle poles, but the sister chromatids remain held together by their connection at the centromere.

Anaphase: Sister chromatids move to opposite spindle poles (Figure 3.7d)

The nearly simultaneous severing of the centromeric connection between the sister chromatids of all chromosomes indicates that anaphase (from the Greek *ana* meaning "up" as in "up toward the poles") is under way. The separation of sister chromatids allows each chromatid to be pulled toward the spindle pole to which it is connected by its kinetochore microtubules; as the chromatid moves toward the pole, its kinetochore microtubules shorten. Because the arms of the chromatids lag behind the kinetochores, metacentric chromatids have a characteristic V shape during anaphase. The connection of sister chromatids to microtubules emanating from opposite spindle poles means that the genetic information migrating toward one pole is exactly the same as its counterpart moving toward the opposite pole.

Telophase: Identical sets of chromosomes are enclosed in two nuclei (Figure 3.7e)

The final transformation of chromosomes and the nucleus during mitosis happens at telophase (from the Greek *telo* meaning "end"). Telophase is like a rewind of prophase. The spindle fibres begin to disperse; a nuclear envelope forms around the group of chromatids at each pole; and one or more nucleoli reappears. The former chromatids now function as independent chromosomes, which decondense (uncoil) and dissolve into a tangled mass of chromatin. Mitosis, the division of one nucleus into two identical nuclei, is over.

Cytokinesis: The cytoplasm divides (Figure 3.7f)

In the final stage of cell division, the daughter nuclei emerging at the end of telophase are packaged into two separate daughter cells. This final stage of division is called cytokinesis (literally "cell movement"). During cytokinesis, the elongated parent cell separates into two smaller independent daughter cells with identical nuclei. Cytokinesis usually begins during anaphase, but it is not completed until after telophase.

The mechanism by which cells accomplish cytokinesis differs in animals and plants. In animal cells, cytoplasmic division depends on a contractile ring that pinches the cell into two approximately equal halves, similar to the way the pulling of a string closes the opening of a bag of marbles (**Figure 3.8a**). Intriguingly, some types of molecules that form the contractile ring also participate in the mechanism responsible for muscle contraction. In plants, whose cells are surrounded by a rigid cell wall, a membrane-enclosed

(a) Cytokinesis in an animal cell

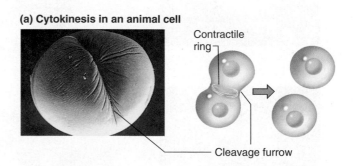

Contractile ring

Cleavage furrow

(b) Cytokinesis in a plant cell

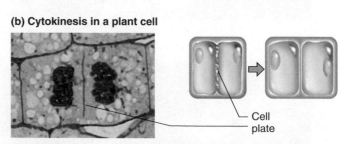

Cell plate

Figure 3.8 Cytokinesis: The cytoplasm divides, producing two daughter cells. (a) In this dividing frog zygote, the contractile ring at the cell's periphery has contracted to form a cleavage furrow that will eventually pinch the cell in two. **(b)** In this dividing onion root cell, a cell plate that began forming near the equator of the cell expands to the periphery, separating the two daughter cells.

disk, known as the cell plate, forms inside the cell near the equator and then grows rapidly outward, thereby dividing the cell in two (**Figure 3.8b**).

During cytokinesis, a large number of important organelles and other cellular components, including ribosomes, mitochondria, membranous structures such as Golgi bodies, and (in plants) chloroplasts, must be parcelled out to the emerging daughter cells. The mechanism accomplishing this task does not appear to predetermine which organelle is destined for which daughter cell. Instead, because most cells contain many copies of these cytoplasmic structures, each new cell is bound to receive at least a few representatives of each component. This original complement of structures is enough to sustain the cell until synthetic activity can repopulate the cytoplasm with organelles.

Sometimes cytoplasmic division does not immediately follow nuclear division, and the result is a cell containing more than one nucleus. An animal cell with two or more nuclei is known as a syncytium. The early embryos of fruit flies are multinucleated syncytia (**Figure 3.9**), as are the precursors of spermatozoa in humans and many other animals. A multinucleate plant tissue is called a coenocyte; coconut milk is a nutrient-rich food composed of coenocytes.

> After mitosis plus cytokinesis, the sister chromatids of every chromosome are separated into two daughter cells. As a result, these two cells are genetically identical to each other and to the original parental cell.

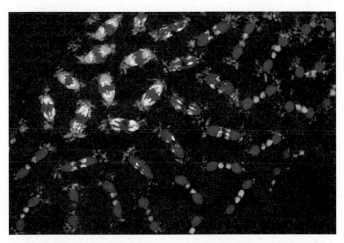

Figure 3.9 If cytokinesis does not follow mitosis, one cell may contain many nuclei. In fertilized *Drosophila* eggs, 13 rounds of mitosis take place without cytokinesis. The result is a single-celled syncytial embryo that contains several thousand nuclei. The photograph shows part of an embryo in which the nuclei are all dividing; chromosomes are in *red*, and spindle fibres are in *green*. Nuclei at the *upper left* are in metaphase, while nuclei toward the *bottom right* are progressively later in anaphase. Membranes eventually grow around these nuclei, dividing the embryo into cells.

Regulatory checkpoints ensure correct chromosome separation

The cell cycle is a complex sequence of precisely coordinated events. In higher organisms, a cell's "decision" to divide depends on both intrinsic factors, such as conditions within the cell that register a sufficient size for division; and signals from the environment, such as hormonal cues or contacts with neighbouring cells that encourage or restrain division. Once a cell has initiated events leading to division, usually during the G_1 period of interphase, everything else follows like clockwork. A number of checkpoints—moments at which the cell evaluates the results of previous steps—allow the sequential coordination of cell-cycle events. Consequently, under normal circumstances, the chromosomes replicate before they condense, and the doubled chromosomes separate to opposite poles only after correct metaphase alignment of sister chromatids ensures equal distribution to the daughter nuclei (**Figure 3.10**).

In one illustration of the molecular basis of checkpoints, even a single kinetochore that has not attached to spindle fibres generates a molecular signal that prevents the sister chromatids of all chromosomes from separating

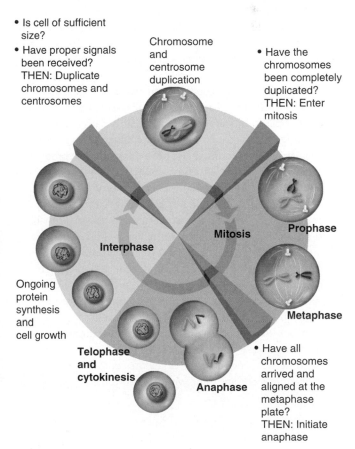

Figure 3.10 Checkpoints help regulate the cell cycle. Cellular checkpoints (*red wedges*) ensure that important events in the cell cycle occur in the proper sequence. At each checkpoint, the cell determines whether prior events have been completed before it can proceed to the next step of the cell cycle. (For simplicity, we show only two chromosomes per cell.)

Metaphase   Anaphase   Metaphase   Anaphase

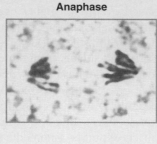

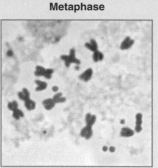

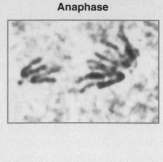

Figure A Metaphase and anaphase chromosomes in a wild-type male fruit fly.

During each cell cycle, the chromosomes participate in a tightly patterned choreography that proceeds through sequential steps, synchronized in both time and space. Through their dynamic dance, the chromosomes convey a complete set of genes to each of two newly forming daughter cells. Not surprisingly, some of the genes they carry encode proteins that direct them through the dance.

A variety of proteins, some assembled into structures such as centrosomes and microtubule fibres, make up the molecular machinery that helps coordinate the orderly progression of events in mitosis. Because a particular gene specifies each protein, we might predict that mutant alleles generating defects in particular proteins could disrupt the dance. Cells homozygous for a mutant allele might be unable to complete chromosome duplication, mitosis, or cytokinesis because of a missing or nonfunctional component. Experiments on organisms as different as yeast and fruit flies have borne out this prediction. Here we describe the effects of a mutation in one of the many *Drosophila* genes critical for proper chromosome segregation.

Although most mistakes in mitosis are eventually lethal to a multicellular organism, some mutant cells may manage to divide early in development. When prepared for viewing under the microscope, these cells actually allow us to see the effects of defective mitosis. To understand these effects, we first present part of a normal mitosis as a basis for comparison. **Figure A** (*left panel*) shows the eight condensed metaphase chromosomes of a wild-type male fruit fly (*Drosophila melanogaster*): two pairs of large metacentric autosomes with the centromere in the centre, a pair of dot-like autosomes that are so small it is not possible to see the centromere region, an acrocentric X chromosome with the centromere very close to

Figure B Metaphase and anaphase chromosomes in a mutant fly. These cells are from a *Drosophila* male homozygous for a mutation in the *zw10* gene. The mutant metaphase cell (*left*) contains extra chromosomes as compared with the wild-type metaphase cell in Figure A. In the mutant anaphase cell (*right*), more chromatids are moving toward one spindle pole than toward the other.

one end, and a metacentric Y chromosome. Because most of the Y chromosome consists of a special form of chromatin known as heterochromatin, the two Y sister chromatids remain so tightly connected that they often appear as one.

Figure B (*left panel*) shows the results of aberrant mitosis in a male fruit fly homozygous for a mutation in a gene called *zw10* that encodes a component of the chromosomal kinetochores. The mutation disrupted mitotic chromosome segregation during early development, producing cells with the wrong number of chromosomes. The problem in chromosome segregation probably occurred during anaphase of the previous cell division.

Figure A (*right panel*) shows a normal anaphase separation leading to the wild-type chromosome complement. Figure B (*right panel*) portrays an aberrant anaphase separation in a *zw10* mutant animal that could lead to an abnormal chromosome complement similar to that depicted in the left panel of the same figure; you can see that many more chromatids are migrating to one spindle pole than to the other.

The smooth unfolding of each cell cycle depends on a diverse array of proteins. Particular genes specify each of the proteins active in mitosis and cytokinesis, and each protein makes a contribution to the coordinated events of the cell cycle. As a result, a mutation in any of a number of genes can disrupt the meticulously choreographed mechanisms of cell division.

at their centromeres. This signal makes the beginning of anaphase dependent on the prior proper alignment of all the chromosomes at metaphase. As a result of multiple cell-cycle checkpoints, each daughter cell reliably receives the right number of chromosomes.

Breakdown of the mitotic machinery can produce division mistakes that have crucial consequences for the cell. Improper chromosome segregation, for example, can cause serious malfunction or even the death of daughter cells. As the **Fast Forward** box **"How Gene Mutations Cause**

Errors in Mitosis" explains, gene mutations that disrupt mitotic structures such as the spindle, kinetochores, or centrosomes are one source of improper segregation. Other problems occur in cells where the normal restraints on cell division, such as checkpoints, have broken down. Such cells may divide uncontrollably, leading to tumour formation. We present the details of cell-cycle regulation, checkpoint controls, and cancer formation in Chapter 16.

3.3   Meiosis: Cell Divisions That Halve Chromosome Number

During the many rounds of cell division within an embryo, most cells either grow and divide via the mitotic cell cycle just described, or they stop growing and become arrested in G_0. These mitotically dividing and G_0-arrested cells are the so-called somatic cells whose descendants continue to make up the vast majority of each organism's tissues throughout the lifetime of the individual. Early in the embryonic development of animals, however, a group of cells is set aside for a different fate. These are the germ cells: cells destined for a specialized role in the production of gametes. Germ cells arise later in plants, during floral development instead of during embryogenesis. The germ cells become incorporated in the reproductive organs—ovaries and testes in animals; ovaries and anthers in flowering plants—where they ultimately undergo meiosis, the special two-part cell division that produces gametes (eggs and sperm or pollen) containing half the number of chromosomes as other body cells.

The union of haploid gametes at fertilization yields diploid offspring that carry the combined genetic heritage of two parents. Sexual reproduction therefore requires the alternation of haploid and diploid generations. If gametes were diploid rather than haploid, the number of chromosomes would double in each successive generation such that in humans, for example, the children would have 92 chromosomes per cell, the grandchildren 184, and so on. Meiosis prevents this lethal, exponential accumulation of chromosomes.

In meiosis, the chromosomes replicate once but the nucleus divides twice

Unlike mitosis, meiosis consists of two successive nuclear divisions, logically named division I of meiosis and division II of meiosis, or simply meiosis I and meiosis II. With each round, the cell passes through a prophase, metaphase, anaphase, and telophase, followed by cytokinesis. In meiosis I, the parent nucleus divides to form two daughter nuclei; in meiosis II, each of the two daughter nuclei divides, resulting in four nuclei (**Figure 3.11**). These four nuclei—the final products of meiosis—become partitioned in four separate daughter cells because cytokinesis occurs after both rounds of division. The chromosomes duplicate at the start of meiosis I, but they do not duplicate in meiosis II, which explains why the gametes contain half the number of chromosomes found in other body cells. A close look at each round of meiotic division reveals the mechanisms by which each gamete comes to receive one full haploid set of chromosomes.

During meiosis I, homologues pair, exchange parts, and then segregate

The events of meiosis I are unique among nuclear divisions (see **Figure 3.12**, meiosis I). The process begins with the replication of chromosomes, after which each one consists of two sister chromatids. A key to understanding meiosis I is the observation that the centromeres joining these chromatids remain intact throughout the entire division, rather than splitting as in mitosis.

As the division proceeds, homologous chromosomes align across the cellular equator to form a coupling that ensures proper chromosome segregation to separate nuclei. Moreover, during the time homologous chromosomes face each other across the equator, the maternal and paternal chromosomes of each homologous pair may exchange parts, creating new combinations of alleles at different genes along the chromosomes. Afterward, the two homologous chromosomes, each still consisting of two sister chromatids connected at a single, unsplit centromere, are pulled to opposite poles of the spindle. As a result, it is homologous chromosomes (rather than sister chromatids as in mitosis) that segregate into different daughter cells at the conclusion of the first meiotic division. With this overview in mind, let us take a closer look at the specific events of meiosis I, bearing in mind that we analyze a dynamic, flowing sequence of cellular events by breaking it down somewhat arbitrarily into the easily pictured, traditional phases.

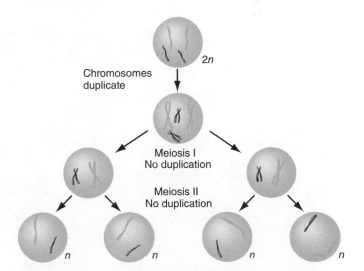

Figure 3.11 An overview of meiosis: The chromosomes replicate once, while the nuclei divide twice. In this figure, all four chromatids of each chromosome pair are shown in the same shade of the same colour. Note that the chromosomes duplicate before meiosis I, but they do not duplicate between meiosis I and meiosis II.

Meiosis: One Diploid Cell Produces Four Haploid Cells

Meiosis I: A reductional division

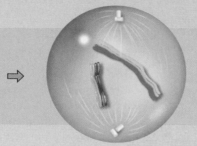

Prophase I: Leptotene

1. Chromosomes thicken and become visible, but the chromatids remain invisible.
2. Centrosomes begin to move toward opposite poles.

Prophase I: Zygotene

1. Homologous chromosomes enter *synapsis*.
2. The *synaptonemal complex* forms.

Prophase I: Pachytene

1. Synapsis is complete.
2. *Crossing-over*, genetic exchange between nonsister chromatids of a homologous pair, occurs.

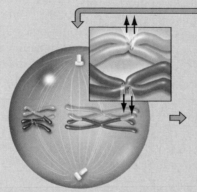

Metaphase I

1. Tetrads line up along the *metaphase plate.*
2. Each chromosome of a homologous pair attaches to fibres from opposite poles.
3. Sister chromatids attach to fibres from the same pole.

Anaphase I

1. The centromere does not divide.
2. The chiasmata migrate off chromatid ends.
3. Homologous chromosomes move to opposite poles.

Meiosis II: An equational division

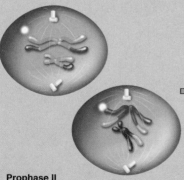

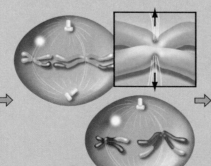

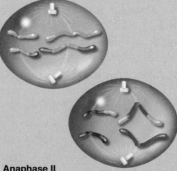

Prophase II

1. Chromosomes condense.
2. Centrioles move toward the poles.
3. The nuclear envelope breaks down at the end of prophase II (not shown).

Metaphase II

1. Chromosomes align at the metaphase plate.
2. Sister chromatids attach to spindle fibres from opposite poles.

Anaphase II

1. Centromeres divide, and sister chromatids move to opposite poles.

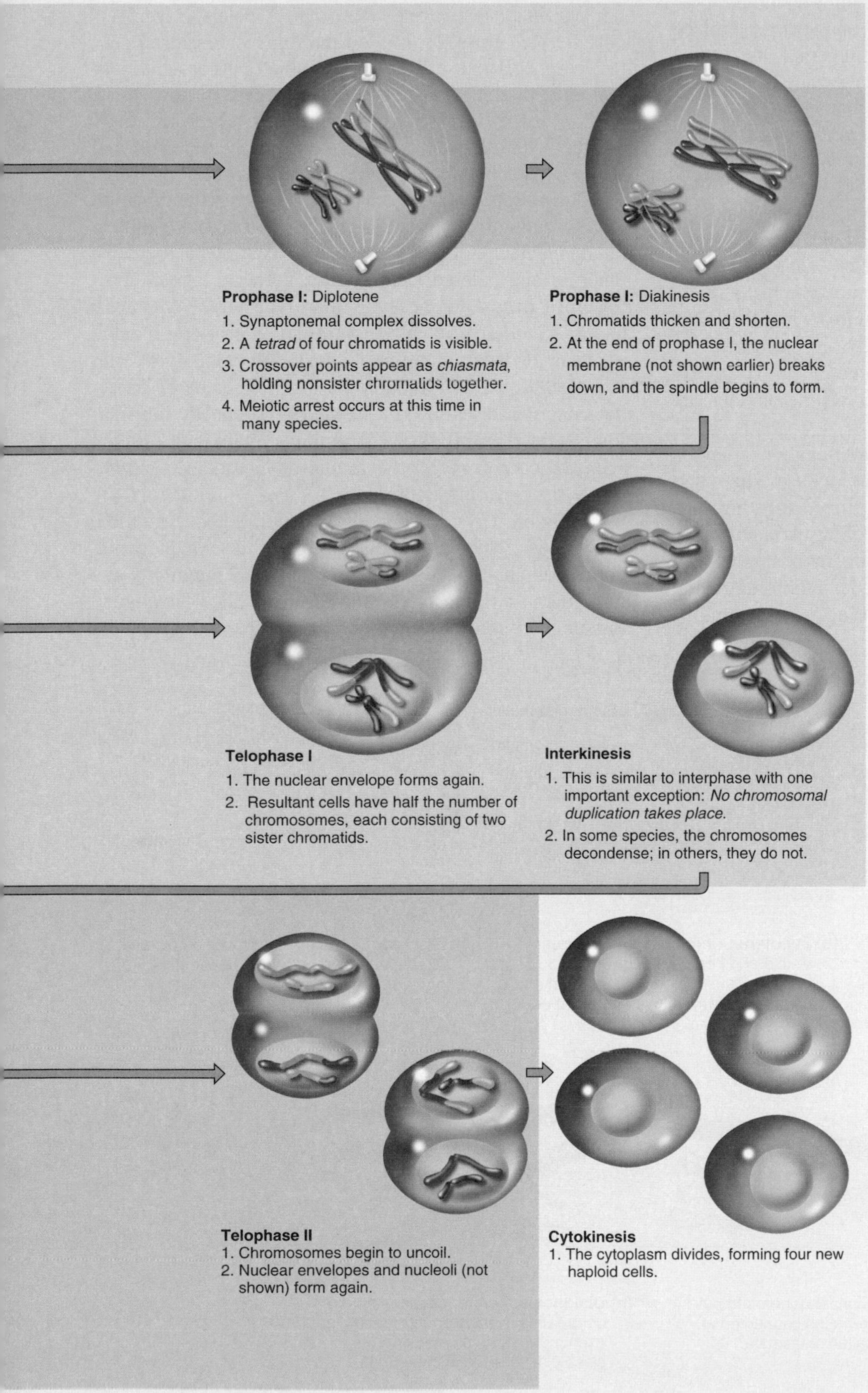

Figure 3.12 To aid visualization of the chromosomes, the figure is simplified in two ways: (1) The nuclear envelope is not shown during prophase of either meiotic division. (2) The chromosomes are shown as fully condensed at zygotene; in reality, full condensation is not achieved until diakinesis.

Prophase I: Diplotene
1. Synaptonemal complex dissolves.
2. A *tetrad* of four chromatids is visible.
3. Crossover points appear as *chiasmata*, holding nonsister chromatids together.
4. Meiotic arrest occurs at this time in many species.

Prophase I: Diakinesis
1. Chromatids thicken and shorten.
2. At the end of prophase I, the nuclear membrane (not shown earlier) breaks down, and the spindle begins to form.

Telophase I
1. The nuclear envelope forms again.
2. Resultant cells have half the number of chromosomes, each consisting of two sister chromatids.

Interkinesis
1. This is similar to interphase with one important exception: *No chromosomal duplication takes place.*
2. In some species, the chromosomes decondense; in others, they do not.

Telophase II
1. Chromosomes begin to uncoil.
2. Nuclear envelopes and nucleoli (not shown) form again.

Cytokinesis
1. The cytoplasm divides, forming four new haploid cells.

Prophase I: Homologues condense and pair, and crossing-over occurs

Among the critical events of prophase I is the condensation of chromatin, the pairing of homologous chromosomes, and the reciprocal exchange of genetic information between these paired homologues. Figure 3.12 shows a generalized view of prophase I; however, research suggests that the exact sequence of events may vary in different species. These complicated processes can take many days, months, or even years to complete. For example, in the female germ cells of several species, including humans, meiosis is suspended at prophase I until ovulation (as discussed further in Section 3.4).

Leptotene (from the Greek for "thin" and "delicate") is the first definable substage of prophase I, the time when the long, thin chromosomes begin to thicken (see **Figure 3.13a** for a more detailed view). Each chromosome has already duplicated prior to prophase I (as in mitosis) and thus consists of two sister chromatids attached at a centromere. At this point, however, these sister chromatids are so tightly bound together that they are not yet visible as separate entities.

Zygotene (from the Greek for "conjugation") begins as each chromosome seeks out its homologous partner, and the matching chromosomes become zipped together in a process known as synapsis. The "zipper" itself is an elaborate protein structure called the synaptonemal complex (see **Figure 3.13b**) that aligns the homologues with remarkable precision, juxtaposing the corresponding genetic regions of the chromosome pair.

Pachytene (from the Greek for "thick" or "fat") begins at the completion of synapsis when homologous chromosomes are united along their length. Each synapsed chromosome pair is known as a bivalent (because it encompasses two chromosomes) or a tetrad (because it contains four chromatids). On one side of the bivalent is a maternally derived chromosome, on the other side a paternally derived one. Because X and Y chromosomes are not identical, they do not synapse completely; there is, however, a small region of similarity (or "homology") between the X and the Y chromosomes known as the pseudoautosomal region (PAR) that allows for a limited amount of pairing.

During pachytene, structures called recombination nodules begin to appear along the synaptonemal complex, and an exchange of parts between nonsister (i.e., between maternal and paternal) chromatids of homologous chromosomes occurs at these nodules (see **Figure 3.13c** for details). Such an exchange is known as crossing-over; it results in the recombination of genetic material. As a result of crossing-over, chromatids may no longer be of purely maternal or paternal origin; however, no genetic information is gained or lost, so all chromatids retain their original size. The current molecular model of meiotic recombination is discussed in Chapter 6.

Diplotene (from the Greek for "twofold" or "double") is signalled by the gradual dissolution of the synaptonemal zipper complex and a slight separation of regions of the homologous chromosomes (see **Figure 3.13d**). The aligned homologous chromosomes of each bivalent nonetheless

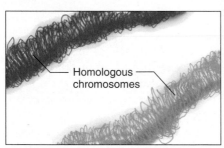

(a) Leptotene: Threadlike chromosomes begin to condense and thicken, becoming visible as discrete structures. Although the chromosomes have duplicated, the sister chromatids of each chromosome are not yet visible under the microscope.

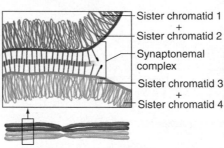

Sister chromatid 1 + Sister chromatid 2
Synaptonemal complex
Sister chromatid 3 + Sister chromatid 4

(b) Zygotene: Chromosomes are clearly visible and begin pairing with homologous chromosomes along the synaptonemal complex to form a bivalent, or tetrad.

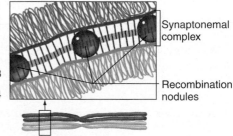

Synaptonemal complex
Recombination nodules

(c) Pachytene: Full synapsis of homologues. Recombination nodules appear along the synaptonemal complex.

(d) Diplotene: The bivalent appears to pull apart slightly but remains connected at crossover sites, called chiasmata.

(e) Diakinesis: Further condensation of chromatids. Nonsister chromatids that have exchanged parts by crossing-over remain closely associated at chiasmata.

Figure 3.13 Prophase I of meiosis at very high magnification.

remain very tightly joined at intervals along their length called chiasmata (singular, *chiasma*), which represent the sites where crossing-over occurred.

Diakinesis (from the Greek for "double movement") is accompanied by further condensation of the chromatids. Because of this chromatid thickening and shortening, it can now clearly be seen that each tetrad consists of four separate chromatids; or viewed in another way, that the two homologous chromosomes of a bivalent are each composed of two sister chromatids held together at a centromere (see **Figure 3.13e**). Nonsister chromatids that have undergone crossing-over remain closely associated at chiasmata. The end of diakinesis is analogous to the prometaphase of mitosis: The nuclear envelope breaks down, and the microtubules of the spindle apparatus begin to form.

> During prophase I, homologous chromosomes pair, and recombination occurs between nonsister chromatids of the paired homologues.

Metaphase I: Paired homologues attach to spindle fibres from opposite poles

During mitosis, each sister chromatid has a kinetochore that becomes attached to microtubules emanating from opposite spindle poles. During meiosis I, the situation is different. The kinetochores of sister chromatids fuse, so that each chromosome contains only a single functional kinetochore. The result of this fusion is that sister chromatids remain together throughout meiosis I because no oppositely directed forces exist that can pull the chromatids apart. Instead, during metaphase I (Figure 3.12, meiosis I), it is the kinetochores of homologous chromosomes that attach to microtubules from opposite spindle poles. As a result, in chromosomes aligned at the metaphase plate, the kinetochores of maternally and paternally derived chromosomes face opposite spindle poles, positioning the homologues to move in opposite directions. Because each bivalent's alignment and hookup is independent of that of every other bivalent, the chromosomes facing each pole are a random mix of maternal and paternal origin.

> The essence of the first meiotic division is the arrangement of chromosomes at metaphase I. The kinetochores of homologous chromosomes are connected to fibres from opposite spindle poles. The homologues are held together by chiasmata.

Anaphase I: Homologues move to opposite spindle poles

At the onset of anaphase I, the chiasmata joining homologous chromosomes dissolve, which allows the maternal and paternal homologues to begin to move toward opposite spindle poles (see Figure 3.12, meiosis I). Note that in the first meiotic division, the centromeres do not divide as they do in mitosis. Thus, from each homologous pair, one chromosome consisting of two sister chromatids joined at their centromere segregates to each spindle pole.

Recombination through crossing-over plays an important role in the proper segregation of homologous chromosomes during the first meiotic division. The chiasmata, in holding homologues together, ensure that their kinetochores remain attached to opposite spindle poles throughout metaphase. When recombination does not occur within a bivalent, mistakes in hookup and transport may cause homologous chromosomes to move to the same pole, instead of segregating to opposite poles. In some organisms, however, proper segregation of nonrecombinant chromosomes nonetheless occurs through other pairing processes. Investigators do not yet completely understand the nature of these processes and are currently evaluating several models to explain them.

Telophase I: Nuclear envelopes form again

The telophase of the first meiotic division, or telophase I, takes place when nuclear membranes begin to form around the chromosomes that have moved to the poles. Each of the developing daughter nuclei contains one-half the number of chromosomes in the original parent nucleus, but each chromosome consists of two sister chromatids joined at the centromere (see Figure 3.12, meiosis I). Because the number of chromosomes is reduced to one-half the normal diploid number, meiosis I is often called a reductional division.

In most species, cytokinesis follows telophase I, with daughter nuclei becoming enclosed in separate daughter cells. A short interphase then ensues. During this time, the chromosomes usually decondense, in which case they must recondense during the prophase of the subsequent second meiotic division. In some cases, however, the chromosomes simply stay condensed. Most importantly, there is no S phase during the interphase between meiosis I and meiosis II; that is, the chromosomes do not replicate during meiotic interphase. The relatively brief interphase between meiosis I and meiosis II is known as interkinesis.

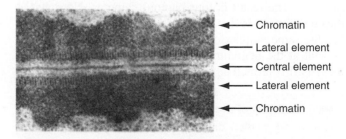

← Chromatin
← Lateral element
← Central element
← Lateral element
← Chromatin

Figure 3.14 The synaptonemal complex brings chromosomes into juxtaposition. Electron micrograph of a synaptonemal complex at pachytene joining nonsister chromatids of homologous chromosomes in *Neottiella rutilans*, a fungal species. The synaptonemal complex consists of three parts: two lateral components and one central element.

During meiosis II, sister chromatids separate to produce haploid gametes

The second meiotic division (meiosis II) proceeds in a fashion very similar to that of mitosis, but because the number of chromosomes in each dividing nucleus has already been reduced by half, the resulting daughter cells are haploid. The same process occurs in each of the two daughter cells generated by meiosis I, producing four haploid cells at the end of this second meiotic round (see Figure 3.12, meiosis II).

Prophase II: The chromosomes condense

If the chromosomes decondensed during the preceding interphase, they recondense during prophase II. At the end of prophase II, the nuclear envelope breaks down, and the spindle apparatus forms again.

Metaphase II: Chromosomes align at the metaphase plate

The kinetochores of sister chromatids attach to microtubule fibres emanating from opposite poles of the spindle apparatus, just as in mitotic metaphase. There are nonetheless two significant features of metaphase II that distinguish it from mitosis. First, the number of chromosomes is one-half that in mitotic metaphase of the same species. Second, in most chromosomes, the two sister chromatids are no longer strictly identical because of the recombination through crossing-over that occurred during meiosis I. The sister chromatids still contain the same genes, but they may carry different combinations of alleles.

Anaphase II: Sister chromatids move to opposite spindle poles

Just as in mitosis, severing of the centromeric connection between sister chromatids allows them to move toward opposite spindle poles during anaphase II.

Telophase II: Nuclear membranes form again, and cytokinesis follows

Membranes form around each of four daughter nuclei in telophase II, and cytokinesis places each nucleus in a separate cell. The result is four haploid gametes. Note that at the end of meiosis II, each daughter cell (i.e., each gamete) has the same number of chromosomes as the parental cell present at the beginning of this division. For this reason, meiosis II is termed an equational division.

> Meiosis consists of two rounds of cell division. The first is a reductional division during which homologues segregate, producing haploid daughter cells. The second is an equational division during which sister chromatids are separated.

Mistakes in meiosis produce defective gametes

Segregational errors during either meiotic division can lead to aberrations, such as trisomies, in the next generation. If, for example, the homologues of a chromosome pair do not segregate during meiosis I (a mistake known as nondisjunction), they may travel together to the same pole and eventually become part of the same gamete. Such an error may at fertilization result in any one of a large variety of possible trisomies. Most autosomal trisomies, as we already mentioned, are lethal *in utero*; one exception is trisomy 21, the genetic basis of the majority of cases of Down syndrome. Like trisomy 21, extra sex chromosomes may also be nonlethal but cause a variety of mental and physical abnormalities, such as those seen in Klinefelter syndrome (see Table 3.1).

In contrast to rare mistakes in the segregation of one pair of chromosomes, some hybrid animals carry nonhomologous chromosomes that can never pair up and segregate properly. **Figure 3.15** shows the two dissimilar sets of chromosomes carried by the diploid cells of a mule. The set inherited from the donkey father contains 31 chromosomes, while the set from the horse mother has 32 chromosomes. Viable gametes cannot form in these animals, so mules are sterile.

Meiosis contributes to genetic diversity

The wider the assortment of different gene combinations among members of a species, the greater the chance that at least some individuals will carry combinations of alleles that allow survival in a changing environment. Two aspects of meiosis contribute to genetic diversity in a population. First, because only chance governs which paternal or maternal homologues migrate to the two poles during the first meiotic division, different gametes carry a different mix of maternal and paternal chromosomes. **Figure 3.16a** shows how two different patterns of homologue migration produce four different mixes of parental chromosomes in the gametes. The amount of potential variation generated by this random independent assortment increases with the number of chromosomes. In *Ascaris*, for example, where $n = 2$ (the chromosome complement shown in Figure 3.16a), the random assortment of homologues could produce only 2^2, or 4 types of gametes. In a human being, however, where $n = 23$, this same mechanism alone could generate 2^{23}, or more than 8 million genetically different kinds of gametes.

A second feature of meiosis, the reshuffling of genetic information through crossing-over during prophase I, ensures an even greater amount of genetic diversity in gametes. Because crossing-over recombines maternally and paternally derived genes, each chromosome in each different gamete could consist of different combinations of maternal and paternal information (**Figure 3.16b**).

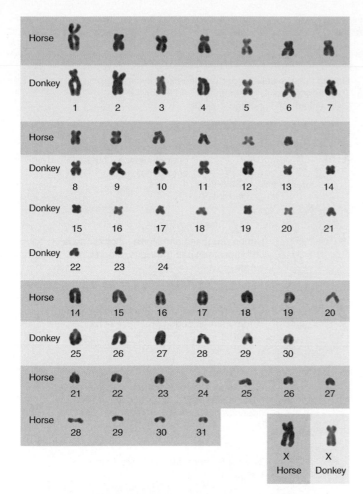

Figure 3.15 Hybrid sterility: When chromosomes cannot pair during meiosis I, they segregate improperly. The mating of a male donkey (*Equus asinus; green*) and a female horse (*Equus caballus; peach colour*) produces a mule with 63 chromosomes. In this karyotype of a female mule, the first 13 donkey and horse chromosomes are homologous and pictured in pairs. Starting at chromosome 14, the donkey and horse chromosomes are too dissimilar to pair with each other during meiosis I.

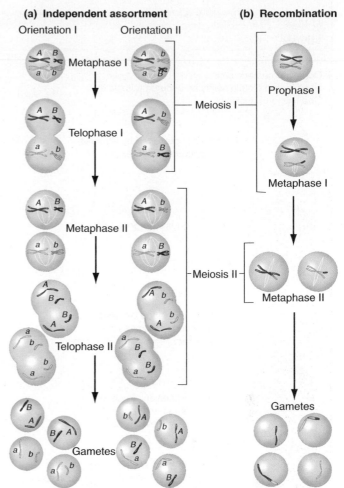

Figure 3.16 How meiosis contributes to genetic diversity. **(a)** The variation resulting from the independent assortment of nonhomologous chromosomes increases with the number of chromosomes in the genome. **(b)** Crossing-over between homologous chromosomes ensures that each gamete is unique.

Of course, sexual reproduction adds yet another means of producing genetic diversity. At fertilization, any one of a vast number of genetically diverse sperm can fertilize an egg with its own distinctive genetic constitution. It is thus not very surprising that, with the exception of identical twins (not accounting for *de novo* or new mutations arising somatically in identical twins naturally), the 7 billion people in the world are all genetically unique.

> Genetic diversity is ensured by the independent assortment of nonhomologous chromosomes and the recombination of homologous chromosomes during meiosis, as well as by the random union of genetically distinct sperm and eggs.

Mitosis and meiosis: A comparison

Mitosis occurs in all types of eukaryotic cells (i.e., cells with a membrane-bounded nucleus) and is a conservative mechanism that preserves the genetic status quo. Mitosis followed by cytokinesis produces growth by increasing the number of cells. It also promotes the continual replacement of roots, stems, and leaves in plants and the regeneration of blood cells, intestinal tissues, and skin in animals.

Meiosis, on the other hand, occurs only in sexually reproducing organisms, in just a few specialized germ cells within the reproductive organs that produce haploid gametes. It is not a conservative mechanism; rather, the extensive combinatorial changes arising from meiosis are one source of the genetic variation that fuels evolution. **Table 3.3** illustrates the significant contrasts between the two mechanisms of cell division.

Table 3.3	Comparing Mitosis and Meiosis

Mitosis	**Meiosis**

Mitosis

Occurs in somatic cells
Haploid and diploid cells can undergo mitosis
One round of division

Mitosis is preceded by S phase (chromosome duplication).

Homologous chromosomes do not pair.

Genetic exchange between homologous chromosomes is very rare.

Sister chromatids attach to spindle fibres from opposite poles during metaphase.

The centromere splits at the beginning of anaphase.

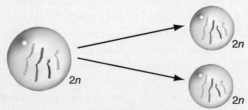

Mitosis produces two new daughter cells, identical to each other and the original cell. Mitosis is thus genetically conservative.

Meiosis

Occurs in germ cells as part of the sexual cycle
Two rounds of division: meiosis I and meiosis II
Only diploid cells undergo meiosis

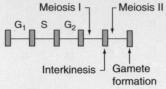

Chromosomes duplicate prior to meiosis I but not before meiosis II.

During prophase of meiosis I, homologous chromosomes pair (synapse) along their length.

Crossing-over occurs between homologous chromosomes during prophase of meiosis I.

Homologous chromosomes (not sister chromatids) attach to spindle fibres from opposite poles during metaphase I.

The centromere does not split during meiosis I.

Sister chromatids attach to spindle fibres from opposite poles during metaphase II.

The centromere splits at the beginning of anaphase II.

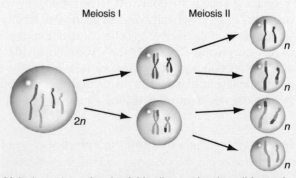

Meiosis produces four haploid cells, one (egg) or all (sperm) of which can become gametes. None of these is identical to each other or to the original cell, because meiosis results in combinatorial change.

3.4 Gametogenesis

In all sexually reproducing animals, the embryonic germ cells (collectively known as the germ line) undergo a series of mitotic divisions that yield a collection of specialized diploid cells, which subsequently divide by meiosis to produce haploid cells. As with other biological processes, many variations on this general pattern have been observed. In some species, the haploid cells resulting from meiosis are the gametes themselves, while in other species, those cells must undergo a specific plan of differentiation to fulfill that function. Moreover, in certain organisms, the four haploid products of a single meiosis do not all become gametes. Gamete formation, or gametogenesis, thus gives rise to haploid gametes marked not only by the events of meiosis per se but also by cellular events that precede and follow meiosis. Here we illustrate gametogenesis with a description of egg and sperm formation in humans. The details of gamete formation in several other organisms appear throughout the book in discussions of specific experimental studies.

Oogenesis in humans produces one ovum from each primary oocyte

The end product of egg formation in humans is a large, nutrient-rich ovum whose stored resources can sustain the early embryo. The process, known as oogenesis (**Figure 3.17**), begins when diploid germ cells in the ovary, called oogonia (singular, *oogonium*), multiply rapidly by mitosis and produce a large number of primary oocytes, which then undergo meiosis.

For each primary oocyte, meiosis I results in the formation of two daughter cells that differ in size, so this division is asymmetric. The larger of these cells, the secondary oocyte, receives over 95 percent of the cytoplasm. The other small sister cell is known as the first polar body.

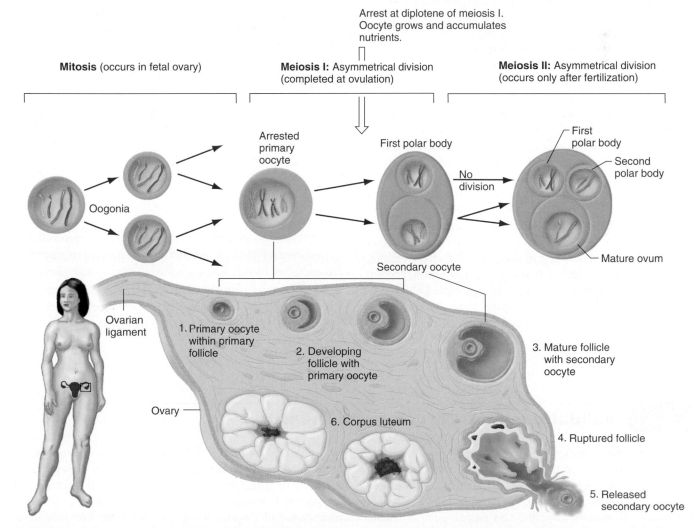

Figure 3.17 In humans, egg formation begins in the fetal ovaries and arrests during prophase of meiosis I. Fetal ovaries each contain about 500 000 primary oocytes arrested in the diplotene substage of meiosis I. If the egg released during a menstrual cycle is fertilized, meiosis is completed. Only one of the three (rarely, four) cells produced by meiosis serves as the functional gamete, or ovum.

During meiosis II, the secondary oocyte undergoes another asymmetrical division to produce a large haploid ovum and a small haploid second polar body. The first polar body usually arrests its development and does not undergo the second meiotic division. However, in a small proportion of cases, the first polar body does divide, producing two haploid polar bodies. The two (or rarely, three) small polar bodies apparently serve no function and disintegrate, leaving one large haploid ovum as the functional gamete. Thus, only one of the three (or rarely, four) products of a single meiosis serves as a female gamete. A normal human ovum carries 22 autosomes and an X sex chromosome.

Oogenesis begins in the fetus. By six months after conception, the fetal ovaries are fully formed and contain about half a million primary oocytes, each arrested in the diplotene substage of prophase I, their homologous chromosomes locked in synapsis. From the onset of puberty, at about age 12, until menopause, some 35–40 years later, most women release one primary oocyte each month (from alternate ovaries), amounting to roughly 480 oocytes released during the reproductive years. The remaining primary oocytes disintegrate during menopause. Although it has been traditionally believed that a girl is born with all the primary oocytes she will ever possess, recent studies have shown that adult human ovaries contain a population of rare oocyte-producing germline stem cells.

At ovulation, a released oocyte completes meiosis I and proceeds as far as the metaphase of meiosis II. If the oocyte is then fertilized (i.e., penetrated by a sperm nucleus), it quickly completes meiosis II. The nuclei of the sperm and ovum then fuse to form the diploid nucleus of the zygote, and the zygote divides by mitosis to produce a functional embryo. In contrast, unfertilized oocytes exit the body during the menses stage of the menstrual cycle.

The long interval before completion of meiosis in oocytes released by women in their 30s, 40s, and 50s may contribute to the observed correlation between maternal age and meiotic segregational errors, including those that produce trisomies. Women in their mid-20s, for example, run a very small risk of trisomy 21; only 0.05 percent of children born to women of this age have Down syndrome. During the later childbearing years, however, the risk rapidly rises; at age 35, it is 0.9 percent of live births, and at age 45, it is 3 percent. You would not expect this age-related increase in risk if meiosis were completed before the mother's birth.

Spermatogenesis in humans produces four sperm from each primary spermatocyte

The production of sperm, or spermatogenesis (**Figure 3.18**), begins in the male testes in germ cells known as spermatogonia. Mitotic divisions of the spermatogonia produce many diploid cells, the primary spermatocytes. Unlike primary oocytes, primary spermatocytes undergo a symmetrical meiosis I, producing two secondary spermatocytes, each of which undergoes a symmetrical meiosis II. At the conclusion of meiosis, each original primary spermatocyte thus yields four equivalent haploid spermatids. These spermatids then mature by developing a characteristic whiplike tail and by concentrating all their chromosomal material in a head, thereby becoming functional sperm. A human sperm, much smaller than the ovum it will fertilize, contains 22 autosomes and *either* an X *or* a Y sex chromosome.

The timing of sperm production differs radically from that of egg formation. The meiotic divisions allowing conversion of primary spermatocytes to spermatids begin only at puberty, but meiosis then continues throughout a man's life. The entire process of spermatogenesis takes about 48–60 days: 16–20 days for meiosis I, 16–20 days for meiosis II, and 16–20 days for the maturation of spermatids into fully functional sperm. Within each testis after puberty, millions of sperm are always in production, and a single ejaculate can contain up to 300 million. Over a lifetime, a man can produce billions of sperm, almost equally divided between those bearing an X and those bearing a Y chromosome.

> Gametogenesis involves mitotic divisions of specialized germline cells that then undergo meiotic divisions to produce gametes. In human females, oocytes undergo asymmetrical meiosis to produce a large ovum and two or three nonfunctional polar bodies. In human males, spermatocytes undergo symmetrical meiosis to produce four sperm.

3.5　Validation of the Chromosome Theory

So far, we have presented two circumstantial lines of evidence in support of the chromosome theory of inheritance. First, the phenotype of sexual identity is associated with the inheritance of particular chromosomes. Second, the events of mitosis, meiosis, and gametogenesis ensure a constant number of chromosomes in the somatic cells of all members of a species over time; one would expect the genetic material to exhibit this kind of stability even in organisms with very different modes of reproduction. Final acceptance of the chromosome theory depended on researchers going beyond the circumstantial evidence to a rigorous demonstration of two key points: (1) that the inheritance of genes corresponds with the inheritance of chromosomes in every detail, and (2) that the transmission of particular chromosomes coincides with the transmission of specific traits other than sex determination.

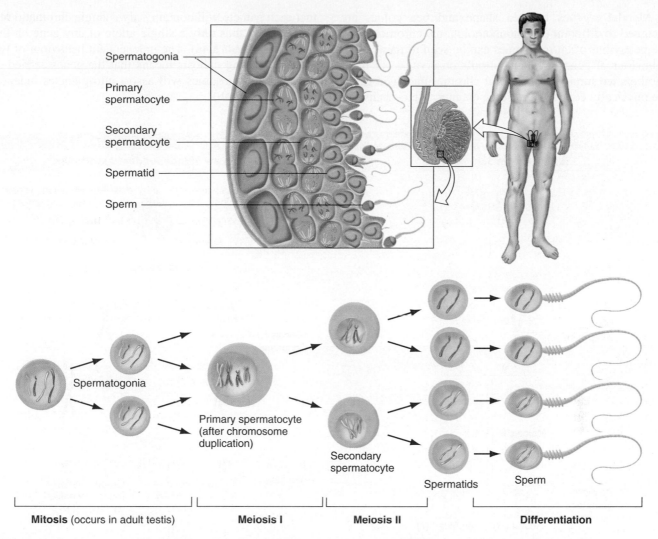

Figure 3.18 Human sperm form continuously in the testes after puberty. Spermatogonia are located near the exterior of seminiferous tubules in a human testis. Once they divide to produce the primary spermatocytes, the subsequent stages of spermatogenesis—meiotic divisions in the spermatocytes and maturation of spermatids into sperm—occur successively closer to the middle of the tubule. Mature sperm are released into the central lumen of the tubule for ejaculation.

Mendel's laws correlate with chromosome behaviour during meiosis

Walter Sutton first outlined the chromosome theory of inheritance in 1902–1903, building on the theoretical ideas and experimental results of Theodor Boveri in Germany, E. B. Wilson in New York, and others. In a 1902 paper, Sutton speculated that "the association of paternal and maternal chromosomes in pairs and their subsequent separation during the reducing division [i.e., meiosis I] . . . may constitute the physical basis of the Mendelian law of heredity." In 1903, he suggested that chromosomes carry Mendel's hereditary units for the following reasons:

1. Every cell contains two copies of each kind of chromosome, and there are two copies of each kind of gene.
2. The chromosome complement, like Mendel's genes, appears unchanged as it is transmitted from parents to offspring through generations.
3. During meiosis, homologous chromosomes pair and then separate to different gametes, just as the alternative alleles of each gene segregate to different gametes.
4. Maternal and paternal copies of each chromosome pair move to opposite spindle poles without regard to the assortment of any other homologous chromosome pair, just as the alternative alleles of unrelated genes assort independently.
5. At fertilization, an egg's set of chromosomes unites with a randomly encountered sperm's set of chromosomes, just as alleles obtained from one parent unite at random with those from the other parent.
6. In all cells derived from the fertilized egg, one-half of the chromosomes and one-half of the genes are of maternal origin, and the other half of paternal origin.

The two parts of **Table 3.4** show the intimate relationship between the chromosome theory of inheritance and Mendel's laws of segregation and independent assortment.

If Mendel's genes for pea shape and pea colour are assigned to different (i.e., nonhomologous) chromosomes, the behaviour of chromosomes can be seen to parallel the behaviour of genes. Walter Sutton's observation of these parallels led him to propose that chromosomes and genes are physically connected in some manner. Meiosis ensures that each gamete will contain only a single chromatid of a bivalent and thus only a single allele of any gene on that chromatid (**Table 3.4a**). The independent behaviour of two bivalents during meiosis means that the genes carried on different chromosomes will assort into gametes independently (**Table 3.4b**).

Table 3.4	How the Chromosome Theory of Inheritance Explains Mendel's Laws
(a) The Law of Segregation	**(b) The Law of Independent Assortment**

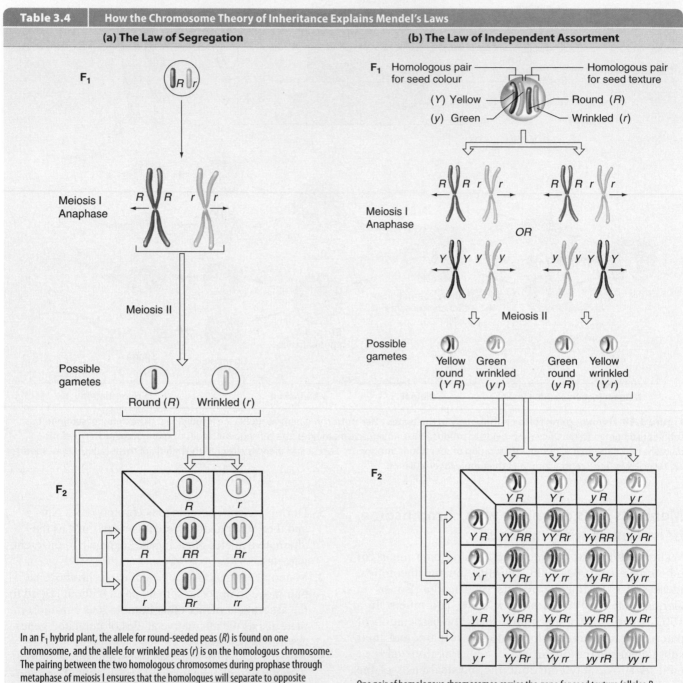

In an F_1 hybrid plant, the allele for round-seeded peas (R) is found on one chromosome, and the allele for wrinkled peas (r) is on the homologous chromosome. The pairing between the two homologous chromosomes during prophase through metaphase of meiosis I ensures that the homologues will separate to opposite spindle poles during anaphase I. At the end of meiosis II, two types of gametes have been produced: half have R, and half have r, but no gametes have both alleles. Thus, the separation of homologous chromosomes at meiosis I corresponds to the segregation of alleles. As the Punnett square shows, fertilization of 50 percent R and 50 percent r eggs with the same proportion of R and r pollen leads to Mendel's 3:1 ratio in the F_2 generation.

One pair of homologous chromosomes carries the gene for seed texture (alleles R and r). A second pair of homologous chromosomes carries the gene for seed colour (alleles Y and y). Each homologous pair aligns at random at the metaphase plate during meiosis I, independently of the other homologous pair. Thus, two equally likely configurations are possible for the migration of any two chromosome pairs toward the poles during anaphase I. As a result, a dihybrid individual will generate four equally likely types of gametes with regard to the two traits in question. As the Punnett square affirms, this independent assortment of traits carried by nonhomologous chromosomes produces Mendel's 9:3:3:1 ratio.

From a review of Figure 3.16, which follows two different chromosome pairs through the process of meiosis, you might wonder whether crossing-over abolishes the clear correspondence between Mendel's laws and the movement of chromosomes. The answer is no. Each chromatid of a homologous chromosome pair contains only one copy of a given gene, and only one chromatid from each pair of homologues is incorporated into each gamete. Because alternative alleles remain on different chromatids even after crossing-over has occurred, alternative alleles still segregate to different gametes as demanded by Mendel's first law. And because the orientation of non-homologous chromosomes is completely random with respect to each other during both meiotic divisions, the genes on different chromosomes assort independently even if crossing-over occurs, as demanded by Mendel's second law.

Specific traits are transmitted with specific chromosomes

The fate of a theory depends on whether its predictions can be validated. Because genes determine traits, the prediction that chromosomes carry genes could be tested by breeding experiments that would show whether transmission of a specific chromosome coincides with transmission of a specific trait. Cytologists knew that one pair of

chromosomes, the sex chromosomes, determines whether an individual is male or female. Would similar correlations exist for other traits?

A gene determining eye colour on the *Drosophila* X chromosome

Thomas Hunt Morgan, an American experimental biologist with training in embryology, headed the research group whose findings eventually established a firm experimental base for the chromosome theory. Morgan chose to work with the fruit fly *Drosophila melanogaster* because it is extremely prolific and has a very short generation time, taking only approximately 11 days to develop from a fertilized egg into a mature adult capable of producing hundreds of offspring. Morgan fed his flies mashed bananas and housed them in empty milk bottles capped with wads of cotton.

In 1910, a white-eyed male appeared among a large group of flies with brick-red eyes. A mutation had apparently altered a gene determining eye colour, changing it from the normal wild-type allele specifying red to a new allele that produced white. When Morgan allowed the white-eyed male to mate with its red-eyed sisters, all the flies of the F₁ generation had red eyes; the red allele was clearly dominant to the white (**Figure 3.19**, cross A).

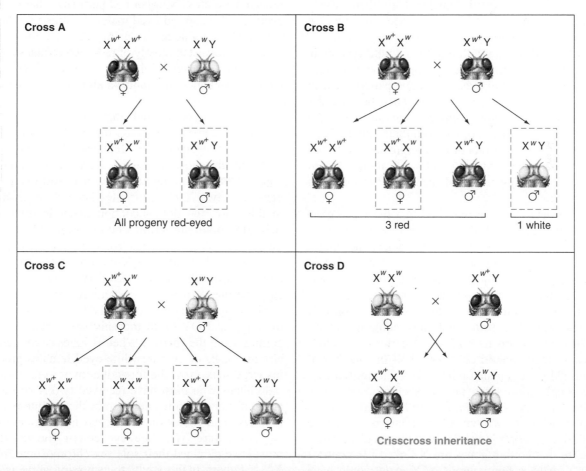

Figure 3.19 A *Drosophila* eye colour gene is located on the X chromosome. X-linkage explains the inheritance of alleles of the *white* gene in this series of crosses performed by Thomas Hunt Morgan. The progeny of crosses A, B, and C outlined with *green dotted boxes* are those used as the parents in the next cross of the series.

Establishing a pattern of nomenclature for *Drosophila* geneticists, Morgan named the gene identified by the abnormal white eye colour, the *white* gene, for the mutation that revealed its existence. The normal wild-type allele of the *white* gene, abbreviated w^+, is for brick-red eyes, while the counterpart mutant *w* allele results in white eye colour. The superscript $+$ signifies the wild type. By writing the gene name and abbreviation in lowercase, Morgan symbolized that the mutant *w* allele is recessive to the wild-type w^+. (If a mutation results in a dominant non-wild-type phenotype, the first letter of the gene name or of its abbreviation is capitalized; thus the mutation known as *Bar* eyes is dominant to the wild-type Bar^+ allele. See the *Guidelines for Gene Nomenclature*, directly following Chapter 24.)

Morgan then crossed the red-eyed males of the F_1 generation with their red-eyed sisters (Figure 3.19, cross B) and obtained an F_2 generation with the predicted 3:1 ratio of red to white eyes. But there was something askew in the pattern: Among the red-eyed offspring, there were two females for every one male, and all the white-eyed offspring were males. This result was surprisingly different from the equal transmission to both sexes of the Mendelian traits discussed in Chapter 2. In these fruit flies, the ratio of various phenotypes was not the same in male and female progeny.

By mating F_2 red-eyed females with their white-eyed brothers (Figure 3.19, cross C), Morgan obtained some females with white eyes, which then allowed him to mate a white-eyed female with a red-eyed wild-type male (Figure 3.19, cross D). The result was exclusively red-eyed daughters and white-eyed sons. The pattern seen in cross D is known as crisscross inheritance because the males inherit their eye colour from their mothers, while the daughters inherit their eye colour from their fathers. Note in Figure 3.19 that the results of the reciprocal crosses red female $\times$ white male (cross A) and white female $\times$ red male (cross D) are not identical, again in contrast with Mendel's findings.

From the data, Morgan reasoned that the *white* gene for eye colour is X-linked; that is, carried by the X chromosome. (Note that while symbols for genes and alleles are italicized, symbols for chromosomes are not.) The Y chromosome carries no allele of this gene for eye colour. Males, therefore, have only one copy of the gene, which they inherit from their mother along with their only X chromosome; their Y chromosome must come from their father. Thus, males are hemizygous for this eye colour gene, because their diploid cells have half the number of alleles carried by the female on her two X chromosomes.

If the single *white* gene on the X chromosome of a male is the wild-type w^+ allele, he will have red eyes and a genotype that can be written $X^{w^+}Y$. (Here we designate the chromosome [X or Y] together with the allele it carries, to emphasize that certain genes are X-linked.) In contrast to an $X^{w^+}Y$ male, a hemizygous X^wY male would have a phenotype of white eyes. Females with two X chromosomes can be one of three genotypes: X^wX^w (white-eyed), $X^wX^{w^+}$ (red-eyed because w^+ is dominant to *w*), or $X^{w^+}X^{w^+}$ (red-eyed).

As shown in Figure 3.19, Morgan's assumption that the gene for eye colour is X-linked explains the results of his breeding experiments. Crisscross inheritance, for example, occurs because the only X chromosome in sons of a white-eyed mother (X^wX^w) must carry the *w* allele, so the sons will be white-eyed. In contrast, because daughters of a red-eyed ($X^{w^+}Y$) father must receive a w^+-bearing X chromosome from their father, they will have red eyes.

> Through a series of crosses, T. H. Morgan demonstrated that the inheritance of a gene controlling eye colour in *Drosophila* was best explained by the hypothesis that this gene lies on the X chromosome.

Support for the chromosome theory from the analysis of nondisjunction

Although Morgan's work strongly supported the hypothesis that the gene for eye colour lies on the X chromosome, he himself continued to question the validity of the chromosome theory until Calvin Bridges, one of his top students, found another key piece of evidence. Bridges repeated the cross Morgan had performed between white-eyed females and red-eyed males, but this time he did the experiment on a larger scale. As expected, the progeny of this cross consisted mostly of red-eyed females and white-eyed males. However, about 1 in every 2000 males had red eyes, and about the same small fraction of females had white eyes.

Bridges hypothesized that these exceptions arose through rare events in which the X chromosomes fail to separate during meiosis in females. He called such failures in chromosome segregation *nondisjunction*. As **Figure 3.20a** shows, nondisjunction would result in some eggs with two X chromosomes and others with none—in this case, the nondisjunction event is referred to as **primary nondisjunction**. Fertilization of these chromosomally abnormal eggs could produce four types of zygotes: XXY (with two X chromosomes from the egg and a Y from the sperm), XXX (with two X's from the egg and one X from the sperm), XO (with the lone sex chromosome from the sperm and no sex chromosome from the egg), and OY (with the only sex chromosome again coming from the sperm). When Bridges examined the sex chromosomes of the rare white-eyed females produced in his large-scale cross, he found that they were indeed XXY individuals who must have received two X chromosomes and with them two *w* alleles from their white-eyed X^wX^w mothers. The exceptional red-eyed males emerging from the cross were XO; their eye colour showed that they must have obtained their sole sex chromosome from their $X^{w^+}Y$ fathers. In this study, transmission of the *white* gene alleles followed the predicted behaviour of X chromosomes during rare meiotic mistakes, indicating that the X

chromosome carries the gene for eye colour. These results also suggested that zygotes with the two other abnormal sex chromosome karyotypes expected from nondisjunction in females (XXX and OY) die during embryonic development and thus produce no progeny.

Because XXY white-eyed females have three sex chromosomes rather than the normal two, Bridges reasoned they would produce four kinds of eggs: XY and X if normal disjunction occurs, or XX and Y if nondisjunction ensues again (termed **secondary nondisjunction** because it transpires in an aneuploid cell that was itself the result of a primary nondisjunction event) (**Figure 3.20b**). You can visualize the formation of these four kinds of eggs by imagining that when the three chromosomes pair and disjoin during meiosis, two chromosomes must go to one pole and one chromosome to the other. With this kind of segregation, only two results are possible: Either one X and the Y go to one pole and the second X to the other (yielding XY and X gametes), or the two X's go to one pole and the Y to the other (yielding XX and Y gametes). The first of these two scenarios occurs much more often because it comes about when the two similar X chromosomes pair with each other, ensuring that they will go to opposite poles during the first meiotic division. The second, less likely possibility happens only if the two X chromosomes fail to pair with each other.

Bridges next predicted that fertilization of these four kinds of eggs by normal sperm would generate an array of sex chromosome karyotypes associated with specific eye colour phenotypes in the progeny. Bridges verified all his predictions when he analyzed the eye colour and sex chromosomes of a large number of offspring. For instance, he showed cytologically that all of the white-eyed females emerging from the cross in Figure 3.20b had two X chromosomes and one Y chromosome, while one-half of the white-eyed males had a single X chromosome and two Y chromosomes. The very small number of males with red eyes had inherited the X chromosome from their father and the Y chromosome from their mother. Bridges's painstaking observations provided compelling evidence that specific genes do in fact reside on specific chromosomes.

X- and Y-linked traits in humans

A person unable to tell red from green would find it nearly impossible to distinguish the rose, scarlet, and magenta in the flowers of a garden bouquet from the delicately variegated greens in their foliage, or to complete a complex electrical circuit by fastening red-clad metallic wires to red ones and green to green. Such a person has most likely inherited some form of red-green colour blindness, a recessive condition that runs in families and affects mostly males. Among Caucasians in North America and Europe,

(a) Nondisjunction in an XX female

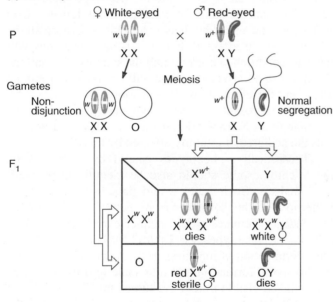

(b) Segregation in an XXY female

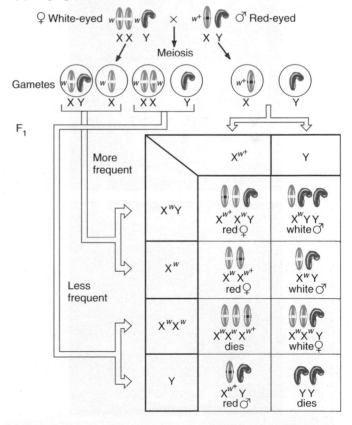

Figure 3.20 Nondisjunction: Rare mistakes in meiosis help confirm the chromosome theory. (a) Rare events of nondisjunction in an XX female produce XX and O eggs. The results of normal disjunction in the female are not shown. XO males are sterile because the missing Y chromosome is needed for male fertility in *Drosophila*. **(b)** In an XXY female, the three sex chromosomes can pair and segregate in two ways, producing progeny with unusual sex chromosome complements.

8 percent of men but only 0.44 percent of women have this vision defect. **Figure 3.21** suggests to readers with normal colour vision what individuals with red-green colour blindness actually see.

In 1911, E. B. Wilson, a contributor to the chromosome theory of inheritance, combined familiarity with studies of colour blindness and recent knowledge of sex determination by the X and Y chromosomes to make the first assignment of a human gene to a particular chromosome. The gene for red-green colour blindness, he asserted, lies on the X because the condition usually passes from a maternal grandfather through an unaffected carrier mother to roughly 50 percent of the grandsons.

Several years after Wilson made this gene assignment, pedigree analysis established that various forms of haemophilia, or "bleeders disease" (in which the blood fails to clot properly), also result from mutations on the X chromosome that give rise to a relatively rare, recessive trait. In this context, rare means "infrequent in the population." The family histories under review, including one following the descendants of Queen Victoria of England (**Figure 3.22a**), showed that relatively rare X-linked traits appear more often in males than in females and often skip generations. The clues that suggest X-linked recessive inheritance in a pedigree are summarized in **Table 3.5**.

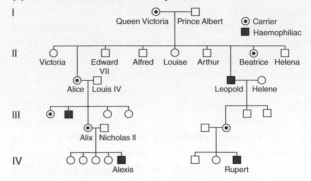

(a) X-linked recessive: Haemophilia

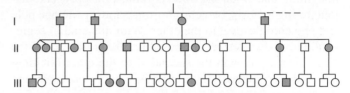

(b) X-linked dominant: Hypophosphataemia

Figure 3.22 X-linked traits may be recessive or dominant.
(a) Pedigree showing inheritance of the recessive X-linked trait haemophilia in Queen Victoria's family. **(b)** Pedigree showing the inheritance of the dominant X-linked trait hypophosphataemia, commonly referred to as vitamin-D-resistant rickets.

Unlike colour blindness and haemophilia, some—although very few—of the known rare mutations on the X chromosome are dominant to the wild-type allele. With such dominant X-linked mutations, more females than males show the aberrant phenotype. This is because all the daughters of an affected male but none of the sons will have the condition, while one-half the sons and one-half the daughters of an affected female will receive the dominant allele and therefore show the phenotype (see Table 3.5). Vitamin-D-resistant rickets, or hypophosphataemia, is an example of an X-linked dominant trait. **Figure 3.22b** presents the pedigree of a family affected by this disease.

Theoretically, phenotypes caused by mutations on the Y chromosome should also be identifiable by pedigree analysis. Such traits would pass from an affected father to all of his sons, and from them to all future male descendants. Females would neither exhibit nor transmit a Y-linked phenotype (see Table 3.5). However, besides the determination of maleness itself, as well as a contribution to sperm formation and thus male fertility, no clear-cut Y-linked visible traits have turned up. The scarcity of known Y-linked traits in humans reflects the fact that the small Y chromosome contains very few genes. Indeed, one would expect the Y chromosome to have only a limited effect on phenotype because normal XX females do perfectly well without it.

Autosomal genes and sexual dimorphism

Not all genes that produce sexual dimorphism (differences in the two sexes) reside on the X or Y chromosomes. Some autosomal genes govern traits that appear in one sex but not the other, or traits that are expressed differently in the two sexes.

Figure 3.21 Red-green colour blindness is an X-linked recessive trait in humans. How the world looks to a person with either normal colour vision (*top*) or a kind of red-green colour blindness known as deuteranopia (*bottom*).

Table 3.5	Pedigree Patterns Suggesting Sex-Linked Inheritance

X-Linked Recessive Trait

1. The trait appears in more males than females since a female must receive two copies of the rare allele to display the phenotype, whereas a hemizygous male with only one copy will show it.

2. The mutation will never pass from father to son because sons receive only a Y chromosome from their father.

3. An affected male passes the X-linked mutation to all his daughters, who are thus unaffected carriers. One-half of the sons of these carrier females will inherit the rare allele and thus the trait.

4. The trait often skips a generation as the mutation passes from grandfather through a carrier daughter to grandson.

5. The trait can appear in successive generations when a sister of an affected male is a carrier. If she is, one-half of her sons will be affected.

6. With the rare affected female, all her sons will be affected and all her daughters will be carriers.

X-Linked Dominant Trait

1. More females than males show the trait.

2. The trait is seen in every generation because it is dominant.

3. All the daughters but none of the sons of an affected male will be affected. This criterion is the most useful for distinguishing an X-linked dominant trait from an autosomal dominant trait.

4. One-half the sons and one-half the daughters of an affected female will be affected.

Y-Linked Trait

1. The trait is seen only in males.

2. All male descendants of an affected man will exhibit the trait.

3. Not only do females not exhibit the trait, they also cannot transmit it.

Sex-limited traits affect a structure or process that is found in one sex but not the other. Mutations in genes for sex-limited traits can influence only the phenotype of the sex that expresses those structures or processes. A curious example of a sex-limited trait occurs in *Drosophila* males homozygous for an autosomal recessive mutation known as *stuck*, which affects the ability of mutant males to retract their penis and release the claspers by which they hold on to female genitalia during copulation. The mutant males have difficulty separating from females after mating. In extreme cases, both individuals die, forever caught in their embrace. Because females lack penises and claspers, homozygous *stuck* mutant females can mate normally. An example of a sex-limited trait in mammalian females is milk production.

Sex-influenced traits show up in both sexes, but expression of such traits may differ between the two sexes because of sex hormone differences. Pattern baldness, a condition in which hair is lost prematurely from the top of the head but not from the sides (**Figure 3.23**), is a sex-influenced trait in humans. Although pattern baldness is a complex trait that can be affected by many genes, an autosomal gene appears to play an important role in certain families. Men in these families who are heterozygous for the balding allele lose their hair while still in their 20s, whereas heterozygous women do not show any significant hair loss. In contrast, homozygotes of both sexes become bald (although the

(a) (b) (c)

Figure 3.23 Male pattern baldness, a sex-influenced trait. **(a)** Prince Philip of England, Duke of Edinburgh. **(b)** Charles, Prince of Wales, and son of Prince Philip. **(c)** Prince William, Duke of Cambridge, and son of Prince Charles. Given the lack of obvious Y-linked visible traits, the father-to-son transmission suggests that the form of male pattern baldness in the royal family is likely determined by an allele of an autosomal gene.

onset of baldness in homozygous women is usually much later in life than in homozygous men). This sex-influenced trait is thus dominant in men, but recessive in women.

The chromosome theory integrates many aspects of gene behaviour

Mendel had assumed that genes are located in cells. The chromosome theory assigned the genes to a specific structure within cells and explained alternative alleles as physically matching parts of homologous chromosomes. In so doing, the theory provided an explanation of Mendel's laws. The mechanism of meiosis ensures that the matching parts of homologous chromosomes will segregate to different gametes (except in rare instances of nondisjunction), accounting for the segregation of alleles predicted by Mendel's first law. Because each homologous chromosome pair aligns independently of all others at meiosis I, genes carried on different chromosomes will assort independently, as predicted by Mendel's second law.

The chromosome theory is also able to explain the creation of new alleles through mutation, a spontaneous or induced change in a particular gene (i.e., in a particular part of a chromosome). If a mutation occurs in the germ line, it can be transmitted to subsequent generations.

Finally, through mitotic cell division in the embryo and after birth, each cell in a multicellular organism receives the same chromosomes—and thus the same maternal and paternal alleles of each gene—as the zygote received from the egg and sperm at fertilization. In this way, an individual's genome—the chromosomes and genes he or she carries—remains constant throughout life.

The idea that genes reside on chromosomes was verified by experiments involving sex-linked genes in *Drosophila* and by the analysis of pedigrees showing X-linked patterns of inheritance in humans. The chromosome theory provides a physical basis for understanding Mendel's laws.

Connections

T. H. Morgan and his students, collectively known as the *Drosophila* group, acknowledged that Mendelian genetics could exist independently of chromosomes. "Why then, we are often asked, do you drag in the chromosomes? Our answer is that because the chromosomes furnish exactly the kind of mechanism that Mendelian laws call for, and since there is an ever-increasing body of information that points clearly to the chromosomes as the bearers of the Mendelian factors, it would be folly to close one's eyes to so patent a relation. Moreover, as biologists, we are interested in heredity not primarily as a mathematical formulation, but rather as a problem concerning the cell, the egg, and the sperm."

The *Drosophila* group went on to find several X-linked mutations in addition to white eyes. One made the body yellow instead of brown, another shortened the wings, and yet another made bent instead of straight body bristles. These findings raised several compelling questions. First, if the genes for all of these traits are physically linked together on the X chromosome, does this linkage affect their ability to assort independently, and if so, how? Second, does each gene have an exact chromosomal address, and if so, does this specific location in any way affect its transmission? In Chapter 4 we describe how the *Drosophila* group and others analyzed the transmission patterns of genes on the same chromosome in terms of known chromosome movements during meiosis, and then used the information obtained to localize genes at specific chromosomal positions.

Essential Concepts

1. Chromosomes are cellular structures specialized for the storage and transmission of genetic material. Genes are located on chromosomes and travel with them during cell division and gamete formation. **[LO1]**

2. In sexually reproducing organisms, somatic cells carry a precise number of homologous pairs of chromosomes, which is characteristic of the species. One chromosome of each pair is of maternal origin; the other, paternal. **[LO3]**

3. Mitosis underlies the growth and development of the individual. Through mitosis, diploid cells produce identical diploid progeny cells. During mitosis, the sister chromatids of every chromosome separate to each of two daughter cells. Before the next cell division, the chromosomes again duplicate to form sister chromatids. **[LO4]**

4. During the first division of meiosis, homologous chromosomes in germ cells segregate from each other. As a result, each gamete receives one member of each matching pair, as predicted by Mendel's first law. **[LO4, LO6]**

5. Also during the first meiotic division, the independent alignment of each pair of homologous chromosomes at the cellular midplane results in the independent assortment of genes carried on different chromosomes, as predicted by Mendel's second law. Mendel's law of independent assortment only applies to genes that reside on different chromosomes or that lie far apart on the same chromosome, as described more extensively in Chapter 4. **[LO4, LO6]**

6. Crossing-over and the independent alignment of homologues during the first meiotic division generate diversity. **[LO4–5]**

7. The second meiotic division generates gametes with a haploid number of chromosomes (n). **[LO4]**

8. Fertilization—the union of egg and sperm—restores the diploid number of chromosomes ($2n$) to the zygote. **[LO2]**

9. The discovery of sex linkage, by which specific genes could be assigned to the X chromosome, provided important support for the chromosome theory of inheritance. Later, the analysis of rare mistakes in meiotic chromosome segregation (nondisjunction) yielded more detailed proof that specific genes are carried on specific chromosomes. **[LO1, LO7]**

Solved Problems

I. In humans, chromosome 16 sometimes has a heavily stained area in the long arm near the centromere. This feature can be seen through the microscope but has no effect on the phenotype of the person carrying it. When such a "blob" exists on a particular copy of chromosome 16, it is a constant feature of that chromosome and is inherited. A couple conceived a child, but the fetus had multiple abnormalities and was miscarried. When the chromosomes of the fetus were studied, it was discovered that it was trisomic for chromosome 16, and that two of the three chromosome 16's had large blobs. Both chromosome 16 homologues in the mother lacked blobs, but the father was heterozygous for blobs. Which parent experienced nondisjunction, and in which meiotic division did it occur?

Answer

This problem requires an understanding of nondisjunction during meiosis. When individual chromosomes contain some distinguishing feature that allows one homologue to be distinguished from another, it is possible to follow the path of the two homologues through meiosis. In this case, because the

fetus had two chromosome 16's with the blob, we can conclude that the extra chromosome came from the father (the only parent with a blobbed chromosome). In which meiotic division did the nondisjunction occur? When nondisjunction occurs during meiosis I, homologues fail to segregate to opposite poles. If this occurred in the father, the chromosome with the blob and the normal chromosome 16 would segregate into the same cell (a secondary spermatocyte). After meiosis II, the gametes resulting from this cell would carry both types of chromosomes. If such sperm fertilized a normal egg, the zygote would have two copies of the normal chromosome 16 and one of the chromosome with a blob. On the other hand, if nondisjunction occurred during meiosis II in the father in a secondary spermatocyte containing the blobbed chromosome 16, sperm with two copies of the blob-marked chromosome would be produced. After fertilization with a normal egg, the result would be a zygote of the type seen in this spontaneous abortion. *Therefore, the nondisjunction occurred in meiosis II in the father.*

II. a. What sex ratio would you expect among the offspring of a cross between a normal male mouse and a female mouse heterozygous for a recessive X-linked lethal gene?

b. What would be the expected sex ratio among the offspring of a cross between a normal hen and a rooster heterozygous for a recessive Z-linked lethal allele?

Answer

This problem deals with sex-linked inheritance and sex determination.

a. Mice have a sex determination system of XX = female and XY = male. A normal male mouse ($X^R Y$) × a heterozygous female mouse ($X^R X^r$) would result in $X^R X^R$, $X^R X^r$, $X^R Y$, and $X^r Y$ mice. The $X^r Y$ mice would die, so there would be a 2:1 ratio of females to males.

b. The sex determination system in birds is ZZ = male and ZW = female. A normal hen ($Z^R W$) × a heterozygous rooster ($Z^R Z^r$) would result in $Z^R Z^R$, $Z^R Z^r$, $Z^R W$, and $Z^r W$ chickens. Because the $Z^r W$ offspring do not live, the ratio of females to males would be 1:2.

III. A woman with normal colour vision whose father was colour-blind mates with a man with normal colour vision.

a. What do you expect to see among their offspring?

b. What would you expect if it were the normal man's father who was colour blind?

Answer

This problem involves sex-linked inheritance.

a. The woman's father has a genotype of $X^{cb}Y$. Because the woman had to inherit an X from her father, she must have an X^{cb} chromosome, but because she has normal colour vision, her other X chromosome must be X^{CB}. The man she mates with has normal colour vision and therefore has an $X^{CB}Y$ genotype. Their children could with equal probability be $X^{CB}X^{CB}$ (normal female), $X^{CB}X^{cb}$ (carrier female), $X^{CB}Y$ (normal male), or $X^{cb}Y$ (colour-blind male).

b. If the man with normal colour vision had a colour-blind father, the X^{cb} chromosome would not have been passed on to him, because a male does not inherit an X chromosome from his father. The man has the genotype $X^{CB}Y$ and cannot pass on the colour-blind allele.

Problems

Vocabulary

1. For each of the terms in the left column, choose the best matching phrase in the right column.

i. meiosis	**1.** X and Y
ii. gametes	**2.** chromosomes that do not differ between the sexes
iii. karyotype	**3.** one of the two identical halves of a replicated chromosome
iv. mitosis	**4.** microtubule organizing centres at the spindle poles
v. interphase	**5.** cells in the testes that undergo meiosis
vi. syncytium	**6.** division of the cytoplasm
vii. synapsis	**7.** haploid germ cells that unite at fertilization
viii. sex chromosomes	**8.** an animal cell containing more than one nucleus
ix. cytokinesis	**9.** pairing of homologous chromosomes
x. anaphase	**10.** one diploid cell gives rise to two diploid cells
xi. chromatid	**11.** the array of chromosomes in a given cell
xii. autosomes	**12.** the part of the cell cycle during which the chromosomes are not visible
xiii. centromere	**13.** one diploid cell gives rise to four haploid cells
xiv. centrosomes	**14.** cell produced by meiosis that does not become a gamete
xv. polar body	**15.** the time during mitosis when sister chromatids separate
xvi. spermatocytes	**16.** connection between sister chromatids

Section 3.1

2. Humans have 46 chromosomes in each somatic cell.

    a. How many chromosomes does a child receive from its father?

    b. How many autosomes and how many sex chromosomes are present in each somatic cell?

    c. How many chromosomes are present in a human ovum?

    d. How many sex chromosomes are present in a human ovum?

3. The figure that follows shows the metaphase chromosomes of a male of a particular species. These chromosomes are prepared as they would be for a karyotype, but they have not yet been ordered in pairs of decreasing size.

    a. How many centromeres are shown?

    b. How many chromosomes are shown?

    c. How many chromatids are shown?

    d. How many pairs of homologous chromosomes are shown?

    e. How many chromosomes on the figure are metacentric? Acrocentric?

    f. What is the likely mode of sex determination in this species? What would you predict to be different about the karyotype of a female in this species?

Section 3.2

4. One oak tree cell with 14 chromosomes undergoes mitosis. How many daughter cells are formed, and what is the chromosome number in each cell?

5. Indicate which of the cells numbered i–v matches each of the following stages of mitosis:

    a. anaphase

    b. prophase

    c. metaphase

    d. G_2

    e. telophase/cytokinesis

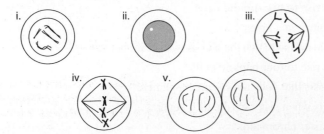

6. a. What are the four major stages of the cell cycle?

    b. Which stages are included in interphase?

    c. What events distinguish G_1, S, and G_2?

7. Answer the questions that follow for each stage of the cell cycle (G_1, S, G_2, prophase, metaphase, anaphase, telophase). If necessary, use an arrow to indicate a change that occurs during a particular cell-cycle stage (e.g., 1 → 2 or yes → no).

    a. How many chromatids comprise each chromosome during this stage?

    b. Is the nucleolus present?

    c. Is the mitotic spindle organized?

    d. Is the nuclear membrane present?

8. Is there any reason that mitosis could not occur in a cell whose genome is haploid?

Section 3.3

9. One oak tree cell with 14 chromosomes undergoes meiosis. How many cells will result from this process, and what is the chromosome number in each cell?

10. Which type(s) of cell division (mitosis, meiosis I, meiosis II) reduce(s) the chromosome number by half? Which type(s) of cell division can be classified as reductional? Which type(s) of cell division can be classified as equational?

11. Complete the following statements using as many of the following terms as are appropriate: mitosis, meiosis I (first meiotic division), meiosis II (second meiotic division), and none (not mitosis nor meiosis I nor meiosis II).

    a. The spindle apparatus is present in cells undergoing _____.

    b. Chromosome replication occurs just prior to _____.

    c. The cells resulting from _____ in a haploid cell have a ploidy of n.

    d. The cells resulting from _____ in a diploid cell have a ploidy of n.

    e. Homologous chromosome pairing regularly occurs during _____.

    f. Nonhomologous chromosome pairing regularly occurs during _____.

    g. Physical recombination leading to the production of recombinant progeny classes occurs during _____.

    h. Centromere division occurs during _____.

    i. Nonsister chromatids are found in the same cell during _____.

12. The five cells shown in figures *a–e* that follow are all from the same individual. For each cell, indicate whether it is in mitosis, meiosis I, or meiosis II. What stage of cell division is represented in each case? What is n in this organism?

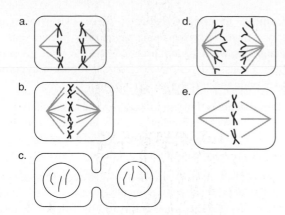

13. One of the first microscopic observations of chromosomes in cell division was published in 1905 by Nettie Stevens. Because it was hard to reproduce photographs at the time, she recorded these observations as *camera lucida* sketches. One such drawing, of a completely normal cell division in the mealworm *Tenebrio molitor*, is shown here. The techniques of the time were relatively unsophisticated by today's standards, and they did not allow her to resolve chromosomal structures that must have been present.

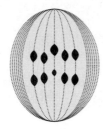

a. Describe in as much detail as possible the kind of cell division and the stage of division depicted in the drawing.

b. What chromosomal structure(s) cannot be resolved in the drawing?

c. How many chromosomes are present in normal *Tenebrio molitor* gametes?

14. A person is simultaneously heterozygous for two autosomal genetic traits. One is a recessive condition for albinism (alleles *A* and *a*); this albinism gene is found near the centromere on the long arm of an acrocentric autosome. The other trait is the dominantly inherited Huntington disease (alleles *HD* and *HD*$^+$). The Huntington gene is located near the telomere of one of the arms of a metacentric autosome. Draw all copies of the two relevant chromosomes in this person as they would appear during metaphase of (a) mitosis, (b) meiosis I, and (c) meiosis II. In each figure, label the location on every chromatid of the alleles for these two genes, assuming that no recombination takes place.

15. Assuming (i) that the two chromosomes in a homologous pair carry different alleles of some genes, and (ii) that no crossing-over takes place, how many genetically different offspring could any one human couple potentially produce? Which of these two assumptions (i or ii) is more realistic?

16. In the moss *Polytrichum commune*, the haploid chromosome number is 7. A haploid male gamete fuses with a haploid female gamete to form a diploid cell that divides and develops into the multicellular sporophyte. Cells of the sporophyte then undergo meiosis to produce haploid cells called spores. What is the probability that an individual spore will contain a set of chromosomes all of which came from the male gamete? Assume no recombination.

17. Is there any reason that meiosis could not occur in an organism whose genome is always haploid?

18. Sister chromatids are held together through metaphase of mitosis by complexes of *cohesin* proteins that form rubber bandlike rings bundling the two sister chromatids. Cohesin rings are found both at centromeres and at many locations scattered along the length of the chromosomes. The rings are destroyed by protease enzymes at the beginning of anaphase, allowing the sister chromatids to separate.

a. Cohesin complexes between sister chromatids are also responsible for keeping homologous chromosomes together until anaphase of meiosis I. With this point in mind, which of the two diagrams that follow (i or ii) properly represents the arrangement of chromatids during prophase through metaphase of meiosis I? Explain your choice.

b. What does your answer to part *a* allow you to infer about the nature of cohesin complexes at the centromere versus those along the chromosome arms? Suggest a molecular hypothesis to explain your inference.

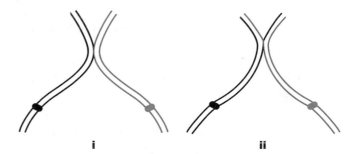

i   ii

Section 3.4

19. Answer the following regarding humans.

a. How many sperm develop from 100 primary spermatocytes?

b. How many sperm develop from 100 secondary spermatocytes?

c. How many sperm develop from 100 spermatids?

d. How many ova develop from 100 primary oocytes?

e. How many ova develop from 100 secondary oocytes?

f. How many ova develop from 100 polar bodies?

20. Somatic cells of chimpanzees contain 48 chromosomes.

How many chromatids and chromosomes are present at (a) anaphase of mitosis, (b) anaphase I of meiosis, (c) anaphase II of meiosis, (d) G$_1$ prior to mitosis, (e) G$_2$

prior to mitosis, (f) G$_1$ prior to meiosis I, and (g) prophase of meiosis I?

How many chromatids or chromosomes are present in (h) an oogonial cell prior to S phase, (i) a spermatid, (j) a primary oocyte arrested prior to ovulation, (k) a secondary oocyte arrested prior to fertilization, (l) a second polar body, and (m) a chimpanzee sperm?

21. In a certain strain of turkeys, unfertilized eggs sometimes develop parthenogenetically to produce diploid offspring. (Females have ZW and males have ZZ sex chromosomes. Assume that WW cells are inviable.) What distribution of sexes would you expect to see among the parthenogenetic offspring according to each of the following models for how parthenogenesis occurs?

a. The eggs develop without ever going through meiosis.

b. The eggs go all the way through meiosis and then duplicate their chromosomes to become diploid.

c. The eggs go through meiosis I, and the chromatids separate to create diploidy.

d. The egg goes all the way through meiosis and then fuses at random with one of its three polar bodies (this assumes the first polar body goes through meiosis II).

22. Female mammals, including women, sometimes develop benign tumours called "ovarian teratomas" or "dermoid cysts" in their ovaries. Such a tumour begins when a primary oocyte escapes from its prophase I arrest and finishes meiosis I within the ovary. (Normally meiosis I does not finish until the primary oocyte is expelled from the ovary upon ovulation.) The secondary oocyte then develops as if it were an embryo, and it implants and develops within the follicle. Development is disorganized, however, and results in a tumour containing a wide variety of differentiated tissues, including teeth, hair, bone, muscle, nerve, and many others. If a dermoid cyst forms in a woman whose genotype is *Aa*, what are the possible genotypes of the cyst?

Section 3.5

23. A system of sex determination known as haplodiploidy is found in honeybees. Females are diploid, and males (drones) are haploid. Male offspring result from the development of unfertilized eggs. Sperm are produced by mitosis in males and fertilize eggs in the females. Ivory eye is a recessive characteristic in honeybees; wild-type eyes are brown.

a. What progeny would result from an ivory-eyed queen and a brown-eyed drone? Give both genotype and phenotype for progeny produced from fertilized and nonfertilized eggs.

b. What would result from crossing a daughter from the mating in part *a* with a brown-eyed drone?

24. Imagine you have two pure-breeding lines of canaries, one with yellow feathers and the other with brown

feathers. In crosses between these two strains, yellow female × brown male gives only brown sons and daughters, while brown female × yellow male gives only brown sons and yellow daughters. Propose a hypothesis to explain these results.

25. Barred feather pattern is a Z-linked dominant trait in chickens. What offspring would you expect from (a) the cross of a barred hen to a nonbarred rooster? (b) the cross of an F$_1$ rooster from part *a* to one of his sisters?

26. Each of the four pedigrees that follow represents a human family within which a genetic disease is segregating. Affected individuals are indicated by filled-in symbols. One of the diseases is transmitted as an autosomal recessive condition, one as an X-linked recessive, one as an autosomal dominant, and one as an X-linked dominant. Assume all four traits are rare in the population.

a. Indicate which pedigree represents which mode of inheritance, and explain how you know.

b. For each pedigree, how would you advise the parents of the chance that their child (indicated by the hexagon shape) will have the condition?

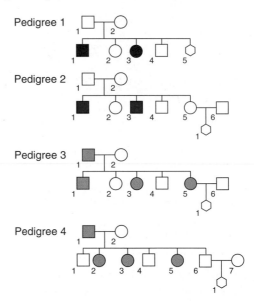

27. In a vial of *Drosophila*, a research student noticed several female flies (but no male flies) with "bag" wings, each consisting of a large liquid-filled blister instead of the usual smooth wing blade. When bag-winged females were crossed with wild-type males, 1/3 of the progeny were bag-winged females, 1/3 were normal-winged females, and 1/3 were normal-winged males. Explain these results.

28. Duchenne muscular dystrophy (DMD) is caused by a relatively rare X-linked recessive allele. It results in progressive muscular wasting and usually leads to death before age 20.

a. What is the probability that the first son of a woman whose brother is affected will be affected?

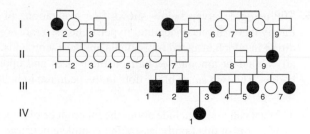

b. What is the probability that the second son of a woman whose brother is affected will be affected, if her first son was affected?

c. What is the probability that a child of an unaffected man whose brother is affected will be affected?

d. An affected man mates with his unaffected first cousin; there is otherwise no history of DMD in this family. If the mothers of this man and his mate were sisters, what is the probability that the couple's first child will be an affected boy? an affected girl? an unaffected child?

e. If two of the parents of the couple in part *d* were brother and sister, what is the probability that the couple's first child will be an affected boy? an affected girl? an unaffected child?

29. The following is a pedigree of a family in which a rare form of colour blindness is found (filled-in symbols). Indicate as much as you can about the genotypes of all the individuals in the pedigree.

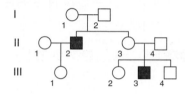

30. In 1995, doctors reported a Chinese family in which retinitis pigmentosa (progressive degeneration of the retina leading to blindness) affected only males. All six sons of affected males were affected, but all of the five daughters of affected males (and all of the children of these daughters) were unaffected.

a. What is the likelihood that this form of retinitis pigmentosa is due to an autosomal mutation showing complete dominance?

b. What other possibilities could explain the inheritance of retinitis pigmentosa in this family? Which of these possibilities do you think is most likely?

31. The pedigree that follows indicates the occurrence of albinism in a group of Hopi Indians, among whom the trait is unusually frequent. Assume that the trait is fully penetrant (all individuals with a genotype that could give rise to albinism will display this condition).

a. Is albinism in this population caused by a recessive or a dominant allele?

b. Is the gene sex-linked or autosomal?

What are the genotypes of the following individuals?

c. individual I-1

d. individual I-8

e. individual I-9

f. individual II-6

g. individual II-8

h. individual III-4

32. When Calvin Bridges observed a large number of offspring from a cross of white-eyed female *Drosophila* to red-eyed males, he observed very rare white-eyed females and red-eyed males among the offspring. He was able to show that these exceptions resulted from nondisjunction, such that the white-eyed females had received two X's from the egg and a Y from the sperm, while the red-eyed males had received no sex chromosome from the egg and an X from the sperm. What progeny would have arisen from these same kinds of nondisjunctional events if they had occurred in the male parent? What would their eye colours have been?

33. In *Drosophila*, a cross was made between a yellow-bodied male with vestigial (not fully developed) wings and a wild-type female (brown body). The F_1 generation consisted of wild-type males and wild-type females. F_1 males and females were crossed, and the F_2 progeny consisted of 16 yellow-bodied males with vestigial wings, 48 yellow-bodied males with normal wings, 15 males with brown bodies and vestigial wings, 49 wild-type males, 31 brown-bodied females with vestigial wings, and 97 wild-type females. Explain the inheritance of the two genes in question based on these results.

34. Consider the following pedigrees from human families containing a male with Klinefelter syndrome (a set of abnormalities seen in XXY individuals; indicated with shaded boxes). In each, *A* and *B* refer to codominant alleles of the X-linked *G6PD* gene. The *phenotypes* of each individual (A, B, or AB) are shown on the pedigree. Indicate if nondisjunction occurred in the mother or father of the son with Klinefelter syndrome for each of the three examples. Can you tell if the nondisjunction was in the first or second meiotic division?

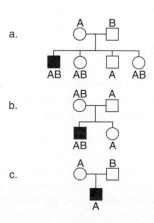

35. The pedigree given below shows five generations of a family that exhibits congenital hypertrichosis, a rare condition in which affected individuals are born with unusually abundant amounts of hair on their faces and upper bodies. The two small black dots in the pedigree indicate miscarriages.

   a. What can you conclude about the inheritance of hypertrichosis in this family, assuming complete penetrance of the trait?

   b. On what basis can you exclude other modes of inheritance?

   c. With how many fathers did III-2 and III-9 have children?

36. In *Drosophila*, the autosomal recessive *brown* eye colour mutation displays interactions with both the X-linked recessive *vermilion* mutation and the autosomal recessive *scarlet* mutation. Flies homozygous for *brown* and simultaneously hemizygous or homozygous for *vermilion* have white eyes. Flies simultaneously homozygous for both the *brown* and *scarlet* mutations also have white eyes. Predict the F_1 and F_2 progeny of crossing the following true-breeding parents:

   a. vermilion females × brown males

   b. brown females × vermilion males

   c. scarlet females × brown males

   d. brown females × scarlet males

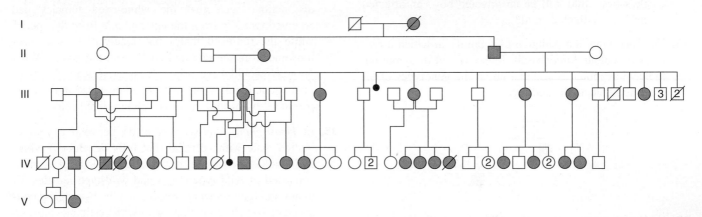

37. Several different antigens can be detected in blood tests. The following four traits were tested for each individual shown:

   ABO type   (I^A and I^B codominant, *i* recessive)

   Rh type   (Rh^+ dominant to Rh^-)

   MN type   (*M* and *N* codominant)

   $Xg^{(a)}$ type   ($Xg^{(a+)}$ dominant to $Xg^{(a-)}$)

   All of these blood type genes are autosomal, except for $Xg^{(a)}$, which is X-linked.

Mother	AB	Rh^-	MN	$Xg^{(a+)}$
Daughter	A	Rh^+	MN	$Xg^{(a-)}$
Alleged father 1	AB	Rh^+	M	$Xg^{(a+)}$
Alleged father 2	A	Rh^-	N	$Xg^{(a-)}$
Alleged father 3	B	Rh^+	N	$Xg^{(a-)}$
Alleged father 4	O	Rh^-	MN	$Xg^{(a-)}$

   a. Which, if any, of the alleged fathers could be the real father?

   b. Would your answer to part *a* change if the daughter had Turner syndrome (the abnormal phenotype seen in XO individuals)? If so, how?

38. In 1919, Calvin Bridges began studying an X-linked recessive mutation causing eosin-coloured eyes in *Drosophila*. Within an otherwise true-breeding culture of eosin-eyed flies, he noticed rare variants that had much lighter cream-coloured eyes. By intercrossing these variants, he was able to make a true-breeding cream-eyed stock. Bridges now crossed males from this cream-eyed stock with true-breeding wild-type females. All the F_1 progeny had red (wild-type) eyes. When F_1 flies were intercrossed, the F_2 progeny were 104 females with red eyes, 52 males with red eyes, 44 males with eosin eyes, and 14 males with cream eyes. Assume this represents an 8:4:3:1 ratio.

   a. Formulate a hypothesis to explain the F_1 and F_2 results, assigning phenotypes to all possible genotypes.

   b. What do you predict in the F_1 and F_2 generations if the parental cross is between true-breeding eosin-eyed males and true-breeding cream-eyed females?

   c. What do you predict in the F_1 and F_2 generations if the parental cross is between true-breeding eosin-eyed females and true-breeding cream-eyed males?

39. As we learned in this chapter, the *white* mutation of *Drosophila* studied by Thomas Hunt Morgan is X-linked and recessive to wild type. When true-breeding white-eyed males carrying this mutation were crossed with true-breeding purple-eyed females, all the F_1 progeny had wild-type (red) eyes. When the F_1 progeny were intercrossed, the F_2 progeny emerged in the ratio 3/8 wild-type females : 1/4 white-eyed males : 3/16 wild-type males : 1/8 purple-eyed females : 1/16 purple-eyed males.

   a. Formulate a hypothesis to explain the inheritance of these eye colours.

   b. Predict the F_1 and F_2 progeny if the parental cross was reversed (i.e., if the parental cross was between true-breeding white-eyed females and true-breeding purple-eyed males).

40. The ancestry of a white female tiger bred in a city zoo is depicted in the pedigree following part *e* of this problem. White tigers are indicated with unshaded symbols. (As you can see, there was considerable inbreeding in this lineage. For example, the white tiger Mohan was mated with his daughter.) In answering the following questions, assume that "white" is determined by allelic differences at a single gene and that the trait is fully penetrant. Explain your answers by citing the relevant information in the pedigree.

a. Could white coat colour be caused by a Y-linked allele?

b. Could white coat colour be caused by a dominant X-linked allele?

c. Could white coat colour be caused by a dominant autosomal allele?

d. Could white coat colour be caused by a recessive X-linked allele?

e. Could white coat colour be caused by a recessive autosomal allele?

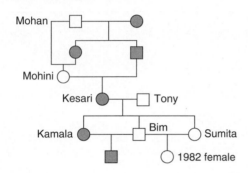

41. The pedigree given below shows the inheritance of various types of cancer in a particular family. Molecular analyses (described in subsequent chapters) indicate that, with one exception, the cancers occurring in the patients in this pedigree are associated with a rare mutation in a gene called *BRCA2*.

a. Which individual is the exceptional cancer patient whose disease is not associated with a *BRCA2* mutation?

b. Is the *BRCA2* mutation dominant or recessive to the normal *BRCA2* allele in terms of its cancer-causing effects?

c. Is the *BRCA2* gene likely to reside on the X chromosome, the Y chromosome, or an autosome? How definitive is your assignment of the chromosome carrying *BRCA2*?

d. Is the penetrance of the cancer phenotype complete or incomplete?

e. Is the expressivity of the cancer phenotype unvarying or variable?

f. Are any of the cancer phenotypes associated with the *BRCA2* mutation sex-limited or sex-influenced?

g. How can you explain the absence of individuals diagnosed with cancer in generations I and II?

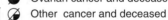

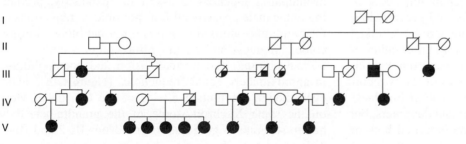

connect For more information on the resources available from McGraw-Hill Ryerson, go to www.mcgrawhill.ca/he/solutions.

Problems   **109**

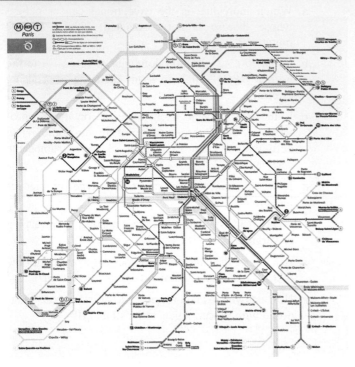

Maps illustrate the spatial relationships of objects, such as the locations of subway stations along subway lines. Genetic maps portray the positions of genes along chromosomes.

In 1928, doctors completed a four-generation pedigree that traced two known X-linked traits: red-green colour blindness and haemophilia A (the more serious X-linked form of "bleeders disease"). The maternal grandfather of the family exhibited both traits, which means that his single X chromosome carried mutant alleles of the two corresponding genes. As expected, neither colour blindness nor haemophilia showed up in his sons and daughters, but two grandsons and one great-grandson inherited both of the X-linked conditions (**Figure 4.1a**). The fact that none of the descendants manifested one of the traits without the other suggests that the mutant alleles did not assort independently during meiosis. Instead they travelled together in the gametes forming one generation and then into the gametes forming the next generation, producing grandsons and great-grandsons with an X chromosome specifying both colour blindness and haemophilia. Genes that travel together more often than expected exhibit **genetic linkage**.

In contrast, another pedigree following colour blindness and a slightly different form of haemophilia, haemophilia B, which also arises from a mutation on the X chromosome, revealed a different inheritance pattern. A grandfather with haemophilia B and colour blindness had four grandsons, but only one of them exhibited both conditions. In this family, the genes for colour blindness and

haemophilia appeared to assort independently, producing in the male progeny all four possible combinations of the two traits—normal vision and normal blood clotting, colour blindness and haemophilia, colour blindness and normal clotting, and normal vision and haemophilia—in approximately equal frequencies (**Figure 4.1b**). Thus, even though the mutant alleles of the two genes were on the same X chromosome in the grandfather, they had to separate to give rise to grandsons III-2 and III-3. This separation of genes on the same chromosome is the result of **recombination**, the occurrence in progeny of new gene combinations not seen in previous generations. (Note that *recombinant progeny* can result in either of two ways: from the recombination of genes on the same

Chapter Outline
4.1 Gene Linkage and Recombination
4.2 The Chi-Square Test and Linkage Analysis
4.3 Recombination: A Result of Crossing-Over During Meiosis
4.4 Mapping: Locating Genes Along a Chromosome
4.5 Mitotic Recombination and Genetic Mosaics

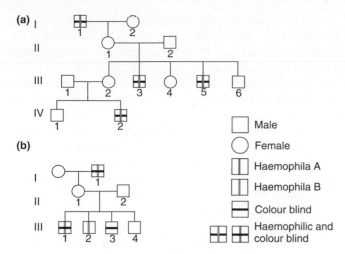

(a)

(b)

Male

Female

Haemophila A

Haemophila B

Colour blind

Haemophilic and colour blind

Figure 4.1 Pedigrees indicate that colour blindness and two forms of haemophilia are X-linked traits. (a) Transmission of red-green colour blindness and haemophilia A. The traits travel together through the pedigree, indicating their genetic linkage. **(b)** Transmission of red-green colour blindness and haemophilia B. Even though both genes are X-linked, the mutant alleles are inherited together in only one of four grandsons in generation III. These two pedigrees indicate that the gene for colour blindness is close to the haemophilia A gene but far away from the haemophilia B gene.

chromosome during gamete formation, discussed in this chapter, or from the independent assortment of genes on nonhomologous chromosomes, previously described in Chapter 3.)

Two important themes emerge as we follow the transmission of genes linked on the same chromosome. The first is that the farther apart two genes are, the greater is the probability of separation through recombination. Extrapolating from this general rule, you can see that the gene for haemophilia A must be very close to the gene for red-green colour blindness, because, as Figure 4.1a shows, the two rarely separate. By comparison, the gene for haemophilia B must lie far away from the colour blindness gene, because, as Figure 4.1b indicates, new combinations of alleles of the two genes occur frequently. A second crucial theme arising from these considerations is that geneticists can use data about how often genes separate during transmission to map the genes' relative locations on a chromosome. Such mapping is a key to sorting out and tracking down the components of complex genetic networks; it is also crucial to geneticists' ability to isolate and characterize genes at the molecular level.

Learning Objectives

1. Discuss the physical basis for genetic linkage.

2. Explain why Mendel's law of independent assortment does not apply to linked genes.

3. Examine how linkage between two genetic loci is determined.

4. Describe when during the cell cycle a crossover occurs, the mechanism, and its potential outcomes.

5. Demonstrate how genetic distance is estimated between two or three linked loci.

6. Construct a simple genetic map based upon data from testcrosses.

4.1   Gene Linkage and Recombination

If people have roughly 20 000 genes but only 23 pairs of chromosomes, most human chromosomes must carry hundreds, if not thousands, of genes. This is certainly true of the human X chromosome: There are approximately 1000 reported protein-encoding genes on this chromosome. This number is likely to grow, at least slightly, as geneticists develop new techniques to analyze the X chromosome's DNA sequence. Moreover, this number does not account for the many genes that do not encode proteins. Recognition that many genes reside on each chromosome raises an important question. If genes on *different* chromosomes assort independently because nonhomologous chromosomes align independently on the spindle during meiosis I, how do genes on the *same* chromosome assort?

Some genes on the same chromosome do not assort independently

We begin our analysis with X-linked *Drosophila* genes because they were the first to be assigned to a specific chromosome. As we outline various crosses, remember

that females carry two X chromosomes, and thus two alleles for each X-linked gene. Males, in contrast, have only a single X chromosome (from the female parent), and thus only a single allele for each of these genes.

We look first at two X-linked genes that determine a fruit fly's eye colour and body colour. These two genes are said to be **syntenic** because they are located on the same chromosome. The *white* gene was previously introduced in Chapter 3; you will recall that the dominant wild-type allele w^+ specifies red eyes, while the recessive mutant allele w confers white eyes. The alleles of the *yellow* body colour gene are y^+ (the dominant wild-type allele for brown bodies) and y (the recessive mutant allele for yellow bodies). To avoid confusion, note that lowercase y and y^+ refer to alleles of the *yellow* gene, while capital Y refers to the Y chromosome (which does not carry genes for either eye or body colour). You should also pay attention to the slash symbol (/), which is used to separate genes found on chromosomes of a pair (either the X and Y chromosomes as in this case, or a pair of X chromosomes or homologous autosomes). Thus $w\ y/Y$ represents the genotype of a male

with an X chromosome bearing w and y, as well as a Y chromosome; phenotypically this male has white eyes and a yellow body.

Detecting linkage by analyzing the gametes produced by a dihybrid

In a cross between a female with mutant white eyes and a wild-type brown body ($w\,y^+/w\,y^+$) and a male with wild-type red eyes and a mutant yellow body ($w^+\,y$/Y), the F_1 offspring are evenly divided between brown-bodied females with normal red eyes ($w\,y^+/w^+\,y$) and brown-bodied males with mutant white eyes ($w\,y^+$/Y) (**Figure 4.2**). Note that the male progeny look like their mother because their phenotype directly reflects the genotype of the single X chromosome they received from her. The same is not true for the F_1 females, who received w and y^+ on the X from their mother and both w^+ and y on the X from their father. These F_1 females are thus dihybrids: With two alleles for each X-linked gene, one derived from each parent, the dominance relations of each pair of alleles determine the female phenotype.

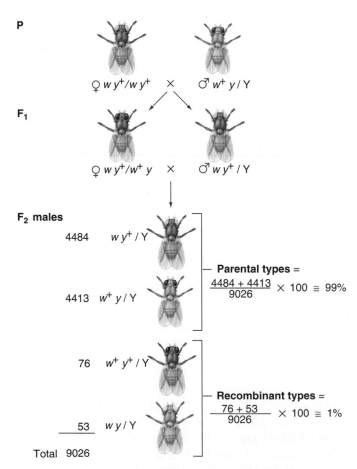

P

$♀\ w\,y^+/w\,y^+$ × $♂\ w^+\,y$/Y

F_1

$♀\ w\,y^+/w^+\,y$ × $♂\ w\,y^+$/Y

F_2 males

4484 $w\,y^+$/Y

4413 $w^+\,y$/Y

— Parental types =
$\dfrac{4484 + 4413}{9026} \times 100 \cong 99\%$

76 $w^+\,y^+$/Y

53 $w\,y$/Y

— Recombinant types =
$\dfrac{76 + 53}{9026} \times 100 \cong 1\%$

Total 9026

Figure 4.2 When genes are linked, parental combinations outnumber recombinant types. Doubly heterozygous $w\,y^+/w^+\,y$ F_1 females produce four types of male offspring. Sons that look like the father ($w^+\,y$/Y) or mother ($w\,y^+$/Y) of the F_1 females are parental types. Other sons ($w^+\,y^+$/Y or $w\,y$/Y) are recombinant types. For these closely linked genes, many more parental types are produced than recombinant types.

Now comes the significant cross for answering our question about the assortment of genes on the same chromosome. If these two *Drosophila* genes for eye and body colour assort independently, as predicted by Mendel's second law, the dihybrid F_1 females should make four kinds of gametes, with four different combinations of genes on the X chromosome—$w\,y^+$, $w^+\,y$, $w^+\,y^+$, and $w\,y$. These four types of gametes should occur with equal frequency; that is, in a ratio of 1:1:1:1. If it happens this way, approximately half of the gametes will be of the two **parental types**, carrying either the $w\,y^+$ allele combination seen in the original female of the P generation or the $w^+\,y$ allele combination seen in the original male of the P generation. The remaining half of the gametes will be of two **recombinant types**, in which reshuffling has produced either $w^+\,y^+$ or $w\,y$ allele combinations not seen in the P generation parents of the F_1 females.

We can see whether the 1:1:1:1 ratio of the four kinds of gametes actually materializes by counting the different types of male progeny in the F_2 generation, as these sons receive their X-linked genes only from their maternal gamete. The bottom part of Figure 4.2 depicts the results of a breeding study that produced 9026 F_2 males. The relative numbers of the four X-linked gene combinations passed on by the dihybrid F_1 females' gametes reflect a significant departure from the 1:1:1:1 ratio expected of independent assortment. By far, the largest numbers of gametes carry the parental combinations $w\,y^+$ and $w^+\,y$. Of the total 9026 male flies counted, 8897, or almost 99 percent, had these genotypes. In contrast, the new combinations $w^+\,y^+$ and $w\,y$ made up little more than 1 percent of the total.

We can explain in one of two ways why the two genes fail to assort independently. Either the $w\,y^+$ and $w^+\,y$ combinations are preferred because of some intrinsic chemical affinity between these particular alleles, or it is the parental combination of alleles the F_1 female receives from one or the other of her P generation parents that shows up most frequently.

Linkage: A prevalence of parental classes of gametes

A second set of crosses involving the same genes but with a different arrangement of alleles explains why the dihybrid F_1 females do not produce a 1:1:1:1 ratio of the four possible types of gametes (see Cross Series B in **Figure 4.3**). In this second set of crosses, the original parental generation consists of red-eyed, brown-bodied females ($w^+\,y^+/w^+\,y^+$) and white-eyed, yellow-bodied males ($w\,y$/Y), and the resultant F_1 females are all $w^+\,y^+/w\,y$ dihybrids. To find out what kinds and ratios of gametes these F_1 females produce, we need to look at the telltale F_2 males.

This time, as cross B in Figure 4.3 shows, $w^+\,y$/Y and $w\,y^+$/Y are the recombinants that account for little more than 1 percent of the total, while $w\,y$/Y and $w^+\,y^+$/Y are the parental combinations, which again add up to almost 99 percent. You can see that there is no preferred association of w^+ and y or of y^+ and w in this cross. Instead, a comparison of the two experiments with these particular

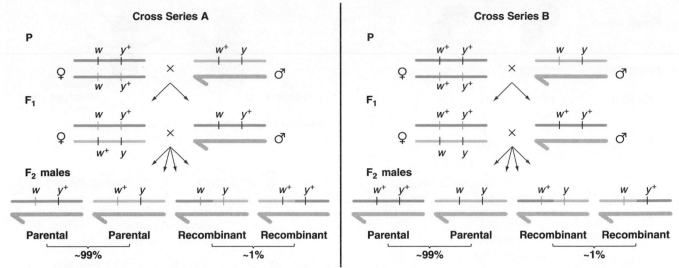

Figure 4.3 **Designations of "parental" and "recombinant" relate to past history.** Figure 4.2 has been redrawn here as **Cross Series A** for easier comparison with **Cross Series B**, in which the dihybrid F_1 females received different allelic combinations of the *white* and *yellow* genes. Note that the parental and recombinant classes in the two cross series are the opposite of each other. The percentages of recombinant and parental types are nonetheless similar in both experiments, showing that the frequency of recombination is independent of the arrangement of alleles.

X chromosome genes demonstrates that the observed frequencies of the various types of progeny depend on how the arrangement of alleles in the F_1 females originated. We have redrawn Figure 4.2 as Cross Series A in Figure 4.3 so that you can make this comparison more directly. Note that in both experiments, it is the **parental classes**—the combinations originally present in the P generation—that show up most frequently in the F_2 generation. The reshuffled **recombinant classes** occur less frequently. It is important to appreciate that the designation of "parental" and "recombinant" gametes or progeny of a dihybrid F_1 female is operational; that is, determined by the particular set of alleles she receives from each of her parents.

When genes assort independently, the numbers of parental and recombinant F_2 progeny are equal, because a dihybrid F_1 individual produces an equal number of all four types of gametes. By comparison, two genes are considered **linked** when the number of F_2 progeny with parental genotypes exceeds the number of F_2 progeny with recombinant genotypes. Instead of assorting independently, the genes behave as if they are often connected to each other. The genes for eye and body colour that reside on the X chromosome in *Drosophila* are an extreme illustration of the linkage concept. The two genes are so tightly coupled that the parental combinations of alleles—$w^+ y$ and $w y^+$ (in Cross Series A of Figure 4.3) or $w^+ y^+$ and $w y$ (in Cross Series B)—are reshuffled to form recombinants in only 1 out of every 100 gametes formed. In other words, the two parental allele combinations of these tightly linked genes are inherited together 99 times out of 100.

Gene-pair-specific variation in the degree of linkage

Linkage is not always this tight. In *Drosophila*, a mutation for miniature wings (*m*) is also found on the X chromosome. A cross of red-eyed females with normal wings

($w^+ m^+/w^+ m^+$) and white-eyed males with miniature wings ($w m/Y$) yields an F_1 generation containing all red-eyed, normal-winged flies. The genotype of the dihybrid F_1 females is $w^+ m^+/w m$. Of the F_2 males, 67.2 percent are parental types ($w^+ m^+$ and $w m$), while the remaining 32.8 percent are recombinants ($w m^+$ and $w^+ m$).

This prevalence of parental combinations among the F_2 genotypes reveals that the two genes are linked: The parental combinations of alleles travel together more often than expected. But compared to the 99 percent linkage between the w and y genes for eye colour and body colour, the linkage of w to m is not that tight. The parental combinations for colour and wing size are reshuffled in roughly 33 (instead of 1) out of every 100 gametes.

Autosomal traits can also exhibit linkage

Linked autosomal genes are not inherited according to the 9:3:3:1 Mendelian ratio expected for two independently assorting genes. Early twentieth-century geneticists William Bateson and Reginald Punnett, for whom the Punnett square was named, were puzzled by the many experimentally observed departures from this ratio, which they could not explain in terms of the gene interactions discussed in Chapter 2. Bateson and Punnett were studying the inheritance of two traits in sweet peas, flower colour and shape of pollen grains. They crossed true-breeding purple-flowered, long pollen plants (*PPLL*) with true-breeding red-flowered, round pollen plants (*ppll*) and, as predicted, the F_1 progeny were all purple-flowered with long pollen (*PpLl*). When the F_1 individuals were crossed to each other, the F_2 progeny were not in the expected 9:3:3:1 ratio (**Figure 4.4**). Bateson and Punnett observed that the two parental phenotypic classes, purple-flowered and long pollen, and red-flowered and round pollen, appeared more often than expected. On the other hand, the two recombinant classes, purple-flowered and round

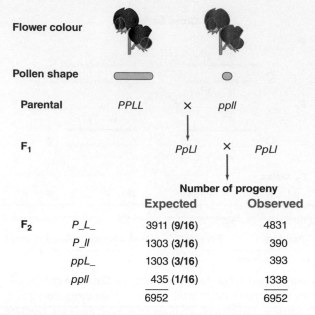

Figure 4.4 Bateson and Punnett observed ratios that were a significant departure from the expected. Crosses between true-breeding purple-flowered, long pollen plants (*PPLL*) and true-breeding red-flowered, round pollen plants (*ppll*) yield F₁ progeny that are all purple-flowered with long pollen (*PpLl*). When dihybrid F₁ individuals are crossed to each other, the F₂ progeny are not in the expected 9:3:3:1 ratio. The numbers of parental-type progeny are greater than expected, while the numbers of recombinant-type progeny are less than expected.

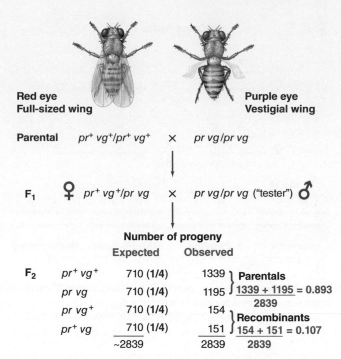

Figure 4.5 Autosomal genes can also exhibit linkage. A testcross shows that the recombination frequency for the eye colour (*pr*) and wing size (*vg*) pair of *Drosophila* genes is 11 percent. Because parentals outnumber recombinants, the *pr* and *vg* genes are genetically linked and must be on the same autosome.

pollen, and red-flowered and long pollen, occurred less frequently than expected. Although Bateson and Punnett could not explain this finding, they suggested that the parental allele combinations, *PL* and *pl*, were somehow physically connected. Bateson and Punnett found it difficult to interpret these unexpected results, because although they knew that individuals receive two copies of each autosomal gene, one from each parent, it was hard to trace which alleles came from which parent.

Thomas Hunt Morgan was also studying the inheritance of two autosomal genes in *Drosophila*. Unlike Punnett and Bateson, however, Morgan set up testcrosses in which one parent (the "tester") was homozygous for the recessive alleles of both genes, enabling him to analyze the gene combinations received from the gametes of the other, dihybrid parent. Morgan's use of the **two-point** (or "two gene") **testcross**, in which a dihybrid F₁ individual is crossed to a true-breeding fly showing the recessive phenotypes, was very important. Since the "tester" fly contributes only recessive alleles to its offspring, the phenotypes of the progeny of a two-point testcross reveal the alleles contributed by the dihybrid parent. Fruit flies, for example, carry an autosomal gene for eye colour (in addition to the X-linked *w* gene); the wild type is once again red, but a recessive mutation in this gene gives rise to purple (*pr*). A second gene on the same autosome helps determine the shape of a fruit fly's wing, with the wild type having full-sized wings and a recessive mutation (*vg*) producing short, crumpled

wings (vestigial). **Figure 4.5** depicts a cross between wild-type flies (*pr⁺ vg⁺/pr⁺ vg⁺*) and purple-eyed flies with vestigial wings (*pr vg/pr vg*). All the F₁ progeny are dihybrid (*pr⁺ vg⁺/pr vg*) and are phenotypically wild type. In a testcross of the F₁ females with *pr vg/pr vg* males, all of the offspring receive the recessive *pr* and *vg* alleles from their father. The phenotypes of the offspring thus indicate the kinds of gametes received from the mother. For example, a purple-eyed fly with wild-type wings would be genotype *pr vg⁺/pr vg*; because we know it received the *pr vg* combination from its father, it must have received *pr vg⁺* from its mother. As Figure 4.5 shows, roughly 89 percent of the testcross progeny in one experiment received parental gene combinations (i.e., allelic combinations transmitted into the F₁ females by the gametes of each of her parents), while the remaining 11 percent were recombinants. Because the parental classes outnumbered the recombinant classes, we can conclude that the autosomal genes for purple-coloured eyes and vestigial wings are linked.

Linkage between two genes can be detected in the proportion of gametes that a dihybrid individual produces. If the numbers of parental-type and recombinant-type gametes are equal, then the two genes are assorting independently. If the parental-type gametes exceed the recombinant form, then the genes are linked.

How do you know from a particular experiment whether two genes assort independently or are genetically linked? At first glance, this question should pose no problem. Discriminating between the two possibilities involves straightforward calculations based on assumptions well supported by observations. For independently assorting genes, a dihybrid F_1 female produces four types of gametes in equal numbers, so one-half of the F_2 progeny are of the parental classes and the other half of the recombinant classes. In contrast, for linked genes, the two types of parental classes by definition always outnumber the two types of recombinant classes in the F_2 generation.

The problem is that, because real-world genetic transmission is based on chance events, even unlinked independently assorting genes can produce deviations from the 1:1:1:1 ratio in a particular study, just as you may easily get 6 heads and 4 tails (rather than the predicted 5 and 5) in 10 tosses of a coin. Thus, if a breeding experiment analyzing the transmission of two genes shows a deviation from the equal ratios of parentals and recombinants expected of independent assortment, can we necessarily conclude the two genes are linked? Is it instead possible that the results represent a statistically acceptable chance fluctuation from the mean values expected of unlinked genes that assort independently? Such questions become more important in cases where linkage is not all that tight, so that even though the genes are linked, the percentage of recombinant classes approaches 50 percent.

The chi-square test evaluates the significance of differences between predicted and observed values

To answer these kinds of questions, statisticians have devised a quantitative measure of the likelihood that an experimentally observed deviation from the predictions of a particular hypothesis could have occurred solely by chance. This measure of the "goodness of fit" between observed and predicted results is a probability test known as the **chi-square test**. The test is designed to account for the fact that the size of an experimental population (the "sample size") is an important component of statistical significance. To appreciate the role of sample size, let us return to the proverbial coin toss before examining the details of the chi-square test.

In 10 tosses of a coin, an outcome of 6 heads (60 percent) and 4 tails (40 percent) is not unexpected because of the effects of chance. However, with 1000 tosses of the coin, a result of 600 heads (60 percent) and 400 tails (40 percent) would intuitively be highly unlikely. In the first case, a change in the results of one coin toss would alter the expected 5:5 ratio to the observed 6:4 ratio. In the second case, 100 tosses would have to change from tails to heads to generate the stated deviation from the predicted 500:500 ratio. Chance events could reasonably, and even

likely, cause one deviation from the predicted number, but not 100.

Two important concepts emerge from this simple example. First, a comparison of percentages or ratios alone will never allow you to determine whether or not *observed* data are significantly different from *predicted* or *expected* values. Second, the absolute numbers obtained are important because they reflect the size of the experiment. The larger the sample size, the closer the observed percentages can be expected to match the values predicted by the experimental hypothesis, *if the hypothesis is correct.* The chi-square test is therefore always calculated with numbers—actual data—and not percentages or proportions.

The chi-square test cannot prove a hypothesis, but it can allow researchers to reject a hypothesis. For this reason, a critical prerequisite of the chi-square test is the framing of a **null hypothesis** "that genes *A* and *B* are *not* linked": a model that might possibly be refuted by the test and that leads to clear-cut numerical predictions. Although contemporary geneticists use the chi-square test to interpret many kinds of genetic experiments, they use it most often to discover whether data obtained from breeding experiments provide evidence for or against the **alternative hypothesis** that two genes are linked. But the problem with the alternative hypothesis that "genes *A* and *B are* linked" is that there is no precise prediction of what to expect in terms of breeding data. The reason is that the frequency of recombinations, as we have seen, varies with each linked gene pair.

In contrast, the null hypothesis "that genes *A* and *B* are *not* linked" gives rise to a precise prediction: that alleles at different genes will assort independently and produce 50 percent parental and 50 percent recombinant progeny. So, whenever a geneticist wants to determine whether two genes are linked, he or she actually tests whether the observed data are consistent with the null hypothesis of no linkage. If the chi-square test shows that the observed data differ significantly from those expected with independent assortment—that is, they differ enough not to be reasonably attributable to chance alone—then the researcher can reject the null hypothesis of no linkage and accept the alternative of linkage between the two genes.

The **Tools of Genetics** box in this chapter presents the general protocol of the chi-square test. The final result of the calculations is the determination of the numerical *probability*—the *p* value—that a particular set of observed experimental results represents a chance deviation from the values predicted by a particular hypothesis. If the probability is high, it is likely that the hypothesis being tested explains the data, and the observed deviation from expected results is considered *insignificant*. If the probability is very low, the observed deviation from expected results becomes *significant*. When this happens, it is unlikely that the hypothesis under consideration explains the data, and the hypothesis can be rejected.

The general protocol for using the chi-square test and evaluating its results can be stated in a series of steps. Two preparatory steps precede the actual chi-square calculation.

1. Use the data obtained from a breeding experiment to answer the following questions:
   a. What is the *total number* of offspring (events) analyzed?
   b. How many different *classes* of offspring (events) are there?
   c. In each class, what is the *number* of offspring (events) *observed*?

2. Calculate how many offspring (events) would be expected for each class if the null hypothesis (here, no linkage) were correct: Multiply the percentage predicted by the null hypothesis (here, 50 percent parentals and 50 percent recombinants) by the total number of offspring. You are now ready for the chi-square calculation.

3. To calculate chi square, begin with one class of offspring. Subtract the expected number from the observed number to obtain the deviation from the predicted value for the class. Square the result, and divide this value by the expected number.

   Do this for all classes and then sum the individual results. The final result is the chi-square (χ^2) value. This step is summarized by the equation

   $$\chi^2 = \Sigma \frac{(Number\ observed - Number\ expected)^2}{(Number\ expected)}$$

   where Σ means "sum of all classes."

4. Next, you consider the **degrees of freedom (df)**. The df is a measure of the number of independently varying parameters in the experiment (see text). The value of degrees of freedom is one less than the number of classes. Thus, if N is the number of classes, then the degrees of freedom (df) = $N - 1$. If there are 4 classes, then there are 3 df.

5. Use the chi-square value together with the df to determine a **p value**: the probability that a deviation from the predicted numbers at least as large as that observed in the experiment would occur by chance. Although the p value is arrived at through a numerical analysis, geneticists routinely determine the value by a quick search through a table of critical χ^2 values for different degrees of freedom, such as **Table 4.1**.

6. Evaluate the significance of the p value. You can think of the p value as the probability that the null hypothesis is true. A value greater than 0.05 indicates that in more than 1 in 20 (or more than 5 percent) repetitions of an experiment of the same size, the observed deviation from predicted values could have been obtained by chance, even if the null hypothesis is actually true; the data are therefore *not significant* for rejecting the null hypothesis. Statisticians have arbitrarily selected the p value of 0.05 as the boundary between accepting and rejecting the null hypothesis. A p value of less than 0.05 means that you can consider the deviation to be *significant*, and you can reject the null hypothesis.

7. Two types of mistakes can be made when rejecting and accepting a null hypothesis. A Type I error occurs when a null hypothesis is falsely rejected, while a Type II error occurs when a null hypothesis is incorrectly accepted. The probability of a Type I error occurring is equivalent to the established p value (less than 5 percent). Scientists can minimize the probability of Type I errors occurring by setting a lower p value, while higher p values can decrease the chances of Type II errors.

Table 4.1	Critical Chi-Square Values						
	p Values						
	Cannot Reject the Null Hypothesis				**Null Hypothesis Rejected**		
	0.99	0.90	0.50	0.10	0.05	0.01	0.001
Degrees of Freedom				χ^2 **Values**			
1	—	0.02	0.45	2.71	3.84	6.64	10.83
2	0.02	0.21	1.39	4.61	5.99	9.21	13.82
3	0.11	0.58	2.37	6.25	7.81	11.35	16.27
4	0.30	1.06	3.36	7.78	9.49	13.28	18.47
5	0.55	1.61	4.35	9.24	11.07	15.09	20.52

Note: χ^2 values that lie in the yellow region of this table allow you to reject the null hypothesis with greater than 95 percent confidence, and for recombination experiments, to postulate linkage.

Applying the chi-square test to linkage analysis: An example

Figure 4.6 depicts two sets of data obtained from testcross experiments asking whether genes *A* and *B* are linked. We first apply the chi-square analysis to data accumulated in the first experiment. The total number of offspring is 50, of which 31 (i.e., 17 + 14) are observed to be parental types and 19 (i.e., 8 + 11) are recombinant types. Dividing 50 by 2, you get 25, the number of parental or recombinant offspring expected according to the null hypothesis of independent assortment (which predicts that parentals = recombinants).

Now, considering first the parental types alone, you square the observed deviation from the expected value, and divide the result by the expected value. After doing the same for the recombinant types, you add the two quotients to obtain the value of chi square.

$$\chi^2 = \frac{(31 - 25)^2}{25} + \frac{(19 - 25)^2}{25} = 1.44 + 1.44 = 2.88$$

You next determine the **degrees of freedom (df)** for this experiment. Degrees of freedom is a mathematical concept that takes into consideration the number of independently varying parameters. For example, if the offspring in an experiment fall into four classes, and you know the total number of offspring as well as the numbers present in three of the classes, then you can directly calculate the number present in the fourth class. Therefore, the degrees of freedom with four classes is one less than the number of classes, or three. When assessing linkage of two genes, there are two classes (parentals and recombinants). Thus, the number of degrees of freedom is 1, and you scan the chi-square table (see Table 4.1) for $\chi^2 = 2.88$ and df = 1. You find by extrapolation that the corresponding *p* value is greater than 0.05 (roughly 0.09). From this

p value you can conclude that it is not possible to reject the null hypothesis on the basis of this experiment, which means that this data set is not sufficient to demonstrate linkage between *A* and *B*.

If you use the same strategy to calculate a *p* value for the data observed in the second experiment, where there are a total of 100 offspring and thus an expected number of 50 parentals and 50 recombinants, you get

$$\chi^2 = \frac{(62 - 50)^2}{50} + \frac{(38 - 50)^2}{50} = 2.88 + 2.88 = 5.76$$

The number of degrees of freedom remains 1, so Table 4.1 arrives at a *p* value greater than 0.01 but less than 0.05. In this case, you can consider the difference between the observed and expected values to be significant. As a result, you can reject the null hypothesis of independent assortment and conclude it is likely that genes *A* and *B* are linked.

Statisticians have arbitrarily selected a *p* value of 0.05 as the boundary between significance and nonsignificance. Values lower than this indicate that there would be less than 5 chances in 100 of obtaining the same results by random sampling if the null hypothesis were true. A *p* value of less than 0.05 thus suggests that the data show major deviations from predicted values significant enough to reject the null hypothesis with greater than 95 percent confidence. More conservative scientists often set the boundary of significance at *p* = 0.01, and they would therefore reject the null hypothesis only if their confidence was greater than 99 percent.

In contrast, *p* values greater than 0.01 or 0.05 do not necessarily mean that two genes are not linked; it may mean only that the sample size is not large enough to provide an answer. With more data, the *p* value normally rises if the null hypothesis of no linkage is correct and falls if there is, in fact, linkage.

Note that in Figure 4.6 all of the numbers in the second set of data are simply double the numbers in the first set, with the percentages remaining the same. Thus, just by doubling the sample size from 50 to 100 individuals, it was possible to go from no significant difference to a significant difference between the observed and the expected values. In other words, the larger the sample size, the less the likelihood that a certain percentage deviation from expected results happened simply by chance. Bearing this in mind, you can see that it is not appropriate to use the chi-square test when analyzing very small samples of less than 10. This creates a problem for human geneticists, because human families produce only a small number of children. To achieve a reasonable sample size for linkage studies in humans, scientists must instead pool data from a large number of family pedigrees.

The chi-square test *does not* prove linkage or its absence. What it *does* do is provide a quantitative measure of the likelihood that the data from an experiment can be

Progeny	Experiment 1	Experiment 2
A B	17	34
a b	14	28
A b	8	16
a B	11	22
Total	50	100

Class	Observed / Expected		Observed / Expected	
Parentals	31	25	62	50
Recombinants	19	25	38	50

Figure 4.6 Applying the chi-square test to see if genes *A* and *B* are linked. The null hypothesis is that the two genes are not linked. For Experiment 1, *p* > 0.05, so it is not possible to reject the null hypothesis. For Experiment 2, with a data set twice the size, *p* < 0.05. Based on this latter result, most geneticists would reject the null hypothesis and conclude with greater than 95 percent confidence that the genes are linked.

explained by a particular hypothesis. The chi-square analysis is thus a general statistical test for significance; it can be used with many different experimental designs and with hypotheses other than the absence of linkage. As long as it is possible to propose a null hypothesis that leads to a predicted set of values for a defined set of data classes, you can readily determine whether or not the observed data are consistent with the hypothesis.

When experiments lead to rejection of a null hypothesis, you may need to confirm an alternative. For instance, if you are testing whether two opposing traits result from the segregation of two alleles of a single gene, you would expect a testcross between an F_1 heterozygote and a recessive homozygote to produce a 1:1 ratio of the two traits in the offspring. If instead, you observe a ratio of 6:4 and the chi-square test produces a p value of 0.009, you can reject the null hypothesis. But you are still left with the question of what the absence of a 1:1 ratio means. There are actually two alternatives: (1) Individuals with the two possible genotypes are not equally viable, *or* (2) more than one gene encodes the trait. The chi-square test cannot tell you which possibility is correct, and you would have to study the matter further. The problems at the end of this chapter illustrate several applications of the chi-square test relevant to genetics.

> Geneticists use the chi-square test to evaluate the probability that differences between predicted results and observed results are due to random sampling error. For linkage analysis, p values of less than 0.05 allow rejection of the null hypothesis that the two genes are not linked.

4.3 Recombination: A Result of Crossing-Over During Meiosis

It is easy to understand how genes that are physically connected on the same chromosome can be transmitted together and thus show genetic linkage. It is not as obvious why all linked genes always show some recombination in a sample population of sufficient size. Do the chromosomes participate in a physical process that gives rise to the reshuffling of linked genes that we call recombination? The answer to this question is of more than passing interest as it provides a basis for gauging relative distances between pairs of genes on a chromosome.

In 1909, the Belgian cytologist Frans Janssens described structures he had observed under the light microscope during prophase of the first meiotic division. He called these structures *chiasmata*; as described in Chapter 3, they seemed to represent regions in which nonsister chromatids of homologous chromosomes cross over each other (review Figure 3.13 in Chapter 3; see **Figure 4.7**). Making inferences from a combination of genetic and cytological data, Thomas Hunt Morgan suggested that the chiasmata observed through the light microscope were sites of chromosome breakage and exchange, resulting in genetic recombination.

Reciprocal exchanges between homologues are the physical basis of recombination

Morgan's idea that the physical breaking and rejoining of chromosomes during meiosis was the basis of genetic recombination seemed reasonable. But although Janssens's chiasmata could be interpreted as signs of the process, before 1930 no one had produced visible evidence that crossing-over between homologous chromosomes actually occurs. The identification of **physical markers**, or cytologically visible abnormalities that make it possible to keep track of specific chromosome parts from one generation to the next, enabled researchers to turn the logical deductions about recombination into facts derived from experimental evidence. In 1931, Harriet Creighton and Barbara McClintock, who studied corn, and Curt Stern, who worked with *Drosophila*, published the results of experiments showing that genetic recombination indeed depends on the reciprocal exchange of parts between maternal and paternal chromosomes. Stern, for example, bred female flies with two different X chromosomes, each containing a distinct physical marker near one of the ends. These same females were also dihybrid for two X-linked **genetic markers**—genes that could serve as points of reference in determining whether particular progeny were the result of recombination.

Figure 4.8 diagrams the chromosomes of these heterozygous females. One X chromosome carried mutations producing carnation eyes (a dark ruby colour, abbreviated *car*) that were kidney-shaped (*Bar*); in addition, this chromosome was marked physically by a visible discontinuity, which resulted when the end of the X chromosome was broken off and attached to an autosome. The other X chromosome had wild-type alleles (+) for both the *car* and

Figure 4.7 Multiple chiasmata are a frequent occurrence in numerous organisms. In this photomicrograph, five chiasmata are observed in a tetrad (or bivalent) of the grasshopper *Chorthippus parallelus*.

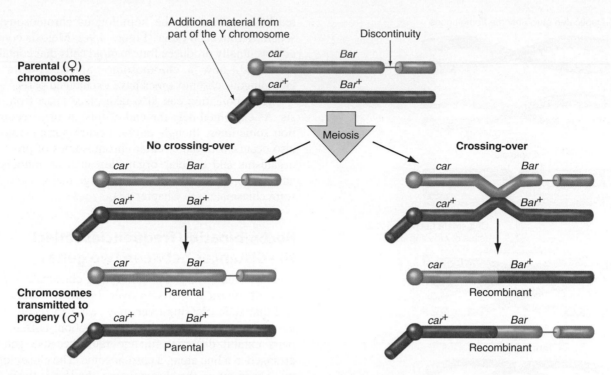

Figure 4.8 Evidence that recombination results from reciprocal exchanges between homologous chromosomes. Genetic recombination between the *car* and *Bar* genes on the *Drosophila* X chromosome is accompanied by the exchange of physical markers observable through the microscope. Note that this depiction of crossing-over is a simplification, as genetic recombination actually occurs after each chromosome has replicated into sister chromatids. Note also that the piece of the X chromosome to the right of the discontinuity is actually attached to an autosome.

the *Bar* genes, and its physical marker consisted of part of the Y chromosome that had become connected to the X-chromosome centromere.

Figure 4.8 illustrates how the chromosomes in these *car Bar/car⁺ Bar⁺* females were transmitted to male progeny. According to the experimental results, all sons showing a phenotype determined by one or the other parental combination of genes (either *car Bar* or *car⁺ Bar⁺*) had an X chromosome that was structurally indistinguishable from one of the original X chromosomes in the mother. In recombinant sons, however, such as those that manifested carnation eye colour and normal eye shape (*car Bar⁺/Y*), an identifiable exchange of the abnormal features marking the ends of the homologous X chromosomes accompanied the recombination of genes. The evidence thus tied an instance of phenotypic recombination to the crossing-over of particular genes located in specifically marked parts of particular chromosomes. This experiment elegantly demonstrated that genetic recombination is associated with the actual reciprocal exchange of segments between homologous chromosomes during meiosis.

Chiasmata mark the sites of recombination

Figure 4.9 outlines what is currently known about the steps of recombination as they appear in chromosomes viewed through the light microscope. Although this low-resolution view may not represent certain details of recombination with complete accuracy, it nonetheless provides a useful frame of reference. In Figure 4.9a, the two homologues of each chromosome pair have already replicated, so there are now two pairs of sister chromatids or a total of four chromatids within each bivalent. In Figure 4.9b, the synaptonemal complex zips together homologous chromosome pairs along their length. The synaptonemal zipper aligns homologous regions of all four chromatids such that allelic DNA sequences are physically near each other (review Figure 3.13b for a detailed depiction). This proximity facilitates crossing-over between homologous sequences; as we will see in Chapter 6, the biochemical mechanism of recombination requires a close interaction of DNAs on homologous chromosomes that have identical, or nearly identical, nucleotide sequences.

In Figure 4.9c, the synaptonemal complex begins to disassemble. Although at least some steps of the recombination process occurred while the chromatids were zipped in synapsis, it is only now that the recombination event becomes apparent. As the zipper dissolves, homologous chromosomes remain attached at chiasmata, the actual sites of crossing-over. Visible under the light microscope, chiasmata indicate where chromatid sections have switched from one molecule to another. In Figure 4.9d, during anaphase I, as the two homologues separate, starting at their centromeres, the ends of the two recombined chromatids pull free of their respective sister chromatids,

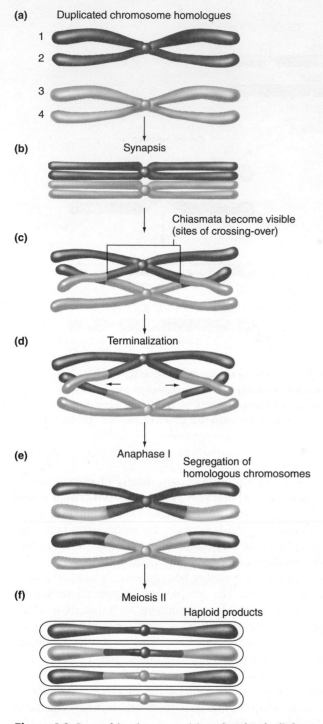

(a) Duplicated chromosome homologues

1
2

3
4

(b) Synapsis

Chiasmata become visible
(sites of crossing-over)

(c)

(d) Terminalization

(e) Anaphase I

Segregation of
homologous chromosomes

(f) Meiosis II

Haploid products

Figure 4.9 Recombination as envisioned under the light microscope. (a) A pair of duplicated homologous chromosomes very early in prophase of meiosis I. **(b)** During leptotene and zygotene of prophase I, the synaptonemal complex helps align corresponding regions of homologous chromosomes, allowing recombination. **(c)** As the synaptonemal complex disassembles during diplotene, homologous chromosomes remain attached at chiasmata. **(d)** and **(e)** The chiasmata terminalize (move toward the chromosome ends), allowing the recombined chromosomes to separate during anaphase and telophase. **(f)** The result of the process is recombinant gametes.

and the chiasmata shift from their original positions toward a chromosome end, or telomere. This movement of chiasmata is known as **terminalization**. When the chiasmata

reach the telomeres, the homologous chromosomes can separate from each other (Figure 4.9e). Meiosis continues and eventually produces four haploid cells that contain one chromatid—now a chromosome—apiece (Figure 4.9f). Homologous chromosomes have exchanged parts.

Recombination can also take place apart from meiosis. As explained near the end of this chapter, recombination sometimes, though rarely, occurs during mitosis. It also occurs with the circular chromosomes of prokaryotic organisms and cellular organelles such as mitochondria and chloroplasts, which do not undergo meiosis and do not form chiasmata (see Chapter 18).

Recombination frequencies reflect the distances between two genes

Thomas Hunt Morgan's belief that chiasmata represent sites of physical crossing-over between chromosomes and that such crossing-over may result in recombination led him to the following logical deduction: Different gene pairs exhibit different linkage rates because genes are arranged in a line along a chromosome. The closer together two genes are on the chromosome, the less is their chance of being separated by an event that cuts and recombines the line of genes. To look at it another way, if we assume for the moment that chiasmata can form anywhere along a chromosome with equal likelihood, then the probability of a crossover occurring between two genes increases with the distance separating them. If this is so, the frequency of genetic recombination also must increase with the distance between genes. To illustrate the point, imagine pinning to a wall 10 inches (25.4 cm) of ribbon with a line of tiny black dots along its length and then repeatedly throwing a dart to see where you will cut the ribbon. You would find that practically every throw of the dart separates a dot at one end of the ribbon from a dot at the other end, while few if any throws separate any two particular dots positioned right next to each other.

Alfred H. Sturtevant, one of Morgan's students, took this idea one step further. He proposed that the percentage of total progeny that were recombinant types, the **recombination frequency (RF)**, could be used as a gauge of the physical distance separating any two genes on the same chromosome. Sturtevant arbitrarily defined one RF percentage point as the unit of measure along a chromosome; later, another geneticist named the unit a **centimorgan (cM)** after T. H. Morgan. Mappers often refer to a centimorgan as a **map unit (m.u.)**. Although the two terms are interchangeable, researchers prefer one or the other, depending on their experimental organism. *Drosophila* geneticists, for example, use map units while human geneticists use centimorgans. In Sturtevant's system, 1% RF = 1 cM = 1 m.u. A review of the two pairs of X-linked *Drosophila* genes we analyzed earlier shows how his proposal works. Since the X-linked genes for eye colour (*w*) and body colour (*y*) recombine in 1.1 percent of F$_2$ progeny, they are 1.1 m.u. apart (**Figure 4.10a**). In contrast, the X-linked genes for eye colour (*w*) and wing

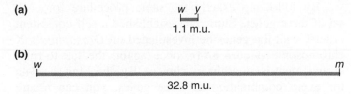

(a)

w y

1.1 m.u.

(b)

w m

32.8 m.u.

Figure 4.10 Recombination frequencies are the basis of genetic maps. (a) 1.1 percent of the gametes produced by a female dihybrid for the genes w and y are recombinant. The recombination frequency (RF) is thus 1.1 percent, and the genes are approximately 1.1 map units (m.u.) or 1.1 centimorgans (cM) apart. **(b)** The distance between the w and m genes is longer: 32.8 m.u. (or 32.8 cM).

Table 4.2	Properties of Linked Versus Unlinked Genes
Linked Genes	
Parentals > recombinants (RF < 50%)	
Linked genes must be syntenic and sufficiently close together on the same chromosome so that they do not assort independently.	
Unlinked Genes	
Parentals = recombinants (RF = 50%)	
Occurs either when genes are on different chromosomes or when they are sufficiently far apart on the same chromosome.	

size (m) have a recombination frequency of 32.8 and are therefore 32.8 m.u. apart (**Figure 4.10b**).

As a unit of measure, the map unit is simply an index of recombination probabilities assumed to reflect distances between genes. According to this index, the y and w genes are much closer together than the m and w genes. Geneticists have used this logic to map thousands of genetic markers to the chromosomes of *Drosophila*, building recombination maps step-by-step with closely linked markers. And as we see next, they have learned that genes very far apart on the same chromosome may appear not to be linked, even though their recombination distances relative to closely linked intervening markers confirm that the genes are indeed on the same chromosome.

Recombination frequencies between two genes never exceed 50 percent

If the definition of linkage is that the proportion of recombinant classes is less than that of parental classes, a recombination frequency of less than 50 percent indicates linkage. But what can we conclude about the relative location of genes if there are roughly equal numbers of parental and recombinant progeny? And does it ever happen that recombinants are in the majority?

We already know one situation that can give rise to a recombination frequency of 50 percent. Genes located on different (i.e., nonhomologous) chromosomes will obey Mendel's law of independent assortment because the two chromosomes can line up on the spindle during meiosis I in either of two equally likely configurations (review Figure 3.16a). A dihybrid for these two genes will thus produce all four possible types of gametes (*AB*, *Ab*, *aB*, and *ab*) with approximately equal frequency. Importantly, experiments have established that genes located very far apart on the same chromosome also show recombination frequencies of approximately 50 percent.

Researchers have never observed statistically significant recombination frequencies between two genes greater than 50 percent, which means that in any cross following two genes, recombinant types are never in the majority. As we explain in more detail later in the chapter, this upper limit of 50 percent on the recombination frequency between two genes results from two aspects of chromosome behaviour during meiosis I. First, multiple crossovers can occur between two genes if they are far apart on the same chromosome; and second, recombination takes place after the chromosomes have replicated into sister chromatids.

For now, simply note that recombination frequencies near 50 percent suggest either that two genes are on different chromosomes or that they lie far apart on the same chromosome. The only way to tell whether the two genes are syntenic (i.e., on the same chromosome) is through a series of matings showing definite linkage with other genes that lie between them. In short, even though crosses between two genes lying very far apart on a chromosome may show no linkage at all (because recombinant and parental classes are equal), you can demonstrate they are on the same chromosome if you can tie each of the widely separated genes to one or more common intermediaries. **Table 4.2** summarizes the relationship between the relative locations of two genes and the presence or absence of linkage as measured by recombination frequencies.

Recombination results from crossing-over of homologues during meiosis I. If two syntenic genes are close together, little chance exists for crossing-over, so the recombination frequency is low. As the distance between syntenic genes increases, the RF increases to a maximum of 50 percent. Thus, genes far enough apart on a single chromosome assort independently, just as do genes on nonhomologous chromosomes.

4.4   Mapping: Locating Genes Along a Chromosome

Maps are images of the relative positions of objects in space. Whether depicting the floor plan of Toronto's Royal Ontario Museum, the layout of the Roman Forum, or the location of cities served by the railways of Europe, maps turn measurements into patterns of spatial relationships that add a new level of meaning to the original data of distances. Maps

that assign genes to locations on particular chromosomes called *loci* (singular **locus**) are no exception. By transforming genetic data into spatial arrangements, maps sharpen our ability to predict the inheritance patterns of specific traits.

We have seen that recombination frequency (RF) is a measure of the distance separating two genes along a chromosome. We now examine how data from many crosses following two and three genes at a time can be compiled and compared to generate accurate, comprehensive gene/chromosome maps.

Comparisons of two-point crosses establish relative gene positions

In his senior undergraduate thesis, Morgan's student A. H. Sturtevant asked whether data obtained from a large number of two-point crosses (crosses tracing two genes at a time) would support the idea that genes form a definite linear series along a chromosome. Sturtevant began by looking at X-linked genes in *Drosophila*. **Figure 4.11a** lists his recombination data for several two-point crosses. Recall that the distance between two genes that yields 1 percent recombinant progeny—an RF of 1 percent—is 1 m.u.

As an example of Sturtevant's reasoning, consider the three genes *w*, *y*, and *m*. If these genes are arranged in a line (instead of a more complicated branched structure, for example), then one of them must be in the middle, flanked on either side by the other two. The greatest genetic distance should separate the two genes on the outside, and this value should roughly equal the sum of the distances separating the middle gene from each outside gene. The data Sturtevant obtained are consistent with this idea, implying that *w* lies between *y* and *m* (**Figure 4.11b**). Note that the left-to-right orientation of this map was selected at random; the map in Figure 4.11b would be equally correct if it portrayed *y* on the right and *m* on the left.

By following exactly the same procedure for each set of three genes, Sturtevant established a self-consistent order for all the genes he investigated on *Drosophila*'s X chromosome (**Figure 4.11c**; once again, the left-to-right arrangement is an arbitrary choice). By checking the data for every combination of three genes, you can assure yourself that this ordering makes sense. The fact that the recombination data yield a simple linear map of gene position supports the idea that genes reside in a unique linear order along a chromosome.

Limitations of two-point crosses

Though of great importance, the pairwise mapping of genes has several shortcomings that limit its usefulness. First, in crosses involving only two genes at a time, it may be difficult to determine gene order if some gene pairs lie very close together. For example, in mapping *y*, *w*, and *m*, 34.3 m.u. separate the outside genes *y* and *m*, while nearly as great a distance (32.8 m.u.) separates the middle *w* from the outside *m* (Figure 4.11b). Before being able to conclude with any confidence that *y* and *m* are truly farther apart, that is, that the small difference between the values of 34.3 and 32.8 is not the result of sampling error, you would have to examine a very large number of flies and subject the data to a statistical test, such as the chi-square test.

A second problem with Sturtevant's mapping procedure is that the actual distances in his map do not always add up, even approximately. As an example, suppose that the locus of the *y* gene at the far left of the map is regarded as position 0 (Figure 4.11c). The *w* gene would then lie near position 1, and *m* would be located in the vicinity of 34 m.u. But what about the *r* gene, named for a mutation that produces rudimentary (very small) wings? Based solely on its distance from *y*, as inferred from the $y \leftrightarrow r$ data in Figure 4.11a, we would place it at position 42.9 (Figure 4.11c). However, if we calculate its position as the sum of all intervening distances inferred from the data in Figure 4.11a, that is, as the sum of $y \leftrightarrow w$ plus $w \leftrightarrow v$ plus $v \leftrightarrow m$ plus $m \leftrightarrow r$, the locus of *r* becomes $1.1 + 32.1 + 4.0 + 17.8 = 55.0$ (**Figure 4.11d**). What can explain this difference, and which of these two values is closer to the truth? Three-point crosses help provide some of the answers.

Three-point crosses provide faster and more accurate mapping

The simultaneous analysis of three markers makes it possible to obtain enough information to position the three genes in relation to each other from just one set of crosses. To describe this procedure, we look at three genes linked on one of *Drosophila*'s autosomes.

A homozygous female with mutations for vestigial wings (*vg*), black body (*b*), and purple eye colour (*pr*) was mated to a wild-type male (**Figure 4.12a**). All the trihybrid F_1 progeny, both male and female, had normal phenotypes for the three characteristics, indicating that the mutations are autosomal recessive. In a testcross of the F_1 females with males having vestigial wings, black body,

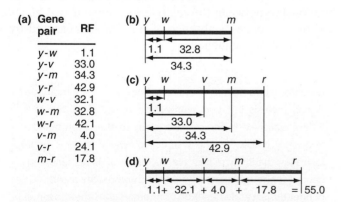

Figure 4.11 Mapping genes by comparisons of two-point crosses. **(a)** Sturtevant's data for the distances between pairs of X-linked genes in *Drosophila*. **(b)** Because the distance between *y* and *m* is greater than the distance between *w* and *m*, the order of genes must be y-w-m. **(c)** and **(d)** Maps for five genes on the *Drosophila* X chromosome. The left-to-right orientation is arbitrary. Note that the numerical position of the *r* gene depends on how it is calculated. The best genetic maps are obtained by summing many small intervening distances as in (d).

(a) Three-point cross results

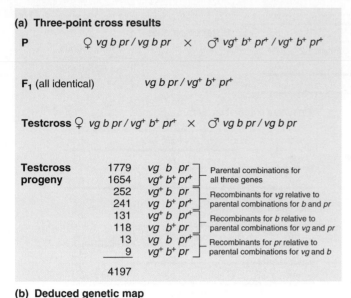

P $♀\ vg\ b\ pr\ /\ vg\ b\ pr$ × $♂\ vg^+\ b^+\ pr^+\ /\ vg^+\ b^+\ pr^+$

F₁ (all identical) $vg\ b\ pr\ /\ vg^+\ b^+\ pr^+$

Testcross $♀\ vg\ b\ pr\ /\ vg^+\ b^+\ pr^+$ × $♂\ vg\ b\ pr\ /\ vg\ b\ pr$

Testcross progeny			
1779	$vg\ b\ pr$	}	Parental combinations for all three genes
1654	$vg^+\ b^+\ pr^+$		
252	$vg^+\ b\ pr$	}	Recombinants for vg relative to parental combinations for b and pr
241	$vg\ b^+\ pr^+$		
131	$vg^+\ b\ pr^+$	}	Recombinants for b relative to parental combinations for vg and pr
118	$vg\ b^+\ pr$		
13	$vg\ b\ pr^+$	}	Recombinants for pr relative to parental combinations for vg and b
9	$vg^+\ b^+\ pr$		
4197			

(b) Deduced genetic map

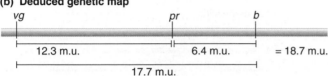

vg — 12.3 m.u. — pr — 6.4 m.u. — b = 18.7 m.u.

17.7 m.u.

Figure 4.12 Analyzing the results of a three-point cross.
(a) Results from a three-point testcross of F₁ females simultaneously heterozygous for vg, b, and pr. **(b)** The gene in the middle must be pr because the longest distance is between the other two genes: vg and b. The most accurate map distances are calculated by summing shorter intervening distances, so 18.7 m.u. is a more accurate estimate of the genetic distance between vg and b than 17.7 m.u.

and purple eyes, the progeny were of eight different phenotypes reflecting eight different genotypes. The order in which the genes in each phenotypic class are listed in Figure 4.12a is completely arbitrary. Thus, instead of $vg\ b\ pr$, one could write $b\ vg\ pr$ or $vg\ pr\ b$ to indicate the same genotype. Remember that at the outset, we do not know the gene order; deducing it is the goal of the mapping study.

In analyzing the data, we look at two genes at a time (recall that the recombination frequency is always a function of a pair of genes). For the pair vg and b, the parental combinations are $vg\ b$ and $vg^+\ b^+$; the nonparental recombinants are $vg\ b^+$ and $vg^+\ b$. To determine whether a particular class of progeny is parental or recombinant for vg and b, we do not care whether the flies are pr or $pr^{\,\text{I}}$. Thus, to the nearest tenth of a map unit, the $vg \leftrightarrow b$ distance, calculated as the percentage of recombinants in the total number of progeny, is

$$\frac{252 + 241 + 131 + 118}{4197} \times 100$$

$$= 17.7 \text{ m.u. } (vg \leftrightarrow b \text{ distance})$$

Similarly, because recombinants for the vg–pr gene pair are $vg\ pr^+$ and $vg^+\ pr$, the interval between these two genes is

$$\frac{252 + 241 + 13 + 9}{4197} \times 100$$

$$= 12.3 \text{ m.u. } (vg \leftrightarrow pr \text{ distance})$$

while the distance separating the b–pr pair is

$$\frac{131 + 118 + 13 + 9}{4197} \times 100$$

$$= 6.4 \text{ m.u. } (b \leftrightarrow pr \text{ distance})$$

These recombination frequencies show that vg and b are separated by the largest distance (17.7 m.u., as compared with 12.3 and 6.4) and must therefore be the outside genes, flanking pr in the middle (**Figure 4.12b**). But as with the X-linked y and r genes analyzed by Sturtevant, the distance separating the outside vg and b genes (17.7) does not equal the sum of the two intervening distances (12.3 + 6.4 = 18.7). In the next section, we learn that the reason for this discrepancy is the rare occurrence of double crossovers.

Correction for double crossovers

Figure 4.13 depicts the homologous autosomes of the F₁ females that are heterozygous for the three genes vg, pr, and b. A close examination of the chromosomes reveals the kinds of crossovers that must have occurred to generate the classes and numbers of progeny observed. In this and subsequent figures, the chromosomes depicted are in late prophase/early metaphase of meiosis I, when there are four chromatids for each pair of homologous chromosomes. As we have suggested previously and demonstrate more rigorously later, prophase I is the stage at which recombination takes place. Note that we call the space between vg and pr "region 1" and the space between pr and b "region 2."

Recall that the progeny from the testcross performed earlier fall into eight groups (review Figure 4.12). Flies in the two largest groups carry the same configurations of genes as did their grandparents of the P generation: $vg\ b\ pr$ and $vg^+\ b^+\ pr^+$; they thus represent the parental classes (Figure 4.13a). The next two groups—$vg^+\ b\ pr$ and $vg\ b^+\ pr^+$—are composed of recombinants that must be the reciprocal products of a crossover in region 1 between vg and pr (Figure 4.13b). Similarly the two groups containing $vg^+\ b\ pr^+$ and $vg\ b^+\ pr$ flies must have resulted from recombination in region 2 between pr and b (Figure 4.13c).

But what about the two smallest groups made up of rare $vg\ b\ pr^+$ and $vg^+\ b^+\ pr$ recombinants? What kinds of chromosome exchange could account for them? Most likely, they result from two different crossover events occurring simultaneously; one in region 1, the other in region 2 (Figure 4.13d). The gametes produced by such double crossovers still have the parental configuration for the outside genes vg and b, even though not one but two exchanges must have occurred.

Because of the existence of double crossovers, the $vg \leftrightarrow b$ distance of 17.7 m.u. calculated in the previous section does not reflect all of the recombination events producing the gametes that gave rise to the observed progeny. To correct for this oversight, it is necessary to adjust the recombination frequency by adding the double crossovers twice, because each individual in the double crossover

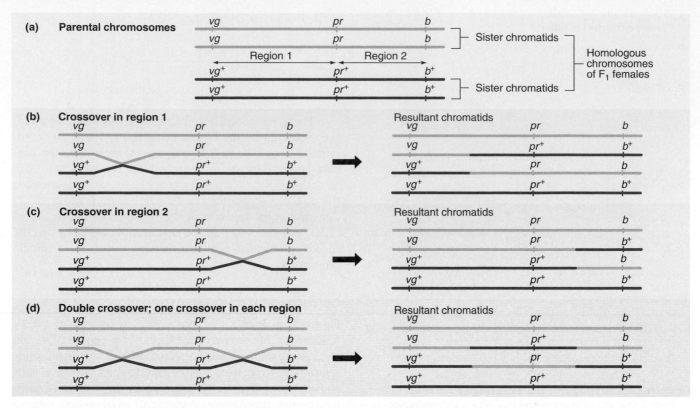

Figure 4.13 Inferring the location of a crossover event. Once you establish the order of genes involved in a three-point cross, it is easy to determine which crossover events gave rise to particular recombinant gametes. Note that double crossovers are needed to generate gametes in which the gene in the middle has recombined relative to the parental combinations for the genes at the ends.

groups is the result of two exchanges between *vg* and *b*. The corrected distance is

$$\frac{252 + 241 + 131 + 118 + 13 + 13 + 9 + 9}{4197} \times 100$$

$$= 18.7 \text{ m.u.}$$

This value makes sense because you have accounted for all of the crossovers that occur in region 1 as well as all of the crossovers in region 2. As a result, the corrected value of 18.7 m.u. for the distance between *vg* and *b* is now exactly the same as the sum of the distances between *vg* and *pr* (region 1) and between *pr* and *b* (region 2).

As previously discussed, when Sturtevant originally mapped several X-linked genes in *Drosophila* by two-point crosses, the locus of the rudimentary wings (*r*) gene was ambiguous. A two-point cross involving *y* and *r* gave a recombination frequency of 42.9, but the sum of all the intervening distances was 55.0 (review Figure 4.11). This discrepancy occurred because the two-point cross ignored double crossovers that might have occurred in the large interval between the *y* and *r* genes. The data summing the smaller intervening distances accounted for at least some of these double crossovers by catching recombinations of gene pairs between *y* and *r*. Moreover, each smaller distance is less likely to encompass a double crossover than a larger distance, so each number for a smaller distance is inherently more accurate.

Note that even a three-point cross like the one for *vg*, *pr*, and *b* ignores the possibility of two recombination

events taking place in, say, region 1. For greatest accuracy, it is always best to construct a map using many genes separated by relatively short distances.

Interference: Fewer double crossovers than expected

In a three-point cross following three linked genes, of the eight possible genotypic classes, the two parental classes contain the largest number of progeny, while the two double recombinant classes, resulting from double crossovers, are always the smallest (see Figure 4.12). We can understand why double crossover progeny are the rarest by looking at the probability of their occurrence. If an exchange in region 1 of a chromosome does not affect the probability of an exchange in region 2, the probability that both will occur simultaneously is the product of their separate probabilities (recall the product rule in Chapter 2, Section 2.2). For example, if progeny resulting from recombination in region 1 alone account for 10 percent of the total progeny (i.e., if region 1 is 10 m.u.) and progeny resulting from recombination in region 2 alone account for 20 percent, then the probability of a double crossover (one event in region 1, the second in region 2) is $0.10 \times 0.20 = 0.02$, or 2 percent. This makes sense because the likelihood of two rare events occurring simultaneously is even less than that of either rare event occurring alone.

If there are eight classes of progeny in a three-point cross, the two classes containing the fewest progeny must have arisen from double crossovers. The numerical

frequencies of observed double crossovers, however, almost never coincide with expectations derived from the product rule. Let us look at the actual numbers from the cross we have been discussing. The probability of a single crossover between vg and pr is 0.123 (corresponding to 12.3 m.u.), and the probability of a single crossover between pr and b is 0.064 (6.4 m.u.). The product of these probabilities is

$$0.123 \times 0.064 = 0.0079 = 0.79\%$$

But the observed proportion of double crossovers (see Figure 4.12) was

$$\frac{13 + 9}{4197} \times 100 = 0.52\%$$

The fact that the number of observed double crossovers is less than the number expected if the two exchanges are independent events suggests that the occurrence of one crossover reduces the likelihood that another crossover will occur in an adjacent part of the chromosome. This phenomenon—of crossovers not occurring independently—is called **chromosomal interference**.

Interference may exist to ensure that every pair of homologous chromosomes undergoes at least one crossover event. It is critical that every pair of homologous chromosomes sustain one or more crossover events because such events help the chromosomes orient properly at the metaphase plate during the first meiotic division. Indeed, homologous chromosome pairs without crossovers often segregate improperly. If only a limited number of crossovers can occur during each meiosis and interference lowers the number of crossovers on large chromosomes, then the remaining possible crossovers are more likely to occur on small chromosomes. This increases the probability that at least one crossover will take place on every homologous pair. Though the molecular mechanism underlying interference is not yet clear, recent experiments suggest that interference is mediated by the synaptonemal complex.

Interference is not uniform and may vary even for different regions of the same chromosome. Investigators can obtain a quantitative measure of the amount of interference in different chromosomal intervals by first calculating a **coefficient of coincidence**, defined as the ratio between the actual frequency of double crossovers observed in an experiment and the number of double crossovers expected on the basis of independent probabilities.

$$\text{Coefficient of coincidence} = \frac{\text{frequency observed}}{\text{frequency expected}}$$

For the three-point cross involving vg, pr, and b, the coefficient of coincidence is

$$\frac{0.52}{0.79} = 0.66$$

The definition of interference itself is

$$\text{Interference} = 1 - \text{coefficient of coincidence}$$

In this case, it is

$$1 - 0.66 = 0.34$$

To understand the meaning of interference, it is helpful to contrast what happens when there is no interference with what happens when it is complete. If interference is 0, the frequency of observed double crossovers equals expected double crossovers, and crossovers in adjacent regions of a chromosome occur independently of each other. If interference is complete (i.e., if interference $= 1$), no double crossovers occur in the experimental progeny because one exchange effectively prevents another. As an example, in a particular three-point cross in mice, the recombination frequency for the pair of genes on the left (region 1) is 20, and for the pair of genes on the right (region 2), it is also 20. Without interference, the expected rate of double crossovers in this chromosomal interval is

$$0.20 \times 0.20 = 0.04, \text{ or } 4\%$$

but when investigators observed 1000 progeny of this cross, they found 0 double recombinants instead of the expected 40.

A method to determine the gene in the middle

The smallest of the eight possible classes of progeny in a three-point cross are the two that contain double recombinants generated by double crossovers. It is possible to use the composition of alleles in these double crossover classes to determine which of the three genes lies in the middle, even without calculating any recombination frequencies. Consider again the progeny of a three-point testcross looking at the vg, pr, and b genes. The F_1 females are $vg\ pr\ b/vg^+\ pr^+\ b^+$. As Figure 4.13d demonstrated, testcross progeny resulting from double crossovers in the trihybrid females of the F_1 generation received gametes from their mothers carrying the allelic combinations $vg\ pr^+\ b$ and $vg^+\ pr\ b^+$. In these individuals, the alleles of the vg and b genes retain their parental associations ($vg\ b$ and $vg^+\ b^+$), while the pr gene has recombined with respect to both the other genes ($pr\ b^+$ and $pr\ b$; $vg\ pr^+$ and $vg^+\ pr$). The same is true for all three-point crosses: In those gametes formed by double crossovers, the gene whose alleles have recombined relative to the parental configurations of the other two genes must be the one in the middle.

Genetic maps of genes along chromosomes can be approximated using data from two-point crosses. Three-point crosses yield more accurate maps because they allow correction for double crossovers as well as estimates of interference (fewer double crossovers than expected). The most accurate maps are constructed with many closely linked genetic markers.

Three-point crosses: A comprehensive example

The technique of looking at double recombinants to discover which gene has recombined with respect to both other genes allows immediate clarification of gene order

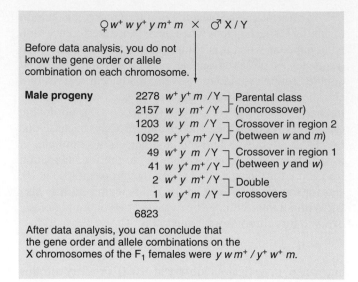

$$♀ w^+ \, w \, y^+ \, y \, m^+ \, m \quad \times \quad ♂ \, X/Y$$

Before data analysis, you do not know the gene order or allele combination on each chromosome.

Male progeny

2278	$w^+ \, y^+ \, m \, / Y$	Parental class
2157	$w \, y \, m^+ \, / Y$	(noncrossover)
1203	$w \, y \, m \, / Y$	Crossover in region 2
1092	$w^+ \, y^+ \, m^+ \, / Y$	(between w and m)
49	$w^+ \, y \, m \, / Y$	Crossover in region 1
41	$w \, y^+ \, m^+ \, / Y$	(between y and w)
2	$w^+ \, y \, m^+ \, / Y$	Double
1	$w \, y^+ \, m \, / Y$	crossovers
6823		

After data analysis, you can conclude that the gene order and allele combinations on the X chromosomes of the F_1 females were $y \, w \, m^+ \, / \, y^+ \, w^+ \, m$.

Figure 4.14 How three-point crosses verify Sturtevant's map. The parental classes correspond to the two X chromosomes in the F_1 female. The genotype of the double recombinant classes shows that w must be the gene in the middle.

even in otherwise difficult cases. Consider the three X-linked genes y, w, and m that Sturtevant located in his original mapping experiment (see Figure 4.11). Because the distance between y and m (34.3 m.u.) appeared slightly larger than the distance separating w and m (32.8 m.u.), he concluded that w was the gene in the middle. But because of the small difference between the two numbers, his conclusion was subject to questions of statistical significance. If, however, we look at a three-point cross following y, w, and m, these questions disappear.

Figure 4.14 tabulates the classes and numbers of male progeny arising from females heterozygous for the y, w, and m genes. Because these male progeny receive their only X chromosome from their mothers, their phenotypes directly indicate the gametes produced by the heterozygous females. In each row of the figure's table, the genes appear in an arbitrary order that does not presuppose knowledge of the actual map. As you can see, the two classes of progeny listed at the top of the table outnumber the remaining six classes, indicating that all three genes are linked to each other. Moreover, these largest groups, which are the parental classes, show that the two X chromosomes of the heterozygous females were $w^+ \, y^+ \, m$ and $w \, y \, m^+$.

Among the male progeny in Figure 4.14, the two smallest classes, representing the double crossovers, have X chromosomes carrying $w^+ \, y \, m^+$ and $w \, y^+ \, m$ combinations, in which the w alleles are recombined relative to those of y and m. The w gene must therefore lie between y and m, verifying Sturtevant's original assessment.

To complete a map based on the $w \, y \, m$ three-point cross, you can calculate the interval between y and w (region 1)

$$\frac{49 + 41 + 1 + 2}{6823} \times 100 = 1.3 \text{ m.u.}$$

as well as the interval between w and m (region 2)

$$\frac{1203 + 1092 + 2 + 1}{6823} \times 100 = 33.7 \text{ m.u}$$

The genetic distance separating y and m is the sum of

$$1.3 + 33.7 = 35.0 \text{ m.u.}$$

Note that you could also calculate the distance between y and m directly by including double crossovers twice, to account for the total number of recombination events detected between these two genes.

$$\text{RF} = \frac{1203 + 1092 + 49 + 41 + 2 + 2 + 1 + 1}{6823} \times 100$$

$$= 35.0 \text{ m.u.}$$

This method yields the same value as the sum of the two intervening distances (region 1 + region 2).

Further calculations show that interference is considerable in this portion of the *Drosophila* X chromosome, at least as inferred from the set of data tabulated in Figure 4.14. The percentage of observed double recombinants was

$$\frac{3}{6823} = 0.000\,44, \text{ or } 0.044\%$$

(rounding to the nearest thousandth of a percent), while the percentage of double recombinants expected on the basis of independent probabilities by the product rule is

$$0.013 \times 0.337 = 0.0044, \text{ or } 0.44\%$$

Thus, the coefficient of coincidence is

$$\frac{0.044}{0.44} = 0.1$$

and the interference is

$$1 - 0.1 = 0.9$$

Do genetic maps correlate with physical reality?

Many types of experiments presented later in this book clearly show that the *order of genes* revealed by recombination mapping corresponds to the order of those same genes along the DNA molecule of a chromosome. In contrast, the *actual physical distances between genes*—that is, the amount of DNA separating them—does not always show a direct correspondence to genetic map distances.

The relationship between recombination frequency and physical distance along a chromosome is not simple. One complicating factor is the existence of double, triple, and even more crossovers. When genes are separated by 1 m.u. or less, double crossovers are not significant because the probability of their occurring is so small ($0.01 \times 0.01 = 0.0001$). But for genes separated by 20, 30, or 40 m.u., the probability of double crossovers skewing the data takes on greater significance. A second confounding factor is the 50 percent limit on the recombination frequency observable in a cross. This limit reduces the precision of RF as a measure

of chromosomal distances. No matter how far apart two genes are on a long chromosome, they will never recombine more than 50 percent of the time. Yet a third problem is that recombination is not uniform, even over the length of a single chromosome: Certain "hotspots" are favoured sites of recombination, while other areas—often in the vicinity of centromeres—are "recombination deserts" in which few crossovers ever take place.

Ever since Morgan, Sturtevant, and others began mapping, geneticists have generated mathematical equations called **mapping functions** to compensate for the inaccuracies inherent in relating recombination frequencies to physical distances. These equations generally make large corrections for RF values of widely separated genes, while barely changing the map distances separating genes that lie close together. This reflects the fact that multiple recombination events and the 50 percent limit on recombination do not confound the calculation of distances between closely linked genes. However, the corrections for large distances are at best imprecise, because mapping functions are based on simplifying assumptions (such as no interference) that are only rarely justified. Thus, the best way to create an accurate map is still by summing many smaller intervals, locating widely separated genes through linkage to common intermediaries. Maps are subject to continual refinement as more and more newly discovered genes are included.

Rates of recombination may differ from species to species. We know this because recent elucidation of the complete DNA sequences of several organisms' genomes has allowed investigators to compare the actual physical distances between genes (in base pairs of DNA) with genetic map distances. They found that in humans, a map unit corresponds on average to about 1 million base pairs. In yeast, however, where the rate of recombination per length of DNA is much higher than in humans, 1 m.u. is approximately 2500 base pairs. Thus, although map units are useful for estimating distances between the genes of an organism, 1 percent RF can reflect very different expanses of DNA in different organisms.

Recombination rates sometimes vary even between the two sexes of a single species. *Drosophila* provides an extreme example: No recombination occurs during meiosis in males. If you review the examples already discussed in this chapter, you will discover that they all measure recombination among the progeny of dihybrid *Drosophila* females. Problem 19 at the end of this chapter shows how geneticists can exploit the absence of recombination in *Drosophila* males to establish rapidly that genes far apart on the same chromosome are indeed syntenic.

Multiple-factor crosses help establish linkage groups

Genes chained together by linkage relationships are known collectively as a **linkage group**. When enough genes have been assigned to a particular chromosome, the terms *chromosome* and *linkage group* become synonymous. If you can demonstrate that gene *A* is linked to gene *B*, *B* to

C, *C* to *D*, and *D* to *E*, you can conclude that all of these genes are syntenic. When the genetic map of a genome becomes so dense that it is possible to show that any gene on a chromosome is linked to another gene on the same chromosome, the number of linkage groups equals the number of pairs of homologous chromosomes in the species. Humans have 23 linkage groups, mice have 20, and fruit flies have 4 (**Figure 4.15**).

The total genetic distance along a chromosome, which is obtained by adding many short distances between genes, may be much more than 50 m.u. For example, the two long *Drosophila* autosomes are both slightly more than 100 m.u. in length (Figure 4.15), while the longest human chromosome is approximately 270 m.u. Recall, however, that even with the longest chromosomes, *pairwise* crosses between genes located at the two ends will not produce more than 50 percent recombinant progeny.

Linkage mapping has practical applications of great importance. For example, the **Fast Forward** box **"Mapping the Cystic Fibrosis Gene"** in this chapter describes how researchers in **Toronto** and the United States used linkage information to locate the gene for this important human hereditary disease. **Table 4.3** also lists selected genetic disorders for which the application of linkage-based approaches has been enormously successful in identifying causal gene loci and specific mutations. A multitude of randomly distributed DNA markers (see the Fast Forward box) completely covering all of the chromosomes is used for gene mapping—this high-density linkage map is called a **saturated gene map** (**Figure 4.16**). A saturated genetic map of the crop species *Brassica napus* (Figure 4.16) was constructed by a research group from the **University of Manitoba** in **Winnipeg**. These genetic maps are not only useful in understanding the genomic architecture of *Brassica napus*, but can also be exploited to identify the economically important genes or traits in this oilseed crop plant.

Table 4.3	Identification of Human Disease-Related Genes by Linkage Mapping	
Disease	**Map Position**	**Gene**
Duchenne muscular dystrophy	Xp21.2	*DMD*
Huntington disease	4p16.3	*HTT*
Polycystic kidney disease 1	16p13.3	*PKD1*
Spherocytosis type 1	8p11.21	*ANK1*
Hereditary nonpolyposis colorectal cancer type 2	3p22.2	*MLH1*
Ataxia-telangiectasia	11q22.3	*ATM*
Marfan syndrome	15q21.1	*FBN1*
Neurofibromatosis type 1	17q11.2	*NF1*
Fragile X mental retardation syndrome	Xq27.3	*FMR1*
Retinoblastoma	13q14.2	*RB1*
Multiple endocrine neoplasia type 1	11q13.1	*MEN1*
Breast-ovarian cancer	17q21.31	*BRCA1*
Haemochromatosis	6p21.3	*HFE*

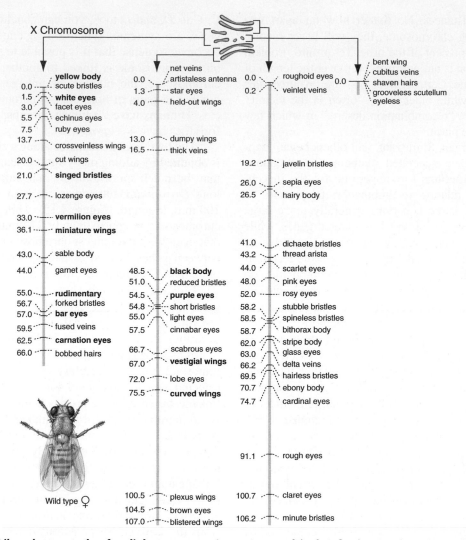

Figure 4.15 _Drosophila melanogaster_ has four linkage groups. A genetic map of the fruit fly, showing the position of many genes affecting body morphology, including those used as examples in this chapter (_highlighted in bold_). Because so many _Drosophila_ genes have been mapped, each of the four chromosomes can be represented as a single linkage group.

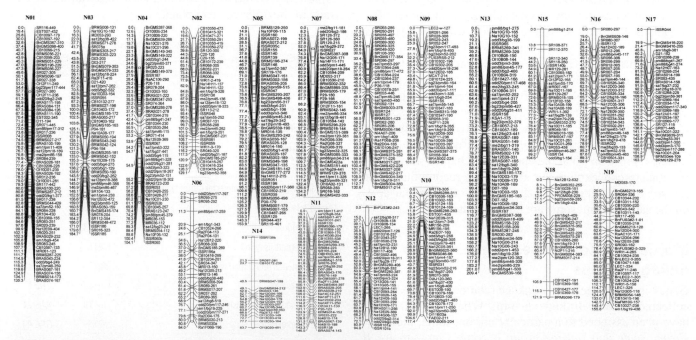

Figure 4.16 A saturated genetic map for the oilseed crop plant _Brassica napus_. A linkage map of _Brassica napus_, the second most important oilseed crop species in the world. This high-density map shows the position of DNA markers covering all of the chromosomes.

For 40 years after the symptoms of cystic fibrosis (CF) were first described in 1938, no molecular clue—no visible chromosomal abnormality transmitted with the disease, no identifiable protein defect carried by affected individuals—suggested the genetic cause of the disorder. As a result, there was no effective treatment for the 1 in 2500 to 3300 Caucasian North Americans born with the disease, most of whom died before they were 30. In 1989, however, geneticists were able to pinpoint a precise chromosomal position, or locus, for the cystic fibrosis gene by combining recently invented techniques for looking directly at DNA with maps constructed by linkage analysis. The discovery was a collaborative effort led by **Drs. Lap-Chee Tsui** and **John Riordan**, researchers at the **Hospital for Sick Children** in **Toronto** and **Dr. Francis Collins**, an investigator at the University of Michigan, Ann Arbor (**Figure A**).

The mappers of the cystic fibrosis gene had faced an overwhelming task. They had searched for a gene that encoded an unknown protein, a gene that had not yet even been assigned to a chromosome. It could have been situated anywhere among the 23 pairs of chromosomes in a human cell. Imagine looking for a close friend you lost track of years ago, who might now be anywhere in the world. You would first have to find ways to narrow the search to a particular continent (the equivalent of a specific chromosome in the gene mappers' search); then to a country (the long or short arm of the chromosome); next to the state or province, county, city, or town, and street (all increasingly narrow bands of the chromosome); and finally, to a house address (the locus itself). Here, we briefly summarize how researchers applied some of these steps in mapping the cystic fibrosis gene.

- A review of many family pedigrees containing first-cousin marriages confirmed that cystic fibrosis is most likely determined by a single gene (*CF*). Investigators collected white blood cells from 47 families with two or more

affected children, obtaining genetic data from 106 patients, 94 parents, and 44 unaffected siblings.

- They next tried to discover if any other trait is reliably transmitted with cystic fibrosis. Analyses of the easily obtainable serum enzyme paroxonase showed that its gene (*PON*) is indeed linked to *CF*. At first, this knowledge was not that helpful, because *PON* had not yet been assigned to a chromosome.

- Then, in the early 1980s, geneticists developed a large series of DNA markers, based on new techniques that enabled them to recognize variations in the genetic material. A **DNA marker** is a piece of DNA of known size, representing a specific locus that comes in identifiable variations. These allelic variations segregate according to Mendel's laws, which means it is possible to follow their transmission as you would any gene's. Chapter 15 explains the discovery and use of DNA markers in the identification of disease-related genes by linkage analysis in greater detail; for now, it is only important to know that they exist and can be identified.

By 1986, linkage analyses of hundreds of DNA markers had shown that one marker, known as *D7S15*, was linked with both *PON* and *CF*. Researchers computed recombination frequencies and found that the distance from the DNA marker to *CF* was 15 cM; from the DNA marker to *PON*, 5 cM; and from *PON* to *CF*, 10 cM. They concluded that the order of the three loci was *D7S15–PON–CF* (**Figure B**). Because *CF* could lie 15 cM in either of two directions from the DNA marker, the area under investigation was approximately 30 cM. And because the human genome consists of roughly 3000 cM, this step of linkage analysis narrowed the search to 1% of the human genome.

Figure A Lap-Chee Tsui (*centre*) with Francis Collins (*left*) and John Riordan on that special day in 1989.

Chromosome 7

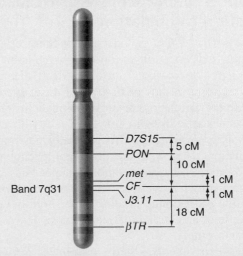

Band 7q31

D7S15 — 5 cM
PON — 10 cM
met — 1 cM
CF — 1 cM
J3.11 — 18 cM
βTR —

Figure B How molecular markers helped locate the gene for cystic fibrosis (*CF*).

(Continued)

4.5   Mitotic Recombination and Genetic Mosaics

The recombination of genetic material is a critical feature of meiosis. It is thus not surprising that eukaryotic organisms express a variety of enzymes (described in Chapter 6) that specifically initiate meiotic recombination. Recombination can also occur during mitosis. Unlike meiosis, however, mitotic crossovers are initiated by mistakes in chromosome replication or by chance exposures to radiation that break DNA molecules, rather than by a well-defined cellular program. As a result, mitotic recombination is a rare event, occurring no more frequently than once in a million somatic cell divisions. Nonetheless, the growth of a colony of yeast cells or the development of a complex multicellular organism involves a multitude of cell divisions such that geneticists can routinely detect these rare mitotic events.

"Twin spots" indicate mosaicism caused by mitotic recombination

In 1936, the *Drosophila* geneticist Curt Stern originally inferred the existence of mitotic recombination from observations of "twin spots" in a few fruit flies. **Twin spots** are adjacent islands of tissue that differ both from each other and from the tissue surrounding them. The distinctive patches arise from homozygous cells with a recessive phenotype growing amid a generally heterozygous cell population displaying the dominant phenotype. In *Drosophila*, the *yellow* (*y*) mutation changes body colour from normal brown to yellow, while the *singed bristles* (*sn*) mutation causes body bristles to be short and curled rather than long and straight. Both of these genes are on the X chromosome.

In his experiments, Stern examined *Drosophila* females of genotype *y sn⁺/y⁺ sn*. These dihybrids were generally wild type in appearance, but Stern noticed that some flies carried patches of yellow body colour, others had small areas of singed bristles, and still others displayed twin spots: adjacent patches of yellow cells and cells with singed bristles (**Figure 4.17**). He assumed that mistakes in the mitotic divisions during the course of fly development could have led to these **mosaic** animals containing tissues of different genotypes. Individual yellow or singed patches could arise from chromosome loss or by mitotic nondisjunction. These errors in mitosis would yield XO cells containing only *y* (but not *y⁺*) or *sn* (but not *sn⁺*) alleles; such cells would show one of the recessive phenotypes.

The twin spots must have a different origin. Stern reasoned that they represented the reciprocal products of mitotic crossing-over between the *sn* gene and the centromere. The mechanism is as follows: During mitosis in a diploid cell, after chromosome duplication, homologous chromosomes occasionally—very occasionally—pair up with each other. While the chromosomes are paired, nonsister chromatids (i.e., one

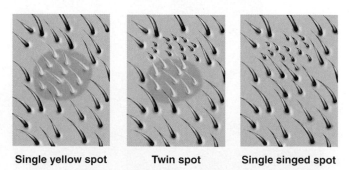

Single yellow spot   Twin spot   Single singed spot

Figure 4.17 Twin spots: A form of genetic mosaicism. In a *y sn⁺/y⁺ sn Drosophila* female, most of the body is wild type, but aberrant patches showing either yellow colour or singed bristles sometimes occur. In some cases, yellow and singed patches are adjacent to each other, a configuration known as *twin spots*.

chromatid from each of the two homologous chromosomes) can exchange parts by crossing-over. The pairing is transient, and the homologous chromosomes soon resume their independent positions on the mitotic metaphase plate. There, the two chromosomes can line up relative to each other in either of two ways (**Figure 4.18a**). One of these orientations would yield two daughter cells that remain heterozygous for both genes and thus be indistinguishable from the surrounding wild-type cells. The other orientation, however, will generate two homozygous daughter cells, one *y sn*⁺*/y sn*⁺, the other *y*⁺ *sn/y*⁺ *sn*. Because the two daughter cells would lie next to each other, subsequent mitotic divisions would produce adjacent patches of *y* and *sn* tissue (i.e., twin spots). Note that if crossing-over occurs between *sn* and *y*, single spots of yellow tissue can form, but a reciprocal singed spot cannot be generated in this fashion (**Figure 4.18b**).

Sectored yeast colonies can arise from mitotic recombination

Diploid yeast cells that are heterozygous for one or more genes exhibit mitotic recombination in the form of **sectors**: portions of a growing colony that have a different genotype than the remainder of the colony. If a diploid yeast cell of genotype *ADE2 / ade2* is placed on a Petri plate, its mitotic descendents will grow into a colony. Usually, such colonies will appear white because the dominant wild-type *ADE2* allele specifies that colour. However, many colonies will contain red sectors of diploid *ade2 / ade2* cells, which arose as a result of mitotic recombination events between the *ADE2* gene and its centromere (**Figure 4.19**). (Homozygous *ADE2 / ADE2* cells will also be produced by the same event, but they cannot be distinguished from heterozygotes because both types of cells are white.) The size of the red sectors indicates when mitotic recombination took place.

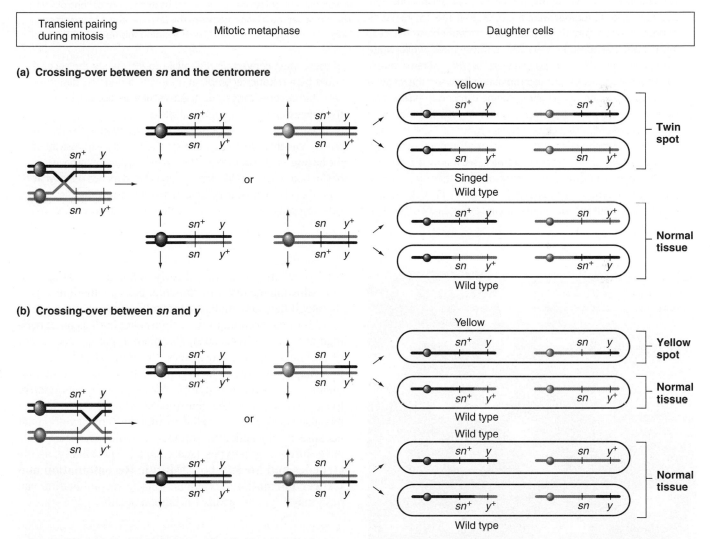

Figure 4.18 Mitotic crossing-over. (a) In a *y sn*⁺*/y*⁺ *sn Drosophila* female, a mitotic crossover between the centromere and *sn* can produce two daughter cells, one homozygous for *y* and the other homozygous for *sn*, that can develop into adjacent aberrant patches (twin spots). This outcome depends on a particular distribution of chromatids at anaphase (*top*). If the chromatids are arranged in the equally likely opposite orientation, only phenotypically normal cells will result (*bottom*). **(b)** Crossovers between *sn* and *y* can generate single yellow patches. However, a single mitotic crossover in these females cannot produce a single singed spot if the *sn* gene is closer to the centromere than the *y* gene.

In humans, some tumours, such as those found in retinoblastoma, may arise from somatic mutations, or as a result of mitotic recombination. Recall from the discussion of expressivity in Chapter 2 that retinoblastoma is the most malignant form of eye cancer. The retinoblastoma gene (*RB*) resides on chromosome 13, where the normal wild-type allele (*RB*$^+$) encodes a protein that regulates retinal cell growth and differentiation. Cells in the eye need at least one copy of the normal wild-type allele to maintain control over cell division. The normal, wild-type *RB*$^+$ allele is thus known as a tumour-suppressor gene.

People with a genetic predisposition to retinoblastoma are born with only one functional copy of the normal *RB*$^+$ allele; their second chromosome 13 carries either a nonfunctional *RB*$^-$ allele or no *RB* gene at all. If a mutagen (such as radiation) or a mistake in gene replication or segregation destroys or removes the single remaining normal copy of the gene in a retinal cell in either eye, a retinoblastoma tumour will develop at that site. In one study of people with a genetic predisposition to retinoblastoma, cells taken from eye tumours were *RB*$^-$ homozygotes, while white blood cells from the same people were *RB*$^+$/*RB*$^-$ heterozygotes. As **Figure A** shows, mitotic recombination between the *RB* gene and the centromere of the chromosome carrying the gene provides one mechanism by which a cell in an *RB*$^+$/*RB*$^-$ individual could become *RB*$^-$/*RB*$^-$. Once a homozygous *RB*$^-$ cell is generated, it will divide uncontrollably, leading to tumour formation.

Only 40 percent of retinoblastoma cases follow the preceding scenario. The other 60 percent occur in people who are born with two normal copies of the *RB* gene. In such people, it takes two mutational events to cause the cancer. The first

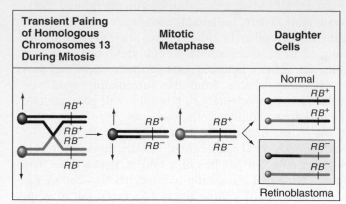

Figure A How mitotic crossing-over can contribute to cancer. Mitotic recombination during retinal growth in an *RB*$^-$/*RB*$^+$ heterozygote may produce an *RB*$^-$/*RB*$^-$ daughter cell that lacks a functional retinoblastoma gene and thus divides out of control. The crossover must occur between the *RB* gene and its centromere. Only the arrangement of chromatids yielding this result is shown.

of these must convert an *RB*$^+$ allele to *RB*$^-$, while the second could be a mitotic recombination event producing daughter cells that become cancerous because they are homozygous for the newly mutant, nonfunctional allele.

Interestingly, the role of mitotic recombination in the formation of retinoblastoma helps explain the variable expressivity of the disease. People born as *RB*$^+$/*RB*$^-$ heterozygotes may have it in one or both eyes (variable expressivity). It all depends on in what cells of the body mitotic recombination (or some other "loss-of-heterozygosity" event that affects chromosome 13) occurs.

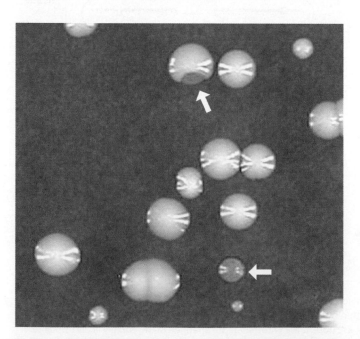

Figure 4.19 Mitotic recombination during the growth of diploid yeast colonies can create sectors. Arrows point to large, red *ade2 / ade2* sectors formed from *ADE2 / ade2* heterozygotes.

If they are large, it happened early in the growth of the colony, allowing the resulting daughter cells a long time to proliferate; if they are small, the recombination occurred later.

Mitotic recombination is significant both as an experimental tool and because of the phenotypic consequences of particular mitotic crossovers. Problem 34 at the end of this chapter illustrates how geneticists use mitotic recombination to obtain information for mapping genes relative to each other and to the centromere. Mitotic crossing-over has also been of great value in the study of development because it can generate animals in which different cells have different genotypes (see Chapter 17). Finally, as the **Genetics and Society** box **"Mitotic Recombination and Cancer Formation"** explains, mitotic recombination can have major repercussions for human health.

Crossing-over can occur in rare instances during mitosis, so that a diploid heterozygous cell can produce diploid homozygous daughter cells. The consequences of mitotic recombination include genetic mosaicism in multicellular organisms and sectoring during the growth of yeast colonies.

Medical geneticists have used their understanding of linkage, recombination, and mapping to make sense of human pedigrees, like the one presented at the beginning of this chapter (see Figure 4.1). The X-linked gene for red-green colour blindness must lie very close to the gene for haemophilia A because the two are tightly linked. In fact, the genetic distance between the two genes is only 3 m.u. The sample size in Figure 4.1a was so small that none of the individuals in the pedigree were recombinant types. In contrast, even though haemophilia B is also on the X chromosome, it lies far enough away from the red-green colour blindness locus that the two genes recombine relatively freely. The colour blindness and haemophilia B genes may appear to be genetically unlinked in a small sample (as in Figure 4.1b), but the actual recombination distance separating the two genes is about 36 m.u. Pedigrees pointing to two different forms of haemophilia, one very closely linked to colour blindness, the other almost not linked at all, provided one of several indications that haemophilia is determined by more than one gene (**Figure 4.20**).

Refining the human chromosome map poses a continuous challenge for medical geneticists. The newfound potential for finding and fitting more and more DNA markers into the map (review the Fast Forward box in this chapter) enormously improves the ability to identify genes that cause disease, as discussed in Chapter 15. Linkage analysis has also been used in conjunction with whole genome sequencing (WGS) and whole exome sequencing (WES) in the discovery of disease-causing mutations (Chapter 19).

Linkage and recombination are universal among life-forms and must therefore confer important advantages to living organisms. Geneticists believe that linkage provides the potential for transmitting favourable combinations of genes intact to successive generations, while recombination produces great flexibility in generating new combinations of alleles. Some new combinations may help a species adapt to changing environmental conditions, whereas the inheritance of successfully tested combinations can preserve what has worked in the past.

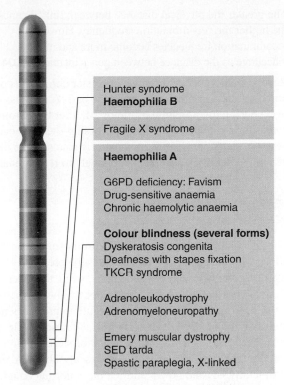

Figure 4.20 A genetic map of part of the human X chromosome.

Thus far, this book has examined how genes and chromosomes are transmitted. As important and useful as this knowledge is, it tells us very little about the structure and mode of action of the genetic material. In the next section (Chapters 5–7), we carry our analysis to the level of DNA, the actual molecule of heredity. In Chapter 5, we look at DNA structure and learn how the DNA molecule carries genetic information. In Chapter 6, we describe how geneticists defined the gene as a localized region of DNA containing many nucleotides that together encode the information to make a protein. In Chapter 7, we examine how the cellular machinery interprets the genetic information in genes to produce the multitude of phenotypes that make up an organism.

Essential Concepts

1. Gene pairs that are close together on the same chromosome are genetically linked because they are transmitted together more often than not. The hallmark of linkage is that the number of parental types is greater than the number of recombinant types among the progeny of dihybrid individuals. **[LO1–3]**

2. The recombination frequencies of pairs of genes indicate how often two genes are transmitted together. For linked genes, the recombination frequency is less than 50 percent. **[LO3]**

3. Gene pairs that assort independently exhibit a recombination frequency of 50 percent, because the number of parental types equals the number of recombinants. Genes may assort independently either because they are on different chromosomes or because they are far apart on the same chromosome. **[LO2–3]**

4. Statistical analysis helps determine whether or not two genes assort independently. The probability value (p) calculated by the chi-square test measures the likelihood that a particular set of data supports the null hypothesis of independent assortment, or no linkage. The lower the p value, the less likely is the null hypothesis, and the more likely the linkage. The chi-square test can also be

used to determine how well the outcomes of crosses fit other genetic hypotheses. **[LO3]**

5. The greater the physical distance between linked genes, the higher the recombination frequency. However, recombination frequencies become more and more inaccurate as the distance between genes increases. **[LO4–5]**

6. Recombination occurs because nonsister chromatids of homologous chromosomes exchange parts (i.e., cross over) during the prophase of meiosis I, after the chromosomes have replicated. Although crossing-over can occur between sister chromatids in some organisms, since the genetic material exchanged is identical, no recombinants

are produced, and thus the effects of these crossover events are not observed in the progeny. **[LO4]**

7. Genetic maps are a visual representation of relative recombination frequencies. The greater the density of genes on the map (and thus the smaller the distance between the genes), the more accurate and useful the map becomes in predicting inheritance. **[LO5–6]**

8. In diploid organisms heterozygous for two alleles of a gene, rare mitotic recombination between the gene and its centromere can produce genetic mosaics in which some cells are homozygous for one allele or the other. **[LO4]**

Solved Problems

I. The *Xg* locus on the human X chromosome has two alleles, a^+ and a. The a^+ allele causes the presence of the Xg surface antigen on red blood cells, while the recessive *a* allele does not allow antigen to appear. The *Xg* locus is 10 m.u. from the *Sts* locus. The *Sts* allele produces normal activity of the enzyme steroid sulfatase, while the recessive *sts* allele results in the lack of steroid sulfatase activity and the disease ichthyosis (scaly skin). A man with ichthyosis and no Xg antigen has a normal daughter with the Xg antigen, who is expecting a child.

a. If the child is a son, what is the probability he will lack the antigen and have ichthyosis?

b. What is the probability that a son would have both the antigen and ichthyosis?

c. If the child is a son with ichthyosis, what is the probability he will have the Xg antigen?

Answer

a. This problem requires an understanding of how linkage affects the proportions of gametes. First designate the genotype of the individual in which recombination during meiosis affects the transmission of alleles: in this problem, the daughter. The X chromosome she inherited from her father (who had ichthyosis and no Xg antigen) must be *sts a*. (No recombination could have separated the genes during meiosis in her father since he has only one X chromosome.) Because the daughter is normal and has the Xg antigen, her other X chromosome (inherited from her mother) must contain the *Sts* and a^+ alleles. Her X chromosomes can be diagrammed as follows:

$$\underline{\qquad sts \qquad\qquad a \qquad}$$

$$\underline{\qquad Sts \qquad\qquad a^+ \qquad}$$

Because the *Sts* and *Xg* loci are 10 m.u. apart on the chromosome, there is a 10 percent recombination frequency. Ninety percent of the gametes will be parental: *sts a* or *Sts* a^+ (45 percent of each type) and 10 percent will be recombinant: *sts* a^+ or *Sts a* (5 percent of each type). The phenotype of a son directly reflects the genotype of the X chromosome from his mother. *Therefore, the probability*

that he will lack the Xg antigen and have ichthyosis (genotype: sts a/Y) is 45/100.

b. *The probability that he will have the antigen and ichthyosis (genotype: sts a^+/Y) is 5/100.*

c. There are two classes of gametes containing the ichthyosis allele: *sts a* (45 percent) and *sts* a^+ (5 percent). If the total number of gametes is 100, then 50 will have the *sts* allele. Of those gametes, 5 (or 10 percent) will have the a^+ allele. *Therefore, there is a 1/10 probability that a son with the sts allele will have the Xg antigen.*

II. *Drosophila* females of wild-type appearance but heterozygous for three autosomal genes are mated with males showing three autosomal recessive traits: glassy eyes, coal-coloured bodies, and striped thoraxes. One thousand progeny of this cross are distributed in the following phenotypic classes:

Wild type	27
Striped thorax	11
Coal body	484
Glassy eyes, coal body	8
Glassy eyes, striped thorax	441
Glassy eyes, coal body, striped thorax	29

a. Draw a genetic map based on this data.

b. Show the arrangement of alleles on the two homologous chromosomes in the parent females.

c. Normal-appearing males containing the same chromosomes as the parent females in the preceding cross are mated with females showing glassy eyes, coal-coloured bodies, and striped thoraxes. Of 1000 progeny produced, indicate the numbers of the various phenotypic classes you would expect.

Answer

A logical, methodical way to approach a three-point cross is described here.

a. Designate the alleles:

t^+ = wild-type thorax	t = striped thorax
g^+ = wild-type eyes	g = glassy eyes
c^+ = wild-type body	c = coal-coloured body

In solving a three-point cross, designate the types of events that gave rise to each group of individuals and the genotypes of the gametes obtained from their mother. (The paternal gametes contain only the recessive alleles of these genes [*t g c*]. They do not change the phenotype and can be ignored.)

Progeny	Number	Type of event	Genotype		
1. wild type	27	single crossover	t^+	g^+	c^+
2. striped thorax	11	single crossover	t	g^+	c^+
3. coal body	484	parental	t^+	g^+	c
4. glassy eyes, coal body	8	single crossover	t^+	g	c
5. glassy eyes, striped thorax	441	parental	t	g	c^+
6. glassy eyes, coal body, striped thorax	29	single crossover	t	g	c

Picking out the parental classes is easy. If all the other classes are rare, the two most abundant categories are those gene combinations that have not undergone recombination. Then there should be two sets of two phenotypes that correspond to a single crossover event between the first and second genes, or between the second and third genes. Finally, there should be a pair of classes containing small numbers that result from double crossovers. In this example, there are no flies in the double crossover classes, which would have been in the two missing phenotypic combinations: glassy eyes, coal body, and striped thorax.

Look at the most abundant classes to determine which alleles were on each chromosome in the female heterozygous parent. One parental class had the phenotype of coal body (484 flies), so one chromosome in the female must have contained the t^+, g^+, and c alleles. (Notice that we cannot yet say in what order these alleles are located on the chromosome.) The other parental class

was glassy eyes and striped thorax, corresponding to a chromosome with the *t*, *g*, and c^+ alleles.

To determine the order of the genes, compare the $t^+ g c^+$ double crossover class (not seen in the data) with the most similar parental class ($t g c^+$). The alleles of *g* and *c* retain their parental associations ($g c^+$), while the *t* gene has recombined with respect to both other genes in the double recombinant class. Thus, the *t* gene is between *g* and *c*.

In order to complete the map, calculate the recombination frequencies between the centre gene and each of the genes on the ends. For *g* and *t*, the nonparental combinations of alleles are in classes 2 and 4, so RF = (11 + 8)/1000 = 19/1000, or 1.9%. For *t* and *c*, classes 1 and 6 are nonparental, so RF = (27 + 29)/1000 = 56/1000, or 5.6%.

The genetic map is

b. The alleles on each chromosome were already determined (*c*, g^+, t^+ and c^+, *g*, *t*). Now that the order of loci has also been determined, the arrangement of the alleles can be indicated.

$$\begin{array}{ccc} c & t^+ & g^+ \\ \hline \hline c^+ & t & g \end{array}$$

c. Males of the same genotype as the starting female ($c\ t^+\ g^+/c^+\ t\ g$) could produce only two types of gametes: parental types $c\ t^+\ g^+$ and $c^+\ t\ g$, because there is no recombination in male *Drosophila*. The progeny expected from the mating with a homozygous recessive female are thus 500 coal body and 500 glassy eyed, striped thorax flies.

Problems

Vocabulary

1. For each of the terms in the left column, choose the best matching phrase in the right column.

i. recombination	**1.** a statistical method for testing the fit between observed and expected results
ii. linkage	**2.** one crossover along a chromosome makes a second nearby crossover less likely
iii. chi-square test	**3.** when two loci recombine in less than 50 percent of gametes
iv. chiasma	**4.** the relative chromosomal location of a gene
v. locus	**5.** the ratio of observed double crossovers to expected double crossovers
vi. coefficient of coincidence	**6.** individual composed of cells with different genotypes
vii. interference	**7.** formation of new genetic combinations by exchange of parts between homologues
viii. mosaic	**8.** structure formed at the spot where crossing-over occurs between homologues

Section 4.1

2. a. A *Drosophila* male from a true-breeding stock with scabrous eyes was mated with a female from a true-breeding stock with javelin bristles. Both scabrous eyes and javelin bristles are autosomal traits. The F_1 progeny all had normal eyes and bristles. F_1 females from this cross were mated with males with both scabrous eyes and javelin bristles. Write all the possible phenotypic classes of the progeny that could be produced from the cross of the F_1 females with the scabrous, javelin males, and indicate for each class whether it is a recombinant or parental type.

b. The cross above yielded the following progeny: 77 scabrous eyes and normal bristles; 76 wild type (normal eyes and bristles); 74 normal eyes and javelin bristles; and 73 scabrous eyes and javelin bristles. Are the genes governing these traits likely to be linked, or do they instead assort independently? Why?

c. Suppose you mated the F_1 females from the cross in part *a* to wild-type males. Why would this cross fail to inform you whether the two genes are linked?

d. Suppose you mated females from the true-breeding stock with javelin bristles to males with scabrous eyes and javelin bristles. Why would this cross fail to inform you whether the two genes are linked?

3. With modern molecular methods it is now possible to examine variants in DNA sequence from a very small amount of tissue like a hair follicle or even a single sperm. You can consider these variants to be "alleles" of a particular site on a chromosome (a "locus"; "loci" in plural). For example, AAAAAAA, AAACAAA, AAAGAAA, and AAATAAA at the same location (call it B) on homologous autosomes in different sperm might be called alleles *1*, *2*, *3*, and *4* of locus *B* (B_1, B_2, etc.). John's genotype for two loci *B* and *D* is B_1B_3 and D_1D_3. John's father was B_1B_2 and D_1D_4, while his mother was B_3B_3 and D_2D_3.

a. What is (are) the genotype(s) of the parental type sperm John could produce?

b. What is (are) the genotype(s) of the recombinant type sperm John could produce?

c. In a sample of 100 sperm, 51 of John's sperm were found to be B_1 and D_1, while the remaining 49 sperm were B_3D_3. Can you conclude whether the *B* and *D* loci are linked, or whether they instead assort independently?

Section 4.2

4. Do the data that Mendel obtained fit his hypotheses? For example, Mendel obtained 315 yellow round, 101 yellow wrinkled, 108 green round, and 32 green wrinkled seeds from the selfing of *Yy Rr* individuals (a total of 556). His hypotheses of segregation and independent assortment predict a 9:3:3:1 ratio in this case. Use the chi-square test to determine whether Mendel's data are significantly different from what he predicted. (The chi-square test did not exist in Mendel's day, so he was not able to test his own data for goodness of fit to his hypotheses.)

5. Two genes control colour in corn snakes as follows:

O– B– snakes are brown, *O– bb* are orange, *oo B–* are black, and *oo bb* are albino. An orange snake was mated to a black snake, and a large number of F_1 progeny were obtained, all of which were brown. When the F_1 snakes were mated to one another, they produced 100 brown offspring, 25 orange, 22 black, and 13 albino.

a. What are the genotypes of the F_1 snakes?

b. What proportions of the different colours would have been expected among the F_2 snakes if the two loci assort independently?

c. Do the observed results differ significantly from what was expected, assuming independent assortment is occurring?

d. What is the probability that differences this great between observed and expected values would happen by chance?

6. A mouse from a true-breeding population with normal gait was crossed to a mouse displaying an odd gait called "dancing." The F_1 animals all showed normal gait.

a. If dancing is caused by homozygosity for the recessive allele of a single gene, what proportion of the F_2 mice should be dancers?

b. If mice must be homozygous for recessive alleles of both of two different genes to have the dancing phenotype, what proportion of the F_2 should be dancers if the two genes are unlinked?

c. When the F_2 mice were obtained, 42 normal and 8 dancers were seen. Use the chi-square test to determine if these results better fit the one-gene model from part *a* or the two-gene model from part *b*.

7. Figure 4.6 applied the chi-square method to test linkage between two genes by asking whether the observed numbers of parental and recombinant classes differed significantly from the expectation of independent assortment that parentals = recombinants. Another possible way to analyze the results from these same experiments is to ask whether the observed frequencies of the four genotypic classes (*A B*, *a b*, *A b*, and *a B*) can be explained by a null hypothesis predicting that they should appear in a 1:1:1:1 ratio. In order to consider the relative advantages and disadvantages of analyzing the data in these two different ways, answer the following:

a. What is the null hypothesis in each case?

b. Which is a more sensitive test of linkage? (Analyze the data in Figure 4.6 by the second method.)

c. How would both methods respond to a situation in which one allele of one of the genes causes reduced viability?

Section 4.3

8. In *Drosophila*, males from a true-breeding stock with raspberry-coloured eyes were mated to females from a true-breeding stock with sable-coloured bodies. In the F_1 generation, all the females had wild-type eye and body colour, while all the males had wild-type eye colour but sable-coloured bodies. When F_1 males and females were mated, the F_2 generation was composed of 216 females with wild-type eyes and bodies, 223 females with wild-type eyes and sable bodies, 191 males with wild-type eyes and sable bodies, 188 males with raspberry eyes and wild-type bodies, 23 males with wild-type eyes and bodies, and 27 males with raspberry eyes and sable bodies. Explain these results by diagramming the crosses, and calculate any relevant map distances.

9. In mice, the dominant allele *Gs* of the X-linked gene *Greasy* produces shiny fur, while the recessive wild-type Gs^+ allele determines normal fur. The dominant allele *Bhd* of the X-linked *Broadhead* gene causes skeletal abnormalities

including broad heads and snouts, while the recessive wild-type Bhd^+ allele yields normal skeletons. Female mice heterozygous for the two alleles of both genes were mated with wild-type males. Among 100 male progeny of this cross, 49 had shiny fur, 48 had skeletal abnormalities, 2 had shiny fur and skeletal abnormalities, and 1 was wild type.

a. Diagram the cross described, and calculate the distance between the two genes.

b. What would have been the results if you had counted 100 female progeny of the cross?

10. $CC\ DD$ and $cc\ dd$ individuals were crossed to each other, and the F_1 generation was backcrossed to the $cc\ dd$ parent. The results were 903 $Cc\ Dd$, 897 $cc\ dd$, 98 $Cc\ dd$, and 102 $cc\ Dd$ offspring.

a. How far apart are the c and d loci?

b. What progeny and in what frequencies would you expect to result from testcrossing the F_1 generation from a $CC\ dd \times cc\ DD$ cross to $cc\ dd$?

11. If the a and b loci are 20 m.u. apart in humans and an $A\ B/a\ b$ woman mates with an $a\ b/a\ b$ man, what is the probability that their first child will be $A\ b/a\ b$?

12. In a particular human family, John and his mother both have brachydactyly (a rare autosomal dominant causing short fingers). John's father has Huntington disease (another rare autosomal dominant). John's wife is phenotypically normal and is pregnant. Two-thirds of people who inherit the Huntington (HD) allele show symptoms by age 50, and John is 50 and has no symptoms. Brachydactyly is 90 percent penetrant.

a. What are the genotypes of John's parents?

b. What are the possible genotypes for John?

c. What is the probability the child will express both brachydactyly and Huntington disease by age 50 if the two genes are not linked?

d. If these two loci are 20 m.u. apart, how will it change your answer to part c?

13. In mice, the autosomal locus coding for the β-globin chain of haemoglobin is 1 m.u. from the albino locus. Assume for the moment that the same is true in humans. The disease sickle-cell anaemia is the result of homozygosity for a particular mutation in the β-globin gene.

a. A son is born to an albino man and a woman with sickle-cell anaemia. What kinds of gametes will the son form, and in what proportions?

b. A daughter is born to a normal man and a woman who has both albinism and sickle-cell anaemia. What kinds of gametes will the daughter form, and in what proportions?

c. If the son in part a grows up and marries the daughter in part b, what is the probability that a child of theirs will be an albino with sickle-cell anaemia?

14. In corn, the allele A allows the deposition of anthocyanin (blue) pigment in the kernels (seeds), while aa plants have yellow kernels. At a second gene, $W-$ produces smooth kernels, while ww kernels are wrinkled. A plant with blue smooth kernels was crossed to a plant with yellow wrinkled kernels. The progeny consisted of 1447 blue smooth, 169 blue wrinkled, 186 yellow smooth, and 1510 yellow wrinkled.

a. Are the a and w loci linked? If so, how far apart are they?

b. What was the genotype of the blue smooth parent? Include the chromosome arrangement of alleles.

c. If a plant grown from a blue wrinkled progeny seed is crossed to a plant grown from a yellow smooth F_1 seed, what kinds of kernels would be expected, and in what proportions?

15. Albino rabbits (lacking pigment) are homozygous for the recessive c allele (C allows pigment formation). Rabbits homozygous for the recessive b allele make brown pigment, while those with at least one copy of B make black pigment. True-breeding brown rabbits were crossed to albinos, which were BB. F_1 rabbits, which were all black, were crossed to the double recessive ($bb\ cc$). The progeny obtained were 34 black, 66 brown, and 100 albino.

a. What phenotypic proportions would have been expected if the b and c loci were unlinked?

b. How far apart are the two loci?

16. Write the number of *different kinds* of phenotypes, excluding gender, you would see among a large number of progeny from an F_1 mating between individuals of identical genotype that are heterozygous for one or two genes (i.e., Aa or $Aa\ Bb$) as indicated. No gene interactions means that the phenotype determined by one gene is not influenced by the genotype of the other gene.

a. One gene; A completely dominant to a.

b. One gene; A and a codominant.

c. One gene; A incompletely dominant to a.

d. Two unlinked genes; no gene interactions; A completely dominant to a, and B completely dominant to b.

e. Two genes, 10 m.u. apart; no gene interactions; A completely dominant to a, and B completely dominant to b.

f. Two unlinked genes; no gene interactions; A and a codominant, and B incompletely dominant to b.

g. Two genes, 10 m.u. apart; A completely dominant to a, and B completely dominant to b; and with recessive epistasis between the genes.

h. Two unlinked duplicated genes (i.e., A and B perform the same function); A and B completely dominant to a and b, respectively.

i. Two genes, 0 m.u. apart; no gene interactions; A completely dominant to a, and B completely dominant to b. (There are two possible answers.)

17. Assume the a and b loci are 40 cM apart and an $AA\ BB$ individual and an $aa\ bb$ individual mate.

a. What gametes will the F_1 individuals produce, and in what proportions? What phenotypic classes and in what

proportions are expected in the F_2 generation (assuming complete dominance for both genes)?

b. If the original cross was *AA bb* $\times$ *aa BB*, what gametic proportions would emerge in F_1? What would be the result in the F_2 generation?

18. A DNA variant has been found linked to a rare autosomal dominant disease in humans and can thus be used as a marker to follow inheritance of the disease allele. In an informative family (in which one parent is heterozygous for both the disease allele and the DNA marker in a known chromosomal arrangement of alleles, and his or her mate does not have the same alleles of the DNA variant), the reliability of such a marker as a predictor of the disease in a fetus is related to the map distance between the DNA marker and the gene causing the disease. Imagine that a man affected with the disease (genotype *Dd*) is heterozygous for the V^1 and V^2 forms of the DNA variant, with form V^1 on the same chromosome as the *D* allele and form V^2 on the same chromosome as *d*. His wife is V^3V^3 *dd*, where V^3 is another allele of the DNA marker. Typing of the fetus by amniocentesis reveals that the fetus has the V^2 and V^3 variants of the DNA marker. How likely is it that the fetus has inherited the disease allele *D* if the distance between the *D* locus and the marker locus is (a) 0 m.u., (b) 1 m.u., (c) 5 m.u., (d) 10 m.u., or (e) 50 m.u.?

Section 4.4

19. In *Drosophila*, the recessive *dp* allele of the *dumpy* gene produces short, curved wings, while the recessive allele *bw* of the *brown* gene causes brown eyes. In a testcross using females heterozygous for both of these genes, the following results were obtained:

wild-type wings, wild-type eyes	178
wild-type wings, brown eyes	185
dumpy wings, wild-type eyes	172
dumpy wings, brown eyes	181

In a testcross using males heterozygous for both of these genes, a different set of results was obtained:

wild-type wings, wild-type eyes	247
dumpy wings, brown eyes	242

a. What can you conclude from the first testcross?

b. What can you conclude from the second testcross?

c. How can you reconcile the data shown in parts *a* and *b*? Can you exploit the difference between these two sets of data to devise a general test for synteny in *Drosophila*?

d. The genetic distance between *dumpy* and *brown* is 91.5 m.u. How could this value be measured?

20. Cinnabar eyes (*cn*) and reduced bristles (*rd*) are autosomal recessive characteristics in *Drosophila*. A homozygous wild-type female was crossed to a reduced, cinnabar male, and the F_1 males were then crossed to the F_1 females

to obtain the F_2. Of the 400 F_2 offspring obtained, 292 were wild type, 9 were cinnabar, 7 were reduced, and 92 were reduced, cinnabar. Explain these results and estimate the distance between the *cn* and *rd* loci.

21. Map distances were determined for four different genes (*MAT*, *HIS4*, *THR4*, and *LEU2*) on chromosome III of the yeast *Saccharomyces cerevisiae*:

HIS4 $\leftrightarrow$ *MAT*	37 cM
THR4 $\leftrightarrow$ *LEU2*	35 cM
LEU2 $\leftrightarrow$ *HIS4*	23 cM
MAT $\leftrightarrow$ *LEU2*	16 cM
MAT $\leftrightarrow$ *THR4*	16 cM

What is the order of these genes on the chromosome?

22. From a series of two-point crosses, the following map distances were obtained for the syntenic genes *A, B, C, D,* and *E* in peas:

$B \leftrightarrow C$	23 m.u.
$A \leftrightarrow C$	15 m.u.
$C \leftrightarrow D$	14 m.u.
$A \leftrightarrow B$	12 m.u.
$B \leftrightarrow D$	11 m.u.
$A \leftrightarrow D$	1 m.u.

Chi-square analysis cannot reject the null hypothesis of no linkage for gene *E* with any of the other four genes.

a. Draw a cross scheme that would allow you to determine the $B \leftrightarrow C$ map distance.

b. Diagram the best genetic map that can be assembled from this data set.

c. Explain any inconsistencies or unknown features in your map.

d. What additional experiments would allow you to resolve these inconsistencies or ambiguities?

23. In *Drosophila*, the recessive allele *mb* of one gene causes missing bristles, the recessive allele *e* of a second gene causes ebony body colour, and the recessive allele *k* of a third gene causes kidney-shaped eyes. (Dominant wild-type alleles of all three genes are indicated with a + superscript.) The three different P generation crosses in the table that follows were conducted, and then the resultant F_1 females from each cross were testcrossed to males that were homozygous for the recessive alleles of both genes in question. The phenotypes of the testcross offspring are tabulated below. Determine the best genetic map explaining all the data.

Parental cross	Testcross offspring of F_1 females	
mb^+ mb^+ e^+ e^+ $\times$ *mb mb e e*	normal bristles, normal body	117
	normal bristles, ebony body	11
	missing bristles, normal body	15
	missing bristles, ebony body	107

Parental cross	Testcross offspring of F₁ females	
$k^+ k^+ e\, e \times k\, k\, e^+ e^+$	normal eyes, normal body	11
	normal eyes, ebony body	150
	kidney eyes, normal body	144
	kidney eyes, ebony body	7
$mb^+\, mb^+\, k^+\, k^+$ $\times mb\, mb\, k\, k$	normal bristles, normal eyes	203
	normal bristles, kidney eyes	11
	missing bristles, normal eyes	15
	missing bristles, kidney eyes	193

24. In the tubular flowers of foxgloves, wild-type coloura-tion is red while a mutation called *white* produces white flowers. Another mutation, called *peloria*, causes the flowers at the apex of the stem to be huge. Yet another mutation, called *dwarf*, affects stem length. You cross a white-flowered plant (otherwise phenotypically wild type) to a plant that is dwarf and peloria but has wild-type red flower colour. All of the F₁ plants are tall with white, normal-sized flowers. You cross an F₁ plant back to the dwarf and peloria parent, and you see the 543 progeny shown in the chart. (Only mutant traits are noted.)

dwarf, peloria	172
white	162
dwarf, peloria, white	56
wild type	48
dwarf, white	51
peloria	43
dwarf	6
peloria, white	5

a. Which alleles are dominant?

b. What were the genotypes of the parents in the original cross?

c. Draw a map showing the linkage relationships of these three loci.

d. Is there interference? If so, calculate the coefficient of coincidence and the interference value.

25. In *Drosophila*, three autosomal genes have the following map:

a. Provide the data, in terms of the expected number of flies in the following phenotypic classes, when $a^+\, b^+\, c^+/a\, b\, c$ females are crossed to $a\, b\, c/a\, b\, c$ males. Assume 1000 flies were counted and that there is no interference in this region.

a^+	b^+	c^+
a	b	c
a^+	b	c
a	b^+	c^+
a^+	b^+	c
a	b	c^+
a^+	b	c^+
a	b^+	c

b. If the cross were reversed, such that $a^+\, b^+\, c^+/a\, b\, c$ males are crossed to $a\, b\, c/a\, b\, c$ females, how many flies would you expect in the same phenotypic classes?

26. A snapdragon with pink petals, black anthers, and long stems was allowed to self-fertilize. From the resulting seeds, 650 adult plants were obtained. The phenotypes of these offspring are listed here.

78	red	long	tan
26	red	short	tan
44	red	long	black
15	red	short	black
39	pink	long	tan
13	pink	short	tan
204	pink	long	black
68	pink	short	black
5	white	long	tan
2	white	short	tan
117	white	long	black
39	white	short	black

a. Using *P* for one allele and *p* for the other, indicate how flower colour is inherited.

b. What numbers of red : pink : white would have been expected among these 650 plants?

c. How are anther colour and stem length inherited?

d. What was the genotype of the original plant?

e. Do any of the three genes show independent assortment?

f. For any genes that are linked, indicate the arrangements of the alleles on the homologous chromosomes in the original snapdragon, and estimate the distance between the genes.

27. Male *Drosophila* expressing the recessive mutations *sc* (*scute*), *ec* (*echinus*), *cv* (*crossveinless*), and *b* (*black*) were crossed to phenotypically wild-type females, and the 3288 progeny listed were obtained. (Only mutant traits are noted.)

653	black, scute, echinus, crossveinless
670	scute, echinus, crossveinless
675	wild type
655	black
71	black, scute
73	scute
73	black, echinus, crossveinless
74	echinus, crossveinless
87	black, scute, echinus
84	scute, echinus
86	black, crossveinless
83	crossveinless
1	black, scute, crossveinless
1	scute, crossveinless
1	black, echinus
1	echinus

a. Diagram the genotype of the female parent.

b. Map these loci.

c. Is there evidence of interference? Justify your answer with numbers.

28. *Drosophila* females heterozygous for each of three recessive autosome mutations with independent phenotypic effects (thread antennae [*th*], hairy body [*h*], and scarlet eyes [*st*]) were testcrossed to males showing all three mutant phenotypes. The 1000 progeny of this testcross were

thread, hairy, scarlet	432
wild type	429
thread, hairy	37
thread, scarlet	35
hairy	34
scarlet	33

a. Show the arrangement of alleles on the relevant chromosomes in the triply heterozygous females.

b. Draw the best genetic map that explains these data.

c. Calculate any relevant interference values.

29. A true-breeding strain of Virginia tobacco has dominant alleles determining leaf morphology (*M*), leaf colour (*C*), and leaf size (*S*). A Carolina strain is homozygous for the recessive alleles of these three genes. These genes are found on the same chromosome as follows:

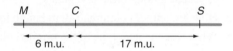

An F_1 hybrid between the two strains is now backcrossed to the Carolina strain. Assuming no interference:

a. What proportion of the backcross progeny will resemble the Virginia strain for all three traits?

b. What proportion of the backcross progeny will resemble the Carolina strain for all three traits?

c. What proportion of the backcross progeny will have the leaf morphology and leaf size of the Virginia strain but the leaf colour of the Carolina strain?

d. What proportion of the backcross progeny will have the leaf morphology and leaf colour of the Virginia strain but the leaf size of the Carolina strain?

30. a. In *Drosophila*, crosses between F_1 heterozygotes of the form *A b/a B* always yield the same ratio of phenotypes in the F_2 progeny regardless of the distance between the two genes (assuming complete dominance for both autosomal genes). What is this ratio? Would this also be the case if the F_1 heterozygotes were *A B/a b*?

b. If you intercrossed F_1 heterozygotes of the form *A b/a B* in mice, the phenotypic ratio among the F_2 progeny would vary with the map distance between the two genes. Is there a simple way to estimate the map distance based on the frequencies of the F_2 phenotypes, assuming rates of recombination are equal in males and females? Could you estimate map distances in the same way if the mouse F_1 heterozygotes were *A B/a b*?

31. The following list of four *Drosophila* mutations indicates the symbol for the mutation, the name of the gene, and the mutant phenotype:

Allele symbol	Gene name	Mutant phenotype
dwp	*dwarp*	small body, warped wings
rmp	*rumpled*	deranged bristles
pld	*pallid*	pale wings
rv	*raven*	dark eyes and bodies

You perform the following crosses with the indicated results:

Cross #1: dwarp, rumpled females × pallid, raven males
→ dwarp, rumpled males and wild-type females

Cross #2: pallid, raven females × dwarp, rumpled males
→ pallid, raven males and wild-type females

F_1 females from cross #1 were crossed to males from a true-breeding *dwarp, rumpled, pallid, raven* stock. The 1000 progeny obtained were as follows:

pallid	3
pallid, raven	428
pallid, raven, rumpled	48
pallid, rumpled	23
dwarp, raven	22
dwarp, raven, rumpled	2
dwarp, rumpled	427
dwarp	47

Indicate the best map for these four genes, including all relevant data. Calculate interference values where appropriate.

Section 4.5

32. A diploid strain of yeast has a wild-type phenotype but the following genotype:

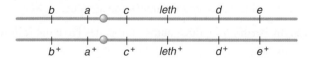

a, *b*, *c*, *d*, and *e* all represent recessive alleles that yield a visible phenotype, and *leth* represents a recessive lethal mutation. All genes are on the same chromosome, and *a* is very tightly linked to its centromere (indicated by a small circle). Which of the following phenotypes could be found in sectors resulting from mitotic recombination in this cell? (1) *a*; (2) *b*; (3) *c*; (4) *d*; (5) *e*; (6) *b e*; (7) *c d*; (8) *c d e*; (9) *d e*; (10) *a b*. Assume that double mitotic crossovers are too rare to be observed.

33. A single yeast cell placed on solid agar will divide mitotically to produce a colony of about 10^7 cells. A haploid yeast cell that has a mutation in the *ade2* gene will produce a red colony; an *ade2+* colony will be white. Some of the colonies formed from diploid yeast cells with a genotype of *ade2+/ade2−* will contain sectors of red within a white colony.

a. How would you explain these sectors?

b. Although the white colonies are roughly the same size, the red sectors within some of the white colonies vary markedly in size. Why? Do you expect the majority of the red sectors to be relatively large or relatively small?

34. Neurofibromas are tumours of the skin that can arise when a skin cell that is originally $NF1^+/NF1^-$ loses the $NF1^+$ allele. This wild-type allele encodes a functional tumour suppressor protein, while the $NF1^-$ allele encodes a nonfunctional protein.

    A patient of genotype $NF1^+/NF1^-$ has 20 independent tumours in different areas of the skin. Samples are taken of normal, noncancerous cells from this patient, as well as of cells from each of the 20 tumours. Extracts of these samples are analyzed by a technique called gel electrophoresis that can detect variant forms of four different proteins (A, B, C, and D) all encoded by genes that lie on the same autosome as $NF1$. Each protein has a slow (S) and a fast (F) form that are encoded by different alleles (e.g., A^S and A^F). In the extract of normal tissue, slow and fast variants of all four proteins are found. In the extracts of the tumours, 12 had only the fast variants of proteins A and D but both the fast and slow variants of proteins B and C; 6 had only the fast variant of protein A but both the fast and slow variants of proteins B, C, and D; and the remaining 2 tumour extracts had only the fast variant of protein A, only the slow variant of protein B, the fast and slow variants of protein C, and only the fast variant of protein D.

a. What kind of genetic event described in this chapter could cause all 20 tumours, assuming that all the tumours are produced by the same mechanism?

b. Draw a genetic map describing these data, assuming that this small sample represents all the types of tumours that could be formed by the same mechanism in this patient. Show which alleles of which genes lie on the two homologous chromosomes. Indicate all relative distances that can be estimated.

c. Another mechanism that can lead to neurofibromas in this patient is a mitotic error producing cells with 45 rather than the normal 46 chromosomes. How can this mechanism cause tumours? How do you know, just from the results described, that none of these 20 tumours is formed by such mitotic errors?

d. Can you think of any other type of error that could produce the results described?

35. In *Drosophila*, the *yellow* (*y*) gene is near the end of the acrocentric X chromosome, while the *singed* (*sn*) gene is located near the middle of the X chromosome. On the wings of female flies of genotype $y\ sn/y^+\ sn^+$, you can very rarely find patches of yellow tissue within which a small subset of cells also have singed bristles.

a. How can you explain this phenomenon?

b. Would you find similar patches on the wings of females having the genotype $y^+\ sn/y\ sn^+$?

![Mc Graw Hill Education] **connect** For more information on the resources available from McGraw-Hill Ryerson, go to www.mcgrawhill.ca/he/solutions.

Problems   **141**

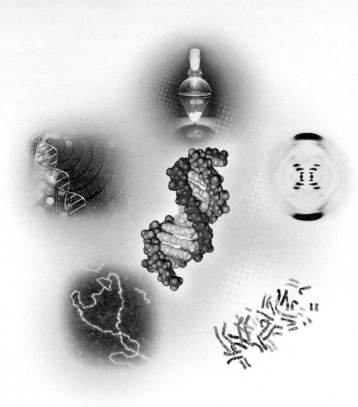

The DNA double helix can be viewed from many unique perspectives (biochemical, biophysical, informatical). An understanding of the many amazing properties of DNA is crucial to appreciating its biological role. *Clockwise from top*: DNA precipitated in ethanol (DNA can be seen as a fluffy white material at the bottom of the flask); X-ray diffraction pattern of B DNA; human karyotype with colour added to differentiate chromosome pairs; atomic force microscopy of a plasmid DNA molecule; an artist's abstract work highlighting the digital nature of DNA; space filling model of B DNA (*centre image*).

In Chapters 2–4, we examined how Mendel used data from breeding experiments to deduce the existence of abstract units of heredity that were later called genes, and how microscopists associated these entities with movements of chromosomes during mitosis and meiosis. These discoveries provided a foundation for predicting the *likelihood* that offspring from defined crosses would express genetically transmitted traits. But in the absence of knowledge about the molecule that carried genetic information, it was impossible to understand anything about the biological processes through which genes determine phenotypes, transmit instructions between generations, and evolve new information. For this reason, we shift our focus to the DNA molecule. The importance of understanding DNA at the molecular level has, perhaps, been most eloquently stated by Francis Crick, the co-discoverer of DNA's double-helical structure and leading theoretician of molecular biology. In 1988, Crick wrote that "Almost all aspects of life are engineered at the molecular level, and without understanding molecules, we can only have a very sketchy understanding of life itself."

As we extend our analysis of the DNA double helix, two general themes emerge. First, DNA's genetic functions flow directly from its molecular structure—the way its atoms are arranged in space. Second, all of DNA's genetic functions depend on specialized proteins that interact with it and "read" the information it carries (because DNA itself is chemically inert). In fact, DNA's lack of chemical reactivity makes it an ideal physical container for long-term maintenance of genetic information in living organisms, as well as their nonliving remains. This last property—combined with modern sequencing technology—has even allowed scientists to analyze DNA sequences from extinct species such as the quagga, a zebra subspecies whose last living member died in captivity in 1883 (**Figure 5.1**).

Chapter Outline
5.1 Evidence That DNA Is the Genetic Material
5.2 The Watson and Crick Double Helix Model of DNA
5.3 DNA Replication

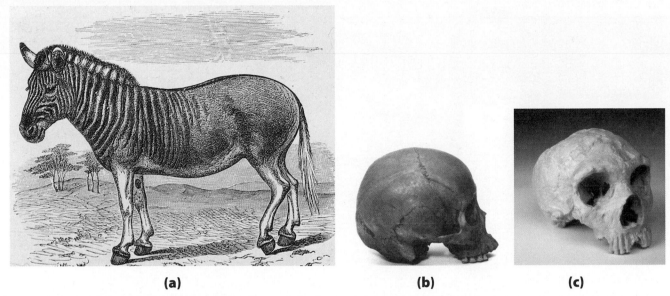

(a)   **(b)**   **(c)**

Figure 5.1 Ancient DNA still carries information. Molecular biologists have successfully extracted and determined the sequence of DNA from **(a)** the remains of a 100-year-old quagga (artist rendition); **(b)** an 8000-year-old human skull; and **(c)** a 38 000-year-old Neanderthal skull. These findings attest to the chemical stability of DNA, the molecule of inheritance.

Learning Objectives

1. Evaluate and appraise the evidence supporting the hypothesis that DNA is the genetic material.

2. Relate DNA's overall molecular structure to the chemical properties of the moieties making up a nucleotide.

3. Relate the molecular structure of DNA to its biological functions.

4. Discriminate between the conservative, semi-conservative, and dispersive models of DNA replication.

5. Summarize the molecular reactions and interactions that drive the process of DNA replication.

5.1   Evidence That DNA Is the Genetic Material

While the importance of the DNA molecule is universally recognized today, only a short time ago (in the early twentieth century) geneticists still did not know that DNA was the genetic material. In fact, it took a cohesive pattern of results, from experiments performed over more than 50 years, to convince the scientific community that DNA was indeed the molecule of heredity. In the following section, we will take you on a historical and scientific journey along the path that led to our current understanding of the biological significance of DNA.

Chemical studies locate DNA in chromosomes

In 1869, Friedrich Miescher extracted a weakly acidic, phosphorus-rich material from the nuclei of human white blood cells and named it "nuclein" (**Figure 5.2**). It was unlike any previously reported chemical compound, and its major component turned out to be DNA. The full chemical name of DNA is <u>d</u>eoxyribo<u>n</u>ucleic <u>a</u>cid. The name reflects three characteristics of the substance: (1) that it is found mainly in cell nuclei; (2) that one of its constituents is a sugar known as deoxyribose; and (3) that it is acidic.

After purifying DNA from the nuclein and performing chemical tests, researchers established that it contained only four distinct chemical building blocks linked in a long chain (**Figure 5.3**). The four individual chemicals belong to a class of compounds known as **nucleotides**; the bonds joining one nucleotide to another are covalent **phosphodiester bonds**; and the linked chain of building block subunits is a type of **polymer**.

A procedure first reported in 1923 made it possible to discover where in the cell DNA resides. Named the Feulgen reaction after its designer, the procedure relies on a chemical called the Schiff reagent, which stains DNA red (**Figure 5.4**). In a preparation of stained cells, the chromosomes redden, while other areas of the cell remain relatively colourless. The reaction shows that DNA is localized almost exclusively within chromosomes.

The finding that DNA was a component of chromosomes did not prove that the molecule had anything to do

Figure 5.2 In 1869, Friedrich Miescher became the first person to isolate DNA. *From left to right:* Friedrich Miescher; a DNA sample isolated by Miescher; Miescher's 1871 publication describing his discovery; and Miescher's laboratory.

with genes. It only suggested the possibility that DNA might be the hereditary material. Another "suspect" in this respect was protein. It was known at the time that in addition to DNA, more than half of a typical eukaryotic chromosome (by weight) comprised protein. Furthermore, because proteins are built of 20 different amino acids, whereas DNA carries just four building-block subunits, many researchers thought proteins had greater potential for diversity and were better suited to serve as the genetic material. These same scientists assumed that even though DNA was an important part of chromosome structure, it was too simple to specify the complexity of genes.

Bacterial transformation implicates DNA as the genetic material

Several studies supported the idea that DNA was the chemical substance carrying genetic information. The most important of these used single-celled bacteria. In a 1923 study of *Streptococcus pneumoniae* bacteria, Frederick Griffith distinguished two bacterial forms: smooth (S) and rough (R). S is the wild type; a mutation in S gives rise to R. From observation and biochemical analysis, Griffith determined that S forms appear smooth because they synthesize a polysaccharide capsule that surrounds pairs of cells. R forms, which arise spontaneously as mutants of S, are unable to make the capsular polysaccharide, and as a result, their colonies appear to have a rough surface (**Figure 5.5**). We now know that the R form lacks an enzyme necessary for synthesis of the capsular polysaccharide. Because the polysaccharide capsule helps protect the bacteria from an animal's immune response, the S bacteria are virulent and kill most laboratory animals exposed to them (**Figure 5.6a**); by contrast, the R forms fail to cause infection (**Figure 5.6b**). In humans, the virulent S forms of *S. pneumoniae* can cause pneumonia.

The phenomenon of transformation

In 1928, Griffith published the astonishing finding that genetic information from dead bacterial cells could somehow be transmitted to live cells. He was working with two types of *S. pneumoniae* bacteria—live R forms and heat-killed S forms. Neither the heat-killed S forms nor the live R forms produced infection when injected into laboratory mice (**Figure 5.6b** and **c**); but a mixture of the two killed the animals (**Figure 5.6d**). Furthermore, bacteria recovered from the blood of the dead animals were living S forms (Figure 5.6d). The ability of a substance to change the genetic characteristics of an organism is known as **transformation**. Something from the heat-killed S bacteria must have transformed the living R bacteria into S. This transformation was permanent and most likely genetic, because all future generations of the bacteria grown in culture were the S form. While undoubtedly a fascinating observation, the key question still remained. What biological molecule from the heat-killed S bacteria transformed the living R bacteria?

DNA as the active agent of transformation

By 1929, two other laboratories had repeated these results, and in 1931, investigators in Oswald T. Avery's laboratory found they could achieve transformation without using any animals at all (i.e., by simply growing R-form bacteria in the presence of components from dead S forms) (**Figure 5.7a**). Avery, who was from Nova Scotia (see the **Focus on Inquiry** box **"Genes Are Made of DNA, not Protein"**) then embarked on a quest that would remain the focus of his work for almost 15 years: "Try to find in that complex mixture, the active principle!" In other words, try to identify the heritable substance in the bacterial extract that induces transformation of harmless R bacteria into pathogenic S bacteria. Avery dubbed the substance he was searching for the "transforming principle" and spent many years trying

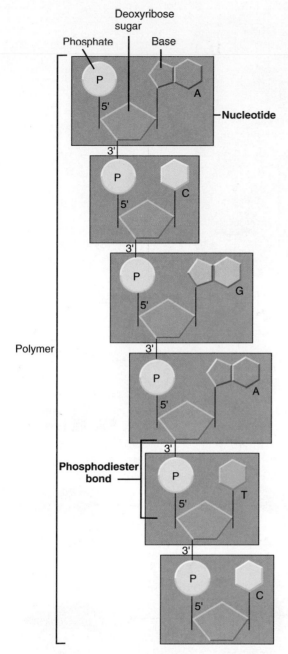

Figure 5.3 **The chemical composition of DNA.** A single strand of a DNA molecule consists of a chain of nucleotide subunits (*blue boxes*). Each nucleotide is made of the sugar deoxyribose (*tan pentagons*) connected to an inorganic phosphate group (*yellow circles*) and to one of four nitrogenous bases (*purple or green polygons*). The phosphodiester bonds that link the nucleotide subunits to each other attach the phosphate group of one nucleotide to the deoxyribose sugar of the preceding nucleotide.

to purify it sufficiently to be able to identify it unambiguously. He and his co-workers eventually prepared a tangible, active transforming principle. In the final part of their procedure, a long whitish wisp materialized from ice-cold alcohol solution and wound around the glass stirring rod to form a fibrous wad of nearly pure principle (**Figure 5.7b**).

Once purified, the transforming principle had to be characterized. In 1944, Avery and two co-workers, Colin MacLeod (also from Nova Scotia) and American, Maclyn

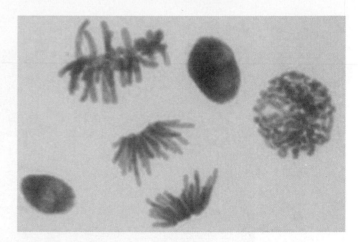

Figure 5.4 **Chromosomes from the onion (*Allum cepa*) visualized using the Feulgen reaction.** Interphase nuclei, as well as nuclei at various stages of mitosis, are visible.

McCarty, published the cumulative findings of experiments designed to determine the transforming principle's chemical composition (**Figure 5.7c**). In these experiments, the purified transforming principle was active at the extraordinarily high dilution of 1 part in 600 million. Although the preparation was almost pure DNA, the investigators nevertheless exposed it to various enzymes to see if some molecule other than DNA could cause transformation. Enzymes that degraded RNA, protein, or polysaccharide had no effect on the transforming principle, but an enzyme that degraded DNA destroyed its activity. The tentative published conclusion was that the transforming principle appeared to be DNA. In a personal letter to his brother, Avery went one step further and confided that the transforming principle "may be a gene."

Despite the paper's abundance of concrete evidence, many within the scientific community still resisted the idea that DNA was the molecule of heredity. They argued that perhaps, Avery's results reflected the activity of contaminants; or perhaps, genetic transformation was not happening at all, and instead, the purified material somehow triggered a physiological switch that transformed bacteria. Unconvinced for the moment, these scientists remained attached to the idea that protein molecules constituted the genetic material.

Viral studies point to DNA, not protein, in replication

Not everyone shared this scepticism. Alfred Hershey and Martha Chase anticipated that they could assess the relative importance of DNA and protein in gene replication by infecting bacterial cells with viruses called **phages**, short for **bacteriophages** (literally "bacteria eaters").

Viruses are the simplest of organisms. By structure and function, they fall somewhere between living cells capable of reproducing themselves and macromolecules such as proteins. Because viruses hijack the molecular machinery of their host cell to carry out growth and replication, they can be very small indeed and contain very few genes.

Figure 5.5 Griffith's demonstration of bacterial transformation. Smooth (S) and rough (R) colonies of *S. pneumoniae*.

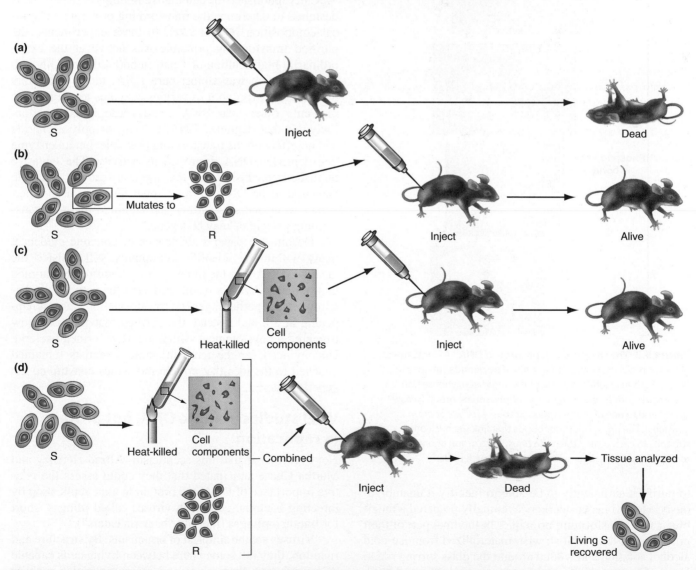

Figure 5.6 Griffith's experiment. (a) S bacteria are virulent and can cause lethal infections when injected into mice. **(b)** Injections of R mutants by themselves do not cause infections that kill mice. **(c)** Similarly, injections of heat-killed S bacteria do not cause lethal infections. **(d)** Lethal infection does result, however, from injections of live R bacteria mixed with heat-killed S strains; the blood of the dead host mouse contains living S-type bacteria.

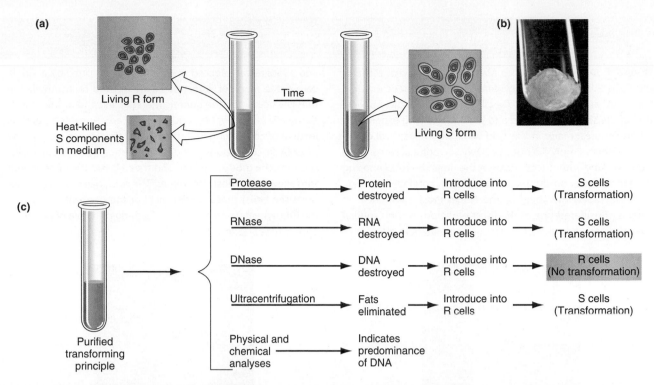

Figure 5.7 The transforming principle is DNA: Experimental confirmation. (a) Bacterial transformation occurs in culture medium containing the remnants of heat-killed S bacteria. Some "transforming principle" from the heat-killed S bacteria is taken up by the live R bacteria, converting (transforming) them into virulent S strains. **(b)** A solution of purified DNA extracted from white blood cells. **(c)** Chemical fractionation of the transforming principle. Treatment of purified DNA with a DNA-degrading enzyme destroys its ability to cause bacterial transformation, while treatment with enzymes that destroy other macromolecules has no effect on the transforming principle.

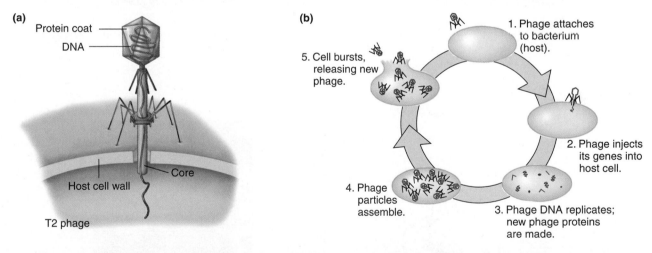

Figure 5.8 Experiments with viruses provide convincing evidence that genes are made of DNA. (a) Bacteriophage T2 structure. The phage particle consists of DNA contained within a protein coat. **(b)** Bacteriophage T2 life cycle. The virus attaches to the bacterial host cell and injects its genes (the DNA) through the bacterial cell wall into the host cell cytoplasm. Inside the host cell, these genes direct the formation of new phage DNA and proteins, which assemble into progeny phages that are released into the environment when the cell bursts.

Each phage particle consists of roughly equal weights of protein and DNA (**Figure 5.8a**). These phage particles can reproduce themselves only after infecting a bacterial cell. Thirty minutes after infection, the cell bursts and hundreds of newly made phages spill out (shown in **Figure 5.8b**). The question is, what substance contains the information used to produce the new phage particles—DNA or protein?

With the invention of the electron microscope in 1939, it became possible to see individual phages, and surprisingly, electron micrographs revealed that the entire phage does not enter the bacterium it infects. Instead, a viral shell—called a *ghost*—remains attached to the outer surface of the bacterial cell wall. Because the empty phage coat remains outside the bacterial cell, one investigator likened phage particles to tiny syringes that bind to the cell

Oswald T. Avery, Colin Munro MacLeod, and Maclyn McCarty provided the first rigorous evidence that the genetic material was DNA, and not protein. Despite their meticulous application of the scientific method and their painstaking experimental procedures, many in the field were highly sceptical of the result. Others simply did not care. Only a handful of researchers were working with nucleic acids at the time, the rest choosing instead to focus on the biological role of proteins. This scepticism included members of the Nobel Prize committee. In what is considered one of the most egregious Nobel "snubs," Avery—even though continuously nominated in the 1940s and 1950s—was blocked from winning the prize by a single committee member who steadfastly refused to acknowledge that DNA could be the genetic material. In spite of the quality of their work, Avery, Macleod, and McCarty remained humble. Instead of concluding with their belief that genes were made of DNA (a conviction confided in a letter from Avery to his brother), the group chose to end their classic 1944 paper with the following modest statement: "The evidence presented supports the belief that a nucleic acid of the deoxyribose type is the fundamental unit of the transforming principle of pneumococcus type III [S form]."

Oswald T. Avery (1937)

Maclyn McCarty (1942)

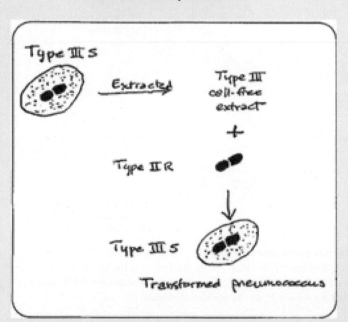

Handdrawn schematic of Avery's model of cell transformation.

Colin MacLeod (1940)

surface and inject the material containing the information needed for viral replication into the host cell.

In their famous Waring blender experiment of 1952, Alfred Hershey and Martha Chase tested the idea that the ghost left on the cell wall is composed of protein, while the injected material consists of DNA (**Figure 5.9**). A type of phage known as T2 served as their experimental system. They grew two separate sets of T2 in bacteria maintained

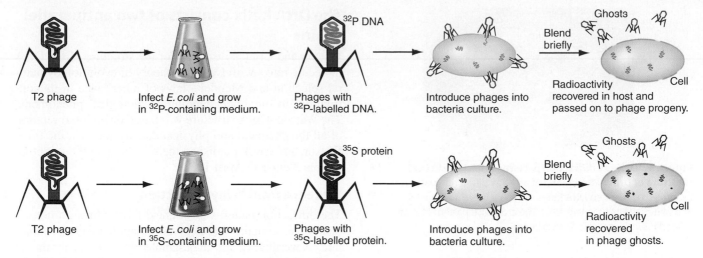

Figure 5.9 The Hershey–Chase Waring blender experiment. T2 bacteriophage particles either with ^{32}P-labelled DNA or with ^{35}S-labelled proteins were used to infect bacterial cells. After a short incubation, Hershey and Chase shook the cultures in a Waring blender and spun the samples in a centrifuge to separate the empty viral ghosts from the heavier infected cells. Most of the ^{35}S-labelled proteins remained with the ghosts, while most of the ^{32}P-labelled DNA was found in the sediment with the T2 gene-containing infected cells.

in two different types of culture media: one that was infused with radioactively labelled phosphorus (^{32}P), and another that was infused with radioactively labelled sulphur (^{35}S). Because proteins incorporate sulphur (but not phosphorus) and DNA incorporates phosphorus (but not sulphur), phages grown on ^{35}S would have radioactively labelled protein while particles grown on ^{32}P would have radioactive DNA. The radioactive tags would serve as markers for the location of each material when the phages infected fresh cultures of bacterial cells.

After exposing one fresh culture of bacteria to ^{32}P-labelled phage and another culture to ^{35}S-labelled phage, Hershey and Chase used a Waring blender to disrupt each one, effectively separating the viral ghosts from the bacteria harbouring the viral genes. Centrifugation of the cultures then separated the heavier infected cells, which ended up in a pellet at the bottom of the tube, from the lighter phage ghosts, which remained suspended in the supernatant solution. Most of the radioactive ^{32}P (in DNA) went to the pellet, while most of the radioactive ^{35}S (in protein) remained in the supernatant. This confirmed that the extracellular ghosts were indeed mostly protein, while the injected viral material specifying production of more phages was mostly DNA. Bacteria containing the radiolabelled phage DNA behaved just as they did when infected with non-labelled

phage, producing and disgorging hundreds of progeny particles. From these observations, Hershey and Chase concluded that phage genes are made of DNA.

The Hershey–Chase experiment, although less rigorous than the Avery project, had an enormous impact. In the minds of many investigators, it confirmed Avery's results and extended them to viral particles. The spotlight was now clearly on DNA.

In the following section, we will begin our in-depth analysis of the DNA molecule. In this way, we can better understand how its many unique properties relate to its biological functions. We will start by analyzing DNA's three-dimensional structure, looking first at the nucleotide building blocks, then at how those subunits are linked together in a polynucleotide chain, and finally, at how two chains associate to form a double helix.

> Experimental evidence in the early to mid-twentieth century pointed to DNA as the genetic material. DNA was identified as a component of chromosomes, was implicated as the agent of bacterial transformation, and was shown to be the information-containing compound that bacteriophages inject into the bacteria they infect.

5.2   The Watson and Crick Double Helix Model of DNA

Under the appropriate experimental conditions, purified molecules of DNA can align alongside each other in fibres to produce an ordered structure. And just as a crystal chandelier scatters light to produce a distinctive pattern on the wall, DNA fibres scatter X-rays to produce a characteristic diffraction pattern (**Figure 5.10**). A knowledgeable X-ray crystallographer can interpret DNA's diffraction pattern to deduce certain aspects of the molecule's three-dimensional

structure. When in the spring of 1951 the 23-year-old James Watson learned that DNA could project a diffraction pattern, he realized that it "must have a regular structure that could be solved in a straightforward fashion."

Nucleotides are the building blocks of DNA

DNA is a long polymer composed of subunits known as nucleotides. Each nucleotide consists of a deoxyribose

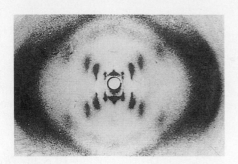

Figure 5.10 X-ray diffraction patterns reflect the helical structure of DNA. Photograph of an X-ray diffraction pattern produced by oriented DNA fibres, taken by Rosalind Franklin and Maurice Wilkins in late 1952. The crosswise pattern of X-ray reflections indicates that DNA is helical.

sugar, a phosphate, and one of four nitrogenous bases. Detailed knowledge of these chemical constituents and the way they combine played an important role in Watson and Crick's model building.

The components of a nucleotide

Figure 5.11 depicts the chemical composition and structure of deoxyribose, phosphate, and the four nitrogenous bases; how these components come together to form a nucleotide; and how phosphodiester bonds link the nucleotides in a DNA chain. Each individual carbon or nitrogen atom in the central ring structure of a nitrogenous base is assigned a number: from 1 to 9 for purines, and 1 to 6 for pyrimidines. The carbon atoms of the deoxyribose sugar are distinguished from atoms within the nucleotide base by the use of primed numbers from $1'$ to $5'$. Covalent attachment of a base to the $1'$ carbon of deoxyribose forms a *nucleoside*. The addition of a phosphate group to the $5'$ carbon forms a complete *nucleotide*.

Information contained in a directional base sequence

Information can be encoded only in a sequence of symbols whose order varies according to the message to be conveyed. Without this sequence variation, there is no potential for carrying information. Because DNA's sugar-phosphate backbone is chemically identical for every nucleotide in a DNA chain, the only difference between nucleotides is in the identity of the nitrogenous base. Thus, if DNA carries genetic information, that information must consist of variations in the sequence of the A, G, T, and C bases. The information constructed from the 4-letter language of DNA bases is analogous to the information built from the 26-letter alphabet of English or French or Italian. Just as you can combine the 26 letters of the alphabet in different ways to generate the words of a book, so too, different combinations of the four bases in very long sequences of nucleotides can encode the information for constructing an organism (see the **Focus on Critical Thinking** box "Information Theory").

The DNA helix consists of two antiparallel chains

Watson and Crick's discovery of the structure of the DNA molecule ranks with Darwin's theory of evolution by natural selection and Mendel's laws of inheritance in its contribution to our understanding of biological phenomena. The Watson–Crick structure was based on an interpretation of all the chemical and physical data available at the time. Watson and Crick published their findings in the scientific journal *Nature* in April 1953.

Evidence from X-ray diffraction

The diffraction patterns of oriented DNA fibres do not, on their own, contain sufficient information to reveal structure. Nevertheless, the photographs do reveal a wealth of structural information to the trained eye. Excellent X-ray images produced by Rosalind Franklin and Maurice Wilkins showed that the molecule is spiral-shaped, or helical; the spacing between repeating units along the axis of the helix is 3.4 angstroms (Å); the helix undergoes one complete turn every 34 Å; and the diameter of the molecule is 20 Å. This diameter is roughly twice the width of a single nucleotide as it is depicted in Figure 5.11, suggesting that a DNA molecule might be composed of two side-by-side DNA chains.

Complementary base-pairing

If a DNA molecule contains two side-by-side chains of nucleotides, what forces hold these chains together? Erwin Chargaff provided an important clue with his data on the nucleotide composition of DNA from various species. Despite large variations in the relative amounts of the bases, the ratio of A to T is not significantly different from 1:1, and the ratio of G to C is the same in every organism (**Table 5.1**). Watson grasped that the roughly 1:1 ratios of

Table 5.1	Chargaff's Data on Nucleotide Base Composition in the DNA of Various Organisms					
	Percentage of Base in DNA				Ratios	
Organism	A	T	G	C	A:T	G:C
Staphylococcus afermentans	12.8	12.9	36.9	37.5	0.99	0.99
Escherichia coli	26.0	23.9	24.9	25.2	1.09	0.99
Yeast	31.3	32.9	18.7	17.1	0.95	1.09
Caenorhabditis elegans*	31.2	29.1	19.3	20.5	1.07	0.96
Arabadopsis thaliana*	29.1	29.7	20.5	20.7	0.98	0.99
Drosophila melanogaster	27.3	27.6	22.5	22.5	0.99	1.00
Honeybee	34.4	33.0	16.2	16.4	1.04	0.99
Mus musculus (mouse)	29.2	29.4	21.7	19.7	0.99	1.10
Human (liver)	30.7	31.2	19.3	18.8	0.98	1.03

*Data for *C. elegans* and *A. thaliana* are based on those for close relative organisms.
Note that even though the level of any one nucleotide is different in different organisms, the amount of A always approximately equals the amount of T, and the level of G is always similar to that of C. Moreover, as you can calculate for yourself, the total amount of purines (A plus G) nearly always equals the total amount of pyrimidines (C plus T).

A Detailed Look at DNA's Chemical Constituents

(a) The separate entities

1. Deoxyribose sugar

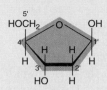

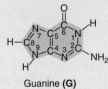

Ribose

2. A phosphate group

3. Four nitrogenous bases

Purines

Adenine **(A)**

Guanine **(G)**

Pyrimidines

Thymine **(T)**

Cytosine **(C)**

(b) Assembly into a nucleotide

1. Attachment of base to sugar

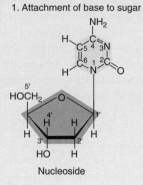

Nucleoside

2. Addition of phosphate

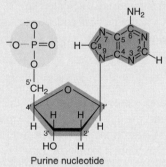

Purine nucleotide

Pyrimidine nucleotide

(c) Nucleotides linked in a directional chain

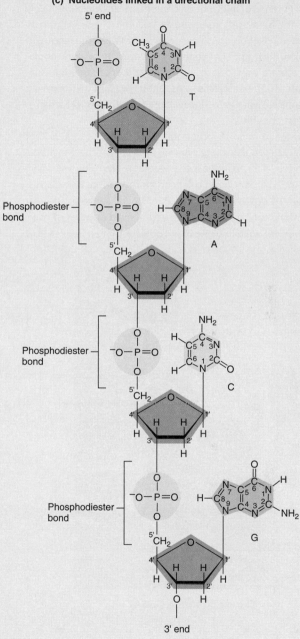

5' end

T

Phosphodiester bond

A

Phosphodiester bond

C

Phosphodiester bond

G

3' end

If one was provided with the genome sequences of four organisms—let us say a virus, a human, a yeast, and a bacterium—and then was asked to rank them (from lowest to highest) in terms of information content, most would intuitively provide the following order: virus < bacteria < yeast < human. What criteria did you use to create this list? Was it the length (in base pairs), the number of genes, the complexity of gene regulation? Is there a way to formally define and quantitate information content? The answer is a resounding yes! In the following paragraphs we will show—in an abstract, mathematical sense—how this is possible.

In order to encode information, one simply requires that there be (1) a set of discrete symbols and (2) that these symbols be encoded in a sequential order. A minimum of two symbols (e.g., 0 and 1) are required to encode information. Using a system having more than two symbols will increase the quantity of information that can be encoded with a fixed number of characters, but will not allow one to encode anything more novel or complex than a simpler binary system. For example, to encode the number one hundred and forty-seven using a decimal system (i.e., using the symbols 0 to 9), you could simply write "147." To encode one hundred and forty-seven using a binary system (i.e., using only 0's and 1's) you could write "10010011." The same quantity is represented; it only took a longer string of characters to encode it using binary. The take-home message here is that having only four symbols in no way limits the ability of DNA to encode complex biological information.

In addition to having discrete symbols, the order in which those symbols appear must also be defined. If one were to write "genetics," each letter would have to appear in a specific position in the string of characters (i.e., "g" in position 1, "e" in position 2 … etc.). If the sequence of letters is not specified, then any number of words (mostly nonsense words like "netsegic") can be written. As you probably noticed, DNA satisfies this criterion. Living things encode information as a string of A's, G's, T's, and C's, which appear in a fixed sequence, as part of their chromosomal complement.

A mathematical equation developed by the founder of information theory, Claude Shannon, that captures these ideas is written as

$$H = L \cdot \log_2 M$$

where L represents the length of the string of characters, M represents the number of discrete symbols, and H can be thought of (at least in this simple scenario) as a measure of the information content in units of "bits." For example, a DNA molecule made up of a string of 68 nucleotides could carry

$$68 \cdot \log_2 4 = 136 \text{ bits of information}$$

Before one dismisses the above discussion as too hypothetical for practically minded geneticists, be reminded of the words of Gottfried Wilhelm Leibniz, one of the co-creators of calculus:

"Without mathematics we cannot penetrate deeply into philosophy. Without philosophy we cannot penetrate deeply into mathematics. Without both we cannot penetrate deeply into anything."

Undoubtedly, the simple ideas presented above cannot in and of themselves solve any biological problems. In order for that to happen, one would need to engage in … more critical thinking!

Claude Shannon (1954)

A to T and of G to C reflect a significant aspect of the molecule's inherent structure.

To explain Chargaff's ratios in terms of chemical affinities between A and T and between G and C, Watson made cardboard cut-outs of the bases in the chemical forms they assume in a normal cellular environment. He then tried to match these up in various combinations, like pieces in a jigsaw puzzle. He knew that the particular arrangement of atoms on purines and pyrimidines play a crucial role in molecular interactions as they can participate in the formation of **hydrogen bonds**: weak electrostatic bonds that result in a partial sharing of hydrogen atoms between reacting groups (**Figure 5.12**). Watson saw that A and T could be paired together such that two hydrogen bonds formed between them. If G and C were similarly paired, hydrogen bonds could also easily connect the nucleotides carrying these two bases. (Watson originally posited two hydrogen bonds between G and C, but there are actually

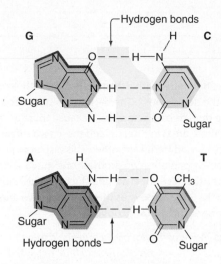

Figure 5.12 Complementary base-pairing. An A on one strand can form two hydrogen bonds with a T on the other strand. G on one strand can form three hydrogen bonds with a C on the other strand. The size and shape of A–T and of G–C base pairs are similar, allowing both to fill the same amount of space between the two backbones of the double helix.

three.) Remarkably, the two pairs (A–T and G–C) had essentially the same shape. This meant that the two pairs could fit in any order between two sugar-phosphate backbones without distorting the structure. It also explained the Chargaff ratios—always equal amounts of A and T and of G and C. Note that both of these base pairs consist of one purine and one pyrimidine. Crick connected the chemical facts with the X-ray data, recognizing that because of the geometry of the base-sugar bonds in nucleotides, the orientation of the bases in Watson's pairing scheme could arise only if the bases were attached to backbones running in opposite directions. **Figure 5.13** illustrates and explains the model Watson and Crick proposed in April 1953: DNA as a double helix.

The double helix may assume alternative forms

Watson and Crick arrived at the double helix model of DNA structure by building models, not by a direct structural determination from the data alone. And even though Watson has written that "a structure this pretty just had to exist," the beauty of the structure is not necessarily evidence of its correctness. At the time of its presentation, the strongest evidence for its correctness was its physical plausibility, its chemical and spatial compatibility with all available data, and its capacity for explaining many biological phenomena.

B DNA and Z DNA

The majority of naturally occurring DNA molecules have the configuration suggested by Watson and Crick. Such molecules are known as **B-form DNA** and spiral to the right (**Figure 5.14a**). DNA structure is, however, more polymorphic than originally assumed. One type, for example, contains nucleotide sequences that cause the DNA to assume what is known as the **Z form**. In this form the helix spirals to the left and the backbone takes on a zigzag shape (**Figure 5.14b**). Researchers have observed many kinds of unusual non-B structures *in vitro* (in the test tube, literally "in glass"), and they speculate that some of these might occur at least transiently in living cells. There is some evidence, for instance, that Z DNA might exist in certain chromosomal regions *in vivo* (in the living organism). Whether the Z form and other unusual conformations have any biological role remains to be determined.

Linear and circular DNA

The nuclear chromosomes of all eukaryotic organisms are long, linear double helixes, but in some instances chromosomes can be circular (**Figures 5.15a** and **b**). These include the chromosomes of prokaryotic bacteria, the chromosomes of organelles such as the mitochondria and chloroplasts that are found inside eukaryotic cells, and the chromosomes of some viruses, including the papovaviruses that can cause cancers in animals and humans. Such circular chromosomes consist of a covalently closed, double-stranded DNA molecule (i.e., a DNA double helix in which the free ends have joined together through the formation of covalent phosphodiester bonds to form a circle).

Single-stranded and double-stranded DNA

In some viruses, the genetic material consists of relatively small, single-stranded DNA molecules. Once inside a cell, the single strand serves as a template for making a second strand, and the resulting double-stranded DNA then governs the production of more virus particles. Examples of viruses carrying single-stranded DNA are bacteriophages φX174 and M13, and mammalian parvoviruses, which are associated with fetal death and spontaneous abortion in humans. In both φX174 and M13, the single DNA strand is in the form of a covalently closed circle; in the parvoviruses, it is linear (**Figures 5.15c** and **d**).

Remarkably, some viruses, including those that cause polio and AIDS, can even use RNA as their genetic material (**Figure 5.16**). How is RNA different than DNA? There are three major chemical differences. First, RNA takes its name from the sugar ribose, which it incorporates instead of the deoxyribose found in DNA (**Figures 5.16a** and **b**). Second, RNA contains the base uracil (U) instead of the base thymine (T); U, like T, base pairs with A (Figure 5.16a). Finally, most RNA molecules are single-stranded and contain far fewer nucleotides than the very long DNA molecules found in nuclear chromosomes. Some completely double-stranded RNA molecules do nonetheless exist. Even within a single-stranded RNA molecule, if folding brings two oppositely oriented regions that carry complementary nucleotide sequences alongside each other, they can form a short double-stranded, base-paired

The Double Helix Structure of DNA

(a) In a leap of imagination, Watson and Crick took the known facts about DNA's chemical composition and physical arrangement in space and constructed a wire-frame model that not only united the evidence but also served as a basis for explaining the molecule's function.

(a)

(b) In the model (shown on the facing page at the left), two DNA chains spiral around an axis with the sugar-phosphate backbones on the outside and pairs of bases (one from each chain) meeting in the middle. Although both chains wind around the helix axis in a right-handed sense, chemically one of them runs 5′ to 3′ upward, while the other runs in the opposite direction of 5′ to 3′ downward. In short, the *two chains are antiparallel.* The base pairs are essentially flat and perpendicular to the helix axis, and the planes of the sugars are roughly perpendicular to the base pairs. As the two chains spiral about the helix axis, they wrap around each other once every ten base pairs, or once every 34 Å. The result is a double helix that looks like a twisted ladder with the two spiralling structural members composed of sugar-phosphate backbones and the rungs consisting of base pairs.

(c) In a space-filling representation of the model (shown on the facing page at the right), the overall shape is that of

a grooved cylinder with a diameter of 20 Å whose axis is the axis of the double helix. The backbones spiral around the axis like threads on a screw, but because there are two backbones, there are two threads, and these two threads are vertically displaced from each other. This displacement of the backbones generates two grooves, one much wider than the other, that also spiral around the helix axis. Biochemists refer to the wider groove as the **major groove** and the narrower one as the **minor groove**.

The *two chains of the double helix are held together by hydrogen bonds between complementary base pairs*; A–T and G–C (see Figure 5.12). Because the overall shapes of these two base pairs are quite similar, either pair can fit into the structure at each position along the DNA. Moreover, each base pair can be accommodated in the structure in two ways that are the reverse of each other: An A purine may be on strand 1 with its corresponding T pyrimidine on strand 2, or the T pyrimidine may be on strand 1 and the A purine on strand 2. The same is true of G and C base pairs.

(d) Interestingly, within the double-helical structure, the spatial requirements of the base pairs are satisfied if and only if each pair consists of one small pyrimidine and one large purine, and even then, only for the particular pairings of A–T and G–C. Pyrimidine–pyrimidine pairs are too small for the structure, and purine–purine pairs are too large. In addition, A–C and G–T pairs do not fit well together; that is, they do not easily form hydrogen bonds. Complementary base-pairing is thus a logical outgrowth of the molecule's steric requirements. Although any one nucleotide pair forms only two or three hydrogen bonds, the sum of these connections between successive base pairs in a long DNA molecule composed of thousands or millions of nucleotides is one basis of the molecule's great chemical stability.

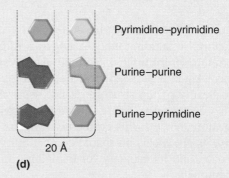

Pyrimidine–pyrimidine

Purine–purine

Purine–pyrimidine

20 Å

(d)

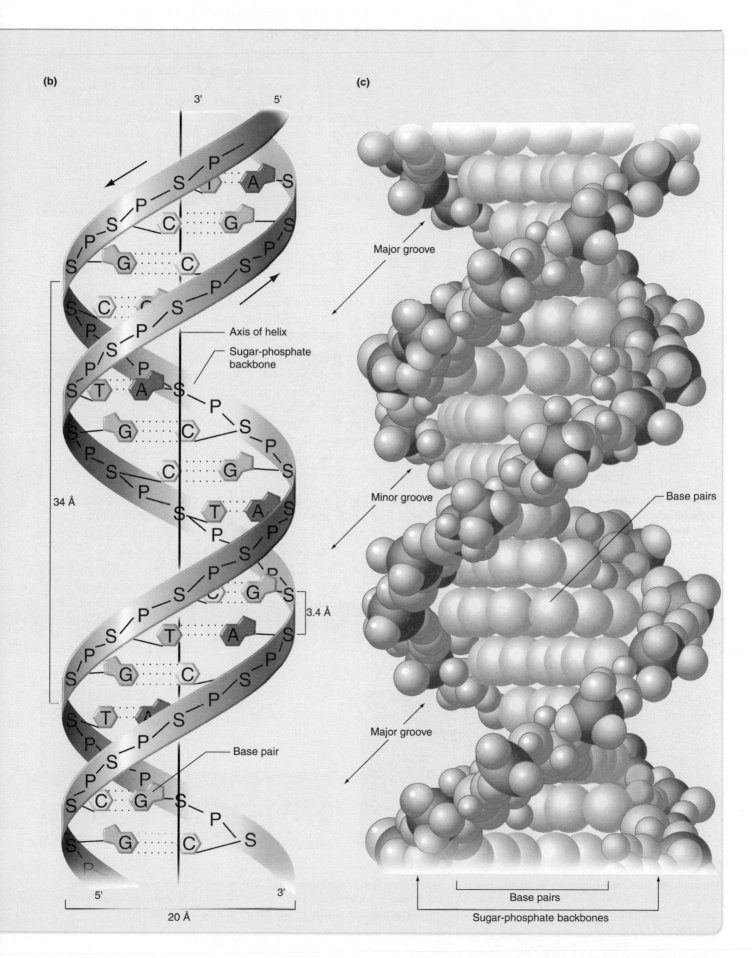

(b)

3' 5'

Axis of helix

Sugar-phosphate
backbone

34 Å

3.4 Å

Base pair

5' 3'

20 Å

(c)

Major groove

Minor groove

Base pairs

Major groove

Base pairs

Sugar-phosphate backbones

(a)

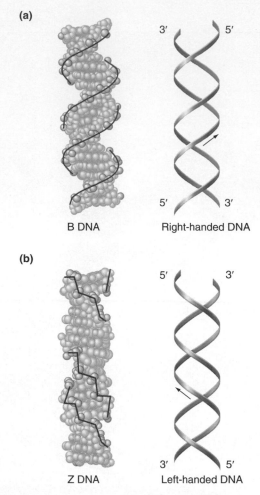

B DNA   Right-handed DNA

(b)

Z DNA   Left-handed DNA

Figure 5.14 Z DNA is one variant of the double helix.
(a) Typical Watson–Crick B-form DNA forms a right-handed helix with a smooth backbone. **(b)** Z-form DNA is left-handed and has an irregular backbone.

stretch within the molecule. This means that, compared with the relatively simple, double-helical shape of a DNA molecule, many RNAs have a complicated structure of short double-stranded segments interspersed with single-stranded loops (**Figure 5.16c**).

DNA structure is the foundation of genetic function

Without sophisticated computational tools for analyzing base sequence, one cannot distinguish bacterial DNA from human DNA. This is because all DNA molecules have the same general chemical properties and physical structure. Proteins, by comparison, are a much more diverse group of molecules with a much greater complexity of structure and function. In his account of the discovery of the double helix, Crick referred to this difference when he said that "DNA is, at bottom, a much less sophisticated molecule than a highly evolved protein and for this reason reveals its secrets more easily."

Four of these "secrets" are embodied in the questions below:

1. How does the molecule carry information?
2. How is that information copied for transmission to future generations?
3. What mechanisms allow the information to change?
4. How does DNA-encoded information govern the expression of phenotype?

The double-helical structure of DNA provides a potential solution to each of these questions, endowing the molecule with the capacity to carry out all the critical functions required of the genetic material. In the remainder of this chapter, we describe how DNA's structure enables it to be copied with great fidelity. The answers to the remaining questions will be discussed in subsequent chapters.

DNA is composed of four nucleotides (A, G, T, and C). Phosphodiester bonds link nucleotides to form a chain with a specific 5'-to-3' polarity. The sequence of nucleotides in a chain specifies genetic information. In the Watson and Crick model for standard B DNA, two antiparallel strands of DNA are held together by the hydrogen bonds of the complementary A–T and C–G base pairs; the two strands are wound around each other in a double helix. DNA can also exist in alternative forms, including Z DNA, circular DNA, and single-stranded DNA.

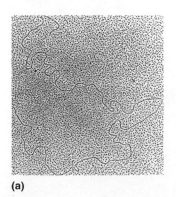

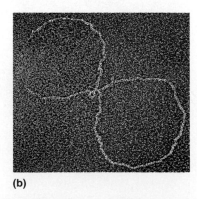

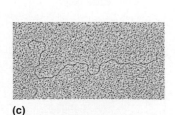

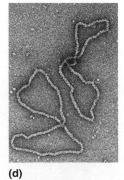

(a)   (b)   (c)   (d)

Figure 5.15 DNA molecules may be linear or circular, double-stranded or single-stranded. These electron micrographs of naturally occurring DNA molecules show **(a)** a fragment of a long, linear double-stranded human chromosome, **(b)** a circular double-stranded papovavirus chromosome, **(c)** a linear single-stranded parvovirus chromosome, and **(d)** circular single-stranded bacteriophage M13 chromosomes.

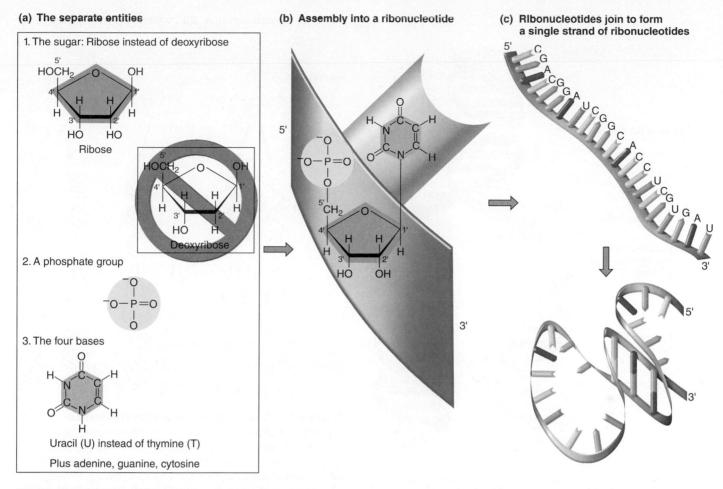

(a) The separate entities

1. The sugar: Ribose instead of deoxyribose

Ribose

Deoxyribose

2. A phosphate group

3. The four bases

Uracil (U) instead of thymine (T)

Plus adenine, guanine, cytosine

(b) Assembly into a ribonucleotide

(c) Ribonucleotides join to form a single strand of ribonucleotides

Figure 5.16 RNA: Chemical constituents and complex folding pattern. (a) and **(b)** Each ribonucleotide contains the sugar ribose, an inorganic phosphate group, and a nitrogenous base. RNA contains the pyrimidine uracil (U) instead of the thymine (T) found in DNA. **(c)** Phosphodiester bonds join ribonucleotides into an RNA chain. Most RNA molecules are single-stranded but are sufficiently flexible so that some regions can fold back and form base pairs with other parts of the same molecule.

5.3 DNA Replication

In one of the most famous understatements in the scientific literature, Watson and Crick wrote at the end of their 1953 paper proposing the double helix model, "It has not escaped our notice that the specific pairing we have postulated immediately suggests a possible copying mechanism for the genetic material." This copying, as we saw in Chapter 3, must precede the transmission of chromosomes from one generation to the next via meiosis, and it is also the basis of the chromosome duplication that occurs prior to each mitosis.

Overview: Complementary base-pairing ensures semiconservative replication

In the process of replication postulated by Watson and Crick, the double helix unwinds to expose the bases in each strand of DNA. Each of the two separated strands then acts as a **template**, or molecular mould, for the synthesis of a new second strand (**Figure 5.17**). The new strand forms as complementary bases align opposite the exposed bases on the parent strand. That is, an A at one position on the original strand signals the addition of a T at the corresponding position on the newly forming strand; a T on the original signifies addition of an A; similarly, G calls for C, and C calls for G, in a process known as **complementary base-pairing**.

Once the appropriate base has aligned and formed hydrogen bonds with its complement, enzymes join the deoxyribose moiety of the nucleotide to the preceding nucleotide by a phosphodiester bond. By repeating this process, nucleotides can be linked together to form the new complementary DNA strand. This mechanism of DNA strand separation and complementary base-pairing followed by the coupling of successive nucleotides yields two "daughter" double helixes that each contain one of the original DNA strands intact (i.e., "conserved") and one completely new strand (**Figure 5.18a**). For this reason, such a pattern of double helix duplication is called **semiconservative replication**: a copying in which one strand

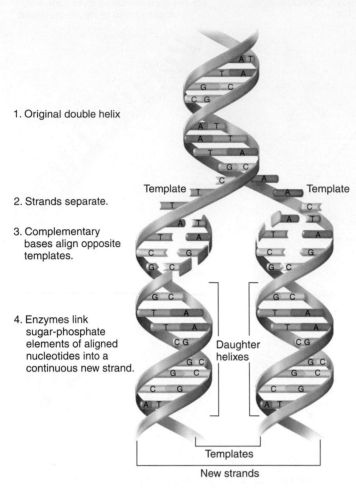

1. Original double helix

2. Strands separate.

3. Complementary bases align opposite templates.

Template   Template

4. Enzymes link sugar-phosphate elements of aligned nucleotides into a continuous new strand.

Daughter helixes

Templates

New strands

Figure 5.17 **The model of DNA replication postulated by Watson and Crick.** Unwinding of the double helix allows each of the two strands to serve as a template for the synthesis of a new strand by complementary base-pairing. The end result: A single double helix becomes transformed into two identical daughter double helixes.

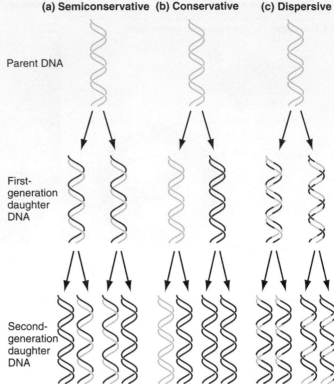

(a) Semiconservative (b) Conservative   (c) Dispersive

Parent DNA

First-generation daughter DNA

Second-generation daughter DNA

Figure 5.18 **Three possible models of DNA replication.** DNA from the original double helix is *blue*; newly made DNA is *magenta*. **(a)** Semiconservative replication (the Watson–Crick model). **(b)** Conservative replication: The parental double helix remains intact; both strands of one daughter double helix are newly synthesized. **(c)** Dispersive replication: At completion, both strands of both double helixes contain both original and newly synthesized material.

of each new double helix is conserved from the parent molecule and the other is newly synthesized.

Watson and Crick's proposal is not the only replication mechanism imaginable. **Figures 5.18b** and **c** illustrate two possible alternatives. With *conservative* replication, one of the two "daughter" double helixes would consist entirely of original DNA strands, while the other helix would consist of two newly synthesized strands. With *dispersive* replication, both "daughter" double helixes would carry blocks of original DNA interspersed with blocks of newly synthesized material. These alternatives are less satisfactory because they do not immediately suggest a mechanism for copying the information in the sequence of bases, and they do not explain the research data (presented below) as well as does semiconservative replication.

Experiments with "heavy" nitrogen verify semiconservative replication

In 1958, Matthew Meselson and Franklin Stahl performed an experiment that confirmed the semiconservative nature

of DNA replication (**Figure 5.19**). The experiment depended on being able to distinguish pre-existing "parental" DNA from newly synthesized daughter DNA. To accomplish this, Meselson and Stahl controlled the isotopic composition of the nucleotides incorporated in the newly forming daughter strands as follows. They grew *E. coli* bacteria for many generations on media in which all the nitrogen was the normal isotope ^{14}N; these cultures served as a control. They grew other cultures of *E. coli* for many generations on media in which the only source of nitrogen was the heavy isotope ^{15}N. After several generations of growth on heavy-isotope medium, essentially all the nitrogen atoms in the DNA of these bacterial cells were labelled with (i.e., contained) ^{15}N. The cells in some of these cultures were then transferred to new medium in which all the nitrogen was ^{14}N. Any DNA synthesized after the transfer would contain the lighter isotope.

Meselson and Stahl isolated DNA from cells grown in the different nitrogen-isotope cultures and then subjected these DNA samples to *equilibrium density gradient centrifugation*, an analytical technique they had just developed. In a test tube, they dissolved the DNA in a solution of the dense salt cesium chloride (CsCl) and spun these solutions at very high speed (about 50 000 revolutions

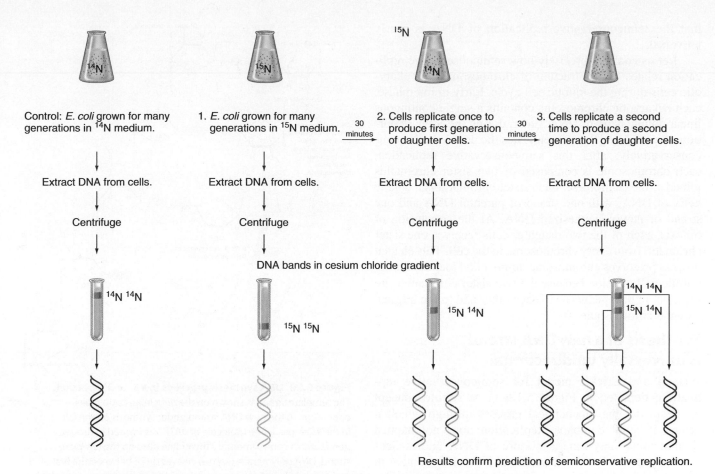

Figure 5.19 How the Meselson-Stahl experiment confirmed semiconservative replication. 1. *E. coli* cells were grown in heavy ^{15}N medium. **2.** and **3.** Some of these cells were transferred to ^{14}N medium and allowed to divide either once or twice. When DNA from each of these sets of cells was prepared and centrifuged in a cesium chloride gradient, the density of the extracted DNA conformed to the predictions of the semiconservative mode of replication, as shown at the *bottom* of the figure, where *blue* indicates heavy original DNA and *magenta* depicts light, newly synthesized DNA. The results are inconsistent with the conservative and dispersive models for DNA replication (compare with **Figures 5.18b** and **c**).

per minute) in an ultracentrifuge. Over a period of two to three days, the centrifugal force (roughly 250 000 times the force of gravity) causes the formation of a stable gradient of CsCl concentrations, with the highest concentration, and thus highest CsCl density, at the bottom of the tube. The DNA in the tube forms a sharply delineated equilibrated band at a position where its own density equals that of the CsCl. Because DNA containing ^{15}N is denser than DNA containing ^{14}N, pure ^{15}N DNA will form a band lower (i.e., closer to the bottom of the tube) than pure ^{14}N DNA (Figure 5.19).

As Figure 5.19 shows, when cells with pure ^{15}N DNA were transferred into ^{14}N medium and allowed to divide once, DNA from the resultant first-generation cells formed a band at a density intermediate between that of pure ^{15}N DNA and that of pure ^{14}N DNA. A logical inference is that the DNA in these cells contains equal amounts of the two isotopes. This finding invalidates the "conservative" model, which predicts the appearance of bands reflecting only pure ^{14}N and pure ^{15}N with no intermediary band. In contrast, DNA extracted from second-generation cells

that had undergone a second round of division in the ^{14}N medium produced two observable bands, one at the density corresponding to equal amounts of ^{15}N and ^{14}N, the other at the density of pure ^{14}N. These observations invalidate the "dispersive" model, which predicts a single band between the two bands of the original generation.

Meselson and Stahl's observations are consistent only with semiconservative replication: In the first generation after transfer from the ^{15}N to the ^{14}N medium, one of the two strands in every daughter DNA molecule carries the heavy isotope label; the other, newly synthesized strand carries the lighter ^{14}N isotope. The band at a density intermediate between that of ^{15}N DNA and ^{14}N DNA represents this isotopic hybrid. In the second generation after transfer, half of the DNA molecules have one ^{15}N strand and one ^{14}N strand, while the remaining half carry two ^{14}N strands. The two observable bands—one at the hybrid position, the other at the pure ^{14}N position—reflect this mix. By confirming the predictions of semiconservative replication, the Meselson–Stahl experiment disproved the conservative and dispersive alternatives. We now know

that the semiconservative replication of DNA is nearly universal.

Let us consider precisely how semiconservative replication relates to the structure of chromosomes in eukaryotic cells during the mitotic cell cycle. Early in interphase, each eukaryotic chromosome contains a single continuous linear double helix of DNA. Later, during the S-phase portion of interphase, the cell replicates the double helix semiconservatively; after this semiconservative replication, each chromosome is composed of two sister chromatids joined at the centromere. Each sister chromatid is a double helix of DNA, with one strand of parental DNA and one strand of newly synthesized DNA. At the conclusion of mitosis, each of the two daughter cells receives one sister chromatid from every chromosome in the cell. This elegant process preserves chromosome number and identity during mitotic cell division because the two sister chromatids are identical in base sequence to each other and to the original parental chromosome.

Synthesis of a new DNA strand is universally unidirectional

Watson and Crick's model for semiconservative replication, depicted in Figure 5.18a, is a simple concept to grasp, but the biochemical process through which it occurs is quite complex. Replication does not happen spontaneously any time a mixture of DNA and nucleotides is present. Rather, it occurs at a precise moment in the cell cycle, depends on a network of interacting regulatory elements, requires considerable input of energy, and involves a complex array of the cell's molecular machinery, including a variety of enzymes. The salient details were deduced primarily by the Nobel laureate Arthur Kornberg and members of his laboratory, who purified individual components of the replication machinery from *E. coli* bacteria. Remarkably, they were eventually able to elicit the reproduction of specific genetic information outside a living cell, in a test tube containing purified enzymes together with DNA template, primer, and nucleotide substrates.

Although the biochemistry of DNA replication was elucidated for a single bacterial species, its essential features are conserved—just like the structure of DNA—within all organisms. The energy required to synthesize every DNA molecule found in nature comes from the high-energy phosphate bonds associated with the four deoxynucleotide triphosphate substrates (dATP, dCTP, dGTP, and dTTP; or dNTP as a general term) that provide bases for incorporation into the growing DNA strand. As shown in **Figure 5.20**, this conserved biochemical feature means that DNA synthesis can proceed only from the hydroxyl group present at the 3′ end of an existing polynucleotide. With energy released from severing the triphosphate arm of a dNTP substrate molecule, the DNA polymerase enzyme catalyzes the formation of a new phosphodiester bond. Once this bond is formed, the enzyme proceeds to

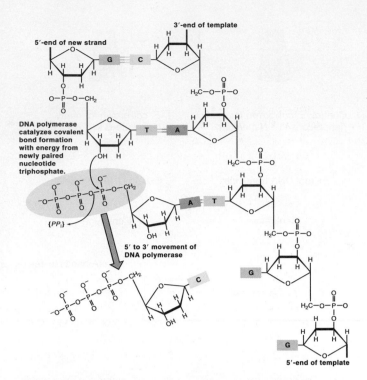

Figure 5.20 DNA synthesis proceeds in a 5′ to 3′ direction. The template strand is shown on the *right* in an antiparallel orientation to the new DNA strand under synthesis on the *left*. In this example, a free molecule of dATP has formed hydrogen bonds with a complementary thymidine base on the template strand. DNA polymerase (*yellow*) cleaves dATP between the first and second phosphate groups, releasing energy to form a covalent phosphodiester bond between the terminal 3′-hydroxyl group on the preceding nucleotide and the first phosphate of the dATP substrate. Pyrophosphate (*PP_i*) is released as a by-product.

join up the next nucleotide brought into position by complementary base-pairing.

The formation of phosphodiester bonds is just one component of the highly coordinated process by which DNA replication occurs inside a living cell. The entire molecular mechanism, illustrated in **Figure 5.21**, has two stages: **initiation**, during which proteins open up the double helix and prepare it for complementary base-pairing, and **elongation**, during which proteins connect the correct sequence of nucleotides on both newly formed DNA double helixes.

DNA replication is a tightly regulated, complex process

DNA replication, which depends in part on DNA polymerase, is complicated by the strict biochemical mechanism of polymerase function. DNA polymerase can lengthen existing DNA chains only by adding nucleotides to the 3′-hydroxy group of the DNA strand. As shown in Figure 5.21, one newly synthesized strand (the *leading strand*) can grow continuously into the opening Y-shaped

The Mechanism of DNA Replication

(a) **Initiation: Preparing the double helix for complementary base-pairing.** A prerequisite of DNA replication is the unwinding of a portion of the double helix, exposing the bases in each DNA strand. These bases may now pair with newly added complementary nucleotides. Initiation begins with the unwinding of the double helix at a particular short sequence of nucleotides known as the **origin of replication**. Each circular *E. coli* chromosome has a single origin of replication. Several proteins bind to the origin, forming a stable complex in which a small region of DNA is unwound and the two complementary strands are separated.

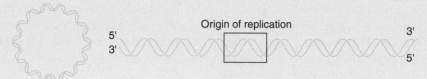

The first of the proteins to recognize and bind to the origin of replication is called the *initiator protein*. A DNA-bound initiator attracts an enzyme called DNA helicase, which catalyzes the localized unwinding of the double helix. The opening up of a region of DNA creates two Y-shaped areas, one at either end of the unwound area, or **replication bubble**. Each Y is called a **replication fork** and consists of the two unwound DNA strands. These single strands will serve as **templates**—molecular moulds—for fashioning new strands of DNA. The molecule is now ready for replication. (Protein molecules are not drawn to scale.)

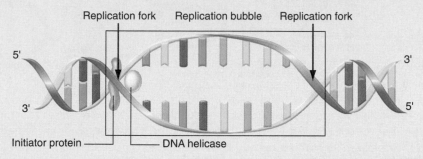

Actual formation of new DNA strands depends on the action of an enzyme complex known as **DNA polymerase III**, which adds nucleotides, one after the other, to the end of a growing DNA strand. DNA polymerase operates according to three strict rules: (1) It can copy only DNA that is unwound and maintained in the single-stranded state; (2) it adds nucleotides only to the end of an existing chain (i.e., it cannot establish the first link in the chain); and (3) it functions in only one direction—5′ to 3′. The requirement for an already existing chain means that something else must prime the about-to-be-constructed chain. That "something else" is RNA. Construction of a very short new strand consisting of a few nucleotides of RNA provides an end to which DNA polymerase can link new nucleotides. This short stretch of RNA is called an RNA **primer**. An enzyme called primase synthesizes the RNA primer at the replication fork, where base-pairing to the single-stranded DNA template takes place. With the double helix unwound and the primer in place, DNA replication can proceed. The third characteristic of DNA polymerase activity (5′ to 3′ only) determines some of the special features of subsequent steps.

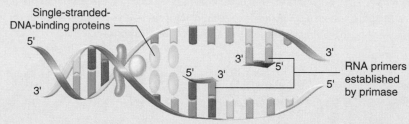

(b) **Elongation: Connecting the correct sequence of nucleotides into a continuous new strand of DNA.** Elongation—the linking together of appropriately aligned nucleotide subunits into a continuous new strand of DNA—is the heart of replication. We have seen that the line-up of bases is determined by complementary base-pairing with the template strand. Thus, the order of bases in the template specifies the order of bases in the newly forming strand. Once complementary base-pairing has determined the next nucleotide to be added, DNA polymerase III catalyzes the joining of this nucleotide to the preceding nucleotide. The linkage of subunits through the formation of phosphodiester bonds is known as **polymerization**.

(Continued)

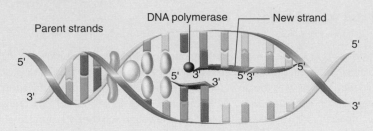

The DNA polymerase III enzyme first joins the correctly paired nucleotide to the 3'-hydroxyl end of the RNA primer, and then it continues to add the appropriate nucleotides to the 3' end of the growing chain. As a result, the DNA strand under construction grows in the 5'-to-3' direction. The new strand is antiparallel to the template strand, so the DNA polymerase molecule actually moves along that template strand in the 3'-to-5' direction.

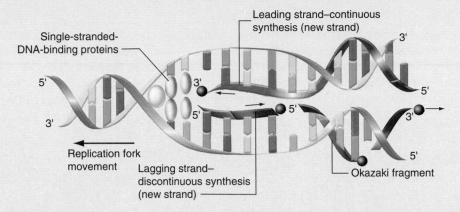

As DNA replication proceeds, helicase progressively unwinds the double helix. DNA polymerase III can then move in the same direction as the fork to synthesize one of the two new chains under construction. The enzyme encounters no problems in the polymerization of this chain—called the **leading strand**—because it can add nucleotides continuously to the growing 3' end as soon as the unravelling fork exposes the corresponding bases on the template strand. The movement of the replication fork, however, presents problems for the synthesis of the second new DNA chain—the **lagging strand**. The polarity of the lagging strand is opposite that of the leading strand, yet as we have seen, DNA polymerase functions only in the 5'-to-3' direction. To synthesize the lagging strand, the polymerase must travel in a direction opposite to that of the replication fork. How can this work?

The answer is that the lagging strand is synthesized *discontinuously* as small fragments of about 1000 bases called **Okazaki fragments** (after two of their discoverers, Reiji and Tuneko Okazaki). DNA polymerase III still synthesizes these small fragments in the normal 5'-to-3' direction, but because the enzyme can add nucleotides only to the 3' end of an existing strand, each Okazaki fragment is initiated by a short RNA primer. The primase enzyme catalyzes formation of the RNA primer for each upcoming Okazaki fragment as soon as the replication fork has progressed a sufficient distance along the DNA. Polymerase then adds nucleotides to this new primer, creating an Okazaki fragment that extends as far as the 5' end of the primer of the previously synthesized fragment. Finally, DNA polymerase I and other enzymes replace the RNA primer of the previously made Okazaki fragment with DNA, and an enzyme known as DNA ligase covalently joins successive Okazaki fragments into a continuous strand of DNA. With the completion of both leading and lagging strands, DNA replication is complete.

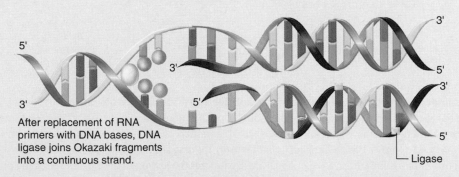

After replacement of RNA primers with DNA bases, DNA ligase joins Okazaki fragments into a continuous strand.

area, but the other new strand (the *lagging strand*) comes into existence only as a series of smaller "Okazaki" fragments (named after their discoverers, Reiji and Tuneko Okazaki). These fragments must be joined together at a second stage of the process.

DNA replication depends on the coordinated activity of many different proteins, including two different DNA polymerases called pol I and pol III (*pol* is short for polymerase). Pol III plays the major role in producing the new strands of complementary DNA, while pol I fills in the gaps between newly synthesized Okazaki segments. Other enzymes contribute to the initiation process. DNA helicase unwinds the double helix. A special group of single-stranded-DNA-binding proteins keep the DNA helix open. An enzyme called primase creates RNA primers to initiate DNA synthesis. The ligase enzyme welds together Okazaki fragments.

It took many years for biochemists and geneticists to discover how the tight collaboration of many proteins drives the intricate mechanism of DNA replication. Today they believe that programmed molecular interactions of this kind underlie most of the biochemical processes that occur in cells. In these processes, a group of proteins, each performing a specialized function, like the workers on an assembly line, cooperate in the manufacture of complex macromolecules.

Integrity and accuracy of genetic information must be preserved

DNA is the sole repository of the vast amount of information required to specify the structure and function of most organisms. In some species, this information may lie in storage for many years and undergo replication many times before it is called on to generate progeny. During this time, the organism must protect the integrity of the information, for even the most minor change can have disastrous consequences (e.g., the production of severe genetic disease or even death). Each organism ensures the informational fidelity of its DNA in three important ways:

- **Redundancy.** Either strand of the double helix can specify the sequence of the other. This redundancy provides a basis for checking and repairing errors arising either from chemical alterations sustained during storage or from malfunctions of the replication machinery.
- **The remarkable precision of the cellular replication machinery.** Evolution has perfected the cellular machinery for DNA replication to the point where errors during copying are exceedingly rare. For example, DNA polymerase has acquired a proofreading ability to prevent unmatched nucleotides from joining a new strand of DNA; as a result, a free nucleotide is attached to a growing strand only if its base is correctly paired with its complement on the parent strand.
- **Enzymes that repair chemical damage to DNA.** The cell has an array of enzymes devoted to the repair of nearly every imaginable type of chemical damage.

All of these safeguards help ensure that the information content of DNA is transmitted intact from generation to generation.

> DNA replication involves many enzymes in a tightly controlled process. The double helix is unwound, and template strands are exposed within the replication bubble, which expands as replication forks progress outward. DNA polymerase can only add nucleotides to the 3' end of a growing chain. As a consequence, one of the two new strands must be formed as a series of Okazaki fragments that are later joined together.

Connections

The Watson–Crick model for the structure of DNA, the single most important biological discovery of the twentieth century, clarified how the genetic material fulfills its primary functions of carrying and accurately reproducing information: Each long, linear or circular molecule carries one of a vast number of potential arrangements of the four nucleotide building blocks (A, T, G, and C).

Unlike its ability to carry information, DNA's capacity for replication is not solely a property of the DNA molecule itself. Rather it depends on the cell's complex enzymatic machinery. But even though replication relies on the complicated orchestration of many different proteins, it occurs with extremely high fidelity. Occasionally, however, errors do occur, providing the genetic basis of evolution. While most errors are detrimental to the organism, a very small percentage of DNA copying errors produce dramatic changes in phenotype without killing the individual. How do such changes affect genes to produce a phenotypic effect? We begin to answer this question in Chapter 8, where we describe how geneticists using mutations as analytical tools demonstrated a correspondence between genes defined in Mendelian terms and specific nucleotide sequences that encode particular proteins.

1. DNA is the nearly universal genetic material. This fact was demonstrated by experiments showing that DNA causes the transformation of bacteria and is the agent of virus production in phage-infected bacteria. **[LO1]**

2. According to the Watson–Crick model, proposed in 1953 and confirmed in succeeding decades, the DNA molecule is a double helix composed of two antiparallel strands of nucleotides; each nucleotide consists of one of four nitrogenous bases (A, T, G, or C), a deoxyribose sugar, and a phosphate. An A on one strand can only pair with a T on the other, and a G can only pair with a C. **[LO2–4]**

3. DNA carries digital information in the sequence of its bases, which may follow one another in any order. Because of the restriction on base-pairing, the information in either strand of a double helix defines the information that must exist in the opposite strand. The two strands are considered complementary. **[LO2–4]**

4. The DNA molecule reproduces by semiconservative replication. In this type of replication, the two DNA strands separate, and the cellular machinery then synthesizes a complementary strand for each. By producing exact copies of the base sequence information in DNA, semiconservative replication allows life to reproduce itself. **[LO2–5]**

Solved Problems

```
5'TAAGCGTAACCCGCTAA      CGTATGCGAAC      GGGTCCTATTAACGTGCGTACAC   3'
3'ATTCGCATTGGGCGATT      GCATACGCTTG      CCCAGGATAATTGCACGCATGTG   5'
```

I. Imagine that the double-stranded DNA molecule shown above was broken at the sites indicated by spaces in the sequence and that, before the breaks were repaired, the DNA fragment between the breaks was reversed. What would be the base sequence of the repaired molecule? Explain your reasoning.

Answer

To answer this question, you need to keep in mind the polarity of the DNA strands involved.

The top strand has the polarity left to right of 5' to 3'. The reversed region must be rejoined with the same polarity. Label the polarity of the strands within the inverting region. To have a 5'-to-3' polarity maintained on the top strand, the *fragment that is reversed must be flipped over*, so the strand that was formerly on the bottom is now on top.

```
5'                                                    3'
TAAGCGTAACCCGCTAA GTTCGCATACG GGGTCCTATTAACGTGCGTACAC
ATTCGCATTGGGCGATT CAAGCGTATGC CCCAGGATAATTGCACGCATGTG
3'                                                    5'
```

II. A new virus has recently been discovered that infects human lymphocytes. The virus can be grown in the laboratory using cultured lymphocytes as host cells. Design an experiment using a radioactive label that would tell you if the virus contains DNA or RNA.

Answer

Use your knowledge of the differences between DNA and RNA to answer this question. RNA contains uracil instead of the thymine found in DNA. *You could set up one culture in which you add radioactive uracil to the medium and a second one in which you add radioactive thymine.* After the viruses have infected cells and produced more new viruses, collect the newly synthesized virus. Determine which culture produced radioactive viruses. If the virus contains RNA, the collected virus grown in media containing radioactive uracil will be radioactive, but the virus grown in radioactive thymine will not be radioactive. If the virus contains DNA, the collected virus from the culture containing radioactive thymine will be radioactive, but the virus from the radioactive uracil culture will not. (You might also consider using radioactively labelled ribose or deoxyribose to differentiate between an RNA- and DNA-containing virus. Technically this does not work as well, because the radioactive sugars are processed by cells before they become incorporated into nucleic acid, thereby obscuring the results.)

III. If you expose a culture of human cells to ³H-thymidine during S phase, how would the radioactivity be distributed over a pair of homologous chromosomes at metaphase? Would the radioactivity be in (a) one chromatid of one homologue, (b) both chromatids of one homologue, (c) one chromatid each of both homologues, (d) both chromatids of both homologues, or (e) some other pattern? Choose the correct answer and explain your reasoning.

Answer

This problem requires application of your knowledge of the molecular structure and replication of DNA and how it relates to chromatids and homologues. DNA replication occurs during S phase, so the ³H-thymidine would be incorporated into the new DNA strands. A chromatid is a replicated DNA molecule, and each new DNA molecule contains one new strand of DNA (semiconservative replication). *The radioactivity would be in both chromatids of both homologues (d).*

Problems

Vocabulary

1. For each of the terms in the left column, choose the best matching phrase in the right column.

i. transformation	**1.** the strand that is synthesized discontinuously during replication
ii. bacteriophage	**2.** the sugar within the nucleotide subunits of DNA
iii. pyrimidine	**3.** a nitrogenous base containing a double ring
iv. deoxyribose	**4.** noncovalent bonds that hold the two strands of the double helix together
v. hydrogen bonds	**5.** Meselson and Stahl experiment
vi. complementary bases	**6.** Griffith experiment
vii. origin	**7.** two nitrogenous bases that can pair via hydrogen bonds
viii. Okazaki fragments	**8.** a nitrogenous base containing a single ring
ix. purine	**9.** a short sequence of bases where unwinding of the double helix for replication begins
x. semiconservative replication	**10.** a virus that infects bacteria
xi. lagging strand	**11.** short DNA fragments formed by discontinuous replication of one of the strands

Section 5.1

2. Griffith, in his 1928 experiments, demonstrated that bacterial strains could be genetically transformed. The evidence that DNA was the "transforming principle" responsible for this phenomenon came later. What was the key experiment that Avery, MacLeod, and McCarty performed to prove that DNA was responsible for the genetic change from rough cells into smooth cells?

3. During bacterial transformation, DNA that enters a cell is not an intact chromosome; instead it consists of randomly generated fragments of chromosomal DNA. In a transformation where the donor DNA was from a bacterial strain that was $a^+ \ b^+ \ c^+$ and the recipient was $a \ b \ c$, 55 percent of the cells that became a^+ were also transformed to c^+, but only 2 percent of the a^+ cells were b^+. Is gene b or c closer to gene a?

4. Nitrogen and carbon are more abundant in proteins than sulphur. Why did Hershey and Chase use radioactive sulphur instead of nitrogen and carbon to label the protein portion of their bacteriophages in their experiments to determine whether parental protein or parental DNA is necessary for progeny phage production?

Section 5.2

5. Imagine you have three test tubes containing identical solutions of purified, double-stranded human DNA. You expose the DNA in tube 1 to an agent that breaks the sugar-phosphate (phosphodiester) bonds. You expose the DNA in tube 2 to an agent that breaks the bonds that attach the bases to the sugars. You expose the DNA in tube 3 to an agent that breaks the hydrogen bonds. After treatment, how would the structures of the molecules in the three tubes differ?

6. What information about the structure of DNA was obtained from X-ray crystallographic data?

7. If 30 percent of the bases in human DNA are A, (a) what percentage are C? (b) What percentage are T? (c) What percentage are G?

8. Which of the following statements are true about double-stranded DNA?

a. A + C = T + G
b. A + G = C + T
c. A + T = G + C
d. A/G = C/T
e. A/G = T/C
f. (C + A) / (G + T) = 1

9. A particular virus with DNA as its genetic material has the following proportions of nucleotides: 20 percent A, 35 percent T, 25 percent G, and 20 percent C. How can you explain this result?

10. When a double-stranded DNA molecule is exposed to high temperature, the two strands separate, and the molecule loses its helical form. We say the DNA has been denatured. (Denaturation also occurs when DNA is exposed to acid or alkaline solutions.)

a. Regions of the DNA that contain many A–T base pairs are the first to become denatured as the temperature of a DNA solution is raised. Thinking about the chemical structure of the DNA molecule, why do you think the A–T-rich regions denature first?

b. If the temperature is lowered, the original DNA strands can reanneal, or renature. In addition to the full double-stranded molecules, some molecules of the type shown here are seen when the molecules are examined under the electron microscope. How can you explain these structures?

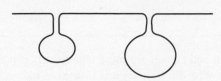

11. A portion of one DNA strand of the human gene responsible for cystic fibrosis is

5′...ATAGCAGAGCACCATTCTG...3′

Write the sequence of the corresponding region of the other DNA strand of this gene, noting the polarity. What do the dots before and after the given sequence represent?

12. The underlying structure of DNA is very simple, consisting of only four possible building blocks.

a. How is it possible for DNA to carry complex genetic information if its structure is so simple?

b. What are these building blocks? Can each block be subdivided into smaller units, and if so, what are they? What kinds of chemical bonds link the building blocks?

c. How does the underlying structure of RNA differ from that of DNA?

13. An RNA virus that infects plant cells is copied into a DNA molecule after it enters the plant cell. What would be the sequence of bases in the first strand of DNA made complementary to the section of viral RNA shown here?

5′ CCCUUGGAACUACAAAGCCGAGAUUAA 3′

14. Bacterial transformation and bacteriophage labelling experiments proved that DNA was the hereditary material in bacteria and in DNA-containing viruses. Some viruses do not contain DNA but have RNA inside the phage particle. An example is the tobacco mosaic virus (TMV) that infects tobacco plants, causing lesions in the leaves. Two different variants of TMV exist that have different forms of a particular protein in the virus particle that can be distinguished. It is possible to reconstitute TMV *in vitro* (in the test tube) by mixing purified proteins and RNA. The reconstituted virus can then be used to infect the host plant cells and produce a new generation of viruses. Design an experiment to show that RNA acts as the hereditary material in TMV.

Section 5.3

15. In Meselson and Stahl's density shift experiments (diagrammed in Figure 5.19), describe the results you would expect in each of the following situations:

a. Conservative replication after two rounds of DNA synthesis on ^{14}N.

b. Semiconservative replication after three rounds of DNA synthesis on ^{14}N.

c. Dispersive replication after three rounds of DNA synthesis on ^{14}N.

d. Conservative replication after three rounds of DNA synthesis on ^{14}N.

16. When Meselson and Stahl grew *E. coli* in ^{15}N medium for many generations and then transferred to ^{14}N medium for one generation, they found that the bacterial DNA banded at a density intermediate between that of pure ^{15}N DNA and pure ^{14}N DNA following equilibrium density centrifugation. When they allowed the bacteria to replicate one additional time in ^{14}N medium, they observed that half of the DNA remained at the intermediate density, while the other half banded at the density of pure ^{14}N DNA. What would they have seen after an additional generation of growth in ^{14}N medium? After two additional generations?

17. If you expose human tissue culture cells to ^{3}H-thymidine just as they enter S phase, then wash this material off the cells and let them go through a second S phase before looking at the chromosomes, how would you expect the ^{3}H to be distributed over a pair of homologous chromosomes? (Ignore the effect recombination could have on this outcome.) Would the radioactivity be in (a) one chromatid of one homologue, (b) both chromatids of one homologue, (c) one chromatid each of both homologues, (d) both chromatids of both homologues, or (e) some other pattern? Choose the correct answer and explain your reasoning.

18. Draw a bidirectional replication fork and label the origin of replication, the leading strands, lagging strands, and the 5′ and 3′ ends of all strands shown in your diagram.

19. As Figure 5.20 shows, DNA polymerase cleaves the high-energy bonds between phosphate groups in nucleotide triphosphates (nucleotides in which three phosphate groups are attached to the 5′-carbon atom of the deoxyribose sugar) and uses this energy to catalyze the formation of a phosphodiester bond when incorporating new nucleotides into the growing chain.

a. How does this information explain why DNA chains grow during replication in the 5′-to-3′ direction?

b. The action of the enzyme DNA ligase in joining Okazaki fragments together is shown in Figure 5.21. Remember that these fragments are connected only after the RNA primers at their ends have been removed. Given this information, infer the type of chemical bond whose formation is catalyzed by DNA ligase and whether or not a source of energy will be required to promote this reaction. Explain why DNA ligase and not DNA polymerase is required to join Okazaki fragments.

20. The bases of one of the strands of DNA in a region where DNA replication begins are shown here. What is the sequence of the primer that is synthesized complementary to the bases in bold? (Indicate the 5′ and 3′ ends of the sequence.)

5′ AGGCCTCG**AATTCGTATAGCTTTCAGA**AAA 3′

21. Replicating structures in DNA can be observed in the electron microscope. Regions being replicated appear as bubbles.

a. Assuming bidirectional replication, how many origins of replication are active in this DNA molecule?

b. How many replication forks are present?

c. Assuming that all replication forks move at the same speed, which origin of replication was activated last?

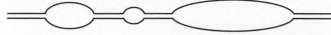

22. Indicate the role of each of the following in DNA replication: (a) helicase, (b) primase, and (c) ligase.

23. Draw a diagram of replication occurring at the end of a double-stranded linear chromosome. Show the leading and lagging strands with their primers. (Indicate the 5′ and 3′ ends of the strands.) What difficulty is encountered in producing copies of both DNA strands at the end of a chromosome?

24. Figure 5.17 depicts Watson and Crick's initial proposal for how the double-helical structure of DNA accounts for DNA replication. Based on our current knowledge, this figure contains a serious error due to oversimplification. Identify the problem with this figure.

25. As we explain in Chapter 14, a DNA synthesizer is a machine that uses automated organic synthesis to create short, single strands of DNA of any given sequence. You have used the machine to create the following three DNA molecules:

(DNA #1) 5′ CTACTACGGATCGGG 3′

(DNA #2) 5′ CCAGTCCCGATCCGT 3′

(DNA #3) 5′ AGTAGCCAGTGGGGAAAAACCCCACTGG 3′

Now you add the DNA molecules either singly or in combination to reaction tubes containing DNA polymerase, dATP, dCTP, dGTP, and dTTP in a buffered solution that allows DNA polymerase to function. For each of the reaction tubes, indicate whether DNA polymerase will synthesize any new DNA molecules, and if so, write the sequence(s) of any such DNAs.

a. DNA #1 plus DNA #3

b. DNA #2 plus DNA #3

c. DNA #1 plus DNA #2

d. DNA #3 only

connect For more information on the resources available from McGraw-Hill Ryerson, go to www.mcgrawhill.ca/he/solutions.

Problems 167

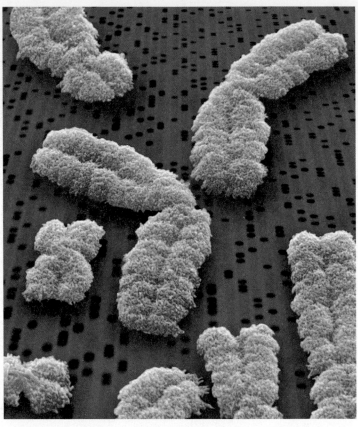

In this collage of images, human metaphase chromosomes (visualized by scanning electron microscopy at 10 000x magnification) are seen overlayed over an autoradiograph of a sequencing gel (see Section 14.6).

In the previous chapter, we carefully examined the molecular structure of DNA. In this chapter, we will examine DNA as it actually exists within the nuclei of eukaryotic cells. DNA molecules do not exist as linear double helixes in isolation; on the contrary, they are carefully packaged with proteins called histones into structures known as chromosomes. As we shall see in this chapter, a chromosome can be viewed as a dynamic organelle for the packaging, replication, segregation, expression, and recombination of the information present in a single molecule of DNA. Each chromosome consists of one DNA molecule combined with a variety of proteins. Flexible DNA-protein interactions condense the chromosome for segregation during mitosis and decondense it for replication or gene expression during interphase. Specific sequences in the chromosomal DNA dictate where spindle attachment occurs for proper segregation; others determine where replication begins. One general theme emerges from our discussion: Chromosomes have a versatile, modular structure that supports a remarkable flexibility of form and function.

Chapter Outline

6.1 Chromosomal DNA and Proteins

6.2 Chromosome Structure and Compaction

6.3 Heterochromatin and Euchromatin

6.4 Chromatin Structure and Gene Expression

6.5 Replication and Segregation of Chromosomes

6.6 Chromosomal Recombination

Learning Objectives

1. Compare (and contrast) naked DNA with chromatin.

2. Deconstruct a eukaryotic chromosome into its constituent parts.

3. Distinguish between heterochromatin and euchromatin.

4. Explain (at the molecular level) why centromeres, telomeres, and origins of replication are required for the duplication and/or segregation of chromosomes.

5. Relate the double-strand-break repair model of meiotic recombination to the process of crossing-over discussed in Chapter 4.

6.1 Chromosomal DNA and Proteins

Each chromosome within a eukaryotic cell nucleus contains one long linear molecule of DNA. This single DNA molecule contains genes composed of coding sequences (exons) interspersed with noncoding sequences (introns) (see the **Fast Forward** box **"What Is a Gene?"**). In addition, substantial stretches of noncoding repetitive DNA can be found concentrated in specific chromosomal regions; for example, at centromeres and telomeres. The repetitive sequences at these locations are critical for the function of the chromosome.

Importantly, these long linear molecules of DNA do not by themselves have the ability to fold up small enough to fit into the cell nucleus. For sufficient compaction, they depend on interactions with two categories of proteins: histones and nonhistone chromosomal proteins. **Chromatin** is the generic term for any complex of DNA and protein found in a cell's nucleus. Chromosomes are the separate pieces of chromatin that behave as a unit during cell division. Chromatin is the same chemical substance that Miescher extracted from the nuclei of white blood cells and named *nuclein* in 1869. Although chromatin is roughly 1/3 DNA, 1/3 histones, and 1/3 nonhistone proteins by weight, it may also contain traces of RNA. Because these RNA bits result mainly from gene expression and are probably unrelated to chromatin structure, we do not include RNA in our discussion of chromatin components.

Histone proteins

Discovered in 1884, **histones** are relatively small proteins with a preponderance of the basic, positively charged amino acids lysine and arginine. The histones' strong positive charge enables them to bind to and neutralize the negatively charged DNA throughout chromatin. Histones make up half of all chromatin protein by weight and are classified into five types of molecules: H1, H2A, H2B, H3, and H4. The last four types, H2A, H2B, H3, and H4, form the core of the most rudimentary DNA packaging unit— the **nucleosome**—and are therefore referred to as **core histones** (we examine the role of these histones in nucleosome structure later).

All five types of histones appear throughout the chromatin of nearly all diploid eukaryotic cells, and they are very similar in all eukaryotes. For example, the H4 proteins of pea plants and cows differ in only two of their 102 amino acid subunits. That histones have changed so little throughout evolution underscores the importance of their contribution to chromatin structure.

Interestingly, specific modifications occur on amino acids located on the exposed tails of histones H3 and H4. The methylation (addition of methyl groups) as well as acetylation (addition of acetyl groups) of these amino acids is important for the functioning and assembly of chromatin. Variations in the acetylation and methylation patterns of different regions of chromatin result in different functions for those regions.

Nonhistone proteins

Fully half of the mass of protein in the chromatin of most eukaryotic cells is not composed of histones. Rather, it consists of hundreds or even thousands of different kinds of nonhistone proteins, depending on the organism. The chromatin of a diploid genome contains from 200 to 2 000 000 molecules of each kind of nonhistone protein. Not surprisingly, this large variety of proteins fulfills many different functions, only a few of which have been defined to date. Some nonhistone proteins play a purely structural role, helping to package DNA into more complex structures. The proteins that form the structural backbone, or **scaffold**, of the chromosome fall in this category (**Figure 6.1**). Others, such as DNA polymerase, are active in replication. Still others are active in chromosome segregation; for example, the motor proteins of kinetochores help move chromosomes along the spindle apparatus and thus expedite the transport of chromosomes from parent to daughter cells during mitosis and meiosis (**Figure 6.2**).

By far the largest class of nonhistone proteins foster or regulate gene expression. Mammals carry 5 000 to 10 000 different proteins of this kind. By interacting with DNA, these proteins influence when, where, and at what rate genes give rise to their protein products.

In a typical chromosome, the chromatin is about one-third DNA and two-thirds histone and nonhistone protein. Four of the five histone proteins in eukaryotic cells, termed the core histones, make up the nucleosomes involved in DNA packaging. The nonhistone proteins include structural components, enzymes such as DNA polymerase, and motor proteins involved in chromosome segregation.

In an abstract sense, a gene can be defined as the basic unit of biological information and heredity. The existence of genes was first deduced by Gregor Mendel as a result of his breeding experiments with pea plants. At that time, a gene could only be thought of as an intangible "entity" that was responsible for the transmission of a hereditary characteristic. Similarly, alleles could only be thought of as alternate "forms" of a gene; these forms being responsible for determining phenotype according to the laws of segregation and independent assortment. Thanks to the contributions of dozens of scientists working in the twentieth century, it became possible to provide a more

"concrete" definition. In a practical sense, a gene can be defined as a segment of DNA in a discrete region of a chromosome that serves as a unit of function by encoding a particular protein (or sometimes simply an RNA molecule like a ribosomal or transfer RNA; see Section 1.4). In eukaryotes the DNA segment encoding a protein is often not contiguous. The sequences within a gene that code for protein are called **exons**, while the noncoding sequences within a gene are called **introns** (see below). In Chapter 7 we will take a detailed look at how DNA sequence information in genes is extracted from the chromosome to produce proteins with specific biological functions.

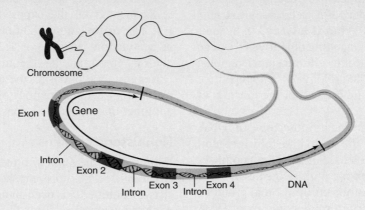

6.2 Chromosome Structure and Compaction

Stretched out into a thin, straight thread, the DNA of a single human cell would be approximately 2 m in length. This is of course much longer than the cell itself, whose dimensions are measured in fractions of millimetres. Several levels of compaction enable the DNA to fit inside the cell (see **Table 6.1**). First, the winding of DNA around histones forms small nucleosomes. Next, tight coiling gathers the nucleosomes together into higher-order structures. Other levels of compaction, which researchers do not yet understand, produce the metaphase chromosomes observable in the microscope.

The nucleosome is the fundamental unit of chromosome packaging

The electron micrograph of chromatin in **Figure 6.3** shows long, nub-studded fibres bursting from the nucleus of a chick red blood cell. The nucleosomes resemble beads on a string, with the beads having a diameter of about 100 Å and the string a diameter of about 20 Å ($1 \text{ Å} = 10^{-10} \text{ m} = 0.1 \text{ nm}$). The 20 Å string is DNA. **Figure 6.4** illustrates how DNA wraps around histone cores to form the chromatin fibre's "beads-on-a-string" structure.

Each bead is a nucleosome containing roughly 160 base pairs (bp) of DNA wrapped around a core composed of eight histones—two each of H2A, H2B, H3, and H4, arranged as shown in the figure. The 160 bp of DNA wrap twice around this core octamer. An additional 40 bp of "**linker**" DNA connects one nucleosome with the next.

Histone H1 lies outside the core, apparently associating with DNA where the DNA enters and leaves the nucleosome. When investigators use specific reagents to remove H1 from the chromatin, some DNA unwinds from each nucleosome, but the nucleosomes do not fall apart; about 140 bp remain wrapped around each core.

One can crystallize the nucleosome cores and subject the crystals to X-ray diffraction analysis. The pictures led to the model of nucleosome structure just described and also indicated that the DNA does not coil smoothly around the histone core (**Figure 6.5**). Instead, depending on the specific sequence, it bends sharply at some positions and barely at all at others. For this reason, base sequence helps dictate preferred nucleosome positions along the DNA.

Duplication of the basic nucleosomal structure occurs in conjunction with DNA replication. Synthesis of the four basic histone proteins increases during S phase of the cell

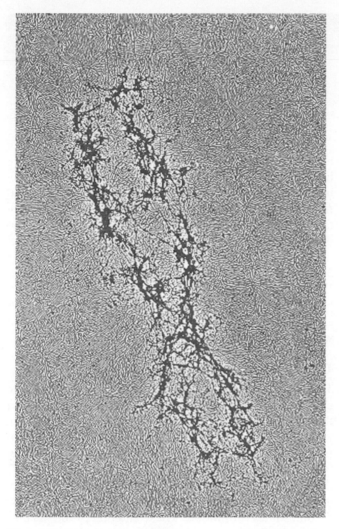

Figure 6.1 Chromosome scaffold. Some nonhistone proteins form the chromosome scaffold. When the condensed human metaphase chromosome in this picture was gently treated with detergents to remove the histones and some of the nonhistone proteins, a dark scaffold composed of some of the remaining nonhistone proteins, and in the shape of the two sister chromatids, became visible. Loops of DNA freed by the detergent treatment surround the scaffold.

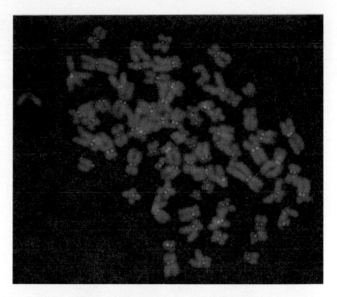

Figure 6.2 Centromere proteins. Some nonhistone proteins power chromosome movements along the spindle during cell division. In this figure, chromosomes are stained in *blue* and a nonhistone protein known as CENP-E is stained in *red*. CENP-E is located at the centromeres of each duplicated chromosome and plays a major role in moving separated sister chromatids toward the spindle poles during anaphase.

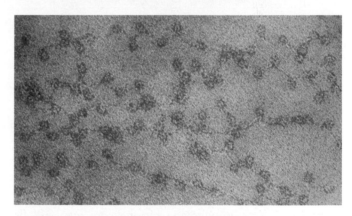

Figure 6.3 Electron microscopy of nucleosomes. In the electron microscope, nucleosomes look like beads on a string.

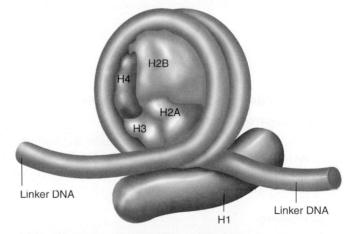

Figure 6.4 Nucleosome structure. The DNA in each nucleosome wraps twice around a nucleosome core composed of two molecules each of H2A, H2B, H3, and H4. One molecule of histone H1 associates with the DNA as it enters and leaves the nucleosome.

cycle to ensure sufficient histone proteins are available to be incorporated onto the newly replicated DNA. Additional proteins mediate the assembly of nucleosomes. Special regulatory mechanisms tightly coordinate DNA and histone synthesis so that both occur at the appropriate time.

The spacing and structure of nucleosomes affect genetic function. The nucleosomes of each chromosome are not evenly spaced, but they do have a well-defined arrangement along the chromatin. This arrangement is transmitted with high fidelity from parent to daughter cells. The spacing of nucleosomes along the chromosome is critical because DNA in the regions between nucleosomes is readily available for interactions with proteins that initiate gene expression, DNA replication, and further DNA compaction. The way in which DNA is wound around a nucleosome also plays a role in determining whether

Table 6.1	Different Levels of Chromosome Compaction	
Mechanism	**Status**	**What It Accomplishes**
Nucleosome	Confirmed by crystal structure	Condenses naked DNA 7-fold to a 100 Å fibre
Supercoiling	Hypothetical model (although the 300 Å fibre predicted by the model has been seen in the electron microscope)	Causes additional 6-fold compaction, achieving a 40- to 50-fold condensation relative to naked DNA
Radial Loop–Scaffold	Hypothetical model (preliminary experimental support exists for this model)	Through progressive compaction of 300 Å fibre, condenses DNA to rodlike mitotic chromosome that is 10 000 times more compact than naked DNA

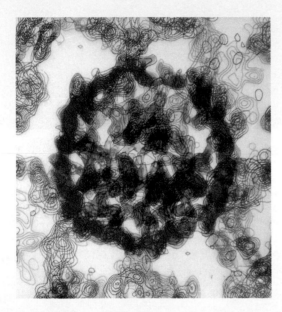

Figure 6.5 X-ray crystallography of a nucleosome. The structure of the nucleosome as determined by X-ray crystallography: In this overhead view of the core particle, you can see that the DNA (*orange*) actually bends sharply at several places as it wraps around the core histone octamer (*blue* and *turquoise*).

certain proteins interact with specific DNA sequences. This is because some DNA sequences in the nucleosome, despite their proximity to the histone core, can still be recognized by nonhistone-binding proteins.

Packaging into nucleosomes condenses naked DNA about sevenfold. With this condensation, the 2 m of DNA in a diploid human genome shortens to approximately 0.25 m in length. This is still much too long to fit in the nucleus of even the largest cell. Thus, additional compaction is required.

Higher-order packaging condenses chromosomes further

The details of chromosomal condensation beyond the nucleosome remain unknown, but researchers have proposed several models to explain the different levels of compaction (see Table 6.1).

Supercoiling

One model of compaction beyond nucleosomal winding proposes that the 100 Å nucleosomal chromatin supercoils into a 300 Å superhelix, achieving a further sixfold chromatin condensation. Support for this model comes in part from electron microscope images of 300 Å fibres that contain about six nucleosomes per turn (**Figure 6.6a**). Whereas the 100 Å fibre is one nucleosome in width, the 300 Å fibre looks three beads wide. Removal of some H1 from a 300 Å fibre causes it to unwind to 100 Å. Adding back the H1 reinstates the 300 Å fibre. Although electron microscopists can actually see the 300 Å fibre, they do not know its exact structure. Higher levels of compaction are even less well understood.

The radial loop–scaffold model

This model proposes that nonhistone proteins bind to chromatin every 60–100 kilobase pairs (kb) and tether the nucleosome-studded 300 Å fibre into structural loops (**Figure 6.6b**). Evidence that nonhistone proteins fasten these loops comes from chemical manipulations in which the removal of histones does not cause the chromatin to unfold completely. A complex of proteins known as **condensins** may act to further condense chromosomes for mitosis. These and other proteins may gather the loops into daisylike rosettes (**Figure 6.6c**) and then compress the rosette centres into a compact bundle. A range of nonhistone proteins thus form the condensation scaffold depicted in Figure 6.1. This proposal of looping and gathering is known as the **radial loop–scaffold model** of compaction.

To visualize how this model achieves condensation, imagine a long piece of string. To shorten it, you knot it at intervals to form loops (separated by straight stretches); the knots are at the base of each loop. To shorten the string still further, you clip together sets of knots. Finally, you pin together all the clips. In this image, the knots, clips, and pins function as the condensation scaffold.

The radial loop–scaffold model offers a simple explanation of progressive chromosome compaction from interphase to metaphase chromosomes. At interphase, the nucleosome-studded chromatin forms many structural loops, which are anchored together in rosettes in some areas. This initial looping and gathering compresses the genetic material sufficiently to fit into the nucleus and to allow the placement of each chromosome in a distinct region or territory within the nucleus. As the chromosomes enter prophase of mitosis, looping and gathering increase, and bundling through protein cross-ties begins. By

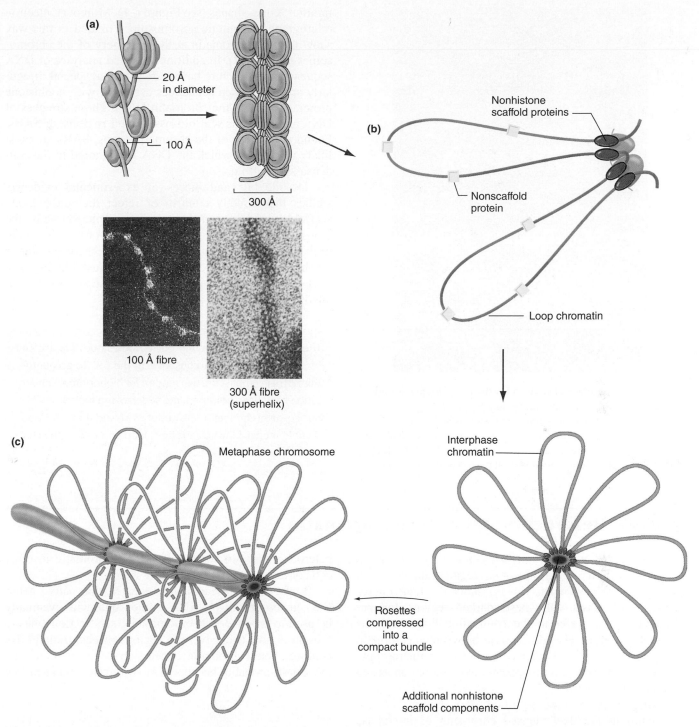

Figure 6.6 Models of higher-order packaging. (a) Electron micrographs contrasting the 100 Å fibre (*left*) with the 300 Å fibre (*right*). The line drawings show the probable arrangement of nucleosomes (with *green cores*) in these structures. **(b)** The radial loop–scaffold model for yet higher levels of compaction. According to this model, the 300 Å fibre is first drawn into loops, each including 60–100 kb of DNA (*purple*), that are tethered at their bases by nonhistone scaffold proteins (*brown* and *orange*). **(c)** Additional nonhistone proteins might gather several loops together into daisylike rosettes and then compress the rosette centres into a compact bundle.

metaphase, this looping, gathering, and bundling achieves a 250-fold compaction of the roughly 40-fold-compacted 300 Å fibre, giving rise to the highly condensed, rodlike shapes we refer to as mitotic chromosomes.

Several pieces of biochemical and micrographic evidence support the radial loop–scaffold model. For example, metaphase chromosomes from which experimenters have extracted all the histones still maintain their

Figure 6.7 Experimental support for the radial loop–scaffold model. A close-up of the image in Figure 6.1, this electron micrograph shows long DNA loops emanating from the protein scaffold at the *bottom* of the picture. The two ends of each DNA loop appear to attach to adjacent locations in the protein scaffold. Note that there are only loops—not ends—at the *top* of the photograph.

familiar X-like shapes (see Figure 6.1). Moreover, electron micrographs of mitotic chromosomes treated in this way show loops of chromatin at the periphery of the chromosomes (**Figure 6.7**). In addition, detailed analyses of DNA sequence and structure have indicated that special, irregularly spaced AT-rich sequences associate with nonhistone proteins to define the chromatin loops. These stretches of DNA are known as **scaffold-associated regions**, or **SARs**. Found at the base of the chromatin loops, SARs are most likely the sites at which the DNA is anchored to the condensation scaffold.

Despite bits and pieces of experimental evidence, studies that directly confirm or reject the radial loop–scaffold model have not yet been completed. Thus, the loops and scaffold concept of higher-order chromatin packaging remains a hypothesis. The hypothetical status of this higher-order compaction model contrasts sharply with nucleosomes, which are entities that investigators have isolated, crystallized, and analyzed in detail.

> The winding of DNA around nucleosomes does not account for the small size of chromosomes in the cell. To account for this further compaction, two models for higher-order condensation have been proposed: the supercoiling model, in which nucleosomal chromatin coils about itself; and the radial loop–scaffold model, in which enzymes loop and gather the chromatin into condensed structures.

6.3 Heterochromatin and Euchromatin

In cells stained with certain DNA-binding chemicals, a small proportion of chromosomal regions appear much darker than others (when viewed under the light microscope). Geneticists call these darker regions **heterochromatin**; they refer to the contrasting lighter regions as **euchromatin**. The distinction between euchromatin and heterochromatin also appears in electron microscopy, where the heterochromatin appears much more condensed than the euchromatin.

Microscopists first identified dark-staining heterochromatin in the decondensed chromatin of interphase cells, where it tends to localize at the periphery of the nucleus. Even highly compacted metaphase chromosomes show the differential staining of heterochromatin versus euchromatin (**Figure 6.8**). Most of the heterochromatin in highly condensed chromosomes is found in regions flanking the centromere, but in some animals, heterochromatin forms in other regions of the chromosomes. In *Drosophila*, the entire Y chromosome, and in humans, most of the Y chromosome, is heterochromatic. Chromosomal regions that remain condensed in heterochromatin at most times in all cells are known as **constitutive heterochromatin**. This is in contrast to **facultative heterochromatin**—regions of chromosomes (or even whole chromosomes) that are

heterochromatic in some cells and euchromatic in other cells of the same organism.

Importantly, research has revealed that active genes (i.e., genes producing RNA copies that will eventually be used to produce a specific protein; see Section 1.4) are present almost exclusively in regions of euchromatin. By contrast, heterochromatin appears to be inactive for the most part, probably because it is so tightly packaged that

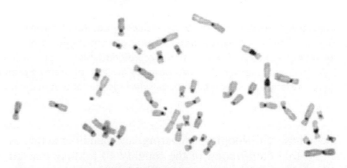

Figure 6.8 Stained heterochromatin. In this image, human metaphase chromosomes were stained by a technique that darkens the constitutive heterochromatin, most of which localizes to regions surrounding the centromere.

Murray Barr (1908–1995; pictured at *left*) was born in Belmont, Ontario in 1908. Educated at University of Western Ontario in London, he returned there—after serving in World War II—to both teach and conduct his groundbreaking research. Murray developed staining techniques that allowed him to identify a small, densely staining chromatin body in the interphase nucleus of female cells. This "Barr" body represented the female X chromosome that has been inactivated early in development (pictured at *right*). The Barr body was never observed in normal male interphase cells. Barr bodies could thus be used to distinguish male from female cells, as well as for diagnosing sex chromosome abnormalities such as Turner syndrome (in which individuals have only one X chromosome and no Barr bodies) and Klinefelter syndrome (in which XXY individuals—who appear outwardly male—have one Barr body). The identification of Barr bodies was also used to detect males fraudulently

competing as females at the 1968 Olympic Games! Dr. Barr received a number of awards, including the Joseph P. Kennedy Award (presented by then United States president John F. Kennedy) and the Gairdner Award of Merit. He was made an Officer of the Order of Canada in 1968 and was posthumously inducted into the Canadian Medical Hall of Fame in 1998.

 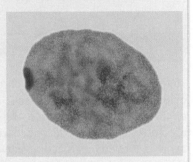

the enzymes required for the production of RNA cannot access the correct DNA sequences. The formation of Barr bodies in mammalian females (see below) clearly illustrates the correlation between heterochromatin formation and a loss of gene activity.

Heterochromatin formation inactivates an X chromosome in cells of female mammals

In both fruit flies and mammals, normal males have one copy of the X chromosome, while females have two. The two sexes, however, require equal amounts of most proteins encoded by genes on the X chromosome. To compensate for the discrepancy in the dose of X-linked genes, male *Drosophila* double the rate at which they express the genes on their single X. Mammals, however, have a different control mechanism for dosage compensation: the random inactivation of all but one X chromosome in each of the female's somatic cells. The inactive X chromosomes are observable in interphase cells as darkly stained heterochromatin masses. Geneticists call these densely staining X chromosomes **Barr bodies** after Murray Barr (born in Belmont, Ontario and educated at the University of Western Ontario), the cytologist who discovered them (see the **Focus on Enquiry** box **"The Discovery of the Barr Body"**). The inactive X chromosome in female mammals is an example of facultative heterochromatin. Here, a whole X chromosome becomes completely heterochromatic in some cells, while remaining euchromatic in others.

An XX female has one Barr body in each somatic cell. The X chromosome that remains genetically active in these cells decondenses and stains as expected during interphase. XY male cells do not contain Barr bodies. As a result, normal male and female mammals have the same number of active X chromosomes. Females with XXX or XXXX karyotypes can survive because they have two or

three Barr bodies and only a single active X. Barr bodies are not restricted to females, but their presence does require more than one X chromosome per cell. Cells from XXY Klinefelter males also contain a Barr body.

X chromosome mosaicism

The "decision" that determines which X chromosome in each cell becomes a Barr body occurs at random in the early stages of development and is inherited by the descendants of each cell. In humans, for example, two weeks after fertilization, when an XX female embryo consists of 500–1000 cells, one of the X chromosomes in each cell condenses to a Barr body. Each embryonic cell "decides" independently which X it will be. In some cells, it is the X inherited from the mother; in others, it is the X inherited from the father.

Once the determination is made, it is clonally perpetuated so that all of the millions of cells descended by mitosis from a particular embryonic cell condense the same X chromosome to a Barr body, thereby inactivating it. Female mammals are thus a mosaic of cells containing either a maternally or a paternally derived inactivated X chromosome.

In an individual female heterozygous for an X-linked gene, some cells express one allele, while other cells express the alternative. In females heterozygous for an X-linked mutation that would be lethal in a male, the relation between the two populations of activated X cells can make the difference between life and death. If the X chromosome carrying the wild-type allele is active in a high enough proportion of cells in which expression of the gene is required, then the individual will survive; if not, the individual will die.

The difference between life and death in this scenario is an excellent example of an "**epigenetic**" phenomenon (from the Greek *epi* meaning "above" or "over"). That is,

a phenomenon in which gene function is *not* determined by variation in DNA sequence. In the scenario described above, viability is determined not by changes in the underlying DNA sequence of the X-linked gene but, instead, by the random nature of X chromosome inactivation in the developing embryo (see Problem 14 in the end-of-chapter problem set for another example involving calico cats). Epigenetics (and epigenomics) will be discussed further in Chapters 11 and 20.

Euchromatin is associated with active genes, while heterochromatin is largely inactive. Dosage compensation ensures that the products of genes on the X chromosome are present in the proper amounts, regardless of gender. In cells of female mammals, one X chromosome randomly undergoes inactivation to become a heterochromatic Barr body.

6.4 Chromatin Structure and Gene Expression

Heterochromatic regions of chromosomes were the first evidence that more compacted DNA is less active. Cells express their genes mainly during interphase when the chromosomes have decondensed, or decompacted—but even the relatively decompacted euchromatic interphase chromatin requires further unwinding to expose the DNA inside nucleosomes for expression. Both the spacing and the structure of nucleosomes affect genetic function. Looking at the level of individual bases in the chromosomes, geneticists have found that nucleosomes of each chromosome are not evenly spaced. They do, however, have a well-defined arrangement along the chromatin.

The position of nucleosomes can be observed at the molecular level. Chromosomal regions from which nucleosomes have been eliminated are experimentally recognizable through their hypersensitivity to cleavage by the enzyme DNase. When one subjects the chromosomal DNA to DNase, the hypersensitive (DH) sites appear in the promoter regions of genes that are active (a promoter is a special DNA sequence near the beginning of a gene that plays a part in initiating gene expression) (**Figure 6.9**). Studies of chromatin structure show that the promoters of most inactive genes are wrapped in nucleosomes. A complex of proteins, referred to as a **remodelling complex**, remove these promoter-blocking nucleosomes or reposition them in relation to the gene and help prepare a gene for expression.

As cells differentiate to perform roles that require the synthesis of specific proteins, patterns of chromatin compaction and decompaction change to allow expression of the appropriate genes. Highly active genes are made accessible in euchromatin. Once established, these patterns of gene decompaction persist in ensuing generations of cells. Thus, because of slight differences in packaging, different areas of chromatin unwind for expression during interphase in different cell types.

Chromatin structure not only varies along a chromosome, but is also dynamic throughout the life of a cell (and of an organism). Three major mechanisms can regulate chromatin patterns. First, modifications to histones (additions of methyl or acetyl groups) are signals for other proteins to interact and cause changes in the level of compaction. Second, as is the case for active genes, the pattern of nucleosomes in chromatin can be altered by remodelling complexes that change the accessibility of DNA sequences. Third, as we will describe later in this chapter, variants of histone proteins can become incorporated into nucleosomes and cause different structures to form.

The position of nucleosomes is a major factor in the control of gene expression. Nucleosomes effectively block a gene's promoter region, and remodelling complexes that remove or reposition nucleosomes can alter the accessibility of DNA sequences.

6.5 Replication and Segregation of Chromosomes

The process by which chromosomes replicate and become segregated into daughter cells is a fascinating field of study requiring an in-depth understanding of both genetics and cell biology. Although the stages can be readily viewed under the light microscope, the mechanics at the molecular level have only recently been revealed.

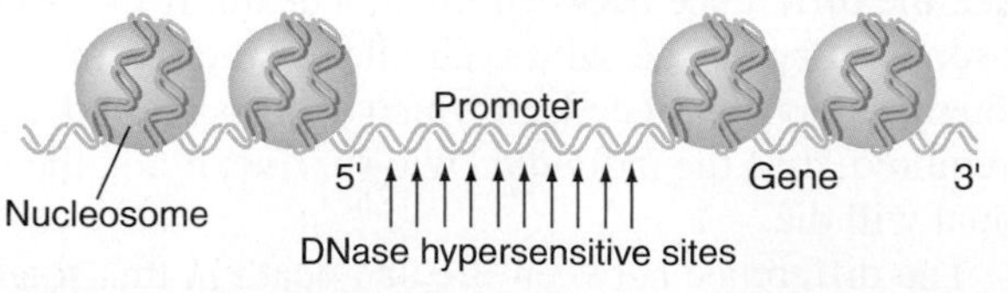

Duplication of chromosomal DNA requires starting points and special ends

As the chromosomes decondense for copying during replication, certain DNA sequences that do not encode proteins regulate the timing and accuracy of the process. Some of these sequences serve as origins of replication that signal where and when the DNA double helix opens up to form

Figure 6.9 DNase hypersensitive sites. Promoters of active genes are exposed for DNase digestion.

replication forks; others function as telomeres that protect the ends of individual chromosomes from progressive decay.

Origins of replication

During replication, the enzyme DNA polymerase assembles a new string of nucleotides according to a DNA template, linking about 50 nucleotides per second in a typical human cell. If there were only one origin of replication, it would take the polymerase about 800 hours (a little more than a month) to copy the 130 million base pairs in an average human chromosome. But the length of the cell cycle in actively dividing human tissues is much shorter, some 24 hours, and S phase (the period of DNA replication) occupies only about a third of this time. Eukaryotic chromosomes meet these time constraints through multiple origins of replication that can function simultaneously.

Most mammalian cells carry approximately 10 000 origins strategically positioned among the chromosomes. As you saw in Chapter 5, each origin of replication binds proteins that unwind the two strands of the double helix, separating them to produce two mirror-image replication forks. Replication then proceeds in two directions (bidirectionally), going one way at one fork and the opposite way at the other, until the forks run into adjacent forks. As replication opens up a chromosome's DNA, a replication bubble becomes visible in the electron microscope, and with many origins, many bubbles appear (**Figure 6.10**).

The DNA running both ways from one origin of replication to the endpoints, where it merges with DNA from adjoining replication forks, is called a **replication unit**, or **replicon**. As yet unidentified controls tie the number of active origins to the length of S phase. In *Drosophila*, for example, early embryonic cells replicate their DNA in less than 10 minutes. To complete S phase in this short a time, their chromosomes use many more origins of replication than are active later in development when S phase is 6–10 times longer. Thus, all origins of replication are not necessarily active during all the mitotic divisions that create an organism.

The 10 000 origins of replication scattered throughout the chromatin of each mammalian cell nucleus are separated from each other by 30–300 kb of DNA, which suggests that there is at least one origin of replication per loop of DNA. Origins of replication in yeast (known as autonomously replicating sequences, or ARSs) can be isolated by their ability to permit replication of plasmids (small, circular double-stranded DNA molecules) in yeast cells. ARSs are capable of binding to the enzymes that initiate replication. They consist of an AT-rich region of DNA adjacent to special flanking sequences (**Figure 6.11**). By digesting interphase chromatin with DNase I, an enzyme that fragments the chromatin only at points where the DNA is not protected inside a nucleosome, investigators have determined that origins of replication are accessible regions of DNA devoid of nucleosomes.

Telomeres: The ends of linear chromosomes

The linear chromosomes of eukaryotic cells terminate at both ends in protective caps called telomeres (**Figure 6.12**). Composed of DNA associated with proteins, these caps contain no genes, but are crucial in preserving the structural integrity of each chromosome. Chromosomes unprotected by telomeres fuse end to end, producing entities with two centromeres. During anaphase of mitosis, if the two centromeres are pulled in opposite directions, the DNA between them will rupture, resulting in broken chromosomes that segregate poorly and eventually disappear from the daughter cells.

Cells must thus preserve their telomeres to maintain the normal genetic complement of chromosomes. But the replication of telomeres is problematic. As you saw in Chapter 5, DNA polymerase, a key component of the replication machinery, functions only in the 5′-to-3′ direction and can add nucleotides only to the 3′ end of an existing chain. With these constraints, the enzyme on its own cannot possibly replicate some of the nucleotides at the 5′ ends of the two DNA strands (one of which is in the telomere at one end of the chromosome, the other of which is in the telomere at the other end). In short, DNA polymerase can reconstruct the 3′ end of each newly made DNA strand in

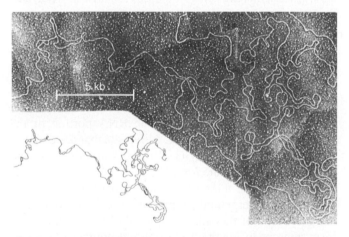

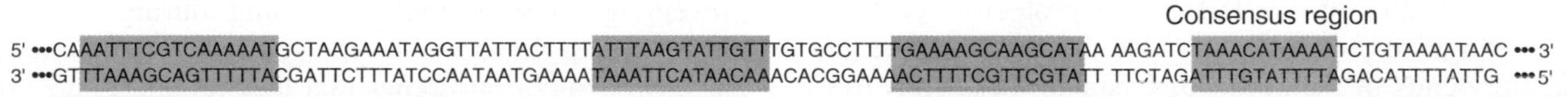

Figure 6.10 Eukaryotic chromosomes have multiple origins of replication. Electron micrograph and diagrammatic interpretation (*lower left*) of a region of replicating DNA from a *Drosophila* embryo. Many origins of replication are active at the same time, creating multiple replicons.

Consensus region

5′ •••CAAATTTCGTCAAAAATGCTAAGAAATAGGTTATTACTTTTATTTAAGTATTGTTTGTGCCTTTTGAAAAGCAAGCATAA AAGATCTAAACATAAAATCTGTAAAATAAC •••3′
3′ •••GTTTAAAGCAGTTTTTACGATTCTTTATCCAATAATGAAAATAAATTCATAACAAACACGGAAAACTTTTCGTTCGTATT TTCTAGATTTGTATTTTAGACATTTTATTG •••5′

Figure 6.11 DNA sequence of an origin of replication. Structure of the yeast origin of replication, ARS1 (the first ARS to be characterized). The *rose-coloured-boxed* sequence is the AT-rich consensus region found in all ARS elements. The *blue boxes* are the flanking sequences close to the ARS1 consensus region that promote function.

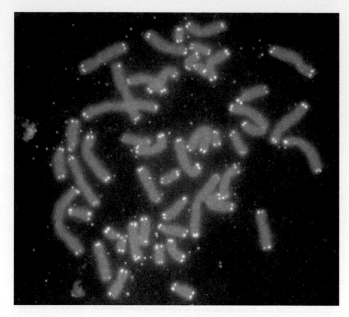

Figure 6.12 Telomeres protect the ends of eukaryotic chromosomes. Human telomeres light up in *yellow* upon *in situ* hybridization with fluorescent probes that recognize the base sequence TTAGGG. The telomeres of humans and many other species contain many repeats of this 6-bp motif.

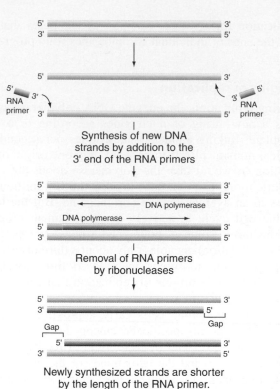

Newly synthesized strands are shorter by the length of the RNA primer.

Figure 6.13 Replication at the ends of chromosomes. Even if an RNA primer at the 5′ end can begin synthesis of a new strand, a gap will remain when ribonucleases eventually remove the primer. The requirement of DNA polymerase for a primer on which to continue polymerization means that the enzyme cannot fill this gap, so newly synthesized strands would always be shorter than parental strands if DNA polymerase were the only player in the production of new ends. In this figure, both parental DNA strands are in *light blue*, the two newly synthesized strands in *dark blue*, and the RNA primers in *red*.

a chromosome, but not the 5′ end (**Figure 6.13**). If left to its own devices, the enzyme would fail to fill in an RNA primer's length of nucleotides at the 5′ end of every new chromosomal strand with each cell cycle. As a result, the chromosomes in successive generations of cells would become shorter and shorter, losing crucial genes as their DNA diminished.

Telomeres and an enzyme called **telomerase** provide a countermeasure to this limitation of DNA polymerase. Telomeres consist of particular repetitive DNA sequences. Human telomeres are composed of the base sequence TTAGGG repeated 250–1500 times. The number of repeats varies with the cell type. Sperm have the longest telomeres. The same exact TTAGGG sequence occurs in the telomeres of all mammals as well as in birds, reptiles, amphibians, bony fish, and many plant species. Some much more distantly related organisms also have repeats in their telomeres, but with slightly different sequences. For example, the telomeric repeat in the chromosomes of the ciliate *Tetrahymena* is TTGGGG. The close conservation of these repeated sequences across phyla suggests that they perform a vital function that emerged in the earliest stages of the evolutionary line leading to eukaryotic organisms, long before dinosaurs roamed Earth.

Telomere DNA helps maintain and replicate chromosome ends by binding two types of proteins: protective proteins and telomerase. The bound protective proteins, which recognize the single-stranded TTAGGG sequences at the very ends of a chromosome, shield these ends from unwanted fusion or degradation, as explained shortly. When the proteins are dislodged, the telomere attracts telomerase, which can also bind to the single-stranded

TTAGGG sequence. The bound enzyme extends the telomere, roughly restoring it to its original length.

Telomerase is an unusual enzyme consisting of protein in association with RNA. Because of this mix, it is called a *ribonucleoprotein*. The RNA portion of the enzyme contains 3′ AAUCCC 5′ repeats that are complementary to the 5′ TTAGGG 3′ repeats in telomeres, and they serve as a template for adding new TTAGGG repeats to the end of the telomere (**Figure 6.14**). In many cells, including the perpetually reproducing cells of yeast and the germ cells of humans, some kind of feedback mechanism appears to maintain the optimal number of repeats at the telomeres. In human somatic cells, low telomerase activity results in the progressive shortening of telomeres as cells divide.

Many studies and observations have shown that telomeres are critical to chromosome function. In addition to preventing chromosomal shortening during replication, the telomeres maintain the integrity of the chromosomal ends. Broken chromosomes that lack telomeres are recognized as defective by the cellular DNA repair machinery, which often remedies the situation by putting the broken ends back together, restoring the telomeres. Sometimes,

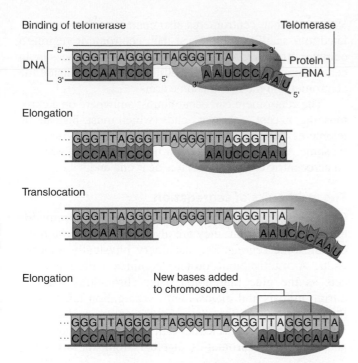

Figure 6.14 How telomerase extends telomeres. Telomerase binds to the ends of chromosomes because of complementarity between the 3′ AAUCCC 5′ repeats of telomerase RNA (*red*) and the TTAGGG repeats of telomeres. Telomerase RNA 3′ AAUCCC 5′ repeats serve as templates for adding TTAGGG repeats to the ends of telomeres. After a telomere has acquired a new repeat, the telomerase enzyme moves (translocates) to the newly synthesized end, allowing additional rounds of telomere elongation.

however, the unprotected, broken, nontelomeric ends are subject to inappropriate repair resulting in chromosome fusion, or they may attract enzymes that degrade the chromosome entirely. Both fusion and degradation disrupt chromosome number and function. Thus, even though they normally carry no genes, telomeres contain information essential to the duplication, segregation, and stability of chromosomes.

Telomerase activity and cell proliferation

The activity of telomerase in normal yeast cells ensures the full reconstruction of each chromosome's ends with each DNA replication. In studies where researchers deleted the yeast gene for telomerase, the telomeres shortened at the rate of about 3 bp per generation, and after significant loss of telomeric length, the chromosomes began to break, and the yeast cells died. Telomerase activity, it seems, endows normal yeast cells with the potential for immortality; given the proper conditions and continual telomere reconstruction, the cells can reproduce forever.

In humans, the telomerase gene is part of every cell's genome. Germline cells, which maintain their chromosomal ends through repeated rounds of DNA replication, express the gene, as do some stem cells; but many normal somatic cells, which have a finite life span, express very little telomerase. In these differentiated somatic cells, the telomeres shorten slightly with each cell division. This

shortening helps determine how many times a particular cell is able to divide. In culture, most somatic cells, after dividing for 30–50 generations, show signs of senescence and then die.

Tumour cells are somatic cells gone awry that continue to divide indefinitely. In contrast to normal somatic cells, many human tumour cells that become immortal exhibit high telomerase activity. Cells isolated from human ovarian tumours, for example, express the telomerase enzyme and maintain stable telomeres; cells from normal ovarian tissue do not. Oncologists hypothesize from these and other observations that expression of telomerase in cancerous human cells may keep those cells from losing their telomeres and eventually dying, and thereby perpetuate tumours. Because high telomerase activity is a characteristic of many tumour cells, pharmaceutical companies are developing cancer treatment drugs that inhibit telomerase activity.

Chromosome duplication includes reproduction of chromatin structure

DNA replication is only one step in chromosome duplication. The complex process also includes the synthesis and incorporation of histone and nonhistone proteins to regenerate tissue-specific chromatin structure. Researchers speculate that the process works something like this:

- Before DNA synthesis can take place, the chromatin fibre must unwind.
- Next, as DNA replication proceeds, newly formed DNA must associate with histones, either pre-existing histones or recently synthesized histones that have just made their way to the nucleus.
- The synthesis and transport of histones must be tightly coordinated with DNA synthesis because the nascent DNA becomes incorporated into nucleosomes within minutes of its formation. Proteins that mediate assembly of nucleosomes have been identified in several organisms, including yeast and humans.
- Finally, the nucleosomal DNA must interact in specific ways with a variety of proteins to produce the same compacted pattern as before.

An exception to the exact replication of compaction patterns occurs in differentiating cells. Changes in available nuclear proteins produce slightly different folding patterns that promote the expression of different genes. Studies with mammalian cells have shown that some hormones can induce changes in gene expression if and only if they are present during chromatin replication.

Segregation of condensed chromosomes depends on centromeres

When cell nuclei divide at mitosis or meiosis II, the two chromatids of each replicated chromosome must separate from one another at anaphase so that each daughter cell receives one, and only one, chromatid from each chromosome. At meiosis I, homologous chromosomes must pair

and segregate so that each daughter cell receives one, and only one, chromosome from each homologous pair. A complex of proteins called **cohesin** holds sister chromatids together after replication and before anaphase. The centromeres of eukaryotic chromosomes ensure this precise distribution during different kinds of cell division by serving as segregation centres.

Characteristics of centromeres

Centromeric constrictions arise because centromeres are contained within blocks of simple, repetitive noncoding sequences, known as **satellite DNAs**, which have a very different chromatin structure and different higher-order packaging than other chromosomal regions.

There are many different kinds of satellite DNA, each consisting of short sequences 5–300 bp long, repeated in tandem (thousands or millions of times) to form large arrays. The predominant human satellite, "α-satellite", is a noncoding sequence 171 bp in length; it is present in a block of tandem repeats extending over a megabase of DNA in the centromeric region of each chromosome.

Various human centromeres also contain sequences unrelated to α-satellite, which give their centromeric regions a complex structure. Although most satellite sequences lie in centromeric regions, some satellites are found outside the centromere on the chromosome arms.

The centromere can occur almost anywhere on a chromosome, except at the very ends (which must, instead, be telomeres). As previously described, in a metacentric chromosome, the centromere is at or near the middle, while in an acrocentric chromosome, it is near one end.

The mechanics of segregation

Centromeres are the sites that hold sister chromatids together. In addition, they are also the sites to which the chromosome segregation machinery physically attaches itself. A multisubunit protein complex called cohesin acts as the glue that holds sister chromatids together during mitosis and meiosis until segregation takes place (**Figure 6.15a**). After the chromosomes replicate in S phase of the cell cycle, cohesin proteins associate with and hold sister chromatids together along the arms and

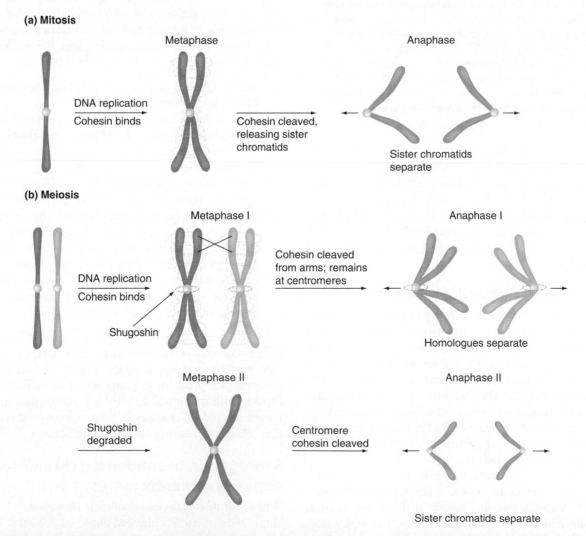

Figure 6.15 Cohesin action in mitosis and meiosis. (a) During mitosis, cohesin holds sister chromatids together through metaphase. Cleavage of cohesin releases sister chromatids so they can segregate at anaphase. **(b)** In meiosis I, cohesin is cleaved from the chromatid arms but is protected by Shugoshin and remains at the centromere to hold sister chromatids together until anaphase II.

in the centromere region. One hypothesis proposes that cohesin encircles the two helixes of the sister chromatids to keep them together.

When the cell enters mitosis, cohesin is lost from the chromosome arms by enzymatic cleavage of the cohesin subunits, but the cohesin complex remains at the centromere. At anaphase the centromeric cohesin is cleaved, and the sister chromatids separate. Mutations in any of cohesin's subunits result in chromosome segregation errors. If a cell expresses a mutant cohesin that is noncleavable, the number of segregation errors increases. The cohesin proteins have been conserved throughout the evolution of eukaryotes.

The cohesin complexes in meiosis behave differently than complexes in mitosis. During meiosis I, homologous chromosomes first pair and then separate, but sister chromatids must stay together. How is this achieved? First, the cohesin found in meiosis contains different proteins than that found in mitosis. At anaphase of meiosis I, the cohesin along the arms of the sister chromatids is cleaved, to allow the resolution of the meiotic crossovers on the chromosome arms (**Figure 6.15b**). Remember that the meiotic crossovers hold homologues together so they pair at the beginning of meiosis.

A meiosis-specific protein, called Shugoshin (meaning "guardian spirit" in Japanese), protects the cohesin at the centromere from cleavage during meiosis I by interacting with the unique centromere components of meiotic cohesin. Upon entering meiosis II, Shugoshin is removed, and the centromere cohesin can now be cleaved at anaphase II, allowing sister chromatids to segregate to opposite poles. Evidence for the role of cohesin in meiosis is found in mutants in several organisms. In mice, for example, a defect in a major cohesin subunit leads to segregation errors and results in infertility.

In addition to holding sister chromatids together, centromeres contribute to proper chromosome segregation through elaboration of a **kinetochore**: a specialized structure composed of DNA and proteins that is the site at which chromosomes attach to the spindle fibres (**Figure 6.16**). Some of the kinetochore proteins are motor proteins that help power chromosome movement during mitosis and meiosis.

During mitosis, a kinetochore develops late in prophase on each sister chromatid, at the part of the centromere that faces one or the other cellular pole. By prometaphase, the kinetochores on the two sister chromatids attach to spindle fibres emanating from centrosomes at opposite poles of the cell. Although it is not yet clear what ensures this bipolar attachment, it appears that kinetochores somehow measure the tension arising when sister chromatids that are connected through their centromere are pulled in opposite directions. At the beginning of anaphase, the cohesin complex is split, freeing the sister chromatids to migrate toward opposite poles (with the assistance of the motor proteins in the kinetochore).

Analysis of centromere structure

Investigators can exploit the centromere's role in chromosome segregation to isolate and then analyze the exact chromosomal regions that make up a centromere. If removal of a DNA sequence disrupts chromosome segregation and reinsertion of that same sequence restores stable transmission, the sequence must be part of the centromere.

In the yeast *S. cerevisiae*, centromeres consist of two highly conserved nucleotide sequences, each only 10–15 bp long, separated by approximately 90 bp of AT-rich DNA (**Figure 6.17**). Evidently, a short stretch of roughly 120 nucleotides is sufficient to specify a centromere in this organism. The centromere sequences of different yeast chromosomes are so closely related that the centromere of one chromosome can substitute for that of another. This indicates that while all centromeres play the same role in chromosome segregation, they do not help distinguish one chromosome from another.

The centromeres of higher eukaryotic organisms are much larger and more complex than those of yeast. In these multicellular organisms, the centromeres lie buried in a considerable amount of darkly staining, highly condensed chromatin, which makes it difficult to discover which specific DNA sequences are critical to centromere function.

The kinetochores in higher eukaryotes attach to many spindle fibres instead of just one, as in yeast. Researchers think that these complex kinetochores are likely to consist

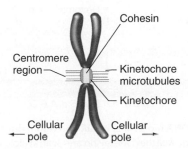

Figure 6.16 Structure of centromeres in higher organisms. Centromeres hold sister chromatids together and contain information for the construction of a kinetochore (*gold*), the structure that allows the chromosome to bind to spindle fibres. Cohesin (*yellow*) binds the sister chromatids together in the centromere region.

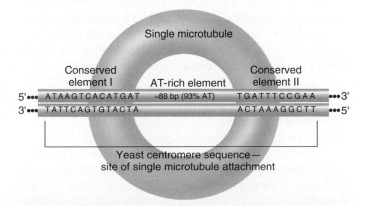

Figure 6.17 Yeast centromeres. Structure and DNA sequence organization of yeast centromeres.

of repeating structural subunits, with each subunit responsible for attachment to one fibre.

Histone variants at centromeres

We mentioned that centromeric DNA consists of repetitive sequences known as satellite DNAs. In higher eukaryotes, the central core of each centromere is composed of unique chromatin that is not readily available for recombination and gene expression. Surrounding this core are regions of heterochromatin interspersed with euchromatin. In all eukaryotes examined, the histone H3 protein has been replaced by a histone variant called CENP-A in the central core. This protein is very similar to histone H3 in its C-terminal region, but different from H3 in its N-terminal portion. The specialized chromatin in the centromere core marks this region for the attachment of the kinetochore protein complexes that are necessary for chromosome segregation.

- Telomeres, maintained by the enzyme telomerase, keep the ends of chromosomes intact. Telomerase activity has been linked to a cell's life span; loss of telomeres with successive cell division leads to chromosome degradation and cell death. In tumour cells, telomerase activity is abnormally high.

- Centromeres play a critical role in chromatid and chromosome segregation. The protein cohesin holds sister chromatids together; during mitosis, it is lost from the arms but is retained at the centromere until anaphase. Centromeric cohesin is protected during meiosis I, keeping chromatids together; protection is removed during meiosis II so that chromatids can segregate.

- Kinetochores are specialized structures that form on each face of a centromere; they contain motor proteins that power chromosome movement.

6.6 Chromosomal Recombination

In Chapter 4 we learned that *recombination* could result not only from independent assortment, but also from reciprocal exchanges between homologous chromosomes. Independent assortment produces gametes carrying new allelic combinations of genes on different chromosomes. Crossing-over, however, can generate new allelic combinations of linked genes. Recombination is crucial for generating genomic diversity in sexually reproducing species as it allows for new combinations of *already existing alleles*. This type of diversity increases the chances that at least some offspring of a mating pair will inherit a combination of alleles best suited for survival and reproduction in a changing environment. Thus, the ability to "recombine" represents one of the most important features of chromosomal behaviour.

Historically, geneticists have used the term "recombination" to indicate the production of new combinations of alleles by any means, including independent assortment. But in the remainder of this chapter, we use **recombination** more narrowly to mean the generation of new allelic combinations—through genetic exchange between homologous chromosomes. In this detailed discussion of the mechanism of recombination, we refer to the products of crossing-over as **recombinants**: chromosomes that carry a mix of alleles derived from different homologues.

During recombination, DNA molecules break and rejoin

When viewed through the light microscope, recombinant chromosomes bearing physical markers appear to result from two homologous chromosomes breaking and exchanging parts as they rejoin (review Figures 4.8 and 4.9). Because the recombined chromosomes, like all other chromosomes, are composed of one long DNA molecule, a logical expectation is that they should show some physical signs of this breakage and rejoining at the molecular level.

Experimental evidence of breaking and rejoining

To evaluate this hypothesis, researchers selected a bacterial virus, **lambda**, as their model organism. Lambda had a distinct experimental advantage for this particular study: It is about half DNA, so the density of the whole virus reflects the density of its DNA.

The experimental technique they used was similar in principle to the one in which Meselson and Stahl monitored changes in DNA density following DNA replication, only in this case, the researchers used the change in DNA density to look at recombination (**Figure 6.18**). They grew two strains of bacterial viruses that were genetically marked to keep track of recombination, one in medium with a heavy

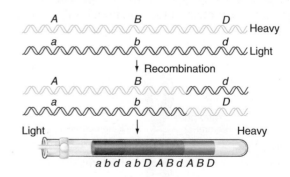

Figure 6.18 DNA molecules break and rejoin during recombination. Matthew Meselson and Jean Weigle infected *E. coli* cells with two different genetically marked strains of bacteriophage lambda previously grown in the presence of heavy (^{13}C and ^{15}N) or light (^{12}C and ^{14}N) isotopes of carbon and nitrogen. They then spun the progeny bacteriophages released from the cells on a CsCl density gradient. The genetic recombinants had densities intermediate between the heavy and light parents.

isotope, the other in medium with a light isotope. They then infected the same bacterial cell with the two viruses under conditions that permitted little if any viral replication. With this type of co-infection, recombination could occur between "heavy" and "light" viral DNA molecules.

After allowing time for recombination and the repackaging of viral DNA into virus particles, the experimenters isolated the viruses released from the lysed cells and analyzed them on a density gradient. Those viruses that had not participated in recombination formed bands in two distinct positions, one heavy and one light, as expected. Those viruses that had undergone recombination, however, migrated to intermediate densities, which corresponded to the position of the recombination event. If the recombinant derived most of its alleles and hence most of its chromosome from a "heavy" DNA molecule, its density was skewed toward the gradient's heavy region; by comparison, if it derived most of its alleles and chromosome from a "light" DNA molecule, it had a density skewed toward the light region of the gradient. These experimental results demonstrated that recombination at the molecular level results from the breakage and rejoining of DNA molecules.

Heteroduplexes at the sites of recombination

Recall that chiasmata, which are visible under the light microscope, indicate where chromatids from homologous chromosomes have crossed over, or exchanged parts (review Figure 4.9). A 100 000-fold magnification of the actual site of recombination within a DNA molecule would reveal the breakage, exchange, and rejoining that constitute the molecular mechanism of crossing-over according to the lambda study. Although current technology does not yet allow us to distinguish base sequences under the microscope, a variety of molecular and genetic procedures do allow us to make inferences regarding the process.

The data obtained using these procedures provide the following two clues about the mechanism of recombination. First, the products of recombination are almost always in exact register, with not a single base pair lost or gained. Geneticists originally deduced this from observing that recombination usually does not cause mutations; today, we know this to be true from analyses of DNA sequence. Second, the two strands of a recombinant DNA molecule do not break and rejoin at the same location on the double helix. Instead, the breakpoints on each strand can be offset from each other by hundreds or even thousands of base pairs. The segment of the DNA molecule located between the two breakpoints is called a **heteroduplex region** (from the Greek *hetero* meaning "other" or "different") (**Figure 6.19**). This name applies not only because one strand of the double helix in this region is of maternal origin, while the other is paternal, but also because the pairing of maternal and paternal strands may produce mismatches in which bases are not complementary. In most organisms, the DNA sequences of the maternal and paternal homologues differ at roughly 1 in every 1000 base pairs, so mismatches are relatively frequent. Within

a heteroduplex, these mismatches prevent proper pairing at the mismatched base pairs, but double helix formation can still occur along the neighbouring complementary nucleotides.

Mismatched heteroduplex molecules do not persist for long. The same DNA repair enzymes that operate to correct mismatches during replication can move in to resolve them during recombination. The outcome of the repair enzymes' work depends on which strand they correct. For example, a repaired G–T mismatch could become either G–C or A–T.

The heteroduplex region of a DNA molecule that has undergone crossing-over has one breakpoint on each strand of the double helix (**Figure 6.19a**). Beyond the heteroduplex region, *both* strands of one DNA molecule have been replaced by both strands of its homologue. There is, however, an alternative type of heteroduplex region in which the initiating and resolving cuts are on the same DNA strand (**Figure 6.19b**). With this type of heteroduplex, only one short segment of one strand has traded places with one short segment of a homologous nonsister strand. Like the first type of heteroduplex, a short heteroduplex arising from a single-strand exchange may also contain one or a few mismatches.

In either type of heteroduplex, mismatch repair may alter one allele to another. For example, if the original homologues carried the *A* allele in one segment of two sister chromatids, and the *a* allele in the corresponding segment of the other pair of sister chromatids, the *A:a* ratio of alleles would be 2:2. Mismatch repair might change that *A:a* allele ratio from 2:2 to 3:1 (i.e., three *A* alleles for every one *a* allele) or 1:3 (one *A* allele for every three *a* alleles; **Figure 6.19c**). Any deviation from the expected 2:2 segregation of parental alleles is known as **gene conversion**, because one allele has been converted to the other.

Although the unusual ratios resulting from gene conversion occur in many types of organisms, geneticists have studied them most intensively in yeast. Interestingly, these observations indicate that gene conversion is associated with crossing-over about 50 percent of the time, but for the other 50 percent of the time, it is an isolated event not associated with a crossover between flanking markers. As we see later, both outcomes derive from the same proposed molecular intermediate, which may or may not lead to a crossover.

Crossing-over at the molecular level: A model

A variety of experimental observations provide the framework for a detailed model of crossing-over during meiosis. First, analysis has shown that only two of the four meiotic products from a single cell are affected by any individual recombination event. One member of each pair of sister chromatids remains unchanged. This provides evidence that recombination occurs during meiotic prophase, after completion of DNA replication. Second, the observation that recombination occurs only between homologous regions and is highly accurate, that is, in exact register,

(a) Heteroduplex region of a recombinant molecule

Heteroduplex region

Breakpoints

(b) Heteroduplex region of a noncrossover molecule

Heteroduplex region

Breakpoints

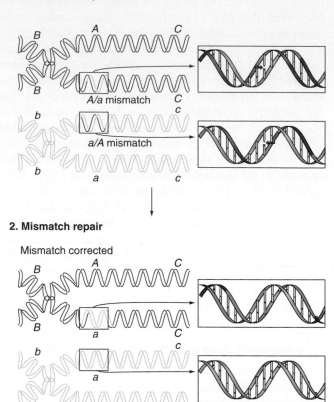

(c) Gene conversion

1. Initial meiotic products

A/a mismatch

a/A mismatch

2. Mismatch repair

Mismatch corrected

Figure 6.19 Heteroduplex regions occur at sites of genetic exchange. (a) A heteroduplex region lies between portions of a chromosome derived from alternative parental homologues after crossing-over. **(b)** A heteroduplex region left behind after an aborted crossover attempt: Sequences from the same parental molecule are found on both sides of the heteroduplex region. The heteroduplexes depicted in (a) and (b) are two alternative products of the same molecular intermediate (as shown in Figure 6.20). **(c)** Gene conversion. **1.** An aborted crossover during meiosis leaves behind two heteroduplex regions with mismatched bases. **2.** DNA repair enzymes eliminate the mismatches, converting both heteroduplexes into the *a* allele.

suggests an important role for base-pairing between complementary strands derived from the two homologues. Third, the observation that crossover sites are often associated with heteroduplex regions further supports the role of base-pairing in the recombination process; it also implies that the process is initiated by single-strand exchange between nonsister chromatids. Finally, the observation of heteroduplex regions associated with gene conversion in the absence of crossing-over indicates that not all recombination events lead to crossovers.

The current molecular model for meiotic recombination derives almost entirely from results obtained in experiments on yeast. Researchers have found, however, that the protein Spo11, which plays a crucial role in initiating meiotic recombination in yeast, is homologous to the Dmc1 protein essential for meiotic recombination in nematodes, plants, fruit flies, and mammals. This finding suggests that the mechanism of recombination presented in detail in **Figure 6.20**—and known as the "double-strand-break repair model"—has been conserved throughout the evolution of eukaryotes. In the figure, we focus on the two nonsister chromatids involved in a single recombination event and show the two nonrecombinant chromatids only at the beginning of the process. These two nonrecombinant chromatids, depicted in the outside positions in Figure 6.20, Step 1, remain unchanged throughout recombination.

Only cells undergoing meiosis express the Spo11 protein, which is responsible for a rate of meiotic recombination several orders of magnitude higher than that found in mitotically dividing cells. Meiotic recombination begins when Spo11 makes a double-strand break in one of the four chromatids. In yeast, where meiotic double-strand breaks have been mapped, it is clear that Spo11 has a preference for some genomic sequences over others, resulting in "hot spots" for crossing-over.

Unlike meiotic cells, mitotic cells do not usually initiate recombination as part of the normal cell-cycle program; instead, recombination in mitotic cells is a consequence of environmental damage to the DNA. X-rays and ultraviolet light, for example, can cause either double-strand breaks or

A Model of Recombination at the Molecular Level

Step 1 Double-strand-break formation. During meiotic prophase, the meiosis-specific Dmc1 protein makes a double-strand break on one of the chromatids by breaking the phosphodiester bonds between adjacent nucleotides on both strands of the DNA.

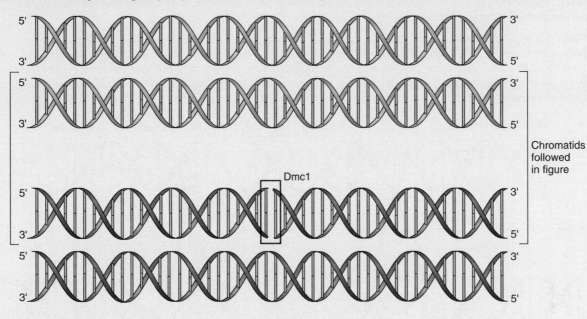

Chromatids followed in figure

Dmc1

Step 2 Resection. The 5′ ends on each side of the break are degraded to produce two 3′ single-stranded tails.

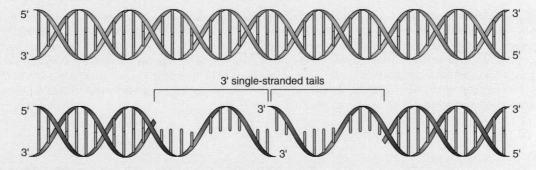

3′ single-stranded tails

Step 3 First strand invasion. One single-stranded tail is recognized and bound by Dmc1 (*orange ovals*). Dmc1 also binds to a double helix in the immediate vicinity. It plays a major role in the ensuing steps of the process, although many other enzymes collaborate with it. Their combined efforts open up the Dmc1-bound double helix, promoting its invasion by the single displaced tail from the other

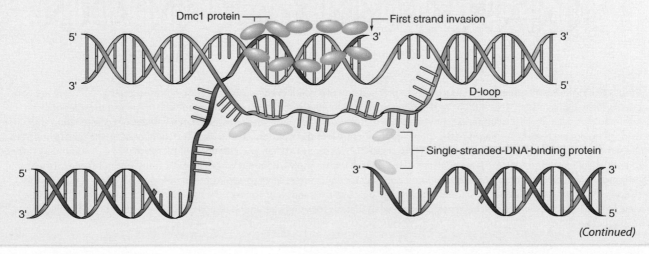

Dmc1 protein

First strand invasion

D-loop

Single-stranded-DNA-binding protein

(Continued)

duplex. Dmc1 then moves along the double helix, prying it open in front and releasing it to snap shut behind. With Dmc1 as its guide, the invading strand scans the base sequence it passes in the momentarily unwound stretches of DNA duplex. As soon as it finds a complementary sequence of sufficient length, it becomes immobilized by dozens of hydrogen bonds and forms a stable heteroduplex. Meanwhile, the strand displaced by the invading tail forms a D-loop (for displacement loop), which is stabilized by binding of the single-stranded-DNA-binding (SSB) protein that played a similar role in DNA replication (see Figure 5.21). D-loops have been observed in electron micrographs of recombining DNA.

Step 4 Formation of a double Holliday junction. New DNA synthesis (indicated by the *dotted string* below) enlarges the D-loop until the single-stranded bases on the displaced strand can form a complementary base pair with the 3′ tail on the nonsister chromatid. New DNA synthesis from this tail re-creates the DNA duplex on the bottom chromatid. The 5′ end on the right side of the break is then connected to the 3′ end of the invading strand. The resulting X structures are called Holliday junctions after Robin Holliday, the scientist who first proposed them.

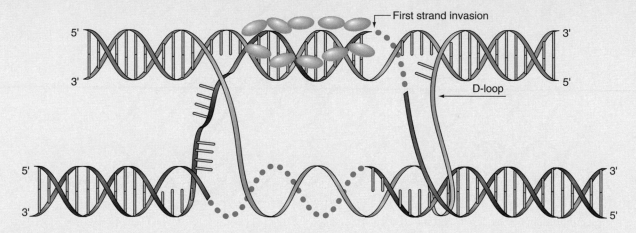

Step 5 Branch migration. The next step, branch migration, results from the tendency of both invading strands to "zip up" by base-pairing along the length of their newly formed complementary strands. The DNA double helixes unwind in front of this double zippering action, and two newly created heteroduplex molecules rewind behind it. The branches of the two ends of the heteroduplex region (where strands from the two homologous chromosomes cross) move in the direction of the arrows. Branch migration thus lengthens the heteroduplex region of both DNA molecules from tens of base pairs to hundreds or thousands. Because the two invading strands began their scanning from complementary bases at slightly different points on the homologous chromatids, branch migration produces two heteroduplex regions that are somewhat different in length.

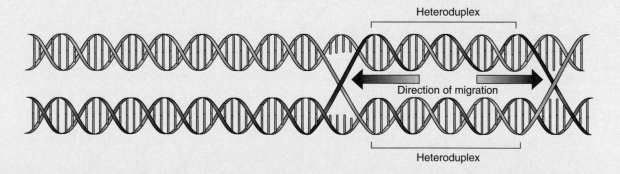

Step 6 The Holliday intermediate. For meiosis to proceed, the two interlocked nonsister chromatids must disengage. There are two equally likely paths to such a resolution. To distinguish these alternative resolutions, we have modified the view of the interlocked intermediate structure. In this figure, we show only one of the two Holliday intermediates associated with each recombination shift. By pushing out each of the four arms of the interlocked structure into the X pattern shown here and then rotating one set of arms from the same original chromatid 180°, we obtain the "isomerized cross-strand exchange configuration" pictured in Step 7, commonly referred to as the "Holliday intermediate." It is important to realize that this is simply a different way of looking at the structure for explanatory purposes. In reality, there is no preferred conformation of chromatid arms relative to each other in this small, localized region. Rather, the arms are free to move about at random, constrained only by the strands that connect the two DNA molecules to each other. The view of the Holliday intermediate, however, clearly reveals that the four single-stranded regions all play an equal role in holding the structure together.

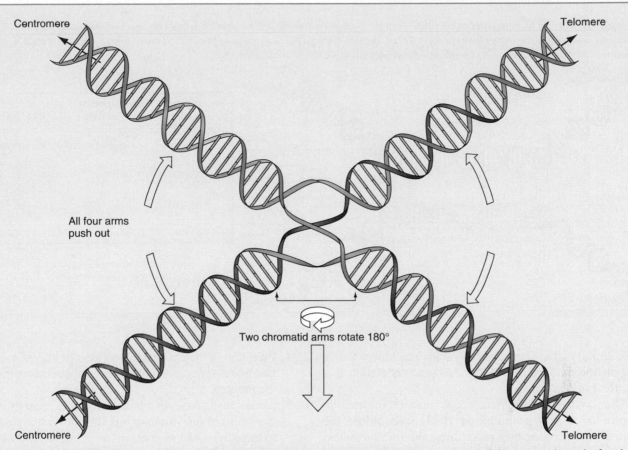

Centromere   Telomere

All four arms
push out

Two chromatid arms rotate 180°

Centromere   Telomere

Step 7 Alternative resolutions. If endonucleases make a horizontal cut (as in this illustration) across a Holliday intermediate, the freed centromeric and telomeric strands of both homologue 1 and homologue 2 can become ligated. In contrast, if the endonucleases make a vertical cut across a strand from homologue 1 and homologue 2, the newly freed strand from the centromeric arm of homologue 1 can now be ligated to the freed strand from the telomeric arm of homologue 2. Likewise, the telomeric strand from homologue 1 can now be ligated to the centromeric strand from homologue 2. This leads to crossing-over between two homologues. However, the resolution of the second Holliday intermediate will determine whether an actual crossing-over event is consummated.

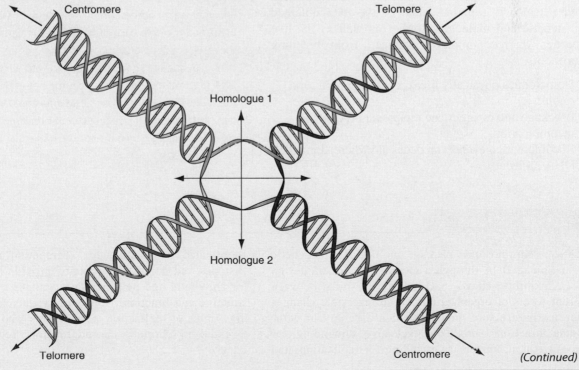

Centromere   Telomere

Homologue 1

Homologue 2

Telomere   Centromere

(Continued)

Step 8 Probability of crossover occurring. Because there are two Holliday junctions, both must be resolved. Resolution of both Holliday junctions in the same plane results in a noncrossover chromatid. For a crossover to occur, the two Holliday junctions must be resolved in opposite planes. (Chromatids are shown in the initial configuration of Step 6.)

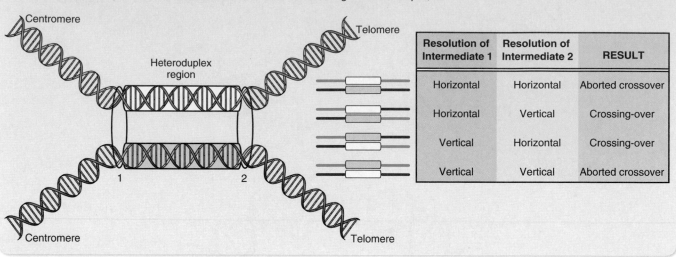

Resolution of Intermediate 1	Resolution of Intermediate 2	RESULT
Horizontal	Horizontal	Aborted crossover
Horizontal	Vertical	Crossing-over
Vertical	Horizontal	Crossing-over
Vertical	Vertical	Aborted crossover

single-strand nicks. The cell's enzymatic machinery works to repair the damaged DNA site, and recombination is a side effect of this process.

The double-strand-break repair model of meiotic recombination was proposed in 1983, well before the direct observation of any recombination intermediates. Since that time, scientists have seen—at the molecular level—the formation of double-strand breaks, the resection of those breaks to produce 3′ single-strand tails, and intermediate recombination structures in which single strands from two homologues have invaded each other. The double-strand-break repair model has become established because it explains much of the data obtained from genetic and molecular studies as well as the five properties of recombination deduced from breeding experiments:

1. Homologues physically break, exchange parts, and rejoin.
2. Breakage and repair create reciprocal products of recombination.
3. Recombination events can occur anywhere along the DNA molecule.

4. Precision in the exchange—no gain or loss of nucleotide pairs—prevents mutations from occurring during the process.
5. Gene conversion—in which a small segment of information from one homologous chromosome transfers to the other—can give rise to an unequal yield of two different alleles. Fifty percent of gene conversion events are associated with crossing-over between flanking markers, but an equal 50 percent are not associated with crossover events.

Recombination occurs when homologous DNA molecules break and rejoin to each other. When breakpoints are offset, the result is a double-stranded DNA heteroduplex region containing a paternal strand base-paired with a maternal strand. DNA repair of mismatches within the heteroduplex can alter the Mendelian 1:1 ratio of allele transmission and explain gene conversion. The double-strand-break repair model provides one possible scenario for the molecular mechanism of crossing-over.

Connections

Eukaryotic chromosomes package and manage the genetic information in DNA through a modular chromatin design whose flexibility allows back-and-forth shifts between different levels of organization. These reversible changes in chromatin structure reliably sustain a variety of chromosome functions, producing selective unwinding for gene expression, universal unwinding for replication, and coordinated compaction for segregation and transport. Histones and nonhistone proteins provide the framework for chromatin and help regulate changes in chromosome structure and function. Noncoding sequences that specify the origins of replication, centromeres, and telomeres are essential to chromosome duplication, segregation, and integrity.

Although the faithful function, replication, and transmission of chromosomes underlie the perpetuation of life within each species, chromosomal changes do occur. In Chapter 9, we examine broader chromosomal rearrangements that produce different numbers of chromosomes, reshuffle genes between nonhomologous chromosomes, and reorganize the genes of a single chromosome. These large-scale modifications, by altering the genetic content of a genome, provide some of the important variations that fuel evolution.

Essential Concepts

1. Each chromosome consists of one long molecule of DNA compacted by histone and nonhistone proteins. The five types of histones—H1, H2A, H2B, H3, and H4—are essential to the establishment of generalized chromosome structure. **[LO1–2]**

2. DNA-protein interactions create reversible levels of compaction. The naked DNA wraps around the four core histones to form nucleosomes, which are secured by H1. **[LO1–2]**

3. Models of higher-order compaction suggest that some sort of supercoiling condenses the nucleosomal fibre to a shorter but wider fibre. Nonhistone proteins then anchor this fibre to form loops. In metaphase chromosomes, higher levels of compaction condense the DNA 10 000-fold. **[LO1–2]**

4. Extremely condensed chromosomal areas appear as darkly staining heterochromatin under the microscope, as contrasted with the lighter euchromatin. The extreme condensation of heterochromatin is associated with the silencing of gene expression. **[LO2–3]**

5. Specific changes in the acetylation and methylation of histones H3 and H4 occur in regions of altered chromatin. The highly condensed heterochromatin of centromeres attracts specific protein complexes necessary for proper chromosome segregation. Chromatin is less compacted in active regions. As cells differentiate, patterns of decompaction are established that allow expression of genes specific to a particular type of cell. **[LO2–3]**

6. Origins of replication are sites accessible for the binding of proteins that initiate DNA replication. In eukaryotic chromosomes, many origins of replication ensure timely replication. **[LO1, LO4]**

7. Telomeres, composed of repetitive base sequences, protect the ends of chromosomes, ensuring their integrity. The enzyme telomerase helps reconstruct the complete telomere with each cell division. In normal cells, telomerase activity becomes reduced over time; in tumour cells, this enzyme's activity may remain high. **[LO1, LO4]**

8. Centromeres, which appear as constrictions in metaphase chromosomes, ensure proper segregation by holding sister chromatids together until anaphase of mitosis and meiosis II. Kinetochores, found on the faces of a centromere, properly attach sister chromatids to spindle fibres and act as motors for chromatid separation. **[LO1, LO4]**

9. Recombination arises from a highly accurate cellular mechanism that includes the base-pairing of homologous strands of nonsister chromatids. Recombination generates new combinations of alleles in sexually reproducing organisms. **[LO5]**

Solved Problems

I. Mouse geneticist Mary Lyon proposed in 1961 that all but one copy of the X chromosome were inactivated in mammals.

   a. What cytological finding supports the Lyon hypothesis?

   b. What is a genetic result that supports the Lyon hypothesis?

Answer

This question requires an understanding of the experimental observations on X inactivation.

   a. Microscopic examination of cells produces cytological evidence. *The number of Barr bodies seen in the cells of individuals with different numbers of X chromosomes supports the hypothesis.* For example, cells from XX females have one Barr body, whereas cells from an XXX female have two Barr bodies.

   b. *Examination of the phenotype of cells from a female heterozygous for two different alleles of an X-linked gene produces genetic evidence that supports the Lyon hypothesis.* In these females, some cells have the phenotype associated with one allele, while other cells have the phenotype associated with the other allele.

Vocabulary

1. For each of the terms in the left column, choose the best matching phrase in the right column.

i. telomere	**1.** site at the base of chromatin loops that anchor loops to the scaffold
ii. kinetochore	**2.** origin of replication in yeast
iii. nucleosome	**3.** repetitive DNA found near the centromere in higher eukaryotes
iv. ARS	**4.** specialized structure at the end of a linear chromosome
v. satellite DNA	**5.** complexes of DNA and protein in the eukaryotic nucleus
vi. chromatin	**6.** small basic proteins that bind to DNA and form the core of the nucleosome
vii. SAR	**7.** complex of DNA and proteins where spindle fibres attach to a chromosome
viii. histones	**8.** beadlike structure consisting of DNA wound around histone proteins

Section 6.1

2. Many proteins other than histones are found associated with chromosomes. What roles do these nonhistone proteins play? Why are there more different types of nonhistone than histone proteins?

Section 6.2

3. What difference is there in the compaction of chromosomes during metaphase and interphase?

4. What is the role of the core histones in compaction compared with the role of histone H1?

5. a. About how many molecules of histone H2A would be required in a typical human cell just after the completion of S phase, assuming an average nucleosome spacing of 200 bp?

   b. During what stage of the cell cycle is it most crucial to synthesize new histone proteins?

   c. The human genome contains 60 histone genes, with 10–15 genes of each type (H1, H2A, H2B, H3, and H4). Why do you think the genome contains multiple copies of each histone gene?

6. The enzyme micrococcal nuclease can cleave phosphodiester bonds on single- or double-stranded DNAs, but DNA that is bound to proteins is protected from digestion by micrococcal nuclease. When chromatin from eukaryotic cells is treated for a short period of time with micrococcal nuclease and then the DNA is extracted and analyzed by electrophoresis and ethidium bromide staining, the pattern shown in lane A on the gel illustrated here is found. Treatment for a longer time results in the pattern shown in lane B, and treatment for yet more time yields that shown in lane C. Interpret these results.

7. Histone H1 appears to play an important role in the formation of the 300 Å fibre, while the other histone proteins do not appear to participate. Why do you think this is true?

8. Chromosome assembly factor (CAF-1) is a complex of proteins that was identified biochemically in extracts from human cells. The sequence of amino acids in the proteins was identified.

   a. How could you use these data to look for homologous genes in yeast?

   b. As a geneticist, why would you find it advantageous to identify the yeast genes to further your understanding of chromatin assembly?

9. The histone proteins H3 and H4 are modified in predictable and consistent ways that are conserved across species. One of the modifications is addition of an acetyl group to the twelfth lysine in the H4 protein. If you were a geneticist working in yeast and had a clone of the H4 gene, what could you do to test whether the acetylation at this specific lysine was necessary for the functioning of chromatin?

Section 6.3

10. For each of the following pairs of chromatin types, which is the most condensed?

   a. 100 Å fibre or 300 Å fibre

   b. 300 Å fibre or DNA loops attached to a scaffold

   c. euchromatin or heterochromatin

   d. interphase chromosomes or metaphase chromosomes

11. How many Barr bodies are present in humans with the following karyotype?

   a. an XX female

   b. an XY male

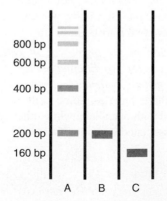

c. an XX male (known as an exceptional male*)

d. an XXY male

e. an XXXX female

f. an XO female

12. A pair of twin sisters were believed to be identical until one was diagnosed with Duchenne muscular dystrophy, an X-linked trait. Her sister did not have the disease. Does this finding mean they are definitely not identical twins (derived from fertilization of one egg)? Why or why not?

13. Females with the genotype $X^{CB} X^{cb}$ are rarely colour blind although some have only partial colour vision. Speculate on why this is true.

14. In cats, the dominant *O* allele of the X-linked *orange* gene is required to produce orange fur; the recessive *o* allele of this gene yields black fur.

a. Tortoiseshell cats have coats with patches of orange fur alternating with patterns of black fur. Approximately 90 percent of all tortoiseshell cats are females. What type of crosses would be expected to produce female tortoiseshell cats?

b. Suggest a hypothesis to explain the origin of male tortoiseshell cats.

c. Calico cats (most of which are females) have patches of white, orange, and black fur. Suggest a hypothesis for the origin of calico cats.

15. In marsupials like the opossum or kangaroo, X inactivation selectively inactivates the paternal X chromosome.

a. Predict the possible coat colour phenotypes of the progeny of both sexes if a female marsupial homozygous for a mutant allele of an X-linked coat color gene was mated with a male hemizygous for the alternative wild-type allele of this gene.

b. Predict the possible coat colour phenotypes of the progeny of both sexes if a male marsupial hemizygous for a mutant allele of an X-linked coat colour gene was mated with a female homozygous for the alternative wild-type allele of this gene.

c. Why are the terms "recessive" and "dominant" not useful in describing the alleles of X-linked coat colour genes in marsupials?

d. Why would marsupials heterozygous for two alleles of an X-linked coat colour gene not have patches of fur of two different colours as did the tortoiseshell cats described in Problem 14?

Section 6.4

16. Which of the following would be suggested by a DNase hypersensitive site?

a. No transcription occurs in this region of the chromosome.

b. The chromatin is in a more open state than a region without the hypersensitive site.

17. Give an example of a chromosomal structure or function affected by the following mechanisms for regulating chromatin structure.

a. methylation of histones

b. new histones in the nucleosome

Section 6.5

18. a. What DNA sequences are commonly found at human centromeric regions?

b. What functions do the two centromere-associated complexes, cohesin and the kinetochore, play in chromosome mechanics?

19. The human genome contains about 3 billion base pairs. During the first cell division after fertilization of a human embryo, S phase is approximately 3 hours long. Assuming an average DNA polymerase rate of 50 nucleotides/s over the entire S phase, what is the minimum number of origins of replication you would expect to find in the human genome?

20. Give at least three examples of types of mutations that would disrupt the process of mitotic chromosome segregation. That is, explain in what DNA structures or in genes encoding what kinds of proteins you would find these segregation-disrupting mutations.

21. The mitotic cell divisions in the early embryo of *D. melanogaster* occur very rapidly (every 8 minutes).

a. If there was one bidirectional origin in the middle of each chromosome, how many nucleotides would DNA polymerase have to add per second to replicate all the DNA in the longest chromosome (66 Mb) during the 8-minute early embryonic cell cycles? (Assume that replication occurs during the entire cell division cycle.)

b. In fact, many origins of replication are active on each chromosome during the early embryonic divisions and are spaced approximately 7 kb apart. Calculate the average rate (per second) at which DNA polymerase adds complementary nucleotides to a growing chain in the early *Drosophila* embryo.

22. The enzyme telomerase consists of protein and an RNA containing a template sequence that directs the addition of an end sequence appropriate for the species. Telomere sequences (TTGGGG) from the ciliated protozoan *Tetrahymena* were cloned onto the ends of a linear YAC, which was then transformed into yeast. The YAC survived as a linear piece of DNA but the YAC now had TGGTGG sequences at the very ends in addition to TTGGGG. Why do you think these sequences were added?

23. The CENP-B and CENP-A proteins are both involved in centromere function. CENP-B is a nonessential protein but CENP-A is essential for viability.

---

* In an exceptional XX male, one of the two X chromosomes has all genes normally found on the X chromosome plus a small region from the Y chromosome that carries the testes-determining factor required for male development.

a. In yeast, if you had a mutant containing a temperature-sensitive allele of CENP-B and another mutant containing a temperature-sensitive allele of CENP-A, what phenotype would you expect for viability and what phenotype for chromosome loss for each of the mutants when you raised the temperature?

b. Describe a test you could use to assay chromosome loss in these mutants.

24. One of the unique proteins found in the meiotic cohesion complex is Rec8. This protein, expressed only during meiosis, is not cleaved during meiosis I but is cleaved during meiosis II to finally allow sister chromatids to segregate. Scientists hypothesized that a protein protects the Rec8 protein from cleavage and degradation during meiosis I. To identify such a protein, researchers first produced the Rec8 protein in mitotically dividing yeast cells. In these cells, Rec8 was cleaved during mitosis and the cells suffered no harmful effects. To find a protein that protects Rec8 from cleavage, researchers then expressed other proteins in the cell expressing Rec8 mitotically and were able to identify Shugoshin that protects Rec8 from degradation. What effect do you think expressing Shugoshin had on the mitotically dividing cells expressing Rec8? What phenotype would the cells show?

Section 6.6

25. Bacterial cells were co-infected with two types of bacteriophage lambda: One carried the c^+ allele and the other the c allele. After the cells lysed, progeny bacteriophages were collected. When a single such progeny bacteriophage was used to infect a new bacterial cell, it was observed in rare cases that some of the resulting progeny were c^+ and others were c. Explain this result.

26. DNA fingerprinting, a technique that will be described in Chapter 15, can show whether two different samples of DNA come from the same individual. One form of DNA fingerprinting relies on chromosome regions called microsatellites, which contain many repeats of a short sequence (e.g., CACACACA). The number of repeats is highly variable from individual to individual in a population. Scientists have suggested that this variability could result from recombination. Use the double-strand-break model, including strand invasion, to explain how a microsatellite could gain or lose repeats during recombination.

CHAPTER 7

Gene Expression: The Flow of Information from DNA to RNA to Protein

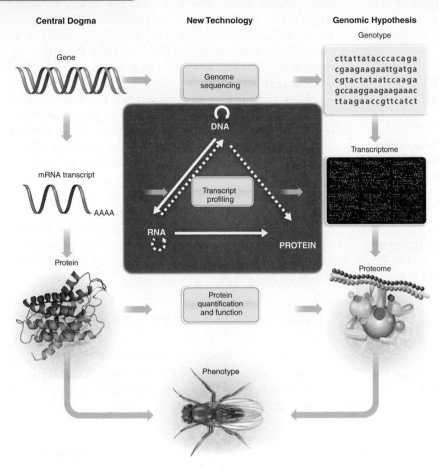

The central dogma of molecular biology as described by Francis Crick in 1970 (*white text on blue background at centre*). Superimposed upon this schematic is a more modern interpretation of the dogma from the early twenty-first century. This latter interpretation reflects the influence of post-genomic ideas and technologies. We will learn more about these topics in Unit II of this text.

In this chapter, we describe the cellular mechanisms that govern gene expression. As intricate as some of the details may appear, the general scheme of gene expression is elegant and straightforward: *Within each cell, genetic information flows from DNA to RNA to protein.* This statement was set forward as the "Central Dogma" of molecular biology by Francis Crick in 1957. As Crick explained, "Once 'information' has passed into protein, it cannot get out again."

The Central Dogma maintains that genetic information flows in two distinct stages (**Figure 7.1**). If you think of genes as instructions written in the language of nucleic acids, the cellular machinery first transcribes the instructions written in the DNA dialect to the same instructions written in the RNA dialect. The conversion of DNA-encoded information to its RNA-encoded equivalent is known as **transcription**. The product of transcription is a **transcript**. In prokaryotes, the RNA transcript serves directly as a **messenger RNA** (mRNA). In eukaryotes the RNA transcript must be processed to become an mRNA (see Section 7.2).

In the second stage of gene expression, the cellular machinery translates mRNA into its polypeptide equivalent in the language of amino acids. This decoding of nucleotide information to a sequence of amino acids is known as **translation**. Translation depends on (1) molecular workbenches called **ribosomes**, which are composed of

Chapter Outline
7.1   The Genetic Code
7.2   Transcription: From DNA to RNA
7.3   Translation: From mRNA to Protein
7.4   Differences in Gene Expression Between Prokaryotes and Eukaryotes
7.5   A Comprehensive Example: Computerized Analysis of Gene Expression in *C. elegans*
7.6   The Effect of Mutations on Gene Expression and Gene Function
7.7   How the Code Was "Cracked"

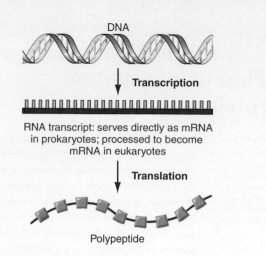

DNA

Transcription

RNA transcript: serves directly as mRNA in prokaryotes; processed to become mRNA in eukaryotes

Translation

Polypeptide

Figure 7.1 Gene expression: The flow of genetic information from DNA via RNA to protein. In transcription, the enzyme RNA polymerase copies DNA to produce an RNA transcript. In translation, the cellular machinery uses instructions in mRNA to synthesize a polypeptide (following the rules of the genetic code).

proteins and **ribosomal RNAs (rRNAs)**; (2) the **genetic code**, a "dictionary" that defines each amino acid in terms of specific sequences of three nucleotides; and (3) **transfer**

RNAs (tRNAs), small RNA adaptor molecules that place specific amino acids at the correct position in a growing polypeptide chain.

The Central Dogma does not explain the behaviour of all genes. As Crick himself realized, a large subset of genes is transcribed into RNAs that are never translated into proteins. For example, the genes encoding rRNAs and tRNAs belong to this group. In addition, scientists later found that certain viruses contain an enzyme that can reverse the DNA-to-RNA flow of information by copying RNA to DNA in a process called **reverse transcription**.

Four general themes emerge from our discussion of gene expression. First, the pairing of complementary bases is key to the transfer of information from DNA to RNA and from RNA to protein. Second, the polarities (directionality) of DNA, RNA, and polypeptides help guide the mechanisms of gene expression. Third, like DNA replication and recombination, gene expression requires an input of energy and the participation of specific proteins and macromolecular assemblies, such as ribosomes. Finally, changes in DNA sequence that alter genetic information, or obstruct the flow of its expression, can have dramatic effects on phenotype.

Learning Objectives

1. Relate the genetic code to the concepts of (i) gene function, (ii) mutation, and (iii) Mendel's laws.

2. Describe the molecular mechanisms responsible for the flow of information from DNA to RNA to protein.

3. Compare and contrast gene expression in prokaryotes and eukaryotes.

4. Illustrate how a typical eukaryotic gene is organized.

5. Summarize the properties of the genetic code and explain how these properties were inferred through genetic analysis.

7.1 The Genetic Code

A code is a system of symbols that equates information in one language with information in another. A useful analogy for the genetic code is the Morse code, which uses dots and dashes to transmit messages over radio or telegraph wires. Various groupings of the dot-dash symbols represent the 26 letters of the English alphabet. Because there are many more letters than the two symbols (dot or dash), groups of one, two, three, or four dots/dashes in various combinations represent individual letters. For example, the symbol for C is dash dot dash dot (– · – ·), the symbol for O is dash dash dash (– – –), D is dash dot dot (– · ·), and E is a single dot (·). Because anywhere from one to four symbols specify each letter, the Morse code requires a symbol for "pause" (in practice, a short interval of time) to signify where one letter ends and the next begins.

Triplet codons of nucleotides represent individual amino acids

The language of nucleic acids is written in four nucleotides (A, G, C, and T in the DNA dialect; A, G, C, and U in the

RNA dialect), while the language of proteins is written in amino acids. The first hurdle to be overcome in deciphering how sequences of nucleotides can determine the order of amino acids in a polypeptide is to determine how many amino acid "letters" exist. Over lunch one day at a local pub, Watson and Crick produced the now accepted list of 20 amino acids that are genetically encoded by DNA or RNA. They created the list by analyzing the known amino acid sequences of a variety of naturally occurring polypeptides. Amino acids that are present in only a small number of proteins or in only certain tissues or organisms did not qualify as standard building blocks; Crick and Watson correctly assumed that such amino acids arise when proteins undergo modification after their synthesis. By contrast, amino acids that are present in most (though not necessarily all) proteins made the list. The question then became, how can four nucleotides encode 20 amino acids?

Like the Morse code, the four nucleotides encode 20 amino acids through specific groupings of A, G, C, and T or A, G, C, and U. Researchers initially arrived at the number of letters per grouping by deductive reasoning and

later confirmed their guess by experiment. They reasoned that if only one nucleotide represented an amino acid, there would be information for only four amino acids: A would encode one amino acid; G, a second amino acid; and so on. If two nucleotides represented each amino acid, there would be $4^2 = 16$ possible combinations of doublets.

Of course, if the code consisted of groups containing one *or* two nucleotides, it would have $4 + 16 = 20$ groups and could account for all the amino acids, but there would be nothing left over to signify the pause required to denote where one group ends and the next begins. Groups of three nucleotides in a row would provide $4^3 = 64$ different triplet combinations, more than enough to code for all the amino acids. If the code consisted of doublets and triplets, a signal denoting a pause would once again be necessary. But a triplets-only code would require no symbol for "pause" if the mechanism for counting to three and distinguishing among successive triplets was very reliable.

Although this kind of reasoning generates a hypothesis, it does not prove it. As it turned out, however, the experiments described later in this chapter did indeed demonstrate that groups of three nucleotides represent all 20 amino acids. Each nucleotide triplet is called a **codon**. Each codon, designated by the bases defining its three nucleotides, specifies one amino acid. For example, GAA is a codon for glutamic acid (Glu), and GUU is a codon for valine (Val). Because the code comes into play only during the translation part of gene expression, that is, during the decoding of messenger RNA to polypeptide, geneticists usually present the code in the RNA dialect of A, G, C, and U, as depicted in **Figure 7.2**. When speaking of genes, they can substitute T for U to show the same code in the DNA dialect.

If you knew the sequence of nucleotides in a gene or its transcript as well as the sequence of amino acids in the corresponding polypeptide, you could then deduce the genetic code without understanding how the underlying cellular machinery actually works. Although techniques for determining both nucleotide and amino acid sequences are available today, this was not true when researchers were trying to crack the genetic code in the 1950s and 1960s. At that time, they could establish a polypeptide's amino acid sequence, but not the nucleotide sequence of DNA or RNA. Because of their inability to read a nucleotide sequence, they used an assortment of genetic and biochemical techniques to fathom the code. They began by examining how DNA sequence changes in a single gene affected the amino acid sequence of the gene's polypeptide product.

First letter	Second letter: U	Second letter: C	Second letter: A	Second letter: G	Third letter
U	UUU Phe	UCU Ser	UAU Tyr	UGU Cys	U
	UUC	UCC	UAC	UGC	C
	UUA Leu	UCA	UAA Stop	UGA Stop	A
	UUG	UCG	UAG Stop	UGG Trp	G
C	CUU Leu	CCU Pro	CAU His	CGU Arg	U
	CUC	CCC	CAC	CGC	C
	CUA	CCA	CAA Gln	CGA	A
	CUG	CCG	CAG	CGG	G
A	AUU Ile	ACU Thr	AAU Asn	AGU Ser	U
	AUC	ACC	AAC	AGC	C
	AUA	ACA	AAA Lys	AGA Arg	A
	AUG Met	ACG	AAG	AGG	G
G	GUU Val	GCU Ala	GAU Asp	GGU Gly	U
	GUC	GCC	GAC	GGC	C
	GUA	GCA	GAA Glu	GGA	A
	GUG	GCG	GAG	GGG	G

Figure 7.2 The genetic code: 61 codons represent the 20 amino acids, while 3 codons signify stop. To read the code, find the first letter in the *left column*, the second letter along the *top*, and the third letter in the *right column*; this reading corresponds to the 5′-to-3′ direction along the mRNA.

In this way, they were able to use the abnormal (genes with anomalous DNA sequences) to understand the normal (the general relationship between genes and polypeptides).

"The "cracking" of the genetic code represented one of the greatest accomplishments of the twentieth century. It also demonstrated the power of the experimental approach (called genetic analysis) that was used in these experiments. The deciphering of the code will be discussed in detail at the end of the chapter. To begin our discussion, however, we provide a comprehensive description of the mechanisms by which cells read the code and transform it into protein.

Geneticists reasoned on theoretical grounds that codons composed of three nucleotides would provide the simplest mechanism by which genes could encode the 20 amino acids commonly found in proteins.

7.2 Transcription: From DNA to RNA

Transcription is the process by which the polymerization of ribonucleotides, guided by complementary base-pairing, produces an RNA transcript of a gene. The template for the RNA transcript is an individual strand of the DNA double helix that defines the gene.

RNA polymerase synthesizes a single-stranded RNA copy of a gene

Figure 7.3 depicts the basic components of transcription and illustrates key events in the process as it occurs in the

Transcription in Bacterial Cells

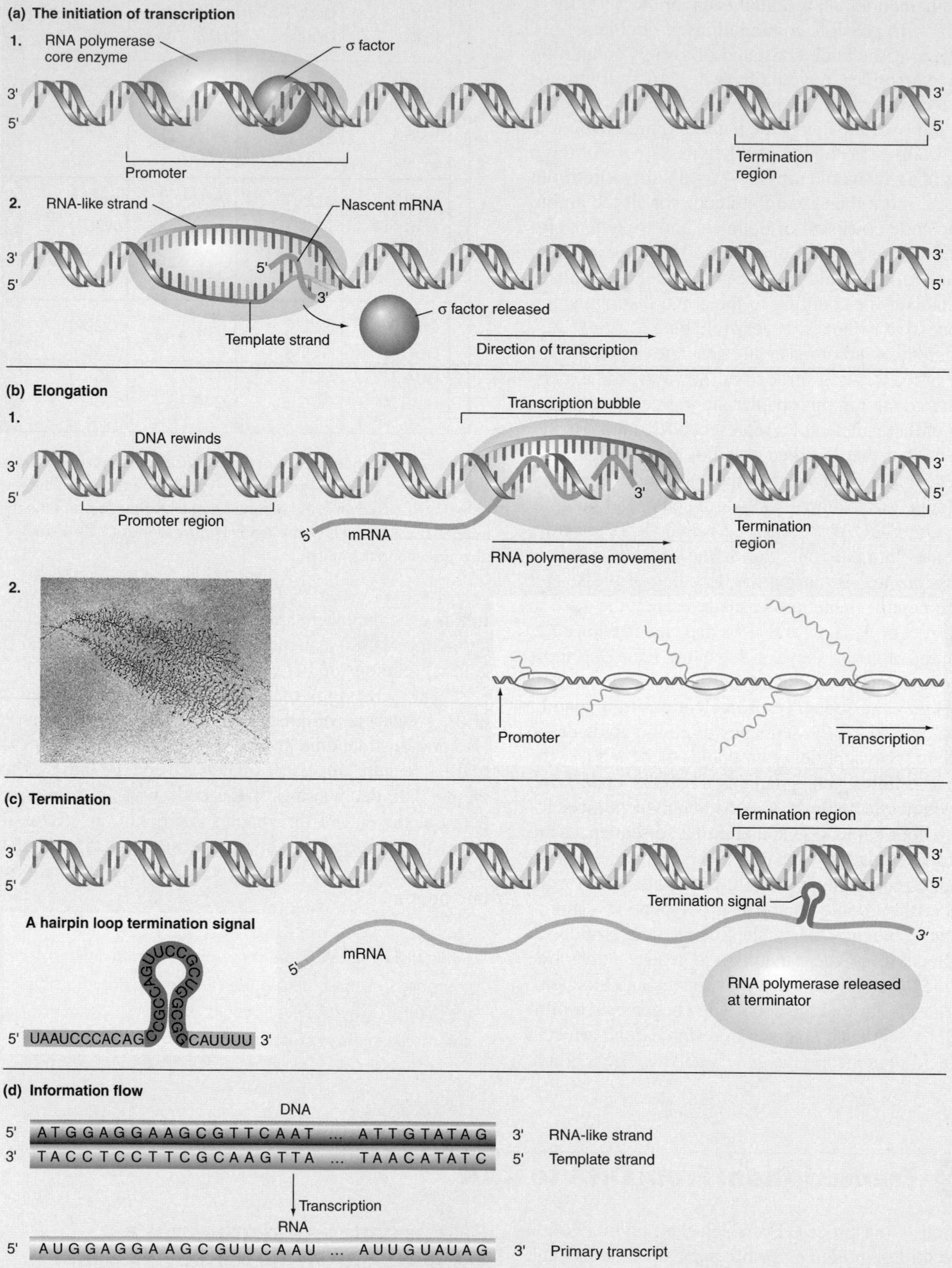

(a) The initiation of transcription

1. RNA polymerase core enzyme   σ factor

Promoter

Termination region

2. RNA-like strand   Nascent mRNA

Template strand

σ factor released

Direction of transcription

(b) Elongation

1.

DNA rewinds

Transcription bubble

Promoter region

mRNA

RNA polymerase movement

Termination region

2.

Promoter

Transcription

(c) Termination

Termination region

Termination signal

A hairpin loop termination signal

mRNA

RNA polymerase released at terminator

5' UAAUCCCACAG CCGCCAGUUCCGCUGGCGCG CAUUUU 3'

(d) Information flow

DNA

5' ATGGAGGAAGCGTTCAAT ... ATTGTATAG 3'   RNA-like strand

3' TACCTCCTTCGCAAGTTA ... TAACATATC 5'   Template strand

Transcription

RNA

5' AUGGAGGAAGCGUUCAAU ... AUUGUAUAG 3'   Primary transcript

(a) The Initiation of Transcription

1. *RNA polymerase binds to double-stranded DNA at the beginning of the gene to be copied.* RNA polymerase recognizes and binds to **promoters**, specialized DNA sequences near the beginning of a gene where transcription will start. Although specific promoters vary substantially, all promoters in *E. coli* contain two characteristic short sequences of 6–10 nucleotide pairs that help bind RNA polymerase (**Figure 7.4**). In bacteria, the complete RNA polymerase (the *holoenzyme*) consists of a core enzyme, plus a σ (sigma) subunit involved only in initiation. The σ subunit reduces RNA polymerase's general affinity for DNA but simultaneously increases RNA polymerase's affinity for the promoter. As a result, the RNA polymerase holoenzyme can hone in on a promoter and bind tightly to it, forming a so-called *closed promoter complex*.

2. *After binding to the promoter, RNA polymerase unwinds part of the double helix, exposing unpaired bases on the template strand.* The complex formed between the RNA polymerase holoenzyme and an unwound promoter is called an *open promoter complex*. The enzyme identifies the template strand and chooses the two nucleotides with which to initiate copying. Guided by base-pairing with these two nucleotides, RNA polymerase aligns the first two ribonucleotides of the new RNA, which will be at the 5′ end of the final RNA product. The DNA transcribed into the 5′ end of the mRNA is often called the *5′ end of the gene* and is the end where the process of translation will begin. RNA polymerase then catalyzes the formation of a phosphodiester bond between the first two ribonucleotides. Soon thereafter, the RNA polymerase releases the σ subunit. This release marks the end of initiation.

(b) Elongation: Constructing an RNA Copy of the Gene

1. *When the σ subunit separates from the RNA polymerase, the enzyme loses its enhanced affinity for the promoter sequence and regains its strong generalized affinity for any DNA.* These changes enable the core enzyme to leave the promoter and yet remain bound to the gene. The core enzyme now moves along the chromosome, unwinding the double helix to expose the next single-stranded region of the template. The enzyme extends the RNA by linking a ribonucleotide positioned by complementarity with the template strand to the 3′ end of the growing chain. As the enzyme extends the mRNA in the 5′-to-3′ direction, it moves in the antiparallel 3′-to-5′ direction along the DNA template strand.

   The region of DNA unwound by RNA polymerase is called the **transcription bubble**. Within the bubble, the nascent RNA chain remains base-paired with the DNA template, forming a DNA-RNA hybrid. However, in those parts of the gene behind the bubble that have already been transcribed, the DNA double helix forms again, displacing the RNA, which hangs out of the transcription complex as a single strand with a free 5′ end.

2. *Once an RNA polymerase has moved off the promoter, other RNA polymerase molecules can move in to initiate transcription.* If the promoter is very strong, that is, if it can rapidly attract RNA polymerase, the gene can undergo transcription by many RNA polymerases simultaneously. Here we show an electron micrograph and an artist's interpretation of simultaneous transcription by several RNA polymerases. As you can see, the promoter for this gene lies very close to where the shortest RNA is emerging from the DNA.

   Geneticists often use the direction travelled by RNA polymerase as a reference when discussing various features within a gene. If, for example, you started at the 5′ end of a gene at point A and moved along the gene in the same direction as RNA polymerase to point B, you would be travelling in the **downstream** direction. If, by contrast, you started at point B and moved in the opposite direction to point A, you would be travelling in the **upstream** direction.

(c) Termination: The End of Transcription

RNA sequences that signal the end of transcription are known as **terminators**. There are two types of terminators: *intrinsic terminators,* which cause the RNA polymerase core enzyme to terminate transcription on its own, and *extrinsic terminators,* which require proteins other than RNA polymerase—particularly a polypeptide known as *rho*—to bring about termination. All terminators, whether intrinsic or extrinsic, are specific sequences in the mRNA that are transcribed from specific DNA regions. Terminators often form **hairpin loops** in which nucleotides within the mRNA pair with nearby complementary nucleotides. Upon termination, RNA polymerase and a completed RNA chain are both released from the DNA.

(d) The Product of Transcription Is a Single-Stranded Primary Transcript

The RNA produced by the action of RNA polymerase on a gene is a single strand of nucleotides known as a *primary transcript*. The bases in the primary transcript are complementary to the bases between the initiation and termination sites in the template strand of the gene. The ribonucleotides in the primary transcript include a start codon, the codons that specify the remaining amino acids of the polypeptide, and a stop codon.

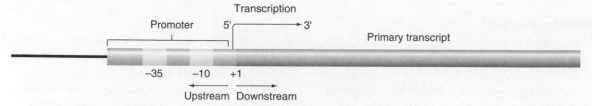

(a) Transcription initiation signals in bacteria

(b) Strong *E. coli* promoters

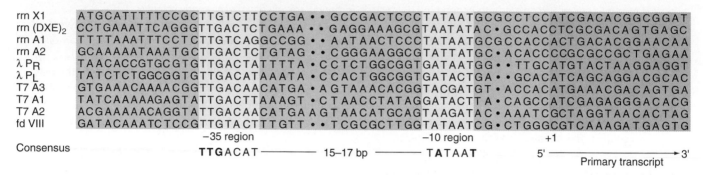

Figure 7.4 The promoters of ten different *E. coli* genes. Only the sequence of the RNA-like strand is shown; numbering starts at the first transcribed nucleotide (+1). **(a)** Most promoters are upstream of the start point of transcription. **(b)** All promoters in *E. coli* share two different short stretches of nucleotides (*yellow*) that are essential for recognition of the promoter by RNA polymerase. The most common nucleotides at each position in each stretch constitute a *consensus sequence*; invariant nucleotides within the consensus are in **bold**.

bacterium *E. coli*. This figure divides transcription into successive phases of *initiation*, *elongation*, and *termination*. The following four points are of particular importance:

1. The enzyme **RNA polymerase** catalyzes transcription.
2. DNA sequences near the beginning of genes, called **promoters**, signal to RNA polymerase where to begin transcription. As seen in Figure 7.4, most bacterial gene promoters share two short regions that have almost identical nucleotide sequences. These are the sites at which RNA polymerase makes particularly strong contact with the promoter.
3. RNA polymerase adds nucleotides to the growing RNA polymer in the 5′-to-3′ direction. The chemical mechanism of this nucleotide-adding reaction is similar to the formation of phosphodiester bonds between nucleotides during DNA replication (review Figure 5.20), with one exception: Transcription uses ribonucleotide triphosphates (ATP, CTP, GTP, and UTP) instead of deoxyribonucleotide triphosphates. Hydrolysis of the high-energy bonds in each ribonucleotide triphosphate provides the energy needed for elongation.
4. Sequences in the RNA products, known as **terminators**, tell RNA polymerase where to stop transcription.

As you examine Figure 7.3, bear in mind that a gene consists of two antiparallel strands of DNA, as mentioned earlier. One—the *RNA-like strand*—has the same polarity and sequence (except for T instead of U) as the emerging RNA transcript. The second—the *template strand*—has the opposite polarity and a complementary sequence that enables it to serve as the template for making the RNA

transcript. When geneticists refer to the sequence of a gene, they usually mean the sequence of the RNA-like strand.

Although the transcription of all genes in all organisms roughly follows the general scheme shown in Figure 7.3, important variations can be found in the details. For example, the transcription of different genes in bacteria can be initiated by alternative sigma (σ) factors. In eukaryotes, promoters are more complicated than those in bacteria, and there are three different kinds of RNA polymerase that can transcribe different classes of genes. Chapters 10 and 11 describe how prokaryotic and eukaryotic cells exploit these and other variations to control when, where, and to what level a given gene is expressed. Finally, the **Genetics and Society** box **"HIV and Reverse Transcription"** describes how the AIDS virus uses an exceptional form of transcription, known as **reverse transcription**, to construct a double strand of DNA from an RNA template.

The result of transcription is a single strand of RNA known as a **primary transcript** (see Figure 7.3d). In prokaryotic organisms, the RNA produced by transcription is the actual messenger RNA that guides protein synthesis. In eukaryotic organisms, by contrast, most primary transcripts undergo *processing* in the nucleus before they migrate to the cytoplasm to direct protein synthesis. This processing has played a fundamental role in the evolution of complex organisms.

In eukaryotes, RNA processing after transcription produces a mature mRNA

Some RNA processing in eukaryotes modifies only the 5′ or 3′ ends of the primary transcript, leaving the information content of the rest of the mRNA untouched.

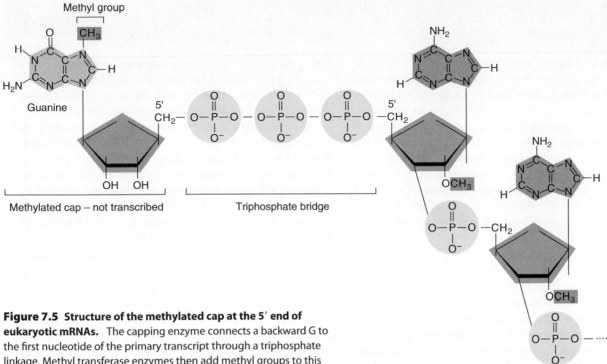

Figure 7.5 Structure of the methylated cap at the 5′ end of eukaryotic mRNAs. The capping enzyme connects a backward G to the first nucleotide of the primary transcript through a triphosphate linkage. Methyl transferase enzymes then add methyl groups to this G and to one or two of the nucleotides first transcribed from the DNA template.

Other processing deletes blocks of information (i.e., the introns) from the middle of the primary transcript, so the content of the mature mRNA is related, but not identical, to the complete set of DNA nucleotide pairs in the original gene.

Adding a 5′ methylated cap and a 3′ poly-A tail

The nucleotide at the 5′ end of a eukaryotic mRNA is a G in reverse orientation from the rest of the molecule; it is connected through a triphosphate linkage to the first nucleotide in the primary transcript. This "backward G" is not transcribed from the DNA. Instead, a special *capping enzyme* adds it to the primary transcript after polymerization of the transcript's first few nucleotides. Enzymes known as *methyl transferases* then add methyl ($-CH_3$) groups to the backward G and to one or more of the succeeding nucleotides in the RNA, forming a so-called **methylated cap** (**Figure 7.5**).

Like the 5′ methylated cap, the 3′ end of most eukaryotic mRNAs is not encoded directly by the gene. In a large majority of eukaryotic mRNAs, the 3′ end consists of 100–200 A's, referred to as a **poly-A tail** (**Figure 7.6**). Addition of the tail is a two-step process. First, a ribonuclease cleaves the primary transcript to form a new 3′ end; cleavage depends on the sequence AAUAAA, which is found in poly-A-containing mRNAs 11–30 nucleotides upstream of the position where the tail is added. Next the enzyme *poly-A polymerase* adds A's onto the 3′ end exposed by cleavage.

Unexpectedly, both the methylated cap and the poly-A tail are critical for the efficient translation of the mRNA into protein, even though neither helps specify an amino

acid. Recent data indicate that particular *eukaryotic translation initiation factors* bind to the 5′ cap, while *poly-A binding protein* associates with the tail at the 3′ end of the mRNA. The interaction of these proteins shapes the mRNA molecule into a circle. This circularization both enhances the initial steps of translation and stabilizes the mRNA in the cytoplasm by increasing the length of time it can serve as a messenger.

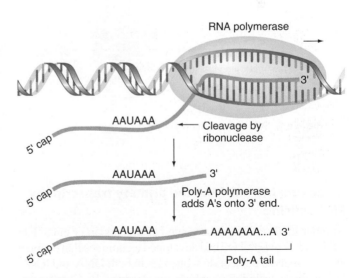

Figure 7.6 How RNA processing adds a tail to the 3′ end of eukaryotic mRNAs. A ribonuclease recognizes AAUAAA in a particular context of the primary transcript and cleaves the transcript 11–30 nucleotides downstream to create a new 3′ end. The enzyme poly-A polymerase then adds 100–200 A's onto this new 3′ end.

The AIDS-causing human immunodeficiency virus (HIV) is the most intensively analyzed virus in history. From laboratory and clinical studies spanning more than a decade, researchers have learned that each viral particle is a rough-edged sphere consisting of an outer envelope enclosing a protein matrix, which, in turn, surrounds a cut-off cone-shaped core (**Figure A**). Within the core lies an enzyme-studded genome: two identical single strands of RNA associated with many molecules of an unusual DNA polymerase known as **reverse transcriptase**.

During infection, the AIDS virus binds to and injects its cone-shaped core into cells of the human immune system (**Figure B**). It next uses reverse transcriptase to copy its RNA genome into double-stranded DNA molecules in the cytoplasm of the host cell. The double helixes then travel to the nucleus where another enzyme inserts them into a host chromosome. Once integrated into a host-cell chromosome, the viral genome can do one of two things. It can commandeer the host cell's protein synthesis machinery to make hundreds of new viral particles that bud off from the parent cell, taking with them part of the cell membrane and sometimes resulting in the host cell's death. Alternatively, it can lie latent inside the host chromosome, which then copies and transmits the viral genome to two new cells with each cell division.

The events of this life cycle make HIV a **retrovirus**: an RNA virus that after infecting a host cell copies its own single strands of RNA into double helixes of DNA, which a viral enzyme then integrates into a host chromosome. RNA viruses that are not retroviruses simply infect a host cell and then use the cellular machinery to make more of themselves, often killing the host cell in the process. The viruses that cause hepatitis A, many types of the common cold, and rabies are this latter type of RNA virus. Unlike retroviruses, they are not transmitted by cell division to a geometrically growing number of new cells.

Reverse transcription, the foundation of the retroviral life cycle, is inconsistent with the one-way, DNA-to-RNA-to-protein flow of genetic information. Because it was so unexpected, the phenomenon of reverse transcription encountered great resistance in the scientific community when first reported by Howard Temin of the University of Wisconsin and David Baltimore, then of MIT. Now, however, it is an established fact. Reverse transcriptase is a remarkable DNA polymerase that can construct a DNA polymer from either an RNA or a DNA template.

In addition to its comprehensive copying abilities, reverse transcriptase has another feature not seen in most DNA polymerases: inaccuracy. As we saw in Chapter 5, normal DNA polymerases replicate DNA with an error rate of one mistake in every million nucleotides copied. Reverse transcriptase, however, introduces one mutation in every 5000 incorporated nucleotides.

HIV uses this capacity for mutation, in combination with its ability to integrate its genome into the chromosomes of

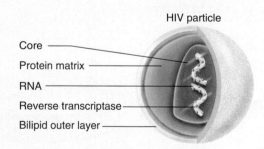

HIV particle

Core
Protein matrix
RNA
Reverse transcriptase
Bilipid outer layer

Figure A Structure of the AIDS virus.

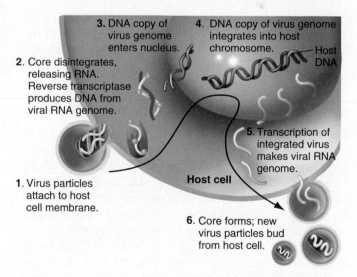

2. Core disintegrates, releasing RNA. Reverse transcriptase produces DNA from viral RNA genome.

3. DNA copy of virus genome enters nucleus.

4. DNA copy of virus genome integrates into host chromosome.
Host DNA

5. Transcription of integrated virus makes viral RNA genome.

1. Virus particles attach to host cell membrane.

Host cell

6. Core forms; new virus particles bud from host cell.

Figure B Life cycle of the AIDS virus.

Removing introns from the primary transcript by RNA splicing

Another kind of RNA processing became apparent in the late 1970s, after researchers had developed techniques that enabled them to analyze nucleotide sequences in both DNA and RNA. Using these techniques, which we describe in Chapter 14, they began to compare eukaryotic genes with the mRNAs derived from them. Their expectation was that just as in prokaryotes, the DNA nucleotide sequence of a gene's RNA-like strand would be identical to the RNA nucleotide sequence of the messenger RNA (with the exception of U replacing T in the RNA). Surprisingly, they found that the DNA nucleotide sequences of many eukaryotic genes are much longer than their corresponding mRNAs, suggesting that RNA transcripts, in addition to receiving a methylated cap and a poly-A tail, undergo extensive internal processing.

An extreme example of the difference in length between primary transcript and mRNA is seen in the human gene for dystrophin (**Figure 7.7**). Abnormalities in the dystrophin gene underlie the genetic disorder of

immune-system cells, to gain a tactical advantage over the immune response of its host organism. Cells of the immune system seek to overcome an HIV invasion by multiplying in response to the proliferating viral particles. The numbers are staggering. Each day of infection in every patient, from 100 million to a billion HIV particles are released from infected immune-system cells. As long as the immune system is strong enough to withstand the assault, it may respond by producing as many as 2 billion new cells daily. Many of these new cells produce antibodies targeted against proteins on the surface of the virus.

But just when an immune response wipes out those viral particles carrying the targeted protein, virions (entire viruses) incorporating new forms of the protein resistant to the current immune response make their appearance. After many years of this complex chase, capture, and destruction by the immune system, the changeable virus outruns the host's immune response and gains the upper hand. Thus, the intrinsic infidelity of HIV's reverse transcriptase, by enhancing the virus's ability to compete in the evolutionary marketplace, increases its threat to human life and health.

This inherent mutability has undermined two potential therapeutic approaches toward the control of AIDS: drugs and vaccines. Some of the antiviral drugs available in Canada for treatment of HIV infection—AZT (zidovudine), ddC (dideoxycytidine), and ddI (dideoxyinosine)—block viral replication by interfering with the action of reverse transcriptase. Each drug is similar to one of the four nucleotides, and when reverse transcriptase incorporates one of the drug molecules rather than a genuine nucleotide into a growing DNA polymer, the enzyme cannot extend the chain any further. However, the drugs are toxic at high doses and thus can be administered only at low doses that do not destroy all viral particles. Because of this limitation and the virus's high rate of mutation, mutant reverse transcriptases soon appear that work even in the presence of the drugs.

Similarly, researchers are having trouble developing safe, effective vaccines. Because HIV infects cells of the immune system and a vaccine works by stimulating immune-system cells to multiply, some of the vaccines tested so far actually increase the activity of the virus; others have only a weak effect on viral replication. Moreover, if it were possible to

Figure C Dr. Yong Kang.

produce a vaccine that could generate a massive immune response against one, or even several HIV proteins at a time, such a vaccine might be effective for only a short while—until enough mutations have built up to make the virus resistant.

One approach to vaccine creation aimed at circumventing this problem has been developed by University of Western Ontario researcher, Dr. Yong Kang (**Figure C**). While previous researchers created vaccines based on the expression of a single protein subunit of the whole virus, Dr. Kang has developed a "killed whole virus" vaccine. By developing methods to create large quantities of inactivated virus—genetically modified to ensure they cannot cause infection—Dr. Kang and his team can use the whole virus (composed of 19 different proteins) as an immunogen. The potential of the immune system to develop a response to all 19 HIV proteins (instead of just one, or a few) will make it much more difficult for the virus to develop mutations conferring resistance. Phase I clinical trials were successfully completed in November 2012. As expected the vaccine was shown to be safe for human use. Furthermore, in one patient, Dr. Kang measured a 32-fold increase in the level of antibodies fighting the virus. In another patient, a tenfold increase was observed. While preliminary, these results are extremely encouraging. The vaccine is currently undergoing phase II human clinical trials.

Duchenne muscular dystrophy (DMD). The dystrophin gene is 2.5 million nucleotides—or 2500 kilobases (kb)—long, whereas the corresponding mRNA is roughly 14 000 nucleotides, or 14 kb, in length. Obviously the gene contains DNA sequences that are not present in the mature mRNA. Those regions of the gene that do end up in the mature mRNA are scattered throughout the 2500 kb of DNA.

Exons and Introns. Sequences found in both a gene's DNA and the mature messenger RNA are called **exons** (for "expressed regions"). The sequences found in the DNA of the gene but not in the mature mRNA are known as **introns** (for "intervening regions"). Introns interrupt, or separate, the exon sequences that actually end up in the mature mRNA.

The *DMD* gene has more than 80 introns; the mean intron length is 35 kb, but 1 intron is an amazing 400 kb long. Other genes in humans generally have many fewer introns, while a few have none—and the introns range from 50 bp to over 100 kb. Exons, in contrast, vary in size from 50 bp to a few kilobases; in the *DMD* gene, the mean

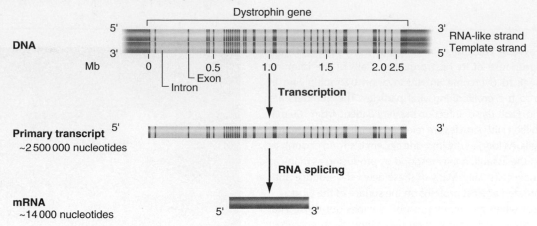

Splicing removes introns from a primary transcript.

Figure 7.7 The human dystrophin gene: An extreme example of RNA splicing. Though the dystrophin gene is 2500 kb (or 2.5 Mb) long, the dystrophin mRNA is only 14 kb long. More than 80 introns are removed from the 2500 kb primary transcript to produce the mature mRNA (which is not drawn to scale).

exon length is 200 bp. The greater size variation seen in introns compared with exons reflects the fact that introns do not encode polypeptides and do not appear in mature mRNAs. As a result, fewer restrictions exist on the sizes and base sequences of introns.

Mature mRNAs must contain all of the codons that are translated into amino acids, including the initiation and termination codons. In addition, mature mRNAs have sequences at their 5′ and 3′ ends that are not translated, but that nevertheless play important roles in regulating the efficiency of translation. These sequences, called the **5′ and 3′ untranslated regions** (5′ and 3′ **UTRs**), are located just after the methylated cap and just before the poly-A tail, respectively. Excepting the cap and tail themselves, all of the sequences in a mature mRNA, including all of the codons and both UTRs, must be transcribed from the gene's exons. Introns can interrupt a gene at any location, even between the nucleotides making up a single codon. In such a case, the three nucleotides of the codon are present in two different (but successive) exons.

How do cells make a mature mRNA from a gene whose coding sequences are interrupted by introns? The answer is that cells first make a primary transcript containing all of a gene's introns and exons, and then they remove the introns from the primary transcript by **RNA splicing**, the process that deletes introns and joins together successive exons to form a mature mRNA consisting only of exons (Figure 7.7). Because the first and last exons of the primary transcript become the 5′ and 3′ ends of the mRNA, while all intervening introns are spliced out, a gene must have one more exon than it does introns. To construct the mature mRNA, splicing must be remarkably precise. For example, if an intron lies within a codon, splicing must remove the intron and reconstitute the codon without disrupting the *reading frame* (the partitioning of groups of three nucleotides from a fixed starting point) of the mRNA. (You will learn more about reading frames in Section 7.7).

The Mechanism of RNA Splicing. **Figure 7.8** illustrates how RNA splicing works. Three types of short sequences within the primary transcript—**splice donors**, **splice acceptors**, and **branch sites**—help ensure the specificity of splicing. These sites make it possible to sever the connections between an intron and the exons that precede and follow it, and then to join the formerly separated exons.

The mechanism of splicing involves two sequential cuts in the primary transcript. The first cut is at the splice donor site, at the 5′ end of the intron. After this first cut, the new 5′ end of the intron attaches, via a novel 2′–5′ phosphodiester bond, to an A at the branch site located within the intron, forming a so-called *lariat structure*. The second cut is at the splice acceptor site, at the 3′ end of the intron; this cut removes the intron. The discarded intron is degraded, and the precise splicing of adjacent exons completes the process of intron removal.

SnRNPs and the Spliceosome. Splicing normally requires a complicated intranuclear machine called the **spliceosome**, which ensures that all of the splicing reactions take place in concert (**Figure 7.9**). The spliceosome consists of four subunits known as small nuclear ribonucleoproteins, or snRNPs (pronounced "snurps"). Each snRNP contains one or two small nuclear RNAs (snRNAs) 100–300 nucleotides long, associated with proteins in a discrete particle. Certain snRNAs can base-pair with the splice donor and splice acceptor sequences in the primary transcript, so these snRNAs are particularly important in bringing together the two exons that flank an intron. Given the complexities of spliceosome structure, it is remarkable that a few primary transcripts can splice themselves without the aid of a spliceosome or any additional factor. These rare primary transcripts function as **ribozymes**—RNA molecules that can act as enzymes and catalyze a specific biochemical reaction.

It might seem strange that eukaryotic genes incorporate DNA sequences that are spliced out of the mRNA before

(a) Short sequences dictate where splicing occurs.

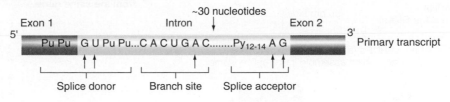

(b) Two sequential cuts remove the intron.

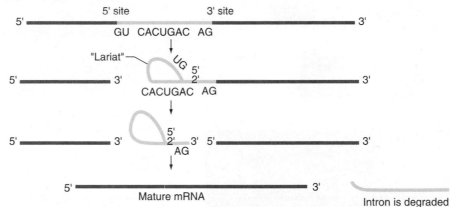

Figure 7.8 How RNA processing splices out introns and joins adjacent exons. (a) Three short sequences within the primary transcript determine the specificity of splicing. (1) The splice donor site occurs where the 3′ end of an exon abuts the 5′ end of an intron. In most splice donor sites, a GU dinucleotide (*arrows*) that begins the intron is flanked on either side by a few purines (Pu; i.e., A or G). (2) The splice acceptor site is at the 3′ end of the intron where it joins with the next exon. The final nucleotides of the intron are always AG (*arrows*) preceded by 12–14 pyrimidines (Py; i.e., C or U). (3) The branch site, which is located within the intron about 30 nucleotides upstream of the splice acceptor, must include an A (*arrow*) and is usually rich in pyrimidines. **(b)** Two sequential cuts, the first at the splice donor site and the second at the splice acceptor site, remove the intron, allowing precise splicing of adjacent exons.

translation and thus do not encode amino acids. No one knows exactly why introns exist. One hypothesis proposes that they make it possible to assemble genes from various exon building blocks, which encode modules of protein function. This type of assembly would allow the shuffling of exons to make new genes, a process that appears to have played a key role in the evolution of complex organisms. The exon-as-module proposal is attractive because it is easy to understand the selective advantage of the potential for exon shuffling. Nevertheless, it remains a hypothesis without proof. There is no hard evidence for or against the hypothesis, and introns may have become established through means that scientists have yet to imagine.

Alternative splicing: Different mRNAs from the same primary transcript

Normally, RNA splicing joins together the splice donor and splice acceptor at the opposite ends of an intron, resulting in removal of the intron and fusion of two successive— and now adjacent—exons. For some genes, however, RNA splicing during development is regulated so that at certain times or in certain tissues, some splicing signals may be ignored. As an example, splicing may occur between the splice donor site of one intron and the splice acceptor site of a different intron downstream. Such **alternative splicing** produces different mRNA molecules that may encode related proteins with different—though partially overlapping—amino acid sequences and functions. In

effect then, alternative splicing can tailor the nucleotide sequence of a primary transcript to produce more than one kind of polypeptide. Alternative splicing largely explains how the approximately 21 000 genes in the human genome can encode the hundreds of thousands of different proteins estimated to exist in human cells.

In mammals, alternative splicing of the gene encoding the antibody heavy chain determines whether the antibody proteins become embedded in the membrane of the B lymphocyte that makes them or are instead secreted into the blood. The antibody heavy-chain gene has eight exons and seven introns; exon number 6 has a splice donor site within it. To make the membrane-bound antibody, all exons except for the right-hand part of number 6 are joined to create an mRNA encoding a hydrophobic (water-hating, lipid-loving) C terminus (**Figure 7.10a**). For the secreted antibody, only the first six exons (including the right part of 6) are spliced together to make an mRNA encoding a heavy chain with a hydrophilic (water-loving) C terminus. These two kinds of mRNAs formed by alternative splicing thus encode slightly different proteins that are directed to different parts of the body.

A rare and unusual strategy of alternative splicing, seen in *C. elegans* and a few other eukaryotes, is **trans-splicing**, in which the spliceosome joins an exon of one gene with an exon of another gene (**Figure 7.10b**). Special nucleotide sequences in the RNAs make trans-splicing possible.

Spliceosome components

Five snRNAs
(small nuclear RNAs) + ~50 proteins

Four snRNPs (small nuclear ribonucleoproteins), which assemble into a spliceosome

Proteins

snRNA

Figure 7.9 Splicing is catalyzed by the spliceosome. (*Top*) The spliceosome is assembled from four snRNP **subunits**, each of which contains one or two snRNAs and several proteins. (*Bottom*) A view of three spliceosomes in the electron microscope.

(a) Alternative splicing produces two different mRNAs from the same gene.

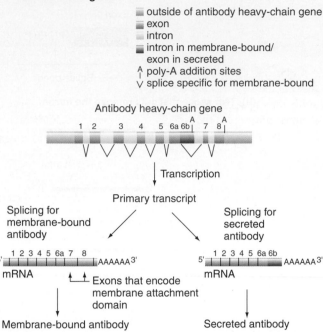

- outside of antibody heavy-chain gene
- exon
- intron
- intron in membrane-bound/ exon in secreted
- ⋀ poly-A addition sites
- ⋁ splice specific for membrane-bound

Antibody heavy-chain gene

Transcription

Primary transcript

Splicing for membrane-bound antibody

Splicing for secreted antibody

5′ 1 2 3 4 5 6a 7 8 AAAAAAA 3′
mRNA

Exons that encode membrane attachment domain

5′ 1 2 3 4 5 6a 6b AAAAAAA 3′
mRNA

Membrane-bound antibody

Secreted antibody

(b) Trans-splicing combines exons from different genes.

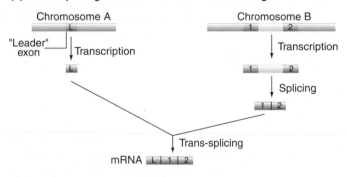

Chromosome A

"Leader" exon

Transcription

Chromosome B

Transcription

Splicing

Trans-splicing

mRNA

Figure 7.10 Different mRNAs can be produced from the same primary transcript. (a) Alternative splicing of the primary transcript for the antibody heavy chain produces mRNAs that encode different kinds of antibody proteins. **(b)** Rare trans-splicing events combine exons from different genes into one mature mRNA.

- RNA polymerase, the key enzyme of transcription, recognizes the promoter at the beginning of a gene and then uses complementary base-pairing with the DNA template strand to add RNA nucleotides to the 3′ end of the growing transcript. When RNA polymerase detects a terminator, it dissociates from both the DNA and the transcript.

- In eukaryotic cells, RNA processing follows transcription to generate an mRNA. Processing steps include additions of a methylated cap to the RNA's 5′ end and a poly-A tail to the 3′ end, as well as the removal of introns from the primary transcript by splicing. Alternative splicing of exons can yield different mRNAs from the same primary transcript.

7.3 Translation: From mRNA to Protein

Translation is the process by which the sequence of nucleotides in a messenger RNA directs the assembly of the correct sequence of amino acids in the corresponding polypeptide. Translation takes place on ribosomes that coordinate the movements of transfer RNAs carrying specific amino acids with the genetic instructions of an mRNA. As we examine the cell's translation machinery, we first describe the structure and function of tRNAs and

ribosomes; we then explain how these components interact during translation.

Transfer RNAs mediate the translation of mRNA codons to amino acids

No obvious chemical similarity or affinity exists between the nucleotide triplets of mRNA codons and the amino acids they specify. Rather, *transfer RNAs (tRNAs)* serve as adaptor molecules that mediate the transfer of information from nucleic acid to protein.

The structure of tRNA

Transfer RNAs are short, single-stranded RNA molecules 74–95 nucleotides in length. Several of the nucleotides in tRNAs contain modified bases produced by chemical alterations of the principal A, G, C, and U nucleotides (**Figure 7.11a**). Each tRNA carries one particular amino acid, and cells must have at least one tRNA for each of the 20 amino acids specified by the genetic code. The name of a tRNA reflects the amino acid it carries. For example, tRNA^{Gly} carries the amino acid glycine.

As **Figure 7.11b** shows, it is possible to consider the structure of a tRNA molecule on three levels.

1. The nucleotide sequence of a tRNA constitutes the primary structure.
2. Short complementary regions within a tRNA's single strand can form base pairs with each other to create a characteristic cloverleaf shape; this is the tRNA's secondary structure.
3. Folding in three-dimensional space creates a tertiary structure that looks like a compact letter L.

At one end of the L, the tRNA carries an **anticodon**: three nucleotides complementary to an mRNA codon specifying the amino acid carried by the tRNA (Figure 7.11b). The anticodon never forms base pairs with other regions of the tRNA; it is always available for base-pairing with its complementary mRNA codon. As with other complementary base sequences, during pairing at the ribosome, the strands of anticodon and codon run antiparallel to each other. For example, if the anticodon is 3′ CCU 5′, the complementary mRNA codon is 5′ GGA 3′, specifying the amino acid glycine.

At the other end of the L, where the 5′ and 3′ ends of the tRNA strand are found, enzymes known as **amino-acyl-tRNA synthetases** connect the tRNA to the amino acid that corresponds to the anticodon (**Figure 7.12**). These enzymes are extraordinarily specific, recognizing unique features of a particular tRNA—despite its general structural similarities with all other tRNAs—while also recognizing the corresponding amino acid.

Aminoacyl-tRNA synthetases are, in fact, the only molecules that read the languages of both nucleic acid and protein. They are thus the actual molecular translators. At least one aminoacyl-tRNA synthetase exists for each of the 20 amino acids, and like tRNA, each synthetase functions with only one amino acid. Figure 7.12 shows the two-step

(a) Some tRNAs contain modified bases.

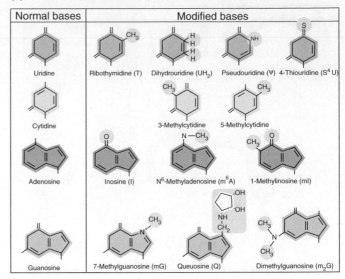

(b) Each tRNA has a primary, secondary, and tertiary structure.

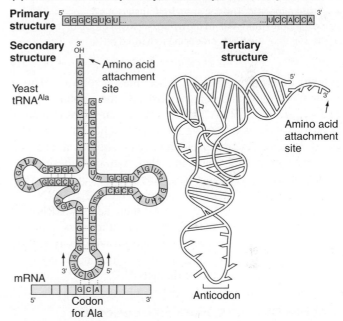

Figure 7.11 tRNAs mediate the transfer of information from nucleic acid to protein. (a) Many tRNAs contain modified bases produced by chemical alterations of A, G, C, and U. **(b)** The primary structures of tRNA molecules fold to form characteristic secondary and tertiary structures. The anticodon and the amino acid attachment site are at opposite ends of the L-shaped structure.

process that establishes the covalent bond between an amino acid and the 3′ end of its corresponding tRNA. A tRNA covalently coupled to its amino acid is called a **charged tRNA**. The bond between the amino acid and tRNA contains substantial energy that is later used to drive peptide bond formation.

The critical role of base-pairing between codon and anticodon

While attachment of the appropriate amino acid charges a tRNA, the amino acid itself does not play a significant

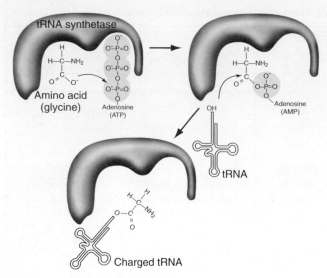

Figure 7.12 Aminoacyl-tRNA synthetases catalyze the attachment of tRNAs to their corresponding amino acids. The aminoacyl-tRNA synthetase first activates the amino acid, forming an AMP-amino acid. The enzyme then transfers the amino acid's carboxyl group from AMP to the hydroxyl (−OH) group of the ribose at the 3' end of the tRNA, producing a charged tRNA.

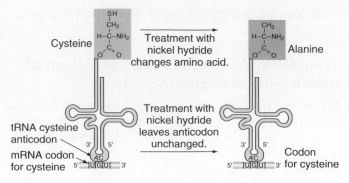

Figure 7.13 Base-pairing between an mRNA codon and a tRNA anticodon determines which amino acid is added to a growing polypeptide. A tRNA with an anticodon for cysteine, but carrying the amino acid alanine, adds alanine whenever the mRNA codon for cysteine appears.

role in determining where it becomes incorporated in a growing polypeptide chain. Instead, the specific interaction between a tRNA's anticodon and an mRNA's codon makes that decision. A simple experiment illustrates this point (**Figure 7.13**). Researchers can subject a charged tRNA to chemical treatments that, without altering the structure of the tRNA, change the amino acid it carries. One treatment replaces the cysteine carried by tRNA^{Cys} with alanine. When investigators then add the tRNA^{Cys} charged with alanine to a cell-free translational system, the system incorporates *alanine* into the growing polypeptide wherever the mRNA contains a cysteine codon complementary to the anticodon of the tRNA^{Cys}.

Wobble: One tRNA, more than one codon

Although at least one kind of tRNA exists for each of the 20 amino acids, cells do not necessarily carry tRNAs with anticodons complementary to all of the 61 possible codon triplets in the genetic code. *E. coli*, for example, makes 79 different tRNAs containing 42 different anticodons. Although several of the 79 tRNAs in this collection obviously have the same anticodon, $61 - 42 = 19$ of 61 potential anticodons are not represented. Thus 19 mRNA codons will not find a complementary anticodon in the *E. coli* collection of tRNAs. How can an organism construct proper polypeptides if some of the codons in its mRNAs cannot locate tRNAs with complementary anticodons?

The answer is that some tRNAs can recognize more than one codon for the amino acid with which they are charged. That is, the anticodons of these tRNAs can interact with more than one codon for the same amino acid, in

keeping with the degenerate nature of the genetic code. Although researchers do not fully understand this "promiscuous" base-pairing between codons and anticodons, Francis Crick spelled out a few of the rules that govern it.

Crick reasoned that the 3' nucleotide in many codons adds nothing to the specificity of the codon. For example, 5' GGU 3', 5' GGC 3', 5' GGA 3', and 5' GGG 3' all encode glycine (review Figure 7.2). It does not matter whether the 3' nucleotide is U, C, A, or G as long as the first two letters are GG. The same is true for other amino acids encoded by four different codons, such as valine, where the first two bases must be GU, but the third base can be U, C, A, or G.

For amino acids specified by two different codons, the first two bases of the codon are, once again, always the same, while the third base must be either one of the two purines (A or G) or one of the two pyrimidines (U or C). Thus, 5' CAA 3' and 5' CAG 3' are both codons for glutamine; 5' CAU 3' and 5' CAC 3' are both codons for histidine. If Pu stands for either purine and Py stands for either pyrimidine, then CAPu represents the codons for glutamine, while CAPy represents the codons for histidine.

In fact, the 5' nucleotide of a tRNA's anticodon can often pair with more than one kind of nucleotide in the 3' position of an mRNA's codon (recall that after base-pairing, the bases in the anticodon run antiparallel to the bases in the codon). A single tRNA charged with a particular amino acid can thus recognize several or even all of the codons for that amino acid. This flexibility in base-pairing between the 3' nucleotide in the codon and the 5' nucleotide in the anticodon is known as **wobble** (**Figure 7.14a**). The combination of normal base-pairing at the first two positions of a codon with wobble at the third position clarifies why multiple codons for a single amino acid usually start with the same two letters.

Crick's "wobble rules" (see **Figure 7.14b**) delimit what kind of flexibility in base-pairing is consistent with the genetic code. For example, methionine (Met) is specified by a single codon (5' AUG 3'). As a result, Met-specific tRNAs must have a C at the 5' end of their anticodons

(a)

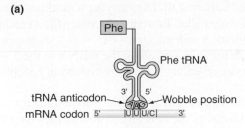

(b)

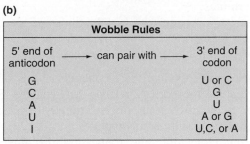

Figure 7.14 Wobble: Some tRNAs recognize more than one codon for the amino acid they carry. (a) The G at the 5' end of the anticodon shown here can pair with either U or C at the 3' end of the codon. **(b)** The chart shows the pairing possibilities for other nucleotides at the 5' end of an anticodon; I = inosine.

(5' CAU 3'), because this is the only nucleotide at that position that can base-pair with the G at the 3' end of the Met codon. By contrast, a single isoleucine-specific tRNA with the modified nucleotide inosine (I) at the 5' position of the anticodon can recognize all three codons (5' AUU 3', 5' AUC 3', and 5' AUA 3') for isoleucine.

Ribosomes are the sites of polypeptide synthesis

Ribosomes facilitate polypeptide synthesis in various ways. First, they recognize mRNA features that signal the start of translation. Second, they help ensure accurate interpretation of the genetic code by stabilizing the interactions between tRNAs and mRNAs; without a ribosome, codon-anticodon recognition, mediated by only three base pairs, would be extremely weak. Third, they supply the enzymatic activity that links the amino acids in a growing polypeptide chain. Fourth, by moving 5' to 3' along an mRNA molecule, they expose the mRNA codons in sequence, ensuring the linear addition of amino acids. Finally, ribosomes help end polypeptide synthesis by dissociating both from the mRNA directing polypeptide construction and from the polypeptide product itself.

The structure of ribosomes

In *E. coli*, ribosomes consist of three different **ribosomal RNAs (rRNAs)** and 52 different ribosomal proteins (**Figure 7.15a**). These components associate to form two different ribosomal subunits called the 30S subunit and the 50S subunit (with S designating a coefficient of sedimentation related to the size and shape of the subunit; the 30S

(a) A ribosome has two subunits composed of RNA and protein.

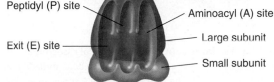

(b) Different parts of a ribosome have different functions.

Figure 7.15 The ribosome: Site of polypeptide synthesis. (a) A ribosome has two subunits, each composed of rRNA and various proteins. **(b)** The small subunit initially binds to mRNA. The large subunit contributes the enzyme peptidyl transferase, which catalyzes the formation of peptide bonds. The two subunits together form the A, P, and E tRNA-binding sites.

subunit is smaller than the 50S subunit). Before translation begins, the two subunits exist as separate entities in the cytoplasm. Soon after the start of translation, they come together to reconstitute a complete ribosome. Eukaryotic ribosomes have more components than their prokaryotic counterparts, but they still consist of two dissociable subunits.

Functional domains of ribosomes

The small 30S subunit is the part of the ribosome that initially binds to mRNA. The larger 50S subunit contributes an enzyme known as **peptidyl transferase**, which catalyzes formation of the peptide bonds joining adjacent amino acids (**Figure 7.15b**). Both the small and the large subunits contribute to three distinct tRNA-binding areas known as the **aminoacyl (or A) site**, the **peptidyl (or P) site**, and the **exit (or E) site**. Finally, other regions of the ribosome distributed over the two subunits serve as points of contact for some of the additional proteins that play a role in translation.

Using X-ray crystallography and electron microscopy, researchers have recently gained a remarkably detailed view of the complicated structure of the ribosome. **Figure 7.16** shows the large subunit of a bacterial ribosome; the small subunit was computationally removed for better visualization of the charged tRNAs occupying the A and P sites. With this illustration, you can see that the tRNAs occupy

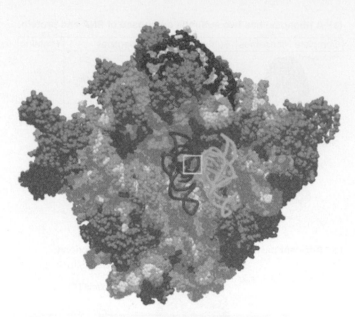

Figure 7.16 The large subunit of a bacterial ribosome. Various ribosomal proteins are *lavender*, 23S rRNA is in *gold* and *white*, and 5S rRNA is *maroon* and *white*. The tRNA in the A site is *green*; the tRNA in the P site is *red*; no tRNA is shown in the E site. The superimposed box shows the location where new peptide bonds are formed.

most of the space in the central part of the ribosome, while the various ribsosomal proteins are studded around the exterior. Surprisingly, no proteins are found close to the region between the two tRNAs where peptide bonds are formed. This finding supports the conclusions of biochemical experiments that peptidyl transferase is actually a function of the 50S subunit's rRNA rather than any protein component of the ribosome; in other words, the rRNA acts as a *ribozyme* that joins amino acids together.

Ribosomes and charged tRNAs collaborate to translate mRNAs into polypeptides

As was the case for transcription, translation consists of three phases: an **initiation** phase that sets the stage for polypeptide synthesis; **elongation**, during which amino acids are added to a growing polypeptide; and a **termination** phase that brings polypeptide synthesis to a halt and enables the ribosome to release a completed chain of amino acids. **Figure 7.17** illustrates the details of the process, focusing on translation as it occurs in bacterial cells. As you examine the figure, note the following points about the flow of information during translation.

- The first codon to be translated—the initiation codon—is an AUG set in a special context at the 5′ end of the gene's reading frame (*not* precisely at the 5′ end of the mRNA).

- Special initiating tRNAs carrying a modified form of methionine called formylmethionine (fMet) recognize the initiation codon.
- The ribosome moves along the mRNA in the 5′-to-3′ direction, revealing successive codons in a stepwise fashion.
- At each step of translation, the polypeptide grows by the addition of the next amino acid in the chain to its C terminus.
- Translation terminates when the ribosome reaches a UAA, UAG, or UGA nonsense codon at the 3′ end of the gene's reading frame.

These points explain the biochemical basis of colinearity; that is, the correspondence between the 5′-to-3′ direction in the mRNA and the N-terminus-to-C-terminus direction in the resulting polypeptide.

During elongation, the translation machinery adds about 2–15 amino acids per second to the growing chain. The speed is higher in prokaryotes and lower in eukaryotes. At these rates, construction of an average-size 300-amino-acid polypeptide (from an average-length mRNA that is somewhat longer than 1000 nucleotides) could take as little as 20 seconds or as long as 2.5 minutes.

Several details have been left out of Figure 7.17 so that you can concentrate on the flow of information during translation. In particular, this figure does not depict the important roles played by protein translation factors, which help shepherd mRNAs and tRNAs to their proper locations on the ribosome. Some translation factors also carry GTP to the ribosome, where hydrolysis of the high-energy bonds in the GTP helps power certain molecular movements (such as translocation of the ribosome along the mRNA).

Processing after translation can change a polypeptide's structure

Protein structure is not irrevocably fixed at the completion of translation. Several different processes may subsequently modify a polypeptide's structure. Cleavage may remove amino acids, such as the N-terminal fMet, from a polypeptide (**Figure 7.18a**), or it may generate several smaller polypeptides from one larger product of translation (**Figure 7.18b**). In the latter case, the larger polypeptide made before it is cleaved into smaller polypeptides is often called a **polyprotein**. The addition of chemical constituents, such as phosphate groups, methyl groups, or even carbohydrates, to specific amino acids may also modify a polypeptide after translation (**Figure 7.18c**). Such cleavages and additions are known as **posttranslational modifications**. Posttranslational changes to a protein can be very important. For example, the biochemical function of many enzymes directly depends on the addition (or sometimes removal) of phosphate groups.

Translation of mRNAs on Ribosomes

(a) Initiation: Setting the stage for polypeptide synthesis The first three nucleotides of an mRNA do not serve as the first codon to be translated into an amino acid. Instead, a special signal indicates where along the mRNA translation should begin. In prokaryotes, this signal is called the **ribosome-binding site**, and it has two important elements. The first is a short sequence of six nucleotides—usually 5′ . . . AGGAGG . . . 3′—named the **Shine–Dalgarno box** after its discoverers. The second element in an mRNA's ribosome-binding site is the triplet 5′ AUG 3′, which serves as the initiation codon.

A special initiator tRNA, whose 5′ CAU 3′ anticodon is complementary to AUG, recognizes an AUG preceded by the Shine–Dalgarno box of a ribosome-binding site. The initiator tRNA carries ***N*-formylmethionine (fMet)**, a modified methionine whose amino end is blocked by a formyl group. The specialized fMet tRNA functions only at an initiation site. An AUG codon located within an mRNA's reading frame is recognized by a different tRNA that is charged with an unmodified methionine. This tRNA cannot start translation.

During initiation, the 3′ end of the 16S rRNA in the 30S ribosomal subunit binds to the mRNA's Shine–Dalgarno box (*not shown*), the fMet tRNA binds to the mRNA's initiation codon, and a large 50S ribosomal subunit associates with the small subunit to round out the ribosome. At the end of initiation, the fMet tRNA sits in the P site of the completed ribosome. Proteins known as **initiation factors** (*not shown*) play a transient role in the initiation process.

In eukaryotes, the small ribosomal subunit binds first to the methylated cap at the 5′ end of the mature mRNA. It then migrates to the initiation site—usually the first AUG it encounters as it scans the mRNA in the 5′-to-3′ direction. The initiator tRNA in eukaryotes carries unmodified methionine (Met) instead of fMet.

Initiation phase

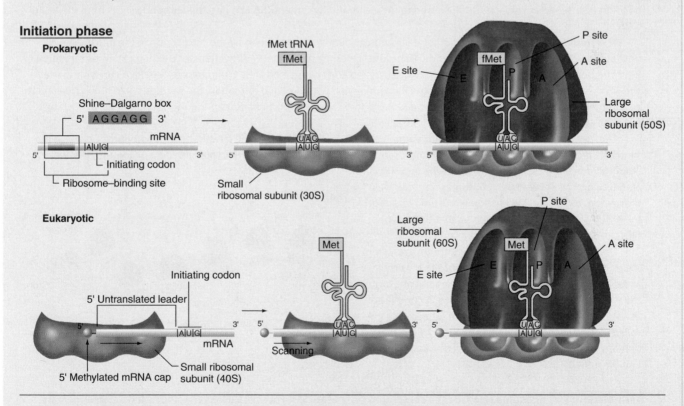

(b) Elongation: The addition of amino acids to a growing polypeptide Proteins known as **elongation factors** (*not shown*) usher the appropriate tRNA into the A site of the ribosome. The anticodon of this charged tRNA must recognize the next codon in the mRNA. The ribosome simultaneously holds the initiating tRNA at its P site and the second tRNA at its A site so that peptidyl transferase can catalyze formation of a peptide bond between the amino acids carried by the two tRNAs. As a result, the tRNA at the A site now carries two amino acids. The N terminus of this dipeptide is fMet; the C terminus is the second amino acid, whose carboxyl group remains covalently linked to its tRNA.

Following formation of the first peptide bond, the ribosome moves, exposing the next mRNA codon. The ribosome's movement requires the help of elongation factors and an input of energy. As the ribosome moves, the initiating tRNA, which

(Continued)

Elongation phase

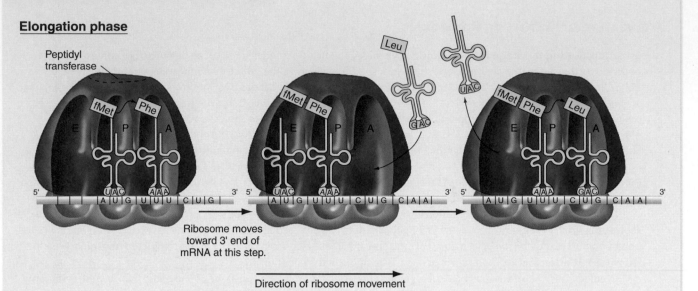

Peptidyl transferase

Ribosome moves toward 3' end of mRNA at this step.

Direction of ribosome movement

no longer carries an amino acid, is transferred to the E site, and the other tRNA carrying the dipeptide shifts from the A site to the P site.

The empty A site now receives another tRNA, whose identity is determined by the next codon in the mRNA. The uncharged initiating tRNA is bumped off the E site and leaves the ribosome. Peptidyl transferase then catalyzes formation of a second peptide bond, generating a chain of three amino acids connected at its C terminus to the tRNA currently in the A site. With each subsequent round of ribosome movement and peptide bond formation, the peptide chain grows one amino acid longer. Note that each tRNA moves from the A site to the P site to the E site (excepting the initiating tRNA, which first enters the P site).

Because the elongation machinery adds amino acids to the C terminus of the lengthening polypeptide, polypeptide synthesis proceeds from the N terminus to the C terminus. As a result, fMet in prokaryotes (Met in eukaryotes), the first amino acid in the growing chain, will be the N-terminal amino acid of all finished polypeptides prior to protein processing. Moreover, the ribosome must move along the mRNA in the 5'-to-3' direction so that the polypeptide can grow in the N-to-C direction.

Once a ribosome has moved far enough away from the mRNA's ribosome-binding site, that site becomes accessible to other ribosomes. In fact, several ribosomes can work on the same mRNA at one time. A complex of several ribosomes translating from the same mRNA is called a **polyribosome**. This complex allows the simultaneous synthesis of many copies of a polypeptide from a single mRNA.

Polyribosome

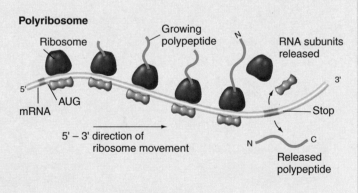

(c) **Termination: The ribosome releases the completed polypeptide** No normal tRNAs carry anticodons complementary to the three nonsense (stop) codons UAG, UAA, and UGA. Thus, when movement of the ribosome brings a nonsense codon into the ribosome's A site, no tRNAs can bind to that codon. Instead, proteins called **release factors** recognize the termination codons and bring polypeptide synthesis to a halt. The tRNA specifying the C-terminal amino acid releases the completed polypeptide, the same tRNA as well as the mRNA separate from the ribosome, and the ribosome dissociates into its large and small subunits.

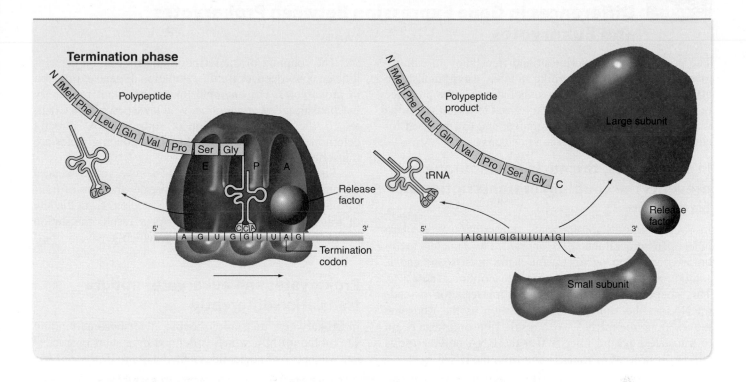

Termination phase

Polypeptide

Release factor

Termination codon

Polypeptide product

tRNA

Large subunit

Release factor

Small subunit

(a) Cleavage may remove an amino acid.

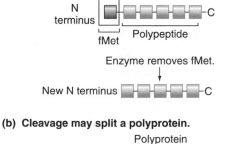

N terminus

fMet

Polypeptide

Enzyme removes fMet.

New N terminus

(b) Cleavage may split a polyprotein.

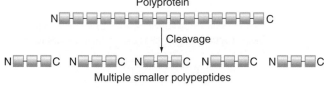

Polyprotein

N ▭▭▭▭▭▭▭▭▭▭▭▭▭▭▭▭ C

Cleavage

N▭▭▭C N▭▭▭C N▭▭▭C N▭▭▭C N▭▭▭C

Multiple smaller polypeptides

(c) Addition of chemical constituents may modify a protein.

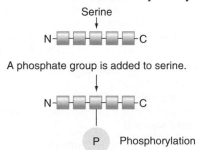

Serine

N-▭▭▭▭▭-C

A phosphate group is added to serine.

N-▭▭▭▭▭-C

P Phosphorylation

Figure 7.18 Posttranslational processing can modify polypeptide structure. Cleavage may remove an amino acid from the N terminus of a polypeptide **(a)** or split a larger *polyprotein* into two or more smaller functional proteins **(b)**. **(c)** Chemical reactions may add a phosphate or other functional group to an amino acid in the polypeptide.

- Transfer RNAs mediate the relationship between codons in the mRNA and the amino acids in the polypeptide product. At one end of the L-shaped tRNA molecule are the three nucleotides of the anticodon that can base-pair with complementary codons. At the other end of the L, the proper amino acid is covalently coupled to the tRNA by a specific aminoacyl-tRNA synthetase enzyme.

- "Wobble" refers to the observation that the nucleotide at the 5′ position of a tRNA's anticodon can often pair with different nucleotides at the 3′ position of an mRNA codon. Wobble explains why alternative codons for a single amino acid usually start with the same two nucleotides.

- The ribosome is a complex made of various proteins and rRNAs at which polypeptide synthesis takes place. The large and small subunits of the ribosome together form three binding sites (A, P, and E) for tRNA molecules.

- Ribosomes initiate mRNA translation at AUG initiation codons. During elongation, the ribosome moves along the mRNA in the 5′-to-3′ direction, while tRNAs base-paired with mRNA codons move through the ribosome's A, P, and E sites. The ribosome's peptidyl transferase forms peptide bonds between successive amino acids. Translation terminates at stop codons in the mRNA.

Differences in Gene Expression Between Prokaryotes and Eukaryotes

The processes of transcription and translation in eukaryotes and prokaryotes are similar in many ways but also are affected by certain differences, including (1) the presence of a nuclear membrane in eukaryotes, (2) variations in the way in which translation is initiated, and (3) the need for additional transcript processing in eukaryotes.

In eukaryotes, the nuclear membrane prevents the coupling of transcription and translation

In *E. coli* and other prokaryotes, transcription takes place in an open intracellular space undivided by a nuclear membrane; translation occurs in the same open space and is sometimes coupled directly with transcription (**Table 7.1**). This coupling is possible because transcription extends mRNAs in the same 5'-to-3' direction as the ribosome moves along the mRNA. As a result, ribosomes can begin to translate a partial mRNA that the RNA polymerase is still in the process of transcribing from the DNA.

The coupling of transcription and translation has significant consequences for the regulation of gene expression in prokaryotes. For example, in an important regulatory mechanism called *attenuation*, which we describe in Chapter 10, the rate of translation of some mRNAs directly determines the rate at which the corresponding genes are transcribed into these mRNAs.

Such coupling cannot occur in eukaryotes because the nuclear envelope physically separates the sites of transcription and RNA processing in the nucleus from the site of translation in the cytoplasm. As a result, translation in eukaryotes can affect the rate at which genes are transcribed only in more indirect ways.

Prokaryotes and eukaryotes initiate translation differently

In prokaryotes, translation begins at a ribosome-binding site on the mRNA, which is defined by a short characteristic sequence of nucleotides called a *Shine–Dalgarno box*

Table 7.1	Differences Between Prokaryotes and Eukaryotes in the Details of Gene Expression	
	Prokaryotes	**Eukaryotes**
Overview	1. No nucleus. Transcription and translation take place in the same cellular compartments, and translation is often coupled to transcription.	1. Nucleus separated from the cytoplasm by a nuclear membrane. Transcription takes place in the nucleus, while translation occurs in the cytoplasm. Direct coupling of transcription and translation is not possible.
	2. Genes are not divided into exons and introns.	2. The DNA of a gene consists of exons separated by introns; the exons are defined by posttranscriptional splicing, which deletes the introns.
Transcription	1. One RNA polymerase consisting of five subunits.	1. Several kinds of RNA polymerase, each containing ten or more subunits; different polymerases transcribe different genes.
	2. Primary transcripts are the actual mRNAs; they have a triphosphate start at the 5' end and no tail at the 3' end.	2. Primary transcripts undergo processing to produce mature mRNAs that have a methylated cap at the 5' end and a poly-A tail at the 3' end.
Translation	1. Unique initiator tRNA carries formylmethionine.	1. Initiator tRNA carries methionine.
	2. mRNAs have multiple ribosome-binding sites and can thus direct the synthesis of several different polypeptides.	2. mRNAs have only one start site and can thus direct the synthesis of only one kind of polypeptide.
	3. Small ribosomal subunit immediately binds to the mRNA's ribosome-binding site.	3. Small ribosomal subunit binds first to the methylated cap at the 5' end of the mature mRNA and then scans the mRNA to find the ribosome-binding site.

adjacent to an initiating AUG codon (Figure 7.17a). There is nothing to prevent an mRNA from having more than one ribosome-binding site, and, in fact, many prokaryotic messages are **polycistronic**: They contain the information of several genes (sometimes referred to as *cistrons*; see Chapter 10), each of which can be translated independently starting at its own ribosome-binding site (Table 7.1).

In eukaryotes, by contrast, the small ribosomal subunit first binds to the methylated cap at the 5' end of the mature mRNA and then migrates to the initiation site. This site is almost always the first AUG codon encountered by the ribosomal subunit as it moves along (or "scans") the mRNA in the 5'-to-3' direction (see Figure 7.17a and Table 7.1). The mRNA region between the 5' cap and the initiation codon is sometimes referred to as either the *5' untranslated region (5' UTR)* or the *5' untranslated leader.* Because of this scanning mechanism, initiation in eukaryotes takes place at only a single site on the mRNA, and each mRNA contains the information for translating only a single kind of polypeptide.

Another translational difference between prokaryotes and eukaryotes is in the composition of the initiating tRNA. In prokaryotes, as already mentioned, this tRNA carries a modified form of methionine known as *N*-formyl-methionine, while in eukaryotes, it carries an unmodified

methionine (see Table 7.1). Thus, immediately after translation, eukaryotic polypeptides all have Met (instead of fMet) at their N termini. Posttranslational cleavage events in both prokaryotes and eukaryotes often create mature proteins that no longer have N-terminal fMet or Met (see Figure 7.18a).

Eukaryotic mRNAs require more processing than prokaryotic mRNAs

Table 7.1 reviews other important differences in gene structure and expression between prokaryotes and eukaryotes. In particular, introns interrupt eukaryotic, but not prokaryotic genes such that the splicing of a primary transcript is necessary for eukaryotic gene expression. Other types of RNA processing that occur in eukaryotes but not prokaryotes add a methylated cap and a poly-A tail to the 5' and 3' ends of the mRNAs, respectively.

> Because mRNA in eukaryotes must leave the nucleus for translation, transcription and translation cannot be coupled, as in prokaryotes. Eukaryotic mRNAs also initiate translation at a single site, rather than at multiple ribosome-binding sites. Finally, in eukaryotes, additional processing steps, including splicing, are required to form mature mRNAs.

7.5 A Comprehensive Example: Computerized Analysis of Gene Expression in *C. elegans*

Caenorhabditis elegans is a soil-living roundworm about 1 mm in length (**Figure 7.19**). Feeding on bacteria, it grows from fertilized egg to adult (either hermaphrodite or male) in just three days. Each hermaphrodite produces between 250 and 1000 progeny. Because of its small size, short life cycle, and capacity for prolific reproduction, *C. elegans* is an ideal subject for genetic analysis.

Using the techniques discussed in Chapter 19, geneticists have determined the precise sequence of nearly all of the 100 million base pairs in the haploid genome of the tiny nematode *C. elegans.* Using their knowledge of gene structure and gene expression, they have also programmed computers to locate the sequences within the genome likely to be genes. Their programs include instructions to search for possible exons by looking for **open reading frames (ORFs)**: strings of amino-acid-encoding nucleotide triplets uninterrupted by in-frame nonsense (stop) codons. Possible introns, on the other hand, are identified by programming the algorithm to recognize likely splice donor and splice acceptor sites.

Once the computer has retrieved regions likely to be genes, the researchers ask it to use the genetic code to predict the amino acid sequences of the polypeptides encoded by these genes. Finally, they scan computerized databases for similar amino acid sequences in the polypeptides of other organisms. If they find a similar sequence in a polypeptide of known function in another organism, they can

Figure 7.19 An adult *C. elegans* roundworm.

conclude that the *C. elegans* version of the polypeptide probably has a parallel function.

Geneticists now know many characteristics of the *C. elegans* genome

Investigators have discovered that the *C. elegans* genome contains roughly 20 000 genes, of which approximately 15 percent encode components of the worm's gene-expression machinery. Many of these gene-expression components are proteins. For example, more than 60 genes encode proteins that function as parts of the ribosome workbench, while more than 300 genes encode **transcription factors**: DNA-binding proteins that regulate transcription.

By contrast, a large contingent of expression-related genes produce RNAs that are not translated into protein. There are 659 tRNA genes in the *C. elegans* genome, about 100 rRNA genes, and 72 genes for spliceosomal RNAs. The

relatively high numbers of RNA-encoding genes reflect the fact that the genome contains several identical or near-identical copies of these untranslated genes. For example, even though there are 72 spliceosomal RNA genes, there are only five different kinds of spliceosomal RNAs.

Computerized predictions based on genomic DNA sequences alone are valuable but not infallible tools. Computer programs are currently very good at predicting the introns and exons and the primary amino acid sequence of genes encoding proteins that are well conserved in evolution. But certain details of the transcription and translation of these genes cannot be established without isolating and characterizing their corresponding mRNAs. For example, although the computer can accurately locate the protein-coding exons of a gene, the gene may contain additional exons and introns at its 5′ or 3′ ends that are more difficult for the computer to find. Similarly, without biochemical analysis of the gene's RNA products, it is not possible to know whether alternative splicing of the gene's primary transcript produces different mRNAs.

A *C. elegans* collagen gene illustrates principles of gene structure

Using techniques described in Chapters 14 and 15, researchers have obtained both the genomic DNA and the mRNA sequences for many *C. elegans* genes. These data allow an examination of the structure of these genes in nucleotide-by-nucleotide detail. One of these genes encodes a particular type of collagen protein. This single-polypeptide protein is a component of the hard cuticle that surrounds and protects the worm. Related forms of collagen occur in all multicellular animals. In vertebrates, collagen is the most abundant protein, found in bones, teeth, cartilage, tendons, and other tissues.

Figure 7.20 shows a diagram of the collagen gene together with the sequence of the corresponding DNA, the primary RNA transcript, the mature mRNA, and the polypeptide product. As you can see, the gene's structural features include three exons and two introns, as well as the signals that allow transcription, RNA processing, and translation into collagen. Note that the ATG-initiated reading

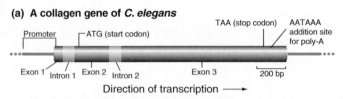

(a) A collagen gene of *C. elegans*

Promoter   ATG (start codon)   TAA (stop codon)   AATAAA addition site for poly-A

Exon 1   Intron 1   Exon 2   Intron 2   Exon 3   200 bp

Direction of transcription ⟶

(b) Sequence of a *C. elegans* collagen gene, mRNA, and polypeptide

RNA-like strand 5′... ACAACACTAGGTATAAAGCGGAAGTGGTGGCTTTAAAAT CACTTGGCTTCTAAAGTCCAGTGACAG GTAAG
Template strand 3′... TGTTGTGATCCATATTTCGCCTTCACCACCGAAATTT AGTGAACCGAAGATTTCAGGTCACTGTC CATTC
mRNA   5′ cap–CACUUGGCUUCUAAAGUCCAGUGACAG

GTTCTCGTTACTTCCGTCTCGATTACTAAGATTTGATTACTTTTAG AAAAATGACCGAAGATCCAAAGCAGATTGCCCAGGAGACTGAG...
CAAGAGCAATGAAGGCAGAGCTAATGATTCTAAACTAATGAAAATC TTTTTACTGGCTTCTAGGTTTCGTCTAACGGGTCCTCTGACTC...
                                               AAAAAUGACCGA AGAUCCAAAGCAGAUUGCCCAGG AGACUGAG...
Polypeptide   Met Thr Glu Asp Pro Lys Gln Ile Ala Gln Glu Thr Glu...

GTTGAATTCTGCCAACACAGATCAAATGGACTTTGGGATGAGTATAAGAGA GTATGTTTTTTTTGTTGAATAATTTTAATTTTAGTTAAATGTTT
CAACTTAAGACGGTTGTGTCTAGTTTACCTGAAACCCTACTCATATTCTCT CATACAAAAAAAACAACTTATTAAAATTAAAATCAATTTACAAA
GUUGAAUUCUGCCAACACAGAUCAAAUGGACUUUGGGAUGAGUAUAAGAGA
Val GluPhe Cys Gln His ArgSer Asn Gly Leu Trp Asp Glu Tyr Lys Arg

GATTTCAG TTCCAAGGAGTTTCTGGAGTTGAAGGACGTATCAAGAGAGACGCATATCACCGTAGCCTCGGAGTTTCTGGTGCTTCCCGC
CTAAAGTC AAGGTTCCTCAAAGACCTCAACTTCCTGCATAGTTCTCTCTGCGTATAGTGGCATCGGAGCCTCAAAGACCACGAAGGGCG
         UUCCAAGG AGUUUCUGGAGUUGAAGGACGUAUCAAGAGAG ACGCAUAUCACCGUAGCCUCGGAG UUUCUGGUGCUUCCCGC
         Phe Gln Gly Val Ser Gly Val Glu Gly Arg Ile Lys Arg Asp Ala Tyr His ArgSer Leu Gly Val SerGly Ala Ser Arg

AAGGCTCGTCGTCAATCTTATGGAAATGACGCTGCTGTCGGAGGATTCGGTGGATCATCTGGAGGATCATGCTGCTCATGCGGATCT...
TTCCGAGCAGCAGTTAGAATACCTTTACTGCGACGACAGCCTCCTAAGCCACCTAGTAGACCTCCTAGTACGACGAGTACGCCTAGA...
AAGGCUCGUCGUCAAUCUUAUGGAAAUGACGCUGCUGUCGGAGGAUUCGGUGGAUCAUCUGGAGGAUCAUGCUGCUCAUGCGGAUCU...
Lys Ala Arg Arg Gln Ser Tyr Gly Asn Asp Ala Ala Val Gly Gly Phe Gly Gly SerSer Gly Gly Ser Cys Cys Ser Cys Gly Ser...

CCAGGACAAGCTGGAGCACCAGGACAAGATGGAGAGAGTGGATCCGAGGGAGCTTGCGATCACTGCCCACCACCACGTACCGCTCCA
GGTCCTGTTCGACCTCGTGGTCCTGTTCTACCTCTCTCACCTAGGCTCCCTCGAACGCTAGTGACGGGTGGTGGTGCATGGCGAGGT
CCAGGACAAGCUGGAGCACCAGGACAAGAUGGAGAGAGUGGAUCCGAGGGAGCUUGCGAUCACUGCCCACCACCACGUACCGCUCCA
Pro Gly Gln Ala Gly Ala Pro Gly Gln Asp Gly Glu Ser Gly Ser Glu Gly Ala Cys Asp His Cys Pro Pro Pro Arg Thr Ala Pro

GGATATTAAGCGCTTCAATGACATCTCATTTGATTATCTCTGCTTTATCTCATTTGTATGTTTTGTGTATGAAAAACGAACACACTTAGAATAG
CCTATAATTCGCGAAGTTACTGTAGAGTAAACTAATAGAGACGAAATAGAGTAAACATACAAAACACATACTTTTTGCTTGTGTGAATCTTATC
GGAUAUUAAGCGCUUCAAUGACAUCUCAUUUGAUUAUCUCUGCUUUAUCUCAUUUGUAUGUUUUGUGUAUGAAAAACGAACACACUUAGAAUAG
Gly Tyr Stop

TGGAATAAATGATTTCATTACAAATTTGAAAT TG AATAAGACAAATGTGAAATGAAAGTATAAAAGAAAATGAGAGAC...3′
ACCTTATTTACTAAAGTAATGTTTAAACTTTAAC TTATTCTGTTTACACTTTACTTTCATATTTTCTTTTACTCTCTG...5′
UGG AAUAAAUGAUUUCAUUACAAAUUUGAAAUUGAAAAAAAAAA...
(poly (A))

Figure 7.20 Expression of a *C. elegans* gene for collagen. **(a)** Landmarks in the collagen gene. **(b)** Comparison of the sequence of the collagen gene's DNA with the sequence of nucleotides in the mature mRNA (*purple*) pinpoints the start of transcription, the location of exons (*red*) and introns (*green*), and the position of the AAUAAA poly-A addition signal (*underlined in purple*). Translation of the mRNA according to the genetic code determines the amino acids of the protein product.

frame for the protein begins only in the second exon. The reason is that the entire first exon and the first four nucleotides of the second exon correspond to the 5' untranslated region (5' UTR). Similarly, the third exon contains both amino-acid-specifying codons, as well as sequences transcribed into an untranslated region near the 3' end of the mature mRNA (the 3' UTR) just upstream of the poly-A tail.

The general structure of the collagen gene is similar to the structure of most eukaryotic genes. This is because the basic pattern of gene expression has remained substantially the same throughout evolution, even though the details, such as gene length, exon number, and the spacing or size of the untranslated 5' and 3' ends, vary from gene to gene and from organism to organism.

Gene expression in *C. elegans* involves trans-splicing and polycistronic transcripts

The sequencing of the 100 million nucleotides in the *C. elegans* genome not only led to the identification of 20 000 genes but also helped reveal some uncommon features in the way the worm expresses its genes. In rare instances, worms use trans-splicing to create an mRNA from the primary transcripts of two different genes (review Figure 7.10b); before observing trans-splicing in the nematode, researchers had seen it mainly in trypanosomes, the single-celled protozoans that cause sleeping sickness.

Like bacteria, *C. elegans* transcribes some groups of adjacent genes as one long polycistronic primary transcript; it is one of the very few eukaryotic organisms in which researchers have observed this predominantly prokaryotic phenomenon. Polycistronic transcripts are permissible in *C. elegans* because they are processed by trans-splicing into mature mRNAs for individual genes.

> In the post-genomic era, it is possible to sequence the entire genome of an organism and use this information to make predictions regarding the number of encoded genes.

7.6 The Effect of Mutations on Gene Expression and Gene Function

We have seen that the information in DNA is the starting point of gene expression. The cell transcribes that information into mRNA and then translates the mRNA information into protein. Heritable changes that alter the sequence of nucleotide pairs (i.e., mutations) may modify any of the steps or products of gene expression and thus have the potential to affect phenotype. In this section we describe the nature of different classes of these mutations and describe how they affect gene expression. In Section 7.7 we will see how mutation was used as a genetic "tool" by scientists to crack the genetic code.

Mutations in a gene's coding sequence may alter the gene product

Because of the nature of the genetic code, mutations in a gene's amino-acid-encoding exons generate a range of repercussions (**Figure 7.21a**).

Silent mutations

One consequence of the code's degeneracy is that some mutations, known as **silent mutations**, can change a codon into a mutant codon that specifies exactly the same amino acid. The majority of silent mutations change the third nucleotide of a codon, the position at which most codons for the same amino acid differ. For example, a change from GCA to GCC in a codon would still yield alanine in the protein product. Because silent mutations do not alter the amino acid composition of the encoded polypeptide, such mutations have no effect on any of the phenotypes influenced by the gene.

Missense mutations

Mutations that change a codon into a mutant codon that specifies a different amino acid are called **missense mutations**. If the substituted amino acid has chemical properties similar

(a) Types of mutation in a gene's coding sequence

Wild-type mRNA	5' GCU	GGA	GCA	CCA	GGA	CAA	GAU	GGA 3'
Wild-type polypeptide	N Ala	Gly	Ala	Pro	Gly	Gln	Asp	Gly C

Silent mutation
GCU GGA GC**C** CCA GGA CAA GAU GGA
Ala Gly Ala Pro Gly Gln Asp Gly

Missense mutation
GCU GGA GCA CCA **AGA** CAA GAU GGA
Ala Gly Ala Pro **Arg** Gln Asp Gly

Nonsense mutation
GCU GGA GCA CCA GGA **UAA** GAU GGA
Ala Gly Ala Pro Gly **Stop**

Frameshift mutation
GCU GGA GC**C** ACC AGG ACA AGA UGG A
Ala Gly Ala **Thr Arg Thr Arg Trp**

(b) Mutations outside the coding sequence

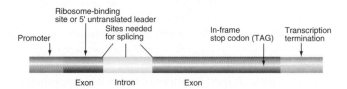

Figure 7.21 How mutations in a gene can affect its expression.
(a) Mutations in a gene's coding sequences. *Silent mutations* do not alter the protein's primary structure. *Missense mutations* replace one amino acid with another. *Nonsense mutations* shorten a polypeptide by replacing a codon with a stop signal. *Frameshift mutations* result in a change in reading frame downstream of the addition or deletion.
(b) Mutations outside the coding region can also disrupt gene expression.

to the one it replaces, then it may have little or no effect on protein function. Such substitutions are **conservative**. For example, a mutation that alters a GAC codon for aspartic acid to a GAG codon for glutamic acid is a conservative substitution because both amino acids have acidic R groups.

By contrast, **nonconservative** missense mutations that cause substitution of an amino acid with very different properties are likely to have more noticeable consequences. A change of the same GAC codon for aspartic acid to GCC, a codon for alanine (an amino acid with an uncharged, nonpolar R group), is one such nonconservative substitution.

The effect on phenotype of any missense mutation is difficult to predict because it depends on how a particular amino acid substitution changes a protein's structure and function.

Nonsense mutations

Mutations known as **nonsense mutations** change an amino-acid-specifying codon to a premature stop codon. Nonsense mutations therefore result in the production of proteins smaller than those encoded by wild-type alleles of the same gene. The shorter, *truncated proteins* lack all amino acids between the amino acid encoded by the mutant codon and the C terminus of the normal polypeptide. The mutant polypeptide will be unable to function if it requires the missing amino acids for its activity.

Frameshift mutations

Frameshift mutations result from the insertion or deletion of nucleotides within the coding sequence (the series of codons specifying the amino acids of the gene product). As discussed earlier, if the number of extra or missing nucleotides is not divisible by 3, the insertion or deletion will skew the reading frame downstream of the mutation. As a result, frameshift mutations cause unrelated amino acids to appear in place of amino acids critical to protein function, destroying or diminishing polypeptide function. (You will learn more about frameshift mutations in Section 7.7.)

Mutations outside the coding sequence can also alter gene expression

Mutations that produce a variant phenotype are not restricted to alterations in codons. Because gene expression depends on several signals other than the actual coding sequence, changes in any of these critical signals can disrupt the process (see **Figure 7.21b**).

We have seen that promoters and termination signals in the DNA of a gene instruct RNA polymerase where to start and stop transcription. Changes in the sequence of a promoter that make it hard or impossible for RNA polymerase to recognize the site diminish or prevent transcription. Mutations in a termination signal can diminish the amount of mRNA produced and thus the amount of gene product.

In eukaryotes, most primary transcripts have splice acceptor sites, splice donor sites, and branch sites that allow splicing to join exons together with precision in the mature mRNA. Changes in a splice acceptor or donor site can obstruct splicing. In some cases, the result will be the absence of mature mRNA and thus no polypeptide. In other cases, the splicing errors can yield aberrantly spliced mRNAs that encode altered forms of the protein.

Mature mRNAs have ribosome-binding sites and in-frame stop codons indicating where translation should start and stop. Mutations affecting a ribosome-binding site would lower the affinity of the mRNA for the small ribosomal subunit; such mutations are likely to diminish the efficiency of translation and thus the amount of polypeptide product. Mutations in a stop codon would produce longer than normal proteins that might be unstable or nonfunctional.

Most mutations that affect gene expression reduce gene function

Mutations affect phenotype by changing either the amino acid sequence of a protein or the amount of the protein produced. Any mutation inside or outside a coding region that reduces or abolishes protein activity in one of the many ways previously described is a **loss-of-function mutation**.

Recessive loss-of-function alleles

Loss-of-function alleles that completely block the function of a protein are called **null**, or **amorphic**, mutations (**Figure 7.22**). Such mutations either prevent synthesis of the protein or promote synthesis of a protein incapable of carrying out its function. For example, a deletion of an entire gene would by definition be a null allele. In an A^+/a heterozygote, in which allele a is recessive to wild-type allele A^+, the A^+ allele would generate functional protein, while the null a allele would not. If the amount of protein produced by the single A^+ allele (usually, though not always, half the amount produced in an A^+/A^+ cell) is above the threshold amount sufficient to fulfill the normal biochemical requirements of the cell, the phenotype of the A^+/a heterozygote will be wild type. For the large number of genes that function in this way, A^+/A^+ cells actually make more than twice as much of the protein needed for the normal phenotype.

A **hypomorphic mutation** is a loss-of-function mutation that produces either much less of a protein or a protein with very weak but detectable function (Figure 7.22). In a B^+/b heterozygote, where b is a hypomorphic allele recessive to wild-type allele B^+, the amount of protein activity

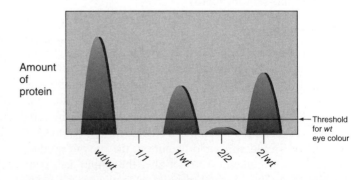

Figure 7.22 Why some mutant alleles are recessive. Researchers subjected fly extracts to "rocket" immunoelectrophoresis to quantify the amount of an enzyme called xanthine dehydrogenase. Flies need only 10 percent of the enzyme produced in wild-type strains (*wt/wt*) to have normal eye colour. Null allele 1 and hypomorphic allele 2 are recessive to wild type because 1/*wt* or 2/*wt* heterozygotes have enough enzyme for normal eye colour.

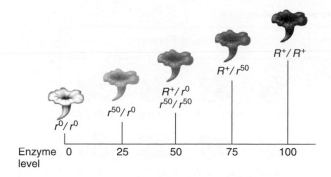

Enzyme level: 0 25 50 75 100

r^0/r^0 r^{50}/r^0 R^+/r^0 R^+/r^{50} R^+/R^+
r^{50}/r^{50}

Figure 7.23 **When a phenotype varies continuously with levels of protein function, incomplete dominance results.**

will be somewhat greater than half the amount in a B^+/B^+ cell. Usually, this is enough activity to fulfill the normal biochemical requirements of the cell. Most hypomorphic mutations are detectable only in homozygotes, and only if the reduction in protein amount or function is sufficient to cause an abnormal phenotype.

Incomplete dominance

Some combinations of alleles generate phenotypes that vary continuously with the amount of functional gene product, giving rise to incomplete dominance. For example, loss-of-function mutations in a single pigment-producing gene can generate a red-to-white spectrum of flower colours, with the white resulting from the absence of an enzyme in a biochemical pathway (**Figure 7.23**). Consider three alleles of the gene encoding this enzyme: R^+ specifies a high, wild-type amount of the enzyme; r^{50} generates half the normal amount of the same enzyme (or the full amount of an altered form that has half the normal level of activity); and r^0 is a null allele. R^+/r^0 heterozygotes produce pink flowers whose colour is halfway between red and white because one-half the R^+/R^+ level of enzyme activity is not enough to generate a full red. Combining R^+ or r^0 with the r^{50} allele produces pigmentation intermediate between red and pink or between pink and white.

Rare dominant loss-of-function alleles

With phenotypes that are exquisitely sensitive to the amount of functional protein produced, even a relatively small change of twofold or less can cause an alteration in phenotype. For example, a heterozygote for a null loss-of-function mutation that generates only half the normal amount of functional gene product may look completely different from the wild type.

The T locus in mice has just such a mutation, with an easy-to-visualize dominant phenotype (**Figure 7.24a**). Mice require the wild-type protein product of the T-locus gene during embryogenesis for the normal development of the posterior portion of the spinal cord and tail. Embryos heterozygous for a null mutation at the T locus produce only half the normal amount of the T-determined protein, and they mature into viable offspring that are normal in all respects except for the absence of the distal two-thirds of

their tail. The severely shortened tail reflects the embryo's sensitivity to the level of T-gene product available during morphogenesis; half the normal amount of T protein is below the threshold needed for normal development.

Geneticists use the term **haploinsufficiency** to describe situations in which one wild-type allele does not provide enough of a gene product. Only a minority of phenotypes are so sensitive to the amount of a particular protein. Thus, as described earlier, null and hypomorphic alleles usually produce phenotypes that are recessive to wild type.

In another mechanism leading to dominance, some alleles of genes encode subunits of multimers that block the activity of the subunits produced by normal alleles. Such blocking alleles cause a loss of function of the gene product in the organism, and are called **dominant negative**, or **antimorphic**, alleles. Consider, for example, a gene encoding a polypeptide that associates with three other identical polypeptides in a four-subunit enzyme. All four subunits are products of the same gene. If a dominant mutant allele D directs the synthesis of a polypeptide that can still assemble into aggregates but whose presence in the multimer—even as one subunit out of four—abolishes enzyme function, the chance of a heterozygote producing a multimer composed solely of functional wild-type d^+ subunits is 1 in 16: $(1/2)^4 = 1/16 = 6.25\%$ (**Figure 7.24b**). As a result, total enzyme activity in D/d^+ heterozygotes is far less than that seen in wild-type d^+/d^+ homozygotes. Dominant negative mutations can also affect subunits in multimers composed of more than one type of polypeptide. The *Kinky* allele at the *fused* locus in mice is an example of such a dominant negative mutation (**Figure 7.24c**).

Unusual gain-of-function alleles are almost always dominant

Because there are many ways to interfere with a gene's ability to make sufficient amounts of active protein, the large majority of mutations in most genes are loss-of-function alleles. However, rare mutations that enhance a protein's function or even confer a new activity on a protein produce **gain-of-function alleles**. Because a single such allele by itself can produce sufficient excess protein to alter phenotype, these unusual gain-of-function mutations are almost always dominant to wild-type alleles.

A **hypermorphic mutation** is a gain-of-function mutation that generates either more protein than the wild-type allele or the same amount of a more efficient protein. A hypermorphic mutation in the rhodopsin gene produces a rhodopsin protein that is activated whether or not light is present, resulting in constant, low-level stimulation of rhodopsin in the photoreceptor cells that detect black and white. These cells, known as rod cells, function primarily at night. People with the mutation can still see in bright daylight, but they have congenital night blindness. The blindness probably arises because the constant rhodopsin stimulation prevents adaptation of the rod cells to the very low light intensities present at night. A very rare class of

(a) Haploinsufficiency

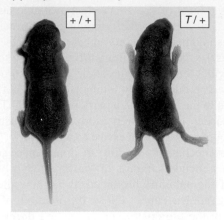

(b) Dominant negative mutations

Functional Enzyme	Nonfunctional Enzyme			
d⁺d⁺d⁺d⁺	d⁺d⁺d⁺D	d⁺d⁺D d⁺	d⁺ D d⁺d⁺	D d⁺d⁺d⁺
	d⁺d⁺D D	d⁺D D d⁺	d⁺ D d⁺D	D d⁺D d⁺
	D d d⁺D	D D d⁺d⁺	d⁺ D D D	D d⁺D D
	D D d⁺D	D D D d⁺	D D D D	

D = dominant mutant subunit
d⁺ = wild-type subunit

(c) *Kinky*: A dominant negative mutation

(d) A result of ectopic expression

Figure 7.24 Why some mutant alleles are dominant. (a) Mice heterozygous for a null mutation of the *T* locus (*T*/+) have tails shorter than wild type (+/+). **(b)** With proteins composed of four subunits encoded by a single gene, a dominant negative mutant may inactivate 15 out of every 16 multimers. **(c)** The *Kinky* allele in mice is a dominant negative mutation that causes a kink in the tail. **(d)** A neomorphic dominant mutation in the fly *Antennapedia* gene causes ectopic expression of a leg-determining gene in structures that normally produce antennae. The photograph at *left* shows two legs growing out of the head; a normal fly head is shown at *right*.

dominant gain-of-function alleles arises from **neomorphic mutations** that generate a novel phenotype. Some neomorphic mutations produce proteins with a new function, while others cause genes to produce the normal protein but at an inappropriate time or place. A striking example of inappropriate protein production is the *Drosophila* gene *Antennapedia*, active during embryonic and larval stages. Normally, the gene makes its protein product in tissues destined to become legs; the protein ensures that these tissues develop into legs and not, for example, head structures such as antennae. Dominant mutations of the gene cause production of the protein in the head region of the animal, where the *Antennapedia* gene is not normally active. Here, the misplaced protein causes tissues that would normally develop into antennae to develop into legs (**Figure 7.24d**).

Production of a protein outside of its normal place or time is called **ectopic expression**.

The effects of a mutation can be difficult to predict

As previously noted, most mutations constitute loss-of-function alleles. This is because many changes in amino acid sequence are likely to disrupt a protein's function, and because most alterations in gene regulatory sites, such as promoters, will make those sites less efficient. Nonetheless, rare mutations at almost any location in a gene can result in a gain of function.

Consider, for example, a hypothetical protein composed of (1) an N-terminal functional (e.g., an enzymatic)

Table 7.2	Mutations Classified by Their Effects on Protein Function				
	Loss-of-Function			**Gain-of-Function**	
Mutation Type	Hypomorphic (leaky)	Amorphic (null)	Antimorphic (dominant negative)*	Hypermorphic	Neomorphic (ectopic expression)
Occurrence	Common	Common	Rare	Rare	Rare
Possible Dominance Relations	Usually recessive to wild type Can be incompletely dominant if phenotype varies continuously with gene product Can be dominant in cases of haploinsufficiency		Usually dominant or incompletely dominant	Usually dominant or incompletely dominant	Usually dominant or incompletely dominant

*Some scientists, focusing on the protein encoded by the mutant allele rather than on the total level of active protein in the cell, classify antimorphic alleles as gain-of-function mutations.

domain, and (2) a C-terminal regulatory domain. Furthermore, imagine that the regulatory domain negatively regulates the functional portion of the protein (except under particular conditions where the enzymatic activity is required by the cell). A nonsense mutation that removes the amino acids needed for this negative regulation would create a hypermorphic allele (i.e., the protein would be active all the time, not just when needed). In another example, the *Antennapedia* mutation shown in Figure 7.24d results from an unusual alteration in the gene's promoter that causes *Antennapedia* to be transcribed in the wrong tissues of the animal.

Even when you know how a mutation affects gene function, you cannot always predict whether the mutation will be dominant or recessive to wild type (**Table 7.2**). Although most loss-of-function mutations are recessive and almost all gain-of-function mutations are dominant, exceptions to these generalizations exist. The reason is that dominance relations between the wild-type and mutant alleles of genes in diploid organisms depend on how drastically a mutation influences protein production or activity, and how thoroughly phenotype depends on the normal wild-type level of the protein.

Mutations in genes encoding the molecules that govern expression may have global effects

Gene expression depends on an astonishing number and variety of macromolecules (**Table 7.3**). A separate gene encodes the subunits of each macromolecule. The genes for all the proteins are transcribed and translated the same as any other gene. The genes for all the rRNAs, tRNAs, and snRNAs are transcribed but *not* translated. Many mutations in these genes have a dramatic effect on phenotype.

Lethal mutations affecting the machinery of gene expression

Mutations in the genes encoding molecules that implement gene expression, such as ribosomal proteins or rRNAs, are often lethal because such mutations adversely affect the synthesis of all proteins in a cell. Even a 50 percent reduction in the amount of some of the proteins enumerated in Table 7.3 can have severe repercussions. In *Drosophila*,

for example, null mutations in many of the genes encoding the various ribosomal proteins are lethal when homozygous. This same mutation in a heterozygote causes a dominant *Minute* phenotype in which the slow growth of cells delays the fly's development.

Mutations in tRNA genes that can suppress mutations in protein-coding genes

If more than one gene encoded the same molecule with a role in gene expression, a mutation in one of these genes would not necessarily be lethal and might even be useful. Bacterial geneticists have found, for example, that mutations in certain tRNA genes can suppress the effect of a nonsense mutation in other genes. The tRNA-gene mutations that have this

Table 7.3	The Cellular Components of Gene Expression
Function	**Cellular Components**
*Transcription**	Core RNA polymerase Sigma subunit Rho factor
Splicing and RNA Processing	snRNAs Protein components of spliceosomes Additional splicing factors Capping enzyme Methyl transferases Poly-A polymerase
Translation	mRNAs tRNAs Aminoacyl-tRNA synthetases rRNAs Protein components of ribosomes Translation factors
Protein Processing	Deformylases Amino peptidases Proteases Methylases Hydroxylases Glycosylases Kinases Phosphatases

* For simplicity, we list here only proteins from prokaryotic organisms involved in transcription. The cellular components needed for transcription in eukaryotic organisms are more complex; for example, eukaryotes have three different kinds of RNA polymerase, each made of numerous subunits.

effect give rise to **nonsense suppressor tRNAs**. Consider, for instance, an otherwise wild-type *E. coli* population with an in-frame UAG nonsense mutation in the tryptophan synthetase gene. All cells in this population make a truncated, nonfunctional form of the tryptophan synthetase enzyme and are thus unable to synthesize tryptophan (**Figure 7.25a**). These *trp⁻* cells are referred to as tryptophan auxotrophs; that is, they are unable to synthesize tryptophan and thus can grow only if the amino acid is added to the growth medium. This is in contrast to *trp⁺* cells that can make their own tryptophan and are thus able to grow in media lacking the amino acid. Interestingly, subsequent exposure of these *trp⁻* auxotrophs to mutagens generates some *trp⁺* cells that carry two mutations: one is the original tryptophan synthetase nonsense mutation, the second is a mutation in the gene that encodes a tRNA for the amino acid tyrosine. Evidently, the mutation in the tRNA gene suppresses the effect of the nonsense mutation, restoring the function of the tryptophan synthetase gene.

As **Figure 7.25b** illustrates, the basis of this nonsense suppression is that the tRNA^Tyr mutation changes an anticodon that recognizes the codon for tyrosine to an anticodon complementary to the UAG stop codon. The mutant tRNA can therefore insert tyrosine into the polypeptide at the position of the in-frame UAG nonsense mutation, allowing the cell to make at least some full-length enzyme. Similarly, mutations in the anticodons of other tRNA genes can suppress UGA or UAA nonsense mutations.

Cells with a nonsense-suppressing mutation in a tRNA gene can survive only if two conditions coexist with the mutation. First, the cell must have other tRNAs that recognize the same codon as the suppressing tRNA recognized before mutation altered its anticodon. Without such tRNAs, the cell has no way to insert the proper amino acid in response to that codon (in our example, the codon for tyrosine). Second, the suppressing tRNA must have only a weak affinity for the stop codons normally found at the ends of mRNA coding regions. If this were not the case, the suppressing tRNA would wreak havoc in the cell, producing a whole array of aberrant polypeptides that are longer than normal. One way cells guard against this possibility is that for many genes, termination depends on two stop codons in a row. Because a suppressing tRNA's chance of inserting an amino acid at both of these codons is very low, only a small number of extended proteins arise.

- Silent mutations have no effect on the encoded polypeptide or on phenotype. The phenotypic consequences of missense, nonsense, and frameshift mutations depend upon how the specific changes in amino acid sequence influence the function of the gene product.
- Most mutations outside the coding sequence, such as in promoters and transcription termination signals, affect the amount but not the nature of the protein product.

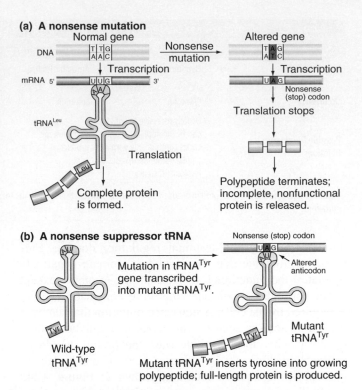

(a) A nonsense mutation

Complete protein is formed.

Translation stops

Polypeptide terminates; incomplete, nonfunctional protein is released.

(b) A nonsense suppressor tRNA

Wild-type tRNA^Tyr

Mutation in tRNA^Tyr gene transcribed into mutant tRNA^Tyr.

Nonsense (stop) codon

Altered anticodon

Mutant tRNA^Tyr

Mutant tRNA^Tyr inserts tyrosine into growing polypeptide; full-length protein is produced.

Figure 7.25 Nonsense suppression. (a) A nonsense mutation that generates a stop codon causes production of a truncated, nonfunctional polypeptide. **(b)** A second, nonsense-suppressing mutation in a tRNA gene causes addition of an amino acid in response to the stop codon, allowing production of a full-length polypeptide.

Rare exceptions include mutations that lead to incorrectly spliced mRNAs, or that convert a stop codon into a codon for an amino acid.

- Most loss-of-function (null or hypomorphic) mutations are recessive because half the normal amount of gene product is usually sufficient for a wild-type phenotype. Exceptions occur when intermediate levels of gene products cause intermediate phenotypes (incomplete dominance), when half the amount of gene product yields an abnormal phenotype (haploinsufficiency), or when a mutant polypeptide blocks the action of the wild-type polypeptide (dominant negative alleles).
- Rare gain-of-function mutations, which are typically dominant, include hypermorphic mutations that generate greater protein function than normal, and neomorphic mutations that either produce proteins with new functions or express normal proteins inappropriately (ectopic expression).
- Mutations altering the genes involved in gene expression are often lethal. Important exceptions include mutations in tRNA genes that can suppress nonsense mutations in protein-coding genes. The suppressing tRNAs insert amino acids into the growing polypeptide chains in response to premature stop codons in the mRNAs.

The deciphering of the genetic code represents one of the greatest scientific achievements of the twentieth century. It involved both rigorous experimentation and highly abstract reasoning. In the final section of this chapter we describe the imaginative and elegant experiments that allowed the code to be broken. As you read, pay close attention not only to the final results and conclusions, but also to the logic and reasoning employed by the researchers. In this way you will develop a better appreciation of the skill and imagination needed to generate meaningful answers to well-thought-out scientific questions.

A gene's nucleotide sequence is colinear with the amino acid sequence of the encoded polypeptide

As you know, DNA is a linear molecule with base pairs following one another down the intertwined chains. Proteins, by contrast, have complicated three-dimensional structures. Even so, if unfolded and stretched out from N terminus to C terminus, proteins have a one-dimensional, linear structure—a specific sequence of amino acids. If the information in a gene and its corresponding protein are colinear, the consecutive order of bases in the DNA from the beginning to the end of the gene would stipulate the consecutive order of amino acids from one end to the other of the outstretched protein.

In the 1960s, Charles Yanofsky began comparing maps of mutations within a gene to the particular amino acid substitutions that resulted. He began by generating a large number of *trp⁻* auxotrophic mutants in *E. coli* that carried mutations in the *trpA* gene for a subunit of the enzyme tryptophan synthetase. He next made a fine-structure recombinational map of these mutations. Yanofsky then purified and determined the amino acid sequence of the mutant tryptophan synthetase subunits. As **Figure 7.26a** illustrates, his data showed that the order of mutations mapped within the DNA of the gene by recombination was indeed colinear with the positions of the amino acid substitutions occurring in the resulting mutant proteins. In spite of this colinearity in order, distances on the genetic map (measured in map units) do not exactly reflect the number of amino acids between the amino acid substitutions. The reason is that recombination as seen on this very high resolution map does not occur with an equal probability at every base pair within the gene.

By carefully examining the results of his analysis, Yanofsky deduced key features of the relationship between nucleotides and amino acids, in addition to his confirmation of the existence of colinearity.

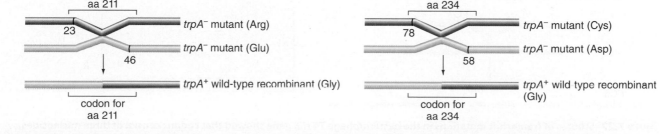

Figure 7.26 Mutations in a gene are colinear with the sequence of amino acids in the encoded polypeptide. (a) The relationship between the genetic map of *E. coli*'s *trpA* gene and the positions of amino acid substitutions in mutant tryptophan synthetase proteins. **(b)** Codons must include two or more base pairs. When two mutant strains with different amino acids at the same position were crossed, recombination could produce a wild-type allele.

Evidence that a codon is composed of more than one nucleotide

Yanofsky observed that different **point mutations** (changes in only one nucleotide pair) may affect the same amino acid. In one example shown in Figure 7.26a, mutation #23 changed the glycine (Gly) at position 211 of the wild-type polypeptide chain to arginine (Arg), while mutation #46 yielded glutamic acid (Glu) at the same position. In another example, mutation #78 changed the glycine at position 234 to cysteine (Cys), while mutation #58 produced aspartic acid (Asp) at the same position. In both cases, Yanofsky also found that recombination could occur between the two mutations that changed the identity of the same amino acid; such recombination would produce a wild-type tryptophan synthetase gene (**Figure 7.26b**). Because the smallest unit of recombination is the base pair, two mutations capable of recombination—in this case, in the same codon because they affect the same amino acid—must be in different (although nearby) nucleotides. Thus, a codon must contain more than one nucleotide.

Evidence that each nucleotide is part of only one codon

As Figure 7.26a illustrates, each of the point mutations in the tryptophan synthetase gene characterized by Yanofsky

alters the identity of only a single amino acid. This is also true of the point mutations examined in many other genes, such as the human genes for rhodopsin and haemoglobin (see Chapter 8). Because point mutations change only a single nucleotide pair, and most point mutations affect only a single amino acid in a polypeptide, each nucleotide in a gene must influence the identity of only a single amino acid. In contrast, if a nucleotide were part of more than one codon, a mutation in that nucleotide would affect more than one amino acid.

Non-overlapping triplet codons are set in a reading frame

Although the most efficient code to specify 20 amino acids requires three nucleotides per codon, more complicated scenarios are possible. But in 1955, Francis Crick and Sydney Brenner obtained convincing evidence for the triplet nature of the genetic code in studies of mutations in the bacteriophage T4 *rIIB* gene (**Figure 7.27**). They induced the mutations with proflavin, an intercalating mutagen that can insert itself between the paired bases stacked in the centre of the DNA molecule (**Figure 7.27a**). Their assumption was that proflavin would act like other mutagens, causing single-base substitutions. If this were true, it would be possible to generate revertants through treatment with other mutagens that might restore the wild-type DNA sequence.

(a) The mutagen proflavin can insert between two base pairs.

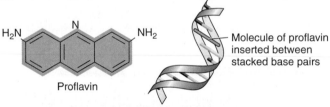

Proflavin

Molecule of proflavin inserted between stacked base pairs

(b) Consequences of exposure to proflavin

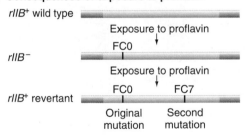

rIIB⁺ wild type

Exposure to proflavin

FC0

rIIB⁻

Exposure to proflavin

FC0 FC7

rIIB⁺ revertant

Original mutation Second mutation

(c) rIIB⁺ revertant X wild type yields rIIB⁻ recombinants.

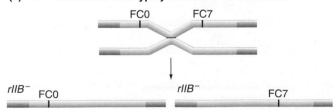

FC0 FC7

rIIB⁻ FC0   rIIB⁻ FC7

(d) Different sets of mutations generate either a mutant or a normal phenotype.

Proflavin-induced Mutations (+) insertion (−) deletion	Phenotype
− or +	Mutant
− − or + +	Mutant
− − − − or − − − − − or + + + + or + + + + +	Mutant
− +	Wild type
− − − or − − − − − − or + + + or + + + + +	Wild type

Figure 7.27 Studies of frameshift mutations in the bacteriophage T4 *rIIB* gene showed that codons consist of three nucleotides. **(a)** The mutagen proflavin slips between adjacent base pairs, eventually causing a deletion or insertion. **(b)** Treatment with proflavin produces a mutation at one site (FC0). A second proflavin exposure results in a second mutation (FC7) within the same gene, which suppresses FC0. **(c)** When the revertant is crossed with a wild-type strain, crossing-over separates the two *rIIB⁻* mutations FC0 and FC7. The reversion to an *rIIB⁺* phenotype was thus the result of intragenic suppression. **(d)** Evidence for a triplet code.

Surprisingly, genes with proflavin-induced mutations did not revert to wild type upon treatment with other mutagens known to cause nucleotide substitutions. Only further exposure to proflavin caused proflavin-induced mutations to revert to wild type (**Figure 7.27b**). Crick and Brenner had to explain this observation before they could proceed with their phage experiments. With keen insight, they correctly guessed that proflavin does not cause base substitutions; instead, it causes insertions or deletions. This hypothesis explained why base-substituting mutagens could not cause reversion of proflavin-induced mutations; it was also consistent with the structure of proflavin. By intercalating between base pairs, proflavin would distort the double helix and thus interfere with the action of enzymes that function in the repair, replication, or recombination of DNA. The result would be the deletion or addition of one or more nucleotide pairs to the DNA molecule.

Evidence for a triplet code

Crick and Brenner began their experiments with a particular proflavin-induced *rIIB⁻* mutation they called FC0. They next treated this mutant strain with more proflavin to isolate an *rIIB⁺* revertant (Figure 7.27b). By recombining this revertant with wild-type bacteriophage T4, Crick and Brenner were able to show that the revertant's chromosome actually contained two different *rIIB⁻* mutations (**Figure 7.27c**). One was the original FC0 mutation; the other was the newly induced FC7. Either mutation by itself yields a mutant phenotype, but their simultaneous occurrence in the same gene yielded an *rIIB⁺* phenotype. Crick and Brenner reasoned that if the first mutation was the deletion of a single base pair, represented by the symbol (−), then the counteracting mutation must be the insertion of a base pair, represented as (+). The restoration of gene function by one mutation cancelling another in the same gene is known as **intragenic suppression**. On the basis of this reasoning, they went on to establish T4 strains with different numbers of (+) and (−) mutations in the same chromosome. **Figure 7.27d** tabulates the phenotypes associated with each combination of proflavin-induced mutations.

In analyzing the data, Crick and Brenner assumed that each codon is a trio of nucleotides and that for each gene there is a single starting point. This starting point establishes a **reading frame**: the partitioning of groups of three nucleotides such that the sequential interpretation of each triplet generates the correct order of amino acids in the resulting polypeptide chain. If codons are read in order from a fixed starting point, one mutation will counteract another if the two are equivalent mutations of opposite signs—that is, (−) and (+). In such a case, each insertion compensates for each deletion, and this counterbalancing restores the reading frame (**Figure 7.28a**). The gene would only regain its wild-type activity, however, if the portion of the polypeptide encoded between the two mutations of opposite sign is not required for protein function, because in the double mutant, this region would have an improper amino acid sequence.

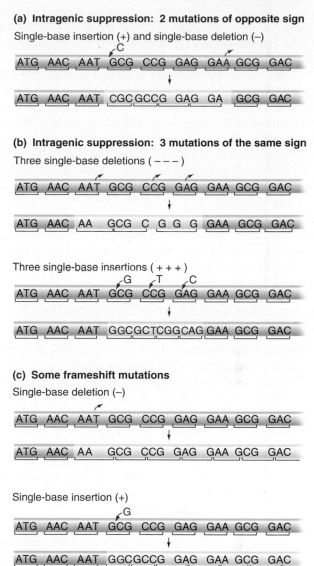

correct triplet
incorrect triplet

(a) Intragenic suppression: 2 mutations of opposite sign

Single-base insertion (+) and single-base deletion (−)

(b) Intragenic suppression: 3 mutations of the same sign

Three single-base deletions (− − −)

Three single-base insertions (+ + +)

(c) Some frameshift mutations

Single-base deletion (−)

Single-base insertion (+)

Figure 7.28 Codons consist of three nucleotides read in a defined reading frame. The phenotypic effects of proflavin-induced frameshift mutations depend on whether the reading frame is restored and whether the part of the gene with an altered reading frame specifies an essential or nonessential region of the polypeptide.

Similarly, if a gene sustains three changes of the same sign (or multiples of three), the encoded polypeptide can still function, because the mutations do not alter the reading frame for the majority of amino acids (**Figure 7.28b**). The resulting polypeptide will, however, have one extra or one fewer amino acid than normal (designated by three plus signs (+) or three minus signs (−), respectively), and the region encoded by the part of the gene between the first and the last mutations will not contain the correct amino acids.

By contrast, a single nucleotide inserted into or deleted from a gene alters the reading frame and thereby affects the identity of not only one amino acid but of all

other amino acids beyond the point of alteration (**Figure 7.28c**). Changes that alter the grouping of nucleotides into codons are called *frameshift mutations*: They shift the reading frame for all codons beyond the point of insertion or deletion, almost always abolishing the function of the polypeptide product.

A review of the evidence tabulated in Figure 7.27d supports all these points. A single (−) or a single (+) mutation destroyed the function of the *rIIB* gene and produced an *rIIB⁻* phage. Similarly, any gene with two base changes of the same sign (− − or + +) or with four or five insertions or deletions of the same sign (e.g., + + + +) also generated a mutant phenotype. However, genes containing three or multiples of three mutations of the same sign (e.g., + + + or − − − − − −), as well as genes containing a (+ −) pair of mutations, generated *rIIB⁺* wild-type individuals. In these last examples, intragenic suppression allowed restitution of the reading frame and thereby restored the lost or aberrant genetic function produced by other frameshift mutations in the gene.

Evidence that most amino acids are specified by more than one codon

As Figure 7.28a illustrates, intragenic suppression occurs only if, in the region between two frameshift mutations of opposite sign, a gene still dictates the appearance of amino acids—even if these amino acids are not the same as those appearing in the normal protein. If the frameshifted part of the gene encodes instructions to stop protein synthesis by introducing, for example, a triplet that does not correspond to any amino acid, then wild-type polypeptide production will not continue. The reason is that polypeptide synthesis would stop before the compensating mutation could re-establish the correct reading frame.

The fact that intragenic suppression occurs as often as it does suggests that the code includes more than one codon for some amino acids. Recall that there are 20 common amino acids but $4^3 = 64$ different combinations of three nucleotides. If each amino acid corresponded to only a single codon, there would be $64 − 20 = 44$ possible triplets not encoding an amino acid. These noncoding triplets would act as "stop" signals and prevent further polypeptide synthesis. If this happened, more than half of all frameshift mutations (44/64) would cause protein synthesis to stop at the first codon after the mutation, and the chances of extending the protein each amino acid farther down the chain would diminish exponentially. As a result, intragenic suppression would rarely occur. However, we have seen that many frameshift mutations of one sign can be offset by mutations of the other sign. The distances between these mutations, estimated by recombination frequencies, are in some cases large enough to code for more than 50 amino acids, which would be possible only if most of the 64 possible triplet codons specified amino acids. Thus, the data of Crick and Brenner provide strong support for the idea that the genetic code is **degenerate**: Two or more nucleotide triplets specify most of the 20 amino acids (see the genetic code in Figure 7.2).

Cracking the code: Which codons represent which amino acids?

Although the genetic experiments just described allowed remarkably prescient insights about the nature of the genetic code, they did not establish a correspondence between specific codons and specific amino acids. The discovery of messenger RNA and the development of techniques for synthesizing simple messenger RNA molecules had to occur first, so that researchers could manufacture simple proteins in the test tube.

The discovery of messenger RNAs

In the 1950s, researchers exposed eukaryotic cells to amino acids tagged with radioactivity and observed that protein synthesis incorporating the radioactive amino acids into polypeptides takes place in the cytoplasm, even though the genes for those polypeptides are sequestered in the cell nucleus. From this discovery, they deduced the existence of an intermediate molecule, made in the nucleus and capable of transporting DNA sequence information to the cytoplasm, where it can direct protein synthesis. RNA was a prime candidate for this intermediary information-carrying molecule.

Because of RNA's potential for base-pairing with a strand of DNA, one could imagine the cellular machinery copying a strand of DNA into a complementary strand of RNA in a manner analogous to the DNA-to-DNA copying of DNA replication. Subsequent studies in eukaryotes using radioactive uracil, a base found only in RNA, showed that although the molecules are synthesized in the nucleus, at least some of them migrate to the cytoplasm. Among those RNA molecules that migrate to the cytoplasm are the messenger RNAs, or mRNAs, depicted in Figure 7.1. They arise in the nucleus from the transcription of DNA sequence information and then move (after processing) to the cytoplasm, where they determine the proper order of amino acids during protein synthesis.

Using synthetic mRNAs and *in vitro* translation

Knowledge of mRNA served as the framework for two experimental breakthroughs that led to the deciphering of the genetic code. In the first, biochemists obtained cellular extracts that, with the addition of mRNA, synthesized polypeptides in a test tube. They called these extracts "*in vitro* translational systems". The second breakthrough was the development of techniques enabling the synthesis of artificial mRNAs containing only a few codons of known composition. When added to *in vitro* translational systems, these simple, synthetic mRNAs directed the formation of very simple polypeptides.

In 1961, Marshall Nirenberg and Heinrich Matthaei added a synthetic poly-U (5′...UUUUUUUUUUUU...3′) mRNA to a cell-free translational system derived from *E. coli*. With the poly-U mRNA, phenylalanine (Phe) was the only amino acid incorporated into the resulting polypeptide (**Figure 7.29a**). Because UUU is the only possible triplet in

(a) Poly-U mRNA encodes polyphenylalanine.

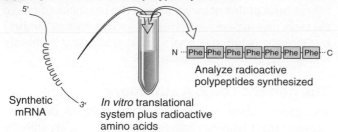

5'

Synthetic mRNA   3'   *In vitro* translational system plus radioactive amino acids

N ···Phe–Phe–Phe–Phe–Phe–Phe–Phe···C

Analyze radioactive polypeptides synthesized

(b) Analyzing the coding possibilities

Synthetic mRNA	Polypeptides Synthesized
	Polypeptides with one amino acid
poly-U UUUU ...	Phe-Phe-Phe ...
poly-C CCCC ...	Pro-Pro-Pro ...
poly-A AAAA ...	Lys-Lys-Lys ...
poly-G GGGG ...	Gly-Gly-Gly ...
Repeating dinucleotides	**Polypeptides with alternating amino acids**
poly-UC UCUC ...	Ser-Leu-Ser-Leu ...
poly-AG AGAG ...	Arg-Glu-Arg-Glu ...
poly-UG UGUG ...	Cys-Val-Cys-Val ...
poly-AC ACAC ...	Thr-His-Thr-His ...
Repeating trinucleotides	**Three polypeptides each with one amino acid**
poly-UUC UUCUUCUUC ...	Phe-Phe.... and Ser-Ser.... and Leu-Leu....
poly-AAG AAGAAGAAG ...	Lys-Lys.... and Arg-Arg.... and Glu-Glu....
poly-UUG UUGUUGUUG ...	Leu-Leu.... and Cys-Cys.... and Val-Val....
poly-UAC UACUACUAC ...	Tyr-Tyr.... and Thr-Thr.... and Leu-Leu....
Repeating tetranucleotides	**Polypeptides with repeating units of four amino acids**
poly-UAUC UAUCUAUC ...	Tyr-Leu-Ser-Ile-Tyr-Leu-Ser-Ile...
poly-UUAC UUACUUAC ...	Leu-Leu-Thr-Tyr-Leu-Leu-Thr-Tyr...
poly-GUAA GUAAGUAA ...	none
poly-GAUA GAUAGAUA ...	none

Figure 7.29 How geneticists used synthetic mRNAs to limit the coding possibilities. (a) Poly-U mRNA generates a poly-phenylalanine polypeptide. **(b)** Polydi-, polytri-, and polytetra-nucleotides encode simple polypeptides. Some synthetic mRNAs, such as poly-GUAA, contain stop codons in all three reading frames and thus specify the construction only of short peptides.

poly-U, UUU must be a codon for phenylalanine. In a similar fashion, Nirenberg and Matthaei showed that CCC encodes proline (Pro), AAA is a codon for lysine (Lys), and GGG encodes glycine (Gly) (**Figure 7.29b**).

The chemist Har Gobind Khorana later made mRNAs with repeating dinucleotides, such as poly-UC (5'...UCUCUCUC...3'), repeating trinucleotides, such as poly-UUC, and repeating tetranucleotides, such as poly-UAUC, and used them to direct the synthesis of slightly more complex polypeptides. As Figure 7.29b shows, his results limited the coding possibilities, but some ambiguities remained. For example, poly-UC encodes the poly-peptide N...Ser-Leu-Ser-Leu-Ser-Leu...C in which serine and leucine alternate with each other. Although the mRNA contains only two different codons (5' UCU 3' and 5' CUC 3'), it is not obvious which corresponds to serine and which to leucine.

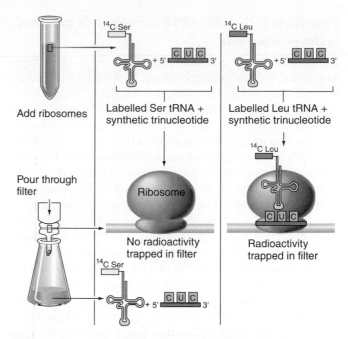

Add ribosomes

Pour through filter

Labelled Ser tRNA + synthetic trinucleotide

Labelled Leu tRNA + synthetic trinucleotide

Ribosome

No radioactivity trapped in filter

Radioactivity trapped in filter

Figure 7.30 Cracking the genetic code with mini-mRNAs. Nirenberg and Leder added trinucleotides of known sequence, in combination with tRNAs charged with a radioactive amino acid, to an *in vitro* extract containing ribosomes. If the trinucleotide specified the radioactive amino acid, the amino-acid-bearing tRNA formed a complex with the ribosomes that could be trapped on a filter. The experiments shown here indicate that the codon CUC specifies leucine, not serine.

Nirenberg and Philip Leder resolved these ambiguities in 1965 with experiments in which they added short, synthetic mRNAs only three nucleotides in length to an *in vitro* translational system containing one radioactive amino acid and 19 unlabelled amino acids, all attached to tRNA molecules. They then poured through a filter the mixture of synthetic mRNAs and translational systems containing a tRNA-attached, radioactively labelled amino acid (**Figure 7.30**). tRNAs carrying an amino acid normally go right through a filter. If, however, a tRNA carrying an amino acid binds to a ribosome, it will stick in the filter, because this larger complex of ribosome, amino-acid-carrying tRNA, and small mRNA cannot pass through the filter. Nirenberg and Leder could thus use this approach to see which small mRNA caused the entrapment of which radioactively labelled amino acid. For example, they knew from Khorana's earlier work that CUC encoded either serine or leucine. When they added the synthetic triplet CUC to an *in vitro* system where the radioactive amino acid was serine, this tRNA-attached amino acid passed through the filter, and the filter thus emitted no radiation. But when they added the same triplet to a system where the radioactive amino acid was leucine, the filter lit up with radioactivity, indicating that the radioactively tagged leucine attached to a tRNA had bound to the ribosome-mRNA complex and gotten stuck in the filter. CUC thus encodes leucine, not serine. Nirenberg and Leder used this technique to determine all the codon–amino acid correspondences shown in the genetic code table (see Figure 7.2).

Polarities: 5′-to-3′ in mRNA corresponds to N-to-C in the polypeptide

In studies using synthetic mRNAs, when investigators added the six-nucleotide-long 5′ AAAUUU 3′ to an *in vitro* translational system, the product N-Lys-Phe-C emerged, but no N-Phe-Lys-C appeared. Because AAA is the codon for lysine and UUU is the codon for phenylalanine, this means that the codon closest to the 5′ end of the mRNA encoded the amino acid closest to the N terminus of the corresponding polypeptide. Similarly, the codon nearest the 3′ end of the mRNA encoded the amino acid nearest the C terminus of the resulting polypeptide.

To understand how the polarities of the macromolecules participating in gene expression relate to each other, remember that although the gene is a segment of a DNA double helix, only one of the two strands serves as a template for the mRNA. This strand is known as the **template strand**. The other strand is the **RNA-like strand**, because it has the same polarity and sequence (written in the DNA dialect) as the RNA. Note that some scientists use the terms *sense strand* or *coding strand* as synonyms for the RNA-like strand; in these alternative nomenclatures, the template strand would be the *antisense strand* or the *noncoding strand*. **Figure 7.31** diagrams the respective polarities of a gene's DNA, the mRNA transcript of that DNA, and the resulting polypeptide.

Nonsense codons and polypeptide chain termination

Although most of the simple, repetitive RNAs synthesized by Khorana were very long and thus generated very long polypeptides, a few did not. These RNAs had signals that stopped construction of a polypeptide chain. As it turned out, three different triplets—UAA, UAG, and UGA—do not correspond to any of the amino acids. When these codons appear in-frame, translation stops. As an example of how investigators established this fact, consider the case of poly-GUAA (review Figure 7.29b). This mRNA will not generate a long polypeptide, because in all possible reading frames, it contains the **stop codon** UAA.

The three stop codons that terminate translation are also known as **nonsense codons**. For historical reasons, researchers often refer to UAA as the *ochre* codon, UAG

as the *amber* codon, and UGA as the *opal* codon. The historical basis of this nomenclature is the last name of one of the early investigators—Bernstein—which means "amber" in German; ochre and opal derive from their similarity with amber as semiprecious materials.

The genetic code: A summary

The genetic code is a complete, unabridged dictionary equating the 4-letter language of the nucleic acids with the 20-letter language of the proteins. The following list summarizes the code's main features:

1. *Triplet codons:* As written in Figure 7.2, the code shows the 5′-to-3′ sequence of the three nucleotides in each mRNA codon; that is, the first nucleotide depicted is at the 5′ end of the codon.
2. The codons are *non-overlapping*. In the mRNA sequence 5′ GAAGUUGAA 3′, for example, the first three nucleotides (GAA) form one codon; nucleotides 4 through 6 (GUU) form the second; and so on. Each nucleotide is part of only one codon.
3. The code includes three *stop*, or *nonsense*, *codons*: UAA, UAG, and UGA. These codons do not encode an amino acid and thus terminate translation.
4. The code is *degenerate*, meaning that more than one codon may specify the same amino acid. The code is nevertheless unambiguous, because each codon specifies only one amino acid.
5. The cellular machinery scans mRNA from a fixed starting point that establishes a *reading frame*. As we see later, the nucleotide triplet AUG, which specifies the amino acid methionine, serves in certain contexts as the **initiation codon**, marking where in an mRNA the code for a particular polypeptide begins.
6. *Corresponding polarities:* Moving from the 5′ to the 3′ end of an mRNA, each successive codon is sequentially interpreted into an amino acid, starting at the N terminus and moving toward the C terminus of the resulting polypeptide.
7. Mutations may modify the message encoded in a sequence of nucleotides in three ways. *Frameshift mutations* are nucleotide insertions or deletions that alter the genetic instructions for polypeptide construction by changing the reading frame. *Missense mutations* change a codon for one amino acid to a codon for a different amino acid. *Nonsense mutations* change a codon for an amino acid to a stop codon.

The effects of mutations on polypeptides helped verify the code

The experiments that cracked the genetic code by assigning codons to amino acids were all *in vitro* studies using cell-free extracts and synthetic mRNAs. A logical question thus arose: Do living cells construct polypeptides according to the same rules? Early evidence that they do came from studies analyzing how mutations actually affect the amino acid composition of the polypeptides encoded by a gene.

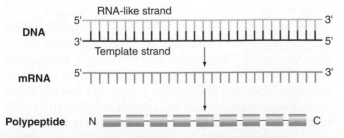

Figure 7.31 Correlation of polarities in DNA, mRNA, and polypeptide. The template strand of DNA is complementary to both the RNA-like DNA strand and the mRNA. The 5′-to-3′ direction in an mRNA corresponds to the N-terminus-to-C-terminus direction in the polypeptide.

Most mutagens change a single nucleotide in a codon. As a result, most missense mutations that change the identity of a single amino acid should be single-nucleotide substitutions, and analyses of these substitutions should conform to the code. Yanofsky, for example, found two *trp*⁻ auxotrophic mutations in the *E. coli* tryptophan synthetase gene that produced two different amino acids (arginine, or Arg, and glutamic acid, or Glu) at the same position—amino acid 211—in the polypeptide chain (**Figure 7.32a**). According to the code, both of these mutations could have resulted from single-base changes in the GGA codon that normally inserts glycine (Gly) at position 211.

Even more telling were the *trp*⁺ revertants of these mutations subsequently isolated by Yanofsky. As Figure 7.32a illustrates, single-base substitutions could also explain the amino acid changes in these revertants. Note that some of these substitutions restore Gly to position 211 of the polypeptide, while others place amino acids such as Ile, Thr, Ser, Ala, or Val at this site in the tryptophan synthetase

molecule. The substitution of these other amino acids for Gly at position 211 in the polypeptide chain is compatible with (i.e., largely conserves) the enzyme's function.

Yanofsky obtained better evidence yet that cells use the genetic code *in vivo* by analyzing proflavin-induced frameshift mutations of the tryptophan synthetase gene (**Figure 7.32b**). He first treated populations of *E. coli* with proflavin to produce *trp*⁻ mutants. Subsequent treatment of these mutants with more proflavin generated some *trp*⁺ revertants among the progeny. The most likely explanation for the revertants was that their tryptophan synthetase gene carried both a single-base-pair deletion and a single-base-pair insertion (− +). Upon determining the amino acid sequences of the tryptophan synthetase enzymes made by the revertant strains, Yanofsky found that he could use the genetic code to predict the precise amino acid alterations that had occurred by assuming the revertants had a specific single-base-pair insertion and a specific single-base-pair deletion.

Yanofsky's results helped confirm not only amino acid codon assignments but other parameters of the code as well. His interpretations make sense only if codons do not overlap and are read from a fixed starting point with no pauses or commas separating the adjacent triplets.

The genetic code is almost, but not quite, universal

We now know that virtually all cells alive today use the same basic genetic code. One early indication of this uniformity was that a translational system derived from one organism could use the mRNA from another organism to convert genetic information to the encoded protein. Rabbit haemoglobin mRNA, for example, when injected into frog eggs or added to cell-free extracts from wheat germ, directs the synthesis of rabbit haemoglobin proteins. More recently, comparisons of DNA and protein sequences have revealed a perfect correspondence according to the genetic code between codons and amino acids in almost all organisms examined.

Conservation of the genetic code

The universality of the code is an indication that it evolved very early in the history of life. Once it emerged, it remained constant over billions of years, in part because evolving organisms would have little tolerance for change. A single change in the genetic code could disrupt the production of hundreds or thousands of proteins in a cell—from the DNA polymerase that is essential for replication to the RNA polymerase that is required for gene expression to the tubulin proteins that compose the mitotic spindle—and such a change would therefore be lethal.

Exceptional genetic codes

Researchers were thus quite amazed to observe a few exceptions to the universality of the code. In some species of the

(a) **Altered amino acids in *trp*⁻ mutations and *trp*⁺ revertants**

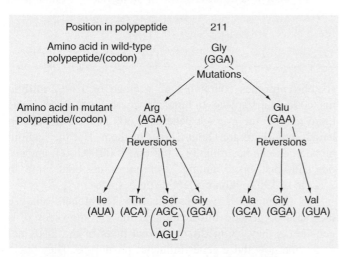

(b) **Amino acid alterations that accompany intragenic suppression**

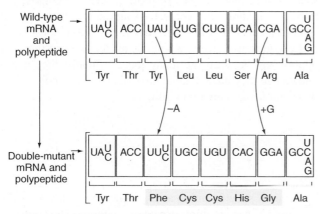

Figure 7.32 Experimental verification of the genetic code.
(a) Single-base substitutions can explain the amino acid substitutions of *trp*⁻ mutations and *trp*⁺ revertants. **(b)** The genetic code predicts the amino acid alterations (*yellow*) that would arise from single-base-pair deletions and suppressing insertions.

single-celled eukaryotic protozoans known as ciliates, the codons UAA and UAG, which are nonsense codons in most organisms, specify the amino acid glutamine; in other ciliates, UGA, the third stop codon in most organisms, specifies cysteine. These ciliates use the remaining nonsense codons as stop codons.

Other systematic changes in the genetic code exist in mitochondria, the semiautonomous, self-reproducing organelles within eukaryotic cells that are the sites of ATP formation. Each mitochondrion has its own chromosomes and its own apparatus for gene expression (which we describe in detail in Chapter 18). In the mitochondria of yeast, CUA specifies threonine instead of leucine.

It may be that ciliates and mitochondria tolerated these changes in the genetic code because the alterations affected very few proteins. For instance, the nonsense codon UGA might have found only infrequent use in one kind of primitive ciliate, so its switch to a "sense" codon would not have made a tremendous difference in protein production. Similarly, mitochondria might have survived a few changes in the code because they synthesize only a handful of proteins.

- Comparison of recombination maps of mutations with the amino acid sequences of mutant polypeptides established colinearity; the order of nucleotides in the gene corresponds to the order of amino acids in the polypeptide. Further analysis demonstrated that a single codon must contain more than one nucleotide, and that each nucleotide in the gene helps encode only a single amino acid.

- Work with frameshift mutations in the bacteriophage T4 *rIIB* gene established that (1) codons consist of three adjacent nucleotides, (2) each gene has a specific starting point to set a reading frame for triplets, and (3) the genetic code is degenerate, with some amino acids specified by more than one codon.

- The addition of synthetic mRNAs to *in vitro* translation systems allowed biochemists to determine which codons specify which amino acids.

- The effects of specific mutations on the amino acid sequence of the encoded polypeptide are consistent with the genetic code table shown in Figure 7.2.

Connections

Our knowledge of gene expression enables us to redefine the concept of a gene. A gene is not simply the DNA that is transcribed into the mRNA codons specifying the amino acids of a particular polypeptide. Rather, *a gene is all the DNA sequences needed for expression of the gene into a polypeptide product.* A gene therefore includes the promoter sequences that govern where transcription begins and, at the opposite end, signals for the termination of transcription. A gene also must include sequences dictating where translation of the mRNA starts and stops. In addition to all of these features, eukaryotic genes contain introns that are spliced out of the primary transcript to make the mature mRNA. Because of introns, most eukaryotic genes are much larger than prokaryotic genes.

Even with introns, a single gene carries only a very small percentage of the nucleotide pairs in the chromosomes that make up a genome. The average gene in *C. elegans* is about 4000 nucleotide pairs in length, and there are roughly 20 000 genes. The worm's haploid genome, however, contains approximately 100 million nucleotide pairs distributed among six chromosomes containing an average of 16–17 million nucleotide pairs apiece. In humans, where genes tend to have more introns, the average gene is 16 000 nucleotide pairs in length, and there are about 21 000 of them. But the haploid human genome has roughly 3 billion (3 000 000 000) nucleotide pairs distributed among 23 chromosomes containing an average of 130 million nucleotide pairs apiece.

In Chapter 19, we describe how researchers analyze the mass of genetic information in the chromosomes of a genome as they try to discover what parts of the DNA are genes and how those genes influence phenotype. They begin their analysis by breaking the DNA into pieces of manageable size, making many copies of those pieces to obtain enough material for study, and characterizing the pieces down to the level of nucleotide sequence. They then try to reconstruct the DNA sequence of an entire genome by determining the spatial relationship between the many pieces. Finally, they use the knowledge they have obtained to examine the genomic variations that make individuals unique.

Essential Concepts

1. Gene expression is the process by which cells convert the DNA sequence of a gene to the RNA sequence of a transcript, and then decode the RNA sequence as the amino acid sequence of a polypeptide. [LO1–4]

2. The nearly universal genetic code consists of 64 codons, each one composed of three nucleotides. Sixty-one codons specify amino acids, while three—UAA, UAG, and UGA—are nonsense or stop codons. The code is degenerate because more than one codon can specify each amino acid except methionine and tryptophan. The codon AUG in the context of a ribosome-binding site is the initiation codon; it establishes the reading

frame that groups nucleotides into non-overlapping codon triplets. **[LO1]**

3. Transcription is the first stage of gene expression. During transcription, RNA polymerase synthesizes a single-stranded primary transcript from a DNA template. In initiation, RNA polymerase binds to the promoter sequence of the DNA and unwinds the double helix to expose bases for pairing. During elongation, the enzyme extends the RNA in the 5′-to-3′ direction by catalyzing bond formation between successively aligned nucleotides. Termination occurs when terminator sequences in the RNA cause RNA polymerase to dissociate from the DNA. **[LO2]**

4. In prokaryotes, the primary transcript is the messenger RNA (mRNA). In eukaryotes, RNA processing after transcription produces a mature mRNA. RNA processing adds a methylated cap to the 5′ end and a poly-A tail to the 3′ end of eukaryotic mRNA. An important aspect of processing is RNA splicing, during which the spliceosome removes introns from the primary transcript and joins together the remaining exons. Alternative splicing allows production of different mRNAs from the same primary transcript. **[LO2–3]**

5. Translation occurs when the cell synthesizes protein according to instructions in the mRNA. This process takes place on ribosomes, which are composed of protein and ribosomal RNA (rRNA). Ribosomes have three binding sites for transfer RNAs (tRNAs)—A, P, and E sites—and they also supply the ribozyme known as peptidyl transferase, which catalyzes formation of peptide bonds between amino acids. **[LO5]**

6. Individual aminoacyl-tRNA synthetases connect the correct amino acids to their corresponding tRNAs; a tRNA carrying an amino acid is said to be charged. Each charged tRNA has an anticodon complementary to the mRNA codon specifying the amino acid the tRNA carries. Because of wobble, some tRNA anticodons recognize more than one mRNA codon. **[LO2]**

7. To initiate translation, the small subunit of the ribosome binds to a ribosome-binding site on the mRNA that includes the AUG initiation codon. Special tRNAs carry the amino acid fMet in prokaryotes or Met in eukaryotes to the ribosomal P site. This amino acid becomes the N terminus of the growing polypeptide. After initiation has begun, a charged tRNA complementary to the next codon of the mRNA enters the A site of the ribosome. **[LO2]**

8. During elongation, the carboxyl group of the amino acid connected to a tRNA at the ribosome's P site becomes bonded to the amino acid carried by the tRNA at the A site. The ribosome then travels three nucleotides toward the 3′ end of the mRNA. The 5′-to-3′ direction in the mRNA thus corresponds to the N-terminus-to-C-terminus direction in the polypeptide under construction. **[LO2]**

9. Termination occurs when the ribosome encounters a nonsense (stop) codon. The ribosome then releases the mRNA and disconnects the completed polypeptide from the tRNA to which it was attached. **[LO2]**

10. Posttranslational processing may alter a polypeptide by adding or removing chemical constituents or by cleaving the polypeptide into smaller molecules. **[LO2]**

11. Mutations in a gene may modify the message encoded in a sequence of nucleotides. Silent mutations usually change the third letter of a codon and have no effect on polypeptide production. Missense mutations change the codon for one amino acid to the codon for another amino acid. Nonsense mutations change a codon for an amino acid to a stop codon. Frameshift mutations change the reading frame of a gene, altering the identity of all subsequent amino acids. **[LO1, LO4–5]**

12. Mutations outside coding sequences that alter signals required for transcription, mRNA splicing, or translation can modify gene expression by altering the amount, time, or place of protein production. **[LO1, LO4–5]**

13. Loss-of-function mutations reduce or completely block gene expression. Most loss-of-function alleles are recessive to wild-type alleles, but in haploinsufficiency, half the normal gene product is not enough for a normal phenotype, so the mutant allele is dominant to wild type. Certain loss-of-function alleles can have dominant effects by disrupting the function of wild-type protein subunits in a complex. **[LO5]**

14. Rare gain-of-function mutations cause either increased protein production or synthesis of a protein with enhanced activity. Some gain-of-function alleles confer a novel function on a gene; one example is ectopic expression, in which the gene product is made in the wrong tissue or at the wrong time in development. Most gain-of-function mutations are dominant. **[LO5]**

15. Mutations in genes that encode molecules of the gene-expression machinery are often lethal. Exceptions include mutations in tRNA genes that suppress nonsense mutations in polypeptide-encoding genes. **[LO5]**

Solved Problems

I. A geneticist examined the amino acid sequence of a particular protein in a variety of *E. coli* mutants. The amino acid in position 40 in the normal enzyme is glycine. The following table shows the substitutions the geneticist found at amino acid position 40 in six mutant forms of the enzyme.

mutant 1	cysteine
mutant 2	valine
mutant 3	serine
mutant 4	aspartic acid
mutant 5	arginine
mutant 6	alanine

Determine the nature of the base substitution that must have occurred in the DNA in each case. Which of these mutants would be capable of recombination with mutant 1 to form a wild-type gene?

Answer

To determine the base substitutions, use the genetic code table (see Figure 7.2). The original amino acid was glycine, which can be encoded by GGU, GGC, GGA, or GGG. Mutant 1 results in a cysteine at position 40; Cys codons are either UGU or UGC. A change in the base pair in the DNA encoding the first position in the codon (a G–C to T–A transversion) must have occurred, and the original glycine codon must therefore have been either GGU or GGC. Valine (in mutant 2) is encoded by GUN (with N representing any one of the four bases), but assuming that the mutation is a single-base change, the Val codon must be either GUU or GUC. The change must have been a G–C to T–A transversion in the DNA for the second position of the codon. To get from glycine to serine (mutant 3) with only one base change, the GGU or GGC would be changed to AGU or AGC, respectively. There was a transition (G–C to A–T) at the first position. Aspartic acid (mutant 4) is encoded by GAU or GAC, so the DNA of mutant 4 is the result of a G–C to A–T transition at position 2. Arginine (mutant 5) is encoded by CGN, so the DNA of mutant 5 must have undergone a G–C to C–G transversion at position 1. Finally, alanine (mutant 6) is encoded by GCN, so the DNA of mutant 6 must have undergone a G–C to C–G transversion at position 2. Mutants 2, 4, and 6 affect a base pair different from that affected by mutant 1, so they could recombine with mutant 1.

In summary, the sequence of nucleotides on the RNA-like strand of the wild-type and mutant genes at this position must be

wild type	5′ G G T/C 3′
mutant 1	5′ T G T/C 3′
mutant 2	5′ G T T/C 3′
mutant 3	5′ A G T/C 3′
mutant 4	5′ G A T/C 3′
mutant 5	5′ C G T/C 3′
mutant 6	5′ G C T/C 3′

II. The double-stranded circular DNA molecule that forms the genome of the SV40 virus can be denatured into single-stranded DNA molecules. Because the base composition of the two strands differs, the strands can be separated on the basis of their density into two strands designated W(atson) and C(rick). When each of the purified preparations of the single strands was mixed with mRNA from cells infected with the virus, hybrids were formed between the RNA and DNA. Closer analysis of these hybridizations showed that RNAs that hybridized with the W preparation were different from RNAs that hybridized with the C preparation. What does this tell you about the transcription templates for the different classes of RNAs?

Answer

An understanding of transcription and the polarity of DNA strands in the double helix are needed to answer this question. *Some genes use one strand of the DNA as a template; others use the opposite strand as a template.* Because of the different polarities of the DNA strands, one set of genes would be transcribed in a clockwise direction on the circular DNA (using say the W strand as the template), and the other set would be transcribed in a counterclockwise direction (with the C strand as template).

III. Geneticists interested in human haemoglobins have found a very large number of mutant forms. Some of these mutant proteins are of normal size, with amino acid substitutions, while others are short, due to deletions or nonsense mutations. The first extra-long example was named Hb Constant Spring, in which the β globin has several extra amino acids attached at the C-terminal end. What is a plausible explanation for its origin? Is it likely that Hb Constant Spring arose from failure to splice out an intron?

Answer

An understanding of the principles of translation and RNA splicing are needed to answer this question. Because there is an extension on the C-terminal end of the protein, *the mutation probably affected the termination (nonsense) codon rather than affecting splicing of the RNA.* This could have been a base change or a frame shift or a deletion that altered or removed the termination codon. The information in the mRNA beyond the normal stop codon would be translated until another stop codon in the mRNA was reached. A splicing defect could explain Hb Constant Spring only in the more unlikely case that an incorrectly spliced mRNA would encode a protein much longer than normal.

Problems

Vocabulary

1. For each of the terms in the left column, choose the best matching phrase in the right column.

i. codon	**1.** removing base sequences corresponding to introns from the primary transcript
ii. colinearity	**2.** UAA, UGA, or UAG
iii. reading frame	**3.** the strand of DNA that has the same base sequence as the primary transcript
iv. frameshift mutation	**4.** a transfer RNA molecule to which the appropriate amino acid has been attached
v. degeneracy of the genetic code	**5.** a group of three mRNA bases signifying one amino acid

vi. nonsense codon	**6.** most amino acids are not specified by a single codon
vii. initiation codon	**7.** using the information in the nucleotide sequence of a strand of DNA to specify the nucleotide sequence of a strand of RNA
viii. template strand	**8.** the grouping of mRNA bases in threes to be read as codons
ix. RNA-like strand	**9.** AUG in a particular context
x. intron	**10.** the linear sequence of amino acids in the polypeptide corresponds to the linear sequence of nucleotide pairs in the gene
xi. RNA splicing	**11.** produces different mature mRNAs from the same primary transcript
xii. transcription	**12.** addition or deletion of a number of base pairs other than three into the coding sequence
xiii. translation	**13.** a sequence of base pairs within a gene that is not represented by any bases in the mature mRNA
xiv. alternative splicing	**14.** the strand of DNA having the base sequence complementary to that of the primary transcript
xv. charged tRNA	**15.** using the information encoded in the nucleotide sequence of an mRNA molecule to specify the amino acid sequence of a polypeptide molecule
xvi. reverse transcription	**16.** copying RNA into DNA

Section 7.1

2. Match the hypothesis from the left column to the observation from the right column that gave rise to it.

a. An intermediate messenger exists between DNA and protein.

b. The genetic code is non-overlapping.

c. The codon is more than one nucleotide.

d. The genetic code is based on triplets of bases.

e. Stop codons exist and terminate translation.

f. The amino acid sequence of a protein depends on the base sequence of an mRNA.

1. Two mutations affecting the same amino acid can recombine to give wild type.

2. One- or two-base deletions (or insertions) in a gene disrupt its function; three-base deletions (or insertions) are often compatible with function.

3. Artificial messages containing certain codons produced shorter proteins than messages not containing those codons.

4. Protein synthesis occurs in the cytoplasm, while DNA resides in the nucleus.

5. Artificial messages with different base sequences gave rise to different proteins in an *in vitro* translation system.

6. Single-base substitutions affect only one amino acid in the protein chain.

3. How would the artificial mRNA 5'. . GUGUGUGU . . 3' be read according to each of the following models for the genetic code?

a. two-base, not overlapping

b. two-base, overlapping

c. three-base, not overlapping

d. three-base, overlapping

e. four-base, not overlapping

Section 7.2

4. Describe the steps in transcription that require complementary base-pairing.

5. The coding sequence for gene *F* is read from left to right on the following figure. The coding sequence for gene *G* is read from right to left. Which strand of DNA (top or bottom) serves as the template for transcription of each gene?

```
5' ┌──────────────┐          ┌──────────────┐  3'
   │   Gene F     │          │   Gene G     │
3' └──────────────┘          └──────────────┘  5'
```

6. If you mixed the mRNA of a human gene with the genomic DNA for the same gene and allowed the RNA and DNA to form a hybrid, what would you be likely to see in the electron microscope? Your figure should include hybridization involving both DNA strands (template and RNA-like) as well as the mRNA.

Section 7.3

7. Describe the steps in translation that require complementary base-pairing.

8. Locate as accurately as possible the listed items that are shown on the following figure. Some items are not

shown. (a) 5′ end of DNA template strand; (b) 3′ end of mRNA; (c) ribosome; (d) promoter; (e) codon; (f) an amino acid; (g) DNA polymerase; (h) 5′ UTR; (i) centromere; (j) intron; (k) anticodon; (l) N terminus; (m) 5′ end of charged tRNA; (n) RNA polymerase; (o) 3′ end of uncharged tRNA; (p) a nucleotide; (q) mRNA cap; (r) peptide bond; (s) P site; (t) aminoacyl-tRNA synthetase; (u) hydrogen bond; (v) exon; (w) 5′ AUG 3′; (x) potential "wobble" interaction.

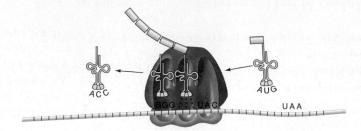

9. Consider the figure for the previous problem (#8).

    a. Which process is being represented?

    b. What is the next building block to be added to the growing chain in the figure? To what end of the growing chain will this building block be added? How many building blocks will there be in the chain when it is completed?

    c. What other building blocks have a known identity?

    d. What details could you add to this figure that would be different in a eukaryotic cell versus a prokaryotic cell?

Section 7.4

10. In prokaryotes, a s earch for genes in a DNA sequence involves scanning the DNA sequence for long open reading frames (i.e., reading frames uninterrupted by stop codons). What problem can you see with this approach in eukaryotes?

11. The yeast gene encoding a protein found in the mitotic spindle was cloned by a laboratory studying mitosis. The gene encodes a protein of 477 amino acids.

    a. What is the minimum length in nucleotides of the protein-coding part of this yeast gene?

    b. A partial sequence of one DNA strand in an exon containing the middle of the coding region of the yeast gene is given here. What is the sequence of nucleotides of the mRNA in this region of the gene? Show the 5′ and 3′ directionality of your strand.

        5′ GTAAGTTAACTTTCGACTAGTCCAGGGT 3′

    c. What is the sequence of amino acids in this part of the yeast mitotic spindle protein?

12. The sequence of a complete eukaryotic gene encoding the small protein Met-Tyr-Arg-Gly-Ala is shown here. All of the written sequences on the template strand are transcribed into RNA.

5′ CCCCTATGCCCCCCTGGGGGAGGATCAAAACACTTACCTGTACATGGC 3′
3′ GGGGATACGGGGGGGACCCCCTCCTAGTTTTGTGAATGGACATGTACCG 5′

a. Which strand is the template strand? Which direction (right to left or left to right) does RNA polymerase move along the template as it transcribes this gene?

b. What is the sequence of the nucleotides in the processed mRNA molecule for this gene? Indicate the 5′ and 3′ polarity of this mRNA.

c. A single-base mutation in the gene results in synthesis of the peptide Met-Tyr-Thr. What is the sequence of nucleotides making up the mRNA produced by this mutant gene?

13. Using recombinant DNA techniques (which will be described in Chapter 14), it is possible to take the DNA of a gene from any source and place it on a chromosome in the nucleus of a yeast cell. When you take the DNA for a human gene and put it into a yeast cell chromosome, the altered yeast cell can make the human protein. But when you remove the DNA for a gene normally present on yeast mitochondrial chromosomes and put it on a yeast chromosome in the nucleus, the yeast cell cannot synthesize the correct protein, even though the gene comes from the same organism. Explain. What would you need to do to ensure that such a yeast cell could make the correct protein?

14. a. The genetic code table shown in Figure 7.2 applies both to humans and to *E. coli*. Suppose that you have purified a piece of DNA from the human genome containing the entire gene encoding the hormone insulin. You now transform this piece of DNA into *E. coli*. Why can *E. coli* cells containing the human insulin gene actually not make insulin?

    b. Pharmaceutical companies have actually been able to obtain *E. coli* cells that make human insulin; such insulin can be purified from the bacterial cells and used to treat diabetic patients. How were the pharmaceutical companies able to create such "bacterial factories" for making insulin?

Section 7.5

15. Arrange the following list of eukaryotic gene elements in the order they would appear in the genome and in the direction travelled by RNA polymerase along the gene. Assume the gene's single intron interrupts the open reading frame. Note that some of these names are abbreviated and thus do not distinguish between elements in DNA versus RNA. For example, "splice donor site" is an abbreviation for "DNA sequences transcribed into the splice donor site" because splicing takes place on the gene's RNA transcript, not on the gene itself. Geneticists often use this kind of shorthand for simplicity, even though it is imprecise. (a) splice donor site; (b) 3′ UTR; (c) promoter; (d) stop codon; (e) nucleotide to which methylated cap is added; (f) initiation codon; (g) transcription terminator; (h) splice acceptor site; (i) 5′ UTR; (j) poly-A addition site; (k) splice branch site.

16. Consider the list of eukaryotic gene elements in the previous problem (#15).

a. Which of the element names in the list are abbreviated? (That is, which of these elements actually occur in the gene's primary transcript or mRNA rather than in the gene itself?)

b. Which of the elements in the list are found partly or completely in the first exon of this gene (or the RNA transcribed from this exon)? In the intron? In the second exon?

Section 7.6

17. Do you think each of the following types of mutations would have very severe effects, mild effects, or no effect at all?

a. Nonsense mutations occurring in the sequences encoding amino acids near the N terminus of the protein

b. Nonsense mutations occurring in the sequences encoding amino acids near the C terminus of the protein

c. Frameshift mutations occurring in the sequences encoding amino acids near the N terminus of the protein

d. Frameshift mutations occurring in the sequences encoding amino acids near the C terminus of the protein

e. Silent mutations

f. Conservative missense mutations

g. Nonconservative missense mutations affecting the active site of the protein

h. Nonconservative missense mutations not in the active site of the protein

18. Null mutations are valuable genetic resources because they allow a researcher to determine what happens to an organism in the complete absence of a particular protein. However, it is often not a trivial matter to determine whether a mutation represents the null state of the gene.

a. Geneticists sometimes use the following test for the "nullness" of an allele in a diploid organism: If the abnormal phenotype seen in a homozygote for the allele is identical to that seen in a heterozygote where one chromosome carries the allele in question and the homologous chromosome is known to be completely deleted for the gene, then the allele is null. What is the underlying rationale for this test? What limitations might there be in interpreting such a result?

b. Can you think of other methods to determine whether an allele represents the null state of a particular gene?

19. The following is a list of mutations that have been discovered in a gene that has more than 60 exons and encodes a very large protein of 2532 amino acids. Indicate whether or not each mutation could cause a detectable change in the size or the amount of mRNA and/or a detectable change in the size or the amount of the protein product. (Detectable changes in size or amount must be greater than 1 percent of normal values.) What kind of change would you predict?

a. Lys576Val (changes amino acid 576 from lysine into valine)

b. Lys576Arg

c. AAG576AAA (changes codon 576 from AAG to AAA)

d. AAG576UAG

e. Met1Arg (there are at least two possible scenarios for this mutation)

f. promoter mutation

g. one base-pair insertion into codon 1841

h. deletion of codon 779

i. IVS18DS, G–A, + 1 (this mutation changes the first nucleotide in the eighteenth intron of the gene, causing exon 18 to be spliced to exon 20, thus skipping exon 19)

j. deletion of the poly-A addition site

k. G-to-A substitution in the 5′ UTR

l. insertion of 1000 base pairs into the sixth intron (this particular insertion does not alter splicing)

20. Consider further the mutations described in the previous problem (#19).

a. Which of the mutations could be null mutations?

b. Which of the mutations would be most likely to result in an allele that is recessive to wild type?

c. Which of the mutations could result in an allele dominant to wild type? What mechanism(s) could explain this dominance?

21. When 1 million cells of a culture of haploid yeast carrying a *met⁻* auxotrophic mutation were plated on Petri plates lacking methionine (met), five colonies grew. You would expect cells in which the original *met⁻* mutation was reversed (by a base change back to the original sequence) to grow on media lacking methionine, but some of these apparent reversions could be due to a mutation in a different gene that somehow suppresses the original *met⁻* mutations. How would you be able to determine if the mutations in your five colonies were due either to a precise reversion of the original *met⁻* mutation or to the generation of a suppressor mutation in a gene on another chromosome?

22. a. What are the differences among null, hypomorphic, hypermorphic, dominant negative, and neomorphic mutations?

b. For each of these kinds of mutations, would you predict they would be dominant or recessive to a wild-type allele in producing a mutant phenotype?

23. A mutant *B. adonis* bacterium has a nonsense suppressor tRNA that inserts glutamine (Gln) to match a UAG (but not other nonsense) codons.

a. What is the anticodon of the suppressing tRNA? Indicate the 5′ and 3′ ends.

b. What is the sequence of the template strand of the wild-type tRNA^Gln-encoding gene that was altered to produce the suppressor, assuming that only a single-base-pair alteration was involved?

c. What is the *minimum* number of *tRNA*^Gln genes that could be present in a wild-type *B. adonis* cell? Describe the corresponding anticodons.

24. You are studying mutations in a bacterial gene that codes for an enzyme whose amino acid sequence is known. In the wild-type protein, proline is the fifth amino acid from the amino terminal end. In one of your mutants with nonfunctional enzyme, you find a serine at position number 5. You subject this mutant to further mutagenesis and recover three different strains. Strain A has a proline at position number 5 and acts just like wild type. Strain B has tryptophan at position number 5 and also acts like wild type. Strain C has no detectable enzyme function at any temperature, and you cannot recover any protein that resembles the enzyme. You mutagenize strain C and recover a strain (C-1) that has enzyme function. The second mutation in C-1 responsible for the recovery of enzyme function does not map at the enzyme locus.

    a. What is the nucleotide sequence in both strands of the wild-type gene at this location?

    b. Why does strain B have a wild-type phenotype? Why does the original mutant with serine at position 5 lack function?

    c. What is the nature of the mutation in strain C?

    d. What is the second mutation that arose in C-1?

25. Another class of suppressor mutations, not described in the chapter, are mutations that suppress missense mutations.

    a. Why would bacterial strains carrying such missense suppressor mutations generally grow more slowly than strains carrying nonsense suppressor mutations?

    b. What other kinds of mutations can you imagine in genes encoding components needed for gene expression that would suppress a missense mutation in a protein-coding gene?

26. Yet another class of suppressor mutations not described in the chapter are mutations in tRNA genes that can suppress frameshift mutations. What would have to be true about a tRNA that could suppress a frameshift mutation involving the insertion of a single base pair?

27. There is at least one nonsense-suppressing tRNA known that can suppress more than one type of nonsense codon.

    a. What is the anticodon of such a suppressing tRNA?

    b. What stop codons would it suppress?

    c. What are the amino acids most likely to be carried by this nonsense-suppressing tRNA?

Section 7.7

28. An example of a portion of the T4 *rIIB* gene in which Crick and Brenner had recombined one + and one − mutation is shown here. (The RNA-like strand of the DNA is shown.)

```
wild type    5′ AAA AGT CCA TCA CTT AAT GCC 3′
mutant       5′ AAA GTC CAT CAC TTA ATG GCC 3′
```

    a. Where are the + and − mutations in the mutant DNA?

    b. What alterations in amino acids occurred in this double mutant, which produces wild-type plaques?

c. How can you explain the fact that amino acids are different in the double mutant compared with the wild-type sequence, yet the phage is wild type?

29. In the *HbS* allele (sickle-cell allele) of the human β-globin gene, the sixth amino acid in the β-globin chain is changed from glutamic acid to valine. In *HbC*, the sixth amino acid in β globin is changed from glutamic acid to lysine. What would be the order of these two mutations within the map of the β-globin gene?

30. The following diagram describes the mRNA sequence of part of the *A* gene and the beginning of the *B* gene of phage φX174. In this phage, there are some genes that are read in overlapping reading frames. For example, the code for the *A* gene is used for part of the *B* gene, but the reading frame is displaced by one base. Shown here is the single mRNA with the codons for proteins A and B indicated.

```
aa       5 6 7 8 9 10 11 12 13 14 15 16
A        AlaLysGluTrpAsnAsnSerLeuLysThrLysLeu
mRNA     GCUAAAGAAUGGAACAACUCACUAAAAACCAAGCUG
B            MetGluGlnLeuThrLysAsnGlnAla
aa           1 2 3 4 5 6 7 8 9
```

Given the following amino acid (aa) changes, indicate the base change that occurred in the mRNA and the consequences for the other protein sequence.

    a. Asn at position 10 in protein A is changed to Tyr.

    b. Leu at position 12 in protein A is changed to Pro.

    c. Gln at position 8 in protein B is changed to Leu.

    d. The occurrence of overlapping reading frames is very rare in nature. When it does occur, the extent of the overlap is not very long. Why do you think this is the case?

31. The amino acid sequence of part of a protein has been determined:

N . . . Gly-Ala-Pro-Arg-Lys . . . C

A mutation has been induced in the gene encoding this protein using the mutagen proflavin. The resulting mutant protein can be purified and its amino acid sequence determined. The amino acid sequence of the mutant protein is exactly the same as the amino acid sequence of the wild-type protein from the N terminus of the protein to the glycine in the preceding sequence. Starting with this glycine, the sequence of amino acids is changed to the following:

N . . . Gly-His-Gln-Gly-Lys . . . C

Using the amino acid sequences, one can determine the sequence of 14 nucleotides from the wild-type gene encoding this protein. What is this sequence?

32. When the artificial mRNA 5′ . . . UCUCUCUC . . . 3′ was added to an *in vitro* protein synthesis system, investigators found that proteins composed of alternating leucine and serine were made. What experiments were done to determine whether leucine was specified by CUC and serine by UCU, or vice versa?

33. Identify all the amino-acid–specifying codons where a point mutation (a single-base change) could generate a nonsense codon.

34. Translate all the sequences shown in Figure 7.28, assuming that in each case the RNA-like strand of the gene is depicted.

35. A particular protein has the amino acid sequence

    N . . . Ala-Pro-His-Trp-Arg-Lys-Gly-Val-Thr . . . C

    within its primary structure. A geneticist studying mutations affecting this protein discovered that several of the mutants produced shortened protein molecules that terminated within this region. In one of them, the His became the terminal amino acid.

    a. What DNA single-base change(s) would cause the protein to terminate at the His residue?

    b. What other potential sites do you see in the DNA sequence encoding this protein where mutation of a single base pair would cause premature termination of translation?

36. In studying normal and mutant forms of a particular human enzyme, a geneticist came across a particularly interesting mutant form of the enzyme. The normal enzyme is 227 amino acids long, but the mutant form was 312 amino acids long, having that extra 85 amino acids as a block in the middle of the normal sequence. The inserted amino acids do not correspond in any way to the normal protein sequence. What are possible explanations for this phenomenon? How would you distinguish among them?

37. How many possible open reading frames (frames without stop codons) are there that extend through the following sequence?

    5'. . . CTTACAGTTTATTGATACGGAGAAGG. . .3'
    3'. . . GAATGTCAAATAACTATGCCTCTTCC. . .5'

38. a. In Figure 7.26, the physical map (the number of base pairs) is not exactly equivalent to the genetic map (in map units). Explain this apparent discrepancy.

    b. In Figure 7.26, which region shows the highest rate of recombination, and which the lowest?

39. The sequence of a segment of mRNA, beginning with the initiation codon, is given here, along with the corresponding sequences from several mutant strains.

    ```
    Normal     AUGACACAUCGAGGGGUGGUAAACCCUAAG. . .
    Mutant 1   AUGACACAUCCAGGGGUGGUAAACCCUAAG. . .
    Mutant 2   AUGACACAUCGAGGGUGGUAAACCCUAAG. . .
    Mutant 3   AUGACGCAUCGAGGGGUGGUAAACCCUAAG. . .
    Mutant 4   AUGACACAUCGAGGGGUUGGUAAACCCUAAG. . .
    Mutant 5   AUGACACAUUGAGGGGUGGUAAACCCUAAG. . .
    Mutant 6   AUGACAUUUACCACCCCUCGAUGCCCUAAG. . .
    ```

    a. Indicate the type of mutation present in each and translate the mutated portion of the sequence into an amino acid sequence in each case.

    b. Which of the mutations could be reverted by treatment with proflavin?

40. You identify a proflavin-generated allele of a gene that produces a 110-amino-acid polypeptide rather than the usual 157-amino-acid protein. After subjecting this mutant allele to extensive proflavin mutagenesis, you are able to find a number of intragenic suppressors located in the part of the gene between the sequences encoding the N terminus of the protein and the original mutation but no suppressors located in the region between the original mutation and the sequences encoding the usual C terminus of the protein. Why do you think this is the case?

connect® For more information on the resources available from McGraw-Hill Ryerson, go to www.mcgrawhill.ca/he/solutions.

Problems   235

The black jaguar cub on the *left* has a mutation that results in increased production of the pigment melanin, compared with the spotted jaguar cub on the *right*.

Human chromosome 3 consists of approximately 220 million base pairs and carries 1000–2000 genes (**Figure 8.1**). Somewhere on the long arm of the chromosome resides the gene for rhodopsin, a light-sensitive protein active in the rod cells of our retinas. The rhodopsin gene determines perception of low-intensity light. People who carry the normal, wild-type allele of the gene see well in a dimly lit room and on the road at night. One simple change—a mutation—in the rhodopsin gene, however, diminishes light perception just enough to lead to night blindness. Other alterations in the gene cause the destruction of rod cells, resulting in total blindness. Medical researchers have so far identified more than 30 mutations in the rhodopsin gene that affect vision in different ways.

The case of the rhodopsin gene illustrates some very basic questions. Which of the 220 million base pairs on chromosome 3 make up the rhodopsin gene? How are the base pairs that comprise this gene arranged along the chromosome? How can a single gene sustain so many

mutations that lead to such divergent phenotypic effects? In this chapter, we describe the ingenious experiments performed by geneticists during the 1950s and 1960s as they examined the relationships among mutations, genes, chromosomes, and phenotypes in an effort to understand, at the molecular level, what genes are and how they function.

We can recognize three main themes from the elegant work of these investigators. The first is that mutations are

Chapter Outline
8.1 Mutations: Primary Tools of Genetic Analysis
8.2 What Mutations Tell Us About Gene Structure
8.3 What Mutations Tell Us About Gene Function
8.4 A Comprehensive Example: Mutations That Affect Vision

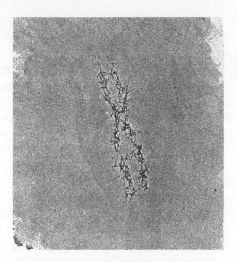

Figure 8.1 The DNA of each human chromosome contains hundreds to thousands of genes. The DNA of this human chromosome has been spread out and magnified 50 000×. No topological signs reveal where along the DNA the genes reside. The darker, chromosome-shaped structure in the middle is a scaffold of proteins to which the DNA is attached.

heritable changes in base sequence that can affect phenotype. The second is that physically, a gene is usually a specific protein-encoding segment of DNA in a discrete region of a chromosome. (We now know that some genes encode various kinds of RNA that do not get translated into protein.) Third, a gene is not simply a bead on a string, changeable only as a whole and only in one way, as some had believed. Rather, genes are divisible, and each gene's subunits—the individual nucleotide pairs of DNA—can mutate independently and can recombine with each other.

Knowledge of what genes are and how they work deepens our understanding of Mendelian genetics by providing a biochemical explanation for how genotype influences phenotype. One mutation in the rhodopsin gene, for example, causes the substitution of one particular amino acid for another in the construction of the rhodopsin protein. This single substitution changes the three-dimensional structure of rhodopsin and thus the protein's ability to absorb photons, ultimately altering a person's ability to perceive light.

Learning Objectives

1. Distinguish between forward and reverse mutations.
2. Examine the various terms used to identify different types of mutations.
3. Explain how spontaneous mutation rate is measured.
4. Differentiate between the various types of spontaneous and induced mutations and how they occur.
5. Distinguish between the various types of DNA repair mechanisms.
6. Examine the possible consequences of genetic mutations.
7. Compare germline and somatic mutations.
8. Analyze how mutations are used to study gene structure and function in a variety of organisms.

8.1 Mutations: Primary Tools of Genetic Analysis

We saw in Chapter 2 that genes with one common allele are *monomorphic*, while genes with several common alleles in natural populations are *polymorphic*. The term **wild-type allele** has a clear definition for monomorphic genes, where the allele found on the large majority of chromosomes in the population under consideration is wild type. In the case of polymorphic genes, the definition is less straightforward. Some geneticists consider all alleles with a frequency of 1 percent or greater to be wild type, while others describe the many alleles present at appreciable frequencies in the population as *common variants* and reserve "wild-type allele" for use only in connection with monomorphic genes.

Mutations are heritable changes in DNA base sequences

A mutation that changes a wild-type allele of a gene (regardless of the definition) to a different allele is called a **forward mutation**. The resulting novel mutant allele can be either recessive or dominant to the original wild type. Geneticists often diagram forward mutations as $A^+ \rightarrow a$ when the mutation is recessive and as $b^+ \rightarrow B$

when the mutation is dominant. Mutations can also cause a novel mutant allele to revert back to wild type ($a \rightarrow A^+$, or $B \rightarrow b^+$) in a process known as **reverse mutation**, or **reversion**. In this chapter, we designate wild-type alleles, whether recessive or dominant, with a plus sign (+).

Mendel originally defined genes by the visible phenotypic effects—yellow or green, round or wrinkled—of their alternative alleles. In fact, the only way he knew that genes existed at all was because alternative alleles for seven particular pea genes had arisen through forward mutations. Close to a century later, knowledge of DNA structure clarified that such mutations are heritable changes in DNA base sequence. DNA thus carries the potential for genetic change in the same place it carries genetic information—the sequence of its bases.

Mutations may be classified by how they change DNA

A **substitution** occurs when a base at a certain position in one strand of the DNA molecule is replaced by one of the other three bases (**Figure 8.2a**); after DNA replication,

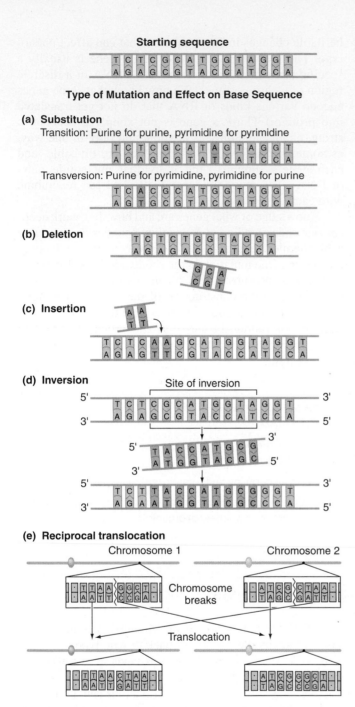

Starting sequence

Type of Mutation and Effect on Base Sequence

(a) Substitution

Transition: Purine for purine, pyrimidine for pyrimidine

Transversion: Purine for pyrimidine, pyrimidine for purine

(b) Deletion

(c) Insertion

(d) Inversion

Site of inversion

(e) Reciprocal translocation

Chromosome 1   Chromosome 2

Chromosome breaks

Translocation

Figure 8.2 Mutations classified by their effect on DNA.

the microscope when they observe chromosomes in the context of a karyotype, such as that shown in Figure 3.4 in Chapter 3.

More complex mutations include **inversions**, 180° rotations of a segment of the DNA molecule (**Figure 8.2d**), and **reciprocal translocations**, in which parts of two non-homologous chromosomes change places (**Figure 8.2e**). Large-scale DNA rearrangements, including megabase deletions and insertions as well as inversions and translocations, cause major genetic reorganizations that can change either the order of genes along a chromosome or the number of chromosomes in an organism. We discuss these **chromosomal rearrangements**, which affect many genes at a time, in Chapter 9. In this chapter, we focus on mutations that alter only one gene at a time.

Only a small fraction of the mutations in a genome actually alter the nucleotide sequences of genes in a way that affects gene function. By changing one allele to another, these mutations modify the structure or amount of a gene's protein product, and the modification in protein structure or amount influences phenotype. All other mutations either alter genes in a way that does not affect their function or change the DNA between genes. We discuss mutations without observable phenotypic consequences in Chapter 15; such mutations are very useful for mapping genes and tracking differences between individuals. In the remainder of this chapter, we focus on those mutations that have an impact on gene function and thereby influence phenotype.

> Mutations—heritable changes in DNA base sequences—include substitutions, deletions, insertions, inversions, and translocations.

Spontaneous mutations occur at a very low rate

Mutations that modify gene function happen so infrequently that geneticists must examine a very large number of individuals from a formerly homogeneous population to detect the new phenotypes that reflect these mutations. In one ongoing study, dedicated investigators have monitored the coat colours of millions of specially bred mice and discovered that on average, a given gene mutates to a recessive allele in roughly 11 out of every 1 million gametes (**Figure 8.3**). Studies of several other organisms have yielded similar results: an average spontaneous rate of $2-12 \times 10^{-6}$ mutations per gene per gamete.

Looking at the mutation rate from a different perspective, you could ask how many mutations there might be in the genes of an individual. To find out, you would simply multiply the rate of $2-12 \times 10^{-6}$ mutations per gene times 20 000, a current estimate of the number of genes in the human genome, to obtain an answer of between $0.04-0.24$ mutations per haploid genome. This very rough calculation would mean that, on average, one new mutation affecting phenotype could arise in every 6–30 human gametes.

a new base pair will appear in the daughter double helix. Substitutions can be subdivided into *transitions*, in which one purine (A or G) replaces the other purine, or one pyrimidine (C or T) replaces the other; and *transversions*, in which a purine changes to a pyrimidine, or vice versa.

Other types of mutations produce more complicated rearrangements of DNA sequence. A **deletion** occurs when a block of one or more nucleotide pairs is lost from a DNA molecule; an **insertion** is just the reverse—the addition of one or more nucleotide pairs (**Figures 8.2b and c**). Deletions and insertions can be as small as a single base pair or as large as megabases (i.e., millions of base pairs). Researchers can see the larger changes under

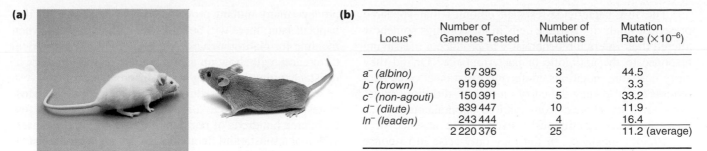

(b)

Locus*	Number of Gametes Tested	Number of Mutations	Mutation Rate ($\times 10^{-6}$)
a^- *(albino)*	67 395	3	44.5
b^- *(brown)*	919 699	3	3.3
c^- *(non-agouti)*	150 391	5	33.2
d^- *(dilute)*	839 447	10	11.9
ln^- *(leaden)*	243 444	4	16.4
	2 220 376	25	11.2 (average)

* Mutation is from wild type to the recessive allele shown.

Figure 8.3 Rates of spontaneous mutation. (a) Mutant mouse coat colours: albino (*left*), brown (*right*). **(b)** Mutation rates from wild-type to recessive mutant alleles for five coat colour genes. Mice from highly inbred wild-type strains were mated with homozygotes for recessive coat colour alleles. Progeny with mutant coat colours indicated the presence of recessive mutations in gametes produced by the inbred mice.

Different genes, different mutation rates

Although the average mutation rate per gene is $2-12 \times 10^{-6}$, this number masks considerable variation in the mutation rates for different genes. Experiments with many organisms show that mutation rates range from less than 10^{-9} to more than 10^{-3} per gene per gamete. More recent studies in humans have calculated the mutation rate as $1-4 \times 10^{-8}$ per site per generation, with the male germ line being more mutagenic than the female. Variation in the mutation rate of different genes within the same organism reflects differences in gene size (larger genes are larger targets that sustain more mutations) as well as differences in the susceptibility of particular genes to the various mechanisms that cause mutations (described later in this chapter).

Estimates of the average mutation rates in bacteria range from 10^{-8} to 10^{-7} mutations per gene per cell division. Although the units here are slightly different than those used for multicellular eukaryotes (because bacteria do not produce gametes), the average rate of mutation in gamete-producing eukaryotes still appears to be considerably higher than that in bacteria. The main reason is that numerous cell divisions take place between the formation of a zygote and meiosis, so mutations that appear in a gamete may have actually occurred many cell generations before the gamete formed. In other words, there are more chances for mutations to accumulate. Some scientists speculate that the diploid genomes of multicellular organisms allow them to tolerate relatively high rates of mutation in their gametes because a zygote would have to receive recessive mutations in the same gene from both gametes for any deleterious effects to occur. In contrast, a bacterium would be affected by just a single mutation that disrupted its only copy of the gene.

Gene function: Easy to disrupt, hard to restore

In the mouse coat colour study, when researchers allowed brother and sister mice homozygous for a recessive mutant allele of one of the five mutant coat colour genes to mate with each other, they could estimate the rate of reversion by examining the F_1 offspring. Any progeny expressing the dominant wild-type phenotype for a particular coat

colour, of necessity, carried a gene that had sustained a reverse mutation. Calculations based on observations of several million F_1 progeny revealed a reverse mutation rate ranging from $0-2.5 \times 10^{-6}$ per gene per gamete; the rate of reversion varied somewhat from gene to gene. In this study, then, the rate of reversion was significantly lower than the rate of forward mutation, most likely because there are many ways to disrupt gene function, but there are only a few ways to restore function once it has been disrupted. The conclusion that the rate of reversion is significantly lower than the rate of forward mutation holds true for most types of mutation. In one extreme example, deletions of more than a few nucleotide pairs can never revert, because DNA information that has disappeared from the genome cannot spontaneously reappear.

Although estimates of mutation rates are extremely rough, they nonetheless support three general conclusions: (1) Mutations affecting phenotype occur very rarely; (2) different genes mutate at different rates; and (3) the rate of forward mutation (a disruption of gene function) is almost always higher than the rate of reversion.

Spontaneous mutations arise from many kinds of random events

Because spontaneous mutations affecting a gene occur so infrequently, it is very difficult to study the events that produce them. To overcome this problem, researchers turned to bacteria as the experimental organisms of choice. It is easy to grow many millions of individuals and then rapidly search through enormous populations to find the few that carry a novel mutation. In one study, investigators spread wild-type bacteria on the surface of agar containing sufficient nutrients for growth as well as a large amount of a bacteria-killing substance, such as an antibiotic or a bacteriophage. Although most of the bacterial cells died, a few showed resistance to the bactericidal substance and continued to grow and divide. The descendants of a single resistant bacterium, produced by many rounds of binary fission, formed a mound of genetically identical cells called a **colony**.

The few bactericide-resistant colonies that appeared presented a puzzle. Had the cells in the colonies somehow altered their internal biochemistry to produce a life-saving response to the antibiotic or bacteriophage? Or did they carry heritable mutations conferring resistance to the bactericide? And if they did carry mutations, did those mutations arise by chance from random spontaneous events that take place continuously, even in the absence of a bactericidal substance, or did they only arise in response to environmental signals (in this case, the addition of the bactericide)?

The fluctuation test

In 1943, Salvador Luria and Max Delbrück devised an experiment to examine the origin of bacterial resistance (**Figure 8.4**). According to their reasoning, if bacteriophage-resistant colonies arise in direct response to infection by bacteriophages, separate suspensions of bacteria containing equal numbers of cells will generate similar, small numbers of resistant colonies when spread in separate Petri plates on nutrient agar suffused with phages. By contrast, if resistance arises from mutations that occur spontaneously even when the phages are not present, then different liquid cultures, when spread on separate Petri plates, will generate very different numbers of resistant colonies. The reason is that the mutation conferring resistance can, in theory, arise at any time during the growth of the culture. If it happens early, the cell in which it occurs will

produce many mutant progeny prior to Petri plating; if it happens later, there will be far fewer mutant progeny when the time for plating arrives. After plating, these numerical differences will show up as fluctuations in the numbers of resistant colonies growing in the different Petri plates.

The results of this **fluctuation test** were clear: Most plates supported zero to a few resistant colonies, but a few harboured hundreds of resistant colonies. From this observation of a substantial fluctuation in the number of resistant colonies in different Petri plates, Luria and Delbrück concluded that bacterial resistance arises from mutations that exist before exposure to bacteriophage. After exposure, however, the bactericide in the Petri plate becomes a selective agent that kills off nonresistant cells, allowing only the pre-existing resistant ones to survive. **Figure 8.5** illustrates how researchers used another technique, known as *replica plating*, to demonstrate even more directly that the mutations conferring bacterial resistance occur before the cells encounter the bactericide that selects for their resistance.

These key experiments showed that bacterial resistance to phages and other bactericides is the result of mutations, and these mutations do not arise in particular genes as a directed response to environmental change. Instead, mutations occur spontaneously as a result of random processes that can happen at any time and hit the genome at any place. Once such random changes occur, however, they usually remain stable. If the resistant mutants of the Luria–Delbrück experiment, for example, were grown for

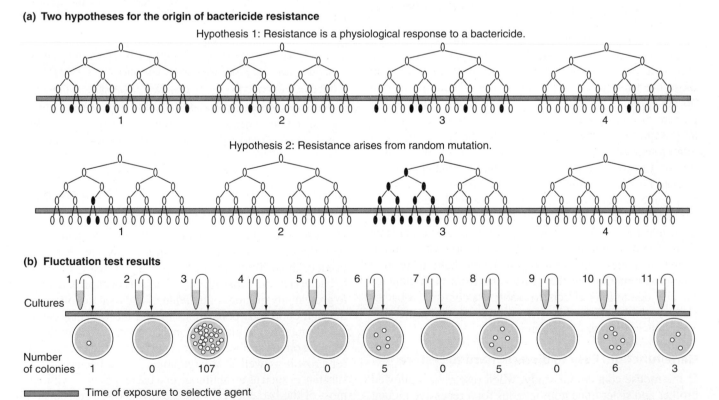

(a) Two hypotheses for the origin of bactericide resistance

Hypothesis 1: Resistance is a physiological response to a bactericide.

Hypothesis 2: Resistance arises from random mutation.

(b) Fluctuation test results

Cultures

Number of colonies

1	2	3	4	5	6	7	8	9	10	11
1	0	107	0	0	5	0	5	0	6	3

Time of exposure to selective agent

Figure 8.4 The Luria–Delbrück fluctuation experiment. (a) Hypothesis 1: If resistance arises only after exposure to a bactericide, all bacterial cultures of equal size should produce roughly the same number of resistant colonies. Hypothesis 2: If random mutations conferring resistance arise before exposure to bactericide, the number of resistant colonies in different cultures should vary (fluctuate) widely. **(b)** Actual results showing large fluctuations suggest that mutations in bacteria occur as spontaneous mistakes independent of exposure to a selective agent.

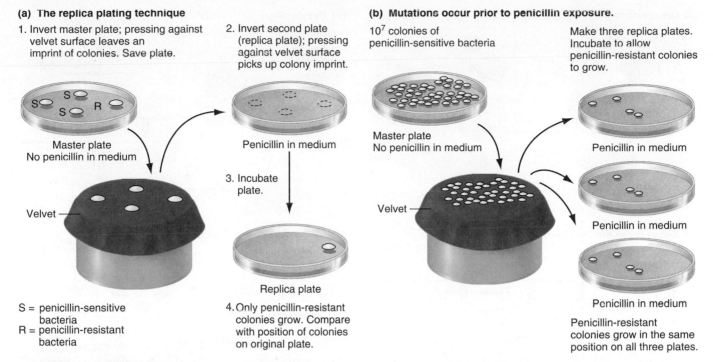

(a) The replica plating technique

1. Invert master plate; pressing against velvet surface leaves an imprint of colonies. Save plate.

2. Invert second plate (replica plate); pressing against velvet surface picks up colony imprint.

Master plate
No penicillin in medium

Penicillin in medium

3. Incubate plate.

Velvet

Replica plate

S = penicillin-sensitive bacteria
R = penicillin-resistant bacteria

4. Only penicillin-resistant colonies grow. Compare with position of colonies on original plate.

(b) Mutations occur prior to penicillin exposure.

10^7 colonies of penicillin-sensitive bacteria

Make three replica plates. Incubate to allow penicillin-resistant colonies to grow.

Master plate
No penicillin in medium

Penicillin in medium

Velvet

Penicillin in medium

Penicillin in medium

Penicillin-resistant colonies grow in the same position on all three plates.

Figure 8.5 Replica plating verifies that bacterial resistance is the result of pre-existing mutations. **(a)** Pressing a *master plate* onto a velvet surface transfers some cells from each bacterial colony onto the velvet. Pressing a *replica plate* onto the velvet then transfers some cells from each colony onto the replica plate. Investigators track which colonies on the master plate are able to grow on the replica plate (here, only penicillin-resistant ones). **(b)** Colonies on a master plate without penicillin are sequentially transferred to three replica plates with penicillin. Resistant colonies grow in the same positions on all three replicas, showing that some colonies on the master plate had multiple resistant cells before exposure to the antibiotic.

many generations in medium that did not contain bacteriophages, they would nevertheless remain resistant to this bactericidal virus.

We now describe some of the many kinds of random events that cause mutations; later, we discuss how cells cope with the damage.

> Luria and Delbrück's fluctuation test showed that mutations in bacteria conferring resistance to bacteriophages occur prior to exposure to the phages and are caused by random, spontaneous events.

Natural processes that alter DNA

Chemical and physical assaults on DNA are quite frequent. Geneticists estimate, for example, that the hydrolysis of a purine base, A or G, from the deoxyribose-phosphate backbone occurs 1000 times an hour in every human cell. This kind of DNA alteration is called **depurination** (**Figure 8.6a**). Because the resulting *apurinic site* cannot specify a complementary base, the DNA replication process sometimes introduces a random base opposite the apurinic site, causing a mutation in the newly synthesized complementary strand three-quarters of the time.

Another naturally occurring process that may modify DNA's information content is **deamination**: the removal of an amino ($-NH_2$) group. Deamination can change cytosine

to uracil (U), the nitrogenous base found in RNA but not in DNA. Because U pairs with A rather than G, deamination followed by DNA replication may alter a C–G base pair to a T–A pair in future generations of DNA molecules (**Figure 8.6b**); such a C–G to T–A change is a transition mutation.

Other assaults include naturally occurring radiation such as cosmic rays and X-rays, which break the sugar-phosphate backbone (**Figure 8.6c**); ultraviolet light, which causes adjacent thymine residues to become chemically linked into thymine–thymine dimers (**Figure 8.6d**); and oxidative damage to any of the four bases (**Figure 8.6e**). All of these changes alter the information content of the DNA molecule.

Mistakes during DNA replication

If the cellular machinery for some reason incorporates an incorrect base during replication, for instance, a C opposite an A instead of the expected T, then during the next replication cycle, one of the daughter DNAs will have the normal A–T base pair, while the other will have a mutant G–C. Careful measurements of the fidelity of replication *in vivo*, in both bacteria and human cells, show that such errors are exceedingly rare, occurring less than once in every 10^9 base pairs. That is equivalent to typing this entire book 1000 times while making only one typing error. Considering the complexities of helix unwinding, base-pairing, and polymerization, this level of accuracy is amazing. How do cells achieve it?

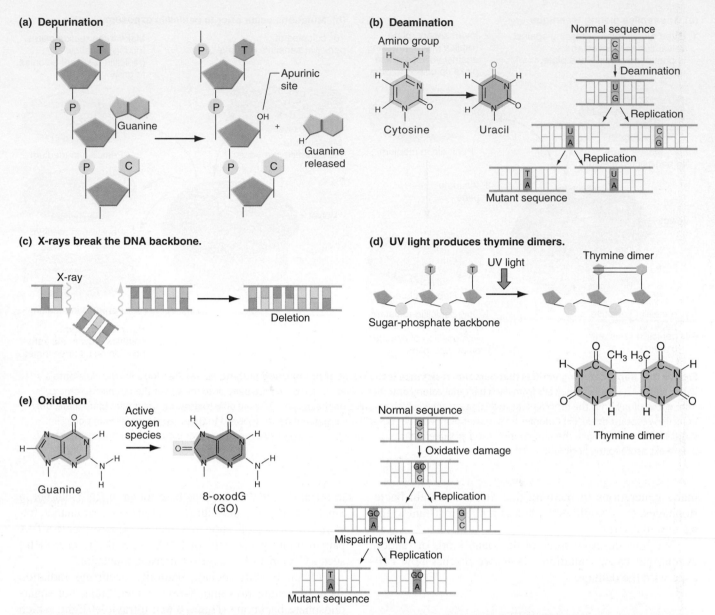

(a) Depurination

Apurinic site

Guanine

Guanine released

(b) Deamination

Amino group

Cytosine → Uracil

Normal sequence

C/G

↓ Deamination

U/G

↘ Replication

U/A   C/G

↓ Replication

T/A   U/A

Mutant sequence

(c) X-rays break the DNA backbone.

X-ray

Deletion

(d) UV light produces thymine dimers.

UV light → Thymine dimer

Sugar-phosphate backbone

Thymine dimer

(e) Oxidation

Guanine → 8-oxodG (GO)

Active oxygen species

Normal sequence

G/C

↓ Oxidative damage

GO/C

↘ Replication

GO/A   G/C

Mispairing with A

↙ Replication

T/A   GO/A

Mutant sequence

Figure 8.6 How natural processes can change the information stored in DNA. **(a)** In depurination, the hydrolysis of A or G bases leaves a DNA strand with an unspecified base. **(b)** In deamination, the removal of an amino group from cytosine (C) initiates a process that causes a transition after DNA replication. **(c)** X-rays break the sugar-phosphate backbone and thereby split a DNA molecule into smaller pieces, which may be spliced back together improperly. **(d)** Ultraviolet (UV) radiation causes adjacent T's to form dimers, which can disrupt the readout of genetic information. **(e)** Both irradiation and normal aerobic respiration cause the formation of *free radicals* (such as oxygen molecules with an unpaired electron) that can alter individual bases. Here, the pairing of the altered base GO with A creates a transversion that changes a G–C base pair to T–A.

The replication machinery minimizes errors through successive stages of correction. In the test tube, DNA polymerases replicate DNA with an error rate of about one mistake in every 10^6 bases copied. This rate is about 1000-fold worse than that achieved by the cell. Even so, it is impressively low and is only attained because polymerase molecules provide, along with their polymerization function, a proofreading/editing function in the form of a nuclease that is activated whenever the polymerase makes a mistake. This nuclease portion of the polymerase molecule, called the *3'-to-5' exonuclease*, recognizes a mispaired base and excises it, allowing the polymerase to copy the nucleotide correctly on the next try (**Figure 8.7**). Without its nuclease

portion, DNA polymerase would have an error rate of one mistake in every 10^4 bases copied, so its editing function improves the fidelity of replication 100-fold. DNA polymerase *in vivo* is part of a replication system including many other proteins that collectively improve on the error rate another 10-fold, bringing it to within about 100-fold of the fidelity attained by the cell.

The 100-fold higher accuracy of the cell depends on a backup system called *methyl-directed mismatch repair* that notices and corrects residual errors in the newly replicated DNA. We present the details of this repair system later in the chapter when we describe the various ways in which cells attempt to correct mutations once they occur.

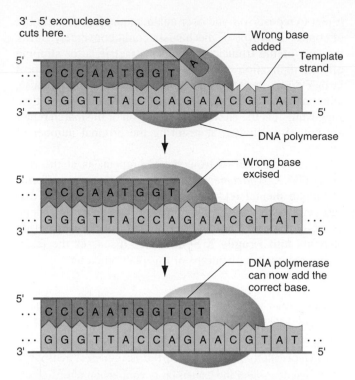

Figure 8.7 DNA polymerase's proofreading function. If DNA polymerase mistakenly adds an incorrect nucleotide at the 3′ end of the strand it is synthesizing, the enzyme's 3′-to-5′ exonuclease activity removes this nucleotide, giving the enzyme a second chance to add the correct nucleotide.

Unequal crossing-over and transposable elements

Some mutations arise from events other than chemical and physical assaults or replication errors. Erroneous recombination is one such mechanism. For example, in **unequal crossing-over**, two closely related DNA sequences that are located in different places on two homologous chromosomes can pair with each other during meiosis. If recombination takes place between the mispaired sequences, one homologous chromosome ends up with a duplication (a kind of insertion), while the other homologue sustains a deletion. As **Figure 8.8a** shows, some forms of red-green colour blindness arise from deletions and duplications in the genes that enable us to perceive red and green wavelengths of light; these reciprocal informational changes are the result of unequal crossing-over.

Another notable mechanism for altering a DNA sequence involves the units of DNA known as **transposable elements (TEs)**. TEs are DNA segments several hundred to several thousand base pairs long that move (or "transpose" or "jump") from place to place in the genome. If a TE jumps into a gene, it can disrupt the gene's function and cause a mutation. Certain TEs frequently insert themselves into particular genes and not others; this is one reason that mutation rates vary from gene to gene. Although some TEs move by making a copy that becomes inserted into a different chromosomal location while the initial version stays put (**replicative transposition**), other TE types actually leave their original position when they move (**conservative transposition**) (**Figure 8.8b**). Mutations caused by TEs that transpose by this second mechanism are exceptions to the general rule that the rate of reversion is lower than the rate of forward mutation. This is because TE transposition can occur relatively frequently, and when it is accompanied by excision of the TE, the original sequence and function of the gene are restored. Chapter 9 discusses additional genetic consequences of TE behaviour.

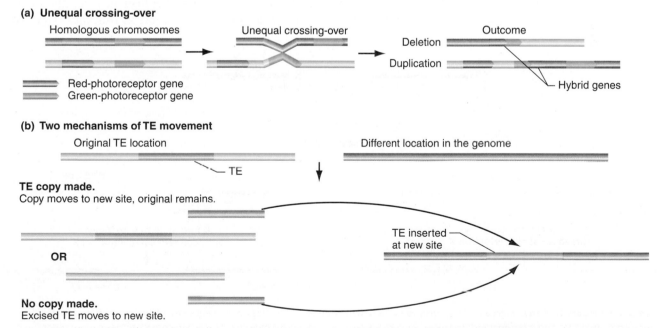

Figure 8.8 How unequal crossing-over and the movement of transposable elements (TEs) change DNA's information content.
(a) If two nearby regions contain a similar DNA sequence, the two homologous chromosomes may pair out of register during meiosis and produce gametes with either a deletion or a reciprocal duplication. Colour blindness in humans can result from unequal crossing-over between the nearby and highly similar genes for red and green photoreceptors. **(b)** TEs move around the genome. Some TEs copy themselves before moving, while others are excised from their original positions during transposition. Insertion of a TE into a gene often has phenotypic consequences.

Unstable trinucleotide repeats

In 1992, a group of molecular geneticists discovered an unusual and completely unexpected type of mutation in humans: the excessive amplification of a CGG base triplet normally repeated only a few to 50 times in succession. If, for example, a normal allele of a gene carries five consecutive repetitions of the base triplet CGG (i.e., CGGCGGCGGCGGCGG on one strand), an abnormal allele resulting from mutation could carry 200 repeats in a row. Further investigations revealed that repeats of several trinucleotides—CAG, CTG, and GAA, in addition to CGG—can be unstable such that the number of repeats often increases or decreases in different cells of a single individual. Instability can also occur during the production of gametes, resulting in changes in repeat number from one generation to the next. The expansion and contraction of trinucleotide repeats has now been found not only in humans but in many other species as well.

The rules governing trinucleotide repeat instability appear to be quite complicated, but one general feature is that the larger the number of repeats at a particular location, the higher the probability that expansion and contraction will occur. Usually, tracts with less than 30–50 repetitions of a triplet change in size only infrequently, and the mutations that do occur cause only small variations in the repeat number. Larger tracts involving hundreds of repeats change in size more frequently, and they also exhibit more variation in the number of repetitions.

Researchers have not yet determined the precise mechanism of triplet repeat amplification. One possibility is that regions with long trinucleotide repeats form unusual DNA structures that are hard to replicate because they force the copying machinery to slip off, then hop back on, slip off, then hop back on. Such stopping and starting may produce a replication "stutter" that causes synthesis of the same triplet to repeat over and over again, expanding the number of copies. This type of mechanism could conversely shrink the size of the trinucleotide repeat tract if, after slipping off, the replication machinery restarts copying at a repeat farther down the template sequence. Whatever the cause, mutations of long trinucleotide stretches occur quite often, suggesting that the enzymes for excision or mismatch repair are not very efficient at restoring the original number of repeats.

The expansion of trinucleotide repeats is at the root of *fragile X syndrome*, one of the most common forms of human mental retardation, as well as Huntington disease and many other disorders of the nervous system. The **Genetics and Society** box **"Unstable Trinucleotide Repeats and Fragile X Syndrome"** discusses the fascinating medical implications of this phenomenon.

> Many naturally occurring mechanisms can generate spontaneous mutations. These include chemical or radiation assaults that modify DNA bases or break DNA chains, mistakes during DNA replication or recombination, the movement of transposable elements, and the expansion or contraction of unstable trinucleotide repeats.

Mutagens induce mutations

Mutations make genetic analysis possible, but most mutations appear spontaneously at such a low rate that researchers have looked for controlled ways to increase their occurrence. H. J. Muller, an original member of Thomas Hunt Morgan's *Drosophila* group, first showed that exposure to a dose of X-rays higher than the naturally occurring level increases the mutation rate in fruit flies (**Figure 8.9**).

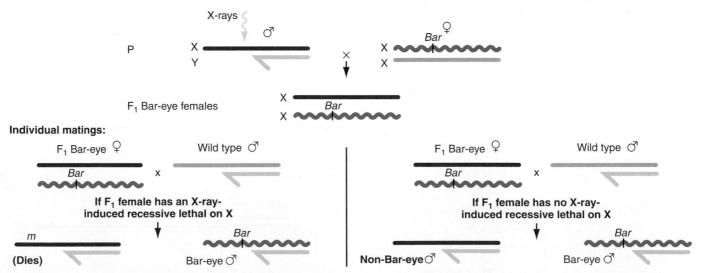

Figure 8.9 Exposure to X-rays increases the mutation rate in *Drosophila*. F₁ females are generated that have an irradiated paternal X chromosome (*red line*), and a *Bar*-marked "balancer" maternal X chromosome (*wavy blue line*). These two chromosomes cannot recombine because the balancer chromosome has multiple inversions (as explained in Chapter 9). Single F₁ females, each with a single X-ray-exposed X chromosome from their father, are then individually mated with wild-type males. If the paternal X chromosome in any one F₁ female has an X-ray-induced recessive lethal mutation (*m*), she can produce only Bar-eyed sons (*left*). If the X chromosome has no such mutation, this F₁ female will produce both Bar-eyed and non-Bar-eyed sons (*right*).

Expansions of the base triplet CGG cause a heritable disorder known as *fragile X syndrome*. Adults affected by this syndrome manifest several physical anomalies, including an unusually large head, long face, large ears, and in men, large testicles. They also exhibit moderate to severe mental retardation. Fragile X syndrome has been found in men and women of all races and ethnic backgrounds. The fragile X mutation is, in fact, a leading genetic cause of mental retardation worldwide, second only to the trisomy 21 that results in Down syndrome.

Specially prepared karyotypes of cells from people with fragile X symptoms reveal a slightly constricted, so-called fragile site near the tip of the long arm of the X chromosome (**Figure A**). The long tracts of CGG trinucleotides, which make up the fragile X mutation, apparently produce a localized constricted region that can even break off in some karyotype preparations. Geneticists named the fragile X disorder for this specific pinpoint of fragility more than 20 years before they identified the mutation that gives rise to it. This cytogenetic test was used to diagnose fragile X syndrome in the past, although it was not always accurate. In the 1990s, identification of the gene that caused the disorder allowed for more accurate testing by Southern blot or PCR analysis (Chapter 15).

The gene in which the fragile X mutation occurs is called *FMR-1* (for fragile-X-associated mental retardation). Near one end of the gene, different people carry a different number of repeats of the sequence CGG, and geneticists now have the molecular tools to quantify these differences. Normal alleles contain 5–54 of these triplet repeats, while the *FMR-1* gene in people with fragile X syndrome contains 200–4000 repeats (**Figure B1**). The rest of the gene's base sequence is the same in both normal and abnormal alleles.

The triplet repeat mutation that underlies fragile X syndrome has a surprising transmission feature. Alleles with a full-blown mutation are foreshadowed by *premutation alleles* that carry an intermediate number of repeats—more than 50 but fewer than 200 (Figure B1). Premutation alleles do not themselves generate fragile X symptoms in most carriers, but they show significant instability and thus forecast the risk of genetic disease in a carrier's progeny. The greater the number of repeats in a premutation allele, the higher the risk of disease in that person's children. For example, if a woman carries a premutation allele with 60 CGG repeats, 17 percent of her offspring run the risk of exhibiting fragile X syndrome. If she carries a premutation allele with 90 repeats, close to 50 percent of her offspring will show symptoms. Interestingly, the expansion of *FMR-1* premutation alleles has some as-yet-unexplained relation to the parental origin of the repeats. Whereas most male carriers transmit their *FMR-1* allele with only a small change in the number of repeats, many women with premutation alleles bear children with 250–4000 CGG repeats in their *FMR-1* gene (**Figure B2**). One possible explanation is that whatever conditions generate fragile X mutations occur most readily during oogenesis.

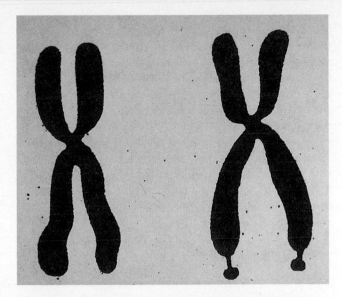

Figure A A karyotype reveals a fragile X chromosome. The fragile X site is seen on the bottom of both chromatids of the X chromosome at the *right*.

1. Effect of (CGG) repeat number

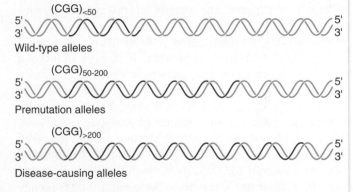

$(CGG)_{<50}$

5'
3'

Wild-type alleles

$(CGG)_{50-200}$

5'
3'

Premutation alleles

$(CGG)_{>200}$

5'
3'

Disease-causing alleles

2. A fragile X pedigree

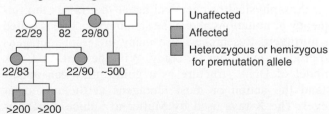

22/29   82   29/80

22/83   22/90   ~500

\>200   >200

☐ Unaffected
■ Affected
■ Heterozygous or hemizygous for premutation allele

Figure B Amplification of CGG triplet repeats correlates with the fragile X syndrome. **1.** *FMR-1* genes in unaffected people generally have fewer than 50 CGG repeats. Unstable premutation alleles have between 50 and 200 repeats. Disease-causing alleles have more than 200 CGG repeats. **2.** A fragile X pedigree showing the number of CGG repeats in each chromosome. Fragile X patients are almost always the progeny of mothers with premutation alleles.

(Continued)

The CGG trinucleotide repeat expansion underlying fragile X syndrome has interesting implications for genetic counselling. Thousands of possible alleles of the *FMR-1* gene exist, ranging from the smallest "normal" allele isolated to-date, with five triplet repeats, to the largest abnormal allele so-far isolated, with roughly 4000 repeats. The relation between genotype and phenotype is clear at both ends of the triplet-repeat spectrum: Individuals whose alleles contain less than 55 repeats are unaffected, while people with an allele carrying more than 200 repeats are almost always moderately to severely retarded. With an intermediate number of repeats, however, expression of the mental retardation phenotype is highly variable, depending to an unknown degree on chance, the environment, and modifier genes.

This range of variable expressivity leads to an ethical dilemma: Where should medical geneticists draw the line in their assessment of risk? Prospective parents with a family history of mental retardation may consult with a counsellor to determine their options. The counsellor would first test the parents for fragile X premutation alleles. If the couple is expecting a child, the counsellor would also want to analyze the fetal cells directly by amniocentesis or CVS, to determine whether the fetus carries an expanded number of CGG repeats in its *FMR-1* gene. If the results indicate the presence of an allele in the middle range of triplet repeats, the counsellor will have to acknowledge the unpredictability of outcomes. The prospective parents' difficult decision of whether or not to continue the pregnancy will then rest on the very shaky ground of an inconclusive, overall evaluation of risk.

Muller exposed male *Drosophila* to increasingly large doses of X-rays and then mated these males with females that had one X chromosome containing an easy-to-recognize dominant mutation causing Bar eyes. This X chromosome (called a *balancer*) also carried chromosomal rearrangements known as inversions that prevented it from crossing-over with other X chromosomes (Chapter 9 explains the details of this phenomenon.) Some of the F_1 daughters of this mating were heterozygotes carrying a mutagenized X from their father and a *Bar*-marked X from their mother. If X-rays induced a recessive lethal mutation anywhere on the paternally derived X chromosome, then these F_1 females would be unable to produce non-Bar-eyed sons. Thus, simply by noting the presence or absence of non-Bar-eyed sons, Muller could establish whether a mutation had occurred in any of the more than 1000 genes on the X chromosome that are essential for *Drosophila* viability. He concluded that the greater the X-ray dose, the greater the frequency of recessive lethal mutations.

Any physical or chemical agent that raises the frequency of mutations above the spontaneous rate is called a **mutagen**. Researchers use many different mutagens to produce mutations for study. With the Watson–Crick model of DNA structure as a guide, they can understand the action of most mutagens at the molecular level. The X-rays used by Muller to induce mutations on the X chromosome, for example, can break the sugar-phosphate backbones of DNA strands, sometimes at the same position on the two strands of the double helix. Multiple double-strand breaks produce DNA fragmentation, and the improper stitching back together of the fragments can cause inversions, deletions, or other rearrangements (see **Figure 8.6c**).

Another molecular mechanism of mutagenesis involves mutagens known as **base analogues**, which are so similar in chemical structure to the normal nitrogenous bases that the replication machinery can incorporate them into DNA (**Figure 8.10a**). Because a base analogue may have pairing properties different from those of the base it replaces, it can cause base substitutions on the complementary strand synthesized in the next round of DNA replication. Other chemical mutagens generate substitutions by directly altering a base's chemical structure and properties (**Figure 8.10b**). Again, the effects of these changes become fixed in the genome when the altered base causes incorporation of an incorrect complementary base during a subsequent round of replication.

Yet another class of chemical mutagens consists of compounds known as **intercalators**: flat, planar molecules that can sandwich themselves between successive base pairs and disrupt the machinery for replication, recombination, or repair (**Figure 8.10c**). The disruption may eventually generate deletions or insertions of a single base pair. Examples of intercalating agents include tobacco smoke and ethidium bromide, the latter a fluorescent compound that is commonly used in molecular biological laboratory procedures such as agarose gel electrophoresis.

> Scientists use mutagens such as X-rays, base analogues, and intercalators to increase the frequency of mutation as an aid to genetic research.

DNA repair mechanisms minimize mutation

Natural environments expose genomes to many kinds of chemicals or radiation that can alter DNA sequences; furthermore, the side effects of normal DNA metabolism within cells, such as inaccuracies in DNA replication or the movement of transposable elements, can also be mutagenic. Cells have evolved a variety of enzymatic systems

Type of Mutagen	Chemical Action of Mutagen
(a) Replace a base: Base analogues have a chemical structure almost identical to that of a DNA base.	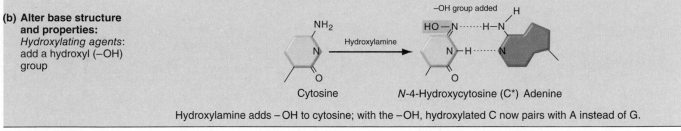 5-Bromouracil–normal state, behaves like thymine Adenine 5-Bromouracil–rare state, behaves like cytosine Guanine 5-Bromouracil: almost identical to thymine. Normally pairs with A; in transient state, pairs with G.

(b) Alter base structure and properties:
Hydroxylating agents: add a hydroxyl (–OH) group

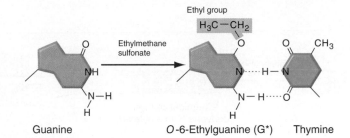

Cytosine *N*-4-Hydroxycytosine (C*) Adenine

Hydroxylamine adds – OH to cytosine; with the –OH, hydroxylated C now pairs with A instead of G.

Alkylating agents: add ethyl (–CH₂–CH₃) or methyl (–CH₃) groups

Guanine *O*-6-Ethylguanine (G*) Thymine

Ethylmethane sulfonate adds an ethyl group to guanine or thymine. Modified G pairs with T above, and modified T pairs with G (not shown).

Deaminating agents: remove amine (–NH₂) groups

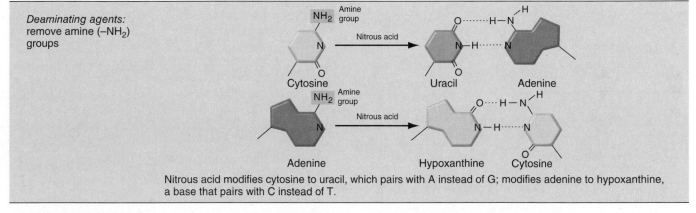

Cytosine Uracil Adenine

Adenine Hypoxanthine Cytosine

Nitrous acid modifies cytosine to uracil, which pairs with A instead of G; modifies adenine to hypoxanthine, a base that pairs with C instead of T.

(c) Insert between bases:
Intercalating agents

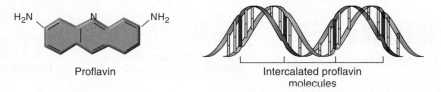

Proflavin Intercalated proflavin molecules

Proflavin intercalates into the double helix. This disrupts DNA metabolism, eventually resulting in deletion or addition of a base pair.

Figure 8.10 How mutagens alter DNA. (a) Base analogues incorporated into DNA may pair aberrantly, allowing the addition of incorrect nucleotides to the opposite strand during replication. **(b)** Some mutagens alter the structure of bases such that they pair inappropriately in the next round of replication. **(c)** Intercalating agents are roughly the same size and shape as a base pair of the double helix. Their incorporation into DNA produces insertions or deletions of single base pairs.

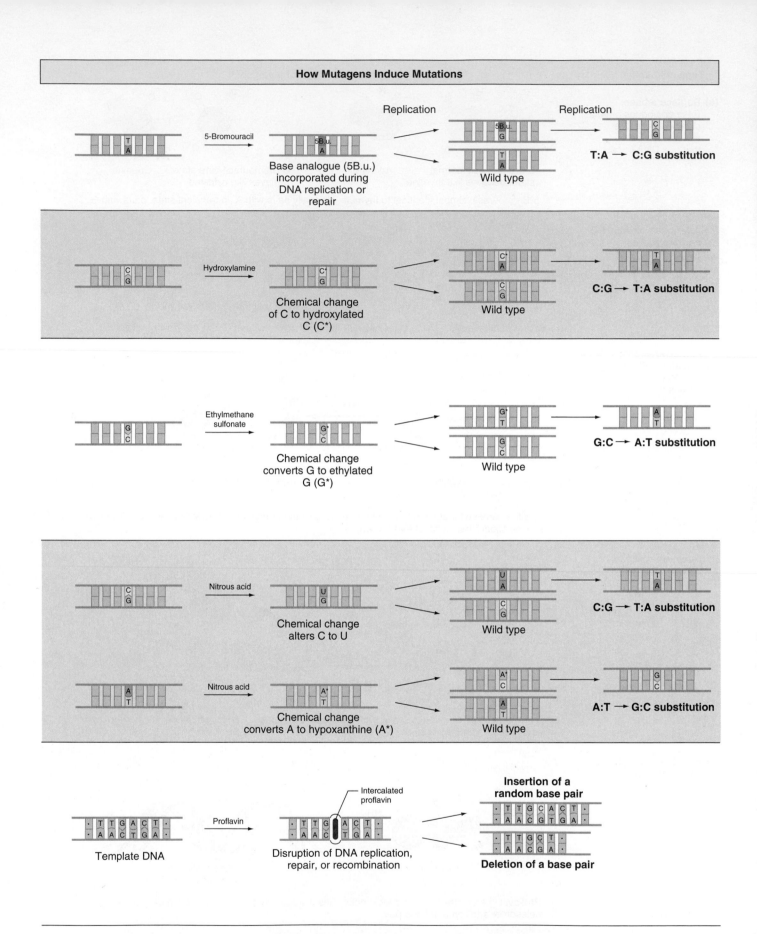

Figure 8.10 How mutagens alter DNA. (*Continued*)

that locate and repair damaged DNA and thereby dramatically diminish the high potential for mutation. The combination of these repair systems must be extremely efficient, because the rates of spontaneous mutation observed for almost all genes are very low.

Reversal of DNA base alterations

If methyl or ethyl groups were mistakenly added to guanine (as in Figure 8.10b), *alkyltransferase* enzymes can remove them so as to re-create the original base. Other enzymes remedy other base structure alterations. For example, the enzyme *photolyase* recognizes the thymine–thymine dimers produced by exposure to ultraviolet light (review Figure 8.6d) and reverses the damage by splitting the chemical linkage between the thymines.

Interestingly, the photolyase enzyme works only in the presence of visible light. In carrying out its DNA repair tasks, it associates with a small molecule called a *chromophore* that absorbs light in the visible range of the spectrum; the enzyme then uses the energy captured by the chromophore to split thymine–thymine dimers. Because it does not function in the dark, the photolyase mechanism is called *light repair*, or *photorepair*.

Removal of damaged bases or nucleotides

Many repair systems use the general strategy of *homology-dependent repair* in which they first remove a small region from the DNA strand that contains the altered nucleotide, and then use the other strand as a template to resynthesize the removed region. This strategy makes use of one of the great advantages of the double-helical structure: If one strand sustains damage, cells can use complementary base-pairing with the undamaged strand to re-create the original sequence.

Base excision repair is one homology-dependent mechanism. In this type of repair, enzymes called *DNA glycosylases* cleave an altered nitrogenous base from the sugar of its nucleotide, releasing the base and creating an apurinic or apyrimidinic (AP) site in the DNA chain (**Figure 8.11**). Different glycosylase enzymes cleave specific damaged bases. Base excision repair is particularly important in the removal of uracil from DNA (recall that uracil often results from the natural deamination of cytosine; review Figure 8.6b). In this repair process, after the enzyme *uracil-DNA glycosylase* has removed uracil from its sugar, leaving an AP site, the enzyme *AP endonuclease* makes a nick in the DNA backbone at the AP site. Other enzymes (known as *DNA exonucleases*) attack the nick and remove nucleotides from its vicinity to create a gap in the previously damaged strand. DNA polymerase fills in the gap by copying the undamaged strand, restoring the original nucleotide in the process. Finally, DNA ligase seals up the backbone of the newly repaired DNA strand.

Nucleotide excision repair (**Figure 8.12**) removes alterations that base excision cannot repair, because the cell lacks a DNA glycosylase that recognizes the problem base. Nucleotide excision repair depends on enzyme complexes containing more than one protein molecule. In *E. coli*, these

1. Deaminated DNA with uracil

2. Glycosylase removes uracil, leaving an AP site. *Uracil released*

3. AP endonuclease cuts backbone to make a nick at the AP site.

4. DNA exonucleases remove nucleotides near the nick, creating a gap.

5. DNA polymerase synthesizes new DNA to fill in the gap.

6. DNA ligase seals the nick.

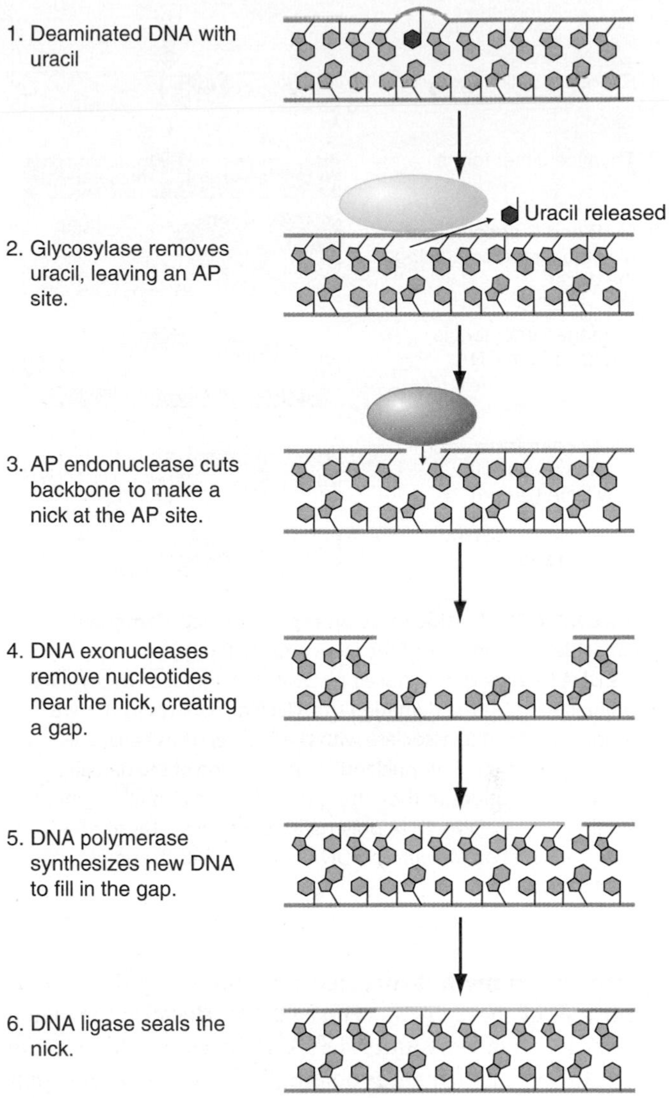

Figure 8.11 Base excision repair removes damaged bases. Glycosylase enzymes remove aberrant bases [like uracil (*red*), formed by the deamination of cytosine], leaving an AP site. AP endonuclease cuts the sugar-phosphate backbone, creating a nick. Exonucleases extend the nick into a gap, which is filled in with the correct information (*green*) by DNA polymerase. DNA ligase reseals the corrected strand.

complexes are made of two out of three possible proteins: UvrA, UvrB, and UvrC. One of the complexes (UvrA + UvrB) patrols the DNA for irregularities, detecting lesions that disrupt Watson–Crick base-pairing and thus distort the double helix (such as thymine–thymine dimers that have not been corrected by photorepair). A second complex (UvrB + UvrC) cuts the damaged strand in two places that flank the damage. This double-cutting excises a short region of the damaged strand and leaves a gap that will be filled in by DNA polymerase and sealed with DNA ligase.

Correction of DNA replication errors

DNA polymerase is remarkably accurate in copying DNA, but the DNA replication system still makes about 100 times more mistakes than most cells can tolerate. A backup repair

1. Exposure to UV light.

2. Thymine dimer forms.

3. UvrB and C endonucleases nick the strand containing the dimer.

4. Damaged fragment is released from DNA.

5. DNA polymerase fills in the gap with new DNA (*green*).

6. DNA ligase seals the repaired strand.

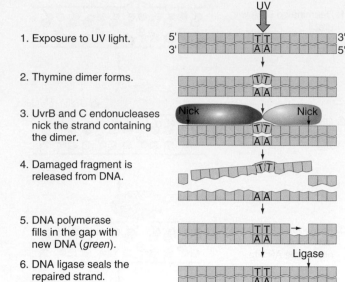

Figure 8.12 Nucleotide excision repair corrects damaged nucleotides. A complex of the UvrA and UvrB proteins (*not shown*) scans DNA for distortions caused by DNA damage, such as thymine–thymine dimers. At the damaged site, UvrA dissociates from UvrB, allowing UvrB (*red*) to associate with UvrC (*blue*). These enzymes nick the DNA exactly four nucleotides to one side of the damage and seven nucleotides to the other side, releasing a small fragment of single-stranded DNA. DNA polymerases then resynthesize the missing information (*green*), and DNA ligase reseals the now-corrected strand.

system called **methyl-directed mismatch repair** corrects almost all of these errors (**Figure 8.13**). Because mismatch repair is active only *after* DNA replication, this system needs to solve a difficult problem. Suppose that a G–C pair has been copied to produce two daughter molecules, one of which has the correct G–C base pair, the other an incorrect G–T. The mismatch repair system can easily recognize the incorrectly matched G–T base pair because the improper base-pairing distorts the double helix, resulting in abnormal bulges and hollows. But how does the system know whether to correct the pair to a G–C or to an A–T?

Bacteria solve this problem by placing a distinguishing mark on the parental DNA strands at specific places: Everywhere the sequence GATC occurs, the enzyme *adenine methylase* puts a methyl group on the A (**Figure 8.13a**). Shortly after replication, the old template strand bears the methyl mark, while the new daughter strand—which contains the wrong nucleotide—is as yet unmarked (**Figure 8.13b**). A pair of proteins in *E. coli*, called MutL and MutS, detect and bind to the mismatched nucleotides. MutL and MutS direct another protein, MutH, to nick the newly synthesized strand of DNA at a position across from the nearest methylated GATC; MutH can discriminate the newly synthesized strand because its GATC is *not* methylated (**Figure 8.13c**). DNA exonucleases then remove all the nucleotides between the nick and a position just beyond the mismatch, leaving a gap on the new, unmethylated strand (**Figure 8.13d**). DNA polymerase can now resynthesize the information using the

(a) Parental strands are marked with methyl groups.

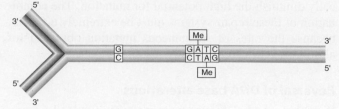

(b) MutS and MutL recognize mismatch in replicated DNA.

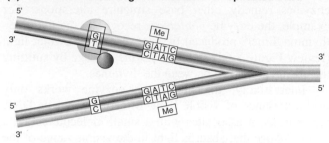

(c) MutL recruits MutH to GATC; MutH makes a nick (*short arrow*) in strand opposite methyl tag.

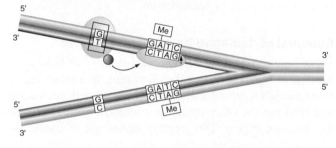

(d) DNA exonucleases (*not shown*) excise DNA from unmethylated new strand.

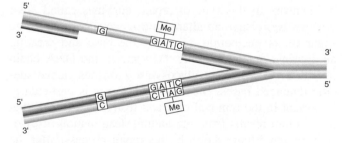

(e) Repair and methylation of newly synthesized DNA strand.

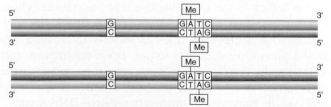

Figure 8.13 In bacteria, methyl-directed mismatch repair corrects mistakes in replication. Parental strands are in *light blue* and newly synthesized strands are *purple*. The MutS protein is *green*, MutL is *dark blue*, and MutH is *yellow*. (See text for details.)

old, methylated strand as a template, and DNA ligase then seals up the repaired strand. With the completion of replication and repair, enzymes mark the new strand with methyl groups so that its parental origin will be evident in the next round of replication (**Figure 8.13e**).

Eukaryotic cells also have a mismatch correction system, but we do not yet know how this system distinguishes templates from newly replicated strands. Unlike prokaryotes, GATCs in eukaryotes are not tagged with methyl groups, and eukaryotes do not seem to have a protein closely related to MutH. One potentially interesting clue is that the MutS and MutL proteins in eukaryotes associate with DNA replication factors; perhaps these interactions might help MutS and MutL identify the strand to be repaired.

> Cells contain many enzymatic systems to repair DNA. The most accurate systems take advantage of complementary base-pairing, using the undamaged strand as a template to correct the damaged DNA strand. Some examples are base or nucleotide excision repair systems, and mismatch repair systems.

Error-prone repair systems: A last resort

The repair systems just described are very accurate in repairing DNA damage because they are able to replace damaged nucleotides with a complementary copy of the undamaged strand. However, cells sometimes become exposed to levels or types of mutagens that they cannot handle with these high-fidelity repair systems. Strong doses of UV light, for example, might make more thymine–thymine dimers than the cell can fix. Any unrepaired damage has severe consequences for cell division: The DNA polymerases normally used in replication will stall at such lesions, so the cells cannot proliferate. Although these cells can initiate emergency responses that may allow them to survive and divide despite the stalling, their ability to proceed in such circumstances comes at the expense of introducing new mutations into the genome.

One type of emergency repair in bacteria, called the **SOS system** (after the Morse code distress signal), relies on error-prone (or "sloppy") DNA polymerases. These sloppy DNA polymerases are not available for normal DNA replication; they are produced only in the presence of DNA damage. The damage-induced, error-prone DNA polymerases are attracted to replication forks that have become stalled at sites of unrepaired, damaged nucleotides. There they add random nucleotides to the strand being synthesized opposite the damaged bases. The SOS polymerase enzymes thus allow the cell with damaged DNA to divide into two daughter cells, but because the sloppy polymerases restore the proper nucleotide only 1/4 of the time, the genomes of these daughter cells carry new mutations. In bacteria, the mutagenic effect of many mutagens either depends on, or is enhanced by, the SOS system.

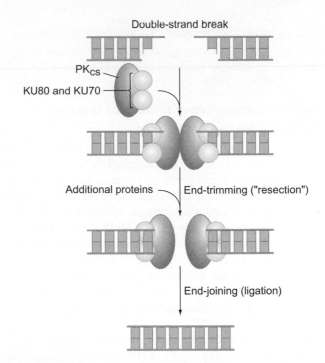

Figure 8.14 Repair of double-strand breaks by nonhomologous end-joining. The proteins KU70, KU80, and PK$_{CS}$ bind to DNA ends and bring them together. Other proteins (*not shown*) trim the ends so as to remove any single-stranded regions, and then ligate the two ends together. This mechanism may result in the deletion of nucleotides and is thus potentially mutagenic.

Another kind of emergency repair system deals with a particularly dangerous kind of DNA lesion: *double-strand breaks*, in which both strands of the double helix are broken at nearby sites (**Figure 8.14**). Recall from Chapter 6 that double-strand breaks occur as the first step in meiotic recombination. We do not consider this type of double-strand break here because the mechanism of recombination repairs them with high fidelity and efficiency using complementary base-pairing (review Figure 6.20). However, double-strand breaks can also result from exposure to high-energy radiation such as X-rays (Figure 8.6c) or highly reactive oxygen molecules. If left unrepaired, these breaks can lead to a variety of potentially lethal chromosome aberrations, such as large deletions, inversions, or translocations.

Cells can restitch the ends formed by such double-strand breaks using a mechanism called **nonhomologous end-joining**, which relies on a group of three proteins that bind to the strand ends and bring them close together (**Figure 8.14**). After binding, these proteins recruit other proteins that cut back (or "resect") any overhanging nucleotides on the ends that do not have a complementary nucleotide to pair with, and then join the two ends together. Because of the resection step, nonhomologous end-joining can result in the loss of DNA and is thus error-prone. Evidently, the mutagenic effects of nonhomologous end-joining are less deleterious to the cell than genomic injuries caused by unrepaired double-strand breaks.

Health consequences of mutations in genes encoding DNA repair proteins

Although differences of detail exist between the DNA repair systems of various organisms, DNA repair mechanisms appear in some form in virtually all species. For example, humans have six proteins with amino acid compositions that are about 25 percent identical with that of the *E. coli* mismatch repair protein MutS. DNA repair systems are thus very old and must have evolved soon after life emerged roughly 3.5 billion years ago. Some scientists believe DNA repair became essential when plants first started to deposit oxygen into the atmosphere, because oxygen favours the formation of free radicals that can damage DNA.

The many known human hereditary diseases associated with the defective repair of DNA damage reveal how crucial these mechanisms are for survival. In one example, the cells of patients with *Xeroderma pigmentosum* lack the ability to conduct nucleotide excision repair; these people are homozygous for mutations in one of seven genes encoding enzymes that normally function in this repair system. As a result, the thymine–thymine dimers caused by ultraviolet light cannot be removed efficiently. Unless these people avoid all exposure to sunlight, their skin cells begin to accumulate mutations that eventually lead to skin cancer (**Figure 8.15**). In another example, researchers have recently learned that hereditary forms of colorectal cancer in humans are associated with mutations in human genes that are closely related to the *E. coli* genes encoding the mismatch repair proteins MutS and MutL. Chapter 16 discusses the fascinating connections between DNA repair and cancer in more detail.

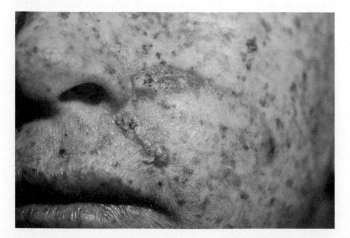

Figure 8.15 Skin lesions in a xeroderma pigmentosum patient. This heritable disease is caused by the lack of a critical enzyme in the nucleotide excision repair system.

Mutations have consequences for species evolution as well as individual survival

"The capacity to blunder slightly is the real marvel of DNA. Without this special attribute, we would still be anaerobic bacteria and there would be no music." In these two sentences, the eminent medical scientist and self-appointed "biology watcher" Lewis Thomas acknowledges that changes in DNA are behind the phenotypic variations that are the raw material on which natural selection has acted for billions of years to drive evolution. The wide-ranging variation in the genetic makeup of the human population—and other populations as well—is, in fact, the result of a balance among (1) the continuous introduction of new mutations; (2) the loss of deleterious mutations because of the selective disadvantage they impose on the individuals that carry them; and (3) the increase in frequency of rare mutations that either provide a selective advantage to the individuals carrying them or that spread through a population by other means.

In sexually reproducing multicellular organisms, only **germline mutations** that can be passed on to the next generation play a role in evolution. Germline mutations are genetic alterations that occur in cells that give rise to gametes (sperm and oocytes); if a mutant gamete contributes to the formation of a zygote, all of the cells of the zygote will carry the mutation. **Somatic mutations**, genetic alterations that occur after conception in non-germline cells, are passed to the progeny of the mutated cells through mitosis; these mutations affect all cells descended from the mutated cell, but cannot be passed on to the next generation. Mutations in somatic cells, nevertheless, can still have an impact on the well-being and survival of individuals. For example, individuals who spend much of their time tanning in the sun, or smoke tobacco (exposure to second-hand smoke included), might experience mutations in their skin cells or lung cells, respectively. The consequences of such mutations are experienced only by those individuals. The skin or lung cells may develop a problem or defect as a result of the mutation, but because the mutation occurred only in a skin or lung cell (a somatic cell), it would not be passed on to succeeding generations. For instance, somatic mutations in genes that help regulate the cell cycle may lead to cancer, but are not transmitted to the next generation. Health Canada tries to identify potential cancer-causing agents (known as carcinogens) by using the **Ames test** to screen for chemicals that cause mutations in bacterial cells (**Figure 8.16**). This test, invented by Bruce Ames, asks whether a particular chemical can induce histidine$^+$ (his^+) revertants of a special histidine$^-$ (his^-) mutant strain of the bacterium *Salmonella typhimurium*. The his^+ revertants can synthesize all the histidine they need from simple compounds in their environment, whereas the original his^- mutants cannot make histidine, so they can survive only if histidine is supplied.

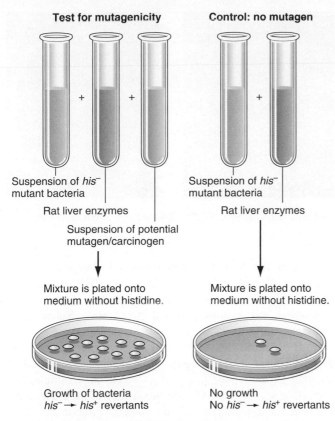

Test for mutagenicity

Control: no mutagen

Suspension of *his⁻* mutant bacteria

Rat liver enzymes

Suspension of potential mutagen/carcinogen

Suspension of *his⁻* mutant bacteria

Rat liver enzymes

Mixture is plated onto medium without histidine.

Mixture is plated onto medium without histidine.

Growth of bacteria *his⁻* → *his⁺* revertants

No growth No *his⁻* → *his⁺* revertants

Figure 8.16 The Ames test identifies potential carcinogens. A compound to be tested is mixed with cells of a *his⁻* strain of *Salmonella typhimurium* and with a solution of rat liver enzymes (which can sometimes convert a harmless compound into a mutagen). Only *his⁺* revertants grow on a Petri plate without histidine. If this plate (*left*) has more *his⁺* revertants than a control plate (also without histidine), containing unexposed cells (*right*), the compound is considered mutagenic and a potential carcinogen. The rare revertants on the control plate represent the spontaneous rate of mutation.

The advantage of the Ames test is that only revertants can grow on Petri plates that do not contain histidine, so it is possible to examine large numbers of cells from an originally *his⁻* culture to find the rare *his⁺* revertants induced by the chemical in question. To increase the sensitivity of mutation detection, the *his⁻* strain used in the Ames test system contains a second mutation that inactivates the nucleotide excision repair system and thereby prevents the ready repair of mutations caused by the potential mutagen, and a third mutation causing defects in the cell wall that allows tested chemicals easier access to the cell interior.

Because most agents that cause mutations in bacteria should also damage the DNA of higher eukaryotic organisms, any mutagen that increases the rate of mutation in bacteria might be expected to cause cancer in people and other mammals. Mammals, however, have complicated metabolic processes capable of inactivating hazardous chemicals. Other biochemical events in mammals can create a mutagenic substance from nonhazardous chemicals. To simulate the action of mammalian metabolism, toxicologists often add a solution of rat liver enzymes to the chemical under analysis by the Ames test (Figure 8.16). Because this simulation is not perfect, Health Canada technicians ultimately assess whether bacterial mutagens identified by the Ames test can cause cancer in rodents by including the agents in test animals' diets.

> Mutations are the ultimate source of variation within and between species. Although some mutations confer a selective advantage, most are deleterious. DNA repair systems help keep mutations to a low level that balances organisms' need to evolve with their need to avoid damage to their genomes.

What Mutations Tell Us About Gene Structure

The science of genetics depends absolutely on mutations because we can track genes in crosses only through the phenotypic effects of their mutant variants. In the 1950s and 1960s, scientists realized they could also use mutations to learn how DNA sequences along a chromosome constitute individual genes. These investigators wanted to collect a large series of mutations in a single gene and analyze how these mutations are arranged with respect to each other. For this approach to be successful, they had to establish that various mutations were, in fact, in the same gene. This was not a trivial exercise, as illustrated by the following situation.

Early *Drosophila* geneticists identified a large number of X-linked recessive mutations affecting the normally red wild-type eye colour (**Figure 8.17**). The first of these to be discovered produced the famous white eyes studied by Morgan's group. Other mutations caused a whole palette of hues to appear in the eyes: darkened shades such as garnet and ruby; bright colours such as vermilion, cherry, and coral; and lighter pigmentations known as apricot, buff, and carnation. This wide variety of eye colour phenotypes posed a puzzle: Were the mutations that caused them multiple alleles of a single gene, or did they affect more than one gene?

Complementation testing reveals whether two mutations are in a single gene or in different genes

Researchers commonly define a gene as a functional unit that directs the appearance of a molecular product that, in turn, contributes to a particular phenotype. They can use this definition to determine whether two mutations are in the same gene or in different genes. If two homologous chromosomes in an individual each carries a mutation recessive to wild type, a wild-type phenotype will

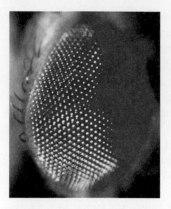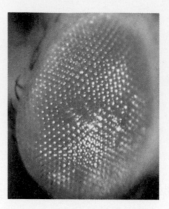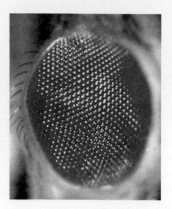

Figure 8.17 *Drosophila* **eye colour mutations produce a variety of phenotypes.** Flies carrying different X-linked eye colour mutations. From the *left*: ruby, white, and apricot; a wild-type eye is at the *far right*.

result if the mutations are in different genes (as described in Chapter 2). The normal phenotype occurs because almost all recessive mutations disrupt a gene's function (as explained in Chapter 7). The dominant wild-type alleles on each of the two homologues can make up for, or **complement**, the defect in the other chromosome by generating enough of both gene products to yield a normal phenotype (**Figure 8.18a**, *left*).

In contrast, if the recessive mutations on the two homologous chromosomes are in the same gene, no wild-type allele of that gene exists in the individual and neither mutated copy of the gene will be able to perform the normal function. As a result, no complementation will occur and no normal gene product will be made, so a mutant phenotype will appear (**Figure 8.18a**, *right*). Ironically, a collection of mutations that do *not* complement each other is known as a **complementation group**. Geneticists often use "complementation group" as a synonym for "gene" because the mutations in a complementation group all affect the same unit of function, and thus, the same gene.

A simple test based on the idea of a gene as a unit of function can determine whether or not two mutations are alleles of the same gene. You simply examine the phenotype of a heterozygous individual in which one homologue of a particular chromosome carries one of the recessive mutations and the other homologue carries the other recessive mutation. If the phenotype is wild type, the mutations cannot be in the same gene. This technique is known as **complementation testing**. For example, because a female *Drosophila* heterozygous for garnet and ruby (*garnet ruby⁺ / garnet⁺ ruby*) has wild-type brick-red eyes, it is possible to conclude that the mutations causing garnet and ruby colours complement each other and are therefore in different genes.

Complementation testing has, in fact, shown that garnet, ruby, vermilion, and carnation pigmentation are governed by separate genes. But chromosomes carrying mutations yielding white, cherry, coral, apricot, and buff phenotypes fail to complement each other. These mutations therefore make up different alleles of a single gene. *Drosophila* geneticists named this gene the *white*, or *w*,

gene after the first mutation observed. The *white* gene encodes a protein that is a member of the ABC transporter superfamily; it functions in the transport of guanine and tryptophan eye colour pigment precursors. Researchers designate the wild-type allele of the gene as w^+ and the various mutations as w^1 (the original white-eyed mutation discovered by T. H. Morgan, often simply designated as *w*), w^{cherry}, w^{coral}, $w^{apricot}$, and w^{buff}. As an example, the eyes of a $w^1 / w^{apricot}$ female are a dilute apricot colour; because the phenotype of this heterozygote is not wild type, the two mutations are allelic. **Figure 8.18b** illustrates how researchers collate data from many complementation tests in a **complementation table**. Such a table helps visualize the relationships among a large group of mutants.

In *Drosophila*, mutations in the *w* gene map very close together in the same region of the X chromosome, while mutations in other eye colour genes lie elsewhere on the chromosome (**Figure 8.18c**). This result suggests that genes are not disjointed entities with parts spread out from one end of a chromosome to another; each gene, in fact, occupies only a relatively small, discrete area of a chromosome. Studies defining genes at the molecular level have shown that most genes consist of 1000–20 000 contiguous base pairs (bp). In humans, among the shortest genes are the roughly 500-bp-long genes that govern the production of histone proteins, while the longest gene so-far identified is the Duchenne muscular dystrophy (*DMD*) gene, which has a length of more than 2 million nucleotide pairs. All known human genes fall somewhere between these extremes. To put these figures in perspective, an average human chromosome is approximately 130 million base pairs in length.

> The complementation test looks at the phenotype of individuals simultaneously heterozygous for two different recessive mutations. A mutant phenotype indicates that the mutations fail to complement each other; that is, they are in the same gene (complementation group). A wild-type phenotype indicates the mutations complement each other, and thus are in different genes.

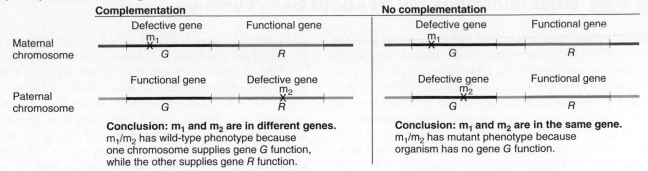

(a) Complementation testing

| Complementation | No complementation |

Maternal chromosome — Defective gene m₁ ... Functional gene

Conclusion: m₁ and m₂ are in different genes.
m₁/m₂ has wild-type phenotype because one chromosome supplies gene *G* function, while the other supplies gene *R* function.

Conclusion: m₁ and m₂ are in the same gene.
m₁/m₂ has mutant phenotype because organism has no gene *G* function.

(b) A complementation table: X-linked eye colour mutations in *Drosophila*

Mutation	white	garnet	ruby	vermilion	cherry	coral	apricot	buff	carnation
white	−	+	+	+	−	−	−	−	+
garnet		−	+	+	+	+	+	+	+
ruby			−	+	+	+	+	+	+
vermilion				−	+	+	+	+	+
cherry					−	−	−	−	+
coral						−	−	−	+
apricot							−	−	+
buff								−	+
carnation									−

(c) Genetic map: X-linked eye colour mutations in *Drosophila*

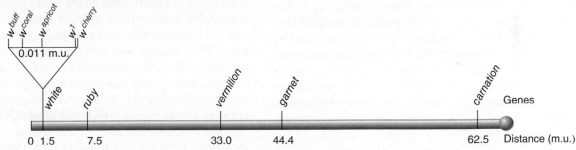

Figure 8.18 Complementation testing of *Drosophila* eye colour mutations. (a) A heterozygote has one mutation (m₁) on one chromosome and a different mutation (m₂) on its homologue. If the mutations are in different genes, the heterozygote will be wild type; the mutations complement each other (*left*). If both mutations affect the same gene, the phenotype will be mutant; the mutations do not complement each other (*right*). Complementation testing makes sense only when both mutations are recessive to wild type. **(b)** This complementation table reveals five complementation groups (five different genes) for eye colour. A "+" indicates mutant combinations with wild-type eye colour; these mutations complement and are thus in different genes. Several mutations fail to complement (−) and are thus alleles of one gene, *white*. **(c)** Recombination mapping shows that mutations in different genes are often far apart, while different mutations in the same gene are very close together.

"Hot spots" of mutation

Some sites within a gene spontaneously mutate more frequently than others and as a result are known as **hot spots**. The existence of hot spots suggests that certain nucleotides can be altered more readily than others. Treatment with mutagens also turns up hot spots, but because mutagens have specificities for particular nucleotides, the highly mutable sites that turn up with various mutagens are often at different positions in a gene than the hot spots resulting from spontaneous mutation.

Nucleotides are chemically the same whether they lie within a gene or in the DNA between genes, and the molecular machinery responsible for mutation and recombination does not discriminate between those nucleotides that are *intragenic* (within a gene) and those that are *intergenic* (between genes). The main distinction between DNA within and DNA outside a gene is that the array of nucleotides composing a gene has evolved a function that determines phenotype. Next, we describe how geneticists discovered what that function is.

> The mechanisms governing mutation and recombination do not discriminate between nucleotide pairs within or outside of genes; however, the nucleotide pairs within a gene together comprise a unit of function that contributes to phenotype.

8.3   What Mutations Tell Us About Gene Function

Mendel's experiments established that an individual gene can control a visible characteristic, but his laws do not explain how genes actually govern the appearance of traits. Investigators working in the first half of the twentieth century carefully studied the biochemical changes caused by mutations in an effort to understand the genotype–phenotype connection.

In one of the first of these studies, conducted in 1902, the British physician Dr. Archibald Garrod showed that a human genetic disorder known as *alkaptonuria* is determined by the recessive allele of an autosomal gene. Garrod analyzed inheritance patterns and performed biochemical analyses on family members with and without the trait. The urine of people with alkaptonuria turns black upon exposure to air. Garrod found that a substance known as homogentisic acid, which blackens upon contact with oxygen, accumulates in the urine of alkaptonuria patients. Alkaptonuriacs excrete all of the homogentisic acid they ingest, while people without the condition excrete no homogentisic acid in their urine even after ingesting the substance.

From these observations, Garrod concluded that people with alkaptonuria are incapable of metabolizing homogentisic acid to the breakdown products generated by normal individuals (**Figure 8.19**). Because many biochemical reactions within the cells of organisms are catalyzed by enzymes, Garrod hypothesized that lack of the enzyme that breaks down homogentisic acid is the cause of alkaptonuria. In the absence of this enzyme, the acid accumulates

and causes the urine to turn black on contact with oxygen. He called this condition an "inborn error of metabolism."

Garrod studied several other inborn errors of metabolism and suggested that all arose from mutations that prevented a particular gene from producing an enzyme required for a specific biochemical reaction. In today's terminology, the wild-type allele of the gene would allow production of functional enzyme (in the case of alkaptonuria, the enzyme is homogentisic acid oxidase), whereas the mutant allele would not. Because the single wild-type allele in heterozygotes generates sufficient enzyme to prevent accumulation of homogentisic acid and thus the condition of alkaptonuria, the mutant allele is recessive.

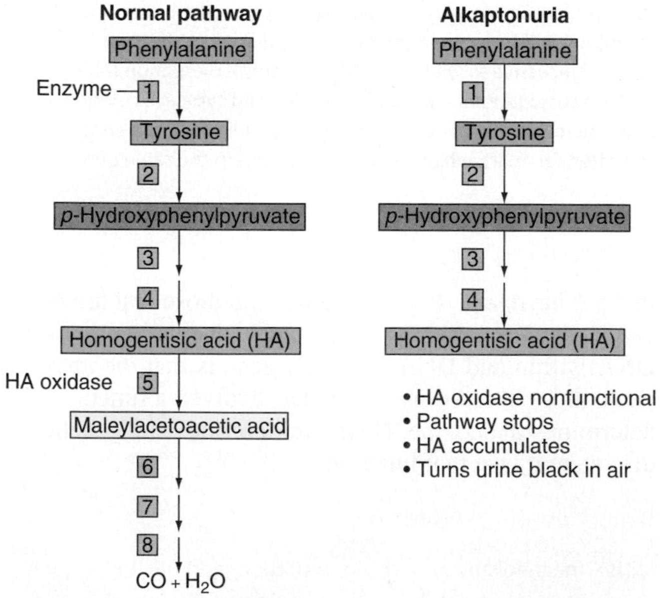

A gene contains the information for producing a specific enzyme: The one gene, one enzyme hypothesis

In the 1940s, George Beadle and Edward Tatum carried out a series of experiments on the bread mould *Neurospora crassa* that demonstrated a direct relation between genes and the enzymes that catalyze specific biochemical reactions. Their strategy was simple. They first isolated a number of mutations that disrupted synthesis of the amino acid arginine, a compound needed for *Neurospora* growth. They next hypothesized that different mutations blocked different steps in a particular **biochemical pathway**: the orderly series of reactions that allows *Neurospora* to obtain simple molecules from the environment and convert them step-by-step into successively more complicated molecules culminating in arginine.

Experimental evidence for "one gene, one enzyme"

Figure 8.20a illustrates the experiments Beadle and Tatum performed to test their hypothesis. They first obtained a set of mutagen-induced mutations that prevented *Neurospora* from synthesizing arginine. Cells with any one of these mutations were unable to make arginine and could therefore grow on a minimal medium containing salt and sugar only if it had been supplemented with arginine. A nutritional mutant microorganism that requires supplementation with substances not needed by wild-type strains is known as an **auxotroph**. The cells just mentioned were arginine auxotrophs. (In contrast, a cell that does not require addition of a substance is a **prototroph** for that factor. In a more general meaning, *prototroph* refers to a wild-type cell that can grow on minimal medium alone.) Recombination analyses located the auxotrophic arginine-blocking mutations in four distinct regions of the genome, and complementation tests showed that each of the four regions correlated with a different complementation group. On the basis of these results, Beadle and Tatum concluded that at least four genes support the biochemical pathway

Figure 8.19 Alkaptonuria: An inborn error of metabolism. The biochemical pathway in humans that degrades phenylalanine and tyrosine via homogentisic acid (HA). In alkaptonuria patients, the enzyme HA hydroxylase is not functional so it does not catalyze the conversion of HA to maleylacetoacetic acid. As a result, HA, which oxidizes to a black compound, accumulates in the urine.

(a) Isolation of arginine auxotrophs

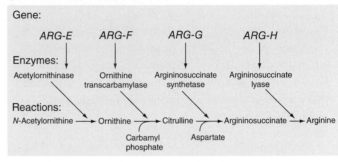

1. X-rays → Wild type → Mutagenized conidia → Crossed with opposite wild type → Fruiting bodies (Asci) → Ascospores dissected and transferred; one to each culture tube.

2. Tubes of complete medium inoculated with single ascospores. → Complete medium
   Germination, production of conidia

3. Conidia from each culture tested on minimal medium. → Minimal medium
   No growth = nutritional mutant

4. Conidia from cultures that fail to grow on minimal medium are tested on minimal medium supplemented with individual amino acids.

Glycine, Leucine, Arginine, Valine, Tyrosine, Proline, Glutamic acid, Asparagine, Serine, Cysteine

Addition of arginine restores growth, reveals arginine auxotroph.

(b) Growth response if nutrient is added to minimal medium

Mutant strain	Supplements				
	Nothing	Ornithine	Citrulline	Arginino-succinate	Arginine
Wild type: Arg^+	+	+	+	+	+
Arg-E^-	−	+	+	+	+
Arg-F^-	−	−	+	+	+
Arg-G^-	−	−	−	+	+
Arg-H^-	−	−	−	−	+

(c) Inferred biochemical pathway

Gene: ARG-E, ARG-F, ARG-G, ARG-H

Enzymes: Acetylornithinase, Ornithine transcarbamylase, Argininosuccinate synthetase, Argininosuccinate lyase

Reactions: N-Acetylornithine → Ornithine → Citrulline → Argininosuccinate → Arginine
(Carbamyl phosphate, Aspartate)

Figure 8.20 Experimental support for the "one gene, one enzyme" hypothesis. (a) Beadle and Tatum mated an X-ray-mutagenized strain of *Neurospora* with another strain, and they isolated haploid ascospores that grew on complete medium. Cultures that failed to grow on minimal medium were nutritional mutants. Nutritional mutants that could grow on minimal medium plus arginine were *arg⁻* auxotrophs. **(b)** The ability of wild-type and mutant strains to grow on minimal medium supplemented with intermediates in the arginine pathway. **(c)** Each of the four *ARG* genes encodes an enzyme needed to convert one intermediate to the next in the pathway.

for arginine synthesis. They named the four genes *ARG-E*, *ARG-F*, *ARG-G*, and *ARG-H*.

They next asked whether any of the mutant *Neurospora* strains could grow in minimal medium supplemented with any of three known intermediates (ornithine, citrulline, and arginosuccinate) in the biochemical pathway leading to arginine, instead of with arginine itself. This test would identify *Neurospora* mutants able to convert the intermediate compound into arginine. Beadle and Tatum compiled a table describing which arginine auxotrophic mutants were able to grow on minimal medium supplemented with each of the intermediates (**Figure 8.20b**).

Interpretation of results: Genes encode enzymes

On the basis of these results, Beadle and Tatum, along with their collaborator Norman Horowitz, proposed a model of how *Neurospora* cells synthesize arginine (**Figure 8.20c**). In the linear progression of biochemical reactions by which a cell constructs arginine from the constituents of minimal medium, each intermediate is both the product of one step and the substrate for the next. Each reaction in the precisely ordered sequence is catalyzed by a specific enzyme, and the presence of each enzyme depends on one of the four *ARG* genes. A mutation in one gene blocks the pathway at a particular step because the cell lacks the corresponding enzyme and thus cannot make arginine on its

own. Supplementing the medium with any intermediate that occurs beyond the blocked reaction restores growth because the organism has all the enzymes required to convert the intermediate to arginine. Supplementation with an intermediate that occurs before the missing enzyme does not work because the cell is unable to convert the intermediate into arginine.

Each mutation abolishes the cell's ability to make an enzyme capable of catalyzing a certain reaction. By inference, then, each gene controls the synthesis or activity of an enzyme, or as stated by Norman Horowitz, one gene, one enzyme. Of course, the gene and the enzyme are not the same thing; rather, the sequence of nucleotides in a gene contains information that somehow encodes the structure of an enzyme molecule.

Although the analysis of the arginine pathway studied by Beadle and Tatum was straightforward, studies of biochemical pathways are not always so easy to interpret. Some biochemical pathways are not linear progressions of stepwise reactions. For example, a branching pathway occurs if different enzymes act on the same intermediate to convert it into two different end-products. If the cell requires both of these end-products for growth, a mutation in a gene encoding any of the enzymes required to synthesize the intermediate would make the cell dependent on supplementation with both end-products. A second

possibility is that a cell might employ either of two independent, parallel pathways to synthesize a needed end-product. In such a case, a mutation in a gene encoding an enzyme in one of the pathways would be without effect. Only a cell with mutations affecting both pathways would display an aberrant phenotype.

Even with nonlinear progressions such as these, careful genetic analysis can reveal the nature of the biochemical pathway on the basis of Beadle and Tatum's insight that genes encode proteins.

> Beadle and Tatum found that mutations in a single complementation group (i.e., a single gene) disrupted one particular enzymatic step of a known biochemical pathway, while mutations in other genes disrupted other steps. They concluded that each gene specifies a different enzyme ("one gene, one enzyme").

Genes specify the identity and order of amino acids in polypeptide chains

Although the one gene, one enzyme hypothesis was a critical advance in understanding how genes influence phenotype, it is an oversimplification. Not all genes govern the construction of enzymes active in biochemical pathways. Enzymes are only one class of the molecules known as proteins, and cells contain many other kinds of proteins. Among the other types are proteins that provide shape and rigidity to a cell, proteins that transport molecules in and out of cells, proteins that help fold DNA into chromosomes, and proteins that act as hormonal messengers. Genes direct the synthesis of all proteins, enzymes and nonenzymes alike. Moreover, as we see next, genes actually determine the construction of polypeptides, and because some proteins are composed of more than one type of polypeptide, more than one gene determines the construction of such proteins.

Proteins: Linear polymers of amino acids linked by peptide bonds

To review the basics, proteins are polymers composed of building blocks known as **amino acids**. Cells use mainly 20 different amino acids to synthesize the proteins they need. All of these amino acids have certain basic features, encapsulated by the formula $NH_2-CHR-COOH$ (**Figure 8.21a**). The $-COOH$ component, also known as *carboxylic acid*, is, as the name implies, acidic; the $-NH_2$ component, also known as an *amino group*, is basic. The R refers to side chains that distinguish each of the 20 amino acids (**Figure 8.21b**). An R group can be as simple as a hydrogen atom (in the amino acid glycine) or as complex as a benzene ring (in phenylalanine). Some side chains are relatively neutral and nonreactive, others are acidic, and still others are basic.

During protein synthesis, a cell's protein-building machinery links amino acids by constructing covalent **peptide bonds** that join the –COOH group of one amino acid to the $-NH_2$ group of the next (**Figure 8.21c**). A pair of amino acids connected in this fashion is a **dipeptide**; several amino acids linked together constitute an **oligopeptide**. The amino acid chains that make up proteins contain hundreds to thousands of amino acids joined by peptide bonds and are known as **polypeptides**. Proteins are thus linear polymers of amino acids. Like the chains of nucleotides in DNA, polypeptides have a chemical polarity. One end of a polypeptide is called the **N terminus** because it contains a free amino group that is not connected to any other amino acid. The other end of the polypeptide chain is the **C terminus**, because it contains a free carboxylic acid group.

Mutations can alter amino acid sequences

Each protein is composed of a unique sequence of amino acids. The chemical properties that enable structural proteins to give a cell its shape, or enzymes to catalyze specific reactions are a direct consequence of the identity, number, and linear order of amino acids in the protein.

If genes encode proteins, then at least some mutations could be changes in a gene that alter the proper sequence of amino acids in the protein encoded by that gene. In the mid-1950s, Vernon Ingram began to establish what kinds of changes particular mutations cause in the corresponding protein. Using recently developed techniques for determining the sequence of amino acids in a protein, he compared the amino acid sequence of the normal adult form of haemoglobin (HbA) with that of haemoglobin in the bloodstream of people homozygous for the mutation that causes sickle-cell anaemia (HbS). Remarkably, he found only a single amino acid difference between the wild-type and mutant proteins (**Figure 8.22a**). Haemoglobin consists of two types of polypeptides: a so-called α (alpha) chain and a β (beta) chain. The sixth amino acid from the N terminus of the β chain is glutamic acid in normal individuals but valine in sickle-cell patients.

Ingram thus established that a mutation substituting one amino acid for another had the power to change the structure and function of haemoglobin and thereby alter the phenotype from normal to sickle-cell anaemia (**Figure 8.22b**). We now know that the glutamic acid–to-valine change affects the solubility of haemoglobin within the red blood cell. At low concentrations of oxygen, the less-soluble sickle-cell form of haemoglobin aggregates into long chains that deform the red blood cell (Figure 8.22a).

Because people suffering from a variety of inherited anaemias also have defective haemoglobin molecules, Ingram and other geneticists were able to determine how a large number of different mutations affect the amino acid sequence of haemoglobin (**Figure 8.22c**). Most of the altered haemoglobins have a change in only one amino acid. In various patients with anaemia, the alteration is generally in different amino acids, but occasionally, two independent mutations result in different substitutions for the same amino acid. Geneticists use the term *missense mutation* (which you learned about in Chapter 7) to describe a genetic alteration that causes the substitution of one amino acid for another.

(a) Generic amino acid structure

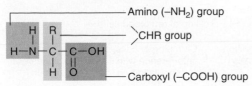

Amino (–NH₂) group

CHR group

Carboxyl (–COOH) group

(c) Peptide bond formation

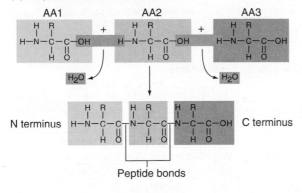

N terminus

C terminus

Peptide bonds

(b) Amino acids with nonpolar R groups

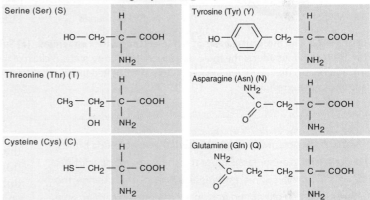

Amino acids with uncharged polar R groups

Amino acids with basic R groups

Amino acids with acidic R groups

Figure 8.21 Proteins are chains of amino acids linked by peptide bonds. (a) Amino acids contain a basic amino group (–NH₂), an acidic carboxylic acid group (–COOH), and a 〉CHR moiety, where R stands for one of the 20 different side chains. **(b)** Amino acids commonly found in proteins, arranged according to the properties of their R groups. **(c)** One molecule of water is lost when a covalent amide linkage (a peptide bond) is formed between the –COOH of one amino acid and the –NH₂ of the next amino acid. Polypeptides such as the tripeptide shown here have polarity; they extend from an N terminus (with a free amino group) to a C terminus (with a free carboxylic acid group).

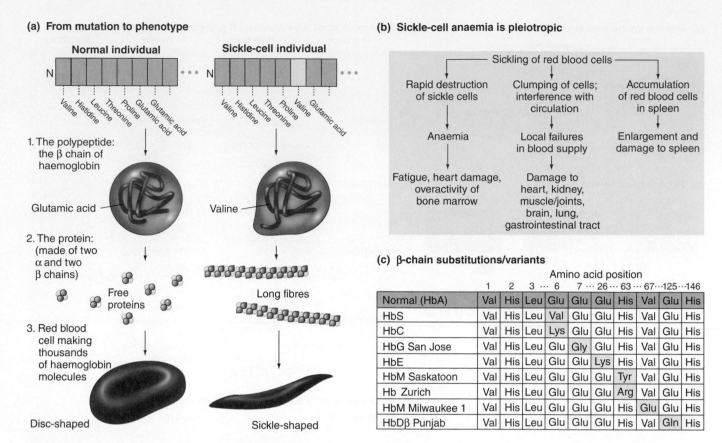

(a) From mutation to phenotype

Normal individual — Sickle-cell individual

N ... | N ...

Valine, Histidine, Leucine, Threonine, Proline, Glutamic acid, Glutamic acid

Valine, Histidine, Leucine, Threonine, Proline, Valine, Glutamic acid

1. The polypeptide: the β chain of haemoglobin

Glutamic acid — Valine

2. The protein: (made of two α and two β chains)

Free proteins — Long fibres

3. Red blood cell making thousands of haemoglobin molecules

Disc-shaped — Sickle-shaped

(b) Sickle-cell anaemia is pleiotropic

Sickling of red blood cells

| Rapid destruction of sickle cells | Clumping of cells; interference with circulation | Accumulation of red blood cells in spleen |

| Anaemia | Local failures in blood supply | Enlargement and damage to spleen |

| Fatigue, heart damage, overactivity of bone marrow | Damage to heart, kidney, muscle/joints, brain, lung, gastrointestinal tract | |

(c) β-chain substitutions/variants

Amino acid position

	1	2	3 ···	6	7 ···	26 ···	63 ···	67 ···	125 ···	146
Normal (HbA)	Val	His	Leu	Glu	Glu	Glu	His	Val	Glu	His
HbS	Val	His	Leu	Val	Glu	Glu	His	Val	Glu	His
HbC	Val	His	Leu	Lys	Glu	Glu	His	Val	Glu	His
HbG San Jose	Val	His	Leu	Glu	Gly	Glu	His	Val	Glu	His
HbE	Val	His	Leu	Glu	Glu	Lys	His	Val	Glu	His
HbM Saskatoon	Val	His	Leu	Glu	Glu	Glu	Tyr	Val	Glu	His
Hb Zurich	Val	His	Leu	Glu	Glu	Glu	Arg	Val	Glu	His
HbM Milwaukee 1	Val	His	Leu	Glu	Glu	Glu	His	Glu	Glu	His
HbDβ Punjab	Val	His	Leu	Glu	Glu	Glu	His	Val	Gln	His

Figure 8.22 The molecular basis of sickle-cell and other anaemias. (a) Substitution of glutamic acid with valine at the sixth amino acid from the N terminus affects the three-dimensional structure of the β chain of haemoglobin. Haemoglobins incorporating the mutant β chain form aggregates that cause red blood cells to sickle. **(b)** Red blood cell sickling has many phenotypic effects. **(c)** Other mutations in the β-chain gene also cause anaemias.

> Proteins are polymers of amino acids linked by peptide bonds; protein chains are polar because they have chemically distinct N and C termini. Some mutations in genes can change the identity of a single amino acid in a protein; such amino acid substitutions can disrupt the protein's function.

Primary, secondary, and tertiary protein structures

Despite the uniform nature of protein construction—a line of amino acids joined by peptide bonds—each type of polypeptide folds into a unique three-dimensional shape. The linear sequence of amino acids within a polypeptide is its **primary structure**. Each unique primary structure places constraints on how a chain can arrange itself in three-dimensional space. Because the R groups distinguishing the 20 amino acids have dissimilar chemical properties, some amino acids form hydrogen bonds or electrostatic bonds when brought into proximity with other amino acids. Nonpolar amino acids, for example, may become associated with each other by interactions that "hide" them from water in localized hydrophobic regions. As another example, two cysteine amino acids can form covalent disulfide bridges (−S−S−) through the oxidation of their −SH groups.

All of these interactions (**Figure 8.23a**) help stabilize the polypeptide in a specific three-dimensional conformation. The primary structure (**Figure 8.23b**) determines three-dimensional shape by generating localized regions with a characteristic geometry known as **secondary structure** (**Figure 8.23c**). Primary structure is also responsible for other folds and twists that together with the secondary structure produce the ultimate three-dimensional **tertiary structure** of the entire polypeptide (**Figure 8.23d**). Normal tertiary structure—the way a long chain of amino acids naturally folds in three-dimensional space under physiological conditions—is known as a polypeptide's *native configuration*. Various forces, including hydrogen bonds, electrostatic bonds, hydrophobic interactions, and disulfide bridges, help stabilize the native configuration.

It is worth repeating that primary structure—the sequence of amino acids in a polypeptide—directly determines secondary and tertiary structures. The information required for the chain to fold into its native configuration is inherent in its linear sequence of amino acids. In one example of this principle, many proteins unfold, or become **denatured**, when exposed to urea and mercaptoethanol or to increasing heat or pH. These treatments disrupt the interactions that normally stabilize the secondary and tertiary structures. When conditions return to normal, many proteins spontaneously refold into their native configuration

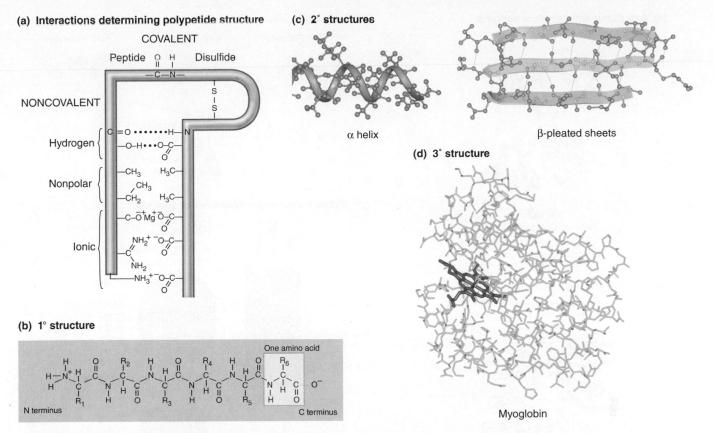

(a) Interactions determining polypeptide structure

COVALENT

Peptide   Disulfide

NONCOVALENT

Hydrogen

Nonpolar

Ionic

(c) 2° structures

α helix

β-pleated sheets

(d) 3° structure

Myoglobin

(b) 1° structure

One amino acid

N terminus

C terminus

Figure 8.23 Levels of polypeptide structure. (a) Covalent and noncovalent interactions determine the structure of a polypeptide. **(b)** A polypeptide's primary (1°) structure is its amino acid sequence. **(c)** Localized regions form secondary (2°) structures such as α helices and β-pleated sheets. **(d)** The tertiary (3°) structure is the complete three-dimensional arrangement of a polypeptide. In this portrait of myoglobin, the iron-containing haem group, which carries oxygen, is *red*, while the polypeptide itself is *green*.

without help from other agents. No other information beyond the primary structure is needed to achieve the proper three-dimensional shape of such proteins.

Quaternary structure: Multimeric proteins

Certain proteins, such as the rhodopsin that promotes black-and-white vision, consist of a single polypeptide. Many others, however, such as the lens crystallin protein, which provides rigidity and transparency to the lenses of our eyes, or the haemoglobin molecule described earlier, are composed of two or more polypeptide chains that associate in a specific way (**Figure 8.24a** and **b**). The individual polypeptides in an aggregate are known as *subunits*, and the complex of subunits is often referred to as a *multimer*. The three-dimensional configuration of subunits in a multimer is a complex protein's **quaternary structure**.

The same forces that stabilize the native form of a polypeptide (i.e., hydrogen bonds, electrostatic bonds, hydrophobic interactions, and disulfide bridges) also contribute to the maintenance of quaternary structure. As Figure 8.24a shows, in some multimers, the two or more interacting subunits are identical polypeptides. These identical chains are encoded by one gene. In other multimers, by contrast, more than one kind of polypeptide makes up the protein (Figure 8.24b). The different polypeptides in these multimers are encoded by different genes.

Alterations in just one kind of subunit, caused by a mutation in a single gene, can affect the function of a multimer. The adult haemoglobin molecule, for example, consists of two α and two β subunits, with each type of subunit determined by a different gene—one for the α chain and one for the β chain. A mutation in the *Hbβ* gene resulting in an amino acid switch at position 6 in the β chain causes sickle-cell anaemia. Similarly, if several multimeric proteins share a common subunit, a single mutation in the gene encoding that subunit may affect all the proteins simultaneously. An example is an X-linked mutation in mice and humans that incapacitates several different proteins all known as interleukin (IL) receptors. Because all of these receptors are essential to the normal function of immune-system cells that fight infection and generate immunity, this one mutation causes the life-threatening condition known as X-linked severe combined immune deficiency (XSCID; **Figure 8.24c**).

The polypeptides of complex proteins can assemble into extremely large structures capable of changing with the needs of the cell. For example, the microtubules that make up the spindle during mitosis are gigantic assemblages of mainly two polypeptides: α tubulin and β tubulin (**Figure 8.24d**). The cell can organize these subunits into very long hollow tubes that grow or shrink as needed at different stages of the cell cycle.

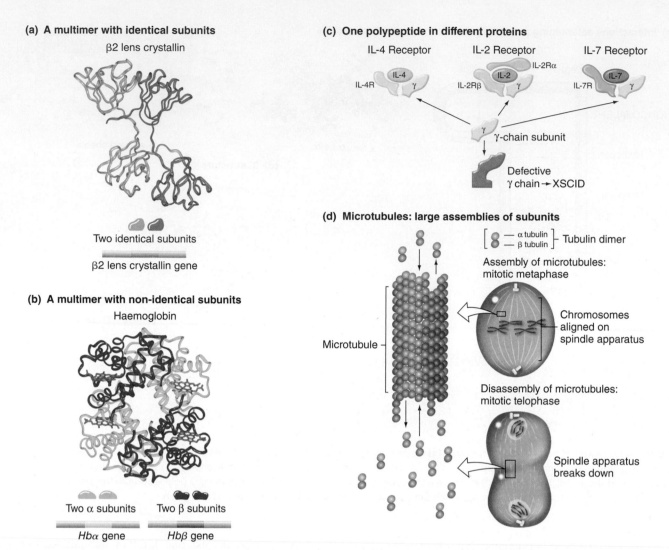

(a) **A multimer with identical subunits**

β2 lens crystallin

Two identical subunits

β2 lens crystallin gene

(b) **A multimer with non-identical subunits**

Haemoglobin

Two α subunits   Two β subunits

Hbα gene   *Hbβ* gene

(c) **One polypeptide in different proteins**

IL-4 Receptor   IL-2 Receptor   IL-7 Receptor

IL-4   IL-2   IL-2Rα   IL-7

IL-4R   γ   IL-2Rβ   γ   IL-7R   γ

γ   γ-chain subunit

Defective
γ chain → XSCID

(d) **Microtubules: large assemblies of subunits**

— α tubulin
— β tubulin   Tubulin dimer

Assembly of microtubules:
mitotic metaphase

Microtubule

Chromosomes
aligned on
spindle apparatus

Disassembly of microtubules:
mitotic telophase

Spindle apparatus
breaks down

Figure 8.24 Multimeric proteins. (a) β2 lens crystallin contains two copies of one kind of subunit; the two subunits are the product of a single gene. The peptide backbones of the two subunits are shown in different shades of *purple*. **(b)** Haemoglobin is composed of two different kinds of subunits, each encoded by a different gene. **(c)** Three distinct protein receptors for the immune-system molecules called interleukins (ILs; *purple*). All contain a common gamma (γ) chain (*yellow*), plus other receptor-specific polypeptides (*green*). A mutant γ chain blocks the function of all three receptors, leading to XSCID. **(d)** One α-tubulin and one β-tubulin polypeptide associate to form a tubulin dimer. Many tubulin dimers form a single microtubule. The mitotic spindle is an assembly of many microtubules.

One gene, one polypeptide

Because more than one gene governs the production of some multimeric proteins, and because not all proteins are enzymes, the "one gene, one enzyme" hypothesis is not broad enough to define gene function. A more accurate statement is "one gene, one polypeptide": Each gene governs the construction of a particular polypeptide. As seen in Chapter 7, even this reformulation does not encompass the function of all genes, as some genes undergo alternative splicing, and a few genes in all organisms do not determine the construction of proteins; instead, they encode RNAs that are not translated into polypeptides.

Beadle and Tatum's experiments were based on the concept that if each gene encodes a different polypeptide and if each polypeptide plays a specific role in the development, physiology, or behaviour of an organism, then a mutation in the gene will block a biological process

(like arginine synthesis in *Neurospora*) in a characteristic way. Other scientists soon realized they could use this approach to study virtually any interesting problem in biology. In the **Fast Forward** box **"Using Mutagenesis to Look at Biological Processes,"** we describe how one biologist found a large group of mutations that disrupted the assembly of bacteriophage T4 particles. By carefully studying the phenotypes caused by these mutations, he inferred the complex pathway that produces an entire bacteriophage. In the **Tools of Genetics** box **"Site-directed Mutagenesis,"** an innovative genetic engineering technique developed and refined by Canadian researcher and Nobel Laureate Dr. Michael Smith, and its relevance to the scientific and medical community, is discussed.

Knowledge about the connection between genes and polypeptides enabled geneticists to analyze how different mutations in a single gene can produce different phenotypes. If each amino acid has a specific effect on the

Geneticists can use mutations to dissect complicated biological processes into their protein components. To determine the specific, dedicated role of each protein, they introduce mutations into the genes encoding the protein. The mutations knock out, or delete, functional protein either by preventing protein production altogether or by altering it such that the resulting protein is nonfunctional. The researchers then observe what happens when the cell or organism attempts to perform the biological process without the deleted protein.

In the 1960s, Robert Edgar set out to delineate the function of the proteins determined by all the genes in the T4 bacteriophage genome. After a single viral particle infects an *E. coli* bacterium, the host cell stops producing bacterial proteins and becomes a factory for making only viral proteins. Thirty minutes after infection, the bacterial cell lyses, releasing 100 new viral particles. The head of each particle carries a DNA genome 200 000 base pairs in length that encodes at least 120 genes.

Edgar's experimental design was to obtain many different mutant bacteriophages, each containing a mutation that inactivates one of the genes essential for viral reproduction. By analyzing what went wrong with each type of mutant during the infective cycle, he would learn something about the function of each of the proteins produced by the T4 genome.

There was just one barrier to implementing this plan. A mutation that prevents viral reproduction by definition makes the virus unable to reproduce and therefore unavailable for experimental study. The solution to this dilemma came with the discovery of *conditional lethal mutants*: viruses, microbes, or other organisms carrying mutations that are lethal to the organism under one condition but not another. One type of conditional lethal mutant used by Edgar was temperature sensitive; that is, the mutant T4 phage could reproduce only at low temperatures. The mutations causing temperature sensitivity changed one amino acid in a polypeptide such that the protein was stable and functional at a low temperature but became unstable and nonfunctional at a higher temperature. Temperature-sensitive mutations can occur in almost any gene. Edgar isolated thousands of conditional lethal bacteriophage T4 mutants, and using complementation studies, he discovered that they fall into 65 complementation groups. These complementation groups defined 65 genes whose function is required for bacteriophage replication.

Edgar next studied the consequences of infecting bacterial cells under *restrictive conditions*; that is, under conditions in which the mutant protein could not function. For the temperature-sensitive mutants, the restrictive condition was high temperature. He found that mutations in 17 genes prevented viral DNA replication and concluded that these 17 genes contribute to that process. Mutations in most of the other 48 genes did not impede viral DNA replication but were necessary for the construction of complete viral particles.

Electron microscopy showed that mutations in these 48 genes caused the accumulation of partially constructed viral particles. Edgar used the incomplete particles to plot the path of viral assembly. As **Figure A** illustrates, three subassembly lines—one for the tail, one for the head, and one for the tail fibres—come together during the assembly of the viral

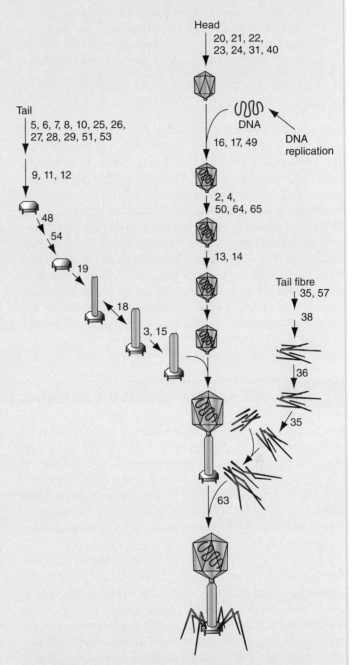

Figure A Steps in the assembly of bacteriophage T4. Robert Edgar determined what kinds of phage structures formed in bacterial cells infected with mutant T4 phage at restrictive temperatures. As an example, a cell infected with a phage carrying a temperature-sensitive mutation in gene 63 filled up with normal-looking phage that lacked tail fibres, and with normal-looking tail fibres. Edgar concluded that gene 63 encodes a protein that allows tail fibres to attach to otherwise completely assembled phage particles.

(Continued)

product. Once the heads are completed and filled with DNA, they attach to the tails, after which attachment of the fibres completes particle construction. It would have been very difficult to discern this trilateral assembly pathway by any means other than mutagenesis-driven genetic dissection.

Between 1990 and 1995, molecular geneticists determined the complete DNA sequence of the T4 genome, and then using the genetic code dictionary (described in Chapter 7), translated that sequence into coding regions for proteins. In addition to the 65 genes identified by Edgar, another 55 genes became evident from the sequence. Edgar did not find these genes because they are not essential to viral reproduction under the conditions used in the laboratory. The previously unidentified genes most likely play important roles in the T4 life cycle outside the laboratory, perhaps when the virus infects hosts other than the *E. coli* strain normally used in the laboratory, or when the virus grows under different environmental conditions and is competing with other viruses.

three-dimensional structure of a protein, then changing amino acids at different positions in a polypeptide chain can alter protein function in different ways. For example, most enzymes have an active site that carries out the enzymatic task, while other parts of the protein support the shape and position of that site. Mutations that change the identity of amino acids at the active site may have more serious consequences than those affecting amino acids outside the active site. Some kinds of amino acid substitutions, such as replacement of an amino acid having a basic side chain with an amino acid having an acidic side chain, would be more likely to compromise protein function than would substitutions that retain the chemical characteristics of the original amino acid.

Some mutations do not affect the amino acid composition of a protein but still generate an abnormal phenotype. As discussed in Chapter 7, such mutations change the amount of normal polypeptide produced by disrupting the biochemical processes responsible for decoding a gene into a polypeptide.

Most (but not all) genes specify the amino acid sequence of a polypeptide; a protein is composed of one or more polypeptides. The primary amino acid sequences of the constituent polypeptides determine a protein's three-dimensional structure and thus its function.

8.4 A Comprehensive Example: Mutations That Affect Vision

Researchers first described anomalies of colour perception in humans close to 200 years ago. Since that time, they have discovered a large number of mutations that modify human vision. By examining the phenotype associated with each mutation and then looking directly at the DNA alterations inherited with the mutation, they have learned a great deal about the genes influencing human visual perception and the function of the proteins they encode.

Using human subjects for vision studies has several advantages. First, people can recognize and describe variations in the way they see, from trivial differences in what the colour red looks like, to not seeing any difference between red and green, to not seeing any colour at all. Second, the highly developed science of psychophysics provides sensitive, non-invasive tests for accurately defining and comparing phenotypes. One diagnostic test, for example, is based on the fact that people perceive each colour as a mixture of three different wavelengths of light—red, green, and blue—and the human visual system can adjust ratios of red, green, and blue light of different intensities to match an arbitrarily chosen fourth wavelength such as yellow. The mixture of wavelengths does not combine to form the fourth wavelength; it just appears that way to the eye. A person with normal vision, for instance, will select a well-defined proportion of red and green lights to match a particular yellow, but a person who cannot tell red from green will permit any proportion of these two colour lights to make the same match. Finally, because inherited variations in the visual system rarely affect an individual's life span or ability to reproduce, mutations generating many of the new alleles that change visual perception remain in a population over time.

Cells of the retina carry light-sensitive proteins

People perceive light through neurons in the retina at the back of the eye (**Figure 8.25a**). These neurons are of two types: rods and cones. The rods, which make up 95 percent of all light-receiving neurons, are stimulated by weak light over a range of wavelengths. At higher light intensities, the rods become saturated and no longer send meaningful information to the brain. This is when the cones take over, processing wavelengths of bright light that enable us to see colour.

The cones come in three forms—one specializes in the reception of red light, a second in the reception of green, and a third in the reception of blue. For each photoreceptor

Tools of Genetics Site-directed Mutagenesis

Dr. Michael Smith (1932–2000) (**Figure A**), born in Blackpool, England, was a distinguished molecular biologist and chemist at the University of British Columbia (UBC) in Vancouver. He was a co-recipient of the Nobel Prize in Chemistry in 1993 for the development of the revolutionary genetic engineering technique known as **site-directed mutagenesis**.

This method is based on the idea that targeted mutations can be created at specific sites in a genome. Using chemically synthesized oligonucleotides (relatively short fragments of nucleic acids, the building blocks of DNA) in a primer extension technique with DNA polymerase, Dr. Smith developed and perfected a method for selectively engineering point, deletion, and insertion mutations in genes (**Figure B**). Before the development of site-directed mutagenesis, there was no method available to induce mutations at specific genomic sites. Geneticists would generate an assortment of different mutants by exposing organisms to radiation or chemical mutagens, and then select the mutant that they desired. This was a random process and it could take a long time to generate the right mutant. Site-directed mutagenesis has thus become one of the fundamental tools of molecular biology and biotechnology, allowing researchers to gain powerful insights into the structure and function of genes, RNA, and proteins, and enabling the development of new diagnostic tests and novel therapeutics for genetic diseases. An inductee of the Canadian Medical Hall of Fame, Dr. Smith has also received a long list of other honours, including Companion of the Order of Canada, Fellow of the Royal Society of Canada and London, and recipient of the Gairdner Foundation International Award. Described as a humanitarian, he donated his $500 000 Nobel Prize award to support schizophrenia research and science education. He then successfully challenged the British Columbia and federal governments to match his generous donation. Dr. Smith's pre-eminent career and steadfast support resulted in the creation of new facilities and resources in British Columbia, and helped establish the province as a leading centre for health research.

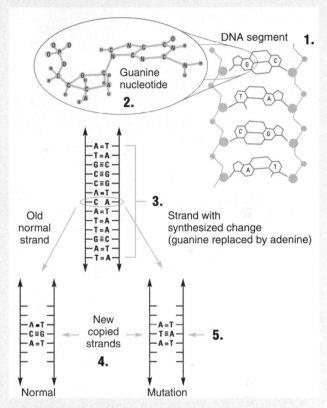

Figure B Site-directed mutagenesis resulting in a point mutation (base-pair substitution). 1. A small DNA segment showing the deoxyribose sugar-phosphate backbone (*light grey*) and the different bases or nucleotides that pair together—adenine-thymine (AT) and cytosine-guanine (CG). **2.** The nucleotide guanine. **3.** Michael Smith's Nobel Prize-winning idea: incorporation of a synthetic genomic fragment, called an oligonucleotide, into the DNA chain, using chemical reagents that disrupt and reform the DNA. In order to induce a mutation, a guanine is replaced by an adenine in this chemically synthesized oligonucleotide. The nucleotides flanking the adenine act as a kind of address that allows this synthesized DNA segment to match up to the appropriate location on the other DNA strand. **4.** When this new, modified DNA is inserted back into an organism, after normal division and reproductive processes, one-half of the DNA will generate an exact copy of the original gene, resulting in a "normal" organism. The other half of the DNA, containing the synthetic oligonucleotide, will produce a mutation, because a thymine residue has now replaced a cytosine (remember, A normally pairs with T, and G combines with C). The resulting mutant organism may have a new phenotype.

Figure A Dr. Michael Smith.

(a) Photoreceptor-containing cells

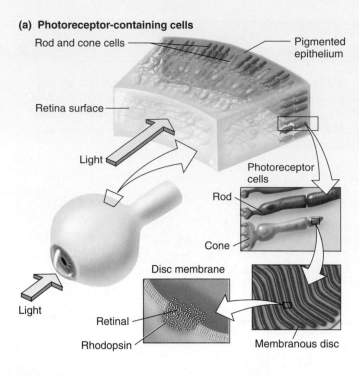

(b) Photoreceptor proteins

Rhodopsin protein

Blue-receiving protein

Green-receiving protein

Red-receiving protein

(c) Red/green pigment genes

X chromosomes from normal individuals:

(d) Evolution of visual pigment genes

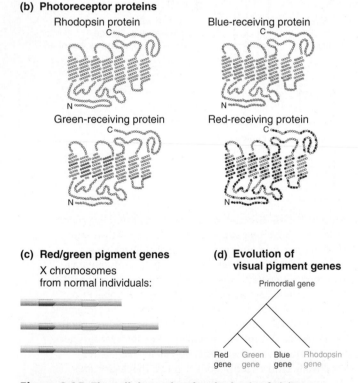

Primordial gene

Red gene   Green gene   Blue gene   Rhodopsin gene

Figure 8.25 The cellular and molecular basis of vision.
(a) Rod and cone cells in the retina carry membrane-bound photoreceptors. **(b)** The photoreceptor in rod cells is rhodopsin. The blue, green, and red receptor proteins in cone cells are related to rhodopsin. **(c)** One red photoreceptor gene and one to three green photoreceptor genes are clustered on the X chromosome. **(d)** The genes for rhodopsin and the three colour receptors probably evolved from a primordial photoreceptor gene through three gene duplication events followed by divergence of the duplicated copies.

cell, the act of reception consists of absorbing photons from light of a particular wavelength, transducing information about the number and energy of those photons to electrical signals, and transmitting the signals via the optic nerve to the brain.

Four related proteins with different light sensitivities

The protein that receives photons and triggers the processing of information in rod cells is **rhodopsin**. (You learned about this protein briefly in Chapter 7.) It consists of a single polypeptide chain containing 348 amino acids that snakes back and forth across the cell membrane (**Figure 8.25b**). One lysine within the chain associates with retinal, a carotenoid pigment molecule that actually absorbs photons. The amino acids in the vicinity of the retinal constitute rhodopsin's active site; by positioning the retinal in a particular way, they determine its response to light. Each rod cell contains approximately 100 million molecules of rhodopsin in its specialized membrane. As you learned at the beginning of this chapter, the gene governing the production of rhodopsin is on chromosome 3.

The protein that receives and initiates the processing of photons in the blue cones is a relative of rhodopsin, also consisting of a single polypeptide chain containing 348 amino acids and also encompassing one molecule of retinal. Slightly less than half of the 348 amino acids in the blue-receiving protein are the same as those found in rhodopsin; the rest are different and account for the specialized light-receiving ability of the protein (**Figure 8.25b**). The gene for the blue protein is on chromosome 7.

Similarly related to rhodopsin are the red- and green-receiving proteins in the red and green cones. These are also single polypeptides associated with retinal and embedded in the cell membrane, although they are both slightly larger at 364 amino acids in length (**Figure 8.25b**). Like the blue protein, the red and green proteins differ from rhodopsin in nearly half of their amino acids; they differ from each other in only four amino acids out of every hundred. Even these small differences, however, are sufficient to differentiate the light sensitivities of the two types of cones and confer on them distinct spectral sensitivities. The genes for the red and green proteins both reside on the X chromosome in a tandem head-to-tail arrangement. Most individuals have one red gene and one to three green genes on their X chromosomes (**Figure 8.25c**).

Evolution of the rhodopsin gene family

The similarity in structure and function among the four rhodopsin proteins suggests that the genes encoding these polypeptides arose by duplication of an original photoreceptor gene and then divergence through the accumulation of many mutations. Many of the mutations that promoted the ability to see colour must have provided selective advantages to their bearers over the course of evolution. The red and green genes are the most similar, differing by less than five nucleotides out of every 100. This suggests

they diverged from each other only in the relatively recent evolutionary past. The less-pronounced amino acid similarity of the red or green proteins with the blue protein, and the even lower relatedness between rhodopsin and any colour photoreceptor, reflect earlier duplication and divergence events (**Figure 8.25d**).

> Duplication and divergence (through mutation) of an ancestral rhodopsin-like gene have produced four specialized genes encoding rhodopsin and the blue, red, and green photoreceptor proteins.

How mutations in the rhodopsin gene family affect the way we see

Mutations in the genes encoding rhodopsin and the three colour photoreceptor proteins can alter vision through many different mechanisms. These mutations range from point mutations that change the identity of a single amino acid in a single protein to larger aberrations resulting from unequal crossing-over that can increase or decrease the number of photoreceptor genes.

Mutations in the rhodopsin gene

At least 29 different single nucleotide substitutions in the rhodopsin gene cause an autosomal dominant vision disorder known as *retinitis pigmentosa* that begins with an early loss of rod function, followed by a slow progressive degeneration of the peripheral retina. **Figure 8.26a** shows the location of the amino acids affected by these mutations. These amino acid changes result in abnormal rhodopsin proteins that either do not fold properly or, once folded, are unstable. Although normal rhodopsin is an essential structural element of rod cell membranes, these nonfunctional mutant proteins are retained in the body of the cell, where they remain unavailable for insertion into the membrane. Rod cells that cannot incorporate enough rhodopsin into their membranes eventually die. Depending on how many rod cells die, partial or complete blindness ensues.

Other mutations in the rhodopsin gene cause the far less serious condition of night blindness (Figure 8.26a). These mutations change the protein's amino acid sequence so that the threshold of stimulation required to trigger the vision cascade increases. With the changes, very dim light is no longer enough to initiate vision.

Mutations in the cone-cell pigment genes

Vision problems caused by mutations in the cone-cell pigment genes are less severe than those caused by similar defects in the rod cells' rhodopsin genes. Most likely, this difference occurs because the rods make up 95 percent of a person's light-receiving neurons, while the cones comprise only about 5 percent. Some mutations in the blue gene on chromosome 7 cause *tritanopia*, a defect in the ability to discriminate between colours that differ only in the amount of blue light they contain (**Figures 8.26b** and **8.27**). Mutations

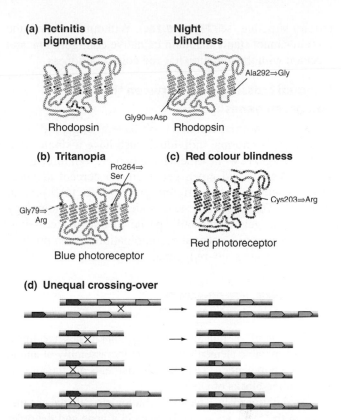

Figure 8.26 How mutations modulate light and colour perception.
(a) Amino acid substitutions (*black dots*) that disrupt rhodopsin's three-dimensional structure result in retinitis pigmentosa. Other substitutions diminishing rhodopsin's sensitivity to light cause night blindness. **(b)** Substitutions in the blue pigment can produce tritanopia (blue colour blindness). **(c)** Red colour blindness can result from particular mutations that destabilize the red photoreceptor. **(d)** Unequal crossing-over between the red and green genes can change gene number and create genes encoding hybrid photoreceptor proteins.

Figure 8.27 How the world looks to a person with tritanopia. (Compare with Figure 3.21 in Chapter 3.)

in the red gene on the X chromosome can modify or abolish red protein function and, as a result, the red cone cells' sensitivity to light. For example, a change at position 203 in the red-receiving protein from cysteine to arginine disrupts one of the disulfide bonds required to support the protein's

tertiary structure (see **Figure 8.26c**). Without that bond, the protein cannot stably maintain its native configuration, and a person with the mutation has red colour blindness.

Unequal crossing-over between the red and green genes

People with normal colour vision have a single red gene; some of these normal individuals also have a single adjacent green gene, while others have two or even three green genes. The red and green genes are 96 percent identical in DNA sequence; the different green genes, 99.9 percent identical. The proximity and high degree of homology make these genes unusually prone to unequal crossing-over. A variety of unequal recombination events produce DNA containing no red gene, no green gene, various combinations of green genes, or hybrid red-green genes (see **Figure 8.26d**). These different DNA combinations account for the large majority of the known aberrations in red-green colour perception, with the remaining abnormalities stemming from point mutations, as described earlier. Because the accurate perception of red and green depends on the differing ratios of red and green light processed, people with no red or no green gene perceive red and green as the same colour (see Figure 3.21 in Chapter 3).

> We see the way we do in part because four genes direct the production of four photoreceptor polypeptides in the rod and cone cells of the retina. Mutations that alter those polypeptides or their amounts change our perception of light or colour.

Connections

Careful studies of mutations showed that genes are linear arrays of mutable elements that direct the assembly of amino acids in a polypeptide. The mutable elements are the nucleotide building blocks of DNA.

As mentioned in Chapter 7, biologists call the parallel between the sequence of nucleotides in a gene and the order of amino acids in a polypeptide **colinearity**. This colinearity arises from base-pairing, a genetic code, specific enzymes, and macromolecular assemblies like ribosomes that guide the flow of information from DNA through RNA to protein.

Although the faithful function, replication, and transmission of chromosomes underlie the perpetuation of life within each species, chromosomal changes do occur. We have already described two mechanisms of change: mutation of individual nucleotides (this chapter) and homologous recombination, which exchanges bases between homologues (Chapters 3, 4, and 6). In Chapter 9, we examine broader chromosomal rearrangements that produce different numbers of chromosomes, reshuffle genes between nonhomologous chromosomes, and reorganize the genes of a single chromosome. These large-scale modifications, by altering the genetic content of a genome, provide some of the important variations that fuel evolution.

Essential Concepts

1. Mutations are alterations in the nucleotide sequence of the DNA molecule that occur by chance and modify the genome at random. Mutations in single-celled organisms or in the germ line of multicelluar organisms can be transmitted from generation to generation when DNA replicates. **[LO1–2, LO7]**

2. Mutations that affect phenotype occur naturally at a very low rate. Forward mutations usually occur more often than reversions. **[LO1]**

3. The agents of spontaneously occurring mutations include chemical hydrolysis, radiation, and mistakes during DNA replication. **[LO4]**

4. Mutagens raise the frequency of mutation above the spontaneous rate. The Ames test screens for mutagenic chemicals. **[LO3–4]**

5. Cells have evolved a number of enzyme systems that repair DNA and thus minimize mutations. **[LO5–6]**

6. Mutations are the raw material of evolution. Although some mutations may confer a selective advantage, many are harmful. Somatic mutations can cause cancer and other illnesses in individuals. **[LO6–7]**

7. Mutations within a single gene usually fail to complement each other. The concept of a complementation group thus defines the gene as a unit of function. A gene is composed of a linear sequence of nucleotide pairs in a discrete, localized region of a chromosome. **[LO8]**

8. The function of most genes is to specify the linear sequence of amino acids in a particular polypeptide (one gene, one polypeptide). The sequence determines the polypeptide's three-dimensional structure, which in turn determines its function. Mutations can alter the amino acid sequence and thus change protein function in many ways. **[LO6, LO8]**

9. Each protein consists of one, two, or more polypeptides. Proteins composed of two or more different subunits are encoded by two or more genes. **[LO8]**

10. The rhodopsin gene family provides an example of how the processes of gene duplication followed by gene divergence mutation can lead to evolution of functional refinements, such as the emergence of accurate systems for colour vision. **[LO6, LO8]**

I. Mutations can often be reverted to wild type by treatment with mutagens. The type of mutagen that will reverse a mutation gives us information about the nature of the original mutation. The mutagen EMS almost exclusively causes transitions; proflavin is an intercalating agent that causes insertion or deletion of a base; ultraviolet (UV) light causes single-base substitutions. Cultures of several *E. coli met⁻* mutants were treated with three mutagens separately and spread onto a plate lacking methionine to look for revertants. (In the chart, − indicates that no colonies grew, and + indicates that some *met⁺* revertant colonies grew.)

	Mutagen treatment		
Mutant number	EMS	Proflavin	UV light
1	+	−	+
2	−	+	−
3	−	−	−
4	−	−	+

a. Given the results, what can you say about the nature of the original mutation in each of the strains?

b. Experimental controls are designed to eliminate possible explanations for the results, thereby ensuring that data are interpretable. In the experiment described, we scored the presence or absence of colonies. How do we know if colonies that appear on plates are mutagen-induced revertants? What else could they be? What control would enable us to be confident of our revertant analysis?

Answer

To answer this question, you need to understand the concepts of mutation and reversion.

a. Mutation 1 is reverted by the mutagen that causes transitions, *so mutation 1 must have been a transition.* Consistent with this conclusion is the fact the UV light can also revert the mutation and the intercalating agent proflavin does not cause reversion. *Mutation 2 is reverted by proflavin and therefore must be either an insertion or a deletion of a base.* The other two mutagens do not revert mutation 2. Mutation 3 is not reverted by any of these mutagenic agents. It is therefore not a single-base substitution, a single-base insertion, or a single-base deletion. *Mutation 3 could be a deletion of several bases or an inversion.* Mutation 4 is reverted by UV light, so it is a single-base change, but it is not a transition, since EMS did not revert the mutation. *Mutation 4 must be a transversion.*

b. *The colonies on the plates could arise by spontaneous reversion of the mutation.* Spontaneous reversion should occur with lower frequency than mutagen-induced reversion. The important control here is to *spread each mutant culture without any mutagen treatment onto selective media to assess the level of spontaneous reversion.*

II. Imagine that ten independently isolated recessive lethal mutations (l^1, l^2, l^3, etc.) map to chromosome 7 in mice. You perform complementation testing by mating all pairwise combinations of heterozygotes bearing these lethal mutations, and you score the absence of complementation by examining pregnant females for dead fetuses. A + in the chart means that the two lethals complemented, and dead embryos were not found. A − indicates that dead embryos were found, at the rate of about one in four conceptions. (The crosses between heterozygous mice would be expected to yield the homozygous recessive showing the lethal phenotype in 1/4 of the embryos.) The lethal mutation in the parental heterozygotes for each cross are listed across the top and down the left side of the chart (i.e., l^1 indicates a heterozygote in which one chromosome bears the l^1 mutation and the homologous chromosome is wild type).

	l^1	l^2	l^3	l^4	l^5	l^6	l^7	l^8	l^9	l^{10}
l^1	−	+	+	+	+	−	−	+	+	+
l^2		−	+	+	+	+	+	+	+	−
l^3			−	−	−	+	+	−	−	+
l^4				−	−	+	+	−	−	+
l^5					−	+	+	−	−	+
l^6						−	−	+	+	+
l^7							−	+	+	+
l^8								−	−	+
l^9									−	+
l^{10}										−

How many genes do the ten lethal mutations represent? What are the complementation groups?

Answer

This problem involves the application of the complementation concept to a set of data. There are two ways to analyze these results. You can focus on the mutations that do complement each other, conclude that they are in different genes, and begin to create a list of mutations in separate genes. Alternatively, you can focus on mutations that do not complement each other and therefore are alleles of the same genes. The latter approach is more efficient when several mutations are involved. For example, l^1 does not complement l^6 and l^7; these three alleles are in one

complementation group. l^2 does not complement l^{10}; they are in a second complementation group. l^3 does not complement l^4, l^5, l^8, or l^9, so they form a third complementation group. *There are three complementation groups.* (Note also that for each mutant, the cross between individuals carrying the same alleles resulted in no complementation, because the homozygous recessive lethal was generated.) *The three complementation groups consist of (1) l^1, l^6, l^7; (2) l^2, l^{10}; and (3) l^3, l^4, l^5, l^8, l^9.*

III. W, X, and Y are the intermediates (in that order) in a biochemical pathway whose product is Z. Z^- mutants are found in five different complementation groups. *Z1* mutants will grow on Y or Z but not W or X. *Z2* mutants will grow on X, Y, or Z. *Z3* mutants will only grow on Z. *Z4* mutants will grow on Y or Z. Finally, *Z5* mutants will grow on W, X, Y, or Z.

a. Order the five complementation groups in terms of the steps they block.

b. What does this genetic information reveal about the nature of the enzyme that carries out the conversion of X to Y?

Answer

This problem requires that you understand complementation and the connection between genes and enzymes in a biochemical pathway.

a. A biochemical pathway represents an ordered set of reactions that must occur to produce a product. This problem gives the order of intermediates in a pathway for producing product Z. The lack of any enzyme along the way will cause the phenotype of Z^-, but the block can occur at different places along the pathway. If the mutant grows when given an intermediate compound, the enzymatic (and hence gene) defect must be before production of that intermediate compound. The *Z1* mutants that grow on Y or Z (but not on W or X) must have a defect in the enzyme that produces Y. *Z2* mutants have a defect prior to X; *Z3* mutants have a defect prior to Z; *Z4* mutants have a defect prior to Y; *Z5* have a defect prior to W. *The five complementation groups can be placed in order of activity within the biochemical pathway as follows:*

$$\xrightarrow{Z5} W \xrightarrow{Z2} X \xrightarrow{Z1,\ Z4} Y \xrightarrow{Z3} Z$$

b. Mutants *Z1* and *Z4* affect the same step, but because they are in different complementation groups, we know they are in different genes. *Mutations Z1 and Z4 are probably in genes that encode subunits of a multisubunit enzyme that carries out the conversion of X to Y.* Alternatively, there could be a currently unknown additional intermediate step between X and Y.

Vocabulary

1. For each of the terms in the left column, choose the best matching phrase in the right column.

   i. an A–T base pair in the wild-type gene is changed to a G–C pair

   ii. an A–T base pair is changed to a T–A pair

   iii. the sequence AAGCTTATCG is changed to AAGCTATCG

   iv. the sequence AAGCTTATCG is changed to AAGCTTTATCG

   v. the sequence AACGTTATCG is changed to AATGTTATCG

   vi. the sequence AACGTCACACACACATCG is changed to AACGTCACATCG

   vii. the gene map in a given chromosome arm is changed from *bog-rad-fox1-fox2-try-duf* (where *fox1* and *fox2* are highly homologous, recently diverged genes) to *bog-rad-fox1-fox3-fox2-try-duf* (where *fox3* is a new gene with one end similar to *fox1* and the other similar to *fox2*)

   viii. the gene map in a chromosome is changed from *bog-rad-fox1-fox2-try-duf* to *bog-rad-fox2-fox1-try-duf*

   ix. the gene map in a given chromosome is changed from *bog-rad-fox1-fox2-try-duf* to *bog-rad-fox1-mel-qui-txu-sqm*

   1. transition

   2. base substitution

   3. transversion

   4. inversion

   5. translocation

   6. deletion

   7. insertion

   8. deamination

   9. X-ray irradiation

   10. intercalator

   11. unequal crossing-over

2. What explanations can account for the pedigree of the very rare trait shown below? Be as specific as possible. How might you be able to distinguish between these explanations?

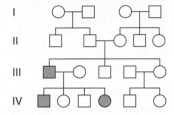

3. The DNA sequence of a gene from three independently isolated mutants is given here. Using this information, what is the sequence of the wild-type gene in this region?

mutant 1	ACCGTAATCGACTGGTAAACTTTGCGCG
mutant 2	ACCGTAGTCGACCGGTAAACTTTGCGCG
mutant 3	ACCGTAGTCGACTGGTTAACTTTGCGCG

4. Among mammals, measurements of the rate of generation of autosomal recessive mutations have been made almost exclusively in mice, while many measurements of the rate of generation of dominant mutations have been made both in mice and in humans. Why do you think there has been this difference?

5. Over a period of several years, a large hospital kept track of the number of births of babies displaying the trait achondroplasia. Achondroplasia is a very rare autosomal dominant condition resulting in dwarfism with abnormal body proportions. After 120 000 births, it was noted that there had been 27 babies born with achondroplasia. One physician was interested in determining how many of these dwarf babies result from new mutations and whether the apparent mutation rate in his area was higher than normal. He looked up the families of the 27 dwarf births and discovered that four of the dwarf babies had a dwarf parent. What is the apparent mutation rate of the achondroplasia gene in this population? Is it unusually high or low?

6. Suppose you wanted to study genes controlling the structure of bacterial cell surfaces. You decide to start by isolating bacterial mutants that are resistant to infection by a bacteriophage that binds to the cell surface. The selection procedure is simple: Spread cells from a culture of sensitive bacteria on a Petri plate, expose them to a high concentration of phages, and pick the bacterial colonies that grow. To set up the selection you could (1) spread cells from a single liquid culture of sensitive bacteria on many different plates and pick every resistant colony *or* (2) start many different cultures, each grown from a single colony of sensitive bacteria, spread one plate from each culture, and then pick a single mutant from each plate. Which method would ensure that you are isolating many independent mutations?

7. In a genetics lab, Kim and Maria infected a sample from an *E. coli* culture with a particular virulent bacteriophage. They noticed that most of the cells were lysed, but a few survived. The survival rate in their sample was about 1×10^{-4}. Kim was sure the bacteriophage induced the resistance in the cells, while Maria thought that resistant mutants probably already existed in the sample of cells they used. Earlier, for a different experiment, they had spread a dilute suspension of *E. coli* onto solid medium in a large Petri dish, and, after seeing that about 10^5 colonies were growing up, they had replica-plated that plate onto three other plates. Kim and Maria decided to use these plates to test their theories. They pipette a suspension of the bacteriophage onto each of the three replica plates. What should they see if Kim is right? What should they see if Maria is right?

8. The pedigree below shows the inheritance of a completely penetrant, dominant trait called amelogenesis imperfecta that affects the structure and integrity of the teeth. DNA analysis of blood obtained from affected individuals III-1 and III-2 shows the presence of the same mutation in one of the two copies of an autosomal gene called *ENAM* that is not seen in DNA from the blood of any of the parents in generation II. Explain this result, citing Figure 3.18 and Figure 8.4. Do you think this type of inheritance pattern is rare or common?

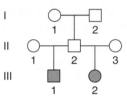

9. A wild-type male *Drosophila* was exposed to a large dose of X-rays and was then mated to an unirradiated female, one of whose X chromosomes carried both a dominant mutation for the trait *Bar* eyes and several inversions. Many F_1 females from this mating were recovered who had the *Bar*, multiply inverted X chromosome from their mother, and an irradiated X chromosome from their fathers. (The inversions ensure that viable offspring of these F_1 females will not have recombinant X chromosomes, as explained in Chapter 9.) After mating to normal males, most F_1 females produced *Bar* and wild-type sons in equal proportions. There were three exceptional F_1 females, however. Female A produced as many sons as daughters, but half of the sons had *Bar* eyes, and the other half had white eyes. Female B produced half as many sons as daughters, and all of the sons had *Bar* eyes. Female C produced 75 percent as many sons as daughters. Of these sons, 2/3 had *Bar* eyes, and 1/3 had wild-type eyes. Explain the results obtained with each exceptional F_1 female.

10. A wild-type *Drosophila* female was mated to a wild-type male that had been exposed to X-rays. One of the F_1 females was then mated with a male that had the following recessive markers on the X chromosome: *yellow* body (*y*), *crossveinless* wings (*cv*), *cut* wings (*ct*), *singed* bristles (*sn*), and *miniature* wings (*m*). These markers are known to map in the order $y-cv-ct-sn-m$.

The progeny of this second mating were unusual in two respects. First, there were twice as many females as males. Second, while all of the males were wild type in phenotype, 1/2 of the females were wild type, and the other 1/2 exhibited the *ct* and *sn* phenotypes.

a. What did the X-rays do to the irradiated male?

b. Draw the X-chromosome pair present in a progeny female fly produced by the second mating that was phenotypically *ct* and *sn*.

c. If the *ct* and *sn* female fly whose chromosomes were drawn in part *b* was then crossed to a wild-type male, what phenotypic classes would you expect to find among the progeny males?

11. In the experiment shown in Figure 8.9, H. J. Muller first performed a control in which the P generation males were not exposed to X-rays. He found that 99.7 percent of the individual F_1 Bar-eyed females produced some male progeny with Bar eyes and some with wild-type (non-Bar) eyes, but 0.3 percent of these females produced male progeny that were all wild type.

a. If the average spontaneous mutation rate for *Drosophila* genes is 3.5×10^{-6} mutations/gene/gamete, how many genes on the X chromosome can be mutated to produce a recessive lethal allele?

b. As of the year 2010, analysis of the *Drosophila* genome had revealed a total of 2283 genes on the X chromosome. Assuming the X chromosome is typical of the genome, what is the fraction of genes in the fly genome that is essential for survival?

c. Muller now exposed male flies to a specific high dosage of X-rays and found that 12 percent of F_1 Bar-eyed females produced male progeny that were all wild type. What does this new information say?

12. Figure 8.10 shows examples of base substitutions induced by the mutagens 5-bromouracil, hydroxylamine, ethylmethane sulfonate, and nitrous acid. Which of these mutagens cause transitions, and which cause transversions?

13. So-called *two-way mutagens* can induce both a particular mutation and (when added subsequently to cells whose chromosomes carry this mutation) a reversion of the mutation that restores the original DNA sequence. In contrast, *one-way mutagens* can induce mutations but not exact reversions of these mutations. Based on Figure 8.10), which of the following mutagens can be classified as one-way and which as two-way?

a. 5-bromouracil

b. hydroxylamine

c. ethylmethane sulfonate

d. nitrous acid

e. proflavin

14. In 1967, J. B. Jenkins treated wild-type male *Drosophila* with the mutagen ethylmethane sulfonate (EMS) and mated them with females homozygous for a recessive mutation called *dumpy* that causes shortened wings. He found some

F_1 progeny with two wild-type wings, some with two short wings, and some with one short wing and one wild-type wing. When he mated single F_1 flies with two short wings to *dumpy* homozygotes, he surprisingly found that only about 1/3 of these matings produced any short-winged progeny.

a. Explain these results in light of the mechanism of action of EMS shown in Figure 8.10.

b. Should the short-winged progeny of the second cross have one or two short wings? Why?

15. Aflatoxin B_1 is a highly mutagenic and carcinogenic compound produced by certain fungi that infect crops such as peanuts. Aflatoxin is a large, bulky molecule that chemically bonds to the base guanine to form the aflatoxin-guanine "adduct" that is pictured below. (In the Figure, the aflatoxin is *orange*, and the guanine base is *purple*.) This adduct distorts the DNA double helix and blocks replication.

a. What type(s) of DNA repair system is (are) most likely to be involved in repairing the damage caused by exposure of DNA to aflatoxin B_1?

b. Recent evidence suggests that the adduct of guanine and aflatoxin B_1 can attack the bond that connects it to deoxyribose; this liberates the adduced base, forming an apurinic site. How does this new information change your answer to part *a?*

Aflatoxin-guanine adduct

16. When a particular mutagen identified by the Ames test is injected into mice, it causes the appearance of many tumours, showing that this substance is carcinogenic. When cells from these tumours are injected into other mice not exposed to the mutagen, almost all of the new mice develop tumours. However, when mice carrying mutagen-induced tumours are mated to unexposed mice, virtually all of the progeny are tumour-free. Why can the tumour be transferred horizontally (by injecting cells) but not vertically (from one generation to the next)?

17. When the *his⁻* *Salmonella* strain used in the Ames test is exposed to substance X, no *his⁺* revertants are seen. If, however, rat liver supernatant is added to the cells along with substance X, revertants do occur. Is substance X a potential carcinogen for human cells? Explain.

Section 8.2

18. Imagine that you caught a female albino mouse in your kitchen and decided to keep it for a pet. A few months later, while vacationing in Guam, you caught a male albino mouse and decided to take it home for some

interesting genetic experiments. You wonder whether the two mice are both albino due to mutations in the same gene. What could you do to find out the answer to this question? Assume that both mutations are recessive.

19. Plant breeders studying genes influencing leaf shape in the plant *Arabidopsis thaliana* identified six independent recessive mutations that resulted in plants that had unusual leaves with serrated rather than smooth edges. The investigators started to perform complementation tests with these mutants, but some of the tests could not be completed because of an accident in the greenhouse. The results of the complementation tests that could be finished are shown in the table that follows.

	1	2	3	4	5	6
1	−	+	−		+	
2		−				−
3			−	−		
4				−		
5					−	+
6						−

a. Exactly what experiment was done to fill in individual boxes in the table with a + or a − ? What does + represent? What does − represent? Why are some boxes in the table filled in *green*?

b. Assuming no complications, what do you expect for the results of the complementation tests that were not performed? That is, complete the table above by placing a + or a − in each of the blank boxes.

c. How many genes are represented among this collection of mutants? Which mutations are in which genes?

20. In humans, albinism is normally inherited in an autosomal recessive fashion. Figure 2.39c in Chapter 2 shows a pedigree in which two albino parents have several children, none of whom is an albino.

a. Interpret this pedigree in terms of a complementation test.

b. It is very rare to find examples of human pedigrees such as Figure 2.39c that could be interpreted as a complementation test. This is because most genetic conditions in humans are rare, so it is highly unlikely that unrelated people with the same condition would mate. In the absence of complementation testing, what kinds of experiments could be done to determine whether a particular human disease phenotype can be caused by mutations at more than one gene?

c. Complementation testing requires that the two mutations to be tested both be recessive to wild type. Suppose that two dominant mutations cause similar phenotypes. How could you establish whether these mutations affected the same gene or different genes?

21. You found five T4 *rII⁻* mutants that will not grow on *E. coli* K(λ). You mixed together all possible combinations of two mutants (as indicated in the following chart),

added the mixtures to *E. coli* K(λ), and scored for the ability of the mixtures to grow and make plaques (indicated as a + in the chart).

	1	2	3	4	5
1	−	+	+	−	+
2		−	−	+	−
3			−	+	−
4				−	+
5					−

a. How many genes were identified by this analysis?

b. Which mutants belong to the same complementation groups?

22. The *rosy* (*ry*) gene of *Drosophila* encodes an enzyme called xanthine dehydrogenase. Flies homozygous for *ry* mutations exhibit a rosy eye colour. Heterozygous females were made that had ry^{41} *Sb* on one homologue and *Ly* ry^{564} on the other homologue, where ry^{41} and ry^{564} are two independently isolated alleles of *ry*. *Ly* (*Lyra* [narrow] wings) and *Sb* (*Stubble* [short] bristles) are dominant markers to the left and right of *ry*, respectively. These females are now mated to males homozygous for ry^{41}. Out of 100 000 progeny, eight have wild-type eyes, *Lyra* wings, and *Stubble* bristles, while the remainder have rosy eyes.

a. What is the order of these two *ry* mutations relative to the flanking genes *Ly* and *Sb*?

b. What is the genetic distance separating ry^{41} and ry^{564}?

Section 8.3

23. In a certain species of flowering plants with a diploid genome, four enzymes are involved in the generation of flower colour. The genes encoding these four enzymes are on different chromosomes. The biochemical pathway involved is as follows (the figure shows that either of two different enzymes is sufficient to convert a blue pigment into a purple pigment):

$$\text{white} \rightarrow \text{green} \rightarrow \text{blue} \begin{smallmatrix}\rightarrow\\\rightarrow\end{smallmatrix} \text{purple}$$

A true-breeding green-flowered plant is mated with a true-breeding blue-flowered plant. All of the plants in the resultant F₁ generation have purple flowers. F₁ plants are allowed to self-fertilize, yielding an F₂ generation. Show genotypes for P, F₁, and F₂ plants, and indicate which genes specify which biochemical steps. Determine the fraction of F₂ plants with the following phenotypes: white flowers, green flowers, blue flowers, and purple flowers. Assume the green-flowered parent is mutant in only a single step of the pathway.

24. In corn snakes, the wild-type colour is brown. One autosomal recessive mutation causes the snake to be orange, and another causes the snake to be black. An orange snake was crossed to a black one, and the F₁ offspring were all brown. Assume that all relevant genes are unlinked.

a. Indicate what phenotypes and ratios you would expect in the F$_2$ generation of this cross if there is one pigment pathway, with orange and black being different intermediates on the way to brown.

b. Indicate what phenotypes and ratios you would expect in the F$_2$ generation if orange pigment is a product of one pathway, black pigment is the product of another pathway, and brown is the effect of mixing the two pigments in the skin of the snake.

25. In each of the following cross schemes, two true-breeding plant strains are crossed to make F$_1$ plants, all of which have purple flowers. The F$_1$ plants are then self-fertilized to produce F$_2$ progeny as shown here.

Cross	Parents	F$_1$	F$_2$
1	blue × white	all purple	9 purple : 4 white : 3 blue
2	white × white	all purple	9 purple : 7 white
3	red × blue	all purple	9 purple : 3 red : 3 blue : 1 white
4	purple × purple	all purple	15 purple : 1 white

a. For each cross, explain the inheritance of flower colour.

b. For each cross, show a possible biochemical pathway that could explain the data.

c. Which of these crosses is compatible with an underlying biochemical pathway involving only a single step that is catalyzed by an enzyme with two dissimilar subunits, both of which are required for enzyme activity?

d. For each of the four crosses, what would you expect in the F$_1$ and F$_2$ generations if all relevant genes were tightly linked?

26. The intermediates A, B, C, D, E, and F all occur in the same biochemical pathway. G is the product of the pathway, and mutants 1 through 7 are all G^-, meaning that they cannot produce substance G. The following table shows which intermediates will promote growth in each of the mutants. Arrange the intermediates in order of their occurrence in the pathway, and indicate the step in the pathway at which each mutant strain is blocked. A + in the table indicates that the strain will grow if given that substance, whereas an O means lack of growth.

	Supplements						
Mutant	A	B	C	D	E	F	G
1	+	+	+	+	+	O	+
2	O	O	O	O	O	O	+
3	O	+	+	O	+	O	+
4	O	+	O	O	+	O	+
5	+	+	+	O	+	O	+
6	+	+	+	+	+	+	+
7	O	O	O	O	+	O	+

27. The following noncomplementing *E. coli* mutants were tested for growth on four known precursors of thymine, A–D.

	Precursor/product				
Mutant	A	B	C	D	Thymine
9	+	−	+	−	+
10	−	−	+	−	+
14	+	+	+	−	+
18	+	+	+	+	+
21	−	−	−	−	+

a. Show a simple linear biosynthetic pathway of the four precursors and the end-product, thymine. Indicate which step is blocked by each of the five mutations.

b. What precursor would accumulate in the following double mutants: 9 and 10? 10 and 14?

28. The pathways for the biosynthesis of the amino acids glutamine (Gln) and proline (Pro) involve one or more common intermediates. Auxotrophic yeast mutants numbered 1–7 are isolated that require either glutamine or proline or both amino acids for their growth, as shown in the following table (+ means growth; − no growth). These mutants are also tested for their ability to grow on the intermediates A–E. What is the order of these intermediates in the glutamine and proline pathways, and at which point in the pathway is each mutant blocked?

Mutant	A	B	C	D	E	Gln	Pro	Gln +Pro
1	+	−	−	−	+	−	+	+
2	−	−	−	−	−	−	+	+
3	−	−	+	−	−	−	−	+
4	−	−	−	−	+	−	−	+
5	−	−	+	+	−	−	−	+
6	+	−	−	−	−	−	+	+
7	−	+	−	−	−	+	−	+

29. Mutations in an autosomal gene in humans cause a form of haemophilia called von Willebrand disease (vWD). This gene specifies a blood plasma protein cleverly called von Willebrand factor (vWF). vWF stabilizes factor VIII, a blood plasma protein specified by the wild-type haemophilia A gene. Factor VIII is needed to form blood clots. Thus, factor VIII is rapidly destroyed in the absence of vWF.

Which of the following might successfully be employed in the treatment of bleeding episodes in haemophiliac patients? Would the treatments work immediately or only after some delay needed for protein synthesis? Would the treatments have only a short-term or a prolonged effect? Assume that all mutations are null (i.e., the mutations result in the complete absence of the protein encoded by the gene) and that the plasma is cell-free.

a. transfusion of plasma from normal blood into a vWD patient

b. transfusion of plasma from a vWD patient into a different vWD patient

c. transfusion of plasma from a haemophilia A patient into a vWD patient

d. transfusion of plasma from normal blood into a haemophilia A patient

c. transfusion of plasma from a vWD patient into a haemophilia A patient

f. transfusion of plasma from a haemophilia A patient into a different haemophilia A patient

g. injection of purified vWF into a vWD patient

h. injection of purified vWF into a haemophilia A patient

i. injection of purified factor VIII into a vWD patient

j. injection of purified factor VIII into a haemophilia A patient

30. In 1952, an article in the *British Medical Journal* reported interesting differences in the behaviour of blood plasma obtained from several individuals who suffered from X-linked recessive haemophilia. When mixed together, the cell-free blood plasma from certain combinations of individuals could form clots in the test tube. For example, the following table shows whether ($+$) or not ($-$) clots could form in various combinations of plasma from four individuals with haemophilia:

1 and 1	$-$	2 and 3	$+$
1 and 2	$-$	2 and 4	$+$
1 and 3	$+$	3 and 3	$-$
1 and 4	$+$	3 and 4	$-$
2 and 2	$-$	4 and 4	$-$

What do these data tell you about the inheritance of haemophilia in these individuals? Do these data allow you to exclude any models for the biochemical pathway governing blood clotting?

31. Adult haemoglobin is a multimeric protein with four polypeptides, two of which are α globin and two of which are β globin.

a. How many genes are needed to define the structure of the haemoglobin protein?

b. If a person is heterozygous for wild-type alleles and alleles that would yield amino acid substitution variants for both α globin and β globin, how many different kinds of haemoglobin protein would be found in the person's red blood cells and in what proportion? Assume all alleles are expressed at the same level.

32. Refer to Figure A in the Fast Forward box in this chapter. For each part that follows, describe what structures Robert Edgar would have seen in the electron microscope if he examined extracts of *E. coli* cells infected with the indicated temperature-sensitive mutant strains of bacteriophage T4 under restrictive conditions.

a. A strain with a mutation in gene 19

b. A strain with a mutation in gene 16

c. Simultaneous infection with two mutant strains, one in gene 13 and the other in gene 14. The polypeptides produced by genes 13 and 14 associate with each other to form a multimeric protein that governs one step of phage head assembly (see Figure A).

d. A strain whose genome contains mutations in both genes 15 and 35

Section 8.4

33. In addition to the predominant adult haemoglobin, HbA, which contains two α-globin chains and two β-globin chains ($\alpha_2\beta_2$), there is a minor haemoglobin, HbA$_2$, composed of two α and two δ chains ($\alpha_2\delta_2$). The β- and δ-globin genes are arranged in tandem and are highly homologous. Draw the chromosomes that would result from an event of unequal crossing-over between the β and δ genes.

34. Most mammals, including "New World" primates such as marmosets (a kind of monkey), are *dichromats*: they have only two kinds of rhodopsin-related colour receptors. "Old World" primates such as humans and gorillas are *trichromats* with three kinds of colour receptors. Primates diverged from other mammals roughly 65 million years ago (Myr), while Old World and New World primates diverged from each other roughly 35 Myr.

a. Using this information, define on Figure 8.25d the time span of any events that can be dated.

b. Some New World monkeys have an autosomal colour receptor gene and a single X-linked colour receptor gene. The X-linked gene has three alleles, each of which encodes a photoreceptor that responds to light of a different wavelength (all three wavelengths are different from that recognized by the autosomal colour receptor). How is colour vision inherited in these monkeys?

c. About 95 percent of all light-receiving neurons in humans and other mammals are rod cells containing rhodopsin, a pigment that responds to low-level light of many wavelengths. The remaining 5 percent of light-receiving neurons are cone cells with pigments that respond to light of specific wavelengths of high intensity. What does this suggest about the lifestyle of the earliest mammals?

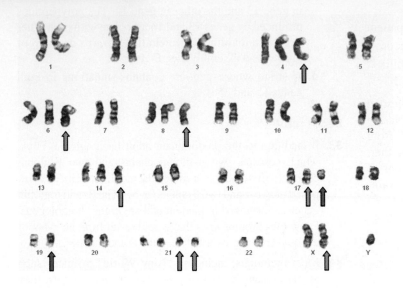

Large-scale chromosomal changes can be detected by karyotype analysis. An acute lymphocytic leukemia (ALL) patient's bone marrow cell karyotype is shown here. The *red arrows* point to a balanced nonreciprocal translocation between chromosomes 17 and 19 [t(17;19)(q22;p13)]. An extra X chromosome is observed, as well as trisomies of chromosomes 4, 6, 8, 14, and 17, along with tetrasomy for chromosome 21 (*blue arrows*).

During the early days of genome sequencing in the 1990s, studies comparing the human genome with that of the laboratory mouse (*Mus musculus*) revealed a surprising evolutionary paradox: At the DNA level, there is a close similarity of nucleotide sequence across hundreds of thousands of base pairs; but at the chromosomal level, mouse and human karyotypes bear little resemblance to each other. These early genomic analyses focused considerable effort on the sequencing of regions encompassing more than 2000 kb of mouse and human DNA containing a complex of genes that encode immune response proteins known as T-cell receptors. Comparisons of the corresponding mouse and human regions show that the nucleotide sequences of the T-cell receptor genes are similar (though not identical) in the two species, as are the order of the genes and the relative positions of a variety of noncoding sequences (of unknown function) along the chromosome. Comparisons of mouse and human Giemsa-stained karyotypes, however, reveal no conservation of banding patterns between the 20 mouse and 23 human chromosomes.

Data for resolving this apparent paradox emerged with the 2002 publication of the nearly complete mouse genome sequence, which researchers could compare with the human genome sequence completed a year earlier. The data showed that each mouse chromosome consists of pieces of different human chromosomes, and vice versa. For example, mouse chromosome 1 contains large blocks of sequences found on human chromosomes 1, 2, 5, 6, 8, 13, and 18 (portrayed in different colours in **Figure 9.1a**). These blocks represent **syntenic segments** in which the identity, order, and transcriptional direction of the genes are almost exactly the same in the two genomes. In principle, scientists could "reconstruct" the mouse genome by breaking the human genome into 342 fragments, each an average length of about 16 Mb, and pasting these fragments together in a different order. Figure 9.1a illustrates this process in detail for mouse chromosome 1; **Figure 9.1b** shows the syntenic relationships between the entire mouse and human genomes at lower resolution. Because a 16 Mb fragment would occupy no more than one or two bands of a stained chromosome, this level of conservation is not visible in karyotypes. It does, however, show up in the sequence of a smaller genomic region, such as that encoding the T-cell receptors.

These findings contribute to our understanding of how complex life-forms evolved. Although mice and humans diverged from a common ancestor about 65 million years ago, the DNA sequence in many regions of the two genomes is very similar. It is thus possible to hypothesize that the mouse and human genomes evolved through a series of approximately 300 reshaping events during which the chromosomes broke apart and the resulting fragments resealed end-to-end in novel ways. After each event, the newly rearranged chromosomes somehow became fixed in the genome of the emerging species.

Chapter Outline
9.1 Rearrangements of DNA Sequences
9.2 Transposable Genetic Elements
9.3 Rearrangements and Evolution: A Speculative Comprehensive Example
9.4 Changes in Chromosome Number
9.5 Emergent Technologies: Beyond the Karyotype

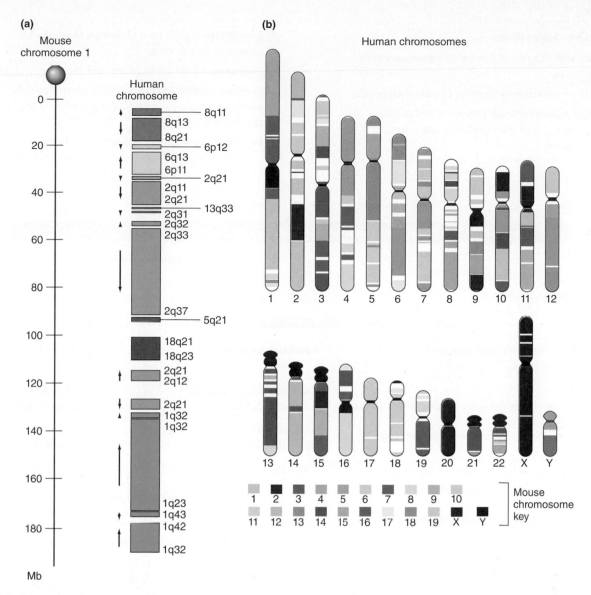

Figure 9.1 Comparing the mouse and human genomes. (a) Mouse chromosome 1 contains large blocks of sequences found on human chromosomes 1, 2, 5, 6, 8, 13, and 18 (portrayed in different colours). *Arrows* indicate the relative orientations of sequence blocks from the same human chromosome. **(b)** Human chromosomes, with segments containing at least two genes whose order is conserved in the mouse genome. Each colour corresponds to a particular mouse chromosome. Centromeres; subcentromeric heterochromatin of chromosomes 1, 9, and 16; and the repetitive short arms of 13, 14, 15, 21, and 22 are in *black*.

Both nucleotide sequence differences and differences in genome organization thus contribute to dissimilarities between the species.

In this chapter, we examine two types of events that reshape genomes: (1) **rearrangements**, which reorganize the DNA sequences within one or more chromosomes, and (2) changes in chromosome number involving losses or gains of entire chromosomes or sets of chromosomes (**Table 9.1**). Rearrangements and changes in chromosome number may affect gene activity or gene transmission by altering the position, order, or number of genes in a cell. Such alterations often, but not always, lead to a genetic imbalance that is harmful to the organism or its progeny.

We can identify two main themes underlying the observations of chromosomal changes. First, karyotypes generally remain constant within a species, not because rearrangements and changes in chromosome number occur infrequently (they are, in fact, quite common), but because the genetic instabilities and imbalances produced by such changes usually place individual cells or organisms and their progeny at a selective disadvantage. Second, despite selection against chromosomal variations, related species almost always have different karyotypes, with closely related species (such as chimpanzees and humans) diverging by only a few rearrangements and more distantly related species (such as mice and humans) diverging by a larger number of rearrangements. These observations suggest there is significant correlation between karyotypic rearrangements and the evolution of new species.

Learning Objectives

1. Explain the causes and possible consequences of deletions.

2. Analyze how deletions can be used to map genes.

3. Compare and contrast the causes and possible consequences of insertions, inversions, and translocations.

4. Examine the significance of Barbara McClintock's experiments with corn.

5. Differentiate DNA transposons from retrotransposons.

6. Compare aneuploidy, monoploidy, and polyploidy, and the causes and consequences of these changes in chromosome number.

7. Evaluate the agricultural value of polyploids.

Table 9.1	Chromosomal Rearrangements and Changes in Chromosome Number

Deletion: Removal of a segment of DNA

Duplication: Increase in the number of copies of a chromosomal region

Inversion: Half-circle rotation of a chromosomal region

Translocations:
Nonreciprocal: Unequal exchanges between nonhomologous chromosomes

Reciprocal: Parts of two nonhomologous chromosomes trade places

Transposition: Movement of short DNA segments from one position in the genome to another

Euploidy: Cells that contain only complete sets of chromosomes

Diploidy (2n/2x): Two copies of each homologue

Monoploidy (n/x): One copy of each homologue

Polyploidy: More than the normal diploid number of chromosome sets

Triploidy (3n/3x): Three copies of each homologue

Tetraploidy (4n/4x): Four copies of each homologue

Aneuploidy: Loss or gain of one or more chromosomes producing a chromosome number that is not an exact multiple of the haploid number

Monosomy (2n − 1)

Trisomy (2n + 1)

Tetrasomy (2n + 2)

All chromosomal rearrangements alter DNA sequence. Some do so by removing or adding base pairs. Others relocate chromosomal regions without changing the number of base pairs they contain. This chapter focuses on heritable rearrangements that can be transmitted through the germ line from one generation to the next, but it also explains that the genomes of somatic cells can undergo changes in nucleotide number or order. For example, the **Fast Forward** box **"Programmed DNA Rearrangements and the Immune System"** describes how the normal development of the human immune system depends on non-inherited, programmed rearrangements of the genome in somatic cells.

Deletions remove material from the genome

We saw in Chapter 8 that deletions remove one or more contiguous base pairs of DNA from a chromosome. They may arise from errors in replication, from faulty meiotic or mitotic recombination, and from exposure to X-rays or other chromosome-damaging agents that break the DNA backbone (**Figure 9.2a**). Here we use the symbol *Del* to

(a) DNA breakage may cause deletions.

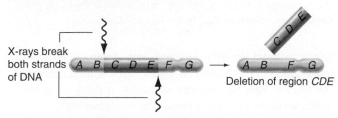

(b) Detecting deletions using PCR

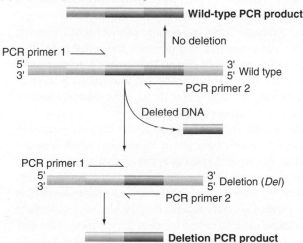

Figure 9.2 Deletions: Origin and detection. (a) When a chromosome sustains two double-strand breaks, a deletion will result if the chromosomal fragments are not properly religated. **(b)** One way to detect deletions is by polymerase chain reaction (PCR). The two PCR primers shown will amplify a larger PCR product from wild-type DNA than from DNA with a deletion.

designate a chromosome that has sustained a deletion. However, many geneticists, particularly those working on *Drosophila*, prefer the term *deficiency* (abbreviated as *Df*) to deletion.

Small deletions often affect only one gene, whereas large deletions can generate chromosomes lacking tens or even hundreds of genes. In higher organisms, geneticists usually find it difficult to distinguish small deletions affecting only one gene from point mutations; they can resolve such distinctions only through analysis of the DNA itself. For example, deletions can result in smaller restriction fragments or polymerase chain reaction (PCR) products, whereas most point mutations would not cause such changes (**Figure 9.2b**). Larger deletions are sometimes identifiable because they affect the expression of two or more adjacent genes. Very large deletions are visible at the relatively low resolution of a karyotype, showing up as the loss of one or more bands from a chromosome.

Lethal effects of homozygosity for a deletion

Because many of the genes in a genome are essential to an individual's survival, homozygotes (*Del/Del*) or hemizygotes (*Del/Y*) for most deletion-bearing chromosomes do not survive. In rare cases where the deleted chromosomal region is devoid of genes essential for viability, however, a deletion hemi- or homozygote may survive. For example, *Drosophila* males hemizygous for an 80-kb deletion including the *white* (*w*) gene survive perfectly well in the laboratory; lacking the w^+ allele required for red eye pigmentation, they have white eyes.

Detrimental effects of heterozygosity for a deletion

Usually, the only way an organism can survive a deletion of more than a few genes is if it carries a nondeleted wild-type homologue of the deleted chromosome. Such a *Del/+* individual is known as a *deletion heterozygote*. Nonetheless, the missing segment cannot be too large, as heterozygosity for very large deletions is almost always lethal. Even small deletions can be harmful in heterozygotes. Newborn humans heterozygous for a relatively small deletion from the short arm of chromosome 5 have *cri du chat* syndrome (from the French for "cry of the cat"), so named because the symptoms include an abnormal cry reminiscent of a mewing kitten. The syndrome also leads to mental retardation.

Why should heterozygosity for a deletion have harmful consequences when the *Del/+* individual has at least one wild-type copy of all of its genes? The answer is that changes in **gene dosage**—the number of times a given gene is present in the cell nucleus—can create a **genetic imbalance**. This imbalance in gene dosage alters the amount of a particular protein relative to all other proteins, and this alteration can have a variety of phenotypic effects.

The human immune system is a marvel of specificity and diversity. It includes close to a trillion B lymphocytes, specialized white blood cells that make more than a billion different varieties of antibodies (also called *immunoglobulins*, or *Igs*). Each B cell, however, makes antibodies against only a single bacterial or viral protein (called an *antigen* in the context of the immune response). The binding of antibody to antigen helps the body attack and neutralize invading pathogens.

One intriguing question about antibody responses is how a genome containing only about 20 000 (2×10^4) genes can encode a billion (10^9) different types of antibodies. The answer is that programmed gene rearrangements, in conjunction with somatic mutations and the diverse pairing of polypeptides of different sizes, can generate roughly a billion binding specificities from a much smaller number of genes. To understand the mechanism of this diversity, it is necessary to know how antibodies are constructed and how B cells come to express the antibody-encoding genes determining specific antigen-binding sites.

The Genetics of Antibody Formation Produce Specificity and Diversity

All antibody molecules consist of a single copy or multiple copies of the same basic molecular unit. Four polypeptides make up this unit: two identical light chains, and two identical heavy chains. Each light chain is paired with a heavy chain (**Figure A**). Each light and each heavy chain has a constant (C) domain and a variable (V) domain. The C domain of the heavy chain determines whether the antibody falls into one of five major classes (designated IgM, IgG, IgE, IgD, and IgA), which influence where and how an antibody functions. For example, IgM antibodies form early in an immune response and are anchored in the B-cell membrane; IgG antibodies emerge later and are secreted into the blood serum. The C domains of the light and heavy chains are not involved in determining the specificity of antibodies. Instead, the V domains of light and heavy chains come together to form the antigen-binding site, which defines an antibody's specificity.

The DNA for all domains of the heavy chain resides on chromosome 14 (**Figure B**). This heavy-chain gene region consists of more than 100 V-encoding segments, each preceded by a promoter, several D (for diversity) segments, several J (for joining) segments, and nine C-encoding segments preceded by an enhancer (a short DNA segment that aids in the initiation of transcription by interacting with the promoter; see Chapter 11 for details). In all germline cells and in most somatic cells, including the cells destined to become B lymphocytes, these various gene segments lie far apart on the chromosome. During B-cell development, however, somatic rearrangements juxtapose random, individual V, D, and J segments together to form the particular variable region that will be transcribed. These rearrangements also place the newly formed variable region next to a C segment and its enhancer, and they further

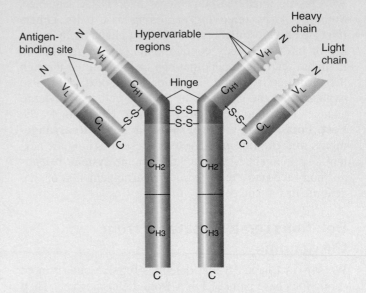

Figure A **How antibody specificity emerges from molecular structure.** Two heavy chains and two light chains held together by disulfide (–S–S–) bonds form the basic unit of an antibody molecule. Both heavy and light chains have variable (V) domains near their N termini, which associate to form the antigen-binding site. "Hypervariable" stretches of amino acids within the V domains vary extensively between antibody molecules. The remainder of each chain is composed of a C (constant) domain; that of the heavy chain has several subdomains (C_{H1}, hinge, C_{H2}, and C_{H3}).

bring the promoter and enhancer into proximity, allowing transcription of the heavy-chain gene. RNA splicing removes the introns from the primary transcript, making a mature mRNA encoding a complete heavy-chain polypeptide.

The somatic rearrangements that shuffle the V, D, J, and C segments at random in each B cell permit expression of one, and only one, specific heavy chain. Without the rearrangements, antibody gene expression cannot occur. Random somatic rearrangements also generate the actual genes that will be expressed as light chains. The somatic rearrangements allowing the expression of antibodies thus generate an enormous diversity of binding sites through the random selection and recombination of gene elements.

Several other mechanisms add to this diversity. First, each gene's DNA elements are joined imprecisely, which is perpetrated by cutting and splicing enzymes that sometimes trim DNA from, or add nucleotides to, the junctions of the segments they join. This imprecise joining helps create the hypervariable regions shown in Figure A. Next, random somatic mutations in a rearranged gene's V region increase the variation of the antibody's V domain. Finally, in every B cell, two copies of a specific H chain that emerged from random DNA rearrangements combine with two copies of a specific L chain that also emerged from random DNA rearrangements to create molecules with a specific, unique binding site. The fact that any light chain can

Heavy-chain gene region

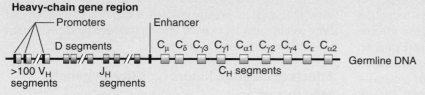

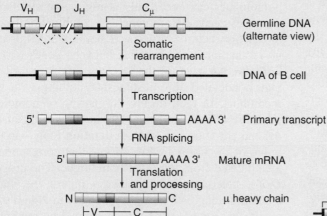

pair with any heavy chain exponentially increases the potential diversity of antibody types. For example, if there were 10^4 different light chains and 10^5 different heavy chains, there would be 10^9 possible combinations of the two.

Mistakes by the Enzymes That Carry Out Antibody Gene Rearrangements Can Lead To Cancer

RagI and RagII are enzymes that interact with DNA sequences in antibody genes to help catalyze the rearrangements just described. In carrying out their rearrangement activities, however, the enzymes sometimes make a mistake that results in a reciprocal translocation between human chromosomes 8 and 14. After this translocation, the enhancer of the chromosome 14 heavy-chain gene lies in the vicinity of the unrelated *c-myc* gene from chromosome 8. Under normal circumstances, *c-myc* generates a transcription factor that turns on other genes active in cell division, at the appropriate time and rate in the cell cycle. However, the translocated antibody-gene enhancer accelerates expression of *c-myc*, causing B cells containing the translocation to divide out of control. This uncontrolled B-cell division leads to a cancer known as Burkitt's lymphoma (**Figure C**).

Thus, although programmed gene rearrangements contribute to the normal development of a healthy immune system, misfiring of the rearrangement mechanism can promote disease.

Chapter 13 describes the evolution of the gene families that encode antibodies and other immune-system proteins.

Figure B The heavy-chain gene region on chromosome 14. The DNA of germline cells (as well as all non-antibody-producing cells) contains more than 100 V_H segments, about 20 D segments, 6 J_H segments, and 9 C_H segments (*top*). Each V_H and C_H segment is composed of two or more exons, as seen in the alternate view of the same DNA on the next line. In B cells, somatic rearrangements bring together random, individual V_H, D, and J_H segments. The primary transcript made from the newly constructed heavy-chain gene is subsequently spliced into a mature mRNA. The μ heavy chain translated from this mRNA is the type of heavy chain found in IgM antibodies. Later in B-cell development, other rearrangements (*not shown*) connect the same V-D-J variable region to other C_H segments such as C_δ, allowing the synthesis of other antibody classes.

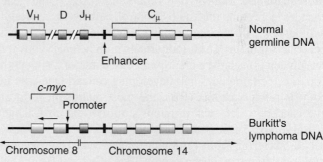

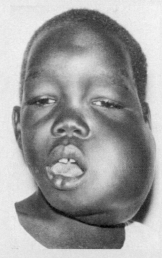

Figure C Misguided translocations can help cause Burkitt's lymphoma. In DNA from this Burkitt's lymphoma patient, a translocation brings transcription of the *c-myc* gene (*green*) under the control of the enhancer adjacent to C_μ. As a result, B cells produce abnormally high levels of the c-myc protein. Apparently, the RagI and RagII enzymes have mistakenly connected a J_H segment to the *c-myc* gene from chromosome 8, instead of to a D segment.

For some rare genes, the normal diploid level of gene expression is essential to individual survival; fewer than two copies of such a gene results in lethality. In *Drosophila*, a single dose of the locus known as *Triplolethal* (*Tpl⁺*) is lethal in an otherwise diploid individual. For certain other genes, the phenotypic consequences of a decrease in gene dosage are noticeable but not catastrophic. For example, *Drosophila* containing only one copy of the wild-type *Notch* gene have visible wing abnormalities but otherwise seem to function normally (**Figure 9.3**). In contrast with these unusual examples, diminishing the dosage of most genes produces no obvious change in phenotype. There is a catch, however. Although a single dose of any one gene may not cause substantial harm to the individual, the genetic imbalance resulting from a single dose of many genes at the same time can be lethal. Humans, for example, cannot survive, even as heterozygotes, with deletions that remove more than about 3 percent of any part of their haploid genome.

Another answer to the question of why heterozygosity for a deletion can be harmful is that with only one remaining wild-type copy of a gene, a cell is more vulnerable to subsequent mutation of that remaining copy. If the gene encodes a protein that helps control cell division, a cell without any wild-type protein may divide out of control and generate a tumour. Thus, individuals born heterozygous for certain deletions have a greatly increased risk of losing both copies of certain genes and developing cancer. One case in point is retinoblastoma (RB), the most malignant form of eye cancer, which was previously introduced in Chapter 4. Karyotypes of normal, noncancerous tissues from many people suffering from retinoblastoma reveal heterozygosity for deletions on chromosome 13. Cells from the retinal tumours of these same patients have a mutation in the remaining copy of the *RB* gene on the nondeleted chromosome 13. Chapter 16, "Somatic Mutation and the Genetics of Cancer," explains in detail how deletion of certain chromosomal regions greatly increases the risk of cancer and how researchers have used this knowledge to locate genes whose mutant forms cause cancer.

Effects of deletion heterozygosity on genetic map distances

Because recombination between maternal and paternal homologues can occur only at regions of similarity, map distances derived from genetic recombination frequencies in deletion heterozygotes will be aberrant. For example, no recombination is possible between genes *C*, *D*, and *E* in **Figure 9.4** because the DNA in this region of the normal, nondeleted chromosome has nothing with which to recombine. In fact, during the pairing of homologues in prophase of meiosis I, the "orphaned" region of the nondeleted chromosome forms a **deletion loop**—an unpaired bulge of the normal chromosome that corresponds to the area deleted from the other homologue. The progeny of a *Del/+* heterozygote will always inherit the markers in a deletion loop as a unit (*C*, *D*, and *E* in **Figure 9.4**). As a result, these genes cannot be separated by recombination, and the map distances between them, as determined by the phenotypic classes in the progeny of a *Del/+* individual, will be zero. In addition, the genetic distance between loci on either side of the deletion (such as between markers *B* and *F* in **Figure 9.4**) will be shorter than expected because fewer crossovers can occur between them.

"Uncovering" genes in deletion heterozygotes

A deletion heterozygote is, in effect, a hemizygote for genes on the normal, nondeleted chromosome that are missing from the deleted chromosome. If the normal chromosome carries a mutant recessive allele of one of these genes, the individual will exhibit the mutant phenotype. This phenomenon is sometimes called **pseudodominance**. In *Drosophila*, for example, the *scarlet* (*st*) eye colour mutation is recessive to wild type. However, an animal heterozygous for the *st* mutation and a deletion that removes the *scarlet* gene (*st/Del*) will have bright scarlet eyes, rather than wild-type, dark red eyes. In these circumstances, the deletion "uncovers" (i.e., reveals) the phenotype of the recessive mutation (**Figure 9.5**).

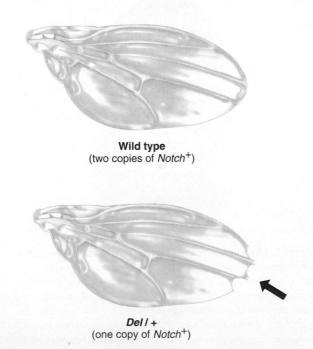

Wild type
(two copies of *Notch⁺*)

Del / +
(one copy of *Notch⁺*)

Figure 9.3 Heterozygosity for deletions may have phenotypic consequences. Flies carrying only one copy of the *Notch⁺* gene instead of the normal two copies have abnormal wings.

Normal homologue

Homologue with a deletion

Deletion loop

Figure 9.4 Deletion loops form in the chromosomes of deletion heterozygotes. During prophase of meiosis I, the undeleted region of the normal chromosome has nothing with which to pair and forms a deletion loop. No recombination can occur within the deletion loop. In this simplified figure, each line represents two chromatids.

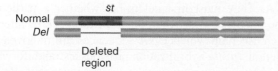

Normal *st*

Del

Deleted region

Figure 9.5 In deletion heterozygotes, pseudodominance shows that a deletion has removed a particular gene. A fly of genotype *st/Del* displays the recessive scarlet eye colour. The deletion has thus "uncovered" the scarlet (*st*) mutation.

Geneticists can use pseudodominance to determine whether a deletion has removed a particular gene. If the phenotype of a recessive-allele/deletion heterozygote is mutant, the deletion has uncovered the mutated locus; the gene thus lies inside the region of deletion. In contrast, if the trait determined by the gene is wild type in these heterozygotes, the deletion has not uncovered the recessive allele, and the gene must lie outside the deleted region. You can consider this experiment as a complementation test between the mutation and the deletion: The uncovering of a mutant recessive phenotype demonstrates a lack of complementation because neither chromosome can supply wild-type gene function.

Using deletions to locate genes

Geneticists can use deletions that alter chromosomal banding patterns to map genes relative to specific regions of metaphase chromosomes. A deletion that results in the loss of one or more bands from a chromosome and also uncovers the recessive mutation of a particular gene places that gene within the missing chromosomal segment.

The greater the number of distinguishable bands in a chromosome, the greater the accuracy of gene localization by this strategy. For this reason, specialized giant chromosomes found in the salivary gland cells of *Drosophila* larvae are a prized mapping resource. The interphase chromosomes in these cells go through ten rounds of replication without ever entering mitosis. As a result, the sister chromatids never separate, and each chromosome consists of 2^{10} (= 1024) double helixes. In addition, because the homologous chromosomes in the somatic cells of *Drosophila* remain tightly paired throughout interphase, pairs of homologues form a cable of double thickness containing 2048 double helixes of DNA (1024 from each homologue). These giant chromosomes consisting of many identical chromatids lying in parallel register are called **polytene chromosomes** (**Figure 9.6a**).

When stained and viewed under the light microscope, *Drosophila* polytene chromosomes have an irregular fine-grain banding pattern in which denser dark bands alternate with lighter interbands. The chromatin of each dark band is roughly ten times more condensed than the chromatin of the lighter interbands (**Figure 9.6b**). Scientists do not yet understand the functional significance of bands and interbands. One possibility is that the bands represent units of transcriptional regulation containing genes activated at the same time. In any event, the precisely reproducible banding patterns of polytene chromosomes provide a detailed physical

(a) Banding pattern of *Drosophila* polytene chromosomes

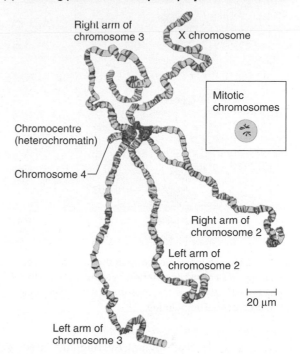

Right arm of chromosome 3

X chromosome

Mitotic chromosomes

Chromocentre (heterochromatin)

Chromosome 4

Right arm of chromosome 2

Left arm of chromosome 2

20 µm

Left arm of chromosome 3

(b) Alignment of chromatids in polytene chromosomes

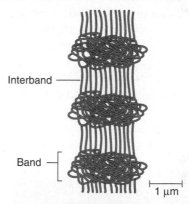

Interband

Band

1 µm

Figure 9.6 Polytene chromosomes in the salivary glands of *Drosophila* larvae. **(a)** A drawing of the banding pattern seen in polytene chromosomes. The inset shows the relative size of normal mitotic chromosomes. Note that the homologous polytene chromosomes are paired along their lengths. **(b)** A hypothetical model showing how the 1024 chromatids of each polytene chromosome are aligned in register, with the chromatin in the bands being more condensed than the chromatin of the interbands.

guide to gene mapping. *Drosophila* polytene chromosomes collectively carry about 5000 bands that range in size from 3 kb to approximately 150 kb; investigators designate these bands by numbers and letters of the alphabet.

Because homologous polytene chromosomes pair with each other, deletion loops form in the polytene chromosomes of deletion heterozygotes (**Figure 9.7**). Scientists can pinpoint the region of the deletion by noting which bands are present in the wild-type homologue but missing in the deletion. If researchers find that a small deletion removing only a few polytene chromosome bands

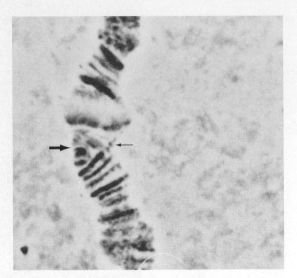

Figure 9.7 Deletion loops also form in the paired polytene chromosomes of *Drosophila* deletion heterozygotes. The *thick arrow* points to the wild-type chromosome; the corresponding region is missing from the *Del* homologue.

uncovers a gene or that several overlapping larger deletions affect the same gene, they can assign the gene to one or a small number of bands, often representing less than 100 kb of DNA. **Figure 9.8** shows how geneticists used this strategy to assign three genes to regions containing only one or two polytene chromosome bands on the *Drosophila* X chromosome.

Geneticists can use deletions analyzed at even higher levels of resolution to help locate genes on cloned fragments

of DNA. They must first determine whether a particular deletion uncovers a recessive allele of the gene of interest and then ascertain which DNA sequences are removed by the deletion. *In situ* hybridization provides a straightforward way to show whether a particular DNA sequence is part of a deletion. Suppose you are trying to determine whether a small segment of the *Drosophila* X chromosome in the vicinity of the *white* gene has been deleted. You could use purified DNA fragments as probes for *in situ* hybridization to polytene chromosomes prepared from female flies heterozygous for various deletions in this region of their X chromosomes. If a probe hybridizes to a *Del* chromosome, the deletion has not completely removed that particular fragment of DNA; lack of a hybridization signal on a *Del* chromosome, however, indicates that the fragment has been deleted (**Figure 9.9**).

Geneticists can also localize deleted regions by asking whether particular bands are removed from human mitotic chromosomes, but because bands in these chromosomes that contain less than 5 Mb of DNA cannot be detected visually, the resolution of this method is much lower than is possible with *Drosophila* polytene chromosomes. As the final section of this chapter on "Emergent Technologies: Beyond the Karyotype" illustrates, new techniques nonetheless allow human geneticists to determine the molecular extent of deletions in human chromosomes. Once this

(a) *In situ* **hybridization of the *white* gene to wild-type polytene chromosomes**

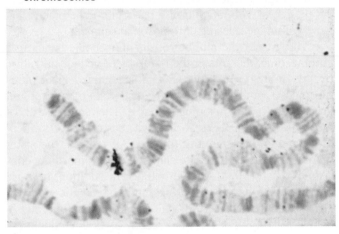

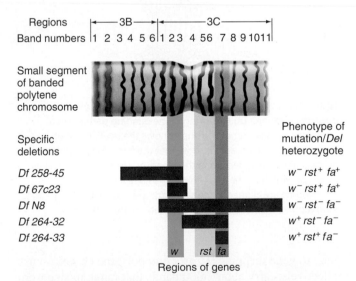

Figure 9.8 Using deletions to assign genes to bands on *Drosophila* polytene chromosomes. *Red bars* show the bands removed by various deletions; for example, *Df 258-45* eliminates bands 3B3–3C3. Complementation experiments determined whether these deletions uncovered the *white* (*w*), *roughest* (*rst*), or *facet* (*fa*) genes. For instance, *w/Df 258-45* females have white eyes, so the *w* gene is removed by this deletion. The *w* gene must lie within bands 3C2–3 (*green*) because that is the region common to the deletions that uncover *w*. Similarly, *rst* must be in bands 3C5–6 (*yellow*) and *fa* in band 3C7 (*purple*).

(b) Characterizing deletions with *in situ* hybridization to polytene chromosomes

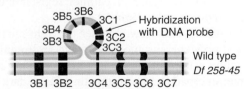

Figure 9.9 *In situ* hybridization as a tool for locating genes at the molecular level. (a) *In situ* hybridization of a probe containing the *white* gene to a single band (3C2) near the tip of the wild-type *Drosophila* X chromosome. **(b)** A particular labelled probe hybridizes to the wild-type chromosome but not to the deletion chromosome in a *Df 258-45/+* heterozygote. The *Df 258-45* deletion thus lacks DNA homologous to the probe.

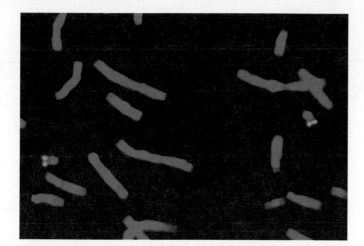

Figure 9.10 Diagnosing DiGeorge syndrome by fluorescence *in situ* hybridization (FISH) to human metaphase chromosomes. The *green* signal is a control probe that identifies both chromosome 22's. The *red* signal is a fluorescent probe from region 22q11, which is deleted in one of the chromosome 22's in DiGeorge syndrome patients. These homologous metaphase chromosomes do not pair with each other and thus do not form a deletion loop.

information is available, *in situ* hybridization to human mitotic chromosomes serves as a useful tool to diagnose whether individuals have genetic diseases associated with heterozygosity for particular deletions. **Figure 9.10** shows an application of this strategy to the diagnosis of DiGeorge syndrome, which accounts for approximately 5 percent of all congenital heart malformations.

Homozygosity or even heterozygosity for deletions can be lethal or harmful; the effects depend on the size of the deletion and the identity of the deleted genes. In deletion heterozygotes, deletions reveal or "uncover" recessive mutations on the intact homologue because the phenotype is no longer masked by the presence of a dominant wild-type allele. Geneticists can use these properties of deletions to map and identify genes.

Duplications add material to the genome

Duplications increase the number of copies of a particular chromosomal region. In **tandem duplications**, repeats of a region lie adjacent to each other, either in the same order or in reverse order (**Figure 9.11a**). In **nontandem** (or *dispersed*) **duplications**, the two or more copies of a region are not adjacent to each other and may lie far apart on the same chromosome or on different chromosomes. Duplications arise by chromosomal breakage and faulty repair, unequal crossing-over, or errors in DNA replication (**Figure 9.11b**). In this book, we use *Dp* as the symbol for a chromosome carrying a duplication.

Most duplications have no obvious phenotypic consequences and can be detected only by cytological or molecular means. Sufficiently large duplications, for example, show up as repeated bands in metaphase or polytene chromosomes. During prophase of meiosis I in heterozygotes for such

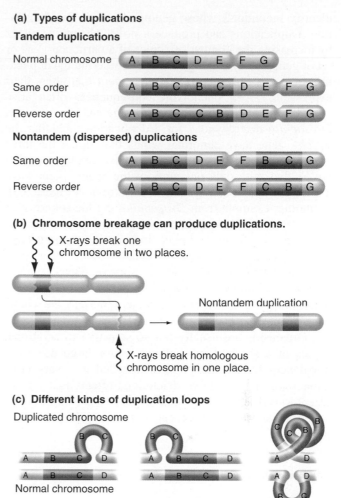

Figure 9.11 Duplications: Structure, origin, and detection. **(a)** In tandem duplications, the repeated regions lie adjacent to each other in the same or in reverse order. In nontandem duplications, the two copies of the same region are separated. **(b)** In one scenario for duplication formation, X-rays break one chromosome twice and its homologue once. A fragment of the first chromosome inserts elsewhere on its homologue to produce a nontandem duplication. **(c)** Duplication loops form when chromosomes pair in duplication heterozygotes (*Dp/+*). During prophase I, the duplication loop can assume different configurations. A single line represents two chromatids in this simplified diagram.

duplications (*Dp/+*), the repeated bands form a **duplication loop**—a bulge in the *Dp*-bearing chromosome that has no similar region with which to pair in the unduplicated normal homologous chromosome. Duplication loops can occur in several alternative configurations (**Figure 9.11c**). Such loops also form in the polytene chromosomes of *Drosophila* duplication heterozygotes, where the pattern of the bands in the duplication loops is a repeat of that seen in the other copy of the same region elsewhere on the chromosome.

Phenotypic effects of duplications

Although duplications are much less likely to affect phenotype than are deletions of comparable size, some duplications do have phenotypic consequences for visible traits or for survival. Geneticists can use such phenotypes to

identify individuals whose genomes contain the duplication. Duplications can produce a novel phenotype either by increasing the number of copies of a particular gene or set of genes, or by placing the genes bordering the duplication in a new chromosomal environment that alters their expression. These phenotypic consequences often arise even in duplication heterozygotes (*Dp/+*). For example, *Drosophila* heterozygous for a duplication including the *Notch*$^+$ gene have abnormal wings that signal the three copies of *Notch*$^+$ (**Figure 9.12a**); we have already seen that *Del/+* flies with only one copy of the *Notch*$^+$ gene have a different kind of wing abnormality (review Figure 9.3). In another example from *Drosophila*, the locus known as *Triplolethal* (*Tpl*$^+$) is lethal when present in one or three doses in an otherwise diploid individual (**Figure 9.12b**). Thus, heterozygotes for a *Tpl* deletion (*Del/+*) or for a *Tpl*$^+$ duplication (*Dp/+*) do not survive. Heterozygotes carrying one homologue deleted for the locus and the other homologue duplicated for the locus (*Del/Dp*) are viable because they have two copies of *Tpl*$^+$.

Organisms are usually not so sensitive to additional copies of a single gene; but just as for large deletions, imbalances for the many genes included in a very large duplication have additive deleterious effects that jeopardize survival. In humans, heterozygosity for duplications covering more than 5 percent of the haploid genome is most often lethal.

Unequal crossing-over between duplications

In individuals homozygous for a tandem duplication (*Dp/Dp*), homologues carrying the duplications occasionally pair out of register during meiosis. **Unequal crossing-over**, that is, recombination resulting from such out-of-register pairing, generates gametes containing increases to three and reciprocal decreases to one in the number of copies of the duplicated region. In *Drosophila*, tandem duplication of several polytene bands near the X chromosome centromere produces the Bar phenotype of kidney-shaped eyes. *Drosophila* females homozygous for the Bar-eye duplication produce mostly Bar-eye progeny. Some progeny, however, have wild-type eyes, whereas other progeny have double-Bar eyes that are even smaller than Bar eyes (**Figure 9.13**). The genetic explanation is that flies with wild-type eyes carry X chromosomes containing only one copy of the region in question, flies with Bar eyes have X chromosomes containing two copies of the region, and flies with double-Bar eyes have X chromosomes carrying three copies. Unequal crossing-over in females homozygous for double-Bar chromosomes can yield progeny with even more extreme phenotypes associated with four or five copies of the duplicated region. Duplications in homozygotes thus allow for the expansion and contraction of the number of copies of a chromosomal region from one generation to the next.

(a) Duplication heterozygosity can cause visible phenotypes.

Wild-type wing: two copies of *Notch*$^+$ gene   Three copies of *Notch*$^+$ gene   Aberrant wing veins

(b) For rare genes, survival requires exactly two copies.

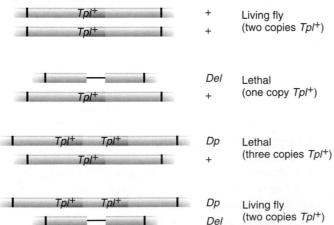

Tpl$^+$ / *Tpl*$^+$	+ / +	Living fly (two copies *Tpl*$^+$)
Del / *Tpl*$^+$	*Del* / +	Lethal (one copy *Tpl*$^+$)
Tpl$^+$ *Tpl*$^+$ / *Tpl*$^+$	*Dp* / +	Lethal (three copies *Tpl*$^+$)
Tpl$^+$ *Tpl*$^+$ /	*Dp* / *Del*	Living fly (two copies *Tpl*$^+$)

Figure 9.12 The phenotypic consequences of duplications. (a) Duplication heterozygotes (*Dp/+*) have three copies of genes contained in the duplication. Flies with three copies of the *Notch*$^+$ gene have aberrant wing veins. This phenotype differs from that caused by only one copy of *Notch*$^+$ (see **Figure 9.3**). **(b)** In *Drosophila*, three copies or one copy of *Tpl*$^+$ are lethal.

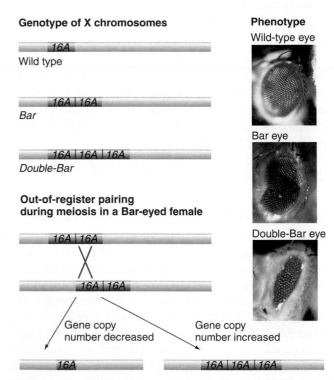

Genotype of X chromosomes   **Phenotype**

Wild-type eye

16A
Wild type

16A | 16A
Bar

Bar eye

16A | 16A | 16A
Double-Bar

Double-Bar eye

Out-of-register pairing during meiosis in a Bar-eyed female

16A | 16A

16A | 16A

Gene copy number decreased   Gene copy number increased

16A   *16A | 16A | 16A*

Figure 9.13 Unequal crossing-over can increase or decrease copy number. Duplication of the X-chromosome polytene region 16A causes Bar eyes. Unequal pairing and crossing-over during meiosis in females homozygous for this duplication produce chromosomes that have either one copy of region 16A (conferring normal eyes) or three copies of 16A (causing the more abnormal double-Bar eyes).

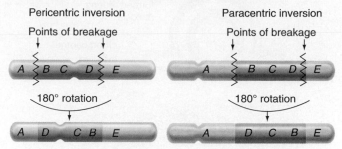

(a) Chromosome breakage can produce inversions.

Pericentric inversion

Points of breakage

Paracentric inversion

Points of breakage

180° rotation

180° rotation

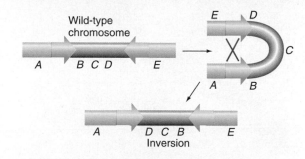

(b) Intrachromosomal recombination can also cause inversions.

Wild-type chromosome

Inversion

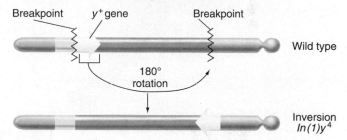

(c) Inversions can disrupt gene function.

Breakpoint y^+ gene Breakpoint

Wild type

180° rotation

Inversion $In(1)y^4$

Figure 9.14 Inversions: Origins, types, and phenotypic effects. **(a)** Inversions can arise when chromosome breakage produces a DNA segment that rotates 180° before it reattaches. When the rotated segment includes the centromere, the inversion is *pericentric*; when the rotated segment does not include the centromere, the inversion is *paracentric*. **(b)** If a chromosome has two copies of a sequence in reverse orientation, rare intrachromosomal recombination can give rise to an inversion. **(c)** An inversion can affect phenotype if it disrupts a gene. Here, the inversion $In(1)y^4$ inactivates the *y* (*yellow*) gene by dividing it in two.

A duplication heterozygote has three copies of a particular chromosomal region, even though the remainder of the genome is diploid. The resulting genetic imbalance can have harmful or even lethal effects, depending on the size of the duplication and the identity of the duplicated genes. Unequal crossing-over between homologous chromosomes bearing the same duplicated region can lead to increases and reciprocal decreases in the number of copies of that region.

Inversions reorganize the DNA sequence of a chromosome

The half-circle rotation of a chromosomal region known as an **inversion (*In*)** can occur when radiation produces two double-strand breaks in a chromosome's DNA. The breaks release a middle fragment, which may turn 180° before religation to the flanking chromosomal regions, resulting in an inversion (**Figure 9.14a**). Inversions may also result from rare crossovers between related DNA sequences present in two positions on the same chromosome in inverted orientation (**Figure 9.14b**), or they may arise by the action of transposable genetic elements (discussed later in this chapter). Inversions that include the centromere are **pericentric**, while inversions that exclude the centromere are **paracentric** (see Figure 9.14a).

Phenotypic effects of inversions

Most inversions do not result in an abnormal phenotype, because even though they alter the order of genes along the chromosome, they do not add or remove DNA and therefore do not change the identity or number of genes. Geneticists can detect some inversions that do not affect phenotype, especially those that cause cytologically visible changes in banding patterns or those that suppress recombination in heterozygotes (as described later) and thereby change the expected results of linkage analysis. In natural populations, however, many inversions that do not affect phenotype go undetected.

If one end of an inversion lies within the DNA of a gene (**Figure 9.14c**), a novel phenotype can occur. Inversion following an intragenic break separates the two parts of the gene, relocating one part to a distant region of the chromosome, while leaving the other part at its original site. Such a split disrupts the gene's function. If that function is essential to viability, the inversion acts as a recessive lethal mutation, and homozygotes for the inversion will not survive.

Inversions can also produce unusual phenotypes by moving genes residing near the inversion breakpoints to chromosomal environments that alter their normal expression. For example, mutations in the *Antennapedia* gene of *Drosophila* that transform antennae into legs (review Figure 7.24d) are inversions that place the gene in a new regulatory environment, next to sequences that cause it to be transcribed in tissues where it would normally remain unexpressed. Inversions that reposition genes normally found in a chromosome's euchromatin to a position near a region of heterochromatin can also produce an unusual phenotype; spreading of the heterochromatin may inactivate the gene in some cells, even if it is still expressed in others, leading to **position-effect variegation (PEV)**.

Inversion heterozygosity and crossover suppression

Individuals heterozygous for an inversion (*In/+*) are *inversion heterozygotes*. In such individuals, when the chromosome carrying the inversion pairs with its homologue at meiosis, formation of an **inversion loop** allows the tightest possible alignment of homologous regions. In an inversion loop, one chromosomal region rotates

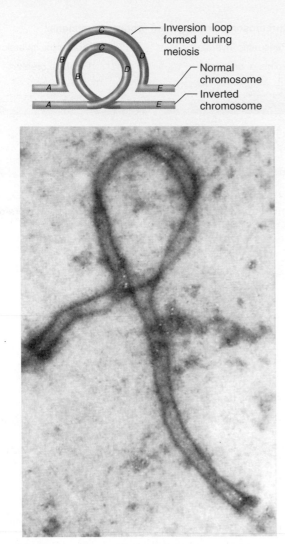

Figure 9.15 Inversion loops form in inversion heterozygotes.
To maximize pairing during prophase of meiosis I in an inversion heterozygote (*In/+*), homologous regions form an inversion loop. (*Top*) Simplified diagram in which one line represents a pair of sister chromatids. (*Bottom*) Electron micrograph of an inversion loop during meiosis I in an *In/+* mouse.

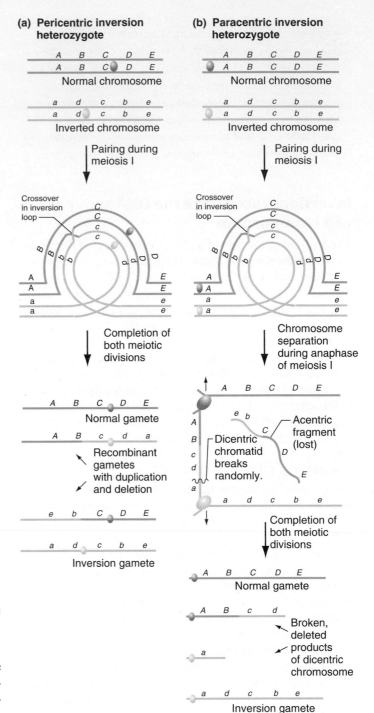

Figure 9.16 Why inversion heterozygotes do not produce recombinant progeny. Throughout this figure, each line represents one chromatid, and different shades of *green* indicate the two homologous chromosomes. **(a)** The chromatids formed by recombination within the inversion loop of a pericentric inversion heterozygote are genetically unbalanced. **(b)** The chromatids formed by recombination within the inversion loop of a paracentric inversion heterozygote are not only genetically unbalanced but also contain two or no centromeres, instead of the normal one.

to conform to the similar region in the other homologue (**Figure 9.15**). Crossing-over within an inversion loop produces aberrant recombinant chromatids whether the inversion is pericentric or paracentric.

If the inversion is pericentric and a single crossover occurs within the inversion loop, each recombinant chromatid will have a single centromere—the normal number—but will carry a duplication of one region and a deletion of a different region that are reciprocal (**Figure 9.16a**). Gametes carrying these recombinant chromatids will have an abnormal dosage of some genes. After fertilization, zygotes created by the union of these abnormal gametes with normal gametes will die because of genetic imbalance.

If the inversion is paracentric and a single crossover occurs within the inversion loop, the recombinant chromatids will be unbalanced not only in gene dosage but also in centromere number (**Figure 9.16b**). One crossover product will be an **acentric fragment** lacking a

centromere, whereas the reciprocal crossover product will be a **dicentric chromatid** with two centromeres. Because the acentric fragment without a centromere cannot attach to the spindle apparatus during the first meiotic division, the

cell cannot package it into either of the daughter nuclei; as a result, this chromosome is lost and will not be included in a gamete. By contrast, at anaphase of meiosis I, opposing spindle forces pull the dicentric chromatid toward both spindle poles at the same time with such strength that the dicentric chromatid breaks at random positions along the chromosome. These broken chromosome fragments are deleted for many of their genes. This loss of the acentric fragment and breakage of the dicentric chromatid results in genetically unbalanced gametes, which at fertilization will produce lethally unbalanced zygotes that cannot develop beyond the earliest stages of embryonic development. Consequently, no recombinant progeny resulting from a crossover in a paracentric inversion loop survive. Any surviving progeny are nonrecombinants.

In summary, whether an inversion is pericentric or paracentric, crossing-over within the inversion loop of an inversion heterozygote has the same effect—formation of recombinant gametes that after fertilization prevent the zygote from developing. Because only gametes containing chromosomes that did not recombine within the inversion loop can yield viable progeny, inversions act as **crossover suppressors**. This does not mean that crossovers do not occur within inversion loops, but simply that there are no recombinants among the viable progeny of an inversion heterozygote.

Geneticists use crossover suppression to create **balancer chromosomes**, which contain multiple, overlapping inversions (both pericentric and paracentric), as well as a marker mutation that produces a visible dominant phenotype (**Figure 9.17**). The viable progeny of a *Balancer/+* heterozygote will receive either the balancer or the chromosome of normal order (+), but they cannot inherit a recombinant chromosome containing parts of both. Researchers can distinguish these two types of viable progeny by the presence or absence of the dominant marker phenotype. Geneticists often generate balancer heterozygotes to ensure that a chromosome of normal order, along with any mutations of interest it may carry, is transmitted to the next generation unchanged by recombination. To help create genetic stocks, the marker in most balancer chromosomes not only causes a dominant visible phenotype, but it also acts as a recessive lethal mutation that prevents the survival of balancer chromosome homozygotes. The *Drosophila* portrait on Connect discusses this and other significant uses of balancer chromosomes in genetic analysis.

> Although inversions do not add or remove DNA, they can alter phenotype if they disrupt a gene or alter its expression. In inversion heterozygotes, recombination within the inversion loop yields genetically unbalanced gametes that produce inviable zygotes. Geneticists can take advantage of this property to create balancer chromosomes that are useful in the production of genetic lines of known composition.

Translocations change the position of chromosomal segments

Translocations are large-scale mutations in which part of one chromosome becomes attached to another region of the same chromosome (**nonreciprocal intrachromosomal translocation**) or to a nonhomologous chromosome (**nonreciprocal interchromosomal translocation**). When parts of two nonhomologous chromosomes trade places, it is known as a **reciprocal translocation** (**Figure 9.18a**). It results when two breaks, one in each of two chromosomes, yield DNA fragments that do not religate to their chromosome of origin; rather, they switch places and become attached to the other chromosome. Depending on the positions of the breaks and the sizes of the exchanged fragments, the translocated chromosomes may be so different from the original chromosomes that the translocation is visible in a cytological examination (**Figure 9.18b**).

Robertsonian translocations are an important type of cytologically visible reciprocal translocation arising from breaks at or near the centromeres of two acrocentric chromosomes (**Figure 9.19**). The reciprocal exchange of broken parts generates one large metacentric chromosome and one very small chromosome containing few, if any, genes. This tiny chromosome may subsequently be lost from the organism. Robertsonian translocations are named after W. R. B. Robertson, who in 1911 was the first to suggest that during evolution, metacentric chromosomes may arise from the fusion of two acrocentrics.

Phenotypic effects of reciprocal translocations

Most individuals bearing reciprocal translocations are phenotypically normal because they have neither lost nor

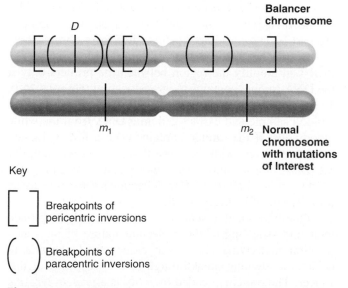

Figure 9.17 Balancer chromosomes are useful tools for genetic analysis. Balancer chromosomes carry both a dominant marker *D* as well as inversions (*brackets*) that prevent the balancer chromosome from recombining with an experimental chromosome carrying mutations of interest (m_1 and m_2). A parent heterozygous for the balancer and experimental chromosomes will transmit either the balancer or the experimental chromosome, but not a recombinant chromosome, to its surviving progeny.

(a) Two chromosome breaks can produce a reciprocal translocation.

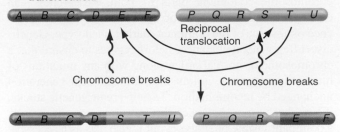

(b) Chromosome painting reveals a reciprocal translocation.

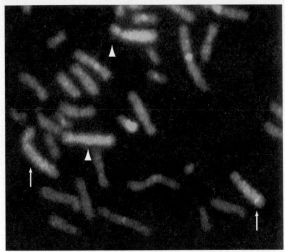

Figure 9.18 Reciprocal translocations are exchanges between nonhomologous chromosomes. (a) In a reciprocal translocation, the region gained by one chromosome is the region lost by the other chromosome. **(b)** Karyotype of a human genome containing a translocation. The two translocated chromosomes are stained both red *and* green (*arrows*). Two normal, nontranslocated chromosomes are stained entirely red *or* entirely green (*arrowheads*), indicating that this person is heterozygous for the translocation.

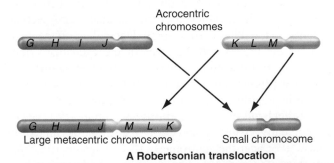

A Robertsonian translocation

Figure 9.19 Robertsonian translocations can reshape genomes. In a Robertsonian translocation, reciprocal exchanges between two acrocentric chromosomes generate a large metacentric chromosome and a very small chromosome. The latter may carry so few genes that it can be lost without ill effect.

gained genetic material. As with inversions, however, if one of the translocation breakpoints occurs within a gene, that gene's function may change or be destroyed. Or if the translocation places a gene normally found in the euchromatin of one chromosome near the heterochromatin of

the other chromosome, normal expression of the gene may cease in some cells, giving rise to position-effect variegation.

Several kinds of cancer are associated with translocations in somatic cells. In normal cells, genes known as *proto-oncogenes* help control cell division. Translocations that relocate these genes can turn them into tumour-producing *oncogenes* whose protein products have an altered structure or level of expression that leads to runaway cell division. For example, in almost all patients with chronic myelogenous leukaemia (CML), a type of cancer caused by overproduction of certain white blood cells, the leukaemic cells have a reciprocal translocation between chromosomes 9 and 22 (**Figure 9.20**). The breakpoint in chromosome 9 occurs within an intron of a proto-oncogene called *c-abl*; the breakpoint in chromosome 22 occurs within an intron of the *bcr* gene. After the translocation, parts of the two genes are adjacent to one another. During transcription, the RNA-producing machinery runs these two genes together, creating a long primary transcript. After splicing, the mRNA is translated into a fused protein in which 25 amino acids at the N terminus of the *c-abl*-determined protein are replaced by about 600 amino acids from the *bcr*-determined protein. The activity of this fused protein releases the normal controls on cell division, leading to leukaemia. (See the Fast Forward box in this chapter for another example of a translocation-induced cancer called Burkitt's lymphoma.)

Medical practitioners can exploit the rearrangement of DNA sequences that accompany cancer-related translocations for diagnostic and therapeutic purposes. To confirm a diagnosis of myelogenous leukaemia, for example, they first obtain a blood sample from the patient, and they then use a pair of PCR primers derived from opposite sides of the breakpoint—one synthesized from the appropriate part of chromosome 22, the other from chromosome 9—to carry out a PCR on DNA from the blood cells. The PCR will amplify the region between the primers only if the DNA sample contains the translocation (**Figure 9.20b**). To monitor the effects of chemotherapy, they again obtain a blood sample and extract genomic DNA from the white blood cells. If the sample contains even a few malignant cells, a PCR test with the same two primers will amplify the DNA translocation from those cells, indicating the need for more therapy. PCR thus becomes a sensitive assay for this type of leukaemic cell.

Pharmaceutical researchers have recently exploited their understanding of the molecular nature of the translocation underlying chronic myelogenous leukaemia to achieve a stunning breakthrough in the treatment of this cancer. The protein encoded by *c-abl* is a *protein tyrosine kinase*, an enzyme that adds phosphate groups to tyrosine amino acids on other proteins. This enzyme is an essential part of the set of signals that dictate cell growth and division. Normal cells closely regulate the activity of the *c-abl* protein, blocking its function most of the time but activating it in response to stimulation by growth factors in the environment. By contrast, the fused protein encoded

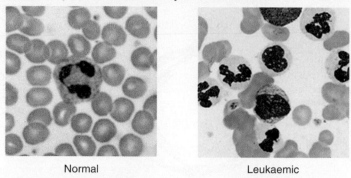

(a) Leukaemia patients have too many white blood cells.

Normal

Leukaemic

(b) The genetic basis for chronic myelogenous leukaemia

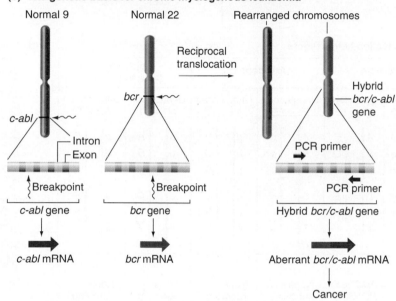

Figure 9.20 How a reciprocal translocation helps cause one kind of leukaemia. (a) Uncontrolled divisions of large, dark-staining white blood cells in a leukaemia patient (*right*) produce a higher ratio of white to red blood cells than that in a normal individual (*left*). **(b)** A reciprocal translocation between chromosomes 9 and 22 contributes to chronic myelogenous leukaemia. This rearrangement makes an abnormal hybrid gene composed of part of the *c-abl* gene and part of the *bcr* gene. The hybrid gene encodes an abnormal fused protein that disrupts controls on cell division. *Black arrows* indicate PCR primers that will generate a PCR product only from DNA containing the hybrid gene.

by *bcr/c-abl* in cells carrying the translocation is not amenable to regulation. It is always active, even in the absence of growth factor, and this leads to runaway cell division. Pharmaceutical companies have developed a drug called Gleevec® that specifically inhibits the enzymatic activity of the protein tyrosine kinase encoded by *bcr/c-abl*. In clinical trials, 98 percent of participants experienced a complete disappearance of leukaemic blood cells and the return of normal white cells. This drug is now the standard treatment for chronic myelogenous leukaemia and is a model for new types of cancer treatments that home in on cancer cells without hurting healthy ones.

Diminished fertility and pseudolinkage in translocation heterozygotes

Translocations, like inversions, produce no significant genetic consequences in homozygotes if the breakpoints do not interfere with gene function. During meiosis in a translocation homozygote, chromosomes segregate normally according to Mendelian principles (**Figure 9.21a**). Even though the genes have been rearranged, both haploid sets of chromosomes in the individual have the same rearrangement. As a result, all chromosomes will find a single partner with which to pair at meiosis, and there will be no deleterious consequences for the progeny.

In translocation heterozygotes, however, certain patterns of chromosome segregation during meiosis produce genetically unbalanced gametes that at fertilization become deleterious to the zygote. In a translocation heterozygote, the two haploid sets of chromosomes do not carry the same arrangement of genetic information. As a result, during prophase of the first meiotic division, the translocated chromosomes and their normal homologues assume a crosslike configuration in which four chromosomes, rather than the normal two, pair to achieve a maximum of synapsis between similar regions (**Figure 9.21b**). To keep track of the four chromosomes participating in this crosslike structure, we

(a) Segregation in a translocation homozygote

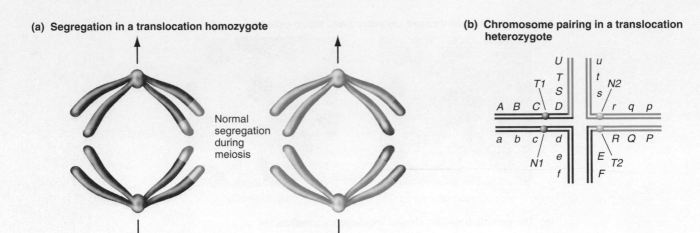

Normal segregation during meiosis

(b) Chromosome pairing in a translocation heterozygote

(c) Segregation in a translocation heterozygote

Segregation pattern	Alternate		Adjacent-1		Adjacent-2 (less frequent)	
	Balanced N1 + N2	Balanced T1 + T2	Unbalanced T1 + N2	Unbalanced N1 + T2	Unbalanced N1 + T1	Unbalanced N2 + T2
Gametes	a b c d e f	p q r s t u	A B C D S T U	P Q R E F	A B C D S T U	p q r s t u	a b c d e f	P Q R E F	a b c d e f	A B C D S T U	p q r s t u	P Q R E F
Type of progeny when mated with normal abcdefpqrstu homozygote	abcdef pqrstu	ABCDEF PQRSTU	None surviving	None surviving	None surviving	None surviving

(d) Semisterility in corn

Figure 9.21 The meiotic segregation of reciprocal translocations. (In all parts of this figure, each *bar* or *line* represents one chromatid.) **(a)** In a translocation homozygote (*T/T*), chromosomes segregate normally during meiosis I. **(b)** In a translocation heterozygote (*T/+*), the four relevant chromosomes assume a cruciform (crosslike) configuration to maximize pairing. The alleles of genes on chromosomes in the original order (*N1* and *N2*) are shown in *lowercase*; the alleles of these genes on the translocated chromosomes (*T1* and *T2*) are in *uppercase letters*. **(c)** Three segregation patterns are possible in a translocation heterozygote. Only the alternate segregation pattern gives rise to balanced gametes. **(d)** This semisterile ear of corn comes from a plant heterozygous for a reciprocal translocation. It has fewer kernels than normal because unbalanced ovules are aborted.

denote the chromosomes carrying translocated material with a *T* and the chromosomes with a normal order of genes with an *N*. Chromosomes *N1* and *T1* have homologous centromeres found in wild type on chromosome 1; *N2* and *T2* have centromeres found in wild type on chromosome 2.

During anaphase of meiosis I, the mechanisms that attach the spindle to the chromosomes in this crosslike configuration still usually ensure the disjunction of homologous centromeres, bringing homologous chromosomes to opposite spindle poles (i.e., *T1* and *N1* go to opposite poles, as do *T2* and *N2*). Depending on the arrangement of the four chromosomes on the metaphase plate, this normal disjunction of homologues produces one of two equally likely patterns of segregation (**Figure 9.21c**). In the **alternate segregation pattern**, the two translocation chromosomes (*T1* and *T2*) go to one pole, while the two normal chromosomes (*N1* and *N2*) move to the opposite pole. Both kinds of gametes resulting

from this segregation (*T1*, *T2* and *N1*, *N2*) carry the correct haploid number of genes, and the zygotes formed by union of these gametes with a normal gamete will be viable. By contrast, in the **adjacent-1 segregation pattern**, homologous centromeres disjoin so that *T1* and *N2* go to one pole, while *N1* and *T2* go to the opposite pole. As a result, each gamete contains a large duplication (of the region found in both the normal and the translocated chromosome in that gamete) and a correspondingly large deletion (of the region found in neither of the chromosomes in that gamete), which make them genetically unbalanced. Zygotes formed by union of these gametes with a normal gamete are usually not viable.

Because of the unusual cruciform pairing configuration in translocation heterozygotes, nondisjunction of homologous centromeres occurs at a measurable but low rate (approximately 4 percent of the time). This nondisjunction produces an **adjacent-2 segregation pattern** in which the homologous

centromeres *N1* and *T1* go to the same spindle pole, while the homologous centromeres *N2* and *T2* go to the other spindle pole (**Figure 9.21c**). The resulting genetic imbalances are lethal after fertilization to the zygotes containing them.

Thus, of all the gametes generated by translocation heterozygotes, only those arising from alternate segregation, which account for slightly less than half the total, can produce viable progeny when crossed with individuals who do not carry the translocation. As a result, the fertility of most translocation heterozygotes, that is, their capacity for generating viable offspring, is diminished by at least 50 percent. This condition is known as **semisterility**. Corn plants illustrate the correlation between translocation heterozygosity and semisterility. The demise of genetically unbalanced ovules produces gaps in the ear where kernels would normally appear (**Figure 9.21d**); in addition, genetically unbalanced pollen grains are abnormally small (not shown).

The semisterility of translocation heterozygotes undermines the potential of genes on the two translocated chromosomes to assort independently. Mendel's second law requires that all gametes resulting from both possible metaphase alignments of two chromosomal pairs produce viable progeny. But as we have seen, in a translocation heterozygote, only the alternate segregation pattern yields viable progeny in outcrosses; the equally likely adjacent-1 pattern and the rare adjacent-2 pattern do not. Because of this, genes near the translocation breakpoints on the nonhomologous chromosomes participating in a reciprocal translocation exhibit **pseudolinkage**: They behave as if they are linked.

Figure 9.21c illustrates why pseudolinkage occurs in a translocation heterozygote. In the figure, lowercase *a b c d e f* represent the alleles of genes present on normal chromosome 1 (*N1*), and *p q r s t u* are the alleles of genes on a nonhomologous normal chromosome 2 (*N2*). The alleles of these genes on the translocated chromosomes *T1* and *T2* are in uppercase. In the absence of recombination, Mendel's law of independent assortment would predict that genes on two different chromosomes will appear in four types of gametes in equal frequencies; for example, *a p*, *A P*, *a P*, and *A p*. But alternate segregation, the only pattern that can give rise to viable progeny, produces only *a p* and *A P* gametes. Thus, in translocation heterozygotes such as these, the genes on the two nonhomologous chromosomes act as if they are linked to each other.

Translocations and gene mapping

In humans, approximately 1 of every 500 individuals is heterozygous for some kind of translocation. While most such people are phenotypically normal, their fertility is diminished because many of the zygotes they produce abort spontaneously. As we have seen, this semisterility results from genetic imbalances associated with gametes formed by adjacent-1 or adjacent-2 segregation patterns. But such genetic imbalances are not inevitably lethal to the zygotes. If the duplicated or deleted regions are very small, the unbalanced gametes generated by these modes of segregation may produce children.

An important example of this phenomenon is seen among individuals heterozygous for certain reciprocal translocations involving chromosome 21, such as the Robertsonian translocation shown in **Figure 9.22**. These

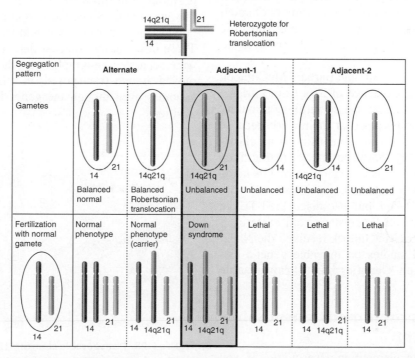

Figure 9.22 How translocation Down syndrome arises. In heterozygotes for a translocation involving chromosome 21, such as 14q21q (a Robertsonian translocation between chromosomes 21 and 14), adjacent-1 segregation can produce gametes with two copies of part of chromosome 21. If such a gamete unites with a normal gamete, the resulting zygote will have three copies of part of chromosome 21. Depending on which region of chromosome 21 is present in three copies, this tripling may cause Down syndrome. (In the original translocation heterozygote, the small, reciprocally translocated chromosome [14p21p] has been lost.)

people are phenotypically normal but produce some gametes from the adjacent-1 segregation pattern that have two copies of a part of chromosome 21 near the tip of its long arm. At fertilization, if a gamete with the duplication unites with a normal gamete, the resulting child will have three copies of this region of chromosome 21. A few individuals affected by Down syndrome have, in this way, inherited a third copy of only a small part of chromosome 21. These individuals with **translocation Down syndrome** provide evidence that the entirety of chromosome 21 need not be present in three copies to generate the phenotype.

Geneticists are now mapping the chromosome 21 regions duplicated in translocation Down syndrome patients to find the one or more genes responsible for the syndrome. Although chromosome 21 is the smallest human autosome, it nevertheless contains an estimated 350 genes, most of them in the 43 million base pairs of its long arm. The mapping of genes relative to the breakpoints of one or more such translocations considerably simplifies the task of identifying those genes that in triplicate produce the symptoms of Down syndrome. One way to locate which parts of chromosome 21 are responsible for Down syndrome is to obtain cloned chromosome 21 sequences from the Human Genome Project and then use these clones as FISH (fluorescence *in situ* hybridization) probes for the genome of the translocation Down syndrome patient. If the probe lights up the translocation chromosome as well as the two normal copies of chromosome 21, it identifies a region of the genome that is of potential importance to the syndrome.

> In a reciprocal translocation, parts of two nonhomologous chromosomes trade places without any net loss or gain of DNA. As with inversions, reciprocal translocations can alter phenotype if they disrupt a gene or its expression. Translocation heterozygotes produce genetically unbalanced gametes from two of three possible meiotic segregation patterns; the result is semisterility and pseudolinkage.

9.2 Transposable Genetic Elements

Large deletions and duplications, as well as inversions and translocations, are major chromosomal reorganizations visible at the relatively low resolution of a karyotype. Small deletions and duplications are lesser chromosomal reorganizations that reshape genomes without any visible effect on karyotype. Another type of cytologically invisible sequence rearrangement with a significant genomic impact is **transposition**: the movement of small segments of DNA—entities known as **transposable elements (TEs)**—from one position in the genome to another.

Marcus Rhoades in the 1930s and Barbara McClintock in the 1950s inferred the existence of TEs from intricate genetic studies of corn. At first, the scientific community did not appreciate the importance of their work because their findings did not support the conclusion from classical recombination mapping that genes are located at fixed positions on chromosomes. Once the cloning of TEs made it possible to study them in detail, geneticists not only acknowledged their existence, but also discovered TEs in the genomes of virtually all organisms, from bacteria to humans. In 1983, Barbara McClintock received the Nobel Prize for her insightful studies on movable genetic elements (see the **Focus on Genetics** box "**Transposable Elements in Corn**").

Molecular studies confirmed transposable element movement

Copia is a transposable element in *Drosophila*. If you examined the polytene chromosomes from two strains of flies isolated from different geographical locations, you would find in general that the chromosomes appear identical. A probe derived from the *white* gene for eye colour, for example, would hybridize to a single site near the tip of the X chromosome in both strains (review Figure 9.9a). However, a probe including the *copia* TE would hybridize to 30–50 sites scattered throughout the genome, and the positions of *in situ* hybridization would not be the same in the two strains. Some sites would be identical in the two polytene sets, but others would be different (**Figure 9.23**). These observations suggest that since the time the strains were separated geographically, the *copia* sequences have moved around (transposed) in different ways in the two genomes even though the genes have remained in fixed positions.

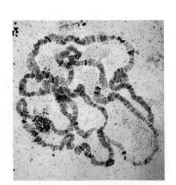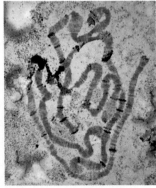

Figure 9.23 Transposable elements (TEs) can move to many locations in a genome. A probe for the *copia* TE hybridizes to multiple sites (*black bands* superimposed over the *blue* chromosomes) that differ in two different fly strains.

Barbara McClintock (1902–1992) (**Figure A**) was a pioneering scientist in an era where the scientific accomplishments of female researchers were regarded with much scepticism. Her work on the cytogenetics of maize (corn) allowed her to develop the theory that there were genes (which she called "controlling elements") that could move around and between chromosomes. These transposable or mobile elements were first discovered in the 1940s and 1950s when McClintock found a transposable element in one strain of corn, which she called **Ds** (**Dissociation**). When inserted at a particular location on chromosome 9, referred to by McClintock as a **mutable site**, this *Ds* element could often cause chromosomal breaks at that position. But this could occur only in the presence of another unlinked genetic element that "activated" this chromosomal break; she named this element **Ac** (**Activator**). She further found that in the presence of *Ac*, *Ds* could jump to other chromosomal locations. At some of these locations (and in the presence of *Ac*), *Ds* would now cause chromosomal breakage at the new position (**Figure B**). Since this breakage of chromosome 9 could occur in numerous cells, which continue to divide and grow as the kernel becomes larger, variable-sized sectors or patches of cells with recessive corn kernel phenotypes were observed: *c* (colourless), *sh* (shrunken), and *wx* (waxy). At other chromosomal positions, it appeared that *Ds* (in the presence of *Ac*) and *Ac* on its own could cause new mutations that were unstable, as shown by their spotted expression in kernels (**Figure C** and Figure 9.24b). Interestingly, the position

Figure A Barbara McClintock: Discoverer of transposable elements.

of the *Ac* element seemed to be very different in various strains of corn. Although Barbara McClintock first published her significant discovery of transposable elements in the 1950s, it was only until the late 1970s and early 1980s that her work gained the recognition that it deserved, when improvements in molecular techniques were made allowing other scientists to confirm her findings. She was consequently awarded the Nobel Prize in Physiology or Medicine in 1983 for her groundbreaking research, which shed new light on the workings of DNA.

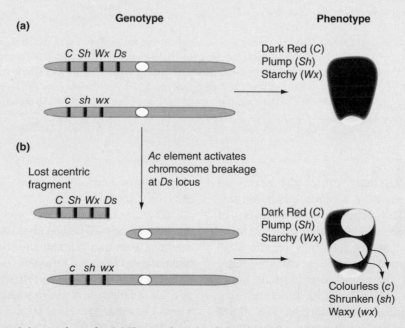

Figure B The sectoring trait in corn kernels. **(a)** The corn kernel is dark red in colour, plump, and starchy due to the presence of a dominant allele at each of the *C*, *Sh*, and *Wx* genes. **(b)** The presence of an *Ac* (*Activator*) element results in chromosome breakage at the *Ds* locus, resulting in an acentric chromosomal fragment containing the *C*, *Sh*, and *Wx* alleles that is lost during mitotic division, and a corn kernel with colourless, shrunken, and waxy patches or sectors.

(Continued)

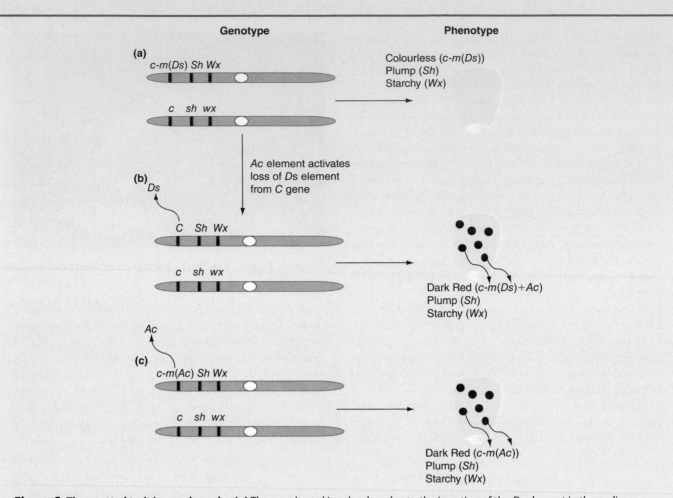

Figure C The spotted trait in corn kernels. (a) The corn kernel is colourless due to the insertion of the *Ds* element in the coding region of the *C* gene (***c-mutable* (*Ds*)**), which inactivates gene function. **(b)** The presence of an *Ac* (*Activator*) element in some kernel cells results in loss or excision of the *Ds* element from the *C* gene (***c-m*utable (*Ds*) + *Ac***), restoring *C* gene function in those particular cells and their mitotic descendants, and resulting in a dark red-spotted kernel phenotype. **(c)** Even in the absence of a *Ds* element, kernels have a spotted phenotype when the *Ac* element, inserted in the coding region of the *C* gene, is excised or lost (***c-mutable* (*Ac*)**).

Any segment of DNA that evolves the ability to move from place to place within a genome is by definition a transposable element, regardless of its origin or function. TEs need not be sequences that do something for the organism; indeed, many scientists regard them primarily as "selfish" parasitic entities carrying only information that allows their self-perpetuation. Some TEs, however, appear to have evolved functions that help their host. In one interesting example, TEs maintain the length of *Drosophila* chromosomes. *Drosophila* telomeres, in contrast to those of most organisms, do not contain TTAGGG repeats that are extendable by the telomerase enzyme (see Figure 6.14). Certain TEs in flies, however, combat the shortening of chromosome ends that accompanies every cycle of replication by jumping with high frequency into DNA very near

chromosome ends. As a result, chromosome size stays relatively constant.

Most transposable elements in nature range from 50 bp to approximately 10 000 bp (10 kb) in length. A particular TE can be present in a genome anywhere from one to hundreds of thousands of times. *Drosophila melanogaster*, for example, harbours approximately 80 different TEs, each an average of 5 kb in length, and each present an average of 50 times. These TEs constitute $80 \times 50 \times 5 = 20\,000$ kb, or roughly 12.5 percent of the 160 000-kb *Drosophila* genome. Mammals carry two major classes of TEs: **LINEs**, or long interspersed elements; and **SINEs**, or short interspersed elements. The human genome contains approximately 20 000 copies of the main human LINE—*L1*—which is up to 6.4 kb in length. The human genome also carries 300 000 copies of the main human SINE—*Alu*—which is 0.28 kb in length (**Figure 9.24a**).

(a) *Alu* SINEs in the human genome

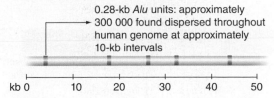

0.28-kb *Alu* units: approximately 300 000 found dispersed throughout human genome at approximately 10-kb intervals

kb 0   10   20   30   40   50

(b) TEs cause mottling in corn.

Figure 9.24 TEs in human and corn genomes. (a) The human genome carries about 300 000 copies of the 0.28 kb *Alu* retrotransposon, the major human SINE. **(b)** Movements of a DNA transposon mottles corn kernels when the transposon jumps into or out of genes that influence pigmentation.

These two TEs alone thus constitute roughly 7 percent of the 3 000 000-kb human genome. Because some TEs exist in only one or a few closely related species, it is probable that some elements arise and then disappear rather frequently over evolutionary time. Chapter 13 describes the evolutionary origins of LINEs and SINEs.

Classification of TEs on the basis of how they move around the genome distinguishes two groups. **Retrotransposons** transpose via reverse transcription of an RNA intermediate. The *Drosophila copia* elements and the human SINEs and LINEs just described are retrotransposons. Retrotransposons undergo **replicative transposition**; that is, they move by making a copy that becomes inserted into a different chromosomal site while the initial TE stays in its original position. This mechanism of transposition is also referred to as "copy-and-paste" transposition. Most **DNA transposons**, however, which move their DNA directly without the requirement of an RNA intermediate, undergo **conservative transposition**; that is, they are removed from their original site and transferred to a new target site in a "cut-and-paste" type of mechanism (see Figure 8.8b). The genetic elements discovered by Barbara McClintock in corn responsible for mottling the kernels are DNA transposons that move via conservative transposition (Figure 9.24b and Focus on Genetics box). Some biologists use the term "transposon" in the broader sense to refer to all TEs. In this book, we reserve it for the direct-movement class of genetic elements, and we use "transposable elements (TEs)" to indicate all DNA segments that move about in the genome, regardless of the mechanism.

Studies in corn and *Drosophila* revealed the existence of transposable genetic elements (TEs): small segments of DNA that can move around, and accumulate in, the genome. TEs can be subdivided according to their mode of transposition. Retrotransposons move via RNA intermediates, whereas DNA transposons move directly without first being transcribed into RNA.

Retrotransposons move via RNA intermediates

The transposition of a retrotransposon begins with its transcription by RNA polymerase into an RNA that encodes a reverse-transcriptase-like enzyme. This enzyme, like the reverse transcriptase made by the AIDS-causing HIV described in the Genetics and Society box in Chapter 7, can copy RNA into a single strand of cDNA and then use that single DNA strand as a template for producing double-stranded cDNA. Many retrotransposons also encode polypeptides other than reverse transcriptase.

Some retrotransposons have a poly-A tail at the 3′ end of the RNA-like DNA strand, a configuration reminiscent of mRNA molecules (**Figure 9.25a**). Other retrotransposons end in *long terminal repeats* (*LTRs*): nucleotide sequences repeated in the same orientation at both ends of the element (Figure 9.25a). The structure of this second type of retrotransposon is similar to the integrated DNA copies of RNA tumour viruses (known as retroviruses), suggesting that retroviruses evolved from this kind of retrotransposon, or vice versa. In support of this notion, researchers sometimes find retrotransposon transcripts enclosed in virus-like particles.

The structural parallels between retrotransposons, mRNAs, and retroviruses, as well as the fact that retrotransposons encode a reverse-transcriptase-like enzyme, prompted investigators to ask whether retrotransposons move around the genome via an RNA intermediate. Experiments in yeast helped confirm that they do. In one study, a copy of the *Ty1* retrotransposon found on a yeast plasmid contained an intron in one of its genes; after transposition into the yeast chromosome, however, the intron was not there (**Figure 9.25b**). Because removal of introns occurs only during mRNA processing, researchers concluded that the *Ty1* retrotransposon passes through an RNA intermediate during transposition.

The mechanisms by which various retrotransposons move around the genome resemble each other in general outline but differ in detail. **Figure 9.25c** outlines what is known of the process for the better understood LTR-containing retrotransposons. As the figure illustrates, one outcome of transposition via an RNA intermediate is that the original copy of the retrotransposon remains in place while the new copy inserts in another location. With this mode of transmission, the number of copies can increase rapidly with time. Human LINEs and SINEs, for example, occur in tens of thousands or even hundreds of thousands of copies within

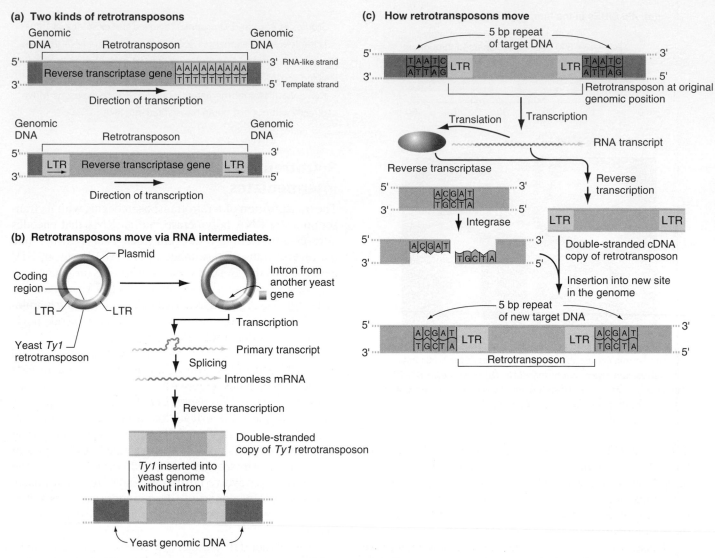

Figure 9.25 Retrotransposons: Structure and movement. (a) Some retrotransposons have a poly-A tail at the end of the RNA-like DNA strand (*top*); others are flanked on both sides by long terminal repeats (LTRs; *bottom*). **(b)** Researchers constructed a plasmid bearing a *Ty1* retrotransposon that contained an intron. When this plasmid was transformed into yeast cells, researchers could isolate new insertions of *Ty1* into yeast genomic DNA. The newly inserted *Ty1* did not have the intron, which implies that transposition involves splicing of a primary transcript to form an intronless mRNA. **(c)** The reverse-transcriptase-like enzyme synthesizes double-stranded retrotransposon cDNA in a series of steps. Insertion of this double-stranded cDNA into a new genomic location (*blue*) involves a staggered cleavage of the target site by integrase that leaves "sticky ends"; polymerization to fill in the sticky ends produces two copies of the 5 bp target site.

the genome. Other retrotransposons, however, such as the *copia* elements found in *Drosophila*, do not proliferate so profusely and exist in much more moderate copy numbers of 30–50. Currently unknown mechanisms may account for these differences by regulating the rate of retrotransposon transcription or by limiting the number of copies through selection at the level of the whole organism.

> Retrotransposons encode a reverse transcriptase enzyme that copies processed retrotransposon RNA (without introns) into complementary DNA; this DNA can insert into a new location in the genome. Because movement of retrotransposons involves an RNA intermediate, the number of copies in the genome can potentially increase rapidly.

Movement of transposon DNA is catalyzed by transposase enzymes

A hallmark of DNA transposons—TEs whose movement does not involve an RNA intermediate—is that their ends are inverted repeats of each other; that is, a sequence of base pairs at one end is present in mirror image at the other end (**Figure 9.26a**). The inverted repeat is usually 10–200 bp long.

DNA between the transposon's inverted repeats commonly contains a gene encoding a **transposase**, a protein that catalyzes transposition through its recognition of those repeats. As Figure 9.26a illustrates, the steps resulting in transposition include excision of the transposon from its original genomic position and integration into a new location. The double-stranded break at the transposon's

(a) Transposon structure

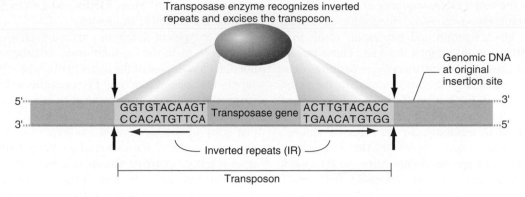

Transposase enzyme recognizes inverted repeats and excises the transposon.

Genomic DNA at original insertion site

5'
3'

GGTGTACAAGT
CCACATGTTCA

Transposase gene

ACTTGTACACC
TGAACATGTGG

3'
5'

Inverted repeats (IR)

Transposon

(b) How *P* element transposons move

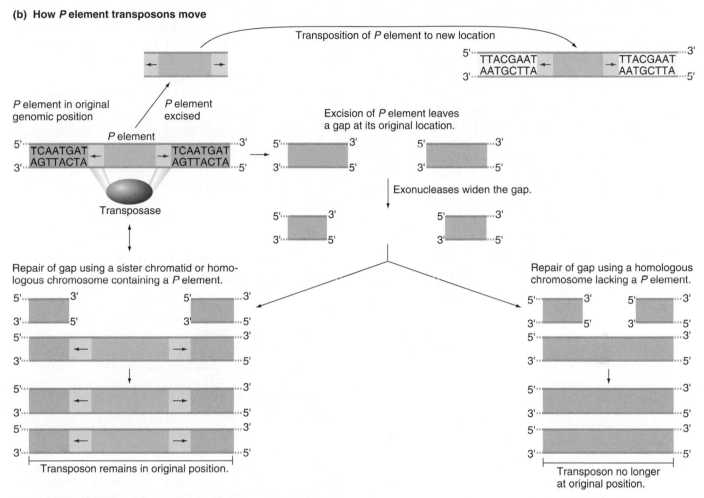

Transposition of *P* element to new location

5'·····TTACGAAT ← → TTACGAAT·····3'
3'·····AATGCTTA → ← AATGCTTA·····5'

P element in original genomic position

P element excised

Excision of *P* element leaves a gap at its original location.

P element

5'·····TCAATGAT ← → TCAATGAT·····3'
3'·····AGTTACTA → ← AGTTACTA·····5'

Transposase

Exonucleases widen the gap.

Repair of gap using a sister chromatid or homologous chromosome containing a *P* element.

Repair of gap using a homologous chromosome lacking a *P* element.

Transposon remains in original position.

Transposon no longer at original position.

Figure 9.26 DNA Transposons: Structure and movement. (a) Most DNA transposons contain inverted repeats at their ends (*light green; red arrows*) and encode a transposase enzyme that recognizes these inverted repeats. The transposase cuts at the borders between the transposon and adjacent genomic DNA, and it also helps the excised transposon integrate at a new site. **(b)** Transposase-catalyzed integration of *P* elements creates a duplication of 8 bp present at the new target site. A gap remains when transposons are excised from their original position. After exonucleases widen the gap, cells repair the gap using related DNA sequences as templates. Depending on whether the template contains or lacks a *P* element, the transposon will appear to remain or to be excised from its original location.

excision site is repaired in different ways in different cases. **Figure 9.26b** shows two of the possibilities. In *Drosophila*, after excision of a transposon known as a *P element*, DNA exonucleases first widen the resulting gap and then repair it using either a sister chromatid or a homologous chromosome as a template. If the template contains the

P element and DNA replication is completely accurate, repair will restore a *P* element to the position from which it was excised; this will make it appear as if the *P* element remained at its original location during transposition (Figure 9.26b, *left*). If the template does not contain a *P* element, the transposon will be lost from the original site after

transposition (Figure 9.26b, *right*). Since the transposase enzyme cleaves the target DNA sequence at staggered recognition sites, transposase-catalyzed integration of all TEs and subsequent DNA ligation and gap repair result in a **target-site duplication** that flanks the TEs. These target-site duplications, also known as *direct repeats*, are identical nucleotide sequences that are oriented in the same direction and repeated. Each type of transposable element differs in the length of its target-site duplication.

Some strains of *D. melanogaster* are called "P strains" because they harbour many copies of the *P* element; "M strains" of the same species do not carry the *P* element at all. Virtually all commonly used laboratory flies are M strains, whereas many flies isolated from natural populations since 1950 are P strains. Because Thomas Hunt Morgan and co-workers in the early part of the twentieth century isolated the flies that have proliferated into most current laboratory strains, these observations suggest that *P* elements did not enter *D. melanogaster* genomes until around 1950. The prevalence of *P* elements in many contemporary natural populations attests to the rapidity with which transposable elements can spread once they enter a species' genome.

Interestingly, the mating of male flies from P strains with females from M strains causes a phenomenon called **hybrid dysgenesis**, which creates a series of defects including sterility of offspring, mutation, and chromosome breakage. One of the more interesting effects of hybrid dysgenesis is to promote the movement of *P* elements to new positions in the genome. Since elevated levels of transposition can foster many kinds of genetic changes, some geneticists speculate that hybrid dysgenesis-like events involving various transposons in different species had a strong impact on evolution. The *Drosophila* portrait on Connect provides more information on the molecular mechanisms underlying hybrid dysgenesis and the ways in which fly geneticists use this phenomenon to introduce new genes into *Drosophila*.

> DNA transposons encode transposase enzymes that recognize the inverted repeats at the ends of the transposon DNA. These enzymes then catalyze the movement of the transposons without the involvement of an RNA intermediate.

Genomes often contain defective copies of transposable elements

Many copies of TEs sustain deletions either as a result of the transposition process itself (e.g., incomplete reverse transcription of a retrotransposon RNA) or as a result of events following transposition (e.g., faulty repair of a site from which a *P* element was earlier excised). If a deletion removes the promoter needed for transcription of a retrotransposon, that copy of the element cannot generate the RNA intermediate for future movements. If the deletion removes one of the inverted repeats at one end of a transposon, transposase will be unable to catalyze transposition

of that element. Such deletions create defective TEs unable to transpose again. Most SINEs and LINEs in the human genome are defective in this way.

Other types of deletions create defective elements that are unable to move on their own, but they can move if nondefective copies of the element elsewhere in the genome supply the deleted function. For example, a deletion inactivating the reverse transcriptase gene in a retrotransposon or the transposase gene in a DNA transposon would "ground" that copy of the element at one genomic location if it is the only source of the essential enzyme in the genome. If reverse transcriptase or transposase were provided by other copies of the same element in the genome, however, the defective copy could move. Defective TEs that require the activity of nondeleted copies of the same TE for movement are called **nonautonomous elements**; the nondeleted copies that can move by themselves are **autonomous elements**. In corn, the *Ds* element is a nonautonomous element, while the *Ac* element is autonomous. Similarly, SINEs are nonautonomous elements, while LINEs are autonomous.

> Deleted, defective copies of TEs that can still transpose are called nonautonomous elements. The movement of non-autonomous elements requires that the genome also contains nondefective copies (autonomous elements) that can supply reverse transcriptase or transposase enzymes.

Transposable elements can disrupt genes and alter genomes

Geneticists usually consider TEs to be segments of "selfish DNA" that exist for their own sake. However, the movement of TEs may have profound consequences for the organization and function of the genes and chromosomes of the organisms in which they are maintained.

Gene mutations caused by TEs

Insertion of a TE near or within a gene can affect gene expression and change phenotype. We now know that the wrinkled pea mutation first studied by Mendel resulted from insertion of a TE into the gene for a starch-branching enzyme. In *Drosophila*, a large percentage of spontaneous mutations, including the w^1 mutation discovered by T. H. Morgan in 1910, are caused by insertion of TEs (**Figure 9.27**). Surprisingly, in light of the large numbers of LINEs and SINEs in human genomes, only a handful of mutant human phenotypes are known to result from insertion of TEs. Among these is a B-type haemophilia caused by *Alu* insertion into a gene encoding clotting factor IX; recall that *Alu* is the main human SINE.

A TE's effect on a gene depends on what the element is and where it inserts within or near the gene (**Figure 9.27**). If an element lands within a protein-coding exon, the additional DNA may shift the reading frame or supply an in-frame stop codon that truncates the polypeptide. If the element falls in an intron, it could diminish the efficiency of splicing. Some

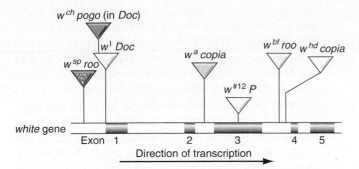

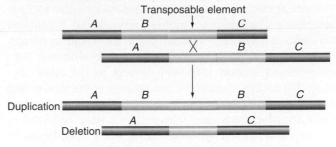

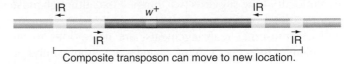

Figure 9.27 TEs can cause mutations on insertion into a gene. Many spontaneous mutations in the *white* gene of *Drosophila* arise from insertions of TEs such as *copia*, *roo*, *pogo*, or *Doc*. The resultant eye colour phenotype (indicated by the colour in the *triangles*) depends on the element involved and where in the *white* gene it inserts.

of these inefficient splicing events might completely remove the element from the gene's primary transcript; this would still allow some—but less than normal—synthesis of functional polypeptide. TEs that land within exons or introns may also provide a transcription stop signal that prevents transcription of gene sequences downstream of the insertion site. Finally, insertions into regions that regulate transcription, such as promoters, can influence the amount of gene product made in particular tissues at particular times during development. Some transposons insert preferentially into the upstream regulatory regions of genes, and some even prefer specific types of genes, such as tRNA genes.

Chromosomal rearrangements caused by TEs

Retrotransposons and DNA transposons can trigger spontaneous chromosomal rearrangements other than transpositions in several ways. Sometimes, deletion or duplication of chromosomal material adjacent to the transposon occurs as a mistake during the transposition event itself. In another mechanism, if two copies of the same TE occupy nearby but not identical sites in homologous chromosomes, the two copies of the TE in heterozygotes carrying both types of homologue may pair with each other and cross over (**Figure 9.28a**). The recombination resulting from this unequal crossover would produce one chromosome deleted for the region between the two TEs and a reciprocal homologue with a tandem duplication of the same region. The duplication associated with the *Bar* mutation in *Drosophila* (review Figure 9.13) probably arose in this way.

Gene relocation due to transposition

When two copies of a transposon occur in nearby but not identical locations on the same chromosome, the inverted

Figure 9.28 How TEs generate chromosomal rearrangements and relocate genes. (a) If a TE (*pink*) is found in slightly different locations on homologous chromosomes (here on opposite sides of segment *B*), unequal crossing-over produces reciprocal deletions and duplications. **(b)** If two copies of a DNA transposon are nearby on the same chromosome, transposase can recognize the outermost inverted repeats (IRs), creating a composite transposon that allows intervening genes such as w^+ (*red*) to jump to new locations.

repeats of the transposons are positioned such that an inverted version of the sequence at the 5′ end of the copy on the left will exist at the 3′ end of the copy to its right (**Figure 9.28b**). If transposase acts on this pair of inverted repeats during transposition, it allows the entire region between them to move as one giant transposon, mobilizing and relocating any genes the region contains. Some composite transposons carry as much as 400 kb of DNA. In prokaryotes, the capacity of two TEs to relocate the intervening genes helps mediate the transfer of drug resistance between different strains or species of bacteria, as will be discussed in Chapter 18.

> The movement of TEs has three main genetic consequences: (1) mutation of a gene due to TE insertion within or near the gene; (2) chromosomal rearrangements either caused by unequal crossing-over between copies of the same TE, or generated as a by-product of the transposition process; and (3) relocation of genes between two nearby transposons on the same chromosome.

<table>
<tr><td>**9.3**</td><td></td></tr>
</table>

9.3 Rearrangements and Evolution: A Speculative Comprehensive Example

We saw at the beginning of this chapter that roughly 300 chromosomal rearrangements could reshape the human genome to a form that resembles the mouse genome.

Many of these rearrangements are transpositions and translocations that could construct a new chromosome from large blocks of sequences that were on different

chromosomes in an ancestral organism. Figure 9.1a provides clear evidence that these reorganizations also include inversions. For example, mouse chromosome 1 contains two adjacent syntenic segments that are found in human chromosome 6, but in a reshuffled order, with one segment turned around 180° with respect to the other segment. Direct DNA sequence comparison of the mouse and human genomes further indicates that deletions, duplications, translocations, and transpositions have occurred in one or the other lineage since humans and mice began to diverge from a common ancestor 65 million years ago.

The occurrence of these various rearrangements over evolutionary time suggests two things. First, although most chromosomal variations, including single-base changes and chromosomal rearrangements, are deleterious to an organism or its progeny, a few changes are either neutral or provide an advantage for survival and manage to become fixed in a population. Second, some rearrangements almost certainly contribute to the processes underlying speciation. Although we still do not know enough to understand how any particular rearrangement that distinguishes the human from the mouse genome may have provided a survival advantage or otherwise helped guide speciation, it is nonetheless useful to consider in a general way how chromosomal rearrangements might contribute to evolution.

Deletions A small deletion that moves a coding sequence of one gene next to a promoter or other regulatory element of an adjacent gene may rarely allow expression of a protein at a novel time in development or in a novel tissue. If the new time or place of expression is advantageous to the organism, the deletion might become established in the genome.

Duplications An organism cannot normally tolerate mutations in a gene essential to its survival, but duplication would provide two copies of the gene. If one copy remained intact to perform the essential function, the other would be free to evolve a new function. The genomes of most higher plants and animals, in fact, contain many **gene families**—sets of closely related genes with slightly different functions, that most likely arose from a succession of gene duplication events. In vertebrates, some *multigene families* have hundreds of members.

Inversions Suppose one region of a chromosome has three mutations that together greatly enhance the reproductive fitness of the organism. In heterozygotes where one homologue carries the mutations and the other does not, recombination could undo the beneficial linkage. If, however, the three mutations are part of an inversion, crossover suppression will ensure that they remain together as they spread through the population.

Translocations On the tiny volcanic island of Madeira off the coast of Portugal in the Atlantic Ocean, two populations of the common house mouse (*Mus musculus*) are in the process of becoming separate species because of translocations that have led to reproductive isolation. The mice live in a few narrow valleys separated by steep mountains. Geneticists have found

that populations of mice on the two sides of these mountain barriers have very different sets of chromosomes because they have accumulated different sets of Robertsonian translocations (**Figure 9.29**). Mice in one Madeira population, for example, have a diploid number (2*n*) of 22 chromosomes, whereas mice in a different population on the island have 24; for most house mice throughout the world, 2*n* = 40. (Recall from Figure 9.19 that Robertsonian translocations can reduce chromosome number if the small chromosome that results from a translocation is lost.)

The hybrid offspring of matings between individuals of these two populations are completely sterile or infertile because chromosomal complements that are so different cannot properly segregate at meiosis. Thus, reproductive isolation has reinforced the already established geographical isolation, and the two populations are close to becoming two separate species. What is remarkable about this example of speciation is

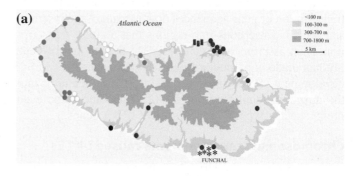

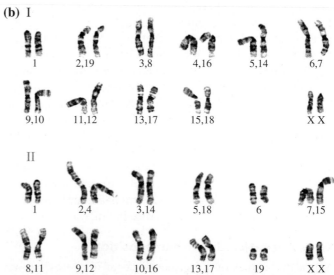

Figure 9.29 Rapid chromosomal evolution in house mice on the island of Madeira. (a) Distribution of mouse populations with different sets of Robertsonian translocations (indicated by circles of different colours). **(b)** Karyotypes of female mice from two different populations. The karyotype I at the *top* is from the population shown with *red dots* in part (a); the karyotype II at the *bottom* is from the population indicated by *green dots*. Robertsonian translocations are indicated by numbers separated by a comma (e.g., 2,19 is a Robertsonian translocation between chromosomes 2 and 19 of the standard mouse karyotype).

that mice were introduced into Madeira by Portuguese settlers only in the fifteenth century. This means that the varied and complicated sets of Robertsonian translocations that contributed to speciation became fixed in the different populations in less than 600 years.

Transpositions Movement of TEs may cause novel mutations, a small proportion of which might be selected for because they are advantageous to the organism. TEs can also help generate potentially useful duplications and inversions.

> Rearrangements and transpositions alter DNA sequences and thus provide raw material for evolutionary change. Duplicated genes can diverge by mutation to acquire different functions. The reduced fertility of heterozygotes for inversions and translocations can contribute to reproductive isolation of populations and thus promote speciation.

9.4 Changes in Chromosome Number

We have seen that in peas, *Drosophila*, and humans, normal diploid individuals carry a $2n$ complement of chromosomes, where n is the number of chromosomes in the gametes. All the chromosomes in the haploid gametes of these diploid organisms are different from one another. In this section, we examine two types of departure from chromosomal diploidy found in eukaryotes: (1) aberrations in usually diploid species that generate cells or individuals whose genomes contain one to a few chromosomes more or less than the normal $2n$, for example, $2n + 1$ or $2n - 1$; and (2) species whose genomes contain complete but nondiploid sets of chromosomes, for example, $3n$ or $4n$.

Aneuploidy is the loss or gain of one or more chromosomes

Individuals whose chromosome number is not an exact multiple of the haploid number (n) for the species are **aneuploids** (review Table 9.1). Individuals lacking one chromosome from the diploid number ($2n - 1$) are **monosomic**, whereas individuals having one chromosome in addition to the normal diploid set ($2n + 1$) are **trisomic**. Organisms with four copies of a particular chromosome ($2n + 2$) are **tetrasomic**.

Deleterious effects of aneuploidy for autosomes

Monosomy, trisomy, and other forms of aneuploidy create a genetic imbalance that is usually deleterious to the organism. In humans, monosomy for any autosome is generally lethal, but medical geneticists have reported a few cases of monosomy for chromosome 21, one of the smallest human chromosomes. Although born with severe multiple abnormalities, these monosomic individuals survived for a short time beyond birth. Similarly, trisomies involving a human autosome are also highly deleterious. Individuals with trisomies for larger chromosomes, such as 1 and 2, are almost always aborted spontaneously early in pregnancy. Trisomy 18 causes Edwards syndrome, and trisomy 13 causes Patau syndrome; both phenotypes include gross developmental abnormalities that typically result in early death.

The most frequently observed human autosomal trisomy, trisomy 21, results in Down syndrome. As one of the shortest human autosomes, chromosome 21 contains only about 1.5 percent of the DNA in the human genome. Although there is considerable phenotypic variation among Down syndrome individuals, traits such as mental retardation and skeletal abnormalities are usually associated with the condition. Many Down syndrome babies die in their first year after birth from heart defects and increased susceptibility to infection. We saw earlier (in the discussion of translocations) that some people with Down syndrome have three copies of only part of, rather than the entire, chromosome 21. It is thus probable that genetic imbalance for only a few genes may be a sufficient cause of the condition. Unfortunately, as of early 2013, scientists had not yet been able to identify any of these genes unambiguously with a particular Down syndrome phenotype.

Dosage compensation through X-chromosome inactivation

Although the X chromosome is one of the longest human chromosomes and contains 5 percent of the DNA in the genome, individuals with X chromosome aneuploidy, such as XXY males, XO females, and XXX females, survive quite well compared with aneuploids for the larger autosomes. The explanation for this tolerance of X-chromosome aneuploidy is that X-chromosome inactivation equalizes the expression of most X-linked genes in individuals with different numbers of X chromosomes. As we saw in Chapter 6, X-chromosome inactivation represses expression of most genes on all but one X chromosome in a cell. As a result, even if the number of X chromosomes varies, the amount of protein generated by most X-linked genes remains constant. Human X-chromosome aneuploidies are nonetheless not without consequence. XXY men have *Klinefelter syndrome*, and XO women have *Turner syndrome*. The aneuploid individuals affected by these syndromes are usually infertile and display skeletal abnormalities, leading in the XXY men to unusually long limbs and in the XO women to unusually short stature.

If X inactivation were 100 percent effective, we would not expect to see even the relatively minor abnormalities of Klinefelter syndrome, because the number of functional X chromosomes—one—would be the same as in XY individuals. One explanation is that during X inactivation,

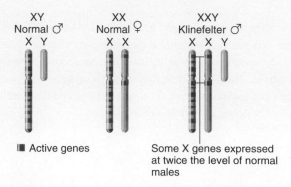

XY
Normal ♂
X Y

XX
Normal ♀
X X

XXY
Klinefelter ♂
X X Y

■ Active genes

Some X genes expressed
at twice the level of normal
males

Figure 9.30 Why aneuploidy for the X chromosome can have phenotypic consequences. X-chromosome inactivation does not affect all genes on the X chromosome. As a result, in XXY Klinefelter males, a few X-chromosome genes are expressed inappropriately at twice their normal level.

several genes near the telomere and centromere of the short arm of the human X chromosome escape inactivation and thus remain active. As a result, XXY males make twice the amount of protein encoded by these few genes as XY males (**Figure 9.30**).

The reverse of X inactivation is *X reactivation*; it occurs in the oogonia, the female germline cells that develop into the oocytes that undergo meiosis (review Figure 3.17). Reactivation of the previously inactivated X chromosomes in the oogonia ensures that every mature ovum (the gamete) receives an active X. If X reactivation did not occur, half of a woman's eggs (those with inactive X chromosomes) would be incapable of supporting development after fertilization. The phenomenon of X reactivation in the oogonia might help explain the infertility of women with Turner syndrome. With X reactivation, oogonia in XX females have two functional doses of X chromosome genes; but the corresponding cells in XO Turner women have only one dose of the same genes and may thus undergo defective oogenesis.

Meiotic nondisjunction

How does aneuploidy arise? Mistakes in chromosome segregation during meiosis produce aneuploids of different types, depending on when the mistakes occur. If homologous chromosomes do not separate (i.e., do not disjoin) during the first meiotic division, two of the resulting haploid gametes will carry both homologues, and two will carry neither. Union of these gametes with normal gametes will produce aneuploid zygotes, half monosomic, half trisomic (**Figure 9.31a**, *left*). By contrast, if meiotic nondisjunction occurs during meiosis II, only two of the four resulting gametes will be aneuploid (**Figure 9.31a**, *right*).

Abnormal *n* + 1 gametes resulting from nondisjunction in a cell that is heterozygous for alleles on the nondisjoining chromosome will be heterozygous if the nondisjunction happens in the first meiotic division, but they will be homozygous if the nondisjunction takes place in the second meiotic division. (We assume here that no recombination has occurred between the heterozygous gene in question and the centromere, as would be the case

for genes closely linked to the centromere.) It is possible to use this distinction to determine when a particular nondisjunction occurred (Figure 9.31a). The nondisjunction events that give rise to Down syndrome, for example, occur much more frequently in mothers (90 percent) than in fathers (10 percent). Interestingly, in women, such nondisjunction events occur more often during the first meiotic division (about 75 percent of the time) than during the second. By contrast, when the nondisjunction event leading to Down syndrome takes place in men, the reverse is true.

Recently obtained data show that many meiotic nondisjunction events in humans result from problems in meiotic recombination. By tracking DNA markers, clinical investigators can establish whether recombination took place anywhere along chromosome 21 during meioses that created *n* + 1 gametes. In approximately one-half of Down syndrome cases caused by nondisjunction during the first meiotic division in the mother (i.e., in about 35 percent of all Down syndrome cases), no recombination occurred between the homologous chromosome 21's in the defective meioses. This result makes sense because chiasmata, the structures associated with crossing-over, hold the maternal and paternal homologous chromosomes together in a bivalent at the metaphase plate of the first meiotic division (review Feature Figure 3.12). In the absence of recombination and thus of chiasmata, there is no mechanism to ensure that the maternal and paternal chromosomes will go to opposite poles at anaphase I. The increase in the frequency of Down syndrome children that is associated with increasing maternal age may therefore reflect a decline in the effectiveness of the mother's machinery for meiotic recombination.

If an aneuploid individual survives and is fertile, the incidence of aneuploidy among his or her offspring will generally be extremely high. This is because half of the gametes produced by meiosis in a monosomic individual lack the chromosome in question, while half of the gametes produced in a trisomic individual have an additional copy of the chromosome (Figure 9.31b).

Mitotic nondisjunction and chromosome loss

As a zygote divides many times to become a fully formed organism, mistakes in chromosome segregation during the mitotic divisions accompanying this development may, in rare instances, augment or diminish the complement of chromosomes in certain cells. In **mitotic nondisjunction**, the failure of two sister chromatids to separate during mitotic anaphase generates reciprocal trisomic and monosomic daughter cells (**Figure 9.32a**). Other types of mistakes, such as a lagging chromatid not pulled to either spindle pole at mitotic anaphase, result in a **chromosome loss** that produces one monosomic and one diploid daughter cell (**Figure 9.32b**).

In a multicellular organism, aneuploid cells arising from either mitotic nondisjunction or chromosome loss may survive and undergo further rounds of cell division, producing clones of cells with an abnormal chromosome count (see the introductory figure at the beginning of this

(a) Nondisjunction can occur during either meiotic division.

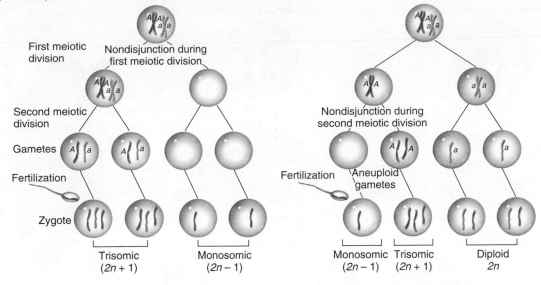

First meiotic division — Nondisjunction during first meiotic division

Second meiotic division

Gametes

Fertilization

Zygote

Trisomic (2n + 1)   Monosomic (2n − 1)

Nondisjunction during second meiotic division

Fertilization — Aneuploid gametes

Monosomic (2n − 1)   Trisomic (2n + 1)   Diploid 2n

(b) Aneuploids beget aneuploid progeny.

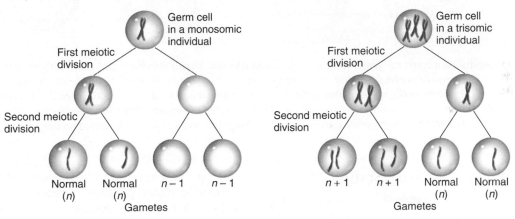

Germ cell in a monosomic individual

First meiotic division

Second meiotic division

Normal (n)   Normal (n)   n − 1   n − 1
Gametes

Germ cell in a trisomic individual

First meiotic division

Second meiotic division

n + 1   n + 1   Normal (n)   Normal (n)
Gametes

Figure 9.31 Aneuploidy is caused by problems in meiotic chromosome segregation. (a) If trisomic progeny inherit two different alleles (*A* and *a*) of a centromere-linked gene from one parent, the nondisjunction occurred in meiosis I (*left*). If the two alleles inherited from one parent are the same (*A* and *A*; or *a* and *a*), the nondisjunction occurred during meiosis II (*right*). **(b)** Because aneuploids carry chromosomes that have no homologue with which to pair, aneuploid individuals frequently produce aneuploid progeny.

chapter). Nondisjunction or chromosome loss occurring early in development will generate larger aneuploid clones than the same events occurring later in development. The side-by-side existence of aneuploid and normal tissues results in a **mosaic** organism whose phenotype depends on what tissue bears the aneuploidy, the number of aneuploid cells, and the specific genes on the aneuploid chromosome. Many examples of mosaicism involve the sex chromosomes. If an XX *Drosophila* female loses one of the X chromosomes during the first mitotic division after fertilization, the result is a **gynandromorph** composed of equal parts male and female tissue (**Figure 9.32c**).

Interestingly, in humans, many Turner syndrome females are mosaics carrying some XX cells and some XO cells. These individuals began their development as XX zygotes, but with the loss of an X chromosome during the embryo's early mitotic divisions, they acquired a clone of XO cells. Similar mosaicism involving the autosomes

also occurs. For example, physicians have recorded several cases of mild Down syndrome arising from mosaicism for trisomy 21. In people with Turner or Down mosaicism, the existence of some normal tissue appears to ameliorate the condition, with the individual's phenotype depending on the particular distribution of diploid versus aneuploid cells.

Aneuploidy for autosomes is usually deleterious, but organisms can better tolerate aneuploidy for sex chromosomes because of dosage compensation mechanisms such as X-chromosome inactivation. Rare events of meiotic nondisjunction can produce aneuploid gametes and thus aneuploid organisms. Rare mistakes in mitosis, including mitotic nondisjunction and chromosome loss, can generate a mosaic organism that has cells with different karyotypes.

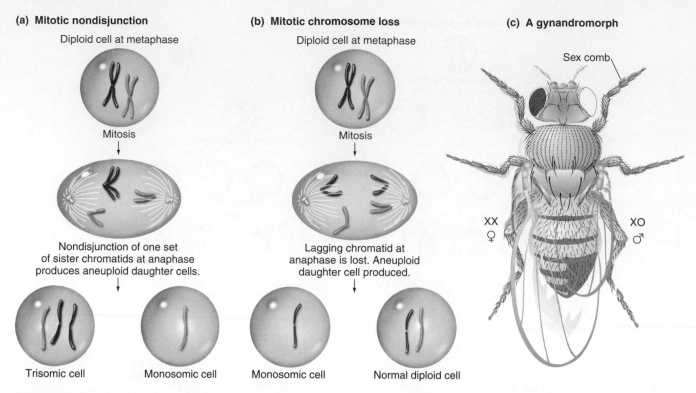

(a) Mitotic nondisjunction

Diploid cell at metaphase

Mitosis

Nondisjunction of one set
of sister chromatids at anaphase
produces aneuploid daughter cells.

Trisomic cell   Monosomic cell

(b) Mitotic chromosome loss

Diploid cell at metaphase

Mitosis

Lagging chromatid at
anaphase is lost. Aneuploid
daughter cell produced.

Monosomic cell   Normal diploid cell

(c) A gynandromorph

Sex comb

XX ♀   XO ♂

Figure 9.32 Mistakes during mitosis can generate clones of aneuploid cells. Mitotic nondisjunction **(a)** or chromosome loss during mitosis **(b)** can create monosomic or trisomic cells that can divide to produce aneuploid clones. **(c)** If an X chromosome is lost during the first mitotic division of an XX *Drosophila* zygote, one daughter cell will be XX (female), while the other will be XO (male). Such an embryo will grow into a gynandromorph. Here, the zygote was $w^+ m^+ / w m$, so the XX half of the fly (*left*) has red eyes and normal wings; loss of the $w^+ m^+$ X chromosome gives the XO half of the fly (*right*) white eyes (*w*), miniature wings (*m*), and a male-specific sex comb on the front leg.

Some euploid species are not diploid

In contrast to aneuploids, **euploid** cells contain only complete sets of chromosomes. Most euploid species are diploid, but some euploid species are **polyploids** that carry three or more complete sets of chromosomes (see Table 9.1). When speaking of polyploids, geneticists use the symbol **n** or **x** to indicate the **basic chromosome number**; that is, the number of different chromosomes that make up a single complete set. Triploid species, which have three complete sets of chromosomes, are then $3n$ or $3x$; tetraploid species with four complete sets of chromosomes are $4n$ or $4x$; and so forth. For diploid species, the number of chromosomes in the gametes is designated as n or x—because each gamete contains a single complete set of chromosomes. Commercially grown bread wheat however, a polyploid species, has a total of 42 chromosomes: 6 nearly (but not wholly) identical sets each containing 7 different chromosomes. Commercial bread wheat is thus a hexaploid ($6x$ or $6n = 42$). But each triploid gamete has one-half the total number of chromosomes, so $3n = 21$. Another form of euploidy, in addition to polyploidy, exists in **monoploid** (n or x) organisms, which have only one set of chromosomes.

Monoploidy and polyploidy are rarely observed in animals. Among the few examples of monoploidy are some species of ants and bees in which the males are monoploid, whereas the females are diploid. Males of these species develop *parthenogenetically* from unfertilized eggs. These monoploid males produce gametes through a modified meiosis that in some unknown fashion ensures distribution of all the chromosomes to the same daughter cell during meiosis I; the sister chromatids then separate normally during meiosis II. Polyploidy in animals normally exists only in species with unusual reproductive cycles, such as hermaphroditic earthworms, which carry both male and female reproductive organs, and goldfish, which are parthenogenetically tetraploid species. The ciliated protozoan *Tetrahymena thermophila* exhibits nuclear dimorphism, with a diploid germline micronucleus and a polyploid somatic macronucleus. In *Drosophila*, it is possible, under special circumstances, to produce triploid and tetraploid females, but never males. In humans, polyploidy is always lethal, usually resulting in spontaneous abortion during the first trimester of pregnancy.

Monoploid organisms

Botanists can produce monoploid plants experimentally by special treatment of germ cells from diploid species that have completed meiosis and would normally develop into pollen. The treated cells divide into a mass of tissue known as an *embryoid*. Subsequent exposure to plant hormones enables the embryoid to develop into a plant (**Figure 9.33a**). Monoploid plants may also arise from rare spontaneous events in a large natural population. Most monoploid

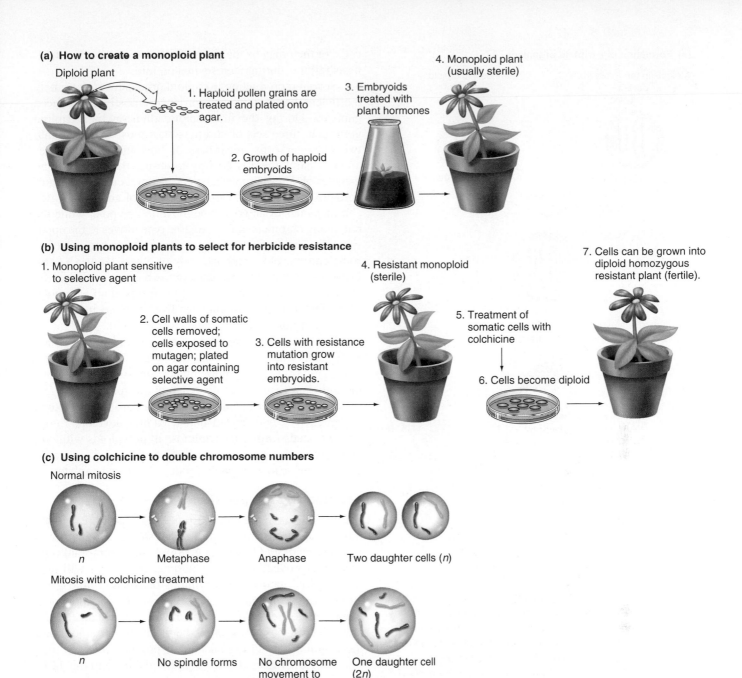

(a) How to create a monoploid plant

Diploid plant

1. Haploid pollen grains are treated and plated onto agar.

2. Growth of haploid embryoids

3. Embryoids treated with plant hormones

4. Monoploid plant (usually sterile)

(b) Using monoploid plants to select for herbicide resistance

1. Monoploid plant sensitive to selective agent

2. Cell walls of somatic cells removed; cells exposed to mutagen; plated on agar containing selective agent

3. Cells with resistance mutation grow into resistant embryoids.

4. Resistant monoploid (sterile)

5. Treatment of somatic cells with colchicine

6. Cells become diploid

7. Cells can be grown into diploid homozygous resistant plant (fertile).

(c) Using colchicine to double chromosome numbers

Normal mitosis

n   Metaphase   Anaphase   Two daughter cells (*n*)

Mitosis with colchicine treatment

n   No spindle forms   No chromosome movement to poles of cell   One daughter cell (2*n*)

Figure 9.33 The creation and use of monoploid plants. (a) Under certain conditions, haploid pollen grains can grow into haploid embryoids. When treated with plant hormones, haploid embryoids grow into monoploid plants. **(b)** Researchers select monoploid cells for recessive traits such as herbicide resistance. They then grow the selected cells into a resistant embryo, which (with hormone treatment) eventually becomes a mature, resistant monoploid plant. Treatment with colchicine doubles the chromosome number, creating diploid cells that can be grown in culture with hormones to make a homozygous herbicide-resistant diploid plant. **(c)** Colchicine treatment prevents formation of the mitotic spindle and also blocks cytokinesis, generating cells with twice the number of chromosomes. *Blue, red,* and *green* colours denote nonhomologous chromosomes.

plants, no matter how they originate, are infertile. Because the chromosomes have no homologues with which to pair during meiosis I, they are distributed at random to the two spindle poles during this division. Rarely do all chromosomes go to the same pole, and if they do not, the resulting gametes are defective as they lack one or more chromosomes. The greater the number of chromosomes in the genome, the lower the likelihood of producing a gamete containing all of them.

Despite such gamete-generating problems, monoploid plants and tissues are of great value to plant breeders. They make it possible to visualize normally recessive traits directly, without crosses to achieve homozygosity. Plant researchers can also introduce mutations into individual monoploid cells; select for desirable phenotypes, such as resistance to herbicides; and use hormone treatments to grow the selected cells into monoploid plants **(Figure 9.33b)**. They can then convert monoploids of their

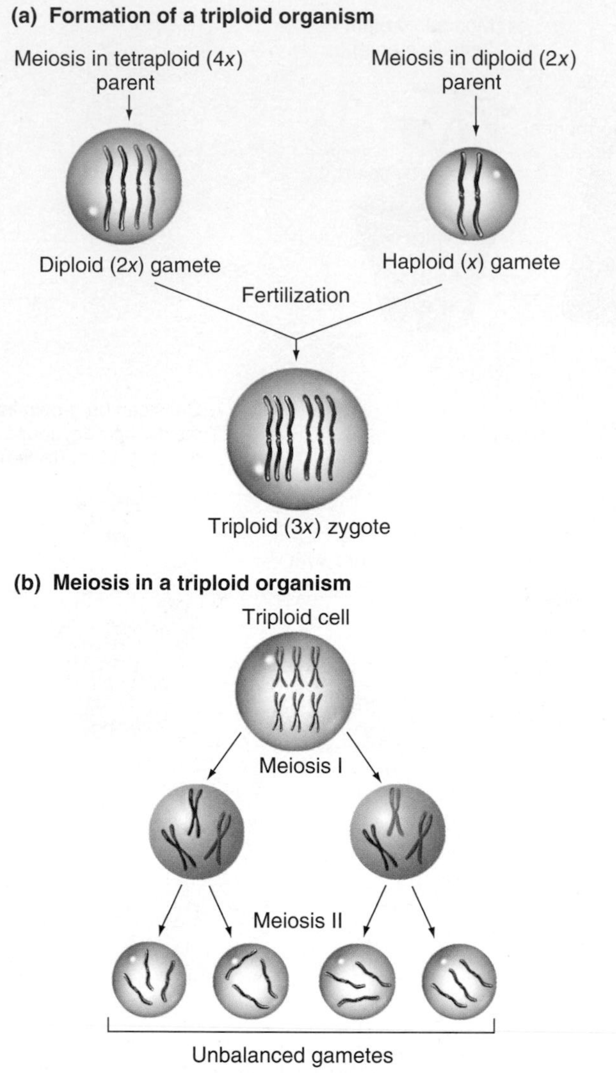

(a) Formation of a triploid organism

Meiosis in tetraploid (4x) parent

Meiosis in diploid (2x) parent

Diploid (2x) gamete

Haploid (x) gamete

Fertilization

Triploid (3x) zygote

(b) Meiosis in a triploid organism

Triploid cell

Meiosis I

Meiosis II

Unbalanced gametes

Figure 9.34 The genetics of triploidy. (a) Production of a triploid (x = 3) from fertilization of a haploid gamete by a diploid gamete. Nonhomologous chromosomes are either *blue* or *red*. **(b)** Meiosis in a triploid produces unbalanced gametes because meiosis I produces two daughter cells with unequal numbers of any one type of chromosome. If x is large, balanced gametes with equal numbers of all the chromosomes are very rare.

choice into homozygous diploid plants by treating tissue with *colchicine*, an alkaloid drug obtained from the autumn crocus. By binding to tubulin—the major protein component of the spindle—colchicine prevents formation of the spindle apparatus. In cells without a spindle, the sister chromatids cannot segregate after the centromere splits, so there is often a doubling of the chromosome set following treatment with colchicine (**Figure 9.33c**). The resulting diploid cells can be grown into diploid plants that will express the desired phenotype and produce fertile gametes.

Triploid organisms

Triploids (3n or 3x) result from the union of haploid (x) and diploid (2x) gametes (**Figure 9.34a**). The diploid gametes may be the products of meiosis in tetraploid (4x) germ

cells, or they may be the products of rare spindle or cytokinesis failures during meiosis in a diploid.

Sexual reproduction in triploid organisms is extremely inefficient because meiosis produces mostly unbalanced gametes. During the first meiotic division in a triploid germ cell, three sets of chromosomes must segregate into two daughter cells; regardless of how the chromosomes align in pairs, there is no way to ensure that the resulting gametes obtain a complete, balanced x or 2x complement of chromosomes. In most cases, at the end of anaphase I, two chromosomes of any one type move to one pole, while the remaining chromosome of the same type moves to the opposite pole. The products of such a meiosis have two copies of some chromosomes and one copy of others (**Figure 9.34b**). If the number of chromosomes in the basic set is large, the chance of obtaining any balanced gametes at all is remote. Thus, fertilization with gametes from triploid individuals does not produce many viable offspring.

It is possible to propagate some triploid species, such as bananas and watermelons, through asexual reproduction. The fruits of triploid plants are seedless because the unbalanced gametes do not function properly in fertilization, or, if fertilization occurs, the resultant zygote is so genetically unbalanced that it cannot develop. Either way, no seeds form. Like triploids, all polyploids with odd numbers of chromosome sets (such as 5x or 7x) are sterile because they cannot reliably produce balanced gametes.

Tetraploidy and speciation

During mitosis, if the chromosomes in a diploid (2x) tissue fail to separate after replication, the resulting daughter cells will be tetraploid (4x; **Figure 9.35a**). If such tetraploid cells arise in reproductive tissue, subsequent meioses will produce diploid gametes. Rare unions between diploid gametes produce tetraploid organisms. Self-fertilization of a newly created tetraploid organism will produce an entirely new species, because crosses between the tetraploid and the original diploid organism will produce infertile triploids (review Figure 9.34a). Tetraploids made in this fashion are **autopolyploids**, a kind of polyploid that derives all its chromosome sets from the same species.

Maintenance of a tetraploid species depends on the production of gametes with balanced sets of chromosomes. Most successful tetraploids have evolved mechanisms ensuring that the four copies of each group of homologues pair two-by-two to form two **bivalents**—pairs of synapsed homologous chromosomes (**Figure 9.35b**). Because the chromosomes in each bivalent become attached to opposite spindle poles during meiosis I, meiosis regularly produces gametes carrying two complete sets of chromosomes. The mechanism requiring that each chromosome pair with only a single homologue suppresses other pairing possibilities, such as a 3:1, which cannot guarantee equivalent chromosome segregation.

Tetraploids, with four copies of every gene, generate unusual Mendelian ratios. For example, even if there are only two alleles of a gene (say, *A* and *a*), five different genotypes are possible: *A A A A, A A A a, A A a a, A a a a,*

(a) Generation of tetraploid (4x) cells

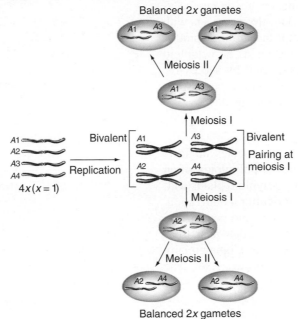

Diploid (2x) interphase cell (x = 2)

Mitotic metaphase in 2x cell

Defective mitosis, chromosomes remain in same cell

Tetraploid (4x) cell

(b) Pairing of chromosomes as bivalents

Balanced 2x gametes

A1 A3   A1 A3

Meiosis II

A1 A3

Meiosis I

A1 A2 A3 A4
4x (x = 1)

Bivalent   A1   A3   Bivalent

Replication   A2   A4   Pairing at meiosis I

Meiosis I

A2 A4

Meiosis II

A2 A4   A2 A4

Balanced 2x gametes

(c) Gametes formed by *A A a a* tetraploids

Chromosomes	Pairing			Gametes Produced by Random Spindle Attachment
1. _A_ 2. _A_ 3. _a_ 4. _a_	1 ↑ _A_ 3 ↑ _a_ 2 _A_ 4 _a_	**or**	1 ↑ _A_ 4 ↑ _a_ 2 _A_ 3 _a_	1 + 3 _A a_ **or** 1 + 4 _A a_ 2 + 4 _A a_ 2 + 3 _A a_
	1 ↑ _A_ 2 ↑ _A_ 3 _a_ 4 _a_	**or**	1 ↑ _A_ 4 ↑ _a_ 3 _a_ 2 _A_	1 + 2 _A A_ **or** 1 + 4 _A a_ 3 + 4 _a a_ 2 + 3 _A a_
	1 ↑ _A_ 2 ↑ _A_ 4 _a_ 3 _a_	**or**	1 ↑ _A_ 3 ↑ _a_ 4 _a_ 2 _A_	1 + 2 _A A_ **or** 1 + 3 _A a_ 3 + 4 _a a_ 2 + 4 _A a_

Total:
2 (*A A*) : 8 (*A a*) : 2 (*a a*)
= 1 (*A A*) : 4 (*A a*) : 1 (*a a*)

Figure 9.35 The genetics of tetraploidy. (a) Tetraploids arise from a failure of chromosomes to separate into two daughter cells during mitosis in a diploid. **(b)** In successful tetraploids, the pairing of chromosomes as bivalents generates genetically balanced gametes. **(c)** Gametes produced in an *A A a a* tetraploid heterozygous for two alleles of a centromere-linked gene, with orderly pairing of bivalents. The four chromosomes can pair to form two bivalents in three possible ways. For each pairing scheme, the chromosomes in the two pairs can assort in two different orientations. If all possibilities are equally likely, the expected genotype frequency in a population of gametes will be 1 (*A A*) : 4 (*A a*) : 1 (*a a*).

and *a a a a*. If the phenotype depends on the dosage of *A*, then five phenotypes, each corresponding to one of the genotypes, will appear. The segregation of alleles during meiosis in a tetraploid is similarly complex. Consider an *A A a a* heterozygote in which the *A* gene is closely linked to the centromere, and the *A* allele is completely dominant. What are the chances of obtaining progeny with the recessive phenotype, generated by only the *a a a a* genotype? As **Figure 9.35c** illustrates, if during meiosis 1, the four chromosomes carrying the gene align at random in bivalents

along the metaphase plate, the expected ratio of gametes is $2\ (A\ A) : 8\ (A\ a) : 2\ (a\ a) = 1\ (A\ A) : 4\ (A\ a) : 1\ (a\ a)$. The chance of obtaining $a\ a\ a\ a$ progeny during self-fertilization is thus $1/6 \times 1/6 = 1/36$. In other words, because A is completely dominant, the ratio of dominant to recessive phenotypes, determined by the ratio of $A - - -$ to $a\ a\ a\ a$ genotypes is 35:1. The ratios will be different if the gene is not closely linked to the centromere or if the dominance relationship between the alleles is not so simple.

New levels of polyploidy can arise from the doubling of a polyploid genome. Such doubling occurs on rare occasions in nature; it also results from controlled treatment with colchicine or other drugs that disrupt the mitotic spindle. The doubling of a tetraploid genome yields an octaploid ($8x$). These higher-level polyploids created by successive rounds of genome doubling are autopolyploids because all of their chromosomes derive from a single species.

Polyploids in agriculture

Roughly one out of every three known species of flowering plants is a polyploid, and because polyploidy often increases plant size and vigour, many polyploid plants with edible parts have been selected for agricultural cultivation. Most commercially grown alfalfa, coffee, and peanuts are tetraploids ($4x$). MacIntosh apple and Bartlett pear trees that produce giant fruits are also tetraploids. Commercially grown strawberries are octaploids ($8x$) (**Figure 9.36**). The evolutionary success of polyploid plant species may stem from the fact that polyploidy, like gene duplication, provides additional copies of genes; while one copy continues to perform the original function, the others can evolve new functions. As you have seen, however, the fertility of polyploid species requires an even number of chromosome sets.

Polyploidy can arise not only from chromosome doubling, but also from crosses between members of two species, even if they have different numbers of chromosomes. Hybrids in which the chromosome sets come from two or more distinct, though related, species are known as **allopolyploids**. In crosses between octaploids and tetraploids, for example, fertilization unites tetraploid and diploid gametes to produce hexaploid progeny. Fertile allopolyploids arise only rarely, under special conditions, because chromosomes from the two species differ in shape, size, and number, so they cannot easily pair with

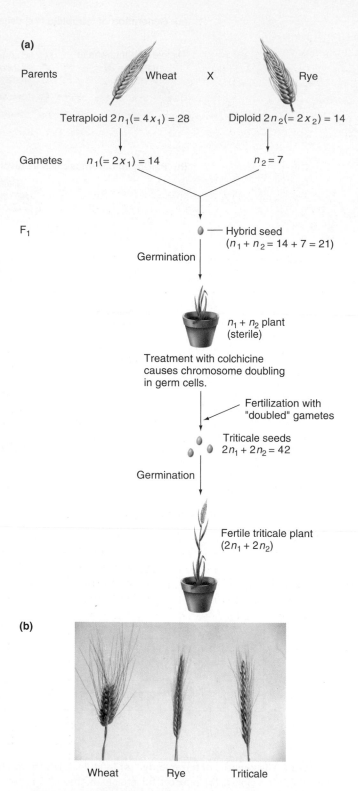

Figure 9.37 Amphidiploids in agriculture. (a) Plant breeders cross wheat with rye to create allopolyploid triticales. Because this strain of wheat is tetraploid, x_1 (the number of chromosomes in the basic wheat set) is one-half n_1 (the number of chromosomes in a wheat gamete). For diploid rye, $n_2 = x_2$. Note that the F_1 hybrid between wheat and rye is sterile because the rye chromosomes have no pairing partners. Doubling of chromosome numbers by colchicine treatment of the F_1 hybrid corrects this problem, allowing regular pairing. **(b)** A comparison of wheat, rye, and triticale grain stalks.

Figure 9.36 Many polyploid plants are larger than their diploid counterparts. A comparison of octaploid (*left*) and diploid (*right*) strawberries.

each other. The resulting irregular segregation creates genetically unbalanced gametes such that the hybrid progeny will be sterile. Chromosomal doubling in germ cells, however, can restore fertility by creating a pairing partner for each chromosome. Organisms produced in this manner are termed **amphidiploids** if the two parental species were diploids; they contain two diploid genomes, each one derived from a different parent. As the following illustrations show, it is hard to predict the characteristics of an amphidiploid or other allopolyploids.

A cross between cabbages and radishes, for example, leads to the production of amphidiploids known as *Raphanobrassica*. The gametes of both parental species contain 9 chromosomes; the sterile F_1 hybrids have 18 chromosomes, none of which has a homologue. Chromosome doubling in the germ cells after treatment with colchicine, followed by union of two of the resulting gametes, produces a new species: a fertile *Raphanobrassica* amphidiploid carrying 36 chromosomes—a full complement of 18 (9 pairs) derived from cabbages and a full complement of 18 (9 pairs) derived from radishes. Unfortunately, this amphidiploid has the roots of a cabbage plant and leaves resembling those of a radish, so it is not agriculturally useful.

By contrast, crosses between tetraploid (or hexaploid) wheat and diploid rye have led to the creation of several allopolyploid hybrids with agriculturally desirable traits from both species (**Figure 9.37**). Some of the hybrids combine the high yields of wheat with rye's ability to adapt to unfavourable environments. Others combine wheat's high level of protein with rye's high level of lysine; wheat protein does not contain very much of this amino acid, an essential ingredient in the human diet. The various hybrids between wheat and rye form a new crop known as triticale. Some triticale strains produce nutritious grains that already appear in breads sold in health food stores. Plant breeders are currently assessing the usefulness of various triticale strains for large-scale agriculture.

Organisms with odd numbers of chromosome sets are generally infertile because during meiosis, some or all of their chromosomes do not have pairing partners. Chromosome doubling can produce new, fertile polyploid species of plants with even numbers of chromosome sets. In autopolyploids, all the chromosomes originally came from a single ancestral species, but in allopolyploids, the chromosomes were derived from two different ancestral species.

9.5 Emergent Technologies: Beyond the Karyotype

Two main problems occur when searching for chromosomal rearrangements and changes in chromosome number by karyotype analysis. First, it is a tedious procedure that depends on highly trained technicians to identify chromosomal alterations under the microscope. Because of the subjective nature of the analysis, mistakes can reduce the accuracy of results. Second, even in the hands of the best technicians, there is a limit to the viewing resolution. Even under optimal circumstances, it is not possible to detect deletions or duplications of less than 5 Mb in human karyotypes. Human populations no doubt have many chromosomes with as yet undetected smaller deletions or duplications.

To overcome the limitations of karyotype analysis, researchers have developed a microarray-based hybridization protocol that can scan the genome for deletions, duplications, and aneuploidy with much greater resolution, very high accuracy, and much greater throughput and without the need for a subjective determination of the result. The technique is called *comparative genomic hybridization (CGH)* or sometimes *virtual karyotyping*.

The protocol works as follows (**Figure 9.38**). First, a series of 20 000 BAC clones with DNA inserts averaging 150 kb that collectively represent the entire human genome are spotted onto a microarray. These BAC clones were characterized in the course of the Human Genome Project.

Next, genomic DNA from a control sample with a normal genome content is labelled with a yellow fluorescent dye, while the genomic DNA from the test sample is labelled with a red fluorescent dye. The two genomic DNA samples are mixed together in equal amounts, denatured, and applied to the microarray as a probe. After hybridization is complete and unhybridized material is washed away, the fluorescence emission from each microarray dot is analyzed automatically by a machine designed for this task.

If the genomic region probed with a particular BAC clone is present in two copies in the test sample, then the ratio of red to yellow dyes on that dot will be 1:1. However, if a particular genomic region is duplicated or deleted from one homologue in the test sample, the ratio of red to yellow will be 1.5:1 or 0.5:1, respectively. An example of this analysis is shown in Figure 9.38.

CGH provides a powerful clinical tool to detect any type of aneuploidy or any deletion or duplication of 50 kb or more anywhere in the genome. Clinicians can use it in conjunction with amniocentesis, CVS, or preimplantation genetic analysis. They can also use CGH to screen tissue biopsies for cancerous cells that have deleted or duplicated regions containing oncogenes or tumour suppressor genes. The technique thus holds great promise for the detection of new genes that contribute to the genesis of cancer.

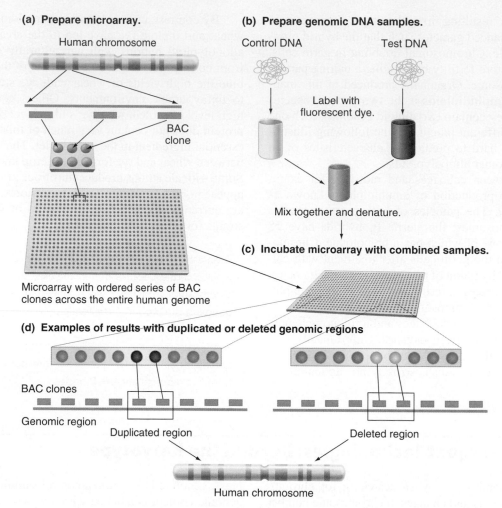

(a) Prepare microarray.

Human chromosome

BAC clone

Microarray with ordered series of BAC clones across the entire human genome

(b) Prepare genomic DNA samples.

Control DNA   Test DNA

Label with fluorescent dye.

Mix together and denature.

(c) Incubate microarray with combined samples.

(d) Examples of results with duplicated or deleted genomic regions

BAC clones

Genomic region

Duplicated region   Deleted region

Human chromosome

Figure 9.38 Comparative genomic hybridization detects duplications, deletions, and aneuploidy. (a) BAC clones representing the human genome are spotted in order onto a microarray. **(b)** The genomic sample to be tested is labelled with one colour dye (here, *red*), and the control genome sample is labelled with a second colour dye (*yellow*). **(c)** The two samples are mixed together, denatured, and then incubated on the microarray. **(d)** Automated analysis of each spot on the microarray detects the ratio of the two dyed probes that hybridize. *Orange* indicates a 1:1 ratio; other colours indicate deletion (0.5:1 ratio; *yellow*) or duplication (1.5:1 ratio; *red*) of BAC clone sequences in the test sample.

Connections

The detrimental consequences of most changes in chromosome organization and number cause considerable distress in humans (**Table 9.2**). Approximately 4 of every 1000 individuals has an abnormal phenotype associated with aberrant chromosome organization or number. Most of these abnormalities result from either aneuploidy for the X chromosome or trisomy 21. By comparison, about 10 people per 1000 suffer from an inherited disease caused by a single-gene mutation.

The incidence of chromosomal abnormalities among humans would be much larger were it not for the fact that many embryos or fetuses with abnormal karyotypes abort spontaneously early in pregnancy. Fully 15 percent to 20 percent of recognized pregnancies end with detectable spontaneous abortions; and half of the spontaneously aborted fetuses show chromosomal abnormalities, particularly trisomy, sex-chromosome monosomy, and triploidy. These figures almost certainly underestimate the rate of spontaneous abortion caused by abnormal chromosomal variations, because embryos carrying aberrations

for larger chromosomes, such as monosomy 2 or trisomy 5, may abort so early that the pregnancy goes unrecognized.

But despite all the negative effects of chromosomal rearrangements and changes in chromosome number, a few departures from normal genome organization survive to become instruments of evolution by natural selection.

As we see in Chapter 18, chromosomal rearrangements occur in bacteria as well as in eukaryotic organisms. In bacteria, transposable elements catalyze many of the changes in chromosomal organization. Remarkably, the reshuffling of genes between different DNA molecules in the same cell catalyzes the transfer of genetic information from one bacterial cell to another.

In Chapters 10 and 11, we examine the molecular mechanisms that regulate gene expression in prokaryotes and regulate nuclear gene expression in eukaryotes. We see again some unifying principles in prokaryotes and eukaryotes but also unique solutions suited to the structure and function of these different types of cells.

Table 9.2	Aneuploidy in the Human Population	
Chromosomes	Syndrome	Frequency at Birth
Autosomes		
Trisomic 21	Down	1/700
Trisomic 13	Patau	1/5000
Trisomic 18	Edwards	1/10 000
Sex chromosomes, females		
XO, monosomic	Turner	1/5000
XXX, trisomic		
XXXX, tetrasomic		1/700
XXXXX, pentasomic		
Sex chromosomes, males		
XYY, trisomic	Normal	1/10 000
XXYY, tetrasomic		
XXY, trisomic	Klinefelter	1/500
XXXY, tetrasomic		
XXXXY, pentasomic		
XXXXXY, hexasomic		

About 0.4 percent of all babies born have a detectable chromosomal abnormality that generates a detrimental phenotype.

Essential Concepts

1. Rearrangements reorganize the DNA sequences within genomes. The results are subject to natural selection, and thus rearrangements serve as instruments of evolution. **[LO1, LO3]**

2. Deletions remove DNA from a chromosome. Homozygosity for a large deletion is usually lethal, but even heterozygosity for a large deletion can create a deleterious genetic imbalance. Deletions may uncover recessive mutations on the homologous chromosome and are thus useful for gene mapping. **[LO1–2]**

3. Duplications add DNA to a chromosome. The additional copies of genes can be a major source of new genetic functions. Homozygosity or heterozygosity for duplications causes departures from normal gene dosage that are often harmful to the organism. Unequal crossing-over between duplicated regions expands or contracts the number of gene copies and may lead to multigene families. **[LO3]**

4. Inversions alter the order, but not the number, of genes on a chromosome. They may produce novel phenotypes by disrupting the activity of genes near the rearrangement breakpoints. Inversion heterozygotes exhibit crossover suppression because progeny formed from recombinant gametes are genetically unbalanced. **[LO3]**

5. In reciprocal translocations, parts of two chromosomes trade places without the loss or gain of chromosomal material. Translocations may modify the function of genes at or near the translocation breakpoints.

Heterozygosity for translocations in the germ line results in semisterility and pseudolinkage. **[LO3]**

6. Transposable elements (TEs) are short, mobile segments of DNA that reshape genomes by generating mutations, causing chromosomal rearrangements, and relocating genes. **[LO4–5]**

7. Aneuploidy, the loss or gain of one or more chromosomes, creates a genetic imbalance. Mistakes in meiosis produce aneuploid gametes, whereas mistakes in mitosis generate aneuploid clones of cells. Autosomal aneuploidy is usually lethal. Sex chromosome aneuploidy is better tolerated because of dosage compensation mechanisms. **[LO6]**

8. Euploid organisms contain complete sets of chromosomes. Organisms with three or more sets of chromosomes are polyploids. In autopolyploidy, all chromosome sets are derived from the same species; in allopolyploidy, chromosome sets come from two or more distinct, though related, species. **[LO6]**

9. Monoploids (with only a single complete chromosome set) as well as polyploids containing odd numbers of chromosome sets are sterile because the chromosomes cannot pair properly during meiosis I. **[LO6]**

10. Polyploids having even numbers of chromosome sets can be fertile if proper chromosome segregation occurs. Amphidiploids, which are allopolyploids produced by chromosome doubling of genomes derived from different diploid parental species, are often fertile and are sometimes useful in agriculture. **[LO6–7]**

I. *Drosophila* males from a true-breeding wild-type stock were irradiated with X-rays and then mated with females from a true-breeding stock carrying the following recessive mutations on the X chromosome: yellow body (*y*), crossveinless wings (*cv*), cut wings (*ct*), singed bristles (*sn*), and miniature wings (*m*). These markers are known to map in the following order:

$$y - cv - ct - sn - m$$

Most of the female progeny of this cross were phenotypically wild type, but one female exhibited *ct* and *sn* phenotypes. When this exceptional *ct sn* female was mated with a male from the true-breeding wild-type stock, there were twice as many females as males among the progeny.

a. What is the nature of the X-ray-induced mutation present in the exceptional female?

b. Draw the X chromosomes present in the exceptional *ct sn* female as they would appear during pairing in meiosis.

c. What phenotypic classes would you expect to see among the progeny produced by mating the exceptional *ct sn* female with a normal male from a true-breeding wild-type stock? List males and females separately.

Answer

To answer this problem, you need to think first about the effects of different types of chromosomal mutations in order to deduce the nature of the mutation. Then you can evaluate the consequences of the mutation on inheritance.

a. Two observations indicate that *X-rays induced a deletion mutation*. The fact that two recessive mutations are phenotypically expressed in the exceptional female suggests that a deletion was present on one of her X chromosomes that uncovered the two mutant alleles (*ct* and *sn*) on the other X chromosome. Second, the finding that there were twice as many females as males among the progeny of the exceptional female is also consistent with a deletion mutation. Males who inherit the deletion-bearing X chromosome from their exceptional mother will be inviable (because other essential genes are located in the region that is now deleted), but sons who inherit a nondeleted X chromosome will survive. On the other hand, all of the exceptional female's daughters will be viable: Even if they inherit a deleted X chromosome from their mother, they also receive a normal X chromosome from their father. As a result, there are half as many male progeny as females from the cross of the exceptional female with a wild-type male.

b. During pairing, *the DNA in the normal (nondeleted) X chromosome will loop out* because there is no homologous region in the deletion chromosome. In the simplified drawing of meiosis I that follows, each line represents both chromatids comprising each homologue.

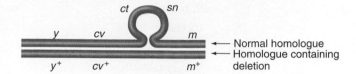

c. *All daughters of the exceptional female will be wild type* because the father contributes wild-type copies of all the genes. Each of the surviving sons must inherit a nondeleted X chromosome from the exceptional female. Some of these X chromosomes are produced from meioses in which no recombination occurred, but other X chromosomes are the products of recombination. *Males can have any of the genotypes listed here and therefore the corresponding phenotypes*. All contain the *ct sn* combination because no recombination between homologues is possible in this deleted region. Some of these genotypes require multiple crossovers during meiosis in the mother and will thus be relatively rare.

y	*cv*	*ct*	*sn*	*m*
+	+	*ct*	*sn*	+
+	*cv*	*ct*	*sn*	*m*
y	+	*ct*	*sn*	+
y	*cv*	*ct*	*sn*	+
+	+	*ct*	*sn*	*m*
+	*cv*	*ct*	*sn*	+
y	+	*ct*	*sn*	*m*

II. One of the X chromosomes in a particular *Drosophila* female had a normal order of genes but carried recessive alleles of the genes for yellow body colour (*y*), vermilion eye colour (*v*), and forked bristles (*f*), as well as the dominant X-linked Bar eye mutation (*B*). Her other X chromosome carried the wild-type alleles of all four genes, but the region including y^+, v^+, and f^+ (but not B^+) was inverted with respect to the normal order of genes. This female was crossed to a wild-type male in the cross diagrammed here.

The cross produced the following male offspring:

y	v	f	B	48
y^+	v^+	f^+	B^+	45
y	v	f	B^+	11
y^+	v^+	f^+	B	8
y	v^+	f	B	1
y^+	v	f^+	B^+	1

a. Why are there no male offspring with the allele combinations $y\,v\,f^+$, $y^+\,v^+\,f$, $y\,v^+\,f^+$, or $y^+\,v\,f$, (regardless of the allele of the Bar eye gene)?

b. What kinds of crossovers produced the $y\,v\,f\,B^+$ and $v^+\,y^+\,f^+\,B$ offspring? Can you determine any genetic distances from these classes of progeny?

c. What kinds of crossovers produced the $y^+\,v\,f^+\,B^+$ and $y\,v^+\,f\,B$ offspring?

Answer

To answer this question, you need to be able to draw and interpret pairing in inversion heterozygotes. Note that this inversion is paracentric.

a. During meiosis in an inversion heterozygote, a loop of the inverted region is formed when the homologous genes align. In the following simplified drawing, each line represents both chromatids comprising each homologue.

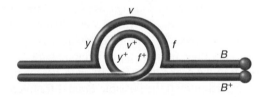

If a single crossover occurs within the inversion loop, a dicentric and an acentric chromosome are formed. Cells containing these types of chromosomes are not viable. The resulting allele combinations from such single crossovers are not recovered. The four phenotypic classes of missing male offspring would be formed by single crossovers between the y and v or between the v and f genes in the female inversion heterozygote and therefore are not recovered.

b. *The $y\,v\,f\,B^+$ and $y^+\,v^+\,f^+\,B$ offspring are the result of single crossover events outside of the inversion loop, between the end of the inversion (just to the right of f on the preceding diagram) and the B gene. This region is approximately 16.7 m.u. in length (19 recombinants out of 114 total progeny).*

c. *The $y^+\,v\,f^+\,B^+$ and $y\,v^+\,f\,B$ offspring would result from two crossover events within the inversion loop, one between the y and v genes and the other between the v and f genes. You should note that these could be*

either two-strand or three-strand double crossovers, but they could not be four-strand double crossovers.

III. In maize trisomics, $n + 1$ pollen is not viable. If a dominant allele at the B locus produces purple colour instead of the recessive phenotype bronze and a $B\,b\,b$ trisomic plant is pollinated by a $B\,B\,b$ plant, what proportion of the progeny produced will be trisomic and have a bronze phenotype?

Answer

To solve this problem, think about what is needed to produce trisomic bronze progeny: three b chromosomes in the zygote. The female parent would have to contribute two b alleles, because the $n + 1$ pollen from the male is not viable. What kinds of gametes could be generated by the trisomic $B\,b\,b$ purple female parent, and in what proportion? To track all the possibilities, rewrite this genotype as $B\,b_1\,b_2$, even though b_1 and b_2 have identical effects on phenotype. In the trisomic female, there are three possible ways the chromosomes carrying these alleles could pair as bivalents during the first meiotic division so that they would segregate to opposite poles: B with b_1, B with b_2, and b_1 with b_2. In all three cases, the remaining chromosome could move to either pole. To tabulate the possibilities as a branching diagram:

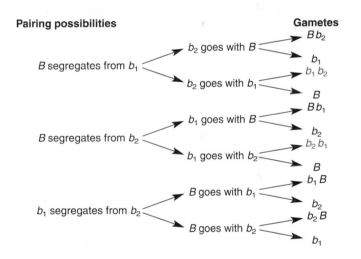

Of the 12 gamete classes produced by these different possible segregations, only the two classes written in red contain the two b alleles needed to generate the bronze ($b\,b\,b$) trisomic zygotes. There is thus a $2/12 = 1/6$ chance of obtaining such gametes.

Although segregation in the $B\,B\,b$ male parent is equally complicated, remember that males cannot produce viable $n + 1$ pollen. The only surviving gametes would thus be B and b, in a ratio ($2/3$ B and $1/3$ b) that must reflect their relative prevalence in the male parent genome. *The probability of obtaining trisomic bronze progeny from this cross is therefore the product of the individual probabilities of the appropriate $b\,b$ gametes from the female parent (1/6) and b pollen from the male parent (1/3): $1/6 \times 1/3 = 1/18$.*

Vocabulary

1. For each of the terms in the left column, choose the best matching phrase in the right column.

i. reciprocal translocation	**1.** lacking one or more chromosomes or having one or more extra chromosomes
ii. gynandromorph	**2.** movement of short DNA elements
iii. pericentric	**3.** having more than two complete sets of chromosomes
iv. paracentric	**4.** exact exchange of parts of two nonhomologous chromosomes
v. euploids	**5.** excluding the centromere
vi. polyploidy	**6.** including the centromere
vii. transposition	**7.** having complete sets of chromosomes
viii. aneuploids	**8.** mosaic combination of male and female tissue

Section 9.1

2. For each of the following types of chromosomal aberrations, tell (i) whether an organism heterozygous for the aberration will form any type of loop in the chromosomes during prophase I of meiosis; (ii) whether a chromosomal bridge can be formed during anaphase I in a heterozygote, and if so, under what condition; (iii) whether an acentric fragment can be formed during anaphase I in a heterozygote, and if so, under what condition; (iv) whether the aberration can suppress meiotic recombination; and (v) whether the two chromosomal breaks responsible for the aberration occur on the same side or on opposite sides of a single centromere, or if the two breaks occur on different chromosomes.

a. reciprocal translocation

b. paracentric inversion

c. small tandem duplication

d. Robertsonian translocation

e. pericentric inversion

f. large deletion

3. In flies that are heterozygous for either a deletion or a duplication, there will be a looped-out region in a preparation of polytene chromosomes. How could you distinguish between a deletion or a duplication using polytene chromosome analysis?

4. For the following types of chromosomal rearrangements, would it theoretically ever be possible to obtain a perfect reversion of the rearrangement? If so, would such revertants be found only rarely, or would they be relatively common?

a. a deletion of a region including five genes

b. a tandem duplication of a region including five genes

c. a pericentric inversion

d. a Robertsonian translocation

e. a mutation caused by a transposable element jumping into a protein-coding exon of a gene

5. Four strains of *Drosophila* were constructed in which one autosome contained recessive mutant alleles of the four genes *rolled eyes*, *thick legs*, *straw bristles*, and *apterous wings*, and the homologous autosome contained one of four different deletions (deletions 1–4). The phenotypes of the flies were as follows:

Deletion	Phenotype
1	rolled eyes, straw bristles
2	apterous wings, rolled eyes
3	thick legs, straw bristles
4	apterous wings

Whole-genome DNA was prepared from the flies. The DNA was digested to completion with the restriction enzyme *Bam*HI, run on an agarose gel, and transferred to nitrocellulose filters. The filters were then probed with a 20-kb cloned piece of wild-type genomic DNA obtained by partially digesting the plasmid clone with *Bam*HI (so the ends of the probe were *Bam*HI ends, but the piece was not digested into all the possible *Bam*HI fragments). The results of this whole-genome Southern blot are shown below. Dark bands indicate fragments present twice in the diploid genome; light bands indicate fragments present once in the genome.

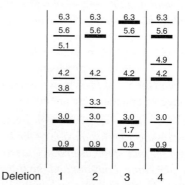

a. Make a map of the *Bam*HI restriction sites in this 20-kb part of the wild-type *Drosophila* genome, indicating distances in kilobases between adjacent *Bam*HI sites. (*Hint:* The genomic DNA fragments in wild type are 6.3, 5.6, 4.2, 3.0, and 0.9 kb long.)

b. On your map, indicate the locations of the genes.

6. A series of chromosomal mutations in *Drosophila* were used to map the *javelin* gene, which affects bristle shape, and *henna*, which affects eye pigmentation. Both the *javelin* and *henna* mutations are recessive. A diagram of region 65 of the *Drosophila* polytene chromosomes is shown here.

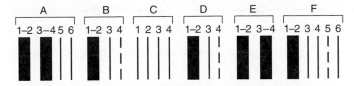

The chromosomal breakpoints for six chromosome rearrangements are indicated in the following table. (For example, deletion A has one breakpoint between bands A2 and A3 and the other between bands D2 and D3.)

Breakpoints in region 65			
Deletions	A	A2–3;	D2–3
	B	C2–3;	E4–F1
	C	D2–3;	F4–5
	D	D4–E1;	F3–4
Breakpoints			
Inversions	A	Band 65A6	Band 82A1
	B	Band 65B4	Band 98A3

Flies with a chromosome containing one of these six rearrangements (deletions or inversions) were mated to flies homozygous for both *javelin* and *henna*. The phenotypes of the heterozygous progeny (i.e., *rearrangement/javelin, henna*) are shown here.

		Phenotypes of F₁ flies
Deletions	A	javelin, henna
	B	henna
	C	wild type
	D	wild type
Inversions	A	javelin
	B	wild type

Using these data, what can you conclude about the cytogenetic location for the *javelin* and *henna* genes?

7. Human chromosome 1 is a large, metacentric chromosome. A map of a cloned region from near the telomere of chromosome 1 is shown here. Three probe DNAs (A, B, and C) from this region were used for *in situ* hybridization to human mitotic metaphase chromosome squashes made with cells obtained from individuals with various genotypes. The breakpoints of chromosomal rearrangements in this region are indicated on the map. The black bars for deletions (*Del*) 1 and 2 represent DNA that is deleted. The breakpoints of inversions (*Inv*) 1 and 2 not shown in the figure are near, but not at, the centromere. For each of the following genotypes, draw chromosome 1 as it would appear after *in situ* hybridization. An example is shown in the following figure for hybridization of probe A to the two copies of chromosome 1 in wild type (+/+).

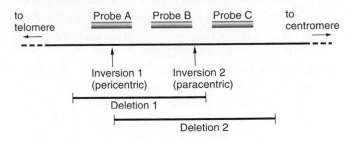

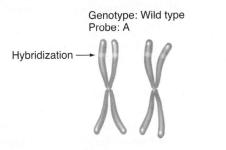

a. genotype: *Del1/Del2*; probe: B
b. genotype: *Del1/Del2*; probe: C
c. genotype: *Del1/ +*; probe: A
d. genotype: *Inv1/ +*; probe: A
e. genotype: *Inv2/ +*; probe: B
f. genotype: *Inv2/Inv2*; probe: C

8. Genes *a* and *b* are 21 m.u. apart when mapped in highly inbred strain 1 of corn and 21 m.u. apart when mapped in highly inbred strain 2. But when the distance is mapped by testcrossing the F₁ progeny of a cross between strains 1 and 2, the two genes are only 1.5 m.u. apart. What arrangement of genes *a* and *b* and any potential rearrangement breakpoints could explain these results?

9. The partially recessive, X-linked z^1 mutation of the *Drosophila* gene *zeste* (*z*) can produce a yellow (zeste) eye colour only in flies that have two or more copies of the wild-type *white* (*w*) gene. Using this property, tandem duplications of the w^+ gene called w^{+R} were identified. Males with the genotype $z^1 w^{+R}$/Y thus have zeste eyes. These males were crossed to females with the genotype $y z^1 w^{+R} spl/y^+ z^1 w^{+R} spl^+$. (These four genes are closely linked on the X chromosome, in the order given in the genotype, with the centromere to the right of all these genes: y = yellow bodies; y^+ = tan bodies; *spl* = split bristles; spl^+ = normal bristles.) Out of 81 540 male progeny of these females, the following exceptions were found:

Class A 2430 yellow bodies, zeste eyes, wild-type bristles

Class B 2394 tan bodies, zeste eyes, split bristles

Class C 23 yellow bodies, wild-type eyes, wild-type bristles

Class D 22 tan bodies, wild-type eyes, split bristles

a. What were the phenotypes of the remainder of the 81 540 males from the first cross?

b. What events gave rise to progeny of classes A and B?

c. What events gave rise to progeny of classes C and D?

d. On the basis of these experiments, what is the genetic distance between *y* and *spl*?

10. Three strains of *Drosophila* (Bravo, X-ray, and Zorro) are obtained that are homozygous for three variant forms of a particular chromosome. When examined in salivary gland polytene chromosome spreads, all chromosomes have the same number of bands in all three strains. When genetic mapping is performed in the Bravo strain, the following map is obtained (distances in map units).

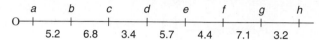

Bravo and X-ray flies are now mated to form Bravo/X-ray F$_1$ progeny, and Bravo flies are also mated with Zorro flies to form Bravo/Zorro F$_1$ progeny. In subsequent crosses, the following genetic distances were found to separate the various genes in the hybrids:

	Bravo/X-ray	Bravo/Zorro
a–b	5.2	5.2
b–c	6.8	0.7
c–d	0.2	< 0.1
d–e	< 0.1	< 0.1
e–f	< 0.1	< 0.1
f–g	0.65	0.7
g–h	3.2	3.2

a. Draw a map showing the relative order of genes *a* through *h* in the X-ray and Zorro strains. Do not show distances between genes.

b. In the original X-ray homozygotes, would the physical distance between genes *c* and *d* be greater than, less than, or approximately equal to the physical distance between these same genes in the original Bravo homozygotes?

c. In the original X-ray homozygotes, would the physical distance between genes *d* and *e* be greater than, less than, or approximately equal to the physical distance between these same genes in the original Bravo homozygotes?

11. In the following group of figures, the *pink lines* indicate an area of a chromosome that is inverted relative to the normal (*black line*) order of genes. The diploid chromosome constitution of individuals 1–4 is shown. Match the individuals with the appropriate statement(s) that follow. More than one diagram may correspond to the following statements, and a diagram may be a correct answer for more than one question.

a. An inversion loop would form during meiosis I and in polytene chromosomes.

b. A single crossover involving the inverted region on one chromosome and the homologous region of the

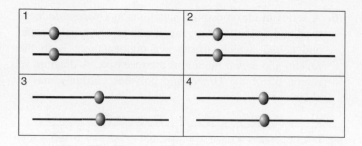

other chromosome would yield genetically unbalanced gametes.

c. A single crossover involving the inverted region on one chromosome and the homologous region of the other chromosome would yield an acentric fragment.

d. A single crossover involving the inverted region yields four viable gametes.

12. In *Drosophila*, the gene for cinnabar eye colour is on chromosome 2, and the gene for scarlet eye colour is on chromosome 3. A fly homozygous for both recessive *cinnabar* and *scarlet* alleles (*cn/cn*; *st/st*) is white-eyed.

a. If male flies (containing chromosomes with the normal gene order) heterozygous for *cn* and *st* alleles are crossed to white-eyed females homozygous for the *cn* and *st* alleles, what are the expected phenotypes and their frequencies in the progeny?

b. One unusual male heterozygous for *cn* and *st* alleles, when crossed to a white-eyed female, produced only wild-type and white-eyed progeny. Explain the likely chromosomal constitution of this male.

c. When the wild-type F$_1$ females from the cross with the unusual male were backcrossed to normal *cn/cn*; *st/st* males, the following results were obtained:

wild type	45%
cinnabar	5%
scarlet	5%
white	45%

Diagram a genetic event at metaphase I that could produce the rare cinnabar or scarlet flies among the progeny of the wild-type F$_1$ females.

13. In the following figure, *black* and *pink lines* represent nonhomologous chromosomes. Which of the figures matches the descriptions below? More than one diagram may correspond to the statements, and a diagram may be a correct answer for more than one question.

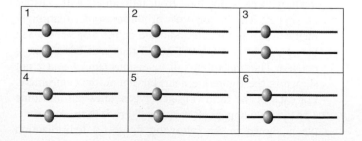

a. gametes produced by a translocation heterozygote

b. gametes that could not be produced by a translocation heterozygote

c. genetically balanced gametes produced by a translocation heterozygote

d. genetically unbalanced gametes that can be produced (at any frequency) by a translocation heterozygote

14. A proposed biological method for insect control involves the release of insects that could interfere with the fertility of the normal resident insects. One approach is to introduce sterile males to compete with the resident fertile males for matings. A disadvantage of this strategy is that the irradiated sterile males are not very robust and can have problems competing with the fertile males. An alternate approach that is being tried is to release laboratory-reared insects that are homozygous for several translocations. Explain how this strategy will work. Be sure to mention which insects will be sterile.

15. Semisterility in corn, as seen by unfilled ears with gaps due to abortion of approximately half the ovules, is an indication that the strain is a translocation heterozygote. The chromosomes involved in the translocation can be identified by crossing the translocation heterozygote to a strain homozygous recessive for a gene on the chromosome being tested. The ratio of phenotypic classes produced from crossing semisterile F_1 progeny back to a homozygous recessive plant indicates whether the gene is on one of the chromosomes involved in the translocation. For example, a semisterile strain could be crossed to a strain homozygous for the yg mutation on chromosome 9. (The mutant has yellow-green leaves instead of the wild-type green leaves.) The semisterile F_1 progeny would then be backcrossed to the homozygous yg mutant.

   a. What types of progeny (fertile or semisterile, green or yellow-green) would you predict from the backcross of the F_1 to the homozygous yg mutant if the gene was not on one of the two chromosomes involved in the translocation?

   b. What types of progeny (fertile or semisterile, green or yellow-green) would you predict from the backcross of the F_1 to the homozygous mutant if the yg gene is on one of the two chromosomes involved in the translocation?

   c. If the yg gene is located on one of the chromosomes involved in the translocation, a few fertile, green progeny and a few semisterile, yellow-green progeny are produced. How could these relatively rare progeny classes arise? What genetic distance could you determine from the frequency of these rare progeny?

16. **a.** Among the selfed (intermated) progeny of a semisterile corn plant that is heterozygous for a reciprocal translocation, what ratio do you expect for progeny plants with normal fertility versus those showing semisterility? In this problem, ignore the rare gametes produced by adjacent-2 segregation.

   b. Among the selfed progeny of a particular semisterile corn plant heterozygous for a reciprocal translocation, the ratio of fertile to semisterile plants was 1:4. How can you explain this deviation from your answer to part a?

17. A *Drosophila* male is heterozygous for a translocation between an autosome originally bearing the dominant mutation *Lyra* (shortened wings) and the Y chromosome; the other copy of the same autosome is *Lyra$^+$*. This male is now mated with a true-breeding, wild-type female. What kinds of progeny would be obtained, and in what proportions?

18. The figure *below* portrays human chromosome 21 in *blue* and chromosome 14 in *red*. The *arrows* represent the 5'-to-3' orientations of various PCR primers. If primer A is one of the two primers used, what is the other primer you could employ to diagnose the presence of a Robertsonian translocation (14q21q) that might be involved in translocation Down syndrome? (That is, which numbered primer, in conjunction with primer A, would produce a PCR product from the genomic DNA of individuals with the translocation, but no PCR product from genomic DNA lacking the translocation?)

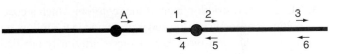

19. Solved Problem I in Chapter 5 shows the genesis of a small chromosomal inversion. Assuming that 11-bp-long primers can be used for the polymerase chain reaction (even though ordinarily longer primers are needed), give the sequences of the two 11-bp primers that could be used to generate the longest PCR product that would indicate the presence of the inversion. (That is, this pair of primers would produce a PCR product from the genomic DNA of individuals with the inversion, but not from wild-type genomic DNA.)

Section 9.2

20. Explain how transposable elements can cause movement of genes that are not part of the transposable element.

21. In the 1950s, Barbara McClintock found a transposable element in corn she called *Ds* (*Dissociation*). When inserted at a particular location, this element could often cause chromosomal breaks at that site, but these breaks occurred only in the presence of another unlinked genetic element she called *Ac* (*Activator*). She found further that in the presence of *Ac*, *Ds* could jump to other chromosomal locations. At some of these locations (and in the presence of *Ac*), *Ds* would now cause chromosomal breakage at the new position; at other positions, it appeared that *Ds* could cause new mutations that were unstable as shown by their patchy, variegated expression in kernels. Interestingly, the position of the *Ac* element seemed to be very different in various strains of corn. Explain these results in terms of our present-day understanding of transposons.

22. Gerasimova and colleagues in the former Soviet Union characterized a mutation in the *Drosophila* cut wing (*ct*) gene called *ct^{MR2}*, which is associated with the insertion of a transposable element called *gypsy*. This allele is very unstable: Approximately 1 in 100 of the progeny of flies bearing *ct^{MR2}* show new *ct* variants. Some of these are *ct^+* revertants, whereas others appear to be more severe alleles of *ct* with stronger effects on wing shape. When the *ct^+* revertants themselves are mated, some of the *ct^+* alleles appear to be stable (no new *ct* mutants appear), whereas others are highly unstable (many new mutations appear). What might explain the generation of stable and unstable *ct^+* revertants as well as the stronger *ct* mutant alleles?

23. In sequencing a region of the human genome, you have come across a segment of about 200 A nucleotides. You suspect that the sequence preceding the A residues may have been moved here by a transposition event mediated by reverse transcriptase. If the adjacent sequence is in fact a retrotransposon, you might expect to find other copies in the genome. How could you determine if other copies of this DNA exist in the genome, and whether this DNA is indeed transposable?

24. The *Eco*RI restriction map of the region in which a coat-colour gene in mice is located is presented in the following diagram. The left-most *Eco*RI site is arbitrarily labelled 0 and the other distances in kilobases are given relative to this coordinate. Genomic DNA was prepared from one wild-type mouse and ten mice homozygous for various mutant alleles. This genomic DNA is digested with *Eco*RI, fractionated on agarose gels, and then transferred to nitrocellulose filters. The filters were probed with the radioactive DNA fragment indicated by the *purple* bar, extending from coordinate 2.6 kb to coordinate 14.5. The resultant autoradiogram is shown schematically.

Assume that each of the mutations 1–10 is caused by one and only one of the events on the following list. Which event corresponds to which mutation?

a. a point mutation exactly at coordinate 6.8

b. a point mutation exactly at coordinate 6.9

c. a deletion between coordinates 10.1 and 10.4

d. a deletion between coordinates 6.7 and 7.0

e. insertion of a transposable element at coordinate 6.2

f. an inversion with breakpoints at coordinates 2.2 and 9.9

g. a reciprocal translocation with another chromosome with a breakpoint at coordinate 10.1

h. a reciprocal translocation with another chromosome with a breakpoint at coordinate 2.4

i. a tandem duplication of sequences between coordinates 7.2 and 9.2

j. a tandem duplication of sequences between coordinates 11.3 and 14.3

Section 9.3

25. In the figure at the bottom of this page, the *top* and *bottom lines* represent chromosomes 4 and 12 of the yeast *Saccharomyces cerevisiae* (*Scer* 4 and *Scer* 12). *Numbers* refer to specific genes, and the *red arrows* represent the direction and extent of transcription. The *middle line* is the sequence of a region from chromosome 1 from a different, but related yeast species called *Kluyveromyces waltii* (*Kwal* 1), with genes indicated in *light blue*. Homologies (close relationships in DNA sequence) are shown as *lines* joining chromosomes of the two species.

a. What is the meaning of the two *K. waltii* genes filled in *dark purple*?

b. Based on these data, formulate a hypothesis to explain the genesis of the part of the *S. cerevisiae* genome illustrated in the figure.

26. Two possible models have been proposed to explain the potential evolutionary advantage of gene duplications. In the first model, one of the two duplicated copies retains the same function as the ancestral gene, leaving the other copy to diverge through mutation to fulfill a new biochemical function. In the second model, both copies can diverge rapidly from the ancestral gene, so that both can acquire new properties. Considering your answer to Problem 25, and given that both the *S. cerevisiae* and *K. waltii* genomes have been completely sequenced, how could you determine which of these two models better represents the course of evolution?

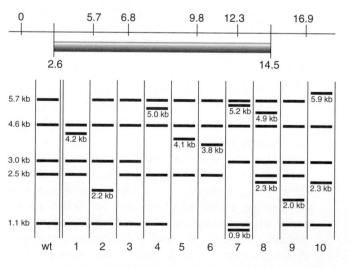

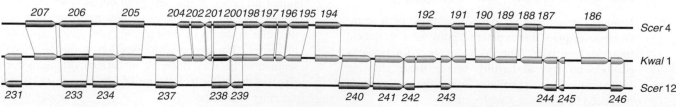

27. The number of chromosomes in the somatic cells of several oat varieties (*Avena* species) are as follows: 14 in sand oats (*Avena strigosa*); 28 in slender wild oats (*Avena barata*); and 42 in cultivated oats (*Avena sativa*).

   a. What is the basic chromosome number (*x*) in *Avena*?

   b. What is the ploidy for each of the different species?

   c. What is the number of chromosomes in the gametes produced by each of these oat varieties?

   d. What is the number of chromosomes (*n*) in each species?

28. Common red clover, *Trifolium pratense*, is a diploid with 14 chromosomes per somatic cell. What would be the somatic chromosome number of the following?

   a. A trisomic variant of this species

   b. A monosomic variant of this species

   c. A triploid variant of this species

   d. An autotetraploid variant

29. Somatic cells in organisms of a particular diploid plant species normally have 14 chromosomes. The chromosomes in the gametes are numbered from 1 through 7. Rarely, zygotes are formed that contain more or fewer than 14 chromosomes. For each of the zygotes below, (i) state whether the chromosome complement is euploid or aneuploid; (ii) provide terms that describe the individual's genetic makeup as accurately as possible; and (iii) state whether or not the individual will likely develop through the embryonic stages to make an adult plant, and if so, whether or not this plant will be fertile.

   a. 11 22 33 44 5 66 77

   b. 111 22 33 44 555 66 77

   c. 111 222 333 444 555 666 777

   d. 1111 2222 3333 4444 5555 6666 7777

30. Genomes A, B, and C all have basic chromosome numbers (*x*) of nine. These genomes were originally derived from plant species that had diverged from each other sufficiently far back in the evolutionary past that the chromosomes from one genome can no longer pair with the chromosomes from any other genome. For plants with the following kinds of euploid chromosome complements, (i) state the number of chromosomes in the organism; (ii) provide terms that describe the individual's genetic makeup as accurately as possible; and (iii) state whether or not it is likely that this plant will be fertile, and if so, give the number of chromosomes (*n*) in the gametes.

   a. AABBC

   b. BBBB

   c. CCC

   d. BBCC

   e. ABC

   f. AABBCC

31. Fred and Mary have a child with Down syndrome. A probe derived from chromosome 21 was used to identify RFLPs in Fred, Mary, and the child (thicker bands indicate signals of twice the intensity). Explain what kind of nondisjunction events must have occurred to produce the child if the child's RFLP pattern looked like that in lanes A, B, C, or D of the following figure:

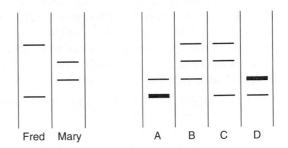

32. *Uniparental disomy* is a rare phenomenon in which only one of the parents of a child with a recessive disorder is a carrier for that trait; the other parent is homozygous normal. By analyzing DNA polymorphisms, it is clear that the child received both mutant alleles from the carrier parent but did not receive any copy of the gene from the other parent.

   a. Diagram at least two ways in which uniparental disomy could arise. (*Hint:* These mechanisms all require more than one error in cell division, explaining why uniparental disomy is so rare.) Is there any way to distinguish between these mechanisms to explain any particular case of uniparental disomy?

   b. How might the phenomenon of uniparental disomy explain rare cases in which girls are affected with rare X-linked recessive disorders but have unaffected fathers, or other cases in which an X-linked recessive disorder is transmitted from father to son?

   c. If you were a human geneticist and believed one of your patients had a disease syndrome caused by uniparental disomy, how could you establish that the cause was not instead mitotic recombination early in the patient's development from a zygote?

33. Human geneticists interested in the effects of abnormalities in chromosome number often karyotype tissue obtained from spontaneous abortions. About 35 percent of these samples show autosomal trisomics, but only about 3 percent of the samples display autosomal monosomies. Based on the kinds of errors that can give rise to aneuploidy, would you expect that the frequencies of autosomal trisomy and autosomal monosomy should be more equal? Why or why not? If you think the frequencies should be more equal, how can you explain the large excess of trisomies as opposed to monosomies?

34. Among adults with Turner syndrome, it has been found that a very high proportion are genetic mosaics. These are of two types: In some individuals, the majority of cells are 45, XO, but a minority of cells are 46, XX. In other

Turner individuals, the majority of cells are 45, XO, but a minority of cells are 46, XY. Explain how these somatic mosaics could arise.

35. The *Drosophila* chromosome 4 is extremely small; there is virtually no recombination between genes on this chromosome. You have available three differently marked chromosome 4's: one has a recessive allele of the gene *eyeless* (*ey*), causing very small eyes; one has a recessive allele of the *cubitus interruptus* (*ci*) gene, which causes disruptions in the veins on the wings; and the third carries the recessive alleles of both genes. *Drosophila* adults can survive with two or three, but not with one or four, copies of chromosome 4.

    a. How could you use these three chromosomes to find *Drosophila* mutants with defective meioses causing an elevated rate of nondisjunction?

    b. Would your technique allow you to discriminate nondisjunction occurring during the first meiotic division from nondisjunction occurring during the second meiotic division?

    c. What progeny would you expect if a fly recognizably formed from a gamete produced by nondisjunction were testcrossed to a fly homozygous for a chromosome 4 carrying both *ey* and *ci*?

    d. Geneticists have isolated so-called *compound 4th chromosomes* in which two entire chromosome 4's are attached to the same centromere. How can such chromosomes be used to identify mutations causing increased meiotic nondisjunction? Are there any advantages relative to the method you described in part *a*?

36. An allotetraploid species has a genome composed of two ancestral genomes, A and B, each of which has a basic chromosome number (*x*) of seven. In this species, the two copies of each chromosome of each ancestral genome pair only with each other during meiosis. Resistance to a pathogen that attacks the foliage of the plant is controlled by a dominant allele at the *F* locus. The recessive alleles F^a and F^b confer sensitivity to the pathogen, but the dominant resistance alleles present in the two genomes have slightly different effects. Plants with at least one F^A allele are resistant to races 1 and 2 of the pathogen regardless of the genotype in the B genome, and plants with at least one F^B allele are resistant to races 1 and 3 of the pathogen regardless of the genotype in the A genome. What proportion of the self-progeny of an F^A F^a F^B F^b plant will be resistant to all three races of the pathogen?

37. You have haploid tobacco cells in culture and have made transgenic cells that are resistant to herbicide. What

would you do to obtain a diploid cell line that could be used to generate a new fertile herbicide-resistant plant?

38. Chromosomes normally associate during meiosis I as bivalents (a pair of synapsed homologous chromosomes) because chromosome pairing involves the synapsis of the corresponding regions of two homologous chromosomes. However, Figure 9.21b shows that in a heterozygote for a reciprocal translocation, chromosomes pair as quadrivalents (i.e., four chromosomes are associated with each other). Quadrivalents can form in other ways: For example, in some autotetraploid species, chromosomes can pair as quadrivalents rather than as bivalents.

    a. How could quadrivalents actually form in these autotetraploids, given that chromosomal regions synapse in pairs? To answer this question, diagram such a quadrivalent.

    b. How can these autotetraploid species generate euploid gametes if the chromosomes pair as quadrivalents rather than bivalents?

    c. Could quadrivalents form in an amphidiploid species? Discuss.

39. Using karyotype analysis, how could you distinguish between autopolyploids and allopolyploids?

Section 9.5

40. The accompanying figure on the bottom of this page shows a virtual karyotype obtained from a line of tumour cells derived from a human leukaemia. The left-to-right direction for each chromosome corresponds to the orientation of that chromosome from the telomere of the small arm to the telomere of the long arm. Every coloured dot corresponds to a different short region of the chromosome analyzed by a microarray technique similar to that shown in Figure 9.38.

    a. Do the data indicate the existence of aneuploidy or any chromosomal rearrangements within the genome of the tumour cell?

    b. What do you think you would see if you did a virtual karyotype of a cell line derived from normal, non-leukaemic cells from the same person?

    c. Are there any kinds of chromosomal rearrangements that could not be detected by this virtual karyotyping method?

    d. What do these data say about genes that might be responsible for the leukaemia?

    e. Do these data tell us anything about the dosage of genes needed for the viability of individual cells?

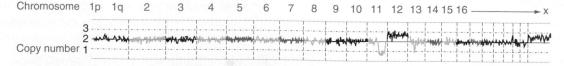

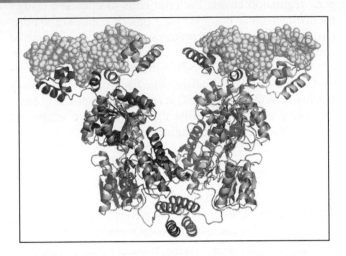

The tetrameric Lac repressor protein (LacI) in *E. coli* consists of two dimers (*blue/grey* and *pink/green*), each bound to different *lac* operon DNA regulatory "operator" elements (*yellow*). Each dimer consists of two identical subunits, with each of the subunits binding to a specific operator DNA sequence. Binding of LacI to specific sites in the *lac* operon results in transcriptional silencing.

Among the many types of bacteria that thrive in sewage water is the species *Vibrio cholerae*, the cause of the life-threatening diarrheal disease cholera. The last worldwide cholera pandemic began in 1961. Today, cholera is nearly absent from areas of the globe where secure sanitation systems are in place, but epidemics of the disease still devastate human populations in regions where sewage treatment and water purification programs are inadequate or nonexistent.

When a person drinks water contaminated by disease-causing *V. cholerae*, the bacteria enter the digestive tract. Soon after, the bacteria encounter the "perilous" environment of the stomach, whose acidity kills the majority of them. The bacteria respond to this hostile environment by curtailing production of several proteins they will use later but do not need for passage through the stomach. Only a large initial *V. cholerae* population ensures that at least a small group of cells will survive to exit the stomach and enter the small intestine.

Upon arrival in the small intestine, these survivors come face to face with a thick mucus that coats their ultimate target—the intestinal epithelial cells. To penetrate this protective mucous layer, the *V. cholerae* cells navigate by chemotaxis and by using flagella (**Figure 10.1**). They also make and secrete proteases that ease their passage by degrading the protein component of mucus.

When the bacteria at last reach their destination, they stop fabricating flagellin (the chief protein component of flagella) and begin production of several virulence proteins, including a pilus (by which they attach to epithelial cells of the small intestine) and a potent toxin that is the actual agent of cholera. Mutant bacteria that produce no toxin do not generate symptoms of disease.

The toxin secreted by cholera bacteria causes chloride ions (Cl^-) to leak from the intestinal cells. To reestablish the osmotic balance, these same cells secrete water. Symptoms of this ionic disruption and fluid flow are watery diarrhea and severe dehydration, which can lead to death within a few hours. The most effective life-saving therapy is oral rehydration: administration of an electrolyte solution consisting of glucose, table salt (NaCl), sodium bicarbonate ($NaHCO_3$), and potassium chloride (KCl) dissolved in purified water. Once toxin production is in full swing, antibiotics without oral rehydration are of little benefit.

The story of *V. cholerae* infection illustrates two key aspects of the life of a unicellular prokaryote: direct contact with their external environment and the ability to respond to changes in that environment by changes in gene expression. This coordinated control of gene expression in a bacterial cell is an example of **prokaryotic gene regulation**, the subject of this chapter. Prokaryotes regulate gene expression by activating, increasing, diminishing, or preventing the transcription and translation of specific genes or groups of genes. *V. cholerae* bacteria that have entered a human host respond to rapid changes in external conditions in part by diminishing or abolishing the production of proteins not required for survival (thereby conserving energy and nutrients) and in part by initiating or increasing

Chapter Outline	
10.1	Overview of Prokaryotic Gene Regulation
10.2	The Regulation of Gene Transcription
10.3	Attenuation of Gene Expression: Termination of Transcription
10.4	Global Regulatory Mechanisms
10.5	A Comprehensive Example: The Regulation of Virulence Genes in *V. cholerae*

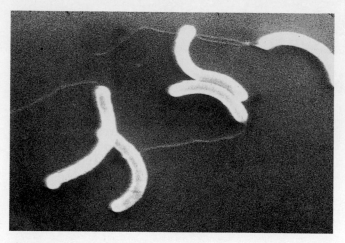

Figure 10.1 *Vibrio cholerae* **bacteria.** *V. cholerae* invade cells in the intestine.

synthesis of proteins required in new environments (such as proteases when they contact mucus, and toxin and other virulence proteins in the vicinity of intestinal epithelial cells). Bacteria attune their gene-function controls in a coordinated way. These cells do not waste energy making unnecessary proteins. In fact, many aspects of prokaryotic gene regulation enable bacterial cells to conserve energy by distinguishing housekeeping proteins that are synthesized continuously from proteins required only in specific situations.

One overarching theme emerges from our discussion. In unicellular organisms like bacteria, the regulatory mechanisms that turn genes on and off in response to environmental conditions enable the organisms to adapt and survive in a constantly changing world.

Learning Objectives

1. Recognize that regulation of gene expression occurs at many levels in prokaryotes.

2. Compare the roles of lactose and allolactose in gene regulation.

3. Explain what an operon is.

4. Diagram and explain the role of the various genetic and molecular elements (e.g., promoter, operator, repressor, inducer, cAMP receptor protein (CRP)) required for the coordinated regulation of the *lac* operon.

5. Distinguish between the various players involved in positive and negative regulation of the *lac* operon.

6. Compare and contrast the regulation of the *lac* and *trp* operons.

7. Differentiate among the various protein factors necessary for regulating the *E. coli* heat-shock response.

8. Evaluate how computational analysis can aid in the identification of regulatory proteins and their DNA-binding sites.

10.1 Overview of Prokaryotic Gene Regulation

We saw in Chapter 7 that *gene expression* is the production of proteins according to instructions encoded in DNA. During gene expression, the information in DNA is transcribed into RNA, and the RNA message is translated into a string of amino acids.

RNA polymerase is the key enzyme for transcription

To begin the process of gene expression in prokaryotes, RNA polymerase transcribes a gene's DNA into RNA. RNA polymerase participates in all three phases of transcription: initiation, elongation, and termination. Initiation requires a special subunit of RNA polymerase—the sigma (σ) subunit—in addition to the two alpha (α), one beta (β), and one beta prime (β') subunits that make up the core enzyme (**Figure 10.2**). When bound to the core enzyme, the σ subunit recognizes and binds specific DNA sequences at the promoter; in its free form, σ does not bind DNA because the σ DNA-binding site is obscured by its own C-terminal tail. The full RNA polymerase—core enzyme plus σ—when bound to the promoter, functions as a complex that both initiates transcription by unwinding the

DNA and begins polymerization of bases complementary to the DNA template strand.

The switch from initiation to elongation requires the movement of RNA polymerase away from the promoter and the release of σ. Elongation continues until the RNA polymerase encounters a signal in the RNA sequence that triggers termination (Figure 10.2). Two types of termination signals are found in prokaryotes: Rho-dependent and Rho-independent. In Rho-dependent termination, a protein factor called Rho (ρ) recognizes a sequence in the newly transcribed mRNA and terminates transcription by binding to the RNA and pulling it away from the RNA polymerase enzyme. In Rho-independent termination, a sequence of about 20 bases in the RNA, with a run of six or more U's at the end, forms a secondary structure, known as a *stem loop*, which serves as a signal for the release of RNA polymerase from the completed RNA.

Translation in prokaryotes begins before transcription ends

Because there is no membrane enclosing the bacterial chromosome, translation of the RNA message into

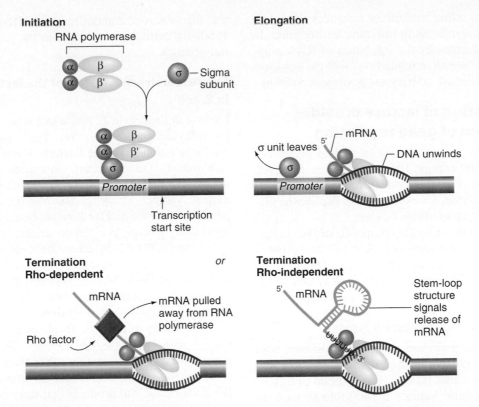

Figure 10.2 Role of RNA polymerase. The core RNA polymerase enzyme plus sigma factor bind to a promoter sequence to initiate transcription. RNA polymerase then moves along the DNA to elongate the transcript, leaving sigma factor behind. Transcription terminates when Rho factor recognizes a sequence on the mRNA, or a stem loop (Rho-independent signal) forms in the mRNA, causing release of the enzyme and message.

a polypeptide can begin while mRNA is still being transcribed. Ribosomes bind to special initiation sites at the 5′ end of the reading RNA frame while transcription of downstream regions of the RNA is still in progress. Signals for the initiation and termination of translation are distinct from signals for the initiation and termination of transcription. Because prokaryotic mRNAs are often polycistronic, that is, contain the information of several genes, ribosomes can initiate translation at several positions along a single mRNA molecule. (See Figure 7.17 for a review of how ribosomes, tRNAs, and translation factors mediate the initiation, elongation, and termination phases of mRNA translation to produce a polypeptide that grows from its N terminus to its C terminus, according to instructions embodied in the sequence of mRNA codons.)

Regulation of expression can occur at many steps

Many levels of control determine the amount of a particular polypeptide in a bacterial cell at any one time. Some controls affect an aspect of transcription: the binding of RNA polymerase to the promoter, the shift from transcriptional initiation to elongation, or the release of the mRNA at the termination of transcription. Other controls are posttranscriptional and determine the stability of the mRNA after its synthesis, the efficiency with which ribosomes recognize the various translational initiation sites along the mRNA, or the stability of the polypeptide product.

As we see next, the critical step in the regulation of most bacterial genes is the binding of RNA polymerase to DNA at the promoter. The other potential points of control, while sometimes important in the expression of certain genes, serve more often to fine-tune the amount of protein produced.

> In prokaryotes, RNA polymerase is the key enzyme in transcription. Translation begins before transcription ends in these organisms, and regulation of gene expression can occur at many different points in this process.

10.2 The Regulation of Gene Transcription

Researchers delineated the principles of gene regulation in prokaryotes through studies of various metabolic pathways in *Escherichia coli*. In this section, we focus our attention on regulation of the lactose utilization genes in *E. coli* because genetic and molecular experimentation in this system established a fundamental principle of gene regulation: The binding of regulatory proteins to DNA targets controls transcription. The DNA binding of these

regulatory proteins either inhibits or enhances the effectiveness of RNA polymerase in initiating transcription. In our discussion, we consider the inhibition of RNA polymerase activity as "negative regulation" and the enhancement of RNA polymerase activity as "positive regulation."

E. coli's utilization of lactose provides a model system of gene regulation

Proliferating *E. coli* can use any one of several sugars as a source of carbon and energy. If given a choice, however, they prefer glucose. *E. coli* grown in medium containing both glucose and lactose, for example, will deplete the glucose before gearing up to utilize lactose.

Lactose is a complex sugar composed of two monosaccharides: glucose and galactose. A membrane protein, lactose (or lac) permease, transports lactose in the medium into the *E. coli* cell. There, the enzyme β-galactosidase splits the lactose into galactose and glucose (**Figure 10.3**).

Induction of gene expression by lactose

The two proteins lac permease and β-galactosidase, both required for lactose utilization, are present at very low levels in cells grown without lactose. The addition of lactose to the bacterial medium induces a 1000-fold increase in the production of these proteins. The process by which a specific molecule stimulates synthesis of a given protein is known as **induction**. The molecule responsible for stimulating production of the protein is called the **inducer**. In the regulatory system under consideration, lactose modified to a derivative known as *allolactose* is the inducer of the genes for lactose utilization.

How lactose in the medium induces the simultaneous expression of the proteins required for its utilization

was the subject of a major research effort in the 1950s and 1960s—a period some refer to as the golden era of bacterial genetics.

The research advantages of the lactose system in *E. coli*

Lactose utilization in *E. coli* was a wise choice as a model for studying gene expression. The possibility of culturing large numbers of the bacteria made it easy to isolate rare mutants. Once isolated, the mutations responsible for the altered phenotypes could be located by mapping techniques. Another advantage was that the lactose utilization genes are not essential for survival because the bacteria can grow using glucose as a carbon source. In addition, there is a striking 1000-fold difference between lactose utilization protein levels in induced and uninduced cells. This makes it easy to see the difference between the mutant and wild-type states, and it also allows the identification of mutants that have partial—not just all-or-none—effects.

The ability to measure levels of expression was critical for many of these experiments. To this end, chemists synthesized compounds other than lactose, such as *o*-nitrophenyl-galactoside (ONPG), that could be split by β-galactosidase into products that were easy to assay. One product of ONPG splitting has a yellow colour, whose intensity is proportional to the amount of product made and thus reflects the level of activity of the β-galactosidase enzyme. A spectrophotometer can easily measure the amount of cleaved yellow product in a sample. Another substrate of the β-galactosidase enzyme that produces a colour change upon cleavage is X-Gal, whose cleavage produces a blue substance; as we will see in a later chapter, X-Gal is often used to indicate whether a piece of DNA has been cloned into plasmid vectors containing parts of the β-galactosidase gene (see Figure 14.7 in Chapter 14).

The operon theory explains how a single substance can regulate several clustered genes

Jacques Monod (**Figure 10.4**), a man of diverse interests, was a catalyst for research on the regulation of lactose utilization. A political activist and a chief of French Resistance

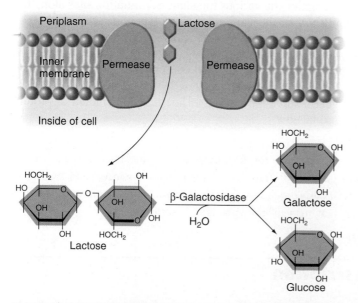

Figure 10.3 Lactose utilization in an *E. coli* cell. Lactose passes through the membranes of the cell via an opening formed by the lactose permease protein. Inside the cell, β-galactosidase splits lactose into galactose and glucose.

Figure 10.4 Jacques Monod. A key scientist in discovering principles of gene regulation, Jacques Monod was also a talented musician, philosopher, and political activist.

operations during World War II, he was also a fine musician and esteemed writer on the philosophy of science. Monod led a research effort centred at the Pasteur Institute in Paris, where scientists from around the world came to study enzyme induction. Results from many genetic studies led Monod and his close collaborator François Jacob to propose a model of gene regulation known as the **operon theory**, which suggested that a single signal can simultaneously regulate the expression of several genes that are clustered together on a chromosome and involved in the same process. They reasoned that because these genes form a cluster, they can be transcribed together into a single mRNA, and thus anything that regulates the transcription of this mRNA will affect all the genes in the cluster. Clusters of genes regulated in this way are called **operons**. We first summarize the theory itself and then describe key experiments that influenced Jacob and Monod's thinking, as well as data that supported components of their theory.

Figure 10.5 presents the players in the theory and how they interact to achieve the coordinate regulation of the genes for lactose utilization. As shown, the structural genes (*lacZ*, *lacY*, and *lacA*) encoding proteins needed for lactose utilization, together with two regulatory elements—the promoter (*P*) and the operator (*O*)—make up the *lac* operon: a single DNA unit enabling the simultaneous regulation of the three structural genes in response to environmental changes. Molecules that interact with the operon include the repressor, which binds to the operon's operator, and the inducer, which when present, binds to the repressor and prevents it from binding to the operator.

Jacob and Monod's theory was remarkable because the authors were working with a very abstract sense of the molecules in the bacterial cell: The Watson–Crick model of DNA structure was only eight years old, mRNA had only recently been identified, and the details of transcription had not yet been described. In 1961, the details of information flow from DNA to RNA to protein were still being established and knowledge of protein roles in the cell was limited. For example, although Monod was a biochemist with a special interest in allostery and its effects, the repressor itself was a purely conceptual construct; at the time of publication, it had not yet been isolated, and it was unknown whether it was RNA or protein. Jacob and Monod thus made a major leap in understanding to propose the theory.

We now know that a key concept of the theory—that proteins bind to DNA to regulate gene expression—holds true for the positive as well as the negative regulation of the *lac* operon. It also applies to many prokaryotic genes outside the *lac* operon and to eukaryotic genes as well. One example of positive transcriptional regulation in the *E. coli* bacteriophage lambda (λ) (a bacterial virus) involves specific transcriptional anti-termination proteins, studied by Canadian researcher Dr. Jack Greenblatt (see the **Focus on Genetics** box **"Transcriptional Regulation in Bacteriophage Lambda"**). In the next section, we look at some of the experiments that suggested how the presence of lactose induces expression of the genes required for its own utilization.

> In induction, a molecule (inducer) stimulates production of a protein. In *E. coli*, three structural genes (*lacZ*, *lacY*, and *lacA*) encode proteins for utilizing lactose; these plus the promoter and operator make up the *lac* operon, a single unit in which the genes are regulated simultaneously. The inducer in this system is allolactose, which forms when lactose is present.

Genetic analysis identifies the roles of the *lac* genes

In proposing the operon theory, Jacob and Monod took enzyme induction, which most of their contemporaries considered a biochemical problem, and used genetic analysis to develop a molecular model explaining how environmental changes could provoke changes in gene activity. On the way to developing the operon theory of gene regulation, Monod and his collaborators isolated many different Lac⁻ mutants; that is, bacterial cells unable to utilize lactose.

Complementation analysis

Using complementation analysis, the researchers showed that the cells' inability to break down lactose resulted from mutations in two genes: *lacZ*, which encodes β-galactosidase, and *lacY*, which encodes lac permease. They also discovered a third *lac* gene, *lacA*, which encodes a transacetylase enzyme that adds an acetyl (CH_3CO) group to lactose and other β-galactoside sugars. Genetic mapping showed that the three genes appear on the bacterial chromosome in a tightly linked cluster, in the order *lacZ-lacY-lacA* (**Figure 10.6**). Because the *lacA* gene product is not required for the breakdown of lactose, most studies of lactose utilization have focused on *lacZ* and *lacY*.

Evidence for a repressor protein

Mutations in another gene named *lacI* produce **constitutive mutants** that synthesize β-galactosidase and lac permease even in the absence of lactose. Constitutive mutants synthesize certain enzymes irrespective of environmental conditions. The existence of these constitutive mutants suggested that *lacI* encodes a negative regulator, or **repressor**. Cells would need such a repressor to prevent expression of *lacY* and *lacZ* in the absence of inducer. In constitutive mutants, however, a mutation in the *lacI* gene generates a defect in the repressor protein that prevents it from carrying out this negative regulatory function.

The historic PaJaMo experiment—named after Arthur Pardee (a third collaborator), Jacob, and Monod—provided further evidence that *lacI* indeed encodes this hypothetical

Figure 10.6 Lactose utilization genes in *E. coli*. Three genes, *lacZ*, *lacY*, and *lacA*, are involved in lactose metabolism in *E. coli*.

The Lactose Operon in *E. coli*

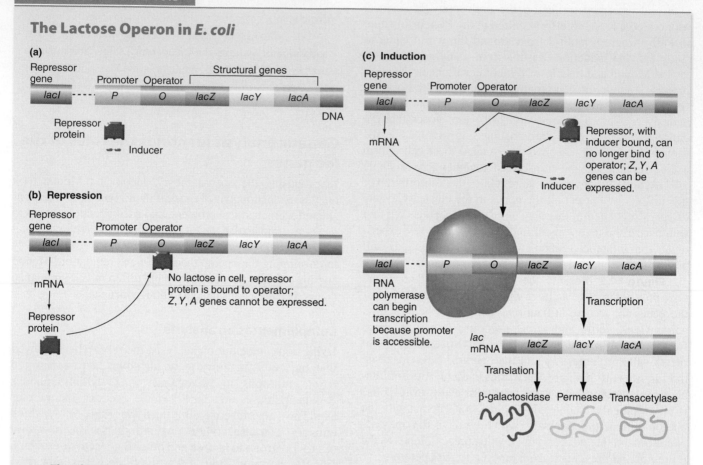

a. The players

Bacteria utilize lactose in an energy-efficient way via the coordination of various elements, including the following:

1. A closely linked cluster of three structural genes—*lacZ*, *lacY*, and *lacA*; these encode the enzymes active in splitting lactose into glucose and galactose.

2. A promoter site, from which RNA polymerase initiates transcription of a polycistronic mRNA. The promoter acts in *cis*, affecting the expression of only downstream structural *lac* genes on the same DNA molecule.

3. A *cis*-acting DNA operator site lying very near the *lac* operon promoter on the same DNA molecule. The three structural genes together with the promoter and the operator constitute the *lac operon*.

4. A *trans*-acting repressor that can bind to the operator. The repressor is encoded by the *lacI* gene, which is separate from the operon and is unregulated. After synthesis, the repressor diffuses through the cytoplasm and binds with its target.

5. An inducer that prevents the repressor's binding to the operator. Although early experimenters thought lactose was the inducer, we now know that the inducer is actually allolactose, a molecule derived from and thus related to lactose.

How the Players Interact to Regulate the Lactose Utilization Genes

b. Repression

In the absence of lactose, the repressor binds to the DNA of the operator, and this binding prevents transcription. The repressor thus serves as a negative regulatory element. RNA polymerase can still access and bind to the *lac* operon promoter, even with the bound repressor, but it cannot transcribe the structural genes of the operon.

c. Induction

1. When lactose is present, allolactose, an inducer derived from the sugar, binds to the repressor. This binding changes the shape of the repressor, making it unable to bind to the operator.

2. With the release of the repressor from the operator, RNA polymerase can access the *lac* operon promoter and initiate transcription of the three lactose utilization genes into a single polycistronic mRNA.

Dr. Jack Greenblatt (**Figure A**) is a Professor in the Banting and Best Department of Medical Research in Toronto and the Department of Molecular Genetics at the University of Toronto. He obtained his B.Sc. in Physics at McGill University and received his Ph.D. in Biophysics from Harvard University before joining the University of Toronto as a faculty member in 1977. A distinguished scientist, Dr. Greenblatt has received numerous awards and prestigious honours, including being named a Fellow of the Royal Society of Canada. The first person to purify a transcriptional anti-termination protein, specifically the anti-termination protein encoded by gene N in bacteriophage lambda (λ), his experiments are also among the earliest to demonstrate the importance of protein–protein interactions during transcription regulation. N protein is required to prevent premature termination of bacteriophage λ DNA transcription by *E. coli* RNA polymerase at Rho-dependent and Rho-independent termination sites in the λ early operons (P_L and P_R). Elongation factors encoded by the *E. coli nusA, nusB, and nusG* genes (Nus = <u>N</u> <u>u</u>tilization <u>s</u>ubstance), as well as ribosomal protein S10, are also necessary to prevent early termination through their direct or indirect interaction with RNA polymerase core enzyme and/or N protein. Moreover, NusG protein facilitates anti-termination by sequestering termination factor Rho. Modification of RNA polymerase by N protein is mediated by an N utilization (nut) sequence, present in each λ early operon and required for the

Figure A Jack Greenblatt.

read-through of transcription termination signals, allowing for transcription of essential downstream genes. The nut site consists of two motifs—a seven-nucleotide boxA sequence followed by a stem-loop boxB motif (**Figure B**).

A principal member of the Proteomics Research Centre at the University of Toronto, Dr. Greenblatt has utilized modern proteomics and genomics technologies to investigate protein–RNA and protein–protein interactions for the bacterium *E. coli*, the budding yeast *Saccharomyces cerevisiae,* mice, and humans, with a goal to identify genes and proteins implicated in human disease.

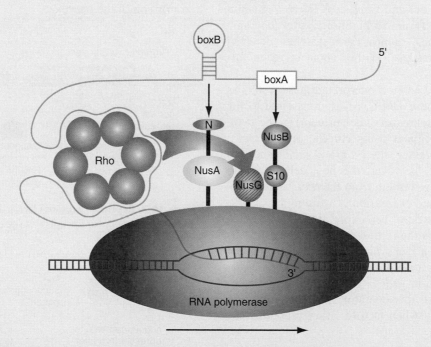

Figure B Model for transcriptional anti-termination of bacteriophage lambda (λ) DNA. N protein binds to NusA and the nut (N utilization) boxB motif to prevent early termination of bacteriophage λ transcription by *E. coli* RNA polymerase. *E. coli* NusA, along with NusB, NusG, and S10, bind directly or indirectly to RNA polymerase to impede premature termination. NusG sequesters termination factor Rho, and NusB also binds to the nut boxA sequence.

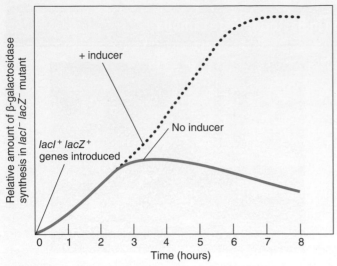

Figure 10.7 The PaJaMo experiment. When DNA carrying *lacI⁺ lacZ⁺* genes was introduced into a *lacI⁻ lacZ⁻* cell, β-galactosidase was synthesized from the introduced *lacZ⁺* gene initially; but as repressor (made from the introduced *lacI⁺* copy of the gene) accumulates, the synthesis of β-galactosidase stops and only residual β-galactosidase is seen. If inducer is added (*dotted line*), the synthesis of β-galactosidase resumes.

negative regulator of the *lac* genes. Matings in which the chromosomal DNA of a donor cell is transferred into a recipient cell served as the basis of the PaJaMo study. The researchers transferred the *lacI⁺* and *lacZ⁺* alleles into a cytoplasm devoid of LacI and LacZ proteins in a medium containing no lactose (**Figure 10.7**). Shortly after the transfer of the *lacI⁺* and *lacZ⁺* genes, the researchers detected synthesis of β-galactosidase. Within about an hour, this synthesis stopped.

Pardee, Jacob, and Monod interpreted these results as follows. When the donor DNA is first transferred to the recipient, there is no repressor (LacI protein) in the cytoplasm because the recipient cell's chromosome is *lacI⁻*. In the absence of repressor, the *lacY* and *lacZ* genes are expressed. Over time, however, the host cell begins to make the LacI repressor protein from the *lacI⁺* gene introduced by the mating and expression is again repressed.

On the basis of the described experiments, Monod and company proposed that the repressor protein prevents further transcription of *lacY* and *lacZ* by binding to a hypothetical **operator site**: a DNA sequence near the promoter of the lactose utilization genes. They suggested that the binding of repressor to this operator site blocks the promoter and occurs only when lactose is not present in the medium.

How the inducer acts to trigger enzyme synthesis

In the final step of the PaJaMo experiment, the researchers added the lactose inducer to the culture medium. With this addition, the synthesis of β-galactosidase resumed.

Their interpretation of this result was that the inducer binds to the repressor. This binding changes the shape of the repressor so that it can no longer bind to DNA. When the inducer is removed from the environment, the repressor, free of inducer, reverts to its DNA-bindable shape. Proteins that undergo reversible changes in conformation when bound to another molecule are called **allosteric proteins.** The binding of inducer to repressor causes an allosteric effect that abolishes the repressor's ability to bind the operator. In this sequence of events, the inducer is an effector molecule that releases repression without itself binding to the DNA.

A repressor with an inducer attached cannot bind to the DNA of the operator, whereas a repressor without an inducer attached can. When the repressor is bound to the operator, RNA polymerase cannot recognize the promoter, so transcription does not occur.

The repressor's two distinct binding domains

If the repressor protein interacts with both the operator and the inducer, what outcome would you predict for mutations that disrupt one of these interactions without affecting the other? Biochemical studies showed that the constitutive *lacI⁻* mutations we discussed earlier produce defects in the repressor's ability to bind DNA (**Figure 10.8**). A different type of *lacI* mutant cannot undergo induction. Researchers designate the non-inducible mutations as *lacIˢ* or superrepressor (**Figure 10.9**). The *lacIˢ* mutants, although they cannot bind inducer, can still bind to DNA and repress transcription of the operon.

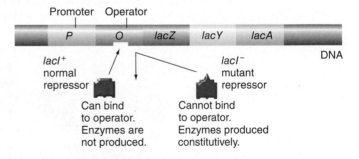

Figure 10.8 Repressor mutant (*lacI⁻*). In the *lacI⁻* mutants, the repressor cannot bind to the operator site and therefore cannot repress the operon.

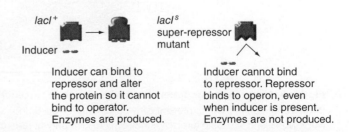

Figure 10.9 Super-repressor mutant *lacIˢ*. In super-repressor mutants, LacIˢ binds to the operator but cannot bind the inducer, so the repressor cannot be removed from the operator and genes are continually repressed.

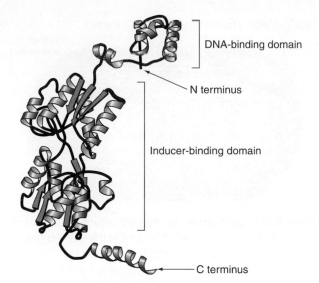

Figure 10.10 Domains of repressor protein. X-ray crystallographic data enable the construction of a model of repressor structure that shows a region to which operator DNA binds and another region to which inducer binds.

The mapping of large numbers of these two types of mutations by DNA sequencing has shown that $lacI^-$ missense mutations, which generate proteins incapable of binding DNA, are clustered in the codons that determine the amino (N) terminus of the repressor, while the $lacI^s$ mutations, which generate proteins incapable of binding inducer, cause amino acid alterations throughout much of the rest of the repressor. Subsequent structural analyses of the repressor protein confirmed what these mutational analyses suggest: The repressor protein has at least two separate domains, one that binds to DNA, and another that binds inducer (**Figure 10.10**).

> The repressor has separate regions, or domains, correlating with two functions that were uncovered through mutations. Defects in either domain, as well as the presence or absence of the inducer, can affect repressor function.

Operator mutants

While mutations in the DNA-binding domain of the repressor can erase repressor activity, mutations that alter the specific nucleotide sequence of the operator recognized by the repressor can have the same effect (**Figure 10.11**).

Mutant operator (o^c)

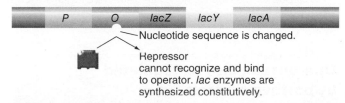

Figure 10.11 Operator mutants. The repressor cannot recognize the altered DNA sequence in the $lacO^c$ mutant site, so it cannot bind and repress the operon.

When mutations change the nucleotide sequence of the operator, the repressor is unable to recognize and bind to the site; the resulting phenotype is the constitutive synthesis of the lactose utilization proteins. Researchers have isolated constitutive mutants whose genetic defects map to the *lac* operator site, which is adjacent to the *lacZ* and *lacY* genes. They call the constitutive operator DNA alterations $lacO^c$ mutations.

Proteins act in *trans*, DNA sites act in *cis*

How can one distinguish the constitutive operator ($lacO^c$) mutants from the previously described constitutive $lacI^-$ mutants, considering that both prevent repression? The answer is found in a *cis/trans* test.

Elements that act in ***trans*** can diffuse through the cytoplasm and act at target DNA sites on any DNA molecule in the cell. Elements that act in ***cis*** can influence only the expression of adjacent genes on the same DNA molecule. Studies of partial diploids in which a second copy of the *lac* genes was introduced helped distinguish mutations in the operator site ($lacO^c$), which act in *cis*, from mutations in *lacI*, which encodes a protein that acts in *trans*.

The partial diploids were made using F′ plasmids that carry a few chromosomal bacterial genes. When F′ *lac* plasmids are present in a bacterium, the cell has two copies of the lactose utilization genes—one on the plasmid and one on the bacterial chromosome. Using F′ *lac* plasmids, Monod's group could create bacterial strains with diverse combinations of regulatory ($lacO^c$ and *lacI*) and structural gene (*lacZ* and *lacY*) mutations. The phenotype of these partially diploid cells allowed Monod and his collaborators to determine whether particular constitutive mutations were in the genes that produce diffusible, *trans*-acting proteins or at *cis*-acting DNA sites that affect only genes on the same molecule.

In one experiment, Monod and colleagues used a $lacI^-$ Z^+ bacterial strain that was constitutive for β-galactosidase production because it could not synthesize repressor (**Figure 10.12**). The introduction of an F′ $lacI^+$ Z^- plasmid into this strain created a partial

F′ $lacI^+$ o^+ Z^- plasmid in $lacI^-$ o^+ Z^+ bacteria

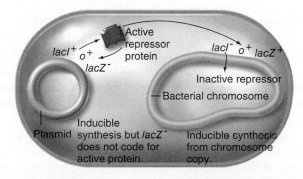

Figure 10.12 $LacI^+$ protein acts in *trans*. Repressor protein, made from the $lacI^+$ gene on the plasmid, can diffuse in the cytoplasm and bind to the operator on the chromosome as well as to the operator on the plasmid.

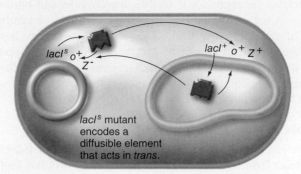

F' *lacI^s Z^−* plasmid in *lacI^+ Z^+* bacteria

lacI^s o^+
Z^−

lacI^+ o^+ Z^+

lacI^s mutant encodes a diffusible element that acts in *trans.*

Figure 10.13 LacI^s protein acts in *trans*. The super-repressor encoded by *lacI^s* on the plasmid diffuses and binds to operators on both the plasmid and chromosome to repress the *lac* operon even if the inducer is present.

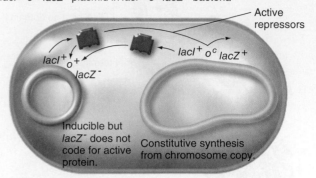

F' *lacI^+ o^+ lacZ^−* plasmid in *lacI^+ o^c lacZ^+* bacteria

Active repressors

lacI^+ o^+
lacZ^−

lacI^+ o^c lacZ^+

Inducible but *lacZ^−* does not code for active protein.

Constitutive synthesis from chromosome copy.

Figure 10.14 lacO^c acts in *cis*. The *lacO^c* constitutive mutation affects only the operon of which it is a part. In this cell, only the chromosomal copy will be expressed constitutively.

diploid that was phenotypically wild type with respect to β-galactosidase expression: both repressible in the absence of lactose and inducible in its presence. Its capacity for repression and subsequent induction indicated that *lacI^+* is dominant to *lacI^−* and that the LacI protein produced from the *lacI^+* gene on the plasmid can bind to the operator on the bacterial chromosome. Thus, the product of the *lacI* gene is a *trans*-acting protein able to diffuse inside the cell and bind to any operator site it encounters, regardless of the chromosomal location of the operator.

In a second experiment, introduction of a *lacI^s* plasmid into a *lacI^+* strain of bacteria that was both repressible and inducible created bacteria that were still repressible but were no longer inducible (**Figure 10.13**). This effect occurred because the mutant LacI^s repressor, while still able to bind to the operator, could no longer bind inducer. The non-inducible repressor was dominant to the wild-type repressor because after awhile, the mutant repressor, unable to bind inducer, occupied all the operator sites and blocked all *lac* gene transcription in the cell.

In a third set of experiments, the researchers used *lacI^+ lacO^c lacZ^+* bacteria that were constitutive for lactose utilization because the wild-type repressor they produced could not bind to the altered operator (**Figure 10.14**). Introduction of an F' *lacI^+ lacO^+ lacZ^−* plasmid did not change this state of affairs—the cells remained constitutive for β-galactosidase production. The explanation is that the *lacO^+* operator on the plasmid had no effect on the *lacZ^+* gene on the chromosome DNA because the operator DNA acts only in *cis*. Because it was able to influence gene expression only of the *lacZ^−* gene on its own DNA molecule, the wild-type operator on the plasmid could not override the mutant chromosomal operator to allow repression of genes on the bacterial chromosome.

A general rule derived from these experiments is that if a gene encodes a diffusible element—usually a protein—that can bind to target sites on any DNA molecule in the cell, whichever allele of the gene is dominant will override any other allele of that gene in the cell (and therefore act in *trans*). If a mutation is *cis*-acting, it affects only the expression of adjacent genes on the same DNA molecule; it does

this by altering a DNA site, such as a protein-binding site, rather than by altering a protein-encoding gene.

Coordinate expression of *lac* genes as a single mRNA

Many of the experiments that led to an understanding of the *lac* genes focused on the expression of *lacZ* because the level of β-galactosidase is easy to measure. But Monod and co-workers also showed that the repression and induction of *lacY* and *lacA* occur in tandem with the repression and induction of *lacZ*.

Observation of the coordinate expression of the genes for lactose utilization led to the proposal that the three genes are transcribed as part of the same polycistronic mRNA. Although researchers in the 1960s hypothesized that RNA was the intermediate between DNA and protein, they had not yet demonstrated the existence of such an intermediary for any gene. In the 1970s, however, biochemical studies showed that RNA polymerase initiates transcription of the tightly linked *lac* gene cluster from a single promoter. During transcription, the polymerase produces a single polycistronic mRNA containing the *lac* gene information in the order 5'-*lacZ-lacY-lacA*-3'. As a result, mutations in the promoter (which must be located just upstream of *lacZ*) affect the transcription of all three genes.

> The *lacO^c* constitutive operator has the same effect as the *lacI^−* constitutive mutant. These can be distinguished by the *cis/trans* test: An altered protein acts in *trans*, whereas an altered operator sequence acts in *cis*. Experiments in the 1970s verified that the *lac* genes are transcribed as a cluster that has a single promoter.

Operons can also be regulated by positive controls

In focusing on the repression and subsequent induction of the genes for lactose utilization, the Jacob and Monod model did not address a key question: Why do *E. coli* cells grown in a

medium containing both glucose and lactose not produce high levels of Lac proteins? We know that glucose is a better carbon source and should be used preferentially, but how is this achieved? If the lactose is present, why does it not act as an inducer? Answers to these questions emerged from molecular studies carried out long after publication of the theory. These studies showed that transcriptional initiation at the *lac* operon is a complex event. In addition to the release of repression, initiation depends on a positive regulator protein that assists RNA polymerase in the start-up of transcription. Without this assistance, the polymerase does not open up the double helix very efficiently. As we see next, the presence of glucose indirectly blocks the function of this positive regulator.

Positive regulation of the *lac* operon by CRP

Inside bacterial cells, the small nucleotide known as cAMP (cyclic adenosine monophosphate) binds to a protein called cAMP receptor protein, or CRP. The binding of cAMP to CRP enables CRP to bind to DNA in the regulatory region of the *lac* operon, and this DNA binding of CRP increases the ability of RNA polymerase to transcribe the *lac* genes (**Figure 10.15**). Thus, CRP functions as a positive regulator that enhances the transcriptional activity of RNA polymerase at the *lac* promoter, while cAMP is an effector whose binding to CRP enables CRP to bind to DNA near the promoter and carry out its regulatory function.

Glucose indirectly controls the amount of cAMP in the cell by decreasing the activity of adenyl cyclase, the enzyme that converts ATP into cAMP. Thus, when glucose is present, the level of cAMP remains low; when glucose is absent, cAMP synthesis increases. As a result, when glucose is present in the culture medium, there is little cAMP available to bind to CRP and therefore little induction of the *lac* operon, even if lactose is present in

the culture medium. The overall effect of glucose in preventing *lac* gene transcription is known as **catabolite repression**, because the presence of a preferred catabolite (glucose) represses transcription of the operon.

In addition to functioning as a positive regulator of the *lac* operon, the CRP–cAMP complex increases transcription in several other catabolic gene systems, including the *gal* operon (whose protein products help break down the sugar galactose) and the *ara* operon (contributing to the breakdown of the sugar arabinose). As you would expect, these other catabolic operons are also sensitive to the presence of glucose, exhibiting a low level of expression when glucose is present and cAMP is in short supply. Mutations in the gene encoding CRP that alter the DNA-binding domain of the protein reduce transcription of the *lac* and other catabolic operons. The binding of the CRP–cAMP complex is an example of a global regulatory strategy in response to limited glucose in the environment.

Positive regulation of the *araBAD* operon by AraC

There are several instances where positive regulators increase transcription of genes in only one pathway. AraC, for example, is a positive regulatory protein specific for all the arabinose genes involved in the breakdown of the sugar arabinose. Three arabinose structural genes, *araB*, *araA*, and *araD*, appearing on the chromosome in that order, constitute an operon (*araBAD*) that is regulated as a single transcription unit. Like the genes for lactose utilization, the arabinose genes are induced when their substrate (arabinose) is present. Evidence that the AraC protein is a positive regulator of the *araBAD* operon came from studies in which *araC⁻* mutants did not express high levels of these three arabinose genes in either the presence or absence of arabinose (**Figure 10.16**). The mutations were recessive,

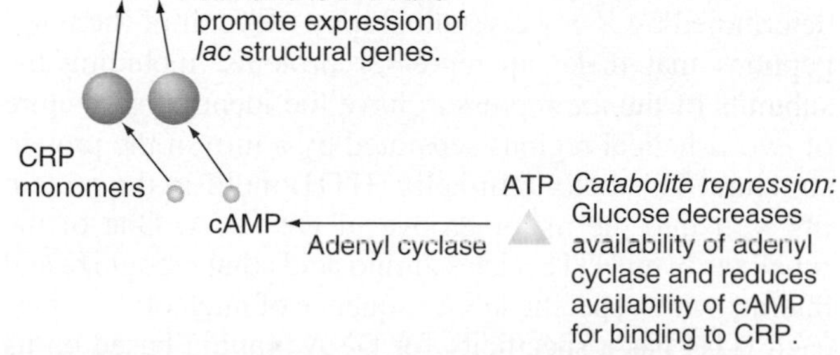

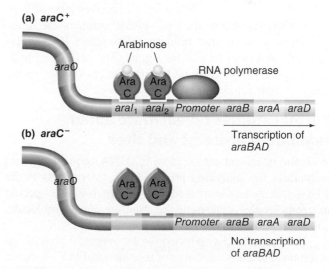

Figure 10.15 Positive regulation by CRP–cAMP. High-level expression of the *lac* operon requires that a positive regulator, the CRP–cAMP complex, be bound to the promoter region.

Figure 10.16 AraC is a positive regulator. Expression of the arabinose genes in *E. coli* requires the AraC protein to be bound next to the promoter. In an *araC⁻* mutant, the defective protein cannot bind, and RNA polymerase will not transcribe the genes.

loss-of-function mutations. When the loss of function of a regulatory protein results in little or no expression of the regulated genes, the protein must be a positive regulator. (By contrast, loss of function of a negative regulator causes constitutive production of the operon's gene products.)

> Positive regulators can be identified when their loss-of-function mutations lead to little or no expression of the regulated genes. Positive regulation occurs when CRP, formed when glucose is absent, enhances the transcriptional activity of RNA polymerase at the *lac* promoter and the promoters of many other catabolic operons. As another example, the AraC protein increases transcription of arabinose genes in the *araBAD* operon.

How DNA-binding proteins control initiation of operon transcription: A summary

In bacteria, the initiation of transcription by RNA polymerase is under the control of regulatory genes whose products bind to specific DNA sequences in the vicinity of the promoter. The binding of negative regulatory proteins prevents the initiation of transcription; the binding of positive regulators assists the initiation of transcription. Regulation of the *lac* operon depends on at least two proteins: the repressor (a negative regulator) and CRP (a positive regulator). Maximum induction of the *lac* operon occurs in media containing lactose but lacking glucose. Under these conditions, the repressor binds inducer and becomes unable to bind to the operator, while CRP complexed with cAMP binds to a site near the promoter to assist RNA polymerase in the initiation of transcription.

Operons that function in the breakdown of other sugars are also under the control of negative and positive regulators. Transcription of the arabinose operon, for example, which is induced in the presence of arabinose, receives a boost from two positive regulators: the CRP–cAMP complex and AraC.

Thus, proteins that bind to DNA affect RNA polymerase's ability to transcribe a gene. The activity of multiple regulators that respond to different cues increases the range of gene regulation.

Further studies reveal more about regulatory proteins and sites

With the development of cloning, DNA sequencing, and techniques for analyzing protein-DNA interactions in the 1970s, researchers increased their ability to isolate specific macromolecules, determine the structure of each molecule, and analyze the interactions between molecules.

In 1966, scientists purified the *lac* repressor protein and determined that it is a tetramer of four identical *lacI*⁻-encoded subunits, with each subunit containing an inducer-binding domain as well as a domain that recognizes and binds to DNA. (Note that we use the term "domains" for the functional parts of proteins but the term "sites" for the DNA sequences with which a protein's DNA-binding domain interacts.)

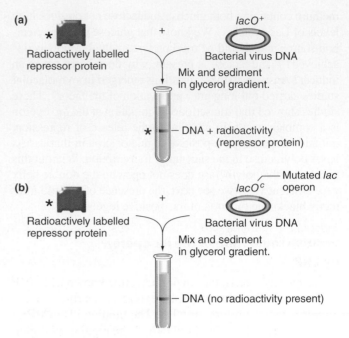

Figure 10.17 **The *lac* repressor binds to operator DNA.** A radioactive tag is attached to the *lac* repressor protein so it can be followed in the experiment. **(a)** When repressor protein from *lacI*⁺ cells was purified and mixed with DNA containing the *lac* operator (on bacterial virus DNA), the protein co-sedimented with the DNA. **(b)** When wild-type repressor was mixed with DNA containing a mutant operator site, no radioactivity sedimented with the DNA.

A radioactively labelled repressor protein and a bacterial virus DNA that contained the *lac* operon were used to show that the repressor binds to operator DNA. When researchers combined the labelled protein and viral DNA and centrifuged the mixture in a glycerol gradient, the radioactive protein co-sedimented with the DNA (**Figure 10.17**). If the viral DNA contained a *lac* operon that had a *lacO^c* mutation, the protein did not co-sediment with the DNA, because it could not bind to the altered operator site. Subsequent sequence analysis of the isolated DNA revealed that the *lac* operator is about 26 bp in length, and it includes the first nucleotides used as a template for the mRNA.

Helix-turn-helix proteins

We can predict a protein's secondary structures—such as α helixes and β sheets—by comparing the amino acid sequence of a newly isolated protein with sequences of proteins whose secondary structures have already been determined by X-ray crystallography. Several of the polypeptides that make up repressor proteins, including the subunits of the *lac* repressor, have the identifiable feature of two α-helical regions separated by a turn in the protein structure. This helix-turn-helix (HTH) motif in the protein fits well into the major groove of the DNA. One of the α helixes in an HTH carries amino acids that recognize and interact with a specific DNA sequence of nucleotides; thus, each HTH has a specificity for DNA binding based on its sequence of amino acids (**Figure 10.18**).

To examine specificity of protein recognition, scientists used cloned DNA to construct made-to-order

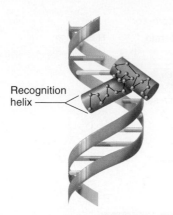

Figure 10.18 DNA recognition sequences by helix-turn-helix motif. A protein motif that has the shape of a helix-turn-helix (*helixes shown here inside a cylindrical shape*) fits into the major groove of the DNA helix. Specific amino acids within the helical region of the protein recognize a particular base sequence in the DNA.

mutations in the gene encoding the repressor of a bacterial virus known as 434. The 434 repressor binds to DNA of the 434 viral DNA that has integrated into the bacterial genome and prevents transcription and production of viral particles. After predicting that a region of the α helix of the 434 repressor recognizes the DNA of its specific operator site, researchers altered the DNA sequence of the gene region encoding this α helix so that it now encoded most of the amino acids in the corresponding α helix of the repressor for another bacterial virus P22 (**Figure 10.19**). The resulting hybrid 434-P22 repressor protein, encoded by the altered gene, contained a P22 α helix that recognized the P22 operator *in vivo*. Binding of the hybrid repressor to the P22 prophage DNA operator region that had integrated into the bacterial host genome shut down transcription of most P22 viral proteins and prevented subsequent infection by the P22 virus. This experiment showed that specific amino acids in this α helix determine the binding specificity of the repressor protein. The P22-like α helix in the hybrid protein is sufficient to convert the binding specificity of the 434 repressor to that of the P22 repressor.

The HTH motif is found in hundreds of DNA-binding proteins. Surprisingly, more than 20 different DNA-binding proteins in bacteria are very similar to the LacI repressor, not only in the HTH DNA-binding domain but throughout much of the protein. This group of repressors is known as the LacI repressor family of proteins. Their structural similarity suggests that they evolved from a common ancestral gene whose duplication and divergence produced a family of transcriptional repressor proteins with similar overall structures but unique recognition regions. The uniqueness of their DNA-recognition regions means that they interact with different operators to regulate different groups of genes.

Rotational symmetry of binding sites

In the 1970s, geneticists studying gene regulation developed new *in vitro* techniques to determine where regulatory proteins bind to the DNA. Purified proteins that bind to fragments of DNA protect the region to which they bind from digestion by enzymes such as DNase I that break the phosphodiester bonds between nucleotides. If a sample of DNA, labelled at one end of one strand and bound by a purified protein, is partially digested with DNase I, the enzyme will cleave phosphodiester bonds in at least some DNA molecules in the sample, except for those phosphodiester bonds that are in regions protected by the bound protein. Gel electrophoresis of the DNA and autoradiography reveal bands at positions corresponding to the cleavage between each base, except in the region where bound protein protected the DNA. Portions of the gel without bands are thus "footprints," indicating the nucleotides of the DNA fragment that were protected by the DNA-binding protein (**Figure 10.20**).

Identified in this way, many of the DNA sequences to which a negative or positive regulator protein binds exhibit rotational symmetry; that is, their two DNA strands have an almost identical sequence when read in the 5′-to-3′ direction on both strands (these sequences are usually not perfect palindromes). An example of such symmetry is in the *lac* operon's CRP-binding site whose sequence is

5′ TGTGAGTTAGCTCACA 3′

3′ ACACTCAATCGAGTGT 5′

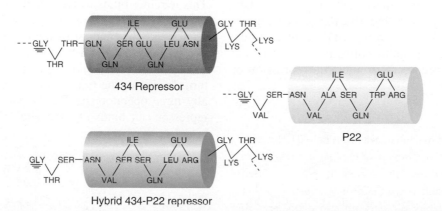

Figure 10.19 Changing amino acids in the recognition sequence. The amino acids inside the recognition helix for phages 434 and P22 and for the hybrid 434-P22 repressor. The amino acids shown in *red* in the hybrid repressor helix are ones that were modified to be like those of the P22 repressor.

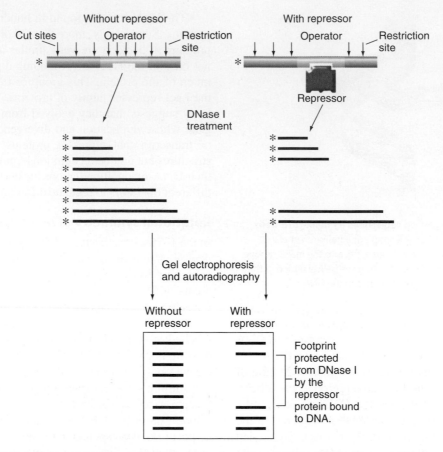

Figure 10.20 DNase footprint shows where proteins bind. DNase footprint establishes the region to which a protein binds. A partial digestion with DNase I produces a series of fragments. If a protein is bound to DNA, DNase cannot digest at sites covered by the protein. Gel electrophoresis of digested products shows which products were not generated and indicates where the protein binds.

Multiple subunits in regulatory proteins

Most regulatory proteins that bind to DNA exist as oligomers composed of two to four polypeptide subunits. The regulatory proteins, which are present in very low numbers in the cell, gain an advantage from this multimeric form. Because each polypeptide subunit of an oligomer has a DNA-binding domain, an assembled oligomer has multiple DNA-binding domains. If the sites to which an oligomeric protein can bind are clustered in a gene's regulatory region, many contacts can be established between the protein and the regulatory region. By increasing the stability of protein-DNA interactions, these multiple binding domains collectively produce the strength of binding necessary to maintain repression or activate transcription.

Many regulatory proteins contain helix-turn-helix (HTH) motifs that allow them to fit into the major groove on DNA, where one of the helixes can recognize a nucleotide sequence. Often these sequences exhibit a symmetry when read in either direction. Most regulatory proteins are also oligomeric, so that the assembled oligomer has multiple binding domains, allowing many contacts.

CRP and *lac* repressor binding

CRP binds to DNA as a dimer at a sequence with rotational symmetry, with one monomer of CRP binding to each side of the sequence (**Figure 10.21**). Thus, CRP-binding sites (such as the site in the *lac* operon whose sequence was just shown) actually consist of two recognition sequences, each able to bind one subunit of the CRP dimer.

The *lac* repressor exists as a tetramer with each of its four subunits containing a DNA-binding HTH motif. This tetramer binds to two operators located far apart on the DNA, with each operator containing two recognition sequences. The binding of the tetrameric repressor to the two operators causes a loop of DNA to form between the two operator sites (**Figure 10.22**); formation of the loop, in turn, facilitates the two-position binding. There are actually three operator sites in the *lac* operon to which the repressor can bind: O_1 (the site originally identified by

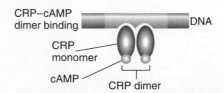

Figure 10.21 CRP–cAMP dimer. CRP–cAMP binds as a dimer to a regulatory region.

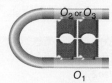

Figure 10.22 *lac* repressor tetramer binds to two sites. The *lac* repressor is a tetramer. For simplicity we previously showed a single repressor object binding to one operator site, but in reality, there are two identical LacI subunits that bind to each operator site. Two of the subunits bind to the sequence in one operator site (O_1), and the other two subunits bind to a second operator (either O_2 or O_3).

lacO^c mutations), O_2, and O_3. Site O_1 has the strongest binding affinity for the repressor, and two subunits of the tetramer always bind at this site. The other two subunits bind at either O_2 or O_3. The distance between operator sites—multiples of ten bases—allows repressor binding to the same side of the helix and thus formation of the loop.

Mutations in *either O_2 or O_3* have very little effect on repression. By contrast, mutations in *both O_2 and O_3* make repression 50 times less effective. The conclusion is that for maximal repression, all four of the repressor's subunits must bind DNA simultaneously. Binding at four recognition sequences (in two operator sites) increases the stability of the protein-DNA interactions. In fact, the DNA binding of the *lac* repressor is so efficient that only ten repressor tetramers per cell are sufficient to maintain repression.

DNA looping and the mechanism of AraC action

Looping first came to light in work on AraC, the regulatory protein that helps control the arabinose operon described previously. AraC functions as a dimer. As we have seen, in the presence of the inducer arabinose, AraC is a positive regulator that helps initiate transcription of the *araBAD* operon. Unexpectedly, in the absence of arabinose, AraC acts as a repressor. In this capacity, the AraC dimer binds to two sites—*araO* and *araI*—that are 194 nucleotide pairs apart (**Figure 10.23**). In one set of experiments analyzing the binding to two sites with concomitant looping of DNA, researchers altered the distance between *araO* and *araI* by inserting several base pairs. The introduction of 11 or 31 bp—alterations that are close to integral changes in the number of turns of the double helix (a full turn of the helix = 10.5 bp)—had little effect on repression. The introduction of 5, 15, or 24 bp, however, noticeably reduced repression. These results suggest that the orientation of the binding sites is a significant variable; only when two sites have an orientation that puts them on the same side of the helix can a dimer bind simultaneously to both and cause formation of a DNA loop.

What enables the AraC dimer to function as both an activator and a repressor? The answer most likely involves allostery and AraC's different binding affinities for recognition sequences at *araI* and *araO*. The *araI* site contains two recognition sequences (*araI$_1$* and *araI$_2$*) to which AraC can bind; the *araO* site contains one recognition sequence. The AraC dimer is an allosteric protein whose structure changes with the binding of inducer (arabinose). One hypothesis of how this allosteric change alters function is that in the absence

(a) No arabinose present

AraC dimer binding at both *araO* and *araI* sites. No *araBAD* genes are transcribed.

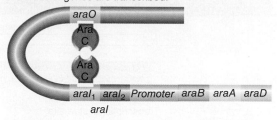

(b) Arabinose present

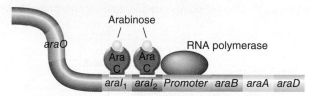

Figure 10.23 AraC acts as both a repressor and an activator. The AraC protein can bind to sites *araI$_1$*, *araI$_2$*, and *araO*. **(a)** When no arabinose is present, the binding of AraC to the *araO* and *araI$_1$* sites causes looping of the DNA and prevents RNA polymerase from transcribing the genes. **(b)** When arabinose (inducer) is present, AraC binds to *araI$_1$* and *araI$_2$* but not to *araO*. RNA polymerase interacts with AraC at the *araI* sites and transcribes the genes.

of arabinose, the size and shape of the protein unbound to inducer allow the AraC dimer to bind to two recognition sequences—*araO* and one of the two sites within *araI*—at the same time; this double binding prevents AraC from binding in a way that would enable it to assist RNA polymerase in the initiation of transcription. When arabinose is present, binding to the inducer changes the shape of the AraC dimer. In this inducer-bound conformation, the regulatory molecule does not bind to *araI* and *araO* at the same time; instead, the inducer–dimer complex binds exclusively to recognition sequences in *araI* (Figure 10.23b). When bound to DNA at only this site, AraC's positive regulatory domain is free to interact with RNA polymerase and increase transcription.

Regulatory proteins and RNA polymerase

Many negative regulators, such as the *lac* repressor, prevent initiation by blocking the functional binding of RNA polymerase. For example, the O_1 operator site to which the *lac* repressor tightly binds consists of 27 nucleotides centred 11 bp downstream from the transcriptional start site. The operator thus includes part of the region where RNA polymerase has to bind to initiate transcription (**Figure 10.24**). When repressor is bound to the operator, its presence on the DNA prevents RNA polymerase from binding in the way needed to initiate transcription.

Positive regulators, by contrast, usually establish a physical contact with RNA polymerase that enhances the enzyme's ability to initiate transcription (**Figure 10.25**). For several positive regulators, researchers have identified points of contact between the regulator and the α, β, or σ subunits of RNA polymerase. Although RNA polymerase will bind to a promoter in the absence of a positive regulator,

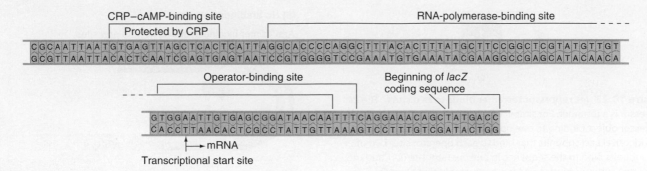

CRP–cAMP-binding site
Protected by CRP
RNA-polymerase-binding site

CGCAATTAATGTGAGTTAGCTCACTCATTAGGCACCCCAGGCTTTACACTTTATGCTTCCGGCTCGTATGTTGT
GCGTTAATTACACTCAATCGAGTGAGTAATCCGTGGGGTCCGAAATGTGAAATACGAAGGCCGAGCATACAACA

Operator-binding site

Beginning of *lacZ*
coding sequence

GTGGAATTGTGAGCGGATAACAATTTCAGGAAACAGCTATGACC
CACCTTAACACTCGCCTATTGTTAAAGTCCTTTGTCGATACTGG

→ mRNA
Transcriptional start site

Figure 10.24 Regulatory-protein-binding sites overlap. The *lac* repressor bound to the operator prevents RNA polymerase from binding. The binding sites for RNA polymerase and repressor (determined by DNase digestion experiments) show that there is overlap between the two sites.

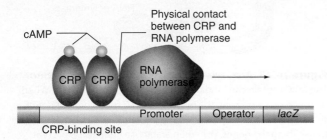

Figure 10.25 CRP-cAMP interaction. The CRP–cAMP complex contacts RNA polymerase directly to help in transcription initiation.

it is less likely to unwind DNA and begin polymerization than when it receives assistance from a positive regulator.

> Negative regulators, such as the tetrameric *lac* repressor, prevent initiation by blocking RNA polymerase's binding to the initiator. Positive regulators, such as CRP, enhance initiation through physical contact with RNA polymerase. AraC can act as a positive regulator or as a repressor; it has different binding affinities in the presence or absence of arabinose.

Fusion of a *lacZ* gene to regulatory regions as a reporter allows functional assessments

Extensive molecular knowledge of the *lacZ* gene and assays to measure its expression have enabled its use as a "reporter" gene in the study of a large variety of regulatory regions in both prokaryotes and eukaryotes. A **reporter gene** is a protein-encoding gene whose expression in the cell is quantifiable by sensitive and reliable techniques of protein detection.

Measuring gene expression

Fusion of the coding region of the reporter gene to *cis*-acting regulatory regions (including promoters and operators) of other genes creates a DNA molecule that enables researchers to assess the activity of the regulatory elements by monitoring the amount of reporter gene product appearing in the cell. For example, with the fusion of gene *X*'s regulatory region to the *lacZ* gene, one can assess the activity of the regulatory elements of gene *X* by monitoring the level

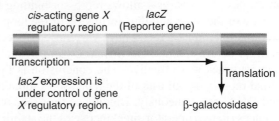

Figure 10.26 *lacZ* gene fused to regulatory region. The *lacZ* structural gene can be fused to a regulatory region of gene *X*. Expression of β-galactosidase will be dependent on signals in the regulatory region to which *lacZ* is fused.

of β-galactosidase expression. With this fusion molecule, conditions that induce expression of gene *X* will generate β-galactosidase (**Figure 10.26**).

Identifying regulatory sites

The use of reporter genes makes it possible to identify the DNA sites necessary for regulation as well as the genes and signals involved in that regulation. For example, you could mutate gene *X*'s control region *in vitro* and then transform the gene *X-lacZ* fusion molecule back into bacterial cells and look for mutations that disrupt a particular aspect of control (as measured by levels of *lacZ*); this protocol would identify *cis*-acting sites important for the regulation. Or, you could mutate the bacteria themselves and then introduce a reporter fusion molecule into the mutated cells and look for changes in level of *lacZ* expression as measured by blue or white colony colour; this protocol would identify *trans*-acting genes.

Identifying sets of genes regulated by the same stimulus

Reporter fusion molecules not only provide the basis for analyzing the regulation of one specific gene, they also make it possible to identify many genes regulated by the same stimulus. To this end, researchers can use transposition to insert the *lacZ* gene without its regulatory region at various sites around the bacterial chromosome. Introducing the *lacZ* reporter into a population of cells generates a collection of *E. coli* cells, some of which contain *lacZ* fused to genes and their regulatory regions. On exposure of this collection to a stimulus such as UV light, the cells containing

lacZ fused to the regulatory regions of genes induced by UV light will produce β-galactosidase (**Figure 10.27**). Researchers identified a set of genes activated by exposure to DNA-damaging agents by this method.

Controlling gene expression

In addition to using *lacZ* as a reporter gene, geneticists studying gene regulation can use their extensive knowledge of the *lac* operon regulatory region to construct recombinant molecules carrying genes whose expression can be controlled. For example, by fusing the *lac* operon control DNA to a human gene expressed in *E. coli*, they could cause overproduction of the human protein in response to an induction cue. The ability to control expression of a foreign gene is important because it provides a way to ensure that protein production is not turned on until cells have proliferated to a high density. The culture will thus contain many cells making the desired foreign protein; and even if the protein has deleterious effects on the growth of *E. coli*, the culture can still grow to high density before addition of the inducer. The production in *E. coli* of human proteins, such as human growth hormone, human insulin, and other pharmacologically useful proteins, is based on this strategy (**Figure 10.28**).

> A reporter gene is a protein-encoding gene that has been fused to other genes as an insertion. The presence or absence of its product can then be used as a reliable indicator (reporter) of expression of those genes. *lacZ* fusions have been widely used to study gene expression.

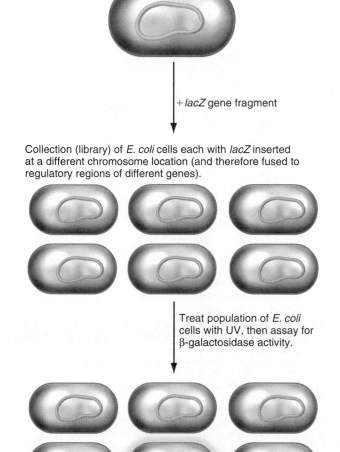

Collection (library) of *E. coli* cells each with *lacZ* inserted at a different chromosome location (and therefore fused to regulatory regions of different genes).

Treat population of *E. coli* cells with UV, then assay for β-galactosidase activity.

This cell exhibits expression of β-galactosidase after UV treatment. Gene to which *lacZ* is fused is regulated by UV light.

Figure 10.27 *lacZ* introduced into a population of *E. coli* cells. Creating a collection of *lacZ* insertions in the chromosome. The *lacZ* gene without its promoter integrates randomly around the chromosome. If *lacZ* integrates within a gene in the orientation of transcription, *lacZ* expression will be controlled by that gene's regulatory region. The library of clones created can be screened to identify insertions in genes regulated by a common signal.

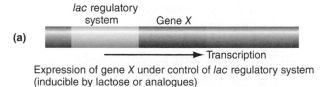

(a)

Transcription

Expression of gene *X* under control of *lac* regulatory system (inducible by lactose or analogues)

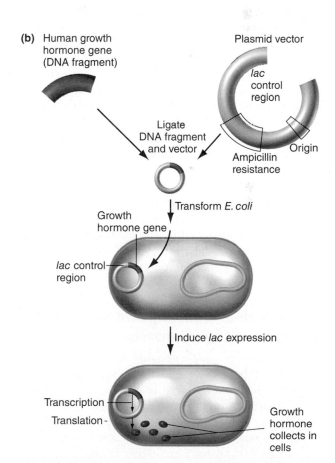

Figure 10.28 Use of fusions to overproduce a gene product. (a) The *lac* regulatory region can be fused to gene *X* to control expression of genes. **(b)** The gene encoding human growth hormone is cloned next to the *lac* control region and transformed into *E. coli*. Conditions that induce *lac* expression will cause expression of growth hormone that can be purified from the cells.

10.3 Attenuation of Gene Expression: Termination of Transcription

In bacteria, the multiple genes of both catabolic and anabolic pathways are clustered together and co-regulated in operons. We have seen that regulators of the catabolic *lac* and *ara* operons respond to the presence of lactose or arabinose, respectively, by inducing gene expression. By contrast, regulators of anabolic operons respond to the presence of the pathway's end-product by shutting down the genes of the operon that encode proteins needed to manufacture the end-product.

There are many anabolic bacterial operons involved in the synthesis of amino acids. A well-studied example is the *E. coli* tryptophan (*trp*) operon, a group of five structural genes—*trpE*, *trpD*, *trpC*, *trpB*, and *trpA*—required for construction of the amino acid tryptophan. Maximal expression of the *trp* genes occurs when tryptophan is absent from the growth medium.

Tryptophan activates a repressor of the *trp* operon

The *trp* operon is regulated by a protein repressor that is the product of the *trpR* gene. In contrast to the *lac* operon, where allolactose functions as an inducer that prevents the repressor from binding to the operator, tryptophan functions as a corepressor: an effector molecule whose binding to the actual TrpR repressor protein allows the negative regulator to bind to DNA and inhibit transcription of the genes in the operon. The binding of tryptophan to the TrpR repressor causes an allosteric alteration in the repressor's shape, and only with this alteration can the TrpR protein bind to the operator site (**Figure 10.29**). Mutations in the *trpR* gene that change either the protein's tryptophan-binding domain or its DNA-binding domain destroy the TrpR repressor's ability to bind DNA, and they result in the constitutive expression of the *trp* genes even when tryptophan is present in the growth medium. The TrpR-mediated repression of the *trp* operon is critical, but it is only one of the regulatory components controlling expression of the *trp* genes in *E. coli*.

Termination of transcription fine-tunes regulation of the *trp* operon

One would expect *trpR⁻* mutants to show constitutive expression of their *trp* genes. With or without tryptophan in the medium, if there is no repressor to bind at the operator, RNA polymerase would have uninterrupted access to the *trp* promoter. Surprisingly, studies show that the *trp* genes of *trpR⁻* mutants are not completely de-repressed (i.e., turned on) when tryptophan is present in the growth medium. As **Table 10.1** shows, the removal of tryptophan from a medium in which *trpR⁻* mutants are growing causes expression of the *trp* genes to increase threefold. What control mechanism is responsible for this repressor-independent change in *trp* operon expression?

Alternative transcripts, different outcomes

In a series of elegant experiments analyzing transcription of the *trp* operon, Charles Yanofsky and co-workers found that initiation at the *trp* promoter can produce two alternative transcripts (Figure 10.29 and **Figure 10.30a**). Sometimes initiation at the promoter leads to transcription of a truncated mRNA about 140 bases long from a short DNA region immediately preceding the first *trp* structural gene (*trpE*); this pre-gene DNA region is called a **leader sequence**, and the RNA transcribed from it is the *RNA leader*. At other times, transcription continues beyond the end of the leader sequence to produce a full operon-length transcript. In analyzing why some mRNAs terminate before they can transcribe the structural *trp* genes, while others do not, the researchers discovered **attenuation**: control of gene expression by premature termination of

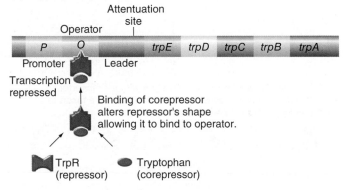

(a) Tryptophan present

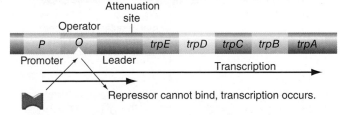

(b) Tryptophan not present

Figure 10.29 Tryptophan acts as a corepressor. (a) When tryptophan is available, it binds to the *trp* repressor, causing the molecule to change shape so that the repressor can bind to the operator of the *trp* operon and repress transcription. **(b)** When tryptophan is not available, the repressor cannot bind to the operator, and the tryptophan biosynthetic genes are expressed.

Table 10.1	Expression of *trp* Operon in *trpR⁺* and *trpR⁻* Strains	
	With Tryptophan* (%)	**Without Tryptophan (%)**
trpR⁺	8	100
trpR⁻	33	100

*In the growth medium

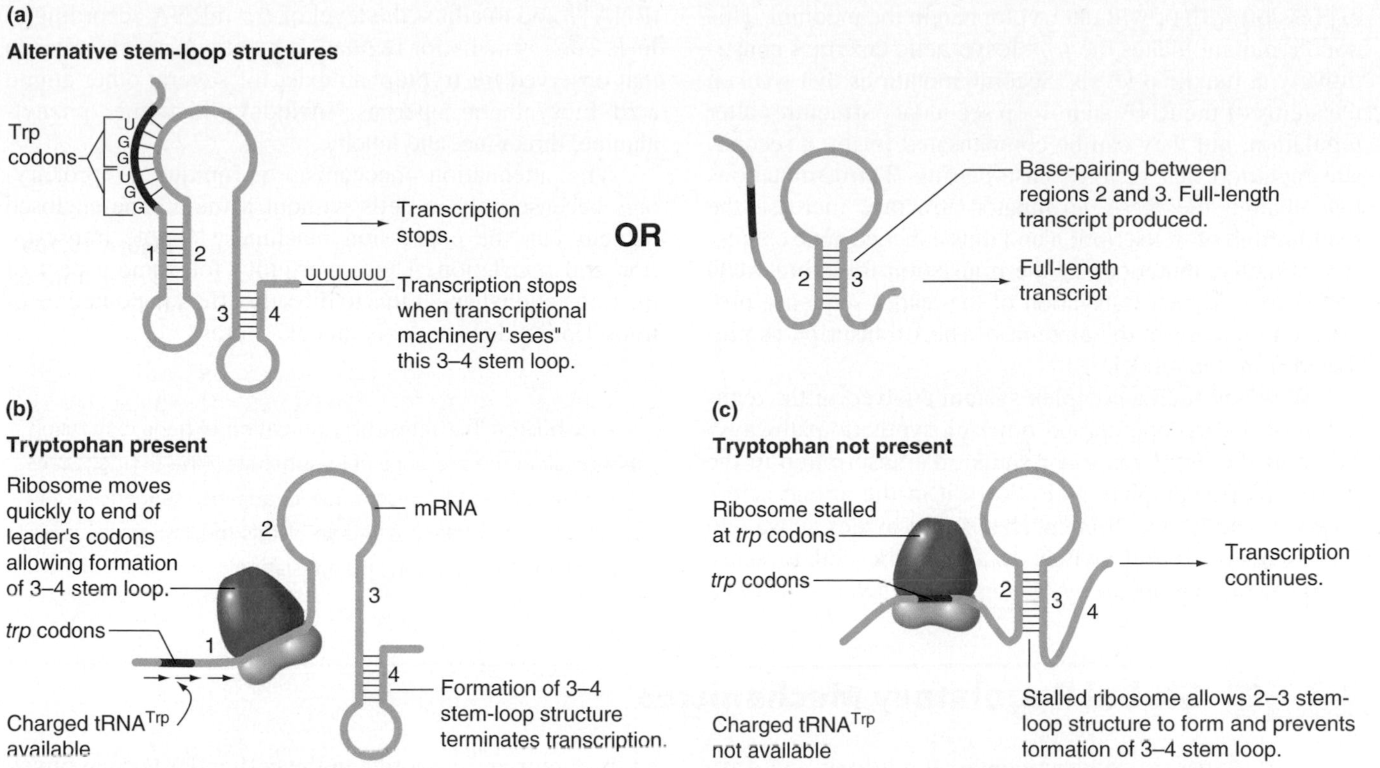

(a)
Alternative stem-loop structures

Trp codons

U
G
G
U
G
G

1 2

3 4

UUUUUUU

Transcription stops.

Transcription stops when transcriptional machinery "sees" this 3–4 stem loop.

OR

Base-pairing between regions 2 and 3. Full-length transcript produced.

2 3

Full-length transcript

(b)
Tryptophan present

Ribosome moves quickly to end of leader's codons allowing formation of 3–4 stem loop.

mRNA

2

3

trp codons

1

4

Charged tRNA^Trp available

Formation of 3–4 stem-loop structure terminates transcription.

(c)
Tryptophan not present

Ribosome stalled at trp codons

trp codons

2 3

4

Transcription continues.

Charged tRNA^Trp not available

Stalled ribosome allows 2–3 stem-loop structure to form and prevents formation of 3–4 stem loop.

Figure 10.30 Attenuation in the tryptophan operon of *E. coli*. (a) Stem loops form by complementary base-pairing in the *trp* leader RNA. Two different secondary structures are possible in the mRNA from the *trp* operon. Stem loops using complementary base pairs between regions 1 and 2 will enable the formation also of the stem-loop 3–4, which is a termination signal for RNA polymerase. Base-pairing between regions 2 and 3 leads to a stem loop that prevents formation of stem-loop 3–4. In the early portion of the transcript, there are two codons for tryptophan. **(b)** When tryptophan is present, the ribosome follows quickly along the transcript, preventing stem-loop 2–3 from forming. Stem-loop 3–4 can form and transcription is terminated. **(c)** If tryptophan is absent, the ribosome stalls at the *trp* codons, allowing formation of stem-loop 2–3, preventing stem-loop 3–4 formation. Transcription continues.

transcription. Whether or not transcription terminates prematurely depends on how the translation machinery reads the secondary structure of the RNA leader.

The RNA leader can fold into two different stable conformations, each one based on the complementarity of bases in the same molecule of RNA. The first structure contains two stem-loop structures with regions 1 and 2 associated by base-pairing and regions 3 and 4 similarly associated by base-pairing. When the transcriptional machinery "sees" the 3–4 stem-loop configuration, which has seven U's at the end, it stops transcription, producing a short, "attenuated" RNA. The alternative RNA structure forms by base-pairing between regions 2 and 3. In this conformation, the leader RNA does not display the 3–4 stem-loop termination signal, and as a result, the transcription machinery reads right through it to produce a full-length transcript that includes the structural-gene sequences.

How a small region of the leader determines the outcome

The early translation of a short portion of the RNA leader (while transcription of the rest of the leader is still taking place) determines which of the two alternative RNA

structures forms. That key portion of the RNA leader includes a short open reading frame containing 14 codons, two of which are *trp* codons (**Figure 10.30b**). When tryptophan is present, the ribosome moves quickly past the *trp* codons in the RNA leader and proceeds to the end of the leader's codons, allowing formation of the stem-loop 3–4 structure and preventing formation of the alternative RNA structure. As we have seen, this 3–4 RNA structure causes the termination of transcription and hence the attenuation (i.e., lessening) of *trp* gene expression. In the absence of tryptophan, the ribosome stalls at the two *trp* codons in the RNA leader because of the lack of charged tRNA^Trp in the cell. The 2–3 stem-loop structure depicted in **Figure 10.30c** is then able to form, and its formation prevents formation of the 3–4 stem-loop RNA structure recognized by the transcriptional terminator. As a result, transcription proceeds through the leader into the structural genes.

How do we know these secondary RNA structures exist *in vivo* and that the translation of the leader RNA plays a significant role in their formation? Several experiments support this model of attenuation. First, deletion of the complete leader sequence results in the loss of control by attenuation; in *trpR*⁻ mutants that also do not contain the leader sequence, there is no difference in *trp*

expression with or without tryptophan in the medium. This double mutant makes the *trp* biosynthetic enzymes constitutively at maximal levels. Second, mutations that weaken the stems of the RNA stem-loop secondary structures alter regulation, but they can be compensated for by a second-site mutation that restores base-pairing. Third, mutations that change the RNA terminator structure increase the readthrough of transcription and thus enhance gene expression. Finally, mutations of the translation-initiating AUG codon that prevent translation of the leader sequence produce an increase in the amount of short transcript, as predicted from the model.

Why has such a complex system evolved in the regulation of the *trp* operon and other biosynthetic pathways? Whereas the TrpR repressor shuts off transcription in the presence of tryptophan and allows it in the amino acid's absence, the attenuation mechanism provides a way to fine-tune this off/on switch. It allows the cell to sense the level of tryptophan by "reading" the level of charged

tRNA^{Trp} and to adjust the level of *trp* mRNA accordingly. In *E. coli*, systems for regulation by attenuation similar to that observed for tryptophan exist for several other amino acid biosynthetic operons, including histidine, phenylalanine, threonine, and leucine.

The attenuation mechanism is unique to prokaryotes because only in cells without a membrane-enclosed nucleus can the expression machinery couple transcription and translation. The opportunity for some aspect of the translational apparatus to directly affect the outcome of transcription does not exist in eukaryotes.

> The *trp* operon illustrates the attenuation of gene expression as a result of the presence of its substrate. Fine-tuning is provided by stalling of transcription at a leader sequence. Only prokaryotes can take advantage of this mechanism because they couple transcription and translation.

10.4 Global Regulatory Mechanisms

Dramatic shifts in environmental conditions can trigger the expression of sets of genes or operons dispersed around the chromosome. The absence of glucose, we have already seen, increases the expression of several catabolic operons that are at least partially controlled by a common factor, the CRP–cAMP complex. A group of genes whose expression is regulated by the same regulatory proteins is called a **regulon**. Another example of such global regulation is *E. coli*'s response to heat shock, which results from exposure to extremely high temperatures (up to 45°C).

An alternative sigma (σ) factor mediates *E. coli*'s global response to heat shock

At high temperatures, most proteins denature or aggregate, or both. In *E. coli*, exposure to high temperature induces the expression of several proteins that alleviate heat-shock-related damage. The induced proteins include those that recognize and degrade aberrant proteins as well as so-called *chaperone proteins*, which assist in the refolding of other proteins and also prevent their aggregation.

E. coli's induction of the proteins that combat heat shock is a highly conserved stress response. Organisms as different as bacteria, flies, and plants induce similar proteins, notably the chaperones, in response to high temperatures.

Conditional mutants and the global mechanism

Conditional lethal *E. coli* mutants in which high temperatures do not induce transcription of the heat-shock genes provided critical evidence for the global regulatory mechanism. These conditional lethal mutants have a defect in the *rpoH* gene that encodes an alternative RNA polymerase sigma factor known as σ^{32}. The normal housekeeping

sigma factor, σ^{70}, is active in the cell under normal physiological conditions. By contrast, the alternative σ^{32} can function at high temperatures; it also recognizes different promoter sequences than those recognized by σ^{70}. Genes induced by heat shock contain nucleotide sequences in their promoters that are recognized by σ^{32} (**Figure 10.31**). σ^{32} mediates the heat-shock response by binding to the core RNA polymerase, thereby allowing the polymerase to initiate transcription of the genes encoding the heat-shock proteins. The RNA polymerase σ^{32} holoenzyme is relatively resistant to heat inactivation compared with the heat-sensitive σ^{70}-dependent RNA polymerase. As a result, when temperatures rise, genes with a σ^{32} promoter undergo transcription, while genes with a σ^{70} promoter do not.

Levels of the σ^{32} protein and the σ^{32} RNA polymerase holoenzyme increase immediately after heat shock. Several factors cause this increase in σ^{32} activity:

- An increase in the transcription of the *rpoH* gene
- An increase in the translation of σ^{32} mRNA stemming from greater stability of the *rpoH* mRNA
- An increase in the stability and activity of the σ^{32} protein. Chaperones DnaJ/K bind to and inhibit σ^{32} under normal physiological conditions. When the temperature rises, these proteins bind to the large

σ^{70} recognizes this promoter sequence.

> T T G A C A 16–18 bp T A T A A T

σ^{32} recognizes this promoter sequence.

> C T T G A A 13–15 bp C C C C A T N T

Figure 10.31 Sigma factor recognition sequences. Base sequences recognized by σ^{32} and σ^{70}. (The *N* indicates that any base can be found at this position.)

number of cellular proteins that become denatured, leaving σ^{32} free to associate with RNA polymerase.

- The inactivity of σ^{70} at high temperatures. Because of this inactivity, σ^{70} does not compete with σ^{32} in forming the RNA polymerase holoenzyme.

How alternative sigma factors are transcribed at high temperatures

Given that high temperatures render σ^{70} inactive, what enables the transcription of σ^{32} during heat shock? The *rpoH* gene, which encodes σ^{32}, has a promoter sequence that is recognized by σ^{70} and used for transcription at lower temperatures. However, at high temperatures, another sigma factor, σ^{24}, recognizes a different promoter sequence at *rpoH* and transcribes the *rpoH* gene from that promoter (**Figure 10.32**). Although σ^{24} is always present in the cell, its own transcription (mediated by the σ^{24} holoenzyme) increases with heat shock and the appearance of denatured proteins.

The RpoS sigma factor

Another sigma factor, RpoS, that is active during many different stress conditions, is also activated during heat shock. The RpoS protein was originally identified as a sigma factor that becomes active as growth of *E. coli* slows. The *rpoS* gene is transcribed throughout growth in *E. coli*, but translation is inhibited during normal growth. A secondary structure forms in the long leader sequence of the *rpoS* mRNA and blocks access of the message to the ribosome for translation (**Figure 10.33**). When *E. coli* is under stress (lack of nutrients, heat shock, or other stresses), a small RNA, *dsrA*, that has complementarity to 20 nucleotides of the *rpoS* mRNA, binds to the message and prevents the secondary structure from forming, thereby allowing translation to occur. Thus the gene encoding the *rpoS* gene, the product of which acts to regulate transcription, is regulated at the translation step in gene expression.

Global regulation by alternative sigma factors

The induction of new sets of genes in many different bacteria is often achieved in bacteria by the turn of alternative sigma factors. By coordinating the transcription of sets of genes in response to cues from the environment, the alternative sigma factors contribute to the control of such complex processes as sporulation, the synthesis of flagella, and nitrogen fixation (see the **Genetics and Society** box "**Nitrogen Fixation and Gene Regulation**").

Genome analysis reveals that the genomes of bacterial species contain several related but slightly different sigma factor genes. For example, the bacterium *Bacillus subtilis*, under the adverse conditions of nutrient deprivation (such as nitrogen or carbon starvation), uses a cascade of sigma factors, induced in a temporal order, to turn on successive sets of genes needed to form spores. With the proper expression of these genes, the bacterial cell becomes a metabolically inert spore able to withstand heat, aridity, extreme cold, toxic chemicals, and radiation.

> Sudden environmental changes can trigger gene expression, as is shown by the heat-shock proteins of *E. coli*. Alternative sigma factors are present that can recognize different promoter sequences and complex with the core RNA polymerase as the temperature rises. Many bacterial species have evolved similar strategies for dealing with environmental change.

Microarrays provide a tool for studying genes regulated in a global response

Microarrays are an important new tool for microbial geneticists studying cellular responses to changing environmental conditions. The cellular responses to these conditions often involve a global change in gene expression that is measurable by microarray analysis of mRNA isolated from cultures of cells grown under different conditions.

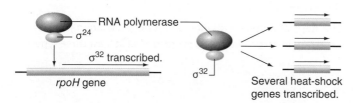

Figure 10.32 Alternative sigma factor in the heat-shock response. At high temperature, the *rpoH* gene (encoding σ^{32}) is transcribed. The σ^{32} interacts with RNA polymerase and transcribes the heat-shock genes.

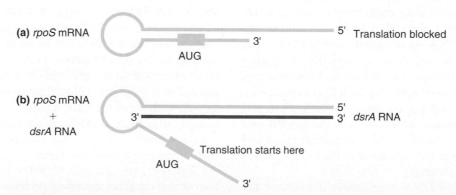

Figure 10.33 Translational control of rpoS. (a) Translation of *rpoS* mRNA is blocked by secondary structure (base-pairing) that occurs in the RNA. **(b)** *dsrA* RNA (*brown*) base-pairs with *rpoS* RNA, freeing the translation start site.

Nitrogen, an essential component of amino acids, chlorophylls, and nucleic acids, is a growth-limiting plant nutrient—the more nitrogen available, the faster most plants grow. However, although gaseous nitrogen (N_2) makes up 78 percent of Earth's atmosphere, plants cannot use nitrogen in this form. They can use only nitrogen that has been *fixed*; that is, converted to ammonia (NH_3) or another nitrogen-containing compound.

Plants obtain fixed nitrogen from three main sources: (1) the decayed organic matter in soils, which releases nitrate and ammonium; (2) the activity of nitrogen-fixing bacteria, which fix atmospheric N_2 into ammonium and other biologically available forms of nitrogen; and (3) inorganic nitrogen fertilizers. The last 50 years has seen a tenfold increase in the application of inorganic fertilizer. This excessive use of fertilizer has produced runoff that increases the mineral content of rivers and coastal waters, which has led to algal blooms and a depletion of oxygen in aquatic ecosystems.

In an attempt to reduce the amount of inorganic fertilizers used in agriculture, many scientists are studying how bacteria fix nitrogen. Several types of bacteria are agents of nitrogen fixation. These bacteria may be free-living cells (such as those in the genus *Klebsiella*) or plant symbionts (such as those in the genus *Rhizobium*). Of the many symbiotic rhizobial species, each one is able to form a working relationship with only one or a few plants, mainly legumes like peas, beans, and alfalfa. For hundreds of centuries, farmers made practical use of the nitrogen-fixing abilities of rhizobial bacteria via the rotation of crops.

Bacteria–Plant Interactions Lead to Nitrogen Fixation

In the symbiotic relationship that develops between *R. meliloti*, a small heterotrophic nitrogen-fixing bacterium, and alfalfa, bacterial genes produce the enzymatic machinery for nitrogen fixation, while the plants provide a low-oxygen environment that allows the nitrogen-fixation enzymes to function. A series of communications between plant and bacteria lead to dramatic changes in both the anatomy of the plant and the structure of the bacteria that enable symbiosis.

Alfalfa's secretion of flavonoids triggers the events leading to nitrogen fixation. *R. meliloti* responds to this environmental signal by expressing *nodulation* (*nod*) genes, whose protein products are enzymes active in the synthesis of liposaccharides known as Nod factors. Release of the Nod factors from the rhizobial cells elicits a curling of root hairs and cell division in the meristem, which lead to the formation of root nodules in the alfalfa plant (**Figure A**). The bacteria now navigate by chemotaxis to the alfalfa host nodules and penetrate the host's root cortex with the help of a gelatinous filament secreted by the plant itself.

Once inside the plant, the *R. meliloti* enter root cells, where they divide and differentiate into *bacteroids*, cells that produce nitrogenase, an enzyme complex that catalyzes the conversion of N_2 to NH_3. The host plant monitors the concentration of oxygen in the area of the nodule where the bacteroids thrive to ensure that it is much lower than in the surrounding plant cells or soil. A nearly anaerobic environment in the nodule is crucial to the survival and function of nitrogenase. The alfalfa plant uses the nitrogen fixed by *R. meliloti* as its source of nitrogen and, in return, provides the bacteria with photosynthetic products and amino acids.

The Genetic Components and Mechanisms That Mediate Nitrogen Fixation in Rhizobial Bacteria

The steps of nodule formation and nitrogen fixation just outlined require the coordinated expression of at least three types

For example, to study changes in gene expression when lactose is substituted for glucose in the extrinsic environment (growth medium), scientists grow one culture of *E. coli* in medium containing glucose as a carbon source and another culture in medium containing lactose as a carbon source. They then isolate RNA from each culture and synthesize labelled DNA complementary to the collection of RNAs from the two different cultures. By comparing the hybridization of the two sets of cDNAs to microarrays containing oligonucleotide spots for each *E. coli* gene, they can see which genes are turned on and which are turned off. In the lactose-treated cells, one would expect to see an increase in the mRNA of the lactose operon genes and therefore an increase in hybridization to the *lacZ*, *Y*, and *A* gene spots on the microarray.

Experiments using DNA microarrays to compare the gene expression patterns of cells grown in media containing as their carbon source glucose, glycerol, succinate, or alanine provide an interesting glimpse into the cellular response to poorer energy sources. The cells grown on glycerol, succinate, or alanine not only turned on the few genes specifically required to use the poorer carbon source, but they also turned on a hierarchy of large sets of additional genes. Cells grown on glycerol or succinate showed increased expression of 40 genes; cells grown on alanine turned on 188 genes, including the set of 40 that were expressed in the glycerol- and succinate-grown cultures. These 40 genes included those of the stress response as well as those of the CRP regulon. Recall that when glucose is not present, cAMP levels rise, and the CRP–cAMP complex binds and increases expression of other genes, including several catabolic genes.

Unexpectedly, the cells grown on poorer carbon sources also turned on genes encoding proteins for motility and for the transport of many compounds in addition to the compound in their medium. The cellular response to carbon

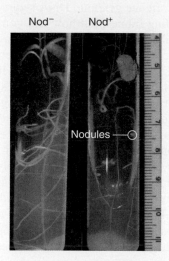

Figure A *R. meliloti's* **release of Nod factors induces the formation of root nodules in alfalfa.** Nod⁻ mutants cannot form nodules.

Nod⁻   Nod⁺

Nodules ──○

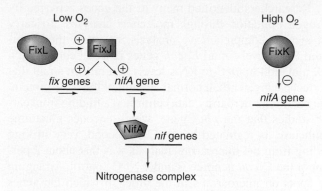

Low O₂   High O₂

FixL → FixJ   FixK

fix genes   *nifA* gene   *nifA* gene

NifA   *nif* genes

Nitrogenase complex

Figure B How environmental signals influence the expression of *nif* genes in *R. meliloti*. Low oxygen activates the *nif* genes via RNA polymerase associated with a σ^{54} factor. A regulatory cascade promotes expression of the *nifA* gene only under appropriate conditions. Plus (+) and minus (−) signs indicate the turning on or off, respectively, of genes.

of *R. meliloti* genes: *nod* genes, which elicit the early steps of nodule formation; *fix* genes, which contribute to the development and metabolism of bacteroids and are essential to nitrogen fixation; and *nif* genes, which encode the polypeptide subunits of the nitrogenase complex. The FixL protein in the membrane senses the O₂ concentration and activates the transcription factor FixJ, which then turns on expression of other *fix* genes as well as *nifA*. The *nif* genes carry a special type of promoter that RNA polymerase recognizes only when the polymerase is associated with a specific σ factor called σ^{54} factor. Initiation of transcription at the σ^{54}-dependent promoters of the *nif* genes depends on NifA, an activator protein responsive to environmental signals that reflect the concentration of oxygen (**Figure B**).

This brief description of the process of nitrogen fixation by *R. meliloti* gives some idea of the complex layering of gene regulation that ensures nitrogen fixation moves forward under favourable conditions but comes to a halt under conditions of too much oxygen. The coordinate expression of these bacterial genes contributes to these bacterial species' ability to respond to environmental signals and become nitrogen-generating symbionts in their host plants.

sources other than glucose seems to be the expression of genes that allow the cell to search out and use any alternative energy source. Motility genes, for example, might be turned on to allow the cell to move about in search of food. Another global transcription change is that RNA polymerase transcribes rRNA at a lower rate in the cells with poorer carbon sources, because these cells channel their energy into the search for and use of an alternative carbon source.

Mutants allow the study of more specific response mechanisms

The experiments previously described, although they enhanced understanding of how genetics controls bacterial physiology, were not specific enough to reveal changes in gene expression related only to the change from glucose to an alternative carbon source. Environmental changes often produce a general physiological reaction, and to get around this experimental difficulty, investigators use bacterial strains with mutations in specific genes—namely, the regulatory genes that serve as the main on/off switch for numerous genes in the pathway they are analyzing. Instead of treating cells to two different growth conditions and measuring RNA levels under the different conditions, they use microarrays to compare RNA levels in a wild-type culture to RNA levels in a culture of cells containing a mutation in the key regulatory gene. Note that the two cultures can be grown under the same environmental conditions because the mutation itself simulates a different condition. For example, a *lacI⁻* cell behaves as if lactose is in the medium, even if the mutant is grown in a glucose medium.

Microbiologists have successfully identified genes that *E. coli* expresses specifically in response to nitrogen limitation. Under the best of circumstances, *E. coli* uses ammonia as its source of nitrogen. A lack of ammonia in the external environment, however, activates a master

control gene called *ntrC*. The NtrC protein, in turn, activates many genes whose expression enables *E. coli* to use sources of nitrogen other than ammonia.

Researchers identified many of the genes activated by nitrogen limitation through molecular analyses. To carry out a more comprehensive analysis, however, they compared RNA levels of all *E. coli* genes in a cell containing a null mutation in *ntrC* to RNA levels in a mutant strain that produced a greater than normal amount of the NtrC protein. The resulting microarray data confirmed a finding from earlier studies that the *glnA* gene, which encodes glutamine synthetase, is regulated by NtrC. A second, very striking finding from the microarray analysis was that about 2 percent of the *E. coli* genome is under NtrC control. Arranging the spots on microarrays in the order in which the genes occur in the genome makes it easier to identify co-regulated adjacent genes that might form an operon.

This type of microarray analysis has revealed several additional genes that are regulated by NtrC. The additional genes are involved in the scavenging of proteins and nitrogen through the transport of nitrogen-containing compounds from the cell wall into the cell and the breakdown of amino acids. NtrC's regulation of these genes has been confirmed by other types of analysis. Microarrays have thus provided an avenue for uncovering changes in gene expression that investigators can confirm and expand on through other methodologies.

> Mutations in regulatory genes combined with microarray analysis allow a further refinement of findings about genes that operate under stress. The environmental conditions can be kept the same in such experiments because the mutation effectively creates the stress environment for the substance in question.

Computer analysis can identify regulatory proteins and their DNA-binding sites

A goal in the post-genomic era is to identify the complete set of proteins that regulate transcription in an organism, as well as their DNA-binding sites and the genes they regulate. This information will help researchers discover molecular targets for controlling cell proliferation and the production of harmful or helpful metabolites; it will also make it possible to model how a cell works. A first step in uncovering the regulatory machinery in bacteria is to identify operons. It is easy to correlate genomic DNA sequence with potential open reading frames, but how can you find genes that are co-transcribed? Because operons have a single promoter for several genes, you can look at clusters of genes that have promoter sequences before the first but not the subsequent genes. You can also look for genes with almost no separation between them. Co-transcribed genes have little space between them in the genome because no nucleotides are needed to regulate the expression of each gene separately. Computer experts have developed algorithms that search for one promoter for several closely spaced genes, as well as for transcription termination signals. While not perfect, these algorithms appear to be good operon predictors because known operons are among the results. The predictions can be further assessed by comparative species analysis. With genes that are co-transcribed and regulated as an operon in one species, a homologous set of adjacent genes in another species is also likely to be co-transcribed as an operon. As with all *in silico* analyses, predictions made with computational tools should be tested experimentally.

Genes encoding regulatory proteins can be identified in the genome by searching for sequences encoding DNA-binding motifs, such as HTH. Of 314 putative regulator proteins identified by the presence of transcription factor domains in *E. coli*, 248 contained the HTH motif. Comparative genomic analyses by computer can help identify the genes that encode these regulatory proteins.

When a set of co-regulated genes is present in different species, the proteins that regulate these genes are often conserved, as are their DNA-binding sites. Researchers can use information from organisms such as *E. coli*, in which extensive genetic and biochemical analyses have defined regulatory pathways, to discover regulatory components in less-understood bacteria.

> Although computerized searches of genomes can reveal potential DNA sites for regulatory proteins, these must be verified experimentally. Nevertheless, computer analysis provides a starting point for further research.

10.5 A Comprehensive Example: The Regulation of Virulence Genes in *V. cholerae*

The principles and mechanisms of gene regulation in *E. coli* apply to gene regulation in other prokaryotes as well, including the bacteria *V. cholerae*, which we described at the beginning of this chapter. As we saw, these bacterial agents of cholera are able to sense changes in their environment and transmit signals about those changes to regulators. These regulators then initiate, enhance, diminish, or repress the expression of various genes as the bacteria pass through the stomach, colonize the intestine, and finally produce a toxin. Of particular interest to epidemiologists and medical practitioners seeking to prevent or treat the symptoms of cholera are the genes bestowing virulence.

lacZ reporters help identify regulators of toxin production

To understand the regulation of the genes for virulence, researchers first cloned the two genes that encode the

polypeptide subunits of cholera toxin, *ctxA* and *ctxB*, which are transcribed and regulated together as an operon. They next made a *ctxA-lacZ* reporter gene fusion molecule that could detect changes in regulation of the operon through changes in levels of β-galactosidase expression. *lacZ* would be expressed and β-galactosidase produced when the cholera toxin promoter was being used, and no β-galactosidase would be produced when the promoter was shut off. They then cut *V. cholerae* genomic DNA into pieces and cloned these into a vector that would replicate in *E. coli*. With the construction of these tools, they were able to perform experiments in *E. coli* cells, which are more amenable than *V. cholerae* to some types of genetic manipulation.

To isolate a gene that regulates expression of the *ctx* operon, they transformed *E. coli* cells already containing the *ctxA-lacZ* fusion molecule with clones containing *V. cholerae* DNA. A clone that contains a gene encoding a positive cholera toxin regulatory protein should turn on expression of the *lacZ* fusion molecule in *E. coli*. Clones that turned on expression contained the regulatory *toxR* gene, which encodes a membrane protein (ToxR) with an N-terminal end in the cytoplasm and a C-terminal end in the periplasm (the space between the inner and outer membranes of the bacterium). In *V. cholerae*, *toxR⁻* mutants do not induce virulence, and the *toxR⁻* mutation is recessive, as you would predict of a positive regulator.

Different fusions reveal genes regulated by ToxR and ToxT

To determine what genes ToxR regulates other than those in the *ctx* operon, researchers fused the *toxR* gene to a constitutive promoter, and they introduced this fusion molecule into a collection of *V. cholerae* strains in which copies of the *lacZ* gene had randomly inserted around the chromosome. Those colonies expressing β-galactosidase (as shown by the blue colour resulting from the splitting of the X-Gal substrate) contained *lacZ* genes adjacent to a promoter region regulated by *toxR*.

In *V. cholerae*, these *lacZ* fusion genes must have been regulated by ToxR (at least indirectly) because all bacteria in the study contained the *toxR* gene fused to a constitutive promoter and thus were constitutive synthesizers of ToxR. However, when transferred into *E. coli*, this collection of genes was not regulated by ToxR. Something required to make these genes respond to ToxR was present in the *V. cholerae* genome but was missing in *E. coli*. The lack of direct regulation by ToxR in *E. coli* triggered a search that culminated in the identification of an intermediate regulatory gene named *toxT*. The ToxT protein is a transcriptional activator that carries out its function by binding to the promoters of many genes, including *ctx* and the other virulence genes. While either ToxR or ToxT can activate the *ctx* genes that produce toxin, ToxT alone activates the additional virulence genes, which encode pili and other proteins that enable the bacteria to colonize the small intestine.

ToxT is a major regulator of several virulence genes, but how is it regulated? Mutations in the *tcpP* gene lead to loss of *toxT* transcription, as do mutations in *toxR*. Analyses of the promoter region of *toxT* showed that TcpP binds to the *toxT* promoter close to the transcription start point, while ToxR binds further upstream. This upstream binding suggests that ToxR helps recruit the TcpP protein to the promoter, where TcpP acts as the positive regulator of *toxT*. Both ToxR and TcpP are membrane-bound, with the N terminal region in the cytoplasm available to bind to DNA and the C terminus in the periplasmic space able to receive environmental signals about the location of the bacterium in the body. Expression of cholera toxin and the pilus is induced when the bacteria have reached the intestine after passing through the bile-laden stomach. Information about the environment is probably transmitted through the activation of *tcpP*, as the transcription of this gene is temperature- and pH-dependent. By comparison, *toxR* is transcribed independently of both temperature and pH.

A model of virulence regulation includes a cascade of regulators

On the basis of these studies, researchers proposed the following model (**Figure 10.34**):

- ToxR is a positive regulator of the *ctx* genes and acts as an auxiliary factor in the regulation of ToxT.
- ToxT is a positive regulator of the many virulence genes that make up the virulence regulon in *V. cholerae*.
- Maximal transcription of *toxT* requires the TcpP regulator, with assistance from ToxR.
- The sensing of environmental change is a part of the regulator gene activation process that is mediated by TcpP.

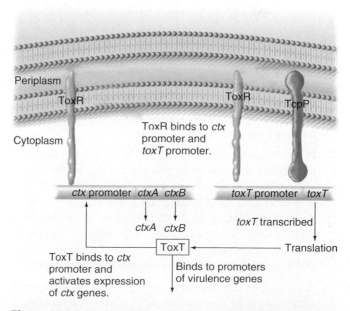

Figure 10.34 Model for how *V. cholerae* regulates genes for virulence. In the cytoplasm, ToxR interacts with the promoter of the *ctxA* and *ctxB* genes. ToxR and TcpP both bind to the *toxT* gene. ToxT in turn regulates the expression of many other virulence genes.

Experiments in which investigators monitored gene expression in the animal (mouse) disease model for *V. cholerae* confirmed the requirement of ToxR as a positive regulator of *ctxA* expression during pathogenesis in the animal. In these animal studies, when *V. cholerae* strains containing the *ctx* reporter gene fusion were injected into mice, expression of *ctxA-lacZ* occurred in the *toxR*$^+$ strain but not in the *toxR*$^-$ mutant. Interestingly, in the mouse model, the specifics of some of the other regulatory pathways for the *V. cholerae* pathogenesis genes did not coincide with the results obtained from the isolated bacterial cultures. From these studies, we can conclude that studying gene expression in pathogenic strains in culture, where it is easier to manipulate genes and measure their expression, provides a valuable first analysis. Once the potential

regulators and pathways have been identified, experiments to detect gene expression during pathogenesis in model animal systems provide critical tests of the models.

Several intriguing questions remain about the regulatory system that controls the expression of the virulence genes in *V. cholerae*. What is the signal that makes the cholera bacteria stop swimming and start to colonize (i.e., adhere to the cells of) the small intestine? What molecular events differentiate swimming and colonization? Why is there a cascade (ToxR, ToxT) of regulatory factors?

Answers to these questions will help scientists complete the picture of how *V. cholerae* generate disease. With a better understanding of pathogenesis, they will be able to devise more effective treatments for cholera as well as measures to prevent it.

Connections

Regulation in prokaryotes depends on the binding of regulatory proteins to specific DNA segments in the vicinity of a gene or group of genes. The existence of these regulatory elements adds another notch to the concept of the gene. Most geneticists would say that a gene consists of the nucleotides that specify amino acids in the gene's protein product or the ribonucleotides in the gene's RNA product, as well as the regulatory elements that influence the gene's transcription.

Some of the ways in which bacteria regulate their genes are available to eukaryotes as well. For example, both types

of organisms can use diffusible regulatory proteins to start or stop transcription. By contrast, eukaryotes cannot regulate transcription by the attenuation mechanism described for the *trp* operon because their nuclear membrane prevents access to the growing transcript by the translational machinery. However, eukaryotic cells, with their larger genomes, have evolved many other mechanisms of gene regulation that go beyond those found in prokaryotic systems. In Chapter 11, we examine the special regulatory needs of eukaryotes and some of the solutions they have evolved.

Essential Concepts

1. Most mechanisms of gene regulation in prokaryotes block or enhance the initiation of transcription. Later steps in gene expression are potential targets for fine-tuning the amount of gene products that accumulate in cells. **[LO1]**

2. Many types of coordinated gene regulation result from the clustering of genes into operons that are transcribed into a single polycistronic mRNA from a single promoter. **[LO3]**

3. In the *lac* operon model proposed by Jacob and Monod, the binding of a repressor protein (encoded by the *lacI* gene) to the DNA *operator* prevents transcription of the structural genes *lacZ*, *lacY*, and *lacA* in the absence of the inducer allolactose. When allolactose is present, its binding to the repressor induces expression of the structural genes by causing the repressor to change its shape and lose its ability to bind to the operator. **[LO2, LO4]**

4. A critical, general principle emerges from the *lac* operon studies: Regulatory genes usually encode *trans*-acting regulatory proteins that interact with *cis*-acting regulatory DNA elements located near the promoter

(such as the operator). Negative regulatory proteins prevent or diminish the rate of transcription, while positive regulatory proteins enhance transcription. **[LO4–5]**

5. Catabolite repression regulates certain catabolic operons by preventing CRP, a positive regulator, from binding to the operons' promoter region in the presence of high concentrations of glucose. **[LO4–5]**

6. Many regulatory proteins, both positive and negative, contain a helix-turn-helix motif, function as oligomers that bind to more than one DNA site, and interact with RNA polymerase to prevent or assist its function. **[LO5]**

7. The binding of repressor proteins to operators can be influenced by either inducers (as for the *lac* repressor) or corepressors (as for the *trp* operon). **[LO6]**

8. Attenuation, a form of fine-tuning for operons involved in the biosynthesis of amino acids, is based on premature termination of mRNA transcription. The termination, in turn, is determined by the intracellular concentration of tRNAs charged with the amino acid produced by the enzyme products of the structural genes in the operon. **[LO6]**

9. Cells can express different sets of genes at different times or under different conditions by using alternative sigma factors or by producing novel RNA polymerases that recognize different classes of promoters. **[LO7]**

10. DNA sequences from many bacterial species are raw data that can be analyzed computationally to identify regulatory features in the bacterial genome. **[LO8]**

Solved Problems

I. In the galactose operon in *E. coli*, a repressor, encoded by the *galR* gene, binds to an operator site, *O*, to regulate expression of three structural genes, *galE*, *galT*, and *galK*. Expression is induced by the presence of galactose in the media. For each of the strains listed, would the cell show constitutive, inducible, or no expression of each of the structural genes? (Assume that *galR⁻* is a loss-of-function mutation.)

a. *galR⁻ galO⁺ galE⁺ galT⁺ galK⁺*

b. *galR⁺ galOᶜ galE⁺ galT⁺ galK⁺*

c. *galR⁻ galO⁺ galE⁺ galT⁺ galK⁻/ galR⁺ galO⁺ galE⁻ galT⁺ galK⁺*

d. *galR⁻ galOᶜ galE⁺ galT⁺ galK⁻/ galR⁺ galO⁺ galE⁻ galT⁺ galK⁺*

Answer

This problem requires an understanding of how regulatory sites and proteins that bind to regulatory sites behave. To predict expression in these strains, look at each copy of the operon individually, and then assess what effect alleles present in the other copy of the operon could have on the expression. After doing that for each copy of the operon, combine the results.

a. The *galR* gene encodes a repressor, so the lack of a *galR* gene product would lead to constitutive expression of *galE, T*, and *K*.

b. The *galOᶜ* mutation is an operator-site mutation. By analogy with the *lac* operon, the designation *galOᶜ* indicates that repressor cannot bind and there is constitutive expression of *galE, T*, and *K*.

c. The first copy of the operon listed has a *galR⁻* mutation. Alone, this would lead to constitutive synthesis of *galE* and *galT*. (*galK* is mutant, so there will not be constitutive expression of this gene.) The other copy is wild type for the *galR* gene, so it produces a repressor that can act in *trans* on both copies of the operon, overriding the effect of the *galR⁻* mutation. Overall, there will be *inducible expression of the three gal genes*.

d. The first copy of the operon contains a *galOᶜ* mutation, leading to constitutive synthesis of *galE* and *galT*. The other copy has a wild-type operator, so it is inducible, but neither operator has effects on the other copy of the operon. The net result is constitutive *galE* and *galT* and inducible *galK* expression.

II. The *araI* site is required for induction of *araBAD*. *I⁻* mutants do not express *araBAD*. In an *I⁻* mutant, a second mutation arose that resulted in constitutive arabinose synthesis. A Southern blot using a probe from the regulatory region and early part of the *araB* gene showed a very different set of restriction fragments than were seen in the starting strain. Based on the altered restriction pattern and constitutive expression, propose a hypothesis about the nature of the second mutation.

Answer

To answer this question, you need to consider how changes in restriction patterns could arise, what effects they could have, and what is necessary to get expression. The fact that the experiment began with a strain that lacked the inducing site, *I*, and there is constitutive synthesis mean that the normal regulation is lacking. Constitutive synthesis could result from a *deletion that fused the araBAD genes to another promoter* (one that is on under the growth conditions used). A deletion would lead to a different pattern of restriction fragments that could be observed in the Southern hybridization analysis.

III. Bacteriophage λ, after infecting *E. coli*, can take one of two routes. It will either produce many progeny that are released by lysis of the cell (lytic growth), or the phage DNA will integrate into the chromosome because transcription from the major phage promoters of the phage will have been shut down. The repressor protein cI, encoded by phage λ, binds to two operator regions to shut down expression, and, therefore, no phages are produced. Mutations in the *cI* gene that destroy the binding ability of the repressor lead to the lytic type of life cycle exclusively; that is, all cells infected by the phage will burst and release progeny phages. Another type of mutation gives the same phenotype—lytic growth only. Such mutations, called λ*vir*, arise at a much lower frequency than the *cI* mutations (about 1 in 10¹² compared with 1 in 10⁶ for *cI* mutants). What do you think these mutations are, and why are they less frequent than *cI* mutations?

Answer

This problem requires an understanding of the types of regulatory mutations that can affect negative regulation. The lack of negative regulation (by *cI*) in the life cycle leads to the lytic cycle of growth only. Such mutations could be either in the gene encoding the negative regulator or in the site to which the repressor binds. You were told that the *cI* mutations are defects in the gene encoding the repressor. The λ*vir* mutations could be mutations in the site to which the repressor binds, but because the repressor has to bind to two sites, *there must be two mutations in a λvir mutant. Therefore, these would arise less frequently* (at a frequency predicted for two independent mutational events combined: 1 in 10⁶ × 1 in 10⁶, or 1 in 10¹²).

Vocabulary

1. For each of the terms in the left column, choose the best matching phrase in the right column.

i. induction		**1.**	glucose prevents expression of catabolic operons
ii. repressor		**2.**	protein undergoes a reversible conformational change
iii. operator		**3.**	often fused to regulatory regions of genes whose expression is being monitored
iv. allostery		**4.**	stimulation of protein synthesis by a specific molecule
v. operon		**5.**	site to which repressor binds
vi. catabolite repression		**6.**	gene regulation involving premature termination of transcription
vii. reporter gene		**7.**	group of genes transcribed into one mRNA
viii. attenuation		**8.**	negative regulator

Section 10.1

2. The following statement occurs early in this chapter: "The critical step in the regulation of most bacterial genes is the binding of RNA polymerase to DNA at the promoter." Why might it be advantageous for bacteria to regulate the expression of their genes at this step?

3. One of the main lessons of this chapter is that several bacterial genes are often transcribed from a single promoter into a large multigene transcript. The region of DNA containing the set of genes that are co-transcribed, along with all of the regulatory elements that control the expression of these genes, is called an *operon*.

a. Which of the mechanisms in the list below could explain differences in the levels of the mRNAs for different operons?

b. Which of the mechanisms in the list below could explain differences in the levels of the protein products of different genes in the same operon?

i. Different promoters might have different DNA sequences.

ii. Different promoters might be recognized by different types of RNA polymerase.

iii. The secondary structures of mRNAs might differ so as to influence the rate at which they are degraded by ribonucleases.

iv. In an operon, some genes are farther away from the promoter than other genes.

v. The translational initiation sequences at the beginning of different open reading frames in an operon might result in different efficiencies of translation.

vi. Proteins encoded by different genes in an operon might have different stabilities.

4. All mutations that abolish function of the Rho termination protein in *E. coli* are conditional mutations. What does this tell you about the *rho* gene?

Section 10.2

5. The promoter of an operon is the site to which RNA polymerase binds to begin transcription. Some base changes in the promoter result in a mutant site to which RNA polymerase cannot bind. Would you expect mutations in the promoter that prevent binding of RNA polymerase to act in *trans* on another copy of the operon on a plasmid in the cell, or only in *cis* on the copy immediately adjacent to the mutated site?

6. You are studying an operon containing three genes that are co-transcribed in the order *hupF*, *hupH*, and *hupG*. Diagram the mRNA for this operon, showing the location of the 5′ and 3′ ends, all open reading frames, translational start sites, stop codons, transcription termination signals, and any regions that might be in the mRNA but do not serve any of these functions.

7. You have isolated a protein that binds to DNA in the region upstream of the promoter sequence of the *sys* gene. If this protein is a positive regulator, which of the following would be true?

a. Loss-of-function mutations in the gene encoding the DNA-binding protein would cause constitutive expression.

b. Loss-of-function mutations in the gene encoding the DNA-binding protein would result in little or no expression.

8. You have isolated two different mutants (*reg1* and *reg2*) causing constitutive expression of the *emu* operon (*emu1 emu2*). One mutant contains a defect in a DNA-binding site, and the other has a loss-of-function defect in the gene encoding a protein that binds to the site.

a. Is the DNA-binding protein a positive or negative regulator of gene expression?

b. To determine which mutant has a defect in the site and which one has a mutation in the binding protein, you decide to do an analysis using F′ plasmids. Assuming you can assay levels of the Emu1 and Emu2 proteins, what results do you predict for the two strains (*i* and *ii*) (see descriptions below) if *reg2* encodes the regulatory protein and *reg1* is the regulatory site?

i. F′ *reg1⁻reg2⁺emu1⁻emu2⁺/ reg1⁺reg2⁺ emu1⁺emu2⁻*

ii. F′ *reg1⁺reg2⁻emu1⁻emu2⁺/ reg1⁺reg2⁺ emu1⁺emu2⁻*

c. What results do you predict for the two strains (*i* and *ii*) if *reg1* encodes the regulatory protein and *reg2* is the regulatory site?

9. Bacteriophage λ, after infecting a cell, can integrate into the chromosome of the cell if the repressor protein, cI, binds to and shuts down phage transcription immediately. (A strain containing a bacteriophage integrated in the chromosome is called a lysogen.) The alternative fate is the production of many more viruses and lysis of the cell. In a mating, a donor strain that is a lysogen was crossed with a lysogenic recipient cell and no phages were produced. However, when the lysogen donor strain transferred its DNA to a non-lysogenic recipient cell, the recipient cell burst, releasing a new generation of phages. Why did mating with a non-lysogenic cell result in phage growth and release but infection of a lysogenic recipient did not?

10. Mutants were isolated in which the constitutive phenotype of a missense *lacI* mutation was suppressed. That is, the operon was now inducible. These mapped to the operon but were not in the *lacI* gene. What could these mutations be?

11. For each of the *E. coli* strains containing the *lac* operon alleles listed, indicate whether the strain is inducible, constitutive, or unable to express β-galactosidase and permease.

a. $I^+ O^+ Z^- Y^+ / I^+ O^c Z^+ Y^+$

b. $I^+ O^+ Z^+ Y^+ / I^- O^c Z^+ Y^-$

c. $I^+ O^+ Z^- Y^+ / I^- O^c Z^+ Y^-$

d. $I^- P^- O^+ Z^+ Y^- / I^+ P^+ O^c Z^- Y^+$

e. $I^s O^+ Z^+ Y^+ / I^- O^+ Z^+ Y^-$

12. For each of the growth conditions listed, what proteins would be bound to *lac* operon DNA? (Do not include RNA polymerase.)

a. glucose

b. glucose + lactose

c. lactose

13. For each of the following mutant *E. coli* strains, plot a 30-minute time course of concentration of β-galactosidase, permease, and acetylase enzymes grown under the following conditions. For the first 10 minutes, no lactose is present; at 10′ lactose becomes the sole carbon source. Plot concentration on the *y*-axis, time on the *x*-axis. (Do not worry about the exact units for each protein on the *y*-axis.)

a. $I^- P^+ O^+ Z^+ Y^+ A^+ / I^+ P^+ O^+ Z^- Y^+ A^+$

b. $I^- P^+ O^c Z^+ Y^+ A^- / I^+ P^+ O^+ Z^- Y^+ A^+$

c. $I^s P^+ O^+ Z^+ Y^+ A^+ / I^- P^+ O^+ Z^- Y^+ A^+$

d. $I^- P^- O^+ Z^+ Y^+ A^+ / I^- P^+ O^c Z^+ Y^- A^+$

e. $I^- P^+ O^+ Z^- Y^+ A^+ / I^- P^- O^c Z^+ Y^- A^+$

14. Maltose utilization in *E. coli* requires the proteins encoded by genes in three different operons. One operon includes the genes *malE*, *malF*, and *malG*; the second includes *malK* and *lamB*; and the genes in the third operon are *malP* and *malQ*. The MalT protein is a positive regulator

that regulates the expression of all three operons; expression of the *malT* gene itself is catabolite-sensitive.

a. What phenotype would you expect to result from a loss-of-function mutation in the *malT* gene?

b. Do you expect the three maltose operons to contain binding sites for CRP (cAMP receptor protein)? Why or why not?

In order to infect *E. coli*, bacteriophage λ binds to the maltose transport protein LamB (also known as the λ receptor protein) that is found in the outer membrane of the bacterial cell. The synthesis of LamB is induced by maltose in the medium via expression of the MalT protein, as described above.

c. List the culture conditions under which wild-type *E. coli* cells would be sensitive to infection by bacteriophage λ.

d. *E. coli* cells that are resistant to infection by bacteriophage λ have been isolated. List the types of mutations in the maltose regulon that λ-resistant mutants could contain.

15. Clones of three adjacent genes involved in arginine biosynthesis have been isolated from a bacterium. If these three genes together make up an operon, what result do you expect when you use the DNA from each of these genes as probes in a Northern analysis? What result do you expect if the three genes do not make up an operon?

16. Given the following data, explain which strains and growth conditions are important for reaching the following conclusions.

a. Arabinose induces coordinate expression of the *araBAD* genes (encoding kinase, isomerase, and epimerase).

b. The *araC* gene encodes a positive regulator of *araBAD* expression.

Genotype	Arabinose in medium	Kinase	Isomerase	Epimerase
1. $C' B' A' D'$	no	−	−	−
2. $C' B' A' D'$	yes	+	+	+
3. $C^- B' A' D'$	no	−	−	−
4. $C^- B' A' D'$	yes	−	−	−

17. Seven *E. coli* mutants were isolated. The activity of the enzyme β-galactosidase produced by cells containing each mutation alone or in combination with other mutations was measured when the cells were grown in medium supplemented with different carbon sources.

	Glycerol	Lactose	Lactose + Glucose
Wild type	0	1000	10
Mutant 1	0	10	10
Mutant 2	0	10	10
Mutant 3	0	0	0
Mutant 4	0	0	0
Mutant 5	1000	1000	10
Mutant 6	1000	1000	10
Mutant 7	0	1000	10
F′ lac from mutant 1/mutant 3	0	1000	10

	Glycerol	Lactose	Lactose + Glucose
F' lac from mutant 2/mutant 3	0	10	10
Mutants 3 + 7	0	1000	10
Mutants 4 + 7	0	0	0
Mutants 5 + 7	0	1000	10
Mutants 6 + 7	1000	1000	10

Assume that each of the seven mutations is one, and only one, of the genetic lesions in the following list. Identify the type of alteration each mutation represents.

a. super-repressor

b. operator deletion

c. nonsense (amber) suppressor tRNA gene (assume that the suppressor tRNA is 100 percent efficient in suppressing amber mutations)

d. defective CRP–cAMP binding site

e. nonsense (amber) mutation in the β-galactosidase gene

f. nonsense (amber) mutation in the repressor gene

g. defective *crp* gene (encoding CRP)

18. Cells containing mutations in the *crp* gene (encoding the positive regulator CRP) are Lac⁻, Mal⁻, Gal⁻, etc. To find suppressors of the *crp* mutation, cells were screened to find those that were both Lac⁺ and Mal⁺.

a. What types of suppressors would you expect to get using this screen compared with a screen for Lac⁺ only?

b. All suppressors isolated were mutant in the gene for the α subunit of RNA polymerase. What hypothesis could you propose based on this analysis?

19. Six strains of *E. coli* (mutants 1–6) that had one of the following mutations affecting the *lac* operon were isolated.

   i. deletion of *lacY*

   ii. *lacO^c* mutation

   iii. missense mutation in *lacZ*

   iv. inversion of the *lac* operon (but not an inversion of the *lacI* gene)

   v. super-repressor mutation

   vi. inversion of *lacZ, Y*, and *A* but not *lacI, P, O*

a. Which of these mutations would prevent the strain from utilizing lactose?

b. The entire *lac* operon (including the *lacI* gene and its promoter) from each of the six *E. coli* strains was cloned into a plasmid vector containing an ampicillin resistance gene. Each recombinant plasmid was transformed into each of the six strains to create partial diploids. In analysis of these strains, mutant 1 was found to carry a deletion of *lacY*, so this strain corresponds to mutation *i* in the list above. Which of the other types of mutations would be expected to complement mutant 1 in these partial diploids so as to allow lactose utilization?

c. In the analysis described in part *b*, each strain was plated on ampicillin media in which lactose is the only carbon source. (Ampicillin was included to ensure maintenance of the plasmid.) Growth of the transformants is scored below (+ = growth, − = no growth). Syntheses of β-galactosidase and permease are required for growth on this medium. Results of this merodiploid analysis are shown here. Which mutant bacterial strain (1–6) contained each of the alterations (*i–vi*) listed previously?

	1	2	3	4	5	6
1	−	+	−	+	−	+
2	+	−	−	+	−	+
3	−	−	−	+	−	+
4	+	+	+	+	−	+
5	−	−	−	−	−	+
6	+	+	+	+	+	+

20. The following data are from a DNaseI footprinting experiment in which either RNA polymerase or a repressor protein was added to a labelled DNA fragment, and then the complex was digested with DNaseI. DNA sequencing reactions were also performed on the same DNA so the bases that were protected by proteins binding could be identified. (Notice that DNaseI does not cut after each base in the DNA fragment.)

a. What is the sequence of the DNA in this fragment?

b. Mark on the sequence the region where the repressor binds.

c. Mark on the sequence the region where RNA polymerase binds.

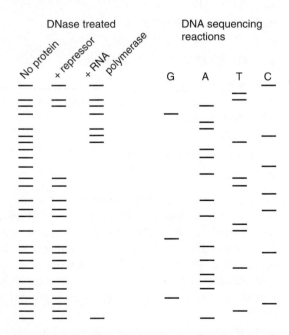

21. a. The original constitutive operator mutations in the *lac* operon were all base changes in O_1. Why do you think mutations in O_2 or O_3 were not isolated in these screens?

b. Explain how a mutagen that causes small insertions could produce an O^c mutation.

c. Would the O^c mutation described above in part *b* be sensitive to LacI^s? Why or why not?

22. In an effort to determine the location of an operator site for a negatively regulated gene, you have made a series

of deletions within the regulatory region. The extent of each deletion is shown by the line underneath the sequence, and the resulting expression from the operon (i = inducible; c = constitutive; $-$ = no expression is also indicated).

```
. . . GGATCTTAGCCGGCTAACATGATAAATATAA . . .
. . . CCTAGAATCGGCCGATTGTACTATTTATATT . . .
1   i   ─────
2   −   ──────────────
3   c   ───────
4   −   ────────
5   c   ─────
```

a. What can you conclude from these data about the location of the operator site?

b. Why do you think deletions 2 and 4 show no expression?

23. An operon fusion consists of a regulatory region cloned next to the coding region of the genes of an operon. A gene fusion consists of a regulatory region of a gene such as *lacZ* and the DNA encoding the first amino acids of the β-galactosidase protein cloned next to the coding region of another gene. What additional feature do you have to consider to create a functional gene fusion that is not necessary for an operon fusion?

Section 10.3

24. a. How many ribosomes are required (at a minimum) for the translation of *trpE* and *trpC* from a single transcript of the *trp* operon?

b. How would you expect deletion of the two tryptophan codons in the RNA leader to affect expression of the *trpE* and *trpC* genes?

25. The following is a sequence of the leader region of the *his* operon mRNA in *Salmonella typhimurium*. What bases in this sequence could cause a ribosome to pause when histidine is limiting (i.e., when there is very little of it) in the medium?

5′ AUGACACGCGUUCAAUUUAAACACCACCAUCAUCACCAUCA
UCCUGACUAGUCUUUCAGGC 3′

26. For each of the *E. coli* strains that follow, indicate the effect of the genotype on expression of the *trpE* and *trpC* genes in the presence and absence of tryptophan. (In the wild type [R^+ P^+ O^+ att^+ $trpE^+$ $trpC^+$], *trpC* and *trpE* are fully repressed in the presence of tryptophan and are fully induced in the absence of tryptophan.)

R = repressor gene; R^n product cannot bind tryptophan; R^- product cannot bind operator

O = operator; O^- cannot bind repressor

att = attenuator; att^- is a deletion of the attenuator

P = promoter; P^- is a deletion of the *trp* operon promoter

$trpE^-$ and $trpC^-$ are null (loss-of-function) mutations

a. R^+ P^- O^+ att^+ $trpE^+$ $trpC^+$

b. R^- P^+ O^l att^l $trpE^+$ $trpC^+$

c. R^n P^+O^+ att^+ $trpE^+$ $trpC^+$

d. R^- P^l O^l att^- $trpE^+$ $trpC^+$

e. R^+ P^+ O^- att^+ $trpE^+$ $trpC^-$ / R^- P^+ O^+ att^+ $trpE^-$ $trpC^+$

f. R^+ P^- O^+ att^+ $trpE^+$ $trpC^-$ / R^- P^+ O^+ att^+ $trpE^-$ $trpC^+$

g. R^+ P^+ O^- att^- $trpE^+$ $trpC^-$ / R^- P^+ O^- att^+ $trpE^-$ $trpC^+$

27. A molecular geneticist is investigating an operon by measuring the amount of expression of the four structural genes (A, B, C, and D) produced in wild-type and mutant bacterial cells after the addition of compound Z to a minimal medium. An additional protein (E) is of very small size (less than 20 amino acids) and cannot be measured by the same analytical system employed for the other proteins. Several of the mutations are nonsense mutations that have an effect on the genes transcribed after them in the operon. In addition to stopping translation of the gene in which the mutations lie, these so-called *nonsense polar mutations* prevent the expression of genes downstream of the mutation. (For example, in the *lac* operon, some *lacZ* nonsense mutations can result in no expression of *lacY* and *lacA*.) The investigator has also obtained mutations in two other sites, *F* and *G*, closely linked to *A–D*. The graphs shown are all semilogarithmic. The percentage of maximal possible expression for a particular protein is plotted on the y-axis, while the x-axis coordinate is time. Compound Z is added at the point specified by the *arrow*.

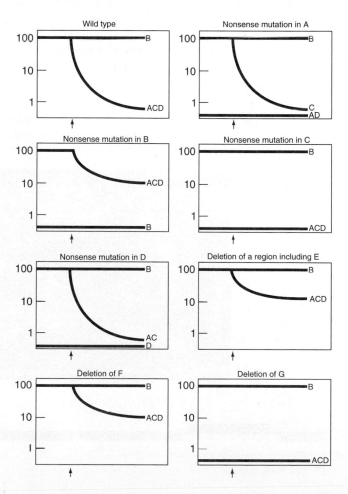

a. Is this operon likely to be involved in a pathway of biosynthesis or a pathway of degradation? Is the operon inducible or repressible?

b. For each of the conditions graphed, state if genes *A–D* are constitutive, completely repressible, partly repressible, or not expressed.

c. Construct a map of this operon. Indicate the relative positions of genes *A, B, C, D,* and *E* as well as the sites *F* and *G*. List possible functions for all the genes and sites. (Possible functions for the genes and sites *A–G*: promoter, operator, enzyme structural gene, CRP-binding site, *crp* gene, attenuator region. This list is not necessarily all inclusive.)

28. The previous problem (#27) introduced the concept of polar mutations in bacterial operons: Nonsense mutations in a "proximal" gene nearer the promoter of the operon can abolish the expression of a "distal" gene in the same operon that is farther from the promoter. Essentially all polar mutations are nonsense mutations; missense mutations do not have this property.

a. Suggest a model to explain why nonsense but not missense mutations might exhibit polarity.

b. Interestingly, in strains that simultaneously carry a polar mutation and a nonsense suppressor mutation in a tRNA gene, the expression both of the gene with the nonsense mutation and of the distal genes in the operon can be restored. However, in strains with both a polar mutation and a loss-of-function mutation in the gene encoding the Rho transcription termination factor, expression of the distal genes can be restored but that of the gene with the nonsense mutation cannot. How might these results influence your model for the underlying cause of polarity?

Section 10.4

29. Many genes whose expression is turned on by DNA damage have been isolated. Loss-of-function mutations in the *lexA* gene leads to expression of many of these genes, even when there has been no DNA damage. Would you hypothesize that LexA protein is a positive or negative regulator? Why?

30. In 2005, Frederick Blattner and his colleagues found that *E. coli* have a global transcriptional program that helps them "forage" for better sources of carbon. Many genes, including genes needed for bacterial motility, are turned on in response to poorer carbon sources so that the bacteria can search for better nutrition. You now want to search for genes that regulate this response. How could you use *lacZ* fusions to try to identify such regulatory genes?

31. To find genes that are turned on or off in response to changes in osmolarity (the total concentration of solutes in solution), you grow a culture of *E. coli* in a medium with high osmolarity and another culture in a medium with low osmolarity. You now want to perform a DNA microarray analysis.

a. What would you use as your probe(s) for the DNA array analysis?

b. What nucleic acids would you spot on the DNA array? How many spots should the DNA array contain?

c. It is possible that osmotic changes may induce a general stress response that may be seen with other stresses as well (e.g., heat shock). How could you distinguish the genes that might be involved in a general stress response from those that are specific for the osmolarity change?

32. Figure A below shows the results of a recent microarray analysis measuring the relative abundance of mRNAs for all of the genes of bacteriophage T4 as a function of time after the infection of *E. coli* cells. The genes can be subdivided into three main classes: early genes (*blue*) transcribed almost immediately after infection, middle genes (*green*) transcribed somewhat later, and late genes (*red*) transcribed later still. Figure B depicts a 10-kb region of

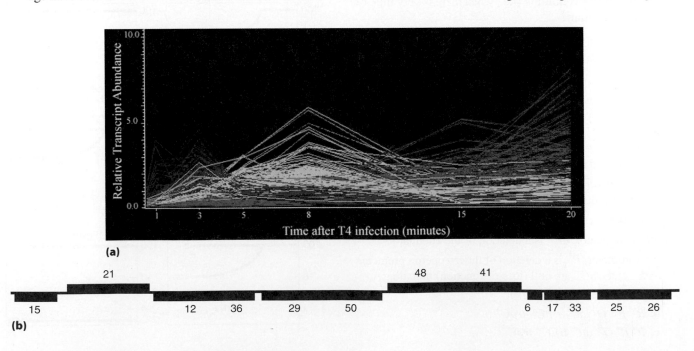

(a)

(b)

the 170-kb T4 genome showing the extent of several genes, each indicated by a number and classified by the same colour scheme. *Boxes* above the *black line* indicate genes transcribed from left to right, while those below the line indicate genes transcribed in the opposite direction.

a. What is the minimal number of promoters in the 10-kb region depicted in Figure B? Which genes could be transcribed as part of the same operon(s)?

b. The *e* gene of bacteriophage T4 encodes an enzyme called endolysin, which helps lyse the *E. coli* host cell to release progeny bacteriophage particles. Would you expect *e* to be an early, middle, or late class gene? Explain your reasoning.

33. Several T4 genes participate in regulating the bacteriophage T4 life cycle described in the preceding problem (#32) and in Figures A and B above. The product of the *motA* gene is a protein that binds to DNA near the promoters for middle genes, enabling the *E. coli* RNA polymerase core enzyme to recognize these promoters.

The gene *asiA* encodes an "anti-σ factor" that associates with *E. coli* σ^{70} and disrupts its function. The protein encoded by *regA* is a ribonuclease that specifically destroys early mRNAs. T4 gene *55*'s product is a σ factor required for recognition of late promoters by the *E. coli* RNA polymerase core enzyme.

a. Of the genes described above (*motA*, *asiA*, *regA*, and *55*), which are likely to be early, middle, or late?

b. What class of T4 genes (early, middle, late) has promoters recognized by the *E. coli* σ^{70} RNA polymerase holoenzyme?

c. What happens to the transcription of genes in the host *E. coli* chromosome as T4 infection progresses?

d. Predict the results of loss-of-function mutations in the *motA*, *asiA*, and *55* genes on the transcription of early, middle, and late mRNAs as well as the mRNAs for host *E. coli* genes.

e. What aspect of Figure A is explained by the function of the RegA ribonuclease?

Mc Graw Hill Education **connect** For more information on the resources available from McGraw-Hill Ryerson, go to www.mcgrawhill.ca/he/solutions.

These inbred female mice are genetically identical. Both have the same genotype at the *agouti* locus (A^{vy}/a); their phenotypic dissimilarities are due to epigenetic differences. The mouse on the *left* is obese, diabetic, prone to cancer development, and has yellow fur and a shorter life expectancy. The mouse on the *right* exhibits the normal wild-type agouti coat colour phenotype, is thinner, healthier, and lives longer than the mouse on the *left*. This variation in phenotypic characteristics between the two mice is the result of differences in methylation at the A^{vy} locus.

When a *Drosophila* male courts a *Drosophila* female, he sings a species-specific song and dances an ancient dance. If successful, his instinctive behaviours culminate in mating. The male senses a female's presence by visual and tactile cues as well as by the pheromones she produces. After orienting himself at a precise angle with respect to his prospective mate, he taps his partner's abdomen with his foreleg and then performs his song by stretching out his wings and vibrating them at a set frequency; when the song is over, he begins to follow the female. If she is unreceptive (perhaps because she has recently mated with another male), she will run away, but if she is receptive, she will let him overtake her. When he does, he licks her genitals with his proboscis, curls his abdomen, mounts the female, and copulates with her for about 20 minutes.

Various mutations in a gene called *fruitless* produce behavioural changes that prevent the male from mating properly. Some mutant alleles alter the song to an unfamiliar melody. Others diminish the male's ability to distinguish females from males; male flies with this mutation court each other. Still others reduce the male's ability to court either sex. Finally, some mutations create lethal null alleles that cause male flies to die just before they emerge from the pupal case, showing that the *fruitless* gene has other functions in addition to its effects on courtship and mating.

Cloned in 1996, *fruitless* is a large gene—roughly 150 kb in length—encoding a product that can regulate transcription at multiple promoters. Some forms of the fruitless protein are sex-specific, appearing only in males or only in females. Other fruitless proteins are expressed at low levels in many kinds of cells in both sexes.

The male version of the fruitless protein, although very similar to the female protein, has an extra 101 amino acids at its N terminus, and this addition almost certainly determines the observed differences in male and female behaviour. Remarkably, the male-specific fruitless mRNA is synthesized in only a few hundred of the tens of thousands of neurons that make up the male *Drosophila*'s nervous system. Most of these *fruitless*-expressing cells are located near motor neurons that control either wing movements (and thus possibly the song) or abdominal movements (and thus possibly the abdominal curling that immediately precedes mating). Work on the *Drosophila fruitless* gene has provided strong evidence that differences in gene expression, and not just differences in alleles, can directly influence complex behaviours.

In this chapter, we see that **eukaryotic gene regulation**—the control of gene expression in the cells of eukaryotes—depends on an array of interacting regulatory elements that turn genes on and off in the right places at the right times. During the embryonic development of multicellular eukaryotic organisms such as *Drosophila* or humans, gene regulation controls not only the elaboration of sex-related characteristics and behaviours but also the

Chapter Outline
11.1 Overview of Eukaryotic Gene Regulation
11.2 Control of Transcription Initiation
11.3 Chromatin Structure and Epigenetic Effects
11.4 Regulation After Transcription
11.5 A Comprehensive Example: Sex Determination in *Drosophila*

differentiation of tissues and organs, as well as the precise positioning of these tissues and organs. Some of the regulatory elements are specific DNA sequences in the vicinity of the gene to be regulated; others are DNA-binding proteins encoded by genes located elsewhere in the genome; and still others are microRNA molecules (miRNAs) that use the specificity of base-pairing to down-regulate, or curtail, specific gene expression after transcription.

In contrast to the theme of environmental adaptation found in the unicellular prokaryotes, the theme in multicellular eukaryotes appears to be maintenance of homeostasis of the organism. But even in unicellular eukaryotes, the mechanisms of regulation are different from those of prokaryotes. In eukaryotic gene regulation, we see a larger, more complex set of interactions than is found in prokaryotes, although some basic principles are shared between them. In addition, since many of the biological functions in multicellular organisms arise from the regulated interactions of large networks of genes, and each gene in a network has multiple potential points of regulation, the possibilities for regulatory refinement are enormous.

Learning Objectives

1. Distinguish between the functions and chromosomal positions of promoters and enhancers.

2. Examine the roles of *trans*-acting transcription factors in the regulation of eukaryotic gene expression.

3. Diagram and explain the role of activators and repressors in the modulation of gene expression.

4. Compare and contrast the epigenetic mechanisms of chromatin modification that result in gene silencing or activation.

5. Analyze the causes and consequences of genomic imprinting.

6. Explain how RNA splicing, RNA interference, translation efficiency, and chemical modification of the gene product can alter gene function.

11.1 Overview of Eukaryotic Gene Regulation

As you explore the intricacies of eukaryotic gene regulation, bear in mind the key similarities and differences between eukaryotes and prokaryotes (**Table 11.1**). In both types of cells, transcriptional regulation can occur through the attachment of DNA-binding proteins to specific DNA sequences that are in the vicinity of the transcription unit itself; and several polypeptide motifs appear in many DNA-binding proteins in prokaryotes and eukaryotes. However, additional levels of complexity are both necessary and possible for controlling expression in eukaryotes for several reasons.

- Eukaryotic genomes contain far more DNA than do those of prokaryotes, making it challenging for proteins to locate binding sequences.
- Chromatin structure makes DNA unavailable to transcription machinery.
- Additional RNA processing events occur.
- Transcription occurs in the nucleus, but translation takes place in the cytoplasm.

In addition to these basic differences between eukaryotes and prokaryotes, multicellular eukaryotes must be able

Table 11.1	Key Regulatory Differences Between Eukaryotes and Prokaryotes		
Characteristic		**Prokaryote**	**Eukaryote**
Control of transcription through specific DNA-binding proteins		Yes	Yes
Re-utilization of same DNA-binding motifs by different DNA binding proteins		Yes	Yes
Activator proteins		Yes	Yes
Repressor proteins		Yes	Yes
Specificity of binding to DNA by regulatory protein		Specific	Highly specific
Affinity of binding		Strong	Very strong
Role played by chromatin structure		No	Yes
Coordinate control achieved with operons		Yes	Rare
Differential splicing		No	Yes
Attenuation		Yes	No
mRNA processing		No	Yes
Differential polyadenylation		No	Yes
Differential transport of RNA from nucleus to cytoplasm		No	Yes

to use gene regulation to control cellular differentiation and the complex interactions of various types of differentiated cells within tissues and organs. Molecular biologists have traditionally assumed that eukaryotic gene expression was regulated predominantly at the point of transcriptional initiation. However, there are many more steps in the process leading to an active product beyond transcription initiation. Recall that gene expression is defined by the production of an active gene product (**Figure 11.1**). Transcript processing (including splicing), export of mRNA from the nucleus, translatability of the message, localization of the protein product in specific organelles in the cell, and modifications to the protein are all activities that can be regulated and that affect the amount of final active product.

With the discovery of alternative splicing in the late 1970s, followed by the progressive appreciation that post-transcriptional mechanisms are powerful means for gene activity control, the view that transcription was the primary process regulating gene expression radically changed. As a result, modern models for gene activity control visualize a complex and highly dynamic gene expression network within which the different biochemical machines responsible for transcription, splicing, and other molecular processes share some components and are therefore tightly coupled. In spite of all this, it is still true that important decisions concerning the amount of gene product in the cell are made during the initiation of transcription, when RNA polymerase starts to make a primary transcript, or RNA copy, of a gene's template strand.

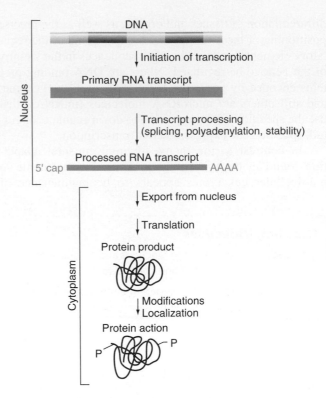

Figure 11.1 Gene expression in eukaryotes. Gene expression involves transcription and mRNA processing in the nucleus, then translation and modifications in the cytoplasm to produce an active protein.

11.2 Control of Transcription Initiation

Three types of RNA polymerases transcribe genes in eukaryotes. RNA polymerase I (pol I) transcribes genes that encode the major RNA components of ribosomes (rRNAs). RNA polymerase II (pol II) transcribes genes that encode all proteins and microRNAs. RNA polymerase III (pol III) transcribes genes that encode the tRNAs as well as certain other, small RNA molecules. We focus on the major transcription activity that produces proteins: pol II transcription.

RNA polymerase II transcribes all protein-encoding genes

During transcription in eukaryotes, RNA polymerase II catalyzes the synthesis of a single-stranded RNA molecule—known as the primary transcript—that is complementary in base sequence to a gene's DNA template strand.

Most of the primary transcripts produced by pol II undergo further processing to generate mRNAs (see Figures 7.5, 7.6, and 7.8). During mRNA formation, introns are spliced out. In addition, ribonuclease cleaves pol II-transcribed primary transcripts to form a new 3′ end, to which the enzyme poly-A polymerase adds a poly-A

tail; and the chemical modification of the 5′ end of the transcript produces a "5′ GTP cap," which protects the molecule from degradation.

Measurements of mRNA and protein levels in eukaryotic cells have revealed that, usually, the more mRNA of a gene that accumulates in the cell, the greater the production of that gene's protein product; however, in many cases this correlation is not perfect because regulatory mechanisms affect RNA processing and translation.

The *cis*-acting regulatory regions: Promoters and enhancers

Although each of the regulatory regions of the thousands of pol II-transcribed genes in a eukaryotic genome is unique, they all contain two kinds of essential DNA sequences. The **promoter** is always very close to the gene's protein-coding region. It includes an initiation site, where transcription begins. The initiation site is most often a "TATA" box, consisting of roughly seven nucleotides of the sequence T–A–T–A–(A or T)–A–(A or T), but it can also be an "initiator" box that is located downstream of the initiation site. Binding of RNA polymerase to the TATA or initiation

box allows a basal level of transcription (**Figure 11.2**). **Enhancers** are regulatory sites that can be quite distant—up to tens of thousands of nucleotides away—from the promoter. Binding of proteins to enhancers augments or represses basal levels of transcription.

Identifying promoters and enhancers

The sequences that make up the promoter and enhancer sites for specific genes are identified using reporter gene fusions, as described in Chapter 10. Recall that reporter constructs are DNA molecules synthesized in the laboratory to contain a gene's postulated regulatory region, but with a "reporter" coding region inserted in place of the gene's own coding region (**Figure 11.3a**). Investigators can systematically identify promoters and enhancers by altering reporter constructs through *in vitro* mutagenesis across a presumed regulatory region and then reintroducing the reporter constructs into the genome by transformation. Cells transformed with the reporter construct "report" the presence or absence of regulatory elements.

In assembling a reporter construct for this purpose, scientists replace the coding region of the gene whose regulation they are studying with the coding region of an easily identifiable product (the "reporter"), such as β-galactosidase or green fluorescent protein (GFP). (Recall that β-galactosidase produces a blue colour in the presence of a substrate known as X-Gal [see Figure 14.7]; similarly, GFP fluoresces green when exposed to light of a particular wavelength.) Reporter constructs are particularly valuable for looking at mutations that affect gene expression rather than the amino acid composition of the gene's polypeptide product. Mutations that alter the amount of reporter synthesized help define the elements necessary for a gene's regulation (**Figure 11.3b**).

In contrast to prokaryotes, eukaryotes have three RNA polymerases. All transcription of genes that yield proteins is performed by RNA polymerase II (pol II). The *cis*-acting regulatory regions include promoters, which are found near the coding region and contain an initiation site, and enhancers, which may be distant and act to increase or repress the basal level of transcription.

---

cis-acting elements

Figure 11.2 *cis*-acting elements. *cis*-acting regulatory elements are regions of DNA sequence that lie nearby on the same DNA molecule as the gene they control. Promoter elements typically lie directly adjacent to the gene that they control. Enhancers that regulate expression can sometimes lie thousands of base pairs away from a gene.

The *trans*-acting proteins control transcription initiation

The binding of proteins to a gene's promoter and enhancer (or enhancers) controls the rates of transcriptional initiation. Different types of proteins bind to each of the *cis*-acting regulatory regions: *Basal factors* bind to the promoter; *activators* and *repressors* bind to the enhancers (**Figure 11.4**). Additional regulatory proteins may interact with these regulators bound to sites on the DNA. The proteins that regulate transcription initiation are collectively known as transcription factors.

Just as with *cis*-acting regions, *trans*-acting elements can be identified using reporter gene fusions (**Figure 11.5**). Mutations that alter the level of expression of a reporter,

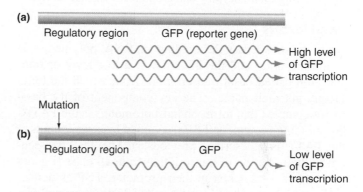

Figure 11.3 Identifying *cis*-acting sites. (a) A fusion between a gene's regulatory region and the GFP gene provides an easy way to monitor levels of transcription. **(b)** Base changes (mutations) that reduce transcription (and therefore the level of green fluorescent protein) identify regulatory sites.

trans-acting gene products

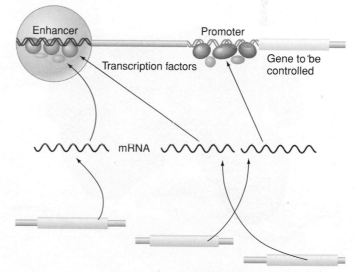

trans-acting genetic elements

Figure 11.4 *trans*-acting factors. *trans*-acting genetic elements encode products called transcription factors that interact with *cis*-acting elements, either directly through DNA binding or indirectly through protein-protein interactions.

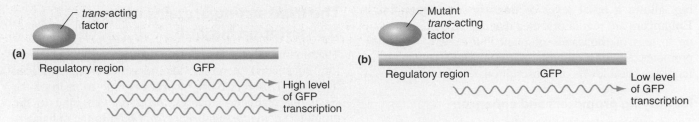

Figure 11.5 Identifying *trans*-acting factors. (a) *trans*-acting factors bind to regulatory regions (enhancers) to increase transcription. **(b)** A *trans*-acting mutation that reduces transcription identifies a regulatory protein.

and that map far from the target gene or reporter construct, are likely to reside in *trans*-acting elements. Biochemical procedures can be used to isolate proteins that bind *in vitro* to *cis*-acting DNA sequences. Once researchers identify a *trans*-acting element, they can clone it for further study.

Basal factors

Basal factors assist the binding of RNA polymerase II to the promoter and the initiation of a low level of transcription called basal transcription (from which the basal factors get their name). The key component of the basal factor complex that forms on most promoters is the TATA-box-binding protein, or **TBP** (so named because it binds to the TATA box described previously). The TBP is essential to the initiation of transcription from all class II genes that have a TATA box in their promoter. TBP associates with several other basal factors called TBP-associated factors, or **TAFs** (**Figure 11.6**). The complex of basal factors binds to the proximal promoter in an ordered pathway of assembly. Once the complex has formed, basal transcription is initiated. Researchers have determined the structure of the TBP–TAF complex on the DNA at the TATA box and find there is a sharp bend in the DNA at the TATA box, induced by TBP.

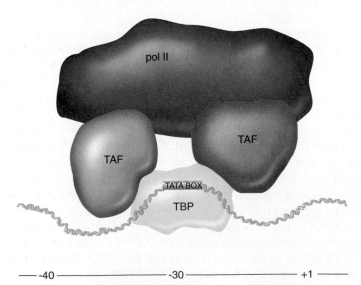

— -40 ——————— -30 ——————— +1 —

Figure 11.6 Basal factors bind to promoters of all protein-encoding genes. Schematic representation of the binding of the TATA-box-binding protein (TBP) to the promoter DNA, the binding of two TBP-associated factors (TAFs) to TBP, and the binding of RNA polymerase (pol II) to these basal factors.

The primary sequence and three-dimensional structure of the basal factors are highly conserved in all eukaryotes, from yeast to humans. This level of evolutionary conservation in sequence and structure underlies a high level of functional conservation across the eukaryotes, which has in turn facilitated the biochemical purification of some of these factors. For example, researchers isolated yeast TBP—the first basal factor to be purified—through its ability to substitute for mammalian TBP *in vitro*.

Activators

Although similar sets of basal factors bind to all the promoters of the tens of thousands of genes in the eukaryotic genome, a cell can transcribe different genes into widely varying amounts of mRNA. This enormous range of transcriptional regulation occurs through the binding of different transcription factors to enhancer elements associated with different genes. When regulatory transcription factors bind to an enhancer element, they can interact directly or indirectly with basal factors at the promoter in a three-dimensional protein/DNA complex to cause an increase in transcriptional activity (**Figure 11.7**). Due to their ability to increase transcriptional activity, these factors are called **activators**. Researchers have already identified hundreds of eukaryotic activators, and it is likely that each eukaryotic genome encodes several thousand of them. At the mechanistic level, transcriptional activator proteins bound at their target sites on DNA can increase RNA synthesis via three different, but not mutually exclusive, systems:

1. They could stimulate the "recruitment" of the basic transcription machinery (such as the RNA polymerase itself and/or some of its associated factors) to core promoter sequences by directly interacting with the components of this machinery.
2. They could stimulate the activity of the basal factors already bound to the promoter.
3. Activators could facilitate the changes in chromatin structure that allow higher transcription levels.

Domains Within Activators. To carry out their function on selected subsets of genes (and not to all genes), transcriptional activator proteins must (a) bind to enhancer DNA in a sequence-specific way, and after binding, they must (b) be able to interact with other proteins to activate transcription. Two structural domains within the activator protein—the DNA-binding domain and the transcription-activator domain—mediate these two biochemical functions.

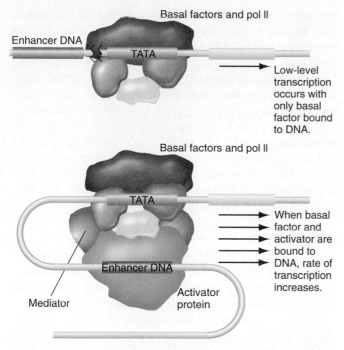

Figure 11.7 **Binding to enhancers increases transcriptional levels.** In the presence of basal factors alone bound to the promoter, low levels of transcription occur. The binding of activator proteins to an enhancer element leads to an increase in transcription beyond the basal level.

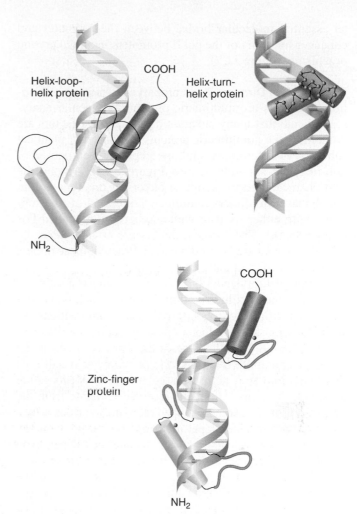

Figure 11.8 **Activator protein domains.** Common motifs found in activator proteins include the helix-loop-helix, helix-turn-helix, and zinc-finger.

A rather small number of protein motifs appear over and over again in the DNA-binding domains of many different activator proteins (**Figure 11.8**). The best characterized of these motifs are the helix-loop-helix and the helix-turn-helix conformations, which are also found in prokaryotic regulators, and the zinc-finger motif, found mostly in eukaryotes. The general function of each of these motifs is to promote binding to the DNA double helix. The proteins fit within or interact within the major groove of DNA. Subtle differences in amino acid sequence among activators can specify high-affinity binding to different DNA sequences associated with different enhancer elements.

Some activators have a third domain that is responsive to specific signals from the environment. An example of activators with this type of domain is the steroid hormone receptors (**Figure 11.9**). Each receptor has a domain that is unique for a particular steroid. The binding of this steroid causes an allosteric change that greatly increases the affinity of the DNA-binding domain of the protein for its target enhancer sequence. Once bound, the hormone–receptor complex activates transcription of its target genes. In the absence of hormone, DNA binding does not occur, and target genes remain unactivated—that is, transcribed only at basal rates. A steroid hormone gene regulation system allows one organ in the human body (a hormone-producing gland) to control gene activity in other organs. There are no universal features of signal response domains.

Proteins and other molecules that play a role in transcriptional activation without binding directly to DNA

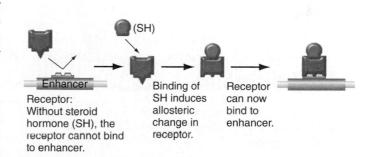

Figure 11.9 **Steroid hormone receptors.** Some activator domains are themselves activated into a DNA-binding conformation through allosteric changes caused by the binding of a steroid hormone molecule to another domain within the activator protein.

are called **coactivators.** The hormone component of a DNA-bound hormone–receptor activation complex is one example of a coactivator. One of the most important protein coactivators is a large multiprotein complex composed of 25–30 proteins called Mediator. The Mediator is considered a central link within the enhancer–pol II promoter pathway. Mediator does not bind to DNA but serves as

an essential molecular bridge between the promoter and enhancer for many of the pol II promoters in all eukaryotic organisms (Figure 11.7).

Formation of Dimers. As in prokaryotes, many transcription regulators are multimeric proteins. Molecular analyses indicate that many eukaryotic transcription factors are homomers (i.e., multimeric proteins composed of identical subunits) or heteromers (multimeric proteins composed of non-identical subunits; review Figure 8.24). Among the best-characterized transcription factors of this type is Jun, which can form dimers (multimers composed of two subunits) with either itself or with another protein called Fos (**Figure 11.10**). The Jun–Jun dimers are *homodimers*; the Jun–Fos dimers are *heterodimers*; and each of these dimers recognizes different enhancer sequences.

Dimerization occurs through yet another transcription factor domain, the **dimerization domain**, which is specialized for specific polypeptide-to-polypeptide interactions. As with other transcription factor domains, certain motifs recur in dimerization domains. One of the most common is the leucine zipper motif (**Figure 11.11**), an amino acid sequence that twirls into an α helix with leucine residues protruding at regular intervals. The motif received its name from the propensity of one leucine zipper motif to interlock like a zipper with a leucine zipper motif on another polypeptide. The ability of two leucine zippers to interlock depends on the specific amino acids that lie between the leucines.

The Jun and Fos polypeptides both contain leucine zippers in their dimerization domains. A Jun leucine zipper can interact with another Jun leucine zipper or with a Fos leucine zipper. But the Fos leucine zipper *cannot* interact with its own kind to form a homodimer. Neither Jun nor Fos alone can bind DNA, so neither can act as a transcription factor as a monomer. Thus, the Jun–Fos transcription factor system can produce only two types of transcription factors: Jun–Jun proteins or Jun–Fos proteins. Both bind to the same enhancer elements, but with different affinities.

The ability to form heterodimers greatly increases the number of potential regulatory transcription complexes a cell can assemble from a set number of gene products. In theory, 100 polypeptides could combine in different ways to form 5000 different transcription factors; with 500 polypeptides, the number jumps to 125 000.

Repressors

Some transcription factors suppress the activation of transcription caused by activator proteins. Any transcription factor that has this effect is considered a **repressor**. Different repressors act in different ways. Some compete with activator proteins for binding to the same enhancer (**Figure 11.12a**). When a repressor binds to an enhancer, it blocks the activator's access to the same sequence.

(a) Competition for binding between repressor and activator proteins

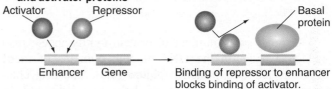

Binding of repressor to enhancer blocks binding of activator.

(b) Quenching

Type I: Repressor binds to and blocks the DNA-binding region of an activator.

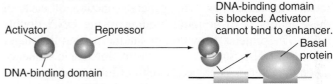

DNA-binding domain is blocked. Activator cannot bind to enhancer.

Type II: Repressor binds to and blocks the activation domain of an activator.

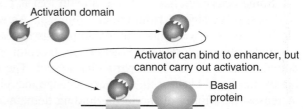

Activator can bind to enhancer, but cannot carry out activation.

Figure 11.12 Repressor proteins act through competition or quenching. (a) Some repressor proteins act by *competing* for the same enhancer elements as activator proteins. But repressor proteins have no activation domain, so when they bind to enhancers, no activation of transcription can occur. **(b)** A second class of repressors act by binding directly to the activator proteins themselves to *quench* activation in one of two ways. Type I quenching is achieved when the repressor prevents the activator from reaching the enhancer. Type II quenching is achieved when the activator can bind to the enhancer, but the repressor prevents the activation domain from binding to basal proteins.

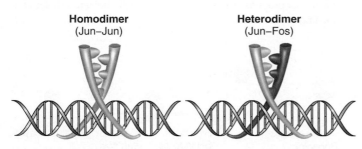

Figure 11.10 Jun–Jun and Jun–Fos dimers. Homodimers contain two identical polypeptides whereas heterodimers contain two different polypeptides.

Leucine zipper

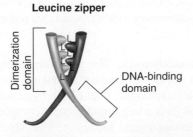

Figure 11.11 Leucine zipper. A common peptide motif present within dimerization domains is the leucine zipper.

The Myc–Max system described in the following section provides an example of this type of activator-repressor competition.

Some repressors operate without binding DNA at all. Instead, in a mechanism called *quenching*, they bind directly to a specific activator (**Figure 11.12b**). In one type of quenching, a repressor binds to and blocks the DNA-binding region of an activator, thereby preventing the activator from attaching to its enhancer. In another type of quenching, a repressor binds to and blocks the activation domain of an activator. These blocked activators still bind to their enhancers, but once bound, they are unable to carry out activation. Quenching polypeptides that operate in this manner are termed **corepressors**. Just like coactivators, corepressors associate indirectly with enhancers through their interaction with DNA-binding proteins.

The repression resulting from both activator-repressor competition and quenching reduces activation, but it has no effect on basal transcription. As in prokaryotes, however, some eukaryotic repressors act directly on the promoter to eliminate almost all transcriptional activity. They can do this by binding to DNA sequences very close to the promoter and thereby blocking RNA polymerase's access to the promoter. Or they can bind to DNA sequences farther from the promoter and then reach over and contact the basal factor complex at the promoter, causing the DNA between the enhancer and promoter to loop out and allow contact between the repressor and the basal factor complex. This second mechanism also denies RNA polymerase access to the promoter and reduces transcription below the basal level.

Whether a transcription factor acts as an activator or a repressor, or has no effect at all, depends not only on the cell type in which it is expressed but also on the gene it is regulating. This is one reason why all *cis*-acting elements bound by either activators or repressors are referred to as enhancers, even though some may actually repress transcription when associated with the appropriate protein.

The specificity of transcription factors can be altered by other molecules in the cell. One example of this phenomenon is observed with the yeast α2 repressor, which helps determine the mating type of *a* cells. Yeast cells can be either haploid or diploid, and haploid cells come in two mating types: α and *a*. In *a* cells, the α2 repressor binds to enhancers that control the activity of a set of *a*-determining genes, whose expression would make the cell type *a*. The binding of the α2 repressor to these *a*-determining genes is one step in the generation of α cells (**Figure 11.13a**). In diploid yeast cells, however, the same α2 repressor plays an additional role. In such cells, expression of the polypeptide known as a1 occurs; the binding of a1 to the α2 repressor alters the repressor's DNA-binding specificity such that α2/a1 now binds to enhancers associated with a set of haploid-specific genes, repressing the expression of those genes (in diploid cells, α2 alone still represses the *a* genes) (**Figure 11.13b**). To summarize, in diploid cells, the α2 repressor maintains the diploid state by repressing haploid-specific genes.

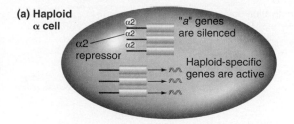

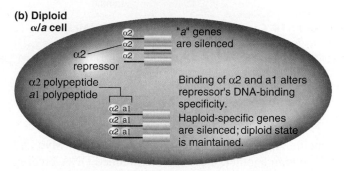

Figure 11.13 The same transcription factors can play different roles in different cells. (a) In haploid α yeast cells, the α2 factor acts to silence the set of "*a*" genes. **(b)** In α/*a* diploid yeast cells, the α2 factor dimerizes with the a1 factor and acts to silence the set of haploid-specific genes.

The *trans*-acting proteins include basal factors that bind to promoters, and activators and repressors that bind to enhancers. Basal factors are responsible for a basal transcription level. Activators have a number of binding domains, and binding at one or more of these domains can change its conformation and thus its affinity to enhancers. Repressors suppress the action of activators by competing for binding at the enhancer or by binding directly with the activator.

The Myc–Max mechanism can activate or repress transcription

The Myc–Max transcription factor system is one in which dimer structure and concentrations of subunits determine whether transcription is activated or repressed. Through the identification of mutations affecting *myc* gene expression in one class of lymphocytes, researchers showed that *myc* plays a critical role in the regulation of cell proliferation. This class of lymphocytes is responsible for Burkitt's lymphoma, a form of cancer. Genetic data suggested that the Myc protein is a transcription factor, but biochemists could find no evidence for this function *in vitro*. Their experiments revealed that even though the Myc polypeptide contains both a helix-loop-helix (HLH) motif and a leucine zipper, it cannot bind to DNA or form homodimers. The apparent contradiction between genetic and biochemical results stymied the scientists who first associated mutations in the *myc* gene with Burkitt's lymphoma and other forms of cancer.

The discovery of the *max* gene product helped resolve this dilemma. Like Myc, Max contains an HLH motif and

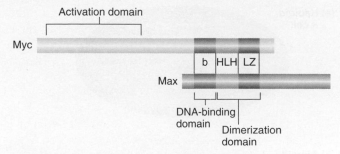

Figure 11.14 Comparative structures of Myc and Max. Linear illustration of the Myc and Max polypeptides and the locations of different domains. The Myc polypeptide has an activation domain, whereas the Max polypeptide does not. Both polypeptides have a DNA-binding domain with a basic amino acid motif and a dimerization domain with adjacent helix-loop-helix (HLH) and leucine zipper (LZ) motifs.

a leucine zipper (**Figure 11.14**). Moreover, both Myc and Max contain another, more recently defined DNA-binding motif called a "basic motif" (because it contains mostly basic amino acids). Unlike Myc, however, the Max polypeptide can form homodimers. When one mixes Max with Myc, heterodimers of the two polypeptides form.

The Myc polypeptide contains an activation domain, but when the molecule is on its own, it cannot bind DNA and thus cannot serve as an activator. The Max polypeptide, on the other hand, can form homodimers and can bind DNA when present without Myc—but Max has no activation domain, so it cannot function as an activator even when it does bind to DNA (**Figure 11.15a**). Only when Myc and Max come together in a heterodimer with both DNA binding and Myc-directed activation does it become possible, and a transcriptional activator is born. Myc–Max heterodimers and Max–Max homodimers both bind to the same enhancer sequences associated with multiple genes that contribute to cell proliferation (**Figure 11.15b**). The binding of a heterodimer results in transcriptional activation, whereas the binding of a homodimer results in transcriptional repression.

One final characteristic of this system is that Myc polypeptides have a much higher affinity for Max polypeptides than Max has for itself. Thus, when Myc and Max are in solution together, the predominant dimer is the heterodimer. With this extensive background, we are ready to see how the cell uses the Myc–Max system to respond rapidly to signals that tell it to proliferate or stop proliferating.

The *max* gene is expressed in all cells at all times, but because its protein product does not carry an activation domain, Max–Max homodimers, when bound to enhancer DNA, inhibit transcription and therefore inhibit cell proliferation. By contrast, the *myc* gene is not universally expressed; the Myc polypeptide is normally synthesized in cells undergoing proliferation but not in cells at rest.

As soon as a cell expresses its *myc* gene, virtually all its Max–Max homodimers convert to Myc–Max heterodimers that bind to the enhancers previously bound by the homodimers. Because the heterodimers include the Myc activation domain, the binding of Myc–Max complexes induces the

(a) Expression of Max monomer alone

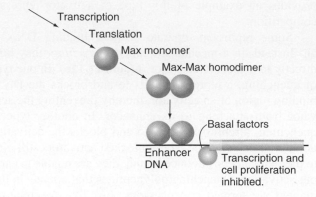

(b) Expression of Myc and Max

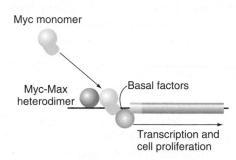

Figure 11.15 Myc–Max system of activation and repression. (a) Gene repression results when a cell makes only the Max polypeptide. **(b)** Gene activation occurs when a cell makes both Myc and Max.

expression of genes required for cell proliferation. Although researchers have not yet characterized all the genes activated by the Myc–Max dimer, they know that the genes guide the cell through its mitotic cycle. Thus, each cell in which *myc* is active divides to produce two daughter cells.

> The Myc–Max system provides a rapid genetic switch for regulating cell division during the cell proliferation and terminal differentiation phases of development. When *myc* is expressed, Myc–Max heterodimers immediately form, serving as transcription activators.

Complex regulatory regions enable fine-tuning of gene expression

In complex multicellular organisms, a large percentage of genes are devoted to transcriptional regulation. Of the estimated 20 000 genes in the human genome, scientists estimate that about 2000 genes encode transcription regulatory proteins. Each gene can have many proteins that regulate its expression, but each regulatory protein may act on many different genes. The number of possible combinations of regulators is staggering and provides the flexibility important for differentiation of cells and development in multicellular eukaryotes. Gene regulation is not just a matter of turning genes on and off, however. It also entails fine-tuning the precise level of transcription—higher or lower in

different cells, and higher or lower in cells of the same tissue but at different stages of development. It also includes mechanisms that allow each cell to modify its program of gene activity in response to constantly changing signals from its neighbours. Organisms accomplish the orchestration of transcription from each of tens of thousands of genes through *cis*-acting regulatory regions that are often far more complex than those we have so far described.

Enhancers and enhanceosomes

A regulatory region may contain a dozen or more enhancer elements, each with the ability to bind different activators and repressors, with varying affinities. At any moment, there may be dozens of transcription factors in the cell whose affinities for DNA or other polypeptides are being modulated by binding to hormones or other molecules. Different sets of these transcription factors compete for different enhancers within the regulatory region. And different sets of coactivators and corepressors compete with each other for binding to different activators or repressors. The biochemical integration of all this information yields a precise level of transcriptional activation or repression.

The term **enhanceosome** is used to describe a multimeric complex of proteins and other small molecules associated with an enhancer element (see Figure 11.4); the multimeric complex of proteins can include activators, coactivators, and other types of transcription factors known as repressors and corepressors. Slight changes in a cell's environment can dispatch signal molecules that cause changes in the balance of transcription factors or in their relative affinities for DNA or for each other. These changes, in turn, lead to the assembly of an altered enhanceosome, which recalibrates gene activity. In short, a large, exquisitely controlled machinery determines the level of primary transcript produced.

Control of the *string* gene in *Drosophila*

The enhancer regions of some class II genes are very large, containing multiple elements that make possible the fine-tuned regulation of a gene. This is particularly true for those genes in multicellular eukaryotes that must be expressed in many different tissues. The *string* gene in *Drosophila* is an example. The gene encodes a protein that activates the fourteenth mitosis of embryonic development. This

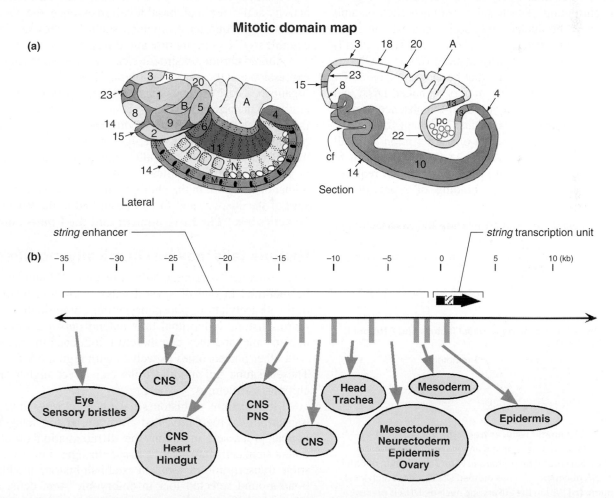

Figure 11.16 *Drosophila string* **gene enhancer. (a)** *Colours* indicate individual mitotic domains during the fourteenth cell cycle of the fruit fly embryo. The cells within each domain divide synchronously, but different domains initiate their divisions at different times. Both lateral and cross-sectional views of the embryo are shown. **(b)** Proteins that bind to the enhancer region in each of these developmental mitotic domains to turn on *string* at the appropriate time are indicated (CNS: central nervous system; PNS: peripheral nervous system).

fourteenth mitosis begins just after membranes simultaneously form around the roughly 6000 nuclei of the giant syncytium that resulted from the first 13 mitoses. What is interesting about the fourteenth mitosis is that cells in different areas, or domains, of the embryo enter this division at different times in an intricate but reproducible temporal pattern (**Figure 11.16a**). Thus, although the cells of each domain go through a fourteenth nuclear division, the time at which that mitosis takes place is different for different domains.

Remarkably, the cells within each domain simultaneously express the *string* gene just before they enter mitosis; in fact, expression of the *string* gene induces their entry into the mitotic cycle. In *string* mutants, all embryonic cells arrest in the G_2 stage of cycle 14 and never undergo mitosis. A roughly 35-kb region upstream of the

Drosophila string gene contains binding sites for many transcription factors known to regulate formation of the *Drosophila* body pattern (**Figure 11.16b**). The complex interaction of these factors ensures that the *string* gene is turned on in the cells of each embryonic domain at the correct time.

An enhanceosome is a large complex of proteins associated with an enhancer; its many components can undergo molecular alteration to fine-tune transcription according to the cell's needs and the environment. The *string* gene in *Drosophila* is an example of how transcription can be timed by binding of different transcription factors to coincide with cell proliferation during development.

11.3 Chromatin Structure and Epigenetic Effects

In Chapter 6 you learned that the DNA of eukaryotic genomes does not float freely in the nucleus but is packaged into chromatin. The basic repeating structural unit of chromatin is the nucleosome, which consists of a ball of histone proteins (two each of H2A, H2B, H3, and H4) around which is wrapped approximately 160 bp of DNA.

In vitro experiments show that basal factors and RNA polymerase readily bind to promoters on naked DNA and initiate high levels of transcription in the absence of activator proteins (**Figure 11.17a**). One significant function of chromatin, then, is the reduction of transcription from all genes to a very low level (**Figure 11.17b**). In contrast to transcriptional modulation in prokaryotes, which requires active repression through the binding of repressors to

cis-acting elements, the normal structure of chromatin in eukaryotes is sufficient by itself to maintain transcriptional activity at the minimal, basal level. In essence, the nucleosomes may sequester promoters, such that they are inaccessible to RNA polymerase and transcription factors.

Altered chromatin structure can cause changes in gene expression and therefore phenotypic changes in a cell or an organism. These changes may be inherited from one generation to the next. These changes are not due to changes in the DNA sequence, but are modifications of the genomic blueprint and are known as **epigenetic** changes. Environmental influences can also alter the epigenome, resulting in gene expression modification. An example of this phenomenon is shown in the photograph of the two mice at the start of the chapter, and further described in the **Focus on Genetics** box "**The Environment and the Epigenome.**"

Histone tails may be chemically modified

Recall that the N-terminal tails of histones H3 and H4 can be modified in one of several ways, including methylation and acetylation, phosphorylation, ubiquitination, and more. Histone N-terminal tails extend outward from the nucleosome, and they can therefore influence interactions with other nucleosomes as well as with regulatory factors. These histone tail modifications can affect higher-order chromatin structures.

Modifications of histones have been closely linked to transcriptional regulation and are required for many biological processes, including the differentiation of pluripotent stem cells into specific tissue lineages. For instance, some transcription factors can establish histone modifications around selected loci in embryonic stem cells, and thus determine lineage-specific gene expression patterns. Thus, histone covalent modifications seem to have crucial roles for the establishment of genetic programs during development.

(a) Naked promoter binds RNA polymerase and basal factors.

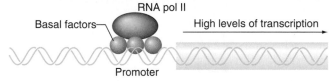

(b) Chromatin reduces binding to basal factors and RNA pol II to very low levels.

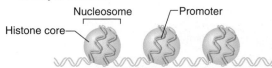

Figure 11.17 Chromatin reduces transcription. (a) DNA molecules containing a promoter and an associated gene can be purified away from chromatin proteins *in vitro*. The addition of basal factors and RNA polymerase to this purified DNA induces high levels of transcription. **(b)** Within the eukaryotic nucleus, DNA is present within chromatin. Promoter regions are generally sequestered within the nucleosome and only rarely bind to basal factors and RNA polymerase. Thus, the chromatin structure maintains basal transcription at very low levels.

At the beginning of this chapter, two mice were shown that were identical at the genetic level, but were phenotypically different. Both of the mice had the same genotype at the **agouti** locus, A^{vy}/a. Agouti is a signalling protein that functions as a regulator of pigmentation in the hair follicle melanocytes. Transcription initiation of the wild-type *agouti* (*A*) allele normally occurs solely in the skin from a developmentally controlled promoter located in exon 2 of the *A* allele. Expression is restricted to the hair follicles during a particular stage of hair growth and leads to the wild-type grey/brown (agouti) coat colour.

The insertion of a murine intracisternal A particle (IAP) retrotransposon approximately 100 kb upstream of the *agouti* gene's transcriptional start site produces the A^{vy} allele. This IAP element contains a cryptic promoter in its proximal end, which promotes *agouti* transcription in a constitutive, ectopic manner and leads to yellow fur colour, obesity, diabetes, and tumour formation (**Figure A**). The IAP retrotransposon can be methylated at CpG sites (and thus transcriptionally silenced) to varying degrees. This allows for the *agouti* gene to be expressed under its normal developmental controls, resulting in a range of coat colours, from yellow (unmethylated) to grey/brown (methylated), in genetically identical A^{vy}/a mice (**Figures A** and **B**). The A^{vy} allele is an example of a **metastable epiallele**—it can be modified in a variable and reversible epigenetic manner, resulting in phenotypic variation even when cells are genetically identical. Thus, expression of the A^{vy} allele

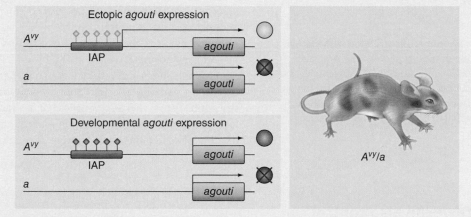

Figure A Epigenetic regulation of the *agouti* gene. In mice carrying an A^{vy} allele, an IAP (intracisternal A particle) retrotransposon is located upstream of the transcription initiation site of the *agouti* gene. The IAP element contains a cryptic promoter that can remain unmethylated at CpG sites (*yellow diamonds*). When unmethylated, the cryptic IAP promoter drives constitutive, ectopic *agouti* expression, resulting in a yellow coat colour phenotype (*top*). Methylation of the IAP element at CpG sites (*red diamonds*) allows normal developmental expression of the *agouti* gene, resulting in a wild-type grey/brown (pseudoagouti) coat colour (*bottom*). A mottled coat colour phenotype occurs in the mouse progeny if methylation of the IAP retrotransposon occurs at a later stage of embryonic development and does not affect all cells (*right*). The *a* allele is not functional and thus cannot regulate pigment production (*box* with a *red cross*).

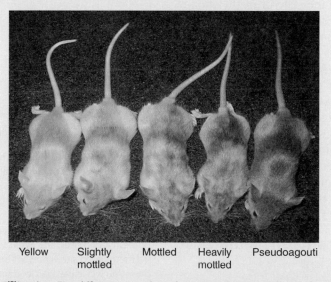

Yellow   Slightly mottled   Mottled   Heavily mottled   Pseudoagouti

Figure B Coat colour variation in A^{vy}/a mice. Five different coat colour phenotypic classes are observed in genetically identical A^{vy}/a littermates.

(Continued)

(c)

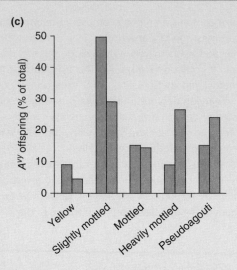

Figure C Coat colour phenotypes correlated with methyl-donor supplementation of maternal diet. In *a/a* female mice mated with male *A^vy^/a* mice, nutritional supplementation with methyl donors results in a shift in coat colour distribution from yellow to pseudoagouti (grey/brown) in the offspring (*tan bars*), compared with that of unsupplemented progeny (*green bars*).

is correlated with the methylation status of the IAP retrotransposon's promoter region.

An interesting result occurs when *a/a* females are mated with male *A^vy^/a* mice under certain conditions. When females are given methyl-donor nutritional supplements (folic acid, vitamin B_{12}, choline, and betaine), or the phytoestrogen genistein, two weeks before mating, and during pregnancy and lactation, the distribution of coat colour in their F_1 generation *A^vy^/a* progeny shifts from yellow to a grey/brown pseudoagouti wild-type colour (**Figure C**). This alteration in coat colour phenotype is the result of an increase in DNA methylation at CpG islands in the IAP promoter region at the *A^vy^* locus. Thus, the diet of a mother during pregnancy can directly influence the adult phenotype of her offspring via changes in the epigenome. These epigenetic modifications can be further inherited in the F_2 generation of offspring through the germ line. This and other evidence suggests that environmental factors during prenatal and early postnatal developmental stages, including nutritional supplements, can result in altered epigenetic and gene expression patterns, affecting adult phenotypes and disease susceptibility risk. These epigenetic changes can potentially be long-term, inherited by offspring in a transgenerational manner, thus affecting the health of future generations.

Meythlation of DNA can also control transcription

Methylation of DNA—the addition of a methyl (CH_3) group—is another common modification associated with transcription changes. Methylation occurs at the fifth carbon of the cytosine base in a CpG dinucleotide pair (see Figure 5.11a).

It is possible to determine the state of methylation of a DNA region by using two restriction enzymes that both cleave at a sequence containing a CG dinucleotide but that have different sensitivities to the methylation of the DNA substrate. For example, *Hpa*II and *Msp*I both cleave at CCGG, but *Hpa*II does not cleave if the middle C of this site is methylated; in contrast, *Msp*I can cleave regardless (**Figure 11.18**). Thus, by digesting genomic DNA with *Hpa*II and *Msp*I and using a specific DNA probe on a Southern blot, you can determine whether a given CCGG sequence is methylated. Although methylation is associated with transcription silencing, we know it cannot be the only mechanism, because some organisms that show silencing, such as yeast, do not contain methylated DNA.

The remodelling of chromatin mediates activation of transcription

Nucleosomes can be repositioned or bumped off the DNA to expose promoter sequences and other regulatory sequences, thereby allowing high levels of transcription.

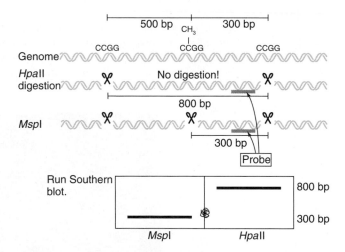

Figure 11.18 Determining the methylation state of DNA. A determination of the methylation status of a DNA region can be made using a pair of restriction enzymes that both recognize the same base sequence, with one being able to digest methylated DNA, while the other cannot. In this example, the restriction site is CCGG, and the enzymes are *Msp*I, which can digest both methylated and unmethylated sites, and *Hpa*II, which cannot digest methylated sites. If a methylated site is present between two unmethylated sites, *Hpa*II digestion will leave a larger fragment than *Msp*I. After electrophoresis and Southern blot analysis with a probe that hybridizes to a sequence on one side of the methylated site, there will be a clearly observable difference in band size.

This activity, known as *chromatin remodelling*, is characteristic of some of the transcriptional activators. Proteins associated together that carry out this function are called remodelling complexes (**Figure 11.19**). The freed DNA becomes much more accessible to basal transcription factors, to RNA polymerase, and probably to enhancer-binding transcription factors, allowing them to interact with DNA sequences to affect transcription.

Chromosomal regions from which the nucleosomes have been eliminated are experimentally recognizable through their hypersensitivity to the enzyme DNase. When one scans a chromosome with the enzyme for the presence of **DNase hypersensitive (DH) sites**, the sites show up at the 5′ ends of genes that are either undergoing transcription or are being prepared for transcription in a later step of cellular differentiation. For example, DH sites appear at the 5′ end of the β-globin gene in human stem cells that are precursors to the haematopoietic cells in which the gene will be activated, but not in cells from other differentiative pathways.

Remodelling by SWI–SNF

Remodelling of chromatin is one way in which gene expression changes occur. One of the best-studied remodelling complexes involves the SWI–SNF proteins in yeast. These proteins form a multisubunit complex that disrupts chromatin structure by removing or repositioning nucleosomes. The resulting chromatin decompaction gives basal factors much greater access to promoter regions, and consequently, transcription rapidly accelerates (**Figure 11.20**).

Chromatin remodelling can expose promoter.

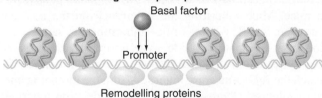

Figure 11.19 Chromatin remodelling. Chromatin remodelling can expose the promoter region. Remodelling proteins cause specific nucleosomes to unravel in specific cells at specific times during differentiation or development. Exposed promoter regions more readily bind basal factors.

SWI–SNF remodelling complex

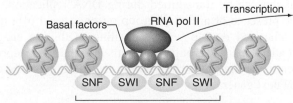

The SWI–SNF multisubunit destabilizes chromatin structure and gives transcription machinery access to the promoter.

Figure 11.20 SWI–SNF remodeller. The SWI–SNF protein complex is a well-characterized remodelling apparatus that functions within yeast cells to expose promoter regions to basal factors, RNA polymerase, and transcriptional activation.

The SWI–SNF complexes use the energy of ATP hydrolysis to alter nucleosome positioning relative to a segment of DNA.

In *Drosophila*, mutations in the gene encoding the ATPase subunit of the SWI-SNF complex impair transcription by RNA polymerase II, suggesting a general role of these remodelling complexes in gene activation. Human cells contain related multisubunit protein complexes that also influence nucleosome position or structure, suggesting that this particular nucleosome-disrupting machinery has been conserved throughout evolution. The SWI–SNF protein complex represents just one of the many that help remodel chromatin at specific chromosomal locations, in specific cells, at particular points of development.

Hypercondensation of chromatin

As described in Chapter 6, many regions of eukaryotic chromosomes, including parts of centromeres and telomeres and all of Barr bodies, are highly condensed into heterochromatic DNA. As a result, most of the genes contained in heterochromatin are transcriptionally inactive or silenced (**Figure 11.21**). We know that the heterochromatic regions are characterized by specific methylation on a lysine in histone H3 and also by methylation of the CpG dinucleotides in DNA.

The heterochromatic state is inherited from one cell generation to the next—a characteristic of an *epigenetic* phenomenon. Epigenetic modifications have phenotypic consequences that are inherited, but without a change in the DNA sequence. Examples described in earlier chapters included X-chromosome inactivation and position-effect variegation in *Drosophila*. In X-chromosome inactivation, most alleles on one chromosome are not expressed, and this can have phenotypic consequences.

Let us look at a specific example in which a human female is heterozygous for the anhidrotic ectodermal dysplasia gene (on the X chromosome), which is required for sweat gland development as well as development of other ectoderm-associated tissues and organs. All cells in the female have the same genotype, but in some cells, the wild-type allele is on the inactivated X chromosome, exposing the mutant phenotype. The phenotype changes

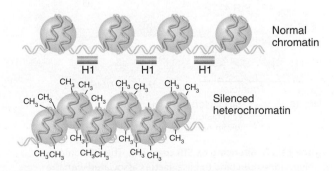

Figure 11.21 Condensed chromatin. Normal chromatin in standard nucleosome conformation can be converted into tightly packed heterochromatin with the addition of methyl groups to a series of cytosine bases within a local DNA region.

without a change in genotype, and this change is inherited in subsequent cell divisions. Therefore, an affected woman may have areas of skin that contain sweat glands while other areas do not, depending on the point at which the X chromosome with the wild-type gene became inactivated.

Insight into the mechanism of silencing comes from studies of mutations that give rise to sterility in yeast. Recall that yeast cells come in two mating types—α and *a*—and the α2 gene product represses certain *a*-determining genes. The chromosomal locus of the α2 gene is known as *MAT* (for mating type). In normally mating yeast cells, there are two additional copies of the *MAT* locus called *HML* and *HMR*; located near the telomeres on each arm of chromosome III, these loci are transcriptionally silent. Mutations that reduce or destroy silencing at these loci cause sterility because they allow the simultaneous expression of α and *a* information. The resulting cells, which behave as diploids, do not mate.

Analysis of these mutations identified the family of *SIR* genes. The SIR polypeptide products associate to form a *trans*-acting complex that mediates silencing by acting at *cis*-acting sites near *HML* and *HMR*. Null mutations that eliminate the activity of any *SIR* gene or mutations that delete a *cis*-acting site abolish silencing. The SIR complex binds to other polypeptides, and these larger complexes interact with histones H3 and H4 (**Figure 11.22**). These interactions with the histones establish a silenced chromosomal domain that remains hidden from the activators and repressors of transcription.

> Alterations to chromatin structure can control transcription. These include additions of chemical groups to histone tails, methylation of DNA at cytosine bases, removal or repositioning of nucleosomes, and hypercondensation of chromatin such as which occurs with the Barr body. Some are epigenetic effects passed on from one cell generation to the next.

SIR complex binds to basal factors and interacts with H3 and H4 components of histones.

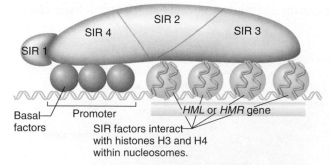

Figure 11.22 Silencing by SIR complex. The SIR complex of polypeptides can bind to basal factors associated with the promoters of the *HML* and *HMR* genes. This binding, in turn, causes the SIR complex to interact with the histones H3 and H4 present in downstream nucleosomes associated with the gene itself. The result is the complete silencing of transcription.

Genomic imprinting results from transcriptional silencing

A major tenet of Mendelian genetics is that the parental origin of an allele—whether it comes from the mother or the father—does not affect its function in the F_1 generation. For the vast majority of genes in plants and animals, this principle still holds true today. Surprisingly, however, experiments and pedigree analyses have uncovered convincing evidence of exceptions to this general rule for some genes in mammals.

The phenomenon in which the expression pattern of a gene depends on the parent that transmits it is known as **genomic imprinting**. In most cases of genomic imprinting, the copy of a gene inherited from one parent is transcriptionally inactive in all or most of the tissues in which the copy from the other parent is active. The term "imprinting" signifies that whatever silences the maternal or paternal copy of an imprinted gene is not encoded in its DNA sequence; rather the "silencer" exercises its effect through some epigenetic alteration of the DNA or chromatin during gametogenesis. With the development of molecular tools able to distinguish between transcripts of a gene from either parental homologue, geneticists observed that expression of a small number of genes—scattered around the genome, but often found in clusters—depends on whether the copy of the gene comes from the female parent or the male parent. The silencing effect is epigenetic and does not involve a change in DNA sequence.

An understanding of the mechanism behind imprinted genes came from studies of the transmission of a deletion in the chromosome 7 insulin-like growth factor gene (*Igf2*) in mice. Mice inheriting the deletion from the paternal side were small, whereas mice inheriting the same deletion from the maternal side were normal size. The simplest explanation of these results is based on a model in which the *Igf2* gene copy inherited from the mother is normally silenced (**Figure 11.23a**). Thus, a deletion inherited from the mother produces no phenotypic effect because the maternal allele is not expressed anyway. If the deletion comes from the father, however, it produces a phenotypic effect because the animal is now unable to make any IGF2 products.

A hypothesis for how this imprint can be maintained from one generation to the next is that the pattern of methylation can be transmitted during DNA replication, with the presence of a methyl group on one strand of a newly synthesized double helix signalling methylase enzymes to add a methyl group to the other strand. For imprinted genes, the imprint is reset during meiosis and passed on to the next generation.

Insulators

Although the biochemical mechanism of genomic imprinting is not yet completely understood, one important component is the methylation of cytosines in CG dinucleotides within the imprinted region (**Figure 11.23b**). The methylated C's silence the gene or genes in the region by

preventing RNA polymerase and other transcription factors from gaining access to the DNA. The methylation pattern is not transient as a response to a short-term stimulus, but is stably inherited.

Further insight into the mechanism of *Igf2* imprinting came from the surprising finding that *H19*, found just 70 kb downstream of *Igf2*, is also imprinted, but in the opposite way. With *H19*, the copy inherited from the father is silenced and the copy inherited from the mother is active in normal mice. A model of how imprinting works at both *H19* and *Igf2* is based on detailed biochemical and genetic studies of a 100-kb region encompassing both genes. Researchers identified an enhancer region downstream of the *H19* gene that can interact with promoters for both genes (**Figure 11.23c**). In the region between the two genes lies another type of transcriptional regulation element called an **insulator**. When an insulator becomes functional, it stops communication between enhancers on one side of it and promoters on the other side. Insulators exist throughout the genome, limiting the chromatin region over which an enhancer can operate. Without insulators, enhancers could wreak havoc in a cell by turning on genes at DNA distances of hundreds of kilobases.

In the *Igf2–H19* region, the insulator DNA becomes functional by binding a protein called CTCF. The binding normally occurs on the maternal chromosome. As a result, the enhancer element on the maternal chromosome can interact only with the promoter of *H19*; this interaction, of course, turns on the *H19* gene. In such a situation, the *Igf2* gene remains unexpressed. On the paternal chromosome, by contrast, both the insulator and the *H19* promoter are methylated. Because methylation of the insulator prevents the binding of CTCF, the insulator is not functional; and without a functional insulator, the enhancer downstream of *H19* can reach over a great distance to activate transcription from the *Igf2* promoter. In addition, methylation of the *H19* promoter suppresses transcription of the paternal *H19* gene. Imprinting of the paternal chromosome by methylation thus turns on transcription of *Igf2* and prevents transcription of *H19*.

This epigenetic imprint remains throughout the life of the mammal, but it is erased and regenerated during each passage of the gene through the germ line into the next generation (**Figure 11.23d**). Some genes receive an imprint in the maternal germ line; others receive it in the paternal germ line. For each gene subject to this effect, imprinting occurs in either the maternal or paternal line, never in both.

Inheritance pattern of imprinted genes

Before the late 1980s, clinical geneticists were accustomed to seeing sex-linked differences in inherited phenotypes related to the sex of the affected individual. With imprinting, however, it is the sex of the parent carrying a mutant allele that counts, and not the sex of the individual inheriting the mutation. After the discovery of imprinting in mice, medical geneticists reanalyzed human pedigrees and determined retrospectively that what appeared to be instances of incomplete penetrance were actually manifestations of imprinting (**Figure 11.23e**). An inactivating mutation in a maternally imprinted gene could pass unnoticed from mother to daughter for many generations (because the maternally derived gene copy is inactive due to imprinting). If, however, the mutation passed from mother to son, the son would have a normal phenotype (having received an active wild-type allele from his father), but the son's children, both boys and girls, would each have a 50 percent chance of receiving a mutant paternal allele and therefore expressing the mutant phenotype resulting from the absence of any gene activity.

Evidence for imprinting as a contributing factor now exists for a variety of human developmental disorders, including the related pair of syndromes known as Prader–Willi syndrome and Angelman syndrome. Children with Prader–Willi syndrome have small hands and feet, underdeveloped gonads and genitalia, a short stature, and mental retardation; they are also compulsive overeaters and obese. Children affected by Angelman syndrome have red cheeks, a large jaw, a large mouth with a prominent tongue, and a happy disposition accompanied by excessive laughing; they also show severe mental and motor retardation. Both syndromes are often associated with small deletions in the q11–13 region of chromosome 15. When the deletions are inherited from the father, the child develops Prader–Willi syndrome; when the deletions come from the mother, the child has Angelman syndrome.

The explanation for this phenomenon is that at least two genes in the region of these deletions are differently imprinted. One gene is maternally imprinted: Children receiving a deleted chromosome from their father and a wild-type (nondeleted) chromosome with an imprinted copy of this gene from their mother exhibit Prader–Willi syndrome because the imprinted, wild-type gene is inactivated. In the case of Angelman syndrome, a different gene in the same region is paternally imprinted: Children receiving a deleted chromosome from their mother and a wild-type, imprinted gene from their father, develop this syndrome.

In the last several years, research in the field of epigenetics has been progressing at an accelerated pace, advancing our knowledge of epigenetic mechanisms of gene regulation and providing a new paradigm for disease aetiology. Dr. Arturas (Art) Petronis, an investigator in the field of epigenetics at the University of Toronto and the Centre for Addiction and Mental Health (CAMH) in Toronto, is featured in the **Genetics and Society** box **"Epigenetics and Complex Disease."**

> Imprinting appears to be accomplished largely by DNA methylation during gametogenesis. In some cases, a functionalized insulator region between two genes selectively interferes with transcription of one or the other of the genes. Some cases of incomplete penetrance have been found to result from imprinting. The inheritance pattern may resemble that of sex-linked alleles, but with generation skipping.

Genomic Imprinting

(a) Deletion of *Igf2* causes mutant phenotype only when transmitted by the father.

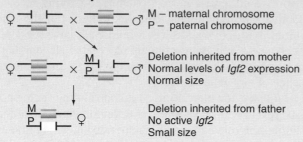

M – maternal chromosome
P – paternal chromosome

Deletion inherited from mother
Normal levels of *Igf2* expression
Normal size

Deletion inherited from father
No active *Igf2*
Small size

The phenotypic effect of an *Igf2* deletion is determined by the parent transmitting the mutant locus. This parent-of-origin effect can be demonstrated in the two-generation cross illustrated here.

(b) Methylation of complementary strands of DNA

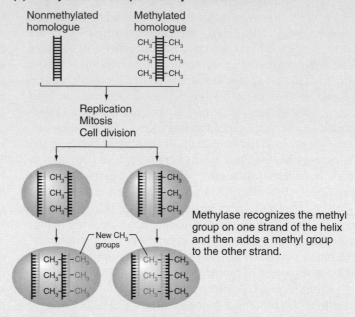

Methylase recognizes the methyl group on one strand of the helix and then adds a methyl group to the other strand.

An epigenetic state of DNA methylation can be maintained across cell generations. This is accomplished by the activity of DNA methylases that recognize methyl groups on one strand of a double helix and respond by methylating the opposite strand.

(c) Methylation of paternally inherited *H19* promoter

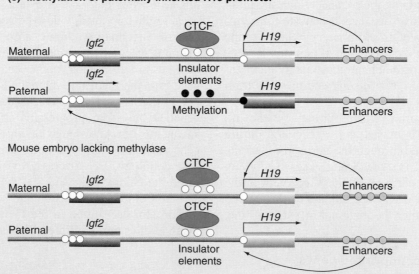

Reciprocal parent-of-origin expression occurs with the *Igf2–H19* gene pair. Only the unmethylated insulator between the two genes can bind to the protein CTCF. On the maternal chromosome, the enhancer only has access to the *H19* promoter. On the paternal chromosome, methylation occurs only at the insulator and the *H19* promoter (indicated with *darkened circles*). This serves the double purpose of blocking transcription of *H19* and allowing access of the enhancer to *Igf2*. In mouse embryos lacking methylase, the paternal chromosome behaves biochemically like the maternal chromosome.

(d) The resetting of genomic imprints during meiosis

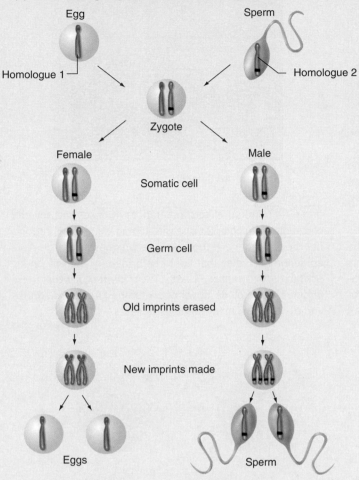

Egg

Sperm

Homologue 1

Homologue 2

Zygote

Female

Male

Somatic cell

Germ cell

Old imprints erased

New imprints made

Eggs

Sperm

Follow the transmission of a pair of homologous chromosomes (homologue 1 from the mother and homologue 2 from the father) from gametes through fertilization and the development of female and male progeny, to meiosis and the creation of a new set of gametes. Maternally imprinted genes are shown in *red*, paternally imprinted genes in *black*. The cellular machinery erases the old imprints and establishes new ones in germs cells during meiosis. Note that in the second generation, one of the chromosomes in both egg (homologue 2) and sperm (homologue 1) will be differently imprinted than the way it was in the first generation.

(e) Genomic imprinting and human disease

Paternal imprinting

Maternal imprinting

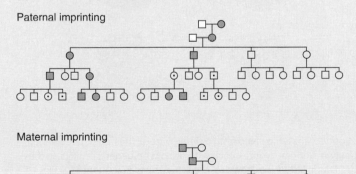

In each pedigree, affected individuals (represented by *filled-in orange circles* and *squares*) are heterozygotes for a deletion removing a gene that has either a paternal or a maternal imprint. In these pedigrees, a *dotted symbol* indicates individuals carrying a deleted chromosome but not displaying the mutant phenotype.

In the last decade or so, there has been an explosion in the field of epigenetics research, resulting in a number of openings at the university level for investigators conducting research in epigenetic mechanisms of gene regulation and the role of epigenetics in disease aetiology. One such researcher is Dr. Arturas (Art) Petronis (**Figure A**), a Professor in the Departments of Psychiatry and Pharmacology at the University of Toronto in Toronto, and a Senior Scientist in the Neuroscience Research Department, and Head of The Krembil Family Epigenetics Laboratory at the Centre for Addiction and Mental Health (CAMH). His research has focused on understanding the role of epigenetic factors in complex, non-Mendelian diseases, such as schizophrenia, diabetes, and bipolar disorder, and has resulted in the development of an epigenetic theory of complex disease. This theory is based on the hypotheses that together with chromosomal DNA sequences, at least some epigenetic factors are passed on from generation to generation and contribute to the heritability of at least some traits. These epigenetic factors have regulatory roles in various genomic activities, resulting in non-Mendelian features such as environmental influences, monozygotic twin discordance (**Figure B**), parental

Figure A Art Petronis.

origin (imprinting) effects, differences in onset and severity of symptoms in complex diseases, to name a few. Dr. Petronis has developed new tools and techniques for large-scale epigenomic studies that could help identify the epigenetic risk factors in complex diseases. The expectation is that these findings will provide novel opportunities for the diagnosis and treatment of many complex diseases.

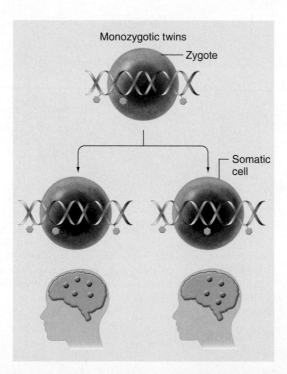

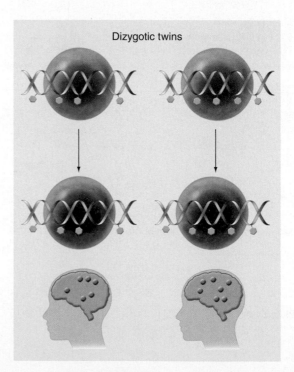

Figure B Epigenetic heritability and twin studies. Since monozygotic twins are derived from a single zygote, their epigenetic profile is initially more similar than that of dizygotic twins, who develop from two different zygotes, each with distinct epigenetic profiles. In the germline (one layer) or somatic cells (multiple layers), DNA methylation at CpG sites are shown as *green hexagons*. Epigenetic changes in monozygotic and dizygotic twins are influenced by stochastic and environmental factors, resulting in further modification of the epigenome and comparable amounts of epigenetic alterations in somatic cells. Given that dizygotic twins, derived from two different zygotes, initially have more epigenetic differences than monozygotic twins, dizygotic twins overall exhibit greater epigenetic variation or discordance in somatic tissues than monozygotic twins. This could explain the substantial phenotypic disparities (*pink circles*) observed between dizygotic twins compared with monozygotic twins.

Gene regulation can take place at any point in the process of gene expression. So far we have mainly discussed the mechanisms that influence rates of transcription; some other systems regulate posttranscriptional events—these include RNA splicing; RNA stability; RNA localization; and protein synthesis, stability, and localization. The regulation of all these processes relies on regulatory proteins as well as on small RNAs.

RNA splicing helps regulate gene expression

One surprise at the completion of genome projects for several complex organisms was that the number of genes was lower than originally anticipated. And yet, multicellular organisms require a large number of proteins for development and for physiological functions of different cell types, tissues, and organs. One way to generate more diversity of proteins to fulfill different needs is to splice primary transcripts into distinct mRNAs that produce different proteins. Alternative splicing is a common feature in eukaryotes and contributes to the mechanisms for regulating gene expression (review Figure 7.10).

One example where we understand the players in regulating RNA splicing and the importance of alternative splicing is found in the regulation of the *Sxl* gene in *Drosophila*. As you will see in the comprehensive example at the end of this chapter, transcription factors in very early female (XX) embryos activate the expression of a key gene called *Sxl* through a promoter called the early promoter (P_e). The cellular machinery splices the resulting transcript to create an mRNA that is translated into the Sxl protein, which is essential to the female-specific developmental program. The *Sxl* gene is not transcribed in early male (XY) embryos, so these embryos do not make the Sxl protein (**Figure 11.24a**).

Later in development, the transcription factors activating the *Sxl* early promoter in females disappear; but to develop as females, these animals still need the Sxl protein. How can they still make the Sxl protein they need? The answer is that later in embryogenesis, the *Sxl* gene in both males and females is transcribed from another promoter—the late promoter (P_L) (**Figure 11.24b**). In males, splicing of the primary *Sxl* transcript generates an RNA that includes an exon (exon 3) containing a stop codon in its reading frame. As a result, this RNA in males is not productive—it does not generate any Sxl protein. In females, however, the Sxl protein previously produced by transcription from the early promoter influences the splicing of the primary transcript initiated at the late promoter. When the earlier-made Sxl protein binds to the later-transcribed RNA, this binding alters the splicing pathway such that exon 3 is no longer part of the final mRNA. Without exon 3, the mRNA is productive—that is, it can be translated to make more Sxl protein. Thus, a small amount of Sxl protein synthesized very early in development establishes a positive feedback loop that ensures more fabrication of Sxl protein later in development.

> Alternative splicing allows transcripts to be combined in different ways, increasing the variety of proteins a genome may produce. Splicing also serves to regulate gene expression by restricting or allowing certain combinations of transcripts.

Some small RNAs are responsible for RNA interference

In the first five years of the twenty-first century, a new family of gene regulators was discovered in the form of small specialized RNAs that prevent the expression of specific genes through complementary base-pairing.

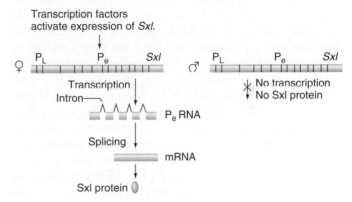

(a) Early embryo

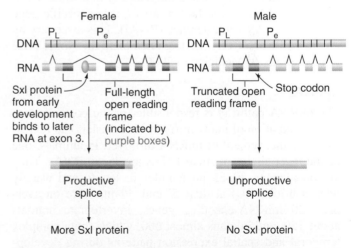

(b) Sxl protein regulates the splicing of its mRNA.

Figure 11.24 Differential RNA splicing in *Drosophila* development. (a) In the early female—but not the male— *Drosophila* embryo, transcriptional activators initiate transcription from the P_e promoter of *Sxl* to produce an mRNA that encodes the Sxl protein. **(b)** Later in development, transcriptional activators that bind the P_L promoter are produced in both male and female animals. When the Sxl protein is present, as it is in females, it causes the splicing apparatus to skip over this exon and splice exon 2 directly to exon 4. The resulting RNA molecule has an intact coding sequence and can be translated into more Sxl protein. This results in a feedback loop that maintains the presence of Sxl protein in females but not in males.

Several classes of small regulatory RNAs have now been described, including **microRNAs (miRNAs)** and **small interfering RNAs (siRNAs)**. New families of small RNAs continue to be discovered, making some of the divisions between classes ever more diffuse. Each small RNA class is generated through distinct pathways, leading to the production of RNAs of slightly different length but always within the range of 21–30 nucleotides.

To exert their functions, each small RNA class requires distinct members of the Argonaute/Piwi protein family with which they form ribonucleoprotein complexes. The complexes are able to recognize nucleic acid targets with perfect or partial complementarity. These small RNAs primarily regulate gene activity at the posttranscriptional level through the regulation of RNA stability or translation, but some recent reports suggest that they may also act at the transcriptional level—for instance, by affecting chromatin structure.

Why were these very important regulators not recognized until recently? The answer probably lies in the intrinsic properties of these molecules. For example, because of their reduced size, they are easily missed by most standard biochemical RNA analysis methods; their short length also makes them very poor targets for inactivation through classical genetic approaches. In addition, given that many miRNAs appear to be able to compensate for each other's function—a phenomenon known as *genetic redundancy*—conventional genetic screens have often failed to expose the biological significance of losing individual small RNAs. The first two miRNAs (*lin-4* and *let-7*) were discovered by genetic experiments in the worm *Caenorhabditis elegans*. Shortly afterward, siRNAs were found in animals, plants, and fungi as key molecular mediators of a phenomenon known as sequence-specific gene silencing, or RNA interference (RNAi). This discovery by Andrew Fire and Craig Mello gained them the Nobel Prize in Physiology or Medicine in 2006.

miRNAs

The miRNA pathway is responsible for the posttranscriptional regulation of many mRNAs via translational repression or enhancement of miRNA turnover. In animals, one of the most abundant small RNAs are the miRNAs. They are on average 20–23 nucleotides in length and usually have a uridine (U) at their 5′ end. Plants have on average 120 miRNA-encoding genes, invertebrate animals about 150, and humans almost 600, which show complex temporal and spatial expression patterns during development. Recently, miRNAs have also been identified in some viruses and green algae, indicating their broad phylogenetic distribution.

Most miRNAs are transcribed by RNA polymerase II from noncoding DNA regions that generate short dsRNA hairpins (**Figure 11.25**). In animal cells, the endoribonuclease Drosha excises the miRNA stem loops from the primary transcript (pri-miRNA) while it is still in the nucleus, releasing an approximately 70-base RNA intermediate (pre-miRNA) (**Figure 11.26**). The pre-miRNA is actively exported to the cytoplasm by a protein complex. Once in the cytoplasm, the pre-miRNA is subsequently processed by the RNAse Dicer to produce a mature miRNA in the form of a double-stranded intermediate. One of the strands in these double-stranded miRNA intermediates—the "guide" strand—is incorporated into ribonucleoprotein complexes that are often referred to as miRNA-induced silencing complexes (miRISCs). The other strand, usually termed miRNA*, is degraded.

The ribonucleotide complexes (miRISCs) containing miRNAs mediate diverse functions depending on the particular Argonaute protein they possess, and on the extent of sequence complementarity between the guide miRNA and the target sequences in mRNA 3′ untranslated regions (3′ UTRs) (**Figure 11.27**). miRNA complexes with perfect complementarity between guide and target RNA cause mRNA cleavage. With less complementarity, the mechanism is often some type of inhibition of translation. The mechanisms used by miRNAs to regulate translational activity are still not fully understood. Recent work shows that miRNAs are able to repress protein expression in at least four distinct manners: (1) co-translational protein degradation, (2) inhibition of translation elongation, (3) premature termination of translation (i.e., ribosome "drop-off"), and (4) inhibition of translation initiation.

siRNAs

In the siRNA pathway, dsRNAs are either produced by transcription of both strands of an endogenous DNA sequence in the genome, or arise from an exogenous source such as a virus. These dsRNAs are the pri-RNAs that are processed by Dicer, and the resulting ssRNA can interfere with expression of a gene containing the complementary sequence. This pathway may also protect the cell from invading dsRNAs produced by viruses by destroying those RNAs.

The siRNA pathway is responsible for detecting exogenous double-stranded RNAs (dsRNAs) and destroying any transcripts derived from the invading RNA, as well as from cellular dsRNA transcripts generated by transcription of both a sense and antisense strand of a gene. In plants, many siRNAs of viral and viroid origin are detected. The plant can destroy RNAs of these invading particles using the siRNA pathway.

Researchers have found the siRNA pathway to be a useful mechanism to selectively shut off expression of targeted genes. To study the function of a specific gene or the effect of loss of the selected gene product, researchers introduce dsRNA of that gene into the cell and expression of the endogenous gene is shut off or knocked down. The double-stranded siRNA is again composed of a guide RNA and a sense strand. The guide RNA will bind to complementary transcripts and mediate mRNA cleavage as with miRNAs. Researchers are excited about the possibility of using RNA interference to treat diseases, and this is detailed in the **Tools of Genetics** box **"RNA Interference and Treatment of Disease."**

(a) Primary transcripts containing miRNA

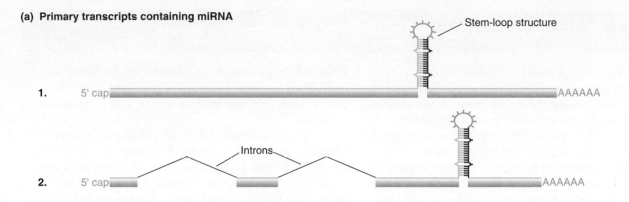

1. 5' cap ————————————————————————— AAAAAA — Stem-loop structure

2. 5' cap —— Introns —— AAAAAA

(b) Predicted pri-miRNA stem loops

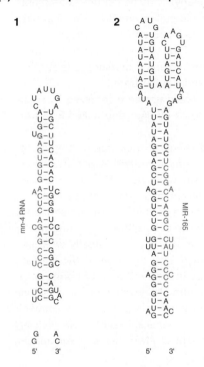

Figure 11.25 microRNA-containing genes. (a) Most primary (pri-) miRNA transcripts do not contain an open reading frame, but some miRNAs are present within the introns of protein-coding mRNAs, as shown in the second example. **(b)** Ribonucleotide sequences and predicted duplex structures of stem loops in different pri-miRNA transcripts. Example 1 is from *C. elegans*; nearly identical homologues of these stem-loop structures have been found in other animals, including flies and mammals. Example 2 is from the plant *Arabidopsis*; a nearly identical homologue has been uncovered in rice and other plants.

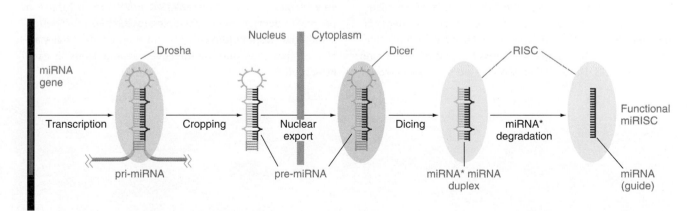

Figure 11.26 miRNA processing. Immediately after transcription, microRNA-containing primary transcripts (pri-miRNAs) are recognized by the nuclear enzyme Drosha, which crops out pre-miRNA stem-loop structures from the larger RNA. The pre-miRNAs undergo active transport from the nucleus into the cytoplasm where they are recognized by the enzyme Dicer. Dicer reduces the pre-miRNA into a short-lived miRNA* miRNA duplex, which is released and picked up by an RNA-induced silencing complex (RISC). RISC eliminates the miRNA* strand from the duplex and becomes a functional and highly specific miRISC.

The recent discovery of RNA interference (RNAi) as a natural process of gene regulation in all eukaryotic cells suggests a new approach toward the development of therapies to combat disease. The general idea is to co-opt the existing cellular RNAi machinery into working with laboratory-designed siRNA molecules that target specific mRNAs from the disease-causing gene for destruction through the mechanism shown in **Figures A1** and **A2**. Among the diseases researchers are currently targeting for RNAi therapy are incurable conditions such Huntington disease, ALS (amyotrophic lateral sclerosis), AIDS, and a variety of cancers. Investigators are also targeting conditions such as hypertension and hypercholesterolaemia, for which current treatments are not specific enough.

The first step in the development of an RNAi therapy involves the design, construction, and experimental validation of an siRNA that can function inside living cells to eliminate disease-causing *target transcripts*, while not affecting the transcripts of any other gene. A well-designed siRNA contains a 21–23-base-long **antisense** sequence that is perfectly complementary to a unique sequence within the target transcript. This antisense sequence must be contained within a longer RNA strand that is itself part of a duplex with a complementary **sense** strand. The duplex structure and extended RNA length are required to allow recognition and binding by Dicer—the first cytoplasmic enzyme in the RNAi processing pathway—which trims the siRNA and passes it on to RISC.

With an automated oligonucleotide synthesizer, researchers can generate a large set of target-specific duplex RNA molecules. These duplex RNAs also have variations in the sequences adjacent to the antisense sequences. Experiments conducted on tissue culture cells can be used to identify which particular siRNAs have the desired properties of high activity and target specificity.

Moving from a therapeutic RNAi model that works well in tissue culture to one that is effective in whole animals—and eventually people—requires the development of a delivery strategy that (1) protects the siRNA sequence from degradation before it reaches cells carrying target transcripts, and (2) guides the siRNA sequence across the plasma membrane into those cells. Naked RNA molecules are not well suited for either task: They are rapidly degraded by RNases present in all bodily fluids, and their negatively charged phosphate groups prevent ready entry into the hydrophobic core of the plasma membrane.

Hans-Peter Vornlocher and his group from Kulmbach, Germany, developed a chemical strategy for overcoming siRNA delivery problems in a mouse model for hypercholesterolaemia, a condition caused by excess low-density lipoproteins (LDLs) in blood serum. The liver protein apolipoprotein B (apoB) functions only in LDL biogenesis and thus presents an excellent target for the development of an LDL-specific RNAi therapy. In tissue culture experiments, Vornlocher and colleagues identified an siRNA with high activity and specificity for mouse apoB transcripts. They then synthesized the two strands of the siRNA, but rather than using them directly, they created chemical modifications at the 5′ and 3′ ends of both strands, as illustrated in Figure A2. The modified siRNA product could not be recognized by serum RNases, which do their work by digesting naked RNA from one end or the other.

Vornlocher's second chemical trick was to employ the lipid cholesterol as the chemical entity that was attached to the 3′ end of the sense strand in the siRNA duplex. Cholesterol not only protects one end of the RNA from degradation, but also tends to incorporate itself into plasma membranes, which facilitates the passage of its siRNA cargo into the cell proper. Evidence that this strategy can actually work in a living animal was obtained after injection of the specially modified anti-apoB siRNA into the tail veins of normal mice. Within 24 hours, their serum LDL levels had fallen by over 50 percent. This remarkable result serves as a proof-of-principle for the use of chemical-based siRNA delivery systems in the development of RNAi therapies for treating chronic human diseases.

So far, the effectiveness of RNAi therapies has been demonstrated only in experimental animals. Although the results are very encouraging, these therapies are not yet ready for use in people. To develop human therapies, researchers must design siRNAs that work in human cells and then conduct full-scale clinical trials to ensure the effectiveness and safety of each RNAi protocol.

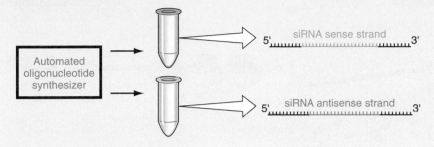

1. *In vitro* synthesis of siRNA

2. Chemically modified siRNA delivery

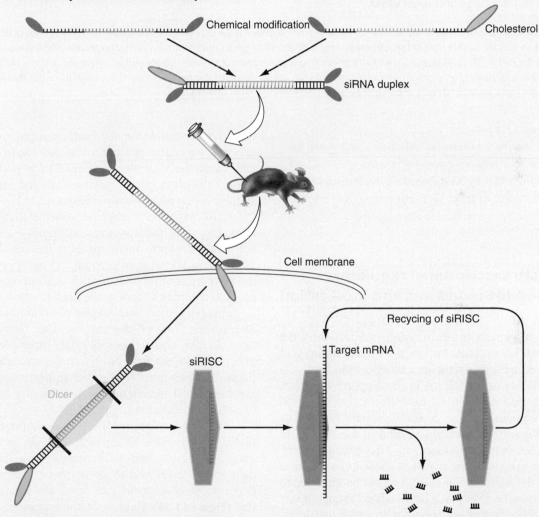

Figure A The development of RNAi therapy. 1. An automated oligonucleotide synthesizer is used to create the antisense and sense strands of a potential siRNA molecule. **2.** The two strands are chemically modified at their 5′ and 3′ ends and then brought together to form an siRNA duplex, which is injected into experimental animals. This chemical modification includes attachment of a cholesterol molecule to the 3′ end of the sense strand. The cholesterol group incorporates itself into the plasma membrane of cells and facilitates entry of the whole siRNA duplex. In the cytoplasm, the RNAi enzyme Dicer recognizes the siRNA, cleaves off its ends, and passes it to RISC. The siRNA-loaded RISC attaches to target mRNA transcripts and destroys them. RISC is then recycled to attack further target mRNAs.

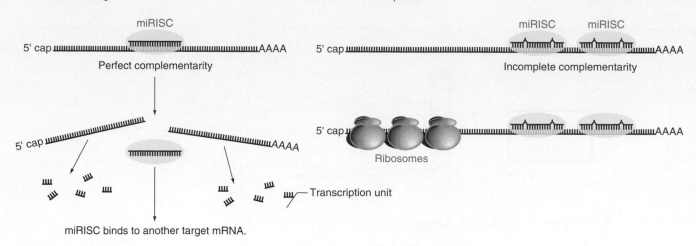

1. mRNA cleavage

2. Translational repression

miRISC binds to another target mRNA.

Figure 11.27 Mechanism of interference. The miRISC can down-regulate gene expression through two different modes of action that are both based on specific binding to a target mRNA. **1.** If the miRNA and its target mRNA contain perfectly complementary sequences, miRISC cleaves the mRNA. The two cleavage products are no longer protected from RNase and are rapidly degraded. **2.** If the miRNA and its target mRNA have only partial complementarity, cleavage does not occur. However, the miRISC remains bound to its target and represses its movement across ribosomes. This mode of down-regulation is less efficient than cleavage.

Small RNAs include a number of subclasses, and more are being discovered. Among these, microRNAs (miRNAs) are incorporated into RNA-induced silencing complexes (miRISCs) that act to repress translation. Small interfering RNAs (siRNAs) detect and destroy foreign double-stranded RNAs.

Other posttranscriptional regulators include half-life indicators and localization markers

The amount of protein made in a cell is affected by the amount of mRNA present. Initiation of transcription and mechanisms to inhibit translation can affect the final outcome, but stability of the mRNA is also important. mRNA contains information about its half-life in the 3′ and 5′ untranslated regions (UTRs). Specific proteins (as well as the miRNAs described earlier) can bind in the 3′ UTR to stabilize mRNA or to cause more rapid degradation.

mRNA also contains information about its localization after leaving the nucleus. Some RNAs must be localized to specific regions where they will be translated at appropriate times, when other factors are present with which they interact. This is especially true in large cells, such as neurons or fertilized eggs, which have polarization of functions.

Proteins may also be modified after translation

The action of a gene is reflected in the activity of its protein product, and a number of posttranslational modifications, including ubiquitination and phosphorylation, affect protein function. Many of these modifications occur extremely rapidly compared with the time it takes to activate gene

transcription and accumulate sufficient protein product for use in a particular process, or to deactivate transcription and await the slow disappearance of a protein product. Thus, cells often rely on posttranslational modification in situations that require a rapid response.

Cells have many enzyme systems that destroy proteins. In one of these systems, ubiquitin—a small, highly conserved protein—functions as a marker. The covalent attachment of chains of ubiquitin to other proteins marks the ubiquitinated proteins for degradation by a large multienzyme complex known as the *proteosome*.

Phosorylation and dephosphorylation often occur in cascades. That is, one protein, after being phosphorylated, is then able to phosphorylate other proteins, which phosphorylate the next protein in the cascade, and so on. Such reactions are found in the transmission of a signal across the cell membranes and eventually to the nucleus (described in detail in Chapter 16).

Another example involving phosphorylation as a regulator of activity is in a process known as *sensitization*; many tissues exposed to hormones for a long time lose their ability to respond to the hormone. An example is the exposure of heart muscle to the stress hormone epinephrine (**Figure 11.28**). Binding of epinephrine to β-adrenergic receptors, located in the plasma membrane of heart muscle cells, normally increases the rate at which the heart contracts. But after several hours of continuous exposure to epinephrine, the heart muscle cells no longer respond in this way. Their sensitization is due to phosphorylation of the β-adrenergic receptors. The phosphorylation does not affect a receptor's ability to bind epinephrine, but it does prevent the receptor from transmitting the hormone signal into the heart muscle cells. The phosphorylation itself depends in large part on the activity of kinase (phosphate-adding) enzymes that phosphorylate the β-adrenergic receptor only

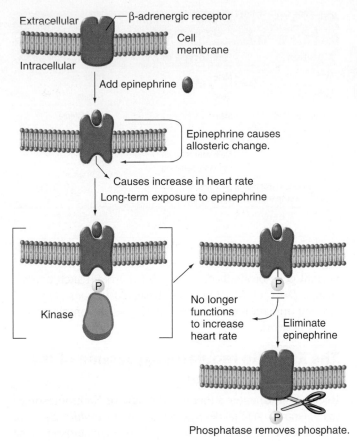

Extracellular

β-adrenergic receptor

Cell membrane

Intracellular

Add epinephrine

Epinephrine causes allosteric change.

Causes increase in heart rate

Long-term exposure to epinephrine

Kinase

No longer functions to increase heart rate

Eliminate epinephrine

Phosphatase removes phosphate.

Figure 11.28 Phosphorylation and desensitization. Covalent phosphorylation of the β-adrenergic receptor has no effect on its binding to epinephrine, but it blocks its downstream function of modulating heart rate.

when the receptor is bound to epinephrine. With the removal of epinephrine from the heart tissue, the kinases no longer act on the receptors, and phosphatase enzymes remove any phosphates already on them. The removal of phosphate from the β-adrenergic receptors eventually restores the heart muscle's ability to respond to new doses of epinephrine.

> Posttranslational modifications may regulate protein function. Ubiquitination targets proteins for breakdown via proteosomes. Phosphorylation and dephosphorylation are responsible for cascade reactions such as which occur in signal transmission, and they also play a role in sensitization of tissues to hormonal signals.

Computer analyses can reveal regulatory mechanisms

Just as with prokaryotes, computer analysis has allowed insight into regulatory mechanisms of eukaryotes. Molecular biologists and biochemists are not yet able to unravel the details of complex regulatory networks, but their knowledge of them increases every day. In addition, the emergence of bioinformatics—a field of science in which biology, computer science, and information technology merge to form a single discipline—promises to facilitate the understanding of complex transcriptional programs.

For example, modern computer programs translate putative open reading frames into *in silico* proteins and recognize motifs within the proteins, such as zinc-finger motifs. These motifs suggest the proteins are transcription factors, and further biochemical analyses can confirm this designation. Sites on *in silico* translated proteins suggest specific posttranslational modifications that could occur and may be important for function.

Possible transcriptional regulatory sites are identified by a global analysis called *phylogenetic footprinting*. In this analysis, genomic sequences of closely related species are compared to find DNA sequences outside of coding regions that are highly conserved between these species. Because noncoding DNA is not usually highly conserved, those sequences that have been conserved suggest important functions such as gene regulation.

Other global analyses of the genome used to analyze transcription factors include the ChIP technology (Chapter 19). Proteins are cross-linked to DNA *in vivo* and the chromosomal DNA is fragmented. The fragments are treated with antibody (Ab) that recognizes a specific transcription factor. The DNA sequence that is precipitated with the Ab and transcription factor can be determined by hybridization to a DNA array or directly sequenced.

A similar process, using the ChIP technology, is being used to identify chromatin patterns and modifications throughout the genome. An Ab that recognizes a specific modification to a histone protein or recognizes a protein that binds to altered chromatin precipitates the protein and its associated DNA. Using this technique, a profile of chromatin modifications can be generated for different types of cells or at different times in development. These approaches are part of the emerging field of epigenomics—the understanding at the global level of the changes in chromatin structure.

11.5 A Comprehensive Example: Sex Determination in *Drosophila*

Male and female *Drosophila* exhibit many sex-specific differences in morphology, biochemistry, behaviour, and function of the germ line (**Figure 11.29**). By examining the phenotypes of flies with different chromosomal constitutions, researchers confirmed that the ratio of X to autosomal chromosomes (X:A) helps determine sex, fertility, and viability (**Table 11.2**). They then carried out genetic experiments that showed that the X:A ratio influences sex through three independent pathways: One determines whether the flies look and act like males or females; another determines whether germ cells develop as eggs or sperm; and a third produces dosage compensation through doubling the rate of transcription of X-linked genes in males. (Note that this strategy of dosage compensation is just the

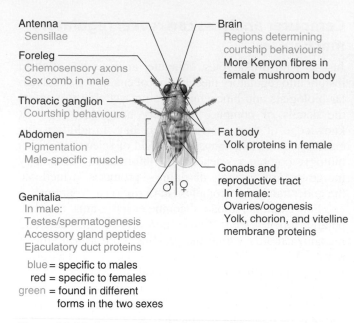

Antenna
Sensillae

Foreleg
Chemosensory axons
Sex comb in male

Thoracic ganglion
Courtship behaviours

Abdomen
Pigmentation
Male-specific muscle

Genitalia
In male:
Testes/spermatogenesis
Accessory gland peptides
Ejaculatory duct proteins

Brain
Regions determining
courtship behaviours
More Kenyon fibres in
female mushroom body

Fat body
Yolk proteins in female

**Gonads and
reproductive tract**
In female:
Ovaries/oogenesis
Yolk, chorion, and vitelline
membrane proteins

blue = specific to males
red = specific to females
green = found in different
forms in the two sexes

Figure 11.29 Sex-specific traits in _Drosophila._ Objects or traits shown in _blue_ are specific to males. Objects or traits shown in _red_ are specific to females. Objects or traits shown in _green_ are found in different forms in the two sexes.

Table 11.2	How Chromosomal Constitution Affects Phenotype in _Drosophila_	
Sex Chromosomes	**X:A**	**Sex Phenotype**
Autosomal Diploids		
X0	0.5	Male (sterile)
XY	0.5	Male
XX	1.0	Female
XXY	1.0	Female
Autosomal Triploids		
XXX	1.0	Female
XYY	0.33	Male
XXY	0.66	Intersex

opposite of that seen in mammals, where the inactivation of one X chromosome in females equalizes the expression of X-linked genes with that in males.)

To simplify this discussion of sex determination in _Drosophila_, we focus on the first-mentioned pathway: the determination of somatic sexual characteristics. An understanding of this pathway emerged from analyses of mutations affecting particular sexual characteristics in one sex or the other. For example, as we saw at the beginning of the chapter, XY flies carrying mutations in the _fruitless_ gene (_fru_) exhibit aberrant male courtship behaviour, whereas XX flies with the same _fruitless_ mutations appear to behave as normal females. **Table 11.3** shows that mutations in other genes also affect the two sexes differently. Clarification of how these mutations influence somatic sex determination came from a combination of genetic experiments (studying, for example, whether one mutation in a

Table 11.3	_Drosophila_ Mutations That Affect the Two Sexes Differently	
Mutation	**Phenotype of XY**	**Phenotype of XX**
_Sx^{fl}*_	Male	Dead
Sxl^{ML}**	Dead	Female
transformer (tra)	Male	Male (sterile)
doublesex (dsx)	Intersex	Intersex
fruitless	Male with aberrant courtship behaviour	Female

*_Sx^{fl}_ is a recessive mutation of _Sex lethal._
**_Sxl^{ML}_ is a dominant mutation of _Sex lethal._

double mutant is epistatic to the other) and molecular biology experiments (in which investigators cloned mutant and normal gene products for analysis). Through such experiments, _Drosophila_ geneticists dissected various stages of sex determination to delineate the following complex regulatory network.

The X:A ratio regulates expression of the _Sex-lethal_ (_Sxl_) gene

Recall from Chapter 3 that it is the ratio of X chromosomes to autosomes (A) that determines sex in _Drosophila._ Since in normal diploids there are two copies of each autosome, the X:A ratio is $2/2 = 1.0$ in a normal XX female and $1/2 = 0.5$ in a normal XY male. In short, when the X:A ratio is 1.0, females develop; when the ratio is 0.5, males develop.

Numerator and denominator elements

Key factors of sex determination are helix-loop-helix proteins encoded by genes on the X chromosome. Sisterless-A (Sis-A) and Sisterless-B (Sis-B) are two such proteins. Referred to as _numerator elements_, these two proteins monitor the X:A ratio through the formation of homodimers containing two of the same kind of subunit, or heterodimers containing two different subunits. The homodimers consist of two numerator elements, whereas the heterodimers are composed of one numerator element and one denominator element. _Denominator elements_ are helix-loop-helix proteins that are encoded by genes on autosomes. Because the number of X chromosomes determines the ratio of numerator homodimers to numerator/denominator heterodimers, the homodimers of numerator elements provide a measure of the X:A ratio (**Figure 11.30**).

The observation that in flies with a greater number of numerator homodimers, transcription of the _Sxl_ gene occurs early in development suggests that numerator subunit homodimers may function as transcription factors that turn on _Sxl._ In this hypothesis, the association of denominator subunits with numerator subunits sequesters the numerator elements in inactive heterodimers that cannot activate transcription. Females produce enough numerator subunits, however, that some remain unbound by denominator elements. Homodimers formed from these free numerator elements act as transcriptional activators of _Sxl_ at the P_e

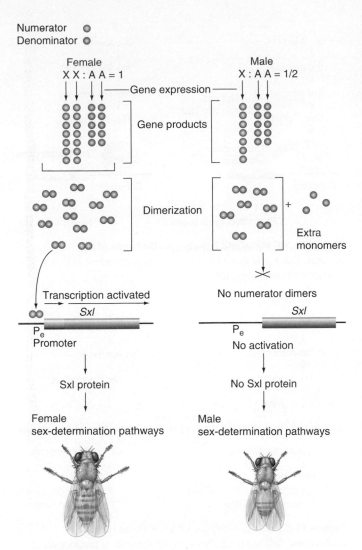

Figure 11.30 The X:A ratio determines the expression of Sxl. Numerator elements are produced by the X chromosome at a slightly higher level than denominator elements are produced by autosomes. When the X:A ratio is 1 (in females), there are too many numerator elements to be occupied by denominators, and those not sequestered can form homodimers, which act as activators of the *Sxl* gene. When the X:A ratio is 1/2 (in males), there are fewer numerators than denominators, and all the numerators become sequestered.

promoter early in development. Males, by contrast, carry only half as many X-encoded numerator subunits; thus, the abundant denominator proteins tie up all the numerator elements, and as a result, there are no free numerator elements in males to turn on the P_e promoter of the *Sxl* gene.

Although this model is likely to be an oversimplification, it suggests how different X:A ratios might activate and repress transcription of the *Sxl* gene.

The action of the Sxl protein in females

The Sxl protein produced early in the development of female embryos participates later in an autoregulatory feedback loop as just described (review Figure 11.24). In this self-regulating system, the Sxl protein catalyzes the synthesis of more of itself through RNA splicing of the P_L-initiated transcript, which results in a productive mRNA.

By contrast, in males where there is no transcription of *Sxl* early in development, activation of the P_L promoter later in embryonic development results in an unproductive *Sxl* transcript containing a stop codon near the beginning of the message. Because no Sxl protein is present to splice out the problem stop codon, this unproductive transcript is not translated to protein—and males thus have no Sxl protein at any point in development.

The effects of *Sxl* mutations

Recessive *Sxl* mutations that produce nonfunctional gene products have no effect in XY males, but they are lethal in XX females (see Table 11.3). The reason is that males, which do not normally express the *Sxl* gene, do not miss its functional product, but females, which depend on the Sxl protein for sex determination, do. The absence of the Sxl protein in females allows the aberrant expression of certain male-specific dosage-compensation genes that increase transcription of genes on the X chromosome—and the hypertranscription of these X-linked genes on two X chromosomes in mutant females proves lethal.

By comparison, rare dominant *Sxl* mutations that allow production of Sxl protein even in XY embryos are without effect in females but lethal to males. In these mutants, the *Sxl* gene product indirectly represses transcription of genes that males need to express for dosage compensation. Without the products of these male-specific dosage-compensation genes, males cannot hypertranscribe X-linked genes and thus do not have enough X-linked gene products to survive.

The Sxl protein triggers a cascade of splicing

In addition to splicing its own transcript, the Sxl protein influences the splicing of RNAs transcribed from other genes. Among these is the *transformer* (*tra*) gene. In the presence of the Sxl protein, the *tra* primary transcript undergoes productive splicing that produces an mRNA translatable to a functional protein. In the absence of Sxl protein, the splicing of the *tra* transcript results in a nonfunctional protein (**Figure 11.31a**).

The cascade continues. The functional Tra protein synthesized only in females, along with another protein encoded by the *tra2* gene (which is transcribed in both males and females), influences the splicing of the *doublesex* (*dsx*) gene's primary transcript. This splicing pathway results in the production of a female-specific Dsx protein called Dsx-F. In males, where there is no Tra protein, the splicing of the *dsx* primary transcript produces the related, but different, Dsx-M protein (**Figure 11.31b**). The N-terminal parts of the Dsx-F and Dsx-M proteins are the same, but the C-terminal parts of the proteins are different.

The Dsx-F and Dsx-M proteins control development of somatic sexual characteristics

Although both Dsx-F and Dsx-M function as transcription factors, they have opposite effects. In conjunction with the

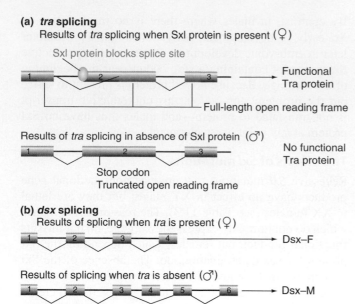

(a) *tra* splicing
Results of *tra* splicing when Sxl protein is present (♀)

Sxl protein blocks splice site

Functional
Tra protein

Full-length open reading frame

Results of *tra* splicing in absence of Sxl protein (♂)

No functional
Tra protein

Stop codon
Truncated open reading frame

(b) *dsx* splicing
Results of splicing when *tra* is present (♀)

Dsx–F

Results of splicing when *tra* is absent (♂)

Dsx–M

Figure 11.31 Regulation by alternative splicing. (a) The presence of *Sxl* alters the splicing of *tra* mRNA. Female transcripts produce functional Tra protein, while male transcripts have a truncated open reading frame and are unable to produce Tra. **(b)** Tra protein, in turn, plays a role in altering the splicing pattern of the *dsx* mRNA. A different Dsx product results in males (Dsx-M) rather than in females (Dsx-F).

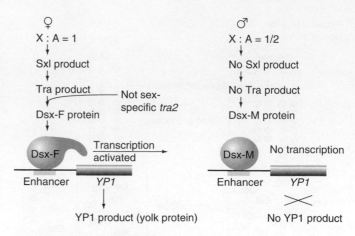

Figure 11.32 Male- and female-specific forms of Dsx protein. Dsx-F acts as a transcriptional activator, whereas Dsx-M acts as a transcriptional repressor.

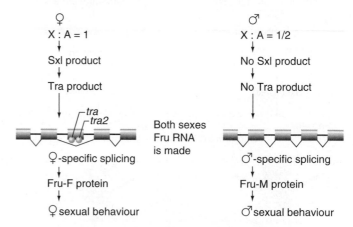

Figure 11.33 The primary *fru* RNA transcript is made in both sexes. Splicing occurs unhindered in males to produce an mRNA, which is translated into the Fru-M protein product. But tra protein (present only in females) causes alternative splicing of the *fru* transcript to produce an alternative mRNA, which encodes an alternative protein product, Fru-F.

protein encoded by the *intersex* (*ix*) gene, Dsx-F functions mainly as a repressor that prevents the transcription of genes whose expression would generate the somatic sexual characteristics of males. Dsx-M, which works independently of the intersex protein, accomplishes the opposite: the activation of genes for the somatic sexual characteristics of males and the repression of genes that determine female somatic sexual characteristics.

Interestingly, the two Dsx proteins can bind to the same enhancer elements, but their binding produces opposite outcomes (**Figure 11.32**). For example, both bind to an enhancer upstream of the promoter for the *YP1* gene, which encodes a yolk protein; females make this protein in their fat body organs and then transfer it to developing eggs. The binding of Dsx-F stimulates transcription of the *YP1* gene in females; the binding of Dsx-M to the same enhancer region inactivates transcription of *YP1* in males, working in conjunction with other transcription factors.

Mutations in *dsx* affect both sexes, because in both males and females, the production of Dsx proteins represses certain genes specific to development of the opposite sex. Null mutations in *dsx* that make it impossible to produce either functional Dsx-F or Dsx-M result in intersexes that cannot repress either certain male-specific or certain female-specific genes.

The Tra and Tra2 proteins also help regulate expression of the *fruitless* gene

We saw at the beginning of this chapter that the courting song and dance of male *Drosophila* are among the

sexual behaviours under the control of the *fruitless* (*fru*) gene. As it turns out, the *fru* primary transcript is another regulatory target of the Tra and Tra2 splicing factors (**Figure 11.33**). In females, whose cells make both Tra and Tra2 proteins, splicing of the *fru* transcript produces an mRNA that encodes a protein we refer to as Fru-F. In males, whose cells carry no Tra protein, alternative splicing of the *fru* transcript generates a related Fru-M protein with 101 additional amino acids at its N terminus. As we mentioned at the beginning of this chapter, these additions almost certainly determine some of the observed differences between male and female behaviour. Because both Fru-F and Fru-M have the zinc-finger motifs characteristic of transcription factors, they probably activate and repress genes whose sex-specific products help generate courting behaviours.

The sex-specific products of *fru* appear in only a few cells in the nervous system, and the location of these neurons is significant. Some are in regions known to help

regulate the courtship song; others are in areas that process chemosensory information from the antennae (perhaps in neurons that receive pheromone signals); still others are in regions that control abdominal movements (suggesting how *fruitless* may influence the male's curling of the abdomen during mating). To understand precisely how changes in gene expression in these few cells control sexual behaviours, *Drosophila* researchers are now trying to discover which genes are the targets of transcriptional regulation by the fruitless protein.

> Sex determination in *Drosophila* illustrates several kinds of gene regulation. X-encoded numerators and autosome-encoded denominators allow assessment of the "femaleness" of a fly, leading to activation of *Sxl* transcription in early female embryos. The Sxl protein then acts as a splicing factor to perpetuate its synthesis in females. Additional splicing cascades result in female- and male-specific versions of Dsx proteins that determine sexual characteristics and behaviour.

Connections

Multiple controls regulate gene activity and function from imprinting, to chromatin remodelling, to the initiation of transcription, to the processing of RNA transcripts, to RNA interference, to the chemical modification of final gene products. At the outset, the regulation of transcription occurs through the interaction of *cis*-acting DNA regions and a variety of transcription factors.

Accurate regulation of gene function is crucial for proper control of development and of the cell cycle. Indeed, a critical network of *cis*-acting control regions, *trans*-acting factors, and protein modifications promotes cell growth, DNA replication, and cell division in response to certain environmental signals, and also delays these events in response to other signals, such as DNA damage.

In Chapter 16, we describe the regulatory network controlling the cell cycle in normal cells and explain how mutations that disrupt one or more aspects of that network can result in cancer.

In Chapters 12 and 13, we shift our focus from the analysis of gene activity in individuals to an analysis of gene transmission in whole populations and an examination, at the molecular level, of how genes and genomes evolve over time. Chapter 12 describes why an understanding of evolution requires knowledge of gene transmission in populations. Chapter 13 then builds on ideas presented throughout the book to reconstruct the molecular strategies by which genes and genomes have evolved throughout the roughly 4 billion years of life on Earth.

Essential Concepts

1. Transcriptional initiation is a critical point in the regulation of gene activity. Analyses of mutations that affect a gene's function without changing the sequence of its product provided insight into this level of regulation. Through these mutations, researchers defined *cis*-acting DNA regulatory elements and *trans*-acting transcription factors. **[LO1–2]**

2. Two types of *cis*-acting regulatory regions—promoters and enhancers—are associated with genes transcribed by RNA polymerase II. The promoters are located at the 5′ end of the gene they influence. Basal factors bind to promoters to allow a low, nonspecific basal level of gene transcription. The enhancers have a more variable location in relation to the genes they control. **[LO1]**

3. The association of transcription factors with enhancer elements can modulate levels of transcriptional initiation. Activation is mediated by transcription factors called activators, which bind to enhancers. Activators can interact with basal factors at the promoter to increase transcription above the basal level. Repressors can compete with activators for enhancer binding or quench the ability of activators to carry out their

function. Activators and repressors that bind directly to DNA often form homodimers and/or heterodimers, which can be a prerequisite for them to function as transcription factors. **[LO3]**

4. The unravelling of the DNA in chromatin is an initial step in activation. Hypercompaction of chromatin domains causes transcriptional silencing by blocking access to the promoter and enhancers of a gene and thereby preventing its activation even in the presence of activator proteins. **[LO4]**

5. Genomic imprinting is an example of epigenetic control over gene expression. Imprinting operates on the copy of a gene received from one parent but not the other. DNA methylation plays a role in the maintenance of imprinting from one mammalian somatic cell generation to the next. **[LO5]**

6. Although the regulation of most genes depends primarily on controls over transcription, in some cases, further regulation down the path to protein production also plays a role. Modulation of gene function can occur through changes in RNA splicing, RNA interference, changes in the efficiency of translation, and chemical modification of the gene product. **[LO6]**

I. You are studying expression of a gene whose protein product is made after UV irradiation. You cloned the gene and made antibody to the protein.

a. If expression is regulated by turning on transcription after UV exposure, what results would you predict from hybridizing a DNA probe to RNA isolated from cells before and after UV irradiation (Northern analysis) and from incubating the antibody to proteins isolated from cells before and after UV treatment (Western analysis)?

b. If expression is regulated by preventing translation, what results would you predict from doing similar Northern and Western analyses?

Answer

To answer this question, you need to consider the consequences of transcriptional and translational regulation on expression and think through what happens experimentally in Northern and Western analyses.

a. If a gene is transcriptionally regulated, the mRNA will not be present in cells that were not exposed to UV. *There will be no hybridizing band in the Northern analysis of mRNA from unexposed cells. The mRNA will be present in cells that have been treated with UV, and there will be a hybridizing band.* Similarly, the protein will only be found in cells that were exposed to UV, and *the antibody will bind to its protein target only in the protein preparation from exposed cells.*

b. If expression is regulated at the translation step, mRNA will be present in the cells whether they have been exposed to UV or not. *Hybridizing bands will be found in both RNA samples. The protein will be present only in those cells that were exposed to UV, so signal will be seen only in the exposed preparation.*

II. The retinoic acid receptor (RAR) is a transcription factor that is similar to steroid hormone receptors. The substance (ligand) that binds to this receptor is retinoic acid. One of the genes whose transcription is activated by retinoic acid binding to the receptor is *myoD*. The diagram at the end of this problem shows a schematic of the RAR protein produced by a gene into which two different 12-base oligonucleotides had been inserted in the sequences encoding sites indicated by a–m. For constructs encoding a–e, oligonucleotide 1 (TTAATTAATTAA) was inserted into the *RAR* gene. For constructs encoding f–m, oligonucleotide 2 (CCGGCCGGCCGG) was inserted into the gene. Each mutant protein was tested for its ability to bind retinoic acid, bind to DNA, and activate transcription of the *myoD* gene. Results are tabulated as follows (the insertion site associated with each mutant protein is indicated with the appropriate letter on the polypeptide map).

NH₂ —f— g —h— i —j— k —l— m— COOH
a   b   c   d   e

Mutant	Retinoic acid binding	DNA binding	Transcriptional activation
A	−	−	−
B	−	−	−
C	−	−	−
D	−	+	+
E	+	+	+
F	+	+	+
G	+	+	−
H	+	+	−
I	+	−	−
J	+	−	−
K	−	+	+
L	−	+	+
M	+	+	+

a. What is the effect of inserting oligonucleotide 1 anywhere in the protein?

b. What is the effect of inserting oligonucleotide 2 anywhere in the protein?

c. Indicate the three protein domains on a copy of the preceding drawing.

Answer

This question involves the concepts of domains within proteins and use of the genetic code to understand effects of oligonucleotide insertions.

a. Oligonucleotide 1 contains a stop codon in any of its three reading frames. This means it will *cause termination of translation of the protein wherever it is inserted.*

b. Oligonucleotide 2 does not contain any stop codons and so will *just add amino acids to the protein.* Because there are 12 bases in the oligonucleotide, it will not change the reading frame of the protein. *Insertion of the oligonucleotide can disrupt the function of a site in which it inserts.*

c. Looking at the data overall, notice that all mutants that are defective in DNA binding are also defective in transcriptional activation, as would be expected for a transcription factor that binds to DNA. The mutants that will be informative about the transcriptional activation domain are those that do not have a DNA-binding defect. Inserts a, b, and c using oligonucleotide 1, which truncates the protein at the site of insertion, are defective in all three activities. The protein must be made at least as far as point d before DNA binding or transcription activation are seen. These two activities must lie

before d. Truncation at d is negative for retinoic acid binding, but the truncation at e does bind to retinoic acid. The retinoic-acid-binding activity must lie before e. Using the oligonucleotide 2 set of insertions, transcriptional activation was disrupted by insertions at sites g and h, indicating that this region is part of the transcriptional domain; i and j insertions disrupted the DNA binding; and k and l insertions disrupted the retinoic acid binding. The minimal endpoints of domains as determined from these data are summarized in the following schematic.

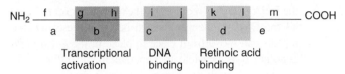

Transcriptional activation — DNA binding — Retinoic acid binding

III. A cDNA clone that you isolated using pituitary gland mRNA from mice was used as a probe against a blot containing RNAs from embryonic heart (EH), adult heart (AH), embryonic pituitary (EP), adult pituitary (AP), and testis (T). The results of the hybridization are shown here.

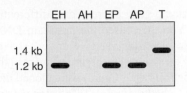

a. What would you conclude about this gene based on the result with AH RNA?

b. How would you explain the result with testis RNA?

Answer

This problem requires an understanding of RNA and transcription.

a. No RNA from adult heart (AH) hybridized with the probe, indicating that the *gene is not transcribed in this tissue.*

b. *A different-sized RNA is seen in the testis sample.* This could be due to *alternate splicing of the transcript or a different start site* in testis compared with other tissues.

Problems

Vocabulary

1. For each of the terms in the left column, choose the best matching phrase in the right column.

i. basal factors	**1.** marks a protein for degradation
ii. transcriptional silencing	**2.** pattern of expression is dependent on which parent transmitted the allele
iii. activators	**3.** multimers of non-identical subunits
iv. imprinting	**4.** heterochromatin
v. RNAi	**5.** multimers of identical subunits
vi. coactivators	**6.** bind to enhancers
vii. homomers	**7.** bind to promoters
viii. heteromers	**8.** bind to activators
ix. ubiquitination	**9.** prevents or reduces gene expression posttranscriptionally

Section 11.1

2. Does each of the following types of gene regulation occur in eukaryotes only? in prokaryotes only? in both prokaryotes and eukaryotes?

a. differential splicing

b. positive regulation

c. chromatin compaction

d. attenuation

e. negative regulation

3. List five events other than transcription initation that can affect the type or amount of active protein produced in a cell.

Section 11.2

4. Which eukaryotic RNA polymerase (RNA pol I, pol II, or pol III) transcribes which genes?

a. tRNAs

b. mRNAs

c. rRNAs

d. miRNAs

5. Which of the following types of fusion gene would you use for which purpose? (The slash indicates the fusion, and the parts of each type of fusion are given in the order in which they would occur.)

Types of fusions:

a. random mouse sequences/*lacZ* gene

b. mouse metallothionein promoter/a mouse gene

Uses:

i. to identify genes turned on in neurons

ii. to turn on expression of a gene by including the metal Zn in the diet

6. You isolated a gene expressed in differentiated neurons in mice. You then fused the upstream DNA and beginning of the gene to *lacZ* (reporter gene) so that you could monitor expression. Different fragments (shown as *dark lines* in the following figure) were cloned next to the *lacZ* gene that lacked a promoter. The clones were introduced into neurons in tissue culture to monitor expression. From the results that follow, which region contains the promoter and which contains an enhancer?

Regulatory region of nerve gene
M M H M   M H

lacZ

lacZ expression   Fragments fused to *lacZ*
0
0
0
5

5
80
5

80

7. In yeast, the GAL4 protein binds to DNA to activate transcription of GAL7 or GAL10. GAL80 represses expression by binding to GAL4 protein and preventing it from binding to DNA. In which gene(s) should you be able to isolate galactose constitutive mutations, and in each case, what characteristics of the protein would the mutation disrupt?

8. A single enhancer site regulates expression of three adjacent genes *GAL1*, *GAL7*, and *GAL10*, but the genes are not co-transcribed as one mRNA. How could you show experimentally that each gene is transcribed separately?

9. Which of the listed motifs is associated with DNA binding, transcription activation, or dimer formation?

  a. zinc finger

  b. helix-loop-helix

  c. leucine zipper

  d. acidic region

  e. helix-turn-helix

10. How could you make a library of genes expressed during sporulation in yeast?

11. *MyoD* is a transcriptional activator that turns on the expression of several muscle-specific genes in human cells. The *Id* gene product inhibits *MyoD* action. How could you determine if *Id* acts by quenching *MyoD* or by blocking access to the enhancer? What differences would you expect to see experimentally?

12. a. Assume that two transcription factors are required for expression of the blue pigmentation genes in pansies. (Without the pigment, the flowers are white.) What phenotypic ratios would you expect from crossing strains heterozygous for each of the genes encoding these transcription factors?

  b. Now assume that either transcription factor is sufficient to get blue colour. What phenotypic ratios would you expect from crossing strains heterozygous for each of the genes encoding these transcription factors?

Section 11.3

13. a. You want to create a genetic construct that will express the β-galactosidase enzyme encoded by *E. coli*'s *lacZ* gene in *Drosophila*. In addition to the *lacZ* coding sequence, what DNA element(s) must you include in order to express this protein in flies if the construct could somehow become integrated into the *Drosophila* genome? Where should such DNA element(s) be located?

  b. In making your construct, you place inverted repeats found at the ends of a particular type of transposable element on either side of the *lacZ* coding region and all of the DNA elements required by the answer to part *a*. Since all the DNA sequences located between these inverted repeats can move from place to place in the *Drosophila* genome, it is possible to generate many different fly strains, each with the construct integrated at a different location in the genome. You treat animals from each strain with a chemical that turns blue in the presence of β-galactosidase. Animals from different strains show different patterns: Some show blue staining in the head, others in the thorax, some show no blue colour, etc. Explain these results and describe a potential use of your construct.

14. In the previous problem (#13), you identified a region that is likely to behave as an enhancer. What experiments could you perform to verify that these DNA sequences indeed share all the characteristics of an enhancer?

15. What experimental evidence indicates that chromatin structure acts to reduce basal levels of transcription?

16. Which of the following would be suggested by a DNase hypersensitive site?

  a. No transcription occurs in this region of the chromosome.

  b. The chromatin is in a more open state than a region without the hypersensitive site.

  c. Transcription terminates at this site.

17. You isolated nuclei from liver cells, treated them with increasing amounts of DNaseI, stopped the reactions, and isolated the DNA from each sample. You next treated the DNAs with the restriction enzyme *Eag*I, electrophoresed the DNAs, transferred them to a blot, and hybridized the blot with a probe from the gene you are studying. With no DNaseI treatment, there was a 20-kb *Eag*I fragment that hybridized with your probe. With trace amounts of DNaseI, two bands of 16 and 4 kb were present. The same DNaseI treatment of DNA from muscle cells produced only a 20-kb fragment. What does this result tell you about the region of DNA?

18. From Northern analysis, you find that the *ADAG* gene is expressed only in the brain. You examine expression in glial and neuronal cell lines (two types of cells in the brain) and find that only glial cells make *ADAG* mRNA. No one has characterized the *cis*- or *trans*-acting elements required for glial-specific expression, so you decide to do so. You make a set of deletions in the regulatory region and fuse these to the *lacZ* gene so you can easily monitor the expression after introducing the clones into tissue culture cells derived from a glial tumour. Deletions beginning at a site upstream of the

gene and extending to base −85 (with the transcription start site considered −1 and bases prior to the start having negative designations) still retain full activity, but a deletion to −75 leaves only 1 percent of the original activity.

a. What do these findings tell you?

You now mix a DNA fragment from this region that is labelled with ^{32}P at the 5′ end of one strand with a purified glial-specific transcription factor. You perform a DNaseI footprinting experiment (as described in Figure 10.20), and obtain the results tabulated below. Lane 1 shows the DNaseI reaction of the labelled DNA alone, Lane 2 is the reaction involving the mixture of DNA and protein. You also analyze the DNA sequence of the same DNA fragment on the same gel.

DNA Sequencing Reactions   **DNaseI Footprinting**

C	T	A	G		1	2

−60

−90

b. What is the sequence of the segment containing the binding site(s) for the glial-specific transcription factor?

c. Additional evidence indicates that the glial-specific factor binds to DNA as a dimer, and each monomer binds the same target sequence. Identify the likely binding sites for the two monomers.

19. Match the gene expression phenomenon with molecular components that modulate each.

    a. transcriptional silencing   **1.** insulator

    b. imprinting   **2.** heterochromatin

    c. restricting the range of enhancer activity   **3.** TBF, TAFs

    d. basal transcription   **4.** methylation

20. Epigenetic changes that are inherited from one generation to the next can involve which of the following?

    a. Methylation of histones

    b. Methylation of DNA

    c. Change in DNA sequence

21. Prader–Willi syndrome is caused by a mutation in a maternally imprinted gene. Answer the following questions as true or false, assuming that the trait is 100 percent penetrant.

    a. Half of the sons of affected males will show the syndrome.

    b. Half of the daughters of affected males will show the syndrome.

    c. Half of the sons of affected females will show the syndrome.

    d. Half of the daughters of affected females will show the syndrome.

22. A boy expresses a mutant phenotype because he has received a mutation in a paternally imprinted gene. From which parent did the boy inherit the mutant allele?

23. The *IGF-2R* gene is autosomal and maternally imprinted. Copies of the gene received from the mother are not expressed, but copies received from the father are expressed. You have found two alleles of this gene that encode two different forms of the IGF-2R protein distinguishable by gel electrophoresis. One allele encodes a 60K blood protein; the other allele encodes a 50K blood protein. In an analysis of blood proteins from a couple named Bill and Joan, you find only the 60K protein in Joan's blood and only the 50K protein in Bill's blood. You then look at their children, Jill and Bill Jr. Jill is producing only the 50K protein, while Bill Jr. is producing only the 60K protein.

    a. With these data alone, what can you say about the *IGF-2R* genotype of Bill Sr. and Joan?

    b. Bill Jr. and a woman named Sara have two children, Pat and Tim. Pat produces only the 60K protein and Tim produces only the 50K protein. With the accumulated data, what can you now say about the genotypes of Joan and Bill Sr.?

24. Assume that the disease illustrated with the pedigree below is due to expression of a rare allele of an autosomal gene that is a paternally imprinted. What would you predict is the genotype of individuals (a) I-1, (b) II-1, and (c) III-2?

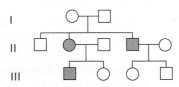

25. Follow the expression of a paternally imprinted gene through three generations. Indicate whether the copy of the gene from the male in generation I is expressed in the germ cells and somatic cells of the individuals listed.

    a. generation I male (I-2) germ cells

    b. generation II daughter (II-2): somatic cells

    c. generation II daughter (II-2): germ cells

    d. generation II son (II-3): somatic cells

    e. generation II son (II-3): germ cells

    f. generation III grandson (III-1): somatic cells

    g. generation III grandson (III-1): germ cells

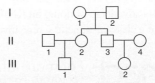

Section 11.4

26. Excluding the possible rare polycistronic message, how can a single mRNA molecule in a eukaryotic cell produce several different proteins?

27. What events occur during processing of a primary transcript?

28. You are studying muscle cells and have found a protein that is only made in this tissue. The data here are from analysis of RNA (Northern blot) using a DNA probe from the gene and analysis of protein (Western blot) using antibody directed against the protein. Is this gene transcriptionally or translationally regulated?

RNA analysis

| Muscle cell RNA | Nerve cell RNA | Skin cell RNA |

Protein analysis

| Muscle cell protein | Nerve cell protein | Skin cell protein |

29. The *hunchback* gene, one of the genes necessary for setting up the anterior-posterior axis of the *Drosophila* embryo, is translationally regulated. The position of the coding region within the transcript is known, and there is additional sequence beyond the coding region at the 5′ and 3′ ends of the mRNA. How could you determine if the sequences at the 5′ or 3′ or both ends are necessary for proper regulation of translation?

30. You isolated a cDNA from skin cells, and when you hybridized that cDNA as a probe with a blot containing mRNAs from skin cells and nerve cells you saw a 1.2 and a 1.3 kb fragment, respectively. How could you explain the different-sized cDNAs?

31. From Northern and Western hybridization, you know that the mRNA and protein produced by a tissue-specific gene are present in brain, liver, and fat cells, but you detect an enzymatic activity associated with this protein only in fat cells. Provide an explanation for this phenomenon.

32. Modern-day geneticists are very excited by the prospect of using RNA interference as a way to find genes involved in various biological processes, such as mitosis or the development of specific body parts like the pancreatic cells that make insulin.

   a. How would you perform an RNAi-based screen to find genes involved in these processes?

   b. What is the advantage of doing an RNAi-based screen as opposed to a classical genetic screen involving mutagens?

   c. What are the disadvantages of performing an RNAi-based screen?

33. You are studying a strain of transgenic mice that express an *E. coli lacZ* reporter gene under the control of *cis*-acting regulatory elements that normally control an interesting gene needed for the early development of mice. Previous evidence from Northern blots indicates that mRNA for the gene of interest can be identified between days 8.5 and 10.5 of gestation. In your strain, staining for β-galactosidase (the protein product of *lacZ*) can be seen from about day 8.75 until at least day 12.

   a. Explain the discrepancy between mRNA and protein expression.

   b. Would you expect β-galactosidase protein expression to indicate more accurately the normal onset of activity for this gene, or the normal cessation of this gene's activity? Explain.

Section 11.5

34. The *Drosophila* gene *Sex lethal* (*Sxl*) is very deserving of its name. Certain alleles have no effect on XY animals, but cause XX animals to die early in development. Other alleles have no effect on XX animals, but XY animals with these alleles die early in development. Thus, some *Sxl* alleles are lethal to females, while others are lethal to males.

   a. Would you expect a null mutation in *Sxl* to cause lethality in males or in females? What about a constitutively active *Sxl* mutation?

   b. Why do *Sxl* alleles of either type cause lethality in a specific sex?

   The gene *transformer* (*tra*) gets its name from "sexual transformation," since some *tra* alleles can change XX animals into sterile males, while other *tra* alleles can change XY animals into normal-appearing females.

   c. Which of these sex transformations would be caused by null alleles of *tra* and which would be caused by constitutively active alleles of *tra*?

   d. XX animals carrying particular alleles of *tra* develop as males, but they are sterile. Why?

   e. In contrast with *Sxl*, null *tra* mutations do not cause lethality either in XX or in XY animals. However, the Sxl protein regulates the production of the Tra protein. Why then do all *tra* mutant animals survive?

   f. Predict the consequences of null mutations in *tra2* on XX and XY animals.

   g. XY males carrying loss-of-function mutations in the *fruitless* (*fru*) gene display aberrant courtship behaviour. Would you predict that either XX or XY animals with wild-type alleles of *fru* but loss-of-function mutations of *tra* would also court abnormally?

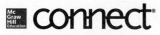 **For more information on the resources available from McGraw-Hill Ryerson, go to www.mcgrawhill.ca/he/solutions.**

CHAPTER **12** **Variation and Selection in Populations**

The enormous range of genetic diversity within tomatoes, sheep, and humans is easy to see.

Tuberculosis (TB) is an ancient and persistent human disease. Bone deformities typical of those produced by the infection are found in Egyptian mummies dating to 2000–4000 BC; and as recently as the mid-nineteenth century, TB was the leading cause of death in Canada (at the time of Confederation in 1867), Europe, and the urban United States. The microbe that causes TB is the bacterium *Mycobacterium tuberculosis*. In humans, populations of *M. tuberculosis* most often infect the lungs and lymph nodes, but sometimes they colonize the bones and skin of a patient (**Figure 12.1a**). *M. tuberculosis* bacteria can spread from person to person through the air when an infected individual exhales bacteria from his or her lungs during coughing.

Beginning in the late nineteenth century, improved sanitation and the quarantine of TB patients led to a steady decline in the incidence and death rates from TB in Canada, Europe, and the United States (**Figure 12.1b**). The introduction of antibiotics during the 1940s and 1950s further reduced TB mortality in those areas, and by the 1960s, many people believed that the disease had been eradicated, at least in North America. But 25 years later, the incidence of TB began to rise in urban areas around the globe. By 2000, TB accounted for more deaths worldwide than any other identifiable infectious disease, claiming close to 3 million lives annually. Three factors contributed to this rapid increase in TB incidence: the emergence of AIDS, which weakens the human immune system; protein deficiencies among the malnourished; and the widespread occurrence of *M. tuberculosis* strains that are resistant to one or more antibiotics.

Despite the cultural, technological, and medical advances of the twentieth century, infectious microorganisms such as *M. tuberculosis* are still among the leading causes of death in many human populations. And in a related arena, populations of plant pests (such as the mites that prey on strawberry plants and almond trees) destroy a substantial fraction of human food supplies.

How do new diseases emerge in human populations? Why do diseases persist in all living organisms? What causes diseases and pests long under control to resurge in frequency and intensity?

One way to answer these questions is to examine genetic variation and its expression as phenotypes within populations of organisms. The scientific discipline that studies what happens in whole populations at the genetic level is known as **population genetics**. It encompasses the evolutionary ideas of Darwin, the laws of Mendel, and the insights of molecular biology.

In this chapter, we explore the nature of genotypic and phenotypic variation within populations and the role of this variation in evolution. We know from Chapter 2 that variation exists in all populations. To begin, we analyze the incidence of genetic diseases, such as cystic fibrosis and retinoblastoma, which are determined by a single gene. Through our analysis, we develop a framework for understanding how the frequency of a disease-causing allele determines the frequency of diseased individuals in a population. We next examine the effects of population size and chance on changes in allele frequency.

Finally, we consider variation in multifactorial (also known as quantitative or complex) traits; that is, in traits determined by two or more genes and their interaction with the environment (see Chapter 2). In fact, most aspects of disease susceptibility are multifactorial.

Chapter Outline
12.1 The Hardy–Weinberg Law: Predicting Genetic Variation in Populations
12.2 Causes of Allele Frequency Changes
12.3 Analyzing Quantitative Variation

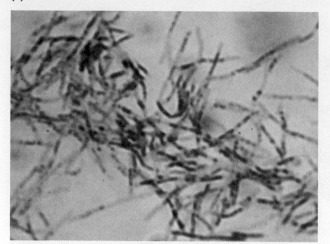

(a)

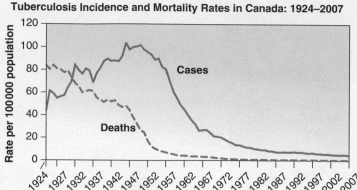

(b)

Tuberculosis Incidence and Mortality Rates in Canada: 1924–2007

Figure 12.1 Tuberculosis in human populations. (a) Photograph of *M. tuberculosis* bacteria colonizing the lungs and bones. **(b)** Case and death rates from tuberculosis in Canada from 1924–2007.

One general theme emerges from our discussion: Population geneticists rely on mathematical models in predicting a population's potential for stasis or change because most of the scientifically useful questions they ask are statistical.

Simple mathematical models not only clarify the questions about frequency of genetic diseases or rate of spread of pathogens, but they also serve as tools for analyzing data and making predictions about future populations.

Learning Objectives

1. Differentiate between population and gene pool.

2. Determine phenotype frequencies, genotype frequencies, and allele frequencies in a population.

3. Explain how the Hardy–Weinberg law was derived and the conditions required for a population to be in Hardy–Weinberg equilibrium.

4. Demonstrate the utility of the Hardy–Weinberg equation in natural populations.

5. Analyze the effects of mutation, genetic drift, and natural selection on allele frequencies in populations.

6. Distinguish between environmental variance and genetic variance and explain how they are used to measure total phenotypic variation in a population.

7. Compare and contrast broad-sense and narrow-sense heritability and how each is measured.

8. Evaluate the importance of narrow-sense heritability in breeding experiments and evolutionary studies.

12.1 The Hardy–Weinberg Law: Predicting Genetic Variation in Populations

Modern genetics began with Mendel's elucidation of formal rules of probability that describe the likelihood of transmission of genes and traits from parents to offspring in controlled breeding experiments. In this section, we describe an extension of Mendel's work that provides researchers with genetic tools for predicting transmission frequencies of traits and alleles in natural populations having an unlimited size.

Population geneticists describe populations with well-defined terms

To population geneticists, a **population** is a group of interbreeding individuals of the same species that inhabit the same space at the same time. An example would be all of the grizzly bears in the Khutzeymateen Valley on the

northern coast of British Columbia, Canada or all of the salmon in the Khutzeymateen River and Inlet on which the grizzly bears feed. The sum total of all alleles carried in all members of a population is that population's **gene pool**. In nature, the genetic makeup of a population changes over time as new alleles arise by mutation or are introduced by immigration, and as rare, pre-existing alleles disappear when all individuals carrying them leave the population or die. Changes in the frequency of alleles within a population are the basis of **microevolution**: alterations of a population's gene pool.

Suppose you were to look at a human population of 20 in which 4 people have blue-coloured eyes because they are homozygous for the *B* allele at a particular "blue eyes" locus, where the alternative allele is *A*. To predict how the number of blue-eyed individuals in the population

will change over time, you need to determine the frequencies of each genotype (homozygous *AA*, heterozygous *AB*, and homozygous *BB*), each phenotype (dark eyes and blue eyes), and each allele (*A* and *B*). Population geneticists define **phenotype frequency** as the proportion of individuals in a population that express a particular phenotype. For our hypothetical population, the phenotype frequencies are 4/20 = 20% blue-eyed (the number of homozygous *BB* individuals expressing the recessive trait) and 16/20 = 80% dark-eyed (the remaining fraction with either *AA* or *AB* genotypes).

Genotype frequencies

Genotype frequency is the proportion of total individuals in a population that carry a particular genotype. To determine genotype frequencies, you simply count the number of individuals of each genotype and divide by the total number of individuals in the population (**Figures 12.2a and 12.2b**). For recessive traits such as blue eyes, it is not possible to distinguish between homozygous dark eyes and heterozygous genotypes: Both give rise to individuals with dark eyes. Thus, the only way to determine genotype

frequencies directly is to use a molecular assay that distinguishes between the different alleles. For our hypothetical population, molecular analyses showed that 12 individuals (of 20) are of genotype *AA*, 4 are of genotype *AB*, and 4 are *BB*. This means that the *AA* genotype frequency is 12/20 = 0.6; the *AB* genotype frequency is 4/20 = 0.2; and the *BB* genotype frequency is also 0.2. Note that these three frequencies (0.6 + 0.2 + 0.2) sum to 1, the totality of genotypes in the whole population.

Allele frequency

The definition of **allele frequency** is the proportion of gene copies in a whole population that are of a given allele type. (Initially, population geneticists used the term "gene frequency" to describe what is now more accurately called "allele frequency.") Because each individual in a population has two copies of each chromosome, the total number of gene copies is two times the number of individuals in the population. Thus, for our hypothetical population of 20 people, there would be 40 gene copies or chromosomes. Of course, both homozygotes and heterozygotes contribute to the frequency of an allele. But homozygotes contribute

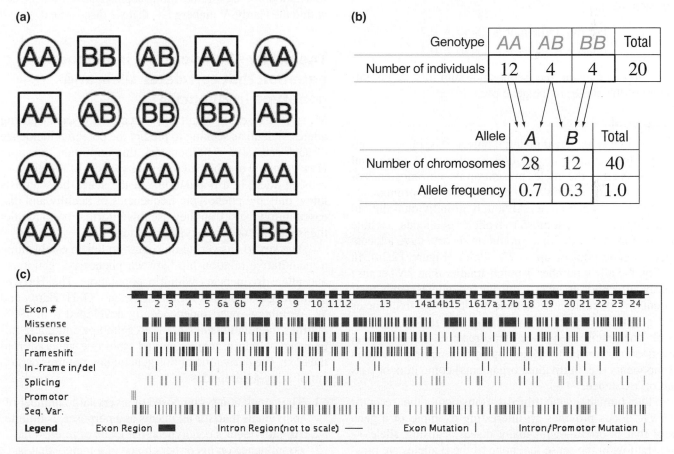

Figure 12.2 From genotype frequencies to allele frequencies. (a) A first-generation population of 20 individuals who are each homozygous or heterozygous for alleles *A* and/or *B* at a locus. **(b)** Whole population numbers for genotypes and alleles at single locus of interest. **(c)** Gene pool of alleles at the *CFTR* locus in European-North American populations. The structure of the *CFTR* gene is not drawn to scale; introns are much larger proportionally than shown. At the level of DNA sequence, thousands of nonfunctional biallelic SNP and InDel variants have been identified at this locus like all others (locations shown in the *Seq. Var.* row). Nonfunctional alleles are generally ignored in studies focused on disease phenotypes. The rows labelled missense, nonsense, frameshift, in-frame in/del, splicing, and promoter show the locations of mutations that affect each of these aspects of gene expression and CFTR protein structure and function.

to the frequency of a particular allele twice, while heterozygotes contribute only once (Figure 12.2b). To find the frequencies of A and B, you first use the number of people with each genotype to compute the number of A and B alleles.

$$12\ AA \rightarrow 24\ \text{copies of } A$$
$$4\ AB \rightarrow 4\ \text{copies of } A$$
$$4\ BB \rightarrow 0\ \text{copies } A$$

Together, $24 + 4 + 0 = 28$ copies of the A allele. Similarly,

$$12\ AA \rightarrow 0\ \text{copies of } B$$
$$4\ AB \rightarrow 4\ \text{copies of } B$$
$$4\ BB \rightarrow 8\ \text{copies of } B$$

Together, $0 + 4 + 8 = 12$ copies of the B allele. Next, you add the 28 A alleles to the 12 B alleles to find the total number of chromosome copies.

$$28 + 12 = 40\ \text{copies of the gene, which is twice the}$$
$$\text{number of people in the population}$$

Finally, you divide the number of each allele by the total number of gene copies to find the proportion, or frequency, for each allele.

$$\text{For the } A \text{ allele, it is } 28/40 = 0.7$$
$$\text{For the } B \text{ allele, it is } 12/40 = 0.3$$

Note that here again, the frequencies sum to 1, representing all of the alleles in the gene pool.

Gene pool

A gene pool is not a real material object. It is, rather, a conceptual term used by population geneticists. A **gene pool** represents all of the alleles present on the chromosomes of all members of a population and the relative prominence or rareness of each allele. Although an individual diploid organism can carry, at most, two alleles at a locus, a whole population of N individuals could, in theory, have a locus-specific gene pool of up to 2N alleles (Figure 12.2a). In reality, the allele number is much smaller than 2N because nearly all children inherit unchanged alleles from their parents. A population is defined by its allele frequencies and genotype frequencies, which together make up a gene pool. The precise frequency measurements of a gene pool are fleeting, constantly changing over time as population components (i.e., individual organisms) come into, or pass out of, existence.

The human cystic fibrosis transmembrane receptor (*CFTR*) locus provides an interesting example of a gene pool that is broad—with many disease-causing alleles—but shallow in the sense that none of these alleles are present at a very high frequency. Over 1900 disease-causing mutations have been identified in the genomes of individuals who express cystic fibrosis (**Figure 12.2c**). The combined frequency of all disease-causing alleles of the *CFTR* gene is 0.02, which is quite small compared with the 0.98 frequency of the functional class of *CFTR* alleles.

One particular mutant allele, called ΔF508 because it deletes amino acid codon number 508, a phenylalanine (F) in normal CFTR polypeptides, accounts for approximately 70 percent of the total disease-causing chromosomes. This value corresponds to an individual allele frequency of 0.014; no other cystic fibrosis disease allele has a frequency of greater than 0.001. In contrast, other loci, like the genetic determinant of eye colour that we will discuss shortly, have allelic forms that predominate in some human populations but are absent from others.

The scientific endeavour of genetics can be divided into subfields based on the unit object that is the focus of a geneticist's attention. To molecular geneticists, the unit entity is "the gene." To formal geneticists, the unit entity is the individual organism, which is defined by genotypes. To population geneticists, the unit entity is the population consisting of a group of interbreeding organisms.

Although an actual gene pool can only be determined empirically by counting all of its constituent alleles (which is nearly impossible to do for a large wild population), population geneticists have developed analytical and computational models for estimating the genetic and phenotypic variations of a population and how they may change over time. The foundation for all of the more sophisticated models lies within the Hardy–Weinberg law that we discuss next.

The Hardy–Weinberg law is a binomial equation that correlates allele and genotype frequencies

Many rare diseases result from recessive disease-causing alleles. Scientists seeking to predict the potential incidence of such recessive conditions ask this important question: How common are the heterozygous carriers of the disease-causing allele in a population? At the start, the scientists know only the phenotypic frequencies of healthy and diseased individuals. Can they use this information to predict the frequency of heterozygous carriers?

The key to answering this question lies in establishing a quantitative relationship between phenotype, genotype, and allele frequencies within a population. The **Hardy–Weinberg law**, named for the two men—G. H. Hardy and W. Weinberg—who independently developed it in 1908, clarifies the relationships between genotype and allele frequencies within a generation and from one generation to the next. The derivation of this general law requires certain simplifying assumptions:

1. The population is composed of a very large number of individuals that, for all intents and purposes, is infinite.
2. An individual's genotype at the locus of interest has no influence on his or her choice of a mate—that is, mating is random.
3. No new mutations appear in the gene pool.
4. No migration takes place into or out of the population.
5. Different genotypes at the locus of interest have no impact on the ability to survive to reproductive age and transmit genes to the next generation.

The assumptions behind the Hardy–Weinberg equilibrium enable the mathematical derivation of an equation for predicting genotype (and thence phenotype) frequencies for a population of diploid individuals. Of course, no actual population is in perfect Hardy–Weinberg equilibrium. All populations are finite; alternative genotypes can make a difference in mating; mutations occur constantly; migration into and out of a population is common; and many genotypes of interest, such as those that cause diseases, affect the ability to survive or reproduce. Nevertheless, even when many of the assumptions of the Hardy–Weinberg equilibrium do not apply, the equation derived on the basis of these assumptions is remarkably robust at providing estimates of genotype and phenotype frequencies in real populations over a limited number of breeding generations. Furthermore, the reverse situation, where allele frequencies are found to be inconsistent with a Hardy–Weinberg equilibrium, can sometimes provide scientists with insight into special biological properties of the locus in question or the population as a whole. Indeed, the equation has always been the most powerful mathematical tool available to population geneticists.

Predicting frequencies from one generation to the next

For a population of sexually reproducing diploid organisms, two steps are needed in translating the genotype frequencies of one generation into the genotype frequencies of the next generation (**Figure 12.3**).

First, if the likelihood that an individual will grow into an adult does not depend on the genotype (i.e., if there is no difference in fitness among individuals), then the allele frequencies in the adults should be the same as in their gametes. For example, if p is the frequency of allele A, and q is the frequency of allele B in the adults, p and q will also be the frequencies of the two alleles in the combination of gametes produced by the whole population of those adults.

Second, the allele frequencies in the gametes can be used to calculate genotype frequencies in the zygotes of the next generation. An enhanced version of the Punnett square, which provides a systematic means of considering all possible combinations of uniting gametes, is the tool of

choice (Figure 12.3). For example, if fertilization is random among individuals with any genotype and if the population of gametes is very large, then the following pattern emerges. Recall that AA zygotes result from fertilization of A-carrying eggs by A-carrying sperm. If p is the frequency of A gametes (eggs and sperm), then, applying the product rule, the frequency of AA zygotes is p (frequency of A eggs) $\times$ p (frequency of A sperm) $= p^2$. Similarly, BB zygotes result from fertilization of B-carrying eggs by B-carrying sperm. If q is the frequency of B gametes (eggs and sperm), the frequency of BB zygotes will be q (the frequency of B eggs) $\times$ q (the frequency of B sperm) $= q^2$. Finally, AB zygotes result either from fertilization of A eggs by B sperm, with a frequency of $p \times q = pq$, or from the fertilization of B eggs by A sperm, occurring at a frequency of $q \times p = pq$. The total frequency of AB zygotes is thus $pq + pq = 2pq$.

The resemblance of the Hardy–Weinberg square shown in Figure 12.3 to the Punnett square that we encountered in the visual representation of formal genetics is not a coincidence. The top and left sides of the Punnett square were divided into sectors representing the frequency of each genetically distinct class of sperm or egg produced by two individual parents. But to population geneticists, individual organisms are not significant. Instead, it is the gametes produced by the population as a whole that matters.

In parallel to the Punnett square, the Hardy–Weinberg square represents a metaphorical mixture of sperm produced by all breeding males along one side, and a mixture of eggs produced by all breeding females along the second side.

To summarize: The genotype frequencies of zygotes arising in a large population of sexually reproducing diploid organisms are p^2 for AA, $2pq$ for AB, and q^2 for BB (see Figure 12.3). These genotype frequencies are known as the Hardy–Weinberg proportions; they exist in populations that satisfy the Hardy–Weinberg assumptions of a large number of individuals, mating at random, with no new mutations, no migration, and no genotype-dependent differences in fitness. Since these genotype frequencies represent the totality of genotypes in the population, they must sum to 1. Thus the binomial equation representing the Hardy–Weinberg proportions is

$$p^2 + 2pq + q^2 = 1 \qquad (12.1)$$

Because we have assumed no differences in fitness, the genotype frequencies of the zygotes will be the genotype frequencies of the adult generation that develops from those zygotes.

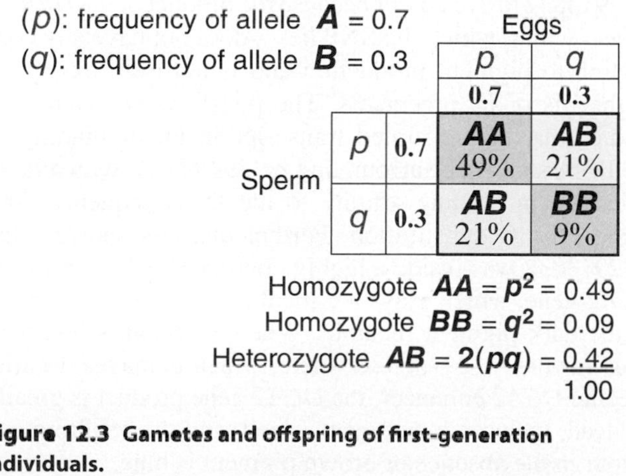

Predicting the frequency of albinism: A case study

This equation thus enables us to use information on genotype and allele frequencies to predict the genotype frequencies of the next generation. Suppose, for example, that in a population of 100 000 people carrying the recessive allele a for albinism, there are 100 aa albinos and 1800 Aa heterozygous carriers. To find what the frequency

(p): frequency of allele A = 0.7
(q): frequency of allele B = 0.3

Eggs

		p 0.7	q 0.3
Sperm	p 0.7	**AA** 49%	**AB** 21%
	q 0.3	**AB** 21%	**BB** 9%

Homozygote **AA** = p^2 = 0.49
Homozygote **BB** = q^2 = 0.09
Heterozygote **AB** = 2(pq) = 0.42
1.00

Figure 12.3 Gametes and offspring of first-generation individuals.

of heterozygous carriers will be in the next generation, you compute the allele frequencies in the parent population.

98 100 *AA* individuals; 1800 *Aa* individuals, and 100 *aa* individuals → 196 200 *A* alleles + 1800 *A* alleles; 1800 *a* alleles + 200 *a* alleles

Out of 200 000 total alleles, the frequency of the *A* allele is

198 000/200 000 = 99/100 = 0.99; thus *p* = 0.99

and the frequency of the *a* allele is

2000/200 000 = 1/100 = 0.01; thus, *q* = 0.01

The Hardy–Weinberg equation for the albino gene in this population is

$$p^2 + 2pq + q^2 = 1$$
$$(0.99)^2 + 2(0.99 \times 0.01) + (0.01)^2 = 1$$
$$0.9801 + 0.0198 + 0.0001 = 1$$

It thus predicts that in the next generation of 100 000 individuals, there will be

100 000 × 0.9801 = 98 010 *AA* individuals

100 000 × 0.0198 = 1980 *Aa* individuals

100 000 × 0.0001 = 10 *aa* individuals

This example shows that in one generation, the genotype frequencies have changed somewhat. A natural question follows: Have the allele frequencies also changed? Recall that the initial frequencies of the *A* and *B* alleles are *p* and *q*, respectively, and that *p* + *q* = 1. You can use the rules for computing allele frequencies from genotype frequencies (see Figure 12.3) to compute the allele frequencies of the next generation. From the Hardy–Weinberg equation, you know that p^2 of the individuals are *AA*, whose alleles are all *A*, and $2pq$ of the individuals are *AB*, 1/2 of whose alleles are *A*. Similarly, q^2 of the individuals are *BB*, whose alleles are all *B*, and $2pq$ of the individuals are *AB*, 1/2 of whose alleles are *B*. If *p* + *q* = 1, then *q* = 1 − *p*, and the frequency of the *A* allele in the next-generation population is

$$p^2 + 1/2[2p(1-p)] = p^2 + p(1-p)$$
$$= p^2 + p - p^2 = p \quad (12.2)$$

Similarly, *p* = 1 − *q*, and the frequency of the *B* allele in the next-generation population is

$$q^2 + 1/2[2q(1-q)] = q^2 + q(1-q)$$
$$= q^2 + q - q^2 = q \quad (12.3)$$

Using these equations to calculate the allele frequencies of *A* and *a* in the second generation of 100 000 individuals, some of whom are albinos, we find

for *p*: 0.9801 + 0.99 − 0.9801 = 0.99
= the frequency of the *A* allele

for *q*: 0.0001 + 0.01 − 0.0001 = 0.01
= the frequency of the *a* allele

These frequencies are the same as those in the previous generation. Thus, even though the genotype frequencies have changed from the first generation to the next, the allele frequencies have not. Note that this is true of both the dominant and the recessive alleles.

Analysis of eye colour shows the power and limitations of Hardy–Weinberg

As an example of dominant and recessive phenotypes that result from different genotypes and alleles at a single locus, let us consider the actual genetics of eye colour in human populations. Until 10 000 years ago, eye colour was not polymorphic—all of our ancestors from this period had brown irises.

Genetics of blue eyes

Blue eyes first appeared between 6000 and 10 000 years ago, probably in a population living near the north shore of the Black Sea, according to anthropological genetic data. Today, the trait predominates in northern European populations, appearing in more than 80 percent of the people living around the Baltic Sea (**Figure 12.4a**). Blue-eyed individuals can be seen throughout most of Europe with diminishing frequencies at lower latitudes. Outside of Europe and the Mediterranean region, the blue eye phenotype is rare, appearing sparsely in some central and northern Asian populations, but essentially absent from sub-Saharan Africa and east Asia.

High-resolution linkage mapping has demonstrated an association of blue eyes with a single-base substitution, from an A to a G, that defines the single nucleotide polymorphism (SNP) locus rs12913832, located in an intron of the *HERC2* gene on chromosome 15 (**Figure 12.4b**). The *rs12913832G* allele is always found as a part of a larger haplotype of conserved SNP alleles across a 50-kb chromosomal region (**Figure 12.4c**).

The extent of the conserved DNA region and the limited distribution of the *G* allele primarily in European populations strongly suggest that this base substitution was a one-time event. All people with blue eyes today can trace themselves back genetically to the single individual in which the mutation occurred.

Scientists were puzzled at first by the association of rs12913832 to eye colour because the product of the *HERC2* gene—within which the SNP lies—does not have any connection to pigment production and is not even expressed in the iris or its precursors. The puzzle was solved with studies that demonstrated transcription factor binding to the DNA sequence surrounding *rs12913832A*, with a large reduction in binding affinity to the DNA sequence containing the G substitution. Further analysis showed that rs12913832 was inside a highly conserved enhancer of the *OCA2* gene, which plays a critical role in the biosynthesis of the dark pigment melanin. When a person is homozygous for the *rs12913832G* allele, which damages the iris-specific *OCA2* enhancer, the *OCA2* gene product is greatly reduced, melanin is not synthesized, and the resulting eye colour in the absence of brown pigment is blue.

Pie diagrams illustrating allele frequencies

Pie diagrams provide an intuitive tool for visualizing allele frequency differences that distinguish one population from another. The two alleles of an SNP locus are represented in contrasting colours that occupy radial portions of a circle, or pie, corresponding to allele frequencies.

Pie diagrams are placed on an appropriate geographical map at the locations of the screened populations (**Figure 12.4d**). To obtain useful data, researchers typically assign people to populations according to the geographical locations where their recent ancestors were born and lived.

By viewing Figure 12.4d, you can see immediately that allele frequencies at the SNP locus rs12913832 are dramatically different in geographically separated populations. The highest frequency of the *rs12913832G* allele in populations screened for this locus is 0.84, associated with a population from northern Finland, whereas all of the Chinese and sub-Saharan African populations do not carry the *G* allele at all.

Use of the Hardy–Weinberg equation with mixed populations

To understand the implications of the Hardy–Weinberg law in the analysis of a population formed from a mixture of previously differentiated populations, imagine that 100 adults from northern Finland and 100 from the Yakut people of eastern Siberia (both men and women from both populations) decide to move to a newly built offshore oil

rig in the Arctic Sea. Imagine further that the 200 men and women on the oil rig marry each other without regard to their ancestry.

Now let us ask two questions and see how the data we have obtained on allele frequencies can be combined with the Hardy–Weinberg law to provide answers. First, we can estimate how many adults on the oil rig have blue eyes. Second, we can estimate both the number of children with blue eyes and the allele frequency in this second generation.

Because the Finnish and Yakut populations have different *rs12913832G* allele frequencies, it is easiest to use the Hardy–Weinberg equation to determine the composition of each separately. We will use p to represent the allele frequency of the brown-eyes-associated *rs12913832A* allele, and q to represent the allele frequency of *rs12913832G*. The data sampled from the Finnish yield $q = 0.84$. Blue eyes are recessive, which means that we can estimate the number of Finnish adults with blue eyes from the genotype frequency for *GG*:

$$q^2 = 0.84 \times 0.84 = 0.71$$

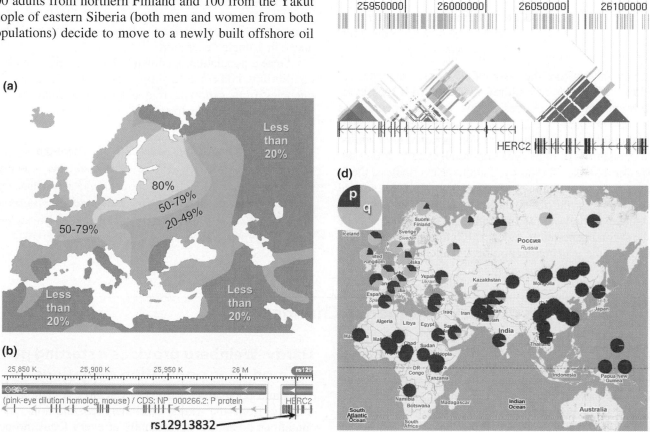

Figure 12.4 Blue-eye population genetics. **(a)** Geographical differences in the proportions of European populations expressing the "blue eyes" phenotype. **(b)** The SNP rs12913832 that controls blue eye colour is located upstream from *OCA2* in an intron of *HERC2*. **(c)** Representation of haplotype structure and coinheritance of SNPs across extended DNA blocks in the *OCA2-HERC2* gene region of chromosome 15. The SNP allele *rs12913832G* is part of a conserved 50-kb chromosomal block that spans the 3′ half of *HERC2*. **(d)** Pie diagrams depict frequencies of alleles *G* and *A* at the SNP locus rs12913832 in different Old World populations.

By multiplying 0.71 by the total number of Finns on the oil rig, we get an estimate for the number of blue-eyed Finns at

$$0.71 \times 100 = 71 \text{ individuals}$$

The frequency of brown-eyed Finnish carriers of the *rs12913832G* allele is calculated as

$$2pq = 2 \times 0.16 \times 0.84 = 0.27$$

which yields 27 carrier individuals on the oil rig.

Similarly, from the Yakut *rs12913832G* allele frequency of 0.10, we obtain

$$q^2 = (0.1)^2 = 0.01$$

which translates into a single Yakut adult with blue eyes, and

$$2pq = 2 \times 0.9 \times 0.1 = 0.18$$

for 18 Yakut adults who are carriers. Together, among the 200 people on the oil rig, 72(71 + 1) will have blue eyes, and 45(27 + 18) will be carriers. With this information, we can calculate the combined *G* allele frequency from a count of chromosomes—two from each blue-eyed person and one from each carrier yields

$$2 \times 72 + 45 = 189 \div 400 \quad \text{(the number of chromosomes is}$$
$$\text{twice the number of people) for}$$
$$\text{an answer of } q = 0.47$$

Now let us use the parental *G* allele frequency to predict the number of children expected with blue eyes. The frequency of blue eyes in this second generation is $q^2 = 0.47 \times 0.47 = 0.22$. If we assume that the population of children is equal in size to the population of adults, then the number of blue-eyed children is $0.22 \times 200 = 44$, far fewer than the 72 blue-eyed adults. Nevertheless, the *G* allele frequency remains unchanged at 0.47. If this second generation intermarries to produce a third generation of 200, the expected number with blue eyes will be the same as the second generation. Indeed, all future generations will have the same expected genotype and allele frequencies.

Properties of populations described by Hardy–Weinberg

The application of Hardy–Weinberg analysis to the study of eye colour in human populations generates two important conclusions. First, even though the proportion of individuals expressing the blue eye phenotype changed dramatically from the first generation of the mixed population to the second generation, no change in allele frequency occurred. The *conservation of allele proportions* principle holds from each generation to the next, as long as the population is sufficiently large, alleles are not lost by mutation or selection, and alleles are not gained by mutation or immigration.

Populations with the same allele frequency do not necessarily have the same genotype or phenotype frequencies. The reason is that a single allele can exist in homozygote or heterozygote genotypes, but a recessive phenotype is only expressed in homozygotes. In the most extreme hypothetical example, a population could have a blue-eyed allele frequency of 0.5 without actually having any people with blue eyes. This situation would arise if everyone in the population is a heterozygote.

Even in this extreme example, the Hardy–Weinberg equilibrium tells us that the Hardy–Weinberg genotype frequencies described by p^2, $2pq$, and q^2, will appear in the very next generation. This is the second significant Hardy–Weinberg implication: A population that is initially stratified because of its founding by individuals from two or more distinct populations will become completely balanced in a single generation.

Once a population is known to be in Hardy–Weinberg equilibrium, it becomes a simple task to predict allele frequencies from genotype frequencies, and genotype and phenotype frequencies from allele frequencies.

In studying populations, geneticists are interested in the frequencies with which different alleles and the resulting genotypes and phenotypes occur. The Hardy–Weinberg equation provides a mathematical method of evaluating these frequencies and predicting them from one generation to the next, based on equilibrium populations.

12.2 Causes of Allele Frequency Changes

If all populations of individuals within the same species were always in Hardy–Weinberg equilibrium, the Hardy–Weinberg equation would lead us to conclude that all allele frequencies would be forever unchanging, and allele frequencies at any particular locus would be equivalent across populations. And yet, as Figure 12.4d shows, different populations of the human species are associated with dramatically different *rs12913832G* allele frequencies spanning the range from 0.0 to 0.84. A similar result is obtained for numerous other loci. What do these observations suggest about the applicability of Hardy–Weinberg to real populations?

Hardy–Weinberg provides a starting point for modelling population deviations

In natural populations, conditions always deviate at least slightly from the Hardy–Weinberg assumptions: New mutations do appear occasionally at every locus, no population is infinite, small groups of individuals sometimes migrate from the main group to become founders of new populations, separate populations can merge together, individuals do not mate at random, and different genotypes do generate differences in rates of survival and reproduction.

And yet, even with these deviations from ideal conditions, the Hardy–Weinberg equation provides remarkably good estimates of allele, genotype, and phenotype frequencies *over the short run*, through one or a few breeding generations of populations of all sizes (except for those so small as to be on the verge of extinction).

Over the *long run*, however, the Hardy–Weinberg equation is rarely applicable for predictive purposes. But it serves a critical role in providing the foundation for both analytical and stochastic methods that do incorporate factors responsible for deviation from equilibrium conditions. With simple modifications to the Hardy–Weinberg equation, or the manner in which it is used, the dynamics of realistic populations may be successfully modelled.

In finite populations, chance plays a critical role

A spontaneous mutation is defined as a variant DNA sequence or chromosomal region in the genome of an individual that is not present in the genomes of either parent. Generally, spontaneous mutations are so infrequent (on the order of 1 in 100 000 to 1 million offspring for each gene) that their impact on a population's allele frequencies can be safely ignored in the short run. Yet, mutations are the source of all alleles at all loci, which means that some mutations do come to matter in the long term.

Mutations that provide a relative benefit to an organism or population are in a rare category of an already rare phenomenon. Many mutations are deleterious, but even more are essentially neutral—that is, the novel allele formed by mutation provides neither a benefit nor harm. The survival or nonsurvival of a neutral mutation is a stochastic phenomenon—it occurs by chance alone.

We now present the modifications to the Hardy–Weinberg law that are required to model the long-term consequences of mutations.

The Monte Carlo simulation

The derivation of the Hardy–Weinberg equation was accomplished by a "sleight-of-hand" extension of Mendel's first law of segregation. Mendel's first law does *not* actually determine which allele a heterozygous parent will transmit to an individual child. Instead, it tells us simply that allele inheritance is like a flip of a coin: A child can receive heads or tails with an equal prior probability. Mendel's law does determine the approximate proportion of a large cohort of offspring that will inherit a particular allele. The larger the cohort population, the more accurate the determination.

The required Hardy–Weinberg sleight-of-hand is the assumption that *precisely* 50 percent of the children in a population inherit each alternative allele from heterozygous parents in each generation. This assumption is only valid as the size of a population approaches infinity. Since no population is infinite, no population truly abides by the Hardy–Weinberg conditions for equilibrium.

To model long-term allele frequency changes in a finite population, the analytical Hardy–Weinberg equation

is replaced by what is known as a **Monte Carlo simulation**. A Monte Carlo simulation is typically performed with a computer program that uses a random-number generator to flip a coin (metaphorically) to choose an outcome for each probabilistic event occurring in a dynamic system defined by predetermined rules of probability.

For population genetics simulations, a specialized Monte Carlo program is initialized with a starting population having a defined number of individuals of each homozygous and heterozygous class. In the simplest case of two alleles at a single locus with no effect on survival, mating, or reproduction, the program sets up matings between individuals chosen randomly with the use of a random-number generator. If a chosen parent is a heterozygote, the program also flips a coin (metaphorically) to decide which allele will be transmitted to a child.

The birth of children is simulated to obtain a total number equal to the population size chosen for the analysis. Then, the first-generation parents are eliminated, and the children are used as the progenitors for the subsequent generation. This process is continued for as many generations as requested by the investigator, or until one allele is lost from the population and every individual is homozygous for the surviving allele. The data are recorded, the program is terminated, and a new "run" is begun with the same initial conditions. A sufficient number of independent Monte Carlo simulations are performed in order for the investigator to get a sense of what outcomes are possible with what probabilities.

Genetic drift

In the first example shown in **Figure 12.5a**, six Monte Carlo simulations have been initialized with populations of just ten individuals who are all set to be heterozygotes; each population has $2 \times 10 = 20$ gene copies (two in each individual), and each allele (A or G) occurs initially with a frequency of 0.5. What allele frequencies are likely to be obtained in the first round of simulated offspring? Because of the particular way in which this example is structured, the Monte Carlo simulation in this first generation is mathematically equivalent to the results obtained from tossing a coin 20 times. As you can see in Figure 12.5a, actual simulations yield first allele (A) frequencies that range from 0.25 (5 heads and 15 tails) to 0.65 (13 heads and 7 tails) with an average of 0.48.

Although the average of populations is not too far from the 0.5 predicted by the Hardy–Weinberg equation, the population-specific values guide each one down a different path of **genetic drift**, defined as a change in allele frequencies as a consequence of the randomness of inheritance from one generation to the next. Genetic drift occurs because the allele frequency in any one generation provides the median for possible allele frequencies in the next generation. So, for example, if one allele has already drifted to a high or low frequency, there is a 50 percent chance that it will go even higher or lower in the following generation.

In four of the six simulated runs shown in Figure 12.5a, genetic drift has culminated in the loss, or extinction, of

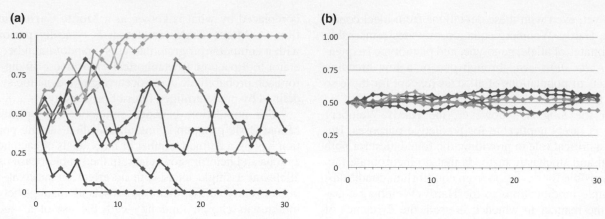

Figure 12.5 Modelling genetic drift in populations of different sizes. (a) Population size is 10. Initial condition is equal numbers of alleles, no selection. **(b)** Population size is 500. Initial condition is equal numbers of alleles, no selection.

one of the two original alleles by generation 18. In each of these instances, a relatively small change in allele frequency from the previous generation caused extinction of one allele, and *fixation* of the remaining allele. A population is considered to be **fixed** at a locus when only one allele has survived and all individuals are homozygous for this allele. At this point, no further changes in allele frequency can occur (in the absence of migration or mutation).

Population size and time to fixation

The results that we would expect to get from a series of coin tosses can provide insight into the effect that the size of a population has on allele frequency changes and time to fixation. Let us start with an experiment that you repeat numerous times, in which you toss four coins in the air and record the frequency of heads each time. The possible frequency results are 0.0, 0.25, 0.5, 0.75, and 1.0. Getting an absolute difference of 4 between the heads and tails count is not so unusual (1/16 + 1/16 = 1/8). If heads represents one allele, and tails the other, then fixation occurs at two frequencies, 0.0 and 1.0, and the probability of fixation in any one experiment is 12.5 percent. If we repeat this experiment with 500 coins in each trial, instead of 4, there is still a good chance that the absolute difference between the heads and tails count may be 4 or more, but an absolute difference of 4 (e.g., 248 heads and 252 tails) translates into relative frequencies of 0.496 and 0.504. Extrapolation from this result leads to an important probabilistic conclusion concerning coin-toss-like models: the larger the sample size, the smaller the typical deviation from a 50:50 ratio.

The impact that population size has on allele frequency dynamics is readily observable in a comparison of the data presented in Figures 12.5a and b. **Figure 12.5b** shows results obtained from six simulations of populations with 500 individuals that are initially all heterozygous. If we follow the lines representing each population, we can see that single generation changes in allele frequency are always relatively small. Because these changes are small, the traditional Hardy–Weinberg equation provides good estimates of allele and genotype frequencies in large populations over the course of a few generations. But a series of

small changes can still add up to large consequences over the long run, and eventually, each of these populations also became fixed for one allele or the other.

In populations with two alleles having equivalent phenotypic effects and present initially at equal frequencies, the median number of generations to fixation is roughly equal to the total number of gene copies in breeding individuals. For a population of 10, the median fixation time is 20 generations; for a population of 500, it is 1000 generations; and for a breeding population of 200 million people, it would be 400 million generations, or 8 billion years (assuming a birth-to-birth generation time of 20 years). Because Earth is expected to perish in flames when our Sun becomes a red giant and expands into our planet's orbit in about 2 to 3 billion years, the implication is that neutral genetic drift is irrelevant to future human evolution.

Founder effect and population bottlenecks

One example of genetic drift in past human populations is the **founder effect**, which occurs when a few individuals separate from a larger population and establish a new one that is isolated from the original. The small number of founders in the new population carries only a fraction of the gene copies from the original population, and by chance, founder allele frequencies can be different.

In Canada, about 8500 individuals from France settled in Quebec between 1608 and 1759, of which there were only 1600 women. These pioneers became isolated for a number of reasons, including geography, culture, religion, and language, and very rarely mixed with other ethnic groups. Furthermore, this colony had a very high birth rate, thus expanding very rapidly. All of these factors contributed to the high relative frequencies among French Canadians of several genetic diseases and disorders that have been attributed to founder effects, including familial hypercholesteraemia and cystic fibrosis.

In the United States, about 200 individuals emigrated from Germany in the early eighteenth century to form the Amish community in eastern Pennsylvania. Since this founding population was completely cut off from Germany and people married only within the group, it was subjected

to genetic drift. Today, the Amish number 14 000 in total, but the population exhibits a much higher incidence of manic-depressive illness than does the larger original European population, most likely because several founders carried alleles producing this disease.

Plant and animal populations are frequently subjected to **population bottlenecks**, which occur when a large proportion of individuals perish, often as a consequence of environmental disturbances. The surviving individuals are essentially equivalent to a founder population.

Consider the perpetuation of allele r, which occurs in 10 percent of the individuals in a population of sunflowers. If a chance occurrence such as a severe summer hailstorm strikes a population of 1 million plants and half die, it is likely that among the 500 000 survivors, roughly 50 000 (10 percent) will still carry the r allele. If the same storm hits a field of just ten plants, only one of which carries the r allele, there is a 50 percent chance that the single plant bearing the r allele will not be among the survivors. The total loss of this one allele from the population would reduce the allele frequency to zero. If, however, the lone individual carrying r does by chance survive, the frequency of allele r would increase to 20 percent.

> Over a few generations, Hardy–Weinberg almost always provides accurate estimates of allele and genotype frequencies. But in the long run, Hardy–Weinberg fails because the randomness of allele segregation causes genetic drift. A Monte Carlo simulation takes account of random segregation and the impact of population size to provide estimates of the number of generations until an allele is lost or becomes fixed.

Natural selection acts on differences in fitness to alter allele frequencies

For many phenotypic traits, including inherited diseases, genotype does influence survival and the ability to reproduce, contrary to the Hardy–Weinberg assumptions. Thus, in real populations, not all individuals survive to adulthood, and some probability always exists that an individual will not live long enough to reproduce. As a result, the genotype frequencies of real populations change as their individual members mature from zygotes to adults.

Fitness and selection

To population geneticists, an individual's relative ability to survive and transmit its genes to the next generation is its **fitness**. But although fitness is an attribute associated with each genotype, it cannot be measured within the individuals of a population; the reason is that each animal with a particular genotype survives and reproduces in a manner greatly affected by chance circumstances. However, by considering all the individuals of a particular genotype together as a group, it becomes possible to measure the relative fitness for that genotype. Thus, for population geneticists, fitness is a statistical measurement only.

Nevertheless, differences in fitness can have a profound effect on the allele frequencies of a population.

Fitness has two basic components: viability and reproductive success. The fitness of individuals possessing variations that help them survive and reproduce in a changing environment is relatively high; the fitness of individuals without those adaptive variations is relatively low. In nature, the process that progressively eliminates individuals whose fitness is lower and chooses individuals of higher fitness to survive and become the parents of the next generation is known as **natural selection**. The mechanisms of selection act independently of any individual; often, interactions between genetically determined phenotypes and environmental conditions are the agents of natural selection. For example, in a hypothetical population of giraffes browsing on the leaves of an ancient savannah, suppose that some of the animals had long necks and some short necks, and that each of these phenotypes resulted from variations in the genes contributing to neck length. If during a long drought, fewer low-hanging leaves were available on the shrubs and trees of the savannah, those long-necked individuals able to reach the higher leaves would have been able to harvest more food. Natural conditions would thus have selected the better adapted, long-necked giraffes—those with the higher fitness—to survive and become the parents of the next generation (**Figure 12.6**). Similarly, in the laboratory, scientists can establish experimental conditions, such as the absence of a nutrient or the presence of an antibiotic, that become the agents of **artificial selection**.

Field studies show that natural selection occurs for phenotypic traits in all natural populations. Until recently, genetic disease and genetic predisposition or resistance to infectious disease were major factors in determining a person's survival and reproduction. As a consequence, the potential for natural selection even in human populations was considerable. Today, the application of vaccines, the availability of antibiotics, and other medical advances have reduced but not eliminated the forces of natural selection. How does selection alter the conclusions that can be drawn from Hardy–Weinberg?

Figure 12.6 Giraffes on the savannah. A visible example of one outcome of natural selection for increased fitness.

Modifications to Hardy–Weinberg

We can see how to apply the Hardy–Weinberg equation in populations undergoing selection with an analysis of the R gene in a population of zygotes in Hardy–Weinberg equilibrium. In this population, the genotype frequencies RR, Rr, and rr are p^2, $2pq$, and q^2, respectively. Now suppose that the viability, that is, the probability of surviving from zygote to adult, depends on genotype, while the second component of fitness—success at productive mating—is independent of genotype. If we define the relative fitness of the three genotypes as w_{RR}, w_{Rr}, and w_{rr}, respectively, the relative frequencies of the three genotypes at adulthood is $p^2 w_{RR}$, $2pq w_{Rr}$, and $q^2 w_{rr}$ (**Figure 12.7**).

The fitness of any one genotype is only defined in relation to the fitness of alternative genotypes, which means that the individual fitness coefficient values (w_{RR}, w_{Rr}, and w_{rr}) are arbitrary. For example, if an experimental result demonstrates a threefold difference in fitness between the RR genotype and the rr genotype, the values of w_{RR} and w_{rr} could be set to 15 and 5, or 6.9 and 2.3, or any of an infinite pair of numbers having a 3:1 ratio.

The fitness-modified Hardy–Weinberg equation is most useful when the fitness coefficients are adjusted so that the three terms sum to 1. In this "normalized" form, each term represents an actual genotype frequency. Normalization is accomplished by calculating the value of a numerical factor that each Hardy–Weinberg term can be divided into so that the terms all add up to 1. The calculation consists simply of setting the sum of the terms in the modified equation to a new variable, designated $\bar{w}$.

$$p^2 w_{RR} + 2pq w_{Rr} + q^2 w_{rr} = \bar{w} \qquad (12.4a)$$

Since $\bar{w}$ represents the sum of the individual fitness values, each multiplied by their relative occurrence in the population, it in fact represents the average fitness of the population.

In populations that satisfy the conditions of the original Hardy–Weinberg equilibrium, when the fitness for each genotype is 1, the value of $\bar{w}$ is also 1. However, when fitness varies from one genotype to another, $\bar{w}$ can also vary, but in a way that can be calculated as long as the value of each of the variables in the equation is known. The modified Hardy–Weinberg equation is thus normalized by dividing each term by $\bar{w}$ such that the new equation becomes

$$\frac{p^2 w_{RR}}{\bar{w}} + \frac{2pq w_{Rr}}{\bar{w}} + \frac{q^2 w_{rr}}{\bar{w}} = 1 \qquad (12.4b)$$

Each term in this normalized equation represents the actual frequency that each genotype will assume in the generation following the one used for the original calculation. Figure 12.7 summarizes this process of calculation.

As an example, let us use the variables p' and q' to represent the frequencies of the R and r alleles in this next generation. Among the gametes produced by the original population, the frequency of allele r will be the result of contributions of r alleles from both Rr and rr adults relative to the number of individuals in the entire adult population. If q' represents the frequency of the r allele in the next-generation adults, then

$$q' = \frac{q^2 w_{rr} + \frac{1}{2}(2pq w_{Rr})}{\bar{w}}$$
$$= \frac{q(q w_{rr} + p w_{Rr})}{\bar{w}} \qquad (12.5)$$

Parental Gametes

Allele	R	r
Frequency	p	q

First-Generation Offspring

	RR	Rr	rr
Genotypes	RR	Rr	rr
Zygote frequency	p^2	$2pq$	q^2
Relative fitness	w_{RR}	w_{Rr}	w_{rr}
Relative frequency after selection	$p^2 w_{RR}$	$2pq w_{Rr}$	$q^2 w_{rr}$
Normalized adult frequency	$p^2 w_{RR}/\bar{w}$	$2pq w_{Rr}/\bar{w}$	$q^2 w_{rr}/\bar{w}$

Gametes from First Generation

Allele	R	r
Frequency	$p' = (p^2 w_{RR} + pq w_{Rr})/\bar{w}$	$q' = (q^2 w_{rr} + pq w_{Rr})/\bar{w}$

Normalization factor $\quad \bar{w} = p^2 w_{RR} + 2pq w_{Rr} + q^2 w_{rr}$

Figure 12.7 Changes in allele frequencies caused by selection. The uncalibrated frequency after selection is calculated by multiplying the zygote frequency by the relative fitness value. Adult frequencies are calibrated through division by the sum of the relative fitness values (w).

Keep in mind,

1. Δq is defined as $q' - q$.
2. A useful identity in moving from Equation 12.5 to Equation 12.6 is $1 - 2q = 1 - q - q = (p - q)$.

Thus, in one generation of selection, the allele frequency of r has changed from q to q'. It is often useful to know the change in allele frequency over one generation of selection. We can represent this change as

$$\Delta q = \frac{pq[q(w_{rr} - w_{Rr}) + p(w_{Rr} - w_{RR})]}{\overline{w}} \quad (12.6)$$

As these equations (12.5 and 12.6) show, selection can cause the frequency of an allele to change from one generation to the next. Equation 12.6 shows that the change in allele frequency resulting from one generation of selection depends on both the allele frequencies and the relative fitness (in this case, the viability component of fitness) of the three genotypes. Note that if the fitness of all genotypes are the same, as in populations at Hardy–Weinberg equilibrium, then the change in q (Δq) = 0; in other words, if there are no genotype-related differences in fitness, there is no possibility of selection, and if there is no possibility of selection, then allele frequencies will be subjected only to genetic drift, as we described in an earlier section.

As an example, let us use Equation 12.6 to look at how a recessive genetic condition, such as thalassaemia, influences the allele frequencies of a population. If the disease, which results from an rr genotype for the R gene, decreases fitness by decreasing the probability of surviving to adulthood, then the fitness of RR and Rr individuals is the same, while the fitness of rr individuals is reduced. Since only the relative values of fitness are important, it is useful to set the values of $w_{RR} = 1$, $w_{Rr} = 1$, and $w_{rr} = 1 - s$, where s is the *selection coefficient* against the rr genotype. This selection coefficient can vary from 0 (no selection against rr) to 1 (rr is lethal, and no rr individuals survive to adulthood). For this example, we can rewrite Equation 12.6 as

$$\Delta q = \frac{pq[q(1 - s - 1) + q(1 - 1)]}{\overline{w}} = \frac{-spq^2}{\overline{w}} \quad (12.7)$$

Equation 12.7 has three interesting features. First, unless there is no selection and $s = 0$, Δq is always negative, and the frequency of the r allele decreases with time.

Second, the rate at which q decreases over time depends on the allele frequencies; in particular, because Δq varies with q^2, the rate at which q decreases diminishes as q becomes smaller. (Recall that because q is always less than 1, $q^2 < q$.) To understand the effect of this correlation between the allele frequency and the rate at which q decreases over time, consider the special case of a lethal recessive disease for which $s = 1$. The dotted line in **Figure 12.8** shows the decrease in allele frequency predicted by Equation 12.7, starting from an initial allele frequency of 0.5. The decrease in allele frequency is rapid at first, and then slows. After ten generations, the predicted frequency of the recessive disease allele is nearly

10 percent, even though the homozygous recessive genotype is lethal. The solid line in Figure 12.8 plots actual data for the decrease in frequency of an autosomal lethal allele in a large experimental population of *Drosophila melanogaster*. In Figure 12.8, the predicted and observed changes in allele frequency match quite closely.

Why is selection less effective as the frequency of a recessive lethal allele moves closer to zero? The answer goes back to our consideration of the frequency of heterozygous carriers of a recessive disease allele. When q is small, individuals homozygous for the disease allele (at a frequency of q^2) are very rare because most copies of the r allele occur in Rr heterozygotes (at a frequency of $2pq$) who do not experience negative selection. (In mathematical terms, the ratio of q^2 to q decreases exponentially for all values of q less than 1.) By contrast, a lethal dominant allele will disappear from a population in a single generation of selection.

The third feature of Equation 12.7 is that it predicts that the allele frequency q should continue to decline, albeit more and more slowly over time, as q moves closer and closer to a value of zero.

A Monte Carlo simulation of natural selection

Modifying the Hardy–Weinberg equation with coefficients of selection overcomes one limitation of the original Hardy–Weinberg equation: the assumption that all possible genotypes are equal in fitness. But the analytical solution of this equation and the derivation of Δq still suffer from a dependence on the assumption of an infinite population. Nevertheless, we can use the modified Hardy–Weinberg equation to develop a similarly modified Monte Carlo simulation to investigate the impact of natural selection on finite populations.

As an example, let us consider a population of 500 individuals in which 499 are homozygous initially for the r allele, and one is heterozygous with an R mutation on one chromosome that provides a slight dominant advantage in survival described by the following selection coefficients: $w_{RR} = 1.0$, $w_{Rr} = 0.98$, and $w_{rr} = 0.98$. These conditions can be modelled with a Monte Carlo approach that randomly eliminates 2 percent of Rr and rr individuals created

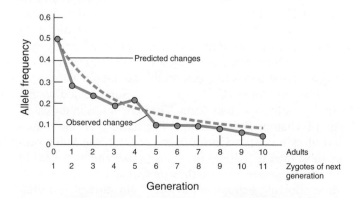

Figure 12.8 Decrease in the frequency of a lethal recessive allele over time. The *dotted line* represents the mathematical prediction. The *blue line* represents the actual data obtained with an autosomal recessive lethal allele.

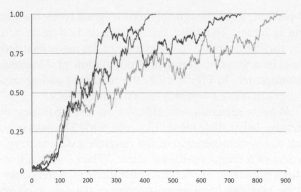

Figure 12.9 Monte Carlo modelling of natural selection. Each coloured data line represents an independent Monte Carlo run on a population of 500 in which a new mutant allele appears in a single individual. In most runs, the mutant allele goes extinct in fewer than 100 generations. The few that survive longer inevitably move to fixation.

in each generation, and replaces them with offspring from a new mating of the parental generation. **Figure 12.9** shows the results of six simulations of this population model. Notice first that in three of them, the new *R* allele never takes off, going extinct within 65 generations. But in the populations where the *R* allele increases in frequency to 0.10, it inevitably moves toward fixation.

This example illustrates several important points concerning the impact that a new mutant allele with a small, yet realistic, fitness advantage can have on a population. First, even though the original copy of the allele provides a selective advantage, it will often go extinct due to chance events of reproduction in the initial generations. But second, if the advantageous allele reaches a threshold frequency level that ensures its survival, it will always eventually increase all the way to fixation. Third, even a fitness advantage of just 2 percent—which is likely to be imperceptible at the individual level—will rise inevitably to fixation.

The fitness of alternative genotypes in different environments

When people migrated out of the East African region in which *H. sapiens* originated, beginning approximately 70 000 years ago, founder populations encountered environmental conditions in Europe and Asia that were distinct from those in Africa. As a result, the relative fitness of alternative alleles at a number of genes became reversed. Among the most obvious changes were differences in allele frequencies at genes that determine skin pigmentation.

The ultraviolet (UV) rays of the Sun provide people with a benefit as well as harm. The benefit lies in the catalysis of vitamin D production. The harm is in the induction of skin cancer. Closer to the equator, the Sun's rays are most intense. Alleles that cause a darkening of the skin are advantageous because they protect against skin cancer while allowing enough ultraviolet light through for vitamin D production. At higher latitudes, where the Sun's rays are less intense, skin cancer is less of a problem, and alleles that lighten the skin allow enough UV penetration for sufficient vitamin D production.

Skin pigmentation is a complex, quantitative trait determined by alleles at many genes, but about a half dozen genes are most influential. One fascinating question concerning our history as a species is whether European and Asian populations derived lighter skin pigmentation from a common ancestral population, or whether the trait evolved separately on the two continents. A mixed answer has been obtained by surveying allele frequencies at multiple pigmentation loci in populations indigenous to different geographical locations around the Old World.

The *KITLG* gene is among the small list with a prominent role in skin pigmentation. As you can see in the pie chart (**Figure 12.10a**), Europeans and Asians share a common SNP variant responsible for a reduction in pigmentation, suggesting that they derived it from a common ancestor after crossing from Africa into the Arabian peninsula and prior to the separation of populations heading northwest and northeast. In contrast, Europeans and Asians independently accumulated variants at two other loci with a role in skin pigmentation (**Figures 12.10b** and **12.10c**), which is evidence that the same selective pressures existed in both populations, which took advantage of different mutations.

Cultural-genetic feedback

Less obvious, at first glance, but just as important to certain populations, has been a genetic response to the cultural innovation of domesticating cattle for milk production. In pre-agricultural societies, only very young children had a diet containing milk from lactating mothers. The lactase enzyme required to digest milk was not needed after weaning, and a DNA regulatory allele that turned it off was advantageous. People living in Turkey domesticated cattle approximately 8000 years ago, and cattle moved with agriculturalists across Europe. European adults who could digest cow's milk gained a survival advantage, and a DNA regulatory variant that maintains lactase production throughout life has increased to high levels. Scientists have identified the DNA change responsible for this allele, which occurred about 5000 years ago. Its advantage to northern Europeans is so great that its frequency is now greater than 90 percent in certain regions.

Eventually, genetic drift will take over and eliminate the mutant allele completely from a finite population.

In some cases, a recessive disease-causing allele remains in a population at a stable frequency, in opposition to expectations from both analytical and Monte Carlo predictions. What maintains these diseases despite continuing selection against them? One answer is that sometimes heterozygotes have a higher fitness than either homozygote, a situation referred to as **balancing selection**.

Balancing selection maintains deleterious alleles in a population

We have seen that sickle-cell anaemia, which includes episodes of severe pain, serious anaemia, and a probability of early death, is a recessive condition resulting from two

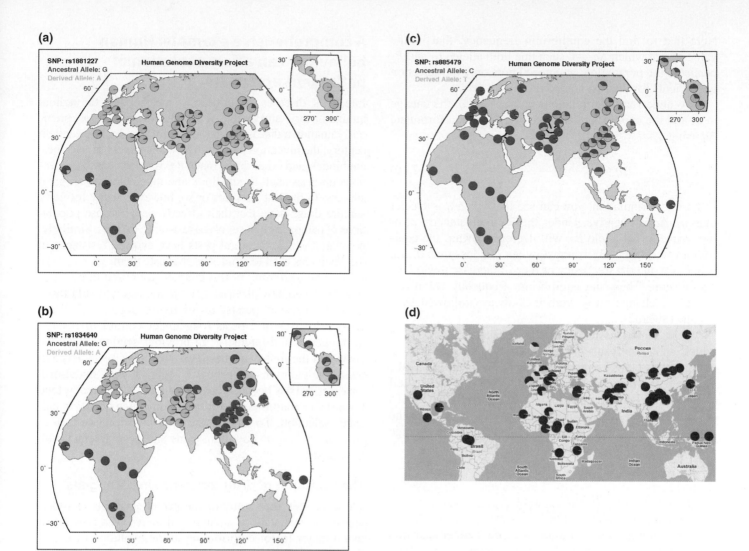

Figure 12.10 Geographical distribution of allele frequencies at skin pigmentation loci. (a) Pie charts showing the distribution of alleles at the *KITLG* locus. **(b)** Pie charts showing the distribution of alleles at the *SLC24A5* locus. **(c)** Pie charts showing the distribution of alleles at the *MC1R* locus. **(d)** Pie charts showing the distribution of alleles at the lactase (*LCT*) locus.

copies of the sickle-cell allele at the *β-globin* locus. The disease allele has not disappeared from several African populations, where it seems to have existed for a very long time.

One clue to the maintenance of the sickle-cell allele in human populations lies in the observation that heterozygotes for the normal and sickle-cell alleles are resistant to malaria. This resistance is due, in part, to the fact that red blood cells infected by the malaria parasite, if they also contain a sickle-cell allele, break open, destroying the parasite as well as the red blood cell itself. By contrast, in cells with two normal haemoglobin alleles, the malaria parasite thrives.

To set up a model of heterozygous advantage, let B_1 represent the normal *β-globin* allele and B_2 stand for the abnormal recessive sickle-cell allele; and for simplicity, assume that B_1B_2 heterozygotes have the maximum relative fitness of 1, while the selection coefficient (representing the selective disadvantage) for B_1B_1 homozygotes is $1 - s_1$, and the selection coefficient for B_2B_2 homozygotes

is $1 - s_2$. We can then represent the changes in allele frequency resulting from selection as

$$\Delta q = \frac{pq(s_1p - s_2q)}{\overline{w}} \qquad (12.8)$$

To maintain both alleles in the population using this equation, Δq must be 0 for some value of q between 0 and 1. The q value at which $\Delta q = 0$ is known as the **equilibrium frequency** of allele B_2. The value of q when $\Delta q = 0$ and both alleles are present occurs when the term inside the parentheses of Equation 12.8 is zero; that is, when

$$s_1p - s_2q = 0$$

Substituting $1 - q$ for p ($p = 1 - q$) and solving this equation for q reveals that the equilibrium frequency of B_2 (represented by q_e) is reached when

$$q_e = \frac{s_1}{s_1 + s_2} \qquad (12.9)$$

Note that to find the equilibrium frequency, that is, the value of q at which $\Delta q = 0$ such that both alleles B_1 and B_2 persist in the population, you need know only the selection coefficients for the two homozygotes.

To understand the relationship between q, the change in q, and the equilibrium frequency q_e, you can formulate Δq using q_e.

$$\Delta q = \frac{-pq(s_1 + s_2)(q - q_e)}{\overline{w}} \quad (12.10)$$

From this formulation, you can see that when q is greater than q_e, Δq is negative. Under these circumstances, q, or the frequency of allele B_2, will decrease toward the equilibrium frequency. By contrast, when q is less than q_e, Δq is positive and the frequency of B_2 will increase toward the equilibrium. Thus, the equilibrium frequency stabilizes, because a change away from it is always followed by a change toward it.

Now, if you assume that the African populations in which sickle-cell anaemia is prevalent are currently at equilibrium relative to their alleles at the β-globin locus, you can use the observed frequency of the sickle-cell allele in these populations to calculate the relative values of the selection coefficients. Field studies show that the actual value of q_e lies between 0.15 and 0.2 for an average value of 0.17. If you plug this number into Equation 12.9, you get

$$0.17 = \frac{s_1}{(s_1 + s_2)}$$

This equation makes it possible to express either selection coefficient in terms of the other. For example, $s_1 = 0.2s_2$.

If you assume that $s_2 = 1$ (i.e., those with the sickle-cell trait never reproduce), as was essentially true before medical advances enabled the survival of children expressing the sickle-cell trait, you will find that $s_1 = 0.2$, which, in turn, means that the relative fitness of the wild-type genotype is 0.8. Recall, however, that we set the fitness of the heterozygote at 1.0. By dividing 1.0 by 0.8, you get 1.25, which represents the relative advantage in fitness that heterozygotes for the sickle-cell allele have over people who do not carry this allele in African populations exposed to malaria. The use of simple statistical methods to calculate this heterozygous advantage demonstrates how medical geneticists can use the tools of population genetics (**Figure 12.11**).

"Fitness" is a measure of relative likelihoods of survival and reproduction due to alternative genotypes at a particular locus in a particular population. More fit genotypes will reproduce more of themselves at each generation, which "drives" an increase in the frequency of the alleles they carry. In some instances, the same genotype can be more fit in one environment, but less fit in another relative to an alternative genotype.

A comprehensive example: Human behaviour can affect evolution of pathogens and pests

Infectious diseases have been a major killer throughout human history, and as the AIDS epidemic illustrates, previously unknown diseases continue to emerge. In the twentieth century, the invention and discovery of a variety of vaccines, antibiotics, and other drugs made it possible to combat infectious diseases such as smallpox and tuberculosis with great success. In the last 30 years or so, however, many formerly surefire drugs have lost their effectiveness because populations of pathogens have evolved resistance to them. Similarly, populations of agricultural pests have evolved resistance to the pesticides used to control or eradicate them.

At the beginning of this chapter, we posed three questions: How do new diseases emerge in human populations? Why do diseases persist in all living organisms? What causes diseases and pests long under control to resurge in frequency and intensity? In previous sections, we have answered the first two questions. New diseases emerge in human populations as a consequence of new mutations. Diseases persist because changes in allele frequency tend toward an evolutionary equilibrium in which mutation balances selection. To answer the third question, we turn to an examination of how pathogens and pests interact with their hosts.

The evolution of drug resistance in pathogens

We have seen that many of the bacterial agents of tuberculosis are resistant to several antibiotics. We now know that a major factor contributing to the evolution of multi-drug-resistant (MDR) TB strains is the failure of patients to complete the lengthy drug regimens required for a cure. The two most widely used drugs for TB, isoniazid and rifampicin, require ingestion for six months and have side effects that include nausea and loss of appetite. However, the symptoms of TB can begin to disappear after only two to four weeks of treatment.

Imagine a TB patient with a persistent cough, shortness of breath, and general weakness. This individual harbours a large, actively growing and dividing population of TB bacteria in his lungs. At first, these bacteria are susceptible to antibiotics, but occasional mutations conferring partial resistance appear at random (**Figure 12.12**). The patient's physician prescribes a six-month course of treatment with the antibiotic isoniazid. After a few weeks of treatment, the bacterial population in the patient's lungs has decreased considerably, and the patient's symptoms have abated, although the negative side effects of the drug continue to cause discomfort. However, the composition of the bacterial population has now changed so that the remaining bacteria are likely to include a high proportion of mutant bacteria possessing partial resistance to the antibiotic.

If the patient continues his course of treatment, the persistent dose of antibiotic will eventually kill all of the bacteria, even those with partial resistance, eliminating the infection.

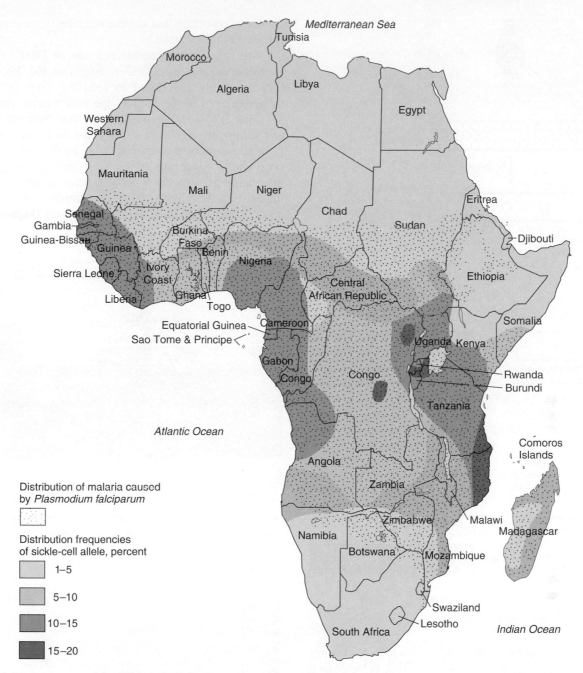

Figure 12.11 Frequency of the sickle-cell allele across Africa where malaria is prevalent.

Distribution of malaria caused by *Plasmodium falciparum*

Distribution frequencies of sickle-cell allele, percent
- 1–5
- 5–10
- 10–15
- 15–20

By contrast, if the patient stops treatment prematurely, the remaining (partially resistant) bacteria will proliferate and within three to four weeks re-establish a large population. Subsequent treatment of the same individual upon relapse or of a new patient to whom the partially resistant bacteria have spread would permit a second cycle of selection. New mutations could then convert the partially resistant bacteria to fully resistant microbes.

It is easy to see how repeated cycles of antibiotic treatment with multiple drugs, coupled with premature cessation of treatment, can promote the evolution of fully resistant bacterial populations, and even bacterial strains resistant to more than one drug, within a single patient. Initially, individual mutant cells that express genes conferring partial resistance increase in frequency; subsequent mutations in some of the mutant bacteria will increase resistance, and incomplete drug dosages will select for the resistant strains.

Several factors contribute to the rapid evolution of resistance in bacterial pathogens. The short generation times—often only a few hours—and rapid rate of reproduction under optimal conditions allow evolution to proceed quickly relative to a human life span. The large population densities typical of bacteria, which may exceed $10^9/cm^3$, ensure that rare resistance-conferring mutations will appear by chance in the population. The strong selection imposed by antibiotics increases the rate of evolution in each generation, unless the bacterial population is entirely eliminated.

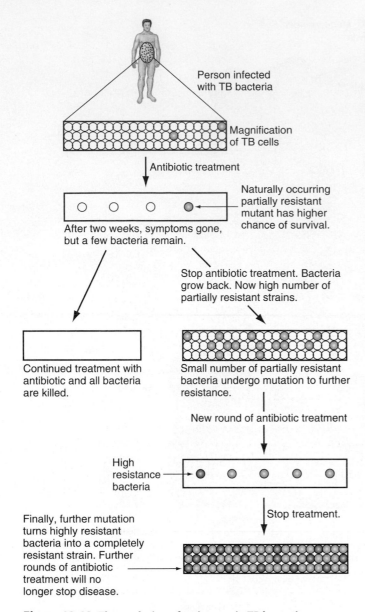

Figure 12.12 The evolution of resistance in TB bacteria.

Labels in figure:
- Person infected with TB bacteria
- Magnification of TB cells
- Antibiotic treatment
- After two weeks, symptoms gone, but a few bacteria remain.
- Naturally occurring partially resistant mutant has higher chance of survival.
- Stop antibiotic treatment. Bacteria grow back. Now high number of partially resistant strains.
- Continued treatment with antibiotic and all bacteria are killed.
- Small number of partially resistant bacteria undergo mutation to further resistance.
- New round of antibiotic treatment
- High resistance bacteria
- Stop treatment.
- Finally, further mutation turns highly resistant bacteria into a completely resistant strain. Further rounds of antibiotic treatment will no longer stop disease.

The large variety of ways by which bacteria can acquire genes also contributes to the rapid evolution of resistance. Many genes for resistance are found on plasmids, and, as you will see in Chapters 14 and 18, the capacity of plasmids to replicate and be transmitted among bacteria allows the amplified expression of resistance genes in bacterial populations. Plasmids also provide a means for the genetic exchange of resistance genes among bacterial populations and species through transformation, conjugation, and transduction. Laboratory studies have demonstrated the ready transfer of plasmid-borne resistance genes to new bacterial species.

The evolution of pesticide resistance

Like infectious bacteria, many agricultural pests spawn large populations because of their short generation times and rapid rates of reproduction. These large, rapidly reproducing populations evolve resistance to the chemical pesticides used to control them via selection for resistance-conferring mutations. Our understanding of this familiar pattern of selection and rapid evolution is most complete for certain insecticides.

The large-scale, commercial use of DDT and other synthetic organic insecticides, begun in the 1940s, was initially highly successful at reducing crop destruction by agricultural pests, such as the boll weevil, and medical pests, such as the mosquitoes that transmit malaria and yellow fever. Within a few years, however, resistance to these insecticides was detectable in the targeted insect populations. Since the 1950s, resistance to every known insecticide has evolved within 10 years of its commercial introduction. By 1984 there were reports of more than 450 resistant species of insects and mites (**Figure 12.13**).

(a)

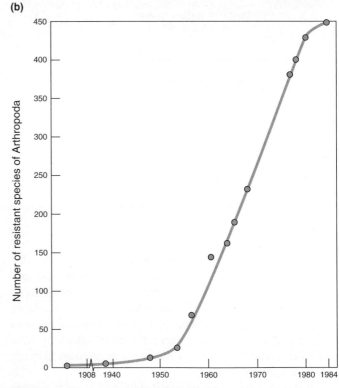

(b)

Figure 12.13 **Increase of insecticide resistance from 1908–1984.**
(a) Insecticide resistance evolved with the aerial spraying of DDT, which began in the 1940s. Within 10 years, resistance to an insecticide becomes widespread as the insects evolve defenses to the insecticides just as they would evolve defenses against infectious bacteria. **(b)** The evolution of resistance among Arthropoda.

Because different populations within a species can become resistant independently of other populations, the number of times insecticide resistance has evolved probably exceeds 1000.

Genetic studies show that insecticide resistance often results from changes in a single gene, and that several significant mechanisms of resistance are similar to those seen in infectious bacteria. DDT, for example, is a nerve toxin in insects. House flies and some mosquitoes develop resistance to DDT from dominant mutations in a single, enzyme-encoding allele. The mutant enzyme detoxifies DDT, rendering it harmless to the insect. As we saw earlier, even at low frequencies, dominant alleles can experience strong selection because of heterozygous advantage. Consider, for example, dominant mutation R (for insecticide resistance), which occurs initially at low frequency in a population. Soon after the mutation appears, most of the R alleles are in SR heterozygotes (in which S is the wild-type susceptibility allele). With the application of insecticide, strong selection favouring SR heterozygotes rapidly increases the frequency of the resistance allele in the population.

A field study of the use of DDT in Bangkok, Thailand, to control *Aedes aegypti* mosquitoes, the carriers of yellow fever, illustrates the rapid evolution of resistance. Spraying of the insecticide began in 1964 (**Figure 12.14**). Within a year, DDT-resistant genotypes emerged and rapidly increased in frequency. By mid-1967, the frequency of resistant RR homozygotes was nearly 100 percent.

The biological balance of resistance and fitness

Since DDT no longer controlled mosquito populations in the region, the Bangkok insecticide program was stopped. The response of the mosquito population to the cessation of spraying was intriguing: The frequency of the R allele decreased rapidly, and by 1969, RR genotypes had virtually disappeared. The precipitous decline of the R allele suggests that in the absence of DDT, the RR genotype produces a lower fitness than the SS genotype. In other words, the homozygous resistance genotype imposes a **fitness cost** on individuals such that in the absence of insecticide, resistance is subject to a negative selection that decreases the frequency of R in the population.

To understand the biological basis of fitness costs, consider how rats evolve resistance to warfarin, a pesticide introduced in the 1940s and 1950s to control small mammals, among other pests. Warfarin interferes with blood clotting by blocking the recycling of vitamin K (a cofactor in the clotting cascade). When a rat ingests warfarin, the inability to form a clot leads to a fatal loss of blood following any internal or external injury. In Europe in the 1960s, the extensive use of warfarin for rat control fuelled the evolution of a single-gene resistance allele in many rat populations. The frequency of resistance, however, did not increase to 100 percent; instead, in most populations, it levelled off at 30 percent to 60 percent. Apparently, some mechanism was maintaining both the R (resistance) and the S (susceptibility) alleles in the presence of warfarin.

Further study showed that in the presence of warfarin, the relative fitness was 0.37 for the SS genotype, 1.0 for the SR genotype, and 0.68 for the RR genotype. The greater relative fitness of the heterozygote is a driver that maintains both the S and R alleles at a particular equilibrium ratio. Further investigation revealed that two phenotypic factors cause the observed differences in fitness.

Both the SR and the RR genotypes are relatively resistant to the effects of warfarin, and thus, they provided higher fitness than the susceptible SS genotype. However, RR homozygotes suffered from a vitamin K deficiency because of the less-efficient vitamin K recycling

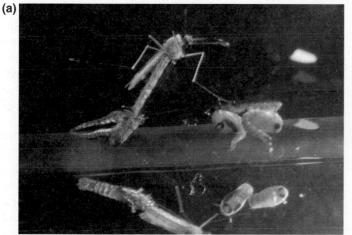

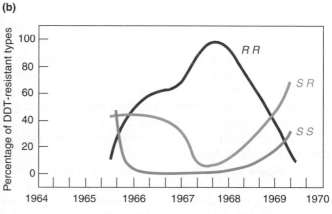

Figure 12.14 How genotype frequencies among populations of *A. aegypti* mosquito larvae changed in response to insecticide.
(a) Mosquitoes and larvae. **(b)** Changing proportions of resistance genotypes of *A. aegypti* (larvae) under selection with DDT (1964–1967), and after selection was relaxed (1968), in a suburb of Bangkok, Thailand.

during blood clotting, and this deficiency reduced the rate of survival when the diet did not contain a large amount of vitamin K. In the absence of warfarin, therefore, *RR* homozygotes had a lower fitness. The biological costs of fitness, which are widespread and occur by various mechanisms, are very likely a major reason why resistance alleles occurred in very low or undetectable frequencies before the routine use of pesticides

Rapid changes in resistance allele frequencies are driven by human use and subsequent disuse of pesticides and antibiotics to control organisms as different as bacteria and rats. The results suggest that fitness benefits often come with fitness costs, and the balance between the two is highly dependent on the environment in which the population lives.

12.3 Analyzing Quantitative Variation

We now examine how population geneticists study quantitative or complex traits. The continuous variation of such traits depends on the number of genes that generate the trait, as well as the genetic and environmental factors that affect the penetrance and expressivity of these genes (see Chapters 2 and 15). One of the goals of quantitative analysis is to discover how much of the variation in a particular trait is the result of genotypic differences among individuals in a population and how much arises from differences in the environment.

Genetic variance can be separated from environmental variance

To sort out the genetic and environmental determinants of phenotypic variation in a population, consider a series of experiments on a population of dandelions, a common weedy plant in lawns and other disturbed areas throughout North America (**Figure 12.15a**). Dandelions have a long tubular stem and a large, yellow composite flowering structure composed of many small individual flowers; each of these flowers can produce a single, tufted, diploid seed, dispersible by the wind. Most dandelion seeds arise from mitotic, rather than meiotic, divisions such that all the seeds from a single plant are genetically identical. Your goal is to compare the influence of genes and environment on the length of the stem at flowering.

To distinguish environmental from genetic effects on phenotypic variation, you need to quantify one variable, say the environment, while controlling for the other one; that is, while holding the genetic contribution steady. You could begin by planting half of the genetically identical seeds on a grassy hillside and allowing them to grow undisturbed until they flower. You then measure the length of the stem of each flowering plant and determine the mean and variance of the distribution of values for this trait in this dandelion population.

As **Figure 12.15b** shows, you find the **mean** by summing the values of all stem lengths and dividing by the number of stems. You then find the **variance** by expressing the stem lengths as plus or minus deviations from the mean, squaring those deviations, and again dividing by the number of stems. Because all members of this population are genetically identical, any observed variation in stem length among individuals should be a consequence of environmental variations, such as different amounts of water and sunlight at different locations on the hillside (if we ignore rare mutations). When represented as a variance from the mean, these observed environmentally determined differences in stem length are called the **environmental variance**, or V_E.

To refine your estimate of environmental variance, you plant the second half of the genetically identical seeds from the single test plant in a controlled greenhouse in which growth conditions are everywhere the same (**Figure 12.15c**). Because environmental conditions are much more similar for all these genetically identical plants, the amount of environmental variance (V_E) among greenhouse plants is much smaller than among hillside plants. In theory, in a perfectly controlled greenhouse, growth conditions would be the same for all plants, the V_E would be zero, and all plants would have identical stem lengths (within measurement error). In reality, there is no such thing as perfect control or a homogeneous environment, and the greenhouse V_E will have some value greater than zero. Nonetheless, the difference between the V_E of the dandelions grown on the hillside and the V_E of dandelions grown in the greenhouse is a measure of the impact of the more diversified hillside environment on the phenotypic variation of stem length.

Even though the greenhouse V_E will have some value greater than zero, for the sake of simplicity in the following discussion, we assume that it is small.

To examine the impact of genetic differences on stem length, you take seeds from many different dandelion plants produced in many different locations, and you plant them in a controlled greenhouse (**Figure 12.15d**). Because you are raising genetically diverse plants in a relatively uniform environment, observed variation in stem length—beyond that found in the genetically identical population—is the result of genetic differences promoting **genetic variance**, or V_G.

Now, to determine the total impact on phenotype of genes and environment, you take the seeds of many different plants from many different locations and grow them on the same hillside (**Figure 12.15e**). For the population of dandelions that grow up from these seeds, the **total phenotype variance** (V_P) in stem length will be the sum of the genetic variance (V_G) and the environmental variance (V_E). The environmental variance is determined directly from the phenotypic variance found in the initial population of genetically identical plants grown on the hillside.

(a)

(b)

Genetically identical plants grown on hillside

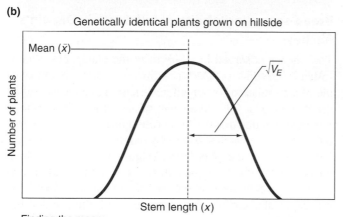

Finding the mean:

Let x_i = the stem length of the plant i in a population of N plants. The mean of stem length, $\bar{x}$, for the population is defined as

$$\bar{x} = \frac{\sum\limits_{i=1}^{N} x_i}{N}$$

Finding the variance:

The variance V_E of stem length for the population is defined as

$$V_E = \frac{\sum\limits_{i=1}^{N} (x_i - \bar{x})^2}{N}$$

(c)

Genetically identical seeds

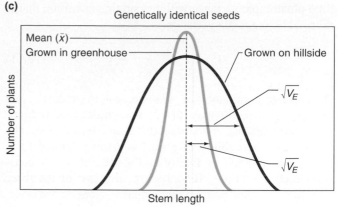

V_E is smaller among greenhouse-grown plants.

(d)

Plants grown in greenhouse

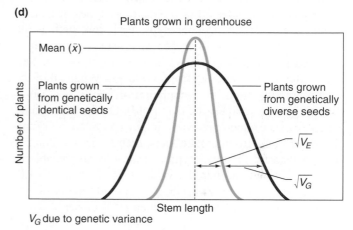

V_G due to genetic variance

(e)

Plants grown on hillside

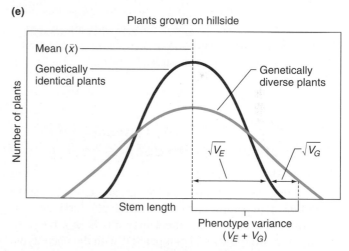

Figure 12.15 Studies of dandelions can help sort out the effects of genes versus the environment. (a) The familiar dandelion (*Taraxacum* sp.) is a useful model when studying population variations. **(b)** Finding the mean and variance of stem length. **(c)** Variance of genetically identical plants grown in a greenhouse and grown on a hillside. **(d)** Genetic variance of plants grown in a greenhouse. **(e)** Phenotype variance of plants grown on a hillside.

It thus becomes possible for you to calculate the genetic variance in the second, mixed population as the difference between the phenotypic variance found in this genetically mixed population and the phenotypic variance found in the genetically identical population. For natural populations of dandelions, both genetic variation among individuals and variation in the environmental conditions experienced by each plant contribute to the total phenotypic variation.

Broad-sense heritability is the proportion of phenotypic variance due to genetic variance

With the ability to determine the relative contributions to phenotypic variation of genes and environment, geneticists have developed a mathematical definition of the **broad-sense heritability** (H^2) of a trait: It is the proportion of

total phenotypic variance within a single generation that is ascribable to genetic variance.

$$H^2 = \frac{V_G}{V_G + V_E} = \frac{V_G}{V_P} \qquad (12.11)$$

Because the amounts of genetic, environmental, and phenotypic variation may differ among traits, among families or populations, and among different environments, the broad-sense heritability of a trait is always defined for a specific population or family and a specific set of environmental conditions. If you know any two of the three variables of total phenotypic variance, genetic variance, and environmental variance, you can find the remaining unknown variance. H^2 is called "broad-sense" heritability because it includes all of the different types of genetic inputs that contribute to phenotypic variation, such as the contributions of individual genes, epistatic interactions between genes, and of course how genes work together. The value of H^2 can range from 0.0 to 1.0. When all of the phenotypic variation in a population or family is attributable to genetic sources, then H^2 is 1.0. On the other hand, when all of the phenotypic variation in a population or family is due to environmental effects, then H^2 is 0.0. Thus, if a large share of the phenotypic variation is attributable to genotypic variation, then the value of H^2 is closer to 1.0, and if a large part of the variation in a population or family is due to environmental sources, then H^2 is closer to 0.0.

> Analysis of a quantitative trait begins by measuring the relative contributions of environmental and genetic variance. Because the combination of these two gives the total phenotypic variance, any two values can be used to calculate the third.

Broad-sense heritability is measured in studies of groups with defined genetic differences

In analyzing the contributions of genes and environment to dandelion stem length, you measured the phenotypic variation among genetically identical individuals in a range of specified environments and compared it with the phenotypic variation among all individuals in the population. Of course, most organisms are not as easy to clone as dandelions. The key to generalizing from the dandelion example is to recognize that genetic clones are simply a special case of the broader notion of genetically related individuals, or genetic relatives, who share certain alleles because they have one or more common ancestors. To quantify this idea, we can define the **genetic relatedness** of two individuals as the average fraction of common alleles at all gene loci that the individuals share because they inherited them from a common ancestor.

To determine the genetic relatedness of two siblings, for example, you simply calculate the probability they received the same allele at any locus from the same parent. If you assume that one sibling received allele *A1* from an

A1/A2 heterozygous parent, the probability that the second sibling received the same allele is 0.5. Because this simple calculation holds for every locus transmitted by both parents, the total genetic relatedness of two siblings is 0.5. With an extension of this probabilistic analysis, we can see that an aunt and niece have 0.25 genetic relatedness, and first cousins 0.125.

If genetic similarity contributes to phenotypic similarity for some trait, it is logical to expect that a pair of close genetic relatives will be more phenotypically similar than a pair of individuals chosen at random from the population at large. Thus, by comparing the phenotypic variation among a well-defined set of genetic relatives with the phenotypic variation of the entire population over some range of environments, it is possible to estimate the broad-sense heritability of a trait.

Broad-sense heritability of bill depth in Darwin's finches

The finches observed by Darwin in the Galápagos Islands (often referred to as "Darwin's finches") provide an example of a population for which geneticists have measured the heritability of a trait under natural conditions in the field. Scientists studied the medium ground finch, *Geospiza fortis*, on the island of Daphne Major by banding many of the individual birds in the population (**Figure 12.16a**). They then measured the depth of the bill for the mother, father, and offspring in each nest on the island and calculated how the bill depth of the offspring correlated with the average bill depth of the mother and father (called the *midparent value*).

The results, depicted in **Figure 12.16b**, show a clear correlation between parents and offspring; parents with deeper bills had offspring with deeper bills, while parents with smaller bill depth had offspring with smaller bill depth. In the figure, the broad-sense heritability of bill depth, as represented by the slope of the line correlating midparent bill depth to offspring bill depth, is 0.82. This means that roughly 82 percent of the variation in bill depth in this population of Darwin's finches is attributable to genetic variation among individuals in the population; the other 18 percent results from variation in the environment. If the environment had had no influence at all on the trait, then the slope of the line representing the broad-sense heritability of bill depth, that is, correlating bill depth in parents with bill depth in offspring, would be 1.0 (**Figure 12.16c**).

Now consider a population in which the bill depth for parents and their offspring is, on average, no more or less similar than the bill depths for any pair of individuals chosen from the population at random. In such a population, there is no correlation between the bill depth trait in parents and in offspring, and a plot of midparent and offspring bill depths produces a circular "cloud" of points (**Figure 12.16d**).

From these examples, you can conclude that phenotypic similarity among genetically related individuals may provide evidence for the broad-sense heritability of a trait. However, conversion of the phenotypic similarity among genetic relatives to a measure of broad-sense heritability

(a)

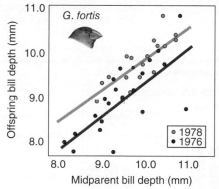

(b) Correlation between parents and offspring

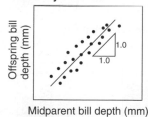

(c) If the heritability were 1.0

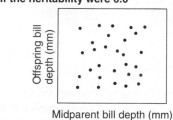

Midparent bill depth (mm)

(d) If the heritability were 0.0

Midparent bill depth (mm)

Figure 12.16 Measuring the broad-sense heritability of bill depth in populations of Darwin's finches. (a) *G. fortis* with bands placed by scientists. **(b)** The correlation between beak size of offspring and their midparent value (the average of the parents' beak size) is 0.90 both in 1976 (*red circles*) and 1978 (*blue circles*), even though the mean beak size increased due to a drought in 1978. This correlation shows constant high heritability in the broad sense independent of environmental change. Note that high broad-sense heritability does *not* mean that a trait is constant: beak size is highly variable (note range of axes) and varies over time (displacement of slopes). **(c)** Results if broad-sense heritability were 1.0. **(d)** Results if broad-sense heritability were 0.0.

depends on a crucial assumption: that the distribution of genetic relatives is random with respect to environmental conditions experienced by the population. In the finch example, we assumed that parents and their offspring do not experience environments that are any more similar than the environments of unrelated individuals.

In nature, however, there may be reasons why genetic relatives violate this assumption by inhabiting similar environments. With finches, for example, all offspring produced by a mother and father during a breeding season normally hatch and grow in a single nest where they receive food from their parents. Because bill depth affects a finch's capacity to forage for food, the amount of feeding in a nest correlates with parental bill depth, for reasons quite distinct from genetic similarities.

One way to reduce the confounding variable of environmental similarity is to remove eggs from the nest of one pair of parents and randomly place them in nests built by other parents in the population; this random relocation of eggs is called **cross-fostering**. In heritability studies of animals that receive parental care, cross-fostering helps randomize environmental conditions. Controlling for both environmental conditions and breeding crosses is a fundamental part of the experimental design of heritability studies carried out on wild and domesticated organisms.

Measuring the broad-sense heritability of complex traits in humans

Mating does not occur at random with respect to phenotypes in human populations, and researchers cannot apply techniques for controlling environmental conditions and breeding crosses to studies of such populations. Nonetheless, in most human societies, family members share similar family and cultural environments. Thus, phenotypic similarity between genetic relatives may result either from genetic similarities or similar environments or, most often, both. How can you distinguish the effects of genetic similarity from the effects of a shared environment?

One way is to study monozygotic, "identical" twins given up for adoption shortly after birth and raised in different families. In such a pair of identical twins, any phenotypic similarity should be the result of genetic similarity. At first glance, then, the study of adopted identical twins eliminates the confounding effects of a similar family environment. Further scrutiny, however, shows that this is often not true. Many pairs of twins are adopted by different genetic relatives; the adoptions often occur in the same geographical region (usually in the same province and even the same city); and families wishing to adopt must satisfy many criteria, including job and financial stability and a certain family size. As a result, the two families adopting a pair of twins are likely to be more similar than a pair of families chosen at random, and this similarity can reduce the phenotypic differences between the twins. A valid scientific study of separated twins must take these factors into consideration.

A related approach is to compare the phenotypic differences between different sets of genetic relatives, particularly different types of twins (**Figure 12.17a**).

For example, monozygotic (MZ) twins, which are the result of a split in the zygote after fertilization, are genetically identical because they come from a single sperm and a single egg; they share all alleles at all loci and thus have a genetic relatedness of 1.0. By contrast, dizygotic (DZ) twins, which are the result of different sperm from a single father fertilizing two different maternal eggs, are like any pair of siblings born at separate times; they have a genetic relatedness of 0.5 (which actually means that their dissimilarity is only 50 percent of the average dissimilarity between two unrelated individuals). Comparing the phenotypic differences between a pair of MZ twins with the phenotypic differences between a pair of DZ twins can help distinguish between the effects of genes and family environment.

Consider a trait in which the differences in phenotype among individuals in the population arise entirely from differences in the environment experienced by each individual; that is, a trait for which the broad-sense heritability is 0.0 (**Figure 12.17b1**). For this trait, you would expect the phenotypic differences among many pairs of MZ twins to be as great as the differences among many pairs of DZ twins. The fact that the MZ twins are more closely related genetically has no effect.

Now consider a trait for which differences in phenotypes among individuals in a population arise entirely from genetic differences; that is, a trait for which the broad-sense heritability is 1.0 (**Figure 12.17b2**). Since MZ twins have a genetic relatedness of 1.0, they always show 100 percent concordance in expression: If one expresses the trait, the other does as well. The concordance of trait expression between unrelated individuals varies based on the commonality of the trait (as shown in Figure 12.17b2). Dizygotic twins would display greater concordance than genetically unrelated individuals, but less than monozygotic twins. In the highly simplified case of a dominant trait caused by an allele at a single autosomal gene, dizygotic twins would show a level of concordance that is halfway between the unrelated value and 100 percent. In reality, nearly all traits are affected by multiple genes that may have dominant, recessive, incompletely dominant, and interacting or epistatic effects, and as mentioned earlier the broad-sense heritabilities of nearly all traits lie between 0.0 and 1.0.

The concept of broad-sense heritability is often used incorrectly by people who are not familiar with its scientific derivation. A measured broad-sense heritability is a statistical value that is only meaningful in the context of a family or population, not an individual. Furthermore, the broad-sense heritability of a trait applies only to a particular family or population in a particular environment, where environment includes every influence on organisms outside of their genomes.

A good example of the impact of environment on broad-sense heritability comes from an analysis of human height. When measured in a prosperous population with modern standards of food production, human height shows a very high broad-sense heritability value, almost 0.9. In contrast, in a poor country where not everyone gets enough nutrition, heritability in the broad sense would be much lower.

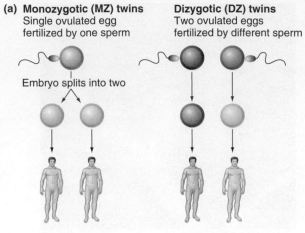

(a) Monozygotic (MZ) twins Single ovulated egg fertilized by one sperm

Dizygotic (DZ) twins Two ovulated eggs fertilized by different sperm

Embryo splits into two

Monozygotic twins 100% genetic identity

Dizygotic twins Decrease in genetic dissimilarity relative to unrelated individuals 25% decrease in genotypic dissimilarity 50% decrease in allelic dissimilarity

(b) Probability that a second child will express a dominant trait that is expressed by a first child

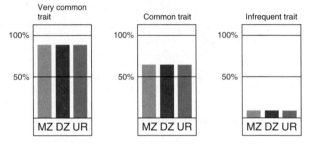

1. Trait with 0.0 heritability

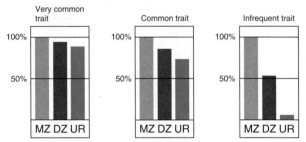

2. Trait with 1.0 heritability

MZ = monozygotic twins, DZ = dizygotic twins, UR = unrelated due to adoption

Figure 12.17 The impact of broad-sense heritability on the concordance of dominant trait expression in two children raised in the same family environment. (a) Monozygotic and dizygotic twins have different genetic origins. **(b)** The frequency with which a second child will share a trait expressed by a first child. **1.** At one extreme are hypothetical traits associated with a broad-sense heritability of 0.0. Irrespective of the frequency of trait expression, no differences would be observed in a comparison of monozygotic twins, dizygotic twins, or situations where one or both children are adopted. **2.** At the opposite extreme are hypothetical traits associated with a broad-sense heritability of 1.0. Monozygotic pairs would be concordant, whereas dizygotic pairs of twins would show a concordance halfway between 100 percent and the concordance found between genetically unrelated children.

The explanation comes from the fact that a person's genome determines their maximum height potential. If their nutrition is sufficient, they will reach this potential; further intake of food will make no difference. However, in some underdeveloped countries, great differences exist in the amount of nutrition that any individual is able to consume. This environmental difference will express itself as an increase in the environmental component of height variance.

> Genetic relatedness is a relative term that describes the degree of genetic similarity between two members of the same extended family that is greater than that expected for two unrelated members of the population. The degree to which genetic relatedness correlates with the expression of a trait provides a measure of the trait's broad-sense heritability.

A trait's narrow-sense heritability determines its potential for evolution

We saw earlier that high values for broad-sense heritability suggest that genotype is important in determining whether the quantitative trait or phenotype arises due to the segregation of particular alleles, however, H^2 does not predict the phenotypes of progeny based on the phenotypes of the parents. Thus, if H^2 is high, the phenotype of an individual is likely attributable to its genotype within that family or population, but even if there is genetic variation in a family or population as measured by H^2, it may not be passed on to the next generation in a predictable manner. Moreover, broad-sense heritability is specific to a particular environment and family or population, so values for H^2 can vary among different families, populations, and environments.

Why is broad-sense heritability not predictive? Well, genetic variation (V_G) encompasses variation due to additive effects (V_A), variation due to dominance effects (V_D), and variation due to epistatic effects (V_I). Both dominance variation (V_D) and epistatic variation (V_I), however, are not transmitted in a predictable manner from parent to offspring. Given that some of the phenotypic variation among the parents is the result of dominance and epistatic interactions between alleles, and since parents pass on their genes but not their genotypes to their progeny, these dominance and epistatic interactions are not transmitted to the offspring: New genotypes, and consequently new dominance and epistatic relationships, are established with each generation. Thus, with dominance and epistasis, the parental phenotypes are not completely heritable.

On the other hand, additive variation (V_A) is the fraction of genetic variation that is predictably transmitted from parent to progeny. With additive gene action, the phenotype or trait value of the heterozygote is exactly intermediate (midway) between the two homozygous parental classes or trait values. Although parents do not transmit their dominance, epistatic, or environmental variations to their progeny, they do pass on their additive variations. Thus, assuming that there are no environmental effects, and the genotype solely determines the phenotype, then gene action is wholly additive and the phenotype is fully heritable. The proportion of total phenotypic variation that is attributable to additive genetic variation is known as the **narrow-sense heritability (h^2)** of the trait.

$$h^2 = \frac{V_A}{V_P} \qquad (12.12)$$

Values for h^2 fall between 0.0 and 1.0. If all of the phenotypic variation is due to additive variation, then h^2 is equal to 1.0. Alternatively, if all of the phenotypic variation in a population or family is attributable to other genetic and environmental effects, then h^2 is 0.0. Thus, if narrow-sense heritability is high, the phenotype of the individual is predictable based on the phenotypes of its parents within that family or population. Like broad-sense heritability though, h^2 is a product of the particular environment, and family or population in which it was calculated; thus an h^2 value from one population, family, or environment may not be useful for another population, family, or environment. Furthermore, although we know that particular additive-effect genes contribute to the phenotype or trait, h^2 cannot tell us what the specific genes are.

Because narrow-sense heritability quantifies the extent to which phenotypic variation among individuals in a family or population is transmitted in a predictable way to their offspring, it is the value considered most important to plant and animal breeders, since it indicates which characteristics can be improved or enhanced by artificial selection. Narrow-sense heritability is also extremely valuable to evolutionary biologists. It was previously shown how the selection of pre-existing mutations generates evolutionary change. Given that the narrow-sense heritability of a multifactorial trait is a measure of the additive genetic component of its phenotypic variation, h^2 quantifies the potential for selection and thus the potential for evolution from one generation to the next. A trait with a high narrow-sense heritability value has a large potential for evolution via selection. Dr. David Coltman, a Professor at the University of Alberta in Edmonton, has been using population and quantitative genetic methods for several years to study the evolution of wildlife populations in Canada such as the bighorn sheep (see the **Focus on Genetics** box "**Population and Quantitative Genetic Approaches in the Wild**").

To grasp the relationship between narrow-sense heritability, selection, and evolution, consider the number of bristles on the abdomens of fruit flies in a laboratory population of *D. melanogaster*. This fruit fly population exhibits some phenotypic variation in the trait of bristle number. If the trait has a high narrow-sense heritability value in the population, the offspring of this original population will closely resemble their parents in bristle number (**Figure 12.18**). If, however, you select as parents of the next generation only those flies among the top 15 percent in bristle number, the average bristle number among these breeders of the next generation will be greater than the average bristle number in the population as a whole. This artificial selection in conjunction with the high narrow-sense heritability of the trait will produce an F_1 generation in which the average bristle number is greater than the average bristle number in the previous generation. In other words, the artificial selection imposed by the experimenter

Dr. David Coltman (**Figure A**) is a professor in the Department of Biological Sciences at the University of Alberta. His research interests include the maintenance of genetic variation in important adaptive traits in natural populations such as bighorn sheep (**Figure B**) and mountain goats. He uses molecular markers, such as SNPs and microsatellites, as tools for population and quantitative genetic analyses. His investigation of evolutionary theory in the wild using longitudinal data and pedigrees has enabled the use of genetic approaches to challenges in conservation and wildlife management. One challenge is the impact of human hunting activities on the evolution and genetic makeup of wildlife populations. His expertise in genetics has also been critical to evolutionary and quantitative genetic studies of traits associated with fitness in bighorn sheep, as well as the generation of a genome map for bighorn sheep.

Figure A David Coltman.

Figure B Bighorn sheep Adult male (*left*) and female (*right*) bighorn sheep from Ram Mountain, Alberta.

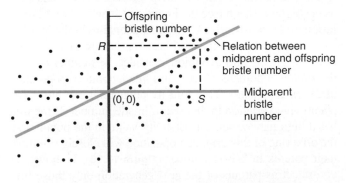

Figure 12.18 Relationship between midparent number of abdominal bristles and bristle number in offspring for a hypothetical laboratory population of *Drosophila*.

will induce an evolutionary change whose magnitude is related to the narrow-sense heritability of the trait. If the h^2 of bristle number were zero, there would be no evolutionary change. (Figure 12.16 also shows the impact of natural selection on a trait with high narrow-sense heritability.)

Additive variation (V_A) is the fraction of genetic variation that is transmitted in a predictable manner from parent to offspring. The proportion of phenotypic variation in a trait that is due to additive genetic variation is known as narrow-sense heritability (h^2). Narrow-sense heritability is the value used by animal and plant breeders, and evolutionary biologists, for programs of artificial selection and predictions of evolution by natural selection.

A mathematical model of the relation between narrow-sense heritability and evolution

Let S represent the average trait value (in this case, the average bristle number) of breeding individuals in the parental population. This value is measured as the difference between the value of this trait for *parents* and the value of the trait in the entire parental population (both breeding and nonbreeding individuals). S is then a measure of the strength of selection on the trait; as such, it is called the **selection differential**. Now let R represent the average trait value in

the offspring of these breeding parents, which is measured as the difference between the trait's value for *offspring* and its value in the entire parental population of breeding and nonbreeding individuals. Used in this way, **R** signifies the **response to selection**; that is, the amount of evolution, or change in mean trait value, resulting from selection.

The narrow-sense heritability of the trait (h^2), as seen in the slope of the line relating parental to offspring trait values in Figure 12.16, determines the relationship between S and R.

$$R = h^2 S \qquad (12.13)$$

In other words, the strength of selection (S) and the narrow-sense heritability of a trait (h^2) directly determine the trait's amount or rate of evolution in each generation. This relationship is the primary reason population and evolutionary geneticists consider the ability to measure narrow-sense heritability so important.

Variations in complex traits arise rapidly because of increased opportunities for change

Geneticists have long used bristle number in *Drosophila* as a model for understanding the variation, selection, and evolution of quantitative traits. Early laboratory studies of selection acting on bristle number showed that the trait has substantial narrow-sense heritability in *Drosophila*, and it evolves rapidly in response to selection for either high or low bristle number (**Figure 12.19**). Two results from these early studies were particularly striking. First, selection can rapidly lead to phenotypes not seen in the original population. After 35 generations of artificial selection for high or low bristle number, no overlap was evident between the unselected and the selected populations. Some of the change in bristle number probably arose from reassortment and changes in frequency of existing alleles, without the appearance of new alleles. However, traits such as bristle number continue to evolve in response to selection for many generations. This observation suggests that new mutation is an additional source of variation in the population.

Experimenters have examined the contribution of mutation to genetic variation in bristle number (and by extension, other quantitative traits) through studies of highly inbred lines of *Drosophila* that at first had low or no genetic variation in bristle number. With these inbred lines,

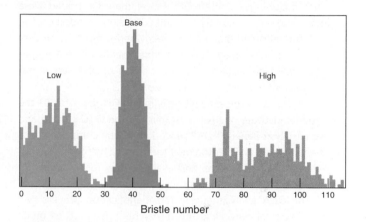

Figure 12.19 Evolution of abdominal bristle number in response to artificial selection in *Drosophila*. Bristle number distributions in different populations under selection for low (*green*) or high (*orange*) values, compared with distributions in a population not subjected to selection (base in *blue*).

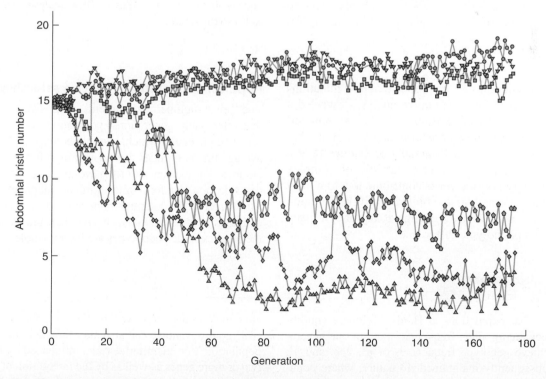

Figure 12.20 The effect of new mutations on mean bristle number in *Drosophila*. The average bristle number in a population under artificial selection for a reduced number over many generations is indicated with the *diamond, pentagon,* and *up-triangle* data points (*green*): Populations not under selection are indicated with the *down-triangle, circle,* and *square* data points (*orange*).

In the wake of the 9/11/01 World Trade Center (WTC) attacks in New York City, the scientific community refined techniques of DNA analysis to work successfully with badly damaged DNA and the lack of any prior tissue samples from some victims. Degraded DNA samples are common to forensic cases and missing persons investigations, and improvements in techniques have extended the range of information recoverable from highly degraded specimens.

The crime scene following the WTC disaster covered 17 acres from which searchers retrieved 19 893 separate body parts, including a single tooth. Fewer than 300 bodies were intact, and only 12 could be identified by sight. In addition, the DNA fragments in many of the recovered samples were very scarce and very small. The traditional means of DNA fingerprinting, which require long, intact pieces of DNA, would not be sufficient to identify the victims of 9/11.

To deal with this difficult situation, in October of 2001 the National Institute of Justice established the WTC Kinship and Data Analysis Panel (KADAP), a working group composed of 25 scientific experts. Over the next three and a half years, KADAP met monthly to develop and evaluate new technologies for the analysis of DNA remains.

The DNA analysis and victim identification process had three main phases. First, the researchers gathered as much information as possible about the missing individuals. Personal effects, such as the victims' hairbrushes, razors, combs, dirty clothes, and toothbrushes, were one source of information; DNA reference samples from family members—including children, spouses, siblings, and parents—were another.

In phase two, these personal effects and family reference samples were subjected to DNA typing tests, including PCR analysis of various microsatellite and SNP alleles, described previously. In some cases, the allele constitution of the missing person could be reconstructed directly from DNA gathered from personal effects. In other cases, a victim's microsatellite and SNP alleles could be inferred only through genetic analysis of surviving children and spouses: Any allele that was present in a child and not in the surviving parent must have come from the victim's genome.

In phase three, the resultant DNA profiles were loaded into computer databases for matching with DNA extracted from the human remains found at the WTC site. Several companies contributed to this third phase, including Bode Technology Group, Celera Genomics, and Orchid GeneScreen. Gene Codes Forensics developed the Mass Fatality Identification System (M-FISys) software to deal with the need for high-throughput analysis.

In February 2005, the New York medical examiner's office announced that it had "exhausted all current technologies" and ended efforts to identify the remains. Of the 2749 people who died, 1100 remained unidentified. In addition, about 10 000 unidentified bone and tissue fragments had not yet been matched with the list of the missing. These will be held in the New York medical examiner's laboratory until identification is possible.

The technologies developed in the aftermath of the WTC disaster were applicable in the aftermath of the tsunami that struck Thailand on December 26, 2004, which killed an estimated 174 000–275 000 people. Thai forensic experts and disaster teams from more than 25 other nations have made positive matches on 2156 bodies, most of them tourists killed while on vacation. These identification efforts, based on continually improving strategies of DNA analysis, will continue well into the future.

---

selection could occur only in the presence of new mutations affecting bristle number. Quantitative analyses revealed a significant selection-driven evolution of bristle number, which means that new mutations affecting bristle number arise in a population at a substantial rate (**Figure 12.20**). These results highlight a key characteristic of complex traits: If many polymorphic genes contribute to a trait, new variation in the trait may arise rapidly even if the mutation rate per gene is low, because a change at any one of many loci can cause a phenotypic difference.

Populations eventually reach a selective plateau

The bristle-number experiments with *Drosophila* showed that after many generations, populations eventually reach a selective plateau at which, even with continued selection, the average bristle number does not change for many more generations. The existence of such evolutionary plateaus suggests that selection can, for a time, eliminate all genetic variation in a trait and that the potential for new mutations allowing further extremes in phenotypes has been exhausted, usually because the most extreme phenotypes are incompatible with viability.

Connections

We have seen that populations at Hardy–Weinberg equilibrium have unchanging allele frequencies, and in one generation, they achieve genotype frequencies of p^2, $2pq$, and q^2, which are subsequently maintained. In nature, where populations are rarely at complete Hardy–Weinberg equilibrium, however, natural selection acts on differences in fitness to alter allele frequencies. New mutations, genetic drift, and heterozygous advantage can also alter the allele frequencies of a population. For quantitative traits influenced by the alleles of two or more genes as well as by the interaction of those alleles with the environment, the narrow-sense heritability of a trait, that is, the proportion of its phenotypic variation attributable

to additive genetic variation, determines its potential for evolution by mutation, selection, and genetic drift.

In Chapter 13, we look at evolution from the point of view of the molecular mechanisms that propel it. In that chapter, we examine the various ways in which changes at the genomic level continually reshuffle the genetic deck to create the ever-changing abundance of life-forms that inhabit Earth.

Essential Concepts

1. A population is defined as a unit entity by its allele frequencies and genotype frequencies, which together make up a gene pool. Although an actual gene pool can only be determined empirically by counting all existing alleles, population geneticists have developed analytical and computational models for estimating the genetic and phenotypic variations of a population and how they may change over time. **[LO1–2]**

2. With the simplifying assumptions of an ideal population of very large size, where individuals mate at random, no new mutations appear, no individuals enter or leave, and there are no genotype-dependent differences in fitness, it becomes possible to derive a simple binomial equation that describes the precise relationships existing between allele, genotype, and phenotype frequencies. This equation, $p^2 + 2pq + q^2 = 1$, is called the Hardy–Weinberg law. **[LO3]**

3. A population satisfying the Hardy–Weinberg assumptions is said to be at Hardy–Weinberg equilibrium. In such a population, allele frequencies remain constant from one generation to the next, and the genotype frequencies of p^2, $2pq$, and q^2 are achieved in one generation, after which they are maintained. **[LO3]**

4. In natural populations, conditions always deviate at least slightly from the Hardy–Weinberg assumptions. And yet, even with these deviations from ideal conditions, the Hardy–Weinberg equation provides remarkably good estimates of allele, genotype, and phenotype frequencies over the *short run*. Over the *long run*, however, the Hardy–Weinberg equation is rarely applicable for predictive purposes. But it serves a critical role in providing the foundation for both analytical and stochastic methods that do incorporate the various factors responsible for deviation from equilibrium conditions. **[LO4]**

5. Monte Carlo simulations provide a computational method for modelling allele frequencies in populations of finite size, which undergo genetic drift. In populations with two alleles having equivalent phenotypic effects and present initially at equal frequencies, the median number of generations to fixation is roughly equal to the total number of gene copies in breeding individuals. Evolution consists of changes in allele frequency over time. Selection acting on genotype-dependent differences in fitness can drive evolution. Selection does not entirely eliminate deleterious recessive alleles from a population. One reason for this is balancing selection. **[LO5]**

6. For quantitative traits, the environmental variance is a measure of the influence of environment on phenotypic variation. Similarly, genetic variance measures the contribution of genes to phenotypic variation. Total phenotype variance is the sum of genetic variance and environmental variance. **[LO6]**

7. Measures of environmental, genetic, and total phenotype variance make it possible to define the broad-sense heritability of a trait as the proportion of total phenotype variance attributable to genetic variance. With traits for which the number and identity of contributing genes remain unknown and genetic clones cannot be obtained, it is possible to correlate phenotypic variation with the genetic relatedness of individuals—that is, the average fraction of common alleles at all genetic loci that the individuals share because they inherited them from a common ancestor—to measure the broad-sense heritability of a trait. **[LO7]**

8. To ascertain the broad-sense heritability of a human trait, population geneticists often turn to studies of twins. The most useful approach is to compare the phenotypic differences between pairs of monozygotic and dizygotic twins. Environmental changes can always influence the degree of heritability. **[LO7]**

9. A trait's narrow-sense heritability determines its predictability and its potential for evolution. In the case of complex traits, opportunities for mutation exist at many loci, therefore allowing relatively rapid changes to occur. Eventually a trait reaches a selective plateau at which no further average variation occurs. **[LO7–8]**

Solved Problems

I. A population called the "founder generation," consisting of 2000 *AA* individuals, 2000 *Aa* individuals, and 6000 *aa* individuals, is established on a remote island. Mating within this population occurs at random, the three genotypes are selectively neutral, and mutations occur at a negligible rate.

   a. What are the frequencies of alleles *A* and *a* in the founder generation?

   b. Is the founder generation at Hardy–Weinberg equilibrium?

   c. What is the frequency of the *A* allele in the second generation (i.e., the generation subsequent to the founder generation)?

   d. What are the frequencies for the *AA*, *Aa*, and *aa* genotypes in the second generation?

e. Is the second generation at Hardy–Weinberg equilibrium?

f. What are the frequencies for the *AA*, *Aa*, and *aa* genotypes in the third generation?

Answer

This question requires calculation of allele and genotype frequencies and an understanding of the Hardy–Weinberg equilibrium principle.

a. To calculate allele frequencies, count the total alleles represented in individuals with each genotype and divide by the total number of alleles.

Number of individuals	Number of *A* alleles	Number of *a* alleles
2000 *AA*	4000	0
2000 *Aa*	2000	2000
6000 *aa*	0	12 000
Total	6000	14 000

The frequency of the A allele (p) = 6000/20 000 = 0.3.
The frequency of the a allele (q) = 14 000/20 000 = 0.7.

b. If a population is at Hardy–Weinberg equilibrium, the genotype frequencies are p^2, $2pq$, and q^2. We calculated in part *a* that $p = 0.3$ and $q = 0.7$ in this population. Therefore,

$$p^2 = 0.09$$

$$2pq = 2(0.3)(0.7) = 0.42$$

$$q^2 = 0.49$$

For a population of 10 000 individuals, the number of individuals with each genotype, if the population were at equilibrium and the allele frequencies were $p = 0.3$ and $p = 0.7$, would be *AA* = 900; *Aa* = 4200; and *aa* = 4900. *The founder population described is therefore not at equilibrium.*

c. Given the conditions of random mating, selectively neutral alleles, and no new mutations, allele frequencies do not change from one generation to the next; $p = 0.3$ *and* $q = 0.7$.

d. *The genotype frequencies for the second generation would be those calculated for part b*, because in one generation the population will go to equilibrium. *AA* = p^2 = 0.09; *Aa* = $2pq$ = 0.42; and *aa* = q^2 = 0.49.

e. *Yes*, in one generation a population not at equilibrium will go to equilibrium if mating is random and there is no selection or significant mutation.

f. *The genotype frequencies will be the same in the third generation as in the second generation.*

II. Two alleles have been found at the X-linked phosphogluco-mutase gene (*Pgm*) in *Drosophila persimilis* populations in California. The frequency of the *Pgm^A* allele is 0.25, while the frequency of the *Pgm^B* allele is 0.75. Assuming the population is at Hardy–Weinberg equilibrium, what are the expected genotype frequencies in males and females?

Answer

This problem requires application of the concept of allele and genotype frequencies to X-linked genes. For X-linked genes, males (XY) have only one copy of the X chromosome, so the genotype frequency is equal to the allele frequency. Therefore, $p = 0.25$ and $q = 0.75$. *The frequency of male flies with genotype* $X^{PgmA} Y$ *is 0.25; the frequency of males with genotype* $X^{PgmB} Y$ *is 0.75.* Three genotypes exist for females: $X^{PgmA} X^{PgmA}$, $X^{PgmA} X^{PgmB}$, and $X^{PgmB} X^{PgmB}$ corresponding to p^2, $2pq$, and q^2. *The frequencies of female flies with these three genotypes are* $(0.25)^2$, $2(0.25)(0.75)$, and $(0.75)^2$; or 0.0625, 0.375, and 0.5625, respectively.

III. Two hypothetical lizard populations found on opposite sides of a mountain in the Arizonan desert have two alleles (A^F, A^S) of a single gene *A* with the following three genotype frequencies:

	$A^F A^F$	$A^F A^S$	$A^S A^S$
Population 1	38	44	18
Population 2	0	80	20

a. What is the allele frequency of A^F in the two populations?

b. Do either of the two populations appear to be at Hardy–Weinberg equilibrium?

c. A huge flood opened a canyon in the mountain range separating populations 1 and 2. They were then able to migrate such that the two populations, which were of equal size, mixed completely and mated at random. What are the frequencies of the three genotypes ($A^F A^F$, $A^F A^S$, and $A^S A^S$) in the next generation of the single new population of lizards?

Answer

This question requires calculation of allele frequencies and genotype frequencies in existing and in newly created populations.

a. The frequency of allele *A* is calculated in the following way:

38 $A^F A^F$ × 2	76 A^F alleles
44 $A^F A^S$ × 1	44 A^F alleles
	120 A^F alleles

120 A^F alleles/200 total alleles = 0.6

b. For population 1, the allele frequencies are $p = 0.6$ and $q = 0.4$. Genotype frequencies when the population is in equilibrium are

$$p^2 = (0.6)^2 = 0.36$$

$$2pq = 2(0.6)(0.4) = 0.48$$

$$q^2 = (0.4)^2 = 0.16$$

For population 1, which consists of 100 individuals, the equilibrium would be 36 $A^F A^F$, 48 $A^F A^S$, and

$16\ A^S A^S$ lizards. *Population 1 does seem to be at equilibrium.* (Sampling error and small population size could lead to slight variations from the expected frequencies.) For population 2, the allele frequency (p) is based solely on the number of A^F alleles from the 80 $A^F A^S$ individuals. The total number of alleles = 200, so the frequency of A^F alleles is 80/200 or 0.4. The genotype frequencies for a population at equilibrium would be

$$p^2 = (0.4)^2 = 0.16$$

$$2pq = 2(0.4)(0.6) = 0.48$$

$$q^2 = (0.6)^2 = 0.36$$

Population 2 does not seem to be at equilibrium.

c. The combination of the two populations of lizards results in one population with the following allele frequencies:

A^F alleles		
$A^F A^F$	38×2	76
$A^F A^S$	44	44
$A^F A^S$	80	80
Total:		200
A^S alleles		
$A^F A^S$	44	44
$A^F A^S$	80	80
$A^S A^S$	18×2	36
$A^S A^S$	20×2	40
Total:		200

The allele frequencies are 200/400, or 0.5, for both p and q. *The genotype frequencies in the next generation will therefore be*

$$p^2 = (0.5)^2 = 0.25$$

$$2pq = 2(0.5)(0.5) = 0.5$$

$$q^2 = (0.5)^2 = 0.25$$

Problems

Vocabulary

1. For each of the terms in the left column, choose the best matching phrase in the right column.

i. fitness

ii. gene pool

iii. fitness cost

iv. allele frequency

v. genotype frequency representation

vi. heterozygote advantage

vii. equilibrium frequency

viii. genetic drift

ix. broad-sense heritability

1. the genotype with the highest fitness is the heterozygote

2. chance fluctuations in allele frequency

3. ability to survive and reproduce

4. proportion of total phenotypic variance representation attributed to genetic variance

5. collection of alleles carried by all members of a population

6. p^2 and q^2

7. p and q

8. the advantage of a particular genotype in one situation is a disadvantage in another situation

9. frequency of an allele at which $\Delta q = 0$

Section 12.1

2. In a certain population of frogs, 120 are green, 60 are brownish green, and 20 are brown. The allele for brown is denoted G^B, while that for green is G^G, and these two alleles show incomplete dominance relative to each other.

a. What are the genotype frequencies in the population?

b. What are the allele frequencies of G^B and G^G in this population?

c. What are the expected frequencies of the genotypes if the population is at Hardy–Weinberg equilibrium?

3. Which of the following populations are at Hardy–Weinberg equilibrium?

Population	AA	Aa	aa
a	0.25	0.50	0.25
b	0.10	0.74	0.16
c	0.64	0.27	0.09
d	0.46	0.50	0.04
e	0.81	0.18	0.01

4. A dominant mutation in *Drosophila* called *Delta* causes changes in wing morphology in *Delta/+* heterozygotes. Homozygosity for this mutation (*Delta/Delta*) is lethal. In a population of 150 flies, it was determined that 60 had normal wings and 90 had abnormal wings.

a. What are the allele frequencies in this population?

b. Using the allele frequencies calculated in part *a*, how many total zygotes must be produced by this population in order for you to count 160 viable adults in the next generation?

c. Given that there is random mating, no migration, and no mutation, and ignoring the effects of genetic drift, what are the expected numbers of the different genotypes in the next generation if 160 viable offspring of the population in part *a* are counted?

d. Is this next generation at Hardy–Weinberg equilibrium? Why or why not?

5. A large, random mating population is started with the following proportion of individuals for the indicated blood types:

0.5 MM

0.2 MN

0.3 NN

This blood-type gene is autosomal and the *M* and *N* alleles are codominant.

a. Is this population at Hardy–Weinberg equilibrium?

b. What will be the allele and genotype frequencies after one generation under the conditions assumed for the Hardy–Weinberg equilibrium?

c. What will be the allele and genotype frequencies after two generations under the conditions assumed for the Hardy–Weinberg equilibrium?

6. A gene called *Q* has two alleles, Q^F and Q^G, that encode alternative forms of a red blood cell protein that allows blood-group typing. A different, independently segregating gene called *R* has two alleles, R^C and R^D, permitting a different kind of blood-group typing. A random, representative population of football fans was examined, and on the basis of their blood typing, the following distribution of genotypes was inferred (all genotypes were equally distributed between males and females):

$Q^F Q^F R^C R^C$	202
$Q^F Q^G R^C R^C$	101
$Q^G Q^G R^C R^C$	101
$Q^F Q^F R^C R^D$	372
$Q^F Q^G R^C R^D$	186
$Q^G Q^G R^C R^D$	186
$Q^F Q^F R^D R^D$	166
$Q^F Q^G R^D R^D$	83
$Q^G Q^G R^D R^D$	83

This sample contains 1480 fans.

a. Is the population at Hardy–Weinberg equilibrium with respect to either or both of the *Q* and *R* genes?

b. After one generation of random mating within this group, what fraction of the *next* generation of football fans will be $Q^F Q^F$ (independent of their *R* genotype)?

c. After one generation of random mating, what fraction of the *next* generation of football fans will be $R^C R^C$ (independent of their *Q* genotype)?

d. What is the chance that the first child of a $Q^F Q^G R^C R^D$ female and a $Q^F Q^F R^C R^D$ male will be a $Q^F Q^G R^D R^D$ male?

7. A population with an allele frequency (*p*) of 0.5 and a genotype frequency (p^2) of 0.25 is at equilibrium. How can you explain the fact that a population with an allele frequency (*p*) of 0.1 and a genotype frequency (p^2) of 0.01 is also at equilibrium?

8. When an allele is dominant, why does it not always increase to produce the phenotype proportion of 3:1 (3/4 dominant : 1/4 recessive individuals) in a population?

9. It is the year 1998, and the men and women sailors (in equal numbers) on the American ship the *Medischol Bounty* have mutinied in the South Pacific and settled on the island of Bali Hai, where they have come into contact with the local Polynesian population. Of the 400 sailors that come ashore on the island, 324 have MM blood type, 4 have the NN blood type, and 72 have the MN blood type. Already on the island are 600 Polynesians between the ages of 19 and 23. In the Polynesian population, the allele frequency of the *M* allele is 0.06, and the allele frequency of the *N* allele is 0.94. No other people come to the island over the next ten years.

a. What is the allele frequency of the *N* allele in the sailor population that mutinied?

b. It is the year 2008, and 1000 children have been born on the island of Bali Hai. If the mixed population of 1000 young people on the island in 1998 mated randomly and the different blood group phenotypes had no effect on viability, how many of the 1000 children would you expect to have MN blood type?

c. In fact, 50 children have MM blood type, 850 have MN blood type, and 100 have NN blood type. What is the observed frequency of the *N* allele among the children?

10. Alkaptonuria is a recessive autosomal genetic disorder associated with darkening of the urine. In the United States, approximately 1 out of every 250 000 people have alkaptonuria.

a. Assuming Hardy–Weinberg equilibrium, estimate the frequency of the allele responsible for this trait.

b. What proportion of people in the American population are carriers for this trait? In this population, what is the ratio of carriers to individuals affected by alkaptonuria?

c. If a woman without alkaptonuria who had a child with this trait with one husband then remarried, what is the chance that a child produced by her second marriage would have alkaptonuria?

d. Alkaptonuria is a relatively benign condition, so there is little selective advantage to individuals with any genotype; as a result, your assumption of Hardy–Weinberg equilibrium in part *a* is reasonable. Could you also use the assumption of Hardy–Weinberg equilibrium to estimate the allele frequencies and carrier frequencies of more severe recessive autosomal conditions such as cystic fibrosis? Explain.

11. The equation $p^2 + 2pq + q^2 = 1$ representing the Hardy–Weinberg proportions examines genes with only two alleles in a population.

   a. Derive a similar equation describing the equilibrium proportions of genotypes for a gene with three alleles. [*Hint:* Remember that the Hardy–Weinberg equation can be written as the binomial expansion $(p + q)^2$.]

   b. A single gene with three alleles (I^A, I^B, and i) is responsible for the ABO blood groups. Individuals with blood type A can be either $I^A I^A$ or $I^A i$; those with blood type B can be either $I^B I^B$ or $I^B i$; people with AB blood are $I^A I^B$, and type O individuals are ii. Among Armenians, the frequency of I^A is 0.360, the frequency of I^B is 0.104, and the frequency of i is 0.536. Calculate the frequencies of individuals in this population with the four possible blood types, assuming Hardy–Weinberg equilibrium.

12. a. Alleles of genes on the X chromosome can also be at equilibrium, but the equilibrium frequencies under the Hardy–Weinberg assumptions must be calculated separately for the two sexes. For a gene with two alleles A and a at frequencies of p and q, respectively, write expressions that describe the equilibrium frequencies for all the genotypes in men and women.

   b. Approximately 1 in 10 000 males in the United States is afflicted with haemophilia, an X-linked recessive condition. If you assume that the population is at Hardy–Weinberg equilibrium, what proportion of American females would be haemophiliacs? About how many female haemophiliacs would you expect to find among the 100 million women living in the United States?

13. In 1927, the ophthalmologist George Waaler tested 9049 schoolboys in Oslo, Norway for red-green colour blindness and found 8324 of them to be normal and 725 to be colour blind. He also tested 9072 schoolgirls and found 9032 that had normal colour vision while 40 were colour blind.

   a. Assuming that the same sex-linked recessive allele c causes all forms of red-green colour blindness, calculate the allele frequencies of c and C (the allele for normal vision) from the data for the schoolboys. (*Note:* Refer to your answer to Problem 12a above.)

   b. Does Waaler's sample demonstrate Hardy–Weinberg equilibrium for this gene? Explain your answer by describing observations that are either consistent or inconsistent with this hypothesis.

   On closer analysis of these schoolchildren, Waaler found that there was actually more than one c allele causing colour blindness in his sample: one kind for the "prot" type (c^p) and one for the "deuter" type (c^d) (protanopia and deuteranopia are slightly different forms of red-green colour blindness). Importantly, some of the "normal" females in Waaler's studies were probably of genotype c^p/c^d. Through further analysis of the 40 colour-blind females, he found that 3 were prot (c^p/c^p) and 37 were deuter (c^d/c^d).

   c. Based on this new information, what is the frequency of the c^p, c^d, and c alleles in the population examined by Waaler? Calculate these values as if the frequencies obey the Hardy–Weinberg equilibrium. (*Note:* Refer to your answer to Problem 11a above.)

   d. Calculate the frequencies of all genotypes among men and women expected if the population is at equilibrium.

   e. Do these results make it more likely or less likely that the population in Oslo is indeed at equilibrium for red-green colour blindness? Explain your reasoning.

14. A new university on a Caribbean island has recruited its 700 faculty members from colleges in France and Kenya. Five hundred came from France and 200 came from Kenya, with equal numbers of men and women in both groups. Upon arrival, you notice that 90 of the French and 75 of the Kenyans express a peculiar trait of rolling their eyes up into their sockets when asked a stupid question. Upon studying this trait, you discover that it is always due to the expression of a dominant allele at a single gene called *Ugh*. Field trips taken to both Kenya and France indicate that the two alleles at the *Ugh* locus are at Hardy–Weinberg equilibrium in both of these separate populations. All of the faculty members arrived on the island single, but after teaching for a few years, they all married other faculty members in a random manner. Among 1000 progeny from these marriages, how many children do you expect will express the eye-rolling phenotype?

15. In *Drosophila*, the vestigial wings recessive allele, *vg*, causes the wings to be very small. A geneticist crossed some true-breeding wild-type males to some vestigial virgin females. The male and female F_1 flies were wild type. He then allowed the F_1 flies to mate and found that 1/4 of the male and female F_2 flies had vestigial wings. He dumped the vestigial F_2 flies into a morgue and allowed the wild-type F_2 flies to mate and produce an F_3 generation.

   a. Give the genotype and allele frequencies among the wild-type F_2 flies.

   b. What will be the frequencies of wild-type and vestigial flies in the F_3?

   c. Assuming the geneticist repeated the selection against the vestigial F_3 flies (i.e., he dumped them in a morgue and allowed the wild-type F_3 flies to mate at random), what will be the frequencies of the wild-type and mutant alleles in the F_4 generation?

   d. Now the geneticist lets all of the F_4 flies mate at random (i.e., both wild-type and vestigial flies mate). What will be the frequencies of wild-type and vestigial F_5 flies?

16. A mouse mutation with incomplete dominance ($t = tailless$) causes short tails in heterozygotes (t^+/t). The same mutation acts as a recessive lethal that causes homozygotes (t/t) to die *in utero*. In a population consisting of 150 mice, 60 are t^+/t^+ and 90 are heterozygotes.

a. What are the allele frequencies in this population?

b. Given that there is random mating among mice, no migration, and no mutation, and ignoring the effects of random genetic drift, what are the expected numbers of the different genotypes in this next generation if 200 offspring are born?

c. Two populations (called Dom 1 and Dom 2) of mice come into contact and interbreed randomly. These populations initially are composed of the following numbers of wild-type (t^+/t^+) homozygotes and tailless (t^+/t) heterozygotes.

	Dom 1	Dom 2
Wild type	16	48
Tailless	48	36

What are the frequencies of the two genotypes in the next generation?

Section 12.2

17. Why is the elimination of a fully recessive deleterious allele by natural selection difficult in a large population and less so in a small population?

18. Would you expect to see a greater Δq from one generation to the next in a population with an allele frequency (q) of 0.2 or in a population with an allele frequency of 0.02? Assume relative fitness is the same in both populations and that the equilibrium frequency for q is 0.01.

19. You have identified an autosomal gene that contributes to tail size in male guppies, with a dominant allele B for large tails and a recessive allele b for small tails. Female guppies of all genotypes have similar tail sizes. You know that female guppies usually mate with males with the largest tails, but the effects of population density and the ratio of the sexes on this preference have not been studied. You therefore place an equal number of males in three tanks. In tank 1, the number of females is twice the number of males. In tank two, the numbers of males and females are equal. In tank 3, there are half as many females as males. After mating, you find the following proportions of small-tailed males among the progeny: tank 1 = 16%; tank 2 = 25%; tank 3 = 30%.

a. In your original population, 25 percent of the males have small tails. Assuming that the allele frequencies in males and females are the same, calculate the frequencies of B and b in your original population.

b. Calculate Δq for each tank.

c. If $w_{BB} = 1.0$, what is w_{Bb} for each tank?

d. If $w_{BB} = 1.0$, is w_{bb} less than, equal to, or greater than 1.0 for each tank?

20. An allele of the *G6PD* gene acts in a recessive manner to cause sensitivity to fava beans, resulting in a haemolytic reaction (lysis of red blood cells) after ingestion of the beans. The same allele also confers dominant resistance to malaria. The heterozygote has an advantage in a region where malaria is prevalent. Will the equilibrium

frequency (q_e) be the same for an African and a North American country? What factors affect q_e?

21. In Europe, the frequency of the CF^- allele causing the recessive autosomal disease cystic fibrosis is about 0.04. Cystic fibrosis causes death before reproduction in virtually all cases.

a. Determine values of fitness (w) and of the selection coefficient (s) for the unaffected, carrier, and affected genotypes.

b. Determine the average fitness at birth of the population as a whole with respect to the cystic fibrosis trait $\overline{w}$ and the expected change in allele frequency over one generation (Δq) when measured at the birth of the next generation.

Now suppose that the mutation rate from CF^+ to CF^- alleles is 1×10^{-6}.

c. What is the expected evolutionary equilibrium frequency ($\hat{q}$) of the CF^- allele? Is this larger or smaller than the observed frequency?

d. Without changing the value of s for the CF^-/CF^- genotype you calculated in part *a*, propose an explanation that might resolve the discrepancy between the observed and expected frequencies of the CF^- allele you noted in part *c*.

Section 12.3

22. How can each of the following be used in determining the role of genetic and/or environmental factors in phenotypic variation in different organisms?

a. genetic clones

b. human monozygotic versus dizygotic twins

c. cross-fostering

23. Which of the following statements would be true of a human trait that has high broad-sense heritability in a population of one country?

a. The phenotypic difference within monozygotic twin pairs would be about the same as the phenotypic differences among members of dizygotic twin pairs.

b. There is very little phenotypic variation between monozygotic twins but high variability between dizygotic twins.

c. The trait would have the same heritability in a population of another country.

24. a. Studies have indicated that for pairs of twins raised in the same family, the environmental similarity for monozygotic (MZ) twins is not significantly different from the environmental similarity for fraternal (dizygotic or DZ) twins. Why is this an important fact for calculations of broad-sense heritability?

b. If you wished to determine the broad-sense heritability of a particular trait in humans, would it be more useful to study MZ or DZ twins? Explain.

25. A study published in 1937 examined the average differences between pairs of twins [either monozygotic (MZ)

or dizygotic (DZ)] and pairs of siblings for three different traits: height, weight, and intelligence quotient (IQ) as measured by the Stanford–Binet test. (The concept of "IQ" is extremely controversial as it is unclear to what extent IQ tests measure native intelligence, but for this problem, consider IQ as a measurable phenotype even if its significance is unknown.) Some of the MZ twins were raised together in the same household (RT), while other MZ twins were raised apart in different families (RA). The results of this study are as follows:

	MZ/RT	MZ/RA	DZ	Siblings
Height	1.7 cm	1.8 cm	4.4 cm	4.5 cm
Weight	1.86 kg	4.49 kg	4.54 kg	4.72 kg
IQ	5.9	8.2	9.9	9.8

a. Which of these three traits appears to have the highest broad-sense heritability? the lowest broad-sense heritability?

b. The Centers for Disease Control and Prevention (CDC) of the National Institutes of Health recently reported that in the United States during the period 1960–2002, the average weight of a 15-year-old boy increased from 135.5 pounds (61.46 kg) to 150.3 pounds (68.17 kg). During the same period, the average height of a 15-year-old boy increased from 67.5 inches (171.5 cm) to 68.4 inches (173.7 cm). How do these statistics match your estimates of broad-sense heritability from part *a*?

26. Two different groups of scientists studying a rare trait in ground squirrels report very different broad-sense heritabilities. What factors influencing heritability values make it possible for both conclusions to be correct?

27. Human geneticists have found the Finnish population to be very useful for studies of a variety of conditions. The Finnish population is small; Finns have extensive church records documenting lineages; and few people have migrated into the population. The frequency of some recessive disorders is higher in the Finnish population than elsewhere in the world; and diseases such as PKU and cystic fibrosis that are common elsewhere do not occur in the Finnish population.

a. How would a population geneticist explain these variations in disease occurrence?

b. The Finnish population is also a source of information for the study of quantitative traits. The genetic basis of schizophrenia is one question that can be explored in this population. What advantage(s) and disadvantage(s) can you imagine for studying complex traits based on the Finnish population structure?

28. Two traits with similar phenotype variance exist in a population. If one trait has two major genes and six minor loci that influence the phenotype, and the second trait has 12 minor loci and no major genes affecting the phenotype, which trait would you expect to respond most consistently to selection? Explain.

29. Two alleles at one locus produce three distinct phenotypes. Two alleles of two genes lead to five distinct phenotypes. Two alleles of six genes lead to 13 distinct phenotypes. (These statements assume that the alleles at any one locus are codominant and that each gene makes an equal contribution to the phenotype.)

a. Derive a formula to express this relationship. (Let n equal the number of genes.)

b. Each of the most extreme phenotypes for a trait determined by two alleles at one locus are found in a proportion of 1/4 in the F_2 generation. If there are two alleles of two genes that determine a trait, each extreme phenotype will be present in the F_2 as 1/16 of the population. In common wheat (*Triticum aestivum*), kernel colour varies from red to white and the genes controlling the colour act additively; that is, both genes and alleles for each gene contribute equally to the colour. A true-breeding red variety is crossed to a true-breeding white variety, and 1/256 of the F_2 have red kernels and 1/256 have white kernels. How many genes control kernel colour in this cross?

30. In a certain plant, leaf size is determined by four independently assorting genes acting additively. Thus, alleles *A*, *B*, *C*, and *D* each adds 4 cm to leaf length and alleles *A'*, *B'*, *C'*, and *D'* each adds 2 cm to leaf length. Therefore, an *A/A*, *B/B*, *C/C*, *D/D* plant has leaves 32 cm long and an *A'/A'*, *B'/B'*, *C'/C'*, *D'/D'* plant has leaves 16 cm long.

a. If true-breeding plants with leaves 32 cm long are crossed to true-breeding plants with leaves 16 cm long, the F_1 will have leaves 24 cm long; that is, *A/A'*, *B/B'*, *C/C'*, *D/D'*. List all possible leaf lengths and their expected frequencies in the F_2 generation produced from these F_1 plants.

b. Now assume that in a randomly mating population, the following allele frequencies occur:

frequency of *A* = 0.9
frequency of *A'* = 0.1
frequency of *B* = 0.9
frequency of *B'* = 0.1
frequency of *C* = 0.1
frequency of *C'* = 0.9
frequency of *D* = 0.5
frequency of *D'* = 0.5

Calculate separately the expected frequency in this population of the three possible genotypes for each of the four genes.

c. What proportion of the plants in the population described in part *b* will have leaves that are 32 cm long?

For more information on the resources available from McGraw-Hill Ryerson, go to www.mcgrawhill.ca/he/solutions.

Problems   425

Evolution and Population Genetics: How Biological Information Varies in Space and Time

CHAPTER 13

Genome Evolution

An artist's interpretation of the evolution of the human glucocorticoid receptor (the ancient form is shown in *blue* and the modern form in *orange*). Glucocorticoid receptors bind to steroid hormones and thus play an important role in development, immunity, and in the control of metabolic processes. As we shall see later in this chapter, evolutionary changes in the transcriptional regulation of steroid hormone receptors may have also played a role in the development of traits that are characteristic of modern humans.

Today, more than 150 years after the publication of *The Origin of Species*, biologists accept Darwin's theory of evolution as a foundation of modern biology. In addition, thanks to Mendel, they also understand the basic principles of heredity. What is more, thanks to Watson and Crick, biologists know, that ultimately, evolution is a process that begins at the molecular level, inside the double helix of DNA. In Chapter 1, you read that the modular construction of genomes had a major impact on the evolution of life. In this chapter, we examine in detail the basic components of evolution at the molecular level: **diversification** into many

variants, followed by **selection** of one or a few variants (i.e., differential reproduction) in a population over many generations.

Chapter Outline
13.1 The Origin and Evolution of Life on Earth
13.2 The Evolution of Genomes
13.3 The Organization of Genomes

Learning Objectives

1. Relate the modular nature of gene regulatory networks to the evolution of species-specific traits.

2. Describe the molecular mechanisms underlying the duplication and diversification of genomic regions.

3. Appraise the role of gene duplication in the evolution of novel traits.

4. Describe the genetic origins of multigene families.

13.1 The Origin and Evolution of Life on Earth

No fundamental law of biochemistry says that all living cells have to be constructed in the way that they are; indeed, an imaginative biochemist could think of an almost infinite number of ways to build a functioning cell based on the laws of chemistry. The observation that the cells of

all plants, animals, fungi, and microorganisms are similar in terms of their subcellular organelles, biochemistry, and genetic processes, suggests that the abundant variety of life-forms alive today descended from a single, original cell. In the remainder of this section we will provide a

brief overview of evolutionary history as it is currently understood. This will be followed by a detailed examination of the mechanisms of genome evolution.

The earliest cells

Scientists agree that planet Earth coalesced some 4.5 billion years ago, and that by 4.2 billion years ago, enormous oceans covered the planet. The first living organisms, consisting of a membrane that surrounded information-replicating and information-executing machinery, evolved about 3.7 billion years ago; these were the precursors of present-day cells. How did this machinery, and the cells in which it was contained, arise from nonliving material on ancient Earth? Scientists continue to speculate about the answer in the absence of direct evidence, but one theory postulates that life began when certain organic molecules became self-replicating. Some arguments suggest RNA as the original replicator; however, many questions remain unanswered with respect to how life may have progressed. In any event, fossil cells laid down 3.5 billion years ago near North Pole, Australia (a small town in an arid, rocky region) provide the earliest evidence of distinct cells (**Figure 13.1**). Once life in cellular form emerged, living organisms evolved into three distinct domains: Archaea, Bacteria, and Eukarya (**Figure 13.2**). Contemporary representatives of Archaea and Bacteria include only single-celled organisms whose genomes carry tightly packed genes.

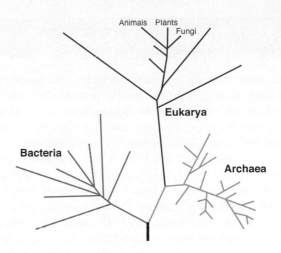

Figure 13.2 Three kingdoms. The distinct branches represent different organisms in each kingdom. The length of the branches is proportional to the times of species divergence.

Complex cells and multicellular organisms

Eukaryotes emerged about 1.4 billion years ago with the symbiotic incorporation of certain single-celled organisms into other single-celled organisms and the complex compartmentalization of the cell's interior, including the segregation of DNA molecules into the nucleus. The incorporated single-celled organisms evolved to become intracellular organelles. The evolution of these relatively complex eukaryotes from the earliest cells thus took almost 2.3 billion years.

About 1 billion years ago, the single-celled ancestors of contemporary plants and animals diverged. The first primitive multicellular organisms appeared 600–900 million years ago. Then 570 million years ago, one of the most remarkable events in the evolution of life occurred: the explosive appearance of a multitude of multicellular organisms, both plants and animals. The multicellular animals are referred to as **metazoans**.

The drama of metazoan evolution

The Burgess shale of southeastern British Columbia provides a rare and fascinating window into the evolutionary history of metazoans. This shale was formed from a mud slide that trapped a wide variety of different organisms in a shallow Cambrian Sea. Events and conditions during and after the slide conspired to achieve the nearly perfect preservation of the three-dimensional structure of the entrapped specimens (see the **Tools of Genetics** box **"The Burgess Shale of Southeastern British Columbia"**).

Three aspects of the Burgess organisms are remarkable. First, they represent a wide variety of very different body plans (**Figure 13.3**). For example, paleobiologists have distinguished 20–30 classes of arthropods in the Burgess sea shale, a striking contrast to the three contemporary classes of arthropods. Second, this emergence of metazoan organisms occurred over a remarkably short (in evolutionary terms) period, perhaps just 20–50 million years.

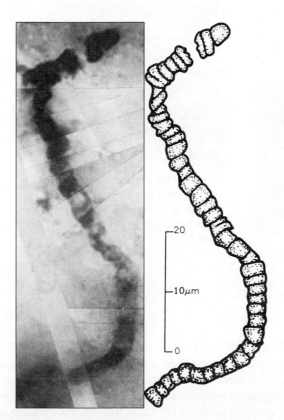

Figure 13.1 Fossilized cells. The oldest fossilized cells.

For those interested in evolutionary biology, the Burgess shale of southeastern British Columbia has proven to be both a useful tool and a treasure trove of information and inspiration. Formed 500 million years ago in what is today Yoho National Park, the Burgess shale contains the well-preserved fossils of organisms dating back to the Cambrian explosion. Remarkably, in addition to hard body parts, fossils found in the shale often have well-preserved soft tissues such as muscle; a most rare and extraordinary find indeed! The area is now a focal point of investigation into evolutionary history. While an exceptional find, one key question remains: What underlying genotypic changes gave rise to this explosion of phenotypic diversity? We will explore this topic in detail later in this chapter.

Jean-Bernard Caron (Department of Earth Sciences, University of Toronto and Curator of Invertebrate Palaeontology, Royal Ontario Museum) is one of the world's foremost experts in the Cambrian explosion; an expertise gained by the scientific examination of the fossils of the Burgess shale (and similar deposits in China). In 2010 Dr. Caron received the Pikaia award from the Canadian Geological Association in recognition for his outstanding research. The awards committee praised him as "an exceptionally innovative and productive young palaeontologist who shows promise for continuing excellence in Canadian paleontological research." To discover more about his research, visit the Virtual Museum of Canada at www.burgess-shale.rom.on.ca.

Figure A Jean-Bernard Caron (centre of photograph) and colleagues examining the Burgess shale.

Figure C Fossil ridge, Yoho National Park, British Columbia.

Figure B The first complete *Anomalocaris* fossil (found in the Burgess shale and now residing in the Royal Ontario Museum, Toronto).

This rapid evolution is an example of **punctuated equilibrium**: the tendency of evolution to proceed through long periods of stasis (lack of change) followed by short periods of explosive change. As we will see later, this rapid change in body plans reflects an equally rapid change in the regulatory networks that control the development of organisms. Third, it seems that all the basic body plans of contemporary organisms initially established themselves in the metazoan explosion. For example, the ancestor of contemporary vertebrates depicted in **Figure 13.3c** probably emerged at about the same time as the ancestor of all contemporary invertebrates.

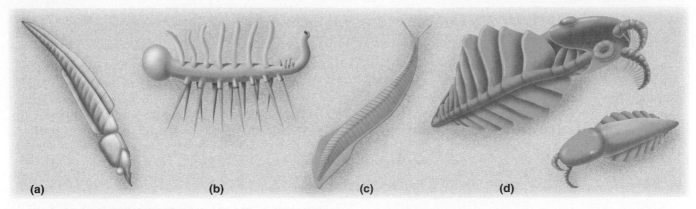

Figure 13.3 Burgess shale organisms. Although all these life-forms are now extinct, fossils have revealed the enormous diversity of their body plans: **(a)** the *Nectocaris*, **(b)** the *Hallucigenia*, **(c)** the *Pikaia*, and **(d)** the *Anomalocaris*. The *Pikaia* contains a notochord, which makes it an ancestor to modern-day vertebrates.

Mass extinctions

The enormous diversity of metazoan body plans that materialized about 500 million years ago has now become tremendously reduced; in part through four to six abrupt extinction events that each destroyed 70 percent to 95 percent of the existing organisms. The most recent example was the global decimation, 65 million years ago, leading to the extinction of the dinosaurs. Many scientists believe this extinction was a consequence of a large meteorite impact in the Yucatan region of present-day Mexico. This impact dramatically changed Earth's climate by propelling enormous amounts of dust into the higher atmosphere. Scientists hypothesize that this thick cloud of dust dispersed and shrouded the globe for several years, preventing solar rays from reaching Earth's surface. The lack of solar energy led to a global drop in temperatures and the demise of all green life, which in turn caused the demise of all large animals, such as dinosaurs, that depended for survival either directly on plants or on animals that ate plants.

Some smaller animals (like mammals' mouse-sized ancestors) presumably survived this long sunless winter because their lesser size allowed them to get by on seeds alone. When the Sun returned, the seeds lying dormant on the ground sprang to life and the world again became an abundantly fertile environment. In the absence of competition from dinosaurs, mammals became the dominant large animal group, diverging into numerous species that could take advantage of all the newly unoccupied ecological niches. Some eventually evolved into our own species.

The evolution of humans

Humans arose from an ancestor common to most contemporary primates 35 million years ago. Humans then diverged from the ancestors of their closest primate relatives, the chimpanzees, about 6 million years ago (**Figure 13.4**). While paleobiologists have not yet sorted out the immediate evolutionary ancestors of *Homo sapiens*, the recent typing of fossil DNA suggests that one previous candidate, the Neanderthal lineage, is not on the direct human evolutionary line.

Figure 13.4 Humans diverged from an ancestor shared with chimpanzees about 6 million years ago. Representatives of primates alive today: **(a)** orangutan, **(b)** gorilla, **(c)** chimpanzee, and **(d)** human.

This fundamental question has fascinated both philosophers and scientists for thousands of years. While previous work on the subject has sparked much lively debate, it is only recently that genetic data emerged to help provide scientific answers. For example, recent genomic analysis by Cory McLean and colleagues (performed in humans, chimpanzees, and macaques) suggests that simple changes in genome architecture can indeed lead to the development of traits that are characteristic of modern humans (**Figure A**). In this

2011 study, researchers examined the sequence of the human genome and discovered a deletion in the regulatory region of a gene, GADD45G, involved in neural function (this regulatory region is present in chimpanzees, macaques, and mice). Interestingly, the deletion "removes a forebrain subventricular zone enhancer near the tumour-suppressor gene, GADD45G, a loss correlated with expansion of specific brain regions in humans." This suggests an evolutionary model where loss of the enhancer promoted expanded human neural cell proliferation and brain development. Interestingly, other human-specific enhancer deletions were found to affect genes involved in steroid hormone signalling. For instance, one such deletion removed an enhancer (from the human androgen receptor gene) responsible for the formation of whiskers (also known as vibrissae) and penile spines, two morphological characteristics observed in chimps, macaques, and mice, but absent in humans. (In **Figure B**, the androgen receptor gene is shown in *black* and the deleted enhancer is represented by a *red triangle*.)

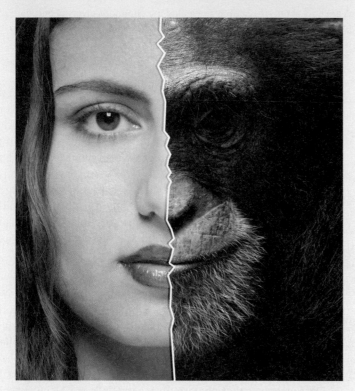

Figure A What genetic variants distinguish humans from other primates?

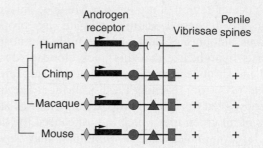

Figure B Structure of the androgen receptor gene in mouse and primates.

Remarkably, on average, the chimpanzee and human genomes are approximately 99 percent similar. Moreover, chimpanzee and human karyotypes are nearly the same. In addition, in every comparison to date, the observed differences between chimpanzee and human DNA sequences have been insignificant in terms of gene function. These data suggest that the evolution of modern humans from a common primate ancestor might be accounted for by a few thousand isolated genetic changes, yet to be uncovered.

Changes in regulatory circuits

Many evolutionary biologists think it likely that these genetic changes occurred in regulatory sequences. Such changes would alter when, and how, master regulatory genes produce transcription factors (and when and how the target genes respond to these regulatory molecules).

For example, the brains of humans and chimps are quite different. The human brain is larger, is far more convoluted (folded), and contains a significantly greater density of neurons (brain cells). Hence, the regulatory networks guiding chimp and human brain development must have diverged strikingly in the 6 million years since the two species split from a common ancestor. These regulatory changes are presumably reflected in modified patterns of transcription-factor-binding sites in the promoter regions of genes that specify brain development (see the **Focus on Inquiry** box "**What Makes Us Human?**"). The rewiring of the regulatory networks in less than 6 million years is a very rapid change in terms of evolutionary time. The same is true of the diversity of body plans generated during the Cambrian explosion of metazoa 570 million years ago.

The idea that evolution occurs primarily because of changes in regulatory networks and not structural genes is supported by the amazing ability of genes from one species to substitute for the absence of homologous genes in other species, in some cases even when the species are as different as yeast and humans. If homologous coding sequences from very different species are functionally indistinguishable, it is reasonable to speculate that species-specific differences in phenotype may arise, to a large degree, from species-specific differences in gene expression.

Multicellular life-forms diversified tremendously during the Cambrian explosion (although many lineages disappeared in mass extinction events). Humans diverged from a primate ancestor about 6 million years ago. What makes humans different from our closest relative, the chimpanzee, may not be the proteins that our genes encode, but rather when, where, and to what level those proteins are expressed during development.

13.2 The Evolution of Genomes

Although Darwin developed his theory of evolution without any knowledge of the molecules that make up living systems, evolution is very much a molecular process that operates on genetic information. This is to say, the variation that initiates each step in the evolutionary process occurs within the genetic material itself (in the form of new mutations). Indeed, new mutations provide a continuous source of variation.

DNA alterations form the basis of genomic evolution

We have seen that mutations arise in several different ways. One is the replacement of individual nucleotides by other nucleotides. Substitutions occurring in a coding region are silent, or *synonymous*, when they have no effect on the amino acid encoded; by contrast, they are *non-synonymous* when the change in nucleotide determines a change in an amino acid or creates a premature termination codon, leading to a truncated gene product. Molecular biologists further distinguish between nonsynonymous changes that cause conservative amino acid changes (e.g., from one acidic amino acid to another) and those that cause nonconservative changes (e.g., from a charged amino acid to a noncharged amino acid). Other gene mutations, arising from errors in replication or recombination, consist of the insertion or deletion of DNA sequences.

Changes can also occur in the order and types of transcription factor binding sites in the promoters of genes; such changes alter the patterns of gene expression. These expression-altering changes may occur very rapidly, which raises the question of whether such regulatory evolution can be explained by single-base mutations followed by selection.

Mutations can be classified according to effect

Different mutations can be *deleterious*, *neutral*, or *favourable* to the organisms that inherit them. In multicellular organisms with large genomes, such as corn or humans, genes and their regulatory sequences make up only a small fraction (less than 3 percent) of the total genetic material. As a result, random mutations occur most often in DNA that plays no role in the development or function of an organism. Such mutations are presumably neutral. It might seem that synonymous mutations within coding regions would also be neutral, but there is some evidence that even changes in the codon used to produce a particular amino acid can provide a minute advantage or disadvantage to the organism, possibly based on the availability of different tRNA molecules and their associated synthetases. While conservative amino acid replacements were once considered neutral, current evidence suggests that they can have an impact on the growth and survival of an organism. Nonconservative amino acid changes and changes such as deletions and insertions involving larger portions of a gene almost certainly have an impact on gene function.

Genetic changes that are truly neutral are unaffected by the agents of selection. They survive or disappear from a population through *genetic drift*, which is the result of chance reproductive events. The smaller the population, the more rapidly genetic drift exerts its effect.

Mutations with only deleterious effects disappear from a population by negative selection—that is, selection against an allele. As you saw in the preceding chapter, however, some mutations (such as those that cause sickle-cell anaemia) are deleterious to homozygotes for the mutation, but advantageous to heterozygotes. Because of this *heterozygous advantage*, selection retains these mutations in a population at a low equilibrium level. However, even deleterious mutations for which there is no heterozygous advantage are eliminated very slowly if they are fully recessive.

Some extremely rare mutations give an organism a significant advantage over other individuals. Because of this advantage, individuals carrying the mutation are more likely to reproduce, and in each succeeding generation, the frequency of the mutation increases. This is positive selection (selection for an allele). Ultimately, the allele that began as a mutation is present in nearly every member of the population on both chromosomes. At this stage, the allele has become *fixed* in the population.

Gene regulatory networks may dominate developmental evolution

Gene regulatory networks that are active in development receive informational signals, integrate and modulate those signals, and then transmit them to protein networks that mediate various aspects of development. For example, a gene network first discovered and analyzed in *Drosophila* and consisting of five transcription factors (NK-2, MEF2, GATA, Hand, and Tbx) controls the expression of patterning and muscle genes responsible for the development of the fruit fly heart. Remarkably, while heart tissue displays drastically different forms in different species, it is clear that this same conserved network of five transcription factors is used to control heart development in organisms ranging from simple chordates to mammals (**Figure 13.5**). Thus, differences in the form of the heart across species are as a result of subtle changes in the "wiring" of this regulatory module, and in the specific downstream muscle and patterning genes that are activated. These differences in wiring can arise from selectable changes to the DNA

of *cis*-control elements or from changes to the amino acid sequences of the transcription factors themselves. Changes to the DNA include (1) the gain or loss of *cis*-control sequences to which the transcription factors bind, (2) repositioning of *cis*-control elements on the regulatory DNA, (3) increases or decreases in a *cis*-control element's binding affinity for a particular transcription factor, and (4) mutations that make a *cis*-control element serve as the binding site for a different transcription factor. These results provide a dramatic example of the modular nature of genome evolution.

A fascinating possibility is that gene regulatory networks in humans may share much of their basic wiring with invertebrates. If this turns out to be true, a detailed study of invertebrate gene regulatory networks could provide powerful insights into the networks controlling development in higher organisms. Moreover, if the subtle rewiring of gene regulatory networks can lead to enormous phenotypic change, it can also explain the evolution of considerable biological complexity (see the **Fast Forward** box **"Network Motifs"**).

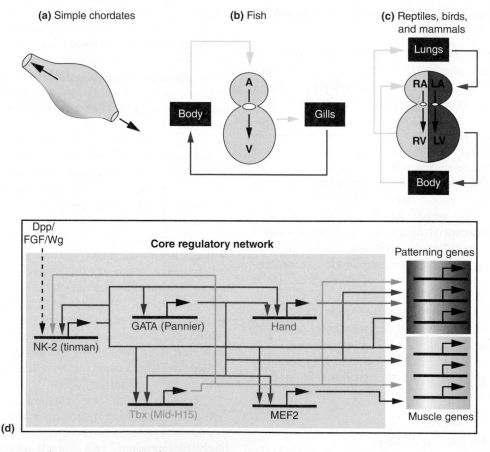

Figure 13.5 Heart structure in organisms ranging from simple chordates to mammals. **(a)** The heart of a simple chordate is tubular and functions as a simple peristaltic pump. **(b)** The fish heart, on the other hand, is much more efficient. It consists of one chamber for receiving blood (atrial chamber) and another separate chamber for pumping blood (ventricular chamber). **(c)** Mammalian hearts are even more complex and consist of four chambers: the right atrial and ventricular chambers (for receiving and pumping deoxygenated blood) and the left atrial and ventricular chambers (for receiving and pumping oxygenated blood). **(d)** Remarkably, this incredible phenotypic diversity is controlled by a core set of evolutionarily conserved transcription factors (NK-2, MEF2, GATA, Tbx, and Hand) that control muscle and patterning genes important to the morphogenesis of cardiac structures.

In addition to highlighting the modularity of genome architecture, research of the last decade has also provided evidence for the existence of "network motifs." A network motif is a pattern of interaction amongst genes/gene-products that occurs significantly more often in "real-world" networks than in randomized networks. The simplest example of a network motif is positive autoregulation, where a transcription factor promotes its own expression by binding to the promoter of the gene encoding the transcription factor. Another very common, but more complex, motif is called a feed-forward loop. This motif is composed of the interaction between three genes/gene-products (**Figure A**). In this motif, transcription factor X promotes the transcription of a second transcription factor, Y. Both X and Y jointly regulate a third gene, Z. The activity of X and Y are "integrated" at the Z promoter through *cis*-acting regulatory sequences. In this case, both X and Y are required to bind to the promoter for efficient transcription of Z (in essence this forms a logical "AND" gate). S_X and S_Y are the inducers of X and Y, respectively. Remarkably, motifs such as these arise again and again over evolutionary time in a wide array of distantly related

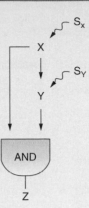

Figure A A network motif.

organisms. Thus, it appears that these motifs or "genetic circuits" have been used repeatedly to tackle similar biological problems over evolutionary history. We will learn more about the properties and functions of these and other network motifs in Chapter 23 (Systems Biology).

An increase in genome size generally correlates with the evolution of complexity

Even though both bacteria and mammals evolved from a common cellular ancestor, the contemporary *Escherichia coli* genome is about 5 Mb in length, while in humans, the genome is about 3000 Mb long. How has evolution fashioned such different genomes from the same original material? The answer lies in the evolutionary potential for increasing the size of the genome through the duplication and diversification of genomic regions and, even more strikingly, through the acquisition of repetitive sequence elements that may represent more than 50 percent of the genome. Note, however, that although the increase in genome sizes from yeast to flies to vertebrates does reflect the increasing complexity of the organisms, there are examples of amoeba, plants, and amphibians with considerably more genomic DNA than humans.

Duplications can occur at random throughout the genome, and the size of the duplication unit can vary from a few nucleotides to the entire genome. When a duplicated segment contains one or more genes, either the original or the duplicated copy of each gene is free to accumulate function-destroying mutations (i.e., diversify) without harm to the organism, because the other "good" copy is still present. With duplications acting as such an important force in evolution, it is critical to understand the two main ways in which they arise.

Duplications resulting from transposition

As we saw in Chapter 9, transposition, the transfer of one copy of a chromosomal sequence from one chromosomal site to another, can occur in various ways: through the direct movement of a DNA sequence; through an RNA intermediate that is copied into a DNA intermediate, leaving the original DNA site intact; or through a DNA intermediate (**Figure 13.6**). When the genomic DNA (rather than its RNA or DNA proxy) moves to a new site, the duplication of genetic material occurs only after the altered chromosome receiving the DNA segregates, together with the unaltered homologue of the chromosome containing the original locus, into an egg or sperm. When the gamete with the duplication unites with a normal gamete, the resulting zygote has three copies of the original locus (**Figure 13.7**). In subsequent generations, the new transposition element may become fixed in the population.

Duplications resulting from unequal crossing-over

Normal crossing-over, or recombination, occurs between equivalent loci on the homologous chromatids present in a synaptonemal complex that forms during the pachytene stage of meiosis. *Unequal crossing-over*, also referred to as *illegitimate recombination*, occurs between nonequivalent loci (review Figure 8.8a). Unequal crossing-over is most often initiated by related sequences located close to each other in the genome. Although the event is unequal in the exchange of nonequivalent segments of DNA, it is still mediated by the sequence similarities at the two separate loci.

So-called **nonhomologous unequal crossing-over** also occurs, although much less often than homologous crossing-over. A nonhomologous crossover may be mediated by at least a short stretch of sequence homology, coding or noncoding, at the crossover's two sites of initiation.

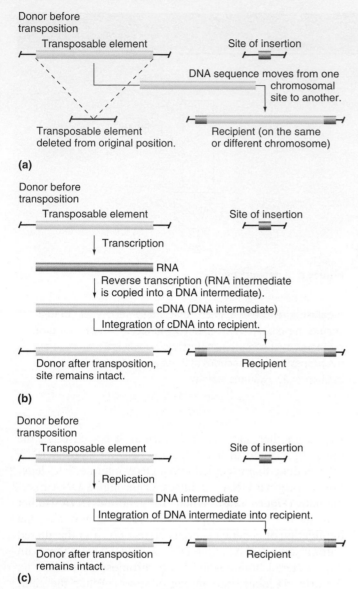

(a)

Donor before transposition

Transposable element   Site of insertion

DNA sequence moves from one chromosomal site to another.

Transposable element deleted from original position.

Recipient (on the same or different chromosome)

(b)

Donor before transposition

Transposable element   Site of insertion

Transcription

RNA

Reverse transcription (RNA intermediate is copied into a DNA intermediate).

cDNA (DNA intermediate)

Integration of cDNA into recipient.

Donor after transposition, site remains intact.

Recipient

(c)

Donor before transposition

Transposable element   Site of insertion

Replication

DNA intermediate

Integration of DNA intermediate into recipient.

Donor after transposition remains intact.

Recipient

Figure 13.6 Duplication by transposition. Transposition may occur by **(a)** excising and reinserting the DNA segment; **(b)** making an RNA copy, which is then converted to a DNA copy for insertion; or **(c)** making a DNA copy for integration. In (b) and (c), the transposon is duplicated.

An initial duplication by unequal crossing-over that produces a two-unit cluster may be either homologous or nonhomologous, but as **Figure 13.8** illustrates, once two units of a related sequence are present in tandem, further rounds of homologous unequal crossing-over between nonequivalent members of the pair readily occur. Thus, small clusters can easily expand to contain three, four, or more copies of an original DNA sequence.

The result of unequal crossing-over between homologous chromosomes is always two reciprocal chromosomal products: one carrying a duplication of the region located between the two crossover sites, and the other carrying a deletion that covers the exact same region. Regions duplicated by unequal crossing-over can vary from a few base pairs to hundreds of kilobases in length, and they may contain no genes, a portion of a gene, a few genes, or many genes.

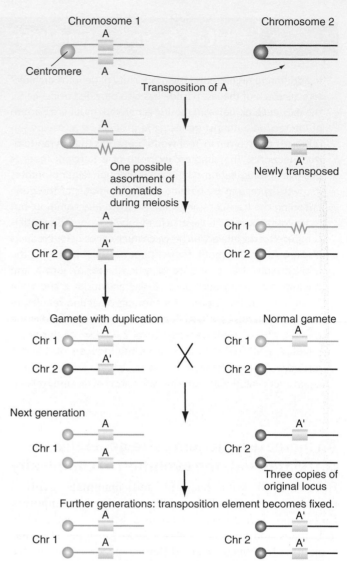

Figure 13.7 Transposition through direct movement of a DNA sequence. A DNA sequence may transpose from one chromosome to a second chromosome in a sex cell. In subsequent generations, this transposed element may become fixed.

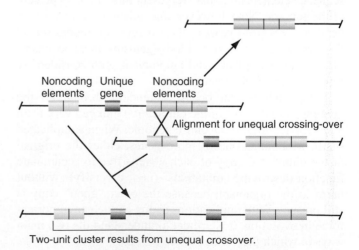

Noncoding elements   Unique gene   Noncoding elements

Alignment for unequal crossing-over

Two-unit cluster results from unequal crossover.

Figure 13.8 Duplications arise from unequal crossing-over. The crossover depicted above involves dispersed repeat elements (*blue boxes*). The *pink boxes* denote a unique gene. In the crossover event, this originally unique gene is deleted in one chromosome and duplicated in the other.

Darwin's evolutionary theory maintains that the diversity and complexity of living organisms are the result of evolution by the natural selection of pre-existing variation encoded by the genome in combination with random drift. Creationism is the belief that certain aspects of life are too complex to have evolved in this way; instead, it maintains that a divine being created the world, including the diversity and complexity of all life from microbes to people. Creationism is a religious belief, not a scientific theory, because the hypothesis at its foundation is not testable.

A battle has been ongoing between creationism and evolution ever since Darwin first proposed his theory of evolution by natural selection. In 1925, John Scopes, a rural high school biology teacher, was sued by the State of Tennessee for teaching evolution in the science classroom. The courtroom battle between two well-known attorneys, William Jennings Bryan for the state and Clarence Darrow for the defence, drew national attention. In the so-called "Monkey Trial," the forces for scientific evolution actually lost, but the decision was later overturned on a technicality.

More recently, the proponents of creationism have cast their argument in a new form known as Intelligent Design (ID). According to this idea, life is too complex to have been created by Darwinian evolution; instead, it must have been created by a higher agent or intelligence. Although not named as such, that intelligence is generally taken to be the Christian God by proponents of ID.

In 2005, the parents of eleven students sued the Dover, Pennsylvania school board over the board's mandate that all science teachers must tell their students that evolution is "just a theory" and that intelligent design literature in the form of a textbook titled *Of Pandas and People* must be made available as an alternative hypothesis. The parents' lawsuit, *Klitzmer v. Dover Area School District*, argued that the mandate to give evolution and intelligent design equal time in the science classroom was inappropriate because intelligent design is not science, but religious belief. On December 20, 2005, Judge John E. Jones III of the federal district court hearing the case issued his decision: a sweeping repudiation of the teaching of intelligent design in the science classroom as an alternative to Darwinian evolution. In his decision, Judge Jones stated, "We find that ID . . . cannot be adjudged a valid, accepted scientific theory. . . . [It] is grounded in theology, not science. . . . It has no place in a science curriculum. . . . The goal of the ID movement is not to encourage critical thought, but to foment a revolution that would supplant evolutionary theory with ID."

Even as the proponents of Intelligent Design base their anti-evolution arguments on a supernatural explanation of how biological complexity emerged, biologists are making great strides toward understanding that complexity scientifically through the systems approach to biology (see Chapter 23).

Evolution of a duplicated gene: New functions or pseudogenes

Duplicated regions, like all other genetic novelties, must originate in the genome of a single individual. At first, the survival of a duplication in at least some animals in each subsequent generation of a population is, most often, a matter of chance. The reason is that the addition of one extra copy of a chromosomal region, including most genes, to the two already present in the diploid genome usually causes no significant harm to the individual. In the terminology of population genetics, the duplicated units are neutral with regard to genetic selection. They are thus subject to genetic drift and inherited at random by some offspring but not others. By chance again, most neutral genetic elements disappear from a population in several generations.

Development of pseudogenes

When a duplicated region that includes a functional gene survives for hundreds or thousands of generations, random mutations in the gene may turn it into a related gene with a different function, or into a nonfunctional *pseudogene.* Some of the mutations generating a pseudogene lead to a loss of regulatory function; others change one or more critical amino acids in the gene product; still others cause premature termination of the growing polypeptide chain, or change the translational reading frame of the gene, or alter the RNA splicing patterns.

Pseudogenes, because they serve no function, are subject to mutation without selection and thus accumulate mutations at a far faster pace than the coding or regulatory regions of a functional gene. Eventually, nearly all pseudogene sequences mutate past a boundary beyond which it is no longer possible to identify the functional genes from which they have been derived. Thus, continuous mutation can turn a once functional sequence into an essentially random sequence of DNA.

Diversification leading to new functionality

Every so often, the accumulation of a set of random mutations in a spare copy of a gene leads to the emergence of a new functional gene that provides benefit and, consequently, selective advantage to the organism in which it resides. Because it provides a selective advantage, the new gene persists in the population. Although its function is usually related to that of the original gene, it almost always has a novel pattern of expression—in time, in

space, or both—which most likely results from alterations in *cis*-regulatory sequences that occur along with codon changes. For example, a duplicated copy of the original human β-globin gene evolved into the myoglobin gene, whose protein product has a higher affinity for oxygen than the haemoglobin molecules composed in part of β-globin polypeptides. The myoglobin gene is active only in muscle cells, while the β-globin gene is expressed only in red blood cell precursors. Thus, duplication, divergence, and selection generated a new gene function from a previously functional gene.

Molecular clocks allow inference of phylogenetic relationships

Before the advent of molecular biology, researchers determined the genealogies of organisms by calculating the rates of evolution in phenotypic traits such as teeth and vertebrae. Then, in the 1950s, Linus Pauling and Emile Zuckerkandl analyzed haemoglobin and cytochrome C protein sequences from different species and noted that the rates of amino acid substitution in each type of protein are similar for various mammalian lineages. On the basis of this observation, they postulated that for a given protein, the rate of evolution is constant across all lineages. They called this idea of a constant rate of change for each type of molecule a **molecular clock**. Its existence enables biologists to determine from molecular data the approximate times when species diverged and then use these dates to reconstruct genealogical phylogenetic trees.

Although the molecular clock hypothesis does not keep perfect time, in many instances where it has been tested, it has produced a reasonably good estimate of the time of divergence between two types of organisms. Today molecular biologists compare the nucleotide sequences of genes as well as the amino acid sequences of proteins to determine phylogenetic divergences.

Scientists use molecular data in various ways to construct **phylogenetic trees** that illustrate the relatedness of homologous genes or proteins. A phylogenetic tree consists of **nodes** and **branches** (**Figure 13.9**). The nodes represent the taxonomic units, which may be species, populations, individuals, or genes, while the branches define the relationship of these units. The branch length suggests the amount of time that has elapsed based on the number of molecular changes that have occurred.

Because different genes accumulate changes at different rates, different types of genes are best suited to the construction of different types of phylogenetic trees. For example, in fibrinopeptides, the major components of wound-responsive blood clots, the exact amino acid sequences are not critical to function. As a result, the fibrinopeptides are not under strong selective pressure, and their genes evolve quickly. In contrast, the ribosomal genes, which are highly conserved, evolve slowly. Thus, the genes encoding fibrinopeptides are useful for looking at recent evolutionary events among very closely related

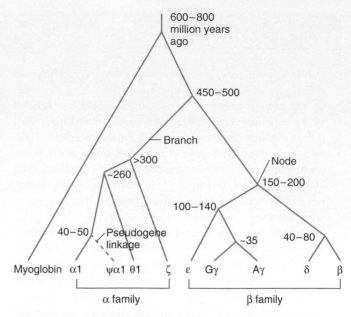

Figure 13.9 A phylogenetic tree. This phylogenetic tree diagrams the evolutionary history of human haemoglobin genes. The *broken line* denotes a pseudogene linkage. Only one of the two α-haemoglobin genes is shown because their date of divergence is so recent.

species, while the ribosomal genes are useful for looking at ancient evolutionary events such as the relationships of phyla to each other. Indeed, phylogenetic trees based on ribosomal genes helped rewrite the most fundamental domain classifications of multicellular organisms (see Figure 13.2).

- Mutations in DNA are the basis of evolution; their effects may be deleterious, neutral, or favourable. Major phenotypic changes appear to result from mutations that affect developmental regulatory networks. Many of these basic networks have been conserved throughout evolution.

- Repeated duplications increase genome size. Transposition moves a copy of a sequence from one chromosomal site to another; unequal crossing-over results in duplication of one region, and deletion of the same region in the reciprocal chromosome.

- In many cases, mutations turn a duplicated gene into a nonfunctional pseudogene. In other cases, mutations may confer a new function on a duplicated gene on which selection can act. If the new function provides a selective advantage, the newly functional gene will be retained.

- Individual proteins have been found to mutate at a constant rate across lineages. Biologists can use these molecular clocks to estimate times of divergence and to construct molecularly based phylogenetic trees. The genes of some proteins evolve more quickly than others, allowing analysis of distant divergences or more recent speciation.

13.3 The Organization of Genomes

With this understanding of the basic mechanisms by which genomes evolve, we turn our attention to the results of genome evolution. Our focus is on the various organizational features of the enormous mammalian genome, which evolved from much simpler bacteria-like genomes through eons of duplication, diversification, and selection.

Molecular geneticists have accumulated data on genome organization by analyzing the genomic sequences of chromosomal DNAs from completed sequences of more than 100 genomes, including those of humans, mice, puffer fish, rice, *Drosophila*, *C. elegans*, *Arabidopsis*, yeast, *E. coli*, and approximately 300 other organisms (mostly microbes). The single-celled organisms exhibit genomes with densely packed genes and few, if any, introns. The mammalian genomes, with their far less densely packed genes, have several distinct features dominating their landscape: genes and families of genes; dispersed repetitive elements constituting more than a third of the genome; simple sequence repeats composed of single nucleotides or di-, tri-, tetra-, pentamers, and so forth; simple repetitive elements serving as a core for centromeres and telomeres; and unique nongene sequences. We now describe how these genomic features could have evolved.

Duplications have created multigene families and gene superfamilies

Four levels of duplication (followed by diversification and selection) have fuelled the evolution of complex genomes. At the lowest level, exons duplicate or shuffle to change the size or function of genes. At the next three increasingly complex levels, entire genes duplicate to create multigene families; multigene families duplicate to produce gene superfamilies; and the entire genome duplicates to double the number of copies of every gene and gene family (**Figure 13.10**). At each of these successively higher levels of organization, the duplication of larger and larger units leads to the hierarchical generation of greater and greater amounts of new information.

Basic components of genes

Genes consist of many different components: exons containing coding regions, 5′ and 3′ untranslated regions, introns, and, finally, associated control regions (**Figure 13.11**). Many of the control regions lie just 5′ of the transcribed region; but some lie far outside the gene and appear to play a role in opening up the chromatin of the gene family locus so that gene expression can proceed at the appropriate time and level. Until all associated regulatory elements have been defined, the boundaries of a genetic locus remain uncertain. Now that we have reviewed the basic structure of a gene, we will next examine how genetic duplication events have shaped the contemporary organization of genomes.

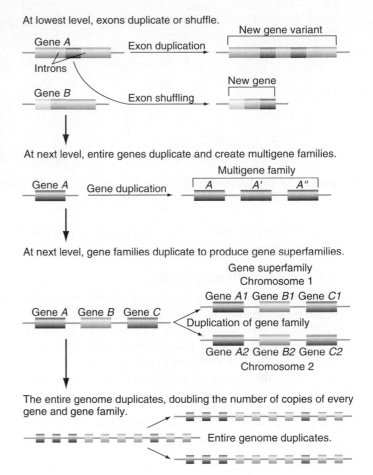

Figure 13.10 Duplications increase genome size. Genome size increases through duplication of exons, genes, gene families, and finally of entire genomes.

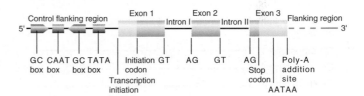

Figure 13.11 The basic structure of a gene. A typical eukaryotic protein-coding gene. The *light green boxes* represent the 5′ and 3′ untranslated mRNA sequences. The small *pink boxes* represent regulatory sequences where transcription factors bind.

The duplication and shuffling of exons

Entirely new genes may arise from **exon shuffling**: the exchange of exons among different genes. Exon shuffling produces mosaic proteins such as tissue plasminogen activator (TPA), a molecule with four domains of three distinct types: kringle (K), growth factor (G), and finger (F). The gene for TPA captured exons governing the synthesis of four domains from the genes for three other proteins: K from the gene for plasminogen, G from the gene for epidermal growth factor, and F from the gene for fibronectin (**Figure 13.12**). Although, the mechanism by which exon

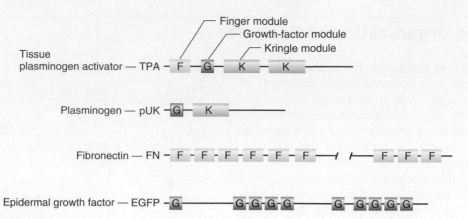

Figure 13.12 Duplication and shuffling of exons. The tissue plasminogen activator (TPA) gene has evolved by the shuffling of exons from the genes for plasminogen, fibronectin, and epidermal growth factor.

shuffling occurs is unknown, it clearly creates new proteins with different combinations of functions.

Creation of a multigene family

A **multigene family** is a set of genes descended by duplication and diversification from one ancestral gene. The members of a multigene family may be either arrayed in tandem (i.e., clustered on the same chromosome) or distributed on different chromosomes (**Figure 13.13**). Unequal crossing-over can expand and contract the number of members in a multigene family cluster (**Figure 13.14**).

Genetic exchange between related DNA elements

The genome contains many places where a flow of genetic information appears to have occurred from one DNA element to other related, but non-allelic elements located nearby or on different chromosomes. Such information flow between related DNA sequences occurs through an alternative outcome of the process responsible for unequal crossing-over. This alternative is known as **intergenic gene conversion** (**Figure 13.15**). In Chapter 6, you learned that for meiosis to proceed, nonsister chromatids engaged in crossing-over must disengage, and that two resolutions via Holliday intermediates are possible (review Figure 6.20, Steps 7 and 8): One is straightforward crossing-over, and the other is gene conversion without crossing-over. The same alternative outcomes can occur with unequal recombination intermediates. The gene conversion outcome of unequal crossing-over allows the transfer of information from one gene to another. In special cases (such as the ribosomal gene family that exists in every eukaryotic species), the flow of information from such intergenic

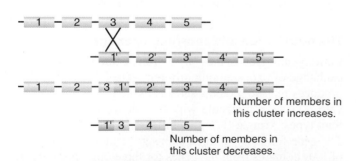

Figure 13.13 Multigene families. Tandem and dispersed multigene families on segments of the indicated chromosomes.

Figure 13.14 Evolution via crossovers within families. A schematic illustration of how a crossover can expand and contract gene numbers in a multigene family.

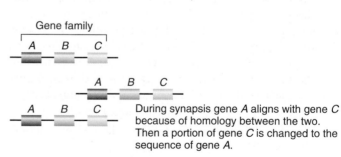

During synapsis gene A aligns with gene C because of homology between the two. Then a portion of gene C is changed to the sequence of gene A.

Note: Gene A remains unchanged, only gene C has a portion (*magenta*) of the gene A sequence.

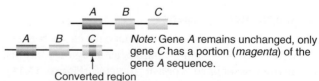

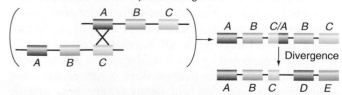

Figure 13.15 Intergenic gene conversion. In intergenic gene conversion, one gene is changed, the other is not.

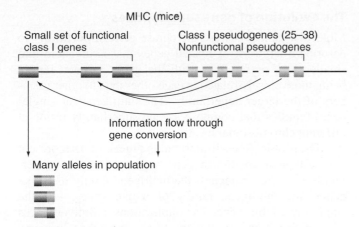

MH C (mice)

Small set of functional class I genes

Class I pseudogenes (25–38)
Nonfunctional pseudogenes

Information flow through gene conversion

Many alleles in population

Figure 13.16 Increasing the number of alleles. How gene conversion events from pseudo-class I genes could increase the polymorphism in functional MHC class I genes in mice.

gene conversion has been so extreme that it has caused all members of a gene family to co-evolve with near identity. And in at least one case—that of the class I genes of the major histocompatibility complex (MHC)—selection has acted on information flow in only one direction, causing information transfer from a series of nonfunctional pseudo-genes to a small subset of just two to three functional genes (**Figure 13.16**). In this unusual case, the pseudogene family members served as a reservoir of genetic information that produced a dramatic increase in the amount of polymorphism (i.e., the number of alleles) in the small number of functional gene members.

Concerted evolution and multigene homogeneity

A few multigene families have evolved under a special form of selective pressure that requires all family members to maintain essentially the same sequence. In these families, the high number of gene copies does not result in variations on a theme; rather it supplies a cell with a large amount of product within a short period of time. Among the gene families with identical elements is the one that produces RNA components of the cell's ribosomes, the one that produces tRNAs, and the one that produces the histones (which must rapidly generate enough protein to coat the new copy of the genome that was replicated during the S phase of the cell cycle).

Each of these gene families consists of one or more clusters of tandem repeats of identical elements. A strong selective pressure maintains the same sequence across all members of each family because all must produce the same product. Optimal functioning of the cell requires that the products of any individual gene be interchangeable in structure and function with the products of all other members of the same family. The problem is that the natural tendency of duplicated sequences is to drift apart over time. How does the genome counteract this natural tendency?

When researchers first compared ribosomal RNA and other gene families in this class, both between and within species, a remarkable picture emerged: Between species, they saw clear evidence of genetic drift, but within species, all sequences appeared essentially equivalent. Thus, it

is not simply that some mechanism suppresses mutational changes in these gene families. Rather, there appears to be an ongoing process of **concerted evolution**, which allows changes in single genetic elements to spread across a complete set of genes in a particular family.

Concerted evolution appears to occur through two different processes. The first is based on the expansion and contraction of gene family size through sequential rounds of unequal crossing-over between homologous sequences (**Figure 13.17**). Selection acts to maintain the absolute size of the gene family within a small range around an optimal mean.

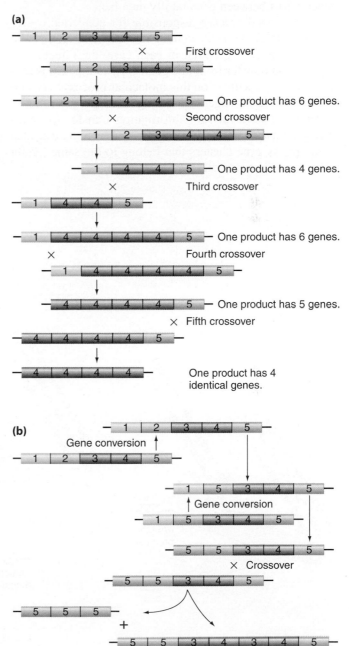

Figure 13.17 Concerted evolution can lead to gene homogeneity. *Boxes* with different colours and numbers represent gene family members with variant DNA sequences. Repeat cycles of unequal crossover events **(a)** or gene conversion **(b)** cause the duplicated genes on each chromosome to become progressively more homogenized.

As the gene family becomes too large, the shorter of the unequal crossover products becomes selected; as the family becomes too small, the longer of the products becomes selected. This cyclic process causes a continuous oscillation around a size mean. However, each contraction results in the loss of divergent genes, whereas each expansion results in the indirect "replacement" of those lost genes with identical copies of other genes in the family. With unequal crossovers occurring at random positions throughout the cluster and with selection acting in favour of the least divergence among family members, this process can act to slow down dramatically the natural tendency of genetic drift between gene family members.

The second process responsible for concerted evolution is intergenic gene conversion between non-allelic family members. Although in each case, the direction of information transfer from one gene copy to the next is random, selection will act on this molecular process to ensure an increase in homogeneity among different gene family members (**Figure 13.17b**). Information transfer (presumably by means of intergenic gene conversion) can also occur across gene clusters that belong to the same family but are distributed to different chromosomes.

The evolution of gene superfamilies

Molecular geneticists use the phrase **gene superfamily** to describe a large set of related genes that is divisible into smaller sets, or families, with the genes in each family being more closely related to each other than to other members of the larger superfamily. The multigene (or single-gene) families that compose a gene superfamily reside at different chromosomal locations.

The globin genes illustrated in **Figure 13.18** represent a prototypical small-size gene superfamily. The superfamily has three branches: the multigene family of β-like genes, the multigene family of α-like genes, and the single myoglobin gene. The duplications and divergences that produced the three superfamily branches occurred early in the evolution of vertebrates; as a result, the three branches of the superfamily are found in all mammals. All functional members of this superfamily play a role in oxygen transport. The products of the α- and β-globin genes are active in red blood cells, while the product of the myoglobin gene transports oxygen in muscle tissue.

The primordial globin gene gave rise to the myoglobin and α-/β-globin precursor genes by gene duplication and transposition. The primordial α- and β-globin

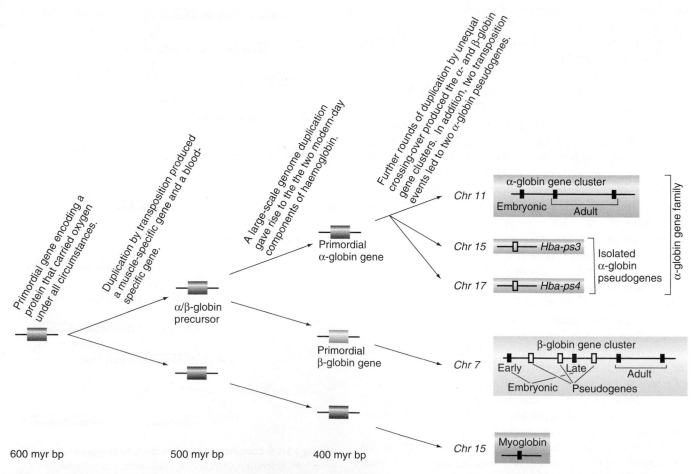

Figure 13.18 Evolution of the mouse globin superfamily. Repeated gene duplication by various mechanisms gave rise to the globin supergene family in mice, with two multigene families (α and β) and one single gene (myoglobin). The α family has both tandemly arrayed and dispersed gene members.

genes probably arose by a large-scale genome duplication (tetraploidization).

The β-like branch of this gene superfamily arose by duplication via multiple unequal crossovers. In the mouse, all the β-like genes are present in a single cluster on chromosome 7 containing four functional genes and three pseudogenes (Figure 13.18). The β-like chains encode similar polypeptides; however, each functions optimally at a different stage of mouse development: one during early embryogenesis, one during a later stage of embryogenesis, and two functioning in the adult.

The α-like branch also evolved by unequal crossovers and divergences that generated a cluster of three genes on mouse chromosome 11: One functions during embryogenesis, and two function in the adult (Figure 13.18). The two adult α-globin genes are virtually identical at the level of DNA sequence, which suggests that the duplication producing them occurred very recently (on the evolutionary time scale). In addition to the primary α-like cluster, there are two nonfunctional α-like genes—pseudogenes—that have dispersed via transposition to locations on chromosomes 15 and 17.

Pseudogenes existing in isolation from their parental families are called *orphons*. Interestingly, the α-globin orphon on mouse chromosome 15 (named *Hba-ps3*) has no introns and thus appears to have arisen through a retrotransposition event involving mRNA copied back to DNA. In contrast, the α-globin orphon on chromosome 17 (*Hba-ps4*) does contain introns and may have arisen by a direct DNA-mediated transposition. The single mouse myoglobin gene on chromosome 15 has no close relatives either nearby or far away. The globin gene superfamily provides a view of the many different mechanisms that the genome can employ to evolve structural and functional complexity.

The mouse *Hox* gene superfamily provides an alternative prototype for the evolution of a gene superfamily (**Figure 13.19**). At first, a single *Hox* gene had evolved to produce a protein product that could bind to DNA enhancer regions and thereby regulate the expression of other genes. Unequal crossover events predating the divergence of insects and vertebrates some 600 million years ago produced a cluster of five related genes encoding DNA-binding proteins that regulated the expression of other genes encoding spatial information (i.e., instructions for the spatial positioning of tissues and organs) in the developing embryo.

The original *Hox* gene family then duplicated en masse and dispersed to four locations—on chromosomes in an ancestor common to all vertebrates. Because of the order of duplication events leading to the superfamily, an evolutionary tree would show that a single gene family within the superfamily has actually splayed out physically across the four gene clusters, as shown in Figure 13.19.

After the en masse duplication that generated the gene superfamily, smaller duplications (by unequal crossing-over) added genes to some of the dispersed clusters and subtracted genes from others, thereby generating differences in gene number and type within a basic framework of homology among the different clusters. Each of the four *Hox* clusters in the mouse superfamily currently contains 9–12 homologous genes.

Repetitive "nonfunctional" DNA families constitute nearly one-half of the genome

Many repetitive nonfunctional DNA families consist of retroviral elements. To review, retroviruses are RNA-containing viruses that can convert their RNA genome into DNA molecules through the viral-associated

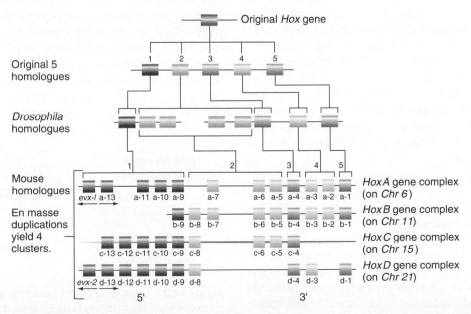

Figure 13.19 Evolution of the *Hox* gene superfamily of mouse and *Drosophila*. This gene superfamily arose by a series of gene duplications. Four multigene families are present in the mouse and one in *Drosophila*.

RNA-dependent DNA polymerase known as reverse transcriptase, which becomes activated upon cell infection. The resulting DNA can integrate itself at random into the host genome, where it becomes a provirus that retains the genetic information of the retroviral genome. Under certain conditions, the provirus can become activated to produce new viral RNA genomes and associated proteins, including reverse transcriptase, that can come together to form new virus particles. These particles are ultimately released from the cell surface by exocytosis. By contrast, many stably integrated retroviral elements appear to be inactive.

Once integrated into a host chromosome, the provirus replicates with every round of host DNA replication, regardless of whether the provirus itself is expressed or silent. Moreover, proviruses that integrate into the germ line, through the sperm or egg genome, segregate along with their host chromosome into the progeny of the host animal and into subsequent generations of animals as well. The genomes of all species of mammals contain inactive integrated proviral elements.

The LINE family: "Selfish DNA"

LINE is an acronym for long interspersed element. These elements, first described in Chapter 9, encode a reverse transcriptase. This enzyme, however, is not required for any normal cellular process of mammals. Geneticists speculate that LINEs represent "selfish DNA" elements (i.e., elements that do not contribute to the reproductive success of the host). The LINEs are one group of these elements.

The LINE family of DNA elements is very old. Homologous families of repetitive elements exist in a wide variety of organisms, including protists and plants. Thus, LINE-related elements, or other DNA elements of a similar nature, are likely to have been the source material that gave rise to retroviruses.

Dispersion to new positions in the germ-line genome presumably begins with the transcription of LINEs in spermatogenic or oogenic cells. The reverse-transcriptase-encoding region on the transcript is translated into an enzyme that preferentially associates with and uses the transcript it came from as a template to produce LINE cDNA sequences; however, the reverse transcriptase often stops before it has made a full-length DNA copy of the RNA transcript. The resulting incomplete cDNA molecules can nevertheless form a second strand to produce truncated double-stranded LINEs that integrate into the genome but remain forever dormant (**Figure 13.20**).

The SINE family

The Alu family in the human genome is an example of a highly repetitive, widely dispersed SINE (short interspersed element) family; SINEs were also described in Chapter 9. Over 500 000 Alu elements are dispersed throughout the human genome. At 300 bp in length, the Alu element is far too short to encode a reverse transcriptase. Nonetheless, like LINEs, Alu and other SINEs are able to disperse themselves throughout the genome by means of an RNA intermediate that undergoes reverse transcription and are

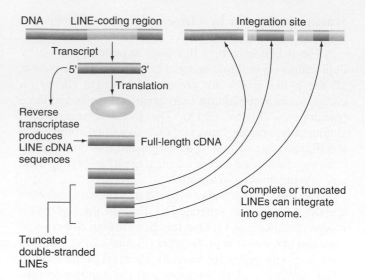

Figure 13.20 Creation of a LINE gene family. A complete LINE sequence can be copied into RNA. It encodes a reverse transcriptase that can make cDNA copies from the RNA. These copies may be complete or truncated and may integrate into other sites on any chromosome.

considered selfish elements. Clearly, SINEs depend on the availability of reverse transcriptase produced elsewhere, perhaps from LINE transcripts or the proviral elements of retroviruses.

All SINEs in the human genome, as well as in other mammalian genomes, appear to have evolved from small cellular RNA species, most often tRNAs, but also the 7S cytoplasmic RNA that is a component of the signal recognition particle (SRP) essential for protein translocation across the endoplasmic reticulum. The defining event in the evolution of a functional cellular RNA into an altered-function, self-replicating SINE is the accumulation of nucleotide changes in the 3′ region that lead to self-complementarity with the propensity to form hairpin loops. Reverse transcriptase can recognize the open end of the hairpin loop as a primer for strand elongation. Because it is likely that the hairpin loops form only rarely among normal cellular RNAs, a cell will preferentially use its SINE transcripts as templates for the production of cDNA molecules that are somehow able to integrate into the genome at random sites (**Figure 13.21**).

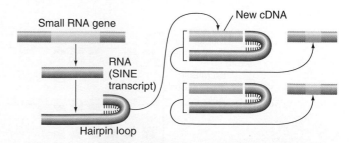

Figure 13.21 Creation of a SINE gene family. SINEs can be transcribed, and because they form 3′ hairpin loops, they can be copied into cDNAs by LINE-encoded reverse transcriptase. These cDNA copies then integrate into the genome.

The potential selective advantage of selfish elements

While SINEs and LINEs may have amplified themselves for selfish purposes, they have had a profound impact on whole-genome evolution. In particular, homologous SINEs or LINEs located near each other can, and will, catalyze unequal but homologous crossovers that result in the duplication of single-copy genes located between the homologous elements. Such duplications initiate the formation of multigene families. In addition, some selfish elements appear to have evolved a regulatory role—acting as enhancers or promoters—through chance insertions next to open reading frames. Thus, selfish elements may confer a selective advantage by facilitating duplication through unequal crossing-over or by becoming regulatory elements.

Tandem repeats

Through large-scale sequencing and hybridization analyses of mammalian genomes, researchers have found tandem repeats of DNA sequences with no apparent function scattered throughout the genome. The size of the repeating units ranges from two to six nucleotides (**simple sequence repeats**) all the way up to 20 kb or more, and the number of tandem repeats varies from two to several hundred. Such sequences, including microsatellites, minisatellites, and macrosatellites, have proven very useful as tags in genome analysis and detection of individual genotype.

The mechanism by which these tandem repeats originate may be different for loci having very short repeat units as compared with loci having longer repeat units. Tandem repeats of short di- or trinucleotides can originate through random changes in nonfunctional sequences. By contrast, the initial duplication of larger repeat units is likely to be a consequence of unequal crossing-over. Once two or more copies of a repeat unit exist in tandem, an increase in the number of repeat units in subsequent generations can occur through unequal crossovers or errors in replication (see Figures 13.8 and 13.14). It is not yet clear whether random mechanisms alone can account for the rich variety of tandem repeat loci in mammalian genomes or whether other selective forces are at play. Either way, tandem repeat loci continue to be highly susceptible to unequal crossing-over and, as a result, tend to be highly polymorphic in overall size.

Centromeres and telomeres contain many repeat sequences

Highly repetitive, noncoding sequences shorter than 200 bp are found in and around centromeres. For example, in the human genome, alphoid, a noncoding sequence 171 bp in length, is present in tandem arrays extending for over a megabase in the centromeric region of each chromosome. In addition, several similar-sized repetitive sequences unrelated to alphoid are found in some centromeres (review Figure 6.17). These regions are sites of interaction with the spindle fibres that segregate chromosomes during meiosis and mitosis. Selection may have acted to retain the thousands of copies of centromeric repeat elements in each centromeric region because they increase the efficiency and/or accuracy of chromosome segregation.

A second type of repeated genomic element with a special location is the hexamer TTAGGG found in the telomeres of all human chromosomes. The six-base unit is repeated in tandem arrays 5–10 kb in length at the ends of human and all other mammalian chromosomes. Selection may have conserved this repeat element because it plays an essential role in maintaining chromosomal length (review Figures 6.13 and 6.14).

- A gene family results from repeated duplication of a single gene and diversification among the copies. A few gene families have maintained homogeneity among members through concerted evolution, which occurs through two processes: selection that acts to keep the size of the gene family within an optimal range; and intergenic gene conversion that increases homogeneity among members.

- A gene superfamily is a collection of related multigene families found at different chromosomal locations. One example is the globin superfamily that contains the myoglobin gene as well as different haemoglobin genes; another is the *Hox* gene superfamily that contains genes involved in the regulation of gene expression.

- Much of the genome consists of "nonfunctional" DNA, including the LINE and SINE families of repeated elements. In contrast, centromeres and telomeres contain highly repetitive noncoding sequences that presumably play a role in chromosome replication.

Connections

A retrospective bird's-eye view of the key events that led to our current understanding of evolution at the molecular level goes something like this. In 1859, Charles Darwin published *The Origin of Species* in which he inferred from the visible evidence of descent through modification that the diverse organisms alive today evolved from a single primordial form, in large part, by a process of natural selection. Several years later,

Mendel published *Experiments on Plant Hybrids* in which he applied the laws of probability to the visible evidence of heredity, inferring the existence of hereditary units that segregate during gamete formation and assort independently of each other (Chapter 2). In the early twentieth century, Thomas Hunt Morgan and co-workers gave Mendel's units of heredity a physical location in the cell, establishing the

chromosomal basis of heredity and showing not only that genes have chromosomal addresses, but also that recombination can separate otherwise linked genes (Chapters 3 and 4). In the 1940s several people, including Oswald Avery, Martha Chase, and Alfred Hershey, showed that the molecule of heredity is DNA. Then in 1953, James Watson and Francis Crick deciphered the structure of DNA and proposed a mechanism by which the molecule replicates (Chapter 5). By the end of the twentieth century, extensive genomic analyses had made it possible to explain how DNA mutates, duplicates, diverges, and is acted on by selection to generate the diversity of life we see around us (Chapters 8, 9, 12, and 13).

Essential Concepts

1. The fossil record as well as living organisms of all levels of complexity provide scientists with a detailed picture of the evolution of complex life from the first cell to human beings. **[LO1–4]**

2. Preliminary studies on the evolution of gene regulatory networks suggest that the "rewiring" of these networks can account, at least in part, for the evolution of biological complexity. **[LO1]**

3. The evolution of organismal complexity generally correlates with an increase in genome size, which occurs through repeated duplications. Some duplications result from transpositions, while others arise from unequal crossing-over. **[LO2–4]**

4. Mutations rendering genes nonfunctional turn many duplicated genes into pseudogenes that over time diverge into random DNA sequences. However, rare advantageous mutations can turn a second copy of a gene into a new functional unit able to survive and spread through positive selection. **[LO3]**

5. The mammalian genome contains genes, multigene families, gene superfamilies, genome-wide repetitive elements; simple sequence repeats; and repetitive elements in centromeres and telomeres. **[LO4]**

6. Complex genomes arose from four levels of duplication followed by diversification and selection: exon duplication to create larger, more complex genes; gene duplication to create multigene families; multigene family duplication to create gene superfamilies; and the duplication of entire genomes. **[LO4]**

7. Genetic exchange between related DNA elements by intergenic gene conversion most often increases the variation among members of a multigene family. Sometimes, however, it can contribute to concerted evolution, which creates a family of nearly identical genes. **[LO4]**

Solved Problems

I. The sequence of two different forms of a gene starting with ATG is shown here. Which of the base differences in the second sequence are synonymous changes and which are nonsynonymous changes?

| Form 1 | ATGTCTCATGGACCCCTTCGTTTG |
| Form 2 | ATGTCTCAAAGACCACATCGTCTG |

Answer

The key to answering this question is understanding the difference between synonymous and nonsynonymous changes in DNA sequence.

Synonymous changes are nucleotide substitutions that do not change the amino acid specified by the DNA sequence. Nonsynonymous changes are nucleotide changes that result in a different amino acid in the protein.

Looking at the amino acids specified by the base sequence:

	Met	Ser	His	Gly	Pro	Leu	Arg	Leu
Form 1	ATG	TCT	CAT	GGA	CCC	CTT	CGT	TTG
	Met	Ser	Gln	Arg	Pro	His	Arg	Leu
Form 2	ATG	TCT	CAA	AGA	CCA	CAT	CGT	CTG

(The base substitutions in form 2 are underlined.) *The first, second, and fourth A substitutions are nonsynonymous changes; the third A substitution and the C substitution are synonymous changes.*

II. What difficulties would arise if you tried to derive a molecular clock rate using a noncoding sequence in some species and a coding sequence in other species?

Answer

To answer this question, you need to think about how molecular clocks are derived and the constraints on base changes in coding and noncoding sequences.

The evolution of coding sequences is restricted by the fact that the gene sequence needs to be maintained for the gene product to function. The sequence of noncoding regions generally can tolerate many base substitutions without selection acting on these sequences. *Therefore, you would expect more substitutions in the noncoding sequence. The result would be an inconsistency in your clock rate* if you are using a coding region in some species and noncoding DNA in other species.

III. If the chromosomes diagrammed below misalign by the pairing of repeated sequences (shown as *solid blocks*) and crossing-over occurs, what will the products be?

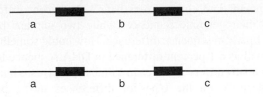

Answer

This question requires an understanding of crossing-over via homology. When you align the homologous repeated sequences out of register, one of the resulting products will have a duplication of the region between the repeats and three copies of the repeated sequence, and the other product will be deleted for the DNA between the repeats and contain only one repeated sequence.

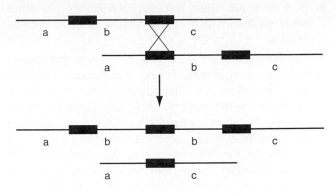

Problems

Vocabulary

1. For each of the terms in the left column, choose the best matching phrase in the right column.

i. ribozymes	**1.** constant rate of change in amino acid sequence
ii. retrotransposition	**2.** sudden explosive evolutionary changes
iii. SINEs	**3.** exchange of pieces of genes
iv. punctuated equilibrium	**4.** RNA molecules that can act as enzymes to catalyze specific chemical reactions
v. molecular clock	**5.** short repeated sequences in human genomes
vi. phylogenetic trees	**6.** RNA intermediate in duplication of genetic material
vii. exon shuffling	**7.** representation of evolutionary relationships

Section 13.1

2. What observations support the unity of life concept that all life-forms evolved from a common ancestor?

Section 13.2

3. Rates of nonsynonymous and synonymous amino acid substitutions for three genes that were compared in humans and mice or rats are shown here. (The rates are expressed as the average number of substitutions per base per year, with the standard deviation given. In each case, the number shown is the rate at which the human and rat sequences have diverged from each other, by either a nonsynonymous or a synonymous substitution.)

a. Why do the rates of nonsynonymous substitutions vary among these genes?

b. Why are the rates of synonymous substitutions similar?

Protein	Nonsynonymous substitutions	Synonymous substitutions
Histone 3	$0.0 \pm 0.0 \times 10^{-9}$	$6.38 \pm 1.10 \times 10^{-9}$
Growth hormone	$1.23 \pm 0.15 \times 10^{-9}$	$4.95 \pm 0.77 \times 10^{-9}$
β globin	$0.8 \pm 0.133 \times 10^{-9}$	$3.05 \pm 0.56 \times 10^{-9}$

4. Synonymous mutations are more prevalent than non-synonymous mutations in most genes. The immunoglobulin (Ig) genes encoding antibody subunits are an exception where nonsynonymous base changes outnumber synonymous changes. Based on the function of the Ig genes, why do you think this might be true?

5. Mutations in the *CF* gene that cause cystic fibrosis are carried by 1 in 20 individuals of Caucasian ancestry. The disease is clearly detrimental, yet the allele is maintained at a relatively high level in the population. What does this paradox suggest about the effect of the mutation?

6. Unequal crossing-over between two copies of a gene can lead to duplication and deletion on the two homologues involved. How could a single gene become duplicated?

7. Human beings have three-colour vision, while most other species of animals have two-colour vision or one-colour vision (i.e., black and white). Three-colour vision is produced by the products of a three-member, cross-hybridizing gene family that encodes light-sensitive pigments active in different ranges of the colour spectrum

(red, green, and blue). What is the most likely molecular explanation for the evolution of three-colour vision in the ancestor to human beings?

8. How do transposition and unequal crossing-over differ based on the location of final copies of the duplicated sequence?

9. You have identified an interesting new gene that appears to be involved in human brain development. You have discovered three cross-hybridizing copies of this gene within the human genome (*A*, *B*, and *C*). In the mouse genome, there is only a single copy of this gene (named *M*) and in the frog *Xenopus*, there is also only a single copy (named *X*). You have sequenced the same 10 000 bp of open reading frame from each of these genes and calculated the number of base-pair differences that exist between different pairs and obtained the following results:

Comparison	Number of base-pair differences
A versus *B*	300
B versus *C*	10
A versus *C*	300
A versus *M*	600
A versus *X*	3000

Assume that a constant rate of evolution has occurred with all members of this gene family, and assume that mice and humans evolved apart 60 million years ago.

a. How long ago did frogs split from the line leading to mice?

b. How many gene duplication events are observable within these data? At what time in the past did each of these duplication events occur?

c. Mapping of the *A*, *B*, and *C* genes shows that *A* and *C* are very closely linked, while *B* assorts independently from either of these genes. With this linkage information, what can you say about the molecular nature of the duplication events that occurred along the evolutionary line leading to human beings?

10. Phylogenetic trees of primates constructed using chromosome mutations (deletions, insertions, inversions, etc.) show the same relationships of species as those constructed using base substitutions within a gene. Which type of genetic alteration would you expect to have the greater impact on the evolution of chimpanzees and humans from a common ancestor? Why?

11. When a phylogenetic tree was constructed using comparisons of glucose-6-phosphate isomerase amino acid sequences from a wide variety of species, including animals, plants, and bacteria, the bacterium *E. coli* was placed on a branch of the phylogenetic tree with a flower. What explanation could there be for this unusual association based on this one protein-coding sequence? (Recall that bacteria and plants are in very different locations on the evolutionary trees derived from analyses of several gene sequences or morphology and physiology.)

12. Humans and chimpanzees have a 1 percent difference in their genomic sequence, while two humans have a 0.1 percent sequence difference. How could something as small as a 1 percent difference in DNA sequence lead to dramatic differences seen in chimpanzees versus humans? Speculate on the types of differences the 1 percent variation may represent.

Section 13.3

13. What is the unit that can be duplicated and modified to form

a. a new gene?

b. a multigene family?

14. It has been hypothesized that in the evolution of vertebrates, there were two successive doublings of the genome (tetraploidization) to produce the vertebrate genome.

a. How does the *Hox* superfamily fit with this hypothesis?

b. Among vertebrates, there is variation in the numbers of genes present within the *Hox* gene family clusters. How could this variation arise?

15. LINEs and SINEs are considered to be selfish DNA, yet they can, in some instances, confer a selective advantage on the organism. What are two ways in which a LINE or SINE can change the genome?

16. How is the size polymorphism of dinucleotide repeated sequences thought to occur?

17. a. What is an example of repetitive noncoding DNA for which we know no physiological function?

b. What is an example of repetitive noncoding DNA for which we know a cellular function?

18. Match the observation on the right with the event that it suggests occurred during evolution.

Event	Evidence in the genome
a. concerted evolution	**1.** genes in different individuals of the same species have 6, 8, or 10 homology units in an immunoglobulin gene family member
b. exons and introns of a gene	**2.** several copies of the same gene that are identical in the genome
c. unequal crossing-over	**3.** poly-A sequences at the end of coding regions in genomic DNA
d. retrotransposition	**4.** blocks of DNA sequence conserved between species separated by nonconserved blocks

 For more information on the resources available from McGraw-Hill Ryerson, go to www.mcgrawhill.ca/he/solutions.

446 **Unit I** **Chapter 13** Genome Evolution

Genetically modified *Gymnocorymbus ternetzi* (commonly known as Black Tetra). These fish, commercially sold as "GloFish," have been modified to express a recombinant gene encoding green fluorescent protein (GFP). In this chapter, we will learn about the techniques used by researchers to both analyze and manipulate DNA. These methodologies have not only made GloFish a reality, but have led to important advances in both medicine and biotechnology.

The vivid red colour of our blood arises from its life-sustaining ability to carry oxygen. This ability, in turn, derives from billions of red blood cells suspended in proteinaceous solution, each one packed with close to 280 million molecules of the protein pigment known as haemoglobin (**Figure 14.1a**). A normal adult haemoglobin molecule consists of four polypeptide chains, two alpha (α) and two beta (β) globins, each surrounding an iron-containing structure known as a haem group (**Figure 14.1b**). The iron atom within the haem sustains a reversible interaction with oxygen, binding it firmly enough to hold it on the trip from lungs to body tissue, but loosely enough to release it where needed. The intricately folded α and β chains protect the iron-containing haems from substances in the cell's interior. Each haemoglobin molecule can carry up to four oxygen atoms, one per haem, and these oxygenated haems impart a scarlet hue to the pigment molecules and thus to the blood cells that carry them.

The genetically determined molecular composition of haemoglobin changes several times during human development, enabling the molecule to adapt its oxygen-transport function to the varying environments of the embryo, fetus, newborn, and adult (**Figure 14.1c**). In the first five weeks after conception, the red blood cells carry *embryonic haemoglobin*, which consists of two α-like zeta (ζ) chains and two β-like epsilon (ϵ) chains. Thereafter, throughout the rest of gestation, the cells contain *fetal haemoglobin*, composed of two α chains and two β-like gamma (γ) chains. Then, shortly before birth, production of *adult haemoglobin*, composed of two α and two β chains, begins to climb. By the time an infant reaches three months of age, almost all of his or her haemoglobin is of the adult type.

Evolution of the various forms of haemoglobin maximized the delivery of oxygen to an individual's cells at different stages of development. The early embryo, which is not yet associated with a fully functional placenta, has the least access to oxygen in the maternal circulation. Both embryonic and fetal haemoglobin evolved to bind oxygen more tightly than adult haemoglobin does; they thus facilitate the transfer of maternal oxygen to the embryo or fetus. All the haemoglobins readily release their oxygen to cells, which have an even lower level of oxygen than any source of the gas. After birth, when oxygen is abundantly available in the lungs, adult haemoglobin, with its more relaxed kinetics of oxygen binding, allows for the most efficient pickup and delivery of the vital gas.

Haemoglobin disorders are the most common genetic diseases in the world and include sickle-cell anaemia, which arises from an altered β chain, and thalassaemia, which results from decreases in the amount of either α- or β-chain production.

The haemoglobin genes lie buried in a diploid human genome containing 6 billion base pairs distributed among

Chapter Outline
14.1 Sequence-specific DNA Fragmentation
14.2 Cloning Fragments of DNA
14.3 Genomic and cDNA Libraries
14.4 Hybridization
14.5 The Polymerase Chain Reaction
14.6 DNA Sequence Analysis
14.7 The Haemoglobin Genes: A Comprehensive Example

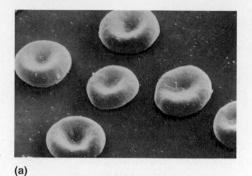

(a)

Figure 14.1 Haemoglobin is composed of four polypeptide chains that change during development. (a) Scanning electron micrograph of adult human red blood cells loaded with haemoglobin. **(b)** Adult haemoglobin consists of two α and two β polypeptide chains, each associated with an oxygen-carrying haem group. **(c)** The haemoglobin carried by red blood cells switches during human development from an embryonic form containing two α-like ξ chains and two β-like ε chains, to a fetal form containing two α chains and two β-like γ chains, and finally to the adult form containing two α and two β chains. In a small percentage of adult haemoglobin molecules, a β-like δ chain replaces the actual β chain. (The α-like chains are represented in *magenta*, and β-like chains are represented in *green*.)

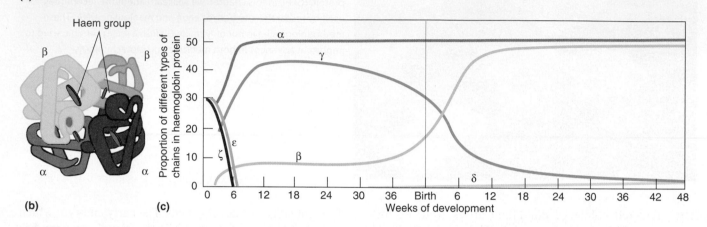

(b)   (c)

46 different strings of DNA (the chromosomes) that range in size from 60 million to 360 million base pairs each. In this chapter, we describe the powerful tools of modern molecular analysis that are used to search through these enormously long strings of information for genes such as the haemoglobin genes (which may be only several thousand base pairs in length). Initially, these tools took advantage of isolated enzymes and biochemical reactions that occur naturally within the simplest life-forms, bacterial cells. But over the last two decades, biologists have collaborated with chemists, engineers, and computer scientists to expand the toolkit to include automated chemical procedures not

found in nature. Researchers now refer to the whole kit of modern tools and reactions as **biotechnology**.

Biotechnology emerged from a technological revolution that began in the mid-1970s, when researchers gained the ability to read the digital information contained within any isolated sequence of DNA. For the first time, the genotypes of organisms could be determined even when they did not express a distinguishable phenotype. Geneticists can use the tools of biotechnology to gather information unobtainable in any other way, or to analyze the results of breeding and cytological studies with greater speed and accuracy than ever before.

Learning Objectives

1. Evaluate the importance of restriction enzymes in recombinant DNA technology.

2. Define the term "molecular cloning".

3. Compare and contrast "genomic" libraries and "cDNA" libraries.

4. List and appraise the advantages/disadvantages of PCR relative to molecular cloning.

5. Describe "Sanger" sequencing and explain how this process has been automated for high-throughput applications.

14.1   Sequence-specific DNA Fragmentation

Every diploid human body cell carries two nearly identical sets of 3 billion base pairs of information. Among these 3 billion base pairs are the 1400 base pairs encoding the β-globin gene, *HBB*. In the late 1960s and early 1970s the sheer volume of information made any attempt to study the genome as a whole a most daunting task. Instead, to reduce the complexity, researchers chose to first cut the

genome into "bite-size" pieces. These pieces contained on the order of hundreds to thousands of base pairs each. Each piece represented a distinct part of the genome. In some cases, the coding sequence of an entire gene (such as the *HBB* gene) would be found within a fragment. In other cases, noncoding sequences (e.g., regulatory elements like promoters or enhancers) would be present. In any event,

this simple strategy allowed geneticists to both analyze and experiment with manageable-sized fragments of DNA and thus to provide insight into the function of the sequences encoded within. In the remainder of this chapter, we will discuss the methodologies used by geneticists to cut, copy, and paste DNA fragments. These methodologies are often collectively referred to as **recombinant DNA technology** since they allow researchers to not only analyze naturally occurring fragments, but to also bring together genetic material from different sources to create novel phenotypes not before seen in nature (like the fluorescent green colour exhibited by recombinant GloFish).

Restriction enzymes fragment the genome at specific sites

Researchers use restriction enzymes to cut the DNA released from the nuclei of cells. These well-defined cuts generate fragments suitable for manipulation and characterization. A **restriction enzyme** recognizes a specific sequence of bases anywhere within the genome and then severs two covalent bonds (one in each strand) in the sugar-phosphate backbone at particular positions within or near that sequence. The fragments generated by restriction enzymes are referred to as **restriction fragments**, and the act of cutting is often called **digestion**.

Restriction enzymes originate in and can be purified from bacterial cells. The enzymes function naturally to protect these prokaryotic cells from viral infection by digesting viral DNA. Bacteria shield their DNA from digestion by their own restriction enzymes through the selective addition of methyl groups ($-CH_3$) to the restriction recognition sites in their DNA. In the test tube, restriction enzymes from bacteria recognize target sequences of 4–8 bp (in DNA isolated from any other organism) and cut the DNA at or near these sites. **Table 14.1** lists the names, recognition sequences, and microbial origins of just ten of the more than 100 commonly used restriction enzymes.

For the majority of these enzymes, the recognition site contains 4–6 base pairs and exhibits a kind of palindromic symmetry in which the base sequences of each of the two DNA strands are identical when read in the 5′-to-3′ direction. Because of this, base pairs on either side of a central line of symmetry are mirror images of each other. Each enzyme always cuts at the same place relative to its specific recognition sequence, and most enzymes make their cuts in one of two ways: either straight through both DNA strands right at the line of symmetry to produce fragments with **blunt ends**, or displaced equally in opposite directions from the line of symmetry by one or more bases to generate fragments with single-stranded ends (**Figure 14.2**). Geneticists often refer to these protruding single strands as **sticky ends**. They are considered "sticky" because they are free to base-pair with a complementary sequence from the DNA of *any* organism cut by the same restriction enzyme (see the **Tools of Genetics** box "**Restriction Enzyme Sites**" and the **Focus on Inquiry** box "**The Discovery of Restriction Enzymes**").

Table 14.1	Ten Commonly Used Restriction Enzymes	
Enzyme	**Sequence of Recognition Site**	**Microbial Origin**
*Taq*I	5′ T C G A 3′ / 3′ A G C T 5′	*Thermus aquaticus* YTI
*Rsa*I	5′ G T A C 3′ / 3′ C A T G 5′	*Rhodopseudomonas sphaeroides*
*Sau*3AI	5′ G A T C 3′ / 3′ C T A G 5′	*Staphylococcus aureus* 3A
*Eco*RI	5′ G A A T T C 3′ / 3′ C T T A A G 5′	*Escherichia coli*
*Bam*HI	5′ G G A T C C 3′ / 3′ C C T A G G 5′	*Bacillus amyloliquefaciens* H.
*Hin*dIII	5′ A A G C T T 3′ / 3′ T T C G A A 5′	*Haemophilus influenzae*
*Kpn*I	5′ G G T A C C 3′ / 3′ C C A T G G 5′	*Klebsiella pneumoniae* OK8
*Cla*I	5′ A T C G A T 3′ / 3′ T A G C T A 5′	*Caryophanon latum*
*Bss*HII	5′ G C G C G C 3′ / 3′ C G C G C G 5′	*Bacillus stearothermophilus*
*Not*I	5′ G C G G C C G C 3′ / 3′ C G C C G G C G 5′	*Nocardia otitidiscaviarum*

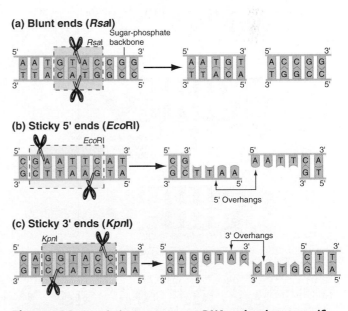

(a) Blunt ends (*Rsa*I)

(b) Sticky 5′ ends (*Eco*RI)

(c) Sticky 3′ ends (*Kpn*I)

Figure 14.2 Restriction enzymes cut DNA molecules at specific locations to produce restriction fragments with either blunt or sticky ends. (a) The restriction enzyme *Rsa*I produces blunt-ended restriction fragments. **(b)** *Eco*RI produces sticky ends with a 5′ overhang. **(c)** *Kpn*I produces sticky ends with a 3′ overhang.

In many types of bacteria, the unwelcome arrival of viral DNA mobilizes minute molecular weapons known as **restriction enzymes**. Each enzyme has the twofold ability to (1) recognize a specific sequence of four to six base pairs anywhere within any DNA molecule, and (2) sever a covalent bond in the sugar-phosphate backbone at a particular position within or near that sequence on each strand. When a bacterium calls up its reserve of restriction enzymes at the first sign of invasion, the ensuing shredding and dicing of selected stretches of viral DNA incapacitates the virus's genetic material and thereby restricts infection.

Since the early 1970s, geneticists have isolated more than 300 types of restriction enzymes and named them for the bacterial species in which they originate. *Eco*RI, for instance, is so named because it was originally isolated from **E**scherichia **co**li strain **R**Y13. Since it was the first enzyme isolated in *E. coli*, its name is ended with the roman numeral **I**. Each restriction enzyme recognizes a different base sequence and cuts the DNA strand at a precise spot in relation to that sequence. *Eco*RI recognizes the sequence 5'. . .GAATTC. . .3' and cleaves between the G and the first A. The DNA of a bacteriophage called lambda (λ), for example, carries the GAATCC sequence recognized by *Eco*RI in five separate places; the enzyme thus cuts the linear lambda DNA at five points, breaking it into six pieces with specific sizes. The DNA of a phage known as φX174, however, contains no *Eco*RI recognition sequences and is not cut by the enzyme.

Figure A illustrates *Eco*RI in action. Note that the recognition sequence in double-stranded DNA is symmetrical; that is, the base sequences on the two strands are identical when each is read in the 5'-to-3' direction. Thus, each time an enzyme recognizes a short 5'-to-3' sequence on one strand, it finds the exact same sequence in the 5'-to-3' direction of the complementary anti-parallel strand. The double-stranded recognition sequence is said to be palindromic; like the phrase "TAHITI HAT" or the number 1881, it reads the same backward and forward. (The analogy is not exact because in English only a single strand of letters or numbers is read in both directions, whereas in the DNA palindrome, reading in opposite directions occurs on opposite strands.)

Restriction enzymes made in other bacteria can recognize different DNA sequences and cleave them in different ways. When the weak hydrogen bonds between the strands dissociate, these cuts leave short, protruding single-stranded flaps known as **sticky** or **cohesive ends**. Like a tiny finger of Velcro®, each flap can stick to—that is, form hydrogen bonds with—a complementary sequence protruding from the end of another piece of DNA.

In the mid-1970s, geneticists took advantage of the activity of restriction enzymes to create DNA molecules composed of restriction fragments derived from different organisms. Such novel DNA fragments (i.e., never before seen in nature) are referred to as **recombinant DNA molecules**. In researchers' hands, the enzymes served as precision scissors that, in effect, revolutionized the study of life and gave birth to recombinant DNA technology. (Although the sticky ends created by restriction enzymes enable two unrelated DNA molecules to come together by base-pairing, another enzyme, known as DNA ligase, is required to stabilize the recombinant molecule. The ligase seals the breaks in the backbones of both strands.)

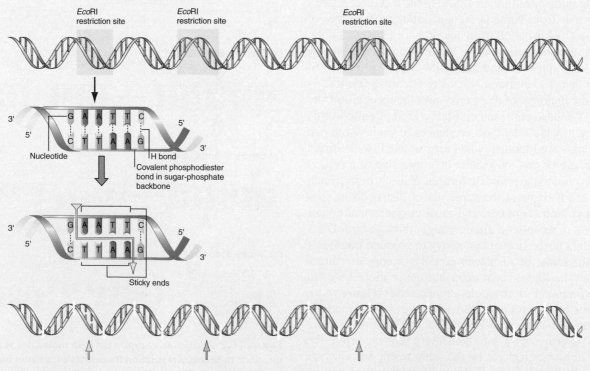

Figure A *Eco*RI in action.

Figure B illustrates one of the many applications of recombinant DNA technology: the splicing of the human gene for insulin into a small circle of DNA known as a plasmid, which can replicate inside a bacterial cell. Here is how it works. EcoRI is added to solutions of plasmids and human genomic DNA, where it cleaves both types of DNA molecules. The cleavage converts the circular plasmids to linear DNAs with EcoRI sticky ends; it also fragments each copy of the human genomic DNA into hundreds of thousands of pieces, all of which terminate with EcoRI sticky ends. When the solutions are then mixed together, the different fragments can adhere to each other in any combination, because of the complementarity of their sticky ends. In one combination, a human genomic fragment containing the insulin gene will become incorporated into a circular DNA molecule after adhering to the two ends of a linearized plasmid. And just as restriction enzymes operate as scissors, DNA ligase acts as a glue that seals the breaks in the DNA backbone by forming new phosphodiester bonds. Investigators can transform bacteria with the recombinant plasmids containing the insulin gene exactly as Avery transformed bacteria with his "transforming principle." The recombinant DNA molecules will enter some cells. When the bacteria copy their own chromosome in preparation for cell division, they will also make copies of any resident plasmids along with all the genes the plasmids contain. In the illustrated example, the plasmid carrying the gene for insulin also carries sequences that can direct its expression into protein. As the bacterial culture grows, so does the number of plasmids carrying sequences that direct the expression of the human insulin gene into protein. Eventually, a population of bacteria grows up in which every cell not only contains a copy of the human gene but also makes the insulin encoded by that gene. With recombinant DNA technology, it thus became possible to provide diabetic patients with a source of safe and inexpensive medicine to treat the symptoms of their disease.

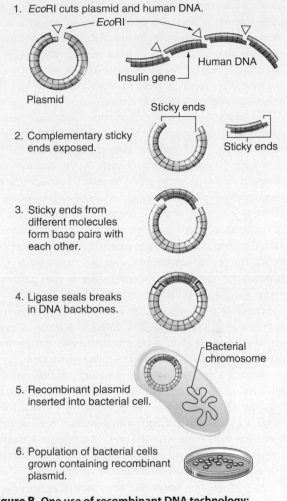

1. EcoRI cuts plasmid and human DNA.

EcoRI

Plasmid

Insulin gene

Human DNA

2. Complementary sticky ends exposed.

Sticky ends

Sticky ends

3. Sticky ends from different molecules form base pairs with each other.

4. Ligase seals breaks in DNA backbones.

Bacterial chromosome

5. Recombinant plasmid inserted into bacterial cell.

6. Population of bacterial cells grown containing recombinant plasmid.

Figure B One use of recombinant DNA technology: Harnessing bacteria to copy the human insulin gene. *E. coli* cells transformed with a recombinant plasmid can become miniature factories for the synthesis of insulin.

Different restriction enzymes produce fragments of different length

The average length of the fragments generated by a particular restriction enzyme can be calculated, and the information used to estimate the approximate number and distribution of recognition sites in a genome. The estimate depends on two simplifying assumptions: (1) that each of the four bases occurs in equal proportions (i.e., that the genome is composed of 25 percent A, 25 percent T, 25 percent G, and 25 percent C); and (2) that the bases are randomly distributed in the DNA sequence. Although these assumptions are never precisely valid, they enable us to determine the average distance between recognition sites of any length by the general formula 4^n, where n is the number of bases in the site (**Figure 14.3**).

Size of restriction enzyme recognition site and fragment length

According to the 4^n formula, *Rsa*I, which recognizes the four-base sequence GTAC, will cut on average once every 4^4, or every 256 bp, creating fragments averaging 256 bp in length. By comparison, the enzyme *Eco*RI, which recognizes the six-base sequence GAATTC, will cut on average once every 4^6, or 4096 bp; because 1000 base pairs = 1 kb pair, researchers often round off this large number to roughly 4.1 kb. Similarly, an enzyme such as *Not*I, which recognizes the eight bases GCGGC-CGC, will cut on average every 4^8 bp, or every 65.5 kb. Note, however, that because the actual distances between restriction sites for any enzyme vary considerably, very few of the fragments produced by the three enzymes

Most of the tools and techniques for cloning and analyzing DNA fragments emerged from studies of bacteria and the viruses that infect them. Molecular biologists had observed, for example, that viruses able to grow abundantly on one strain of bacteria grew poorly on a closely related strain of the same bacteria. While examining the mechanisms of this discrepancy, they discovered restriction enzymes.

To follow the story, one must know that researchers compare rates of viral proliferation in terms of *plating efficiency*: the fraction of viral particles that enter and replicate inside host bacterial cells, causing the cells to lyse and release viral progeny. These progeny go on to infect and replicate inside neighbouring cells, which in turn lyse and release further virus particles. When a Petri dish is coated with a continuous "lawn" of bacterial cells, an active viral infection can be observed as a visibly cleared spot, or plaque, where bacteria have been eliminated. The plating efficiency of lambda virus grown on *E. coli* C is nearly 1.0. This means that 100 original virus particles will cause close to 100 plaques on a lawn of *E. coli* C bacteria.

The plating efficiency of the same virus grown on *E. coli* K12 is only 1 in 10^4, or 0.0001. The ability of a bacterial strain to prevent the replication of an infecting virus, in this case the growth of lambda on *E. coli* K12, is called **restriction**.

Restriction is rarely absolute. Although lambda virus grown on *E. coli* K12 produces almost no progeny (the viruses infect cells but cannot replicate inside them), a few viral particles inside a few cells do manage to proliferate. If their progeny are then tested on *E. coli* K12, the plating efficiency is nearly 1.0. The phenomenon in which growth on a restricting host modifies a virus so that succeeding generations grow more efficiently on that same host is known as **modification**.

What mechanisms account for restriction and modification? Studies following viral DNA after bacterial infection found that during restriction, the viral DNA is broken into pieces and degraded (**Figure A**). When the enzyme responsible for the initial breakage was isolated, it was found to be an endonuclease, an enzyme that breaks the phosphodiester bonds in the viral DNA molecule, usually making double-strand cuts at a specific sequence in the viral chromosome. Because this breakage restricts the biological activity of the viral DNA, researchers called the enzymes that accomplish it *restriction enzymes*. Subsequent studies showed that the small percentage of viral DNA that escapes digestion and goes on to generate new viral particles has been modified by the addition of methyl groups during its replication in the host cell. Researchers named the enzymes that add methyl groups to specific DNA sequences **modification enzymes**.

Biologists have identified a large number of complementary restriction-modification systems in a variety of bacterial strains. Purification of the systems has yielded a mainstay of recombinant DNA technology: the battery of restriction enzymes used to cut DNA *in vitro* for cloning, mapping, and ligation (see Table 14.1).

This example of serendipity in science sheds some light on the debate between administrators who distribute and oversee research funding and scientists who carry out the research. Microbial investigators did not set out to find restriction enzymes; they could not have known these enzymes would be one of their finds. Rather, they sought to understand the mechanisms by which viruses infect and proliferate in bacteria. Along the way, they discovered restriction enzymes and how they work. The politicians and administrators in charge of allocating funds often want to direct research spending to urgent health or agricultural problems, while the scientists in charge of laboratory research call for a broad distribution of funds to all projects investigating interesting biological phenomena. The validity of both views suggests the need for a balanced approach to the funding of research activities.

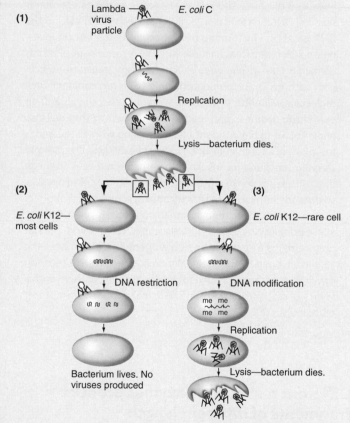

Figure A Operation of the restriction enzyme/modification system in nature. **(1)** *E. coli* strain C does not have a functional restriction enzyme/modification system and is susceptible to infection by the lambda phage. **(2)** In contrast, *E. coli* strain K12 generally resists infection by the viral particles produced from a phage infection of *E. coli* C. This is because *E. coli* K12 makes several restriction enzymes, including *Eco*RI, which cut the lambda DNA molecule before its genes can be expressed. **(3)** However, in rare K12 cells, the lambda DNA is modified by an enzyme that protects its recognition sites from the host cell's restriction enzymes. This modified lambda DNA can now replicate and generate phage particles, which eventually destroy the bacterial cell.

(a) Calculating Average Restriction Fragment Size

1. Probability that a four-base recognition site will be found in a genome =

$$1/4 \times 1/4 \times 1/4 \times 1/4 = 1/256$$

2. Probability that a six-base recognition site will be found =

$$1/4 \times 1/4 \times 1/4 \times 1/4 \times 1/4 \times 1/4 = 1/4096$$

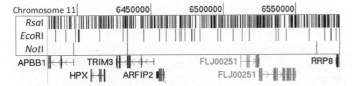

Figure 14.3 The number of base pairs in a recognition site determines the average distance between sites in a genome and thus the size of fragments produced. (a) *Rsa*I recognizes and cuts at a 4-bp site, *Eco*RI cuts at a 6-bp site, and *Not*I cuts at an 8-bp site. **(b)** *Rsa*I, *Eco*RI, and *Not*I restriction sites in a 200-kb region of human chromosome 11, followed by the names and locations of genes in this region. Each *vertical black line* represents a restriction site. As expected, *Rsa*I sites are very common, and *Not*I sites are very rare. *Eco*RI sites are present at an intermediate frequency.

mentioned here will be precisely 65.5 kb, 4.1 kb, or 256 bp in length.

Geneticists often need to produce DNA fragments of a particular length—larger ones to study the organization of a chromosomal region, smaller ones to examine a whole gene, and ones that are smaller still for DNA sequence analysis (i.e., for the determination of the precise order of bases in a DNA fragment). If their goal is 4-kb fragments, they have a range of six-base cutter enzymes to choose from. Exposing the DNA to a six-base cutter for a long enough time gives the restriction enzyme ample opportunity for digestion. The result is a **complete digest** in which the DNA has been cut at every one of the recognition sites it contains.

Different restriction enzymes produce different numbers of fragments

We have seen that the four-base cutter *Rsa*I cuts the genome on average every 4^4 (256) bp. If you exposed the haploid human genome with its 3 billion bp to *Rsa*I for a sufficient time under the appropriate conditions, you would ensure that all of the recognition sites in the genome that could be cleaved would be cleaved, and you would get

$$\frac{3\,000\,000\,000 \text{ bp}}{\sim256 \text{ bp}} = \begin{array}{l} \sim12\,000\,000 \text{ fragments that are} \\ \sim256 \text{ bp in average length} \end{array}$$

By comparison, the six-base cutter *Eco*RI cuts the DNA on average every 4^6 (4096) bp, or every 4.1 kb. If you exposed the haploid human genome with its 3 billion bp, or 3 million kb, to *Eco*RI in the proper way, you would get

$$\frac{3\,000\,000\,000 \text{ bp}}{\sim4100 \text{ bp}} = \begin{array}{l} \sim700\,000 \text{ fragments that are} \\ \sim4.1 \text{ kb in average length} \end{array}$$

And if you exposed the same haploid human genome to the eight-base cutter *Not*I, which cuts on average every 4^8 (65 536) bp, or 65.5 kb, you would obtain

$$\frac{3\,000\,000\,000 \text{ bp}}{\sim65\,500 \text{ bp}} = \begin{array}{l} \sim46\,000 \text{ fragments that are} \\ \sim65.5 \text{ kb in average length} \end{array}$$

Clearly, the larger the recognition site, the smaller the number of fragments generated by enzymatic digestion.

Restriction enzymes were first used to study the very small genomes of viruses such as bacteriophage lambda (λ), whose genome has a length of approximately 48.5 kb, and the animal tumour virus SV40, whose genome has a length of 5.2 kb. We now know that the six-base cutter *Eco*RI digests lambda DNA into five fragments, and the four-base cutter *Rsa*I digests SV40 into 12 fragments. But when molecular biologists first used restriction enzymes to digest these viral genomes, they also needed a tool that could distinguish the different fragments in a genome from each other and determine their sizes. That tool is gel electrophoresis.

Gel electrophoresis distinguishes DNA fragments according to size

Electrophoresis is the movement of charged molecules in an electric field. Biologists use it to separate many different types of molecules; for example, DNA of one length from DNA of other lengths, DNA from protein, or one kind of protein from another. In this discussion, we focus on its application to the separation of DNA fragments of varying length in a gel (**Figure 14.4**). To carry out such a separation, you place a solution of DNA molecules into indentations called wells at one end of a porous gel-like matrix. When you then place the gel in a buffered aqueous solution and set up an electric field between bare wires at either end, the electric field causes all charged molecules in the wells to migrate in the direction of the electrode having an opposite charge. Because all of the phosphate groups in the backbone of DNA carry a net negative charge in a solution near neutral pH, DNA molecules are pulled through a gel toward the wire with a positive charge.

Several variables determine the rate at which DNA molecules (or any other molecules) move during electrophoresis. These variables are the strength of the electric field applied across the gel, the composition of the gel, the charge per unit volume of molecule (known as *charge density*), and the physical size of the molecule. The only one of these variables that actually differs among any set of linear DNA fragments migrating in a particular gel is size. The reason is that all molecules placed in a well are subjected to the same electric field and the same gel matrix, and all DNA molecules have the same charge density (because the charge of all nucleotide pairs is nearly identical). As a result, only differences in size cause different linear DNA molecules to migrate at different speeds during electrophoresis.

Gel Electrophoresis

(a)

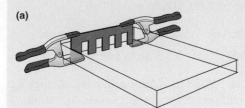

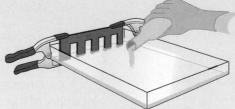

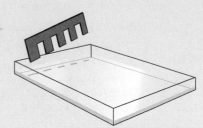

1. Attach a comb to a clear acrylic plate with clamps.

2. Pour heated molten agarose into plate. Allow to cool and harden.

3. Remove comb from gel; shallow wells are left in gel. Remove gel from plate.

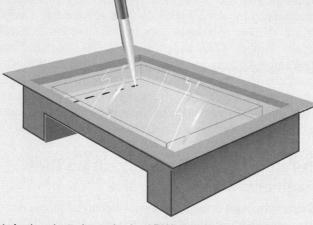

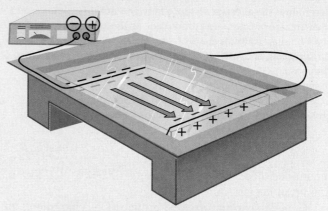

4. A micropipette is used to load DNA samples into each well. Each sample contains a blue dye to make it easier to visualize.

5. Electrode wires are placed along each end of the gel and are attached to a power supply. The current is switched on, and DNA molecules in each sample migrate toward the "+" end of the box (along the paths depicted with *orange arrows*). Electrophoresis continues for 1–20 hours.

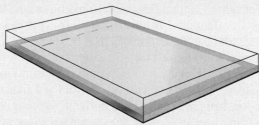

6. Remove gel from gel box. Incubate with ethidium bromide, then wash to remove excess dye.

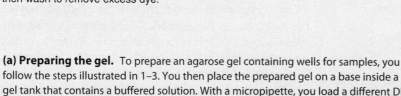

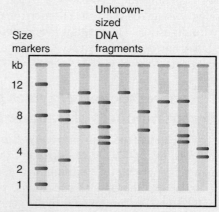

(a) Preparing the gel. To prepare an agarose gel containing wells for samples, you follow the steps illustrated in 1–3. You then place the prepared gel on a base inside a gel tank that contains a buffered solution. With a micropipette, you load a different DNA sample into each well (Step 4). A special "size marker" sample containing DNA fragments of known size is loaded into the first well. You now connect wires at either end of the box to a power supply, turn on the electric current, and allow the fragments to migrate for 1–20 hours. You then remove the gel from the electrophoresis chamber and place it into a box containing a solution of ethidium bromide, a fluorescent dye that will bind tightly to any DNA fragments in the gel. After incubating the gel for several hours, you immerse the gel in water to wash away any unbound dye molecules. Then, with exposure to ultraviolet light, the bound dye absorbs photons in the UV range and gives off photons in the visible red range. The DNA molecules appear as red bands, and a digital image shows the relative positions to which they have migrated in the gel.

To determine the length of a DNA fragment, you chart the mobility of the band composed of that fragment relative to the migration of the size marker bands in the first gel lane.

7. Expose gel to UV light.
DNA molecules will appear as red bands. A photograph of the bands will provide a black-and-white image. The sizes of the bands in the unknown samples can be calibrated by comparison to size markers that have been run in the leftmost lane of the gel.

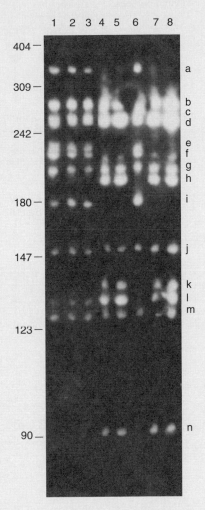

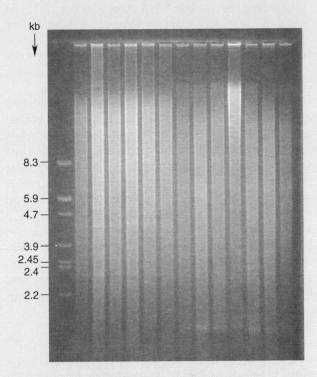

(b) Different types of gels separate different-sized DNA molecules. Gels can be composed of two different types of chemical matrices: polyacrylamide and agarose. Polyacrylamide, which has smaller pores, can separate DNA molecules in the range of 10–500 nucleotides in length; agarose separates only larger molecules. The polyacrylamide gel on the *left* was used to separate restriction fragments formed by the digestion of eight different DNA samples with an enzyme that recognizes a four-base sequence. The fragments *a–n* range in size from 90–340 bases in length. The agarose gel on the *right* was used to separate restriction fragments in a series of complex samples digested with *Eco*RI. The first lane contains size markers ranging in size from 2.2 kb to 8.3 kb. The other lanes contain digested whole genomic DNA from different mice. The mouse genome contains approximately 700 000 unique *Eco*RI restriction fragments, which appear as a smear when stained with ethidium bromide.

With linear DNA molecules, differences in size are proportional to differences in length: the longer the molecule, the larger the volume it occupies as a random coil. The larger the volume a molecule occupies, the less likely it is to find a pore in the gel matrix big enough to squeeze through and the more often it will bump into the matrix. And the more often the molecule bumps into the matrix, the slower will be its rate of migration (also referred to as its mobility). With this background, you can follow the steps of **Figure 14.4a** to determine the length of the restriction fragments in the DNA under analysis.

When electrophoresis is completed, the gel is incubated with a fluorescent DNA binding dye called ethidium bromide. After the unbound dye has been washed away, it is easy to visualize the DNA by placing the gel under an ultraviolet light. The actual size of restriction fragments observed on gels is determined by comparison to the migration distances of known *marker fragments* that are subjected to electrophoresis in an adjacent lane of the gel.

DNA molecules range in size from small fragments of less than 10 bp to whole human chromosomes that have an average length of 130 000 000 bp. No one sizing procedure has the capacity to separate molecules throughout this enormous range. To detect DNA molecules in different size ranges, researchers use a variety of protocols based mainly on two kinds of gels: polyacrylamide (formed by covalent bonding between acrylamide monomers), which is good for distinguishing smaller DNA fragments, and agarose (formed by the noncovalent association of agarose polymers), which is suitable for looking at larger fragments. **Figure 14.4b** illustrates these differences.

Restriction maps provide sequence-specific landmarks in the DNA terrain

Researchers can use restriction enzymes not only as molecular scissors to create unique DNA fragments but also as an analytical tool to create maps of viral genomes and other purified DNA fragments. These maps, called *restriction maps*, show the relative order and distances between multiple restriction sites, which thus act as landmarks along a DNA molecule.

The derivation of a restriction map can be approached in several ways. One of the most commonly used methods involves digestion with multiple restriction enzymes—alone or mixed together—followed by gel electrophoresis to visualize the fragments produced. If the relative arrangement of sites for the various restriction enzymes employed does not create too many fragments, the data obtained can provide enough information to piece together a map showing the position of each restriction site. **Figure 14.5** shows how the process of elimination can be used to infer the arrangement of restriction sites (in a way that is consistent with the results of three sets of digestions using either of two enzymes alone or both enzymes simultaneously).

Figure 14.5 How to infer a restriction map from the sizes of restriction fragments produced by two restriction enzymes.
(a) Divide a purified preparation of cloned DNA into three aliquots; expose the first aliquot to *Eco*RI, the second to *Bam*HI, and the third to both enzymes. **(b)** Now separate by gel electrophoresis the restriction fragments that result from each digestion and determine their sizes in relation to defined markers. **(c)** Finally, use the process of elimination to derive the only arrangement that can account for the results obtained with all three samples.

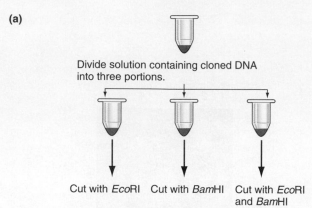

(a) Divide solution containing cloned DNA into three portions.

Cut with *Eco*RI   Cut with *Bam*HI   Cut with *Eco*RI and *Bam*HI

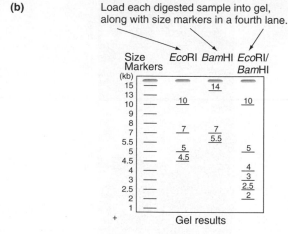

(b) Load each digested sample into gel, along with size markers in a fourth lane.

Gel results

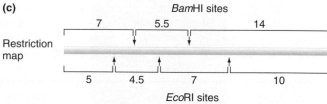

(c) Restriction map

*Bam*HI sites

*Eco*RI sites

- Restriction enzymes recognize specific DNA sequences and cut each strand of DNA at specific locations in or near the target sequence. The result of digesting a particular genome with a particular restriction enzyme is a collection of restriction fragments of defined length and composition.

- Geneticists use enzyme-specific recognition-site size, and time of exposure to the enzyme, to create complete or partial digests of DNA genomes.

- By using different restriction enzymes, scientists can generate different numbers of unique fragments from a single genome. The larger the recognition site, the smaller the number of fragments.

- Gel electrophoresis is used to separate and measure the different lengths of DNA molecules present in a complex solution.

- Restriction maps are constructed through a process of logical deduction based on the sizes of restriction fragments obtained after digestion with two or more restriction enzymes (both separately and together). Restriction fragment sizes are determined by gel electrophoresis.

14.2   Cloning Fragments of DNA

While restriction enzyme digestion and gel electrophoresis provide a means for analyzing simple DNA molecules, the genomes of animals, plants, and even microorganisms are far too large to be analyzed in this way. For example, the *E. coli* genome is approximately 4200 kb or 4.2 Mb in length. An *Eco*RI digestion of this genome would produce approximately 1000 fragments. If you subjected these complex mixtures of DNA fragments to gel electrophoresis, all you would see at the end is a smear rather than discrete bands (review Figure 14.4b). To study any one fragment (e.g., a fragment containing the *HBB* gene) within this complex mixture, it first must be purified away

from all the other fragments and then amplified; that is, used to make many identical copies of the originally purified molecule. Researchers can then apply chemical and physical techniques—including restriction mapping and DNA sequencing—to analyze the isolated DNA fragment.

Scientists now use two strategies to accomplish the purification and amplification of individual fragments: *molecular cloning*, which replicates individual fragments of previously uncharacterized DNA, and the *polymerase chain reaction* (or *PCR*), which can purify and amplify a previously sequenced genomic region (or a transcribed version of it) from any source much more rapidly than cloning. Here we present the protocol for the molecular cloning of DNA. Later in the chapter, we describe PCR.

Molecular cloning is the process that takes a complex mixture of restriction fragments and uses living cells to purify and make many exact replicas of just one fragment at a time. It consists of two basic steps. In the first, DNA fragments that fall within a specified range of sizes are inserted into specialized chromosome-like carriers called *vectors*, which ensure the transport, replication, and purification of individual inserts. In the second step, the combined vector-insert molecules are transported into living cells, and the cells make many copies of these molecules. Because all the copies are identical, the group of replicated DNA molecules is known as a **DNA clone**. DNA clones may be purified for immediate study or stored within cells or viruses as collections of clones known as *libraries* for future analysis. We now describe each step of molecular cloning.

Cloning step 1: Splicing inserts to vectors produces recombinant DNA

On their own, restriction fragments cannot reproduce themselves in a cell. To make replication possible, it is necessary to splice each fragment to a **vector**: a specialized DNA sequence that can enter a living cell, signal its presence to an investigator by conferring a detectable property to the host cell, and provide a means of replication for itself and the foreign DNA inserted into it. A vector must also possess distinguishing physical traits, such as size or shape, by which it can be purified away from the host cell's genome. Several types of vectors are in use and each one behaves as a chromosome capable of accepting foreign DNA inserts and replicating independently of the host cell's genome. The cutting and splicing together of vector and inserted fragment—DNA from two different origins—creates a **recombinant DNA molecule**.

Sticky ends and base-pairing

Two characteristics of single-stranded or "sticky" ends provide a basis for the efficient production of a vector-insert recombinant: the ends are available for base-pairing, and no matter what the origin of the DNA (e.g., bacterial or human), two sticky ends produced with the same enzyme are complementary in sequence. You simply cut the vector with the same restriction enzyme used to generate the fragment of genomic DNA and mix the digested vector and genomic DNAs together in the

presence of DNA ligase. You then allow time for the base-pairing of complementary sticky ends and for the ligase to stabilize the molecule. Certain laboratory "tricks" (discussed later) help prevent two or more genomic fragments from joining with each other rather than with vectors.

Choice of vectors

Available vectors differ from one another in biological properties, carrying capacity, and the type of host they can infect. The simplest vectors are minute circles of double-stranded DNA known as **plasmids** that can gain admission to and replicate in the cytoplasm of many kinds of bacterial cells, independently of the bacterial chromosomes (**Figure 14.6**). The most useful plasmids

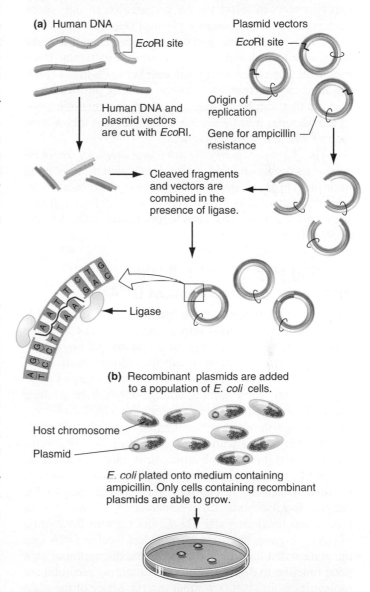

Figure 14.6 Creating recombinant DNA molecules with plasmid vectors. (a) Human genomic DNA is cut with *Eco*RI to produce a mixture of fragments. A plasmid vector is also cut with *Eco*RI at its single *Eco*RI recognition site. The two are mixed together in the presence of the enzyme ligase, which sutures them to each other to form circular recombinant DNA molecules. **(b)** *E. coli* cells transformed with recombinant plasmids are recognized by their growth in the presence of ampicillin.

contain several recognition sites (e.g., one *Eco*RI site, one *Hpa*I site, and so forth). This provides flexibility in the choice of enzymes that can be used to digest the DNA containing the fragment, or fragments, of interest. Exposure to any one of these restriction enzymes opens up the vector at the corresponding recognition site, allowing the insertion of a foreign DNA fragment (Figure 14.6).

Each plasmid vector carries an origin of replication and a gene for resistance to a specific antibiotic. The origin of replication enables it to replicate independently inside a bacterium. The gene for antibiotic resistance confers on the host cell the ability to survive in a medium containing a specific antibiotic; the resistance gene thereby enables experimenters to select for the propagation of only those bacterial cells that contain a plasmid (**Figure 14.7**). Antibiotic resistance genes, and other vector genes that make it possible to pick out cells harbouring a particular DNA molecule, are called **selectable markers**. Plasmids fulfill the final requirement for vectors—ease of purification— because they can be purified away from the genomic DNA of the bacterial host by several techniques that take advantage of size and other differences.

The largest-capacity vectors are *artificial chromosomes*: recombinant DNA molecules formed by combining multiple chromosomal replication and segregation elements of a specific host with a DNA insert. A bacterial artificial chromosome (BAC) can accommodate a DNA insert of 300 kb.

Cloning step 2: Host cells take up and replicate recombinant DNA

Although each type of vector functions in a slightly different way, and can enter only a specific kind of host, the general scheme of entering a host cell and taking advantage of the cellular environment to replicate itself is the same for all. We divide our discussion of this step of cloning into three parts: getting foreign DNA into the host cell; selecting cells that have received a DNA molecule; and distinguishing insert-containing recombinant molecules from vectors without inserts. Figure 14.7 illustrates the three-part process with a plasmid vector containing an origin of replication, the gene for resistance to ampicillin (*amp^r*), and the *E. coli lacZ* gene, which encodes the enzyme β-galactosidase. By constructing the vector with a common restriction site like *Eco*RI right in the middle of the *lacZ* gene, researchers can insert foreign DNA into the gene at that location and then use the disruption of *lacZ* gene function to distinguish insert-containing recombinant molecules from vectors without inserts. Many of the plasmid vectors used today incorporate most if not all of the features depicted in Figure 14.7.

Transformation of host cells

Transformation, as you saw in Chapter 5, is the process by which a cell or organism takes up a foreign DNA molecule, changing the genetic characteristics of that cell or organism. What we now describe is similar to what Avery and his colleagues did in the transformation experiments that determined DNA was the molecule of heredity (see Section 5.1).

First, recombinant DNA molecules are added to a suspension of specially prepared *E. coli*. Under conditions favouring entry, such as suspension of the bacterial cells in a cold CaCl$_2$ solution or treatment of the solution with high-voltage electric shock (a technique known as *electroporation*), the plasmids will enter about 1 in 1000 cells (**Figure 14.7b**). These protocols increase the permeability of the bacterial cell membrane, in essence punching temporary holes through which the DNA gains entry. The probability that any one plasmid will enter any one cell is so low (0.001) that the probability of simultaneous entry of two plasmids into a single cell is insignificant (0.001 × 0.001 = 0.000 001).

Identification and isolation of transformed cells

To identify the 0.1 percent of cells housing a plasmid, the bacteria-plasmid mixture is decanted onto a plate containing agar, nutrients, and ampicillin. Only cells transformed by a plasmid providing resistance to ampicillin will be able to grow and multiply in the presence of the antibiotic. The plasmid's origin of replication enables it to replicate in the bacterial cell independently of the bacterial chromosome; in fact, most plasmids replicate so well that a single bacterial cell may end up with hundreds of identical copies of the same plasmid molecule.

Each viable plasmid-containing bacterial cell will multiply to produce a distinct spot on an agar plate, consisting of a colony of tens of millions of genetically identical cells. The colony as a whole is considered a **cellular clone**. Such clones can be identified when they have grown to about 1 mm in diameter (**Figure 14.7c**). The millions of identical plasmid molecules contained within a colony together make up a **DNA clone**. These plasmid molecules can be isolated for further analysis by simply lysing the *E. coli* cells and then purifying the plasmid DNA away from the other cellular components.

Screening for insert-containing DNA molecules

If prepared under proper conditions, most treated plasmids contain an insert. Some plasmids, however, slip through without one. **Figure 14.7d** shows how the system we are discussing distinguishes cells with only vectors from cells with vectors containing inserts.

The medium on which the transformed, ampicillin-resistant bacteria grow contains, in addition to nutrients and ampicillin, a chemical compound known as X-Gal. This compound serves as a substrate for the reaction catalyzed by the intact β-galactosidase enzyme (the protein encoded by the *lacZ* gene); one product of the reaction is a new, blue-coloured chemical. Cells containing vectors without inserts turn blue because they carry the

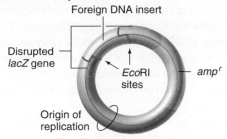

(a) A recombinant plasmid

Foreign DNA insert

Disrupted
lacZ gene

*Eco*RI sites

*amp*ʳ

Origin of replication

(b) Transformation: Foreign DNA enters the host cell.

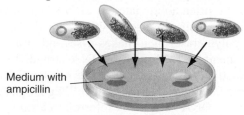

(c) Selecting cells that have received a plasmid

Medium with ampicillin

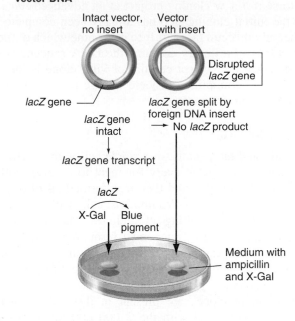

(d) Distinguishing cells carrying recombinant molecules from cells carrying just nonrecombinant vector DNA

Intact vector, no insert

Vector with insert

Disrupted *lacZ* gene

lacZ gene

lacZ gene intact

lacZ gene split by foreign DNA insert
⟶ No *lacZ* product

lacZ gene transcript

lacZ

X-Gal   Blue pigment

Medium with ampicillin and X-Gal

Figure 14.7   How to identify transformed bacterial cells containing plasmids with DNA inserts.   (a) Plasmid vectors are often constructed so that they contain the *E. coli lacZ* gene with a restriction site right in the middle of the gene. If the vector reanneals to itself without inclusion of an insert, the *lacZ* gene will remain uninterrupted; if it accepts an insert, the gene will be interrupted. **(b)** Transformation: When added to a culture of bacteria, plasmids enter about 1 in 1000 cells. **(c)** Only cells transformed by a plasmid carrying a gene for ampicillin resistance will form colonies on Petri plates. **(d)** Cells containing vectors that have reannealed to themselves without the inclusion of an insert will express the uninterrupted *lacZ* gene. The polypeptide product of the gene is β-galactosidase. Reaction of this enzyme with a substrate known as X-Gal produces a molecule that turns the cell blue. Any cells containing recombinant plasmids will not generate active β-galactosidase and will therefore not turn blue.

original intact β-galactosidase gene. Cells containing plasmids with inserts remain colourless, because the interrupted *lacZ* gene does not allow production of functional β-galactosidase enzyme. This process of engineering an insert to interrupt a host gene is termed *insertional inactivation*.

- The first step in the creation of recombinant DNA clones is the ligation of the DNA of interest to a vector, such as a plasmid or an artificial chromosome. Vectors contain one or more origins of replication and a selectable marker, such as an antibiotic resistance gene, so that the recombinant molecule can be replicated and identified.

- The vector component of the recombinant DNA molecule (1) provides a receptacle for the DNA fragment of interest, (2) carries a selectable marker, (3) hijacks the cell's biochemical machinery to replicate the recombinant molecule, (4) provides a means for distinguishing recombinant molecules from vector-only molecules, and (5) can be trimmed away to allow purification of the amplified insert DNA.

14.3   Genomic and cDNA Libraries

Moving step by step from the DNA of any organism to a single purified DNA fragment is a long and tedious process. Fortunately, scientists do not have to return to Step 1 every time they need to purify a new genomic fragment from the same organism. Instead, they can build a **genomic library**: a long-lived collection of cellular clones that contains copies of every sequence in the whole genome inserted into a suitable vector. Like traditional

book libraries, genomic libraries store large amounts of information for retrieval upon request. They make it possible to start a new cloning project at an advanced stage, when the initial cloning step has already been completed and the only difficult task left is to determine which of the many clones in a library contains the DNA sequence of interest. Once the correct cellular or viral clone is identified, it can be amplified to yield a large amount of the desired genomic fragment.

Genomic libraries

If you digested the genome of a single cell with a restriction enzyme and ligated every fragment to a vector with 100 percent efficiency, and then transformed all of these recombinant DNA molecules into host cells with 100 percent efficiency, the resulting set of clones would represent the entire genome in a fragmented form. A hypothetical collection of cellular clones that includes one copy—and one copy only—of every sequence in the entire genome would be a *complete genomic library*.

How many clones are present in this hypothetical library? If you started with the 3 000 000 kb of DNA from a haploid human sperm and reliably cut it into a series of 150-kb restriction fragments, you would generate 3 000 000/150 = 20 000 genomic fragments. If you placed each and every one of these fragments into BAC cloning vectors that were then transformed into *E. coli* host cells, you would create a perfect library of 20 000 clones that collectively carry every locus in the genome. The number of clones in this perfect library defines a **genomic equivalent**. To find the number of clones that constitute one genomic equivalent for any library, you simply divide the length of the genome (here, 3 000 000 kb) by the average size of the inserts carried by the library's vector (in this case, 150 kb).

In real life, it is impossible to obtain a perfect library. Each step of cloning is far from 100 percent efficient, and the DNA of a single cell does not supply sufficient raw material for the process. Researchers must thus harvest DNA from the millions of cells in a particular tissue or organism. If you make a genomic library with this DNA by collecting only one genomic equivalent (20 000 clones for a human library in BAC vectors), then by chance some human DNA fragments will appear more than once, while others will not be present at all. However, by including four to five genomic equivalents, one can produce an average of four to five clones for each locus, and ensure, with 95 percent probability, that any individual locus is present at least once.

cDNA libraries

Often, only the information in a gene's coding sequence is of experimental interest. Thus, it would be advantageous to limit analysis to the gene's exons without having to determine the structure of the introns as well. However, because coding sequences account for a very small percentage of genomic DNA in higher eukaryotes, it is inefficient to look for them in genomic libraries. The solution is to generate **cDNA libraries**, which store sequences copied into DNA from all the RNA transcripts present in a particular cell type, tissue, or organ. Because they are obtained from RNA transcripts, these sequences carry only exon information.

To produce DNA clones from mRNA sequences, researchers rely on a series of *in vitro* reactions that mimic several stages in the life cycle of viruses known as **retroviruses**. Retroviruses carry their genetic information in molecules of RNA. As part of their gene-transmission kit, retroviruses also contain an unusual enzyme known as **RNA-dependent DNA polymerase**, or simply **reverse transcriptase** (review the Genetics and Society box in Chapter 7). After infecting a cell, a retrovirus uses reverse transcriptase to copy its single strand of RNA into a mirror-image-like strand of complementary DNA, often abbreviated as **cDNA**. The reverse transcriptase, which can also function as a DNA-dependent DNA polymerase, then makes a second strand of DNA complementary to this first cDNA strand. Finally, this double-stranded DNA copy of the retroviral RNA chromosome integrates into the host cell's genome. Although the term "cDNA" originally referred to a single strand of DNA complementary to an RNA molecule, it now refers to any DNA—single- or double-stranded—derived from an RNA template.

Suppose you were interested in studying the structure of a mutant β-globin protein. You have already analyzed haemoglobin obtained from a patient carrying this mutation and found that the alteration affects the amino acid structure of the protein itself and not its regulation, so you now need only look at the sequence of the mutant gene's coding region to understand the primary genetic defect. To establish a library enriched for the mutant gene sequence and lacking all the extraneous information, you would first obtain mRNA from the cytoplasm of the patient's red blood cell precursors (**Figure 14.8a**). About 80 percent of the total mRNA in these red blood cells is from the α- and β-haemoglobin genes, so the mRNA preparation contains a much higher proportion of the sequence corresponding to the β-globin (*HBB*) gene than do the genomic sequences found in a cell's nuclear DNA.

The addition of reverse transcriptase to the total mRNA preparation—as well as ample amounts of the four deoxyribonucleotide triphosphates and primers to initiate synthesis—generates single-stranded cDNA bound to the mRNA template (**Figure 14.8b**). The primers used in this reaction would be oligo(dT)—single-stranded fragments of DNA containing about 20 T's in a row—that can bind through hybridization to the poly-A tail at the 3′ end of eukaryotic mRNAs and initiate polymerization of the first cDNA strand. Upon exposure to high temperature, the mRNA-cDNA hybrids separate, or "denature," into single

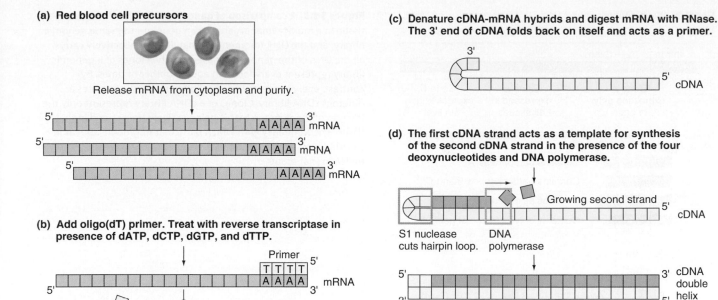

(a) Red blood cell precursors

Release mRNA from cytoplasm and purify.

5' [□□□□□□□□□□□□□□□|A|A|A|A| 3' mRNA

5' [□□□□□□□□□□□□□|A|A|A|A| 3' mRNA

5' [□□□□□□□□□□□□□|A|A|A|A| 3' mRNA

(b) Add oligo(dT) primer. Treat with reverse transcriptase in presence of dATP, dCTP, dGTP, and dTTP.

Primer 5'
[T|T|T|T]
5' [□□□□□□□□□|A|A|A|A| 3' mRNA

5' Growing cDNA
[T|T|T|T]
5' [□□□□□□□□□|A|A|A|A| 3' mRNA
Reverse transcriptase

3' [□□□□□□□□□□□□□□] 5' cDNA
5' [□□□□□□□□□□□□□□] 3' mRNA

(c) Denature cDNA-mRNA hybrids and digest mRNA with RNase. The 3' end of cDNA folds back on itself and acts as a primer.

3'
[□□□□□□□□□□□□□□□□□□□] 5' cDNA

(d) The first cDNA strand acts as a template for synthesis of the second cDNA strand in the presence of the four deoxynucleotides and DNA polymerase.

Growing second strand 5' cDNA

S1 nuclease cuts hairpin loop.   DNA polymerase

5' [□□□□□□□□□□□□□□□□□□] 3' cDNA double helix
3' [□□□□□□□□□□□□□□□□□□] 5'

(e) Insert cDNA into vector.

Figure 14.8 Converting RNA transcripts to cDNA. (a) Obtain mRNA from red blood cell precursors. **(b)** Create a hybrid cDNA-mRNA molecule using reverse transcriptase. **(c)** Heat the mixture to separate mRNA and cDNA strands, and then eliminate the mRNA transcript. The 3' end of the cDNA strands loops around and binds by chance to complementary nucleotides within the same strand, forming the primer for DNA polymerization. **(d)** Create a second cDNA strand complementary to the first. After the reaction is completed, the enzyme S1 nuclease is used to cleave the "hairpin loop" at one end. **(e)** Insert the newly created double-stranded DNA molecule into a vector for cloning.

strands. The addition of an RNase enzyme that digests the original RNA strands leaves intact single strands of cDNA (**Figure 14.8c**). Most of these fold back on themselves at their 3' end to form transient hairpin loops via base-pairing with random complementary nucleotides in nearby sequences in the same strand. These hairpin loops serve as primers for synthesis of the second DNA strand. Now the addition of DNA polymerase, in the presence of the requisite deoxyribonucleotide triphosphates, initiates the production of a second cDNA strand from the just synthesized single-stranded cDNA template (**Figure 14.8d**).

After using restriction enzymes and ligase to insert the double-stranded cDNA into a suitable vector (**Figure 14.8e**) and then transforming the vector-insert recombinants into appropriate host cells, you would have a library of double-stranded cDNA fragments, with the cDNA fragment in each individual clone corresponding to an mRNA molecule in the red blood cells that served as your sample. This library includes only the exons from that part of the genome that the red blood precursors were actively transcribing for translation into protein. For genes expressed infrequently or in very few tissues, you would have to screen many clones of a cDNA library to find the gene of interest. For highly expressed genes, such as the *HBB* gene, you would have to screen only a few clones in a red blood cell precursor library.

Genomic versus cDNA libraries

Figure 14.9 compares genomic and cDNA libraries. The main advantage of genomic libraries is that the genomic clones within them represent all regions of DNA equally and show what the intact genome looks like in the region of each clone. The chief advantage of cDNA libraries is that the cDNA clones reveal which parts of the genome contain the information used in making proteins in specific tissues, as determined from the prevalence of the mRNAs for the genes involved. To gain as much information as possible about a gene's structure and function, researchers rely on both types of libraries.

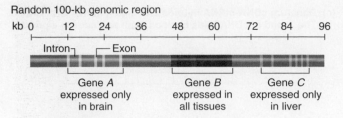

Random 100-kb genomic region

kb 0 12 24 36 48 60 72 84 96

Intron— —Exon

Gene A
expressed only
in brain

Gene B
expressed in
all tissues

Gene C
expressed only
in liver

Clones from a genomic library with 20-kb inserts that are homologous
to this region

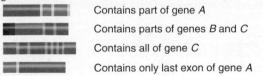

Contains part of gene A

Contains parts of genes B and C

Contains all of gene C

Contains only last exon of gene A

Clones from cDNA libraries

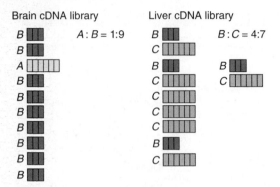

Brain cDNA library

A : B = 1:9

Liver cDNA library

B : C = 4:7

Figure 14.9 A comparison of genomic and cDNA libraries. Every tissue in a multicellular organism can generate the same genomic library, and the DNA fragments in that library collectively carry all the DNA of the genome. On average, the clones of a genomic library represent every locus an equal number of times. By contrast, every tissue in a multicellular organism generates a different cDNA library. Clones of a cDNA library represent only the fraction of the genome that is being actively transcribed in that tissue. The frequency with which particular fragments appear in a cDNA library is proportional to the level of the corresponding mRNA in that tissue.

> A genomic library is a collection of cloned DNA fragments, each of which is equivalent to a portion of an organism's genome. In an ideal library, every region of the genome is represented in an equal number of clones. By contrast, a cDNA library contains only sequences present in the mRNA transcripts of the particular source tissue.

14.4   Hybridization

Once you have collected the hundreds of thousands of human DNA fragments in a genomic or cDNA library, how do you find the gene you wish to study, the proverbial "needle in the haystack"? For example, how would you go about finding a genomic clone containing the *HBB* gene and its surrounding nontranscribed region? One way is to take advantage of hybridization—the natural propensity of complementary single-stranded molecules of DNA or RNA to base-pair and form stable double helixes. Once a β-globin cDNA clone is available, it can be denatured (separated into single strands), linked with a radioactive or fluorescent tag, and then used to probe a whole-genome library that is spread out as a series of colonies on one or more Petri plates. The tagged DNA probe will hybridize with denatured DNA from the genomic clones that contain a complementary sequence. After nonhybridizing probe is washed away, only the tiny number of β-globin-containing clones (among the hundreds of thousands in the library) are tagged by virtue of their hybridization to the probe. Individual cellular clones can then be retrieved from the library and put into culture to produce larger amounts of material in preparation for recovery of the purified DNA insert.

Hybridization has a single critical requirement: The region of complementarity between two single strands must be sufficiently long and accurate to produce a large enough number of hydrogen bonds to generate a cohesive force. Accuracy refers to the percentage of bases within the complementary regions that are actual complements of each other (C–G or A–T). The cohesive force formed by adding together large numbers of hydrogen bonds counteracts the thermal forces that tend to disrupt the double helix. If two single strands form hydrogen bonds between 15 or more contiguous base pairs, the combined force is sufficient. Hybridization can occur between any two single strands of nucleic acid: DNA/DNA, DNA/RNA, or RNA/RNA.

DNA probes are used to screen libraries

DNA probes are purified fragments of single-stranded DNA, 25 to several thousand nucleotides in length, that are subsequently labelled with a radioactive isotope (typically ^{32}P) or a fluorescent dye. DNA probes can be produced from previously cloned fragments of DNA, from purified fragments of DNA amplified by PCR (described in the next section), or from short single strands of chemically synthesized DNA.

The population of the European corn borer (a type of caterpillar) has steadily increased in Canada over the last decade. As a result of this population increase, potato farmers in Prince Edward Island, New Brunswick, and Nova Scotia have lost millions of dollars. These pests are particularly difficult to control since the larvae are able to "bore" into the tissues of the potato plant where they are protected from insecticides. One strategy used to control the corn borer population makes use of genes found in the common soil bacterium *Bacillus thuringiensis kurstaki* (abbreviated *Bt*). *Bt* normally uses these genes to protect itself from being eaten by the same caterpillars that cause so many problems for farmers. About a dozen genes in the *Bt* genome code for crystalline (CRY) polypeptides that function as specialized endotoxins. When a caterpillar ingests the bacteria, the CRY proteins bind to specific intestinal membrane sites and disrupt digestion, leading to the insect's rapid death. CRY binding is highly specific for proteins found only in the larvae of moths and butterflies and not in any vertebrate species. Even high doses of CRY proteins have no toxic or allergenic effects on birds, mammals, reptiles, or amphibians.

In the mid-1980s, agricultural molecular biologists realized they could use the newly developed tools of recombinant DNA technology to create genetically modified (GM) crops that would be resistant to insect infestation, without the need for pesticide applications. Based on the extensive safety record associated with whole-organism *Bt* use and a detailed understanding of the biochemical mechanism of CRY pesticidal action, researchers developed a strategy for creating plants that expressed a *cry* gene within their own cells as follows. They cloned a *cry* family gene named *cry1Ab* into a plasmid vector, cut out the insert with a restriction enzyme, and purified the insert. Next, they ligated a restriction fragment containing a plant gene intron (required to stabilize RNA transcripts) to the 5' end of the coding region. At the 5' end of this joined molecule, they added another restriction fragment containing a promoter from a plant virus; at the 3' end of the construct, they attached a special plant transcription termination signal sequence. They then inserted this four-part *cry*-gene construct into a bacterial plasmid vector resembling the one shown in Figure 14.7, which was used to transform a bacterial culture. Finally, they identified bacterial clones containing the construct based on antibiotic resistance and the absence of *lacZ* production and used these clones to produce a purified DNA insert containing the *cry* gene and associated genetic elements (**Figure A**).

Cells from many different plant species can be grown in Petri dishes where DNA transformation with the recombinant *cry* insert can easily occur. Transformed cells are identified, isolated, and then grown under conditions that allow them to regenerate whole plants. Genetically modified plants containing the *cry* gene were grown commercially for the first time in 1996 (**Figure B**). By 2004, *cry* genes had been used to create insect-resistant canola, cotton, corn, papaya, potato, rice, soybean, squash, sugar beet, tomato, and wheat. These and other genetically modified crops are being grown on over

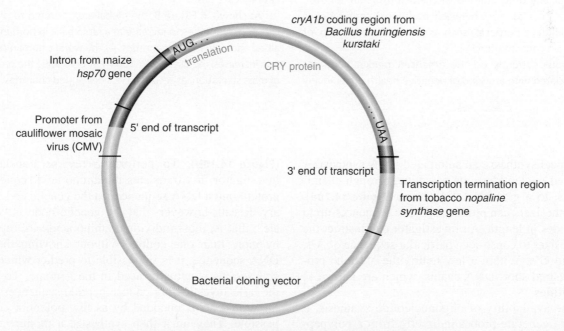

Figure A DNA construct with recombinant gene that can express CRY protein in plants.

(Continued)

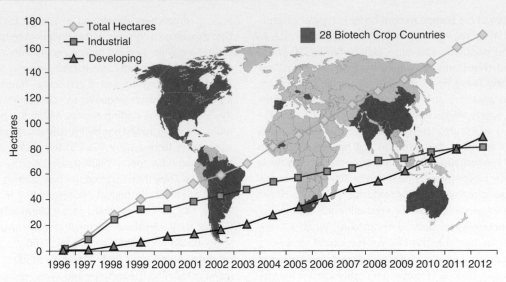

Figure B Global area of biotech crops (1996–2012).

160 million hectares of land around the world, in both industrialized and developing countries.

Before genetically modified plants can be grown commercially, they must pass stringent tests for efficacy and safety on a case-by-case basis. In the case of corn engineered to express the *cry1Ab* gene, the CRY protein product was detected at a level of three parts per 10 million (0.000 000 3) in corn and is nonexistent in extracted corn syrup used for soft-drink production. As expected, no difference could be detected between the GM and non-GM variety in amino acids, vitamins, carbohydrates, or any other nutritional characteristic. (In contrast, large differences *do* exist between traditional corn varieties bred for different purposes, such as pig feed, corn syrup, or direct human consumption.)

GM foods currently on the American market have not been associated with any kind of negative health effect in any person. This does not mean that all future GM plants will be without ill effect and risk free, but risk assessment only makes sense in comparison to the substitute foods that people would eat in the absence of a particular GM product. In some situations, the risk exists for genetically engineered traits to migrate unintentionally into wild plants. Indeed, most scientists take this risk more seriously than alleged health risks. With a scientifically informed regulatory process, the risk of significant eco-harm can be assessed up front and included in the decision to implement, redesign, or reject a particular GM technology on a case-by-case basis.

As shown in Figure B, the global area devoted to planting GM crops continues to increase at a rapid pace in both industrialized and developing countries. As the world's human population increases, the use of GM crops may help solve the problems of mass starvation, especially in less developed countries.

In chemical synthesis, an automated DNA synthesizer adds specified nucleotides, one at a time, through chemical reactions, to a growing DNA strand (**Figure 14.10a**). Modern synthesizers can produce specific sequences up to 200 nucleotides in length. An investigator can instruct the DNA synthesizer to construct a particular sequence of A's, T's, C's, and G's. Within a few hours, the machine produces the desired short DNA chains, which are known as **oligonucleotides**.

With the availability of oligonucleotide synthesis, it is possible to generate probes indirectly from a polypeptide sequence whose corresponding gene-coding sequence is unknown, rather than directly from a known DNA sequence. This process is known as *reverse translation*

(**Figure 14.10b**). To perform a reverse translation, an investigator first translates the amino acid sequence of a protein into a DNA sequence via the genetic code dictionary. Recall, however, that the genetic code is "degenerate"; that is, most individual amino acids are represented by more than one codon. Without knowing the coding DNA sequence, it is impossible to predict which of several codons is actually used in the genome. To simplify the task, investigators choose peptide sequences containing amino acids encoded by as few potential codons as possible. They must then synthesize a mixture of oligonucleotides containing all possible combinations of codons for each amino acid. This is no problem for an automated DNA synthesizer: An investigator can direct the machine

(a)

(b) Synthesizing DNA probes based on reverse translation

Protein sequence	Glu	Asp	Met	Trp	Tyr
↓					
Degenerate coding sequences	GAA	GAT	ATG	TGG	TAT
	GAG	GAC			TAC

Sequences that must be present in the probe

GAAGATATGTGGTAT
GAGGATATGTGGTAT
GAAGACATGTGGTAT
GAGGACATGTGGTAT
GAAGATATGTGGTAC
GAGGATATGTGGTAC
GAAGACATGTGGTAC
GAGGACATGTGGTAC

Figure 14.10 How to make oligonucleotide probes for screening a library. (a) A DNA synthesizer from the lab of Dr. Chris Wilds (Department of Chemistry and Biochemistry, Concordia University). A synthesizer is a machine that automates the addition of specified nucleotides to growing DNA chains, known as oligonucleotides. The bottles contain solutions of A, T, C, and G, along with reagents used in the reactions. **(b)** Reverse translation. An amino acid sequence can be "reverse translated" into a degenerate DNA sequence, which can be programmed into a DNA synthesizer to create a set of oligonucleotides that must include the one present in the actual genomic DNA.

to add in a defined mixture of nucleotides (e.g., A and G) at each ambiguous position in the oligonucleotide. With this indirect method of obtaining a DNA probe, researchers can locate and clone genes even if they have only partial coding information based on the proteins the genes encode.

Hybridization can occur between single strands that are not completely complementary, including related sequences from different species. In general, two single DNA strands that are longer than 50–100 bp will hybridize so long as the extent of their complementarity is more than 80 percent, even though mismatches may appear throughout the resulting hybrid molecule. Imperfect hybrids are less stable than perfect ones, but geneticists can exploit this difference in stability to evaluate the similarity between molecules from two different sources. Hybridization, for example, occurs between the mouse and human genes for the cystic fibrosis protein. Researchers can thus use the human genes to identify and isolate the corresponding mouse sequences and then use these sequences to develop a mouse carrying a defective cystic fibrosis gene. Such a mouse provides a model for cystic fibrosis in a species that, unlike humans, can be used in experimental analysis.

Southern blots allow visualization of rare DNA fragments in complex samples

Researchers use hybridization to screen a library of thousands of clones for particular ones complementary to specific probes. Hybridization with a cloned probe can also provide information about similar DNA regions in a whole-genome sample. The protocol for accomplishing this task combines gel electrophoresis (review Figure 14.4) with the hybridization of DNA probes to DNA targets immobilized on nitrocellulose paper.

Suppose you had a clone of a gene called *H2K* from the mouse major histocompatibility complex (MHC). The *H2K* gene plays a critical role in the body's ability to mount an immune response to foreign cells. You want to know whether other genes in the mouse genome are similar to *H2K* and, by extrapolation, also play a role in the immune response. To get an estimate of the number of *H2K*-like genes that exist in the genome, you could turn to a hybridization technique called the **Southern blot**, named for Edward Southern, the British scientist who developed it. **Figure 14.11** illustrates the details of the technique.

Southern blotting can identify individual *H2K*-like DNA sequences within the uncloned expanse of DNA present in a mammalian genome. Cutting the total genomic DNA with *Eco*RI produces about 700 000 different fragments. When you separate these fragments by gel electrophoresis and stain them with ethidium bromide, all you see is a smear, because it is impossible to distinguish 700 000 fragments spread over a distance of some 10 cm (Figure 14.11). But you can blot the smear of fragments to a nitrocellulose filter paper and probe the resulting blot with a labelled *H2K* clone, which picks out the bands containing the *H2K*-like gene sequences. The result shown in Figure 14.11 is a pattern of approximately two dozen fragments that constitute a series of related MHC genes within the mouse genome. The Southern blot thus makes it possible to start with a very complex mixture and identify the small number of fragments among hundreds of thousands within a whole genome that are related to your original clone.

Southern blotting can also determine the location of one cloned sequence (such as a 1.4-kb human β-globin cDNA sequence from a plasmid vector) within a larger cloned sequence (such as a 30-kb genomic clone containing the β-globin locus and surrounding genome). Suppose you want to use these two clones to discover the location of the

Southern Blot Analysis

Genomic DNA was purified from the tissues of seven mice, and each sample was subjected to digestion with the restriction enzyme *Eco*RI. Digested samples were separated by electrophoresis in an agarose gel, as illustrated in Figure 14.4.

Stain with ethidium bromide to visualize total genomic DNA under UV illumination.

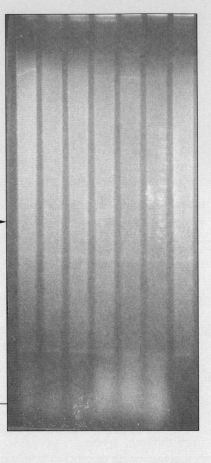

Next you place the gel in a strongly alkaline solution to denature the DNA, and then in a neutralizing solution. You now cover the gel containing the separated DNA restriction fragments with a piece of nitrocellulose filter paper. On top of the filter paper, you place a stack of paper towels, and beneath the gel, a sponge saturated with buffer. Within this setup, the dry paper towels act as a blotter, pulling liquid from the buffer-saturated sponge, through the gel, the nitrocellulose filter, and into the towels themselves. The large DNA molecules do not pass through the filter into the paper towels. Instead, they become trapped at points directly above their locations in the gel, forming a Southern blot: the nitrocellulose filter containing DNA fragments in a pattern that is a replica of their migration pattern in the gel.

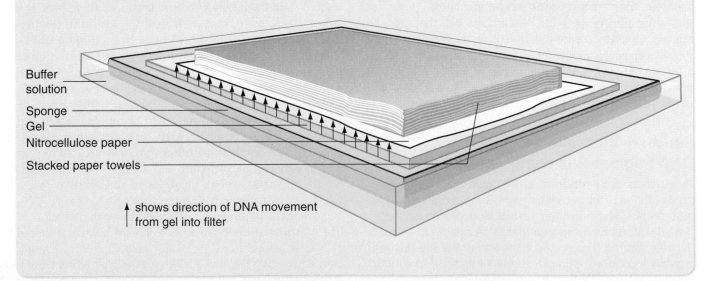

Buffer solution

Sponge

Gel

Nitrocellulose paper

Stacked paper towels

↑ shows direction of DNA movement from gel into filter

The Southern blot is removed from the blotting apparatus, incubated with NaOH to denature the transferred DNA, and then baked and exposed to UV radiation to attach the single-stranded DNA to the blot.

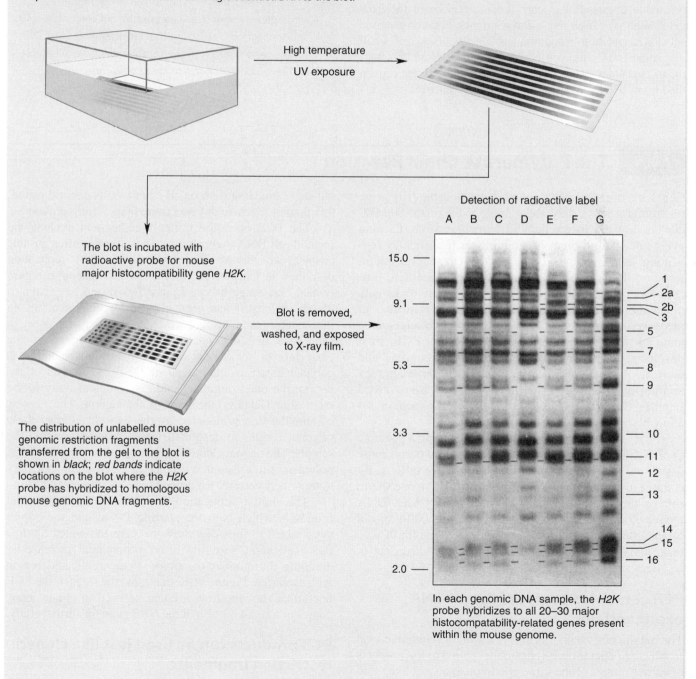

High temperature

UV exposure

Detection of radioactive label

The blot is incubated with radioactive probe for mouse major histocompatibility gene *H2K*.

Blot is removed, washed, and exposed to X-ray film.

The distribution of unlabelled mouse genomic restriction fragments transferred from the gel to the blot is shown in *black*; *red bands* indicate locations on the blot where the *H2K* probe has hybridized to homologous mouse genomic DNA fragments.

A B C D E F G

15.0

9.1

5.3

3.3

2.0

1
2a
2b
3
5
7
8
9
10
11
12
13
14
15
16

In each genomic DNA sample, the *H2K* probe hybridizes to all 20–30 major histocompatability-related genes present within the mouse genome.

gene within the genomic clone as well as to learn which parts of the full gene are exons and which parts are introns. To answer both of these questions in a very straightforward fashion, you would turn to Southern blotting.

First you would use gel electrophoresis of restriction-enzyme-digested DNA from the 30-kb clone to construct a restriction map. Next, you would transfer the restriction fragments from the gel onto a filter paper and probe the filter by hybridization to the labelled β-globin cDNA clone. A system for detecting the location and intensity of the label, which can be either radioactive or fluorescent, then shows the precise restriction fragments that carry coding regions of the *HBB* gene. With very high-resolution restriction maps of genomic subclones and high-resolution gel electrophoresis of small restriction fragments, you could distinguish restriction fragments containing exons from the other fragments that contain the introns of the *HBB* gene or a flanking sequence.

- Hybridization is the process through which complementary DNA strands base-pair to form stable double-helical structures. Hybridization occurs even between strands that have a small number of mismatches. A purified DNA fragment can be tagged and used as a probe to screen genomic or cDNA libraries (derived from any species) for clones containing related DNA sequences. DNA probes can be produced from previously cloned fragments or from synthesized oligonucleotides.

- In the Southern blot technique, restriction fragments from a complex genome are separated by gel electrophoresis and transferred by blotting to a nitrocellulose filter. DNA sequences of interest in the complex genome are identified by hybridization to a tagged DNA probe.

14.5 The Polymerase Chain Reaction

Genes are rare targets in a complex genome: the *HBB* gene, for example, spans only about 1400 of the 3 000 000 000 nucleotide pairs in the haploid human genome. Cloning overcomes the problem of studying such rarities by replicating large amounts of a specific DNA fragment in isolation. But cloning is a tedious, labour-intensive process. Once a sequence is known, or even partially known, molecular biologists now use an alternative method to recover versions of the same sequence from any source material: the **polymerase chain reaction**, or **PCR**. First developed in 1985, PCR is faster, less expensive, and more flexible in application than cloning. From a complex mixture of DNA—like that present in a person's blood sample—PCR can isolate a purified DNA fragment in just a few hours.

PCR is also extremely efficient. In creating a genomic or cDNA library, a large number of cells from one or more tissues are necessary as the source of DNA or mRNA. By contrast, the single copy of a genome present in one sperm cell or the minute amount of severely degraded DNA recovered from the bone marrow of a 30 000-year-old Neanderthal skeleton provides enough material for PCR to make a billion or more copies of a target DNA sequence in an afternoon.

PCR generates copies of target DNA exponentially

The polymerase chain reaction is a kind of reiterative loop in which an operation is repeatedly applied to the products of earlier rounds of the same operation. You can liken it to the operation of an imaginary, generously paying automatic slot machine. You start the machine by inserting a quarter, at which point the handle cranks, and the machine pays out two quarters; it then reinserts those two coins, cranks, and produces four quarters; reinserts the four, cranks, and spits out eight coins, and so on. By the twenty-second round, this fantasy machine delivers more than 4 million quarters.

The PCR operation brings together and exploits the method of DNA hybridization described earlier in this chapter and the essential features of DNA replication described in Chapter 5. Once a specific genomic region (which may range in size from a few dozen base pairs to 25 kb in length) has been chosen for amplification, an investigator uses prior knowledge of the sequence to synthesize two oligonucleotides that correspond to the two ends of the target region. One oligonucleotide is complementary to one strand of DNA at one end of the region; the other oligonucleotide is complementary to the other strand at the other end of the region. The process of amplification is initiated by the hybridization of these oligonucleotides to denatured DNA molecules within the sample. The oligonucleotides act as primers, directing DNA polymerase to create new strands of DNA complementary to those between the two primed sections (**Figure 14.12**).

This initial replication is followed by subsequent rounds in which both the starting DNA, and the copies synthesized in previous steps, become templates for further replication, resulting in an exponential increase by doubling the number of copies of the replicated region with each step. Figure 14.12 diagrams the steps of the PCR operation, showing how it could be used to obtain many copies of a small portion of the *HBB* gene for further study.

PCR products can be used just like cloned restriction fragments

When properly executed, PCR provides all of the highly enriched DNA you could want for unambiguous analyses of many types. PCR products can be labelled to produce hybridization probes or can be sequenced (as described in the next section) to determine the exact genetic information

Polymerase Chain Reaction

Suppose you are a physician and wish to understand the molecular details of an *HBB* gene mutation that results in a novel form of anaemia in one of your patients. To characterize the potentially novel allele, you would turn to PCR. You begin by preparing a small amount of genomic DNA from skin, blood, or other tissue that is easy to obtain from the patient suffering from the novel anaemia. You then synthesize two specific oligonucleotide primers, each a short single-stranded chain of 16–26 nucleotides, whose sequence is chosen from the already known sequence of the wild-type β-globin allele. One of these oligonucleotides (arbitrarily called the "left primer" in the diagram) is equivalent in sequence to a section of DNA along a 5′ strand adjacent to the target region (coloured *blue* in the diagram). The second oligonucleotide, the "right primer," is equivalent to a sequence on the opposite adjacent 5′ strand. As you can see, the target DNA amplified by PCR is that stretch of the genome lying between the two primers.

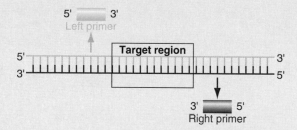

Next you put the patient's genomic DNA in a test tube along with the specially prepared primers, a solution of the four deoxynucleotides, and *Taq* DNA polymerase, a specialized polymerase obtained from *Thermus aquaticus* bacteria living in hot springs. This specialized DNA polymerase remains active at the high temperatures employed during the PCR protocol. Now place the test tube with these components in a machine called a thermal cycler, which repeatedly changes the temperature of incubation according to a preset program with three phases.

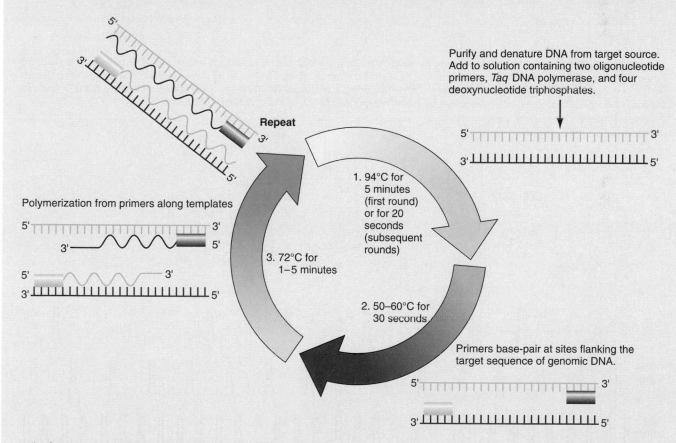

In the first round of a typical program, the cycler follows these steps:

1. The cycler heats the solution to 94°C for 5 minutes. At this temperature, the target DNA separates into single strands.
2. The temperature is next lowered to 50–60°C for 30 seconds to allow the primers to base-pair with complementary sequences in the single-stranded genomic DNA. Specifics of both temperature and timing within these ranges depend on the length and GC:AT ratio of the primer sequences.

(Continued)

3. The thermal cycler then raises the temperature to 72°C, the temperature at which the *Taq* polymerase functions best. Holding the temperature at 72°C for 1–5 minutes (depending on the length of the target sequence) allows DNA polymerization to proceed. At the end of this period, with the completion of DNA synthesis, the first round of PCR is over, and the amount of target DNA has doubled.

To start the next round, the cycler again raises the temperature to 94°C (but this time for only about 20 seconds) to denature the short stretches of DNA consisting of one of the original strands of genomic DNA and a newly synthesized complementary strand initiated by a primer. These short single strands become the templates for the second round of replication because the synthesized primers are able to base-pair to them.

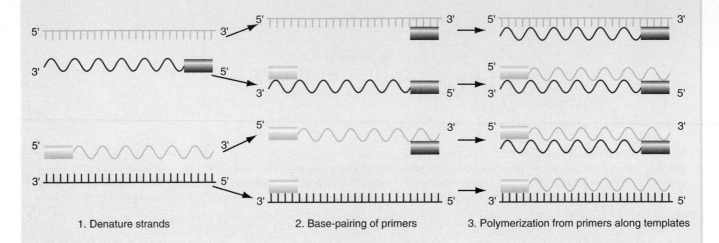

1. Denature strands   2. Base-pairing of primers   3. Polymerization from primers along templates

The machine repeats the cycle again and again, generating an exponential increase in the amount of target sequence: 22 repetitions produce over a million copies of the target sequence; 32 repetitions over a billion. The length of the accumulating DNA strands becomes fixed at the length of the DNA between the 5′ ends of the two primers, as shown. This is because, beginning with round 3, the 5′ end of a majority of templates is defined by a primer that has been incorporated in one strand of a PCR product.

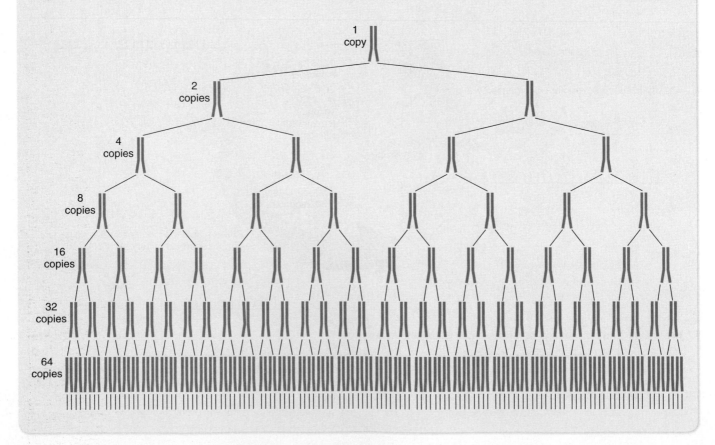

they contain. Because the products are obtained without cloning, it is possible to amplify and learn the sequence of a specific DNA segment in a very short time. In fact, in checking for a particular haemoglobin mutation, one could start with a blood sample and determine a DNA sequence within two days.

As an analytical tool, PCR has advantages over cloning. First, it provides the ultimate in sensitivity: The minimum input is a single DNA molecule. Second, as we have seen, it is very fast, requiring no more than a few hours to generate enough amplified DNA for analysis. PCR is nevertheless unsuitable in certain situations. For example, since the protocol only copies DNA fragments up to 25 kb in length, it cannot amplify larger regions of interest. Furthermore, since the synthesis of PCR primers depends on sequence information from the vicinity of the target region, the protocol cannot serve as the starting point for the analysis of genes or genomic regions that have not yet been cloned and sequenced.

PCR has many uses

PCR is one of the most powerful techniques in molecular biology. Its originator, Kary Mullis, received the 1993 Nobel Prize in Chemistry for his 1985 invention of this tool for genetic analysis. PCR has made molecular analysis an essential component of genotype detection and gene mapping (Chapter 15). In addition, PCR has revolutionized evolutionary studies, enabling researchers to analyze sequences from both living and extinct organisms, and to determine the relatedness between these organisms with greater accuracy than ever before. The study of gene diversity at the nucleotide level in populations has been facilitated tremendously by PCR, and it has greatly simplified the process of monitoring genetic changes in a group over time (see Chapter 12 for the details of population genetics). Finally, PCR has helped bring molecular genetics to many fields outside of traditional genetics. The following example of its use in diagnosing infectious disease provides an inkling of its impact on medicine.

HIV, the virus associated with AIDS, gains entry to a person's body through the bloodstream or lymphatic system, then docks at specific membrane receptors on a few types of white blood cells, fuses with the cell membrane, and releases its RNA chromosome, along with several copies of reverse transcriptase, into the cell. Once inside the cell, the reverse transcriptase copies the RNA to cDNA. The double-stranded DNA copy of the viral genome then integrates itself into the host genome where it can lie latent for up to 10 years. When activated, it directs the cellular machinery to make more viral particles.

Standard tests for HIV detect antibodies to the virus, but it may take several months for the antibodies produced by an infected person's immune system to reach levels that are measurable in the blood. Then, in another few months, when ongoing viral activity inside many types of circulating white blood cells subsides, most of the antibodies may disappear from the circulation. The reason is that once the viral particles have entered the latent state, they are literally in hiding (inside chromosomal DNA) and able to avoid detection by the immune system.

With PCR, it is possible to detect small amounts of virus circulating in the blood or lymph very soon after infection, before antibody production is in full swing. PCR can also detect viral DNA incorporated in the genome of any cell, picking up as few as 1–10 copies of viral DNA per million cells. Thus, with PCR, it becomes possible to diagnose and begin treating HIV infection during the critical period before antibodies reach measurable levels. It also becomes possible to follow the progress of each person's HIV infection and thus to tailor therapies accordingly, using a large dose of certain drugs to combat a large amount of viral activity, but small doses of perhaps other drugs to prevent a small number of cells from emerging from latency.

- PCR is a powerful tool used to isolate and make large quantities of a defined DNA fragment from a complex genome. PCR takes advantage of hybridization and synthetic oligonucleotides. Its starting material can be as small as a single cell. Because amplification is exponential, a single DNA molecule can be copied into trillions of copies in a single day.

- Once a reference DNA sequence has been established, PCR can be used to pick out variant forms of that sequence in any DNA sample. PCR amplifies any sequence between the primers used, even if the order of bases is slightly different, and it allows screening of thousands of samples very quickly.

14.6 DNA Sequence Analysis

Although scientists have known since the 1953 discovery of the double helix that genes and genomes are defined by sequences of base pairs, it was not until the early 1970s that the first specific sequences of genomic DNA were determined directly by chemical methods. The DNA sequences of the library fragments representing an entire genome provide a staggering amount of practical information. Restriction enzyme recognition sites are immediately visible. Open reading frames of genes are recognized and translated amino acid sequences are inferred. These primary polypeptide structures provide information about possible protein structure and function. Furthermore, the comparison of genomic and cDNA sequences immediately shows how a gene is divided into exons and introns and may suggest whether alternative splicing of the gene's primary transcript occurs.

The first DNA sequence to be determined was a 24-base-pair region of the *E. coli* genome that binds to the *lac* repressor:

TGGAATTGTGAGCGGATAACAATT

ACCTTAACACTCGCCTATTGTTAA

It was "a laborious process that took several years," according to Walter Gilbert. The frustration of that experience galvanized Gilbert and his colleague Alan Maxam to invent a general-purpose sequencing method based on the chemical cleavage of DNA molecules at specific nucleotide types. A second technology, developed by Fred Sanger during the same mid-1970s time frame, was based on the enzymatic extension of DNA strands to a defined terminating base. Gilbert and Sanger both won the Nobel Prize for their contribution to DNA sequencing technology. Their techniques have a similar throughput of 500–700 bases obtained in each several-day-long experiment, and a similar accuracy, which approaches 99.9 percent. However, only the Sanger technique was readily amenable to automation.

Sanger sequencing generates sets of nested fragments separated by size

There are two steps to the Sanger method of sequencing. The first step is the generation of a complete series of single-stranded subfragments complementary to a portion of the DNA template under analysis (although both strands of a DNA fragment are present in a typical DNA sample, only one is used as a template for sequencing). Each subfragment differs in length by a single nucleotide from the preceding and succeeding fragments; the graduated set of fragments is known as a *nested array*. A critical feature of the subfragments is that each one is distinguishable according to its terminal 3′ base. Thus, each subfragment has two defining attributes—relative length and one of four possible terminating nucleotides.

In the second step of the sequencing process, biologists analyze the mixture of DNA subfragments through polyacrylamide gel electrophoresis, under conditions that allow the separation of DNA molecules differing in length by just a single nucleotide.

The original Sanger sequencing procedure (illustrated in **Figure 14.13**) begins with the denaturation of the DNA to be sequenced. The single strands are then mixed in

Sanger Sequencing

Begin by mixing the purified, denatured DNA with a labelled oligonucleotide primer that is complementary to a particular site on one strand of the cloned insert. Add DNA polymerase and the four deoxynucleotide triphosphates.

Next divide the mixture into four aliquots and, into each one, add a small amount of a single chain-terminating dideoxyribonucleotide triphosphate abbreviated as "dideoxynucleotide triphosphate" or simply "ddNTP." One aliquot, for example, contains the deoxynucleotides A, T, C, and G spiked with the dideoxynucleotide analogue of T. Polymerization from the primer strand continues until, by chance, the dideoxynucleotide is incorporated.

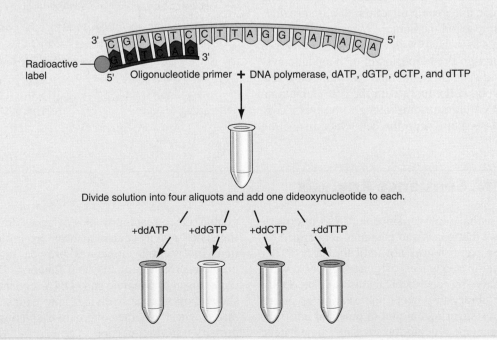

Because a dideoxynucleotide analogue has no oxygen at the 3′ position in the sugar, its incorporation prevents the further addition of nucleotides to the strand and thus terminates a growing chain wherever it becomes incorporated in place of an actual deoxynucleotide. The aliquot that has the dideoxy form of thymidine, for example, will generate a population of DNA molecules that terminate at each of the thymidines in the original template strand under analysis.

Then use electrophoresis on a polyacrylamide gel to separate the fragments in each of the four aliquots according to size. The resolution of the gel is such that you can distinguish DNA molecules that differ in length by only a single base. The appearance of a DNA fragment

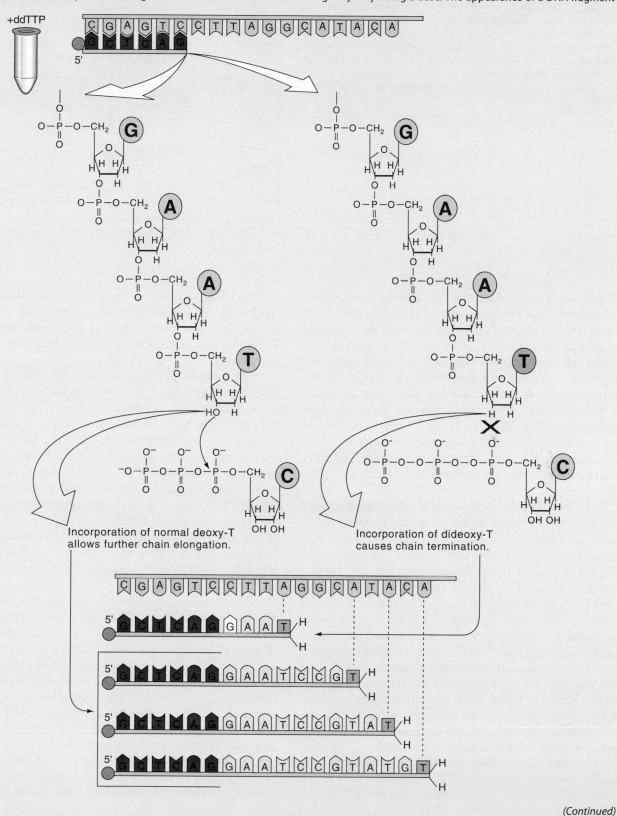

Incorporation of normal deoxy-T allows further chain elongation.

Incorporation of dideoxy-T causes chain termination.

(Continued)

of a particular length demonstrates the presence of a particular nucleotide at that position in the strand.

Suppose, for example, that the aliquot polymerized in the presence of dideoxythymidine shows fragments 32, 35, and 39 bases in length. These fragments indicate that thymidine is present at those positions in the strand of nucleotides. In practice, one does not independently determine the exact lengths of each fragment. Instead, one starts at the bottom of the gel, looks at which of the four lanes has a band in it, records that base, then moves up one position and determines which lane has the next band, and so on. In this way, it is possible to read several hundred bases from a single set of reactions.

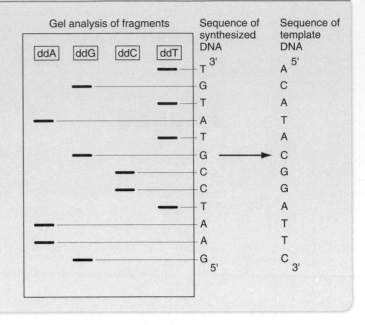

solution with DNA polymerase, the four deoxynucleotide triphosphates, and a radioactively labelled oligonucleotide primer complementary to DNA adjacent to the 3′ end of the template strand under analysis. The solution is next divided into four aliquots. To each one, an investigator adds a small amount of a single type of a nucleotide triphosphate lacking the 3′-hydroxyl group that is critical for the formation of the phosphodiester bonds that lead to chain extension (review Figure 5.20); this nucleotide analogue is called a **dideoxyribonucleotide** (or **dideoxynucleotide**), and it comes in four forms: **ddTTP**, **ddATP**, **ddGTP**, or **ddCTP** (abbreviated even further as ddT, ddA, ddG, and ddC).

In each of the four aliquots, the oligonucleotide primer hybridizes at the same location on the template DNA strand. As a primer, it will supply a free 3′ end for DNA chain extension by DNA polymerase. The polymerase adds nucleotides to the growing strand that are complementary to those of the sample's template strand (i.e., the actual DNA strand under analysis). The addition of nucleotides continues until, by chance, a dideoxynucleotide is incorporated instead of a normal nucleotide. The absence of a 3′-hydroxyl group in the dideoxynucleotide prevents the DNA polymerase from forming a phosphodiester bond with any other nucleotide, ending the polymerization for that new strand of DNA. Next, after allowing enough time for the polymerization of all molecules to reach completion, an investigator releases the templates from the newly synthesized strands by denaturing the DNA at high temperature. Each sample tube now holds a whole collection of single-stranded radioactive DNA chains as well as the nonradioactive single strands of the template DNA. The

lengths of the radioactive chains reflect the distance from the 5′ end of the oligonucleotide primer to the position in the sequence at which the specific dideoxynucleotide present in that particular tube was incorporated into the growing chain.

The samples in the four tubes are now electrophoresed in adjacent lanes on a polyacrylamide gel, and the gel is subjected to a system that detects the presence of the radioactive label. Because the template strands are not labelled, they do not show up. The investigator reads out the sequence of the radioactive strand by starting at the bottom and moving up, determining which lane carries each subsequent band in the ascending series, as shown in Figure 14.13. As you ascend, each band represents a chain that is one nucleotide longer than the chain of the band below. Once the sequence of the newly synthesized DNA is known, it is a simple matter to convert this sequence into the complementary sequence of the template strand under analysis.

To automate the DNA sequencing process, molecular geneticists changed the method of labelling the newly formed complementary DNA strands. Instead of placing a single radioactive label on the primer oligonucleotide, they labelled each of the four chain-terminating dideoxynucleotides with a different colour fluorescent dye. As a result, instead of four separate reactions, all four dideoxynucleotides could be combined in a single reaction mixture that could be analyzed in a single lane on a gel (**Figure 14.14**). A DNA sequencing machine follows the DNA chains of each length in the ascending series through a special detector that can distinguish the different colours associated with each

(a) Automated sequencing

1. Generate nested array of fragments; each with a fluorescent label corresponding to the terminating 3' base.

2. Fragments separated by electrophoresis in a single vertical gel lane.

3. As migrating fragments pass through the scanning laser, they fluoresce. A fluorescent detector records the colour order of the passing bands. That order is translated into sequence data by base-calling software.

(b) Fluorescent bands in a sequencing gel

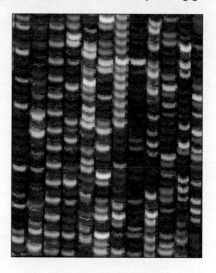

(c) Chromatogram and inferred sequence

CTNGCTTTGGAGAAAGGCTCCATTGNCAATCAAGACACACAGAGGTGTCCTCTTTTTCCCCTGGTCAGCGNCCAGGTACATNGCACCAAGGCTGCGTAGTGAACTTGNCACCAGNCCATGGAC
(For ease of readability, the *yellow* colour of G nucleotides has been replaced by *black*.)

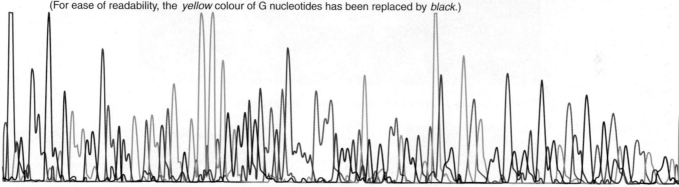

Figure 14.14 Automated sequencing. (a) For automated sequencing, the Sanger protocol is performed with all four fluorescently labelled terminating nucleotides present in a single reaction. At completion of the reaction, DNA fragments terminating at every base in the sequence are present and colour-coded by the identity of the terminating base. Separation by gel electrophoresis is next. As each fragment moves past a laser beam, the colour of the terminal base is detected and recorded. **(b)** Image of a sequencing gel. Each lane displays the sequence obtained with a separate DNA sample and primer. **(c)** The raw data are displayed as peaks of four different colours, called a chromatogram. The base-calling software produces a text sequence of the newly synthesized, complementary DNA strand from left to right, which corresponds to the 5'-to-3' direction. The machine records any ambiguity in the base call as an "N"; ambiguity may be due to a mixture of alleles in the starting sample or technical failure. In large-scale sequencing projects, each genomic region is sequenced multiple times on both strands, which allows resolution of most ambiguities.

terminating dideoxynucleotide. Thus, in each lane of a gel, it is possible to run a different DNA fragment for complete sequence analysis.

The Sanger method remains a very useful technique that is well suited for the small-scale sequencing needs of most molecular biology labs. However, using the Sanger method for genome-scale projects can be prohibitively expensive. In response to these needs, faster and more cost-efficient sequencing methods have been developed. These are often referred to as "next-generation" sequencing technologies. Remarkably, these methods allow researchers to sequence an entire genome in a matter of days, and at a miniscule fraction of the cost of the Sanger method (see

the **Fast Forward** box **"Next-Generation Sequencing"**). These methods will be discussed in Chapter 19, Sequencing and Analyzing Genomes.

> Sanger sequencing begins with the hybridization of a DNA primer to the template DNA under analysis. DNA polymerase extends the primer until, by chance, a particular dideoxynucleotide is incorporated, stopping the polymerization. The result is a nested set of DNA fragments tagged according to their terminating base.

The human genome project was completed in 2001 and took over a decade to complete (at a cost of $3 billion dollars). As part of the effort, hundreds of Sanger sequencing machines were put to work at research centres around the world (**Figure A**). While the publication of the human genome clearly stood out as a landmark in the history of genetics, in other respects it was just a starting point. This is because the sequence data were derived from only a small handful of individuals (i.e., it was a "composite" sequence). To truly understand genetic variation and its role in determining phenotype, thousands upon thousands of personal genome sequences will need to be analyzed. The need to sequence at this scale has precipitated the development of technologies that work faster and are less expensive than the Sanger method. These "next-generation" methods, exemplified by the Roche 454 pyrosequencer (**Figure B**), produce as much data in a day as hundreds of Sanger-type sequencers. We will learn more about how pyrosequencers work, as well as the ramifications of next-generation sequencing technologies in Chapter 19, Sequencing and Analyzing Genomes.

Figure A Research centre with Sanger sequencing machines.

Figure B The Roche 454 pyrosequencer.

14.7 The Haemoglobin Genes: A Comprehensive Example

Geneticists have used the tools of biotechnology and bioinformatics to analyze the clusters of related genes that make up the α- and β-globin loci. Fundamental insights from these studies have helped explain how DNA encodes the instructions for the development of the haemoglobin system, including the changes in globin expression observed during normal development. The studies have also clarified how the globin genes evolved and how a large number of different mutations produce the phenotypes associated with globin-related disorders. In this section,

we will explore the haemoglobin system as revealed by DNA technology.

Haemoglobin genes occur in two clusters on two chromosomes

The α-globin (*HBA*) gene cluster contains five functional genes and spans about 28 kb on chromosome 16 (**Figure 14.15a**). All the genes in the α-gene cluster are oriented in the same direction; that is, they all use the same strand of DNA as the template for transcription. Moving in the 5′-to-3′ direction along the RNA-like strand, the α or α-like genes appear in the order *HBZ, HBM, HBA2, HBA1,* and *HBQ1*. The genes in the β-gene cluster, like those in the α-gene cluster, all have the same orientation. The β-globin (*HBB*) cluster covers 45 kb on chromosome 11 and also contains five functional genes in the order *HBE, HBG2, HBG1, HBD,* and *HBB* (**Figure 14.15b**). Geneticists refer to the chromosomal region carrying all of the *HBA*-like genes as the **α-globin locus** and the region containing the *HBB*-like genes as the **β-globin locus**. Note that the term *locus* signifies a location on a chromosome; that location may be as small as a single nucleotide or as large as a cluster of related genes.

Correlation of globin gene order with timing of expression

The linear organization of the genes in the α- and β-gene clusters reflects the order in which they are expressed during development. For the α-like chains, that temporal order is *HBZ* during the first five weeks of embryonic life, followed by *HBA2* and *HBA1* during fetal and adult life. For the β-like chains, the order is *HBE* during the first five weeks of embryonic life; then *HBG2* and *HBG1* during fetal life; and finally, within a few months of birth, mostly *HBB* but also some δ chains (see Figures 14.1 and 14.15).

The fact that the order of genes on the chromosomes parallels the order of their expression during development suggests that whatever mechanism turns these genes on and off takes advantage of their relative positions. We now understand what that mechanism is: A *locus control region* (or *LCR*) associated with specialized DNA-binding proteins at the 5′ end of each locus works its way down the locus, bending the chromatin back on itself to turn genes on and off in order.

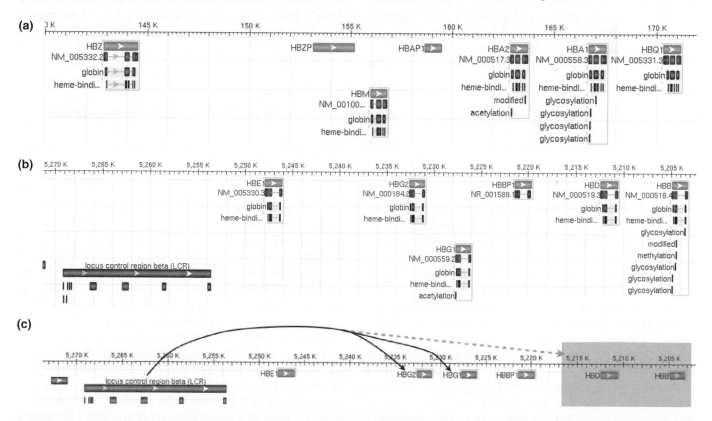

Figure 14.15 The genes for the polypeptide components of human haemoglobin are located in two genomic clusters on two different chromosomes. (a) Schematic representation of the *HBA* gene cluster on chromosome 16. The *HBA* gene homologues are indicated with *green* boxes. Transcripts are shown below active genes by a series of boxes, representing exons, connected with lines representing introns. The translated portions of each transcript are indicated in *red*, and the translation products are represented below transcripts in *black*. Sites of posttranslation modification, including haem-binding, acetylation, and glycosylation, are also shown. The cluster contains five functional genes and two pseudogenes (designated with the appended letter P). **(b)** Schematic representation of the *HBB* gene cluster on chromosome 11; this cluster has five functional genes and one pseudogene. (The pseudogene HBBP is actually transcribed, but the transcript is not translated.) Upstream from both the *HBA* and *HBB* gene clusters lie the locus control regions (LCR) (which is only shown for the β-globin locus here). **(c)** In this example of a mutant chromosome, the adult *HBB* genes β and δ have been deleted; as a result, the LCR cannot switch from activating the fetal genes to activating the adult genes, and the fetal genes remain active in the adult.

Fetal globin expression in adults caused by a deletion

One consequence of a master regulatory element that controls an entire gene complex is seen in a rare medical condition with a surprising prognosis. In some adults, the red blood cell precursors express neither the *HBB* nor the *HBD* genes. Although this should be a lethal situation, these adults remain healthy. Cloning and sequence analysis of the β-globin locus from affected adults show that they have a deletion extending across the *HBB* and *HBD* genes. Because of this deletion, the master regulatory control cannot switch around birth, as it normally would, from γ-globin production to β- and δ-globin production (**Figure 14.15c**). People with this rare condition continue to produce large amounts of fetal γ globin throughout adulthood, and that γ globin is sufficient to maintain a near-normal level of health.

Globin-related diseases result from a variety of mutations

By comparing DNA sequences from affected individuals with those from healthy individuals, researchers have learned that there are two general classes of disorders arising from alterations in the haemoglobin genes. In one class, mutations change the amino acid sequence and thus the three-dimensional structure of the α- or β-globin chain, and these structural changes result in an altered protein whose malfunction causes the destruction of red blood cells. Diseases of this type are known as haemolytic anaemias (**Figure 14.16a**). An example is sickle-cell anaemia, caused by an A-to-T substitution in the sixth codon of the β-globin chain. This simple change in DNA sequence alters the sixth amino acid in the chain from glutamic acid to valine, which, in turn, modifies the form and function of the affected haemoglobin molecules. Red blood cells carrying these altered molecules often have abnormal shapes that cause them to block blood vessels or be degraded.

The second major class of haemoglobin-related genetic diseases arises from DNA mutations that reduce or eliminate the production of one of the two globin polypeptides. The disease state resulting from such mutations is known as thalassaemia, from the Greek words *thalassa* meaning "sea" and *emia* meaning "blood"; the name arose from the observation that a relatively high rate of this blood disease occurs among people who live near the Mediterranean Sea. Several different types of mutation can cause thalassaemia, including those that delete an entire *HBA* or *HBB* gene, those that alter the sequence in regions that are outside the gene but necessary for its regulation, or those that alter the sequence within the gene such that no protein can be produced. The consequence of these changes in DNA sequence is the total absence or a deficient amount of one or the other of the normal haemoglobin chains. Because there are two *HBA* genes (*HBA1* and *HBA2*) that see roughly equal expression beginning a few weeks after conception,

individuals carrying deletions within the α-globin locus may be missing anywhere from one to four copies in total (**Figure 14.16b**). A person lacking only one would be a heterozygote for the deletion of one of two *HBA* genes; a person missing all four would be a homozygote for deletions of both *HBA* genes. The range of mutational possibilities explains the range of phenotypes seen in α-thalassaemia. Individuals missing only one of four possible copies of the α genes are normal, those lacking two of the four have a mild anaemia, and those without all four die before birth.

The fact that the *HBA* genes are expressed early in fetal life explains why the α-thalassaemias are detrimental *in utero*. By contrast, β-thalassaemia major, the disease occurring in people who are homozygotes for most deletions of the single *HBB* gene, also usually results in death, but not until soon after birth. These individuals survive that long because the *HBB* homologue *HBD* is expressed in the fetus (review Figure 14.1).

Comparison of the altered DNA sequences from affected individuals with wild-type sequences from healthy individuals has helped illuminate the sequences necessary for normal haemoglobin expression. In some β-thalassaemia patients, for example, disease symptoms arise from the alteration of a few nucleotides adjacent to the 5′ end of the coding region for the β chain. Data of this type have defined sequences that are important for expression of the β-globin locus. One such segment is the TATA box, a sequence found in many eukaryotic promoters (**Figure 14.17a**). In other thalassaemia patients, the entire α-globin locus and adjacent regulatory segments, including the TATA box, are intact, but a mutation has altered the LCR found far to the 5′ side of all the α-like genes. This LCR is necessary for a high level of tissue-specific expression of all α-like genes in red blood cell progenitors (**Figure 14.17b**). Mutations in the TATA box or the locus control region, depending on how disruptive they are, produce α- or β-thalassaemias of varying severity.

In addition to simply examining naturally occurring variants of the *HBB* and *HBA* genes, it is also possible to experimentally create user-defined alleles using a technique known as site-directed mutagenesis. This method allows researchers to alter the sequence of a gene at a specific nucleotide position and then to study the effects of this change on protein function. This method was first developed and utilized by University of British Columbia researcher, Dr. Michael Smith, to create a G-to-A substitution in a gene of a bacteriophage. Since then Dr. Smith's methods have been used in genetic studies involving enumerable genes and organisms. Dr. Smith was awarded the Nobel Prize in Chemistry for this work in 1993 (review the Tools of Genetics box "Site-directed Mutagenesis" in Chapter 8). As you might have guessed, this technique has also been used extensively to define the relationship between the sequences of the *HBA* and *HBB* genes and the biochemical functions of the encoded proteins (**Figure 14.18**).

(a.1) Major types of structural variants causing haemolytic anaemias

Name	Molecular Basis of Mutation	Change in Polypeptide	Pathophysiological Effect of Mutation	Inheritance
HbS	Single nucleotide substitution	β 6 Glu ↓ Val	Deoxygenated HbS polymerizes→ sickle cells→ vascular occlusion and haemolysis	Autosomal recessive
HbC	Single nucleotide substitution	β 6 Glu ↓ Lys	Oxygenated HbC tends to crystallize→ less deformable cells → mild haemolysis; the disease in HbS:HbC compounds is like mild sickle-cell anaemia	Autosomal recessive
Hb Hammer-smith	Single nucleotide substitution	β 42 Phe ↓ Ser	An unstable Hb → Hb precipitation → haemolysis; also low O₂ affinity	Autosomal dominant

(b.1) Clinical results of various α-thalassaemia genotypes

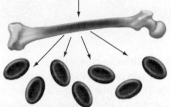

Clinical Condition	Genotype	Number of Functional α Genes	α-Chain Production
Normal	HBAHBA/ HBAHBA	4	100%
Silent carrier	HBAHBA/ HBA—	3	75%
Heterozygous α-thalassaemia— mild anaemia	HBA—/HBA— or HBAHBA/— —	2	50%
HbH (β₄) disease— moderately severe anaemia	HBA—/— —	1	25%
Homozygous α-thalassaemia— lethal	— —/— —	0	0%

(a.2) Basis of sickle-cell anaemia

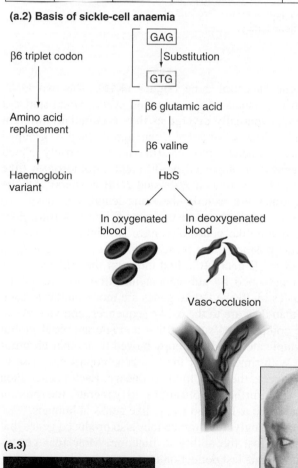

β6 triplet codon → Amino acid replacement → Haemoglobin variant

GAG → Substitution → GTG

β6 glutamic acid → β6 valine → HbS

In oxygenated blood   In deoxygenated blood → Vaso-occlusion

(a.3)

(b.2) A β-thalassaemia patient makes only α globin, not β globin.

Four α subunits combine to make abnormal haemoglobin.

Abnormal haemoglobin molecules clump together, altering shape of red blood cells. Abnormal cells carry reduced amounts of oxygen.

To compensate for reduced oxygen level, medullary cavities of bones enlarge to produce more red blood cells.

The spleen enlarges to remove excessive number of abnormal red blood cells.

— Spleen

Too few red blood cells in circulation result in anaemia.

(b.3)

Figure 14.16 Mutations in the DNA for haemoglobin produce two classes of disease. (a.1) The major types of haemoglobin variants causing haemolytic anaemias. **(a.2)** The basis of sickle-cell anaemia. **(a.3)** Sickling red blood cells appear as crescents among more rounded nonsickling cells. **(b.1)** Thalassaemias associated with deletions in the α-globin polypeptide. **(b.2)** The physiological basis of β-thalassaemia major. **(b.3)** Child suffering from β-thalassaemia major.

(a) Promoter region of the *HBB* gene

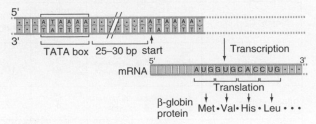

TATA box   25–30 bp start

Transcription

mRNA

Translation

β-globin protein   Met · Val · His · Leu · · ·

(b) Locus control region of the *HBA* locus

α-globin cluster

5' LCR   ζ2   α2 α1 θ'

60   30   20   10 5   0

kb

Figure 14.17 Regulatory regions affecting globin gene expression. (a) Mutations in the TATA box associated with the *HBB* gene can eliminate transcription and cause β-thalassaemia. **(b)** A locus control region is present 25–50 kb upstream of the *HBA* gene cluster. The function of the LCR is to open up the chromatin domain associated with the complete cluster of *HBA* genes. Mutations in the LCR can prevent expression of all the *HBA* genes, resulting in severe α-thalassaemia.

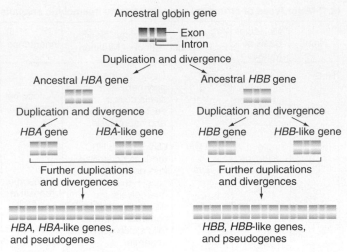

Ancestral globin gene

Exon
Intron

Duplication and divergence

Ancestral *HBA* gene   Ancestral *HBB* gene

Duplication and divergence   Duplication and divergence

HBA gene   *HBA*-like gene   *HBB* gene   *HBB*-like gene

Further duplications and divergences   Further duplications and divergences

HBA, *HBA*-like genes, and pseudogenes   *HBB*, *HBB*-like genes, and pseudogenes

Figure 14.19 Evolution of the globin gene family. Duplication of an ancestral gene followed by divergence of the separate duplication products established the α- and β-globin lineages. Further rounds of duplication and divergence within the separate lineages generated the two sets of genes and pseudogenes of the globin gene family.

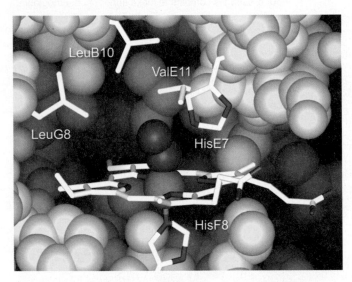

Figure 14.18 Site-directed mutagenesis of haemoglobin. Model of the adult human oxyhaemoglobin α-subunit active site (PDB access code, 1HHO). The schematic shows a space-filling representation of the haem pocket of the α-globin molecule. The oxygen and the haem iron are shown in space-filling representation in *dark red* and *grey*, respectively. Using site-directed mutagenesis researchers have been able to re-engineer the haem pocket. Such genetically engineered haemoglobins can be used as "blood substitutes" and are designed to correct oxygen deficit due to ischaemia.

All of the globin genes can be traced back to a single ancestral DNA sequence

With the use of bioinformatics, researchers can see that all the human globin genes form a closely related group, or gene family, that evolved by duplication and divergence from one ancestral gene (**Figure 14.19**). The two DNA sequence products of a duplication event, which start out identical, eventually diverge as they accumulate different mutations. The members of a gene family may be grouped together on one chromosome (like the very closely related *HBB* genes) or dispersed on different chromosomes (like the less closely related *HBA* and *HBB* clusters). All the β-like genes are exactly the same length and have two introns at exactly the same positions (Figure 14.16b). Four of the five α-like genes also have two introns at exactly the same positions, but these positions are different from those of the β genes (the first intron of the *HBZ* gene has been lengthened by subsequent insertion of DNA). The sequences of all the β-like genes are more similar to each other than they are to the α-like sequences, and vice versa. These comparisons suggest that a single ancestral globin gene duplicated, and one copy moved to another chromosome. With time, one of the two gene copies gave rise to the α lineage, the other to the β lineage. Each lineage then underwent further duplications to generate the present array of three α-like and five β-like genes in humans.

Interestingly, the duplications also produced genes that eventually lost the ability to function. Molecular geneticists made this last deduction from data showing two additional α-like sequences within the α locus and one β-like sequence within the β locus that no longer have the capacity for proper expression. The reading frames are interrupted by frameshifts, missense mutations, and nonsense codons, while regions needed to control the expression of the genes have lost key DNA signals. Sequences that look like, but do not function as a gene are known as **pseudogenes**; they occur throughout all higher eukaryotic genomes.

- Geneticists have found that the haemoglobin genes occur in two clusters in two separate chromosomes. The genes in the two clusters are transcribed in order at different stages of development, explaining how the structure of haemoglobin changes from embryo to adult.

- Disease-causing mutations in the globin genes range from the point mutation that causes sickle-cell anaemia to the variety of mutations that cause thalassaemia, including deletions, frameshifts, and changes to regulatory genes.

Connections

The tools of recombinant DNA technology grew out of an understanding of the DNA molecule and its interaction with the enzymes that operate on DNA in normal cells. Geneticists use the tools singly or in combination to look at DNA directly. Through cloning, hybridization, PCR, and sequencing, they have been able to isolate the genes that encode, for example, the hemoglobin proteins; identify sequences near the genes that regulate their expression; determine the complete nucleotide sequence of each gene; and discover the changes in sequence produced by the hundreds of mutations that affect hemoglobin production. The results give a fascinating and detailed picture of how the nucleotides along a DNA molecule determine protein structure and function and how

mutations in sequence produce far-ranging and varied effects on human health. The methods of classical genetics that we examined in Chapters 2–4 complement those of recombinant DNA technology to produce an integrated picture of genes and genomes at many levels.

In Chapters 19–22 we describe how the use of recombinant DNA technology has expanded from the analysis of single genes and gene complexes to the sequencing and examination of whole genomes. Through the automation of sequencing, and the sophisticated computer analysis of the data, scientific teams have determined the DNA sequence of the entire human genome as well as the genomes of many other organisms.

Essential Concepts

1. An intact eukaryotic genome is too complex for most types of analysis. Geneticists have appropriated the enzymes that normally operate on foreign DNA molecules inside a bacterial cell and used them in the test tube to create the tools of recombinant DNA technology. Restriction enzymes cut DNA at defined sites, ligase splices the pieces together, DNA polymerase makes DNA copies, and reverse transcriptase copies RNA into DNA. **[LO1–3]**

2. Gel electrophoresis provides a method for separating DNA fragments according to their size. When biologists subject a viral genome, plasmid, or small chromosome to restriction digestion and gel electrophoresis, they can observe the resulting DNA fragments by ethidium bromide staining. They then determine the size of the fragments by comparing their migration within the gel with the migration of known marker fragments. **[LO1–4]**

3. New technologies have allowed the cloning of DNA fragments. Restriction fragments and cloning vectors with matching sticky ends can be spliced together to produce recombinant DNA molecules. A cloning vector is a DNA sequence that can enter a host cell, produce a selectable phenotype, and provide a means of replicating and purifying both itself and any DNA to which it is spliced. **[LO1–5]**

4. Once inside a living cell, vector-insert recombinants are replicated during each cell cycle, just as the cell's own chromosomes are. A cellular clone consists of

the millions of cells arising from the consecutive divisions of a single cell. The vector-insert recombinant molecules inside the cells of a clone, often referred to as DNA clones, can be purified by procedures that separate recombinant molecules from host DNA. Restriction enzymes can cut away the insert, which can then undergo purification and processing. **[LO1, LO3–4]**

5. Genomic libraries are random collections of vector-insert recombinants containing DNA fragments of a given species. The most useful libraries carry at least four to five genomic equivalents. cDNA libraries carry DNA copies of the RNA transcripts produced in a particular tissue at a particular time. The clones in a cDNA library represent only that part of the genome transcribed and spliced into mRNA in the cells of a specific tissue, organ, or organism. **[LO3–4]**

6. Hybridization is the process whereby complementary DNA strands form stable double helixes. Hybridization makes it possible to use previously purified DNA fragments as labelled probes. Biologists use such probes to identify clones containing identical or similar sequences within genomic or cDNA libraries. Hybridization can also be used with gel electrophoresis as part of the technique called Southern blotting. Southern blot hybridization allows an investigator to determine the numbers and positions of complementary sequences within isolated DNA fragments or whole genomes of any complexity. **[LO2, LO4]**

7. The polymerase chain reaction (PCR) is a method for the rapid purification and amplification of a single DNA fragment from a complex mixture such as the whole human genome. The DNA fragment to be amplified is defined by a pair of oligonucleotide primers complementary to either end on opposite strands. The PCR procedure operates through a reiterative loop that amplifies the sequence between the primers in an exponential manner. PCR is used in place of cloning to purify DNA fragments whenever sequence information for primers is already available. **[LO5]**

8. Sequencing provides the ultimate description of a cloned fragment. Automation has increased the speed and scope of sequencing. **[LO6]**

Solved Problems

I. The following map of the plasmid cloning vector pBR322 shows the locations of the ampicillin (*amp*) and tetracycline (*tet*) resistance genes as well as two unique restriction enzyme recognition sites, one for *Eco*RI and one for *Bam*HI. You digested this plasmid vector with both *Eco*RI and *Bam*HI enzymes and purified the large *Eco*RI-*Bam*HI vector fragment. You also digested the cellular DNA that you want to insert into the vector with both *Eco*RI and *Bam*HI. After mixing the plasmid vector and the fragments together and ligating, you transformed an ampicillin-sensitive strain of *E. coli* and selected for ampicillin-resistant colonies. If you test all of your selected ampicillin-resistant transformants for tetracycline resistance, what result do you expect, and why?

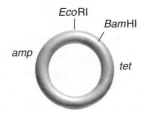

Answer

This problem requires an understanding of vectors and the process of combining DNAs using sticky ends generated by restriction enzymes.

The plasmid must be circular to replicate in *E. coli*, and, in this case, a circular molecule will be formed only if the insert fragment joins with the cut vector DNA. The cut vector will not be able to religate without an inserted fragment because the *Bam*HI and *Eco*RI sticky ends are not complementary and cannot base-pair. All ampicillin-resistant colonies therefore contain a *Bam*HI-*Eco*RI fragment ligated to the *Bam*HI-*Eco*RI sites of the vector. Fragments cloned at the *Bam*HI-*Eco*RI site interrupt and therefore inactivate the tetracycline resistance gene. *All ampicillin-resistant clones will be tetracycline sensitive.*

II. The gene for the human peptide hormone somatostatin (encoding nine amino acids) is completely contained on an *Eco*RI (5′ G^AATTC 3′) fragment, which can be cut out of the larger fragment shown below. (The ^ symbol indicates the site where the sugar-phosphate backbone is cut by the restriction enzyme.)

a. What is the amino acid sequence of human somatostatin?

b. Indicate the direction of transcription of this gene.

c. The first step in synthesizing large amounts of human somatostatin for pharmacological treatments involves constructing a so-called fusion gene. In this fusion construct, the N terminus of the protein encoded by the fusion gene consists of the N-terminal half of the *lacZ* gene (encoding β-galactosidase), while the remainder of the product of the fusion gene is human somatostatin. A family of three plasmid vectors for the construction of such a fusion gene has been created. All of these vectors have an ampicillin resistance gene and part of the *lacZ* gene encoding the first 583 amino acids of the β-galactosidase protein. The *Eco*RI fragment (i.e., the fragment produced by cutting with *Eco*RI) containing human somatostatin can be inserted into the single *Eco*RI restriction site on the vectors. The sequence of three vectors in the vicinity of the *Eco*RI site is shown here. The numbers refer to amino acids in the β-galactosidase protein, with the N-terminal amino acid being number 1. The DNA sequence presented is the same as that of the *lacZ* mRNA (with T's replacing the U's found in RNA). In which of these three vectors must the *Eco*RI fragment containing human somatostatin be inserted to generate a fusion protein with an N-terminal region from β-galactosidase and a C-terminal region from human somatostatin?

582 583
    Gly Asn   *Eco*RI
pWR590-1   5′ GGCAACCGGGCGAGCTCGAATTCG

582 583
    Gly Asn   *Eco*RI
pWR590-2   5′ GGCAACCGGGCGAGCTCGAATTC

582 583
    Gly Asn   *Eco*RI
pWR590-3   5′ GGCAACGGGGCAGCTCGAATTCGA

Answer

This problem requires an understanding of the sticky ends formed by restriction enzyme digestion and the requirement of appropriate reading frames for the production of proteins.

5′ GCCG^AATT CGATCCTATCAACACGAAGTGAAAGTCTTACAACCCATG^AATT CGATTCG 3′

3′ CGGC TTAA^GCTAGGATAGTTGTGCTTCACTTTCAGAATGTTGGGTAC TTAA^GCTAAGC 5′

a. The only complete open reading frame (ATG start codon to a stop codon) is found on the bottom strand (underlined on the 3' to 5' sequence below). *The amino acid sequence is Met-Gly-Cys-Lys-Thr-Phe-Thr-Ser-Cys.*

b. Based on the amino acid sequence determined for part *a*, *the gene must be transcribed from right to left.*

c. The cut site for *Eco*RI is after the G at the 5' end of the *Eco*RI recognition sequence on each strand. For each of the three vectors, the cut will be shifted relative to the reading frame of the *lacZ* gene by one base. The *Eco*RI fragment containing somatostatin can ligate to the vector in two possible orientations, but because we know the sequence on the bottom strand codes for the protein, a fusion protein will be produced only if the fragment is inserted with that coding sequence on the same strand as the vector coding sequences. Consider only this orientation to determine which vector will produce the fusion protein. The *Eco*RI fragment to be inserted into the vector next to the *lacZ* gene has five nucleotides that precede the first codon of the somatostatin gene (see the figure above) and therefore requires one more nucleotide to match the reading frame of the vector. For pWR590-1, the cut results in an in-frame end; pWR590-2 has one base extra beyond the reading frame; pWR590-3 has a two-base extension past the reading frame. *The Eco*RI *fragment must be inserted in the vector pWR590-2 to get somatostatin protein produced.*

III. Imagine you have cloned a 14.7-kb piece of DNA, which contains restriction sites as shown here.

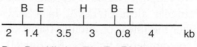

B = *Bam*HI site, E = *Eco*RI site,
H = *Hind*III site

Numbers under the segments represent the sizes of the regions in kilobases (kb). You have labelled the left end of the molecule with ^{32}P. What radioactive bands would you expect to see following electrophoresis if you did a complete digestion with *Bam*HI? *Eco*RI? *Hind*III?

Answer

This problem deals with partial and complete digests and radioactive labelling of fragments. Only the left-most fragment would be seen after complete digestion with any of the three enzymes because only the left end contains radioactivity. *Radioactive bands seen after digestion with BamHI: 2 kb; EcoRI: 3.4 kb; and HindIII: 6.9 kb.*

5' GCCG^AATT CGATCCTATCAACACGAAGTGAAAGTCTTACAACCCATG^AATT CGATTCG 3'

3' CGGC TTAA^GCTAGGAT<u>AGTTGTGCTTCACTTTCAGAATGTTGGGTAC</u> TTAA^GCTAAGC 5'

Problems

Vocabulary

1. For each of the terms in the left column, choose the best matching phrase in the right column.

i. oligonucleotide	**1.** a DNA molecule used for transporting, replicating, and purifying a DNA fragment
ii. vector	**2.** a collection of the DNA fragments of a given species, inserted into a vector
iii. sticky ends	**3.** DNA copied from RNA by reverse transcriptase
iv. recombinant DNA	**4.** stable binding of single-stranded DNA molecules to each other
v. reverse translation	**5.** efficient and rapid technique for amplifying the number of copies of a DNA fragment
vi. genomic library	**6.** computational method for determining the possible sequence of base pairs associated with a particular region of a polypeptide
vii. genomic equivalent	**7.** contains genetic material from two different organisms
viii. cDNA	**8.** the number of DNA fragments that are sufficient in aggregate length to contain the entire genome of a specified organism
ix. PCR	**9.** short single-stranded sequences found at the ends of many restriction fragments
x. hybridization	**10.** a short DNA fragment that can be synthesized by a machine

Section 14.1

2. Approximately how many restriction fragments would result from the complete digestion of the human genome (3×10^9 bases) with the following restriction enzymes? (The recognition sequence for each enzyme is given in parentheses, where N means any of the four nucleotides.)

a. *Sau*3A (^GATC)

b. *Bam*HI (G^GATCC)

c. *Sfi*I (GGCCNNNN^NGGCC)

3. Why do longer DNA molecules move more slowly than shorter ones during electrophoresis?

4. You have a circular plasmid containing 9 kb of DNA, and you wish to map its *Eco*RI and *Bam*HI sites. When you digest the plasmid with *Eco*RI and run the resulting DNA on a gel, you observe a single band at 9 kb. You get the same result when you digest the DNA with *Bam*HI. When you digest with a mixture of both enzymes, you observe two bands, one 6 kb and the other

3 kb in size. Explain these results. Draw a map of the restriction sites.

5. The linear bacteriophage λ genomic DNA has at each end a single-strand extension of 20 bases. (These are "sticky ends" but are not, in this case, produced by restriction enzyme digestion.) These sticky ends can be ligated to form a circular piece of DNA. In a series of separate tubes, either the linear or circular forms of the DNA are digested to completion with *Eco*RI, *Bam*HI, or a mixture of the two enzymes. The results are shown here.

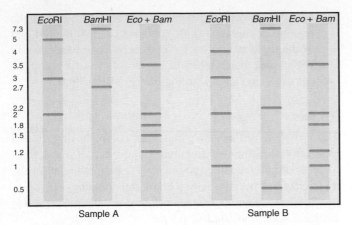

Sample A   Sample B

a. Which of the samples (A or B) represents the circular form of the DNA molecule?

b. What is the total length of the linear form of the DNA molecule?

c. What is the total length of the circular form of the DNA molecule?

d. Draw a restriction map of the linear form of the DNA molecule. Label all restriction enzyme sites as *Eco*RI or *Bam*HI.

6. The following fragments were found after digestion of a circular plasmid with restriction enzymes as noted. Draw a restriction map of the plasmid.

*Eco*RI: 7.0 kb

*Sal*I: 7.0 kb

*Hin*dIII: 4.0, 2.0, 1.0 kb

*Sal*I + *Hin*dIII: 2.5, 2.0, 1.5, 1.0 kb

*Eco*RI + *Hin*dIII: 4.0, 2.0, 0.6, 0.4 kb

*Eco*RI + *Sal*I: 2.9, 4.1 kb

Section 14.2

7. What purpose do selectable markers serve in vectors?

8. Why do geneticists studying eukaryotic organisms often construct cDNA libraries, whereas geneticists studying bacteria almost never do? Why would bacterial geneticists have difficulties constructing cDNA libraries even if they wanted to?

9. A plasmid vector pBS281 is cleaved by the enzyme *Bam*HI (G^GATCC), which recognizes only one site in the DNA molecule. Human DNA is digested with the enzyme *Mbo*I (^GATC), which recognizes many sites in human DNA. These two digested DNAs are now ligated together. Consider only those molecules in which the pBS281 DNA has been joined with a fragment of human DNA. Answer the following questions concerning the junction between the two different kinds of DNA.

a. What proportion of the junctions between pBS281 and all possible human DNA fragments can be cleaved with *Mbo*I?

b. What proportion of the junctions between pBS281 and all possible human DNA fragments can be cleaved with *Bam*HI?

c. What proportion of the junctions between pBS281 and all possible human DNA fragments can be cleaved with *Xor*II (C^GATCG)?

d. What proportion of the junctions between pBS281 and all possible human DNA fragments can be cleaved with *Eco*RII (Pu Pu A ^ T Py Py)? (Pu and Py stand for purine and pyrimidine, respectively.)

e. What proportion of all possible junctions that can be cleaved with *Bam*HI will result from cases in which the cleavage site in human DNA was not a *Bam*HI site in the human chromosome?

Section 14.3

10. Consider three different kinds of human libraries: a genomic library, a brain cDNA library, and a liver cDNA library.

a. Assuming inserts of approximately equal size, which would contain the greatest number of different clones?

b. Would you expect any of these not to overlap the others at all in terms of the sequences it contains? Explain.

c. How do these three libraries differ in terms of the starting material for constructing the clones in the library?

11. As a molecular biologist and horticulturist specializing in snapdragons, you have decided that you need to make a genomic library to characterize the flower colour genes of snapdragons.

a. How many genomic equivalents would you like to have represented in your library to be 95 percent confident of having a clone containing each gene in your library?

b. How do you determine the number of clones that should be isolated and screened to guarantee this number of genomic equivalents?

12. Imagine that you are a molecular geneticist studying a particular gene in which mutations cause a serious human disease. The gene, including its flanking regulatory sequences, spans 200 kb of DNA. The distance from the first to the last coding base is 140 kb, which is divided among 10 exons and 9 introns. The exons contain a total of 9.7 kb, and the introns contain 130.3 kb of DNA.

You would like to obtain the following for your work: (a) an intact clone of the whole gene, including flanking sequences; (b) a clone containing the entire coding sequences but no noncoding sequences; and (c) a clone of exon 3, which is the site of the most common disease-causing mutation in this gene. For each of these clones, describe the source of the human DNA to be inserted into the vector, and decide whether you would use a plasmid vector or a BAC vector. (*Note:* Where possible, it is technically easier to use plasmids than BACs as vectors.) Explain your answers.

13. A 49-bp *Eco*RI fragment containing the somatostatin gene was inserted into the vector pWR590 shown below. The sequence of the inserted fragment is

5′ AATTCGATCCTATCAACACGAAGTGAAAGTCTTACAACCCATGAATTCG 3′

3′ GCTAGGATAGTTGTGCTTCACTTTCAGAATGTTGGGTACTTAAGCTTAA 5′

Distances between adjacent restriction sites in the pWR590 vector are indicated in the diagram *below*. What are the patterns of restriction digests with *Eco*RI (G^AATTC) or with *Mbo*I (^GATC) before and after cloning the somatostatin gene into the vector? (E = *Eco*RI, M = *Mbo*I)

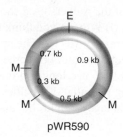

pWR590

14. The *Notch* gene involved in *Drosophila* development is contained within a restriction fragment of *Drosophila* genomic DNA produced by cleavage with the enzyme *Sal*I. The restriction map of this *Drosophila* fragment for several enzymes (*Sal*I, *Pst*I, and *Xho*I) is shown here; numbers indicate the distances between adjacent restriction sites. This fragment is cloned by sticky-end ligation into the single *Sal*I site of a bacterial plasmid vector that is 5.2 kb long. The plasmid vector has no restriction sites for *Pst*I or *Xho*I enzymes.

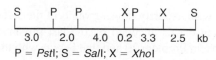

Make a sketch of the expected patterns seen after agarose gel electrophoresis and staining of a *Sal*I digest (alone), of a *Pst*I digest (alone), of a *Xho*I digest (alone), and of the plasmid containing vector and *Drosophila* fragment. Indicate the fragment sizes in kilobases.

15. Your undergraduate research advisor has assigned you a task: Insert an *Eco*RI-digested fragment of frog DNA into an *E. coli* plasmid that carries a *lacZ* gene with an *Eco*RI site in the middle (see Figure 14.7). Your advisor

suggests that after you digest your plasmid with *Eco*RI, you should treat the plasmid with the enzyme alkaline phosphatase. This enzyme removes phosphate groups that may be located at the 5′ ends of DNA strands. You will then add the fragment of frog DNA to the vector and join the two together with the enzyme DNA ligase.

You do not quite follow your advisor's reasoning, so you set up two ligations, one with plasmid that was treated with alkaline phosphatase and the other without such treatment. The ligation mixtures are otherwise identical. After the ligation reactions are completed, you transform a small aliquot (portion) of each ligation into *E. coli* and spread the cells on Petri plates containing both ampicillin and X-gal. The next day, you observe 100 white colonies and one blue colony on the plate transformed with alkaline-phosphatase-treated plasmids and 100 blue colonies and one white colony on the plate transformed with plasmids that had not been treated with the enzyme.

a. Explain the results seen on the two plates.

b. Why was your research advisor's suggestion a good one?

c. Why would you normally treat plasmid vectors with alkaline phosphatase but not the DNA fragments you want to add to the vector?

Section 14.4

16. a. Given the following restriction map of a cloned 10-kb piece of DNA, what size fragments would you see after digesting this linear DNA fragment with each of the enzymes or combinations of enzymes listed? (1) *Eco*RI, (2) *Bam*HI, (3) *Eco*RI + *Hin*dIII, (4) *Bam*HI + *Pst*I, and (5) *Eco*RI + *Bam*HI.

b. What fragments in the last three double digests would hybridize on a Southern blot with a probe made from the 4-kb *Bam*HI fragment?

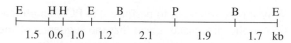

17. You have cloned and characterized a particularly interesting protein-coding gene from the bacterium *Bacillus subtilis*, and you would like to isolate the corresponding, homologous gene from the rare, poorly characterized bacterial species *Beneckea nigripulchritudo* that infects certain shrimp. You decide to make degenerate probes to identify, by hybridization, clones containing this homologous gene. The amount of degeneracy is a potential problem because the more types of different DNA molecules contained in the probe, the worse the signal-to-noise ratio in the hybridization experiment. How can you minimize the degeneracy? Be as specific as possible, mentioning such factors as the length of the probe and the region of DNA you will choose to synthesize by reverse translation.

18. It is possible to use hybridization techniques similar to those described for the Southern blot procedure (Figure 14.11) to identify within a library particular clones homologous to a nucleic acid probe. The idea is to transfer some DNA from the colonies growing on a plate to a nitrocellulose filter, hybridize the filter with the radioactive probe, and then pick cells from the original plate that correspond to the positions of probe hybridization. With this idea in mind, place in an appropriate order the following steps that could be used.

a. Mix together BAC vector DNA and hoot owl DNA with ligase.

b. Expose nitrocellulose paper dics to UV radiation and baking.

c. Extract genomic DNA from hoot owl cells.

d. Visualize labelled DNA fragments.

e. Produce a labelled DNA probe to the *idiosynchratase* gene.

f. Completely digest BAC vector DNA with the restriction enzyme *Hin*dIII.

g. Place nitrocellulose paper dics onto the agar surface to transfer colonies.

h. Incubate DNA probe with nitrocellulose paper dics.

i. Distribute bacteria onto a Petri plate containing agar and nutrients and allow growth into colonies.

j. Partially digest hoot owl genomic DNA with the restriction enzyme *Hin*dIII.

k. Transform bacteria.

Section 14.5

19. Using PCR, you want to amplify an approximately 1-kb exon of the human autosomal gene encoding the enzyme phenylalanine hydroxylase from the genomic DNA of a patient suffering from the autosomal recessive condition phenylketonuria (PKU).

a. Why might you wish to perform this PCR amplification in the first place, given that the sequence of the human genome has already been determined?

b. Calculate the number of template molecules that are present if you set up a PCR using 1 ng (1×10^{-9} g) of chromosomal DNA as the template. Assume that each haploid genome contains only a single gene for phenylalanine hydroxylase and that the molecular weight of a base pair is 660 g/mol. The human genome contains 3×10^9 base pairs.

c. Calculate the number of PCR product molecules you would obtain if you perform 25 PCR cycles and the yield from each cycle is exactly twice that of the previous cycle. What would be the mass of these PCR products taken together?

20. Which of the following set(s) of primers could you use to amplify the target DNA sequence below, which is part of the last protein-coding exon of the *CFTR* gene?

5′ GGCTAAGATCTGAATTTTCCGAG ... TTGGGCAATAATGTAGCGCCTT 3′
3′ CCGATTCTAGACTTAAAAGGCTC ... AACCCGTTATTACATCGCGGAA 5′

a. 5′ GGAAAATTCAGATCTTAG 3′;
   5′ TGGGCAATAATGTAGCGC 3′

b. 5′ GCTAAGATCTGAATTTTC 3′;
   3′ ACCCGTTATTACATCGCG 5′

c. 3′ GATTCTAGACTTAAAGGC 5′;
   3′ ACCCGTTATTACATCGCG 5′

d. 5′ GCTAAGATCTGAATTTTC 3′;
   5′ TGGGCAATAATGTAGCGC 3′

21. Problem 20 raises several interesting questions about the design of PCR primers.

a. PCR is important because it can amplify a single region of DNA from a complex genome. How can you be sure that the two primers you chose as your answer to Problem 20 will amplify only an exon of the *CFTR* gene from a sample of human genomic DNA?

b. The protocol for PCR shown in Figure 14.12 states that each of the primers used should be 16–26 nucleotides long. (i) Why do you think the lower limit would be approximately 16? (ii) The upper limit of 26 nucleotides is not absolute. For some applications of PCR, it is possible to use longer primers, but at the risk of introducing potential difficulties. What complications or disadvantages might be associated with longer primers?

c. Suppose that one of the primers you designed in your answer to Problem 20 had a mismatch with a single base in the genomic DNA of a particular individual. Would you be more likely to obtain a PCR product from this genomic DNA if the mismatch were at the 5′ end or at the 3′ end of the primer? Why?

d. Suppose you wanted to clone the region you amplified in Problem 20 into a plasmid vector with a single site for the restriction enzyme *Eco*RI? How could you modify the PCR primers to produce a PCR product with *Eco*RI sites at both ends?

22. You wish to purify large amounts of the part of the CFTR protein that is encoded by the last protein-coding exon shown in Problem 20 and that begins with the amino acid sequence

N...Leu-Arg-Ser-Glu-Phe-Ser-Glu...C

and ends with the sequence

N...Trp-Ala-Ile-Met (C terminus)

You will start this process by cloning an appropriate PCR product into the pMore vector, part of whose sequence is shown *below*. The pMore vector makes large amounts of maltose binding protein (MBP) when transformed into *E. coli*. The amino acids shown with the vector sequence

correspond to the C-terminal end of MBP. To do the cloning, you will digest both the pMore vector and your PCR product with both the *EcoRI* (G^AATTC) and *SalI* (G^TCGAC) restriction enzymes and then ligate the pieces together. The vector has only a single site for each of these enzymes.

```
5'... AGGATTTCAGAATTCGGATCCTCTAGAGTCGACCTGTAGGGCAA ...3'
   ArgIleSerGluPheGlySerSerArgValAspLeup
```

a. As discussed in Solved Problem II, a fusion protein contains amino acid sequences derived from two or more naturally occurring polypeptides. Describe the fusion protein that will be made when the PCR product is ligated into the vector. What are the orientations of the parts of MBP and CFTR relative to that of the fusion protein?

b. What advantages might there be for cutting the vector and PCR product with two restriction enzymes instead of one?

c. Design PCR primers that will allow you to construct the desired recombinant DNA molecule. Note (i) that the sequence shown in Problem 20 has neither *EcoRI* nor *SalI* sites, (ii) that additional nucleotides can be added to appropriate locations in the PCR primers, and (iii) that restriction enzymes require about five nucleotides on either side of the restriction site for the enzymes to work. This problem is extremely difficult, but will help you integrate a great deal of information about gene structure and recombinant DNA technology.

d. MBP can bind to the sugars amylose and maltose. The last 20 amino acids at the C terminus of MBP are not required for this property. It is also possible to chemically synthesize an amylose resin (beads with covalently bound amylose). How would these facts be helpful in allowing you to purify a large amount of a region of the CFTR protein?

Section 14.6

23. Several of the techniques discussed in this chapter, particularly restriction mapping and methods based on DNA hybridization such as Southern blots, are still often used for studying genes in unusual organisms. However, in the twenty-first century, these techniques are used much more rarely than in the late twentieth century for studying genes in humans or in model organisms such as yeast, *C. elegans*, *Drosophila*, or mice. What has changed with the millennium, and what new techniques have arisen as replacements?

24. Which of the following processes used in biotechnology relies on specific enzymes? What are those enzymes? What is the basis for any of these processes that are not enzyme based?

a. DNA ligation

b. cleavage of DNA at specific sites

c. DNA hybridization

d. DNA sequencing

e. cDNA synthesis

f. PCR

25. a. If you are presented with the following sequencing autoradiogram, what can you say about the sequence of the template strand used in these sequencing reactions?

b. If the template for sequencing is the strand that resembles the mRNA, write out the sequence of the mRNA insofar as it can be determined.

c. Is this portion of the genome likely to be within a coding region? Explain your answer.

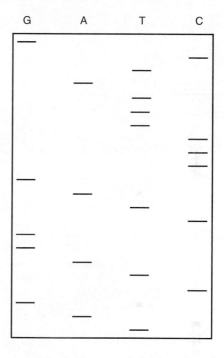

26. You read the following sequence directly from a gel.

5' TCTAGCCTGAACTAATGC 3'

a. Make a drawing that reproduces the autoradiogram from which this sequence was read. How would you know the reading frame if you are reading this short sequence off a gel?

b. Assuming this sequence is from an exon in the middle of a gene, does this newly synthesized strand or the template strand have the same sequence as the mRNA for the gene (except that T's are present instead of U's)? Justify your answer.

c. Using the genetic code table, give the amino acid sequence of the hexapeptide (six amino acids) translated from the 18-base message. Indicate which is the amino-terminal end of the peptide.

27. The following figure portrays a trace derived from the automated sequencing of a certain PCR product produced by the amplification of the genomic DNA from a

particular person's cells. The left-to-right orientation of the peaks on the trace corresponds to smaller-to-larger fragments of DNA. The height of the peaks is unimportant. (red = T; green = A; black = G; purple = C)

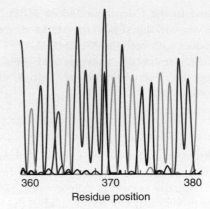

a. What does the green peak at the left end of the trace signify? Be as precise as possible.

b. Write the sequence of DNA revealed by this trace, indicating the 5′-to-3′ orientation.

c. What do you think is meant by "residue position"? That is, what is located at residue position 1?

d. Explain the apparent anomaly at residue position 370.

For more information on the resources available from McGraw-Hill Ryerson, go to www.mcgrawhill.ca/he/solutions.

488   Unit I   Chapter 14   Molecular Analysis of DNA

Genome-wide Analysis of Genetic Variation

With the advent of high-throughput and low-cost DNA sequencing methods, the era of personalized genomics is upon us. But what can your genome sequence reveal? In this chapter, we will begin to answer this question by examining the methods used by scientists to study genetic variation among individual genomes.

A couple whose firstborn suffers from cystic fibrosis learns from the medical diagnosis of that child's symptoms that both parents are carriers of a recessive, disease-producing mutation. Together they run a 25 percent risk of having a second child exhibiting this life-threatening hereditary condition. In the early 1990s, one such couple did not want to take that chance and after genetic counselling decided to try an experimental protocol: *in vitro* fertilization combined with the direct detection *in vitro* of the embryo's genotype, before its placement in the mother's womb.

The procedure took less than a week. At the start, a team of medical workers, including an obstetrician and a reproductive biologist, obtained ten eggs from the woman and fertilized them with sperm from the man. Three days later, after the fertilized eggs had undergone several mitotic divisions to generate embryos with six to ten cells, a research assistant used micropipettes to remove one cell from each embryo (**Figure 15.1**). Because embryos that split naturally at this stage can develop into healthy identical twins, the removal of a single cell would not prevent normal development.

Next, the research assistant used PCR (review Figure 14.12) to amplify from each isolated cell a specific DNA segment from both the maternal and paternal copy of chromosome 7. The amplified segment contained the site of the most common mutation found within the gene responsible for cystic fibrosis (the *CFTR* gene; cystic fibrosis transmembrane conductance regulator). They then used wild-type and mutant CFTR probes (called allele-specific oligonucleotides or ASOs) to detect the genotype of the cells, and by inference, of each embryo.

Chapter Outline
15.1 Genetic Variation Among Individual Genomes
15.2 Single Nucleotide Polymorphisms (SNPs) and Small-Scale-Length Variations
15.3 Deletions or Duplications of a DNA Region
15.4 Positional Cloning: From DNA Markers to Disease-causing Genes
15.5 Complex Traits

On the same day that the embryos were biopsied and genotyped, the doctor, in consultation with the parents, selected for placement in the mother's womb two embryos of known genotype. Of the two embryos they chose, one was a heterozygous carrier and the other a homozygote for the normal allele. The use of two embryos improves the chances of at least one implantation and is a part of many *in vitro* fertilization procedures. As it happened, only the embryo carrying two normal alleles of the *CFTR* gene implanted itself successfully into the woman's uterus.

Nine months later, the mother gave birth to a healthy 3.3 kg (7 lb 3 oz) baby girl. When evaluated by a pediatrician at her four-week checkup, the infant daughter was found to be completely normal in both physical and mental development. At the same time, testing at an independent laboratory confirmed that neither of her homologous chromosome 7's carried the *CFTR* mutation.

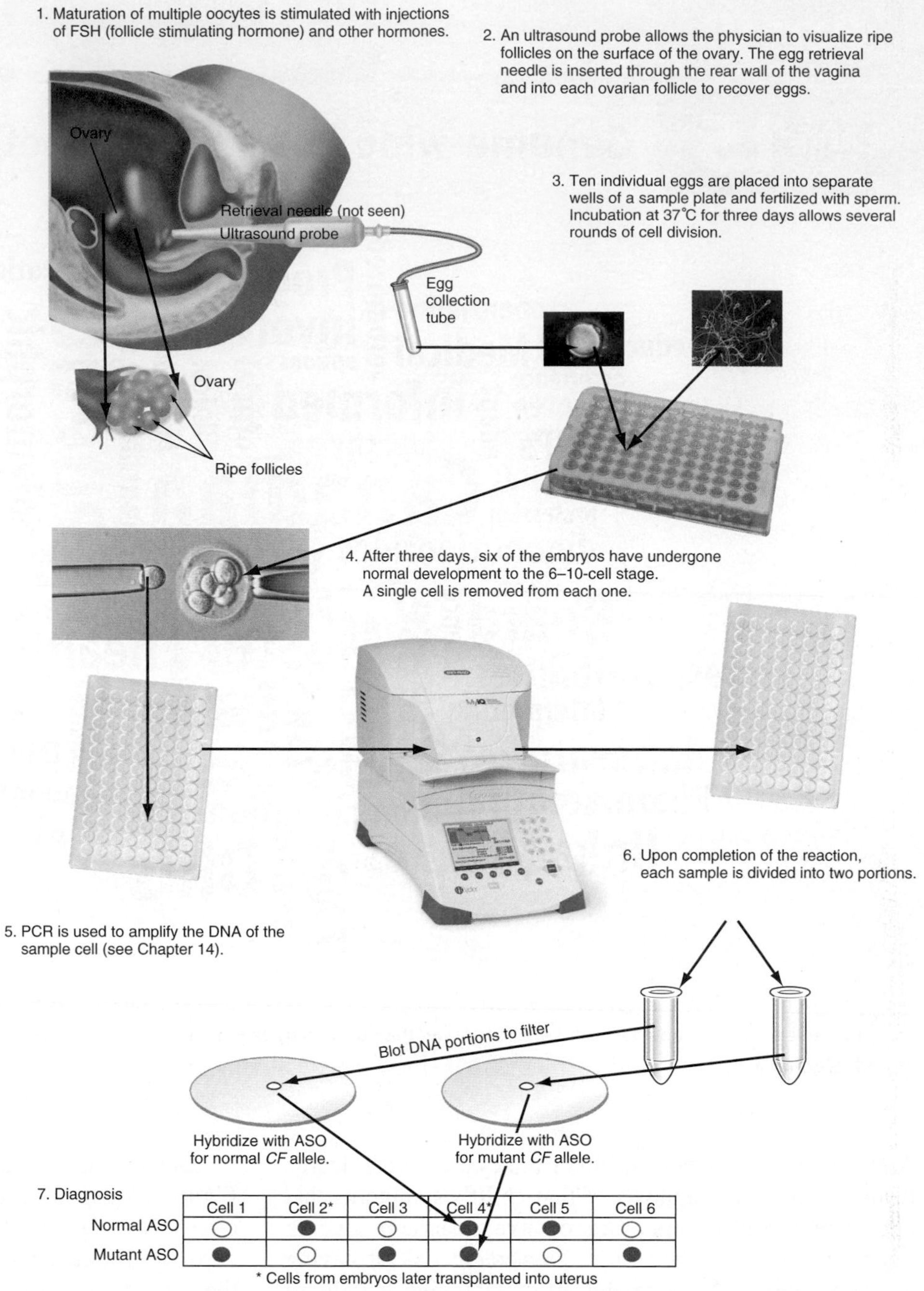

In this chapter, we examine how geneticists use an array of molecular tools to detect DNA differences among individuals. The techniques are sensitive enough to operate on single hair follicles and even single cells from human embryos.

The ability to detect genotype directly at the DNA level has far-reaching consequences. It provides geneticists with tools for mapping and identifying the genes responsible for human diseases (like the cystic fibrosis gene) that were previously defined only by their effect on phenotype. Once the genes and their variant alleles are identified, DNA genotyping can help predict the probability of future disease—*in vitro*, *in utero*, or after birth—or reveal the presence of a silent recessive disease allele in a carrier who shows no evidence of disease. The application of whole-genome analysis to thousands of individuals who have also been evaluated clinically also gives human geneticists the power to uncover common DNA variants that contribute to extremely complex traits, including diabetes and heart disease.

1. Maturation of multiple oocytes is stimulated with injections of FSH (follicle stimulating hormone) and other hormones.

2. An ultrasound probe allows the physician to visualize ripe follicles on the surface of the ovary. The egg retrieval needle is inserted through the rear wall of the vagina and into each ovarian follicle to recover eggs.

3. Ten individual eggs are placed into separate wells of a sample plate and fertilized with sperm. Incubation at 37°C for three days allows several rounds of cell division.

4. After three days, six of the embryos have undergone normal development to the 6–10-cell stage. A single cell is removed from each one.

5. PCR is used to amplify the DNA of the sample cell (see Chapter 14).

6. Upon completion of the reaction, each sample is divided into two portions.

Blot DNA portions to filter

Hybridize with ASO for normal *CF* allele.

Hybridize with ASO for mutant *CF* allele.

7. Diagnosis

	Cell 1	Cell 2*	Cell 3	Cell 4*	Cell 5	Cell 6
Normal ASO	○	●	○	●	●	○
Mutant ASO	●	○	●	●	○	●

* Cells from embryos later transplanted into uterus

Figure 15.1 Preimplantation embryo diagnosis. Plucking one cell from an eight-cell embryo for the direct detection of genotype.

Our discussion of these novel genetic technologies and applications suggests two general themes: (1) the ability to distinguish genotypic differences of all kinds extends our concept of a locus and the alleles that define it, and (2) techniques for the direct detection of genotype provide access to genetic details about human individuals never before available. Who should have access to this information, and how should it be used? (See the **Genetics and Society** box "**Social and Ethical Issues Surrounding Preimplantation Genetic Diagnosis**" at the end of this chapter for a discussion of this topic).

15.1 Genetic Variation Among Individual Genomes

Although Darwin's theory of natural selection predated the rediscovery of Mendel's laws, these two pillars of modern biology were separate intellectual fields until they were first brought together under the rubric of the "modern synthesis" in the 1940s. Consequently, early geneticists worked under the age-old conceptualization of a whole species as a unique and *discrete* entity. This view led to the assumption that an "ideal specimen" of any particular species would carry species-specific "wild-type" alleles at each of the genes in its genome. Alleles that were not wild type were considered mutant. But mutant alleles could only be recognized indirectly as the causative agents of mutant phenotypes.

Extensive allelic variation distinguishes individuals within a species

The first crack in the wild-type/mutant allele dichotomy came during the 1950s and 1960s with the application of gel electrophoresis to study the chemical properties of specific proteins. The results obtained from a variety of species (from *Drosophila* to humans) were consistent and striking—presumed "wild-type" individuals of the same species frequently produced different variant forms of proteins, encoded by variant alleles. A locus with two or more alleles that are each present in more than 1 percent of a species' members is considered to be **polymorphic**, and the alleles of a polymorphic locus are called **genetic variants**, rather than wild type or mutant.

With the advent of DNA cloning and sequencing, and with more recent forays into personal genome sequencing, the staggering degree to which individual human genomes differ from each other has become slowly uncovered. A direct comparison of the genomes of James Watson, co-discoverer of the DNA double helix, and J. Craig Venter, a pioneer of DNA sequencing, reveals 2 021 206 single nucleotide substitutions, of which 5015 cause changes in amino acid sequences of expressed proteins (**Figure 15.2**). Pairwise comparisons of Watson's or Venter's genome with that of an anonymous Chinese man reveals similar differences. The genomes of the two men also differ by small additions or subtractions of genetic material—deletions or insertions—at over 100 000 genomic sites, ranging in size from 2 to 38 896 bp.

Not only is there no such thing as a wild-type human genome, there is also no such thing as a wild-type human

genome length. Polymorphic deletions, insertions, and duplications result in genome lengths that differ by as much as 1 percent in healthy individuals.

The ability to distinguish genotypic differences of all kinds extends our concept of a locus and the alleles that define it. In the early days of genetic analysis, loci and genes were synonymous simply because nonfunctional regions of DNA were invisible (by definition) at the level of phenotype. By contrast, modern geneticists, able to look

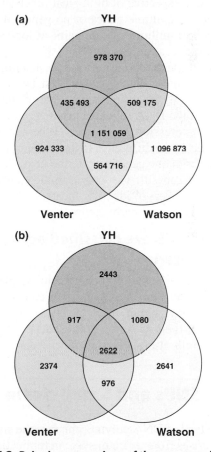

Figure 15.2 Pairwise comparison of three personal genomes. Single nucleotide substitutions in the genomes of J. Craig Venter, James D. Watson, and an anonymous Chinese man (YH). Numbers of unique nucleotide substitutions are shown in non-overlapping portions of each circle. Variants shared by two of the three men are shown in the double-overlap regions. The central three-way overlap indicates numbers of substitutions relative to the human reference sequence. **(a)** Differences across the whole genome. **(b)** Amino-acid-changing substitutions only.

Table 15.1 | Categories of Genetic Variants

	Short Name	Size	Frequency	Total Loci Recorded	Method of Detection			
					DNA Microarray	PCR & Gel Electrophoresis	PCR & ASO Hybridization	DNA Sequence
Single nucleotide polymorphism	SNP	1 bp	1 kb	38 000 000	Yes		Yes	
Insertion/deletion	InDel	2–100 bp	10 kb	200 000	(Yes)	Yes		
Simple sequence repeat	SSR, microsatellite	3–200 bp	30 kb	100 000		Yes		
Copy number variation/copy number polymorphism	CNV/CNP	0.1–1000 kb	3 Mb	8 600	Yes			
Complex variant			500 kb					Yes

at genotype directly, can pick out genotypic differences in both coding and noncoding DNA regions on the basis of changes in DNA sequence alone. As a result, a **locus** is now considered to be any location in the genome that is defined by chromosomal coordinates for the convenience of researchers, irrespective of biological function. A DNA locus can contain multiple genes or no genes; it can be a single base pair or millions of base pairs, as long as it has a defined genomic location and length.

Along with a new definition of a locus comes a new definition of an allele. Since a locus can be any defined segment of DNA in the genome, an allele of that locus is any variation in the DNA sequence itself, even if it has no impact on the expression of any trait. Whether it is functional or not makes no difference in the manner that a locus is transmitted from one generation to the next. We will see that researchers can still use polymorphic nonfunctional loci as genetic markers to identify, locate, isolate, and follow the transmission of nearby genes.

Genetic variants are classified according to several criteria

For the purposes of genotyping, geneticists place polymorphic DNA loci into one of the five categories based on size, frequency within individual genomes, and the method used for their detection (**Table 15.1**). The simplest and most generally useful class of genetic variants are the

single nucleotide polymorphisms called **SNPs** ("snips"). SNPs are particular base positions in the genome where alternative letters of the DNA alphabet commonly distinguish some people from others.

Beyond the first category of SNPs, genetic variants arise in every size and complexity. For convenience, geneticists place them into one of four additional categories: (1) short deletions and insertions called InDels or DIPs; (2) regions of repeating two- or three-base-long units termed simple sequence repeats (SSRs); (3) large regions of duplication or deletion (copy number polymorphisms, CNPs, or copy number variants, CNVs, depending on their frequency of occurrence); and (4) a catch-all category of complex variants that do not fit into any other category. Although the borders between these classes are fuzzy, they serve the purpose of orienting researchers to a sense of what a particular genetic variant looks like.

> The original view of what constitutes a genetic locus has changed as modern technology has revealed the genome at the base-pair level. Widespread genetic variance exists among members of a species, even when the variations do not cause a phenotypic difference. Examples of variants include single nucleotide polymorphisms, short deletions and insertions, simple sequence repeats, and copy number polymorphisms.

15.2 SNPs and Small-Scale-Length Variations

The simplest type of DNA polymorphism is the single-base SNP, which arises from a rare mistake in replication or due to a mutagenic chemical (**Figure 15.3**). SNPs account for the vast majority of the total variation that exists between human genomes, occurring on average once every 1000 bases in any pairwise comparison. The per-base mutation rate is less than 1 in 30 million per generation, which is so low that in nearly all cases, each individual SNP can be traced back to a genomic change that occurred once in a single ancestral genome. This also means that those people

who did not inherit the variant allele have a more ancient allele that was present long before the human species took form.

The origin of human SNPs is determined by comparison with other species

Geneticists can take advantage of the close relationship between the human and chimpanzee genomes to distinguish between the original chimp-shared allele and the

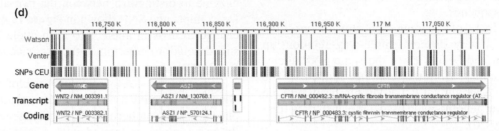

(a)

Allele	Genome	Single-Strand Notation
1	AGCTCTGATAC TCGAGACTATG	AGCTCTGATAC
2	AGCTCCGATAC TCGAGGCTATG	AGCTCCGATAC

(b)

Genotype	Notation
Homozygous T	TT
Heterozygous T	TC
Homozygous C	CC

(c)

Human 1   TTGAC**G**TATAAATGATCTTTATAT**T**TTCAGAAGTC

Human 2   TTGAC**G**TATAAATGATCTTTATAT**C**TTCAGAAGTC

Chimp   TTGAC**A**TATAAATGATCTTTATAT**C**TTCAGAAGTC

(d)

Figure 15.3 **Single nucleotide polymorphisms.** **(a)** Single nucleotide polymorphisms (SNPs) are single-base-pair polymorphisms. Geneticists routinely represent genome sequences as the single strand with a 5′ direction toward the start position of the chromosome. **(b)** SNP genotypes are indicated with a two-letter notation. **(c)** A comparison of two human genomic sequences with the chimp sequence. *Boxes* indicate positions where single-base changes have occurred since divergence of the species. **(d)** The NCBI Sequence Viewer and the UCSC Genome Browser were used to interrogate a 400-kb region of chromosome 7 (from 116 700 001 to 117 100 000) that contains three genes, including *CFTR*, and a small unnamed pseudogene. Three tracks of SNPs are shown. Two were read from the personal genomes of Watson and Venter. The third track shows the positions of all SNPs that have been uncovered in sequences from European subjects.

derived allele (**Figure 15.3c**). With a comparison of the two genomes, it is possible to identify which of the two alleles at a human SNP locus is the original version that was present in the common chimp-human ancestor. In the example shown, two single-base changes have occurred in this small genomic region since the divergence of the two species. One is shared by all human genomes and is thus not polymorphic.

The second base change was from a C in the common chimp-human ancestor to a T in the ancestor of some people but not others. This means that if you and a friend share an allele at an anonymous SNP locus, you both got that allele from the same ancestor (who may have lived thousands or even hundreds of thousands of years ago). The fact that every random pair of human beings on the planet shares many unlinked SNP alleles indicates recent common ancestry for all people.

SNP distributions

Although SNPs in coding sequences can alter the amino acid sequence of a gene product and have a direct impact on phenotype, the vast majority of SNPs appear to be functionally silent. The reason is illustrated with the 400-kb genomic region in **Figure 15.3d**. This particular region is denser than average in transcription units with three functional genes that include the cystic fibrosis transmembrane receptor (*CFTR*), which can mutate to cause cystic fibrosis. But the actual number of base pairs used for protein coding in all three genes is only 6945, which is less than 2 percent of the total. Even if we add in the additional base pairs

involved in gene regulation and splicing, over 95 percent still remains in the nonfunctional category.

No evolutionary advantage or disadvantage is present for mutations at these noncoding, nonregulatory loci. Thus, single-base mutations that occur at nonfunctional sites will not be selected against, and although most are lost just by chance, some will remain and gain frequency in a population. But as described earlier, some do alter coding regions, and some of those are likely to have a phenotypic effect. Functional SNPs will be subject to selective pressures like other functional mutations.

Human SNPs

Although it would be possible in theory for SNPs to exist at a billion or more genomic sites, the predicted number of common human SNPs is much lower for reasons that will be fully elucidated later in the chapter. To date, the analysis of thousands of human genomes has led to the identification of approximately 38 million SNPs, which is likely to represent the majority of those that actually exist. In Figure 15.3d, you can see the distribution of all known SNPs across the 400-kb CFTR region.

Most SNP variation among people is confined to a limited number of positions. This result is seen most clearly in a comparison of the two most fully validated whole-genome sequences: those of James Watson and J. Craig Venter. Approximately 3.3 million SNPs were identified that distinguish the two individuals (as predicted for an average pairwise SNP frequency of once per kilobase). Of the total SNPs found, 82 percent and 85 percent,

respectively, are listed in the SNP database (dbSNP) at NCBI. These two results are remarkably consistent with each other and with prior predictions.

In Figure 15.3 is a display of the SNP differences observed in a comparison of the personal genomes of Watson or Venter against the genome represented in the human reference sequence. You can also get a sense of the similarity of the Watson and Venter genomes to each other. In some large blocks of genome (e.g., 116 740 to 116 830 K), differences of either man with the reference genome are sparse and unique. In these blocks, Watson and Venter are no more related to each other than either is to the reference sequence.

A 20-kb block of genome centring on position 116 880 shows a very different three-way relationship in which Watson and Venter carry the same SNP alleles that differ substantially from the reference sequence. The block patterns of SNP similarity and dissimilarity provide the foundation for genome-wide associations studies (discussed in Chapter 20).

Some SNPs that do not have a direct effect on phenotype lie so close to a disease gene, or other genes influencing significant phenotypic differences, that they can serve as DNA markers: specific DNA loci with identifiable variations. Medical researchers can use such markers to identify and follow phenotypic differences in groups of people.

SNPs can be genotyped with several different molecular methods

Because alleles of a SNP locus are well-defined single-base changes in DNA sequence, they can be distinguished

by a variety of molecular biology protocols that operate upon, or resolve, specific DNA sequences. These protocols include restriction enzyme digestion, gel electrophoresis, Southern blotting, PCR, allele-specific oligonucleotide hybridization, and DNA microarrays.

Southern blot analysis of restriction-site-altering SNPs

By chance, a small proportion of SNPs will eliminate or create a restriction site recognized by a restriction enzyme. When this happens, researchers can use the restriction enzyme to distinguish between the two alleles. Consider, for example, the SNP shown in **Figure 15.4a**. A single-base change from A in allele 1 to G in allele 2 (using single-strand notation) determines the presence or absence of an *Eco*RI restriction site.

Prior to the 1990s, a Southern blot was the tool of choice for detecting a SNP-related restriction site polymorphism. In the Southern blot protocol, genomic DNA from the test samples is treated with *Eco*RI and the digested DNA separated by gel electrophoresis (and then transferred to a filter paper). The resulting Southern blot is hybridized with a DNA probe obtained from the region between the polymorphic restriction site and an adjacent nonpolymorphic restriction site.

The length of the genomic restriction fragment that hybridizes to the probe reveals which version of the polymorphic restriction site is present. The probe in **Figure 15.4b** detects a 3-kb restriction fragment in DNA with SNP allele 1 and a 5-kb restriction fragment in DNA with SNP allele 2. Because the different SNP alleles change the size of the hybridizing restriction fragment detected on

(a)

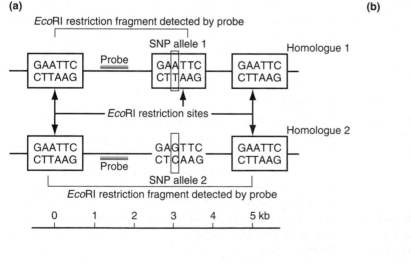

(b)

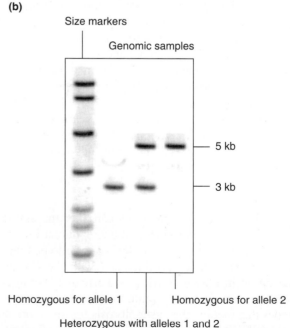

Figure 15.4 Restriction-site-altering SNPs detected by Southern blots. (a) A SNP can create a restriction site polymorphism at an *Eco*RI site. The two SNP alleles will produce different-sized restriction fragments. **(b)** Southern blot analysis and hybridization with the probe can distinguish among the three possible genotypes at this SNP locus.

the Southern blot, this type of polymorphism is called a restriction fragment length polymorphism, or RFLP.

PCR analysis of restriction-site-altering SNPs

Restriction site polymorphisms can be detected much more quickly and cheaply with a PCR-based protocol. This protocol has three steps:

1. Amplification by PCR of a several hundred base-pair region encompassing the SNP.
2. Exposure of the PCR products to the appropriate restriction enzyme.
3. Evaluation of the samples by gel electrophoresis and ethidium bromide staining, followed by a reading of the size of the DNA fragments off the gel.

We illustrate this experimental approach with a solution to the real-life problem of detecting the mutation at the β-globin locus that is responsible for sickle-cell anaemia. Sickle-cell anaemia occurs, as we have seen, when a person carries two copies of a mutant form of the *HBB* gene with a single-base substitution that replaces an A with a T and changes the encoded amino acid from glutamic acid to valine (see Section 14.7). The normal allele is called *A*, and the sickle-cell allele is *S*. Since the sickle-cell mutation also by chance destroys the recognition site of the restriction enzyme *Mst*II (**Figure 15.5a**), it is possible to use PCR and restriction enzyme digestion to detect the mutant allele.

Suppose a carrier couple (both of genotype *AS*) have a child with sickle-cell anaemia (genotype *SS*) and want to know the genotype of the fetus they have recently conceived. Through amniocentesis, their doctor recovers fetal cells from the pregnant woman's womb. He or she next subjects the genomic DNA in this sample, as well as in samples from both parents and the first child, to PCR amplification with primers complementary to sequences on either side of the sickle-cell mutation (Figure 15.5a). The doctor then mixes the restriction enzyme *Mst*II with the PCR products and separates the resulting DNA fragments according to size by gel electrophoresis. It is easy to distinguish DNA derived from the normal allele—which is digested into two fragments by *Mst*II—from the indigestible DNA derived from the mutant allele. **Figure 15.5b** shows the results: The fetus is *AA*, so the younger sibling will neither have sickle-cell anaemia nor carry the sickle-cell trait.

Detection of any SNP with allele-specific oligonucleotide hybridization

Most SNP variants do not alter restriction sites. Fortunately, however, they can be detected by a protocol that exploits differences in hybridization between short oligonucleotide probes and a genomic sequence containing a SNP.

Only with very short probes—oligonucleotides containing around 40 bases—can single-base changes provide a large enough difference to be readily detected. The reason is that for very small DNA molecules—those composed

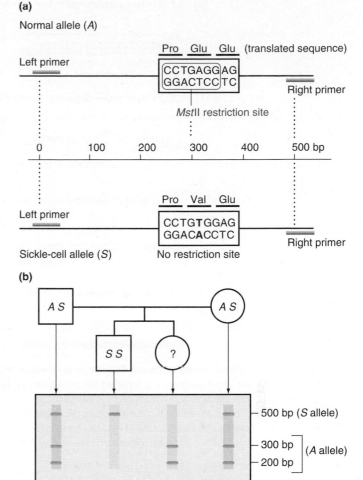

Figure 15.5 Detection of the sickle-cell-causing SNP with PCR. (a) The normal (*A*) and sickle-cell (*S*) alleles at the β-globin locus differ by a single base-pair substitution that changes glutamic acid (Glu) to valine (Val) in the protein product. The base-pair change also eliminates the restriction site *Mst*II. **(b)** PCR amplification of the region containing this SNP, with the primers shown, produces a 500-bp product. Exposure of the normal PCR product to *Mst*II digests this DNA fragment into two smaller fragments, 200 and 300 bp in size; exposure of the mutant PCR product to the restriction enzyme has no effect. Three possible genotypes can be distinguished.

of no more than 60 bp—the length of the molecule itself helps determine whether the double helix remains intact or falls apart. The effective length, and therefore the strength of the hydrogen-bond forces holding together the double helix of a short-probe/short-target DNA hybrid, depends on the longest stretch that does not contain any mismatches. When the two strands do not match exactly, there may not be enough weak hydrogen bonds in a row to hold them together.

If, for example, a 40-base probe hybridizes to a target strand that differs at a single base in the middle of the sequence, the effective length of the resulting double-stranded hybrids is only 20 bp. Since a 20-bp hybrid is significantly less stable than a 40-bp hybrid, one can devise hybridization conditions to select between them—for example, by choosing temperatures under which the perfect hybrids will remain intact, while the imperfect hybrids

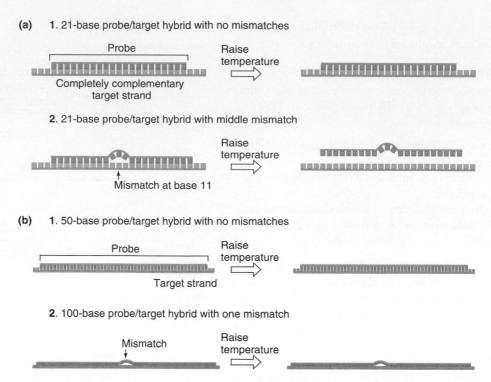

(a) **1.** 21-base probe/target hybrid with no mismatches

Probe

Raise temperature

Completely complementary target strand

2. 21-base probe/target hybrid with middle mismatch

Raise temperature

Mismatch at base 11

(b) **1.** 50-base probe/target hybrid with no mismatches

Probe

Raise temperature

Target strand

2. 100-base probe/target hybrid with one mismatch

Mismatch

Raise temperature

Figure 15.6 Short hybridization probes can distinguish single-base mismatches. (a) Researchers allow hybridization between a short 40-base probe and two different target sequences. **1.** A perfect match between probe and target extends across all 40 bases. When the temperature rises, this hybrid has enough hydrogen bonds to remain intact. **2.** With a single-base mismatch in the middle of the probe, the effective length of the probe-target hybrid is only 20 bases. When the temperature rises, this hybrid falls apart. **(b)** Researchers allow hybridization to occur with a probe of 100 bases. **1.** A perfect match between a 50-base probe and its target bases achieves stability; any extension in the length of the match has no significant effect on temperature dissociation. **2.** Thus, a 100-bp hybrid with one mismatched base is not easily distinguished from a 100-bp hybrid with a perfect match.

will not (**Figure 15.6a**). By comparison, molecules longer than 60 bp can maintain their double helix conformation even with intermittent mismatches. Once a critical number of hydrogen bonds required for double helix stability is achieved, any further increase in the number of these bonds makes no difference (**Figure 15.6b**). Short oligonucleotides of 30–40 bases that hybridize to only one of the two alleles at a SNP locus under appropriate conditions are known as **allele-specific oligonucleotides**, or **ASOs**.

Detection of millions of SNPs simultaneously with DNA microarrays

Rapid advances in DNA microarray (or chip) manufacturing technology (described in more detail in Chapters 1 and 20) have led to an exponential rise in their capacity. Standard microarrays produced by several companies detect SNP alleles at over 2 million loci for a cost of several hundred dollars per sample (this works out to a per-SNP genotyping cost that is a small fraction of a penny).

International cooperation among geneticists has allowed the development of a standardized SNP nomenclature system and a freely accessible public database (http://www.ncbi.nlm.nih.gov/projects/SNP/) where SNP information is compiled. Anyone can now purchase a SNP chip analysis of their own personal genome from several direct-to-consumer companies. With SNP results in hand, consumers can now perform their own comparisons to see

what secrets their genomes hold. **Figure 15.7** provides an example of text output from dbSNP for all SNP loci within the coding regions of the first four *CFTR* exons.

Genetic variation can be caused by subtraction or addition of short sequences

The DNA changes in this category are the result of mutagenic events that expand or contract the length of a DNA region by deleting, duplicating, or inserting genetic material into chromosomes. These changes in genomic length can range in size from one base pair to multiple megabases. In this section, we address genetic variants of just one or a few base pairs. In the next section, we will discuss larger duplications and deletions.

Deletion-insertion polymorphisms (DIPs)

Short insertions or deletions of genetic material that are typically one or a few base pairs in length represent the second most common form of genetic variation in the human genome. These variants are referred to as **InDels** or **DIPs**. A direct comparison of the Venter whole-genome sequence to that of the human reference sequence detects 292 102 DIPs ranging in length from one base pair to 571 base pairs, with a steep decline in relative frequency in relation to length (**Figure 15.8a**).

A visual sense of the density and distribution of DIPs relative to SNPs and SSRs (discussed in the next section) can

	Position			Allele		Variant Function	SNP ID	Allele Frequency
	mRNA	Protein	Codon	Base	Amino Acid			
Exon_1	133	1				start codon		
	156	8	3	A	Lys [K]	synonymous	rs1800071	N.D.
		8	3	G	Lys [K]			
	163	11	1	A	Ile [I]	missense	rs1800072	N.D.
		11	1	G	Val [V]			
Exon_2	204	24	3	A	Leu [L]	synonymous	rs55773134	N.D.
		24	3	G	Leu [L]			
	223	31	1	T	Cys [C]	missense	rs1800073	0.014
		31	1	C	Arg [R]			
	263	44	2	T	Val [V]	missense	rs1800074	N.D.
		44	2	A	Asp [D]			
Exon_3	345	71	3	C	Asn [N]	synonymous	rs1800075	N.D.
		71	3	T	Asn [N]			
	356	75	2	A	Gln [Q]	missense	rs1800076	0.012
		75	2	G	Arg [R]			
Exon_4	485	118	2	C	Ser [S]	frameshift	rs35871908	N.D.
		118	2	-/C	Ser [S]			
	492	120	3	A	Ala [A]	synonymous	rs1800077	N.D.
		120	3	G	Ala [A]			
	545	138	2	C	Pro [P]	missense	rs1800078	N.D.
		138	2	T	Leu [L]			
	575	148	2	C	Thr [T]	missense	rs35516286	0.026
		148	2	T	Ile [I]			

Figure 15.7 Known SNP loci and alleles in the first four exons of the *CFTR* gene. Only SNP loci within the coding regions of the first four *CFTR* exons are shown. Many more SNPs are located in the introns between these exons. Transcript (mRNA) and codon (protein) addresses are indicated for each SNP, along with position within each codon. Both alleles of each SNP locus are indicated. No function is provided for the alleles that appear in the human reference sequence.

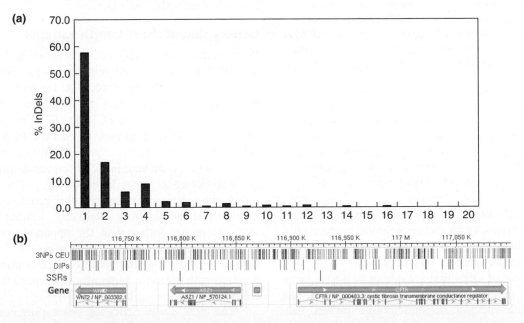

Figure 15.8 Genetic variants defined by changes in DNA length. (a) Size distribution of InDel variations detected between the human reference genome and the Venter genome. **(b)** Distribution of SNPs, DIPs (InDels), and SSR variants in the *CFTR* region.

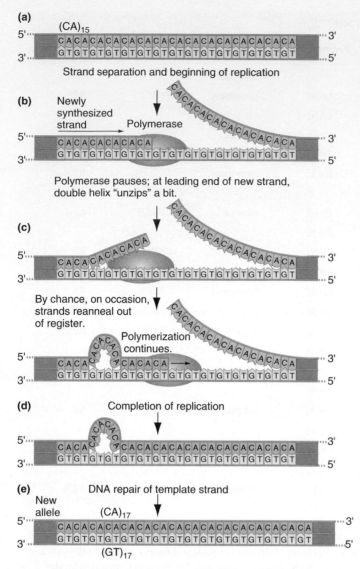

(a)

$(CA)_{15}$

5′ ···| C A C A C A C A C A C A C A C A C A C A C A C A C A C A C A |··· 3′
3′ ···| G T G T G T G T G T G T G T G T G T G T G T G T G T G T G T |··· 5′

Strand separation and beginning of replication

(b) Newly synthesized strand → Polymerase

5′ ···| C A C A C A C A C A C A |··· 3′
3′ ···| G T G T G T G T G T G T G T G T G T G T G T G T G T G T G T |··· 5′

Polymerase pauses; at leading end of new strand, double helix "unzips" a bit.

(c)

5′ ···| C A C A C A C A C A |··· 3′
3′ ···| G T G T G T G T G T G T G T G T G T G T G T G T G T G T G T |··· 5′

By chance, on occasion, strands reanneal out of register.

Polymerization continues.

5′ ···| C A C A | C A C A C A |→ 3′
3′ ···| G T G T G T G T G T G T G T G T G T G T G T G T G T G T G T |··· 5′

(d) Completion of replication

5′ ···| C A C A | C A C A C A C A C A C A C A C A C A |··· 3′
3′ ···| G T G T G T G T G T G T G T G T G T G T G T G T G T G T G T |··· 5′

(e) DNA repair of template strand

New allele $(CA)_{17}$

5′ ···| C A C A C A C A C A C A C A C A C A C A C A C A C A C A C A C A |··· 3′
3′ ···| G T G T G T G T G T G T G T G T G T G T G T G T G T G T G T G T |··· 5′

$(GT)_{17}$

Figure 15.9 Simple sequence repeats (SSRs) are highly polymorphic because of their potential for faulty replication.
(a) An SSR consisting of 15 tandem repeats of the CA dinucleotide sequence. **(b)** Replication of the strands by DNA polymerase moving in the 5′-to-3′ direction. **(c)** Pauses can occur if the required nucleotides are, by chance, not in the vicinity of the polymerase. **(d)** When the required nucleotides become available, the newly synthesized strand reanneals to the template and acts as a primer for further replication. But the new strand may be out of register such that the polymerase begins by adding one or more nucleotides across from a part of the template strand that has already been replicated. **(e)** The resulting DNA molecule will have one or more identical repeats in the newly synthesized strand. DNA repair processes then adjust the template strand to make it the same length as the newly synthesized strand.

be attained in a view of the 400-kb genomic region around *CFTR*, shown in **Figure 15.8b**. While SNP loci occur with a frequency of about one per kilobase, DIPs are distributed at a frequency of about one in every 10 kb of DNA.

The 75 percent of DIPs that are one or two base pairs in length can be detected on DNA microarrays alongside SNPs, with allele-specific oligonucleotide probes that match the presence or absence of the base pairs. Larger DIPs display more of a size differential between alleles

and are amenable to detection by PCR and gel electrophoresis as detailed next for SSRs.

Simple sequence repeats

The genomes of humans and other complex organisms are loaded with loci defined by **simple sequence repeats (SSRs)**. The most common repeating units are one-, two-, or three-base sequences repeated in tandem 15–100 times. Examples of SSRs are AAAAAAAAAAAAAAA or CACACACA-CACACACACACACA. In the mammalian genome, the CA-repeat SSR occurs on average once in every 30 000 bp.

SSRs arise spontaneously from random events that initially produce a short repeated sequence with four to five repeat units. Once a short SSR mutates into existence, however, it can expand into a longer sequence by the process shown in **Figure 15.9**. Unlike SNPs—which are biallelic and do not change after the mutational event that gave rise to them—individual SSR loci often mutate into multiple alleles.

Research has shown that faulty DNA replication is the main mutational mechanism behind SSR polymorphism (Figure 15.9). Because the same short homologous unit (e.g., CA) is repeated over and over again, DNA polymerase may develop a stutter during replication; that is, it may slip and make a second copy of the same dinucleotide, or skip over a dinucleotide. SSRs are thus highly polymorphic in the number of repeats they carry, with many alleles distinguishable at each SSR locus.

New alleles arise at SSR loci at an average rate of one in every thousand gametes. This frequency is much greater than the single nucleotide mutation rate, and results in a large amount of SSR variation among unrelated individuals within a population. At the same time, the rate of SSR mutation is low enough that changes usually do not occur within a few generations of even a large family; because of this, SSRs can serve as relatively stable, highly polymorphic DNA markers in linkage studies of human families, other animals, and plants.

Genotyping of short-length variants

Small variations in the actual size of a locus can be directly and easily distinguished by gel electrophoresis as illustrated in **Figure 15.10**. You begin by using a pair of primers (complementary to sequences on either side of the polymorphism) to PCR amplify the locus from an individual's DNA. You then subject the PCR products to gel electrophoresis to separate DNA fragments according to their size. After staining with ethidium bromide, each allele shows up as a specific band.

A single researcher can use this protocol to genotype hundreds of samples in a single day, without any specialized equipment (other than the apparatus necessary for PCR and gel electrophoresis). The protocol can also be automated by using fluorescently tagged primers in conjunction with the same apparatus that is used for automated DNA sequencing (see Figure 14.14).

The most useful SSRs for genotyping consist of 2 or 3 bp units repeated 15–100 times in a row. These SSRs are highly polymorphic, with multiple alleles that differ in

(a) Determine sequences flanking microsatellites.

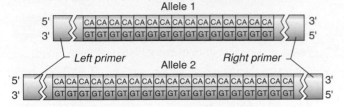

Allele 1

Allele 2

(b) Amplify alleles by PCR.

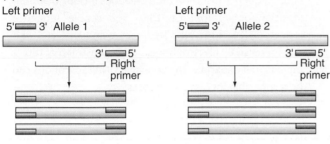

(c) Analyze PCR products by gel electrophoresis.

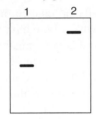

(d) Example of population with three alleles

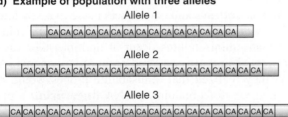

Allele 1

Allele 2

Allele 3

Six diploid genotypes are present in this population.

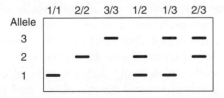

Figure 15.10 Detection of simple sequence repeat (SSR) polymorphisms by PCR and gel electrophoresis. (a) SSR alleles differ in length. Left and right primers are devised based on sequences that flank the SSR locus. **(b)** Genomic DNA is amplified by PCR with primers specific for the SSR locus. **(c)** Gel electrophoresis and ethidium bromide staining distinguish the alleles from each other. **(d)** SSRs are often highly polymorphic with many different alleles present in a population.

2 or 3 bp unit increments. All of the multiple alleles can be detected as different-sized PCR products (Figure 15.10).

SSRs played a crucial role in the development of linkage maps across the genomes of mice, humans, and other

species. The reasons for their widespread use include their frequent appearance in all vertebrate genomes and their extensive polymorphism.

SSRs and disease

A small number of genes naturally contain SSR sequences with triplet repeat units within their coding regions. The propensity of the SSR sequences to change in size from one generation to the next can produce mutant alleles with drastic effects on phenotype. One example of a disease in this category is Huntington disease (HD).

Huntington disease is transmitted as an autosomal dominant mutation. Over 3500 Canadians show one or more symptoms of the disease—involuntary, jerky movements; unsteady gait; mood swings; personality changes; slurred speech; impaired judgment. An additional 17 500 have an affected parent, which gives them a 50:50 chance of carrying and expressing the dominant condition themselves as they age. Although symptoms usually show up between the ages of 30 and 50, the first signs of the disease have appeared in people as young as 2 and as old as 83. Some people with a family history of HD would like to know their genotype before deciding whether to have a family.

In 1993, after 10 years of intensive research, investigators identified and cloned the *HD* gene. With the gene in hand, they were able to uncover the unusual mutation that causes the disease (**Figure 15.11a**). Unlike the vast majority of disease mutations, which result from base-pair

(a) Basic structure of the *HD* gene's coding region

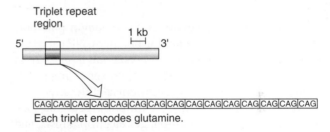

Each triplet encodes glutamine.

(b) Some alleles at the *HD* locus

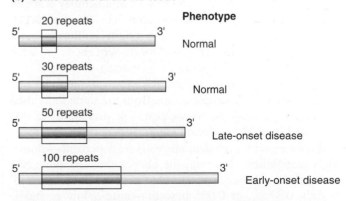

Figure 15.11 Mutations at the Huntington disease locus are caused by expansion of a triplet repeat microsatellite in a coding region. (a) Near the 5′ end of the coding region is a repeating triplet sequence that codes for a string of glutamines. **(b)** Different alleles at the *HD* locus have different numbers of repeating units. Fewer than 34 repeats gives a normal phenotype. As the number of repeats increases beyond 42, the onset of the disease is earlier.

changes or the elimination of genetic information, HD is caused by too much genetic information: an expansion of a CAG trinucleotide repeat in the coding sequence, which translates into a string of glutamine amino acids.

It is possible to detect *HD* alleles directly with the same size-based PCR procedure used to detect other SSR alleles. The normal allele contains up to 34 repeats, while disease-causing alleles carry 42 or more. In general, the greater the number of repeats, the earlier the age of disease onset (**Figure 15.11b**). Those who inherit a disease allele invariably get the disease if they live long enough. Thus, although expressivity, which depends on the number of triplet repeats, is variable, penetrance is complete. Several other diseases caused by triplet repeat expansion have been uncovered, including a variety of neurological disorders, and the fragile X syndrome described in the Genetics and Society box in Chapter 8.

- SNPs are the simplest and most frequent type of stable genetic variation. In a comparison of any two unrelated haploid human genomes, alternative SNP alleles will be found, on average, once in every 1000 base pairs.

- Automated technology allows millions of SNPs to be detected simultaneously. Hybridization, PCR amplification, and the Southern blot technique can allow comparison and identification of individual SNPs.

- Short-sequence insertions and deletions include deletion-insertion polymorphisms (DIPs) and simple sequence repeats (SSRs). These may be detected and analyzed with PCR followed by electrophoresis. SSRs consisting of triplet repeats that expand with successive DNA replications can be responsible for genetic diseases.

15.3 Deletions or Duplications of a DNA Region

Earlier we mentioned that a fourth category of DNA variants consists of duplications or deletions. Included in this category are the short copy number repeats (minisatellites), which are useful for DNA fingerprinting, and the large-scale duplications and deletions referred to as copy number polymorphisms (CNPs) and copy number variants (CNVs).

Minisatellites are ideal for DNA fingerprinting

Minisatellites are a subcategory of DNA length polymorphisms that are at the low end of the broader CNP category. They are defined arbitrarily as repeats having a unit size in the range of 500 bp to 20 kb. The real power of minisatellites lies in the fact that particular minisatellite sequences often occur at a small number of different genomic loci. With restriction enzyme digestion, gel electrophoresis, and Southern blot hybridization using a cross-hybridizing minisatellite probe, researchers can look simultaneously at allelic variation at these multiple unlinked loci (**Figure 15.12**). (This strategy would not work for microsatellites because their core mono-, di-, or trinucleotide sequences are each present thousands of times in the genome.)

How many loci would you have to examine to be certain that two DNA samples come from the same individual (or identical twins) and no one else? The probability of two unrelated individuals having identical genotypes at a locus with two equally prevalent alleles is 37.5 percent—quite a high probability. However, the chance that the same two individuals will be identical at ten such loci, all unlinked, is only 0.375^{10}, or 0.005 percent—quite a low probability. The result of 0.005 percent means there is 1 chance in 20 000 that the two will, by chance, have the same genotype at ten unlinked loci. By extension, if you simultaneously detect genotype at 24 unlinked two-allele loci, the chance of two individuals being the same at all 24 drops to 0.375^{24}, or 1 in 17 billion. Since the total human population is less than

8 billion, there is virtually no chance that two individuals (who are not identical twins) would have the same genotype at all 24 loci. In short, a relatively small number of loci are sufficient to produce a combination genotype pattern that, like a traditional fingerprint, will be unique for each individual (or pair of identical twins) within the species.

How minisatellite comparison generates DNA fingerprints

In 1985, Alec Jeffreys and co-workers made two key findings: (1) each minisatellite locus is highly polymorphic, and (2) many minisatellites occur at multiple sites (usually between 2 and 50) scattered around the genome. As a result, they realized that minisatellite probes would be perfect reagents for obtaining a **DNA fingerprint**: a pattern produced by the simultaneous detection of genotype at a group of unlinked, highly polymorphic loci.

The most useful minisatellite families have 10–20 members per genome. This range of numbers is small enough to allow the resolution of all the loci as individual bands on an autoradiograph, but large enough to provide true fingerprint information. If one fingerprint is not sufficient to resolve the relationship between two different DNA samples, investigators can always obtain data from two, three, or even more minisatellite families.

Figure 15.13 illustrates an interesting example of the utility of DNA fingerprinting. In 1997, scientists from the Roslin Institute in Scotland announced that they had cloned a sheep by injecting a diploid nucleus from an adult udder cell (grown in culture) into an unfertilized egg whose own genetic material had been removed. Initially, many scientists were sceptical of this result and thought that "Dolly" might actually be the result of a fertilization between some contaminating sperm and the egg. If this were the case, Dolly's genome would be unique. Instead, the results of the fingerprint analysis shown in Figure 15.13 demonstrated that

(a) Digest DNA with restriction enzyme that does not cut inside minisatellite.

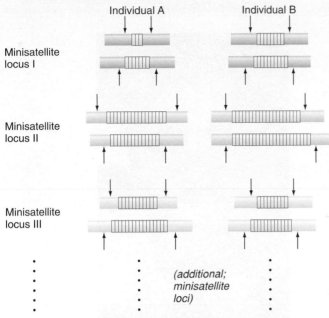

Minisatellite locus I

Minisatellite locus II

Minisatellite locus III

(additional; minisatellite loci)

(b) Run DNA samples on a gel. Perform Southern blotting. Hybridize with probe containing minisatellite sequence.

Individuals

A   B   C   D

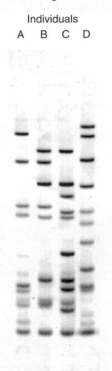

Figure 15.12 Minisatellite analysis provides a broad comparison of whole genomes. (a) Two individuals each carry two alleles at three loci containing the same minisatellite-repeating unit sequence. The *arrows* indicate a restriction site recognized by a particular restriction enzyme. Notice that minisatellite lengths are different both among alleles at a locus and among different loci. **(b)** After restriction enzyme digestion, gel electrophoresis, Southern blotting, and hybridization to the minisatellite probe, researchers can obtain an autoradiograph of the type shown here for four individuals.

Dolly's DNA fingerprint was identical to that of the adult udder cell used to clone her. This established beyond a doubt that her genome was indeed a clone of that cell's DNA.

The uses of DNA fingerprinting

DNA fingerprints are a powerful tool for forensic analysis. Crown attorneys and defence lawyers alike use them to show the likelihood of a suspect's presence at the scene of a crime or to prove the innocence of someone falsely accused (see the **Genetics and Society** box **"Correcting a Miscarriage of Justice"**). More recently, DNA fingerprinting demonstrated that skeletal remains unearthed at Ekaterinburg in the Ural Mountains of Russia belonged to Czar Nicholas II and his family, who were murdered in 1918 during the Bolshevik Revolution. Geneticists established the relationship by comparing

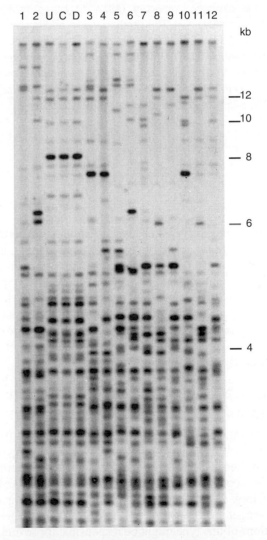

Figure 15.13 DNA fingerprint analysis confirmed that Dolly was cloned from an adult udder cell. Genomic DNA samples were prepared from the donor udder cells (U), from the cell culture prepared from the udder cells (C), from Dolly's blood cells (D), and from control sheep 1–12. The DNA fingerprints of the 12 control sheep are all different from one another and from the cells involved in the Dolly experiment. Dolly's DNA fingerprint is identical to the fingerprints of both the udder cells and the derived cell culture. This result provides very strong evidence that Dolly is a clone of the ewe that donated the udder cells.

Guy Paul Morin (**Figure A**) was only 25 years old when arrested and accused of raping and murdering his 9-year old neighbour, Christine Jessop, in Queensville, Ontario. After a series of trials and appeals, he was finally convicted and imprisoned in 1992. Fortunately, for Mr. Morin, DNA fingerprinting technology (that had not been available during his trials and appeals) conclusively established his innocence. He was acquitted by the Ontario Court of Appeal in 1995. The identity of Christine Jessop's murderer is still unknown.

The testing that proved Morin's innocence was funded, in part, by the Justice for Guy Paul Morin Committee. This small group of volunteers later became the Association in Defence of the Wrongly Convicted (AIDWYC) based in Toronto, Ontario (aidwyc.org). This group reviews, and in some cases, supports claims of innocence in homicide cases. One of their most useful tools is DNA fingerprinting technology. As of 2013, AIDWYC has helped exonerate 14 wrongfully convicted individuals. In five of these cases DNA fingerprinting was crucial in establishing the innocence of the accused.

Figure A   Guy Paul Morin. In 1995, Morin was acquitted of the crime of murder with the help of the Justice for Guy Paul Morin Committee.

DNA from the excavated bones with samples obtained from a number of living relatives of the Romanov family, including Prince Philip, Duke of Edinburgh. This information disproved the claim of Anna Anderson (**Figure 15.14a**) that she was the Grand Duchess Anastasia (**Figure 15.14b**); in three independent analyses, her DNA (obtained from hair and from biopsy samples removed during an examination for cancer years before her death in 1984) did not match that of members of the Romanov line—living or dead.

Large-scale deletions and duplications commonly differentiate human genomes

With advances in DNA microarray technology (see Chapter 1), it became possible to scan individual genomes for the presence of deletions and duplications that were large enough to cover one or more whole genes. Several research groups took advantage of this new technological capability to investigate the possibility that relatively large genomic alterations of this type might be responsible for severe mental diseases that were clearly heritable, but had resisted previous attempts at genetic localization.

Identification of CNPs and CNVs

Researchers unexpectedly discovered an extensive degree of polymorphism within their control population (non-diseased) for the deletion or duplication of relatively large blocks of genetic material that can measure up to 1 Mb in

length without causing disease. This category of genetic variants is referred to as **copy number variants (CNVs)** or **copy number polymorphisms (CNPs)**, depending on whether their frequency in the population is less than or

(a)   **(b)**

Figure 15.14 Anna Anderson (1901–1984) and Anastasia Nikolaevna (1901–1918). (a) Anne Anderson was one of many imposters claiming to be the Grand Duchess Anastasia of Russia. Anna Anderson's fraudulent claim was refuted through DNA fingerprinting analysis. **(b)** The real Anastasia was killed along with her family in 1918 by Bolshevik revolutionaries.

greater than 1 percent. For ease of description, the CNV term will be used to encompass both CNVs and CNPs when they are grouped together.

Unlike SNPs, CNVs can be detected through microarray analysis even if they have not been observed previously. The Affymetrix 6.0 DNA microarray is designed specifically to allow detection of CNVs with several hundred thousand nonpolymorphic oligonucleotide probes (NPOs), which are spaced uniformly across the genome. CNVs are detected as an increase or decrease in hybridization—for duplications and deletions, respectively—across a contiguous set of NPOs (**Figure 15.15a**).

Surprisingly to geneticists, CNVs turn out to be quite common both in their distribution across the genome and in their frequency of occurrence within human populations (**Figure 15.15b**). Over 6000 CNV loci have been identified, and pairwise comparison of any two genomes typically identifies a different allele for several hundred. Over 99 percent of all CNVs are derived from inheritance rather than new mutation.

Copy number variation in the olfactory receptor (*OR*) gene family

One example of copy number variants (CNVs) is the olfactory receptor (*OR*) gene family, which is composed of several hundreds to thousands of members that provide animals with the ability to smell a diverse array of odours. A typical mouse genome carries 1400 *OR* genes distributed at numerous chromosomal sites. In humans, however, a keen sense of smell is no longer as important for survival. As a result, *OR* genes can be lost without consequence, and people typically carry less than 1000 genes. However, individuals vary widely around the mean.

Figure 15.16a shows the variation in copy number among 60 people at 11 representative *OR* loci. One locus, *OR4K2*, varies in copy number from two to six in different genomes, while seven of the eleven loci are completely missing from some individuals. All together, some people can have hundreds of *OR* genes more than others do, resulting in large differences in the abilities of people to distinguish odours.

(a)

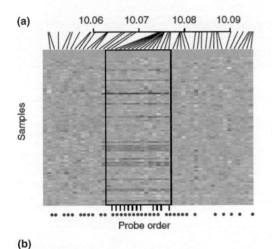

Figure 15.15 Chromosomal locations of CNPs or CNVs identified in multiple individuals. (a) Results of DNA microarray analysis performed on 88 samples, lined up in rows, for adjacent probes across a region of chromosome 4. Each column portrays the intensity of hybridization for a particular probe. *Red* indicates low intensity and *yellow* high intensity. Several samples display evidence of a deletion across the same 15-kb genomic region. **(b)** Representation of a human karyotype. *Blue* and *green* bars represent the locations of CNVs and CNPs.

(b)

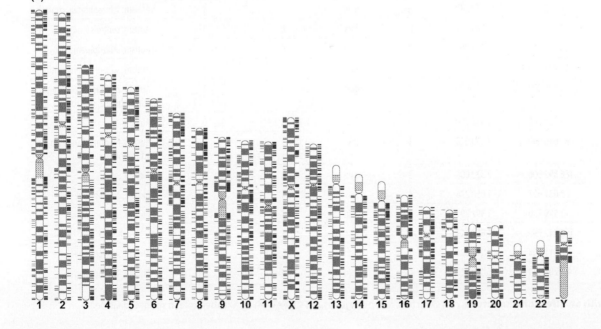

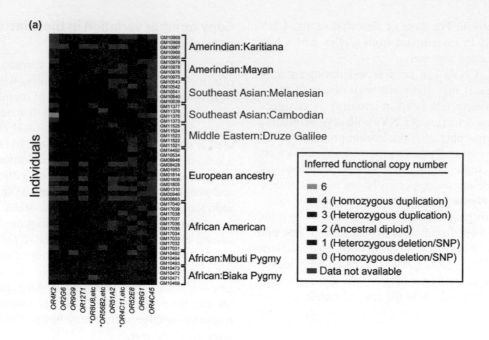

(b)

Chr	Start	Stop	Length	Total CNVs	Type	Diseased	Controls	Diseases
	Chromosomal Location					**Number of People**		
chr15	27 015 263	30 650 000	3 634 737	19	loss	19	0	Schizophrenia
chr15	18 376 200	30 756 771	12 380 571	58	gain	45	13	Autism, mental retardation, schizophrenia, controls
chr22	17 014 900	19 993 127	2 978 227	31	loss	31	0	Autism, mental retardation, schizophrenia
chr22	17 200 000	21 546 762	4 346 762	14	gain	9	5	Autism, schizophrenia, controls
chr1	142 540 000	146 059 433	3 519 433	27	loss	24	3	Autism, schizophrenia, controls
chr1	142 800 580	146 009 436	3 208 856	15	gain	12	3	Autism, mental retardation, schizophrenia, controls
chr22	45 144 027	49 509 153	4 365 126	4	loss	4	0	Autism
chr22	47 572 875	48 323 417	750 542	6	gain	5	1	Autism, schizophrenia, controls
chr16	29 470 951	30 252 473	781 522	11	loss	8	3	Autism, controls
chr16	29 474 810	30 235 818	761 008	7	gain	6	1	Autism, schizophrenia
chr17	14 000 000	15 421 835	1 421 835	7	loss	6	1	Autism, schizophrenia, controls
chr17	12 650 000	15 540 000	2 890 000	5	gain	4	1	Autism, mental retardation, schizophrenia, controls
chr16	60 141 700	61 581 600	1 439 900	4	loss	4	0	Autism
chr11	78 120 000	85 610 000	7 490 000	3	loss	3	0	Autism, mental retardation, schizophrenia
chr2	184 270 000	186 892 000	2 622 000	3	gain	3	0	Autism
chr15	82 573 421	83 631 697	1 058 276	4	loss	4	0	Autism, schizophrenia
chr9	206 456	1 599 250	1 392 794	3	gain	3	0	Autism, schizophrenia
chr3	197 179 156	198 842 299	1 663 143	3	loss	3	0	Schizophrenia
chr16	21 693 739	22 611 363	917 624	5	loss	5	0	Autism, schizophrenia
chr16	80 737 839	82 208 451	1 470 612	4	gain	4	0	Autism, schizophrenia

Figure 15.16 CNVs with an effect on phenotype. (a) Olfactory receptor genes. **(b)** CNVs with a causative role in mental disease.

Copy number variation and mental disease

Serious psychiatric diseases can be devastating to the lives of individuals and their families. Together, the five most frequent of these occur in 5 percent of the population (schizophrenia, bipolar I, bipolar II, autism, and autism spectrum disorder [ASD]). Until recently geneticists struggled to come up with reducible associations of specific genes with these diseases.

Although most CNVs appear not to be associated with strong phenotypic effects, the exceptions are stark. In general, long deletions and duplications—over a megabase in length anywhere in the genome—increase the risk of psychiatric disease to a level of 30 percent. And in particular, CNVs that cover a number of specific genomic regions have been directly associated with autism, schizophrenia, or mental retardation, as indicated in **Figure 15.16b**.

The association of individual SNPs with diseases and traits is usually demonstrable only through large-scale whole-genome-association studies (see Chapter 20). In contrast, the disease-causing potential of a newly discovered CNV can often be evaluated immediately by looking at the NCBI database to determine whether critical genes have been deleted or duplicated. If, for example, a gene with an essential role in brain development or function is deleted or disrupted on one chromosome of a healthy women, and the same gene is deleted or disrupted in her reproductive partner, there is a 25 percent risk of giving birth to a child who has no copies of the essential gene and is likely to be mentally disabled or strongly predisposed to mental illness.

- Modern genetic analysis enables quick comparison of minisatellites—short repeats of 500 bp to 1 kp found at a relatively small number of genomic loci. Analysis of minisatellite genotypes at 10 to 24 of these loci is the basis for DNA fingerprinting, which can be applied to identification of individuals.

- Large-scale deletions and duplications are surprisingly common in the human genome. CNPs and CNVs are present normally in some systems, and they have been implicated in certain diseases, notably mental conditions.

15.4 Positional Cloning: From DNA Markers to Disease-causing Genes

A knowledge of the chromosomal location of individual DNA markers can be of great use to geneticists who want to determine the identity of genes responsible for disease phenotypes and other inherited differences among people, animals, and plants. The chromosomal positions of DNA markers that are linked to these genes can serve as the basis for a process called **positional cloning**.

In a few cases, a causative gene can be discovered without mapping

Medical researchers learned from the analysis of pedigrees like the one shown in **Figure 15.17a**, that haemophilia A is an X-linked recessive trait governed by a single gene. Geneticists were also able to make an educated guess as to the biochemical function of the responsible gene. The function of the wild-type haemophilia A gene, they proposed, is production of a normal clotting factor; mutations that inactivate this factor produce haemophilia A.

Once molecular investigators worked out the details of the blood-clotting cascade (**Figure 15.17b**), they could look for clotting factors in normal individuals that were absent in haemophiliacs (**Figure 15.17c**). In this way, they identified a protein known as factor VIII. They next determined the amino acid sequence of factor VIII and used the genetic code to predict the nucleotide sequence of the corresponding gene. This information allowed them to develop a degenerate oligonucleotide probe that could identify clones of the factor VIII gene (now called the *F8* gene) within genomic libraries (**Figure 15.17d**). When they sequenced this gene from people suffering from haemophilia A, they found mutations with an absolute correlation to the disease phenotype and thereby verified the gene as the causative agent of the disease.

Of the thousands of genes responsible for known human genetic diseases, only a small number can be identified in the manner just outlined. Much more often, making an educated guess about the protein altered by the disease allele is difficult.

Cystic fibrosis, for example, is a recessive autosomal genetic condition inherited by 1 child in every 2500 born from two parents of European descent. But because the trait is recessive, the frequency of unaffected carriers in the population is much greater, about 1 in every 36 people. Many carriers come from families where the disease has never appeared, and so the first birth of a child with the disease can come as a complete shock. Children with the disease have a variety of symptoms arising from abnormally viscous secretions in the lungs, pancreas, sweat glands, and several other tissues. Even with modern medical treatments that combat some symptoms of the disease, half of CF patients in Canada will die before the age of 47 (in 1974 half would die before the age of 23).

Unfortunately, the gross symptoms of cystic fibrosis did not provide insight into the underlying molecular cause

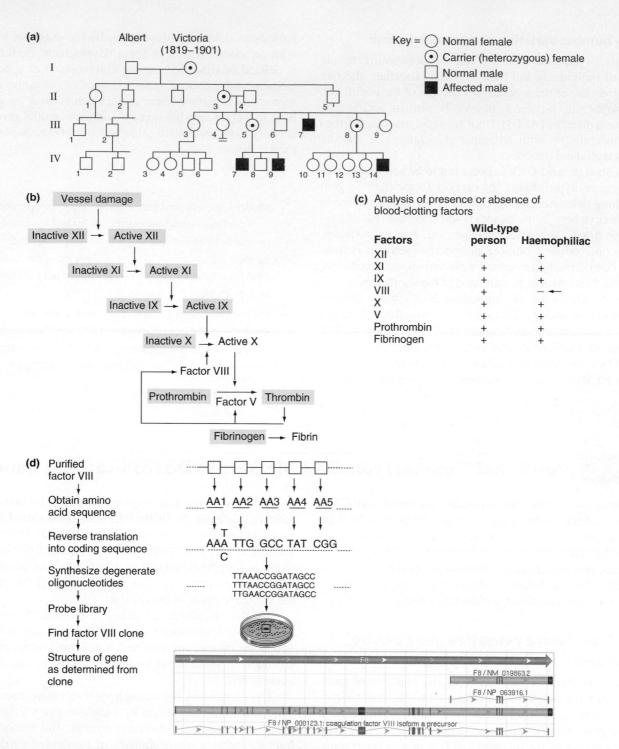

Figure 15.17 How geneticists identified and cloned the haemophilia A gene. **(a)** A pedigree of the royal family descended from Queen Victoria. This family tree uses the standard pedigree symbols. **(b)** The blood-clotting cascade. Vessel damage induces a cascade of enzymatic events that convert inactive factors to active factors. The cascade results in the transformation of fibrinogen to fibrin and the formation of a clot. **(c)** Blood tests can determine whether an active form of each factor involved in the clotting cascade is present. The results of such analyses show that many haemophiliacs, such as those found in Queen Victoria's pedigree, lack an active factor VIII in their blood. **(d)** Researchers purified factor VIII, determined its amino acid sequence, used this information to infer all possible degenerate coding sequences, constructed oligonucleotides for a region with minimal degeneracy, probed a genomic library with these oligonucleotides, and obtained genomic clones of the *F8* gene.

of the disease. Hundreds of proteins contribute to the process of cell secretion, and most of these were still unidentified in the 1980s. Without a way to determine which one was defective in cystic fibrosis, investigators had no simple way to work their way from gene function to the protein's DNA-coding sequence.

In positional cloning, linkage analysis with DNA markers helps identify disease genes

The standard approach for identifying genes associated with human disease is to combine linkage analysis (as described in Chapter 4) with the use of DNA markers

(described earlier in this chapter) to localize the human disease gene to a specific region of chromosomal DNA. Other techniques can then determine which gene—among the small number in this region—contains mutations that correlate with the disease phenotype. This entire protocol is called **positional cloning**.

You learned in Chapter 4 that a simple two-point cross can demonstrate linkage if the two loci under analysis lie close enough together on the same chromosome. You also saw that the frequency of recombination between the two loci provides a direct measure of the distance separating them, as recorded in centimorgans (cM), or map units (m.u.) (geneticists studying humans, mice, and other mammals use the centimorgan unit of measure, which we adopt in this chapter). Finally, you learned that it is possible to integrate multiple pairs of linked loci into a "linkage group" by performing many different two- or three-point crosses with overlapping sets of loci. The **linkage maps** constructed from these crosses depict the distances between loci as well as the order in which they occur on a chromosome.

With the use of SNPs and other polymorphic DNA markers discussed in this chapter, rather than markers defined by phenotype, there is no limit to the number of loci that can be mapped in a single cross or extended human family. In place of a traditional three-point cross, it becomes possible to perform linkage analysis combining thousands, or hundreds of thousands, of DNA loci with the disease locus of interest. If genetic linkage can be demonstrated between a disease trait and one or more previously mapped DNA markers, then the gene responsible for the trait must lie in the same subchromosomal region as those DNA markers.

Discovery of a DNA marker that shows linkage to the disease locus is the first goal of positional cloning. For traits expressed in plants or small animals, it is a simple matter to set up a single testcross for the production of hundreds of offspring that can be easily analyzed to identify the map position of the trait in question. For human traits such as disease phenotypes, directed breeding is not a possibility. Instead, until recently, researchers had no other choice but to try to find many different extended families—each with a large number of children—in which some individuals express the mutant phenotype and others do not.

The mapping of a human disease locus begins with the genotyping of all members of the disease-carrying families for a series of DNA markers ideally spaced along each chromosome. Because the human genome has been sequenced, you already know the chromosomal positions of all DNA markers. The only unknown map position is that of the disease locus. If the number of genotyped individuals is sufficiently large, a simply inherited Mendelian disease locus must show linkage to one or more of these markers. Finding linkage to at least one marker of known position will place the disease locus in a particular subregion of a particular chromosome (**Figure 15.18**).

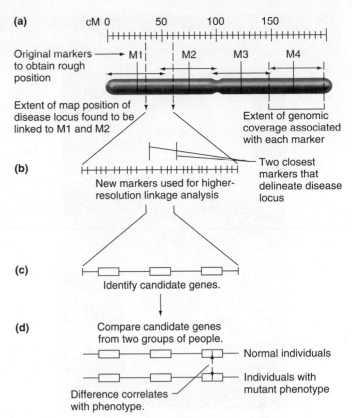

Figure 15.18 Positional cloning: from phenotype to chromosomal location to guilty gene. (a) Diagram of a human chromosome with four markers—M1, M2, M3, and M4—used in the linkage analysis of a disease phenotype. Each marker provides "linkage coverage" of a portion of the chromosome. This suggests that the gene responsible for the disease lies between those markers. **(b)** With this information, an investigator could type additional markers that lie between M1 and M2 to position the disease locus with higher resolution. **(c)** Looking for candidate genes. Analysis of the region between recombination sites (that define the smallest area within which the disease locus can lie) should reveal the presence of candidate genes. **(d)** Finding the correct candidate through comparisons of the structure and expression of each candidate gene in many diseased and nondiseased individuals.

In 1984, the Huntington disease (*HD*) locus became the first human disease gene to be successfully mapped by positional cloning. **Figure 15.19** shows the five-generation, 104-member family pedigree used to demonstrate linkage between a previously mapped DNA marker named G8 and the *HD* locus. Preliminary linkage between the *HD* locus and the G8 marker placed the disease gene on human chromosome 4. Further linkage analysis then narrowed down the map position of *HD* to less than 1000 kb.

The discovery of the *HD* gene was followed soon after by the mapping of the cystic fibrosis gene by Canadian, Lap-Chee Tsui, of the University of Toronto in 1989 (see the **Focus on Inquiry** box **"Cloning of the Cystic Fibrosis (*CF*) Gene—Twenty-five Years Later"**).

Despite the cloning of the *CFTR* gene in 1989 (see Section 15.4 and the Fast Forward box in Chapter 4), CF remains one of Canada's leading killers (on average one Canadian a week dies of CF). CF affects 1 in every 3600 Canadian-born children. Furthermore, 1 in 25 Canadians is a carrier of the defective allele. While Dr. Lap-Chee Tsui was able to identify *CFTR* as the disease-causing locus through positional cloning, this has not yet provided a cure. Although disappointing, there is still good news to report; recent medical advances (based on a detailed understanding of the molecular function of the CFTR protein made possible by Dr. Tsui's work) have dramatically increased life-expectancy of CF patients. At the time of the discovery of the *CFTR* gene in 1989, half of all CF patients would die before the age of 32. Today this has increased to 47 years.

Figure A **Danny Bessette in 2009, holding a copy of the September 8, 1989 issue of *Science* magazine that reported the cloning of the CF gene.** This issue bears Danny's picture on the cover (he was four years old at the time).

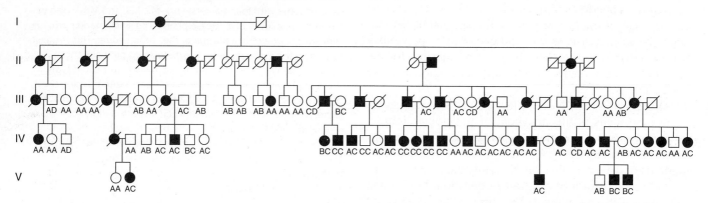

Figure 15.19 **Detection of linkage between the DNA marker G8 and the locus responsible for Huntington disease (HD) was the first step in the cloning of the *HD* gene.** Portion of a large Venezuelan pedigree affected by Huntington disease. For living members of the pedigree, alleles at the G8 marker locus are indicated (A, B, C, and D). It is easy to see the co-transmission of marker alleles with the mutant and wild-type alleles at the *HD* locus. Pedigree analysis shows that the *HD* locus is within 5 cM of the G8 marker.

In some genetic diseases, the affected loci can be identified through inference and by working backward from a defective protein to the DNA sequence in question. Positional cloning allows mapping of disease alleles through comparison with thousands of markers of known position, narrowing the region of the disease locus.

15.5 Complex Traits

In humans, only a small fraction of disease traits follow the simple Mendelian pattern of single-gene inheritance seen in cystic fibrosis, Huntington disease, and haemophilia A. Most common characteristics of human appearance, such as height, skin colour, the shape of the face, hair type, and many essential measures of human physiology have a more complex pattern of inheritance, as described initially in Chapter 2.

Table 15.2	Complexities That Alter Traditional Mendelian Ratios	
Category of Complexity	Problem in Observed Relationship Between Genotype and Phenotype	Changes Necessitated In Mapping Strategy
Incomplete penetrance	Disease genotype can occur in an individual who does not express the disease phenotype	Eliminate nondiseased individuals from analysis
Phenocopy	Disease phenotype can be expressed by an individual who does not have the disease genotype	Limit studies to families that show evidence for inheritance of the trait
Genetic heterogeneity	In different families, different disease genotypes are responsible for the same disease phenotype	Divide complete set of disease-transmitting families into subgroups (based on various parameters such as average age of onset) and perform linkage analysis separately on each subgroup
Polygenic determination	Mutant alleles at more than one locus influence expression of the disease phenotype in a single individual	Search for complex patterns of association between the disease trait and multiple DNA markers

We now review the chief causes of complex inheritance and describe the problems they pose for the linkage mapping and positional cloning protocols described so far for single-gene traits (**Table 15.2**). We then explain how researchers adapt the procedures for analyzing single-gene traits to the more difficult task of identifying, mapping, and characterizing the genes that contribute to complex traits.

With incomplete penetrance, a mutant genotype does not always cause a mutant phenotype

All cancers have a genetic basis; that is, they are the result of mutations in genes that regulate cell proliferation (see Chapter 16 for details). Most of these mutations occur in somatic tissues and are *not* inherited. Cancers arising in this way are termed *sporadic* and account for 90 percent of all breast cancers. A correlation has been found between sporadic breast cancer and several environmental factors, including alcohol consumption. By contrast, about 10 percent of women with breast cancer have inherited an allele that

predisposes them to this condition, as suggested by the observation that their mothers and aunts have a higher than normal incidence of breast cancer as well.

Medical investigators used a positional cloning protocol to map and clone the *BRCA1* (<u>BR</u>east <u>CA</u>ncer <u>1</u>) gene, one of several genes that can cause breast cancer. Significantly, only 66 percent of women who carry a mutant allele at the *BRCA1* locus develop breast cancer by the age of 55. As seen in the first pedigree in **Figure 15.20**, it is possible for a mother to carry a mutant *BRCA1* allele and remain disease-free, while her daughter becomes afflicted with the disease. Thus, although the mutant *BRCA1* allele predisposes a woman to breast cancer, it does not guarantee that the disease phenotype will occur; that is, it is not completely penetrant. By comparison, a disease such as sickle-cell anaemia, in which a mutant genotype always causes a mutant phenotype, is completely penetrant.

The causes of incomplete penetrance vary from trait to trait and from individual to individual. With breast cancer, it seems that chance plays the largest role in determining which predisposed individuals get the disease—through

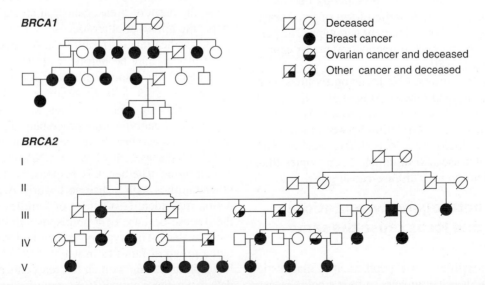

Figure 15.20 Incomplete penetrance and genetic heterogeneity in the inheritance of breast cancer. Both pedigrees in this figure show evidence of the transmission of a dominant mutation with incomplete penetrance that causes breast cancer. Linkage analysis shows that the mutation in the first pedigree resides on chromosome 17, whereas the mutation in the second pedigree is on chromosome 13. The first family is segregating a mutant *BRCA1* allele, while the second family is segregating a mutant *BRCA2* allele.

the accumulation of secondary somatic mutations. With heart disease, the individual's environment—especially diet and amount of exercise—plays a large role in determining whether a predisposing genotype results in a mutant phenotype, and if so, at what age.

Incomplete penetrance hampers linkage mapping and positional cloning for one main reason: Individuals who do not express a mutant phenotype may nevertheless carry a mutant genotype. The simplest solution to this problem is to exclude all nondiseased individuals from the analysis. With age-dependent traits like breast cancer and Huntington disease, such exclusion has meant that in disease-carrying families, the majority of children and adults under the age of 40 could not be included in the analysis. As a result, many more families were required for the studies that led to the mapping and cloning of the genes associated with both diseases.

With variable expressivity, individuals exhibit different degrees of expression of a mutant trait

Variation in gene expression may be in age of onset, phenotypic severity, or any other measurable parameter. Variable expressivity does not normally interfere with genetic analysis, because geneticists can use any degree of mutant phenotype as evidence for the presence of a mutant allele.

Phenocopy describes a disease phenotype that occurs in the absence of any inherited, predisposing mutation

The observation that 3 percent of women who do not carry a mutation at the *BRCA1* locus (or have any family history of the disease) still develop breast cancer by age 55 suggests that the disease can arise entirely from one or more somatic mutations in the breast cells themselves. This form of the disease is considered a **phenocopy** because it is indistinguishable from the inherited form of the disease yet is not caused by an inherited mutant genotype. The percentage of women who develop phenocopy breast cancer rises to 8 percent by age 80.

We have seen that researchers focus on families with a history of disease to map predisposing alleles. If a small but significant fraction of women who develop the disease carry wild-type alleles, the correlation between the inheritance of the disease locus (or a locus-linked marker) and expression of the disease will diminish. Phenocopies thus make it more difficult to map disease-causing loci.

With genetic heterogeneity, mutations at more than one locus cause the same phenotype

Sometimes it is possible to use sophisticated diagnostic techniques to separate what appears to be a single disease into a set of related diseases caused by mutations in different genes. For example, researchers can distinguish insulin-dependent from insulin-independent diabetes on the basis of their different physiological origins. However, even when the limit of disease subdivision has been reached, what appears to be a homogeneous phenotype may still arise from genetic heterogeneity. The seemingly simple disease of thalassaemia is a case in point. Mutations in either the α-globin gene or the β-globin gene can cause the same phenotype: severe reduction or elimination of the functional haemoglobin molecules produced in red blood cell precursors.

Genetic heterogeneity complicates attempts to map disease-causing loci in the following way. Although individual human families usually segregate only a single mutation responsible for a rare disease, most families do not have enough members to provide sufficient data for determining linkage. For this reason, linkage studies in humans almost always combine data from multiple families. But if a disease is heterogeneous, a marker linked to the disease locus in one family may assort independently from a different disease locus in a second family. When data from the two families are combined, the calculated probability of linkage between the marker and the initial disease locus would drop below that obtained with the first family alone.

Genetic heterogeneity is suspected whenever a comprehensive analysis of many families, each with several affected members, fails to map a locus responsible for the disease trait. In such a case, investigators try to divide the complete set of disease-transmitting families into subsets—based on any of several phenotypic parameters—and then combine only the families in each subset for linkage analysis. For example, when researchers combined the data from a large set of breast-cancer-prone families, they found no evidence of linkage to any marker. But when they selected a subset of the families in which the average age of disease onset was less than 47 years, they obtained strong evidence for a disease locus on chromosome 17, named *BRCA1* (**Figure 15.20a**). It is now clear that mutations in *BRCA1* cause the onset of breast cancer at an earlier age than predisposing alleles at other loci.

Classification of families into early-onset and late-onset groups may be helpful with any trait showing age-dependent expression. With traits for which classification by age of onset fails to produce evidence of linkage, classification by other variables, such as severity of the expressed phenotype, may prove helpful.

Once researchers have identified a first locus responsible for a disease, they can use DNA markers at that locus to determine whether it is responsible for the disease in other families with a disease history. A process of elimination may identify a subset of families that must inherit the disease because of a predisposing allele at a different locus or loci. This type of testing and elimination identified a group of families in which *BRCA1* could not be the locus predisposing women to breast cancer. The combined data from these non-*BRCA1* families revealed a second breast cancer locus, on chromosome 13, named *BRCA2* (**Figure 15.20b**). Mutations at *BRCA1* or *BRCA2* account for many inherited breast cancers, but not all.

The range of responses to the issues generated by new reproductive technologies, like preimplantation genetic diagnosis (PGD), shows a diversity of approach based in part on national culture and history. It also reflects international apprehension about the potential for misuse and abuse of the technologies. Here are some of the main concerns.

Which Genetic Variants Should Be Screened?

The couple in our opening story whose firstborn suffered from cystic fibrosis faced a medical problem. PGD could help them have a second child unaffected by the disease. With no cure at present for CF and no therapy that allows for CF-affected people to look forward to a life of normal health or length, this is an example of medically therapeutic screening. Governmental and professional committees in most industrialized countries permit PGD for this purpose, although Germany and Japan ban any use of preimplantation genetic testing for any purpose. And even in countries where PGD is permitted, there is opposition from people who hold a religious belief that all human embryos—even those at the earliest stages of development—have a right to life. Others object to PGD because they think people should not have a right to interfere with the natural process of allele segregation, even if nature selects a lethal childhood disease.

How Should the Tests Be Carried Out?

The couple screening for CF began by consulting a genetic counsellor and then worked with medical practitioners associated with a university laboratory. Most geneticists agree that counselling before a procedure should foster an open discussion of all the issues (including the possibility that, in the case of CF, the tests might give false negatives) and that long-term follow-up should be part of the process. PGD itself, like other forms of genetic testing, should be carried out by highly trained personnel in licensed laboratories that are subject to standards and review.

Who Should Have Access to the Technology?

In Canada the cost of *in vitro* fertilization and PGD testing ranged from $12 000–$15 000 in 2012. Should the government provide tests for people who cannot afford them? How should society decide this issue?

Should Parents Have the Right to Make Any Genetic Decision?

If, for instance, parents decide to forgo PGD and then have a child affected by a genetic disease, should they bear all financial responsibility for the child's care, or does society have an obligation to assist with medical treatment? On the other hand, how should physicians handle a request from prospective parents who wish to select against alleles responsible for minor diseases like myopia (nearsightedness) or late-onset diseases like Huntington disease or Alzheimer's disease? What about selection for alleles that provide a child with a relative advantage such as complete resistance to infection by HIV?

Who Should Have Access to Test Results?

The parents and eventually the child? The parents, the child, and certain community institutions, such as schools? Some combination of these plus commercial enterprises, such as insurance companies and places of employment? (We discuss these same questions of privacy in relation to other types of genetic testing in the Genetics and Society boxes in Chapter 2.)

What Constitutes a Person?

Cultural and religious beliefs, rather than scientific knowledge, are the basis for answers to this question. Some people see PGD as an alternative to abortion that allows a couple to make a decision before pregnancy begins. Others argue that even at the eight-cell stage, a microscopic preimplantation embryo is the equivalent of a human being, and rejection of an embryo is equivalent to killing a human being. The difference between these two positions is in part the result of different religious beliefs about the moral significance of embryonic cells.

Although there are no simple solutions to these complex issues, geneticists around the globe agree on the need for continuous open discussion and oversight of the development of new reproductive technologies.

Polygenic inheritance occurs when two or more genes interact in expression of a phenotype

So far, we have examined ways in which diseases caused by mutations in a single gene are associated with complex patterns of inheritance. But as we saw in Chapter 2, many traits arise from the interaction between two or more genes. Some such polygenic traits are discrete: They either show up or they do not. The occurrence of a heart attack, or myocardial infarction, is a discrete polygenic trait. Other polygenic traits are quantitative: They vary over a continuous range of measurement, from one extreme, through the normal range, to the opposite extreme. Blood sugar levels, cholesterol levels, and depression are examples of quantitative traits. Loci that influence the expression of such quantitative traits are known as **quantitative trait loci**, or **QTLs**. Although extreme values of QTL expression are considered abnormal, the border between normal and abnormal is arbitrary.

Virtually an unlimited number of transmission patterns are possible for polygenic traits. A completely penetrant discrete trait may require mutations at multiple loci to cause

the abnormal phenotype. With other discrete polygenic traits, penetrance may increase as the number of mutant loci increases. With quantitative polygenic traits, the measured degree of expression (expressivity) may vary with the number of mutant loci present in the individual or with the degree to which different mutations at a single locus alter the level of polypeptide production.

Many other factors can complicate the analysis of polygenic traits. Some members of a set of interacting polygenic loci may make a disproportionately large (or small) contribution to the penetrance or expressivity of the trait. Mutations at some loci in a set may be recessive, while mutations at other loci are dominant or codominant. Some traits may arise from a mixture of polygenic and heterogeneous components. For example, one form of a disease may be caused by mutations at loci A, B, C, and D;

a second form, by mutations at B and E; and a third form, by a single mutation at F. The more complex the inheritance pattern of a trait, the more difficult it is to identify the loci involved.

Complex traits are hard to pin down with linkage mapping and positional cloning because of ambiguity in disease effects. Penetrance and expressivity may make it difficult to distinguish a potential disease genotype from a healthy one. In some cases, a disease arises from phenocopy rather than from inheritance, so that pedigrees do not indicate transmitted alleles. Finally, traits affected by mutations in more than a single gene, or that result from interactions of several genes, make loci identification a daunting task.

Connections

In Chapter 14, we described the new tools of biotechnology and bioinformatics that enable geneticists to interpret the DNA sequence in individual genomes. In this chapter, we have examined how researchers apply these molecular tools to determine genotype at the DNA level. Health professionals can use direct genotyping to diagnose hereditary disease; forensic experts can use it to determine the identity and degree of relatedness of DNA samples. Finally, researchers can use direct genotyping to identify and characterize the genes responsible for any inherited trait that differs in its

expression among individuals within a population. With these new approaches, the human species has become a superb system for genetic analysis.

Biologists are fully aware, however, that a static list of genes cannot describe life. Rather, life is a dynamic system of molecular interactions and information processing. The study of these dynamic processes at the level of an organism or discrete biological systems is called "systems biology." In Chapter 23, we explore the experimental methods used by systems biologists and the insight into life provided by these approaches.

Essential Concepts

1. Using a variety of tools, researchers have detected enormous variation in nearly all animals and plants. When two or more alleles exist at a DNA locus, the locus is polymorphic, and the variations themselves are DNA polymorphisms. Polymorphic DNA loci that are useful for genetic studies are known as DNA markers. **[LO1–2]**

2. The four classes of DNA polymorphisms are single nucleotide polymorphisms (SNPs), microsatellites, minisatellites, and deletions/duplications/insertions in nonrepeat loci (InDels or DIPs). **[LO1–2]**

3. Several methods allow genotyping of SNPs: (1) Southern blot analysis of SNPs that eliminate or create a restriction site, (2) PCR analysis of the same type of SNPs, (3) allele-specific oligonucleotide hybridization to find any type of SNP; and (4) DNA microarrays that allow identification of millions of SNPs at once. **[LO1–2]**

4. Deletion and subtraction of short sequences also causes genetic variation. These include deletion-insertion polymorphisms (DIPs) and simple sequence repeats (SSRs). Automated PCR combined with gel electrophoresis can readily distinguish SSRs in the range of 15–300 repeats. SSRs are the cause of certain genetic diseases such as Huntington disease and fragile X syndrome. **[LO1–2]**

5. Copy number variants (CNVs) and copy number polymorphisms (CNPs) are relatively large blocks of genetic material. Some of these have normal function, or at least do not produce detectable phenotypic variation, but others have been implicated in mental disease. **[LO1–3]**

6. Positional cloning identifies the genes responsible for traits whose molecular cause is unknown. To localize a trait-affecting gene to a specific region of chromosomal DNA, researchers combine formal linkage analysis with the use of DNA markers. Researchers catalogue all possible candidate genes in a suspected genomic region to narrow the search. **[LO4]**

7. To identify the one candidate gene that is responsible for the trait of interest, researchers compare groups of phenotypically normal and abnormal individuals. A finding that the gene's DNA sequence or transcript is altered in all individuals exhibiting the mutant trait is strong evidence that the candidate gene is responsible for the trait. **[LO4]**

8. Most common genetically determined trait variation among individuals results from complex interactions that exhibit non-Mendelian inheritance. **[LO5]**

I. The figure shows the pedigree of a family in which a completely penetrant, autosomal dominant disease is transmitted through two generations, together with a corresponding Southern blot with individual pedigree samples digested with *Eco*RI and probed with a DNA fragment that detects a restriction fragment length polymorphism (RFLP). Do the data suggest the existence of genetic linkage between the RFLP locus and the disease locus? If so, what is the estimated genetic distance between the two loci?

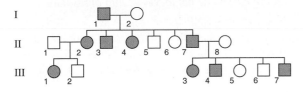

Answer

To solve this problem, you need to understand how DNA polymorphisms can be followed through a pedigree and how they can be tested for linkage to a locus defined by phenotype alone. First, examine the Southern blot pattern to determine what the forms of the DNA polymorphism are. The two segregating DNA alleles in this pedigree are represented by RFLPs having sizes of 8 and 7 kb. Some individuals are heterozygous, carrying both restriction fragments; and some are homozygous, with just the 8-kb fragment or the 7-kb fragment alone.

When two parents have one DNA allele in common, but are different at the second allele, it is possible for a child to inherit the common allele from either or both parents. If the child is homozygous for the allele, then he or she must have received it from both parents. But, if the child has just one copy of the common allele, exclusion analysis can be used to determine which parent had to be the one that transmitted it. For example, children II-5, II-6, and II-7 have a 7-kb allele that could have come only from their father (I-1), because their mother does not carry this allele. By exclusion, their second allele—which is 8 kb—must have come from their mother, even though it is present in the genomes of both parents:

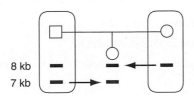

In the second generation set of siblings (II-2 through II-7), inheritance of the paternal 8-kb allele correlates with inheritance of the disease allele in five of six children: II-2, II-3, and II-4 inherit the paternal 8-kb allele along with the disease, and II-5 and II-6 do not receive the paternal 8-kb allele and do not exhibit the disease. The remaining child (II-7) exhibits

the disease but inherited the 7-kb allele from his father. There are two possible explanations for this discrepant individual. First, the RFLP locus could be unlinked to the disease locus, and the six out of seven transmission correlation could be a chance event. Second, the loci could indeed be linked with the II-7 child representing a recombination event that brings the disease locus onto the same chromosome as the 7-kb allele.

To distinguish between these possibilities, you need to examine transmission to the third-generation children using the same logic but with different facts. First, in the third-generation family on the left, the diseased parent (II-2) is homozygous for the 8-kb RFLP allele, and thus, no useful data on linkage can be obtained from her children III-1 and III-2. But useful data can be obtained from the third-generation family on the right. If your hypothesis of linkage is correct, it should now be the 7-kb allele that is transmitted in correlation with the disease allele. Indeed, III-3, III-4, and III-7 receive a paternal 7-kb allele together with the disease allele, and III-5 and III-6 do not receive either the 7-kb allele or the disease allele.

When linkage data are combined from the entire pedigree, you find that there are 11 informative offspring. *In 10 of 11 cases, alleles at the disease locus show co-transmission with alleles at the RFLP locus.* This is evidence for linkage between the two loci. The recombination rate can be estimated as 1/11 = 0.09, which translates into a genetic distance between the disease locus and the DNA marker of 9 cM. Further studies with additional families would be required to confirm this linkage.

II. A clear limitation to gene mapping in humans is that family sizes are small, so it is very difficult to collect enough data to get accurate recombination frequencies. A technique that circumvents this problem begins with the purification of DNA from single sperm cells. (Remember that recombination occurs during meiosis. Analysis of individual sperm in a large population can provide a large data set for linkage studies.) The DNA from single sperm cells can be used for SNP studies. Four pairs of primers were used for PCR amplification of four defined SNP loci from one man's somatic cells and from 21 single sperm that he provided for this research. Each of these primer pairs amplifies a different SNP locus referred to as A, B, C, and D. The four pairs of PCR primers were used simultaneously on each sample of DNA. Each of the amplified DNAs was divided into eight aliquots (identical subsamples), and these aliquots were denatured and spotted onto eight nitrocellulose membrane strips (vertically, as shown in the figure). Each of these strips was then hybridized with a different ASO (allele-specific oligonucleotide). There are two different ASOs for each SNP. For example, ASOs named A1 and A2 detect different alleles at SNP locus A. Black spots indicate that the amplified DNA hybridized to the ASO probe.

a. Based on the results shown, which SNP loci could be X-linked?

b. Which SNP loci could be on the Y chromosome?

c. Which SNP loci must be autosomal and homozygous?

d. Which SNP loci must be autosomal and heterozygous?

e. Do any SNP loci appear to be linked to each other?

f. Ignoring the results from sperm number 21, what is the distance between the two linked SNP loci?

g. How could you map the genomic region defined by SNP locus A?

h. What event could have given rise to sperm number 21?

Answer

For this problem, you need to understand how ASOs detect SNP alleles and the advantages and limitations of SNPs as DNA marker loci. An advantage of SNP analysis by PCR is that the technique is so sensitive that the single alleles present within individual sperm cells can be assayed. An ASO result is either positive or negative. If the result is positive (as indicated by a black dot of hybridization in this example), a tested somatic cell sample can be either homozygous or heterozygous for the corresponding ASO allele; a positive ASO result by itself does not distinguish between these possibilities. If the result is negative, the sample does not contain the ASO allele under analysis, but nothing can be said about the alleles (if any) that are present at the tested locus.

a. Half of the sperm cells will not have an X chromosome and would not be expected to show a positive result with any ASO for any X-linked SNP locus. The other half of the sperm cell will carry the same SNP allele. *Gene C shows this type of pattern.*

b. Similarly, a gene on the Y chromosome would be found in only half the sperm. Again, *gene C is a candidate for a Y-linked gene.*

c. If an individual is homozygous at an autosomal SNP locus, all the sperm from that individual will show hybridization to one ASO for that locus and not any other. *Locus A appears to be homozygous and autosomal.*

d. At a heterozygous SNP locus, one ASO will hybridize to approximately one-half of the sperm samples, and a second ASO will hybridize to those samples that do not hybridize to the first ASO. *SNP loci B and D show this type of pattern.*

e. Alleles at linked loci will segregate together more than 50 percent of the time and would therefore end up in the same sperm. Alleles B1 and D2 are transmitted together more often than not, and the reciprocal alleles B2 and D1 are also transmitted together more often than not. *This result suggests that loci B and D are linked.*

f. Sperm 3, 9, and 18 show evidence of recombination between alleles at the B and D loci. Three out of 20, or 15 percent, are recombinant. *The distance between the B and D loci is therefore 15 cM.*

g. Since SNP locus A is homozygous in this individual, it cannot be mapped in a linkage analysis. But, SNPs are found at an approximate rate of one in a thousand base pairs. *By sequencing several kilobases of genomic DNA around SNP locus A from this individual, you could identify a nearby SNP locus that is heterozygous and that could be mapped in a linkage analysis.*

h. *Sperm sample number 21 could accidentally have two sperm cells rather than one. It is also possible that a single sperm in this sample has accidently received two copies of the chromosome that carries loci B and D through meiotic nondisjunction.*

Problems

Vocabulary

1. For each of the terms in the left column, choose the best matching phrase in the right column.

i. DNA polymorphism

ii. RFLP

iii. ASO

iv. SNP

v. DNA fingerprinting

vi. minisatellite

vii. locus

1. DNA element composed of tandemly repeated identical sequences

2. two different nucleotides appear at the same position in genomic DNA from different individuals

3. location on a chromosome

4. a DNA sequence that occurs in two or more variant forms

5. a short oligonucleotide probe that will hybridize to only one allele at a chosen SNP locus

6. detection of genotype at a number of unlinked highly polymorphic loci using one probe

7. variation in the length of a restriction fragment detected by a particular probe due to nucleotide changes at a restriction site

Section 15.1

2. What advantages do anonymous DNA markers afford for genetic mapping as opposed to traditional allelic markers associated with visible phenotypes? What are the disadvantages of anonymous DNA markers for mapping?

3. Would you characterize the pattern of inheritance of anonymous DNA polymorphisms as recessive, dominant, incompletely dominant, or codominant?

4. Would you be more likely to find SNPs in the protein-coding or in the noncoding DNA of the human genome?

5. Mutations at microsatellite loci occur at a frequency of 1×10^{-3}, which is much higher than the rate of base substitutions at other loci.

   a. What is the nature of microsatellite polymorphisms?

   b. By what mechanism are these polymorphisms generated?

   c. Minisatellites also mutate at a relatively high frequency. Do these mutations occur by the same or a different mechanism?

6. If you were comparing two closely related but non-identical gene sequences from different individuals of the same species, how would you distinguish whether these sequences represented polymorphisms of a single gene or two different paralogous genes?

Section 15.2

7. Each of the following reagents can be used to detect SNPs. Where is the polymorphism located in relation to the probe or primer DNA sequence used in each of these techniques?

   a. allele-specific oligonucleotide (ASO)

   b. primer-extension oligonucleotide

   c. RFLP probe

8. An 18-bp deletion in the *PAX-3* gene causes Waardenburg syndrome (an autosomal dominant condition that is responsible for a small percentage of deafness in humans). What features of this mutation make it amenable to molecular analyses that could not be applied to detection of the mutation in the β-globlin gene responsible for sickle cell anaemia?

9. Given the two allelic sequences shown here and the site at which a single-base polymorphism occurs (underlined in the sequence), what sequences would you use as oligonucleotide probes for ASO analysis of genotype by hybridization? (Assume that ASO probes are usually 19 bp in length.)

allele 1:

5′ GGCATTGCATGCTAACCCTATAAATG<u>C</u>GCTAGGCGTAGTTAGCTGGGAA
    TAAAAAGCT 3′

allele 2:

5′ GGCATTGCATGCTAACCCTATAAATG<u>GG</u>CTAGGCGTAGTTAGCTGGGAA
    TAAAAAGCT 3′

10. The ASO technique was used to determine the genotypes of ten family members with regard to sickle-cell anaemia, as shown here. Each pair of dots represents the results of ASO analysis for the DNA from one person. The upper row represents hybridization with the normal oligonucleotide, and the lower row represents the results of hybridization using the mutant oligonucleotide. The three replications of the assay were incubated at 100°C (upper set), 90°C (middle set), and 80°C (lower set).

   a. Why do the three replications of the same sample set look different?

   b. What are the genotypes of the individuals?

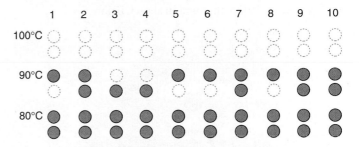

11. Angela and George have one child, and Angela has sickle-cell anaemia. They want to have more children but do not want any of them to suffer from this disease. They also do not want to be in a position of having to abort a fetus, so they elect to have *in vitro* fertilization and embryo screening. Briefly list the steps they must take to accomplish this goal.

Section 15.3

12. DNA fingerprinting can be used to settle cases concerning paternity. In the DNA fingerprint shown, the mother's DNA sample is in lane 1, the daughter is in lane 2, and two samples of men that could be the father are in lanes in 3 and 4. Can you determine from these data if one of the men must be the father?

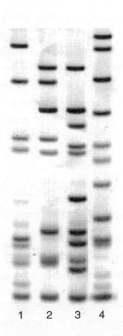

13. The police discovered the body of a woman who had been brutally beaten and raped while working late in her office one evening. They suspected that one of her seven co-workers might be responsible for the crime. To test this possibility, they recovered semen from her vagina and used it to prepare a DNA sample. They also took DNA samples from each of her co-workers. All eight samples were subjected to DNA fingerprinting analysis based on restriction digestion, gel electrophoresis, Southern blotting, and probing with a minisatellite. The results are shown in the photograph shown below.

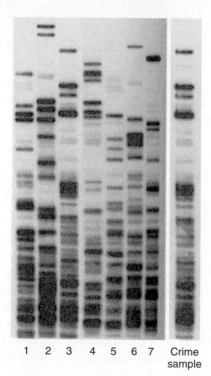

1　2　3　4　5　6　7　Crime sample

Is any one of her co-workers likely to be the perpetrator of the crime? Which one? Estimate how likely it is that this particular person is the perpetrator rather than another person unknown to the authorities.

14. Individuals homozygous for a point mutation, changing an A to a T, in the human β-globin gene develop sickle-cell anaemia. The wild-type gene sequence over this region is shown here (the top strand is the RNA-like coding strand):

```
5′ ATGGTGCACCTGACTCCTGAGGAG 3′
3′ TACCACGTGGACTGAGGACTCCTC 5′
```

and the sickle-cell allele sequence is

```
5′ ATGGTGCACCTGACTCCTGTGGAG 3′
3′ TACCACGTGGACTGAGGACACCTC 5′
```

Design a PCR-based strategy to distinguish the DNA from homozygous wild-type, heterozygous, and homozygous sickle-cell allele individuals. Your strategy should exploit the single A-to-T transversion and should produce an "all or none" response in which amplification will occur with particular sets of primers while no amplification will occur with other sets of primers.

15. The trinucleotide repeat region of the Huntington disease (*HD*) locus in six individuals is amplified by PCR and analyzed by gel electrophoresis as shown in the following figure; the numbers to the right indicate the sizes of the PCR products in bp. Each person whose DNA was analyzed has one affected parent.

a. Which individuals are most likely to be affected by Huntington disease, and in which of these people is the onset of the disease likely to be earliest?

b. Which individuals are least likely to be affected by the disease?

c. Consider the two PCR primers used to amplify the trinucleotide repeat region. If the 5′ end of one of these primers is located 70 nucleotides upstream of the first CAG repeat, what is the maximum distance downstream of the last CAG repeat at which the 5′ end of the other primer could be found?

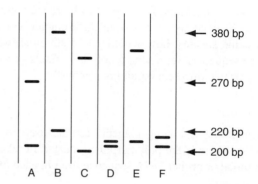

16. Sperm samples were taken from two men just beginning to show the effects of Huntington disease. Individual sperm from these samples were analyzed by PCR for the length of the trinucleotide repeat region in the *HD* gene. In the graphs that follow, the horizontal axes represent the number of CAG repeats in each sperm, and the vertical axes represent the fraction of total sperm of a particular size. The first graph shows the results for a man whose mutant *HD* allele (as measured in somatic cells) contained 62 CAG repeats; the man whose sperm were analyzed in the second graph had a mutant *HD* allele with 48 repeats.

a. What is the approximate CAG repeat number in the HD^+ alleles from both patients?

b. Assuming that these results indicate a trend, what can you conclude about the processes that give rise to mutant *HD* alleles? In what kinds of cells do these processes take place?

c. How do these results explain why approximately 5 percent to 10 percent of Huntington disease patients have no family history of this condition?

d. Predict the results if you performed this same PCR analysis on single blood cells from each of these patients instead of single sperm.

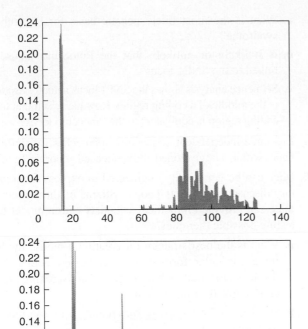

Section 15.4

17. A relatively frequent, completely penetrant recessive disease known as the foul mouth syndrome (FM) has been found to be due to a variety of mutations in the *FM* gene, which has recently been cloned. Analysis of Southern blots of human DNA cleaved with the enzyme *Hpa*I and probed with a radioactively labelled fragment of the *FM* gene has revealed that *Hpa*I does not cleave within the gene itself. However, the positions of *Hpa*I sites surrounding the gene vary among individuals, producing at least three RFLPs in the population: The sizes of the RFLP alleles are 13, 10, and 7 kb. Shown here are two small pedigrees of families in which individuals with the disease are shaded in *black*. Below each pedigree symbol is the corresponding result obtained from the DNA sample in a Southern blot analysis with the FM probe.

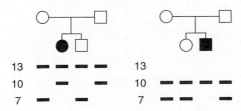

a. Which restriction fragment is associated with the disease mutation in the father shown in the left pedigree?

b. Which restriction fragment is associated with the disease mutation in the mother shown in the right pedigree?

c. If the male child from the left pedigree marries the female child from the right pedigree, what is the probability that their child will be diseased?

d. What is the probability their child will be a carrier?

18. The recessive disease cystic fibrosis displays extensive *allelic heterogeneity*: more than 1900 different mutations of the *CFTR* gene have been shown to be associated with cystic fibrosis worldwide. Approximately 1 in 25 Canadians is a carrier for some mutant allele of *CFTR*. One of these alleles, a deletion of three nucleotides that results in the loss of a single phenylalanine from the encoded protein, accounts for approximately 70 percent of the mutant alleles in populations of western European descent. With these facts in mind, is it feasible and worthwhile to mount a nationwide screening program for cystic fibrosis, and if so, how should this screening program be conducted?

19. When a researcher begins to choose DNA markers for linkage analysis of a disease trait in a particular family, what are the first criteria used in this choice?

20. Imagine that you have identified a SNP marker that lies 1 cM away from a locus causing a rare hereditary autosomal dominant disease. You test additional nearby markers and find one that shows no recombination with the disease locus in the one large family that you have used for your linkage analysis. Furthermore, you discover that all afflicted individuals have a G base at this SNP on their mutant chromosomes, while all wild-type chromosomes have a T base at this SNP. You would like to think that you have discovered the disease locus and the causative mutation but realize you need to consider other possibilities.

a. What is another possible interpretation of the results?

b. How would you go about obtaining additional genetic information that could support or eliminate your hypothesis that the base-pair difference is responsible for the disease?

21. Approximately 3 percent of the population carries a mutant allele at the *CFTR* gene responsible for cystic fibrosis. New disease-causing mutations at this locus arise at a frequency of 1 in 10^4 gametes. A genetic counsellor is examining a family in which both parents are known to be carriers for a *CFTR* mutation. Their first child was born with the disease, and the parents have come to the counsellor to assess whether the new fetus inside the mother is also diseased, is a carrier, or is completely wild type at the *CF* locus. DNA samples from each family member and fetus are tested by PCR and gel electrophoresis for a microsatellite marker within one of the *CFTR* gene's introns. The following results are obtained:

a. What is the probability that the child who will develop from this fetus will exhibit the disease?

b. If this child grows up and gets married, what is the probability that one of her children will be afflicted with the disease?

22. The pedigrees indicated here were obtained with three unrelated families whose members express the same completely penetrant disease caused by a dominant mutation that is linked at a distance of 10 cM from a marker locus with three alleles numbered 1, 2, and 3. The marker alleles present within each live genotype are indicated below the pedigree symbol. The phenotypes of the newly born labelled individuals—A, B, C, and D—are unknown. What is the probability of disease expression in each of these individuals?

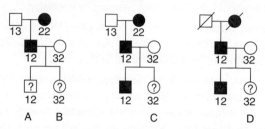

23. One of the difficulties faced by human geneticists is that matings are not performed with a scientific goal in mind, so pedigrees may not always provide desired information. As an example, consider the following matings (W, X, Y, or Z). Which of these matings are informative and which non-informative for testing linkage between anonymous loci A and B? (A1 and A2 are different alleles of locus A, B1 and B2 are different alleles of locus B, etc.) Explain your answer for each mating.

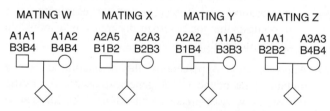

MATING W   MATING X   MATING Y   MATING Z

24. The next disease that you decide to tackle is Pinocchio syndrome, which causes the noses of afflicted individuals to grow larger when they tell a lie. You discover a family in which this disease is segregating as indicated in the following pedigree. You have reason to believe that a SNP locus called SNP1 is linked to the Pinocchio gene, and you test each individual in the family with ASOs that recognize allele 1 or allele 2 at this SNP locus, with the genotyping results shown here.

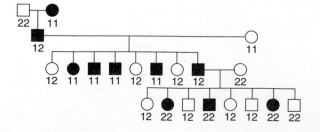

a. What is the most likely genetic basis for Pinocchio syndrome?

b. Is it likely or unlikely that the Pinocchio locus is linked to the SNP1 locus?

c. Sequence analysis shows the SNP1 locus actually resides in the middle of a coding region. How likely is it that this coding region is equivalent to the Pinocchio locus?

25. List three independent conceptual approaches for finding genes within a large cloned and sequenced genomic region.

26. Mice can be genetically engineered to express a hereditary disease that afflicts people. Strains of these mutant mice can provide biomedical scientists with a model for testing possible therapies.

a. What is the best strategy for creating a mouse model for a particular form of hypercholesterolaemia that results from a mutant human gene that overexpresses a cholesterol-forming enzyme?

b. What is the best strategy for creating a mouse model for haemophilia?

27. A rare human disease leads to overgrowth of the heart without any other effect on the afflicted individual. Linkage analysis with DNA markers has been used to map the disease locus to a small chromosomal region. This region has been divided into five DNA fragments (named A through E) that are each labelled and used to probe Northern blots containing a single lane of RNA from one of three tissues—liver, heart, and muscle—taken from nondiseased cadavers. The results are shown here.

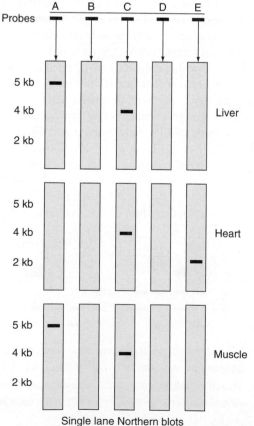

Genomic region linked to disease

Single lane Northern blots

a. Which of the five DNA fragments is likely to contain a gene?

b. How many genes have been identified in this region by Northern blot analysis?

c. Is it possible that there are more genes in this region than those detected here?

d. Which of the five DNA fragments could possibly contain the gene responsible for this disease?

e. Which fragment is most likely to contain the gene?

f. What is your next step in testing this candidate gene as the causative factor for the disease?

Section 15.5

28. You have decided to study another disease trait that is very rare. You have searched far and wide to come up with an extended family in which a number of non-sibling individuals express the disease. The pedigree is as follows:

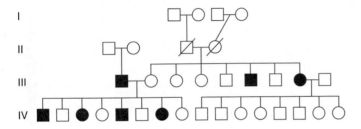

a. What is the most likely genetic basis for the disease?

b. Are there any individuals in the pedigree who *must* carry the disease mutation even though they do not express the disease trait? If so, list those individuals according to their generation number (in Roman numerals) and their number within the generation (counting from left to right across the entire pedigree).

29. You have decided to study another disease trait that is very rare. You have found an extended family in which a number of non-sibling individuals express the disease. One pair of identical twins are indicated with a horizontal line that joins both symbols together and with descent lines that join together at a vertex.

The pedigree is as follows:

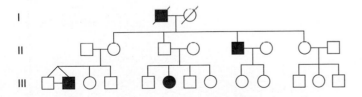

a. What is the most likely genetic basis for the disease?

b. Are there any individuals in the pedigree who *must* carry the disease mutation even though they do not express the disease trait? If so, list those individuals according to their generation number (in Roman numerals) and their number within the generation (counting from left to right across the entire pedigree).

30. Among the most prevalent diseases that afflict human beings is heart disease, which can have a severe impact on quality of life as well as result in premature death. While heart disease mostly afflicts those who are older, 1 or 2 percent of people in their 30s, and even their 20s, suffer from this disease. There are genetic and environmental components to this disease. Use this information to answer the following questions.

a. What strategy might you use to choose families to participate in a linkage study of heart-disease-causing genes?

b. Once you have cloned a gene that you believe plays a role in heart disease, how would you confirm this role?

31. Human chromosome 6 has a region containing several closely linked genes encoding cell surface proteins called human leukocyte antigens. Three of the genes in this region, called *HLA-A*, *HLA-B*, and *HLA-C*, are highly polymorphic: About 25 alleles of *HLA-A*, 50 alleles of *HLA-B*, and 10 alleles of *HLA-C* are known.

a. How many different haplotypes for these three genes are possible in human populations? Note that a "haplotype" is a combination of alleles at multiple tightly linked loci that are transmitted together over many generations.

b. How many diplotypes (i.e., different pairs of haplotypes) are possible?

Now consider the inheritance of HLA alleles in the following family:

	HLA-A		HLA-C		HLA-B	
Father	A23	A25	C2	C4	B7	B35
Mother	A3	A24	C5	C9	B8	D44
Child #1	A24	A25	C4	C5	B7	B8
Child #2	A3	A23	C2	C9	B35	B44
Child #3	A23	A24	C2	C5	B8	B35
Child #4	A3	A25	C4	C9	B7	B44

c. Diagram the two haplotypes in the father and the two haplotypes in the mother. Because the genes are so closely linked, assume none of these children is the result of recombination events in the parents.

d. For tissue transplantation to succeed, it is best that the donated tissue has the same alleles of the three *HLA* genes as the recipient. What is the chance that the next child born to this family (child #5) would be able to serve as a bone marrow donor to child #1 (his sister) with no danger of rejection due to incompatibility between *HLA-A*, *HLA-B*, and *HLA-C* antigens?

32. Canavan disease is a recessive, severe neurodegenerative syndrome usually causing death by the age of 18 months. The frequency of Canavan disease is particularly high in Jewish populations. In an effort to map the gene causing this condition, researchers looked at 10 SNPs (1–10) spaced at roughly 100-kb distances along chromosome 17

in five affected Jewish patients and four unaffected control Jewish individuals. In the table below, each row depicts a single haplotype. G, C, A, and T represent the actual nucleotide at the indicated SNP location.

a. Does the disease-causing mutation appear to be in linkage disequilibrium with any of the SNP alleles? If so, which ones?

b. Where is the most likely location for the Canavan disease gene? About how long is the region to which you can ascribe the gene?

c. How many independent mutations of the Canavan gene are suggested by these data?

d. Suppose that individuals 2–9 are Ashkenazic (whose ancestors lived in the Rhine river basin of Germany and France after the Jews were expelled from Judea in 70 AD) while individual 1 is Sephardic (a non-Ashkenazic Jew). Would these facts provide any information about the history of the mutations causing Canavan disease?

e. For mapping genes by haplotype association, why is it often helpful to focus on certain subpopulations? Does this strategy have any disadvantages?

f. Human chromosome 17 is an autosome, so each person contains two copies of each region along the chromosome. With this in mind, how could the researchers determine any individual haplotype, such as those shown in the table?

Patient	SNP1	SNP2	SNP3	SNP4	SNP5	SNP6	SNP7	SNP8	SNP9	SNP10
1	G	T	G	T	T	T	C	A	G	T
2	A	T	G	T	T	T	C	A	G	T
3	G	T	G	T	T	T	C	A	G	C
4	A	A	G	T	T	T	C	T	C	C
5	G	A	G	C	C	T	G	A	C	C
Control										
6	A	A	G	T	T	T	C	A	G	T
7	G	T	G	G	C	T	G	A	G	T
8	A	T	C	T	C	G	C	T	C	C
9	G	T	C	G	T	G	G	A	C	T

connect For more information on the resources available from McGraw-Hill Ryerson, go to www.mcgrawhill.ca/he/solutions.

520   Unit I   Chapter 15   Genome-wide Analysis of Genetic Variation

CHAPTER **16**   # Somatic Mutation and the Genetics of Cancer

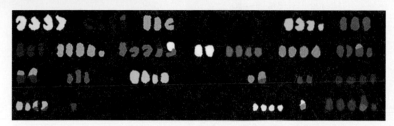

Sporadic ovarian tumours often exhibit chromosomal instability. In this example, spectral karyotype (SKY) analysis of a cancer patient's tumour cells reveals a near-tetraploid chromosome number and other structural chromosome rearrangements, such as translocations, deletions, and duplications. These chromosomal abnormalities are consequences of aberrant mitotic segregation or cytokinesis, cell-cycle checkpoint errors, and faulty DNA replication and repair mechanisms.

We saw in Chapter 8 that a great variety of mutations can occur during the replication and segregation of DNA, including single-base changes, deletions of many bases, inversions, and insertions, as well as translocations between chromosomes. Although these events are rare per gene and per cell division, there are many genes in each cell and many cell divisions (about 10^{13}) between the fertilization of the egg cell and the production of an adult human individual. This means that all cells have some mutations, although not all of them necessarily affect the functions of genes. The mutations that occur between the formation of the fertilized egg and the production of the germ cells are the substrates upon which evolution acts by selecting for advantageous changes and selecting against deleterious ones.

What about the mutations that occur in the somatic cells of the organism? Because these do not appear in the germ cells, they are not transmitted to the next generation. Can they affect the individual in which they occur? If the mutation led to a nonfunctional cell, then it might have little or no consequence because there are so many other cells in the tissue or organism that could compensate. But some mutations lead to abnormal cellular behaviour, and these changes could affect health. Indeed, cancer is usually the result of mutations that occur during the division of somatic cells.

Most of us give little thought to the life-sustaining cycle of cell death and renewal in somatic cells, because despite the continuous comings and goings of our body's cells, we generally remain the same shape and size. But the appearance of an abnormal growth or a set of symptoms diagnosed as cancer startles us into the realization that we take for granted the intricate checks and balances that control cell division and behaviour. These controls enable all cells of the body to function as part of a tightly organized, cooperative society. Cancer results when some cells divide out of control and eventually acquire the ability to spread beyond their prescribed boundaries. Although there are many types of cancer, they all result from excessive and inaccurate cellular proliferation (**Figure 16.1**).

Thus, the body's ability to regulate cell division is a foundation of health.

In this chapter, we describe how mutations cause cancer. We then examine the genes and gene products that control normal cell proliferation, including molecules that control the machinery of cell division, molecules that integrate the repair of DNA damage with progression through the cell cycle, and molecules that relay messages about whether conditions are right for cell division.

Two unifying themes can be inferred from our study of cancer. First, cancer is ultimately a disease of the genes: The multiple phenotypes collectively referred to as cancer all result from mutations in genes that regulate a cell's passage through the cycle of growth and division. Chemicals in the environment that raise the rate of gene mutation increase the probability of cancer incidence. Second, cancer differs in two ways from cystic fibrosis, Huntington disease, and other genetic conditions caused by the inheritance of one or two copies of a single defective gene: (1) although some people inherit mutations that predispose them to cancer, most mutations that lead to cancer occur in the somatic cells of one tissue; and (2) multiple mutations in an array of genes must accumulate over time in the clonal descendants of a single cell before the cancer phenotype appears—this is known as the **multi-hit model** of carcinogenesis. By contrast, the mutations that cause cystic fibrosis and Huntington disease are transmitted through the germ line; thus, the mutant alleles of one particular gene appearing in all cells of all somatic tissues cause the disease in an affected individual.

Chapter Outline
16.1 Overview: Initiation of Division
16.2 Cancer: A Failure of Control over Cell Division
16.3 The Normal Control of Cell Division

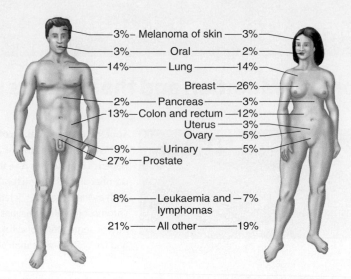

Figure 16.1 The relative percentages of new cancers in Canada that occur at different sites in the bodies of men and women.

3% – Melanoma of skin — 3%
3% — Oral — 2%
14% — Lung — 14%
Breast — 26%
2% — Pancreas — 3%
13% — Colon and rectum — 12%
Uterus — 3%
Ovary — 5%
9% — Urinary — 5%
27% — Prostate

8% — Leukaemia and — 7% lymphomas
21% — All other — 19%

Learning Objectives

1. Describe the two basic types of signalling systems and their molecular components regulating cell division.

2. Summarize the phenotypic changes differentiating tumour cells from normal cells.

3. Examine the multi-hit model of carcinogenesis, the evidence supporting this model, and explain how most cancers arise.

4. Compare and contrast oncogenes and tumour-suppressor genes.

5. Differentiate between the roles of cyclins, cyclin-dependent kinases (CDKs), and other regulatory cell-cycle genes in cell division processes.

6. Distinguish between the various cell-cycle checkpoints required for genomic stability.

16.1 Overview: Initiation of Division

How do cells know when to divide? To function according to the needs of the body as a whole, cells depend on signals sent from one tissue to another. These signals tell them whether to divide, metabolize (i.e., make the products they are programmed to make), or die. The two basic types of signals are extracellular signals and cell-bound signals.

Extracellular signals in the form of steroids, peptides, and proteins act over long or short distances and are collectively known as hormones (**Figure 16.2a**). The thyroid-stimulating hormone (TSH) produced by the brain's pituitary gland, for example, travels through the bloodstream to the thyroid gland, where it stimulates cells to produce another hormone, thyroxine, which in turn increases metabolic rate.

Cell-bound signals, such as the histocompatibility proteins that, like fingerprints, distinguish an individual's cells from all foreign cells and molecules, require direct contact between cells for transmission (**Figure 16.2b**). The macrophages, helper T cells, and antibody-producing B cells of the immune system communicate via cell-bound signals about the presence of viral particles, bacteria, and toxins.

Each signalling system has four components

Although the details can be complex depending on the individual system, both types of signalling systems have four molecular components that control cell division.

- **Growth factors**. These factors are extracellular hormones or cell-bound signals that stimulate or inhibit cell proliferation. Most growth factors deliver their message to specific receptors embedded in the membrane of the receiving cell (**Figure 16.3a**).

- The **receptors** are proteins that have three parts: a signal-binding site outside the cell, a transmembrane segment that passes through the cell membrane, and an intracellular domain that relays the signal (i.e., the binding of growth factor) to proteins inside the cell's cytoplasm.

- These cytoplasmic proteins are known as **signal transducers**. They are responsible for relaying the signal inside the cell.

- The final link is usually a **transcription factor** that activates the expression of specific genes in the

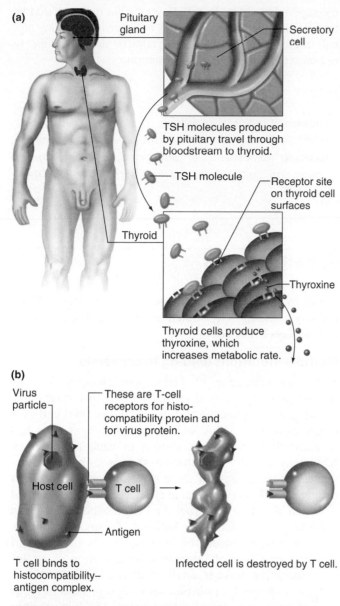

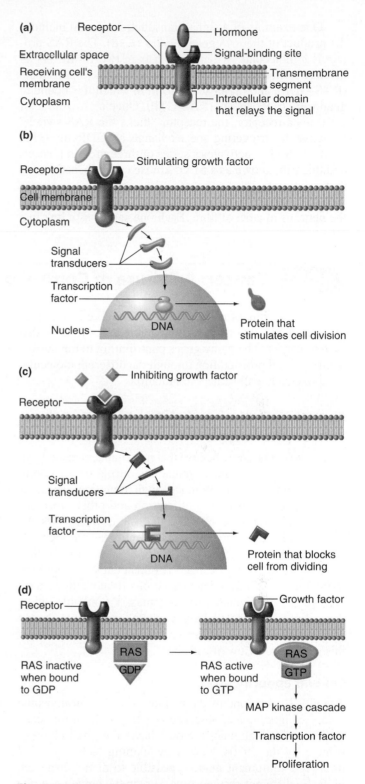

Figure 16.2 Extracellular signals can diffuse from one cell to another or be delivered by cell-to-cell contact. (a) The pituitary gland produces thyroid-stimulating hormone (TSH) that moves through the circulation to the thyroid gland, which produces another hormone, thyroxine, that acts on many cells throughout the body. (b) A killer T cell recognizes its target cell by direct cell-to-cell contact.

nucleus, either to promote or to inhibit cell proliferation (**Figures 16.3b** and **c**). These factors were described in detail in Chapter 11.

Molecular interactions relay a signal

Binding of a growth factor to its specific receptor elicits a cascade of biochemical reactions inside the cell, often involving a large number of molecules. Each molecule in the cascade transmits the receptor's binding-of-messenger signal by activating or inhibiting another molecule. The activation and inhibition of intracellular targets after growth-factor binding is called **signal transduction**.

Figure 16.3 Many hormones transmit signals into cells through receptors that span the cellular membrane. (a) Hormones bind to a specific cell surface receptor. The extracellular surface of the receptor transmits a signal to the intracellular domain of the receptor, which, in turn, interacts with other signalling molecules in the cell either to (b) stimulate growth or (c) inhibit growth. (d) The RAS protein is an intracellular signalling molecule that is induced to exchange a bound GDP (inactive) for a bound GTP (active) when a growth factor binds to the cellular receptor with which RAS interacts.

One example of a signal transduction system includes the product of the *RAS* gene (**Figure 16.3d**). The RAS protein is a molecular switch that exists in two forms: (1) an inactive form in which it is bound to guanosine diphosphate (RAS-GDP), and (2) an active form in which it is bound to guanosine triphosphate (RAS-GTP). Once a growth factor activates a receptor, the receptor "flicks the RAS switch" to active by triggering the exchange of GDP for GTP. Next, RAS-GTP activates a series of three protein kinases, and this trio, known as a **MAP kinase cascade**, activates a transcription factor.

The proteins in a signal transduction system are like the neurons in a nerve fibre: Each one serves as a link in a message-relay chain. In deciding whether or not to divide, the cell, like the brain, combines messages from many signal transduction systems and adjusts its behaviour in response to the integrated information.

> Initiation of cell division involves either extracellular signals or cell-bound signals. Both systems comprise four general molecular components: growth factors, receptors, signal transducers, and transcription factors that promote DNA replication.

16.2 Cancer: A Failure of Control over Cell Division

An understanding of the molecular basis of cell-cycle regulation sheds light on the life-threatening proliferative disease of cancer. The many genes contributing to the normal control of cell proliferation through the different molecular components are all subject to mutations.

Accumulation of mutations results in the cancer phenotype

Cancer biologists now believe that most cancers result from the accumulation of many mutations during the proliferation of somatic cells. When enough mutations accumulate in genes controlling proliferation and other processes within a single clone of cells, that clone overgrows the normal cells that surround it, disseminates through the bloodstream to other parts of the body, and forms a life-threatening tumour, or cancer. (In this chapter, we use the term "tumour" to designate cancerous tissue and the term "growth" to designate a benign mass.) Epidemiological data, clinical studies, and experimental analyses of a range of cell types in a variety of species provide evidence for this gene-based view of cancer.

Cellular abnormalities

Theodor Boveri, one of the architects of the chromosome theory of inheritance, observed as early as 1914 that cells excised from malignant tumours have abnormal chromosomes. By the 1970s, when new staining techniques and improved equipment made it possible to distinguish each of the 23 different chromosome pairs in the human genome by their specific banding patterns, investigators noted that many different chromosomal abnormalities appear in tumour cells. Using tools developed in the 1980s, geneticists confirmed that most tumour cells exhibit karyotypic instability.

Figure 16.4 shows the main characteristics that distinguish tumour cells from normal cells. The cancer phenotype includes uncontrolled cell growth, genomic and karyotypic instability, the potential for immortality, and the ability to invade and disrupt local and distant tissues.

Although no one cancer cell necessarily manifests all the phenotypic changes illustrated in Figure 16.4, each cancer cell displays a number of them.

Multiple mutations leading to conversion

The large catalogue of phenotypic changes seen in tumour cells suggests that many mutations in a number of genes are necessary to convert a normal cell into a cancerous cell. DNA sequencing of tumour cells has revealed thousands of mutations in each tumour, but how many actually contribute to the cancer phenotype is unclear because most of these mutations are different in different tumours of the same type. The number of genes in which mutations can fuel the progression to cancer is quite large—at least a few hundred. To study mutations associated with cancer, researchers initially identify and isolate a mutation of interest using a variety of strategies, including linkage analysis of markers, traditional genetic mapping to a chromosome, and positional cloning (all techniques described in Chapter 15). It is possible to test in mice whether a mutation in a single gene associated with cancer is sufficient to induce a tumour. If the mutation acts in a dominant fashion, researchers insert a copy of the mutant allele into the mouse genome of a fertilized egg; if the mutation is recessive, they delete one copy of the homologous gene from the early embryonic mouse genome and then breed animals homozygous for the deletion.

In gene transfer experiments where a dominant cancer-causing mutation was inserted into a mouse genome under the control of a breast-cell-specific promoter, the transgenic mice produced a few breast tumours (**Figure 16.5a**). Doubly transgenic mice made by breeding these transgenic mice carrying one mutated gene with transgenic mice carrying a different mutated gene implicated in cancer generated more tumours earlier. Even in these mice, however, only a small percentage of the transformed cells proliferated abnormally.

Studies of recessive mutations in the *p53* (also known as *Tp53*) gene point to the same conclusion. Mice with both

copies of the *p53* gene deleted from their genome develop relatively normally. The *p53* mutant mice, however, have shortened life spans and get a variety of tumours more frequently than wild-type mice (**Figure 16.5b**). This experiment shows that the *p53* gene is not essential for development or for normal cell function, but it does play a role in preventing tumour formation. Consequently, deleting both copies of the wild-type gene from all of a mouse's cells does not convert every cell to a tumour cell, but it does increase the probability that at least one cell will become cancerous. The conclusion is that mutations in *p53* are just one of the many genetic changes that may occur in a cell to produce cancer.

Clonal proliferation

Examination of cells from women heterozygous for X-linked alleles provides evidence that cancer originates in a single somatic cell (**Figure 16.6**). Although the random inactivation of one of the two X chromosomes in each cell of a female means that individual cells express only one of the two X-linked alleles, in small samples of normal somatic tissues, one usually finds both alleles expressed. The reason is that most somatic tissues are constructed from many clones of cells.

In contrast to normal tissue, tumours from females invariably express only one allele of an X-linked gene

---

FEATURE FIGURE 16.4

Phenotypic Changes That Distinguish Tumour Cells from Normal Cells

Most Normal Cells   **Many Cancer Cells**

a.1   Autocrine stimulation

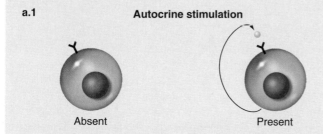

Absent   Present

a.2   Contact inhibition

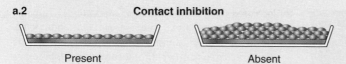

Present   Absent

a.3   Cell death

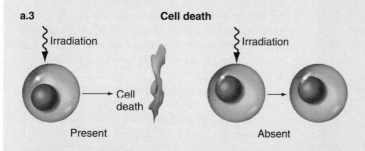

Irradiation   Irradiation

Cell death

Present   Absent

a.4   Gap junctions

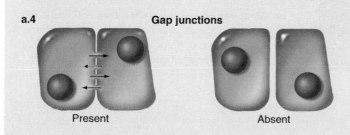

Present   Absent

a. Changes that produce uncontrolled cell growth

1. *Autocrine stimulation.* Most cells "decide" whether or not to divide only after receiving signals from neighbouring cells. Many tumour cells, by contrast, make their own stimulatory signals, in a process known as autocrine stimulation, or they have become insensitive to negative signals.

2. *Loss of contact inhibition.* Normal cells stop dividing when they come in contact with one another, as evidenced by the fact that normal cell types that grow in culture form sheets one-cell thick. Tumour cells, in contrast, climb all over each other to produce piles that are many-cells thick. This change contributes to the disordered array of cells seen in tumours.

3. *Loss of cell death.* Normal cells die when starved of growth factors or when exposed to agents that damage them. Programmed cell death (apoptosis) is activated by the expression of certain genes in the cell; it is probably a safeguard against the early stages of cancer. Most cancer cells are much more resistant than normal cells to programmed cell death.

4. *Loss of gap junctions.* Normal cells connect to their neighbours by small pores, or gap junctions, in their membranes. The gap junctions permit the transfer of small molecules that may be important in controlling cell growth. Most tumour cells have lost these channels of communication.

(Continued)

b.1

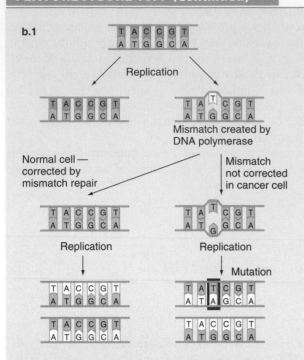

Replication

Mismatch created by
DNA polymerase

Normal cell —
corrected by
mismatch repair

Mismatch
not corrected
in cancer cell

Replication

Replication

Mutation

b. **Changes that produce genomic and karyotypic instability**

1. *Defects in the DNA replication machinery.* Cancer arises most often in cells that have lost the ability to reproduce their genomes faithfully. You saw in Chapters 6 and 8 that cells have elaborate systems for repairing DNA damage; these systems include the enzymatic machinery for mismatch repair and the repair of damage caused by radiation or ultraviolet light. Work on yeast and bacteria has shown that mutant organisms defective in DNA repair have enormously increased rates of mutation. These increased mutation rates often lead to cancer in multicellular organisms.

b.2

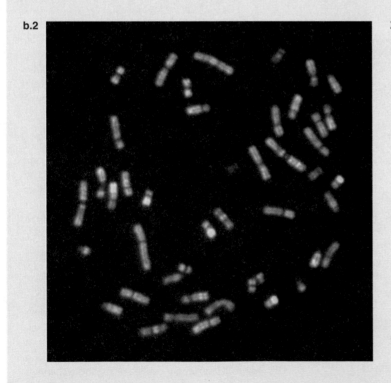

2. *Increased rate of chromosomal aberrations.* Tumour-cell karyotypes often carry gross rearrangements, including broken chromosomes, with some of the pieces rejoined to other chromosomes; multiple copies of individual chromosomes, rather than the normal two; and deletions of large chromosomal segments and of whole chromosomes. Studies have confirmed that the fidelity of chromosome reproduction is greatly diminished in tumour cells. Normal fibroblast cells, for example, have an undetectable rate of gene amplification (an increase in the number of copies of a gene), whereas tumour cells have amplification rates as high as 1 in 100 cells.

Probably only a small fraction of these chromosomal rearrangements lead to cancer; for example, tumours from solid tissues typically carry many chromosomal rearrangements, but most of these aberrations do not recur in all tumours. A few rearrangements, however, regularly appear in specific tumour types. Examples include the translocation between chromosomes 8 and 14 found in patients with certain kinds of lymphoma and the translocation between chromosomes 9 and 22 found in certain types of leukaemias (see Figure 9.20).

c. Changes that produce a potential for immortality

1. *Loss of limitations on the number of cell divisions.* Most normal cells (except for the rare stem cells) die spontaneously after a specifiable number of cell divisions. Tumour cells, by contrast, can divide indefinitely.

2. *Ability to grow in culture.* Cells derived from tumour cells usually grow readily in culture, making cancerous cell lines available for study. Normal cells do not grow well in culture.

3. *Restoration of telomerase activity (not shown).* Most normal human somatic cells do not express the enzyme telomerase, and this lack of telomerase expression prevents them from replicating the repeated sequences in the telomeres at the ends of their chromosomes, contributing to cell aging and death (see Chapter 6). Tumour cells have the ability to express telomerase, a feature that most likely contributes to their immortality.

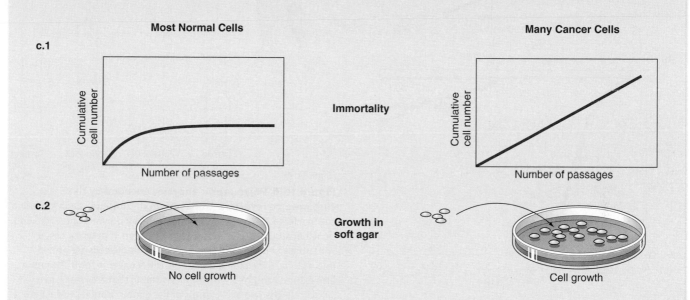

Most Normal Cells   **Many Cancer Cells**

c.1

Cumulative cell number — Number of passages

Immortality

Cumulative cell number — Number of passages

c.2

Growth in soft agar

No cell growth   Cell growth

d. Changes that enable a tumour to disrupt local tissue and invade distant tissues

1. *The ability to metastasize.* Normal cells stay within rigidly defined boundaries. Tumour cells, by comparison, often acquire the capacity to invade surrounding tissues and eventually to travel through the bloodstream to colonize distant tissues. Metastasis—the invasion of other tissues—is a complicated behaviour requiring many genetic changes.

2. *Angiogenesis.* Once the adult human body has developed, new blood vessels do not normally form except to heal a wound. Tumour cells, however, secrete substances that cause blood vessels to grow toward them. The new vessels serve as supply lines through which the tumour can tap new sources of nutrients and as escape routes through which tumour cells can metastasize.

3. *Evasion of immune surveillance (not shown).* The human immune system may recognize cancer cells as foreign and attack them, thereby helping to eliminate tumours even before they become large enough for clinical detection. As evidence, cancer patients often have antibodies and/or killer T cells directed against their cancer cells. Successful tumour cells, however, somehow develop the ability to evade detection by the immune system.

d.1 Metastasis   **d.2** Angiogenesis

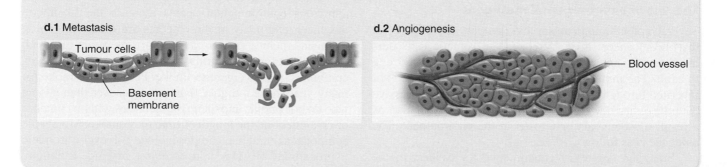

Tumour cells — Basement membrane — Blood vessel

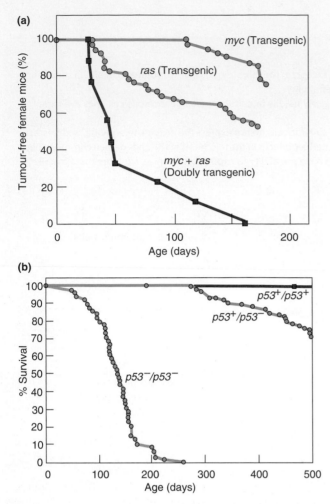

(a)

(b)

Figure 16.5 The percentage of mice still alive as a function of age. **(a)** The activated *myc* oncogene produces tumours more slowly than the *ras* oncogene. Mice containing both oncogenes develop tumours even faster than mice with *ras*. **(b)** Homozygous *p53*$^+$ mice rarely get life-threatening tumours, whereas those heterozygous for a *p53*$^-$ mutation develop tumours late in life. Mice homozygous for the *p53*$^-$ mutation develop tumours early in life.

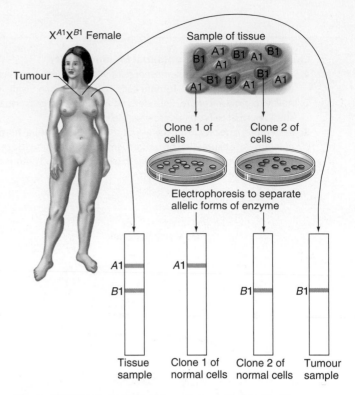

Figure 16.6 Polymorphic enzymes encoded by the X chromosome reveal the clonal origin of tumours. Each individual cell in a female expresses only one form of a polymorphic X-linked gene because of X-chromosome inactivation. A patch of tissue will usually contain both types of cells. If single cells are grown into a clone, they exhibit one or the other enzyme form. A tumour also exhibits only one form, demonstrating that it arose from a single cell. The two allelic forms of the gene's protein product are distinguished by electrophoresis.

(review the discussion of X-chromosome inactivation in Chapter 9). This finding suggests that the cells of each tumour are the clonal descendants of a single somatic cell that sustained a rare mutation.

The role of environmental mutagens

Several epidemiological surveys support the hypothesis that most cancers arise by chance in somatic cells during their division and differentiation from fertilized egg to adult. The mutations that produce these cancers are not inherited through the germ line in a dominant or recessive pattern; rather, they arise sporadically in a population as a result of chemicals or viruses in the environment. The evidence is as follows.

First, the degree of concordance for cancers of the same type among first-degree relatives, such as sisters and brothers or even identical twins, is low for most forms of cancer in the population as a whole (we discuss specific exceptions later). If one sibling or twin gets a cancer, the other usually does not.

Second, although rates for the incidence of specific cancers vary worldwide (**Table 16.1**), when populations migrate from one place to another, their profile of cancer incidence becomes more like that of the people indigenous to the new location. The change in cancer profile often takes decades, suggesting that the environment acts over a long period of time to induce the cancer.

Third, epidemiological studies have established that numerous environmental agents increase the likelihood of cancer, and many of these agents are mutagens. These mutagens include cancer-causing viruses, some of which carry mutant forms of normal genes that control cell proliferation, as well as cigarette smoke. People who smoke for many years have a higher risk of lung cancer than people who do not smoke, and their risk increases with the number of cigarettes and the length of time they smoke. Mutagenic compounds found in cigarettes include polycyclic aromatic hydrocarbons (PAHs), nitrosamines, arsenic, cadmium, cyanide, and several other carcinogens. Tobacco exposure has been linked to tumour-causing mutations in *p53* and *KRAS* (a RAS family member).

Site of Origin of Cancer	High-Incidence Population		Low-Incidence Population	
	Location	*Incidence**	*Location*	*Incidence**
Prostate	Canada (Prince Edward Island)	112	China (Jiashan)	1.4
Lung, trachea, and bronchus	USA (New Orleans, blacks)	97	Peru (Trujillo)	5.9
Breast	Canada (Nova Scotia)	83	Oman (Omani)	14.6
Stomach	Japan (Hiroshima)	80	Kuwait (Kuwaitis)	3.4
Melanoma	Australia (Queensland)	56	Korea (Daejeon)	0.3
Cervix	Zimbabwe (Harare, African)	47	Israel (non-Jews)	2.4
Liver	China (Guangzhou City)	43	Canada (New Brunswick)	2.3
Colon	Canada (Northwest Territories)	40	India (Chennai)	1.9
Bladder	Spain (Tarragona)	38	Chile (Valdivia)	3.8
Nasopharynx	China (Zhongshan)	27	UK (southwestern)	0.4
Oesophagus	France (Calvados)	15	Italy (Syracuse)	1.3
Ovary	UK (Northern Ireland)	15	Portugal (Porto)	4.3
Pancreas	USA (District of Columbia, blacks)	13	Uganda (Kyadondo County)	1.2
Lip	Canada (Newfoundland and Labrador)	7	Japan (Nagasaki)	0.1

*Incidence indicates number of new cases per year per 100 000 population, adjusted for a standardized population age distribution (so as to eliminate effects merely to differences of population age distribution). Figures for cancers of breast, cervix, and ovary are for women; other figures are for men. (Adapted from Curado, M., et al. (eds.) (2007). *Cancer Incidence in Five Continents, 9,* Lyon: International Agency for Research on Cancer.)

Cancer development over time

The data on lung cancer show that decades elapse between the time a population begins smoking and the time that lung cancer begins increasing. Both in Canada and the United States, cancer incidence in men rose dramatically after 1940, roughly two decades after men began frequent smoking; women did not begin frequent smoking until several decades after men, and lung cancer incidence in women did not begin its dramatic increase until after 1960 (**Figure 16.7a**).

Epidemiological data also show that the incidence of cancer rises with age. The prevalence of cancer in older people supports the idea that cancer develops over time as well as the idea that the accumulation of many mutations in the clonal descendants of a somatic cell fuels the progression from normal to cancerous. If you assume that the rate of accumulation of cancer-causing mutations is constant over a lifetime, the slope of a logarithmic curve plotting cancer incidence against age is a measure of the number of mutations required for cancer (**Figure 16.7b**). Interestingly, the data for many types of tumours generate a similar curve in which the evolution of cancer requires six to ten mutations. Thus, the correlation between cancer incidence and aging, as well as the time lag between exposure to carcinogens and the appearance of tumours, suggests that the mutations that produce cancer accumulate over time. However, this simple interpretation is only part of the picture because cells increase their mutation rate at some point during their progression to cancer.

Cancers that run in families

In some families, a specific type of cancer recurs in many members, indicating the inheritance of a predisposition through the germ line. Retinoblastoma is an example of this type of cancer (see the Genetics and Society box in Chapter 4). Half the individuals in families affected by retinoblastoma inherit a mutation in the *RB* gene from one parent. Because all of their somatic cells carry one defective copy of the gene, a mutation in the single remaining wild-type copy of the *RB* gene in the cells that proliferate to produce the retina predisposes these individuals to develop retinal cancer (**Figure 16.8**). People who do not inherit a mutation in the *RB* gene need to experience a mutation in both copies of the gene in the same cell to develop cancer; this type of double hit is very rare. Interestingly, for nearly all common types of cancer that occur sporadically in a population, rare families can be found that exhibit an inherited predisposition to that cancer.

Figure 16.9 summarizes the sequence of events, as scientists now understand them, that ultimately lead to a malignant cell, known as the multi-hit model of carcinogenesis.

> Cancer cells arise from multiple mutations occurring over time in a single cell that then produces a clone of malignant cells. This is known as the multi-hit model of cancer development. Environmental mutagens are responsible for most cancers; inheritance of certain mutations predisposes some families to development of specific cancers.

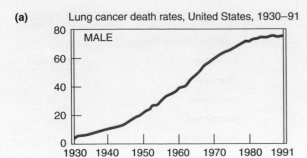

(a) Lung cancer death rates, United States, 1930–91

MALE

FEMALE

Rates are per 100 000 and are age-adjusted to the 1970 U.S. census population.

(b)

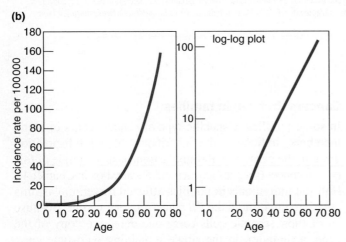

log-log plot

Figure 16.7 **Lung cancer death rates and incidence of cancer with age.** **(a)** Lung cancer death rates in the United States during the twentieth century began increasing rapidly for men in the 1940s and for women in the 1960s. This reflects the fact that smoking became prevalent among men about 20 years before it did among women. **(b)** The incidence of most cancers shows a dramatic increase with age, a result thought to reflect the accumulation of mutations in somatic cells.

Mutations create dominant oncogenic alleles or recessive tumour-suppressor alleles

Research has not only revealed that cancer results from multiple genetic changes in the clonal descendants of one cell, it has also established that the mutations found in tumours are of two general types: those that improperly activate genes (e.g., the genes responsible for promoting cell proliferation) and those that improperly inactivate genes (e.g., the genes responsible for preventing excessive cell proliferation).

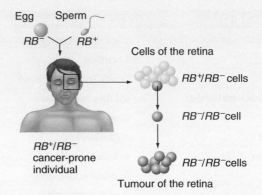

Figure 16.8 **Individuals who inherit one copy of the RB^- allele are prone to cancer of the retina.** During the proliferation of retinal cells, the RB^+ allele is lost or mutated, and cancers grow out of the RB^-/RB^- clone of cells.

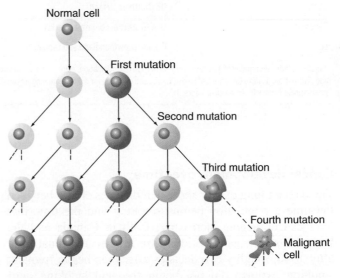

Figure 16.9 **Cancer is thought to arise by successive mutations in a clone of proliferating cells.**

The mutant alleles that lead to cancer are referred to as **cancer genes**, but the term "genes" is a misnomer. All cancer genes are, in fact, mutant alleles of normal genes. When present in all or a subset of cells within an organism, these mutant alleles predispose the individual to develop cancer over a lifetime. Mutant alleles that act dominantly are known as **oncogenes**; in a diploid cell, one mutant oncogenic allele is sufficient to alter the cell phenotype (**Figure 16.10a**). Mutant alleles that act recessively are known as mutant **tumour-suppressor** genes; in a diploid cell, both copies of a tumour-suppressor gene must be mutant to make the cell abnormal (**Figure 16.10b**).

Increased cell proliferation from oncogenes

Two approaches to identifying oncogenes are the study of tumour-causing viruses and the study of tumour DNA itself.

Tumour viruses are useful tools for studying cancer-causing genes, first because they carry very few genes

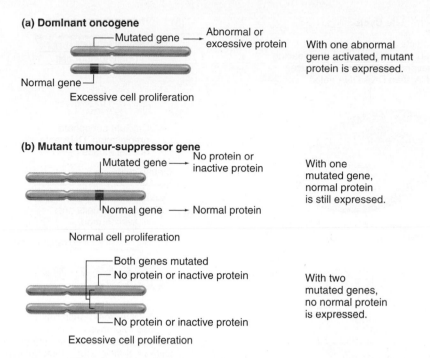

(a) Dominant oncogene

Mutated gene → Abnormal or excessive protein

Normal gene

Excessive cell proliferation

With one abnormal gene activated, mutant protein is expressed.

(b) Mutant tumour-suppressor gene

Mutated gene → No protein or inactive protein

Normal gene → Normal protein

Normal cell proliferation

With one mutated gene, normal protein is still expressed.

Both genes mutated

No protein or inactive protein

No protein or inactive protein

Excessive cell proliferation

With two mutated genes, no normal protein is expressed.

Figure 16.10 Cancer-producing mutations occur in two forms. (a) Dominant gain-of-function mutations generate oncogenes that exhibit abnormal activity or produce an excessive amount of protein. **(b)** Recessive loss of function mutations produce altered tumour suppressors that usually generate little or no phenotype when heterozygous with a wild-type allele, but that affect cell proliferation when a second mutation inactivates the wild-type allele.

themselves, and second because they infect and change cultured cells to tumour cells, which makes it possible to study them *in vitro*. A large number of the viruses that generate tumours in animals are retroviruses whose RNA genome, upon infecting a cell, is copied to cDNA, which then integrates into the host chromosome (review the Genetics and Society box in Chapter 7). Later, during excision from the host chromosome, the virus can pick up copies of host genes. These normal genes change to abnormally activated oncogenes either through mutations that occur during viral propagation or through their placement near powerful promoters and enhancers in the viral genome (**Figure 16.11a**). The oncogenes carried by tumour-producing viruses are thus mutated versions of normal host-cell genes. The wild-type genes that become oncogenes upon mutation are known as **proto-oncogenes**. When a virus carrying one or more oncogenes infects a cell, the oncogenes cause abnormal proliferation that can lead to the accumulation of more mutations and eventually to cancer. The analysis of tumour-causing retroviruses led to the discovery of oncogenes in a variety of species (**Table 16.2**).

Some DNA viruses also carry oncogenes. An example is the human papillomavirus (HPV). HPV infection of a woman's cervical cells is probably the first step in the development of cervical cancer. The papillomavirus carries at least two oncogenes capable of transforming appropriate recipient cells in culture: *E6* and *E7*. The E6 and E7 proteins bind to and inactivate the

normal products of the *p53* and *RB* genes. In addition, only those HPV subtypes whose E6 and E7 proteins bind p53 and RB proteins are associated with cervical cancer in women. Progression of HPV-infected cells to cancer requires additional mutations in genes not yet identified.

Scientists also identify oncogenes by isolating DNA from tumour cells and exposing noncancerous cells in culture to this tumour DNA. Some tumour DNA transforms cultured cells into cells capable of producing tumours (**Figure 16.11b**). For example, the DNA responsible for the transformation of mouse cells by human tumour DNA can be identified by re-isolating the human DNA from the transformed mouse cells with probes for the short interspersed elements known as *Alu* sequences. These sequences appear only in the human genome (review Figure 9.24a). The oncogenes identified in this way, like those discovered in studies of tumour viruses, are oncogenic alleles of normal cellular genes that have mutated to abnormally active forms.

Sometimes the two approaches have identified the same oncogene, for example, *RAS*. The oncogenic forms of the *RAS* gene generate proteins that are always (or constitutively) in the GTP-activated form; therefore, whether or not growth factor is present, a cell carrying a *RAS* oncogene receives signals to divide (**Figure 16.11c**). Like mutated *RAS*, many oncogenes continuously turn on one or more of a cell's many signal transduction systems. They do this by encoding receptors, signal transmitters, and

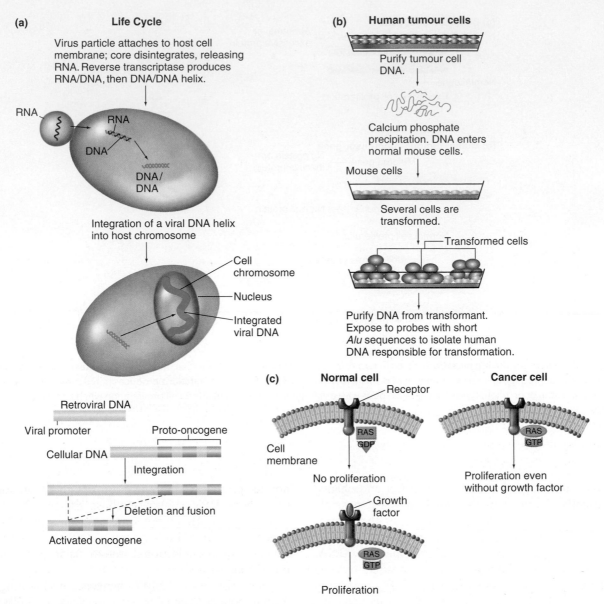

Figure 16.11 Two methods to isolate oncogenes. **(a)** Retroviruses that cause cancer carry a mutant or overexpressed copy of a cellular growth-promoting gene. If the genome of a retrovirus integrates into the host chromosome near a proto-oncogene, the cellular gene may be packaged with the viral genome when the virus leaves the cell. **(b)** DNA isolated from some human cancers is able to transform mouse cells into cancer cells. These cells are found to contain a human oncogene. **(c)** The *RAS* oncogene, a mutant form of the *RAS* proto-oncogene, produces a protein that becomes locked into the GTP-activated form.

Table 16.2	Retroviruses and Their Associated Oncogenes*		
Virus	**Species**	**Tumour**	**Oncogene**
Rous sarcoma	Chicken	Sarcoma	*src*
Harvey murine sarcoma	Rat	Sarcoma and erthyroleukaemia	*H-ras*
Kristen murine sarcoma	Rat	Sarcoma and erthyroleukaemia	*K-ras*
Moloney murine sarcoma	Mouse	Sarcoma	*mos*
FBJ murine osteosarcoma	Mouse	Chondrosarcoma	*fos*
Simian sarcoma	Monkey	Sarcoma	*sis*
Feline sarcoma	Cat	Sarcoma	*sis*
Avian sarcoma	Chicken	Fibrosarcoma	*jun*
Avian myelocytomatosis	Chicken	Carcinoma, sarcoma, and myelocytoma	*myc*
Ableson leukaemia	Mouse	B-cell lymphoma	*abl*

*Retroviruses identified as causative agents of tumours in animals contain oncogenes that were derived from a cellular gene. Adapted from Lewin, *Genetics*, 1e, Oxford University Press, Inc. by permission.

Table 16.3	Oncogenes Are Members of Signal Transduction Systems*		
Name of Oncogene	**Tumour Associations**	**Mechanism of Activation**	**Properties of Gene Product**
hst	Stomach carcinoma	Rearrangement	Growth factor
erbB	Mammary carcinoma, glioblastoma	Amplification	Growth factor receptor
trk	Papillary thyroid carcinoma	Rearrangement	Growth factor receptor
H-ras	Bladder carcinoma	Point mutation	GDP/GTP-binding signalling protein
raf	Stomach carcinoma	Rearrangement	Cytoplasmic serine/threonine kinase
myc	Lymphomas, carcinomas	Amplification, chromosomal translocation	Nuclear transcription factor

*The roles of several oncogene products that are members of the signal transduction pathway and the ways in which they get activated in human cells are shown.

transcription factors that are active with or without growth factor (see **Table 16.3** and the **Focus on Genetics** box "**Cell Signalling Mechanisms and Cancer**").

Enhanced mutation potential in proliferating cells

Like the oncogenic *RAS* gene, many of the oncogenes so-far identified affect cell signalling pathways that tell a cell whether or not to divide. The importance of these genes in generating cancer is not just that they cause cells to proliferate, because an increase in proliferation alone, without other changes, generates benign growths that are not life threatening and can be removed by surgery. Rather, increased proliferation provides a large clone of cells within which further mutations can occur, and these further mutations may eventually lead to malignancy. The more cells that exist in a clone, the more likely that rare mutations will occur in the clone—which already has the potential for rapidly propagating them. Although not all cancer-causing genes are dominant oncogenes, oncogenic mutations have been the easiest to identify for technical reasons.

Increased cell proliferation from mutant tumour-suppressor alleles

Mutant tumour-suppressor genes are recessive alleles of genes whose normal alleles help put cell division on hold, whether in terminally differentiated cells or in cells with DNA damage. Targets for tumour-suppressor mutations include *RB*, *p53*, and *p16*. One wild-type copy of these genes apparently produces enough protein to regulate cell division; the loss of both wild-type copies releases a brake on proliferation (see Figure 16.10b). Researchers have identified dozens of tumour-suppressor genes through the genomic analysis of families with an inherited predisposition to specific types of cancer or through the analysis of specific chromosomal regions that are reproducibly deleted in certain tumour types.

Retinoblastoma provides an example of this identification process. A cancer of the colour-perceiving cone cells in the retina, retinoblastoma is one of several cancers inherited in an autosomal dominant fashion in human families (**Figure 16.12a**). Roughly half the children of a parent with retinoblastoma develop the disease. Retinoblastoma tumours are easy to diagnose and remove before they become invasive. As you saw in Chapter 9, karyotypes of

normal, noncancerous tissues from many people suffering from retinoblastoma reveal heterozygosity for deletions in the long arm of chromosome 13; that is, the patients carry one normal and one partially deleted copy of 13q. Karyotypes of the cancerous retinal cells from some of these same patients show homozygosity for the same chromosome 13 deletions that are heterozygous in the noncancerous cells (**Figure 16.12b**). Although the deletions vary in size and position from patient to patient, they all remove band 13q14.

These observations indicate that band 13q14 includes a gene whose removal contributes to the development of retinoblastoma. *RB* is the symbol for this gene. The heterozygous cells in a patient's normal tissues carry one copy of the gene's wild-type allele (RB^+), and this one copy prevents the cells from becoming cancerous. Tumour cells homozygous for the deletion, however, do not carry any copies of RB^+, and without it, they begin to divide out of control.

Geneticists used their understanding of retinoblastoma inheritance to find the *RB* gene. They cloned DNA carrying the gene by looking for DNA sequences in band 13q14 that were lost in all of the deletions associated with the hereditary condition. They then identified the gene by characterizing a very small deletion that affected only one transcriptional unit—the *RB* gene itself. Analysis of the gene's function showed that it encodes a protein involved (along with many other proteins) in regulating the cell cycle. *RB* thus fits our definition of a tumour-suppressor gene: The protein it determines helps prevent cells from becoming cancerous. Cancer can arise when cells heterozygous for an *RB* deletion lose the remaining functional copy of the gene.

This picture of the genetics of retinoblastoma raises a perplexing question: How can the retinoblastoma trait be inherited in an autosomal dominant fashion if a deletion of the *RB* gene is recessive to the wild-type *RB* allele? At the level of the organism, *RB* deletions are dominant because of the very strong likelihood that in at least one of the hundreds of thousands of retinal cells heterozygous for the deletion, a subsequent genetic event will disable the single remaining *RB* allele, resulting in a mutant cell with no functional tumour-suppressor gene. This one cell then multiplies out of control, eventually generating a clone of cancerous cells (see the Genetics and Society box in Chapter 4).

Geneticists first recognized the recessive *RB* mutation that leads to retinoblastoma through the genomic analysis

Figure A Dr. Tony Pawson.

Dr. Anthony (Tony) Pawson (1952–2013) (**Figure A**), born in Maidstone, England, was a Distinguished Scientist and Apotex Chair in Molecular Oncology at the Samuel Lunenfeld Research Institute of Mount Sinai Hospital in Toronto and a professor at the University of Toronto. His ground-breaking studies have contributed significantly to advances in biomedical research. A global leader in the field of signal transduction (cell signalling research), his investigations and discoveries have led to a greater understanding of the genetic and molecular mechanisms governing cell responses and communication, and the organization and regulation of intracellular signalling pathways and networks. Building on his discovery of the phosphotyrosine-binding Src-homology 2 (SH2) domain, the prototypical modular protein domain that mediates protein-protein interactions and kinase activities during cell signalling processes, he further identified, using genetic and proteomic tools, a number of different protein modules that are crucial for proper signal transduction.

Dr. Pawson received his Master's degree in Biochemistry from the University of Cambridge, studying under Nobel Laureate Tim Hunt. In 1976, he obtained his Ph.D. from King's College, University of London/Imperial Cancer Research Fund, working on retroviral gene expression and replication mechanisms. As a post-doctoral fellow at the University of California at Berkeley, he identified a number of retroviral oncogene products. His studies suggested that tyrosine phosphorylation may play a role in the transformation of normal cells to cancer cells. He continued his work on cancer-causing retroviruses at the University of British Columbia in 1980 as an assistant professor in the Department of Microbiology. There he discovered that the oncogenic protein of the Fujinami sarcoma virus (named Fps) is an active tyrosine kinase that has three different modular domains (including an SH2 domain and kinase domain), similar to Src and other proteins such as Abl (**Figure B**). These domains are critical for their oncogenic activities.

In 1985, he accepted a faculty position at the University of Toronto in the Department of Molecular Genetics, and was one of the founding members of the Samuel Lunenfeld Research Institute of Mount Sinai Hospital, where he investigated the role of the SH2 domain in more detail. This domain is present in a number of different signalling proteins, and binds specifically to growth-inducing receptors that are in their active state. His research allowed him to propose the following cell signalling mechanism: The dimerization and activation of receptor proteins (protein tyrosine kinases), either by growth factors or by tumour-causing mutations, results in the subsequent phosphorylation of these receptors on tyrosine residues. This creates recognition motifs allowing receptor proteins to physically bind the SH2 domains of intracellular proteins, which signal their targets evoking cellular responses (**Figure C**). This signalling network, composed of proteins containing modular

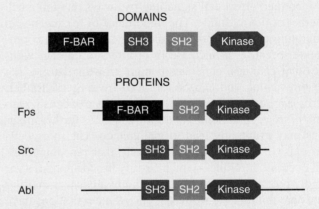

Figure B **Modular protein domains.** Related intracellular proteins, such as Fps, Src, and Abl, have protein domains in common, such as the SH2 and kinase domains.

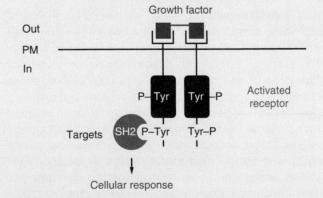

Figure C **Receptor activation results in altered cell behaviour.** Binding of growth factors to the extracellular domains of receptor tyrosine kinases (RTKs) results in their dimerization and subsequent activation of the receptors' cytoplasmic kinase domains. This leads to receptor tyrosine phosphorylation, creating binding sites for intracellular protein targets containing SH2 domains, and allowing for changes in cellular responses. (PM = plasma membrane)

domains, allows cells to respond quickly to external cues. When these receptor proteins are mutated though, cell signalling mechanisms are no longer properly regulated, resulting in increased cell proliferation and tumour formation. Knowledge of cellular communication has had a profound impact on understanding diseases such as cancer, heart disease, and immune-system deficiencies, and has led to the development of novel cancer therapies.

More recently, Dr. Pawson led an international $13 million large-scale project with Genome Canada, with a goal to map protein interactions within human cells in order to determine whether and how modifications in cell signalling networks can cause diseases such as cancer. It is anticipated that the findings from this project will result in novel proteomic and computational technologies, as well as cancer therapeutics.

Dr. Pawson received a number of awards over the course of his career, including the Gairdner Foundation International Award, the AACR/Pezcoller International Award for Cancer Research, the Killam Prize for Health Sciences, the Wolf Prize in Medicine, the Royal Medal from the Royal Society of London, and the Kyoto Prize (the first Canadian scientist to receive this honour). He was also a Fellow of the Royal Societies of London and Canada, an inductee of the Canadian Medical Hall of Fame, and a recipient of the Order of Canada.

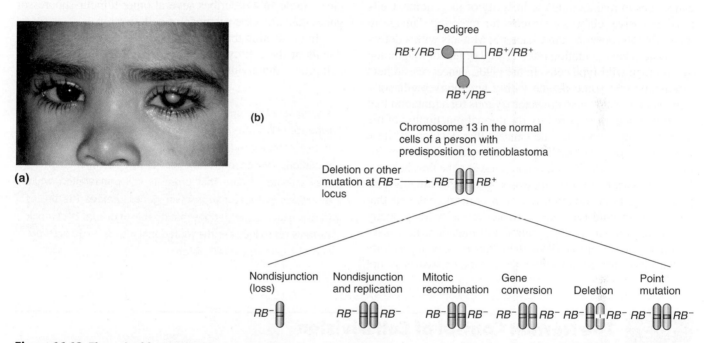

Figure 16.12 The retinoblastoma tumour-suppressor gene. (a) A child with a retinoblastoma tumour in the left eye. **(b)** The RB^- gene is inherited through the germ line as an autosomal recessive mutation. Subsequent changes to the RB^+ allele during somatic divisions generate a clone of cells homozygous or hemizygous for the RB^- allele.

of families inheriting a predisposition to the cancer. More recently, they noted that both copies of the *p16* gene on chromosome 9 are deleted in roughly 75 percent of all melanomas (a malignant skin cancer) and in approximately 85 percent of all gliomas (the most common form of brain cancer). The *p16* gene encodes a protein that binds to and inactivates CDK4. In another example, observations of deletions of both copies of a specific region of chromosome 18 in all colorectal cancers led to identification of the *DCC* (deleted in colorectal cancer) gene.

Many tumour-suppressor mutations occur in genes that control the cell cycle and, with it, the accuracy of genomic replication. It is important to distinguish mutations that determine how the cell cycle is completed from mutations in genes that control proliferation. Alterations in genes that control proliferation result in an enlarged clone of cells, but aside from their increase in number, these cells—if they sustain no further mutations—are normal and thus form a benign growth. By contrast, mutations in genes that control the cell cycle can alter the accuracy with which a cell reproduces its genome. The resulting mutant cells can produce offspring with many more mutations than occur in normal cells, and this increase in the frequency of mutations vastly increases the probability that the cascade of mutations necessary to produce the phenotypic changes of tumour cells will occur.

Because cancer arises most often in cells that have lost the ability to reproduce their genomes faithfully, it seems reasonable to conclude that a cell's primary safeguard against

Table 16.4	Mutant Alleles of These Tumour-Suppressor Genes Decrease the Accuracy of Cell Reproduction*	
Gene	**Normal Function of Gene (if known), or Disease Syndrome Resulting from Mutation**	**Function of Normal Protein Product**
p53	Controls G_1-to-S checkpoint	Transcription factor
RB	Controls G_1-to-S transition	Inhibits a transcription factor
ATM	Controls G_1-to-S phase, and G_2-to-M checkpoint	DNA-dependent protein kinase
BS	Recombinational repair of DNA damage	DNA/RNA ligase
XP	Excision of DNA damage	Several enzymes
hMSH2, hMLH1	Correction of base-pair mismatches	Several enzymes
FA	Fanconi anaemia	Unknown
BRCA1	Repair of DNA breaks	Unknown
BRCA2	Repair of DNA breaks	Unknown

*Many tumour-suppressor genes have been associated with a specific function in the cell cycle necessary for accuracy of cell division.

cancer lies in maintaining the integrity of its genome. Cells have extensive, elaborate systems for repairing damage to their DNA, as described in Chapter 8. Mutants with a defective system have mutation rates several orders of magnitude greater than wild-type cells. In the l990s, cancer researchers discovered that some people with a hereditary predisposition to colorectal cancer are heterozygous for a mutation that inactivates a gene required for the normal functioning of the mismatch repair system; the cancers that develop in these individuals consist of cells that have lost the single remaining wild-type allele. This mismatch repair gene thus behaves like a classical tumour-suppressor gene. Presumably, the greatly increased mutation rate in a homozygous cell that has lost both wild-type alleles makes it easier for progeny cells to accumulate the large number of mutations necessary to produce a cancer cell. Why these cancers develop mainly in the colon or rectum rather than in other tissues is not clear. **Table 16.4** describes several other tumour-suppressor genes that affect the accuracy of cell division.

In the section that follows, we take a closer look at the details of the cell cycle—the process by which a somatic cell grows and divides.

> Accumulated oncogenic and tumour-suppressor mutations produce cells with grossly altered genomes that proliferate both excessively and inaccurately. Dominant, gain-of-function mutations convert proto-oncogenes into oncogenes, which may activate proteins that promote cell proliferation, while mutations in tumour-suppressor genes remove the brakes from cell division. Mutations that disable or disrupt DNA repair systems act to increase the mutational rate as errors accumulate with each round of division.

16.3 The Normal Control of Cell Division

A variety of genes and proteins control the events of the cell cycle. These genes and proteins allow progression to the next stage of the cycle when all is well, but they cause the cellular machinery to slow down when damage to the genome or to the machinery itself requires repair. We now describe the molecules that control cell division.

Cyclins and cyclin-dependent kinases ensure proper timing and sequence of events

Cell division, as you learned in Chapter 3, requires the duplication of chromosomes and other cellular components as well as the precise partitioning of the duplicated elements into two daughter cells. During this complicated process, the cell coordinates the function of hundreds of different proteins. To see how the cell orchestrates the events of cell division, we first review the stages of the cell cycle and then look at some of the proteins that control progression through that cycle.

The four phases of the cell cycle: G_1, S, G_2, and M

To review the cell cycle briefly, G_1 is the "gap" period between the end of mitosis and the DNA synthesis that precedes the next mitosis (**Figure 16.13**). During G_1, the cell grows in size, imports materials to the nucleus, and prepares in other ways for DNA replication. S is the period of DNA synthesis, or replication. G_2 is the "gap" between DNA synthesis and mitosis. During G_2, the cell prepares for division. M, the phase of mitosis, includes the breakdown of the nuclear membrane, the condensation of the chromosomes, their attachment to the mitotic spindle, and the segregation of chromosomes to the two poles; at the completion of mitosis, the cell divides by cytokinesis.

During M, the cell must coordinate the activities of a variety of proteins: those that cause chromosome condensation (see Chapters 3 and 6), tubulins that polymerize to form the mitotic spindle on which the chromosomes move, motor proteins in the kinetochores that power chromosome

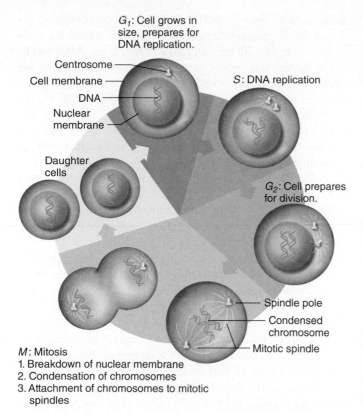

G_1: Cell grows in size, prepares for DNA replication.

Centrosome
Cell membrane
DNA
Nuclear membrane

S: DNA replication

Daughter cells

G_2: Cell prepares for division.

Spindle pole
Condensed chromosome
Mitotic spindle

M: Mitosis
1. Breakdown of nuclear membrane
2. Condensation of chromosomes
3. Attachment of chromosomes to mitotic spindles

Figure 16.13 The cell cycle is the series of events that transpire between one cell division and the next. After division, a cell begins in the G_1 phase, progresses into S phase, where the chromosomes replicate, to the G_2 phase, and to the M phase, where replicated sister chromosomes segregate to daughter cells. In M phase, the nuclear membrane breaks down, centrosomes form the poles of the spindle, and microtubules construct a scaffold on which chromosomes migrate.

movement, proteins that dissolve and form the nuclear membrane again at the beginning and end of mitosis, and others.

Discovery of kinases in yeasts

The budding yeast *Saccharomyces cerevisiae* and the fission yeast *Schizosaccharomyces pombe* have been instrumental in identifying the genes that control cell division. Several properties of both yeast species make them particularly useful for this study. First, both can grow as haploid or diploid organisms. As a result, recessive mutations can be identified in the haploid cells, and then diploid cells can be constructed containing two mutations. These can be allowed to proliferate, and the resulting cell populations tested to determine the number of complementation groups defined by the mutations.

The budding yeast *S. cerevisiae* has yet another property that facilitates cell-cycle analysis. At the beginning of the cell cycle, toward the end of G_1, a new daughter cell arises as a bud on the surface of the mother cell. As the mother cell progresses through the division cycle, the bud grows in size; it is small during S phase and large during mitosis. Bud size thus serves as a marker of progress through the cell cycle (**Figure 16.14a**). One can order cells in an asynchronously cycling population according to

position in the cell cycle by observing the relative si their buds. A normal population of growing yeast cell tains nonbudding cells as well as cells with buds of all sizes.

Isolation of Cell-Cycle Mutants. Mutations that interfere with the cell cycle are lethal; and because cell proliferation depends on successive repeats of the cell cycle, a mutant unable to complete the cell cycle cannot grow into a population of cells (**Figure 16.14b**). Researchers have obtained cell-cycle-defective mutants by isolating cells with temperature-sensitive mutations (see the Fast Forward box in Chapter 8). In these mutants, a protein needed for cell division functions normally at a low permissive temperature, but it loses function at a higher

(a)

(b)

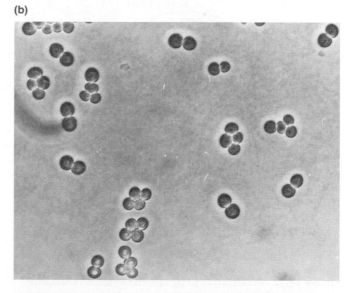

Figure 16.14 A cell-cycle mutant of yeast. (a) Cells of a temperature-sensitive mutant growing at the permissive temperature display buds of all sizes. **(b)** After incubation at the restrictive temperature, the same cells have arrested—all with a large bud. Cells that are early in the cell cycle at the time of the temperature shift arrest in the first cell cycle; these cells have the small buds in (a). Cells that are later in the cell cycle finish the first cell cycle and arrest in the second, producing clumps with two large-budded cells.

restrictive temperature. At the permissive temperature, the mutants grow normally, producing a population of cells for study. A shift to the restrictive temperature causes the temperature-sensitive protein in the mutant population to become nonfunctional; researchers can then study the consequences of its loss.

To isolate temperature-sensitive mutations, investigators expose haploid cells to a mutagen and then plate them at the permissive temperature, allowing them to form colonies (**Figure 16.15**). After the colonies grow up, each from a single mutant cell, the experimenters use replica plating to imprint them on two plates. They incubate one plate at the permissive temperature and the other plate at the restrictive temperature. Cells that have sustained a temperature-sensitive mutation grow at the permissive temperature but not at the restrictive temperature.

Analysis of Cell-Cycle Mutants. With this protocol, researchers have isolated thousands of temperature-sensitive mutations. These mutations could occur in any gene required for cell reproduction. Genes of particular interest are those whose protein product functions at only one stage in the cell cycle. Such mutants were identified in *S. cerevisiae* by observing under the light microscope the shape and behaviour of cells shifted from the permissive to the restrictive

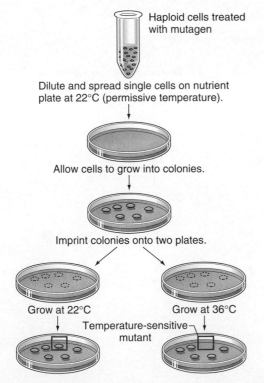

Figure 16.15 The isolation of temperature-sensitive mutants of yeast. Mutations are induced in a culture of haploid cells by exposure to a chemical mutagen. The treated cells are distributed onto solid medium. Each cell proliferates into a colony (clone) of cells, passing on the mutation. Replicas of the colonies are imprinted onto solid medium. One is grown at the permissive temperature (22°C), while the other is grown at the restrictive temperature (36°C). Colonies that grow on the former, but not the latter, carry a temperature-sensitive mutation.

temperatures (see Figure 16.14). A population of cells growing at the permissive temperature includes unbudded cells as well as cells with the full range of bud sizes. After a cell-cycle mutant has grown at the restrictive temperature for about two cycles, however, the cells have a uniform appearance. In the mutant population shown in Figure 16.14, for example, all cells have a single large bud. Moreover, the nuclei (not visible in the figure) are uniformly located at a position between the mother cell and the daughter cell, as if beginning to divide. This uniformity identifies a particular cell-cycle mutant. Other cell-cycle mutants would arrest with different but also uniform morphologies, for example, with all unbudded cells. Thus, mutants that acquire a uniform bud-related morphology at the restrictive temperature are each defective at one stage of the cell cycle.

Further examination of cells transferred from permissive to restrictive temperatures illustrates another property of cell-cycle mutants—a requirement for the normal gene product at a particular stage of the cell cycle. Note that some of the cells in Figure 16.14 formed one cell with a large bud at the restrictive temperature, while others formed two cells, each with a large bud. Note also that the former group all had smaller buds than the latter group at the time of the shift to the restrictive temperature. This observation indicates that cells early in the cell cycle at the moment of temperature shift arrested division in the first cell cycle, while those later in the cell cycle at the time of temperature shift finished the first cycle and became arrested only in the second cell cycle. The point at which a cell acquires the ability to complete a cell cycle is the moment at which the temperature-sensitive protein has fulfilled its function in that cycle.

By analyzing the morphology of buds on cells shifted from permissive to restrictive temperatures and using other methods, yeast geneticists have identified over 100 cell-cycle genes (**Table 16.5**). The significance of the *CDC28* gene in particular became apparent when geneticists identified related genes in other organisms. They found, for example, that the *CDC2* gene in fission yeast controls a step of commitment in that cell. They also learned that in extracts of *Xenopus laevis* (African clawed frog) embryos, the activity of a protein known as MPF (for maturation-promoting factor) controls the rapid early divisions. Sequences of the cloned budding and fission yeast genes revealed that they encode **protein kinases**, enzymes that add phosphate groups to their protein substrates. The *Xenopus* MPF also turned out to be a protein kinase. Moreover, genetic swapping experiments showed that the budding yeast *CDC28* gene and the fission yeast *CDC2* gene can replace each other in either organism, demonstrating that they encode proteins that carry out the same activity; the same is true of the *Xenopus* MPF-encoding gene.

Thus, in three different organisms, genes that seem to be the central controlling element of the cell cycle encode functionally homologous protein kinases. Further work has shown that these kinases are **cyclin-dependent kinases (CDKs)**; that is, they require another protein known as a **cyclin** for their activity.

Table 16.5	Some Important Cell-Cycle and DNA Repair Genes
Genes	**Gene Products and Their Function**
CDKs	Enzymes known as cyclin-dependent protein kinases that control the activity of other proteins by phosphorylating them
CDC28	A CDK discovered in the yeast *Saccharomyces cerevisiae* that controls several steps in the *S. cerevisiae* cell cycle
CDC2	A CDK discovered in the yeast *Schizosaccharomyces pombe* that controls several steps in the *S. pombe* cell cycle; also the designation for a particular CDK in mammalian cells
CDK4	A CDK of mammalian cells important for the G_1-to-S transition
CDK2	A CDK of mammalian cells important for the G_1-to-S transition
Cyclins	Proteins that are necessary for and influence the activity of CDKs
cyclinD	A cyclin of mammalian cells important for the G_1-to-S transition
cyclinE	A cyclin of mammalian cells important for the G_1-to-S transition
cyclinA	A cyclin of mammalian cells important for S phase
cyclinB	A cyclin of mammalian cells important for the G_2-to-M transition
E2F	A transcription factor of mammalian cells important for the G_1-to-S transition
RB	A mammalian protein that inhibits E2F
p21	A protein of mammalian cells that inhibits CDK activity
p16	A protein of mammalian cells that inhibits CDK activity
p53	A transcription factor of mammalian cells that activates transcription of DNA repair genes as well as transcription of *p21*
RAD9	A protein that inhibits the G_2-to-M transition of *S. cerevisiae* in response to DNA damage
E6	A protein of HPV that inhibits p53
E7	A protein of HPV that inhibits Rb

The role of cyclin-dependent kinases

The CDKs are a family of kinases that regulate the transition from G_1 to S and from G_2 to M through phosphorylations that activate or inactivate target proteins. As mentioned, CDKs function only after associating with a cyclin. The cyclin portion of a CDK–cyclin complex specifies which set of proteins a particular CDK phosphorylates; the CDK portion of the complex then performs the phosphorylation (**Figure 16.16a**). One CDK–cyclin complex, for example, activates target proteins required for DNA replication at the onset of the S phase, whereas another CDK–cyclin activates proteins necessary for chromosome condensation and segregation at the beginning of the M phase. The cyclins that guide the CDK phosphorylations appear on cue at each phase of the cell cycle. After they associate with the appropriate CDKs and point out the proper protein targets, they then disappear to make way for the succeeding set of cyclins. The cycle of precisely timed cyclin appearances and disappearances is the result of two mechanisms: gene regulation that turns on and off the synthesis of particular cyclins, and regulated protein degradation that removes the cyclins. As an example, consider the action of one CDK on the **nuclear lamins**, a group of proteins that underlie the inner surface of the nuclear membrane (**Figure 16.16b**). The nuclear lamins provide structural support for the nucleus and may also provide sites for the assembly of proteins that function in DNA replication, transcription, RNA transport, and chromosome structure. During most of the cell cycle, the lamins form an insoluble structural matrix. At mitosis, however, the lamins become soluble, and this solubility allows dissolution of the nuclear membrane into vesicles. Lamin solubility requires phosphorylation; mutant lamins that resist phosphorylation do not become soluble at

mitosis. Thus, one critical mitotic event—dissolution of the nuclear membrane—is most likely triggered by CDK phosphorylation of nuclear lamins.

Genetic studies of yeast provided much of the evidence that CDK–cyclin complexes are key controlling

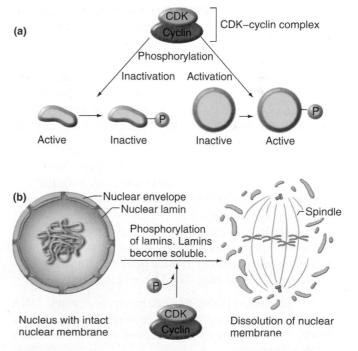

Figure 16.16 The cyclin-dependent kinases (CDK) control the cell cycle by phosphorylating other proteins. (a) A CDK combines with a cyclin and acquires the capacity to phosphorylate other proteins. Phosphorylation of a protein can either inactivate or activate it. **(b)** CDK phosphorylation of the nuclear structure proteins, lamins, is responsible for the dissolution of the nuclear membrane at mitosis.

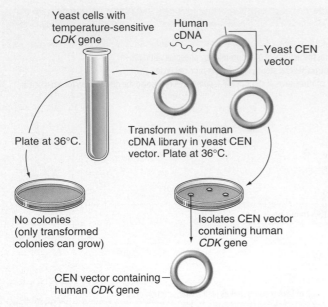

Figure 16.17 Mutant yeast permit the cloning of a human *CDK* gene. A culture of yeast cells containing a temperature-sensitive mutation in the *CDK* gene was transformed with a library composed of human cDNA cloned into a yeast centromere-containing (CEN) vector. The transformed yeast cells were spread on solid medium at the restrictive temperature. Only the rare transformants with a functional copy of the human *CDK* gene were able to grow.

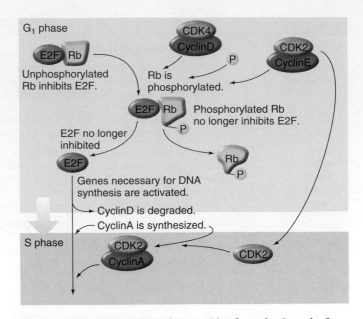

Figure 16.18 CDKs mediate the transition from the G$_1$ to the S phase of the cell cycle. In human cells, CDK4 complexed to cyclinD, and CDK2 complexed to cyclinE, phosphorylate the Rb protein, causing it to dissociate from, and thus activate, the E2F transcription factor. E2F stimulates transcription of many genes needed for DNA replication. At the transition into S phase, cyclinD is destroyed, cyclinA is synthesized, and the CDK2–cyclinA complex activates DNA replication.

agents in all eukaryotic cell cycles. In one series of studies, geneticists used yeast mutants that carry defective CDKs or cyclins to find the corresponding human genes (**Figure 16.17**) and to show that the human CDKs and cyclins can function in yeast in place of the native proteins.

Control of the G$_1$-to-S transition

Cell-cycle investigators have identified many of the molecular events controlling the transition from G$_1$ to S in human cells by analogy with similar events in the cell cycle of yeast. From their analyses, they have pieced together the following scenario.

- The first CDK–cyclin complexes to appear during G$_1$ in humans are CDK4–cyclinD and CDK2–cyclinE (**Figure 16.18**). These complexes initiate the transition to S by a programmed succession of specific phosphorylations, among which are phosphorylations of the protein product of the retinoblastoma (*RB*) gene.
- Unphosphorylated Rb protein inhibits a transcription factor, E2F. Phosphorylated Rb no longer inhibits E2F.

Rb phosphorylation thus indirectly activates DNA synthesis by releasing the brakes on E2F and thereby allowing it to activate the transcription of genes necessary for DNA synthesis.

Control of the G$_2$-to-M transition

Human cells appear to make the transition from G$_2$ to mitosis much as the well-studied cells of the yeast *S. pombe* accomplish the same transition.

- In the yeast, a CDK known as CDC2 (the second C replaces the K for historical reasons) forms a complex with cyclinB. Both the CDC2 kinase and cyclinB are present throughout G$_2$, but phosphorylation of a specific tyrosine residue on the cyclin-dependent kinase (by another protein kinase) keeps it inactive.
- When the time comes to initiate mitosis, a phosphatase enzyme removes the phosphate group from the CDC2 tyrosine; this removal activates the CDK, and the cell enters mitosis (**Figure 16.19**).

Cell-cycle control: A summary

We now have some insight into how a cell is able to replicate its DNA at one time in the cell cycle and segregate its chromosomes at another. The two different phases of the cell cycle are governed by different kinase activities.

During S phase, a CDK is complexed with a cyclin that is specific to S phase. In this complex, it phosphorylates many proteins that lead to a cascade of protein synthesis and activation, and the newly synthesized proteins provide hundreds of activities required for DNA replication. During M phase, a different CDK is complexed with a cyclin specific for M phase. Its activity in this complex leads to the synthesis and activation of hundreds of proteins needed for mitosis. In summary, CDKs and cyclins together set the "state" of the cell: S phase or M phase.

How does the cell change from one state to the other? Among the cellular processes activated by CDKs are those that irreversibly destroy key regulatory proteins, including the cyclins. Thus, as the cell enters either S phase or

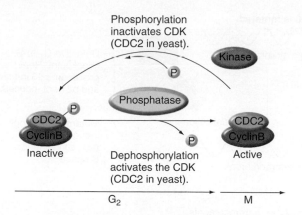

Figure 16.19 CDK activity in yeast is controlled by phosphorylation and dephosphorylation. The CDC2 protein complexed with cyclinB is inactivated prior to mitosis through phosphorylation by a specific kinase and then activated at the onset of mitosis through dephosphorylation by a specific phosphatase.

M phase, it sets in place the end of each phase by removing cyclins and many other proteins whose activities must be limited to either S or M phase.

Just as attachment of a phosphate group activates or deactivates proteins, the covalent attachment of a ubiquitin tag marks proteins for degradation. During S phase, activation of a group of proteins called SCF occurs. The activated SCF adds ubiquitin to proteins such as the S-phase cyclins. During M phase, activation of a group of proteins called APC takes place, and the activated APC adds ubiquitin to proteins such as the M-phase cyclins. Proteins tagged with ubiquitin are rapidly degraded by the cell in the large multiprotein complex, the proteasome. Thus, the cell cycle has an intrinsic ratchet-like mechanism, ensuring that activation of one phase (S or M) leads inevitably to the irreversible end of that phase and elimination of any proteins that could interfere with the next phase.

The discovery of cyclins and cyclin-dependent kinases through experiments with cell-cycle mutants in yeasts has provided insight into the control of cell division. Highly specific cyclins, produced at key points in the cell's cycle, bind with particular CDKs and lead to specialized protein activation. Cyclins produced for one phase are irreversibly marked for destruction at the end of that phase, ensuring that the processes go in only one direction.

Cell-cycle checkpoints ensure genomic stability

Damage to a cell's genome, whether caused by environmental agents or random errors of the cellular machinery as it attempts to replicate and segregate the chromosomes, can cause serious problems for the cell. Damage to the cell-cycle machinery can also cause problems. It is therefore not surprising that elaborate mechanisms have evolved to arrest the cell cycle while repair takes place. These additional controls

are called checkpoints because they check the integrity of the genome and cell-cycle machinery before allowing the cell to continue to the next phase of the cell cycle.

The G₁-to-S checkpoint

When radiation or chemical mutagens damage DNA during G_1, DNA replication is postponed. This postponement allows time for DNA repair before the cell proceeds to DNA synthesis. Replication of the unrepaired DNA could exacerbate the damage; for example, replication over a single-strand nick or gap would produce a double-strand break. In mammals, cells exposed to ionizing radiation or UV light during G_1 delay entry into S phase by activating the *p53* pathway (**Figure 16.20a**). p53 is a transcription factor that induces expression of DNA repair genes as well as expression of the CDK inhibitor known as p21. Like other CDK inhibitors, p21 binds to CDK–cyclin complexes and inhibits their activity; specifically, p21 prevents entry into S phase by inhibiting the activity of CDK4–cyclinD complexes.

Mutations in *p53* disrupt the G_1-to-S checkpoint. One sign of this disruption is a propensity for gene amplification: an increase from the normal two copies to hundreds of copies of a gene. This amplification is visible under the microscope, appearing as an enlarged area within a chromosome known as a *homogeneously staining region* (*HSR*) or as small chromosome-like bodies (called **double minutes**) that lack centromeres and telomeres (**Figure 16.20b**). Normal human cells do not generate gene amplification in culture, but *p53* mutant cells exhibit high rates of such amplification; *p53* mutants also exhibit many types of chromosomal rearrangements. The explanation is as follows: Cells carrying mutations in the *p53* gene most likely have a defective G_1-to-S checkpoint that allows the replication of single-strand nicks. This replication produces double-strand breaks, which, in turn, lead to chromosomal rearrangements. Some of the rearrangements generate gene amplification (**Figure 16.20c**).

Wild-type cells able to produce functional p53 not only arrest in G_1 in the presence of DNA damage; if the damage is great enough, they also "commit suicide" in a process known as **programmed cell death (PCD)**, or **apoptosis**. During apoptosis, the cellular DNA is degraded, and the nucleus condenses. The cell may then be devoured by neighbouring cells or by phagocytes (**Figure 16.20d**). Programmed cell death and the proteins that regulate it—including those that are part of the p53 pathway—appear in multicellular animals from roundworms to humans. It makes sense for multicellular organisms to have a mechanism for eliminating cells that have sustained chromosomal damage. The survival and reproduction of such cells could generate cancers.

The G₂-to-M checkpoint

Damage to DNA during G_2 delays mitosis, allowing time for repair before chromosome segregation (**Figure 16.21a**). Researchers have identified many genes in mammalian and yeast cells that mediate this control. One of these genes is *RAD9*. Whereas wild-type yeast cells can pause to repair as many as 100 double-strand breaks

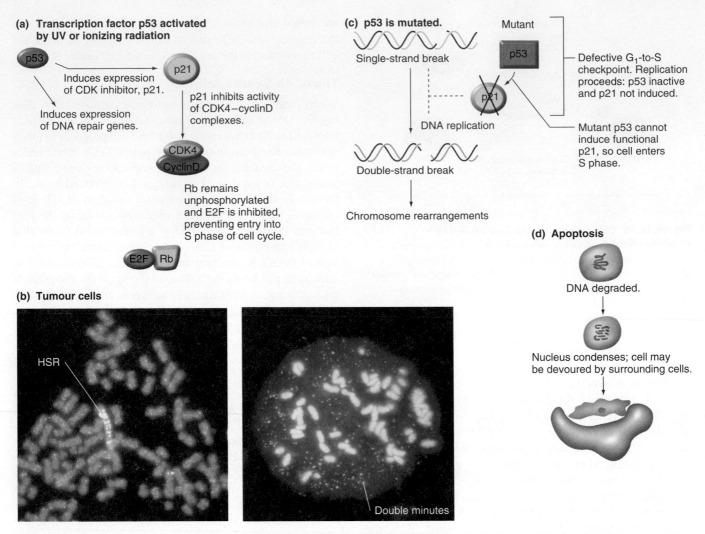

Figure 16.20 Cellular responses to DNA damage. (a) DNA damage activates the *p53* transcription factor, which, in turn, induces expression of the *p21* gene. The p21 protein inhibits CDK activity, producing an arrest of the cell cycle in the G₁ phase. **(b)** Tumour cells exhibit amplified regions of DNA, unlike normal cells, that can appear as homogeneously staining regions (HSRs) within a chromosome, or as double minutes, small pieces of extrachromosomal DNA. **(c)** When the *p53* gene is mutated in cancers, *p21* expression is not induced, cell-cycle progress is not arrested, and cells replicate damaged DNA, producing DNA double-strand breaks from single-strand nicks or gaps. **(d)** DNA damage in normal cells often leads to apoptosis.

before entering mitosis, *RAD9* mutants fail to arrest in G₂ and die as a result of any double-strand breaks that were not repaired before mitosis.

A spindle checkpoint in M

During mitosis, one checkpoint oversees formation of the mitotic spindle and proper engagement of all pairs of sister chromatids (**Figure 16.21b**). Observations of living cells reveal that as chromosomes condense and attach to the spindle, sometimes a single chromosome fails to attach at the expected time. When this happens, the cell does not initiate sister chromatid separation or anaphase chromosome movement until the lagging chromosome attaches to the spindle. Other studies show that in yeast cells exposed to an inhibitor that prevents assembly of a functional spindle, sister chromatids remain firmly attached.

These observations suggest the presence of a checkpoint that prevents chromosome segregation until all chromosomes are properly attached to the spindle. Mutations that eliminate the surveillance of chromosome behaviour during mitosis have helped researchers identify several genes in yeast responsible for this checkpoint.

The necessity of checkpoints

Checkpoints are not essential for cell division. In fact, experiments in mice and other animals demonstrate that mutant cells with one or more defective checkpoints are viable and divide at a normal rate. These mutant cells, however, are much more vulnerable to DNA damage than normal cells.

Knowledge of how checkpoints work hand-in-hand with repair processes clarifies how checkpoints help prevent transmission of three types of genomic instability (described in Chapter 9): chromosome aberrations; aneuploidy (the loss or gain of one or more chromosomes); and changes in ploidy, for example, from 2*n* to

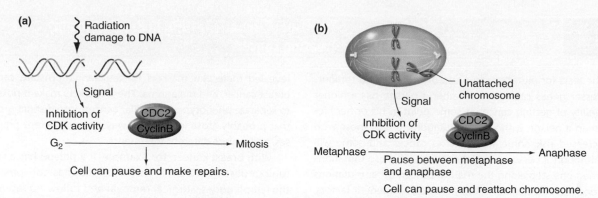

(a)

Radiation damage to DNA

Signal

Inhibition of CDK activity

G₂ ──────────────────────→ Mitosis

Cell can pause and make repairs.

(b)

Unattached chromosome

Signal

Inhibition of CDK activity

Metaphase ──────────────────────→ Anaphase

Pause between metaphase and anaphase

Cell can pause and reattach chromosome.

Figure 16.21 Checkpoints acting at the G₂-to-M cell-cycle transition or during M phase. **(a)** DNA damage, particularly double-strand breaks, induces a signal that inhibits CDK activity, preventing entry into mitosis. **(b)** Spindle damage resulting from the failure of a chromosome to attach to the mitotic spindle generates a signal that inhibits CDK activity and thereby prevents the metaphase-to-anaphase transition.

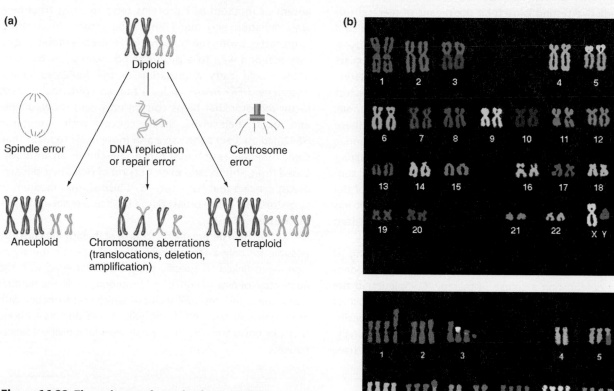

(a)

Diploid

Spindle error

DNA replication or repair error

Centrosome error

Aneuploid

Chromosome aberrations (translocations, deletion, amplification)

Tetraploid

Figure 16.22 Three classes of error lead to aneuploidy in tumour cells. **(a)** Spindle errors can segregate chromosomes incorrectly, resulting in whole-chromosome aneuploidy; DNA replication and/ or repair damage can lead to chromosome aberrations; centrosome errors can result in changes in cell ploidy. **(b)** "Chromosome painting" techniques use fluorescent dyes attached to chromosomal DNA sequences. An appropriate choice of dyes and probes can cause each normal chromosome or chromosome arm to appear relatively homogeneous with a unique colour (*top*), while cancer cell chromosomes reveal many rearrangements and whole-chromosome changes (*bottom*).

4*n* (**Figure 16.22**). Single-strand nicks resulting from oxidative or other types of DNA damage are probably fairly common. A cell normally repairs such nicks to DNA in G₁ before it enters S phase. If the checkpoint coordinating this repair fails to function, however, the copying of single-strand breaks during replication would produce double-strand breaks that could lead to chromosome rearrangements. Chromosome loss or gain can occur if a chromosome fails to attach properly to the spindle. Normally, the M-phase spindle checkpoint recognizes such failures and prevents the initiation of anaphase until the cell has fixed the problem. Cells without a functional checkpoint produce daughter cells carrying too few or too many chromosomes.

Genetic tests for mutations in proto-oncogenes and tumour-suppressor genes can reveal whether a person has a higher probability of getting cancer at some point in his or her lifetime than a person without the mutations. But of those with an increased risk, some will develop cancer and some will not. Although a person who inherits one of these mutations is pushed one step along the road to cancer, other mutations must occur in one clone of cells by chance; nongenetic factors, such as exposure to radiation, influence whether the additional mutations occur. Given this situation, what good is it to learn from a genetic test that you have an increased probability of getting cancer sometime in your life?

Predictive testing is useful if the means of medical surveillance make it possible to detect the cancer to which a mutation predisposes at an early stage. Thus, testing for a genetic predisposition to skin, breast, or colorectal cancers can often lead to increased cancer-specific testing to detect cancers in their earliest stages. A person whose genetic test shows a predisposition to colorectal cancer, for example, could undergo a colonoscopy each year. If one of these colon exams discloses a small cancer, doctors could remove it by surgery or treat it by other means. Predictive testing is not yet useful for some cancers such as pancreatic cancer, because as yet no way to detect small tumours of the pancreas exists. By the time this cancer is identified it has almost always reached an aggressive state and metastasized to other tissues.

Once a cancer has been diagnosed, genetic testing of tumour cells can provide information for making a prognosis and determining a course of therapy. Completion of the Human Genome Project has opened up new possibilities for identifying and tracking the effects on survival of specific cancer-cell mutations. Indeed, comparative microarray analyses (see Chapter 19) of cancerous and normal tissues have revealed molecular markers for leukaemia, prostate cancer, breast cancer, and melanoma. These markers make it possible to separate phenotypically similar cancers into distinct groups that probably arose in different ways, have different prognoses, and require different treatments.

With breast cancer, for example, if a person has a small tumour (less than 2 cm in diameter) that has not spread to the lymph nodes, surgical removal and follow-up radiation (sometimes in conjunction with chemotherapy) usually give the patient a good chance of overcoming the cancer. However, if the tumour cells carry mutations in the *p53* gene, the prognosis is poorer, because breast tumours with absent or mutated p53 proteins tend to resist treatment with radiation and many anticancer drugs. In contrast, large, fast-growing tumours that have already metastasized may respond well to a drug named Herceptin—but only if their cells carry a mutation in the *HER2/neu* proto-oncogene. *HER2/neu* encodes a human epidermal growth factor receptor that helps control how cells grow, divide, and repair themselves. Breast cancers with a mutated *HER2/neu* are very aggressive and more likely to recur than some other types of breast cancer. Herceptin, an antibody-based drug, shrinks and even gets rid of *HER2/neu*-positive breast cancers that have spread; it also shrinks medium to large tumours in the breast tissue itself and reduces the risk of recurrence.

Tests to help determine the course of therapy are currently possible for only a few cancers in which specific mutations have been linked to specific prognoses. However, with the application of new genomic and proteomic tools, the number and scope of such tests will increase, which in conjunction with more precisely targeted drugs, will enable doctors to hone their diagnoses and tailor their treatments for individual cancer patients.

Finally, changes in ploidy can occur if a cell begins S phase before completing mitosis or if a cell fails to replicate or to properly segregate its microtubule-organizing centres, or centrosomes. Checkpoints also recognize these errors, ensuring integration of the centrosome cycle with DNA replication and the formation and function of the mitotic spindle.

> Checkpoints at G_1-to-S and G_2-to-M, along with a spindle checkpoint, help to ensure that cells repair DNA damage or spindle attachments before replication and division proceed. Normally, cells that do not pass a checkpoint undergo apoptosis.

Summary: The accumulation of oncogenic and tumour-suppressor mutations produces cancer cells with grossly altered genomes

Cancer-causing mutations disrupt the normal controls that create a balance between activation and inhibition of cell division. Dominant gain-of-function mutations that change proto-oncogenes to oncogenes may overactivate expression of proteins that promote proliferation. Recessive loss-of-function mutations in tumour-suppressor genes may release the brakes that keep cells from proliferating. Both types of mutations may tip the balance toward excessive and inaccurate cell proliferation.

Mutations that disable one part of a cell's elaborate DNA repair system increase its mutation rate and thus its

likelihood of becoming cancerous. Although no single mutation converts a normal cell to a cancer cell, if a cell has a mutation in one gene that predisposes to cancer, that cell has a higher than normal probability of becoming cancerous because it is already one step along the way. The early mutations in a cell's progression from normal to cancerous may lead to increased proliferation and affect the accuracy of cellular reproduction, allowing the accumulation of several mutations. Other subsequent or simultaneous mutations may enable the abnormally and inaccurately proliferating cells of a single lineage to avoid programmed cell death, evade the immune system, increase formation of blood vessels supplying the abnormal clone, alter the proteins that control tissue architecture, and invade nearby or distant tissues (i.e., metastasize). Environmental factors such as radiation and mutagenic chemicals cause most of the mutations that result in cancer, but rare inherited defects can contribute the first step (see the **Genetics and Society** box **"The Uses of Genetic Testing in Predicting and Treating Cancer"**).

Connections

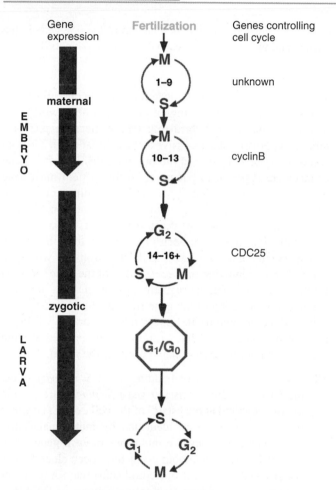

Figure 16.23 Regulation of the cell-cycle changes during _Drosophila_ development. Each step of development has built-in regulators that act as barriers to uncontrolled reproduction of "selfish" cells. Some of these regulators, such as cyclinB and CDC25, are known; others have not yet been identified.

The existence of numerous controls in cell-cycle pathways suggests that evolution has erected many barriers in multicellular animals to the uncontrolled reproduction of "selfish" cells. At the same time, the hundreds of genes contributing to normal cell-cycle regulation provide hundreds of targets for cancer-producing mutations.

Variations on the theme of cell-cycle regulation play a key role in the development of eukaryotic organisms. In _Drosophila_, for example, after fertilization, nuclear division occurs without cell division for the first 13 cycles; during these cycles, the nuclei go through many rapid S and M phases without any intervening G_1 or G_2 (**Figure 16.23**). In cycles 10–13, the synthesis and degradation of cyclinB regulates mitosis. Sometime during cycles 14–16, a G_2 phase appears, and distinct patches of cells with different-length cycles become evident within the embryo. The differences in cycle time between the different cell types are the result of variable G_2 phases. Late in G_2, CDC25 activates cyclin-dependent kinases to control the timing of mitosis. Many tissues stop dividing at cycle 16, but a few continue. In the still-dividing cells, a G_1 phase appears. Some of these cells will arrest in G_1 during larval growth, only to start dividing again in response to signals relayed during metamorphosis, when the larva changes into adult form.

In Chapter 17, we present the basic principles of development and describe how biologists have used genetic analysis in various model organisms to examine development at the cellular and molecular levels.

Essential Concepts

1. The genes and proteins of various signal transduction systems relay signals about whether or not to enter the cell cycle. The four molecular components of these systems include growth factors, receptors for these factors, intracellular transducers that propagate the signal, and transcription factors that begin DNA replication. **[LO1]**

2. Cancer is a genetic disease resulting from the growth of a clone of mutant cells. A cell requires many mutations to become cancerous. Exposure to environmental mutagens probably generates most of these mutations. **[LO2–3]**

3. Many mutations that lead to cancer jeopardize cell-cycle regulation. Mutations in growth factors, receptors,

and other elements of signal transduction pathways can release cells from control by the signals normally required for proliferation. Mutations in CDKs and the proteins that control them may also lead to inappropriate proliferation or genomic instability. The latter may permit rapid evolution of abnormal tumour cells. Mutations in DNA repair and checkpoint controls lead to genomic instability and, often, to loss of the surveillance system that kills aberrant cells by apoptosis. **[LO4–5]**

4. Several genetic pathways help control cell division. The inhibition or activation of CDKs inhibits or activates G_1-to-S and G_2-to-M transitions. The measured synthesis and degradation of different cyclins guide CDKs to the appropriate targets at the appropriate times. Checkpoints that integrate repair of chromosomal damage with events of the cell cycle minimize the replication of damaged DNA. **[LO5–6]**

Solved Problems

I. The addition of growth factors to tissue culture cells stimulates cell division. A number of candidate drugs can be tested for their ability to stop this stimulation of cell division. What do you think the target of these drugs could be?

Answer

This question concerns the regulation of cell division. Growth factors are made by one cell and bind to receptors of another cell to stimulate the cell division cycle. *A drug that binds to receptors* would block access and prevent growth factors from binding. Alternatively, *the drug could bind to the growth factor*, thereby preventing its interaction with the receptor. (These are the most obvious targets. If you are familiar with the signal transduction pathway inside the cell, you might also propose that proteins in this pathway could be targets for drug development.)

II. The *p53* gene has been cloned, and you are using it to analyze DNA in patients in which *p53* defects are involved in the development of their tumours. DNA samples were obtained from normal and tumour tissue of three different cancer patients, digested with *Bam*HI, electrophoresed on an agarose gel, and transferred to filter paper that was probed with a labelled *p53* fragment. Each of the patients inherited a *p53* mutation.

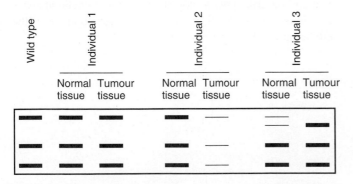

Thin bands indicate half the DNA content of thick bands. DNA from an individual who did not inherit a *p53* mutation is shown in the lane labelled wild type. All wild-type alleles in this study produce the same three fragments. Assuming the model of *p53* acting as a tumour-suppressor gene is correct and that *p53* defects are involved in each of these cancer patients, how

would you describe the genetic makeup of the *p53* gene in the normal and tumour tissue of each of the three patients?

Answer

This question requires knowledge of tumour-suppressor genes. The wild-type *p53* region (as seen in the "wild-type" individual) has three hybridizing bands. Because *p53* is a tumour suppressor, it is recessive at the cellular level, and both copies must be defective in the tumour cells. No observable changes are apparent in patient 1, so this patient must have inherited a point mutation in *p53*, and in the tumour cells, the second copy would also contain a small mutation, thereby inactivating both copies of *p53* in the tumour. In patient 2, a point mutation must have been inherited. In the tumour, the whole region containing *p53* was deleted from the wild-type copy (thereby removing the second copy of the gene), as seen by the loss of restriction fragments. In patient 3, a mutation is evident in one copy of the gene from the altered restriction pattern in the normal tissue, and in the tumour, the wild-type copy of the gene was deleted (probably by gene conversion since the tumour has two mutant copies of the gene).

III. The CDC28 protein of budding yeast *S. cerevisiae* and the CDC2 protein of fission yeast *S. pombe* are protein kinases required at the "start" of the cell cycle. The genes for both proteins were identified by mutational analysis (temperature-sensitive mutations in each gene cause cell-cycle arrest), and both genes have been cloned. How could you determine if one could substitute for the other functionally? (Be sure to mention sources of DNA and genotypes involved.)

Answer

The *CDC28* gene of *S. cerevisiae* could be cloned into a vector and transformed into a temperature-sensitive *cdc2* mutant of *S. pombe*. If CDC28 has the same role (function) as CDC2, the transformed cell will now grow and divide at nonpermissive temperatures. Conversely, the *CDC2* gene of *S. pombe* could be cloned into a vector able to transform a temperature-sensitive *cdc28* mutant of *S. cerevisiae*. If CDC2 of *S. pombe* can substitute for CDC28 of *S. cerevisiae*, the transformed cells would grow and divide at nonpermissive temperatures.

Problems

Vocabulary

1. For each of the terms in the left column, choose the best matching phrase in the right column.

i. growth factor	**1.** mutations in these genes are dominant for cancer formation
ii. tumour suppressor	**2.** programmed cell death genes
iii. cyclin-dependent protein kinases	**3.** series of steps by which a message is transmitted
iv. apoptosis	**4.** proteins that are active cyclically during the cell cycle
v. oncogenes	**5.** control progress in the cell cycle in response to DNA damage
vi. receptor	**6.** mutations in these genes are recessive at the cellular level for cancer formation
vii. signal transduction	**7.** signals a cell to leave G_0 and enter G_1
viii. checkpoints	**8.** cell-cycle enzymes that phosphorylate proteins
ix. cyclins	**9.** protein that binds a hormone

Section 16.1

2. Molecules outside and inside the cell regulate the cell cycle, making it start or stop.

a. What is an example of an external molecule?

b. What is an example of a molecule inside the cell that is involved in cell-cycle regulation?

3. a. Would you expect a cell to continue or stop dividing at a nonpermissive high temperature if you had isolated a temperature-sensitive *RAS* mutant that remained fixed in the GTP-bound form at the nonpermissive temperature?

b. What would you expect if you had a temperature-sensitive mutant in which *RAS* stayed in the GDP-bound form at high temperature?

4. Put the following steps in the correct ordered sequence.

a. kinase cascade

b. activation of a transcription factor

c. hormone binds receptor

d. expression of target genes in the nucleus

e. RAS molecular switch

Section 16.2

5. Mouse tissue culture cells infected with the SV40 virus lose normal growth control and become transformed. If transformed cells are transferred into mice, they grow into tumours. The SV40 protein responsible for this transformation is called T antigen. T antigen has been found to associate with the cellular protein p53. If the *p53* gene fused to a high-level expression promoter is transfected into tissue culture cells, the cells are no longer transformed through infection by SV40.

a. Propose a hypothesis to explain how the high expression of *p53* saves the cells from transformation by T antigen.

b. You have decided to examine the functional domains of the p53 protein by mutagenizing the cDNA, fusing it to the high-level promoter, and transfecting into cells. Results are shown in the following table. How would you explain the effects of mutations 1 and 2 on p53 function?

c. What is the effect of mutation 3 on p53 function?

	Morphology	
p53 construct	Noninfected cells	SV40-infected cells
None	Normal	Transformed
Wild type	Normal	Normal
Mutation 1	Normal	Transformed
Mutation 2	Normal	Transformed
Mutation 3	Normal	Normal

6. What are four characteristics of the cancer phenotype?

7. Amplification of DNA sequences in *p53* mutants can be visualized using electron microscopy.

a. Using a different technique, how could you detect amplification of a specific sequence?

b. How could you detect gross rearrangements (greater than 10 Mb) of chromosomal DNA?

8. Some germline mutations predispose to cancer, yet often environmental factors (chemicals, exposure to radiation) are considered major risks for developing cancer. Are these conflicting views of the cause of cancer or can they be reconciled?

9. The incidence of colorectal cancer in Canada is 12 times higher than it is in India. Differences in diet and/or genetic differences between the two populations may contribute to these statistics. How would you assess the role of each of these factors?

10. Put the following steps in the order appropriate for the positional cloning of *BRCA1*, a gene involved in predisposition to breast cancer.

a. Locate transcripts corresponding to the DNA.

b. Use the physical map to get clones.

c. Determine the tissues in which the transcripts are present.

d. Look for homologous DNA in other organisms.

e. Determine linkage to RFLPs and other molecular markers.

f. Sequence the DNA from affected individuals.

11. Because mutations occur in the development of cancer, researchers suspected that defects in DNA repair machinery might lead to a predisposition to cancer. Place the following steps in appropriate order for following a candidate gene approach to determine if defects in mismatch repair genes lead to cancer.

a. Use molecular markers near the homologous gene to determine if the candidate gene is linked to a predisposition phenotype.

b. Isolate a human homologue of a yeast mismatch repair gene.

c. Compare the DNA sequence of the mismatch repair gene of affected and unaffected persons in a family with predispositions.

d. Determine the map location of the human homologue of the yeast mismatch repair gene.

12. Which of the following events is unlikely to be associated with cancer?

a. mutations of a cellular proto-oncogene in a normal diploid cell

b. chromosomal translocations with breakpoints near a cellular proto-oncogene

c. deletion of a cellular proto-oncogene

d. mitotic nondisjunction in a cell carrying a deletion of a tumour-suppressor gene

e. incorporation of a cellular oncogene into a retrovirus chromosome

13. You have decided to study genetic factors associated with colorectal cancer. An extended family from Morocco in which the disease presents itself in a large percentage of family members at a very early age has come to your attention. (The pedigree is shown below.) In this family, individuals either get colorectal cancer before the age of 16, or they do not get it at all.

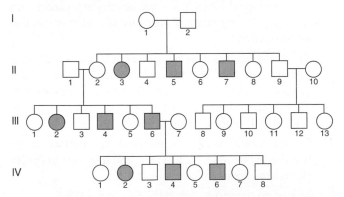

a. Based on the information you have been given, what evidence, if any, suggests an inherited contribution to the development of this disease?

b. You decide to take a medical history of all of the 36 people indicated in the pedigree and discover that a very large percentage drink a special coffee on a daily basis, while the others do not. The only ones who do not drink coffee are individuals numbered I-1,

II-2, II-4, II-9, III-7, III-13, IV-1, and IV-3. Could the drinking of this special coffee possibly play a role in colorectal cancer? Explain your answer.

14. To further understand the basis for colorectal cancer, you find a family from Canada in which two members also get the disease before the age of 16. If there were a dominant inherited mutation segregating in this family, which of the individual(s) would you predict had the mutation in their colon or rectal cells but did not develop the disease?

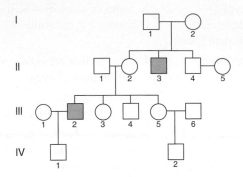

15. You suspect that a very specific point mutation in the *p53* gene is responsible for the majority of *p53* mutations found associated with tumours. Which combination of these techniques would you be most likely to use in developing a simple assay for predictive testing?

a. polymerase chain reaction with oligonucleotide primers flanking the mutation

b. restriction enzyme digestion followed by Southern blot

c. RNA isolation followed by Northern blot

d. hybridization with allele-specific oligonucleotides

16. A 19-year-old female patient is diagnosed with chronic myelogenous leukaemia (CML), whose symptoms are anaemia and internal bleeding due to a massive buildup of leukaemic white blood cells. Karyotype analysis shows that the leukaemic cells of this patient are heterozygous for a reciprocal translocation involving chromosomes 9 and 22. However, none of the normal, nonleukaemic cells of this patient contain the translocation. Which of the following statements is true and which is false?

a. The translocation results in the inactivation (loss-of-function) of a tumour-suppressor gene.

b. The translocation results in the inactivation (loss-of-function) of an oncogene.

c. There is a 50 percent chance that any child of this patient will have CML.

d. This patient is a somatic mosaic in terms of the karyotype.

e. DNA extracted from leukaemic cells of this patient, if taken up by normal mouse tissue culture cells, could potentially transform the mouse cells into cells capable of causing tumours.

f. The normal function of the affected tumour-suppressor gene or proto-oncogene at the translocation breakpoint could potentially block the function of the cyclin proteins that drive the cell cycle forward.

g. This woman is heterozygous for an X-linked gene; the two alleles encode two distinguishable variant forms of the protein product of the gene. If you looked at different normal cells from different parts of her body, some would express exclusively one variant form of the protein, and other normal cells would express exclusively the other variant form.

h. If you examined different leukaemic cells from this patient for the protein described in part *g*, all would express the same variant form of the protein.

i. Two rare events must have occurred to disrupt both copies of the tumour-suppressor gene or proto-oncogene at the translocation breakpoint in the leukaemic cells.

j. A possible treatment of the leukaemia would involve a drug that would turn on the expression of the tumour-suppressor gene or proto-oncogene at the translocation breakpoint in the leukaemic cells.

17. Describe a molecular test to determine if chemotherapy given to the patient described in Problem 16 was completely successful. That is, devise a method to make sure that the patient's blood is now free of leukaemic cells.

18. A generic signalling cascade is shown in the following figure. A growth factor (GF) binds to a growth factor receptor, activating the kinase function of an intracellular domain of the growth factor receptor. One substrate of the growth factor receptor kinase is another kinase, kinase A, which has enzymatic activity only when it is itself phosphorylated by the growth factor receptor kinase. Activated kinase A adds phosphate to a transcription factor. When it is unphosphorylated, the transcription factor is inactive and stays in the cytoplasm. When it is phosphorylated by kinase A, the transcription factor moves into the nucleus and helps turn on the transcription of a *mitosis factor* gene whose product stimulates cells to divide.

a. The following list contains the names of the genes encoding the corresponding proteins. Which of these could potentially act as a proto-oncogene? Which might be a tumour-suppressor gene?

   i. *growth factor*

   ii. *growth factor receptor*

   iii. *kinase A*

   iv. *transcription factor*

   v. *mitosis factor*

Though it is not pictured, the cell in the figure also has a phosphatase, an enzyme that removes phosphates from proteins—in this case, from the transcription factor. This phosphatase is itself regulated by kinase A.

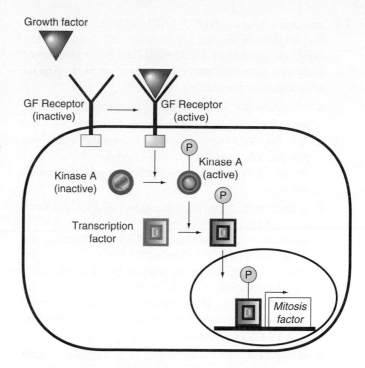

b. What would you expect to be the effect when kinase A adds a phosphate group to the phosphatase? Would this activate the phosphatase enzyme or inhibit it? Explain.

c. Is the *phosphatase* gene likely to be a proto-oncogene or a tumour-suppressor gene?

d. Several mutations are listed below. For each, indicate whether the mutation would lead to excessive cell growth or decreased cell growth if the cell were either homozygous for the mutation or heterozygous for the mutation and a wild-type allele. Assume that 50 percent of the normal activity of all these genes is sufficient for normal cell growth.

   i. A null mutation in the *phosphatase* gene

   ii. A null mutation in the *transcription factor* gene

   iii. A null mutation in the *kinase A* gene

   iv. A null mutation in the *growth factor receptor* gene

   v. A mutation that causes production of a constitutively active growth factor receptor whose kinase function is active even in the absence of the growth factor

   vi. A mutation that causes production of a constitutively active kinase A

   vii. A reciprocal translocation that places the *transcription factor* gene downstream of a very strong promoter

   viii. A mutation that prevents phosphorylation of the *phosphatase* gene

   ix. A mutation that causes the production of a phosphatase that acts as if it is always phosphorylated

19. Are genome and karyotype instabilities consequences or causes of cancer?

20. Neurofibromatosis type 1 (NF1; also known as von Recklinghausen disease) is an inherited dominant disorder. The phenotype usually involves the production of many skin neurofibromas (benign tumours of the fibrous cells that cover the nerves).

a. Is it likely that *NF1* is a tumour-suppressor gene or an oncogene?

b. Are the *NF1* neurofibromatosis-causing mutations that are inherited by affected children from affected parents likely to be loss-of-function or gain-of-function mutations?

c. Neurofibromin, the protein product of *NF1*, has been found to be associated with the RAS protein. RAS is involved in the transduction of extracellular signals from growth factors. The active form of RAS (the form initiating the signal transduction cascade causing proliferation) is complexed with GTP; the inactive form of RAS is complexed with GDP. Would the wild-type neurofibromin protein favour the formation of RAS-GTP or RAS-GDP?

d. Which of the following events in a normal cell from an individual inheriting a neurofibromatosis-causing allele could cause the descendents of that cell to grow into a neurofibroma?

　i. A second point mutation in the allele of *NF1* inherited from the afflicted parent

　ii. A point mutation in the allele of *NF1* inherited from the normal parent

　iii. A large deletion that removes the *NF1* gene from the chromosome inherited from the afflicted parent

　iv. A large deletion that removes the *NF1* gene from the chromosome inherited from the normal parent

　v. Mitotic chromosomal nondisjunction or chromosome loss

　vi. Mitotic recombination in the region between the *NF1* gene and the centromere of the chromosome carrying *NF1*

　vii. Mitotic recombination in the region between the *NF1* gene and the telomere of the chromosome carrying *NF1*

e. The *American Journal of Medical Genetics* published a report in 1999 that certain patients with neurofibromatosis type I who had an affected parent also inherited specific facial anomalies from that parent. Formulate a succinct hypothesis to explain why these patients inherit this additional phenotype, but most other patients with inherited neurofibromatosis I do not.

f. There is a much rarer form of NF1 called segmental NF1. In this form of the disease, neither parent of the patient has any clinical sign of the disease. The tumours in the patient are restricted to one part of the body, like the right leg. Suggest an explanation for the genesis of segmental NF1 and why it is restricted to one part of the body.

Section 16.3

21. During which phase(s) of the cell cycle would the following enzymes or proteins be most active?

a. tubulins in the spindle fibres

b. centromere motor

c. DNA polymerase

d. CDC28 of *S. cerevisiae* or CDC2 of *S. pombe*

22. Conditional mutations are useful for genetic analysis of essential processes. For example, temperature-sensitive cell-cycle mutations in yeast do not divide at 37°C (nonpermissive temperature) but will divide at 30°C. An alternative type of conditional mutation is a cold-sensitive mutation in which the nonpermissive temperature is low (23°C). List the steps you would go through to isolate cold-sensitive cell-cycle mutants of yeast.

23. Many temperature-sensitive yeast mutants that showed defects in the cell cycle were isolated in the 1970s. The mutants that arrested at the unbudded stage were mated with each other to do a complementation analysis. A + sign on the chart indicates that the resulting diploids grew at the high (nonpermissive) temperature. How many complementation groups (i.e., how many genes) are represented by these mutants?

	1	2	3	4	5	6	7	8	9
1	−	+	+	−	−	+	+	+	+
2	+	−	+	+	+	+	+	−	+
3	+	+	−	+	+	−	−	+	−
4	−	+	+	−	−	+	+	+	+
5	−	+	+	−	−	+	+	+	+
6	+	+	−	+	+	−	−	+	−
7	+	+	−	+	+	−	−	+	−
8	+	−	+	+	+	+	+	−	+
9	+	+	−	+	+	−	−	+	−

24. In 1951, a woman named Henrietta Lacks died of cervical cancer. Just before she died, a piece of her tumour was taken and put into culture in a laboratory in an attempt to induce the cells to grow *in vitro*. The attempt succeeded, and the resulting cell line (known as HeLa cells) is still used today in laboratories around the world for studies of various aspects of cell biology. In the cell cycle of typical HeLa cells, G_1 lasts about 11 hours, S lasts about 8 hours, G_2 lasts 4 hours, and mitosis (M) takes about 1 hour.

a. Cultured cells do not typically grow synchronously; that is, the individual cells in a culture are randomly distributed throughout the cell cycle. If you looked through the microscope at a sample of HeLa cells, in approximately what proportion of them would you expect the chromosomes to be visible? (The cells do not split apart completely after cytokinesis, and each joined double cell should be counted as one.)

b. Approximately what proportion would be in interphase?

25. The activity of key cell-cycle regulatory proteins is cyclical, appearing only when needed. What are three ways by which a cell can achieve this cyclical nature of protein activity?

26. True or false?

    a. CDKs phosphorylate proteins in the absence of cyclins.

    b. Degradation of cyclins is required for the cell cycle to proceed.

    c. CDKs are involved in checking for aberrant cell-cycle events.

27. Checkpoints occur at several different times during the cell cycle to check that the DNA content of the cell has not been damaged or altered. Match the defect in a checkpoint with the consequences of that defective checkpoint.

Defective checkpoint	Consequences of defect in checkpoint
a. G_1 to S	1. aneuploidy
b. G_2 to M	2. single-strand nicks get replicated and amplification occurs
c. M	3. unrepaired double-strand breaks result in broken chromosomes

28. Draw a diagram illustrating the accumulation of S-phase and M-phase cyclins during the cell cycle. When are SCF and APC (the protein complexes that add ubiquitin to S-phase and M-phase cyclins, respectively) activated?

29. One of the hallmarks of mitotic anaphase is the separation of sister chromatids. Sister chromatids are held together by a protein complex called "cohesin." Based on your answer to Problem 28, propose a mechanism that would allow sister chromatids to separate during anaphase. How might your proposed mechanism also explain the M-phase checkpoint that prevents sister chromatid separation until all the chromosomes have connected properly to the mitotic spindle?

connect For more information on the resources available from McGraw-Hill Ryerson, go to www.mcgrawhill.ca/he/solutions.

Problems   **551**

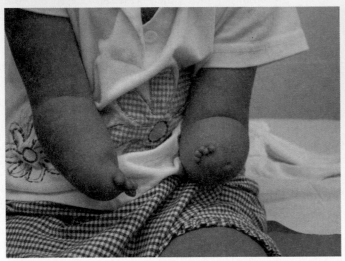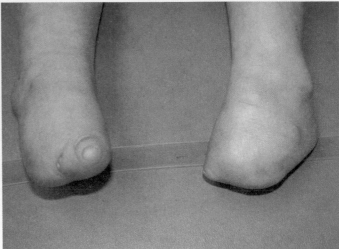

A young patient with Adams–Oliver syndrome, a rare genetic disorder characterized by a combination of developmental abnormalities, including distal limb defects. In this patient, there is extreme shortening of the forearms along with underdeveloped and/or absent fingers (*left*), and absent toes (*right*).

The union of a human sperm and egg (**Figure 17.1a**) initiates the amazing process of development in which a single cell—the fertilized egg—divides by mitosis into trillions of genetically identical cells. These cells differentiate from each other during embryonic development to form hundreds of different cell types. Cells of various types assemble into wondrously complex yet carefully structured systems of organs, including two eyes, a heart, two lungs, and an intricate nervous system. Within a period of three months, the human embryo develops into a fetus whose form anticipates that of the baby who will be born six months later (**Figure 17.1b**). At birth, the baby is already capable of crying, breathing, and eating; and the infant's development does not stop there. New cells form and differentiate throughout a person's growth, maturation, and even senescence (aging).

Biologists now accept that genes direct the cellular behaviours underlying development, but as recently as the 1940s, this idea was controversial. Many embryologists could not understand how cells with identical chromosome sets, and thus the same genes, could form so many different types of cells if genes were the major determinants of development. As we now know, the answer to this riddle is very simple: Not all genes are "turned on" in all tissues. Cells regulate the expression of their genes so that each gene's protein product appears only when and where it is needed. Two central challenges for scientists studying development are to identify which genes are critical for the development of particular cell types or organs; and to figure out how these genes work together to ensure that each is expressed at the right time, in the right place, and in the right amount.

Biologists who use genetics to study how the fertilized egg of a multicellular organism becomes an adult are called **developmental geneticists**. Like other geneticists, they analyze mutations; in this case, mutations that produce developmental abnormalities. An understanding of such mutations helps clarify how normal genes control cell growth, cell communication, and the emergence of specialized cells, tissues, and organs.

Given the great diversity of the 7 billion humans on our planet, it is not hard to find rare individuals carrying mutations that alter their development. For example, pedigree analysis shows that one form of *Adams–Oliver syndrome* (with distal limb reduction) is inherited in families as a Mendelian autosomal recessive trait (see photographs on this page). There are, however, significant ethical and practical limitations on the study of developmental genetics in humans. These include taboos on the deliberate production of mutants, on the experimental manipulation of affected individuals, and on forced matings between

(a) Fertilization

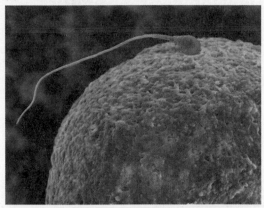

(b) A human fetus three months after fertilzation

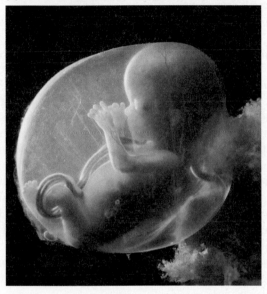

Figure 17.1 Human development. Fertilization of an egg by a sperm **(a)** creates a zygote, which undergoes many rounds of division and cell differentiation to produce a fetus **(b)** by the end of the first trimester of pregnancy.

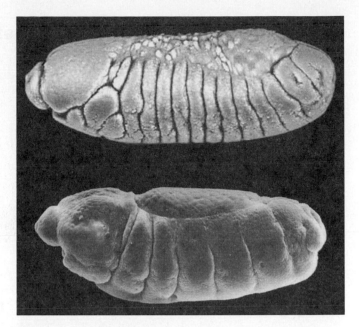

Figure 17.2 Mutations in *Drosophila* genes can affect early development. Wild-type embryo (*top*); embryo homozygous for a mutation in a gene called *ftz* (*bottom*). The mutant embryo has fewer body segments than normal.

individuals with various abnormalities. But one important limitation is not so obvious: Mutations that disrupt the earliest (and to some, the most interesting) stages of development almost always cause the spontaneous abortion of the affected embryo or fetus, often before the mother knows she is pregnant.

As a result, most modern developmental geneticists, even those whose primary interest is in human development, study mutations affecting the development of model organisms more amenable to experimentation. In

Drosophila, for example, only a few dozen genes guide the formation of the early embryo's segmented body plan. Mutations in some of these genes eliminate specific body segments (**Figure 17.2**). Once the embryo has divided into segments that will become parts of the head, thorax, and abdomen, the activation and inactivation of different sets of genes direct the development of specialized structures, such as wings and legs, in each segment.

We examine in this chapter how the single cell of the fertilized egg, or zygote, differentiates into hundreds of cell types. It is impossible to present this complex topic in depth in a single textbook chapter. What we provide here is an overview of the experimental strategies scientists have used to examine this question, along with a synopsis of the major results they have obtained. We can discern two key themes in our exploration of genetics and development. One is that, surprisingly, many genes that control development have been highly conserved through evolution. Thus, for example, the study of a process in *Drosophila* can shed light on events that occur during the development of other animals, including humans. A second theme is that genes themselves are not the only determinants of development because signals that pass between cells, or from the environment to the cells, can strongly influence how these genes function.

Learning Objectives

1. Summarize the advantages of using model organisms to study development.

2. Compare and contrast loss-of-function and gain-of-function mutations and how they are generated.

3. Explain how scientists conduct a functional analysis of genes and their developmental pathways.

4. Distinguish between the different classes of genes and their functions that control early development in *Drosophila*.

5. Evaluate the importance of regulated gene expression and various signalling mechanisms in cellular differentiation.

Throughout the twentieth century, developmental geneticists concentrated their research efforts on a small number of organisms that sampled a range of species from different phyla. The organisms that have contributed most to our understanding of development include the following:

- the yeast *Saccharomyces cerevisiae*
- the plant *Arabidopsis thaliana*
- the fruit fly *Drosophila melanogaster*
- the nematode (roundworm) *Caenorhabditis elegans*
- the mouse *Mus musculus*

Although we focus here on these five eukaryotic organisms, some researchers have made major findings in other model systems, such as corn and the zebrafish. Even prokaryotic organisms and viruses have provided paradigms for tackling certain developmental problems in eukaryotes.

Why study these model organisms?

The five model organisms we discuss in this chapter and in the genetic portraits on Connect are easy to cultivate and rapidly produce large numbers of progeny. Geneticists can thus find rare mutations and study their segregation and behaviour through successive generations. Each organism has attracted a dedicated group of researchers who share information, mutants, and other reagents. Stock centres maintain these mutants and make them available to the whole community of geneticists. Moreover, each model organism's genome has been completely sequenced, and the results have been collated and annotated on computer databases. The completion of these genome projects makes it much easier for geneticists to identify genes whose alteration by mutation produces a phenotypic effect on the organism's development.

In addition to these shared advantages, each model organism also possesses idiosyncratic features that make it valuable for particular types of genetic or developmental analyses. Take yeast, for example. Although *S. cerevisiae* is a single-celled eukaryote, yeast cells signal to each other and differentiate into two mating types using variations of processes involved in the development of multicellular eukaryotes. Because *S. cerevisiae* cells can grow as haploids or diploids, researchers can identify extremely rare mutations in very large populations of haploid cells and then combine mutations in diploid cells for complementation analysis. In another example, the roundworm *C. elegans* is transparent (**Figure 17.3**) and contains an invariant number of somatic cells as an adult—959 in the female/hermaphrodite and 1031 in the male. Because of these unusual properties, researchers can discern the lineage of every cell as the fertilized egg develops into the multicellular adult.

All living forms are related . . .

In the last 150 years, biologists have come to realize that life-forms are related on many levels. For example,

Figure 17.3 The transparency of *C. elegans* facilitates study of the worm's development.

the cells of all eukaryotic organisms have many features in common that are recognizable in the light or electron microscope, such as a nucleus and mitochondria. Moreover, the metabolic pathways by which cells make or degrade organic molecules are virtually identical in all living organisms, and almost all cells use the same genetic code to synthesize proteins. The relatedness of organisms is even visible at the level of the amino acid sequence of individual proteins. For example, over roughly 2 billion years, evolution has conserved the sequence of the histone protein H4, so the H4 proteins of widely divergent species are identical at all but a few amino acids. Most other proteins are not as invariant as H4, but nonetheless, scientists can often trace the evolutionary descent of a protein through the amino acid similarities of its homologues in various species.

Of particular importance to this chapter is the conservation of many basic strategies of development in all multicellular eukaryotes, even in organisms with body plans that look quite different. A graphic example is seen in studies of the genetic control of eye development in fruit flies, mice, and humans. *Drosophila* homozygous for mutations of the *eyeless* (*ey*) gene have either no eyes at all or, at best, very small eyes (**Figure 17.4a**). Mutations in the *Pax6* gene in mice (**Figure 17.4b**) and the *Aniridia* gene in humans also reduce or totally abolish eye formation.

When researchers cloned the *ey*, *Pax6*, and *Aniridia* genes, they found that the amino acid sequences of all three encoded proteins were closely related. This result was surprising because the eyes of vertebrates and insects are so dissimilar: Insect eyes are composed of many facets called ommatidia, whereas the vertebrate eye is a single camera-like organ (see Figure 1.9). Biologists had thus long assumed that the two types of eyes evolved independently. However, the homology of *ey*, *Pax6*, and *Aniridia* suggests instead that the eyes of insects and vertebrates evolved from a single prototypical light-sensing organ whose development required a gene ancestral to *ey* and its mouse and human homologues.

(a)

(b)

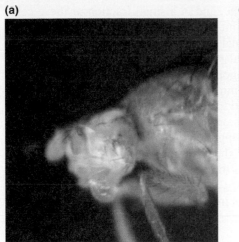
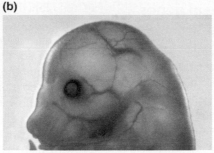
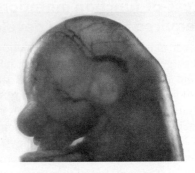

Figure 17.4 The *eyeless/Pax6* gene is critical for eye development. **(a)** Hypomorphic or null mutations of the *eyeless* gene reduce the size of eyes or completely abolish them in adult flies. **(b)** Mutations in the homologous mouse *Pax6* gene also reduce or abolish eye development. *Left:* wild-type mouse fetus. *Right: Pax6* mutant fetus.

... Yet all species are unique

Although the conservation of developmental pathways makes it tempting to conclude that humans are simply large fruit flies, this is obviously not true. Evolution is not only conservative, it is also innovative. Organisms sometimes use disparate strategies to accomplish the same developmental goal.

One example is the difference between the two-cell embryos that form in *C. elegans* and humans upon completion of the first mitotic division in the zygote. If one of the two cells is removed or destroyed in a *C. elegans* embryo at this stage, a complete nematode cannot develop. Because each of the two cells has already received a different set of molecular instructions to guide development, the descendents of one of the cells can differentiate into only certain cell types, and the descendents of the other cell into other types. The situation is very different in humans: If the two embryonic cells are separated from each other, two complete individuals (identical twins) will develop. In fact, as

we saw in Figure 15.1, removal of a cell from a 6–10-cell human embryo has no effect on the development of the remainder of the embryo.

An intrinsic difference exists, therefore, in the way worm and human embryos develop at these early stages. As soon as the *C. elegans* embryo has been formed by mitosis, each cell has already been assigned a specific fate; this pattern of development is often called **mosaic determination**. In contrast, the cells of a human embryo can alter or "regulate" their fates according to the environment, for example, to make up for missing cells; this is called **regulative determination**.

> Genetic studies of development in model organisms often provide key information that can be generalized to all eukaryotes. These studies can also illustrate how evolution has moulded the action of conserved genes to produce diverse developmental programs in different species.

17.2 Using Mutations to Dissect Development

Because proteins are the basic elements of cellular function, biologists can try to understand development by defining the roles played by individual proteins. To do this, they eliminate all copies of a single type of protein from a cell or organism and determine the consequences. From these consequences, they can often infer the function of the normal protein in normal development.

Genetics makes this experimental strategy possible. All an investigator has to do is isolate a mutant cell or organism with a specific, inactivated gene. Such mutants are usually found in the course of genetic screens to look for animals whose development is aberrant in interesting ways. A mutant with an altered gene will lack the wild-type protein encoded by that gene. Careful analysis of the mutant phenotype can then pinpoint what the protein

does in development. As you will see, this basic strategy, although not the only way researchers can harness genetics to study development, is almost always the first and most important step.

Once a gene affecting a developmental process has been identified by a mutation, geneticists try to isolate many additional mutant alleles of that gene. If the mutations affect the function of the corresponding protein in a different way, studies of the phenotypes associated with the various mutations may shed light on the diverse roles the protein plays in the organism. The appendix *Genetic Tools for the Analysis of Development* on Connect discusses in detail the types of genetic screens that researchers can perform to identify genes that are important to development and different kinds of alleles of these genes.

Loss-of-function mutations reveal genes required for normal development

Most mutations are loss-of-function mutations: They disrupt gene function by altering the amino acid sequence (and thus the three-dimensional structure) of the protein product or by interfering with any step of gene expression (transcription, translation, or RNA processing). As a result, such mutations give rise to proteins with diminished (or no) biochemical activity; or they decrease (or stop) production of an otherwise normal protein. We describe four kinds of loss-of-function mutations.

Null mutations

The best way to draw legitimate conclusions about the importance of a protein in development is to study an organism that completely lacks the function provided by that protein. In an analogy, if you tried to ride a bicycle with no chain, you might conclude that the chain is required to move the wheels. If, however, you tried to ride a bicycle whose chain had a damaged link (the equivalent of a partially defective protein) such that the bicycle would move but respond only erratically to your peddling, you might conclude that the chain is not critical to wheel movement and instead affords the cyclist some control over wheel movement. In Chapter 8, we saw that mutations that remove all function act as *null alleles* and that such null alleles are usually (but not always) recessive to wild-type alleles.

It is unfortunately not always easy or even possible to find null mutations in a genetic screen, even if mutagens are employed to increase the mutation rate. However, at least in some model organisms, scientists can use targeted mutagenesis to construct animals bearing null mutations in genes suspected of playing a critical role in development. The idea is to take a cloned gene, use recombinant DNA techniques to destroy its function, and then replace the wild-type gene in the genome with the inactivated cloned copy.

Knockout mice provide key examples of this kind of targeted mutagenesis (**Figure 17.5**). The formation of knockout mice depends on the existence of **embryonic stem (ES) cells**, discovered by Canadian researchers Drs. James Till and Ernest McCulloch. ES cells are undifferentiated cells, originally derived from early embryos that can grow in cell culture and remain undifferentiated (see the **Genetics and Society** box "Stem Cells and Human Cloning"). If these cultured cells are injected into a different early embryo, they can contribute to any and all of the tissues in the mouse that develops from that embryo. Dr. Janet Rossant, a researcher at the University of Toronto, is internationally recognized for her pioneering work in generating knockout mice and stem cell lines (see the **Focus on Genetics** box "Stem Cells and the Genetic Control of Embryonic Development").

To generate knockout mice, scientists first disrupt the cloned gene of interest by inserting foreign DNA into the middle of the gene. They then treat a culture of ES cells with the altered gene. Some of the cells will "take up" the altered cloned gene, and in a small fraction of these cells, homologous recombination will allow the altered gene to replace the original gene (Figure 17.5). Researchers can use one of several strategies to select those rare ES cells in which homologous recombination has occurred. These cells, now containing a null allele in place of a wild-type allele, are allowed to multiply in cell culture, and some of the resulting cells are injected into mouse embryos. Mice with tissues carrying the mutation are then used to begin a series of matings that culminate in the generation of animals homozygous for the null knockout mutation. The article *Mus musculus: Genetic Portrait of the House Mouse* on Connect discusses in more detail the protocols for creating knockout mice.

Hypomorphic mutations

Although the use of null mutations allows investigators to infer the most straightforward explanations for the function of the wild-type protein in development, there are situations in which it is actually more desirable to have a partial loss-of-function (*hypomorphic*) mutant allele. The reason is that many molecules function at multiple times in development. For example, the *wingless* (*wg*) gene in *Drosophila* is needed both for the formation of a proper embryo early in development and for the formation of an adult wing much later in development. An animal homozygous for a null allele of *wg* will die during embryogenesis. Because the animal dies before the wings form, you could not infer from this homozygote that the gene also functions to generate wing structures. In contrast, flies homozygous for a certain hypomorphic allele of the *wg* gene survive to adulthood, but they have no wings. Observing the effects of this allele alone, you would conclude that the gene is involved in wing formation, but you could not infer its role in early development. This example illustrates the importance of obtaining several different mutant alleles of a gene whose function you wish to study.

The ethical controversy over embryonic stem cell research arises from the destruction of the blastocyst when the ES cells are harvested for therapeutic cloning. To most opponents of abortion, this is believed to be the destruction of a human life, whether the embryo was cloned or whether it was left over from *in vitro* fertilization attempts and would eventually be discarded.

Future scientific developments may eventually make the ES cell controversy moot. For example, considerable progress has been reported on techniques to "reprogram" adult stem cells or even adult somatic cells to behave more like pluripotent ES cells via introduction of four transcription factors—Oct3/4, Sox2, c-Myc, and Klf4. These cells are known as **induced pluripotent stem (iPS) cells**. The next few years should help clarify the scientific and political issues surrounding the potential for stem-cell-based therapies.

Conditional mutations

Another way to study genes with effects on diverse developmental processes is to isolate *conditional*

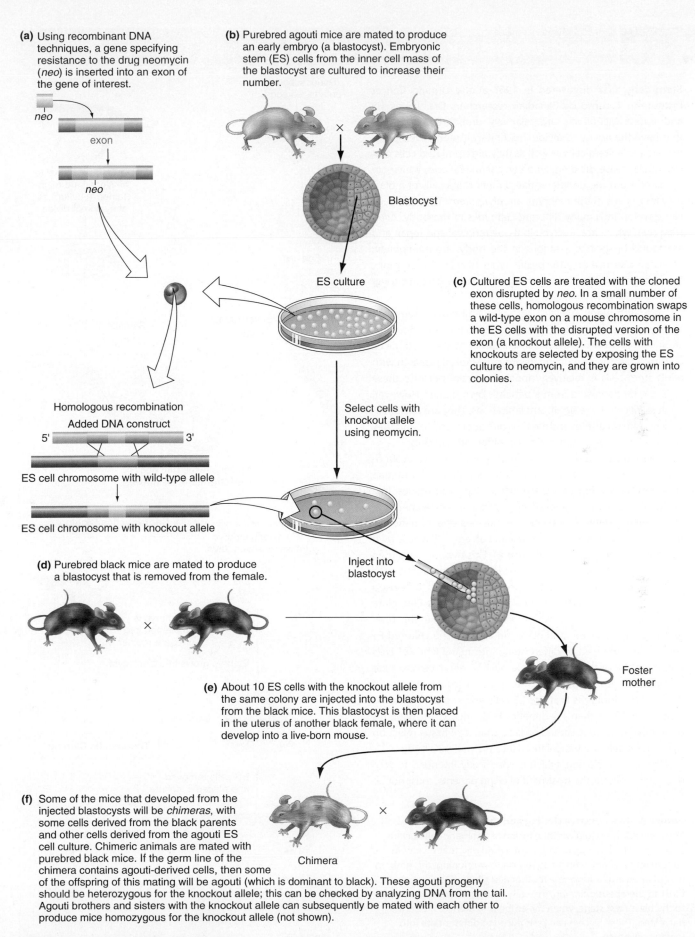

(a) Using recombinant DNA techniques, a gene specifying resistance to the drug neomycin (*neo*) is inserted into an exon of the gene of interest.

neo

exon

neo

(b) Purebred agouti mice are mated to produce an early embryo (a blastocyst). Embryonic stem (ES) cells from the inner cell mass of the blastocyst are cultured to increase their number.

×

Blastocyst

ES culture

(c) Cultured ES cells are treated with the cloned exon disrupted by *neo*. In a small number of these cells, homologous recombination swaps a wild-type exon on a mouse chromosome in the ES cells with the disrupted version of the exon (a knockout allele). The cells with knockouts are selected by exposing the ES culture to neomycin, and they are grown into colonies.

Homologous recombination

Added DNA construct

5'   3'

ES cell chromosome with wild-type allele

ES cell chromosome with knockout allele

Select cells with knockout allele using neomycin.

Inject into blastocyst

(d) Purebred black mice are mated to produce a blastocyst that is removed from the female.

×

Foster mother

(e) About 10 ES cells with the knockout allele from the same colony are injected into the blastocyst from the black mice. This blastocyst is then placed in the uterus of another black female, where it can develop into a live-born mouse.

(f) Some of the mice that developed from the injected blastocysts will be *chimeras*, with some cells derived from the black parents and other cells derived from the agouti ES cell culture. Chimeric animals are mated with purebred black mice. If the germ line of the chimera contains agouti-derived cells, then some of the offspring of this mating will be agouti (which is dominant to black). These agouti progeny should be heterozygous for the knockout allele; this can be checked by analyzing DNA from the tail. Agouti brothers and sisters with the knockout allele can subsequently be mated with each other to produce mice homozygous for the knockout allele (not shown).

Chimera

×

Figure 17.5 Constructing knockout mice.

Stem cells, first discovered in 1963 at the Ontario Cancer Institute in Toronto by Canadian researchers Drs. James Till and Ernest McCulloch, are relatively undifferentiated cells that have the ability to divide indefinitely. Among their progeny are more stem cells as well as fully differentiated cells that eventually cease dividing. *Embryonic stem (ES) cells*, which are obtained from the undifferentiated inner-mass cells of a blastocyst (an early-stage embryo), are *pluripotent*. Their progeny can develop into many different cell types in the body. *Adult stem cells*, which are involved in tissue renewal and repair and are found in specific locations in the body, are *multipotent*: They can give rise only to specific types of cells. For example, haematopoietic stem cells in the bone marrow give rise to an array of red and white blood cells.

Although many investigators value embryonic stem cells because of their pluripotency, research with human embryonic stem cells is controversial because in order to start a stem cell culture, a blastocyst must be destroyed. Medical research with adult stem cells is relatively noncontroversial because these cells can be harvested from a patient's own tissues. However, adult stem cells have significant limitations. They are present in only minute quantities and are thus difficult to isolate, and they can give rise to only certain kinds of differentiated cells.

For medical researchers, the greatest excitement surrounding the use of embryonic stem cells is the potential for human **therapeutic cloning** to replace lost or damaged tissues. In a protocol known as *somatic cell nuclear transfer*, researchers create a cloned embryo by taking the nucleus of a somatic cell from one individual and inserting it into an egg cell whose own nucleus has been removed (**Figure A**). This hybrid egg is then stimulated to begin embryonic divisions by treatment with electricity or certain ions. The embryo is not allowed to develop to term; instead, it is cultured for about five days in a Petri plate to the blastocyst stage, at which point the ES cells in the inner cell mass are collected and placed in culture. The cultured ES cells can be induced to differentiate into many kinds of cells that might be of therapeutic value, such as nerve cells to treat Parkinson disease (Figure A). One of the major advantages of therapeutic cloning is that the ES cells and the differentiated cells derived from them are genetically identical to the patient's own cells. Thus, there should be little chance of tissue rejection when these cells are transplanted into the patient's body.

Therapeutic cloning, which is specifically intended to produce stem cells for the treatment of ailing patients, must not be

Figure A Reproductive cloning and therapeutic cloning.
Both procedures begin with the fusion of a somatic cell nucleus and an enucleated egg, producing a hybrid egg that divides in culture into an early embryo. In reproductive cloning, this embryo is implanted into a surrogate mother and allowed to develop until birth. In therapeutic cloning, the early embryo develops in culture to the blastocyst stage, when the embryonic stem (ES) cells are harvested. These ES cells can be induced to differentiate into various cell types.

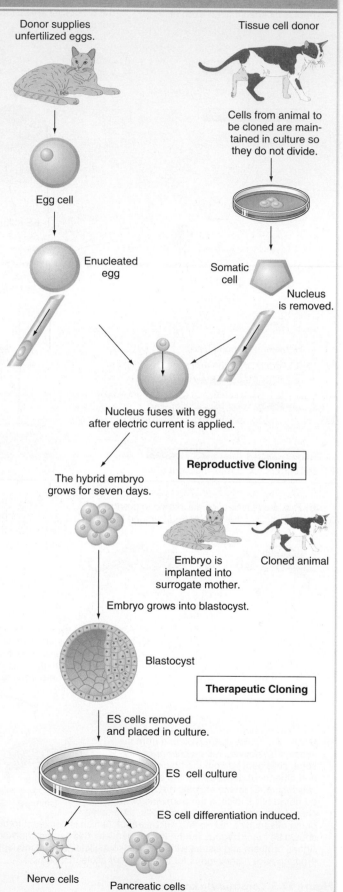

Donor supplies unfertilized eggs.

Tissue cell donor

Egg cell

Cells from animal to be cloned are maintained in culture so they do not divide.

Enucleated egg

Somatic cell

Nucleus is removed.

Nucleus fuses with egg after electric current is applied.

The hybrid embryo grows for seven days.

Reproductive Cloning

Embryo is implanted into surrogate mother.

Cloned animal

Embryo grows into blastocyst.

Blastocyst

Therapeutic Cloning

ES cells removed and placed in culture.

ES cell culture

ES cell differentiation induced.

Nerve cells

Pancreatic cells

confused with **reproductive cloning**, a type of cloning designed to make genetically identical complete organisms. The idea here is to create a cloned embryo by the same method just described for therapeutic cloning. In this case, however, the embryo is implanted into the uterus of a foster mother and allowed to develop to term (Figure A). Reproductive cloning has been successfully performed in several mammalian species, such as sheep and cats, but many cloned animals exhibit puzzling developmental defects such as obesity. No country or group of scientists has yet condoned the reproductive cloning of humans.

FOCUS ON Genetics Stem Cells and the Genetic Control of Embryonic Development

Dr. Janet Rossant **(Figure A)** is a Senior Scientist in the Developmental & Stem Cell Biology Program, the Lombard Insurance Chair in Paediatric Research, and Chief of Research at The Hospital for Sick Children in Toronto. She is also a professor at the University of Toronto. Born in the United Kingdom, Dr. Rossant received her undergraduate and graduate training at the Universities of Oxford and Cambridge, respectively, moving to Canada in 1977 after accepting an academic position at Brock University in St. Catherines, Ontario. In 1985, she moved to Toronto to work at the Samuel Lunenfeld Research Institute at Mount Sinai Hospital, Toronto. Dr. Rossant's research program uses the mouse as a model to understand the genetic control of early development (specifically, how genetically identical cells acquire distinct characteristics during embryogenesis to become a complex organism), with the aim of applying those research findings to human embryo development and the establishment of stem cells. A pioneer of cellular and genetic manipulation techniques, such as the generation of knockout mice via targeted mutagenesis by homologous recombination, the creation of mouse mutants by gene trapping approaches, and the production of mouse chimeras, she has employed these tools to explore the establishment of early embryonic patterning and signalling mechanisms in the mouse. These studies have resulted in the identification of human disease genes, since some of these mutant mice develop diseases similar to those observed in humans. Established mouse models of human disease have allowed for a more thorough investigation of disease characteristics and the effectiveness of various therapeutic strategies.

Dr. Rossant's interest in early embryonic development has also led to her discovery of trophoblast stem (TS) and extra-embryonic endoderm stem (XEN) cells. These novel stem cell types, along with embryonic stem (ES) cells, are made from the three earliest tissue lineages within the blastocyst, the trophectoderm, primitive endoderm, and epiblast, respectively **(Figure B)**. These stem cell lines are being utilized by Dr. Rossant to study the genetic and molecular control of pluripotent and restricted cell lineage formation in the mammalian blastocyst embryo, with the goal of applying this knowledge of development and cell fate decisions to the derivation, maintenance, and differentiation of restricted cell lineage progenitors from human ES and iPS (induced pluripotent stem) cells in order to explore

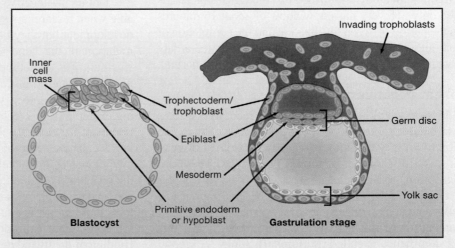

Figure B Cell lineages in the human blastocyst and gastrula embryo. Before implantation, the earliest cell types in the human blastocyst embryo (*left*) are the trophectoderm, epiblast, and primitive endoderm or hypoblast. As gastrulation begins following implantation (*right*), the trophectoderm gives rise to the trophoblast layers of the fetal part of the placenta. The epiblast of the germ disc is formed by the inner cell mass and gives rise to all cell types of the fetus itself and extra-embryonic mesoderm. The primitive endoderm gives rise to the endoderm layers of the yolk sac.

Figure A Dr. Janet Rossant.

(Continued)

mutations that cause a loss of function only under special circumstances. The most commonly studied type of conditional mutation, the temperature-sensitive mutation, produces a protein that is functional at a lower, *permissive temperature* but defective at a higher, *restrictive temperature*. In contrast, the protein product of the wild-type allele functions at both temperatures. It is best if the conditional mutation produces completely nonfunctional protein at the restrictive temperature, but it is sometimes difficult to determine whether the resulting protein is completely nonfunctional or remains partly functional.

Temperature-sensitive mutations have one main experimental advantage. They make it possible to raise an animal at the permissive temperature (which allows the early stages of development to proceed normally) and then to increase the temperature to assess the importance of the gene product at later developmental stages. **Figure 17.6** shows a temperature-shift analysis of a mutant strain of *C. elegans* carrying a temperature-sensitive lethal allele of the *zyg-9* gene, which helps determine the basic polarity of the early embryo. This temperature-shift study established that the ZYG-9 protein is required only in a very narrow window of about 15 minutes during the period between fertilization and the completion of the first mitotic division. If the protein is inactivated at any time outside the 15-minute window, development is normal.

Dominant-negative mutations

Most loss-of-function mutations have recessive effects because heterozygotes have about 50 percent of the gene function of wild-type homozygotes, and this level of gene function is sufficient for a normal phenotype. There are two exceptions to this rule. First, as discussed in Chapter 8, for a small number of developmentally important genes, one wild-type copy is insufficient for normal development. The mutant allele will thus be dominant to wild type; this type of dominance is called **haploinsufficiency**. A second situation in which a loss-of-function mutation can have a dominant effect occurs with so-called **dominant-negative mutants**. Here, the inactive protein encoded by a mutant allele "poisons" or otherwise counteracts the function of the protein encoded by the wild-type allele. Figure 7.23 illustrates one of several ways in which this can occur: In a multimeric protein, the presence of one abnormal subunit

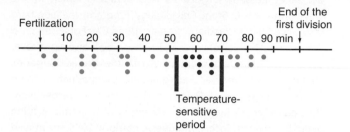

Figure 17.6 Time-of-function analysis. *C. elegans* embryos from mothers homozygous for a temperature-sensitive allele of the *zyg-9* gene develop properly if they are subjected to a short pulse of high temperature starting at any of the times indicated by the *green circles*. They develop incorrectly and subsequently die only if the high temperature begins at one of the times indicated by the *red circles*. (Each circle represents an experiment with one embryo.) These data show that the ZYG-9 protein is needed only during a 15-minute window of development.

might block the function of the protein even if the protein's other subunits are wild type.

Dominant-negative mutations can be particularly valuable when researchers suspect that a gene has an impact on development but they have not yet found a loss-of-function mutation in the gene to test their hypotheses. In such a situation, it is sometimes possible to engineer a dominant-negative mutant transgene *in vitro*, and then introduce this transgene back into a wild-type organism. In an interesting example of this technology, one research group made a dominant-negative fibroblast growth factor receptor (FGFR) in mice. These receptors are normally found on the surface membranes of many cell types (**Figure 17.7a**). One part of the receptor molecule faces the outside of the cell, where it can bind to a molecule called fibroblast growth factor (FGF). FGF is a **ligand**: a molecule involved in cell-to-cell communication that is produced by the cell sending the signal. Binding of the ligand to a receiving cell's receptor alters the behaviour of that cell. The binding of FGF to the extracellular part of FGFR causes several changes in the receptor, including the dimerization of two FGFR subunits and the activation of a kinase in each subunit that adds a phosphate group to the other subunit. This reciprocal phosphorylation, in turn, initiates a complicated intracellular signalling mechanism that changes the receiving cell's developmental fate.

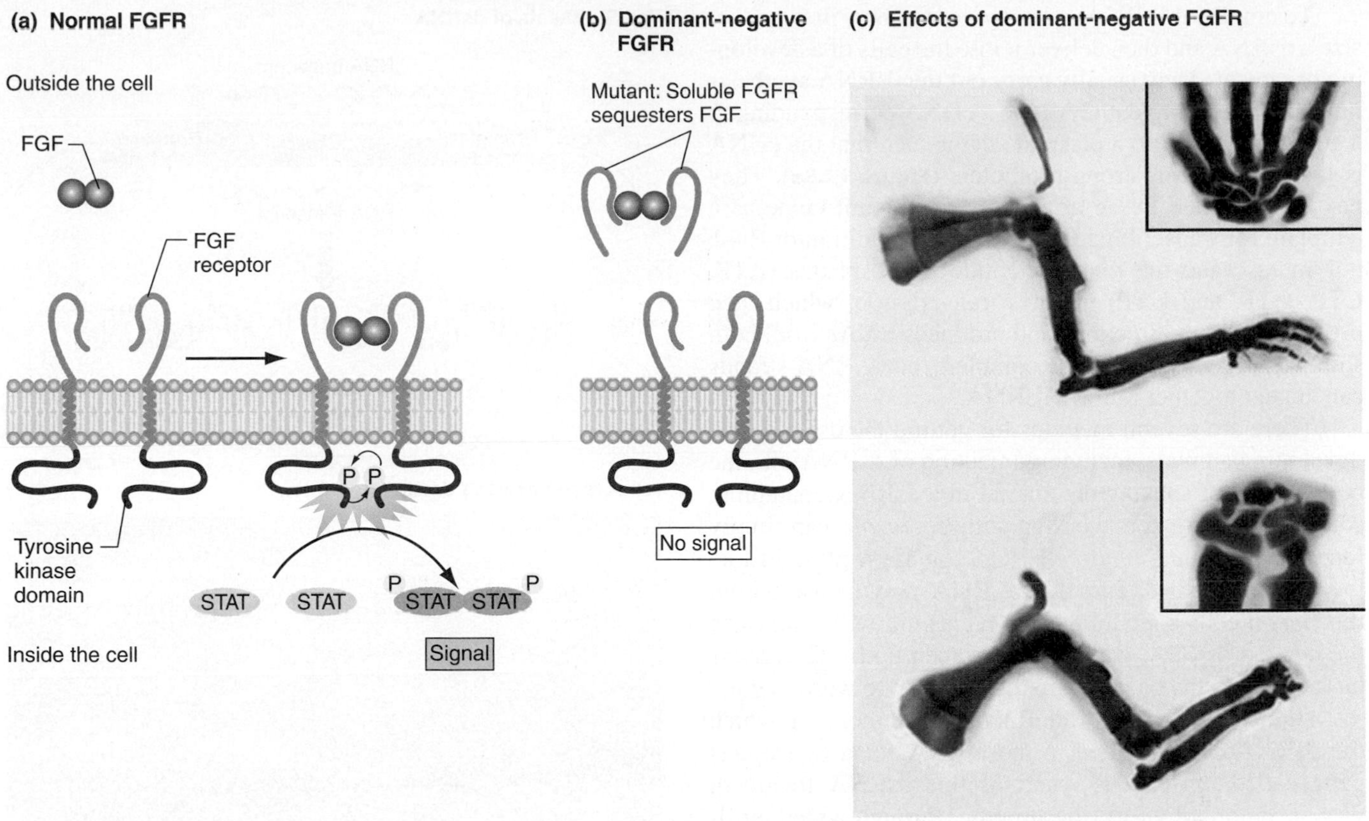

(a) Normal FGFR

Outside the cell

FGF

FGF receptor

Tyrosine kinase domain

Inside the cell

STAT STAT P STAT STAT P

Signal

(b) Dominant-negative FGFR

Mutant: Soluble FGFR sequesters FGF

No signal

(c) Effects of dominant-negative FGFR

Figure 17.7 Engineering a dominant-negative mutation in a mouse fibroblast growth factor receptor (FGFR) gene. (a) A dimer of fibroblast growth factor (FGF) binds two FGFR molecules in the cell membrane. As a result, the protein kinase domains of the two FGFRs phosphorylate (add a phosphate group to) each other; they also phosphorylate and dimerize two STAT molecules, which initiates a signal necessary for development. **(b)** A truncated mutant soluble form of FGFR can bind to FGF, preventing it from binding to normal FGFR in the cell membrane. **(c)** Phenotypic effects. *Top:* wild-type mouse limb. *Bottom:* limb from a mouse engineered to contain the dominant-negative version of the FGFR gene shown in part (b). Note the poor development of the digits in the mutant mouse (*insets*).

To make a dominant-negative FGFR mutant, the investigators synthesized a gene that gives rise to an abnormal form of the receptor. This mutant receptor cannot localize to the cell membrane and is instead secreted out of the cell. The researchers reasoned that the secreted form of FGFR would bind to FGF and thereby prevent the ligand from reaching the normal membrane-bound FGFR (**Figure 17.7b**). When they injected their engineered transgene into early mouse embryos, they observed a number of defects, including problems in limb development (**Figure 17.7c**). These results demonstrated that FGF signalling contributes to the developmental pathway leading to normal limbs.

- Loss-of-function mutations allow researchers to assess the effects of reduced activity or complete absence of a gene product.
- Null mutations (knockouts) reveal the earliest developmental processes influenced by the gene.
- Hypomorphic and conditional mutations allow evaluation of a gene's importance later in development. Dominant-negative mutations are useful when loss-of-function mutations have not yet been found.

RNA interference disrupts gene function without mutations

The genetic screens performed to find mutations in developmentally important genes require a considerable investment of effort and are often subject to unanticipated difficulties. As a result, geneticists have so far identified mutations in only a subset of the genes that play a role in development. But within the last several years, researchers have been able to employ a new strategy to deplete the protein products of specific genes from developing organisms. This strategy, which makes use of *RNA interference* (*RNAi*), is based on the following discovery. When cells ingest or are injected with double-stranded RNA (dsRNA) corresponding to the sequence of a gene's mRNA, the intracellular presence of the dsRNA triggers the degradation of the corresponding mRNA into short fragments. In the absence of intact mRNA, the cell cannot synthesize the protein. The details of RNAi are being worked out, but it appears that many kinds of cells have enzymes that degrade long dsRNAs into shorter dsRNAs roughly 21 nucleotides in length. These shorter dsRNAs then serve as templates for the degradation of homologous mRNAs into similar 21-bp fragments (see Chapter 11 as well as the *C. elegans* portrait on Connect).

To employ this RNAi strategy, researchers first synthesize a dsRNA and then deliver it into the cells of a developing organism. They usually carry out the dsRNA synthesis *in vitro*. For example, they clone a cDNA corresponding to a gene's mRNA into a plasmid vector such that the cDNA is located between strong promoters (**Figure 17.8a**). They next use purified DNA from the recombinant clone as a template for transcribing the cDNA. The addition of RNA polymerase and the four nucleotide triphosphates (ATP, CTP, GTP, and UTP) initiates transcription, which then proceeds in both directions and produces RNAs from both strands of the cDNA. These complementary RNA strands can anneal together to form dsRNA.

There are several methods for getting the dsRNA into developing animals, including injection of dsRNA into the body cavity or soaking the animal in a dsRNA-containing solution. Investigators working with *C. elegans* can simply feed larvae with *E. coli* cells that contain a plasmid like the one shown in Figure 17.8a. RNA polymerase within the *E. coli* cells containing such a plasmid will synthesize the desired dsRNA, which is then taken up by *C. elegans* larval cells as the bacteria are digested in the worm's gut.

Figure 17.8b shows an RNAi experiment in which the dsRNA corresponded to the mRNA for a *C. elegans* gene called *par-1*. The result of this dsRNA treatment was an abnormal vulva (the structure through which fertilized eggs are released) that protruded outside of the animal. Though worm researchers already knew that *par-1* functions very early in development to help establish the anterior-posterior axis of the animal, the results shown in Figure 17.8b showed that the gene also functions later in development in the patterning of the vulva.

RNA interference is an extremely useful technique for creating a **phenocopy** that mimics a loss-of-function mutation. However, a phenocopy is not a true, heritable mutation. Another slight drawback of the RNAi method is that results may vary because they depend on the relative level of dsRNA uptake. Consequently, even if the results of an RNAi experiment provide clues to a developmentally interesting gene's function, it is usually desirable to obtain and study a classical, heritable mutation in the gene as well.

> RNA interference utilizes dsRNA to degrade a corresponding mRNA. Expression of dsRNA within cells, or transfer of dsRNA into an organism's cells by ingestion or injection, can generate loss-of-function phenotypes.

Gain-of-function mutations also identify genes important for development

Mutations that produce too much protein, or proteins with a new function not present in the wild-type protein, are **gain-of-function mutations**. The alleles resulting from gain-of-function mutations are often dominant to the wild-type allele, in contrast with the majority of loss-of-function alleles, which are recessive to wild type.

(a) Synthesis of dsRNA

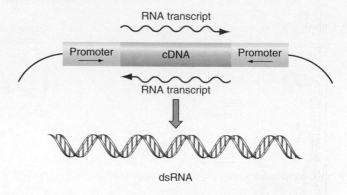

dsRNA

(b) A result of *par-1* dsRNA treatment

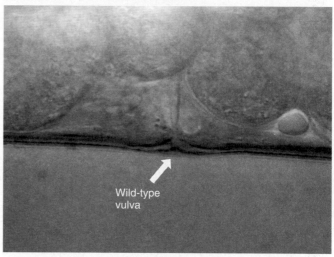

Wild-type vulva

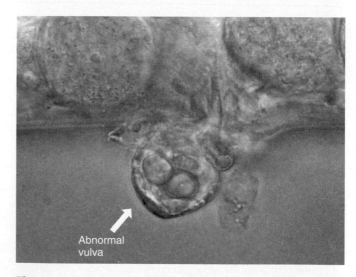

Abnormal vulva

Figure 17.8 RNA interference (RNAi): A tool for studying development. **(a)** How to make double-stranded RNA (dsRNA). A cDNA is cloned between two promoters, allowing transcription from both cDNA template strands. Complementary RNA transcripts will anneal with each other to make dsRNA. **(b)** Abnormal structure of the vulva in *C. elegans* treated with dsRNA for the *par-1* gene. *Top:* wild-type vulva. *Bottom:* protruding vulva in an animal treated with *par-1* dsRNA.

It is hard to understand unambiguously the role of a protein in development from a gain-of-function allele. The reason is that the mutation, rather than taking something away, adds something unusual to the organism, which might behave in an unpredictable way. Nevertheless, gain-of-function mutants can help identify developmentally important genes and clarify the roles they play in development. We now look at two kinds of gain-of-function mutations.

Mutations causing excessive gene activity

Such mutations are rare because they result only from highly specific changes in a gene, in contrast with loss-of-function mutations, which can disrupt gene function in many ways. Nonetheless, there are several ways in which mutations can lead to increased gene activity. One mechanism involves changes to promoters that make the promoters more accessible to transcription factors and RNA polymerase. Other possible scenarios are illustrated by various dominant mutations in *FGFR3*, one of the four genes in mice and humans that encode related yet distinct fibroblast growth factor receptors (review Figure 17.7). Some of these mutations increase the affinity of the FGFR3 receptor protein for its ligand FGF, inappropriately turning on the developmental signal when the concentration of the FGF ligand would normally be too low to accomplish this. Other *FGFR3* mutations allow the developmental signal to be turned on in the absence of FGF. These mutations cause the constitutive (continuous) activation of the phosphate-group-adding kinase domain of FGFR3; they accomplish this by altering a part of the protein that normally blocks the kinase function in the absence of the ligand.

Interestingly, a single amino acid substitution in the FGFR3 protein is sufficient to cause a gain-of-function dominant phenotype through the constitutive activation of the kinase domain. This substitution causes achondroplasia, the most common form of short-limb dwarfism in humans. Researchers have engineered mice with exactly the same amino acid substitution in their homologous *FGFR3* gene. Remarkably, this mutant gene produces what appears to be the same dominant dwarf phenotype seen in human achondroplasia (**Figure 17.9**).

Mutations causing ectopic gene expression

Suppose you suspect that a particular protein plays an important role in initiating the development of some structure like the legs or eyes. If that were true, it might be possible that expression of this protein in tissues in which it is not normally made could lead to the development of legs or eyes in unusual locations in the animal. The expression of a gene at an abnormal place or time is called **ectopic gene expression**.

Rarely, spontaneous mutations cause ectopic expression of genes important to development. An interesting example of such a mutation occurred in *Drosophila* when a chromosomal inversion moved the

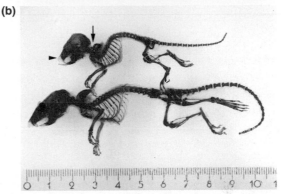

Figure 17.9 Achondroplastic dwarfism in the mouse. (a) The dwarf mouse at the *right* is heterozygous for an *FGFR3* allele with the same amino acid change as that causing achondroplasia in humans. A control littermate is at the *left*. **(b)** Skeletal abnormalities in the dwarf mouse at the *top* include a shorter face, overgrowth of the incisor teeth (*arrowhead*), and improper connection of the head to the spine (*arrow*) as compared with a control littermate at the *bottom*.

Antennapedia gene (normally transcribed in tissues destined to become legs) next to a specific kind of enhancer, which turned the gene on in tissues normally destined to develop into antennae. Animals carrying this *Antennapedia* mutation have legs growing out of their heads in place of antennae (review Figure 7.24). The phenotype of this ectopic mutant shows that the wild-type protein encoded by *Antennapedia* plays a critical role in leg development.

Instead of relying on rare and unpredictable mutations that might cause ectopic gene expression, researchers can now use recombinant DNA technology to make such mutations in a systematic way. They can change the promoter of a cloned gene by adding enhancers or other elements that might cause it to be transcribed at inappropriate places or times, and then introduce this altered gene back into the organism's genome by transformation. One research group placed the *eyeless* gene of *Drosophila* (review Figure 17.4a) under the control of a promoter for a "heat-shock" gene whose transcription in any tissue is turned on by higher than normal temperatures. Flies bearing this recombinant gene that were grown at high temperature made the *eyeless*-encoded

protein throughout their bodies. These animals had eye tissue growing at many different locations, even on their wings and antennae (**Figure 17.10**). This result demonstrates that the Eyeless protein is a master developmental switch that can activate a cellular program causing eye development.

Ectopic eyes also arise when the mouse *Pax6* or the human *Aniridia* gene is expressed in *Drosophila* under the control of the same heat-shock gene promoter. This result means that both elements of the amino acid sequence and the actual function of this master switch have been conserved throughout animal evolution.

Gain-of-function mutations often produce dominant phenotypes that can provide clues about a gene's role in a developmental process. Some genes, when ectopically expressed in tissues different from their normal sites of action, can activate programs of development that alter cellular fates.

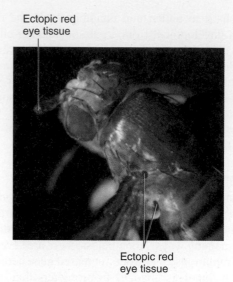

Ectopic red eye tissue

Ectopic red eye tissue

Figure 17.10 Ectopic expression of the *eyeless* gene produces ectopic eye tissue. This fly carries a synthetic *eyeless* gene that is turned on inappropriately. As a result, eye tissue grows in unexpected places, such as at the end of the antennae and on the thorax above the wings.

Analysis of Developmental Pathways

Once you have isolated a comprehensive set of mutations and identified as many as possible of the genes involved in the biological process of interest, the next step is to establish the functions performed by these genes. The ultimate aim of such studies is to discern a *developmental pathway*: a detailed description of how the products of these many genes interact and cooperate with each other to produce a particular outcome in development.

The action of each gene in a pathway must be characterized

Before looking at a complicated pathway as a whole, investigators must first learn as much as possible about each of the genes that comprise it. Specifically, details about the nature of the encoded protein, the location and timing of the gene's expression, the location of the protein product in the organism or in individual cells, and the developmental phenotypes associated with mutations in the gene all help scientists establish a theoretical framework to guide further analysis.

Nature of the encoded protein

With the completion of genome projects for key model organisms, researchers can often identify the mutant gene within a few months of finding the mutation. Once you know the nucleotide sequence of a gene, you automatically know the amino acid sequence of the protein it encodes. You can then use computer programs to search the amino acid sequence for motifs that offer clues to the protein's function. For example, computer programs can often predict whether a protein resembles known membrane-bound

receptors, or whether a protein acts as a kinase that phosphorylates other proteins.

One motif seen in many proteins with developmental significance is the **homeodomain** (**Figure 17.11**). It is found in the proteins encoded by the *eyeless/Pax6* and *Antennapedia* genes discussed earlier in this chapter. The homeodomain is a region of about 60 amino acids that is

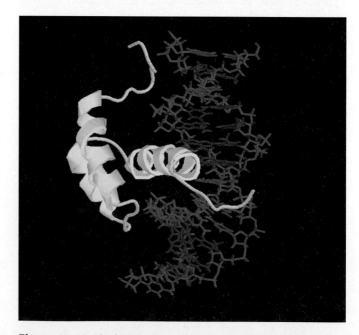

Figure 17.11 The homeodomain: A DNA-binding motif found in many transcription factors that regulate development. The amino acid backbone of a homeodomain (*yellow*) interacts with specific sequences in a DNA double helix (*red* and *blue*).

structurally related to the helix-turn-helix motif of many bacterial regulatory proteins. The homeodomain binds to specific DNA sequences, so its presence suggests that a protein might be a transcription factor.

Location and timing of gene expression

One way to answer the questions of where and when a gene is transcribed is to perform an RNA *in situ* hybridization experiment. To do this, you label cDNA sequences corresponding to the gene's mRNA and then use the labelled cDNA as a probe for the mRNA on preparations of thinly sectioned tissues. Signals where the probe is retained indicate cells containing the gene's mRNA (**Figure 17.12**). Defining the tissues in which the gene is expressed can help formulate hypotheses concerning the gene's role in development. For example, if a mutation in the gene affects the development of a tissue other than that in which the gene is transcribed, you might hypothesize that the gene encodes a signalling molecule like a hormone. Such molecules of cellular communication are made in one tissue but influence the fate of cells in other tissues that contain receptors for the hormone.

Location of the protein product

It is often technically easier to find and evaluate the tissues in which a gene is expressed by following the gene's protein product rather than by using RNA *in situ* hybridization to look for the gene's mRNA. In addition, an mRNA may be found in a tissue that does not contain the protein. This would point to the existence of regulatory controls that prevent translation of the mRNA. Finally, the intracellular localization of a protein often provides clues to its function. For example, concentration of the protein in the nucleus would be consistent with a role as a transcription factor.

Methods to follow a protein usually involve the generation of antibodies against parts of the protein. One way to do this is to use recombinant DNA techniques to construct a fused gene (**Figure 17.13a**). In this construct, part of a cDNA for the gene of interest is cloned downstream of, and in the same reading frame as, part of a gene encoding a protein that can be made at high levels in bacteria. If you transform a plasmid containing this fused gene into *E. coli*, the bacterial cells will make large amounts of a **fusion protein** whose N-terminal amino acids are from the bacterial protein and whose C-terminal amino acids are from the eukaryotic developmental protein. If you inject this fusion protein into rabbits or other animals, they will synthesize antibodies against it. And once you label these antibodies with a fluorescent tag, you can track the tagged antibodies as they react with the corresponding protein of developmental interest in preparations of tissues and cells (**Figure 17.13b**).

A newer way to track a protein is to construct a gene encoding a tagged protein that will itself fluoresce. The idea, illustrated in **Figure 17.13c**, is to synthesize an open reading frame that encodes not only the entire protein of interest, but also (at the protein's N or C terminus) the amino acids

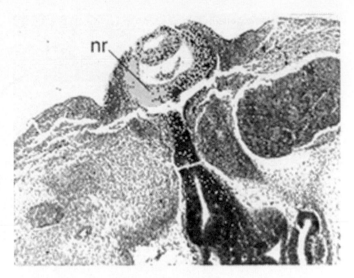

Figure 17.12 *In situ* **hybridization locates cells expressing a gene of interest.** This example shows that mRNA for the *Pax6* gene (*yellow signals*) accumulates in the eye of a human fetus in the seventh week of gestation. Hybridization is specific to the developing neural retina (nr) and the developing eye lens above it.

composing a naturally fluorescent protein from jellyfish called *green fluorescent protein* (*GFP*). When this recombinant gene is reintroduced into the genome by transformation, the organism will make the GFP fusion protein in the same places and at the same times it makes the normal untagged protein. Investigators can keep track of the fusion protein by following GFP fluorescence (**Figure 17.13d**). A major advantage of this approach is that researchers can use it to follow a GFP-tagged protein in living cells or animals, which is generally not possible with tagged antibodies for technical reasons. With the GFP fusion protein, researchers can even record videos that reveal subtle changes in the location of the protein over time.

Developmental phenotypes

Phenotypes may be evaluated in many ways to understand how a mutation impacts particular tissues and the development of the organism as a whole. For example, the morphology of mutant tissues can be examined with increasingly powerful microscopes, and the physiology of these tissues can be analyzed by various biochemical tests.

The importance of such investigations is underlined by the apparently simple question, "What cells or tissues are affected by the loss of gene function?" At first glance, it might seem that this question can be answered by the studies just described to define the location and timing of gene expression. That is, only the cells that make the protein would show the phenotypic effects of mutations that prevent that protein's synthesis. But this "obvious" solution is misleading because cells often communicate with each other to influence developmental decisions. In one simple example, if a gland synthesizes a hormone that circulates through the blood, and the gland can no longer make the hormone, the phenotypic effects might not show up in the

(a) Fusion protein gene in *E. coli*

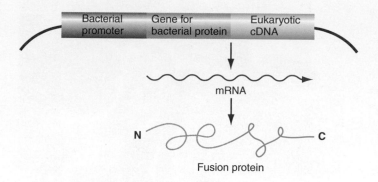

Bacterial promoter | Gene for bacterial protein | Eukaryotic cDNA

↓

mRNA

N — ⟨⟩ — C

Fusion protein

(b) A tissue stained with fluorescent antibodies

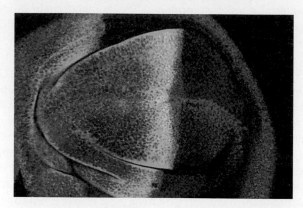

(c) Tagging a protein with GFP

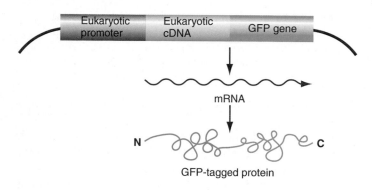

Eukaryotic promoter | Eukaryotic cDNA | GFP gene

↓

mRNA

↓

N — ⟨⟩ — C

GFP-tagged protein

(d) A mouse with a GFP-tagged transgene

Figure 17.13 Using antibodies and GFP tagging to follow the localization of proteins. (a) This synthetic gene encodes a fusion protein that will be made at high levels when transformed into *E. coli* cells. Animals injected with purified fusion protein will make antibodies against the protein of interest. **(b)** A *Drosophila* larval imaginal disc is stained with antibodies against several proteins. Each antibody is tagged with a dye that fluoresces in a particular colour. **(c)** Making a GFP-tagged protein. This recombinant gene encodes a fusion protein that contains GFP at its C terminus. **(d)** This mouse contains a GFP-labelled transgene expressed throughout the skin; the entire mouse becomes fluorescent when illuminated with UV light as at the *bottom*. The same mouse is shown in normal light at the *top*.

gland itself but rather in target cells elsewhere in the body that contain receptors for the hormone.

To address a variety of issues involving communication between cells, developmental geneticists construct **genetic mosaics**: organisms in which some cells (like those in the gland just described) have one genotype, whereas other cells (such as those in the hormone's target tissues) have a different genotype. Researchers can use several techniques to make such genetic mosaics. The technique chosen often depends on the species. *Drosophila* geneticists usually employ mitotic recombination; those working with *C. elegans* use methods based on the loss of small extra chromosomes during mitosis; and investigators studying mice mix embryonic cells from mutant and wild-type strains to make *chimeric mice* with two different cell types. **Chimeras** are genetic mosaics in which cells of different genotype originate from two different individuals. (For more species-specific details concerning the use of genetic mosaics, see the genetic portraits of flies, worms, and mice on Connect.)

Most mosaics are constructed with markers that allow investigators to differentiate between tissues with mutant and wild-type genotypes for the developmental gene. **Figure 17.14a** shows mosaic seedlings of the plant *Arabidopsis* in which blue tissue contains both a marker gene resulting in blue colour and a wild-type gene called *AGAMOUS*$^+$, whereas white tissue lacks the marker gene and is simultaneously mutant for *AGAMOUS*. **Figure 17.14b** diagrams how researchers used such marked mosaics to show that cells from a particular layer of undifferentiated cells (called L2) in the apical meristem send a signal needed for the proper differentiation of cells in a different layer (L1). This signal depends on the presence of a wild-type *AGAMOUS*$^+$ allele in L2 cells. In other words, even *AGAMOUS*$^+$ genotypically wild-type L1 cells develop abnormally if the adjacent L2 cells are mutant for this gene.

Careful analyses of protein structure, patterns of gene and protein expression, and phenotype provide clues about a gene's function in development. The construction

(a) Mosaic seedlings of *Arabidopsis*

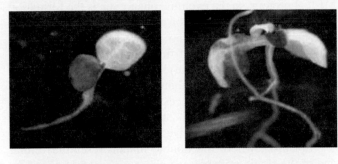

(b) Using mosaics to study cell signalling

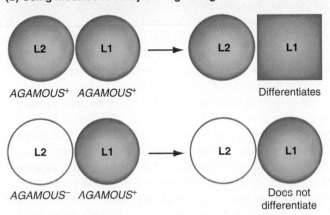

Figure 17.14 Mosaic analysis. (a) In these mosaic seedlings, *blue* tissue contains both a marker gene and the *AGAMOUS⁺* gene, whereas *white* tissue contains neither (it is *AGAMOUS⁻*). **(b)** A signal from blue *AGAMOUS⁺* L2 cells is needed for the proper differentiation of nearby L1 cells. If L2 cells lack the *AGAMOUS⁺* gene (*white*), nearby L1 cells do not differentiate properly, even if they are themselves *AGAMOUS⁺*.

of genetic mosaics is useful for determining whether the product of the gene influences the development of tissues other than those in which the gene is expressed.

The interactions of genes in a pathway must be determined

As you already know, genes do not work in isolation. Instead, complicated biological events demand the coordinated action of many genes. A full description of development from a genetic perspective thus requires not only the identification and analysis of the individual genes that contribute to development, but also the eventual elucidation of how the products of those genes work together.

It is generally easiest to focus first on the interaction of the genes in pairs: How does one gene influence the other, and vice versa? We mention here two of the most common approaches to answering this question. You will see other examples later in the chapter.

Analysis of how one gene affects the expression of another

Once you have defined the tissue distribution and intracellular location of one gene's mRNA or protein, you can

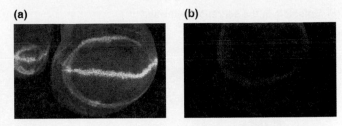

Figure 17.15 A mutation in one gene can affect the expression of another gene. (a) A wild-type *Drosophila* wing imaginal disc stained for Wingless (Wg, *green*) and Vestigial (Vg, *red*) proteins. A thin band of cells expressing both Wg and Vg is *yellow*. **(b)** A *wingless* mutant wing disc stained as in part (a). Not only is there no Wg protein, but Vg protein is made only in a narrow band about two cells wide, not in the broader region about 12 cells wide where it is normally made as in part (a).

ask how mutations in a different gene affect this distribution or localization. For example, consider the *wingless* and *vestigial* genes of *Drosophila*. Certain mutations in either gene result in the loss or reduction in size of the adult wings. The protein products of these genes are expressed only in small, overlapping subsets of the cells within tissues called "imaginal discs" that eventually develop into wings (**Figure 17.15a**). Flies mutant for *wingless* not only fail to make the Wingless protein, but they also fail to produce the Vestigial protein in many of the cells where it would normally be found (**Figure 17.15b**). This suggests that the expression of the *vestigial* gene in those cells is dependent upon *wingless* gene function in adjacent cells.

Analysis of double mutants

If two mutations define successive steps in a biochemical or secretory process, the double mutant will often arrest with the phenotype characteristic of the earliest block in the process; that is, the earlier-acting mutation is *epistatic* to the other. **Figure 17.16a** diagrams an analysis of two loss-of-function mutations in yeast that disrupt the secretion of molecules from the cell. Such secretion is important for many downstream developmental events. The phenotype of the double mutant in this example makes sense. We would expect that a molecule needs to be loaded into secretory vesicles before the fusion of these vesicles with the cell membrane allows the molecule to be secreted from a cell.

It would be incorrect, however, to conclude that an epistatic gene always governs an earlier step than the gene whose mutant phenotype is masked. In developmental or signal transduction pathways, the opposite is true. For example, in *C. elegans* the pathway leading to formation of the vulva includes three genes: *let-60, lin-45,* and *mek-2* (**Figure 17.16b**). The LET-60 protein becomes activated in cells that receive an extracellular signal. Active LET-60 then activates the LIN-45 protein, which subsequently activates the MEK-2 protein; active MEK-2 protein leads through several steps to vulva formation. Gain-of-function *let-60* alleles cause the overactivation of

(a) Epistasis in the secretion pathway

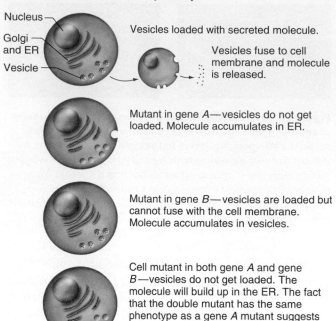

Nucleus

Golgi and ER

Vesicle

Vesicles loaded with secreted molecule.

Vesicles fuse to cell membrane and molecule is released.

Mutant in gene *A*—vesicles do not get loaded. Molecule accumulates in ER.

Mutant in gene *B*—vesicles are loaded but cannot fuse with the cell membrane. Molecule accumulates in vesicles.

Cell mutant in both gene *A* and gene *B*—vesicles do not get loaded. The molecule will build up in the ER. The fact that the double mutant has the same phenotype as a gene *A* mutant suggests that gene *A* acts before gene *B*.

(b) Epistasis in the pathway for vulva formation

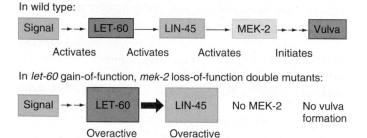

In wild type:

| Signal | → → | LET-60 | → | LIN-45 | → | MEK-2 | → → | Vulva |

Activates   Activates   Activates   Initiates

In *let-60* gain-of-function, *mek-2* loss-of-function double mutants:

| Signal | → → | LET-60 | ⟹ | LIN-45 | No MEK-2 | No vulva formation |

Overactive   Overactive

Figure 17.16 Double-mutant analysis. **(a)** The product of gene *A* helps lead *red* molecules into small, round vesicles. The product of gene *B* allows vesicles to fuse with the cell membrane, causing secretion of the vesicles' contents. Mutations in gene *A* are epistatic to those in gene *B*. **(b)** In this signal transduction pathway for vulva formation, a mutation in the gene controlling a later step is epistatic to a mutation in a gene whose product acts earlier.

LIN-45 protein, which in turn results in too much active MEK-2 protein, so the eventual phenotype is the formation of too many vulvas. Loss-of-function *mek-2* alleles have the opposite phenotype of no vulvas. Double mutant animals (with both the gain-of-function *let-60* allele and the loss-of-function *mek-2* allele) have no vulvas, so the *mek2* mutation is epistatic, even though *let-60* encodes a protein that acts earlier in the process. This result makes sense because if there is no MEK-2 protein, the relative activity of LET-60 cannot affect vulva formation.

Epistasis, where the double mutant resembles either one of the single mutants, is only one of several possibilities for the phenotype of a double mutant. In some cases, mutations in one gene are counteracted by mutations in a second *suppressor gene*, so the phenotype of the double mutant is nearly wild type. In other cases, the effect of mutations in a gene might be worsened by the simultaneous presence of mutations in a different *enhancer gene*.

Whether double-mutant analysis indicates epistasis, suppression, or enhancement, it is dangerous to interpret these results in isolation. Much more information is required. Are the mutations being analyzed loss-of-function or gain-of-function alleles? Does the blockage of each step in the pathway cause a different aberrant outcome, or does the whole pathway have a single output? Do the protein products have recognizable biochemical roles—for example, as transcription factors, or kinases, or hormone receptors? Answering these questions helps illuminate gene interactions.

> Commonly used techniques to explore how genes interact in developmental pathways include (1) determining whether mutations in one gene affect the timing or pattern of expression of another gene, and (2) observing the phenotype of an individual with mutations in two genes (a double mutant). True understanding of a developmental pathway requires integrating the results of many kinds of analysis.

17.4   A Comprehensive Example: Body-Plan Development in *Drosophila*

Studies on the genetic control of the basic body plan of *Drosophila* have revolutionized our understanding of development. Here, we focus on the aspect of this work that explains how the fly's body becomes differentiated and specialized along the *anterior-posterior* (*AP*) *axis*, the line running from the animal's head to its tail.

The research we describe was based on the observation that a fertilized *Drosophila* egg becomes subdivided into several clearly defined segments (review Figure 17.2), each of which eventually has a specific appearance and function. Some segments become parts of the head, others parts of the thorax, and still others, parts of the abdomen.

Scientists designed experiments to answer two fundamental questions about this segmentation. First, how does the developing animal establish the proper number of body segments? And second, how does each body segment "know" what kinds of structures it should form and what role it should play in the animal's biology? Results showed that very early in development, the action of a large group of genes, called the **segmentation genes**, subdivides the body into an array of essentially identical body segments. Later in development, the expression of a different set of genes, called **homeotic genes**, assigns a unique identity to each body segment.

Drosophila embryos become divided into segments

To understand how the segmentation and homeotic genes function, it is helpful to consider some of the basic events that take place in the first few hours of *Drosophila* development (**Figure 17.17**). The egg is fertilized in the uterus as it is being laid, and the meiotic divisions of the oocyte nucleus, which had previously arrested in the metaphase of meiosis I, resume at this time. After fusion of the haploid male and female pronuclei, the diploid zygotic nucleus of the embryo undergoes 13 rounds of nuclear division at an extraordinarily rapid rate, with the average time of mitotic cycles 2 through 9 being only 8.5 minutes.

Nuclear division in early *Drosophila* embryos, unlike most mitoses, is not accompanied by cell division, so the early embryo becomes a multinucleate syncytium. During the first eight division cycles, the multiple nuclei are centrally located in the egg; during the ninth division, most of the nuclei migrate out to the cortex—just under the surface of the embryo—to produce the **syncytial blastoderm**. During the tenth division, nuclei at the posterior pole of the egg are enclosed in membranes that invaginate (creating a pocket by infolding) from the egg cell membrane to form the first embryonic cells; these "pole cells" are the primordial germ cells. At the end of the thirteenth division cycle, about 6000 nuclei are present at the egg cortex.

During the interphase of the fourteenth cycle, membranes in the egg's cortex grow inward between these nuclei, creating an epithelial layer called the **cellular blastoderm** that is one-cell deep (**Figures** 17.17 and **17.18a**). The embryo completes formation of the cellular blastoderm about three hours after fertilization. At the cellular blastoderm stage, no regional differences in cell shape or size are apparent (with the exception of the pole cells at the posterior end). Experiments in which blastoderm cells have been transplanted from one location to another, however, show that despite this morphological uniformity, the segmental identity of the cells has already been determined. Consistent with this finding, molecular studies reveal that most segmentation and homeotic genes function during or even before the cellular blastoderm stage.

Immediately after cellularization, **gastrulation** and establishment of the embryonic germ layers begin. The *mesoderm* forms by invagination of a band of midventral cells that extends most of the length of the embryo. This infolding (the ventral furrow; **Figure 17.18b**) produces an internal tube whose cells soon divide and migrate to produce a mesodermal layer. The *endoderm* forms by distinct invaginations anterior and posterior to the ventral furrow; one of these invaginations is the cephalic furrow seen in Figure 17.18b. The cells of the endodermal infoldings migrate over the yolk to produce the gut. Finally, the

(a) The first three hours after fertilization

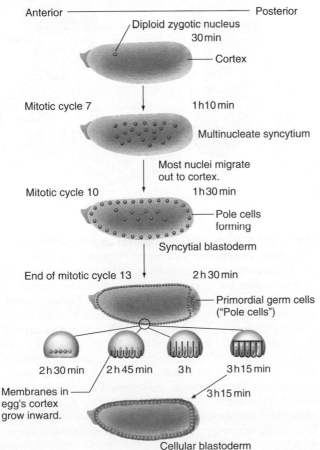

(b) Early embryonic stages in cross section

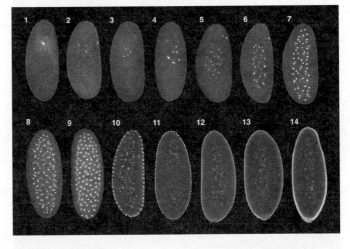

Figure 17.17 Early *Drosophila* development: From fertilization to cellular blastoderm. (a) The zygotic nucleus undergoes 13 very rapid mitotic divisions in a single syncytium. A few nuclei at the posterior end of the embryo become the germline *pole cells*. At the *syncytial blastoderm* stage, the egg surface is covered by a monolayer of nuclei. At the end of the thirteenth division cycle, cell membranes enclose the nuclei at the cortex into separate cells to produce a *cellular blastoderm*. **(b)** Photomicrographs of early embryonic stages stained with a fluorescent dye for DNA.

nervous system arises from neuroblasts that segregate from bilateral zones of the ventral *ectoderm*.

The first visible signs of segmentation are periodic bulges in the mesoderm, which appear about 40 minutes after gastrulation begins. Within a few hours of gastrulation, the embryo is divided into clear-cut body segments that will become the three head segments, three thoracic segments, and eight major abdominal segments of the larva (**Figure 17.18c**). Even though the animal eventually undergoes metamorphosis to become an adult fly, the same basic body plan is conserved in the adult stage (**Figure 17.18d**).

> The first rounds of mitosis in the *Drosophila* embryo produce a syncytial blastoderm. Cell membranes then grow around the thousands of nuclei under the embryonic surface, forming the cellular blastoderm. Some of these cells invaginate toward the middle of the embryo to make a gastrula. Although segmentation is first visible only after gastrulation, the genes responsible for segmentation function even earlier in development.

Segment number is first specified by maternal genes

Very little transcription of genes occurs in the embryonic nuclei between fertilization and the end of the 13 rapid syncytial divisions. Because of this near (but not total) absence of transcription, developmental biologists suspected that formation of the basic body plan initially requires **maternally supplied components** deposited by the mother into the egg during oogenesis. How could they identify the genes encoding these maternally supplied components? Christiane Nüsslein-Volhard and Eric Wieschaus realized that the embryonic phenotype determined by such genes does not depend on the embryo's own genotype; rather, it is determined by the genotype of the mother. They devised genetic screens to identify recessive mutations in maternal genes that influence embryonic development; these recessive mutations are often called **maternal-effect mutations**.

To carry out their screens, Nüsslein-Volhard and Wieschaus established individual balanced stocks for thousands of mutagen-treated chromosomes, and they then examined the phenotypes of embryos obtained from

(a) Cellular blastoderm

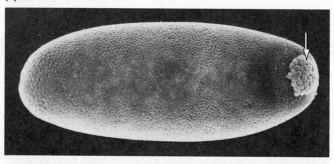

(b) Gastrulation

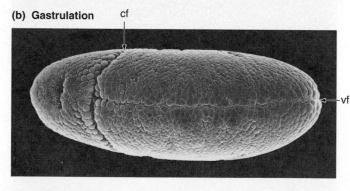

(c) Segmentation

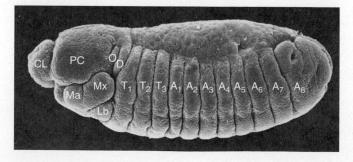

(d) Segment identity is preserved throughout development.

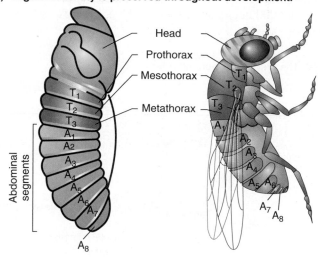

Figure 17.18 ***Drosophila* development after formation of the cellular blastoderm.** **(a)** Scanning electron micrograph of a cellular blastoderm. Individual cells are visible at the periphery of the embryo, and the pole cells at the posterior end can be distinguished (*arrow*). **(b)** A ventral view of some of the furrows that form during gastrulation, roughly four hours after fertilization: vf, ventral furrow; cf, cephalic furrow. **(c)** By ten hours after fertilization, it is clear that the embryo is subdivided into segments. Ma, Mx, and Lb are the three head segments. CL, PC, O, and D refer to nonsegmented regions of the head. The three thoracic segments (labelled T_1, T_2, and T_3) are the prothorax, the mesothorax, and the metathorax, respectively, whereas the abdominal segments are labelled $A_1 - A_8$. **(d)** The identities of embryonic segments (*left*) are preserved through the larval stages and are also retained through metamorphosis into the adult (*right*).

homozygous mutant mothers. They focused their attention on stocks in which homozygous mutant females were sterile, because they anticipated that the absence of maternally supplied components needed for the earliest stages of development would result in embryos so defective that they could never grow into adults. Through these large-scale screens, Nüsslein-Volhard and Wieschaus identified a large number of maternal-effect genes that are required for the normal patterning of the body. For this and other contributions, they shared the Nobel Prize for Physiology or Medicine with Edward B. Lewis—whose work we describe later.

We focus here on two groups of the genes they found. One group is required for normal patterning of the embryo's anterior; the other is required mainly for normal posterior patterning. The genes in these two groups are the first genes activated in the process that determines segment number.

The finding that separate groups of maternal-effect genes control anterior and posterior patterning is consistent with the conclusions of classical embryological experiments. Studies in which polar cytoplasm from the embryo's ends was transplanted, or in which preblastoderm embryos were separated into two halves by constriction of the embryo with a fine thread, suggested that the insect body axis is patterned during cleavage by the interaction of two signalling centres located at the anterior and posterior poles of the egg. In a specific model, Klaus Sander proposed that each pole of the egg produces a different substance, and that these substances form opposing gradients by diffusion. He suggested that the concentrations of these substances then determine the types of structures produced at each position along the body axis.

Molecular characterization of the maternal-effect genes of the anterior and posterior groups indicates that the Sander model for body axis patterning is correct. Substances that define different cell fates in a concentration-dependent manner are known as **morphogens**.

Bicoid: The anterior morphogen

Embryos from mothers homozygous for null alleles of the *bicoid* (*bcd*) gene lack all head and thoracic structures. The protein product of *bcd* is a DNA-binding transcription factor whose transcript is localized at the anterior pole of the egg cytoplasm (**Figure 17.19a**). Translation of the *bcd* transcripts takes place after fertilization. The newly made Bcd protein diffuses from its source at the pole to produce a high-to-low, anterior-to-posterior concentration gradient that extends over the anterior two-thirds of the embryo by the ninth division cycle (**Figure 17.19b**). This gradient determines most aspects of head and thorax development.

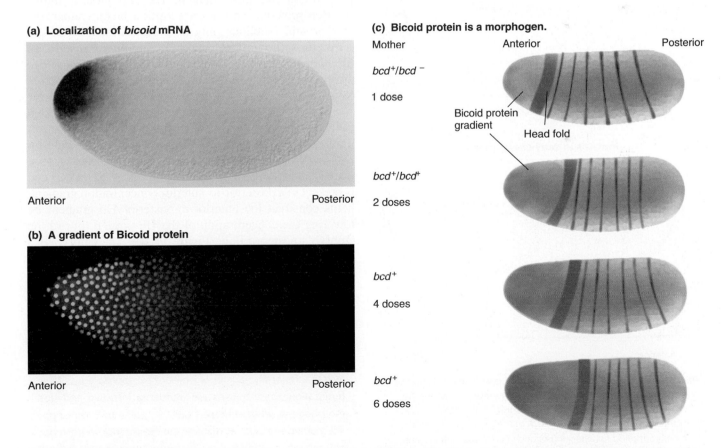

(a) Localization of *bicoid* mRNA

Anterior   Posterior

(b) A gradient of Bicoid protein

Anterior   Posterior

(c) Bicoid protein is a morphogen.

Mother   Anterior   Posterior

bcd⁺/*bcd* ⁻
1 dose

Bicoid protein gradient   Head fold

bcd⁺/*bcd*⁺
2 doses

bcd⁺
4 doses

bcd⁺
6 doses

Figure 17.19 Bicoid is the anterior morphogen. (a) The *bicoid* (*bcd*) mRNA (visualized by *in situ* hybridization in *purple*) concentrates at the anterior tip of the embryo. **(b)** The Bicoid (Bcd) protein (seen by *green* antibody staining) is distributed in a gradient: high at the anterior end and trailing off toward the posterior. The Bcd protein (a transcription factor) accumulates in the nuclei of this syncytial blastoderm embryo. **(c)** The greater the maternal dosage of *bcd*⁺, the higher the concentration of Bcd in the embryo, and the more of the embryo that is devoted to anterior structures. Head structures will develop anterior of the head fold invagination; thoracic and abdominal structures posterior to it.

One of the first lines of evidence that the Bcd protein functions as a morphogen came from experiments in which the maternal dosage of the *bcd* gene varied (**Figure 17.19c**). Mothers that carried only one dose of the *bcd* gene instead of the normal diploid dose incorporated about half the normal amount of *bcd* RNA into their eggs. As a result, translation yielded less Bcd protein, and the Bcd gradient was shallower and shifted to the anterior. In these Bcd-deficient embryos, the thoracic segments developed from more anterior regions than normal, and less of the body was devoted to the head. The opposite effect occurred in mothers carrying extra doses of the *bcd* gene. These and other observations suggested that the level of Bcd protein is a key to the determination of head and thoracic fates in the embryo. Three other genes work with *bcd* in the anterior group of maternal-effect genes; the function of the protein products of these three genes is to localize *bcd* transcripts to the egg's anterior pole.

The Bcd protein itself works in two ways: (1) as a transcription factor that helps control the transcription of genes farther down the regulatory pathway, and (2) as a translational repressor. The target of its repressor activity is the transcript of the *caudal* (*cad*) gene, which also encodes a DNA-binding transcription factor. The *cad* transcripts are uniformly distributed in the egg before fertilization, but because of translational repression by the Bcd protein, translation of these transcripts produces a gradient of Cad protein that is complementary to the Bcd gradient. That is, there is a high concentration of Cad protein at the posterior end of the embryo and lower concentrations toward the anterior (**Figure 17.20**). The Cad protein plays an important role in activating genes expressed later in the segmentation pathway to generate posterior structures.

Nanos: The primary posterior morphogen

The *nanos* (*nos*) RNA is localized to the posterior egg cytoplasm by proteins encoded by other posterior group maternal-effect genes. Like *bcd* RNAs, *nos* transcripts are translated during the cleavage stages. After translation, diffusion produces a posterior-to-anterior Nos protein concentration gradient. The Nos protein, unlike the Bcd protein, is not a transcription factor; rather, the Nos protein functions only as a translational repressor. Its major target is the maternally supplied transcript of the *hunchback* (*hb*) gene, which is deposited in the egg during oogenesis and is uniformly distributed before fertilization.

For development to occur properly, the Hb protein (which is another transcription factor) must be present in a gradient with high concentrations at the embryo's anterior and low concentrations at the posterior. The Nos protein, which represses the translation of *hb* maternal mRNA and is present in a posterior-to-anterior concentration gradient, helps construct the anterior-to-posterior Hb gradient by lowering the concentration of the Hb protein toward the embryo's posterior pole (Figure 17.20). The embryo also has a second mechanism for establishing the Hb protein gradient that functions somewhat later: It transcribes the *hb* gene from zygotic nuclei only in the anterior region (see following).

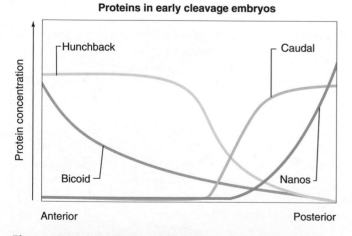

Figure 17.20 Distribution of the mRNA and protein products of maternal-effect genes within the early embryo. *Top:* In the oocyte prior to fertilization, *bicoid* (*bcd*) mRNA is concentrated at the anterior tip and *nanos* (*nos*) mRNA at the posterior tip, whereas maternally supplied *hunchback* (*hb*) and *caudal* (*cad*) mRNAs are uniformly distributed. *Bottom:* In early cleavage stage embryos, the Bicoid (Bcd) and Hunchback (Hb) proteins are found in concentration gradients high at the anterior and lower toward the posterior (A to P), whereas the Nanos (Nos) and Caudal (Cad) proteins are distributed in opposite P-to-A gradients.

Maternal *bcd* and *nos* mRNAs are concentrated respectively at the anterior and posterior poles of *Drosophila* eggs. After fertilization, these mRNAs are translated into Bcd and Nos morphogens, which diffuse from the poles to form oppositely oriented gradients that pattern the anterior-to-posterior embryo axis. Bcd and Nos regulate translation of maternal *cad* and *hb* mRNAs, respectively, generating gradients of Cad and Hb proteins. Bcd, Hb, and Cad are transcription factors that control expression of later-functioning segmentation genes.

Segment number is further specified by zygotic genes

The maternally determined Bcd, Hb, and Cad protein gradients control the spatial expression of zygotic segmentation genes. Unlike the products of maternal-effect genes, whose mRNAs are placed in the egg during oogenesis, the products of zygotic genes are transcribed and translated from DNA in the nuclei of embryonic cells descended from the original zygotic nucleus. The expression of zygotic segmentation genes begins in the syncytial blastoderm stage, a few division cycles before cellularization (roughly cycle 10).

Most of the zygotic segmentation genes were identified in another mutant screen that was carried out in the late 1970s also by Christiane Nüsslein-Volhard and Eric Wieschaus. In this screen, the two *Drosophila* geneticists placed individual ethyl methane sulfonate (EMS)-mutagenized chromosomes into balanced stocks and then examined homozygous mutant embryos from these stocks for defects in the segmentation pattern of the embryo. These embryos were so aberrant that they were unable to grow into adults; thus, the mutations causing these defects would be classified as recessive lethals.

After screening several thousand such stocks for each of the *Drosophila* chromosomes, Nüsslein-Volhard and Wieschaus identified three classes of zygotic segmentation genes: gap genes (14 different genes); pair-rule genes (9 genes in total); and segment polarity genes (about 17 genes). These three classes of zygotic genes fit into a hierarchy of gene expression.

Gap genes

The gap genes are the first zygotic segmentation genes to be transcribed. Embryos homozygous for mutations in the gap genes show a gap in the segmentation pattern caused by an absence of particular segments that correspond to the position at which each gene is transcribed (**Figure 17.21**).

How do the maternal transcription factor gradients ensure that the various gap genes are expressed in their broad zones at the proper position in the embryo? Part of the answer is that the binding sites in the promoter regions of the gap genes have different affinities for the maternal transcription factors. For example, some gap genes are activated by the Bcd protein (the anterior morphogen). Gap genes such as *hb* with low-affinity Bcd-protein-binding sites are activated only in the most anterior regions, where the concentration of Bcd is at its highest; by contrast, genes with high-affinity sites have an activation range extending farther toward the posterior pole.

Another part of the answer is that the gap genes themselves encode transcription factors that can influence the expression of other gap genes. The *Krüppel* (*Kr*) gap gene, for example, appears to be turned off by high amounts of Hb protein at the anterior end of its band of expression; activated within its expression band by Bcd protein in conjunction with lower levels of Hb protein; and turned off at the posterior end of its expression zone by the products of the *knirps* (*kni*) gap gene (**Figure 17.21c**). (Note that the *hb* gene is usually classified as a gap gene, despite the maternal supply of some *hb* RNA, because the protein translated from the transcripts of zygotic nuclei actually plays the more important role.)

Pair-rule genes

After the gap genes have divided the body axis into broad, generalized regions, activation of the pair-rule genes generates more sharply defined sections. These genes encode transcription factors that are expressed in seven stripes in preblastoderm and blastoderm embryos (**Figure 17.22a**). The stripes have a double-segment periodicity; that is, there is one stripe for every two of the fourteen total segments. Mutations in pair-rule genes cause the deletion of similar pattern elements from every alternate segment. For example, larvae mutant for the *fushi tarazu* (*ftz*) gene ("segment deficient" in Japanese) lack parts of abdominal segments A1, A3, A5, and A7 (see Figure 17.2). Mutations in the *even-skipped* gene cause loss of portions of even-numbered abdominal segments.

There are two classes for the typical pair-rule genes: primary and secondary. The striped expression pattern of the three primary pair-rule genes depends on the transcription factors encoded by the maternal-effect genes and the zygotic gap genes. Specific elements within the regulatory region of each pair-rule gene drive the expression of that pair-rule gene within a particular stripe. For example, as **Figures 17.22b** and **c** show, the DNA regulatory region responsible for driving the expression of *even-skipped* (*eve*) in the second stripe contains multiple binding sites for the Bcd protein and the proteins encoded by the gap genes *Krüppel* (*Kr*), *giant* (*gt*), and *hunchback* (*hb*). The transcription of *eve* in this stripe of the embryo is activated by Bcd and Hb, while it is repressed by Gt and Kr. Only in the stripe 2 region are Gt and Kr levels low enough and Bcd and Hb levels high enough to allow for activation of the element driving *eve* expression.

In contrast with the primary pair-rule genes, the five pair-rule genes of the secondary class are controlled by interactions with transcription factors encoded by other pair-rule genes.

Segment polarity genes

Many segment polarity genes are expressed in stripes that are repeated with a single-segment periodicity; that is, there is one stripe per segment (**Figure 17.23a**). Mutations in segment polarity genes cause deletion of part of each segment, often accompanied by mirror-image duplication of the remaining parts. The segment polarity genes thus function to determine anterior-posterior polarity patterns that are repeated in each segment.

The regulatory system that directs the expression of segment polarity genes in a single stripe per segment is quite complex. In general, the transcription factors encoded by pair-rule genes initiate the pattern by directly regulating certain segment polarity genes. Interactions between various cell polarity genes then maintain this periodicity

(a) Zones of gap gene expression

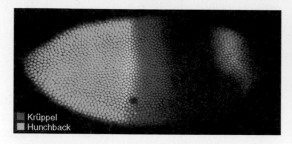

Krüppel
Hunchback

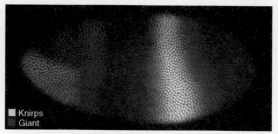

Knirps
Giant

(b) Phenotypes caused by gap gene mutations

T1
T2
T3
A1
A2
A3
A4
A5
A6
A7
A8

normal

A6
A6
A7
A8

Krüppel

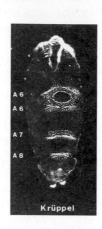

T1
A1
A2
A3
A4
A5
A6
A7
A8

hunchback

T1
T2
T3
A1
A8

knirps

(c) Gap genes: a summary

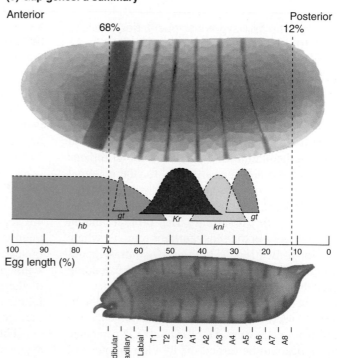

Anterior

68%

Posterior

12%

Egg length (%)

hb   gt   Kr   kni   gt

100 90 80 70 60 50 40 30 20 10 0

Mandibular
Maxillary
Labial
T1 T2 T3 A1 A2 A3 A4 A5 A6 A7 A8

Figure 17.21 Gap genes. (a) Zones of expression of four gap genes (*hunchback* [*hb*], *Krüppel* [*Kr*], *knirps* [*kni*], and *giant* [*gt*]) in late syncytial blastoderm embryos, as visualized with fluorescently labelled antibodies. **(b)** Defects in segmentation caused by mutations in selected gap genes, as seen in late embryos. Only the remaining thoracic and abdominal segments are labelled; the head segments at the anterior end are highly compressed and not labelled. **(c)** Mutation of a particular gap gene results in the loss of segments corresponding to the zone of expression of that gap gene in the embryo.

(a) Distribution of pair-rule gene products

Anterior   Posterior

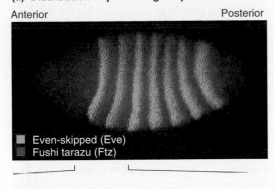

■ Even-skipped (Eve)
■ Fushi tarazu (Ftz)

(b) Proteins regulating *eve* transcription

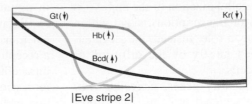

Gt(↓)   Kr(↓)
          Hb(↑)
          Bcd(↑)

|Eve stripe 2|

(c) Upstream regulatory region of *eve*

-1500 base pairs   -800

Kr   Gt   Gt   Kr   Gt   Kr   Kr   Kr
Bcd   Bcd   Bcd   Bcd   Hb   Bcd   Hb   Hb

Figure 17.22 Pair-rule genes. (a) Zones of expression of the proteins encoded by the pair-rule genes *fushi tarazu* (*ftz*) and *even-skipped* (*eve*) at the cellular blastoderm stage. Each gene is expressed in seven stripes. Eve stripe 2 is the second *green* stripe from the *left*. **(b)** The formation of Eve stripe 2 requires activation of *eve* transcription by the Bcd and Hb proteins and repression at its left and right ends by Gt and Kr proteins, respectively. **(c)** The 700-bp upstream regulatory region of the *eve* gene that directs the Eve second stripe contains multiple binding sites for the four proteins shown in part (b).

(a) Distribution of Engrailed protein

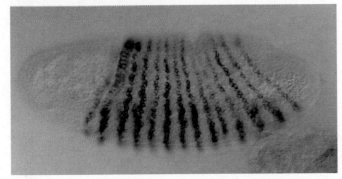

Figure 17.23 Segment polarity genes. (a) Wild-type embryos express the segment polarity gene *engrailed* in 14 stripes. **(b)** The border between a segment's posterior and anterior compartments is governed by the *engrailed* (*en*), *wingless* (*wg*), and *hedgehog* (*hh*) segment polarity genes. Cells in posterior compartments express *en*. The En protein activates the transcription of the *hh* gene, which encodes a secreted protein ligand. Binding of this Hh protein to the Patched receptor in the adjacent anterior cell initiates a signal transduction pathway (through the Smo and Ci proteins) leading to the transcription of the *wg* gene. Wg is also a secreted protein that binds to a different receptor in the posterior cell, which is encoded by *frizzled*. Binding of the Wg protein to this receptor initiates a different signal transduction pathway (including the Dsh, Zw3, and Arm proteins) that stimulates the transcription of *en* and of *hh*. The result is a reciprocal loop stabilizing the alternate fates of adjacent cells at the border.

(b) Segment polarity genes establish compartment borders.

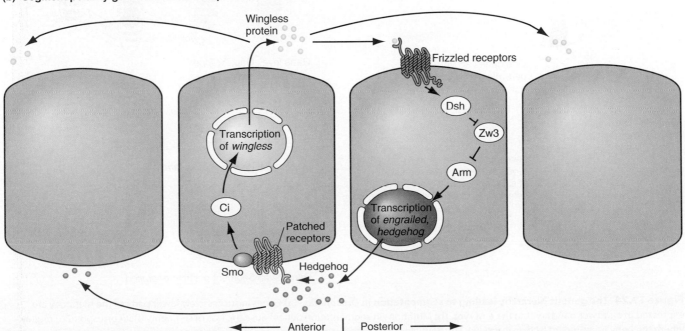

later in development. Significantly, activation of segment polarity genes occurs after cellularization of the embryo is complete, so the diffusion of transcription factors within the syncytium ceases to play a role. Instead, intrasegmental patterning is determined mostly by the diffusion of secreted proteins between cells.

Two of the segment polarity genes, *hedgehog* (*hh*) and *wingless* (*wg*), encode secreted proteins. These proteins, together with the transcription factor encoded by the *engrailed* (*en*) segment polarity gene, are responsible for many aspects of segmental patterning (**Figure 17.23b**). A key component of this control is that a one-cell-wide stripe of cells secreting the Wg protein is adjacent to a stripe of cells expressing the En protein and secreting the Hh protein. The interface of these two types of cells is a self-reinforcing, reciprocal loop. The Wg protein secreted by the more anterior of the two adjacent stripes of cells is required for the continued expression of *hh* and *en* in the adjacent posterior stripe. The Hh protein secreted by the more posterior stripe of cells maintains expression of *wg* in the anterior stripe. Gradients of Wg and Hh proteins made from these adjacent stripes of cells control many aspects of patterning in the remainder of the segment. The products of both *wg* and *hh* appear to function as morphogens; that is, responding cells appear to adopt different fates depending on the concentration of Wg or Hh protein to which they are exposed.

Other segment polarity genes encode proteins involved in **signal transduction pathways** initiated by the binding of Wg and Hh proteins to receptors on cell surfaces. Signal transduction pathways enable a signal received from a receptor on the cell's surface to be converted to a final intracellular regulatory response—usually the activation or repression of particular target genes. The signal transduction pathways initiated by the Wg and Hh proteins determine the ability of cells in portions of each segment to differentiate into the particular cell types characteristic of those locations.

Homologues of the segment polarity genes are key players in many important patterning events in vertebrates. For example, the chicken *sonic hedgehog* gene (related to the fly *hh*) is critical for the initiation of the left-right asymmetry in the early chicken embryo, as well as for the processes that determine the number and polarity of digits produced by the limb buds. The mammalian homologue of *sonic hedgehog* has the same conserved functions.

Summary of segment number specification

The pattern of expression for members of each class of segmentation genes is controlled either by genes higher in the hierarchy or by members of the same class, never by genes of a lower class (**Figure 17.24**). In this regulatory cascade, the maternal-effect genes control the gap and pair-rule genes, the gap genes control themselves and the

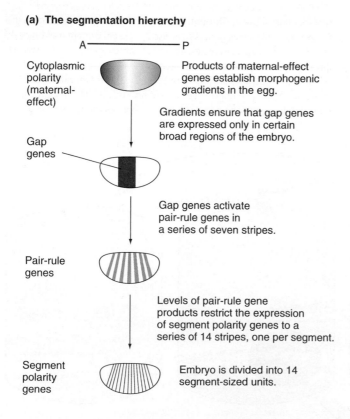

(a) The segmentation hierarchy

A ————————— P

Cytoplasmic polarity (maternal-effect) — Products of maternal-effect genes establish morphogenic gradients in the egg.

Gradients ensure that gap genes are expressed only in certain broad regions of the embryo.

Gap genes

Gap genes activate pair-rule genes in a series of seven stripes.

Pair-rule genes

Levels of pair-rule gene products restrict the expression of segment polarity genes to a series of 14 stripes, one per segment.

Segment polarity genes — Embryo is divided into 14 segment-sized units.

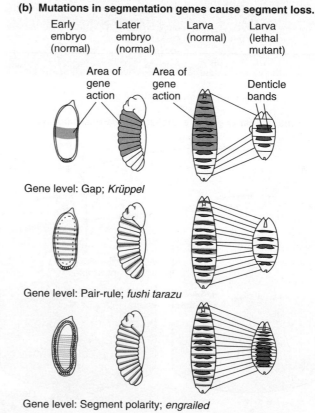

(b) Mutations in segmentation genes cause segment loss.

Early embryo (normal) | Later embryo (normal) | Larva (normal) | Larva (lethal mutant)

Area of gene action | Area of gene action | Denticle bands

Gene level: Gap; *Krüppel*

Gene level: Pair-rule; *fushi tarazu*

Gene level: Segment polarity; *engrailed*

Figure 17.24 The genetic hierarchy leading to segmentation in *Drosophila*. (a) Genes in successively lower parts of the hierarchy are expressed in narrower bands within the embryos. **(b)** Mutations in segmentation genes cause the loss of segments that correspond to regions where the gene is expressed (*shown in yellow*). The denticle bands (*dark brown*) are features that help researchers identify the segments.

pair-rule genes, and the pair-rule genes control themselves and the segment polarity genes. The expression of genes in successively lower parts of the hierarchy is increasingly spatially restricted within the embryo.

The cellular blastoderm looks from the outside like a uniform layer of cells (as seen in Figure 17.18a), but the coordinated action of the segmentation genes has actually already divided the embryo into segment primordia. A few hours after gastrulation, these primordia become distinguishable as clear-cut segments (Figure 17.18c).

> Most of the proteins produced by segmentation genes are transcription factors that control gene expression in the syncytial blastoderm. These factors are hierarchical, acting to restrict transcription of genes of the same or lower classes to increasingly narrow regions. After cellularization, pattern formation also depends upon intercellular communication mediated by secreted proteins.

Segment identity is established by homeotic genes

After the segmentation genes have subdivided the body into a precise number of segments, the homeotic selector (or segment identity) genes help assign a unique identity to each segment. They do this by functioning as master regulators that control the transcription of batteries of genes responsible for the development of segment-specific structures. The homeotic selector genes themselves are regulated by the gap, pair-rule, and segment polarity genes so that at the cellular blastoderm stage, or shortly thereafter, each homeotic gene becomes expressed within a specific subset of body segments. Most homeotic genes then remain active throughout the rest of development, functioning continuously to direct proper segmental specialization.

Mutations in homeotic genes, referred to as **homeotic mutations**, cause particular segments, or parts of them, to develop as if they were located elsewhere in the body. Because some of the mutant homeotic phenotypes are quite spectacular, researchers noticed them very early in *Drosophila* research. In 1915, for example, Calvin Bridges found a mutant he called *bithorax* (*bx*). In homozygotes for this mutation, the anterior portion of the third thoracic segment (T3) develops like the anterior second thoracic segment (T2); in other words, this mutation transforms part of T3 into the corresponding part of T2, as illustrated in **Figure 17.25a**. This mutant phenotype is very dramatic, as T3 normally produces only small club-shaped balancer organs called *halteres*, whereas T2 produces the wings. Another homeotic mutation is *postbithorax* (*pbx*), which affects only posterior T3, causing its transformation into posterior T2. (Note that in this context, *Drosophila* geneticists use the term "transformation" to mean a change of body form.) In the *bx pbx* double mutant, all of T3 develops as T2 to produce the now famous four-winged fly (**Figure 17.25b**).

In the last half of the twentieth century, researchers isolated many other homeotic mutations, most of which map within either of two gene clusters. Mutations affecting segments in the abdomen and posterior thorax lie within a cluster known as the **bithorax complex (BX-C)**; mutations affecting segments in the head and anterior thorax lie within the **Antennapedia complex (ANT-C)** (**Figure 17.26**).

(a) Effects of *bx* or *pbx* mutations

Wing

T2

T3 — Anterior

Haltere — Posterior

Wild type

T2

T3

bithorax mutant

T2

T3

postbithorax mutant

(b) A fly with both *bx* and *pbx* mutations

Figure 17.25 Homeotic transformations. (a) In flies homozygous for the mutation *bithorax* (*bx*), the anterior compartment of T3 (the third thoracic segment that makes the haltere) is transformed into the anterior compartment of T2 (the second thoracic segment that makes the wing). The mutation *postbithorax* (*pbx*) transforms the posterior compartment of T3 into the posterior compartment of T2. **(b)** In a *bx pbx* double mutant, T3 is changed entirely into T2. The result is a four-winged fly.

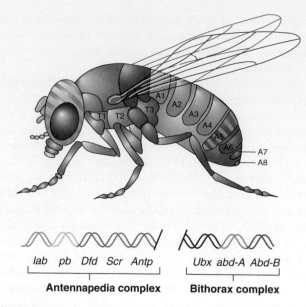

Figure 17.26 Homeotic selector genes. Two clusters of genes on *Drosophila*'s chromosome 3—the Antennapedia complex and the bithorax complex—determine most aspects of segment identity. Interestingly, the order of genes in these complexes is the same as the order of the segments each gene controls.

The bithorax complex

Edward B. Lewis shared the 1995 Nobel Prize for Physiology or Medicine with Christiane Nüsslein-Volhard and Eric Wieschaus for his extensive genetic studies of the BX-C. In his work, Lewis isolated BX-C mutations that, like *bx* and *pbx*, affected the posterior thorax; he also found novel BX-C mutations that caused anteriorly directed transformations of each of the eight abdominal segments. Lewis named mutations affecting abdominal segments *infra-abdominal* (*iab*) mutations, and he numbered these according to the primary segment they affect. Thus, *iab-2* mutations cause transformations of A2 toward A1, *iab-3* mutations cause transformations of A3 toward A2, and so forth.

Researchers initiated molecular studies of the bithorax complex in the early 1980s, and in 15 years, they not only extensively characterized all of the genes and mutations in the BX-C at the molecular level but also completed the sequencing of the entire 315-kb region. **Figure 17.27** summarizes the structure of the complex. A remarkable feature of the BX-C is that mutations map in the same order on the chromosome as the anterior-posterior order of the segments that each mutation affects. Thus, *bx* mutations, which affect anterior T3, lie near the left end of the complex, whereas *pbx* mutations, which affect posterior T3, lie immediately to their right. In turn, *iab-2*, which affects A2, is to the right of *pbx* but to the left of the A3-determining *iab-3*.

Because the *bx*, *pbx*, and *iab* elements are independently mutable, Lewis thought that each was a separate gene. However, the molecular characterization of the region revealed that the BX-C actually contains only three protein-coding genes: *Ultrabithorax* (*Ubx*), which controls the identity of T3; *abdominal-A* (*abd-A*), which controls the identities of A1–A5; and *Abdominal-B* (*Abd-B*), which controls the identities of A5–A8 (Figure 17.27). The expression patterns of these genes are consistent with their roles. *Ubx* is expressed in segments T3–A8 (but most strongly in T3); *abd-A* is expressed in A1–A8 (most strongly in A1–A4); and *Abd-B* is expressed in A5–A8. The *bx*, *pbx*, and *iab* mutations studied by Lewis affect large *cis*-regulatory regions that control the intricate spatial and temporal expression of these genes within specific segments.

The Antennapedia complex

Genetic studies in the early 1980s showed that a second homeotic gene cluster, the Antennapedia complex (ANT-C), specifies the identities of segments in the head and anterior thorax of *Drosophila*. The five homeotic genes of the ANT-C are *labial* (*lab*), which is expressed in the intercalary region; *proboscipedia* (*pb*), expressed in the maxillary and labial segments; *Deformed* (*Dfd*), expressed in the mandibular and maxillary segments; *Sex combs reduced* (*Scr*), expressed in the labial and T1 segments; and *Antennapedia* (*Antp*), expressed mainly in T2, although it is also active at lower levels in all three thoracic and most abdominal segments. (Figure 17.18c shows these head and thoracic segments, whereas Figure 17.26 illustrates the order of the homeotic genes in the ANT-C.) As with the BX-C, the order of genes in the ANT-C is the same (with the exception of *pb*) as the order of segments each controls.

The homeodomain in development and evolution

As researchers started to characterize the genes of the ANT-C and the BX-C at the molecular level, they were surprised to find that all of these genes contained some closely related DNA sequences. Similar sequences were also found in many other genes important for development, such as *bicoid* and *eyeless*, which are located outside the homeotic gene complexes. The region of sequence homology, called the **homeobox**, is about 180 bp in length and is located in the protein-coding part of each gene. The 60 amino acids encoded by the homeobox constitute the *homeodomain*, a region of each protein that can bind to DNA (review Figure 17.11). We now know that almost all proteins containing homeodomains are transcription factors in which the homeodomain is responsible for the sequence-specific binding of the proteins to the *cis*-acting control sites of the genes they regulate.

Surprisingly, however, DNA-binding studies have shown that most of these homeodomains have very similar binding specificities. The homeodomains of the Antp and Ubx proteins, for example, bind essentially the same DNA sequences. Because different homeotic proteins are thought to regulate specific target genes, this lack of DNA-binding specificity seems paradoxical. Much current

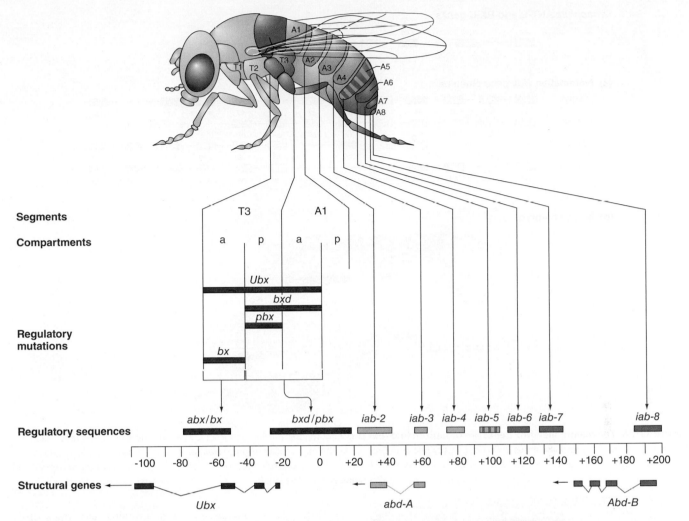

Segments

Compartments

Regulatory mutations

Regulatory sequences

Structural genes

Figure 17.27 The 315-kb bithorax complex region. The complex contains only three homeotic genes: *Ubx*, *abd-A*, and *Abd-B*. Many homeotic mutations such as *bx* and *pbx* affect regulatory regions that influence the transcription of one gene in particular segments. For example, *bx* mutations prevent the transcription of *Ubx* in the anterior compartment of the third thoracic segment, whereas *iab-8* mutations affect the transcription of *Abd-B* in segment A8. Note that the order of these regulatory regions corresponds to the anterior-to-posterior order of segments in the fly.

research is directed toward understanding how the homeotic proteins target specific genes that dictate different segment identities.

The discovery of the homeobox was one of the most important advances in the history of developmental biology because it allowed scientists to isolate by homology many other genes with roles in the development of *Drosophila* and other organisms. In the late 1980s and 1990s, the biological community was astonished to learn that the mouse and human genomes contain clustered homeobox genes called *Hox* genes with clear homologies to the ANT-C and BX-C genes in *Drosophila* (**Figure 17.28**). Remarkably, in all mammals studied to date, the genes within these clusters are arranged in a linear order that reflects their expression in particular regions along the spine of developing mammalian embryos (Figure 17.28). In other words, these gene clusters in mice and humans

are arranged in the genome and are regulated along the anterior-posterior axis in almost exactly the same way as the fly ANT-C and BX-C.

As it turns out, all animal genomes, even those of sponges, the most primitive animals, contain *Hox* genes, so these genes are ancient and have played important (though not necessarily identical) roles in the developmental patterns of all animals. Generally, the more complex the body plan, the more *Hox* genes: Humans and other mammals have four *Hox* gene clusters that together contain 38 *Hox* genes (see Figure 17.28). In just one demonstration that *Hox* genes mediate the developmental fate of specific regions in the body of animals other than *Drosophila*, it has been shown that the malformation of the digits in humans, in a condition called *synpolydactyly*, is caused by mutations in *HoxD13*, one of these 38 *Hox* genes (**Figure 17.29**).

Drosophila ANT-C and BX-C genes

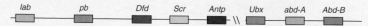

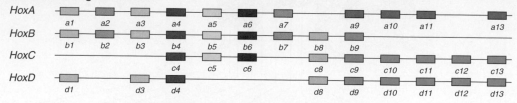

(a) Mammalian *Hox* gene clusters

HoxA
a1 a2 a3 a4 a5 a6 a7 a9 a10 a11 a13

HoxB
b1 b2 b3 b4 b5 b6 b7 b8 b9

HoxC
c4 c5 c6 c8 c9 c10 c11 c12 c13

HoxD
d1 d3 d4 d8 d9 d10 d11 d12 d13

(b) Mouse embryo

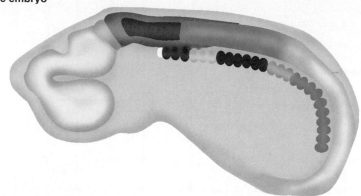

Figure 17.28 The mammalian *Hox* genes are organized into four clusters. **(a)** Mammalian genomes contain multiple homologues of each of the ANT-C and BX-C homeobox genes in *Drosophila*. **(b)** Just as in *Drosophila*, the mammalian (mouse) *Hox* genes in each cluster are arranged in the order that they are expressed along the anterior-posterior axis of the embryo. The *coloured discs* represent somites—precursors of the vertebrae and other structures. The other coloured areas are regions of the central nervous system. The colours represent the *Hox* genes expressed in that tissue.

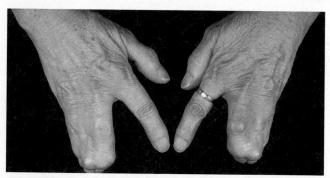

Most homeodomain-containing proteins are transcription factors that often play key roles in the development of multicellular organisms. The genomes of animals contain clusters of *Hox* genes with homeoboxes that encode homeodomains. The particular set of *Hox* genes expressed in a segment or region of the embryo helps dictate its eventual developmental fate.

Figure 17.29 Synpolydactyly caused by mutations in the human *HoxD13* gene.

17.5 How Genes Help Control Development

The previously described analysis of *Drosophila* body-plan development revealed some of the strategies by which genes control the development of multicellular organisms. These strategies form the basic underpinnings for many diverse developmental pathways in many organisms. Here are highlights of the lessons learned from *Drosophila*.

Development requires sequential changes in gene expression

The enormous diversity of cells within the body of a multicellular organism results in a remarkable variety of cell shapes and functions. Even a single tissue, such as the blood, harbours many different kinds of cells (**Figure 17.30**).

How do cells that contain the same genes make such varied developmental decisions and become so different? As you saw in the introduction to this chapter, the reason is simple: Different cell types express different, characteristic subsets of genes. It is easy to understand this point when comparing cells whose function depends on the production of a large amount of a particular gene product. Red blood cells produce copious amounts of haemoglobin, the cone

and rod cells in our eyes synthesize vast numbers of photo-receptor molecules, and certain pancreatic cells produce insulin and secrete it into the bloodstream. But the biochemical differences between cell types are not restricted to the expression of a single key gene. Instead, the differentiation of these various cells requires changes in the expression of many genes. **Figure 17.31** illustrates how complex these developmental patterns of expression can be. It shows that in *Drosophila* the many different proteins necessary for generating the structure of an adult wing are expressed in very precise, partially overlapping subsets of cells in larval imaginal discs that give rise to the wing.

Progressive refinement of cell fate

Differentiation into many types of cells and tissues requires that cells undergo a successive restriction in developmental potential that affects both themselves and their descendents. For example, the two daughter cells of a human zygote can each generate descendents that are able to fulfill any fate in the adult. But later in development, cells must "decide" whether they and their descendents will adopt one kind of fate (say, that of neurons) or a different kind of fate (say, that of epidermal cells). Once a developmental decision is made, a cell and its descendents embark on a pathway of differentiation that excludes them from an alternative fate.

The hierarchical developmental system that determines the number of segments in *Drosophila* embryos provides a clear example (review Figure 17.24). The gap genes such as *Krüppel* are expressed in broad regions covering roughly one-quarter to one-third of the embryo, but later, pair-rule genes are expressed in a fashion that subdivides the regions in which each gap gene was expressed. And later still, the segment polarity genes are expressed in even more sharply defined areas.

The key role of transcriptional regulation

The most efficient point at which protein production can be controlled is at the first step: the initiation of transcription. Indeed, most of the processes that influence cellular fates culminate in decisions to turn on or off the transcription of "target" genes (such as those for haemoglobin) whose expression is important to that cell type (the red blood cell precursor).

We can make this generalization based on three kinds of observations. First, RNA *in situ* hybridization experiments, such as the one shown in Figure 17.12, demonstrate that the mRNA for many developmentally important genes appears only in certain cells at certain times in development. Second, measurements of mRNA levels in many kinds of differentiated cells, using techniques such as microarrays or quantitative PCR, show that the levels of almost all proteins in those cells reflect the abundance of the mRNA encoding that protein. Third, many of the genes that play key roles in developmental decision-making encode proteins that function as transcription factors. For example, *bicoid*, all gap, and most pair-rule genes encode transcription factors. We have also seen that the homeodomain characteristic of

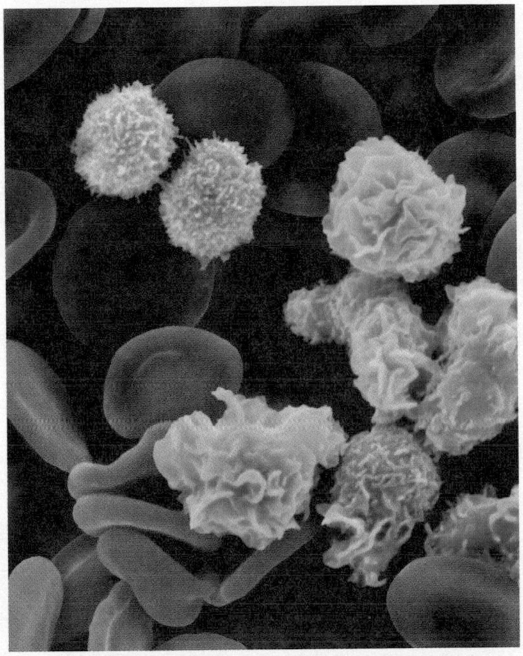

Figure 17.30 Different types of blood cells. The *red* cells are erythrocytes, the oxygen-carrying red blood cells. The cells coloured in *green* are macrophages that ingest and destroy invading microbes. The *yellow* cells are T lymphocytes involved in the immune system. The single *blue* cell is a monocyte, an immature cell that can develop into a macrophage. The colours other than red are computer-generated.

Figure 17.31 Development requires precise control of the expression of many genes. Each imaginal disc was stained with a fluorescently tagged antibody against a different specific protein important for patterning of the wing. Each protein is expressed in a unique set of cells in these imaginal discs.

proteins like the products of the BX-C and ANT-C genes, as well as other genes such as *eyeless/Pax6*, allows these proteins to bind to DNA and thus act as regulatory transcription factors. Hierarchies of transcription factors allow an organism to provide its cells with increasingly specific information that guides them to specific fates.

Posttranscriptional gene regulation and development

Although regulation of transcriptional initiation is the most general strategy by which cells control gene expression during development, it is by no means the only one. The progression from gene to protein involves many subsequent steps, each of which is amenable to regulation. In eukaryotes, a gene's primary transcript has to be spliced into a mature mRNA. This mRNA must be translocated from the nucleus into the cytoplasm, and then it must be translated into a protein. The relative stability of an mRNA or protein can affect its concentration in the cell. And finally, once made, a protein can be altered after translation in ways that affect its activity. A number of molecular mechanisms underlying development exploit each of these steps of gene regulation. You saw one case of developmentally important posttranscriptional regulation in the comprehensive example: Bicoid and Nanos, two proteins encoded by maternal-effect genes in *Drosophila*, act as repressors of translation.

The contributions of both maternal and zygotic genes

The earliest stages of development require not only the regulation of the expression of genes in the developing individual's genome but also the regulation of gene expression in the mother's genome. Before fertilization, the egg in most organisms already contains many of the mRNAs and proteins needed for the earliest stages of development. The egg must load up on these molecules because transcription of the zygotic genome does not begin immediately after fertilization. In *Drosophila*, for example, transcription of zygotic genes does not usually begin in earnest until the embryo contains roughly 6000 cells and has completed some of the earliest steps determining cell fates.

During early embryonic development, the fate of cells becomes increasingly narrowed. This differentiation is based on successive changes in gene expression that are mostly, but not exclusively, regulated at the level of transcription. The earliest steps of development require maternal RNAs and proteins that are expressed from the mother's genome and deposited into the egg during oogenesis.

Development exploits asymmetries

For cells to differentiate into different types, they must either be exposed to different signals from their environment or they must be intrinsically biochemically distinct. Nature has used both strategies to guide the differentiation of cell types.

In some species, the egg is inherently asymmetric, providing a way for cells in the early embryo to receive information about their relative position. The *Drosophila* egg cell (the oocyte), for example, is part of a more complicated structure called an *egg chamber* (**Figure 17.32**). Within the egg chamber, certain cells known as *nurse cells* act as factories that synthesize large amounts of mRNAs and proteins; the nurse cells then deposit these molecules into the oocyte.

The oocyte has an anterior-to-posterior (i.e., head-to-tail) sense of direction in large part because it is connected to the nurse cells only at its anterior end. The mRNA of the *bicoid* gene is transcribed in the nuclei of the nurse cells and then transported into the oocyte. The *bicoid* mRNA, in association with certain proteins that bind to its 3' UTR, appears to become ensnared by microtubules within the egg cell. These microtubules act as tracks along which the mRNA and its associated proteins are transported to the cortex (the cytoplasm just beneath the cell membrane) at the oocyte's anterior end.

In other species, the first asymmetries important for development occur after fertilization. For example, in *C. elegans*, the site at which the sperm enters the egg to effect fertilization defines the posterior end of the embryo. Before fertilization, the egg has no polarity; sperm entry initiates rearrangements of the cytoplasm that establish the anterior-to-posterior axis of the embryo. The asymmetries affecting early mammalian development emerge even later, after four rounds of mitosis have produced a 16-cell embryo.

In organisms like *Drosophila*, certain maternal mRNAs and proteins are concentrated in particular parts of the egg before fertilization. In other organisms, like *C. elegans* or humans, the asymmetries required for differentiation are established after fertilization in response to signals provided by the entering sperm, or to emergent variations in the interactions between embryonic cells.

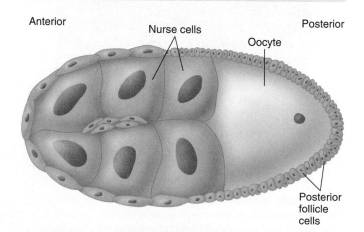

Anterior   Nurse cells   Posterior

Oocyte

Posterior follicle cells

Figure 17.32 A *Drosophila* egg chamber. Large nurse cells at the anterior of the egg chamber synthesize mRNA and proteins and transport them to the oocyte. The nurse cells and the oocyte are surrounded by a layer of *follicle cells.* Cell nuclei are in *purple.*

Cell-to-cell communication is essential for proper development

Construction of a large, complicated multicellular organism depends on more than broad, asymmetric cues such as morphogen gradients. Cells must "talk" to each other to obtain information about their relative positions in the organism. The information obtained from cell-to-cell communication enables cells to refine the decisions that guide their subsequent development.

Cells can communicate with each other either by direct contact or by diffusible factors (usually proteins) released from one cell and received by a second cell. Cell-to-cell communication usually takes place at the surface of the second cell when a ligand made by the signalling cell binds to a receptor embedded in the membrane of the receiving cell. One type of cell-to-cell communication, called *juxtacrine signalling*, takes the form of direct contact. In such signalling, the ligand is a cell surface molecule anchored in the membrane and extending outside of the signalling cell.

Other cellular interactions are mediated by *paracrine factors*: ligands secreted by the signalling cell. Ligands called *hormones*, or *endocrine factors*, circulate throughout the body in the blood and can affect tissues far removed from the gland that produces them. By contrast, some ligands diffuse only over short distances. The reciprocal interactions of *Drosophila* embryonic cells making the Wingless and Hedgehog segment polarity proteins (review Figure 17.23b) illustrate this kind of short-range paracrine signalling. Both Wingless and Hedgehog are secreted by certain cells, and only nearby cells with appropriate receptors can respond to these ligands.

Figure 17.23b emphasizes another feature common to most kinds of cell-to-cell communication: The binding of the ligand to a cell surface receptor initiates a signal transduction pathway that culminates in changes to the transcriptional regulation of suites of genes in the receiving cell's nucleus. Different ligand/receptor combinations activate different signal transduction pathways. For example, in Figure 17.23b, the Smo and Ci proteins participate in the pathway activated by the binding of the Hedgehog ligand to its receptor (Patched), whereas the Dsh, Zw3, and Arm proteins are part of the pathway initiated by the binding of the Wingless ligand to its receptor (Frizzled).

> In juxtacrine signalling, two adjacent cells communicate via a surface ligand on one cell that binds a receptor on the second cell. In paracrine signalling, the signalling cell secretes a hormone that can bind to receptors on cells elsewhere in the body. In either case, binding initiates a signal transduction pathway that alters gene expression in the receiving cell.

Genes explain much, but not everything, about development

Throughout this chapter, we have considered cells in developing multicellular organisms as complex computers. These cellular computers integrate a variety of inputs: the cell's history, its location within the organism, signals from neighbouring cells, and signals from more distant cells. The outputs of the cellular computer are alterations to the transcription of a large suite of target genes, which determine the developmental fate of the cell. The central processors that convert the inputs into the outputs are located near the promoters of the target genes, where assessment of the combinatorial effects of many transcription factors determines the time and rate of target gene transcription.

This reductionist point of view has been remarkably successful in building our understanding of development. We now have lists of many genes that play important roles in development, and we are beginning to fathom how each of these genes works and how they interact with each other. Particularly remarkable in the recent past has been our growing appreciation for the way in which evolution has conserved critical genes and pathways, while at the same time creating new twists that underlie the enormous complexity and diversity of lifeforms on Earth.

Although genes clearly set the ground rules for an individual's development, the same set of genes does not inevitably lead to precisely the same result. Many events in development reflect the strong influence of environmental factors or chance on the execution of the genetic blueprint. For example, the name of the first kitten cloned from an adult cell (**Figure 17.33**) is "cc" for "carbon copy", but this is somewhat of a misnomer. Though the kitten has exactly the same alleles of all genes as the cat that donated the adult cell, the coats of the two animals are dissimilar due to different prenatal environments.

Chance occurrences often influence expression of the genome as well as cellular behaviours. In mammalian

Figure 17.33 "cc" the cloned kitten.

females, for example, the decision of which X chromosome is inactivated in which cell is determined by stochastic (chance) events. Similarly, the choice of which cells adopt particular fates depends on small chance fluctuations in the concentrations of certain ligands and receptors. Finally, the incredibly complex connections between neurons in the developing brain are highly plastic and can be influenced by the environment, particularly through learning.

> Chance events and environmental influences can significantly alter the course of the genetically determined programs underlying development.

Connections

This chapter has presented ample evidence for the conservation of genes that play important roles in development. The *eyeless/Pax6/Aniridia* gene, for example, acts as a master switch to initiate the development of eyes in many types of organisms. Yet the eyes of various species show tremendous differences, from the compound eyes of *Drosophila* to the single camera-like organs of humans.

The themes of conservation and change have been central to our understanding of evolution since Darwin. Evolution creates and then preserves genetic solutions to problems organisms encounter in their development, biochemistry, physiology, and behaviour; but evolution also tinkers with these solutions to produce novel outcomes.

In the next chapter, we see that although there is significant diversity of biological and genetic mechanisms in bacterial organisms and the cellular organelles mitochondria and chloroplasts, the study of these organisms and organelles highlights the unity of genetic phenomena in all forms of living entities.

Essential Concepts

1. Developmental geneticists use model organisms as the basis for studying how a fertilized egg becomes a multicellular adult. The evolutionary relatedness of all organisms often makes it possible to extrapolate from model organisms to all living forms. **[LO1]**

2. A key to the genetic dissection of development is the isolation of a comprehensive set of mutations. Loss-of-function mutations are especially useful in revealing genes whose action is critical for normal development, but gain-of-function mutants can also point to genes that participate in developmental processes. RNA interference provides a method to achieve loss of gene function without mutations. **[LO2]**

3. Researchers first analyze the role of individual genes in development by characterizing the nature of the gene product, the locations in which the gene is transcribed and in which the protein product of the gene accumulates, and the phenotypes associated with mutations in the gene. The construction of genetic mosaics can help determine which cells need to express the gene so that the organism can develop normally. Scientists then examine the interactions of multiple genes affecting the same process to elucidate developmental pathways. **[LO3]**

4. Genetic analysis of the *Drosophila* body plan revealed several basic mechanisms by which genes help control development. A hierarchy of segmentation genes subdivides the body into an array of body segments; the expression of homeotic genes assigns a unique identity to each segment. **[LO4]**

5. Cellular differentiation requires progressive changes in gene expression. These changes usually, but not always, result from decisions concerning the transcription of batteries of genes. The earliest stages of development require control of gene expression in both the maternal and zygotic genomes. **[LO5]**

6. Differentiation requires either that cells have intrinsic differences at the biochemical level or that they are exposed to different information in their environment. Asymmetries in early embryonic development or in the distribution of molecules during cell division can generate intrinsic differences. Cell-to-cell communication, effected by the binding of ligands to receptors and mediated by signal transduction pathways, supplies cells with information about their position in the organism. **[LO5]**

Problems

Vocabulary

1. For each of the terms in the left column, choose the best matching phrase in the right column.

i.	mosaic determination	**1.**	divide the body into identical units (segments)
ii.	regulative determination	**2.**	initiated by the binding of ligand to receptor
iii.	haploinsufficiency	**3.**	individuals with cells of more than one genotype

iv. RNAi	**4.** the fate of early embryonic cells can be altered by the environment
v. ectopic expression	**5.** assign identity to body segments
vi. homeodomain	**6.** substance whose concentration determines cell fates
vii. green fluorescent protein	**7.** suppression of gene expression by double-stranded RNA
viii. genetic mosaics	**8.** when a null allele is dominant to a wild-type allele
ix. segmentation genes	**9.** a DNA-binding motif found in certain transcription factors
x. homeotic genes	**10.** encode proteins that accumulate in unfertilized eggs and are needed for embryo development
xi. morphogen	**11.** early embryonic cells are assigned specific fates
xii. maternal-effect genes	**12.** a gene is turned on in an inappropriate tissue or at the wrong time
xiii. signal transduction pathways	**13.** a tag used to follow proteins in living cells

Section 17.1

2. a. If you were interested in the role of a particular gene in the embryonic development of the human heart, why would you probably study this role in a model organism, and which model organism(s) would you choose?

b. If you were interested in finding new genes that might be required for human heart development, why would you try to find these genes in a model organism, and which model organism(s) would you choose?

3. Early *C. elegans* embryos display mosaic determination, whereas early mouse embryos exhibit regulative determination. Predict the results you would expect if the following treatments were performed on four-cell embryos of each of these two species (assuming these manipulations could actually be performed):

a. A laser is used to destroy one of the four cells (this technique is called *laser ablation*).

b. The four cells of the embryo are separated from each other and allowed to develop.

c. The cells from two different four-celled embryos are fused together to make an eight-celled embryo.

Sections 17.2 and 17.3

Problems 4–7 concern a *Drosophila* gene called *rugose* (*rg*). Adult flies homozygous for recessive mutations in this gene have rough eyes in which the regular pattern of the eye segments called *ommatidia* is disrupted. The scanning electron micrographs below contrast the smooth eyes of wild-type flies on the *left* with the rough eyes of *rugose* mutants on the *right*. The disruption of the eye segment pattern is caused by the absence of one or more so-called *cone cells* from ommatidia; in the wild type, each ommatidium has four cone cells.

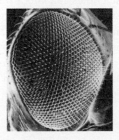

4. In 1932, H. J. Muller suggested a genetic test to determine whether a particular mutation whose phenotypic effects are recessive to wild type is a null (*amorphic*) or hypomorphic allele of a gene. Muller's test was to compare the phenotype of homozygotes for the recessive mutant alleles with the phenotype of a heterozygote in which one chromosome carries the recessive mutation in question and the homologous chromosome carries a deletion for a large region including the gene. In a recent study utilizing Muller's test, investigators examined two mutant alleles of *rugose* named rg^{41} and $rg^{\gamma3}$. The eye phenotypes displayed by flies of several genotypes is indicated in the following table. *Df(1)JC70* is a large deletion that removes *rugose* and several genes to either side of it.

Genotype	Eye surface	Cone cells per ommatidium
wild type	smooth	4
rg^{41} / rg^{41}	mildly rough	2–3
rg^{41} / *Df(1)JC70*	moderately rough	1–2
$rg^{\gamma3}$ / $rg^{\gamma3}$	very rough	0–1
$rg^{\gamma3}$ / *Df(1)JC70*	very rough	0–1

a. Which allele (rg^{41} or $rg^{\gamma3}$) is "stronger" (i.e., which causes the more severe phenotype)?

b. Which allele directs the production of higher levels of functional Rugose protein?

c. How would Muller's test discriminate between a null allele and a hypomorphic allele? Suggest a theoretical explanation for Muller's test. Based on the results shown in the table, is either of these two mutations likely to be a null allele of *rugose*? If so, which one?

5. The molecular identity of the fruit fly *rugose* gene is now known. cDNA clones corresponding to the *rugose* gene mRNA and antibodies that recognize the Rugose protein are also available. Outline several alternatives to the approach described in Problem 4 that might help you decide whether a newly discovered recessive allele of *rugose* is a null or a hypomorphic mutation.

6. In a *Drosophila* population of genotype $rg^{\gamma3}/rg^{\gamma3}$, it was noticed that about 35 percent of fertilized eggs develop into defective embryos that are unable to hatch into

larvae. In contrast, only about 3 percent of fertilized wild-type eggs fail to develop into larvae.

a. In light of this information as well as the data presented in Problem 4, predict which fly tissues at which stages of development require the function of the Rugose protein. (*Note:* The eyes of adult *Drosophila* are not pre-formed in embryos or larvae; instead, they develop from sacs of tissue in larvae called *imaginal discs.*)

b. How could you determine whether the *rugose* gene was expressed in the tissues you predicted in part *a*? Does the expression of the gene in those tissues establish that the Rugose protein plays an essential function there?

7. The *rugose* gene (*rg*) is located about midway between the centromere and telomere of the acrocentric *Drosophila* X chromosome. The *white* gene for eye colour is located near the X chromosome telomere; the dominant *w*⁺ allele specifies red colour in eye cells, whereas *w*⁻ causes eye cells to be white. Mitotic recombination like that shown in Figure 4.18 (Chapter 4) can be induced by exposing *Drosophila* larvae to X-rays.

a. Scientists can use mitotic recombination to create adult flies with mosaic eyes in which some eye cells would be simultaneously homozygous for mutant alleles of *rugose* and *white*, whereas the other cells in the eye would be heterozygous for the mutant and wild-type alleles of both genes. Diagram an arrangement of mutant and wild-type alleles of these two genes that would create such mosaic eyes upon X-ray-induced mitotic recombination.

b. How could you use this system of mitotic recombination to determine whether the lack of the Rugose protein in one ommatidium might affect the proper development of an adjacent ommatidium?

c. Suppose for the sake of argument that all animals homozygous for a true null mutation of *rugose* would die as embryos. How could you use this system of mitotic recombination to determine the effect of a complete lack of the Rugose protein on development of the adult eye?

Problems 8–11 concern a recombinant DNA construct called *myo-2::GFP* that *C. elegans* developmental geneticists have transformed into worms. Worms containing this construct express green fluorescent protein (GFP) in their pharynx, as shown in the following picture. The pharynx is an organ located between the mouth and the gut that grinds up the bacteria *C. elegans* eats so that these bacteria can be used as a food source. The *myo-2::GFP* construct was made by cloning the open reading frame for jellyfish GFP downstream of the promoter for *myo-2*, a gene that is specifically expressed in the muscle cells of the pharynx.

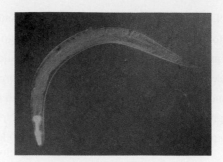

8. a. Explain how you could use worms transformed with *myo-2::GFP* to find mutations that disrupt the structure of the pharynx.

b. Nematodes homozygous for loss-of-function mutations in a gene called *pha-4* have no detectable pharyngeal structures. What do you think will be the fate of these worms?

c. How could you use *myo-2::GFP* to determine if *pha-4* is a master regulatory gene that directs development of the pharynx in a manner similar to the way *Pax6/ eyeless* controls eye development?

9. How could you use the pictured *myo-2::GFP* construct to find out what DNA sequence elements in the *myo-2* gene promoter are required for the pharynx-specific expression of the *myo-2* gene?

10. Suppose you wanted to determine whether a particular gene *X* was important for specification of the pharynx, but mutations in this same gene disrupt embryonic development well before pharyngeal structures appear. How could you use *myo-2::GFP*, the *myo-2* promoter, the DNA sequence of gene *X*, and your knowledge of RNA interference (RNAi) to generate worms that lack gene *X* expression in the pharynx but express gene *X* in all other tissues in which it is expressed in wild-type *C. elegans*?

11. The procedure normally used to transform *C. elegans* involves injection of DNA into the gonads of hermaphrodites. The DNA is incorporated into oocytes, but the injected DNA molecules usually recombine with each other, forming extrachromosomal arrays. These extrachromosomal arrays can be lost during mitosis at a low frequency, producing cells that lack the arrays. How could you use *myo-2::GFP* to create nematodes with mosaic pharynges, such that some cells are homozygous for null mutations of gene *X* while other cells in the same pharynx have gene *X* activity? (Assume that null mutations and the genomic DNA of gene *X* are both available.)

12. Figure 17.5 shows how scientists can knock out any gene in mice using homologous recombination. An alternative and technically much simpler methodology to manipulate the mouse genome is an "add-on" strategy in which DNA is injected into a pronucleus of a fertilized egg, and the injected one-cell embryo is placed into an oviduct of

a receptive female. In this add-on strategy, the injected DNA will integrate into various locations in the genome at random. For each of the following situations, indicate whether it would be preferable to use a knockout or add-on strategy, and explain both your decision and how you would employ the technology of your choice.

a. You want to create a mouse model of a human genetic disease in which a particular missense mutation has a recessive deleterious effect on development.

b. You want to create a mouse model of a human genetic disease in which a particular missense mutation has a dominant deleterious effect on development.

c. You want to explore the potential effects of the ectopic expression of a gene in a tissue in which it is normally not expressed.

d. You want to explore the potential deleterious haplo-insufficient effects of the deletion of a gene.

e. You want to explore the potential deleterious effects of homozygosity for the deletion of a particular gene.

f. You want to explore the potential effects of the absence of gene function associated with the expression of a dominant-negative allele of a gene.

g. You want to suppress the function of a particular gene by RNA interference.

h. You want to find *cis*-acting regulatory sequences that cause a certain gene to be expressed only in particular tissues.

i. You want to prove that a polymorphism you have detected in the DNA of a particular candidate gene is responsible for a specific phenotype of abnormal development seen in mutant animals. (Assume that the mutation actually causing the phenotype is associated with a loss of function, but consider mutations that are recessive or dominant to wild type separately.)

13. As explained in Problem 12, when the "add-on" strategy is used to create transgenic mice, the injected DNA can insert at random into any chromosome. Subsequent matings produce animals homozygous for the transgene insertion, and sometimes an interesting developmental phenotype is generated by the insertion event itself. In one case, after injection of DNA containing the mouse mammary tumour virus (MMTV) promoter fused to the *c-myc* gene, investigators identified a recessive mutation that causes limb deformity. In this mouse, the distal bones were reduced and fused together; the mutation also caused kidney malfunction.

a. The mutant phenotype could be due to insertion of the MMTV/*c-myc* transgene in a particular region of the chromosome or a chance point mutation that arose in the mouse. How could you distinguish between these two possibilities?

b. The mutation in this example was in fact caused by insertion of the transgene. How could you use this transgene insertion as a tag for cloning and identifying a gene important for development?

c. The insertion mutation was mapped to chromosome 2 of mice in a region where a mutation called limb deformity (*ld*) had previously been identified. Mice carrying this mutation are available from a major mouse research laboratory. How could you tell if the *ld* mutation was in the same gene as the transgenic insertion mutation?

Section 17.4

14. Which of the following is not a property of the *hunchback* gene in *Drosophila*?

a. The *hunchback* mRNA is uniformly distributed in the egg by the mother.

b. Transcription of *hunchback* is enhanced by Bicoid (the anterior morphogen).

c. Translation of the *hunchback* mRNA is inhibited by Nanos (the posterior morphogen).

d. The Hunchback protein eventually is distributed in a gradient (anterior high; posterior low).

e. Hunchback protein directs the distribution of *bicoid* mRNA.

15. The *hunchback* gene contains a promoter region, the structural region (the amino-acid-coding sequence), and a 3′ untranslated region (DNA that will be transcribed into sequences appearing at the 3′ end of the mRNA that are not translated into amino acids).

a. What important sequences required to control *hunchback* gene expression are found in the promoter region of *hunchback*?

b. What sequence elements that encode specific protein domains are found in the structural region of *hunchback*?

c. There is another important kind of sequence that turns out to be located in the part of the gene transcribed as the 3′ UTR (untranslated region) of the *hunchback* mRNA. What might this sequence do?

16. How do the segment polarity genes differ in their mode of action from the gap and pair-rule genes?

17. One important demonstration that Bicoid is an anterior determinant came from injection experiments analogous to those done by early embryologists. Injection experiments involve introduction of components such as cytoplasm from an egg or mRNA that is synthesized *in vitro* into the egg by direct injection. Describe injection experiments that would demonstrate that Bicoid is the anterior determinant.

18. In flies developing from eggs laid by a *nanos*⁻ mother, development of the abdomen is inhibited. Flies developing from eggs that have no maternally supplied

hunchback mRNA are normal. Flies developing from eggs laid by a *nanos⁻* mother that also have no maternally supplied *hunchback* mRNA are normal. If there is too much Hunchback protein in the posterior of the egg, abdominal development is prevented.

a. What do these findings say about the function of the Nanos protein and of the *hunchback* maternally supplied mRNA?

b. What do these findings say about the efficiency of evolution?

19. Mutant embryos lacking the gap gene *knirps* (*kni*) are stained at the syncytial blastoderm stage to examine the distributions of the Hunchback and Krüppel proteins. The results of the *knirps⁻* and wild-type embryos stained for the Hunchback and Krüppel proteins are shown schematically on the following figure.

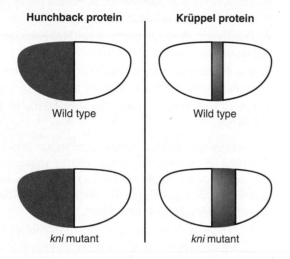

Hunchback protein   **Krüppel protein**

Wild type   Wild type

kni mutant   *kni* mutant

a. Based on these results, what can you conclude about the relationships among these three genes?

b. Would the pattern of Hunchback protein in embryos from a *nanos⁻* mutant mother differ from that shown? If yes, describe the difference and explain why. If not, explain why not.

20. In *Drosophila* with loss-of-function mutations affecting the *Ubx* gene, transformations of body segments are always in the anterior direction. That is, in *bx* mutants, the anterior compartment of T3 is transformed into the anterior compartment of T2, whereas in *pbx* mutants, the posterior part of T3 is transformed into the posterior compartment of T2. In wild type, the *Ubx* gene itself is expressed in T3–A8, but most strongly in T3.

a. The *Abd-B* gene is transcribed in segments A5–A8. Assuming the mode of function of *Abd-B* is the same as that of *Ubx*, what is the likely consequence of homozygosity for a null allele of *Abd-B* (i.e., what segment transformations would you expect to see)?

b. Because *abd-A* is expressed in segments A1–A8, there is some transcription of all three genes of the BX-C (*Ubx*, *abd-A*, and *Abd-B*) in segments A5–A8. Why then are segments A5, A6, A7, and A8 morphologically distinguishable?

c. What segment transformations would you expect to see in an animal deleted for all three genes of the BX-C (*Ubx*, *abd-A*, and *Abd-B*)?

d. Certain *contrabithorax* mutations in the BX-C cause transformations of wing to haltere. Propose an explanation for this phenotype based on the transcription of the *Ubx* gene in particular segments. Do you anticipate that *contrabithorax* mutations would be dominant or recessive to wild type? Explain.

e. During wild-type development, *Antp* is expressed in T1, T2, and T3, but most strongly in T2 and only weakly in T3. In animals with *Ubx* null mutations, *Antp* is expressed at much higher levels in T3 as compared with wild type. In animals with deletions that remove both *Ubx* and *abd-A*, *Antp* is expressed at high levels in T2, T3, and abdominal segments A1–A5. In animals with deletions that remove all three genes of the BX-C, *Antp* is expressed in T2, T3, and abdominal segments A1–A8. Given that the three genes of the BX-C encode proteins with homeodomains, suggest a model that explains how these genes dictate segment identity.

Section 17.5

21. If you were searching for mutations that affect early embryonic development in a model organism that had not been previously studied, why would you need to conduct separate genetic screens for genes encoding maternally supplied components and for genes whose transcription begins only after fertilization? What kinds of screens would you employ in both cases?

22. The unfertilized eggs of *C. elegans* have no predetermined anterior or posterior end. The polarity of the embryo instead depends on the site of sperm entry, which becomes the posterior end. Very soon after fertilization, so-called PAR (for "partitioning") proteins, which are uniformly distributed in unfertilized eggs, become localized to the embryonic cortex (the layer of cytoplasm just under the cell membrane) at one or the other end of the embryo. The following figure shows the distribution of two of these proteins, PAR-2 (which becomes localized to the posterior cortex) and PAR-3 (which goes to the anterior cortex). After the redistribution of these proteins has been achieved, the zygote divides so as to produce a two-cell embryo with an anterior cell and a posterior cell.

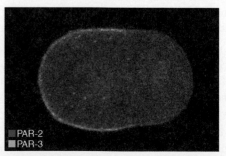

PAR-2
PAR-3

a. How do these findings help explain why the early development of *C. elegans* embryos displays mosaic, rather than regulative, determination?

b. Mutations in the *par-2* or *par-3* genes cause the arrest of development in early embryonic stages. Would you be more likely to find mutations in these genes in screens looking for maternal-effect genes or in screens for zygotic genes?

c. In zygotes produced by hermaphrodite mothers homozygous for loss-of-function *par-2* alleles, the PAR-3 protein is distributed uniformly around the cortex; the same is true of the PAR-2 protein in zygotes made by *par-3* mutant hermaphrodites. What does this information say about the establishment of early polarity in *C. elegans*?

23. At the end of two rounds of mitosis, the *C. elegans* embryo has four cells named ABa, ABp, P₂, and EMS (see the following figure). The ABa and ABp cells are originally developmentally equivalent, but they become different as a result of interactions between ABp and P₂ that involve two proteins called GLP-1 and APX-1, as shown in the figure. Both GLP-1 and APX-1 are membrane bound, with domains that lie outside of the cell. GLP-1 is expressed around the entire surface of ABa and ABp, whereas APX-1 is found at the membrane junction between ABp and P₂ as shown in the figure.

a. The mRNA for *glp-1* is found in all four cells, but the GLP-1 protein is found only in ABa and ABp. In light of your answer to Problem 22, provide an explanation for this observation.

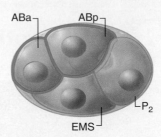

b. Based on the information in the figure, suggest a hypothesis to explain the localization of APX-1 in only one region of the membrane of the P₂ cell.

c. Assuming that the effect of these proteins on the fate of ABp is caused by a signal transduction pathway, which of the two proteins GLP-1 and APX-1 is likely to be a ligand, and which a receptor for this ligand?

d. Describe the effects on the fate of the ABp cell of the following: (i) laser ablation of the P₂ cell; (ii) a null mutation in the *apx-1* gene; (iii) a null mutation in the *glp-1* gene; (iv) a null mutation in a gene encoding a component of the signal transduction pathway initiated by binding of the ligand to its receptor. (Assume here that the mutations in parts *ii–iv* only affect the fate of the ABp cell and not other processes in the early nematode embryo.)

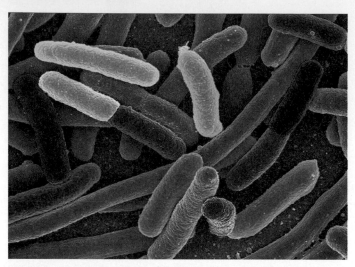

The metabolic engineering of prokaryotic bacteria (like the *E. coli* cells shown at *left*) holds great promise for the creation of advanced biofuels and new pharmaceuticals. For example, genetically engineered *E. coli* cells can be used to degrade lignocellulose from agricultural waste products to synthesize biofuels. In this chapter, we will learn about the genetics of prokaryotic organisms like *E. coli* and see how this fundamental knowledge can be used for both basic and applied research.

Gonorrhea, a sexually transmitted infection of the urogenital tract in men and women, is on the rise in many parts of the world. Caused by the bacterium *Neisseria gonorrhoeae*, the disease is rarely fatal, but in men it can spread from an initial site, usually the urethra, to the prostate gland and the epididymis, diminishing sperm count. In women, it can move from the cervix to the uterine lining and fallopian tubes, leading to sterility. Because infants passing through a gonorrhea-infected birth canal can contract severe eye infections, hospitals routinely treat the eyes of newborns with a few drops of silver nitrate solution or penicillin. Until the late 1970s, a few shots of penicillin were a certain cure for gonorrhea, but by 1995, more than 20 percent of *N. gonorrhoeae* bacteria isolated from patients worldwide were resistant to the drug.

Geneticists now know that the agent of this alarming increase in antibiotic resistance was the transfer of DNA from one bacterium to another. Penicillin-resistant *N. gonorrhoeae* bacteria first appeared in a patient receiving penicillin treatment for the disease. This patient was also fighting an infection caused by another species of bacteria, *Haemophilus influenzae*. Some of the patient's *H. influenzae* bacteria apparently carried a plasmid (a small, circular molecule of double-stranded DNA) that contained a gene encoding penicillinase, an enzyme that destroys penicillin. When the doubly infected patient mounted a specific immune response to *H. influenzae* that degraded these cells, the broken bacteria released their plasmids. Some of the freed circles of DNA entered *N. gonorrhoeae* cells, transforming them to penicillin-resistant bacteria.

The transformed gonorrhea bacteria then multiplied, and successive exposures to penicillin selected for the resistant bacteria. As a result, the patient transmitted penicillin-resistant *N. gonorrhoeae* to subsequent sexual partners. Thus, while penicillin treatment does not create the genes for resistance, it accelerates the spread of those genes. Today in North America, many *N. gonorrhoeae* are simultaneously resistant to penicillin and two other antibiotics—spectinomycin and tetracycline.

In this chapter, we focus first on the remarkable diversity of bacteria, on genetic analysis in bacteria, and on how genome analysis has vastly increased our knowledge of the bacterial world. We also examine the mechanisms by which bacteria transfer genes between cells of the same species, between cells of distantly related species, and between bacterial cells and bacterial viruses.

The second focus of this chapter is the genetics of two organelles, mitochondria and chloroplasts, which are believed to be derived from ancient prokaryotic cells. We describe their genomes, their inheritance from one generation to the next, and human conditions caused by mutations in mitochondrial genes.

One main theme can be found in our exploration of bacterial and organellar genetics: DNA and genes of a single species do not exist in complete isolation. Not only do DNA segments migrate within a genome (e.g., transposable elements), but they also are capable of migration between species, or of migrating from one genome (mitochondria or chloroplast) to another (nucleus).

Learning Objectives

1. Evaluate the utility of bacteria as model organisms.

2. Compare and contrast the makeup of a typical bacterial genome with the makeup of a typical eukaryotic genome.

3. Compare and contrast the molecular mechanisms used by bacteria to transfer genes horizontally.

4. Relate the endosymbiont theory to human genetics.

5. Define and give examples of non-Mendelian inheritance.

18.1 A General Overview of Bacteria

Bacteria are termed prokaryotes because they lack the membrane-bounded nucleus found in eukaryotes. The study of bacteria was of critical importance to the development of the field of genetics. From the 1940s to the 1970s (the era of classical bacterial genetics), virtually everything researchers learned about gene structure, gene expression, and gene regulation came from analyses of bacteria and the *bacteriophages* (bacterial viruses; often abbreviated as *phages*) that infect them. The advent of recombinant DNA technology in the 1970s and 1980s depended on an understanding of genes, chromosomes, and restriction enzymes in bacteria. Many recombinant DNA manipulations of genes from a variety of organisms still rely on bacteria for the development and propagation of genetically engineered molecules.

Bacteria exhibit immense diversity

Bacteria are crucial to the maintenance of Earth's environment. Various species release oxygen into the atmosphere; recycle carbon, nitrogen, and other elements; and digest human and other animal wastes as well as neutralize pesticides and other pollutants, which would otherwise eventually poison the air, soil, and water. In contrast, bacteria also cause hundreds of animal and plant diseases. Even so, harmful species make up a small fraction of all bacteria. Many species, in fact, produce vitamins and other materials essential to the health and survival of humans and other organisms. Two key features of bacteria are their astounding ability to proliferate and their enormous diversity. A human adult carries at least 100 g (roughly a quarter pound) of live bacteria, mainly in the intestines. An estimated 10^{14} bacteria make up those 100 g, a number many thousands of times as great as the number of people on Earth.

Bacterial size and characteristics

The smallest bacteria are about 200 nm (nanometres = billionths of a metre) in diameter. The largest are 500 μm (micrometres = millionths of a metre) in length, which makes them 10 billion times larger in volume and mass than the smallest bacterial cells. These large bacteria are visible without the aid of the microscope. Some bacteria live independently on land, others float freely in aquatic environments, and still others live as parasites or symbionts inside other life-forms.

Although bacteria come in a variety of shapes and sizes and are adapted to a range of habitats, all lack a defined nuclear membrane as well as membrane-bounded organelles, such as the mitochondria and chloroplasts found in eukaryotic cells. Bacterial chromosomes fold to form a dense **nucleoid body** that appears to exclude ribosomes, which function in the surrounding cytoplasm. In most species of bacteria, the membrane is supported by a cell wall composed of carbohydrate and peptide polymers. Some bacteria have, in addition to the cell wall, a thick, mucouslike coating called a *capsule* that helps them resist attack by the immune system. Many bacteria have flagella that propel them toward food or light.

Metabolic diversity

Bacteria have evolved to live in a wide variety of habitats. Some soil bacteria obtain the energy to fuel their metabolism from the chemical ammonia; others obtain their energy from sunlight through photosynthesis. Because of their metabolic diversity, bacteria play essential roles in many natural processes, including the decomposition of materials essential for nutrient cycling. The balance of microorganisms is key to the success of these ecological processes and maintain the environment.

In the cycling of nitrogen, for example, decomposing bacteria break down plant and animal matter rich in nitrogen and produce ammonia (NH_3). Nitrifying bacteria then use this ammonia as a source of energy and release nitrate (NO_3), which some plants can use as is; denitrifying bacteria convert the nitrate not used directly by plants to atmospheric nitrogen (N_2); and nitrogen-fixing bacteria, such as *Rhizobium*, that live in the roots of peas and other leguminous plants convert N_2 to ammonium (NH_4^+), which their host plants can use.

Recently, geneticists and molecular biologists have used microbes to isolate unusual enzymes that carry out natural and industrial processes. They then clone and manipulate the genes encoding these enzymes. One use of unusual microorganisms and their enzymes is the development of bioremediating bacteria that can, for example, break down the hydrocarbons found in oil. In 1989, when the tanker *Exxon Valdez* spilled millions of barrels of oil along the Alaskan coastline, cleanup crews used oil-digesting bacteria in an attempt to revive the environment.

Bacteria must be grown and studied in cultures

Researchers grow bacteria in liquid media (**Figure 18.1**) or on media solidified by agar in a plate, called a Petri dish (**Figure 18.2**). In a liquid medium, the cells of commonly studied species, such as *Escherichia coli* (*E. coli*), grow to a concentration of 10^9 cells per millilitre within a day. In agar-solidified medium, a single bacterium will multiply to a visible colony containing 10^7 or 10^8 cells in less than one day. The ability to grow large numbers of cells is one advantage that has made bacteria, especially *E. coli*, so attractive for genetic studies. Only a relatively small number of bacterial species can be grown in culture in the laboratory; a vast number of species exist only in their native environments.

Genetic studies of bacteria require techniques to count these large numbers of cells and to isolate individual cells of interest. Researchers can use a solid medium to calculate the number of cells in a liquid culture. They begin with sequential dilutions of cells in the liquid medium. They then spread a small sample of the diluted solutions on agar-medium plates and count the number of colonies that form. Although it is still difficult to work with a single bacterial cell, except in very specialized studies, the cells constituting a single colony contain the genetically identical descendants of the one bacterial cell that founded the colony.

Figure 18.2 Bacteria in the laboratory. Bacteria grow as colonies on solid nutrient agar in a Petri dish.

E. coli: A versatile model organism

The most studied and best understood species of bacteria is *E. coli*, a common inhabitant of the intestines of warm-blooded animals (**Figure 18.3**). *E. coli* cells can grow in the complete absence of oxygen—the condition found in the intestines—or in air. The *E. coli* strains studied in the laboratory are not pathogenic, but other strains of the species can cause a variety of intestinal diseases, most of them mild, a few life-threatening.

Because *E. coli* encodes all the enzymes it needs for amino acid and nucleotide biosynthesis, it is a *prototrophic* organism that can grow in *minimal media* containing a single carbon/energy source, such as glucose, and inorganic salts. In a minimal medium, *E. coli* cells divide every hour, doubling their numbers 24 times a day. In a richer, more complex medium containing several sugars and amino acids, *E. coli* cells divide every 20 minutes to produce 72 generations per day. Two days of logarithmic growth at this rate, if unchecked by any limiting factor, would generate a mass of bacteria equal to the mass of Earth.

The rapidity of bacterial multiplication makes it possible to grow an enormous number of cells in a relatively short time and, as a result, to obtain and examine very rare genetic events. For example, wild-type *E. coli* cells are normally sensitive to the antibiotic streptomycin. By spreading a billion wild-type bacteria on an agar-medium plate containing streptomycin, it is possible to isolate a few extremely rare streptomycin-resistant mutants that have arisen by chance among the 10^9 cells. It is not as easy to find and examine such rare events with nonmicrobial

Figure 18.1 Bacterial cultures. Bacteria grow as a suspension of cells in a liquid medium.

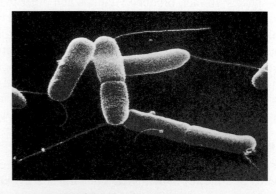

Figure 18.3 *Escherichia coli.* Scanning electron micrograph of *E. coli* (14 000×).

organisms; in multicellular animals, it is almost impossible. For this reason, *E. coli* makes an excellent model organism.

Finding mutations in bacterial genes

Most bacterial genomes carry one copy of each gene and are therefore effectively haploid. The relation between gene mutation and phenotypic variation is thus relatively straightforward; that is, in the absence of a second, wild-type allele for each gene, all mutations express their phenotype.

Bacteria are so small that the only practical way to examine them is in the colonies of cells they form on a Petri dish. Nevertheless, it is still possible to identify many different kinds of mutations, such as the following:

1. Mutations affecting *colony morphology*; that is, whether a colony is large or small, shiny or dull, round or irregular.
2. Mutations conferring *resistance* to bactericidal agents such as antibiotics or bacteriophages.
3. Mutations that create *auxotrophs* unable to grow and reproduce on minimal medium; auxotrophic mutations occur in genes encoding enzymes needed by the bacteria to synthesize relatively complex compounds such as amino acids or nucleotide components. Thus, unless the compound in question is added to the growth medium by the researcher, these bacteria are unable to proliferate.
4. Mutations affecting the *ability of cells to break down and use complicated chemicals* in the environment; for example, the *lacZ* gene in *E. coli* encodes the enzyme β-galactosidase needed to break down the sugar lactose into glucose and galactose. Wild-type cells can grow if lactose, rather than glucose, is the sole source of carbon in the medium, but *lacZ⁻* mutants cannot.
5. Mutations in *essential genes* whose protein products are required for growth; because a null mutation in an essential gene would prevent a colony from growing in any environment, bacteriologists must work with conditional lethal mutations such as temperature-sensitive (*ts*) mutations that allow growth at one temperature, but not at another.

Bacteriologists use different techniques to isolate rare mutations. With mutations conferring resistance to a particular agent, researchers can do a straightforward **selection**; that is, establish conditions in which only the desired mutant will grow. For example, if wild-type bacteria are streaked on a Petri dish containing the antibiotic streptomycin, the only colonies to appear will be streptomycin resistant (Str^r). It is also possible to select for prototrophic revertants of strains carrying auxotrophic mutations by simply plating cells on minimal medium agar, which does not contain the compounds auxotrophs require for growth.

Because the key characteristic of most of the other types of mutants just described is their inability to grow under particular conditions, it is not possible to select for

them directly. Instead, researchers must identify these mutations by a **genetic screen**: an examination of each colony in a population for its phenotype. They can, for example, use a toothpick to transfer cells from a colony growing on minimal medium supplemented with methionine to a Petri plate containing minimal medium without methionine. Failure of those cells to grow on the unsupplemented medium would indicate that the corresponding colony on the original plate is auxotrophic for methionine.

Spontaneous mutations in specific bacterial genes occur very rarely, in 1 in 10^6 to 1 in 10^8 cells, depending on the gene. Therefore, it would be virtually impossible to identify such rare mutations if the phenotype of a million to a hundred million colonies had to be checked through the individual transfer of each one with a toothpick. A number of techniques simplify the process. We describe four.

1. *Replica plating* allows the simultaneous transfer of thousands of colonies from one plate to another (see **Figure 8.5**).
2. *Treatments with mutagens* increase the frequency with which a mutation in a gene appears in the population (see **Figure 8.10**).
3. *Enrichment* increases the proportion of mutant cells in a population by exposing the population to agents that kill wild-type cells. Penicillin is one agent of enrichment; it acts by disrupting the formation of the cell wall in growing cells and thus it kills cells that are dividing, but not cells that are unable to divide. **Figure 18.4** shows how researchers use this property of penicillin to enrich the proportion of auxotrophic cells in a mixed population of auxotrophic and prototrophic cells.
4. *Testing for visible mutant phenotypes* on a Petri plate. In an important example, *E. coli* producing functional β-galactosidase (the product of the *lacZ* gene) cleaves the colourless artificial compound X-Gal, producing a blue product. Thus *lacZ⁺* colonies turn blue on medium containing X-Gal, while *lacZ⁻* colonies remain white (the usual colour of the colonies; review Figure 14.7).

Designation of bacterial alleles

Researchers designate the genes of bacteria by three lowercase, italicized letters that signify something about the function of the gene. For example, genes in which mutations result in the inability to synthesize the amino acid leucine are *leu* genes. In *E. coli*, there are four *leu* genes—*leuA*, *leuB*, *leuC*, and *leuD*—that correspond to the three enzymes (one constructed from two different polypeptides) needed for the synthesis of leucine from other compounds. A mutation in any one of the *leu* genes changes a bacterium from a prototroph to an auxotroph for leucine; that is, into a cell unable to synthesize the amino acid. Such a cell can grow only in media supplemented with leucine.

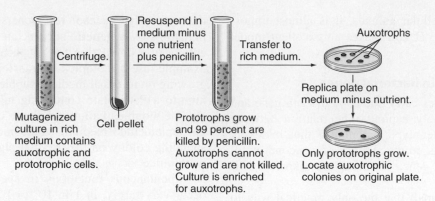

Figure 18.4 Penicillin enrichment for auxotrophic mutants. Penicillin selectively kills growing cells that are making new cell walls, but not bacteria whose growth is arrested. In the absence of nutrients, auxotrophs will not be killed by the penicillin. After enrichment, cells must be screened by replica plating to identify auxotrophs because penicillin does not kill 100 percent of the prototrophs.

Mutations in genes required for the breakdown of a sugar (e.g., the *lacZ* gene) produce cells unable to grow in medium containing only that sugar (lactose) as a source of carbon. Other types of mutations give rise to antibiotic resistance; *str^r* is a mutation producing streptomycin resistance. To designate the alleles of genes present in wild-type bacteria, researchers use a superscript "+": *leu$^+$, str$^+$, lacZ$^+$*. To designate mutant alleles, they use a superscript "−", as in *leuA$^−$* and *lacZ$^−$*, or a superscript description, as in *str^r*.

The phenotype of a bacterium that is wild type or mutant for a particular gene is indicated by the three letters that designate the gene, written, however, with an initial capital letter, no italics, and a superscript of minus, plus, or a one-letter abbreviation: Leu$^−$ (requires leucine for growth); Lac$^+$ (grows on lactose); Str^r (is resistant to streptomycin). A Leu$^−$ *E. coli* strain cannot multiply unless it grows in a medium containing leucine; a Lac$^+$ strain can grow if lactose replaces the usual glucose in the medium; a Str^r strain can grow in the presence of streptomycin.

- Bacteria exhibit tremendous metabolic diversity and play critical roles in Earth's nutrient cycles. Only a few species produce disease. Bacteria generally have a single, circular chromosome of double-stranded DNA and are therefore haploid organisms. A bacterial cell is surrounded by a cell wall, and many species also have a capsule external to the cell wall.

- A single cell of a bacterium such as *E. coli* can grow to a billion cells within a day in a very small culture tube or dish. Because of this rapid generation time, bacteria that can be cultured are ideal for genetic research. The vast numbers generated also allow selection for extremely rare mutants.

18.2 Bacterial Genomes

The essential component of a typical bacterial genome is the **bacterial chromosome**: a single molecule of double-helical DNA arranged in a circle (**Figure 18.5**). This chromosome is 4–5 Mb long in most of the commonly studied species. The circular chromosome of *E. coli*, if broken at one point and laid out in a line, would form a DNA molecule 2.4 nm wide and 1.6 mm long, almost a thousand times longer than the *E. coli* cell in which it is found (**Figure 18.6**). Inside the cell, the long, circular DNA molecule condenses by supercoiling and looping into a densely packed nucleoid body.

During the bacterial cell cycle, each bacterium replicates its circular chromosome and then divides by binary fission into two identical daughter cells, each with its own chromosome. While the majority of bacteria contain a single circular chromosome, there are exceptions. Genomic analyses have shown that some bacteria, such as *Vibrio cholerae* (the cause of the disease cholera), carry two different circular chromosomes that are both essential for viability. Other bacteria contain linear DNA molecules.

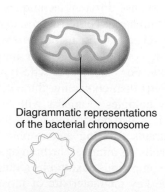

Diagrammatic representations of the bacterial chromosome

Figure 18.5 Chromosomal DNA. Chromosomal DNA is shown either as a double helix or as a single ring in this chapter.

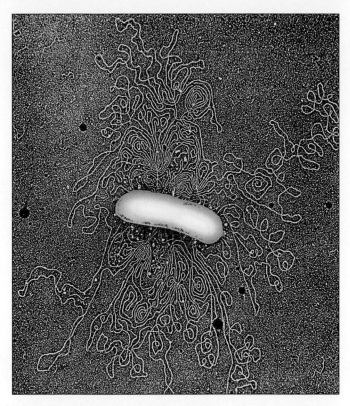

Figure 18.6 E. coli chromosomal DNA. An electron micrograph of an E. coli cell that has been lysed, allowing its chromosome to escape.

The *E. coli* genome has been completely sequenced

In 1997, molecular geneticists completely sequenced the 4.6-million-bp genome of the *E. coli* strain known as K12. From previous genetic work, they knew many of the genes within this genome. In addition, they could identify others because the polypeptides they encode have amino acid sequences similar to those of already sequenced and characterized proteins found in other bacteria (or even in eukaryotic species). Some of the sequences identified as genes, however, encode proteins with functions that have not yet been determined. Such presumed genes are known as **open reading frames**, or **ORFs**; they consist of long stretches of codons in the same reading frame uninterrupted by stop codons.

Close to 90 percent of *E. coli* DNA encodes proteins; on average, every kilobase of the chromosome contains one gene. This contrasts sharply with the human genome, in which less than 5 percent of the DNA encodes proteins and there is roughly one gene every 100 kb. One reason for this discrepancy is that *E. coli* genes have no introns. In addition, there is very little repeated DNA in bacteria, and intergenic regions tend to be very small.

The complete sequence of the *E. coli* genome revealed 4288 genes; surprisingly, the function of 40 percent of these genes remains a mystery at this time. Given the small genome size and the tools developed over the years for genetic analysis in *E. coli*, however, researchers can easily mutate the genes and examine the resulting cells for

phenotypic effects. So far, they have grouped genes whose function is known or has been deduced on the basis of sequence into broad functional classes. The 427 genes that are thought to have a transport function make up the largest class. Other classes include the genes for translation, amino acid biosynthesis, DNA replication, and recombination.

Another interesting feature of the *E. coli* genome is the existence, in eight different locations, of remnants of bacteriophage genomes. The presence of these sequences suggests an evolutionary history of bacteria that includes invasion by viruses on several occasions.

Bacterial genomes contain small insertion sequence (IS) elements

DNA sequence analysis of bacterial genomes also revealed the position of several small transposable elements called **insertion sequence (IS)** elements. These elements, which dot the chromosomes of many types of bacteria, are transposable elements that do not contain selectable markers (such as genes conferring antibiotic resistance). Researchers have identified several distinct elements ranging in length from 700–5000 bp; they named the elements IS1, IS2, IS3, and so forth, with the numbers designating the order of discovery. Like the ends of transposable elements in eukaryotic cells (see Chapter 9), the ends of IS elements are inverted repeats (**Figure 18.7a**); and each IS includes a gene encoding a transposase that initiates transposition by recognizing these mirror-image ends. Because insertion sequence elements can move to other sites on the bacterial chromosome when they transpose, their distribution varies in different strains of a single bacterial species. For example, one strain of *E. coli* may have 15 insertion sequence elements of five different kinds, while a second strain isolated from a different population may have 25 insertion sequence elements, lack one of the types found in the first

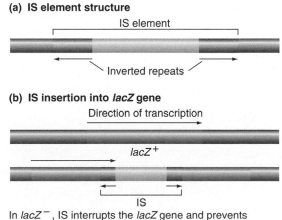

(a) IS element structure

IS element

Inverted repeats

(b) IS insertion into *lacZ* gene

Direction of transcription

lacZ⁺

IS

In *lacZ*⁻, IS interrupts the *lacZ* gene and prevents transcription of the entire gene.

Figure 18.7 IS elements. (a) An IS element showing the inverted repeats at each end. **(b)** Insertion of an IS into a gene. Here, insertion of an IS inactivates the *lacZ* gene because the IS contains a transcription termination signal.

strain, and have a different distribution of IS elements around the chromosome. Some bacterial species, such as *Bacillus subtilis*, carry no insertion sequences.

Insertion sequence elements were first identified in the 1970s as elements that caused inactivation of genes required for galactose metabolism (Gal⁻ mutants) in *E. coli*. When an IS transposes and lands within the coding region of a gene, it generally inactivates the gene (**Figure 18.7b**). We now know that many of the spontaneous mutations isolated in *E. coli* are the result of IS transposition into a gene. Researchers have exploited this ability to cause mutation by using a more complex type of transposable element in bacteria: a **Tn element**. In addition to carrying a gene for transposase, Tn elements contain genes conferring resistance to antibiotics or toxic metals such as mercury. One Tn element known as Tn10 consists of two IS10 elements flanking a gene encoding resistance to tetracycline (**Figure 18.8**). After the introduction of Tn10 into a cell, its transposition into, for example, the *lacZ* gene produces a *lacZ⁻* mutant that is phenotypically both Lac⁻ and Tet^r (resistant to tetracycline).

Because of these effects, the Tn element in the gene is an easily scored genetic marker for mapping experiments and for transferring the disrupted gene to another strain. In addition, because researchers know the sequence of Tn10, they can make a primer corresponding to the end of the Tn10 element and use this primer to begin DNA sequencing and discover the base sequence of the adjacent DNA (**Figure 18.9**). That is, they match the obtained sequence to that of the genome and thereby identify the gene that was mutated by the Tn10 insertion.

Genomic analyses in bacteria have created an information explosion

Although the first complete bacterial genome sequences were reported in the mid- to late-1990s, we now have complete genome data for hundreds of prokaryotic species and partial genomic sequence for thousands of species. The explosion of genome data provides intriguing information about pathogenesis, bacterial evolution, and unusual metabolic pathways and enzymes; it has also stimulated new avenues of inquiry and experimentation.

Microbial ecology and communities

Bacteria that live in extreme and unusual environments (e.g., in the deep sea, mining sites, and in whale carcasses) are often difficult to culture in the laboratory. As a result, we know little about what organisms are present in these challenging environments, what their numbers are, what they do, and how they interact. Rapid DNA sequencing, large-scale PCR amplification, and DNA arrays have opened the door to investigations of certain aspects of microbial ecology, including surveying the composition of microbial communities and the unusual metabolic capabilities of organisms in many settings.

In one recent study, researchers used PCR to amplify microbial DNA from communities in several niches (e.g., in soil and in whale carcasses) using primers for the bacterial 16S rRNA subunit (**Figure 18.10**). The 16S rRNA molecule is found in all bacteria, and yet it shows enough variation from species to species, and even in different strains of a species, to be used as an indicator of the number of types of bacteria present. Researchers amplified 1700 sequences from soil and found 847 distinct types of rRNA sequences. Whole-genome analysis of any one of these distinct bacteria is impossible because cloning and sequencing random fragments would be unlikely to yield overlapping pieces of genomic DNA from the same species. Although the complete genomic sequence of individual bacteria cannot be obtained in these complex communities teeming with different bacteria, much useful information can be learned from cloning random DNA fragments from these environments.

The analysis of genomic DNA from a community or habitat using the types of sampling described is called **metagenomics**. Researchers can learn a tremendous amount about the sheer numbers of distinct organisms present and discover unusual metabolic capabilities through these investigations. In one study, investigators examined ocean microbes by cloning and sequencing random DNA fragments from 200 litres of seawater (Figure 18.10). The sequence data

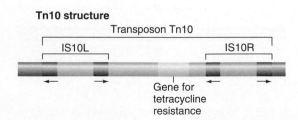

Tn10 structure

Figure 18.8 Transposable elements. The composite transposon Tn10, in which two slightly different IS10's (IS10L and IS10R) flank 7 kb of DNA, including a gene for tetracycline resistance. Because it is flanked by IS10 inverted repeats, Tn10 can be mobilized by the IS10 transposase.

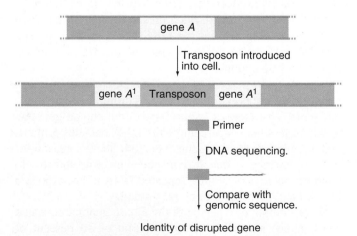

Figure 18.9 Identifying a mutated gene. To identify the gene that was mutated by insertion of a transposon, a primer corresponding to the DNA sequence in the transposon is used for sequencing through the gene *A¹* DNA.

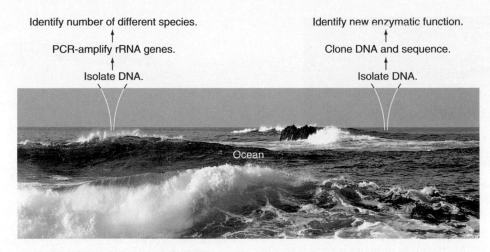

Figure 18.10 New analyses for assessing microbial diversity. Samples from the environment are analyzed to estimate either the number of different microbial species present or to identify new metabolic activities.

indicated a vast diversity of organisms and many new metabolic activities, including several new photosynthetic molecules. Another study examined microbes in indoor air and showed that these species were not simply outdoor microbes that had moved indoors, but bacteria that had adapted to the indoor environment by being able to withstand desiccation and oxidative damage.

While the "shotgun" survey approaches just described have been a focus of recent analyses, study of complete microbial genomes is still a very important tool. For organisms that have been cultured in the laboratory and are recognized as major contributors to global nutrient cycles, complete analysis of the genome is very valuable. Researchers have already sequenced the genomes of several marine cyanobacterial species critical for carbon fixation via photosynthesis. They can now analyze these genomes for their unique properties.

Comparative genome analysis

Much can be learned about the vast microbial world using *comparative genome analysis*—the examination and comparison of different species' genomes. Microbiologists can use comparative genome analysis to explore the similarities between species or among isolates (different strains) of a single species. When complete nucleotide sequences are available for two organisms, base-to-base comparisons of the genomes can be done with computers. However, complete DNA sequences are not necessary for all comparative analyses. An entire cloned genome for one organism can be laid out on a DNA array, which is then hybridized with isolated DNA from other species or strains to identify matches as well as unique sequences.

Researchers can also use the comparative genome approach to study many bacterial functions, including survival at high temperatures and pathogenesis. For example, genes present in thermophilic bacteria, but not in closely related nonthermophilic bacteria, are candidates for further study to determine biochemical functions necessary for survival at high temperatures. Similarly, pathogenic bacteria could be compared with their nonpathogenic relatives to identify candidate genes for pathogenicity.

Genome studies and public health

Many bacteriologists hope that genomic knowledge of pathogenic bacteria will lead to the identification of vaccine candidates. This is an ever more pressing concern as bacterial resistance to antibiotics increases. Genomic analysis could also aid in the discovery of new drug targets. For example, the identification of the genes and gene products essential for growth in a pathogenic species could allow rational drug design—a process by which pharmacologists synthesize compounds that target only those proteins not found in the host species.

Genomic technology and information provide epidemiologists with the ability to unambiguously identify specific bacterial strains. They can use this knowledge to trace the history of an infection. For example, during an outbreak of *Vibrio cholerae*, DNA array analysis using characteristic, specific DNA from several known disease-causing isolates could enable officials to determine whether the outbreak is caused by a previously identified strain or by a newly evolved one. If it turns out to be a new pathogen, investigators could identify key features of the new strain to understand how it evolved.

Plasmids carry additional DNA

Bacteria carry their essential genes—those necessary for growth and reproduction—in their large circular chromosome. In addition, some bacteria carry genes (not needed for growth and reproduction under normal conditions) in smaller circles of double-stranded DNA known as **plasmids** (**Figure 18.11**). Plasmids come in a range of sizes. The smallest are 1000-bp long; the largest are several megabases (Mb) in length. Bacteria usually harbour no more than one extremely large plasmid, but they can house several or even hundreds of copies of smaller DNA circles.

Although plasmids carry genes not normally needed by their bacterial hosts for growth and reproduction, these

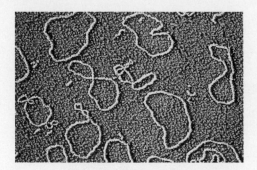

Figure 18.11 Plasmids. Electron micrograph showing circular plasmid DNA molecules.

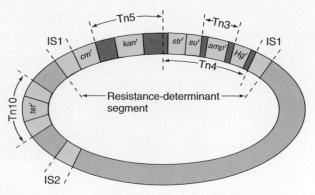

Figure 18.12 Resistance plasmids. Some plasmids contain multiple antibiotic resistance genes (shown in *yellow*: *cm^r* for chloramphenicol, *kan^r* for kanamycin, *str^r* for streptomycin, *su^r* for sulfonamide, *amp^r* for ampicillin, *Hg^r* for mercury, and *tet^r* for tetracycline). Transposons (IS and Tn elements, shown in *tan* and *red*, respectively) facilitate the movement of the antibiotic resistance genes onto the plasmid. Note that many antibiotic resistance genes are located between two IS1 elements, allowing them to transpose as a unit.

same genes may benefit the host cell under certain conditions. For example, the plasmids in many bacterial species carry genes that protect their hosts against toxic metals such as mercury. The plasmids of various soil-inhabiting *Pseudomonas* species encode proteins that allow the bacteria to metabolize chemicals such as toluene, naphthalene, or petroleum products. Since the 1980s, natural and genetically engineered plasmids of this type have become part of the tool kit for cleaning up oil spills and other contaminated sites. Plasmids thus help expand the capabilities of bacteria in nature, and they also provide a rich source of unusual and useful proteins for commercial purposes.

Many of the genes that contribute to pathogenicity reside in plasmids. For example, the toxins produced by *Shigella dysenteriae*, the causative agent of dysentery, are encoded by plasmids. Genes encoding resistance to antibiotics are also often located on plasmids. The plasmid-determined resistance to multiple drugs was first discovered in *Shigella* in the 1970s. Multiple antibiotic resistance is often due to composite IS/Tn elements on a plasmid (**Figure 18.12**). As described later, plasmids can be transferred from one bacterium to another, sometimes even across species. Plasmids thus have terrifying implications for medicine. If resistance plasmids are transferred to new strains of pathogenic bacteria, the new hosts acquire resistance to many antibiotics in a single step. We encountered an example of this potential in the chapter opening story on gonorrhea.

One important group of plasmids allows the bacterial cells that carry them to make contact with another bacterium

and transfer genes—both plasmid and bacterial—to the second cell. We describe this cell-to-cell mating, known as conjugation, in the next section on gene transfer.

- Close to 90 percent of *E. coli* DNA encodes proteins. Insertion sequence (IS) elements, found throughout genomes of many bacteria, are transposable elements that often disrupt gene activity. Complex Tn elements may contain IS elements; Tn elements can be introduced into bacterial genomes for use in mapping and sequencing.

- Metagenomics has revealed an enormous amount of genetic variation among unknown species of bacteria in particular habitats. Comparative genomic studies have yielded data on metabolic capabilities of bacteria and on potential targets for disease therapy. Plasmids are small circles of DNA that can carry genes for antibiotic resistance and other metabolic functions.

18.3 Gene Transfer in Bacteria

Gene transfer from one individual to another plays an important role in the evolution of new variants in nature. Vertical gene transfer, for example, occurs from one generation to the next and is particularly important in organisms utilizing sexual reproduction. By contrast, **lateral gene transfer** (or **horizontal gene transfer**) means that the traits involved are not transferred by inheritance from parents to offspring; rather, they are introduced from unrelated individuals or from different species. Many cases of horizontal gene transfer have come to light through recent molecular and DNA sequencing analyses.

Comparative genomic analysis of many different genes in various bacterial species has shown similarities of genes in species that were thought to be only distantly related. The simplest explanation is that significant transfer of DNA between bacteria has occurred throughout evolution. A close examination of the known mechanisms of DNA transfer helps illuminate this phenomenon. In addition, you will see that researchers can use gene transfer to map genes and to construct bacterial strains with which to test the function and regulation of specific genes.

Bacteria can transfer genes from one strain to another by three different mechanisms: **transformation**, **conjugation**, and **transduction** (**Figure 18.13**). In all three mechanisms, one cell—the **donor**—provides the genetic material for transfer, while a second cell—the **recipient**—receives the material. In transformation, DNA from a donor is added to the bacterial growth medium and is then taken up from that medium by the recipient. In conjugation, the donor carries a special type of plasmid that allows it to come in contact with the recipient and transfer DNA directly. In transduction, the donor DNA is packaged within the protein coat of a bacteriophage and transferred to the recipient when the phage particle infects it. The recipients of a gene transfer are known as **transformants**, **exconjugants**, or **transductants**, depending on the mechanism of DNA transfer that created them.

All bacterial gene transfer is asymmetrical in two ways. First, transfer goes in only one direction, from donor to recipient. Second, most recipients receive 3 percent or less of a donor's DNA; only some exconjugants contain a greater percentage of donor material. Thus, the amount of donor DNA entering the recipient is small relative to the size of the recipient's chromosome, and the recipient retains most of its own DNA. We now examine each type of gene transfer in detail.

In transformation, the recipient takes up DNA that alters its genotype

A few species of bacteria spontaneously take up DNA fragments from their surroundings in a process known as **natural transformation**. The large majority of bacterial species, however, can take up DNA in this way only after laboratory procedures make their cell walls and membranes permeable to DNA in a process known as **artificial transformation**.

Natural transformation

Researchers have studied several species of bacteria that undergo natural transformation, including *S. pneumoniae*, the pathogen in which transformation was discovered by Frederick Griffith (see Chapter 5) and that causes pneumonia in humans; *B. subtilis*, a harmless soil bacterium; *H. influenzae*, a pathogen causing various diseases in humans; and *N. gonorrhoeae*, the microbial agent of gonorrhea.

In one study of natural transformation, investigators isolated *B. subtilis* bacteria with two mutations—*trpC2* and *hisB2*—that made them Trp⁻, His⁻ double auxotrophs. These double auxotrophs served as the recipients in the study; Trp⁺, His⁺ wild-type cells were the donors (**Figure 18.14a**). The experimenters extracted and purified donor DNA and grew the *trpC⁻ hisB⁻* recipients in a suitable medium until the cells became **competent**; that is, able to take up DNA from the medium.

Different bacterial species require different regimens to achieve competence. For *B. subtilis*, competence occurs only in nearly starving cells at very specific times in the growth of the culture. Investigators can starve the cells by growing them in a glucose-salts medium containing a limited amount of tryptophan and histidine. As growth of the culture slows toward the end of the logarithmic phase

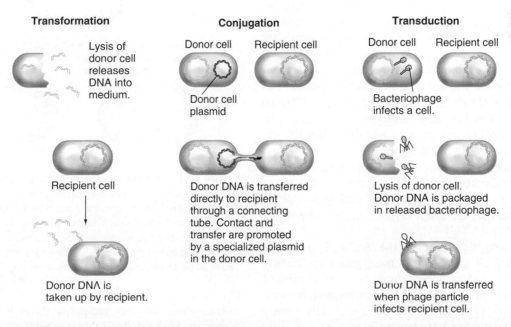

Figure 18.13 Gene transfer in bacteria: An overview. In this figure, and throughout this chapter, the donor's chromosome is *blue*, and the recipient's chromosome is *orange*. In transformation, fragments of donor DNA released into the medium enter the recipient cell. In conjugation, a specialized plasmid (shown in *red*) in the donor cell promotes contact with the recipient and initiates the transfer of DNA. In transduction, DNA from the donor cell is packaged into bacteriophage particles that can infect a recipient cell, transferring the donor DNA into the recipient.

of growth, a fraction of the bacteria—1 to 5 percent in *B. subtilis*—become competent and will take up DNA added to the medium.

When a recipient takes up DNA, only one strand of a fragment of donor DNA enters the cell, while the other strand is degraded (**Figure 18.14b**). The entering strand recombines with the recipient chromosome, producing a transformant when the recipient cell divides.

To observe and count Trp⁺ transformants, researchers decanted the liquid containing newly transformed recipient cells onto Petri dishes containing a simple glucose-salts solid medium lacking tryptophan and containing histidine. Recipient cells that did not take up donor DNA are unable to grow on this medium because it lacks tryptophan, but the Trp^+ transformants can grow and be counted. To select for His^+ transformants, researchers poured the transformation mixture on glucose-salts solid medium lacking histidine and containing tryptophan. In this study, the numbers of Trp^+ and His^+ transformants were equal. In conditions where *B. subtilis* bacteria become highly competent, 10^9 cells will produce approximately 100 000 Trp^+ transformants and 100 000 His^+ transformants.

To discover whether any of the Trp^+ transformants were also His^+, the researchers used sterile toothpicks to transfer colonies of Trp^+ transformants to a glucose-salts solid medium containing neither tryptophan nor histidine. Forty of every 100 Trp^+ transferred colonies grew on this minimal medium, indicating that they were also His^+. Similarly, tests of the His^+ transformants showed that roughly 40 percent are also Trp^+. Thus, in 40 percent of the analyzed colonies, the $trpC^+$ and $hisB^+$ genes had been co-transformed. **Co-transformation** is the simultaneous transformation of two or more genes.

Because donor DNA replaces only a small percentage of the recipient's chromosome during transformation, it might seem surprising that the two *B. subtilis* genes are co-transformed with such high frequency. The explanation is that the *trpC* and *hisB* genes lie very close together on the chromosome and are thus genetically linked. The entire *B. subtilis* chromosome is approximately 4700 kb long. Only genes in the same chromosomal vicinity can be co-transformed; the closer together the genes lie, the more frequently they will be co-transformed. Therefore, although the donor chromosome is fragmented into small pieces of about 20 kb during its extraction for the transformation process, the wild-type $trpC^+$ and $hisB^+$ alleles are so close that they are often together in the same donor DNA molecule.

Sequence analysis shows that the *trpC* and *hisB* genes are only about 7 kb apart. By contrast, genes sufficiently far apart that they cannot appear together on a fragment of donor DNA will almost never be co-transformed, because transformation is so inefficient that recipient cells usually take up only a single DNA fragment.

Transformation usually incorporates a single strand of a linear donor DNA fragment into the bacterial

(a) Donor and recipient genomes

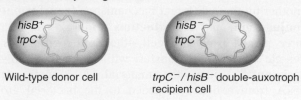

Wild-type donor cell $trpC^-$ / $hisB^-$ double-auxotroph recipient cell

(b) Mechanism of natural transformation

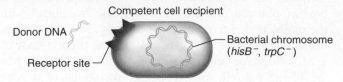

Donor DNA binds to recipient cell at receptor site.

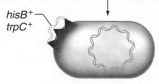

One donor strand is degraded. The admitted donor strand pairs with homologous region of bacterial chromosome. The replaced strand is degraded.

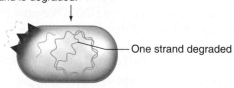

Donor strand is integrated into bacterial chromosome.

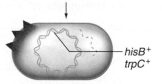

After cell replication, one cell is identical to original recipient; the other carries the mutant genes.

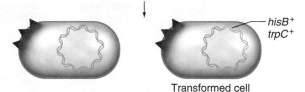

Transformed cell

Figure 18.14 Natural transformation in *B. subtilis*. **(a)** A wild-type donor and a $hisB^-$ $trpC^-$ double-auxotroph recipient. Selection for His^+ and/or Trp^+ phenotypes identifies transformants. **(b)** Mechanism of natural transformation in *B. subtilis*. One strand of a fragment of donor DNA enters the recipient, while the other strand is degraded. The entering strand recombines with the recipient chromosome, producing a transformant when the recipient cell divides.

chromosome of the recipient through recombination. However, if the donor DNA includes plasmids, recipient cells may take up an entire plasmid and acquire the characteristics conferred by the plasmid genes. Bacteriologists

suspect that penicillin-resistant *N. gonorrhoeae*, described in the introduction to this chapter, originated through transformation by plasmids. The donors of the plasmids were *H. influenzae* cells disrupted by the immune defences of a doubly infected patient. The plasmids carried the gene for penicillinase; and thus the recipient *N. gonorrhoeae* bacteria transformed by the plasmids acquired resistance to penicillin.

Artificial transformation

Although the study described above was a laboratory manipulation of natural transformation, researchers have devised many methods to transform bacteria that do not undergo natural transformation. The existence of these techniques was critical for the development of the gene-cloning technology described in Chapter 14. All the methods include treatments that damage the cell walls and membranes of recipient bacteria so that donor DNA can diffuse into the cells. With *E. coli*, the most common treatment consists of suspending the cells in a high concentration of calcium at cold temperature. Under these conditions, the cells become permeable to single- and double-stranded DNA.

Another technique of artificial transformation is *electroporation*, in which researchers mix a suspension of recipient bacteria with donor DNA and then subject the mixture to a very brief high-voltage shock. The shock most likely causes holes to form in the cell membrane. With the proper shocking conditions, recipient cells take up the donor DNA very efficiently. Transformation by electroporation works with most bacteria.

In conjugation, a donor transfers DNA directly to a recipient

In the late 1940s, Joshua Lederberg and Edward Tatum analyzed two *E. coli* strains that were each multiple auxotrophs and made the striking discovery that genes seemed to transfer from one type of *E. coli* cell to the other (**Figure 18.15**). Neither strain could grow on a minimal glucose-salts medium. Strain A required supplementation with methionine and biotin; strain B required supplementation with threonine, leucine, and thiamine (vitamin B_1). Lederberg and Tatum grew the two strains together on supplemented medium. When they then transferred a mixture of the two strains to minimal medium, about 1 in every 10^7 transferred cells proliferated to a visible colony. What were these colonies, and how they did they arise?

More than a decade of further experiments confirmed that Lederberg and Tatum had observed what became known as bacterial conjugation: a one-way DNA transfer from donor to recipient initiated by **conjugative plasmids** in donor strains. Many different plasmids can initiate conjugation because they carry genes that allow them to transfer themselves (and sometimes some of the donor's chromosome) to the recipient.

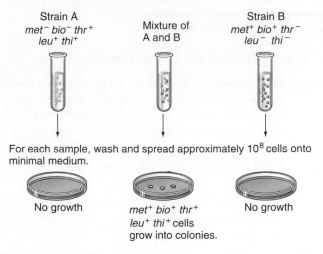

For each sample, wash and spread approximately 10^8 cells onto minimal medium.

No growth

$met^+ bio^+ thr^+$ $leu^+ thi^+$ cells grow into colonies.

No growth

Figure 18.15 Conjugation. Neither of two multiple auxotrophic strains analyzed by Lederberg and Tatum formed colonies on minimal medium. When cells of the two strains were mixed, gene transfer produced some prototrophic cells that did indeed form colonies.

The F plasmid and conjugation

Figure 18.16 illustrates the type of bacterial conjugation initiated by the first conjugative plasmid to be discovered—the **F plasmid** of *E. coli*. Cells carrying an F plasmid are called F^+ cells; cells without the plasmid are F^-. The F plasmid carries many genes required for the transfer of DNA, including genes for formation of an appendage, known as a pilus, by which a donor cell contacts a recipient cell, and a gene encoding an endonuclease that nicks the F plasmid's DNA at a specific site (the origin of transfer).

Once a donor has contacted a recipient cell (lacking the F plasmid) via the pilus, retraction of the pilus pulls the donor and recipient close together. The F plasmid DNA is then nicked, and a single strand moves across a bridge between the two cells. Movement of the F plasmid DNA into the recipient cell is accompanied by synthesis in the donor of another copy of the DNA strand that is leaving. When the donor DNA enters the recipient cell, it forms a circle again and the recipient synthesizes the complementary DNA strand. In this $F^+ \times F^-$ mating, the recipient becomes F^+, and the donor remains F^+. By initiating and carrying out conjugation, the F plasmid acts in bacterial populations the way an agent of sexually transmitted disease acts in human populations. When introduced via a few donor bacteria into a large culture of cells that do not carry the plasmid, the F plasmid soon spreads throughout the entire culture, and all the cells become F^+.

Conjugational transfer of chromosomal genes

The F plasmid contains three different IS elements: one copy of IS2, two copies of IS3, and one copy of the particularly long IS1000. These IS sequences on the F plasmid are identical to copies of the same IS elements found at various

The F Plasmid and Conjugation

a. **The F plasmid contains genes for synthesizing connections between donor and recipient cells.** The F plasmid is a 100-kb-long circle of double-stranded DNA. Host cells that carry it generally have one copy of the plasmid. By analogy with sexual reproduction, researchers think of F$^+$ cells as *male bacteria* because the cells can transfer genes to other bacteria. About 35 percent of F-plasmid DNA consists of genes that control the transfer of the plasmids. Most of these genes encode polypeptides involved in the construction of a structure called the F pilus (plural, pili): a stiff, thin strand of protein that protrudes from the bacterial cell. Other regions of the plasmid carry IS's and genes for proteins involved in DNA replication.

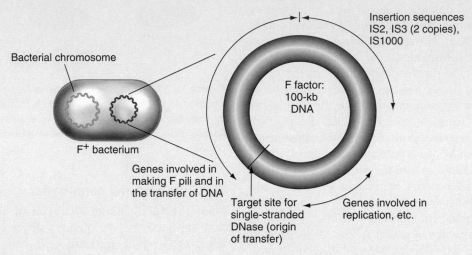

Bacterial chromosome

F$^+$ bacterium

Insertion sequences
IS2, IS3 (2 copies),
IS1000

F factor:
100-kb
DNA

Genes involved in
making F pili and in
the transfer of DNA

Target site for
single-stranded
DNase (origin
of transfer)

Genes involved in
replication, etc.

b. **The process of conjugation.**

1. **The pilus.** An average pilus is 1 μm in length, which is almost as long as the average *E. coli* cell. The distal tip of the pilus consists of a protein that binds specifically to the cell walls of F$^-$ *E. coli* not carrying the F factor.

2. **Attachment to F$^-$ cells (female bacteria).** Because they lack F factors, F$^-$ cells cannot make pili. The pilus of an F$^+$ cell, on contact with an F$^-$ cell, retracts into the F$^+$ cell, drawing the F$^-$ cell closer. A narrow passageway forms through the now adjacent F$^+$ and F$^-$ cell membranes.

3. **Gene transfer: A single strand of DNA travels from the male to the female cell.** Completion of the cell-to-cell corridor signals an endonuclease to cut one strand of the F-plasmid DNA at a specific site (*the origin of transfer*). The F$^+$ cell extrudes the cut strand through the passageway into the F$^-$ cell. As it receives the single strand of F-plasmid DNA, the F$^-$ cell synthesizes a complementary strand. The formerly F$^-$ cell contains a double-stranded F plasmid and is now an F$^+$ cell.

4. **In the original F$^+$ cell, newly synthesized DNA replaces the single strand transferred to the previously F$^-$ cell.** When the two bacteria separate at the completion of DNA transfer and synthesis, they are both F$^+$.

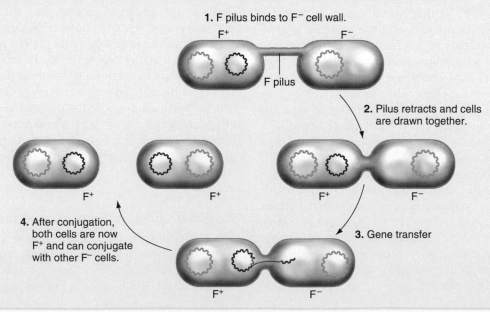

1. F pilus binds to F$^-$ cell wall.

F$^+$ F$^-$

F pilus

2. Pilus retracts and cells
are drawn together.

F$^+$ F$^+$ F$^+$ F$^-$

4. After conjugation,
both cells are now
F$^+$ and can conjugate
with other F$^-$ cells.

3. Gene transfer

F$^+$ F$^-$

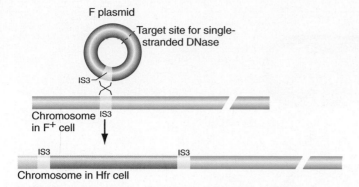

Figure 18.17 **Formation of an Hfr chromosome.** In this figure, the *filled bar* represents both strands of DNA. Recombination between an IS on the F plasmid and the same kind of IS on the bacterial chromosome creates an Hfr chromosome.

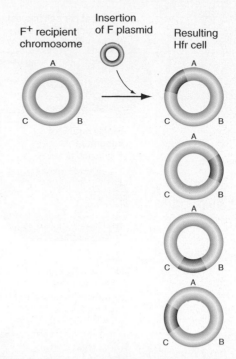

Figure 18.18 **Different Hfr chromosomes.** Recombination can occur between any IS on the F plasmid and any corresponding IS on the bacterial chromosome to create many different Hfr strains.

positions along the bacterial chromosome. In roughly 1 of every 10^5 (100 000) F$^+$ cells, homologous recombination (i.e., a crossover) between an IS on the plasmid and the same IS on the chromosome integrates the entire F plasmid into the *E. coli* chromosome (**Figure 18.17**). Cells whose chromosomes carry an integrated plasmid are called **Hfr** bacteria, because, as we will see, they produce a <u>h</u>igh <u>f</u>requency of <u>r</u>ecombinants for chromosomal genes in mating experiments with F$^-$ strains.

Because the recombination event that results in the F plasmid's insertion into the bacterial chromosome can occur between any of the IS elements on the F plasmid and any of the corresponding IS elements in the bacterial chromosome, geneticists can isolate 20–30 different strains of Hfr cells (**Figure 18.18**). A plasmid that can integrate into the genome is called an **episome**. Various Hfr strains are distinguished by the location and orientation (clockwise or counterclockwise) of the episome with respect to the bacterial chromosome.

During bacterial reproduction, the integrated plasmid of an Hfr cell replicates with the rest of the bacterial chromosome. As a result, the chromosomes in daughter cells produced by cell division contain an intact F plasmid at exactly the same location that the plasmid originally integrated into the chromosome of the parental cell. All progeny of an Hfr cell are thus identical, with the F plasmid inserted into the same chromosomal location and in the same orientation. The integrated F plasmid still has the capacity to initiate DNA transfer via conjugation, but now that it is part of a bacterial chromosome, it can promote the transfer of some or all of that chromosome as well (**Figure 18.19**).

The transfer of DNA from an Hfr cell mated to an F$^-$ cell starts with a single-strand nick in the middle of the integrated F plasmid at the origin of transfer. Very often, the mating process terminates before the entire chromosome is transferred. Once the donor DNA has been transferred to the recipient, recombination occurs between donor DNA and the chromosome in the recipient. Hfr crosses were used for creating genetic maps of the order

of genes by artificially interrupting mating (**Figure 18.20**). Genes closer to the origin of transfer are more likely to get transferred and recombined into the chromosome (**Figures 18.21a** and **b**).

In transduction, a phage transfers DNA from a donor to a recipient

Bacteriophages that infect, multiply in, and kill various species of bacteria are widely distributed in nature. Most bacteria are susceptible to one or more such viruses. During infection, a virus particle may incorporate a piece of the bacterial chromosome and introduce this piece of bacterial DNA into other host cells during subsequent rounds of infection. The process by which viral particles transfer bacterial DNA from one host cell to another is known as *transduction*.

The lytic cycle of phage multiplication

When a bacteriophage injects its DNA into a bacterial cell, the phage DNA takes over the cell's protein synthesis and DNA replication machinery, forcing it to express the phage genes, produce phage proteins, and replicate the phage DNA. The newly produced phage proteins and DNA assemble into phage particles, after which the infected cell bursts, or lyses, releasing 100–200 new viral particles ready to infect other cells. The cycle resulting in cell lysis and release of progeny phage is called the **lytic cycle** of phage multiplication. The population of phage particles released from the host bacteria at the end of the lytic cycle is known as a **lysate**.

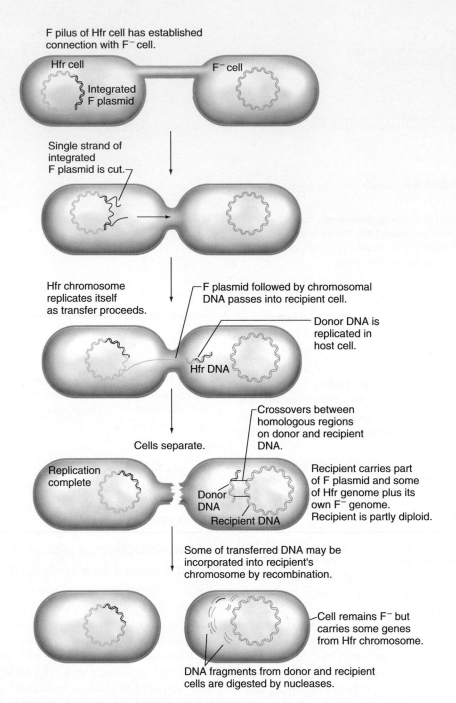

F pilus of Hfr cell has established connection with F⁻ cell.

Hfr cell

F⁻ cell

Integrated F plasmid

Single strand of integrated F plasmid is cut.

Hfr chromosome replicates itself as transfer proceeds.

F plasmid followed by chromosomal DNA passes into recipient cell.

Donor DNA is replicated in host cell.

Hfr DNA

Crossovers between homologous regions on donor and recipient DNA.

Cells separate.

Replication complete

Donor DNA

Recipient DNA

Recipient carries part of F plasmid and some of Hfr genome plus its own F⁻ genome. Recipient is partly diploid.

Some of transferred DNA may be incorporated into recipient's chromosome by recombination.

Cell remains F⁻ but carries some genes from Hfr chromosome.

DNA fragments from donor and recipient cells are digested by nucleases.

Figure 18.19 Gene transfer between Hfr donors and F⁻ recipients. In an Hfr × F⁻ mating, single-stranded DNA is transferred into the recipient, starting with the origin of transfer on the integrated F plasmid. Within the recipient cell, this single-stranded DNA is copied into double-stranded DNA. If mating is interrupted, the recipient cell will contain a double-stranded linear fragment of DNA plus its own chromosome. Genes from the donor are retained in the exconjugant (separated cells) only if they recombine into the recipient's chromosome.

Generalized transduction

Many kinds of bacteriophages encode enzymes that destroy the chromosomes of the host cells. Digestion of the bacterial chromosome by these enzymes sometimes generates fragments of bacterial DNA about the same length as the phage genome, and these phage-length bacterial DNA fragments occasionally get incorporated into phage particles in place of the phage DNA (**Figure 18.22**).

After lysis of the host cell, the phage particles can attach to and inject the DNA they carry into other bacterial cells, thereby transferring genes from the first bacterial strain (the donor) to a second strain (the recipient). Recombination between the injected DNA and the chromosome of the new host completes the transfer. This process, which can result in the transfer of any bacterial gene between related strains of bacteria, is known as **generalized transduction**.

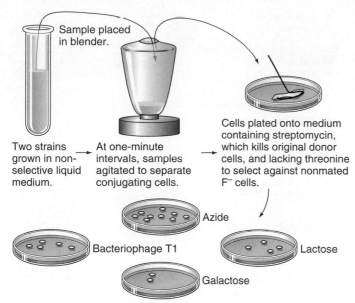

Two strains grown in non-selective liquid medium.

Sample placed in blender.

At one-minute intervals, samples agitated to separate conjugating cells.

Cells plated onto medium containing streptomycin, which kills original donor cells, and lacking threonine to select against nonmated F⁻ cells.

Bacteriophage T1

Azide

Lactose

Galactose

Replica plating transfers each colony to media that select for four donor markers other than streptomycin.

Figure 18.20 Interrupted-mating experiments. Hfr and F⁻ cells were mixed to initiate mating. Samples were agitated at one-minute intervals in a kitchen blender to disrupt gene transfer. Cells were plated onto a medium that contained streptomycin (to kill the Hfr donor cells) and that lacked threonine (to prevent growth of F⁻ cells that had not mated). The phenotypes of the exconjugants for other markers were established by replica plating.

(a) Time of gene transfer

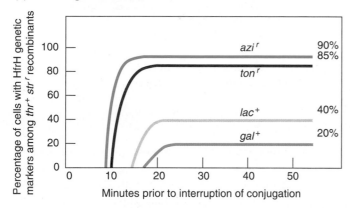

Percentage of cells with HfrH genetic markers among thr^+ str^r recombinants

azi^r — 90%
ton^r — 85%
lac^+ — 40%
gal^+ — 20%

Minutes prior to interruption of conjugation

(b) Map based on mating results

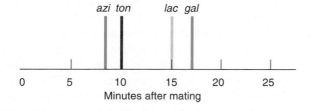

azi ton   lac gal

Minutes after mating

Figure 18.21 Mapping genes. (a) Results of the interrupted-mating experiment. **(b)** Gene order established from the data with positions determined by the time a donor gene first enters the recipient.

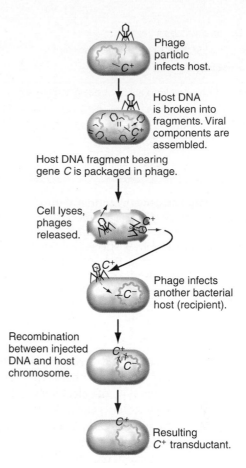

Phage particle infects host.

Host DNA is broken into fragments. Viral components are assembled.

Host DNA fragment bearing gene C is packaged in phage.

Cell lyses, phages released.

Phage infects another bacterial host (recipient).

Recombination between injected DNA and host chromosome.

Resulting C^+ transductant.

Figure 18.22 Generalized transduction. The incorporation of random fragments of bacterial DNA from a donor into bacteriophage particles yields generalized transducing phages. When these phage particles infect a recipient, donor DNA is injected into the recipient's cell. Recombination of donor DNA fragments with the recipient cell chromosome yields transductants.

Mapping genes by generalized transduction

As with co-transformation, two genes close together on the bacterial chromosome may be co-transduced. The frequency of co-transduction depends directly on the distance between the two genes: The closer they are, the more likely they are to appear on the same short DNA fragment and be packaged into the same transducing phage. Two genes that are farther apart than the length of DNA that can be packaged into a single phage particle can never be co-transduced. For bacteriophage P1, a phage often used for generalized transduction experiments with *E. coli*, the maximum separation allowing co-transduction is about 90 kb of DNA, which corresponds to about 2 percent of the bacterial chromosome.

Consider, for example, the three genes—*thyA*, *lysA*, and *cysC*—that all map to a similar region of the *E. coli* chromosome. Where do they lie in relation to one another? You can find out by using a P1 generalized transducing lysate from a wild-type strain to infect a

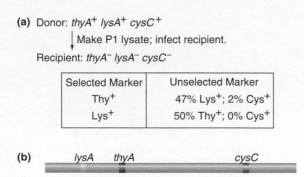

(a) Donor: *thyA⁺ lysA⁺ cysC⁺*

Make P1 lysate; infect recipient.

Recipient: *thyA⁻ lysA⁻ cysC⁻*

Selected Marker	Unselected Marker
Thy⁺	47% Lys⁺; 2% Cys⁺
Lys⁺	50% Thy⁺; 0% Cys⁺

(b)

Figure 18.23 Mapping genes by co-transduction frequencies. (a) A P1 lysate of a *thyA⁺ lysA⁺ cysC⁺* donor is used to infect a *thyA⁻ lysA⁻ cysC⁻* recipient. Either Thy⁺ or Lys⁺ cells are selected and then tested for the unselected markers. **(b)** Genetic map based on the data in part (a). The *thyA* and *cysC* genes were co-transduced at a low frequency, so they must be closer together than *lysA* and *cysC*, which were never co-transduced.

thyA⁻, *lysA⁻*, *cysC⁻* strain and then selecting the transductants for either Thy⁺ or Lys⁺ phenotypes. After replica plating, you test each type of selected transductant for alleles of the two nonselected genes. As the phenotypic data in **Figure 18.23a** indicate, *thyA* and *lysA* are close to each other but far from *cysC*; *lysA* and *cysC* are so far apart that they never appear in the same transducing phage particle; *thyA* and *cysC* are only rarely co-transduced. Thus, the order of the three genes must be *lysA*, *thyA*, *cysC* (**Figure 18.23b**).

Temperate phages

The types of bacteriophages we have discussed so far are **virulent**: After infecting a host, they always enter the lytic cycle, multiplying rapidly and killing the cell. Other types of bacteriophages are **temperate**: Although they can enter the lytic cycle, they can also enter an alternative **lysogenic cycle**, during which their DNA integrates into the host genome and multiplies along with it, doing little or no harm to the host (**Figure 18.24**). The integrated copy of the temperate bacteriophage is called a **prophage**. The integrated prophage replicates along with the chromosome, but does not produce the proteins that lead to production of more virus particles. The choice of lifestyle—lytic or lysogenic—occurs when a temperate phage injects its DNA into a bacterial cell and depends on many factors, including environmental conditions. Normally when temperate phages inject their DNA into host cells, some of the cells undergo a lytic cycle, while others undergo a lysogenic cycle. **Figure 18.25** shows one temperate phage commonly used in research, bacteriophage lambda (λ).

Under certain conditions, it is possible to induce an integrated viral genome to excise from the chromosome, undergo replication, and form new viruses (**Figure 18.26**).

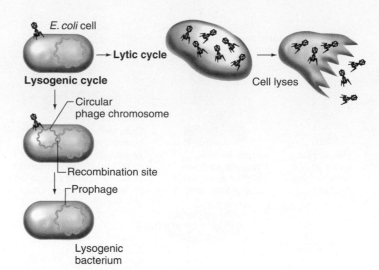

Figure 18.24 Lytic and lysogenic modes of reproduction. Cells infected with temperate bacteriophages (whose chromosomes are shown in *green*) enter either the lytic or lysogenic cycles. In the lytic cycle, phages reproduce by forming new bacteriophage particles that lyse the host cell and can infect new hosts. In the lysogenic cycle, the phage chromosome becomes a prophage incorporated into the host chromosome.

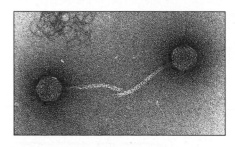

Figure 18.25 Bacteriophage lambda. Electron micrograph of a temperate phage, bacteriophage lambda (λ).

In a small percentage of excision events, some of the bacterial genes adjacent to the site where the bacteriophage integrated may be cut out along with the viral genome and be packaged as part of that genome. Viruses produced by the faulty excision of a lysogenic virus from the bacterial genome are called **specialized transducing phages** (**Figure 18.26b**). During the production of such a phage, bacterial genes may become passengers along with the viral DNA. When the specialized transducing phage then infects other cells, these few bacterial genes may be transferred into the infected cells. The phage-mediated transfer of a few bacterial genes is known as **specialized transduction**. Temperate phages are thought to be a significant vehicle for the lateral transfer of genes from one bacterial strain to another or even from one species to another.

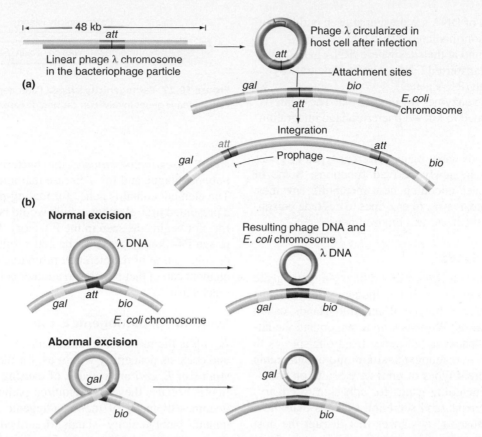

Figure 18.26 Lysogeny and excision. **(a)** Integration of the phage DNA initiates the lysogenic cycle. Recombination between *att* sites on the phage and bacterial chromosomes allows integration of the prophage. **(b)** Errors in prophage excision produce specialized transducing phages. Normal excision produces circles containing only lambda DNA. If excision is inaccurate, adjacent bacterial genes are included in the circles that form and in the resulting bacteriophages. Illegitimate recombination between the prophage and bacterial chromosome forms a circle that lacks some phage genes but has acquired the adjacent *gal* genes.

Comparison of generalized and specialized transduction

Phage particles that act as agents of generalized transduction differ in critical ways from particles that carry out specialized transduction.

1. Generalized transducing phages pick up donor bacterial DNA during the lytic cycle, at the point when DNA is packaged inside a phage protein coat; specialized transducing phages pick up the donor bacterial DNA during the transition from the lysogenic to the lytic cycle.

2. Generalized transducing phages can transfer any bacterial gene or set of genes contained in the right size DNA fragment into the bacterial chromosome; specialized transducing phages can transfer just those genes near the site where the bacteriophage inserted into the bacterial genome.

Lateral gene transfer has significant evolutionary implications

The mechanisms of gene transfer just described were characterized in bacteria that were easy to study genetically.

Understanding these transfer mechanisms facilitated construction of strains needed for genetic dissection of metabolic processes in the cell. In recent years, researchers have uncovered the prevalence of these transfer mechanisms in many bacterial species. The widespread evidence of lateral gene transfer indicates that these mechanisms are very important for rapid adaptation of bacteria to a changing environment and the development of pathogenic strains of bacteria.

Putative gene transfers recognized by genomic analysis may have occurred by any of the mechanisms described. An example of phage-mediated transfer of genes is the presence of the diptheria toxin of *Corynebacterium diphtheriae* on a lysogenic bacteriophage. Toxins in other strains are found on plasmids that could easily be transferred by transformation or by conjugation.

Large segments of DNA (10–200 kb in size), called **genomic islands**, show properties that suggest that they originated from the transfer of foreign DNA into a bacterial cell. Some hallmarks of these genomic islands follow:

- The G + C content of the DNA in the island is different from the G + C content of the rest of the bacterial chromosome.

- Direct repeats of DNA are present at each end (similar to transposon-mediated events).
- Islands are found at the sites where tRNA genes are located (transferred DNA seems to integrate by homology with tRNA genes).
- Islands encode enzymes for integration. These integrases are related to known bacteriophage integration enzymes.

Genomic islands carry many different types of genes that are involved in newly derived functions. Some of these included genes encoding new metabolic enzymes, antibiotic resistance, toxins, or enzymes to degrade poisonous substances in the bacteria's environment.

Pathogenicity islands

Among the significant genomic findings in pathogenic bacteria is the observation that pathogenic determinants are often clustered in a subtype of genomic islands, called **pathogenicity islands**. With such an arrangement, the lateral transfer of a "package" of genes from one species to another can turn a nonpathogenic strain into a pathogenic strain. Many different types of genes are found on pathogenicity islands, including genes for adhesion to eukaryotic (host) cells, toxins, and secretion systems that allow the bacteria to transport substances that disrupt the host cells (**Figure 18.27**).

Most pathogens contain pathogenicity islands. Islands encoding pathogenicity determinants are found in *Vibrio cholerae* strains that cause the disease cholera. Pathogenicity islands in these strains include genes for an enterotoxin that interferes with host-cell function, for invasion proteins that allow the bacteria to make its way through mucus of the intestinal tract, for phage-related integrases, for pilus formation that allows bacteriophages to infect the cell, and many more. Epidemics of cholera are caused by specific strains of *V. cholerae*, and genomic analysis of several of these disease strains reveals variation in the genes present in the pathogenicity islands, although all contain the toxin gene. The severity of an epidemic strain depends on the genes present in the strain.

An intriguing type of pathogenicity island is the type of element called an **integrative and conjugative element (ICE)**. These elements contain features of conjugative plasmids (like the F factor) in addition to the characteristics of genomic islands. ICEs encode an integrase, like a lambda phage, which allows the DNA to integrate or excise from the chromosome. Furthermore, they possess the machinery needed for both conjugation and DNA transfer. Conjugation initiated by ICEs is usually "promiscuous," allowing transfer of DNA between many different species. The gene content of the ICE therefore suggests a mechanism by which some of the pathogenicity islands can be transferred between species.

An ICE in a pathogenic *E. coli* strain contains a 135-kb DNA fragment that is similar to *Yersinia pestis*

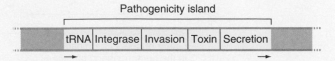

Pathogenicity island

| tRNA | Integrase | Invasion | Toxin | Secretion |

Figure 18.27 Pathogenicity island. Pathogenicity islands can contain many genes involved in causing disease.

and *Y. pseudotuberculosis*, the bacteria responsible for bubonic plague and for a disease that mimics tuberculosis. The element contains genes for mating-pair formation and a presumed oriT element, which would be the site at which transfer begins (as seen in the F factor). The induction of a phage P4-like integrase in the cell results in excision and circularization of the element, providing evidence that this element can in fact transfer to another cell by a conjugative mechanism.

Evolution of pathogenic *E. coli*

E. coli is the most abundant organism in the human colon and coexists peacefully within us, for the most part. Some strains of *E. coli* are capable of causing diarrhea or meningitis because they have acquired pathogenic genes. The genomes of many of these pathogenic strains of *E. coli* contain pathogenicity islands, described above. In recent years, a newly evolved strain, *E. coli* O157:H7, has caused severe illness when people have eaten undercooked, contaminated beef or tainted lettuce.

Genomic analysis indicates that *E. coli* O157:H7 contains a particular type of pathogenicity island found in many pathogenic bacteria. This island encodes proteins that facilitate attachment to epithelial cells, and secretion systems and proteins that cause cytoskeletal changes and loss of fluid. What makes this strain more potent is the presence of a toxin from the bacterium *Shigella* that targets the rRNA of the host cells, stopping protein synthesis in these cells. This toxin acts in several organs, including the kidneys, which get inflamed and may fail, and the intestine, where damage leads to bloody diarrhea. The DNA sequence shows that this toxin gene was transferred into *E. coli* by bacteriophage transduction and became part of the large pathogenicity island. Additional smaller pathogenicity islands are present in the O157:H7 strain, including genes that aid in adhesion to host cells.

The glimpse into prokaryotic history made possible by comparative genome analysis has altered our view of evolution. Geneticists had thought that bacteria started out with a set of genes that slowly evolved through point mutation, deletion, and duplication within the species. But the data showing that some genes or sets of genes in one bacterial species are very similar to those in another species suggest that bacterial genomes have picked up DNA from several different sources during the course of their evolution. Biologists now recognize that lateral gene transfer is a significant evolutionary factor in pathogenicity and many other bacterial functions.

- Bacterial transformation involves the uptake of DNA from the environment that leads to a change in genotype. In natural transformation, cells are able to take up DNA because of certain conditions that make them competent, such as starvation for a nutrient they cannot synthesize. Geneticists can treat bacteria with agents that damage their cell walls, allowing DNA to enter; this technique is termed *artificial transformation*.

- Conjugation transfers a conjugative plasmid, such as the F plasmid of *E. coli*, to another bacterium through direct contact and connection. The F plasmid contains genes for formation of a connecting pilus and for production of an endonuclease that turns the F plasmid into a linear strand. Interrupted mating allows mapping of genes on the F plasmid.

- Transduction is transfer of DNA via a bacteriophage. Short fragments of bacterial DNA are sometimes incorporated into new phage particles and then are released into new cells upon viral infection. In some cases, viruses integrate into the host's DNA; when these viruses later undergo excision, they may take some bacterial DNA with them.

- Genomic islands in bacterial DNA appear to have originated from lateral transfer of foreign DNA. Pathogenicity islands contain a package of genes that confer the ability to create disease. Integrative and conjugative elements (ICEs) contain features of conjugative plasmids plus additional DNA that may confer pathogenicity. An ICE is thought to have been involved in the development of the pathogenic *E. coli* strain O157:H7.

18.4 Bacterial Genetic Analysis

As geneticists learn more about genome structure and the mechanisms of gene transfer, including the transposition of DNA sequences, transduction, and conjugation, they are able to devise ever more clever ways of carrying out genetic analysis. Here we describe how an *E. coli* geneticist might approach the genetic dissection of a biochemical or physiological pathway. Many of the principles we present are applicable to other bacteria in which similar gene transfer mechanisms exist.

Transposons allow manipulation of bacterial genomes

Transposons have played the largest role in simplifying genetic analysis because they can create mutations. The insertion of a transposon into a gene, resulting in the gene's inactivation, is the basis of many mutant screens. Transposons are useful as mutagenic agents because they contain genes for easily selectable antibiotic resistance.

To carry out transposon mutagenesis, geneticists introduce a transposon into a cell as part of a DNA molecule that is not able to replicate on its own inside the cell. The mechanism of gene transfer can be transformation, transduction, or conjugation. For the transposon to be passed on during cell division, it must transpose from the incoming DNA molecule to the bacterial chromosome. By growing cells on a medium containing antibiotics, it is possible to select for those cells in which transposition has occurred. A researcher can then screen the resulting population of cells, which contain transposons at different locations around the chromosome, for the mutant phenotype of interest.

Transposon insertion usually inactivates the gene receiving the insertion, thereby creating a knockout or a null mutation. Such mutations can be useful, but if a gene is essential for a bacterium's survival, it will not be possible to isolate the knockout mutation. For genes encoding essential proteins, conditional mutants (e.g., temperature-sensitive mutants) are isolated. Even for nonessential genes, conditional mutations may be the most informative, because cells grown under permissive conditions can be shifted to nonpermissive conditions and then observed for changes in phenotype. The many useful features of transposon mutagenesis have even been used to help define the minimal set of genes required for life (see the **Fast Forward** box **"Defining the Minimal Gene-Set Required for Life"**).

Reverse genetics provides a way to insert synthetic genes to test function

The sequence analysis of bacterial genomes has led to the identification of genes whose functions are not yet known. One approach to determining the function of such genes is to make a knockout mutation in the chromosomal gene using recombinant DNA techniques and the homologous recombination machinery of bacteriophages. This approach is known as **recombineering**. In a nutshell, a mutant version of the gene is constructed *in vitro* and then introduced into a cell. Then *in vivo* recombination inserts the constructed gene into the chromosome in place of the wild-type copy.

For example, to analyze the function of gene *X*, a defective copy (knockout mutation) of the gene could be created. Once this copy is integrated into a cell and its progeny grown, the phenotype of these cells could be examined. One way to create a defective allele is to insert an antibiotic resistance gene into gene *X*. The antibiotic resistance gene also serves as a selectable marker.

To construct this *in vivo*, 40 bp of the gene *X* sequence (known from the genomic sequence), together with 20 bp of sequence from the drug resistance gene,

The bacterium *Mycoplasma genitalium* possesses the small-est genome of any organism that can grow independently in culture (482 genes; 582 970 base pairs). For this reason, *M. genitalium* is often used in genetic studies aimed at decipher-ing the minimum requirements of life. For example, it has been used to help define the minimal set of genes required for a cell

to show the characteristics of a living organism. In these simple, but insightful experiments, Hamilton O. Smith and colleagues used transposon mutagenesis to identify the *M. genitalium* genes that were indeed essential. The researchers isolated and analyzed transposon-mediated gene disruption mutants that were capable of growth. If a given disruptant was able

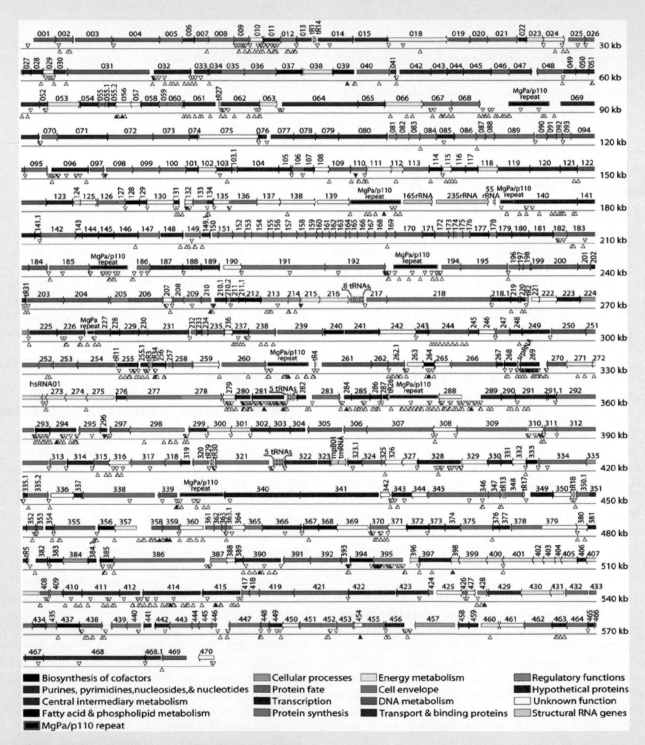

Figure A The essential genes of *Mycoplasma genitalium*.

to grow and form a colony, the researchers could then conclude that the function of the disrupted gene was *not* essential for life. If on the other hand, the researchers were unable to obtain a clone bearing a disruption in a given gene, then they could conclude that this gene *was* essential for life. In the final analysis, the researchers were able to recover only 100 unique gene disruption mutants and thus showed

that 382 of the 482 *M. genitalium* genes were required for life. **Figure A** notes the locations of the recovered transposon insertions (*triangles*) from the study. As you shall see in Chapter 24 (Synthetic Biology), the ability to define a minimal gene-set has important ramifications for those interested in synthesizing novel organisms with user-defined phenotypic characteristics.

are chemically synthesized into one fragment for use as a primer (**Figure 18.28**). Primers are produced for both ends of the gene. PCR amplification using these primers and a fragment containing only the antibiotic gene as the template produces a DNA fragment containing the gene *X* sequence at either end of a complete antibiotic resistance gene. This fragment is used to transform a cell.

The cell to be transformed contains the recombination genes from bacteriophage lambda. These genes are repressed (i.e., are inactive) in the cell at low temperature, but are expressed at high temperature. Raising the temperature causes expression of the recombination functions, and the transformed DNA fragment produced by *in vitro* PCR is recombined into the chromosome using homology to gene *X* at either end of the fragment. The addition of antibiotic to the growth medium allows for the selection of only those cells in which the integration has occurred (i.e., cells in which integration did not occur will not have a copy of the antibiotic resistance gene and thus will not be able to proliferate). Proliferating cells (that did integrate the fragment) can then be analyzed for phenotypes resulting from the "knockout" of the gene.

This approach will work only if the gene is not essential. If the gene is essential, no antibiotic-resistant cells would be expected because of the lethal effects of disrupting the gene's function. But for nonessential genes, the phenotype of the cell containing the knockout mutation can provide clues about the function of the gene.

Genomic and genetic approaches may be combined

Genomic experimentation adds an exciting new dimension to the impressive set of tools developed since the 1950s for the analysis of bacterial life. The marriage of genomic, genetic, and recombinant DNA approaches has led to elegant and innovative experiments. For example, a recent study of *Pseudomonas aeruginosa* used gene transfer techniques involving specially constructed transposons to produce a large-scale library of mutants for further genetic analysis.

P. aeruginosa is an opportunistic pathogen that causes pulmonary infections in immune-compromised and cystic fibrosis patients. The sequence of its large

(a) *in vitro*

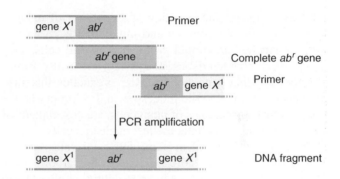

(b) *in vivo*

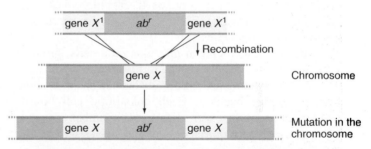

Figure 18.28 Recombineering. (a) A fragment containing the antibiotic resistance gene (*ab^r*) flanked by the *gene X* sequence is produced by PCR amplification. **(b)** The fragment recombines *in vivo* using phage recombination genes.

6.3-Mb genome was determined in 2000. However, since knowledge of a DNA sequence does not immediately indicate what all the genes do, geneticists wanted to produce specific mutations in the genes to study their function. Taking a global approach, they generated a set of isolates, each containing a different gene mutated by insertion of a transposon carrying an antibiotic resistance marker into the open reading frame. To introduce the transposon-carrying DNA into the cell, the researchers mated *P. aeruginosa* with *E. coli* (using the "promiscuous" type of conjugation described earlier). The transposons in *E. coli* jumped into many different places in the *P. aeruginosa* chromosome. Mutants were selected

on the basis of transposon-conferred drug resistance. The genes mutated by transposon insertion could be identified by PCR amplification followed by sequencing from the transposon into the adjacent DNA. The full genomic sequence was the reference material that allowed identification of the disrupted gene.

Using this protocol, the investigators disrupted about 90 percent of the ORFs in the genome and characterized 36 000 mutations. The remaining 10 percent of the ORFs presumably included essential genes that could not be mutated without lethal effects. The mutant library now provides a resource for additional studies on the function of individual *P. aeruginosa* genes and demonstrates the effectiveness of a strategy that is broadly applicable to many other bacterial species.

> Insertion of transposons into bacterial genomes can inactivate genes and may also add genes for antibiotic resistance that serve as markers. Recombineering involves the creation of gene knockouts to observe the effects of null mutations on phenotype. Combined approaches often yield a wealth of information from a single, carefully planned experiment.

18.5 The Genetics of Chloroplasts and Mitochondria

We next consider the genetics of two types of eukaryotic organelles: chloroplasts and mitochondria. Although these organelles are found within eukaryotic cells, they show many characteristics of prokaryotic cells. Biologists believe that mitochondria, the organelles that produce energy for metabolic processes, and chloroplasts, the photosynthetic organelles of plant cells, are descendants of bacteria that fused with the earliest nucleated cells.

Chloroplasts, found in plant and algal cells, capture energy from light and store this energy in carbohydrates (**Figure 18.29**). Chloroplasts have structural similarities to certain cyanobacteria, which are capable of photosynthesis. In corn, each leaf cell contains 40–50 chloroplasts, and each square millimetre of leaf surface carries more than 500 000 of the organelles.

Mitochondria, found in all eukaryotic cells, produce most of the cell's usable energy in the form of ATP molecules. Mitochondria are similar in size and shape to some modern aerobic bacteria. Each eukaryotic cell houses many mitochondria, with the exact number depending on the energy requirements of the cell as well as the chance distribution of mitochondria during cell division. In humans, nerve, muscle, and liver cells each carry more than a thousand mitochondria.

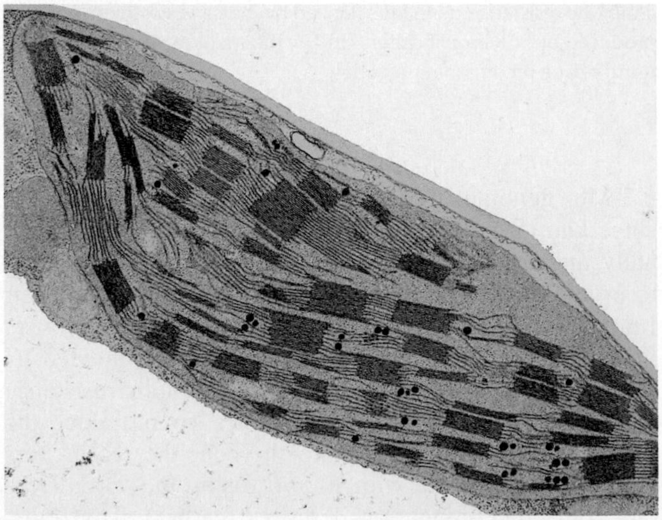

Mitochondria and chloroplasts carry their own DNA

When viewed under the light microscope, cells stained with DNA-specific dyes reveal DNA molecules in the mitochondria and the chloroplasts, as well as in the nucleus. Using methods for purifying mitochondria and chloroplasts, researchers have extracted DNA directly from these organelles and have shown by analyses of base composition and buoyant density that an organism's organellar DNA differs from its nuclear DNA. Although both these organelles replicate and express all the genes in their own DNA, their genomes encode only some of the proteins they require for their activities.

Based in part on these observations, a theory first proposed by Lynn Margulis in 1966, the **endosymbiont theory**, posits that chloroplasts and mitochondria originated when free-living bacteria were engulfed by primitive nucleated cells. Host and guest formed cellular communities in which each member adapted to the group arrangement and derived benefit from it.

The varied genomes of mitochondria contain genes for oxidative processes and unique functions

Mitochondrial DNA (mtDNA) appears in the organelle in highly condensed structures called *nucleoids*. The number of nucleoids in mitochondria varies depending on growth conditions and energy needs of the cell. Variations in the number of mitochondria in a cell and the number of mtDNA molecules within each mitochondrion are regulated by complicated means that researchers do not yet understand.

The replication of mtDNA molecules, as well as the division of the mitochondria, can occur throughout the cell cycle independent of the replication of genomic nuclear DNA (which occurs only during S phase). Interestingly,

Figure 18.29 Chloroplast. Electron micrograph of an isolated chloroplast in a leaf cell of timothy grass (*Phleum pratense*) (11 000×).

Table 18.1	Mitochondrial DNA Sizes
Organism	**Size (kb)**
Plasmodium	6
Yeast	75
Drosophila	18
Pea	110
Human	16.5

which mtDNA molecules undergo replication seems to be determined at random; as a result, some molecules replicate many times in each cell cycle, while others do not replicate at all. This is one cause of the mitotic segregation of mitochondrial genomes discussed later in the chapter.

Mitochondrial variation across species

The size and gene content of mitochondrial DNA vary from organism to organism. The mtDNAs in the malaria parasite, *Plasmodium falciparum*, are only 6 kb in length; those in the free-living nematode *Ascaris suum* are 14.3 kb; those in the muskmelon, *Cucumis melo*, are a giant 2400 kb long. These mtDNA size differences do not necessarily reflect comparable differences in gene content. Although the large mtDNAs of higher plants do contain more genes than the smaller mtDNAs of other organisms, the 75-kb mtDNA of baker's yeast encodes fewer proteins of the respiratory chain than does the 16.5-kb mtDNA of humans. **Tables 18.1** and **18.2** summarize the size and gene content of mtDNAs from organisms representative of plants, animals, and fungi.

Like size and gene content, the shape of mtDNAs varies. Biochemical analyses and mapping studies have shown that the mtDNAs of most species are circular; but the mtDNAs of the ciliated protozoans *Tetrahymena* and *Paramecium*, the alga *Chlamydomonas*, and the yeast *Hansenula* are linear.

Protozoan parasites of the genera *Trypanosoma*, *Leishmania*, and *Crithidia* exhibit mtDNAs that have a highly unusual organization. These single-celled eukaryotic organisms carry a single mitochondrion known as a kinetoplast. Within this structure, the mtDNA exists in one place (contrary to the mtDNA of most other cells) as a large network of 10–25 000 minicircles 0.5–2.5 kb in length interlocked with 50–100 maxicircles 21–31 kb long (**Figure 18.30**). The maxicircles contain most of the genes usually found on mtDNA, while the minicircles play a role in RNA editing, as described later.

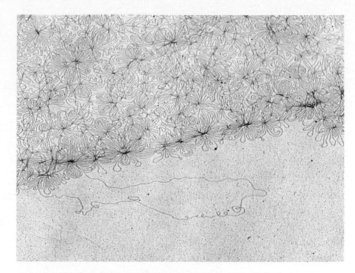

Figure 18.30 Kinetoplast DNA network. In certain protozoan parasites, there is a single mitochondrion, or kinetoplast, that contains a large interlocking network of DNA molecules present in mini- and maxicircles.

Comparisons of mitochondrial genomes

A significant feature of the human mitochondrial genome is the compactness of its gene arrangement. Adjacent genes either abut each other, or slightly overlap. With virtually no nucleotides between them and no introns within them, the genes are packed very tightly. The reason for this compact arrangement is not yet known.

The mitochondrial genome of the yeast *S. cerevisiae* is more than four times longer than human and other animal mtDNAs. Two DNA elements account for the larger size of the yeast mitochondrial genome: long intergenic sequences and introns.

The mtDNA of *M. polymorpha* was the first plant mtDNA to be entirely sequenced. Although it is one of the smallest plant mitochondrial genomes, it is far larger and has many more genes than non-plant mtDNAs. Thus, although mitochondria in different eukaryotic organisms play similar roles in the conversion of food to energy, evolution has produced mtDNAs with an astonishing diversity in the content and organization of their genes. As we see next, mitochondrial evolution has also led to some remarkable variations on the basic mechanisms of gene expression.

RNA editing of mitochondrial DNA transcripts

Researchers discovered the phenomenon of RNA editing in the mitochondria of trypanosomes. As already noted, these

Table 18.2	Comparison of Some Functions Encoded in mtDNA		
Organism	**Oxidative Phosphorylation Genes**	**tRNAs**	**Genome Size (kb)**
Yeast	7	25	75
Marchantia (liverwort)	14	29	186.0
Human	13	22	16.5

protozoan parasites have a single, large mitochondrion—the kinetoplast—which contains much more DNA than the mitochondria of other organisms and which has the DNA arranged as a series of interlocking maxi- and minicircles. DNA sequencing shows that the minicircles carry no protein-encoding genes. The detection of transcripts from maxicircle DNA, however, confirms that these larger circles do carry and express genes.

Surprisingly, the sequencing of maxicircle DNA revealed only short, recognizable gene fragments, instead of whole mitochondrial genes. Furthermore, the sequencing of RNA molecules in the kinetoplast revealed both RNAs that looked like the strange fragments of kinetoplast genes and related RNAs that could encode recognizable mitochondrial proteins. From these observations, investigators concluded that kinetoplastid DNA (kDNA) encodes a precursor (the strange fragment observed) for each mRNA. After transcription, the cellular machinery turns these precursors into functional mRNAs through the insertion or deletion of nucleotides.

The process that converts pre-mRNAs to mature mRNAs is **RNA editing**. Without RNA editing, the pre-mRNAs do not encode polypeptides. Some pre-mRNAs lack a first codon suitable for translation initiation; others lack a stop codon for the termination of translation. RNA editing creates both types of sites, as well as many new codons within the genes.

In addition to the kinetoplasts of trypanosomes, the mitochondria of some plants and fungi carry out RNA editing. The extent of RNA editing varies from mRNA to mRNA and from organism to organism. In trypanosomes, the RNA editing machinery adds or deletes uracils. In plants, the editing adds or deletes cytosines. At present, researchers understand the general mechanism of uracil editing, but not that of cytosine editing. As **Figure 18.31** shows, uracil editing occurs in stages in which enzymes use an RNA template as a guide for correcting the pre-mRNA. The guide RNAs are encoded by short stretches of kDNA on both maxi- and minicircles, and a structure known as an "editosome" is the workbench where the RNA editing takes place.

Mitochondrial exceptions to the "universal" genetic code

As mitochondrial DNA carrying its own rRNA and tRNA genes would suggest, mitochondria have their own distinct translational apparatus. Mitochondrial translation is quite unlike the cytoplasmic translation of mRNAs transcribed from nuclear genes in eukaryotes. Many aspects of the mitochondrial translational system resemble details of translation in prokaryotes. For example, as in bacteria, N-formyl methionine and tRNA^{fMet} initiate translation in mitochondria. Moreover, inhibitors of bacterial translation, such as chloramphenicol and erythromycin, which have no effect on eukaryotic cytoplasmic protein synthesis, are potent inhibitors of mitochondrial protein synthesis.

We stated in Chapter 7 that the genetic code is almost, but not quite, universal. The mtDNA sequences of tRNAs and protein-encoding genes in several species cannot explain the sequences of the resulting proteins in terms of the "universal" code. For example, in human mtDNA, the codon UGA specifies tryptophan rather than stop (as in the standard genetic code); AGG and AGA specify stop instead of arginine; and AUA specifies methionine rather than isoleucine (**Table 18.3**). No single mitochondrial genetic code functions in all organisms, and the mitochondria of higher plants use the universal code. Moreover, while an f-Met-tRNA usually initiates translation in mitochondria by reading AUG or AUA, other triplets, which do not specify methionine, often mark the site of initiation. The genetic codes of mitochondria probably diverged from the universal code by a series of mutations occurring some time after the organelles became established components of eukaryotic cells.

As we see next, chloroplast DNA, although similar in many ways to mtDNA, has some remarkable features of its own.

Table 18.3	Variations in the Genetic Code of Human Mitochondria	
Characteristic	**Universal Code**	**mtDNA Code**
Number of tRNAs	32	22
UGG	Trp	Trp
UGA	Stop	Trp
AGG	Arg	Stop
AGA	Arg	Stop
AUG	Met	Met
AUA	Ile	Met

Altered genetic code. The human mtDNA genetic code is simplified such that a modified U in the tRNA "wobble" position can read all four codons in a codon family (i.e., UUU, UUC, UUA, and UUG). An unmodified U can read both purines, and G can read both pyrimidines. Tryptophan tRNA has a U in the wobble position, so it will read both the traditional UGG codon and the associated UGA stop codon as tryptophan. Similarly, the methionine codon reads both AUG and the associated AUA as methionine. Finally, human mtDNA has only a single arginine tRNA such that two of the six arginine codons (AGG and AGA) now function as stop codons.

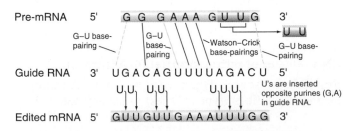

Figure 18.31 RNA editing in trypanosomes. Example of a portion of a pre-mRNA sequence is shown at the *top*. This pre-mRNA forms a double-stranded hybrid with a guide RNA through both standard Watson–Crick A–U and G–C base-pairing, as well as atypical G–U base-pairing. Unpaired G and A bases within the guide RNA initiate the insertion of U's within the pre-mRNA sequence, while unpaired U's in the pre-mRNA are deleted, bringing about the final edited mRNA.

The genomes of chloroplasts include genes for some enzymes of photosynthesis and for gene expression

Chloroplasts occur in plants and algae. The genomes they carry are much more uniform in size than the genomes of mitochondria. Although chloroplast DNAs range in size from 120 to 217 kb, most are between 120 and 160 kb long (**Table 18.4**). Chloroplast DNA (cpDNA) contains many more genes than mtDNA. Like the genes of bacteria and human mtDNA, these genes are closely packed, with relatively few nucleotides between adjacent coding sequences. Like the genes of yeast mtDNA, they contain introns. Most cpDNAs exist as linear and branched forms.

The cpDNA-encoded proteins include many of the molecules that carry out photosynthetic electron transport and other aspects of photosynthesis, as well as RNA polymerase, translation factors, ribosomal proteins, and other molecules active in chloroplast gene expression. The RNA polymerase of chloroplasts is similar to the multisubunit bacterial RNA polymerases. Inhibitors of bacterial translation, such as chloramphenicol and streptomycin, inhibit translation in chloroplasts as they do in mitochondria.

Techniques for introducing genes and DNA fragments into organelles

In the early days of recombinant DNA technology, chloroplast researchers were frustrated by an inability to transfer cloned genes and mutated DNA fragments into organelle genomes. Development of the gene gun and a gene delivery method known as *biolistic transformation* in the late 1980s solved the problem. The basic idea is to coat small (1 μm) metal particles with DNA and then shoot these DNA-carrying "bullets" at cells. Biolistic transformation occurs when a particle lands within a cell, such as a plant protoplast, without killing it, and the DNA is released from the metal. In rare instances, the DNA enters the nucleus or organelles where it may recombine into the genome. If the DNA shot into the cell contains a strong selectable marker, plant geneticists can isolate the rare transformed plant cells in which the released DNA has entered the organelle. The cells may then be cultured as clones to produce a complete plant. A variety of vectors exist for many plant species, and these may contain different selectable markers and sequences that support the expression of introduced genes.

For plants whose chloroplast genome sequence has been determined, organelle transformation and the generation of mutants provide a way to determine the function

of ORFs for which no function has yet been assigned. To explore the function of an ORF, a DNA molecule is constructed containing a selectable chloroplast antibiotic resistance gene within the ORF. This DNA is shot into cells and integrates into the chloroplast genome via homologous recombination (replacing the wild-type ORF with the mutant ORF). Selection for the marker gene increases the proportion of chloroplasts containing the mutant ORF, and this makes it possible to study the phenotype of the transformed cells in culture or in reconstituted plants. Researchers have used this protocol to identify chloroplast genes encoding novel subunits of cytochrome complexes and assembly factors for photosystem genes in tobacco and *Chlamydomonas reinhardtii*.

Potential uses of transformed chloroplasts

Transformation of the chloroplast genome is a suitable mechanism for altering the properties of commercially important crop plants. One goal might be to produce herbicide-resistant plants. The advantage of introducing herbicide resistance into chloroplast DNA instead of nuclear DNA relates to the fact that foreign DNA in the chloroplasts will be inherited maternally, not through the male pollen. The risk that introduced genes will spread to neighbouring plant populations is therefore low.

Chloroplast transformation also makes it possible to make plants into protein-production factories. One could, for example, produce a vaccine in the leaves of an edible plant by incorporating the genes encoding the vaccine into the chloroplast genome. A modified *E. coli* labile-toxin (LT) gene has already been introduced into the chloroplasts of tobacco (the LT toxin protein causes diarrhea). Transformation of the same gene into chloroplasts in edible leaves such as lettuce or spinach would generate an ingestible vaccine that could, in principle, stimulate the human immune system to respond to and eliminate any *E. coli* LT it encountered.

Nuclear and organellar genomes cooperate with one another

The maintenance and assembly of functional mitochondria and chloroplasts depend on gene products from both the organelles themselves and from the nuclear genome (**Figure 18.32**). For example, in most organisms, cytochrome C oxidase, the terminal protein of the mitochondrial electron transport chain, is composed of seven subunits, three of which are encoded by mitochondrial genes whose mRNAs are translated on mitochondrial ribosomes. The remaining four are encoded by nuclear genes whose messages are translated on ribosomes in the cytoplasm. In all organisms, nuclear genes encode the majority of the proteins active in mitochondria and chloroplasts. For example, although mitochondrial genomes carry the rRNA genes, nuclear genomes carry the genes for most (in yeast and plants) or all (in animals) of the proteins in the mitochondrial ribosome. Because mitochondria and chloroplasts do not carry all the genes for the proteins (and in

Table 18.4	Chloroplast DNA Sizes
Organism	**Size (kb)**
Chlamydomonas reinhardtii	196
Marchantia (liverwort)	121
Nicotiana tabacum (tobacco)	156
Oryza sativa (rice)	135

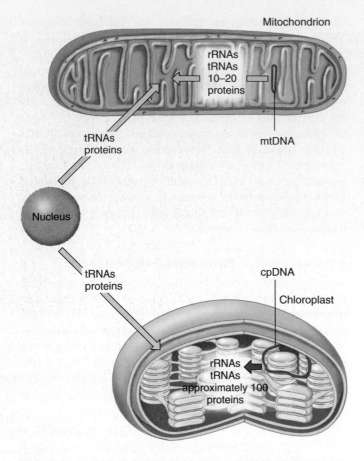

Mitochondrion

rRNAs
tRNAs
10–20
proteins

tRNAs
proteins

mtDNA

Nucleus

tRNAs
proteins

cpDNA

Chloroplast

rRNAs
tRNAs
approximately 100
proteins

Number and genomic location of oxidative phosphorylation genes

| Genomic Location | Number of Polypeptides | | | | | |
| | Electron transport chain | | | | ATP synthase | |
	I	II	III	IV	V	Total
Mitochondrion	7	0	1	3	2	13
Nucleus	≥ 33	4	10	10	10	≥ 67
Total	≥ 40	4	11	13	12	≥ 80

Figure 18.32 Mitochondria and chloroplasts depend on gene products from the nucleus. Although some organelles in some species have many more genes than others, all are dependent on RNA and protein products encoded by nuclear genes. The location of oxidative phosphorylation genes is shown.

some organisms, the tRNAs) they need to function and reproduce, these organelles are *semiautonomous*, requiring the constant provision of proteins (and tRNAs) encoded by nuclear genes.

Gene transfer between an organelle and the nucleus

The symbiotic relationship that developed between organelle and cell allowed loss of some genes from the organelle. Redundant genes could be eliminated, but some genes essential to organelle function were also transferred to the nucleus. Researchers have some understanding of the mechanisms by which this transfer occurred.

In many plants, the mitochondrial genome encodes the *COXII* gene of the mitochondrial electron transport chain; in other plants, the nuclear DNA (nDNA) encodes that same gene; and in several plant species where the nuclear *COXII* gene is functional, the mtDNA still contains a recognizable, but nonfunctional, copy of the gene (i.e., a *COXII* pseudogene). Remarkably, the mtDNA gene contains an intron, while the nuclear gene does not. Geneticists have interpreted this finding to mean that the *COXII* gene transferred from mtDNA to nDNA via an RNA intermediate using reverse transcriptase. The RNA would have lacked the intron, and when the mRNA was copied into DNA by reverse transcriptase and integrated into a chromosome in the nucleus, the resulting nuclear gene also had no intron.

Good evidence also exists for the transfer of many genes at the DNA level. The fact that some plant mtDNAs carry large fragments of cpDNA shows that pieces of cpDNA can move from one organelle to another. Similarly, nonfunctional, intact or partial copies of organellar genes litter the nuclear genomes of eukaryotes. DNA sequencing reveals strong similarities between the organellar and nuclear genes, which means that the nuclear copies are relatively young. This, in turn, suggests that the organelle-to-nucleus transfer of DNA is still going on.

In this evolutionary perspective, the properties of mitochondrial and chloroplast genomes that vary among the organelles of present-day species are probably relatively new. These recently established features include long stretches of cpDNA incorporated in the mtDNAs of many plants, as well as many of the introns in organellar genomes. Some of these introns may have originated in the earliest bacterial symbionts; or they may have been incorporated into the organellar genomes after horizontal transfers between organelle DNAs long after the organelles were established.

The high rate of mutation in mitochondrial DNA

In the 1980s, surveys of DNA sequence variations among individuals of a given species (and between closely related species) showed that the mtDNA of vertebrates evolves almost ten times more rapidly than does nuclear DNA. The higher rate of DNA mutation in mitochondria probably reflects more errors in replication and less efficient repair mechanisms.

Because of mtDNA's high mutation rate, the variation among mitochondrial genomes provides a valuable tool for studying the evolutionary relationships of organisms whose nuclear DNAs are very similar. Conversely, mtDNA variation, because it accumulates so rapidly, is of little value in evaluating the relationships of distant evolutionary relations (here sequence variation data in nuclear genomes would be useful). Sequence analyses of mtDNA have shown that the maternal lineage of all present-day humans, no matter what ethnic group they belong to, traces back to a few female ancestors who lived in Africa some 200 000 years ago.

- Analysis of mtDNA shows wide variation in size, sequence, and number of genes among species. RNA transcribed from mtDNA undergoes editing before it can produce functional polypeptides. The triplet code used by mtDNA is different in some respects from the "universal" DNA code found in nuclear DNA.

- Chloroplast genomes are much more uniform than those of mitochondria. Biolistic techniques allow DNA to be shot into cells, and this DNA integrates most successfully

into chloroplasts. Potential exists for altering plants to produce their own insecticide or even vaccines for human use.

- Mitochondrial and chloroplast functions depend on gene products from both the organelles and the cell's nucleus. Gene transfer has occurred, most likely through an RNA intermediate. Although mtDNA has a high rate of mutation, variation in this genome is useful for studying the evolutionary relationships of closely related species.

18.6 Non-Mendelian Inheritance of Chloroplasts and Mitochondria

As you learned in Chapter 2, Mendel performed reciprocal crosses in which either the male or female plant carried the wild-type or variant allele. His result showed no difference in inheritance based on which parent showed the variant. However, just nine years after the rediscovery of Mendel's laws, plant geneticists reported a perplexing phenomenon that challenged one of Mendel's basic assumptions.

In a 1909 paper, Carl Correns and his colleagues described the results of reciprocal crosses analyzing the transmission of green versus variegated leaves in flowering plants known as four-o'clocks (*Mirabilis jalapa*) (**Figure 18.33**). Fertilization of eggs from a plant with variegated leaves by pollen from a green-leafed plant produced uniformly variegated offspring. Surprisingly, the reciprocal cross—in which the leaves of the mother plant were green and those of the father variegated—did not lead to the same outcome; instead all of the progeny from this cross displayed green foliage. From these results, it appeared that offspring inherit their form of the variegation trait from the mother only. This type of transmission, known as maternal inheritance, challenged Mendel's assumption that maternal and paternal gametes contribute equally to inheritance. Geneticists thus said that the trait in question exhibited **non-Mendelian inheritance**.

Another example of a non-Mendelian trait emerged 40 years later. In 1949, French researchers published studies on the size of yeast colonies in laboratory strains of the single-celled organism *Saccharomyces cerevisiae*. Mitotically dividing cultures of these cells, when grown on plates containing glucose as the source of carbon, produced colonies of two distinctly different sizes. Ninety-five percent of the colonies were large (in French, *grande*); the remaining 5 percent were small (*petite*). Cells from *grande* colonies, when separated and grown on fresh plates containing glucose, yielded some *petite* colonies, but cells from *petite* colonies never generated *grande* colonies. From these observations, the researchers deduced that the founder cells of *petite* colonies arose from frequent mutations—1 in 20 cells—in cells of the *grande* colonies.

The French researchers pursued their study of the genetic basis of this difference in colony size by analyzing various matings using haploid cells from *grande* and *petite* colonies. As described in Chapter 4, the diploid cells formed by mating haploid cells of opposite mating types may, under stressful conditions (e.g., not enough nutrients), enter meiosis. When the French researchers mated *grande* cells of one mating type with *grande* cells of the opposite mating type, the resulting diploids were *grande*; and when these *grande* diploids sporulated via meiosis, each one yielded four *grande* spores (i.e., spores that after germination produced *grande* colonies) and zero *petite* spores.

Figure 18.33 Four-o'clocks. The first example of non-Mendelian inheritance uncovered by geneticists was seen in the flowering plants known as four-o'clocks.

A cross of two cells from *petite* colonies produced only *petite* diploids, which, however, could not sporulate because of their deficiency in respiration. Matings of *grande* with *petite* generated only *grande* diploids, and each sporulation of those diploids yielded four *grande* spores and zero *petites*. This 4:0 ratio consistently replaced the 2:2 ratio predicted by Mendelian genetics.

From these observations, the researchers concluded that a genetic factor necessary for respiratory growth is present in *grande* cells but absent from *petite* cells. They named the factor "rho" (symbolized by "ρ"), and they designated *grande* cells "ρ$^+$" and *petite* cells "ρ$^-$". They also noted that because of the non-Mendelian inheritance pattern, the rho factor did not segregate at meiosis.

Geneticists thought that a connection must exist between the maternal inheritance of leaf variegation in four-o'clocks and the unusual 4:0 inheritance pattern of *grande* and *petite* colony sizes in yeast. Decades of experiments have shown that both traits are determined by genes that do not reside in the nucleus but instead lie in the genomes of non-nuclear organelles. Mutations resulting in leaf variegation occur in cpDNA (in genes that encode proteins active in photosynthesis). Mutations that diminish yeast colony size occur in mtDNA (in regions of the genome that influence the efficiency of a cell's energy use).

Mutations in organellar genes can produce readily detectable whole-organism phenotypes if the altered proteins and RNAs they encode disrupt the production of cellular energy. The cpDNA mutations that cause variegation in four-o'clocks incapacitate proteins essential for photosynthesis. The *petite* mutants form smaller colonies because they are unable to carry out cellular respiration and must obtain the energy they need for survival from the less efficient, anaerobic energy conversion pathway of fermentation. Although most mutations in the genes for these energy-producing systems are lethal in both plants and animals, some organellar gene mutations yield detectable, nonlethal phenotypes that researchers can study genetically. Data from such genetic studies show that the modes of organelle gene transmission vary among organisms. We describe the main modes of transmission from one generation to the next.

Maternal inheritance of differences in wild-type mtDNAs

In a classic experiment documenting maternal inheritance in vertebrates, investigators purified mtDNA from frog eggs, which contain a large number of mitochondria, and used hybridization tests to distinguish the mtDNA of one frog species, *Xenopus laevis*, from the mtDNA of the closely related *Xenopus borealis*. In these tests, probes from *X. laevis* hybridized more efficiently with *X. laevis* mtDNA than with *X. borealis* mtDNA, and vice versa. Because crosses between the two species yield viable progeny, the analysis of F$_1$ mtDNA was one way to trace the inheritance of that DNA in frogs. **Figure 18.34** diagrams the reciprocal crosses and mtDNA typing that formed the basis of the study. The first-generation progeny of both crosses carried mtDNA like that of the maternal parent.

Although the analysis might have missed small contributions from the paternal genome, these *Xenopus* crosses confirmed the predominantly maternal inheritance of mtDNA in these species. They also showed that it is possible to follow pre-existing differences in functionally wild-type mtDNAs in a cross. Since the 1980s, analysis of crosses using DNA polymorphisms confirmed maternal mtDNA inheritance among horses, donkeys, and many other vertebrates.

Maternal inheritance of specific genes in cpDNA

Interspecific crosses tracing biochemically detectable species-specific differences in several chloroplast proteins provided further evidence of maternal inheritance. In the mtDNA inheritance studies just described, the identity of the organelle gene containing the markers was not known—and did not matter. By contrast, in cpDNA studies, researchers identified specific organelle genes through the analysis of proteins.

They began by isolating from tobacco plants (*Nicotiana* species) proteins in which interspecies differences could be distinguished by gel electrophoresis. To determine each protein's mode of inheritance, they evaluated the allele expressed in the offspring of a controlled cross, carefully noting the maternal (ovum) and paternal (pollen) contributions. In one set of experiments, they

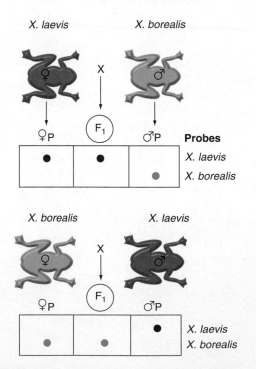

Figure 18.34 Maternal inheritance of *Xenopus* mtDNA. *X. laevis* and *X. borealis* mtDNA can be distinguished by strong hybridization only to probes made from the same species. Reciprocal crosses between two species produce F$_1$ hybrids. Each F$_1$ hybrid retains mtDNA only from its mother.

observed ribulose bisphosphate carboxylase (Rubisco, for short), the first enzyme of photosynthetic carbon fixation in plants and the most abundant protein in tobacco leaves. Rubisco has a 55-kiloDaltons (kDa) large subunit (called LSU) and a 12-kDa small subunit (called SSU). The researchers purified Rubisco from many strains of tobacco plants, digested the purified proteins with trypsin, and analyzed the digests for informative differences. When they followed the inheritance of these differences, they found that LSUs manifested patterns of maternal inheritance, while SSUs showed biparental inheritance. From these results, they hypothesized that a chloroplast gene encodes the LSU polypeptide, while a nuclear gene encodes the SSU.

These studies of the two Rubisco subunits reveal that organelle and nuclear genomes cooperate in specifying even a relatively simple enzyme with only two different subunits. Both genomes contribute essential information for most photosynthetic activities, including those whose elements are much more complex.

The inheritance studies just described followed differences in functionally wild-type organelle genomes. To verify and understand the details of uniparental inheritance, organelle geneticists followed the inheritance of mutations affecting phenotype at both the biochemical and the organismal levels, as described in the section that follows.

LHON: A maternally inherited neurodegenerative disease in humans

Leber's hereditary optic neuropathy, or LHON, is a disease in which flaws in the mitochondria's electron transport chain lead to optic nerve degeneration and blindness (**Figure 18.35**). Family pedigrees show that LHON passes only from mother to offspring. In the late 1980s and early 1990s, a series of molecular studies showed that a G-to-A substitution at nucleotide 11 778 in the human mitochondrial genome is a main cause of the condition. The substitution alters an arginine-specifying codon in the NADH dehydrogenase subunit 4 gene to a histidine codon. The resulting protein product diminishes the efficiency of electron flow down the respiratory transport chain, reducing the cell's production of ATP and causing a gradual decline in cell function and ultimately cell death. Because optic nerve cells have a relatively high requirement for energy, the genetic defect affects vision before it affects other physiological systems.

In other pedigrees of large families, not all offspring show signs of the disease, and not all siblings manifesting the condition have symptoms of the same severity. The random allotment to daughter cells of a large number of mitochondria during mitosis helps explain these observations.

Distribution of Organelles During Mitosis. A diploid cell contains dozens to thousands of organelle DNAs. It is therefore not possible to use the terms "homozygous" and "heterozygous" to describe a cell's complement of mtDNA or cpDNA. Instead, geneticists use the terms "heteroplasmic" and "homoplasmic" to describe the genomic makeup of a cell's organelles. **Heteroplasmic** cells contain a mixture of organelle genomes. **Homoplasmic** cells carry only one type of organelle DNA.

Except for the rare appearance of a new mutation, the mitotic progeny of homoplasmic cells carry a single type of organelle DNA. By contrast, the mitotic progeny of heteroplasmic cells may be heteroplasmic, homoplasmic wild type, or homoplasmic mutant. In most people affected by LHON, for example, the optic nerve cells are homoplasmic for the disease mutation; but in some LHON patients, these optic nerve cells are heteroplasmic. Homoplasmy causes earlier appearance of the disease as well as more severe symptoms.

How Distribution Affects Phenotype. The mitotic segregation of organellar genomes has distinct phenotypic consequences. In a woman whose cells are heteroplasmic for the LHON mutation, some ova may carry a few mitochondria with the LHON mutation and a large number of mitochondria with the wild-type gene for subunit 4 of NADH dehydrogenase; other ova may carry mainly mitochondria with LHON mutations; still others may carry only wild-type organelles. The precise combination depends on the random partitioning of mitochondria during the mitotic divisions that gave rise to the germ line.

After fertilization, as a result of the mitotic divisions of embryonic development, the random segregation of mutation-carrying mitochondria from heteroplasmic cells can produce tissues with completely normal ATP production and tissues of low energy production. If cells homoplasmic for low energy production happen to end up in the optic nerve, LHON will result.

Effects of mutations in chloroplast genomes

In plants where cpDNA mutations would be lethal (due to defects in photosynthesis), heteroplasmy for chloroplast genomes is prevalent. In fact, mitotic segregation of the chloroplasts of heteroplasmic cells explains the transmission of variegation in four-o'clocks.

Most female gametes from a variegated plant are heteroplasmic for mutant and wild-type cpDNAs. Zygotes resulting from fertilization with pollen from a wild-type green plant will develop into variegated progeny. Segregation of wild-type and mutant chloroplasts during F_1 plant development may however generate some female gametes with only mutant cpDNA. Fertilization of these homoplasmic mutant gametes with wild-type pollen produces

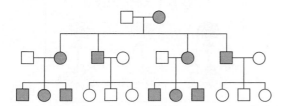

Figure 18.35 LHON pedigree. A hypothetical characteristic pedigree of mitochondrial disease. All offspring of diseased mothers show the disease phenotype, while none of the offspring of diseased fathers show the disease phenotype.

zygotes with only mutant cpDNA; the seedlings that develop from these zygotes cannot carry out photosynthesis and eventually die.

Mechanisms that contribute to uniparental inheritance

Differences in gamete size help explain maternal inheritance in some species. In most higher eukaryotes, the male gamete is much smaller than the female gamete. As a result, the zygote receives a very large number of maternal organelles and, at most, a very small number of paternal organelles.

In some organisms, cells degrade the organelles or the organellar DNA of male gametes. In some plants, the early divisions of the zygote distribute most or all of the paternal organelle genomes to cells that are not destined to become part of the embryo. In certain animals, details of fertilization prevent a paternal cell from contributing its organelles to the zygote. In the pre-vertebrate chordates called tunicates, for example, events of fertilization allow only the sperm nucleus to enter the egg, physically excluding the paternal mitochondria. In some organisms where the complete gametes fuse, the zygote destroys the paternal organelles after fertilization.

The **Genetics and Society** box "**Mitochondrial DNA Tests as Evidence of Kinship in Argentine Courts**" describes how a human rights organization in Argentina used mtDNA sequences as the legal basis for reuniting kidnapped children with their biological families. The maternal inheritance of mitochondria makes it possible to compare and match the DNA of a grandmother and a grandchild.

Some organisms exhibit biparental inheritance of organellar genomes

Although uniparental inheritance of organelles is the norm among most metazoans and plants, single-celled yeast and some plants inherit their organelle genomes from both parents in a **biparental** fashion.

The earliest report of biparental inheritance of organelles is a 1909 description of reciprocal crosses between green and variegated geraniums (*Pelargonium zonale*) (**Figure 18.36**). Unlike what happens in four-o'clocks, both reciprocal crosses yielded green, white,

and variegated seedlings in varying proportions (**Figure 18.37**). Thus, variegated leaves in geraniums are a chimeric condition that results from the chloroplast traits inherited from both parents. Many other plants, as well as yeast, similarly inherit their organelle genomes from both parents.

Principles of non-Mendelian inheritance: A summary

Three features distinguish the non-Mendelian traits encoded by organelle genomes from the Mendelian traits encoded by nuclear genomes.

1. In the inheritance of organelle genomes from one generation to the next, there is a 4:0 segregation of parental alleles, instead of the 2:2 pattern seen for the alleles of nuclear genes.
2. In most organisms, transmission of organelle-encoded traits is uniparental, mainly maternal, although in a few organisms transmission is biparental.
3. With both uniparental and biparental inheritance, when the parents transmit organelles of more than one genotype, mitotic segregation of those genotypes occurs in the offspring. This segregation of genotypes during mitosis is a consequence of the random partitioning of organelles during cell division.

Figure 18.37 Reciprocal crosses show biparental inheritance. Reciprocal crosses between green and variegated geraniums yield the same classes of offspring, indicating that the gene is inherited from both parents.

 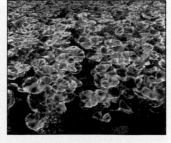

Figure 18.36 Biparental inheritance of variegation in geraniums. Examples of green and variegated *P. zonale* plants.

Between 1976 and 1983, the military dictatorship of Argentina kidnapped, incarcerated, and killed more than 10 000 university students, teachers, social workers, union members, and others who did not support the regime. Many very young children disappeared; and close to 120 babies were born to women in detention centres. In 1977, the grandmothers of some of these infants and toddlers began to hold vigils in the main square of Buenos Aires to bear witness and inform others about the disappearance of their children and grandchildren. They soon formed a human rights group—the "Grandmothers of the Plaza de Mayo."

The grandmothers' goal was to locate the more than 200 grandchildren they suspected were still alive and reunite them with their biological families. To this end, they gathered information from eyewitnesses, such as midwives and former jailers, and set up a network to monitor the papers of children entering kindergarten. They also publicized their work inside Argentina and contacted organizations outside the country, including the United Nations Human Rights Commission and the American Association for the Advancement of Science (AAAS).

What the grandmothers asked of AAAS was help with genetic analyses that would stand up in court. By the time a democracy had replaced the military regime and the grandmothers could argue their legal cases before a relatively impartial court, children abducted at age two or three or born in 1976 were 7–10 years old. Although the grandmothers had compiled an enormous amount of circumstantial evidence, the Argentine courts did not accept such evidence as proof of a young person's identity and biological relatedness. The courts did acknowledge, however, that although the size and other external features of the children had changed, their genes—relating them unequivocally to their biological families—had not. The grandmothers, who had educated themselves about the potential of genetic tests, sought help with the details of obtaining and analyzing such tests. Starting in 1983, the courts agreed to accept their test results as proof of kinship.

In 1983, the best way to confirm or exclude the relatedness of two or more individuals was to compare proteins called human lymphocyte antigens (HLAs). People carry a unique set of HLA markers on their white blood cells, or lymphocytes, and these markers are diverse enough to form a kind of molecular fingerprint. HLA analyses can be carried out even if a child's parents are no longer alive, because for each HLA marker, a child inherits one allele from the maternal grandparents and one from the paternal grandparents. Statistical analyses can establish the probability that a child shares genes with a set of grandparents.

The AAAS put the grandmothers in touch with Mary Claire King, then at the University of California. In the 1980s, King taught Argentine medical workers to analyze the diverse HLA markers on white blood cells. The grandmothers then obtained the HLA types of as many living members as possible of the missing children's families and stored that information in an HLA bank. When a child whom they believed to be one of the missing turned up, they analyzed his or her HLA type and tried to find a match among their data. Depending on the number of different alleles carried and the rarity of the variations, the probability that a tested child belonged to the family claiming him or her on the basis of eyewitness accounts of a birth or abduction varied from 75 percent to 99 percent. As time passed and the "easier" cases had been settled, the limitations of the HLA approach became apparent. By the mid-1980s, for example, there were too few living relatives in some families to establish a reliable match through HLA typing. But the advent of new tools such as PCR and DNA sequencing made it possible to look at DNA directly.

King and two colleagues—C. Orrego and A. C. Wilson—used the new techniques to develop a mtDNA test based on the PCR amplification and direct sequencing of a highly variable noncoding region of the mitochondrial genome. Maternal inheritance and lack of recombination mean that as long as a single maternal relative is available for matching, the approach can resolve cases of disputed relatedness. The extremely polymorphic noncoding region makes it possible to identify grandchildren through a direct match with the mtDNA of only one person—their maternal grandmother, or mother's sister or brother—rather than through statistical calculations assessing data from four people.

To validate their approach, King and colleagues amplified sequences from three children and their three maternal grandmothers without knowing who was related to whom. The mtDNA test unambiguously matched the children with their grandmothers. Thus, after 1989, the grandmothers included mtDNA data in their archives.

Today, the grandchildren—the children of "the disappeared"—have reached adulthood and attained legal independence. Although most of their grandmothers have died, the grandchildren may still discover their biological identity and what happened to their families through the HLA and mtDNA data the grandmothers left behind.

- Non-Mendelian inheritance patterns can result from genes carried in organelles rather than in the nucleus. Organelles are generally inherited from only one parent, usually the female. In humans, certain diseases of the muscles and nervous system have been shown to pass only from mothers to their offspring; these diseases result from faulty mitochondrial genes.

- Some organisms inherit organelles from both parents. Where organellar genomes differ between parents, offspring show a chimeric body pattern. Traits subject to non-Mendelian inheritance exhibit three features: 4:0 segregation of alleles; maternal inheritance of traits in most organisms; and in the case of biparental inheritance, mitotic segregation of genotypes due to random partitioning of organelles.

Some debilitating diseases of the human nervous system pass from mother to daughters and sons, from affected daughters to granddaughters and grandsons, and so on down through the maternal line. The pattern of inheritance suggests the mutations are mitochondrial. Unexpectedly, the symptoms of these diseases vary enormously among family members, even among very close relatives. In addition, some hypotheses link mitochondrial function with the so-called diseases of aging, including Alzheimer's disease (see the **Focus on Inquiry** box **"Understanding the Role of Mitochondria in Human Disease"**).

MERRF has a cluster of symptoms related to mutations in mtDNA

People with a rare inherited condition known as <u>m</u>yoclonic <u>e</u>pilepsy and <u>r</u>agged <u>r</u>ed <u>f</u>ibre disease (MERRF) have a range of symptoms: uncontrolled jerking (the myoclonic epilepsy part of the condition), muscle weakness, deafness, heart problems, kidney problems, and progressive dementia. Affected individuals often have an unusual "ragged" staining pattern in regions of their skeletal muscles, which explains the ragged red fibre part of the condition's name (**Figure 18.38**).

As the pedigree in **Figure 18.39** shows, family members inherit MERRF from their mothers; in the pedigree, none of the offspring of the affected male sibling (II-4) exhibit symptoms of the disease. The family history also reveals individual variations in the number and severity of symptoms. From these two features of transmission, clinical researchers suspected that MERRF results from mutations in the mitochondrial genome.

Molecular analyses confirmed this hypothesis. The mtDNA from patients affected by MERRF carries a

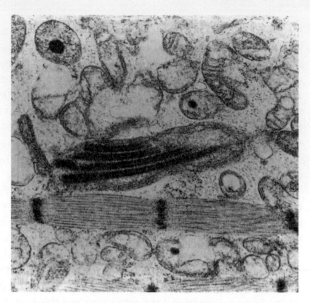

Figure 18.38 Muscle cell of MERRF patient. Transmission electron micrograph of muscle mitochondria from patients expressing MERRF. Mutant mitochondria are highly abnormal, showing paracrystalline arrays and crista degeneration.

mutation in the gene for tRNA^{Lys} or one of the other mitochondrial tRNAs. These tRNA mutations disrupt the synthesis of proteins in multiplexes I and IV of the mitochondrial electron transport chain, thereby decreasing the production of ATP.

In a second large family in which clinicians looked at the mitochondria of muscle cells, an individual carrying 73 percent mutant and 27 percent normal mtDNAs showed no symptoms of MERRF; a relative with 85 percent mutant mtDNA showed no external signs of the disease, but lab

---

FOCUS ON Inquiry   Understanding the Role of Mitochondria in Human Disease

Figure A   Dr. Brian Robinson.

Dr. Brian Robinson (**Figure A**) is Professor of Genetics and Genome Biology at the University of Toronto. His interests lie in understanding the origins of mitochondrial genetic diseases in children; the effects of which can range from mild intolerance to exercise, to severe neurodegeneration. During his long and distinguished career, Dr. Robinson and his colleagues have developed a multitude of tests used in the diagnosis of mitochondrial diseases affecting children. Based on the results of these tests (i.e., identification of the affected gene), medical professionals can make informed decisions regarding the most appropriate form of treatment. You can learn more about Dr. Robinson's research at www.sickkids.ca/Research/AbouttheInstitute/Profiles/robinson-profile.html.

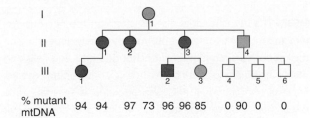

% mutant mtDNA: 94 94 97 73 96 96 85 0 90 0 0

Figure 18.39 Maternal inheritance of the mitochondrial disease MERRF. Pedigree of family showing inheritance of MERRF. The pedigree shows a typical pattern of maternal transmission observed with mitochondrial mutations. The percentage of mutant mtDNA in the cells of individuals varies and corresponds with the severity of the condition (indicated by different colour coding).

tests revealed some muscle tissue abnormalities; and two family members with 98 percent mutant mitochondria showed serious symptoms of MERRF. This suggests that a relatively small percentage of normal mitochondria can have a strong protective effect.

It is likely that many tissues in individuals affected by MERRF are heteroplasmic. Within each person, the ratio of mutant-to-wild-type mtDNA varies considerably from tissue to tissue, and because each tissue has its own energy requirements, even the same ratio can affect different tissues to varying extent (**Figure 18.40**). Muscle and nerve cells have the highest energy needs of all types of cells and are therefore the most dependent on oxidative phosphorylation. Mitochondrial mutations that by chance segregate to these tissues generate the defining features of MERRF.

Mitochondrial mutations may have an impact on aging

Some mutations in mtDNA are inherited through the germ line, while others arise sporadically in somatic cells as a result of random events, such as radiation or chemical mutagens. We have also seen that the rate of somatic mutations is much higher in mtDNA than in nuclear DNA. In part, this rate is a result of DNA-damaging free radicals, which are generated by the mitochondrial oxidative phosphorylation system. In one study, mtDNA accumulated 16 times more oxidative damage than nuclear DNA.

Some researchers focusing on the genetics of aging think that the accumulation of mtDNA mutations over a lifetime results in an age-related decline in oxidative phosphorylation. This decline, in turn, accounts for some of the symptoms of aging, such as decreases in heart and brain function.

Proponents of this hypothesis suggest that individuals born with deleterious mtDNA mutations start life with a diminished capacity for ATP production, and as a result, several of their tissues may cross the threshold from function to nonfunction early or in the middle of life. By comparison, people born with a normal mitochondrial genome start life with a high capacity for ATP production and may

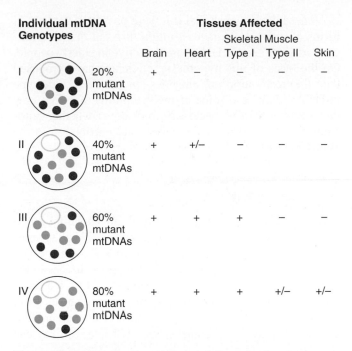

Individual mtDNA Genotypes		Tissues Affected		Skeletal Muscle		
		Brain	Heart	Type I	Type II	Skin
I	20% mutant mtDNAs	+	−	−	−	−
II	40% mutant mtDNAs	+	+/−	−	−	−
III	60% mutant mtDNAs	+	+	+	−	−
IV	80% mutant mtDNAs	+	+	+	+/−	+/−

Figure 18.40 Disease phenotypes and the ratio of mutant-to-wild-type mtDNAs. The proportion of mutant mitochondria determines the severity of the MERRF phenotype and the tissues that are affected (+). Tissues with higher energy requirements (e.g., brain) are least tolerant of mutant mitochondria. Tissues with low energy requirements (e.g., skin) are affected only when the proportion of wild-type mitochondria is greatly reduced.

die before a large number of tissues dip below the required energy threshold.

Evidence in support of an association between mtDNA mutations and aging comes from a variety of studies. In one study, researchers looked at 140 hearts obtained from autopsies and found significant decreases in cytochrome C oxidase, a respiratory enzyme largely encoded by mtDNA. In another study, researchers analyzed a 7.4-kb and a 5-kb deletion in heart and brain mtDNAs in people of different ages. The percentage of hearts that had the 7.4-kb deletion increased with age, and the number of 5-kb deletions increased in normal heart tissue after age 40. Moreover, the 5-kb deletion was absent from the brain tissue of children but present in the brain tissue of adults.

Finally, although biomedical researchers had known for decades that the brain cells of people showing symptoms of Alzheimer's disease (AD) have an abnormally low energy metabolism, they recently discovered that 20 percent to 35 percent of the mitochondria in the brain cells of most AD patients carry mutations in two of their three cytochrome C oxidase genes, which could impair the brain's energy metabolism. To confirm an association between this enzymatic abnormality and AD, the researchers transferred mitochondria from AD patients into normal cultured cells from which they had removed the native mitochondria, and they found that the engineered cells had defective energy production.

These data suggest that if it were possible to assess all forms of mtDNA damage, it might turn out that a significant proportion of mtDNAs are defective in elderly people. On the basis of this hypothesis, clinicians have proposed that the restoration of enzymes encoded by wild-type mtDNA might ease some of the symptoms of aging. Further research will be necessary to discover whether mitochondrial damage makes a significant contribution to the aging process.

> MERRF is a mitochondrial disease affecting the nervous system and muscle tissues. Symptoms vary widely depending on the random segregation of mitochondria into egg cells and their subsequent location by chance in a developing embryo. Because mitochondrial energy production is so vital, some researchers believe that certain conditions of aging result from diminished mitochondrial function.

Connections

The study of bacterial, chloroplast, and mitochondrial genetics underscores the unity of genetic phenomena in all types of living organisms. Double-stranded DNA serves as the genetic material in bacteria and in these organelles, as it does in the nuclear genome of eukaryotes (Chapters 5 and 6). However, we have also seen a remarkable diversity of mechanistic detail in biological processes. Although bacteria do not produce gametes that fuse to become zygotes (Chapter 3), they can exchange genes between different strains through transformation, conjugation, and transduction. These three modes of gene transfer increase the potential for the evolution of prokaryotic genetic material. In a similar way, transfer of DNA from chloroplast and mitochondrial genomes to the nuclear genome has produced a unique symbiosis between the nucleus and organelles.

Essential Concepts

1. Bacteria are prokaryotic cells with no membrane-enclosed nucleus or other cell organelles. The bacterial genome consists of a single circular chromosome in which the genes are tightly packed, with about one gene per kilobase pair. **[LO1–2]**

2. In addition to their chromosome, most bacteria carry plasmids: small circles of double-stranded DNA. Plasmids may include genes that benefit the bacterial host under certain conditions. One important group of plasmids promotes conjugative gene transfer between two bacteria. **[LO2]**

3. Bacterial genomes contain IS and Tn elements. These elements are transposons that can move between sites on any DNA molecule in the cell. **[LO2]**

4. Transformation is a form of gene transfer in which donor DNA that is floating free in the growth medium enters a recipient cell. Conjugation is a second form of gene transfer. It depends on direct cell-to-cell contact between a donor carrying a conjugative plasmid (the F plasmid is one example) or an integrated conjugative element and a recipient lacking such a plasmid or element. Transduction is a third form of gene transfer in bacteria involving the packaging of bacterial donor DNA in the protein coat of a bacteriophage. **[LO3]**

5. The tools of genomics, including comparative genome analyses and DNA arrays, have provided new insights into pathogenesis, evolution, and microbial diversity. **[LO1–3]**

6. According to the endosymbiont theory, mitochondria and chloroplasts evolved from bacteria engulfed by the precursors of eukaryotic cells. The genomes of these organelles have probably lost more than two-thirds of their original bacterial genes in the course of evolution. **[LO4]**

7. Mitochondria and chloroplasts are semiautonomous organelles of energy conversion. They carry their own double-stranded DNA in circular or linear chromosomes whose size and gene content vary from species to species. **[LO4]**

8. Translation in the mitochondria of many species depends on an alternative genetic code. **[LO4]**

9. In most species, organelle genomes show uniparental inheritance, mainly through the maternal line. Cells containing a mixture of organelle genomes are heteroplasmic. Cells carrying only one type of organelle DNA are homoplasmic. The genomes of heteroplasmic cells are not evenly partitioned at mitosis. **[LO5]**

10. Diseases caused by mutation in mtDNA are recognized by maternal inheritance of the disease. The extent of the disease phenotype often depends on the ratio of mutant versus wild-type mitochondria in a cell. **[LO5]**

Solved Problems

I. You have cloned the gene encoding the major protein in the flagella of a new bacterial strain. In screening for mutant bacteria that have a defective flagellar protein, you found mutants at an exceptionally high frequency (1 in 10^3 bacterial cells). You suspect these may have been caused by insertion of a transposable element into the gene. How could you determine if this had occurred?

Answer

One way to determine if the high-frequency mutants result from insertion into the gene is to perform a Southern hybridization. *The cloned gene would be used as a probe to hybridize with DNA from the wild-type and the mutant cells.* If the mutant arises from insertion of a transposable element, the size of fragments containing the interrupted gene will be different from fragments containing the normal gene.

II. Using bacteriophage P22 you performed a three-factor cross in *Salmonella typhimurium*. The cross was between an Arg⁻ Leu⁻ His⁻ recipient bacterium and bacteriophage P22, which was grown on an Arg⁺ Leu⁺ and His⁺ strain. You selected for 1000 Arg⁺ transductants and tested them on several selective media by replica plating. You obtained the following results:

Arg⁺ Leu⁻ His⁻	585
Arg⁺ Leu⁻ His⁺	300
Arg⁺ Leu⁺ His⁺	114
Arg⁺ Leu⁺ His⁻	1

a. What is the order of the three markers?

b. What are the co-transduction frequencies?

Answer

a. The order can be determined by looking at the relative frequencies of each phenotypic class. Arg⁺ Leu⁻ His⁻ is the largest class, with only the *arg* gene transferred. The next largest class is Arg⁺ and His⁺. Therefore, *arg* and *his* are closer to each other than *arg* and *leu*. *The order of the genes is arg-his-leu.*

b. Co-transduction frequency is the percentage of cells that received two markers. For *arg* and *his*, this includes the Arg⁺ Leu⁻ His⁺ cells (300) and Arg⁺ Leu⁺ His⁺ cells (114). The co-transduction frequency of *arg* and *his* is 414/1000 = 41.4%. The co-transduction frequency of *arg* and *leu* is 114 + 1 or 115/1000 = 11.5%.

III. Differential hybridization of a probe to mitochondrial DNA from two *Xenopus* species was the methodology employed to demonstrate maternal inheritance in vertebrates (see Figure 18.34). However, this hybridization technique was not sensitive enough to detect small amounts of paternal DNA. What technique that is more sensitive to small amounts of DNA could be used today? How could you use this technique to determine if paternal mitochondrial DNA was present in the progeny of the interspecies cross?

Answer

Polymerase chain reaction (PCR) is a sensitive technique that detects very small amounts of DNA. Oligonucleotide primers that are specific for each of the mitochondrial DNAs in each of the two different species could be used to determine if paternal DNA is present in the offspring from the interspecies cross.

IV. a. Does the following pedigree suggest mitochondrial inheritance? Why or why not?

b. Is there another mode of inheritance that is consistent with these data?

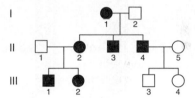

Answer

a. The data presented in this pedigree *are consistent with mitochondrial inheritance because the trait is transmitted by females; the affected males in this family did not transmit the trait; and all of the females' progeny have the trait.*

b. This inheritance pattern is *also consistent with transmission of an autosomal dominant trait.* According to this hypothesis, individuals I-1 and II-2 passed on the dominant allele to all children, but II-4 did not pass on the dominant allele to either child.

Problems

Vocabulary

1. For each of the terms in the left column, choose the best matching phrase in the right column.

i. transformation

ii. conjugation

iii. transduction

iv. lytic cycle

v. lysogeny

vi. episome

vii. auxotroph

1. requires supplements in medium for growth

2. transfer of DNA between bacteria via virus particles

3. small circular DNA molecule that can integrate into the chromosome

4. transfer of naked DNA

5. transfer of DNA requiring direct physical contact

6. integration of phage DNA into the chromosome

7. infection by phages in which lysis of cells releases new virus particles

Section 18.1

2. The unicellular rod-shaped bacterium *E. coli* is approximately 2 μm long and 0.8 μm wide, and has a genome consisting of a single 5.6-Mb circular DNA molecule. The unicellular archaean *Methanosarcina acetivorans* is spherical (coccus-shaped) with a diameter of 3 μm and has a 5.7-Mb circular genome. The unicellular eukaryote *Saccharomyces cerevisiae* is roughly spherical, with a diameter of 5–10 μm. It has a haploid genome of 12 Mb divided among 16 linear chromosomes. Given these descriptions, how could you determine whether a new, uncharacterized microorganism was a bacterium, an archaean, or a eukaryote?

3. A liquid culture of *E. coli* at a concentration of 2×10^8 cells/mL was diluted serially, as shown in the following diagram, and 0.1 mL of cells from the last two test tubes were spread on agar plates containing rich medium. How many colonies do you expect to grow on each of the two plates?

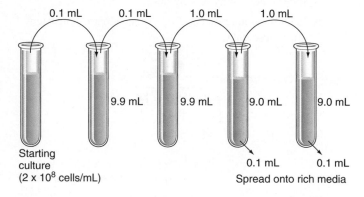

4. Now that the sequence of the entire *E. coli* genome (about 5 Mb) is known, you can determine exactly where a cloned fragment of DNA came from in the genome by sequencing a few bases and matching that data with genomic information.

   a. How many nucleotides of sequence information would you need to determine exactly where a fragment is from?

   b. If you had purified a protein from *E. coli* cells, roughly how many amino acids of that protein would you need to know to establish which gene encoded the protein?

5. Pick out the medium (i, ii, iii, or iv) onto which you would spread cells from a Lac⁻ Met⁻ *E. coli* culture to

   a. select for Lac⁺ cells

   b. screen for Lac⁺ cells

   c. select for Met⁺ cells

      i. minimal medium + glucose + methionine

      ii. minimal medium + glucose (no methionine)

      iii. rich medium + X-Gal

      iv. minimal medium + lactose + methionine

6. Linezolid is a new type of antibiotic that inhibits protein synthesis in several bacterial species by binding to the 50S subunit of the ribosome and inhibiting its ability to participate in the formation of translational initiation complexes. Physicians are particularly interested in this antibiotic for treating pneumonia caused by penicillin-resistant *Streptococcus pneumoniae* (also called "pneumococci"). To explore the mechanisms by which pneumococci can develop resistance to linezolid, you want first to identify linezolid-resistant strains. Next, using one of these strains as starting material, you now want to identify derivatives of these mutants that are no longer tolerant of linezolid.

   a. Outline the techniques you would use to identify linezolid-resistant mutant pneumococci and linezolid-sensitive derivatives of these mutants. In each case, would your techniques involve direct selection, screening, replica plating, enrichment, treating with mutagens, or testing for a visible phenotype?

   b. Suggest possible mutations that could be responsible for the two kinds of phenotypes you will identify. What types of events in the bacterial cells would be altered by the mutations? Can you classify these mutations as loss-of-function or gain-of-function?

Section 18.2

7. DNA sequencing of the entire *H. influenzae* genome was completed in 1995. When DNA from the nonpathogenic strain *H. influenzae Rd* was compared with that of the pathogenic *b* strain, eight genes of the fimbrial gene cluster (located between the *purE* and *pepN* genes) involved in adhesion of bacteria to host cells were completely missing from the nonpathogenic strain. What effect would this have on co-transformation of *purE* and *pepN* genes using DNA isolated from the nonpathogenic versus the pathogenic strain?

8. a. Using the following pieces of technical information, in the order given, explain how you would be able to identify the genes encoding proteins in *E. coli* cells that could bind directly to β-galactosidase: (1) β-galactosidase protein binds very tightly to a resin called APTG-agarose; (2) the 20 amino acids found in proteins vary widely in molecular weight; (3) the enzyme trypsin can cleave proteins into smaller peptides that are in the range of 3–40 amino acids long; the enzyme cleaves in a very predictable and reproducible way (after lysine and arginine amino acids); (4) modern techniques of mass spectrometry can measure the molecular weight of peptides to an accuracy of 0.01 percent; mass spectrometry machines measure the molecular weights of a large number of peptides in a complex mixture at the same time; (5) the entire *E. coli* genome has been sequenced.

   b. Generalize the technique you described in part *a* to identify the genes encoding proteins that bind to any other particular protein in *E. coli*. (*Hint:* Use the fact that β-galactosidase binds to the APTG-agarose in your scheme.)

9. List at least two examples in which bacterial strains have acquired new pathogenicity genes. State both the organism and mode of introduction of the gene.

10. The numbers of IS1 elements in different laboratory strains of *E. coli* vary. There are no recognition sites for the enzyme *Eco*RI in IS1. How could you determine the number of IS1 elements in the two strains *E. coli B* and *E. coli K*?

11. There is usually one copy of the F plasmid per cell in an *E. coli* strain. You suspect you have isolated a cell in which a mutation increases the copy number of F to three to four per cell. (The copy number is determined by hybridization experiments.) How could you distinguish between the possibility that the copy number change was due to a mutation in the F plasmid versus a mutation in a chromosomal gene?

12. Genome sequences show that some pathogenic bacteria contain virulence genes next to bacteriophage genes. Why does this suggest lateral gene transfer, and what would the mechanism of transfer have been?

Section 18.3

13. Bacteria are promiscuous creatures, sharing DNA within and between species by several mechanisms.

   a. What are three general mechanisms of gene transfer in bacteria?

   b. Which type of transfer mechanism(s) can occur using a plasmid?

   c. Which type of transfer mechanism(s) requires a bacteriophage?

   d. Which mechanism(s) require recombination in the recipient to produce new genetically stable cells?

14. In *E. coli*, the genes *purC* and *pyrB* are located halfway around the chromosome from each other. These genes are never co-transformed. Why is this?

15. Genes encoding toxins are often located on plasmids. There has been a recent outbreak in which a bacterium that is usually nonpathogenic is producing a toxin. Plasmid DNA can be isolated from this newly pathogenic bacterial strain and separated from the chromosomal DNA. To determine if the plasmid DNA contains a gene encoding the toxin, you could determine the sequence of the entire plasmid and search for a sequence that looks like other toxin genes previously identified. There is an easier way to determine whether the plasmid DNA carries the gene(s) for the toxin that does not involve DNA sequence analysis. Describe an experiment using this easier method.

16. a. You want to perform an interrupted mating mapping with an Hfr strain that is Pyr$^+$, Met$^+$, Xyl$^+$, Tyr$^+$, Arg$^+$, His$^+$, Mal$^+$, and Str^s. Describe an appropriate bacterial strain to be used as the other partner in this mating.

   b. In an Hfr × F$^-$ cross, the *pyrE* gene enters the recipient in five minutes, but at this time point there are no exconjugants that are Met$^+$, Xyl$^+$, Tyr$^+$, Arg$^+$, His$^+$, or Mal$^+$. The mating is now allowed to proceed for 30 minutes and Pyr$^+$ exconjugants are selected. Of the Pyr$^+$ cells, 32 percent are Met$^+$, 94 percent are Xyl$^-$, 7 percent are Tyr$^+$, 59 percent are Arg$^+$, 0 percent are His$^+$, and 71 percent are Mal$^+$. What can you conclude about the order of the genes?

17. In a cross between an Hfr that has the genotype *ilv$^+$ bgl$^+$ mtl$^+$*, and an F$^-$ that is *ilv$^-$ bgl$^-$ mtl$^-$*, the *ilv* gene is known to be transferred later than *bgl* and *mtl*. To determine the order of *bgl* and *mtl* with respect to *ilv*, ilv$^+$ exconjugants were selected, and these colonies were screened for Bgl and Mtl phenotypes. Based on the following data, what is the order of the three genes?

Exconjugant type	Number of exconjugants
Ilv$^+$ Mtl$^+$ Bgl$^+$	220
Ilv$^+$ Mtl$^-$ Bgl$^-$	60
Ilv$^+$ Mtl$^+$ Bgl$^-$	0
Ilv$^+$ Mtl$^-$ Bgl$^+$	18

18. Starting with an F$^+$ strain that was prototrophic (i.e., had no auxotrophic mutations) and Str^s, several independent Hfr strains were isolated. These Hfr strains were mated to an F$^-$ strain that was Str^r Arg$^-$ Cys$^-$ His$^-$ Ilv$^-$ Lys$^-$ Met$^-$ Nic$^-$ Pab$^-$ Pyr$^-$ Trp$^-$. Interrupted mating experiments showed that the Hfr strains transferred the wild-type alleles in the order listed in the following table as a function of time. The time of entry for the markers within parentheses could not be distinguished from one another.

Hfr strain	Order of transfer →
HfrA	*pab ilv met arg nic (trp pyr cys) his lys*
HfrB	*(trp pyr cys) nic arg met ilv pab lys his*
HfrC	*his lys pab ilv met arg nic (trp pyr cys)*
HfrD	*arg met ilv pab lys his (trp pyr cys) nic*
HfrE	*his (trp pyr cys) nic arg met ilv pab lys*

   a. From these data, derive a map of the relative position of these markers. Indicate with labelled arrows the position and orientation of the integrated F plasmid for each Hfr strain.

   b. To determine the relative order of the *trp, pyr,* and *cys* markers and the distances between them, HfrB was mated with the F$^-$ strain long enough to allow transfer of the *nic* marker, after which Trp$^+$ recombinants were selected. The unselected markers *pyr* and *cys* were then scored in the Trp$^+$ recombinants, yielding the following results:

Number of recombinants	Trp	Pyr	Cys
790	+	+	+
145	+	+	−
60	+	−	+
5	+	−	−

Draw a map of the *trp, pyr,* and *cys* markers relative to each other. (Note that you cannot determine the order relative to the *nic* or *his* genes using these data.) Express map distances between adjacent genes as the frequency of crossing-over between them.

19. Suppose you have two Hfr strains of *E. coli* (HfrA and HfrB), derived from a fully prototrophic streptomycin-sensitive (wild-type) F$^+$ strain. In separate experiments you allow these two Hfr strains to conjugate with an F$^-$ recipient strain (Rcp) that is streptomycin resistant and auxotrophic for glycine (Gly$^-$), lysine (Lys$^-$), nicotinic acid (Nic$^-$), phenylalanine (Phe$^-$), tyrosine (Tyr$^-$), and uracil (Ura$^-$). By using an interrupted mating protocol you determined the earliest time after mating at which each of the markers can be detected in the streptomycin-resistant recipient strain, as shown here.

	Gly$^+$	Lys$^+$	Nic$^+$	Phe$^+$	Tyr$^+$	Ura$^+$
HfrA × Rcp	3	*	8	3	3	3
HfrB × Rcp	8	3	13	8	8	8

(The * indicates that no Lys$^+$ cells were recovered in the 60 minutes of the experiment.)

a. Draw the best map you can from these data, showing the relative locations of the markers and the origins of transfer in strains HfrA and HfrB. Show distances where possible.

b. To resolve ambiguities in the preceding map, you studied co-transduction of the markers by the generalized transducing phage P1. You grew phage P1 on strain HfrB and then used the lysate to infect strain Rcp. You selected 1000 Phe$^+$ clones and tested them for the presence of unselected markers, with the following results:

Number of transductants	Phenotype					
	Gly	Lys	Nic	Phe	Tyr	Ura
600	−	−	−	+	−	−
300	−	−	−	+	−	+
100	−	−	−	+	+	+

Draw the order of the genes as best you can based on the preceding co-transduction data.

c. Suppose you wanted to use generalized transduction to map the *gly* gene relative to at least some of the other markers. How would you modify the co-transduction experiment just described to increase your chances of success? Describe the composition of the medium you would use.

20. In two isolates (one is resistant to ampicillin and the other is sensitive to ampicillin) of a new bacterium, you found that genes encoding ampicillin resistance are being transferred into the sensitive strain. To determine if the gene transfer is transduction or transformation, you treat the mixed culture of cells with DNase. Why would this treatment distinguish between these two modes of gene transfer? Describe the results predicted if the gene transfer is transformation versus transduction.

21. You can carry out matings between an Hfr and F$^-$ strain by mixing the two cell types in a small patch on a plate and then replica plating to selective medium. This methodology was used to screen hundreds of different cells for a recombination-deficient *recA$^-$* mutant. Why is this an assay for RecA function? Would you be screening for a *recA$^-$* mutation in the F$^-$ or Hfr strain using this protocol?

22. Generalized and specialized transduction both involve bacteriophages. What are the differences between these two types of transduction?

Section 18.4

23. Recombineering involves *in vitro* production of mutant DNA to be transferred into a recipient and *in vivo* incorporation into the genome of the recipient. Are the following the *in vitro* or *in vivo* parts of this procedure?

a. primer DNA

b. antibiotics

c. recombination enzymes

d. PCR amplification

24. *Streptococcus parasanguis* is a bacterial species that initiates dental plaque formation by adhering to teeth. To investigate ways to eliminate plaque, researchers constructed a plasmid, depicted in the figure shown, to mutagenize *S. parasanguis*. The key features of this plasmid include *repA-ts* (a temperature-sensitive origin of replication), *kan^r* (a gene for resistance to the antibiotic kanamycin), and the transposon *IS256*. This transposon contains the *erm^r* gene for resistance to the antibiotic erythryomycin and transposes in *S. parasanguis* thanks to a gene encoding a transposase enzyme that moves all DNA sequences located between the transposon's inverted repeats (IRs).

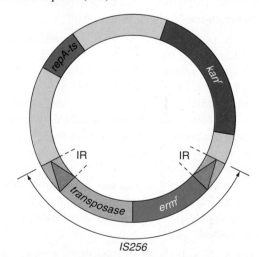

IS256

a. How could the researchers use this plasmid as a mutagen? Consider how they could get the transposon into the bacteria, and how they could identify strains that had new insertions of *IS256* into *S. parasanguis* genes. Your answer should explain why the plasmid has two different antibiotic resistance genes as well as a temperature-sensitive origin of replication.

b. Why would the researchers use this plasmid as a mutagen?

25. Is each of these statements true of chloroplast or mitochondrial genomes, both, or neither?

    a. contain tRNA genes

    b. exist as condensed structures called nucleoids

    c. all genes necessary for function of the organelle are present

    d. vary in size from organism to organism

26. Some genes required for chloroplast function are encoded in the nuclear genome; others are encoded in the chloroplast genome. Nuclear and chloroplast DNAs have different buoyant densities and can therefore be separated from each other by centrifugation based on these differences. There is a small amount of cross-contamination in the separation of nuclear and chloroplast DNAs using this technique. You have just found that a probe for a photosynthetic gene that is present in the chloroplast genome of plants hybridizes to nuclear DNA of a red alga.

    a. Do these results clearly show that the gene of interest is nuclear in the red alga? Why or why not?

    b. What additional DNA hybridization information would allow you to clarify your answer to part *a*?

    c. Assuming this red alga shows uniparental inheritance of chloroplast genes and can be used in reciprocal crosses, design an experiment to confirm the genomic location of the gene discussed in part *a*.

27. "Reverse translation" is a term given to the process of deducing the DNA sequence that could encode a particular protein. Assume that you had the amino acid sequence Trp-His-Ile-Met.

    a. What mammalian nuclear DNA sequence could have encoded these amino acids? (Include all possible variations.)

    b. What mammalian mitochondrial DNA sequence could have encoded these amino acids? (Include all possible variations.)

28. a. Results from hybridization using a probe for the small-subunit gene of the Rubisco protein and a probe for the large-subunit gene of the Rubisco protein to chloroplast (cp) and nuclear DNAs from a green, a red, and a brown alga are shown here. What conclusions would you reach about the location of the small- and large-subunit genes in each of the three types of algae?

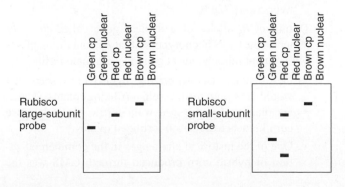

b. When RNA was extracted from the same three algal species and hybridized with a large-subunit Rubisco probe and also with a small-subunit probe, the following results were obtained. What conclusion would you reach about large- and small-subunit gene transcription in red and golden brown algae? Is this consistent with your answer in part *a*?

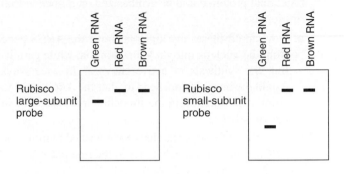

29. Which of the following characteristics of chloroplasts and/or mitochondria make them seem more similar to bacterial cells than to eukaryotic cells?

    a. Translation is sensitive to chloramphenicol and erythromycin.

    b. Alternate codons are used in mitochondria genes.

    c. Introns are present in organelle genes.

    d. DNA in organelles is not arranged in nucleosomes.

30. An example of a cloning vector used for biolistic transformation of chloroplasts is shown in the following diagram. The vector DNA can be prepared in large quantities in *Escherichia coli*. Once "shot" into a chloroplast, the vector DNA integrates into the genome. Match the component of the vector with its function.

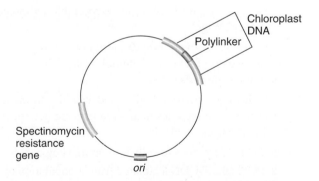

    a. spectinomycin resistance gene

    b. chloroplast DNA

    c. polylinker (multiple restriction sites)

    d. *ori*

    1. homologous DNA that mediates integration

    2. gene used to select chloroplast transformants

    3. sequence for replication in *E. coli*

    4. site at which DNA can be inserted

31. The *Saccharomyces cervevisiae* nuclear gene *ARG8* encodes an enzyme that catalyzes a key step in biosynthesis of the amino acid arginine; this protein is normally synthesized on cytoplasmic ribosomes, but then is transported into mitochondria, where the enzyme conducts its functions. In 1996, T. D. Fox and his colleagues constructed a strain of yeast in which a gene encoding the Arg8 protein was itself moved into mitochondria, where functional protein could be synthesized on mitochondrial ribosomes.

    a. How could these investigators move the *ARG8* gene from the nucleus into the mitochondria, while permitting the synthesis of active enzyme? In what ways would the investigators need to alter the *ARG8* gene to allow it to function in the mitochondria instead of in the nucleus?

    b. Why might these researchers have wished to move the *ARG8* gene into mitochondria in the first place?

Section 18.6

32. Studies distinguishing between uniparental and biparental inheritance of organelles employed a variety of detection methods. Match the system studied with the method used.

    a. *Xenopus laevis* and *X. borealis* mitochondrial DNA   **1.** protein analysis

    b. *Nicotiana* large and small sub-units of Rubisco   **2.** pedigrees

    c. LHON phenotype in humans   **3.** differential hybridization

33. Describe two ways in which the contribution of mitochondrial genomes from male parents is prevented in different species.

34. If a human trait is determined by a factor in the cytoplasm, would an offspring more resemble its mother or its father? Why?

35. Why are very severe mitochondrial or chloroplast mutations usually found in heteroplasmic cells instead of homoplasmic cells?

36. Which of the two methods listed would you choose to determine if organelles in an organism are heteroplasmic or homoplasmic and why?

    a. hybridize probes to cells immobilized on a slide

    b. PCR-amplify DNA isolated from a population of cells

37. In the early 1900s, Carl Correns reported the results of observations he made on the inheritance of leaf colour in the four-o'clock plant *Mirabilus jalapa*. He noticed that on the same plant, some branches contained all green leaves, some branches contained all white leaves, and some branches contained variegated leaves that had patches of green and white tissue.

    a. Explain why some branches have green leaves, some have white leaves, and some have variegated leaves.

Explain why variegated leaves have some patches of white and some patches of green tissue.

    b. When Correns fertilized ovules from a green-leafed branch with pollen from flowers on any type of branch, he found that all the leaves in all of the progeny were green. When he fertilized ovules from a variegated branch with pollen from flowers on any type of branch, 90 percent of the progeny had some branches with green, some with white, and some with variegated leaves. Five percent of the progeny had only green leaves, and the remaining 5 percent of the progeny had white leaves but were severely stunted and died soon after germination. Explain these results. How could ovules from a variegated branch produce progeny with all green or all white leaves? Why did the completely white-leaved plants die early?

    c. Given your answer to part *b*, how could variegated plants have branches with apparently healthy white leaves?

38. A form of male sterility in corn is inherited maternally. Marcus Rhoades first described this cytoplasmic male sterility by crossing female gametes from a male sterile plant with pollen from a male fertile plant. The resulting progeny plants were male sterile.

    a. Diagram the cross, using different colours on lines to distinguish between nuclear and cytoplasmic genomes from the male sterile and male fertile strains.

    b. Female gametes from the male sterile progeny were backcrossed with pollen from the same male fertile parent of the first cross. The process was repeated many times. Diagram the next two generations including possible crossover events.

    c. What was the purpose of the series of backcrosses? (*Hint:* Look at your answer to part *b* and think about what is happening to the nuclear genome.)

39. Plant breeders have long appreciated the phenomenon called *hybrid vigour* or *heterosis*, in which hybrids formed between two inbred strains have increased vigour and crop yield relative to the two parental strains. Starting in the 1930s, seed companies exploited cytoplasmic male sterility (CMS) in corn so that they could cheaply produce hybrid corn seed to sell to farmers. This type of CMS is caused by mutant mitochondrial genomes that prevent pollen formation.

    a. How would CMS aid seed companies in producing hybrid corn seed?

Dominant *Rf* alleles of a nuclear gene called *Restorer* suppress the CMS phenotype, so that *Rf*-containing plants with mutant mitochondrial genomes are male fertile.

    b. Describe a cross-generating hybrid corn seed that would grow into fertile (self-fertilizing) plants. (Farmers planting hybrid seed want fertile plants because corn kernels result from fertilized ovules.)

    c. One of the historical challenges in the commercialization of hybrid corn produced through CMS was the

maintenance of strains with CMS mitochondria: How could the seed companies keep producing male sterile corn plants if they never themselves produced pollen? Suggest a strategy by which they could continue to obtain male sterile plants every breeding season.

d. Are there any potential disadvantages to the use of hybrid corn?

Section 18.7

40. What characteristics in a human pedigree suggest a mitochondrial location for a mutation affecting the trait?

41. The first person in the family represented by the pedigree shown here who exhibited symptoms of the mitochondrial disease MERFF was II-2.

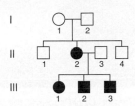

a. What are two possible explanations of why the mother I-1 was unaffected but daughter II-2 was affected?

b. How could you differentiate between the two possible explanations?

42. In 1988, neurologists in Australia reported the existence of identical twins who had developed myoclonic epilepsy in their teens. One twin remained only mildly affected by this condition, but the other twin later developed other symptoms of full-blown MERRF, including deafness, ragged red fibres, and ataxia (loss of the ability to control muscles). Explain the phenotypic dissimilarity in these identical twins.

43. Kearns–Sayre is a disease in which mitochrondrial DNA carries deletions of up to 7.6 kb of the mitochrondrial genome. Although Kearns–Sayre is due to a mitochondrial DNA defect, it does not show maternal inheritance but arises as a new mutation in an individual. The severity of symptoms ranges from mild to severe and affected people can have defects in some tissues but not in others. How can you explain the variation in tissues affected and severity of symptoms? (Assume that the size of the deletion does not contribute to phenotypic differences.)

44. If you were a genetic counsellor and had a patient with MERRF who wanted to have a child, what kind of advice could you give about the chances the child would also have the disease? Are there any tests you could suggest that could be performed prenatally to determine if a fetus would be affected by MERRF?

45. Deletions of various sizes in the mitochondrial genome have been found to increase with age in humans. Which technique would you use to analyze human mtDNA samples for deletions—PCR or gel electrophoresis? Why?

For more information on the resources available from McGraw-Hill Ryerson, go to www.mcgrawhill.ca/he/solutions.

Problems **631**

CHAPTER 19 — Sequencing and Analyzing Genomes

On the *left* stands James Watson, co-discoverer of the structure of DNA and Nobel Laureate. On the *right* sits Ozzy Osbourne, the self-proclaimed "prince of darkness" and unquestioned "heavy metal" legend. What do these two individuals have in common? How have their paths crossed so that their pictures appear here together to open this chapter? The answers to these questions relate to personalized whole-genome sequencing. As described later in this chapter, Watson and Osbourne were among the first human beings to have their genomes sequenced. As we shall see, the ability to generate such individualized genetic information has important implications with respect to the study and understanding of both biology and medicine.

A "**genome**" can be thought of as the total digital information contained within the DNA sequences of an organism's chromosomes. Although the human genome is contained within 46 chromosomes, 99.9 percent of the information in each autosome is the same as in its partner homologue. Thus, each pair of autosomes (22 pairs) is counted only once, together with the X and Y chromosomes, for a total of 24 chromosomes that roughly describe the information content of the human genome. These 24 strings of G's, C's, T's, and A's contain a total of approximately 3 billion nucleotides and range in size from 45 million to 250 million bp. The **Human Genome Project** (found at www.ncbi.nlm.nih.gov/genome/guide/human/) was initiated to sequence and analyze the information content of these 3 billion nucleotides.

Genomics is the branch of biology dedicated to the study of whole genomes (both human and nonhuman). By virtue of its inherent "global" perspective, the practice of genomics involves large-scale, "high-throughput" technologies (see Chapters 20 and 21) that generate vast amounts of data requiring computers (and sophisticated software) to be analyzed. For example, the Human Genome Project was enabled only after the development of (1) very fast and reliable automated DNA sequencing technology, and (2) ingenious software that permitted the capture, storage, and analysis of vast amounts of cloning, mapping, and sequence data. In this chapter, we will explore the tools and ramifications of whole-genome sequencing. While no one claims to have definitive answers, we will attempt to provide an unbiased and objective discussion of the ethical questions arising from living in the post-genomic era.

Chapter Outline
19.1 The Evolution of Whole-Genome Sequencing
19.2 Sequencing and Assembling Entire Genomes
19.3 Bioinformatics: Analyzing Genomes
19.4 Next-Generation Sequencing
19.5 Repercussions of the Human Genome Project

Learning Objectives

1. Identify the goals of the Human Genome Project and evaluate whether or not these goals have been achieved.

2. Compare and contrast the sequencing methodologies used by the publicly funded Human Genome Project and its privately funded competitor, Celera.

3. Define the term "bioinformatics" and relate bioinformatic analysis to both human genetics and evolutionary biology.

4. Compare and contrast "Next-Generation" sequencing methods with the methods used by the Human Genome Project.

5. Discuss the ramifications of the "high-throughput" and "massively parallel" nature of next-generation sequencing technologies.

6. Describe and discuss the ethical ramifications associated with living in the post-genomic era.

19.1 The Evolution of Whole-Genome Sequencing

The first meeting of the Human Genome Project (HGP) took place in Santa Cruz, California, in the spring of 1985. The chancellor of the University of California at Santa Cruz had assembled 12 biologists of diverse backgrounds to explore the idea of starting an institute to sequence the human genome. After two days of heated discussion, the 12 biologists concluded that it would, indeed, be possible to develop the technology required to accomplish this seemingly impossible objective. However, the group was split on whether it would be a good idea for the scientific community.

Two aspects of their discussion were striking. First, the concept of the Human Genome Project introduced the idea of *discovery science*—a new scientific approach to biology. In **discovery science**, one seeks to identify all the elements of a biological system—for instance, the complete sequences of the 24 chromosomes that contain the 3 billion nucleotides of the human genome—and place them in a database to enrich the infrastructure of biology. Discovery science stands in contrast to hypothesis-driven approaches to biology, in which one asks questions and seeks experimental verification of possible answers. Second, it was clear that the Human Genome Project required the development of very fast and reliable ("high-throughput") automated DNA sequencing technology, as well as novel computational tools for analyzing the data.

Not surprisingly, most biologists initially viewed the Human Genome Project with scepticism. They thought the project would not be particularly worthwhile because only approximately 2 percent of the genome codes for proteins; the remaining 98 percent, they argued, must be just "junk." In the mid-1980s, most biologists also believed that the Human Genome Project was not really a scientific endeavour because it was not hypothesis-driven. Many did not understand how the discovery approach to determining the sequence of the human genome would revolutionize the power and potential of genetic and other biological studies. Finally, some viewed the Human Genome Project as "big science" that would inappropriately compete for funds with more fruitful and productive, small-scale, hypothesis-driven science.

In 1988, the United States National Academy of Sciences appointed a committee (half proponents and half opponents) to consider the scientific merits of the Human Genome Project. After a year of vigorous debate and analysis, the committee unanimously endorsed the project, marking a major turning point in its acceptance. The U.S. government-funded Human Genome Project began in 1990 with a projected 15-year time scale and a $3 billion budget for completing the human genome sequence.

A rough sequence draft of the human genome was completed in February 2001; in this "draft," the sequence did not yet have an appropriate level of accuracy (an error rate of 1/10 000), and it had some gaps. An accurate sequence covering 97 percent of the genome was completed in 2003, two years ahead of the originally proposed 2005 finish date. The early finish was in part catalyzed by the 1998 promise of Celera, a private company, to complete a draft of the genome in just three years. The federally supported genome effort reacted by moving its timetable ahead by several years. By April 2013, whole-genome sequences had been completed for 9327 distinct species, including 112 different vertebrates.

The HGP reference sequence is a "composite" of a few individuals

It is important to note that the sequence provided by the HGP was obtained from only a small number of anonymous donors. This is to say, the "reference" HGP sequence is in actuality a "composite" of only a handful of people. While providing a tremendous amount of information regarding overall gene and genome organization, these data do not address the relationship between individual genetic variation and phenotype (Chapter 15). Fortunately, the recent advent of new technologies (that are both faster and less expensive than traditional methods) has allowed complete "individual" genome sequences to be obtained. For example, while the HGP required $3 billion (over an entire decade) to create a composite sequence, modern sequencing technologies can sequence an individual's genome in days for under $10 000. These advancements have led to a rapid increase in the number of sequencing projects and have resulted in vast amounts of individual, personalized genome data (**Figure 19.1**). If these trends continue, the availability of personal genome sequence data—together with all the ethical ramifications that come along with it—is likely to become commonplace in developed countries. In the next section, we will discuss the methodologies used by researchers to obtain whole-genome sequence data.

> The Human Genome Project was a large, multinational scientific endeavour that aimed to provide the complete sequence of the human genome. The project began in 1990 and ended in 2003 and was able to provide an accurate sequence covering 97 percent of the genome at a cost of $3 billion.

19.2 Sequencing and Assembling Entire Genomes

The genomes of microbes and eukaryotes range in size from 700 000 base pairs (700 kb) to more than 100 billion base pairs (gigabase pairs, or Gb) and can be distributed into a single microbial chromosome or into multiple eukaryotic chromosomes. **Table 19.1** gives the genome sizes of representative microbes, plants, and animals. The lungfish,

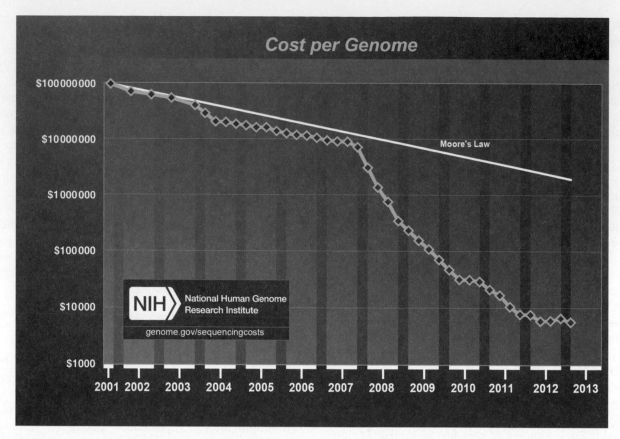

Figure 19.1 The cost of human genome sequencing (2001–2013). Cost is indicated by the plot in *green*. The pace of Moore's law (the observation that computing power doubles every 18 months) is shown in *white*. As you can see, advances in sequencing technology (resulting in lowered cost) easily outpace Moore's law.

Protopterus aethiopicus, has a genome of approximately 130 billion base pairs, and wheat has a genome of 15 billion base pairs. To put these numbers in perspective, the human genome is 200 times larger than the yeast genome and 40 times smaller than the genome of *Protopterus aethiopicus*. Thus, the information content of a genome is not necessarily proportional to the complexity of the organism it defines. The large size of some genomes presents fascinating challenges for their ultimate characterization and analysis. Genomicists face major challenges in dealing with this immense body of data. One of these is how to map sequences accurately. In this section, we consider how high-resolution sequence maps are created.

High-resolution sequence maps

Sequence maps show the order of nucleotides in a cloned piece of DNA. The goal of both the HGP and Celera was the determination of the complete nucleotide sequence for every chromosome in the genome. Two basic strategies were employed: the *hierarchical shotgun approach* (used by the HGP) and the *whole-genome shotgun approach* (used by Celera). The term **shotgun** means that the

Table 19.1	A Comparison of the Developmental Complexity and Genome Features* of Model Organisms					
Organism		**Developmental Complexity**	**Genome Size* (Megabases)**	**No. of Genes**	**No. of Genes per Million bp Sequenced**	**Date Genome Finished**
Type	**Species**					
Bacterium	*Escherichia coli*	1-cell prokaryote	4.64	3244	905	1997
Yeast	*Saccharomyces cerevisiae*	1-cell eukaryote	12.07	6201	483	1996
Worm	*Caenorhabditis elegans*	~1000 cells	100	20 470	205	1998
Fly	*Drosophila melanogaster*	~50 000 cells	180	15 016	83	2000
Mustard weed	*Arabidopsis thaliana*	10^{10} cells	125	25 700	206	2000
Rice (draft)	*Oryza sativa*	5×10^{10} cells	466	39 045	84	2002
Mouse	*Mus musculus*	10^{11} cells	3200	23 786	7	2005
Human	*Homo sapiens*	10^{14} cells	3200	20 687	6	2003

*Haploid genome size, including heterochromatic DNA.

overlapping fragments to be sequenced are randomly generated by shearing (via sonication) or by partial digestion with restriction enzymes.

The hierarchical shotgun sequencing strategy

The publicly funded effort to obtain a draft sequence of the human genome employed the hierarchical shotgun strategy. This strategy first requires that chromosomal DNA be isolated from cells derived from the organism under study. The purified genomic DNA is then treated with restriction enzymes in a reaction that is not allowed to proceed to completion. In this way, not all potential restriction sites are cleaved and large fragments (approximately 250 kb) are generated. These fragments are then cloned into large capacity vectors known as bacterial artificial chromosomes (BACs; see Chapter 18) to create a BAC library in which each genomic DNA sequence is represented (**Figure 19.2**).

The next critical step involves organizing the DNA fragments in the BAC clones into a physical map. This is done through the use of naturally occurring DNA sequences called sequence tagged sites (STSs). An STS is simply a short, unique DNA sequence (200–500 bp in length) that is found only once in the genome. The presence of a particular STS in a given BAC clone can thus be determined by PCR using primers specific to the given STS.

By checking for the presence of STS markers in a given set of BAC clones, one can build what is known as a "tiling path." For example, if two separate BAC clones contain the same STS, then one can conclude that the two clones overlap and thus represent a contiguous region (or path) through the genome. By examining many different STS sequences, it is possible to infer a tiling path that defines a "contig" (a set of overlapping clones representing a much larger contiguous region of a chromosome). The "minimum tiling path" comprises the smallest number of BAC clones encompassing the contig and displaying the minimum amount of overlap. Analyzing only the BAC clones from the minimum tiling path (as opposed to all the BAC clones) greatly reduces the amount of sequencing that needs to be performed and thus aids in reducing the amount of time, money, and effort required to complete the project.

In the final step, the insert of each BAC clone on the minimum tiling path is sheared at random into pieces approximately 2 kb in length and cloned into small plasmid vectors (**Figure 19.3**). The insert of each plasmid

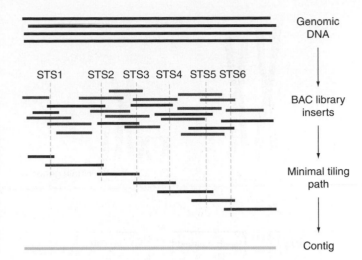

Figure 19.2 Idealized representation of the hierarchical shotgun sequencing strategy. A library is constructed by fragmenting the target genome and cloning the pieces into BAC vectors. The fragments are then organized into a physical map using sequence tagged sites (STSs). By assembling fragments containing the same STS, it is possible to obtain a minimum tiling path (shown in red). The BAC clones containing the red inserts are then sequenced as described in the text.

vector is then sequenced via the Sanger method. For a BAC clone 200 kb long, approximately 1000 plasmids are sequenced. In this way, tenfold coverage of the genome can be achieved (i.e., 1000 plasmids $\times$ 2 kb each = 2000 kb, which is ten times the number of base pairs in the original BAC clone). Sequencing to tenfold coverage greatly limits the chance that any portion of the genome is "missed" (i.e., not sequenced). Next, using sophisticated computer algorithms, the raw sequence data are analyzed for regions of overlap to provide the final assembled sequence.

The whole-genome shotgun sequencing strategy

To obtain its draft sequence of the human genome, Celera employed a distinct strategy referred to as the whole-genome shotgun approach. As part of this method, genomic DNA was randomly sheared three times, first to construct a plasmid library with approximately 2-kb inserts, second to generate a plasmid library with approximately 10-kb inserts, and third to produce a BAC library of approximately 200-kb inserts (**Figure 19.4**). Next, each end of the 2-kb, 10-kb, and 200-kb inserts were sequenced to attain approximately sixfold, threefold, and onefold coverage, respectively.

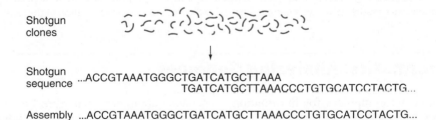

Figure 19.3 The assembly of cloned shotgun sequences to reconstruct the sequence of the contig. BAC clones on the minimum tiling path are sheared at random into pieces approximately 2 kb long and cloned into plasmid vectors. The insert of each plasmid vector is then sequenced via the Sanger method. Using sophisticated computer algorithms, the raw sequence data are assembled into long contiguous strings using regions of overlap.

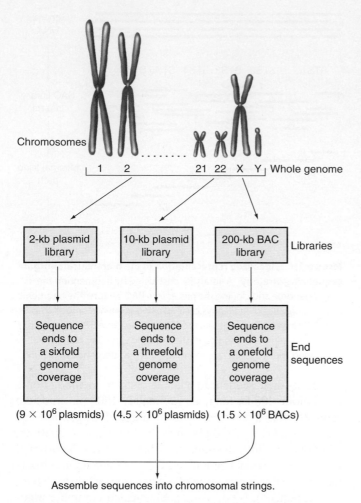

Figure 19.4 **Hypothetical whole-genome shotgun sequencing strategy.** Only three libraries of differing-sized fragments (2 kb, 10 kb, and 200 kb) need to be constructed. The challenge is to assemble these sequences when they include large numbers of repeats.

This "paired-end sequencing" strategy overcomes the problem posed by identical repetitive DNA sequences scattered throughout the genome. These repeat sequences confound assembly since they appear multiple times and thus cannot be definitively assigned a position on a chromosome. For example, if one end of an insert is sequenced and found to contain a unique nonrepetitive sequence, and the other end is sequenced and found to contain a repetitive sequence, then one can unambiguously determine the chromosomal location of the repeat. Since most repetitive sequences are less than 3 kb long, the "paired-end" sequences from the 10-kb and 200-kb libraries are able to span the gaps between regions interrupted by repetitive DNA sequences (**Figure 19.5**). Taken together, the sixfold, threefold, and onefold coverages amount to a tenfold coverage. Finally, a genome-wide shotgun computer program assembles all these sequences into chromosomal strings.

The whole-genome shotgun strategy has several advantages. First, it does not require the construction of a minimal tiling path. Second, the "paired end" strategy aids in the unambiguous assembly of the genome sequences. Third, the whole-genome shotgun strategy relies on only a single highly automated and very mature technology—DNA sequencing. In the end, the Celera effort did incorporate data from the public effort into its own data. Since then, however, genome projects have used only the whole-genome shotgun approach.

> High-resolution sequence maps showing the complete order of nucleotides can be produced in two ways. The hierarchical shotgun sequencing approach generates overlapping BAC clones to generate a minimal tiling path. The whole-genome shotgun strategy shears the genome randomly into segments of known length; these segments are then assembled by a computer program into chromosomal strings.

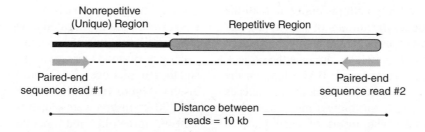

Figure 19.5 **The paired-end sequencing strategy aids in unambiguously assigning repetitive DNA sequences to chromosomes.** In the example above, paired-end sequence read #1 provides a unique "anchor" with which to assign the location of the repetitive sequence.

19.3 Bioinformatics: Analyzing Genomes

The digital language used by computers for information storage and processing is ideally suited to handle the digital code produced by sequencing projects. Keeping pace with the revolution in biological data generation that began in the 1980s was a parallel revolution in information technology. The Internet came into existence along with personal computers that were linked together to establish rapid transmission of electronic sequencing data from one laboratory to another. It was a straightforward task to channel the vast output of DNA sequencing machines directly into electronic storage media, from which sequences were available for analysis and transmission to other scientists.

DNA sequences online

The first official repository for DNA sequences was the GenBank database, established by the National Institutes of Health in 1982. GenBank serves as an open-access, permanent online repository of sequence data generated in molecular biology laboratories from around the world. Individual scientists deposit their sequences electronically, and anyone in the world with an Internet connection can download and analyze them. From its establishment, the GenBank database doubled in size every 18 months, from less than 1 million base pairs initially to a total of over 100 billion base pairs by the beginning of 2013 (**Figure 19.6**).

In 2008, a new generation of nanotechnology-based DNA sequencers provided scientists with the ability to obtain over 100 billion base pairs of sequence data—more than the combined total of global scientific output from 1973 until 2007—in a single experiment (see Section 19.4). As the cost of sequencing continues to drop, and billion-base-pair sequencing experiments become routine, it is no longer feasible for GenBank to act as an all-inclusive repository for the primary sequences generated by the world's scientists.

Hacking the genome

The meaning of DNA sequences must be interpreted through software programs. Initial programs analyzed sequence data for biological landmarks (e.g., restriction-enzyme recognition sites and amino acid sequences encoded in open reading frames). Software was also developed to search for hidden sequence patterns, and to identify statistically significant similarities among different sequences. Results obtained from software-driven studies led to new biological understanding, which was incorporated into more sophisticated computer programs, which led to further understanding, and so on. The integration of biological data and computer analysis gave rise to the new field of bioinformatics.

Bioinformatics provides tools for visualizing functional features of genomes

Bioinformatics is the science of using computational methods to decipher the biological meaning of information contained within organismal systems. Among the most important bioinformatics tools are those that allow researchers to visualize genomic data through graphic presentations constructed on-the-fly for online viewing through a web browser. The National Center for Biotechnology Information (NCBI; at ncbi.nlm.nih.gov) was established in 1988 to oversee GenBank, create additional public databases of biological information, and develop bioinformatic applications for analyzing, systemizing, and disseminating the data. This section provides some examples of bioinformatics tools (developed by scientists at NCBI and elsewhere) that can be accessed through any web browser to visualize publicly available genome data.

The species RefSeq

Comparisons of experimental data involving DNA sequences generated by different laboratories are critically dependent on the use of a universally agreed-upon standard for analysis. This role is played by a species reference sequence, abbreviated as RefSeq. A RefSeq is a single, complete, annotated version of the species genome that is freely available online. A RefSeq need not be derived from a single individual, and it need not contain the most common genetic variants found in species members. Rather, it is

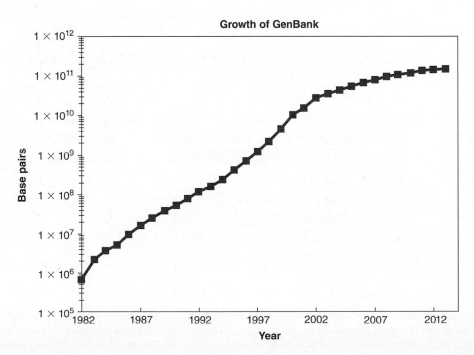

Figure 19.6 Accumulation of genome sequence data (1982–2013). Growth of total sequence data deposited in GenBank. In the 30-year period from 1983 to 2013, GenBank's accumulated data repository grew approximately 60 000-fold to over 100 billion base pairs.

simply an arbitrary, but well-characterized, example against which all newly obtained sequences from that species can be compared. By May 2013, whole-genome RefSeqs had been established for each of 24 656 species, including our own (visit www.ncbi.nlm.nih.gov/RefSeq/).

Visualizing genes

A number of web-based programs have been developed that allow a user to visualize public and private genome data. Among the most popular is the UCSC Genome Browser developed at the University of California, Santa Cruz (genome.ucsc.edu). The UCSC Genome Browser can be used to visualize the genes identified in the human RefSeq (**Figure 19.7**). At different levels of resolution, it becomes possible for a viewer to gain insight into different aspects of human genome organization.

Figure 19.7a depicts the locations of all identified genes along the 158 821 424-bp length of human chromosome 7. The 1503 genes are each represented by a separate blue box indicating location and length. Although very little molecular detail is visible at this resolution, you can see immediately that the density of genes varies enormously

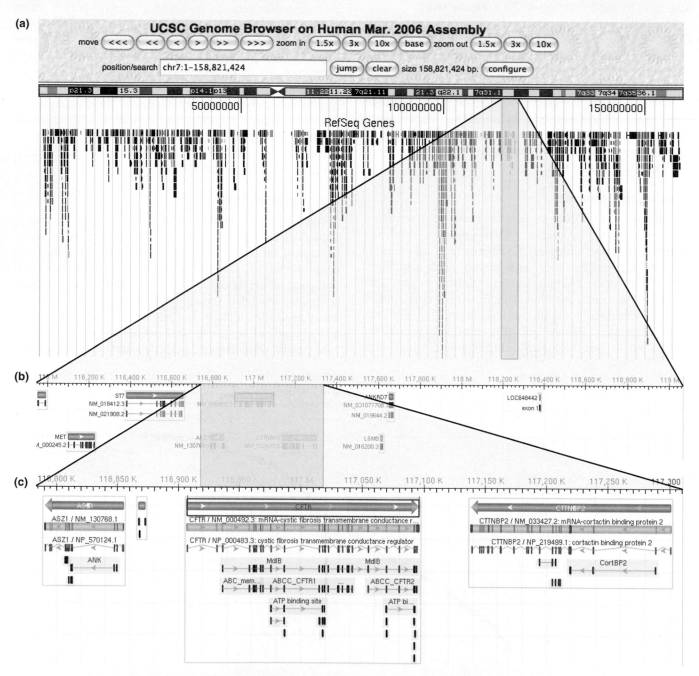

Figure 19.7 Visualizing genes of the human RefSeq genome with the UCSC Genome Browser. (a) Locations of the 1503 genes identified along the 158 821 424-bp length of human chromosome 7. **(b)** A 3-Mb-pair region of chromosome 7 between sequence positions 116 000 001 and 119 000 000, showing the locations and lengths of nine genes labelled on the left with their official names. The genomic region from 117 700 000 to 119 000 000 is a "gene desert." **(c)** Visualization of a 540-kb region of chromosome 7 containing the *CFTR* gene with the NCBI Sequence Viewer.

along the chromosome. Some regions are particularly rich in genes, whereas other regions are "gene deserts." Furthermore, long-range repeating patterns of either gene density or gene sizes are absent.

Variation in gene density is even more apparent when you zoom into a 3-Mb region around the *CFTR* gene at position 117 Mb on the long arm of the chromosome (**Figure 19.7b**). Each gene in the region is now clearly visible as a separate group of vertically extended lines or boxes linked together by a horizontal line; vertical extensions represent exons, and lines represent spliced-out introns. When visible, arrows along an intron indicate the direction of transcription. You can see that nine non-overlapping genes are located in the leftmost 1.7 Mb of the region, whereas none are in the remaining 1.3 Mb. The variation in the lengths of genes is also apparent in this view.

Visualizing gene structure and functional capacity

Transcribed genomic regions, exon-intron structures, and locations of protein-coding regions are best visualized by switching to the NCBI Sequence Viewer (www. ncbi.nlm.nih.gov/nuccore/89161213?content=5&v= 116750000:117350000&report=graph). A 540-kb region around *CFTR* is shown in **Figure 19.7c**, where transcription units are indicated with green bars (containing arrows that indicate the direction of transcription), the exon-intron structure of each gene is shown beneath with blue boxes and connecting lines, the spliced RNA product is indicated with red boxes, and genomic regions corresponding to polypeptide products are in black. Each of the three well-defined genes in this region encodes multiple polypeptides extending across different portions of mature transcripts.

The human genome contains approximately 21 000 genes

The analysis of the HGP sequence with tools like the ones described above has provided striking new insights into gene organization, genome architecture, and the evolution of chromosomes. For example, one of the first surprises to emerge from the human genome sequence was the discovery of just approximately 21 000 genes, a much lower number than expected. A back-of-the-envelope calculation done at the initiation of the Human Genome Project had suggested that there might be 100 000 human genes—approximately 1 gene for every 30 kb. This has not proven to be the case; although the human genome has more genes than the genomes of simpler model organisms, it has not nearly as many as one would expect from the increased complexity (see Table 19.1). This means that mechanisms other than the expression of different germline genes must help generate metazoan (multicellular animal) complexity.

Using whole-genome comparisons to better understand evolution

In addition to helping us understand genome organization/architecture, the availability of sequence data has also allowed researchers to better understand evolutionary history. For example, based on Darwin's model of evolution, one would predict that related species would have related genomes. But how can you tell whether DNA sequences from two sources are similar by chance or by common origin?

To answer this question let us consider a specific, but random, 50-bp sequence and calculate the probability that an independently derived DNA segment could be 100 percent identical, just by chance. The probability of occurrence of any DNA sequence of length n is obtained simply by raising 0.25 (the chance occurrence of the same base at a particular position) to the 50th power (the number of independent chance events required): $(0.25)^{50} = 8 \times 10^{-31}$. For all intents and purposes, this probability is essentially zero, which negates the null hypothesis and tells us that two perfectly matched 50-bp DNA sequences found in nature are almost certainly derived from the same ancestral sequence, rather than by chance.

DNA sequence conservation

A segment of DNA is said to be a *homologue* of a sequence in another species when the two show evidence of derivation from the same DNA sequence in a common ancestor. For perfectly matched sequences that are 50 bp in length or longer, the evidence is clear. But evidence for homology of imperfectly matched DNA regions requires a more sophisticated statistical analysis, a task that is readily performed by specialized bioinformatics programs. When homologues of a DNA sequence are found in many different species, the sequence is said to be *conserved*.

A traditional phylogenetic tree, like the one shown in **Figure 19.8a**, depicts the relatedness of multiple species to each other, with branch points that represent a series of nested common ancestors. When the human genome is compared as a whole with other representative vertebrate species, the percentage of sequence conservation is relatively high for chimps and monkeys, but generally decreases as the elapsed time to a common ancestor increases (**Figure 19.8b**). At a distance of over 400 million years, the fish genome contains only 2 percent of the DNA sequences present in the human genome. In contrast, when comparisons are restricted to human protein-coding sequences, conservation levels remain high—at more than 82 percent—throughout vertebrate evolution.

Functional DNA sequences such as protein-coding regions are subject to loss or lessening of function by at least some mutations. As a result, they evolve more slowly than nonfunctional sequences, which are not similarly constrained by functional requirements. Unconstrained sequence divergence would eventually eliminate all evidence of common ancestry.

Homology mapping of genomes

Using our knowledge of sequence conservation, together with a genome visualization tool, it is now possible to explore DNA sequence conservation directly along the genome, as well as across evolutionary time. An example

(a)

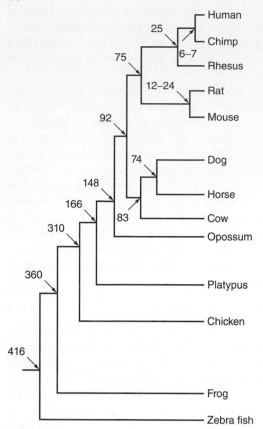

(b)

Scientific Name	Common Name	1	2
Homo sapiens	Human	100%	100%
Pan troglodytes	Chimp	93.9%	96.58%
Macaca mulatta	Rhesus	85.1%	96.31%
Rattus norvegicus	Rat	35.7%	94.47%
Mus musculus	Mouse	37.6%	95.36%
Canis familiaris	Dog	55.4%	95.18%
Equus caballus	Horse	58.8%	92.70%
Bos taurus	Cow	48.2%	94.78%
Monodelphis domestica	Opossum	11.1%	91.43%
Ornithorhynchus anatinus	Platypus	8.2%	86.43%
Gallus gallus	Chicken	3.8%	88.61%
Xenopus tropicalis	Frog	2.6%	87.44%
Danio rerio	Zebra fish	2.0%	82.38%

Figure 19.8 Species relatedness and genome conservation between *Homo sapiens* and other vertebrates. (a) A phylogenetic tree showing branch points at which organisms diverged; the number at each branch point represents millions of years before the present. **(b)** Relatedness of the *H. sapiens* genome to that of other vertebrates is evaluated according to two bioinformatic measures: in **column 1**, the proportion of the complete human genome sequence that is found in the species being compared; and in **column 2**, the proportions of human protein-coding sequences that are found in each vertebrate genome.

of cross-species homology analysis is shown in **Figure 19.9** for a 100-kb region containing the *HOXA* family of genes. The locations and exon-intron structures of the ten human RefSeq genes are displayed in the bottom row. Above this row are homology maps for five representative vertebrate species; conservation of sequence homology is indicated with dark lines or blocks.

As anticipated from whole-genome data, nearly complete conservation of human sequences exists across the entire region in a chimp genome. In other mammals, represented here by the mouse, conservation is also apparent across the entire region, but the pattern is choppy, indicating small regions of conservation interspersed with small regions that are not conserved.

As we move farther across the phylogenetic landscape to frogs and fish, we can more clearly distinguish sequences subject to evolutionary constraints from those

that are not. The coding regions of the *HOXA* genes are all conserved; these genes are critical to proper development of all vertebrates. But in addition, other conserved DNA sequences can be observed at locations between coding regions. Although these sequences do not have coding potential, they may be sites of sequence-specific binding to proteins required for gene regulation or local chromatin structure.

Digital computer technology has proved to be an ideal tool for use with the four-value DNA code. Bioinformatics allows visualization of the functional features of genomes at almost any scale, as well as comparison of genome features. These whole-genome comparisons enable identification of genomic elements conserved by natural selection.

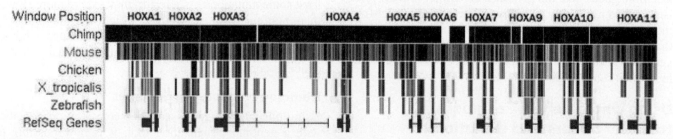

Figure 19.9 Homology map for a 100-kb region of the human genome. Conservation of DNA sequences across a region of chromosome 7 from sequence positions 27 092 501 to 27 192 500 containing the *HOXA* gene family.

19.4 Next-Generation Sequencing

While the Sanger method of sequencing dominated the marketplace for over two decades, recent technological advances have pushed the sequencing envelope even further. These advances have allowed for the design and execution of experiments that could only have been dreamed of a decade earlier. For the most part, these developments were spurred on by the desire for personalized genomic analysis.

As mentioned earlier, the reference human genome is a composite of only a handful of individuals. Thus, questions related to individual genetic variation cannot be answered through the analysis of the reference HGP sequence. For example, one might wonder why some individuals are predisposed to heart disease while others are not. Another might ask why a drug—effective in treating some cancer patients—is at the same time completely ineffective when used on others with the same type of cancer. Yet another might ponder if mental disorders like autism have an underlying genetic basis, and furthermore, whether the critical genetic variants could be revealed by the analysis of the full genome sequences of affected and unaffected individuals. Lastly, and possibly most importantly, you yourself might ask whether the analysis of your genome might provide valuable information regarding your predisposition to disease.

For these questions to be answered—that is, for personalized genomics to become a reality—issues related to the speed and cost of sequencing need to be addressed. Obviously, a $3 billion price tag, together with a decade and a half of effort, makes the HGP strategy impractical with regard to answering the type of questions described above. For these reasons, in the mid-to-late 1990s, private industry began serious efforts to produce technologies that could cheaply, and easily, sequence billions of nucleotides in days or weeks instead of years. The desire for personalized genomics even inspired the Archon Genomics XPRIZE (see the **Focus on Inquiry** box **"The Archon Genomics XPRIZE"**). While the XPRIZE is currently unclaimed, many private groups have made strong headway towards this goal.

The first commercially available "next-" or "second-" generation sequencing platform was produced by Roche/454 Life Sciences. In an effort to prove the effectiveness of their system, the researchers sequenced and published the complete personal genome of Nobel Laureate James Watson (the co-discoverer of the structure of DNA) in 2008. Amazingly, their "pyrosequencing" technology was able to provide Watson's complete sequence in only four months and at a cost of less than $1.5 million (a huge improvement over the $3 billion price tag of the HGP). In the next section, we will describe how this revolutionary new technology provides such huge improvements with respect to the speed and cost of DNA sequencing.

FOCUS ON Inquiry   The Archon Genomics XPRIZE

The Archon Genomics XPRIZE was first unveiled in October of 2006 to foster the development of novel technologies that would result in fast, accurate, and low-cost whole-genome sequencing. The XPRIZE Foundation initially offered a prize of $10 million (sponsored by a Canadian, Stewart Blusson, President of Archon Minerals Ltd.) to the first group capable of sequencing 100 genomes in ten days at a cost of less than $10 000 per genome. More recently, the rules were made even more stringent and ambitious. As of 2011, the prize will be awarded to the team(s) who meet the following criteria: sequence 100 human genomes within 30 days to an accuracy of 1 error per 1 000 000 bases with 98 percent completeness; identify insertions, deletions, and rearrangements; and provide a complete haplotype, at an audited total cost of $1 000 per genome. Furthermore, the genomes to be sequenced will be donated by 100 centenarians in the hopes of providing clues to their amazing longevity (http://genomics.xprize.org/100-over-100). Once claimed, this landmark competition will signal the beginning of the new era of routine whole-genome sequencing and personalized genomics.

XPRIZE launch in New York on October 26, 2011.

Pyrosequencing

Pyrosequencing represents a completely novel and highly sophisticated method of DNA sequencing. Amazingly, the system is based on the detection of light produced as a consequence of the incorporation (via DNA polymerase) of a nucleotide into a strand of DNA (**Figure 19.10**). Unlike a typical Sanger sequencing machine (which can run at most 384 reactions in parallel), the pyrosequencing method can run 1.6 million reactions at the same time! For these reasons "pyrosequencing" is often referred to as being "massively parallel."

The pyrosequencing process begins with the isolation of genomic DNA from the organism under study and its subsequent fragmentation into small pieces (approximately 300–800 bp long). Next, smaller (approximately 40-bp-long) adapter sequences are ligated to the end of each fragment (these act as PCR and sequencing primer binding sites in later stages). The DNA fragments are then attached to small "sequencing" beads only 40 μm in diameter. These beads are coated with millions of copies of a short oligonucleotide that is complementary in sequence to one of the adapters. Thus, DNA fragments are attached to the beads through hybridization of complementary sequences. To begin, each bead is bound to only a single DNA fragment. This is done by mixing a large excess of beads with the fragmented DNA sample. DNA-bound beads are then isolated and placed together in a test tube. The individual beads within the test tube are compartmentalized within tiny aqueous droplets surrounded by an oil sphere. These droplets also contain the enzymes, reagents, and primers needed for PCR (note that the oligos coating the beads can also act as PCR primers). This is referred to as an "emulsion" PCR due to the water-in-oil design. After the PCR cycles are complete, each bead is covered with tens of millions of copies of a single, unique double-stranded DNA fragment.

At this point, the beads are isolated and the DNA strands denatured. The beads (now carrying single-stranded DNA) are placed within the wells of a picotiter plate. Remarkably,

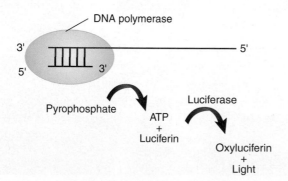

Figure 19.10 How light is produced through pyrosequencing. Incorporation of a complementary nucleotide generates pyrophosphate (PP$_i$). PP$_i$ is converted to ATP by ATP sulfurylase, which is then used to convert luciferin to oxyluciferin in the presence of luciferase. This last reaction produces light in equimolar concentrations to the amount of PP$_i$ produced. (The template DNA is shown in *red*. An annealed primer is shown in *black*. The *green oval* represents DNA polymerase.)

this plate contains over 1.6 million wells (55 μm deep, 44 μm wide) that are designed to accept only a single bead. In addition to the DNA-bound "sequencing" beads, the wells are also packed with smaller "enzyme" beads, which carry with them all the components necessary for the subsequent sequencing reaction (**Figure 19.11a**).

The reaction cycle begins with the incorporation of a dNTP into the polymerizing non-template strand (through the action of DNA polymerase). Each nucleotide incorporated results in the release of a pyrophosphate (PP$_i$) that is converted to ATP with the help of ATP sulfurylase and adenosine 5′ phosphosulphate. The ATP produced then drives the conversion of luciferin to oxyluciferin via the enzyme luciferase. A by-product of this last reaction is light, which is detected by an extremely sensitive digital camera that is able to track light emission in each of the 1.6 million wells of the picotiter plate.

The dNTPs are flowed into the wells in a sequential manner. This is to say, dTTP is first flowed through the wells and the emitted light measured. This is then followed by the flow of dATP and the subsequent measurement of emitted light. Finally dCTP and then dGTP are flowed through, and the emitted light again measured after each flow. After approximately 100 nucleotide flows (i.e., 25 times through the dTTP, dATP, dCTP, dGTP cycle) a flowgram is generated for each individual well and the sequence determined (**Figure 19.11b**). Since the emission of light is proportional to the number of nucleotides incorporated into the growing strand, the measured light can be used to determine how many nucleotides were added after each flow.

For example, if (in a given well) one unit of light is produced after the flow of dTTP, then this would indicate the addition of one T to the sequence. If three units of light are produced after the subsequent flow of dATP, then the next three nucleotides in the sequence can be inferred to be AAA. If no light is detected after the subsequent dCTP flow, then one can infer that no C was incorporated (and that the next letter must be a T, A, or G). As it turns out in this example, the next five bases are G since five units of light are detected after the dGTP flow. Thus (after one flow cycle), we can infer that the sequence of this particular portion of the DNA fragment is 5′ TAAAGGGGG 3′. In this way, the sequence of each bead-bound DNA fragment in each of the 1.6 million wells can be read using the produced flowgrams. Again, sophisticated computer algorithms are then used to assemble the sequences into chromosomal strings using regions of sequence overlap. Therefore, a single run of the Roche/454 pyrosequencing system can produce 1.6 million sequence reads (approximately 400 bp each) in only about eight hours (a total of 640 million bp). In contrast, a single run of a Sanger machine can produce 384 reads of about 600 bp each (a total of only approximately 230 000 bp). In addition, the pyrosequencing method requires no cloning of fragments or generation of BAC or plasmid libraries. The entire process is performed *in vitro* and is both streamlined and automated. The speed, simplicity, automation,

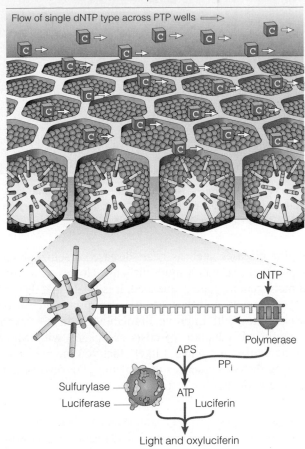

(a)
Roche/454 — Pyrosequencing

1–2 million template beads loaded into PTP wells

Flow of single dNTP type across PTP wells ⟹

dNTP

Polymerase

APS

PP$_i$

Sulfurylase

Luciferase

ATP

Luciferin

Light and oxyluciferin

(b)

Flowgram

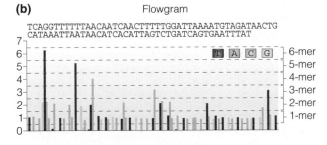

TCAGGTTTTTTAACAATCAACTTTTTGGATTAAAATGTAGATAACTG
CATAAATTAATAACATCACATTAGTCTGATCAGTGAATTTAT

T A C G

6-mer
5-mer
4-mer
3-mer
2-mer
1-mer

Figure 19.11 Pyrosequencing methodology. (a) Sequencing beads (large *yellow spheres*) bound to single-stranded DNA fragments derived from the genome of the organism under study are loaded into the wells of a picotiter plate (PTP). Only one bead is able to fit within a single well. Enzyme beads (small *orange spheres*) coupled to sulfurylase and luciferase are also packed within the wells of the picotiter plate. In the diagram below, dCTP (as well as all other required sequencing reagents) is flowed over the picotiter plate. The light emitted from each well of the plate is measured with a high-resolution digital camera. **(b)** The amount of light generated from the pyrosequencing reaction after each flow of dTTP, dATP, dCTP, or dGTP is recorded for each well, generating a flowgram. The DNA sequence can then be read (left to right) using the intensity of light as a measure of how many nucleotides were incorporated after each flow. In the example, the final sequence is 5′ TCA GGT TTT TTA ACA ATC AAC TTT TTG GAT TAA AAT GTA GAT AAC TGC ATA AAT TAA TAA CAT CAC ATT AGT CTG ATC AGT GAA TTT AT 3′.

and massively parallel nature of this system (and others like it) have revolutionized the field of whole-genome sequencing.

The Nobel Laureate and the "heavy metal" legend

James Watson's personal genome sequence (obtained via pyrosequencing) was published in the journal *Nature* in April of 2008. Furthermore, the generated sequence data were made freely available to the public by its posting to the Internet. You can browse this sequence by going to http://jimwatsonsequence.cshl.edu/cgi-perl/gbrowse/jwsequence/. While the advances in sequencing technology that made the publication of such data possible are indeed impressive, the ability to interpret the biological meaning of the revealed sequence has posed an even greater challenge. Let us start by examining what the analysis of his genome sequence revealed about James Watson.

First, by comparing the sequence data with the Human Gene Mutation Database (a compendium of known human disease alleles), the researchers were able to identify 32 matches (i.e., Dr. Watson possessed 32 mutations known to be associated with a disease phenotype). Of these 32 alleles, 12 were known to cause disease when homozygous (e.g., Cockayne syndrome). Luckily for Dr. Watson, he was heterozygous for all these alleles and thus did not display any mutant phenotypes. The remaining 20 matching alleles were associated with an increased risk for a variety of different diseases. For example, Dr. Watson carries a mutant allele of *BRCA1* that is associated with breast cancer.

In addition, the genome sequence revealed the answer to a medical mystery that had plagued Watson for years. This mystery revolved around the fact that he had had great difficulty in controlling his blood pressure using typically prescribed drugs. Interestingly, the genome sequence revealed that he had genetic variants that made him more sensitive to these drugs (beta blockers). Thus, by simply reducing his prescribed dose, he has since been able to better control this medical condition.

While examples like these demonstrate the power of knowing your own genome sequence, there is also a flip side. For instance, Watson's grandmother died of Alzheimer's disease at the age of 84. Thus, he had a one in four chance of carrying the mutant allele (in a gene called *ApoE4*) associated with the disease. However, unlike in the case of his blood pressure, there was no simple intervention that could be used to counter the effects of the defective allele. Under these circumstances, Watson chose not to be informed as to whether he carried the abnormal *ApoE4* allele. If placed in his shoes, would you have wanted to know? These examples help reveal the positive and negative aspects of living in the post-genomic era.

At the time that Watson's sequence was revealed in 2008, only two individuals possessed personal genomic sequence data: James Watson and J. C. Venter (who had his genome sequenced using traditional methods). Since

that time, the cost of sequencing has decreased to the point where the number of personal genomes sequenced has risen into the thousands. Due to the relatively high cost ($20 000–$80 000 for good quality sequence data), most of these genomes have been sequenced at the request of affluent members of society. Just as in the case of James Watson, the analysis of the genome of one of these individuals, Ozzy Osbourne, illustrates the power of personalized genomics (**Figure 19.12**).

According to Osbourne, he chose to get his genome sequenced for the following reasons: "I was curious, given the swimming pools of booze I've guzzled over the years—not to mention all of the cocaine, morphine, sleeping pills, cough syrup, LSD, Rohypnol . . . you name it—there's really no plausible medical reason why I should still be alive. Maybe my DNA could say why." Remarkably, his genome did in fact reveal possible answers to this question. Interestingly, Osbourne was shown to possess genetic variants that affected how his brain processed dopamine. Based on these variants, researchers concluded that Osbourne had a sixfold increase in his predisposition for alcohol dependence. Furthermore, they also discovered that he possessed variants that allowed him to metabolize recreational drugs better than others. These data thus provide a reasonable hypothesis for explaining how he fell into and survived his drug and alcohol ridden lifestyle. In addition, and perhaps more importantly, Osbourne was also interested in determining if his genome sequence could reveal clues to his Parkinsonian tremors (a condition characterized by involuntary trembling or quivering). Again, the data revealed that variation in the *TTN* gene (a gene involved in muscle function) was likely responsible.

While these examples make it clear that genome sequencing can indeed provide important medical information, its overall predictive value is still quite limited. Though excellent with regard to predicting relatively simple diseases such as CF—which is known to be caused by single gene mutations—predicting the risk of more

Figure 19.12 Genomicist Nathaniel Pearson talks to Sharon Osbourne and Ozzy Osbourne about Ozzy's genomic history on October 29, 2010 in San Diego, California.

complex diseases such as heart disease or schizophrenia has proven to be much more difficult. However, as more and more genomes are sequenced, it is likely that the ability to accurately correlate genetic variation with various disease states will improve dramatically. In any event, the continued reduction of costs associated with whole-genome sequencing make it likely that you the reader will be facing the consequences of knowing your own genome sequence in the not too distant future.

Next-generation sequencing methodologies were developed in the hopes of quickly producing whole-genome sequences at low cost. Pyrosequencing (developed by Roche/454 Life Sciences) is one such method that uses a massively parallel design to produce approximately 640 million bp of sequence in only eight hours. The utility of personalized genomics is illustrated by the individual analysis of the genomes of geneticist James Watson and "heavy metal" legend Ozzy Osbourne.

19.5 Repercussions of the Human Genome Project

The genome sequences of humans and model organisms have transformed all of biology. Knowledge of these sequences enables us to identify and readily access most human genes, and the ability to do this greatly facilitates our understanding of their functions. We can also use the genome sequences to search for the control elements that help regulate the gene's expression. In addition, quick inexpensive whole-genome personal sequencing will soon transform the practice of medicine, moving it from a reactive to a predictive, preventive, and personalized mode.

As exemplified earlier, the inexpensive sequences will serve as one basis of predictive medicine because they will provide access to the DNA polymorphisms underlying human variability. Although most polymorphisms fall outside of genes and do little to change human phenotypes, some are responsible for differences in normal physiology, and others predispose to disease. For example, a single defective copy of the breast cancer 1 gene (*BRCA1*), such as the one carried by James Watson, causes 70 percent of the women who inherit it to be afflicted with breast cancer by the age of 60 (luckily for Watson this risk is much reduced in males). Why only 70 percent? Either environmental factors operate in concert with the defective gene, or other modifying, disease-predisposing genes are present only 70 percent of the time. In either case, a prediction can be made about the future likelihood of disease for individuals carrying the defective gene. In time, physicians will be able to scan the genomes of the young and provide a probabilistic projection of what the future may hold with

Some people argue that genetic information, the naturally occurring raw material of life's evolution, is a common heritage that belongs to everyone. Yet since the mid-1970s, universities and biotechnology companies have sought patents on specific DNA sequences in virtually all types of genomes—plant, human, other animals, bacterial, viral, and plasmid.

Patent examiners evaluate the patentability of a product or process by three criteria: Is it *new*? Is it *non-obvious*? Is it *useful*? In the DNA arena, the courts have made the following interpretations of patent law. Raw materials of nature, such as wild-type DNA in a living organism, are not novel and thus are not patentable; modified products, such as bacterial DNA altered by a synthetic mutation or human DNA in a mouse genome, are novel and thus eligible for patents by the novelty criterion. DNA-based processes that produce a novel material, such as clones of a gene (consisting of a DNA construct in a vector), are also patentable. Publication in a scientific journal makes an item, such as a DNA sequence, obvious and thus unpatentable per se; but a specific use of a published sequence may be non-obvious and thus patentable. A well-defined use might be a particular test for a genetic aberration or a particular process for the manufacture of a therapeutic agent.

Different countries apply the basic tenets of patent law in different ways. In Canada, claims on "isolated" molecules such as a DNA, RNA, or cDNA molecule are patentable, while "higher" life-forms (e.g., plants and animals) and non-isolated molecules that occur in nature, are not patentable. In the United States, the purified form of a gene or protein is patentable because genes and proteins do not exist in nature in purified form. In England, a naturally occurring gene sequence is not patentable no matter what form it is in. And in France, the code on intellectual property declares unpatentable "the human body, its elements and products as well as knowledge of the partial or total structure of a human gene."

The rationale for granting patents is to encourage innovation—the invention of useful contributions to society—by providing a time-limited monopoly to protect an invention from imitation. This commercial protection is given in exchange for the complete disclosure of the information related to a product or process. In fact, in pursuing a patent, a company protects its interests by making available as much information as possible about the modified product or process. Patents protect use for profit; they do not interfere with research or other noncommercial uses of the information in the patent.

Currently, much debate is taking place over the application of patent criteria, originally developed for mechanical or chemical inventions, to life-forms or materials derived from them. This debate identifies several areas of concern.

Openness Versus Secrecy

A company putting a great deal of money into the research and development of what it hopes will be a patentable gene therapy may withhold publication of its data or publish only partial results until its patent application is in the pipeline.

For example, in the 1980s, before the discovery of the cystic fibrosis gene, one gene diagnostics company published an article about markers for the *CF* gene but, in an effort to protect their work, did not include the fact that those markers were on chromosome 7. As it happens, other groups subsequently found closer markers and then the gene. But the question remains regarding to what extent commercial considerations interfere with free exchange of ideas, and if so, whether anything can or should be done about it.

Profitability Versus the Social Good

Most companies consider potential profitability as the basis for pursuing research and development. Some people wonder whether they should also factor into the equation concerns about serving the poor (who cannot pay much for drugs or therapies) and the relatively few patients suffering from uncommon diseases. For example, about 25 million people in Africa are infected with HIV, the virus that causes AIDS. In North America and Europe, AIDS can be managed by triple drug therapy—the simultaneous administration of three drugs that attack AIDS with different mechanisms—but this therapy costs more than $12 000 per year. Most infected individuals in Africa cannot afford the treatment.

India has started to manufacture generic AIDS drugs costing just a few dollars per day, roughly $1100 per year. India can manufacture these drugs inexpensively because the very costly research to identify promising compounds and develop them into drugs was paid for by European and American pharmaceutical companies. In fact, India is violating drug-company patents when it sells the AIDS drugs inexpensively in Africa. Millions of infected people in Africa, however, could never receive appropriate treatment for AIDS without the inexpensive generic drugs manufactured in India. The terrible dilemma faced is the ongoing need for additional research, for which companies must pay, contrasted with the need of suffering people to receive care for their illnesses.

Development and Funding of Socially Useful Research Applications

Because companies must pour large sums of money into developing DNA-based drugs or therapies, testing them on large numbers of people, and bringing them to market, many researchers—at both universities and commercial enterprises—maintain that the survival of biotechnology companies, particularly small ones, depends on patent protection. Without the possibility of a patent, they argue that companies may not be able to afford to use knowledge of, for example, the *CF* gene and the transmembrane conductance regulator it encodes to

(Continued)

regard to a wide variety of diseases. Of course, being able to predict a disease without being able to cure or prevent it leaves physicians in a very uncomfortable position. As the field matures, scientists hope to be able to learn how to circumvent the limitations of these defective genes—with, for example, novel drugs, environmental controls, or other approaches such as stem cell transplants or gene therapy. Preventive measures may be designed to avoid or greatly delay the onset of the disease.

Social, ethical, and legal issues have no simple solutions

The ability to analyze the genomes of individual humans raises a host of pressing questions about the privacy of genetic information, limitations on the use of genetic testing, the patenting of DNA sequences, society's view of older people, the training of physicians, and the extent to which the human genetic engineer should seek to engineer himself or herself.

Because of the complexity of these controversial issues, we have presented individual topics in the **Genetics and Society** essays found throughout this book. In this chapter's essay, **"Patentability of DNA,"** we consider who owns the information being revealed by research and in what ways that information may legally—and ethically—be used. Other social and ethical issues include confidentiality, privacy, genetic diagnosis, and screening.

Another consideration is the impact of longevity that may be brought about through improved predictive/preventive medicine. With more people living productively into their nineties and beyond, societies may need to rethink philosophies of retirement, pension plans, and access to medical care.

Many preventative or remedial measures constitute **somatic gene therapy**, where medical practitioners compensate for a faulty gene by inserting a replacement gene into the affected tissue where the gene is expressed. Somatic gene therapy causes biochemical and physiological changes in the genetically modified tissue or tissues that die with the individual. A potential therapy for cystic fibrosis, for example, involves inserting a wild-type *CFTR* gene into lung cells. This type of genetic engineering is not different in kind from drug therapies aimed at correcting a particular physiological deficiency (e.g., the insulin injections aimed at treating diabetes). The alterations resulting from somatic gene therapy affect only the somatic cells of the individual undergoing the therapy and cannot be transmitted to offspring.

Most of the controversy surrounding genetic engineering stems from the potential for **germline gene therapy**: modifications of the human germ line. Germline gene therapy produces changes in germ cells that are passed on to progeny. Ethical concerns focus on what is and is not appropriate. For example, should parents be able to eliminate a cancer-predisposing gene in their unborn child? Should they be allowed to alter the child's potential for obesity or longevity? Should they have the option of choosing the child's eye colour? (Review the Genetics and Society box in Chapter 15.)

Many geneticists and bioethicists support the idea of somatic gene therapy but oppose germline therapy. At the same time, they urge serious and open discussion of the issues before perfection of the technology overtakes our ability to control its use.

> While new technologies open up novel avenues of medical intervention, ethical questions remain to be resolved. Somatic gene therapy may allow individuals to survive otherwise debilitating diseases. Germline gene therapy is far more controversial.

Connections

One of the triumphs of modern genetics has been the determination of the complete sequence of the human genome and the genomes of many model organisms. With powerful new "shotgun" strategies for sequencing large genomes, genomicists have sequenced more than 9000 different microbe, plant, and animal genomes. The genome projects now under way in laboratories around the world will eventually add thousands of additional genomes to this list.

The high-throughput DNA sequencing platforms developed by the Human Genome Project have catalyzed the

emergence of similar platforms for analyzing mRNA and proteins. Based on these tools, analyses have provided insights into the architecture and evolution of genomes and proteomes. The genetics parts lists made available by the genome sequences are transforming the practice of biology and medicine.

The human genome defines the human species. Indeed, every individual person carries a diploid genome that is 99.9 percent identical to that carried by every other individual. But the flip side of a 99.9 percent identity is a 0.1 percent nonidentity that distinguishes people (other than identical twins) from each other. This 0.1 percent difference translates into 6 million DNA sequence differences, or polymorphisms, that are responsible for all of the inherited ways in which individuals differ from one another. Of critical interest to medical researchers are the specific DNA polymorphisms that either directly alter or indirectly mark the genes that cause or predispose to disease. In Chapter 15 on genome-wide variation, we discussed how researchers identified and used DNA polymorphisms to uncover disease-causing genes and other genes of interest in humans and other species. In Chapter 20, we continue this discussion to include an examination of how genome-wide association studies are used to study complex genetic traits.

Essential Concepts

1. Sequence maps, compiled from the sequences of subclones, provide a readout of every nucleotide in each chromosome. The subclones are derived either from previously mapped large insert clones (hierarchical shotgun approach) or directly from the genome (whole-genome shotgun approach). **[LO1–4]**

2. A variety of major insights have emerged from analyses of human and model organism genomes. The number of human genes, approximately 21 000, is surprisingly low, and an organism's complexity is not an indicator of the number of genes its DNA carries. Many gene sequences have been conserved over evolutionary time, allowing the possibility that gene function can be deduced from known functions of similar genes in other species. **[LO1, LO3]**

3. Comparison of genomes allows investigators to infer evolutionary relationships as well as the function of genes across species. The DNA sequences of microbes, plants, and animals all employ the same genetic code and show a remarkable similarity among many basic biological systems. This affirms the idea that all present life descended from a single common ancestor. **[LO1, LO3]**

4. The Human Genome Project has catalyzed the development of automated DNA sequencing, next-generation sequencing technologies, and personalized whole-genome studies. **[LO1, LO3–6]**

5. The Human Genome Project has brought fundamental change to the disciplines of biology and medicine. Predictive medicine may allow researchers to correlate polymorphisms with disease predisposition and to make health-history predictions for each individual. Preventive medicine may allow identification of defective genes and facilitate development of ways to avoid their limitations with drugs, diet, gene therapy, or stem cell therapy. The potential of predictive/preventive medicine raises social, ethical, and legal questions for which there are no easy answers. **[LO1, LO3–6]**

Solved Problems

I. A physical map of overlapping clones (a contig) was available for an area of a human chromosome containing three genes (and part of a fourth gene) that are transcribed within ovarian tumour tissue. The restriction map of the region is shown *below*; tick marks above the line indicate *Bam*H1 restriction sites, while those below the line indicate *Xho*I sites. Sizes of DNA fragments between adjacent restriction sites are given in kilobases. Individual restriction fragments were purified, made radioactive, and used as probes for Northern blots of poly-A-containing RNA derived from ovarian tumours. In a *Northern blot*, mRNAs are analyzed just as DNA fragments are in a Southern blot (see Chapter 14). The poly-A-containing mixture of mRNAs from a particular cell type is subjected to electrophoresis—separating the mRNAs according to size (smaller mRNA migrate more rapidly than larger ones). The mRNAs are then transferred from the gel to a nitrocellulose filter and a radiolabelled probe for a particular gene (or gene fragment) is hybridized against the mRNAs on the filter. If size standards are also run, the presence and size of particular mRNAs can be established. The resulting autoradiograms are presented below the restriction map. Using these data, characterize the four genes within the contig in the following ways:

a. What is the length of the mRNA for each of the three complete genes?

b. What is the minimum length of the primary transcript for the largest of these RNAs?

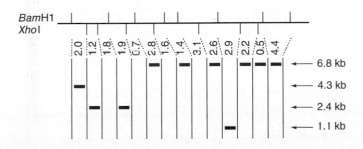

c. What is the minimum number of exons for each gene?

d. What is the minimum number of introns for each gene?

Answer

This problem requires an understanding of primary and processed transcripts and the analysis of RNAs using Northern hybridization.

a. The bands on the Northern blots that hybridize with the probes represent the mRNAs from this region. The three mRNAs corresponding to the genes in this region are *1.1, 2.4, and 6.8 kb in length*. The 4.3-kb transcript comes from a gene that is only partly contained in the DNA used as a probe since the one DNA fragment that hybridizes is only 2.0 kb in length.

b. The bands on the Northern blot represent processed transcripts (introns have been removed). Primary mRNAs are made by copying from contiguous DNA sequences, including those regions that will be removed by splicing. The minimum length of a primary transcript is based on the sizes of the restriction fragments that hybridize with the RNAs and any intervening fragments. *For the 6.8-kb largest mRNA, the minimum size is 2.8 kb + 1.6 kb + 1.4 kb + 3.1 kb + 2.6 kb + 2.9 kb + 2.2 kb + 0.5 kb + 4.4 kb, or 21.5 kb.* We cannot say from these data whether transcription begins before the 2.8-kb fragment or extends beyond the 4.4-kb fragment.

c./d. The minimum number of exons is determined by counting the number of hybridizing fragments or groups of contiguous fragments. This is a minimum number because any of the fragments could contain more than one exon. The nonhybridizing fragments that separate these must contain introns.

Transcript	Minimum number of exons	Minimum number of introns
1.1 kb	one exon	none
2.4 kb	two exons	one
6.8 kb	four exons	three

II. Reverse transcriptase, the enzyme used to synthesize cDNA starting with mRNA as a template, often falls off the template before completely copying the mRNA. When screening for a cDNA clone of a gene, it is therefore not uncommon to isolate partial cDNA clones. What comparison could you make experimentally that would indicate if you had isolated a plasmid clone containing a partial cDNA?

Answer

For this problem, consider the two alternatives: You have the complete cDNA clone or you have a partial cDNA clone. What would be the differences that you could detect experimentally? The cDNA must be as long as its corresponding mRNA to be full length. You need to find out what the full length of the message is to do the comparison. The way to *measure the length of mRNA is to run a Northern gel and hybridize with the cDNA as a probe.* (Another alternative would be to hybridize a fragment from near the 5′ end of the gene to your clone. This would require that you have a fragment that you know is from the 5′ end of the gene.)

Problems

Vocabulary

1. For each of the terms in the left column, choose the best matching phrase in the right column.

i. sequence tagged sites (STSs)

ii. whole-genome shotgun sequencing

iii. contig

iv. physical map

v. RefSeq

vi. pyrosequencing

1. a map showing the order of cloned inserts of DNA

2. a single, complete annotated version of a species genome that is available online

3. a genome sequencing strategy that avoids physical mapping

4. set of overlapping clones representing a larger contiguous region of a chromosome

5. unique DNA sequences that serve as molecular markers

6. a DNA sequencing method based on the detection of light as result of incorporating a nucleotide into a DNA strand

Section 19.1

2. Describe the rationale for funding the Human Genome Project. Discuss the potential pros and cons of this endeavour. Has the HGP fulfilled its promise?

Section 19.2

3. To make a set of clones more suitable for the analysis of DNA sequence, a series of clones was prepared by digesting a BAC clone and subcloning the resulting restriction fragments. The restriction patterns of the inserted fragments are shown next. Arrange these four clones into a physical map, showing the order of the clones and the overlap between them.

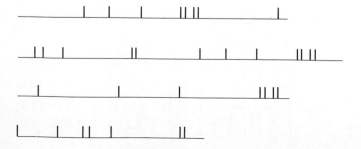

4. During the course of the genome project for the rhesus monkey *Macaca mulatta*, five BAC clones (A–E) forming a single contig were obtained. Researchers determined a short (approximately 500 bp) sequence of monkey DNA from each of the two ends of the BAC clones (i.e., from where the monkey genomic DNA was joined to the BAC vector). The scientists converted these sequences into sequence tagged sites (STSs) by making PCR primers that could amplify the 500 bp of monkey DNA if it were present in any DNA sample. The table below shows which STSs were found in each of the five BACs; each clone of course has the two STSs corresponding to the sequenced monkey DNA at each end.

BAC clone	End STSs	Other STSs
A	1 2	4 5 7 10
B	3 4	2 9
C	5 6	1 8
D	7 8	1 5
E	9 10	2 4

a. Why is it very efficient to determine the sequences of monkey DNA at locations where it is joined to the BAC vector?

b. Diagram a physical map of this region consistent with the data, indicating the relative order of the BAC clones and the location of the STSs.

c. If you wanted to determine the DNA sequence of the entire contig, it would be advantageous to work with the minimal tiling path of the BAC clones. Why? Diagram the minimal tiling path consistent with the data in the table.

d. Estimate the size of the contig in kilobases (kb).

5. In the course of sequencing a genome, a computer is trying to assemble the following six DNA sequences into contigs:

5′ CAAATAGCAGCAAATTACAGCAATATGAAG 3′

5′ AAAATGCCCTAAAGGAAATGAGATTTTTAA 3′

5′ TGATCTCTTCATATTGCTGTAATTTGCTGC   3′

5′ GTAGTATCTCCTTTTAAAAATCTCATTTCC   3′

5′ CAATATGAAGAGATCATACAGTCCACTGAA 3′

5′ TCTCATTTCCTTTAGGGCATTTTCAAATTC   3′

How many contigs are represented by this set of DNA sequences, and what is the sequence of each contig?

6. Repetitive DNA sequences present a challenge to genome projects. Why is this so? What types of repetitive sequences are most problematic? How can whole-genome shotgun sequencing strategies deal with this problem?

7. It is often difficult to find genome-unique PCR primers in certain regions of the genome. Offer two explanations.

8. What are two potential difficulties with the whole-genome shotgun sequencing strategy?

Section 19.3

9. With new information from the Human Genome Project, many new genes will be identified for which the function is not known.

a. What features of the DNA sequence might help you determine the function of a newly identified gene?

b. What are two other types of analysis that would help you learn more about a new gene?

10. Refer to Figure 19.7.

a. What is the significance of the RefSeq genes appearing to "pile up" in the vertical direction on part (a) of the figure?

b. What is the approximate location of the centromere on human chromosome 7?

c. Is the *CFTR* gene located on the short arm or the long arm of human chromosome 7?

d. In which direction is the *CFTR* gene transcribed: toward the centromere, or away from the centromere?

e. What is the approximate number of exons in the *CFTR* gene? Why is this number only an approximation?

11. a. If you found a zinc-finger domain (which facilitates DNA binding) in a newly identified gene, what hypothesis would you make about the gene's function?

b. In another gene, what would a high percentage of similarity throughout the gene with a previously identified gene in the same organism suggest about the origin of the gene?

12. Complete genome sequences indicate that the human genome has roughly 21 000 genes, and the worm (nematode) genome has 19 000 genes. Explain how the human genome can encode a creature enormously more complex than the worm with at most only one-tenth more genes in its genome.

Section 19.4

13. What is meant by the term *next-generation sequencing*? What is the difference between traditional sequencing technologies and "next-generation" methods?

14. Next-generation methods are often referred to as being massively parallel. Why is a massively parallel design so advantageous with regard to the speed and cost of sequencing?

15. Explain how the pyrosequencing method utilizes the measurement of light intensity to quickly and easily sequence whole genomes.

16. Using the flowgram that follows *below* determine the sequence of the template strand (T = orange, A = green, C = blue, G = black).

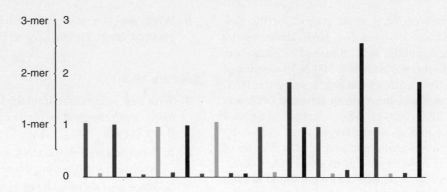

Section 19.5

17. While personalized whole-genome sequencing has great potential to benefit society, there are also many ethical issues that need to be considered. Discuss the pros and cons of living in the post-genomic era. If given the chance, would you want access to your personal genome sequence? Explain.

CHAPTER 20 # Global Genetic Analysis

| Understanding the structure of genomes | Understanding the biology of genomes | Understanding the biology of disease | Advancing the science of medicine | Improving the effectiveness of healthcare |

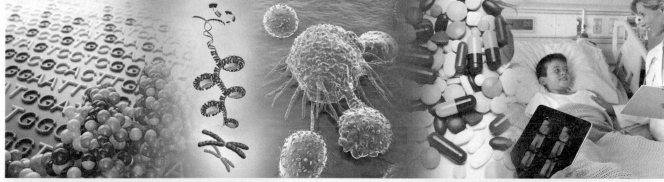

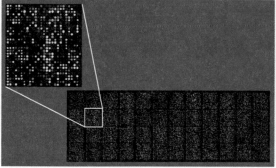

The development of whole-genome sequencing has forever changed the manner in which scientists practise genetics. For instance, it is now possible for researchers—using sequencing data as a foundation—to quickly and easily perform genetic analysis on a global scale (i.e., analyze every gene in the genome in a single experiment). Such analysis is increasingly being used to provide a better understanding of a variety of different biological processes. As we shall see later in this chapter, this type of global genetic analysis, referred to as "genomics," has become an important part of the toolkit of many medical professionals.

The scientific challenges of the Human Genome Project initiated a revolution in the development of sequencing technologies. Similarly, the need for efficient and cost-effective tools to exploit the resultant data has driven the development of powerful, high-throughput platforms that characterize genomic properties on a global (i.e., genome-wide) scale. In this context, a *platform* denotes all of the components needed for the automated acquisition of a set of data. These platforms represent a fusion of the efforts of both research scientists and private industry. In this chapter, we will describe some of the common platforms that have been developed for large-scale genomic analyses and explain how they are used to provide biological insight.

Chapter Outline
20.1 Gene Expression Arrays
20.2 "ChIP-on-Chip" and "ChIP-Seq" Analyses
20.3 Comparative Genome Hybridization
20.4 Genome-wide Association Studies
20.5 Epigenomics

Learning Objectives

1. Compare and contrast traditional genetic analysis with the global analysis discussed in this chapter.

2. Relate nucleic acid hybridization to (i) the measurement of global gene expression levels, and (ii) techniques used to define protein-DNA interactions.

3. Define comparative genome hybridization and illustrate how this methodology can be applied in a clinical setting.

4. Evaluate the ability of genome-wide association studies to correlate genetic variation with complex traits.

5. Define epigenetics. Explain how epigenetic modifications are characterized on a genome-wide level.

Gene Expression Arrays

As we saw in Chapters 10 and 11, the control of transcription is one of the most important ways in which cells regulate gene expression. For this reason, one of the first applications of the fruits of genomic data was the quantitative analysis of mRNA levels. In this section, we will discuss how such analysis is possible and how this methodology can be applied to further our understanding of biology.

Genome annotation

A prerequisite of any "omics" experiment—from gene expression analysis to proteomic analysis—is the accurate annotation of the genome under study. In other words, the raw sequence data from genome sequencing projects must be analyzed so that the "features" of the genome (e.g., protein-coding genes, ribosomal RNAs, noncoding RNAs, promoters, intron/exon junctions, 5′ and 3′ UTRs) are accurately defined. As one might imagine, this cannot be done through the manual inspection of the millions of nucleotides generated by genome sequencing. Instead, sophisticated computer algorithms have been created to recognize the DNA sequence patterns that are characteristic of each of these features.

One of the simplest features to look for is a long stretch of nucleotides that begins with a start codon and ends with a stop codon. If the nucleotide sequence in between is devoid of in-frame stop codons, then one can consider this stretch to be a potential protein-coding gene. In addition to start and stop codons, genes also possess other characteristic sequences that can be used to help determine whether the candidate sequence does indeed define a protein-coding gene (**Figure 20.1**). For example, eukaryotic gene promoters possess highly conserved DNA sequences

(e.g., promoter proximal elements like the "TATA" and "CAAT" boxes) that function to recruit the core transcriptional machinery. Moreover, a consensus sequence called the Kozak signal, GCC(A/G)CCATGG, which encompasses the start codon, can also be used to help define a gene. Also, introns and exons can often be defined by the presence of conserved sequences (splice acceptor/donor sites) required for the splicing of pre-mRNAs. While an in-depth description of these computational and statistical methods is beyond the scope of this text, we can say that the reliability of these algorithms has improved to the point that raw genomic sequence data can be analyzed to define genomic features to a reasonably accurate degree.

Building the DNA microarray

Armed with accurate gene annotation data, it is now possible to exploit whole-genome sequences to build what is known as a DNA microarray. A DNA microarray, sometimes referred to as a "gene chip," is created by attaching (or "spotting") specific DNA fragments—each to a predefined position—onto a solid support such as a small glass slide (**Figure 20.2a**). These spots, themselves composed of millions of copies of a given DNA fragment, are so small that thousands of spots can be placed on the chip. In the photograph that opens this chapter (*bottom left*), you can see such a slide with approximately 40 000 individual spots. These gene chips are created with the aid of robotically controlled pins capable of accurately spotting tiny amounts of DNA at precise locations (**Figure 20.2b**).

As you might have guessed, the DNA in each spot corresponds to the sequence of an annotated gene. Typically, a sequence representing a small stretch of an expressed portion of a gene is chosen to be placed on the chip (**Figure 20.3a**). These small fragments can be generated via PCR using gene-specific primers, or synthesized using specialized DNA synthesis machines. PCR-generated double-stranded DNA fragments are typically a few hundred base pairs long, while the single-stranded oligonucleotides produced by DNA synthesis machines are typically 30–80 nucleotides long. Once placed on the gene chip, each spot of DNA is referred to as a "probe" since, as we shall see in the next section, it allows the researcher to probe for the presence of specific expressed mRNAs.

For studying the fission yeast *Schizosaccharomyces pombe* (a unicellular eukaryote that is often used as a model organism in genetic research), the gene chip would be composed of approximately 10 000 DNA spots. These spots would be composed of DNA fragments representing the expressed portions of all the approximately 5000 *S. pombe* genes (in duplicate). While early arrays could carry only about 12 000 spots, modern methods can place approximately 140 000 spots onto a single glass slide. Other technologies, which use proprietary methods to attach DNA to the solid support, can create slides

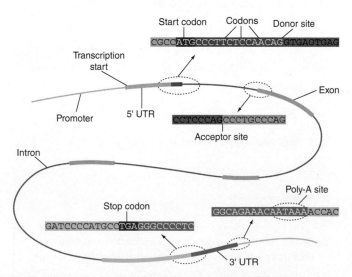

Figure 20.1 Conserved DNA sequence features that can be used to indicate the presence of a gene. Raw genomic sequence data can be analyzed for sequence features like the ones shown in the schematic drawing (e.g., start and stop codons, splice donor/acceptor sites, poly-A sites) to define genes.

Figure 20.2 Constructing a gene chip. (a) The researcher is holding a small glass slide onto which specific DNA fragments have been spotted using a robotic microarray spotter. **(b)** The printhead of a robotic microarray spotter is composed of tiny metallic pins capable of placing nanograms of DNA at predefined locations on a slide.

containing well over 1 000 000 unique DNA fragments (**Figure 20.4**). While these technological advances have been impressive, the underlying strategy that makes this methodology possible is based on the simple hybridization of complementary DNA sequences. In the next section, we will describe how these gene chips are actually used to measure gene expression levels.

Hybridizing to the microarray

DNA fragments placed on the microarray are rationally designed to represent expressed portions of all annotated genes. Thus, cDNAs (derived from mRNAs isolated from

the organism under study) would, upon denaturation, exhibit complementarity to the denatured DNA fragments placed upon the gene chip (see Chapter 14 for a review of cDNA construction). It is this complementarity that is exploited to measure gene expression.

Typically, gene expression experiments are performed in a comparative manner. For example, a researcher might choose to compare the gene expression levels of a group of control cells with an experimental group of cells (e.g., cells derived from normal lung tissue versus cells derived from cancerous lung tissue). In such a scenario, the goal is to identify significant changes in gene expression between the samples. Genes that exhibit significantly different expression are termed "differentially regulated" genes and may provide clues as to the phenotypic differences between the two populations.

Consider a hypothetical example in which the transcriptional response of fission yeast cells to the drug bleomycin

Figure 20.3 Microarray probe design. (a) Probes for gene expression arrays are chosen so that they represent an expressed portion of a gene. In the schematic, the probe matches the sequence of the last exon (*red blocks*) of the gene. **(b)** Probes for intergenic gene chips (intergenic regions are shown in *blue*) are chosen to represent the location of potential DNA-binding sites (e.g., regulatory regions, like promoters). Sometimes intragenic regions such as introns (*blue regions* between the *red blocks*) can also contain binding sites for transcriptional regulators and are also included on these chips.

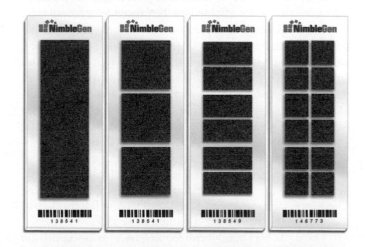

Figure 20.4 NimbleGen gene chips. The private company NimbleGen produces a single gene chip that carries 12 replicates of a grid containing 135 000 DNA fragments (*far right*)!

(induces double-stranded breaks in DNA) is to be characterized. To analyze the transcriptional response, researchers would make use of an experimental group of cells treated with the drug, and a control population of cells not exposed to the drug. By comparing these two groups, researchers will be able to determine if fission yeast cells can recognize the DNA damage and induce the expression of genes involved in the repair of double-stranded breaks. The process works as follows.

First, fission yeast cells are cultured in liquid growth medium and the sample is split into two. One of the two samples (the control group) remains untreated, while the other sample (the experimental group) is cultured in the presence of bleomycin. After a given period of time, mRNAs from both of these populations are purified and converted to cDNAs. It is at this point that each of the samples is labelled with one of two fluorescent dyes: Cy3 (a green fluorophore) or Cy5 (a red fluorophore). Labelling is done by including nucleotides conjugated to Cy3 or Cy5 (usually dCTP) along with normal, nonconjugated dATP, dTTP, and dGTP during synthesis of the cDNA. In this way, the cDNAs generated from each of the samples will—upon exposure to the appropriate wavelength of UV light—fluoresce green (if Cy3 is used) or red (if Cy5 is used). In this example, the control cDNAs are labelled with Cy5, and the experimental cDNAs labelled with Cy3.

Once the red- and green-labelled cDNAs are synthesized, the samples are mixed together in a single test tube. This combined sample is then allowed to hybridize to the gene chip. Typically, the cDNAs are denatured (made single-stranded) and then aliquoted onto the chip so that the sample covers all the spots on the array, allowing for the cDNA molecules to hybridize to the denatured DNA fragments on the chip. The specificity of binding of the cDNA strands (to the spotted DNA on the gene chip) is based upon the complementarity of the sequences. Furthermore, the amount bound to each spot will depend on how much message was present in the original samples. Take for example two hypothetical genes from the fission yeast genome, "X" and "Y." Assume that gene X is a "housekeeping" gene that is not affected by the presence of bleomycin. Gene X would thus be expressed both in the control sample and in the treated sample. In fact, the level of expression would be about equal in both samples. In this case, similar amounts of Cy3- and Cy5-labelled cDNA (derived from the gene X message in the experimental and control samples, respectively) would hybridize to the spot on the chip representing gene X. On the other hand, let us assume that gene Y is a DNA repair gene, and that it is dramatically up-regulated in response to bleomycin. In this scenario, there would be far more Cy3-labelled cDNA derived from the gene Y message in comparison with Cy5-labelled cDNA. Therefore, far more green-labelled cDNA will hybridize to the spot on the array corresponding to gene Y.

Once excess, nonhybridized cDNA is washed off the chip, the microarray is dried and prepared for scanning. In this process, a microarray scanner is used to quantitate the fluorescent cDNA molecules that bind to the gene

chip (**Figure 20.5**). The device first exposes the chip to a specific wavelength of UV light that excites the Cy3 fluorophore. Next, a digital camera captures an image of the chip as the DNA fluoresces green. Subsequently, the chip is exposed to a wavelength of UV light that excites the Cy5 fluorophore and a second image is captured (as the DNA fluoresces red). The two captured images are then overlaid, one on top of the other by computer, to create a single composite image (**Figure 20.6**). This image is then analyzed using computer software to provide quantitative data regarding changes in gene expression.

For instance, if a gene is expressed at similar levels in both the control and experimental samples, then the spot representing that gene on the chip will appear yellow in the composite image. If, on the other hand, a gene is up-regulated in the experimental sample relative to the control, then the spot will appear greener in the composite image. Conversely, if a gene is down-regulated in the experimental sample relative to the control, then the spot representing this gene on the chip would appear more red. Since the position of each gene fragment on the chip is known, one can—with the help of computer software that keeps track of all the spots—calculate the relative amount of expression for every gene in the genome. Thus, in the given example, genes that are both up- and down-regulated in response to DNA damage can be determined. As you can surmise, a similar strategy to compare gene expression between any two samples of interest (e.g., normal versus cancerous tissue, drug-treated versus untreated tissue culture cells, heat-shocked versus non-heat-shocked *Drosophila*) could be used. See the **Tools of Genetics** box **"Diagnosing Cancer"** for how this technology is applied for the diagnosis and treatment of distinct types of cancer.

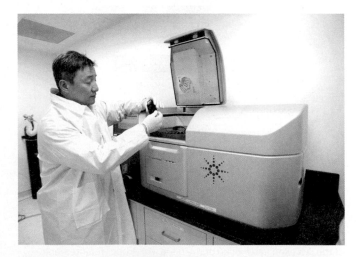

Figure 20.5 A microarray scanner. This tabletop device is able to scan gene chips to measure the fluorescence emitted by both the Cy3 and Cy5 fluorophores. The system also contains a digital camera to record the necessary images. The data are then passed on to computer programs that are able to quantitate the ratio of green to red fluorescence for every DNA spot on the chip.

A key problem in the effective treatment of cancer relates to the proper classification of tumours. If a tumour is classified improperly, the result may be the prescription of treatments that are completely ineffective against the tumour afflicting the patient. For example, acute lymphoblastic leukaemia (ALL) responds well to a mixture of corticosteroids, vincristine, and methotrexate, whereas acute myeloid leukaemia (AML) responds well to a different set of drugs, daunorubicin and cytarabine. Unfortunately, differentiating between these two distinct types of cancers can be difficult, even for experienced medical professionals. Remarkably, advances in gene expression analysis have provided a simple and effective way of distinguishing between ALL and AML. Extensive gene expression analysis of both types of tumours has identified a

small set of genes that can be used as biomarkers to distinguish between the two. In the figure shown, the level of gene expression of these genes (listed on the *right*) is analyzed in a series of tumours derived from different patients. (Each column represents a different tumour isolate.) The level of gene expression is represented by colour; *dark blue* representing low gene expression, pale *blue/pink* representing intermediate gene expression, and *dark red* representing high levels of gene expression, as indicated by the scale at the bottom of the figure. As one can see in the given data, it is possible, based on the pattern of gene expression, to distinguish between these two cancers. Therefore, doctors and patients alike can now rest assured that the most effective treatment regimen will be prescribed.

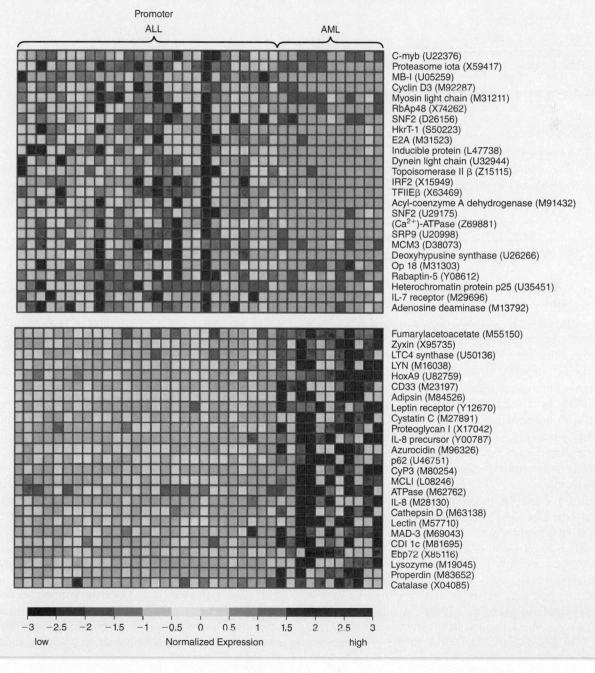

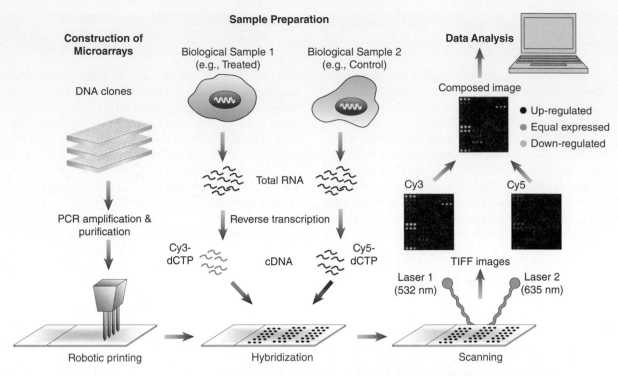

Construction of Microarrays

DNA clones

PCR amplification & purification

Robotic printing

Sample Preparation

Biological Sample 1 (e.g., Treated)

Biological Sample 2 (e.g., Control)

Total RNA

Reverse transcription

Cy3-dCTP   cDNA   Cy5-dCTP

Hybridization

Data Analysis

Composed image

● Up-regulated
● Equal expressed
● Down-regulated

Cy3   Cy5

TIFF images

Laser 1 (532 nm)   Laser 2 (635 nm)

Scanning

Figure 20.6 Schematic representation of the steps required to measure relative changes in gene expression levels.

Gene expression microarrays are composed of a solid support onto which small DNA fragments are robotically spotted. Each individual DNA spot on the chip is made up of millions of copies of a DNA fragment representing an expressed portion of a gene. Thousands of individual DNA spots (representing an entire genome) can be placed on a single gene chip. By competitively hybridizing fluorescently labelled cDNAs (derived from mRNAs isolated from the organism under study) to the gene chip, relative changes in gene expression between a control and an experimentally treated sample can be determined.

20.2 "ChIP-on-Chip" and "ChIP-Seq" Analyses

While gene expression analysis is a powerful tool, it represents but the tip of the iceberg in terms of what may be learned from genomic methods. For example, it was not long after the successful application of global gene expression analysis that the same technologies were extrapolated for alternative purposes. One such technology, also related to the regulation of gene expression, is referred to as ChIP-on-chip analysis. The first "ChIP" in the phrase stands for <u>ch</u>romatin <u>i</u>mmuno<u>p</u>recipitation, while the second "chip" stands for "gene <u>chip</u>." In contrast to expression analysis, which measures gene expression levels, the goal of this technique is to determine the gene targets of any given transcription factor.

Take, for example, the Myc transcription factor. As you know (see Chapter 11), this protein is a critical regulator of cell proliferation and is misregulated in many cancers. If Myc target genes (i.e., the genes in the genome that are transcriptionally regulated by the binding of Myc to their promoters) could be identified, then researchers would be one step closer to understanding its role in cancer progression. Luckily, with the availability of the full human genome sequence, it is now trivial to achieve this goal. In this scenario, the spots on the gene chip are

chosen—not to represent the expressed portion of a gene—but instead to represent regulatory regions; that is, intergenic sequences such as promoters (see Figure 20.3b). Now let us see how these intergenic gene chips can be used to identify transcription-factor-binding sites.

The methodology is dependent on the availability of reagents (such as antibodies) that are capable of specifically binding to the transcription factor of interest. If available, such an antibody could be used to immunoprecipitate the protein of interest from the tissues of the organism under study. In a typical ChIP–chip experiment, the relevant cells are first cultured under conditions where the transcription factor is expressed and active (i.e., bound to the promoters of its target genes). The cells are then treated with the fixative formaldehyde. This results in the transcription factor becoming covalently cross-linked to the DNA to which it is bound. The DNA is then sheared at random into small fragments (less than 1000 bp) by sonication. Then the protein-bound DNA fragments are incubated with the specific antibody. Once the antibody is bound, the protein-DNA complex precipitates out of solution. (All other protein-DNA complexes remain in solution and are removed along with the supernatant.) What is now left are

immunoprecipitated complexes of the transcription factor of interest bound to its target DNA sequences.

To purify and identify the bound DNA, the covalent cross-links are reversed by heating and the DNAs purified from the protein component of the sample. At this point, small linker DNAs (that serve as PCR-primer-binding sites) are ligated to the ends of each fragment. Next, using primers specific to the linker sequences, the DNA fragments are amplified in a PCR using Cy5-conjugated nucleotides. In a complementary experiment, a control sample is prepared in exactly the same fashion (except that no specific antibody is used and the fragments are labelled with Cy3-conjugated nucleotides). This results in the experimental sample being composed of DNA fragments enriched in transcription-factor-binding sites, while the control sample is composed of a random collection of genomic DNA fragments.

As you might have guessed, the identity of the binding sites can be determined by mixing the two samples together and hybridizing the mixture to the intergenic gene chip. After scanning, the spots representing transcription-factor-binding sites will appear red since these sequences are over-represented in the sample. All other spots on the gene chip will show low levels of green fluorescence since the binding sites are relatively under-represented in the control. Again, with the aid of computer software one can determine the identity of the intergenic regions bound by the transcription factor by simply measuring the intensity and colour of the spots. Genes containing such identified binding sites within their promoters (or other regions capable of regulating gene expression) can thus be classified as being candidate target genes of the transcription factor.

With the advent of next-generation sequencing it should also be noted that, while still commonly used, Chip–chip methodologies are slowly being replaced with techniques that rely on direct sequencing of the fragments as opposed to hybridization to a chip. These ChIP-seq methods use the same procedure for the ChIP portion of the experiment (i.e., immunoprecipitation with a specific antibody), but do not identify the binding sites via hybridization to an intergenic gene chip array. Instead, the DNA fragments from the control and experimental samples are individually sequenced by a next-generation sequencing platform (like the Roche/454 pyrosequencer). Since the transcription-factor-binding sites are highly represented in the antibody-treated sample, these fragments will be sequenced many, many times over. Thus, by simply enumerating how many times a given stretch of DNA is sequenced (relative to the control sample), the propensity of any transcription factor for binding any DNA sequence in the genome can be determined.

> Intergenic microarrays together with chromatin immuno-precipitation techniques can be used to identify the genomic-binding sites of any given DNA-binding protein. Such data can be used to help identify the gene targets of transcription factors.

20.3 Comparative Genome Hybridization

In Chapter 9, you learned about karyotype analysis and how it is used to detect chromosomal rearrangements. However, two problems occur when searching for chromosomal rearrangements and changes in chromosome number by karyotype analysis. First, it is a tedious procedure that depends on highly trained technicians to identify chromosomal alterations under the microscope. Because of the subjective nature of the analysis, mistakes can reduce the accuracy of the results. Second, even in the hands of the best technicians, there is a limit to the viewing resolution. Even under optimal circumstances, it is not possible to detect deletions or duplications of less than 5 Mb in human karyotypes. Human populations no doubt have many chromosomes with as yet undetected smaller deletions or duplications.

Again, with the help of whole-genome sequence data and microarray technology, researchers have developed a microarray-based hybridization protocol that can scan the genome for deletions, duplications, and aneuploidy with much greater resolution, very high accuracy, and much greater throughput (and without the need for a subjective determination of the result). The technique is called *comparative genomic hybridization* (*CGH*) or sometimes *virtual karyotyping*.

The protocol works as follows. First, a series of 20 000 BAC clones with DNA inserts averaging 150 kb that (collectively representing the entire human genome) are spotted onto a microarray. These BAC clones were characterized in the course of the Human Genome Project. Next, genomic DNA from a control sample with a normal genome content is labelled with a yellow fluorophore (called Cy3.5, a derivative of Cy3), while the genomic DNA from the test sample is labelled with Cy5. The two genomic DNA samples are mixed together in equal amounts, denatured, and applied to the microarray. After hybridization is complete and unhybridized material is washed away, the fluorescence emission from each microarray spot is analyzed automatically by a microarray scanner.

If the genomic region probed with a particular BAC clone is present in two copies in the test sample, then the ratio of red to yellow dyes on that dot will be 1:1. However, if a particular genomic region is duplicated or deleted from one homologue in the test sample, the ratio of red to yellow will be 1.5:1 or 0.5:1, respectively. An example of this analysis is shown in **Figure 20.7**.

CGH provides a powerful clinical tool to detect any type of aneuploidy or any deletion or duplication of 50 kb or more anywhere in the genome. Clinicians can use it in conjunction with amniocentesis or preimplantation genetic

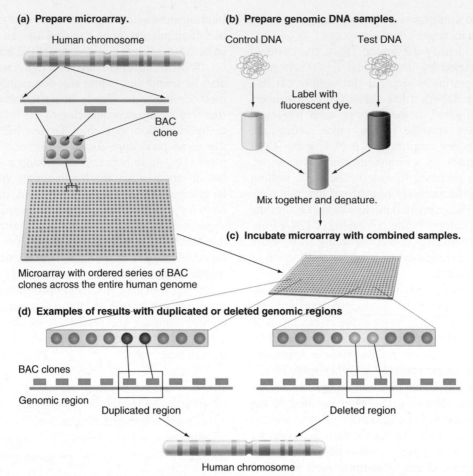

(a) Prepare microarray.

Human chromosome

BAC clone

Microarray with ordered series of BAC clones across the entire human genome

(b) Prepare genomic DNA samples.

Control DNA

Test DNA

Label with fluorescent dye.

Mix together and denature.

(c) Incubate microarray with combined samples.

(d) Examples of results with duplicated or deleted genomic regions

BAC clones

Genomic region

Duplicated region

Deleted region

Human chromosome

Figure 20.7 Comparative genomic hybridization detects duplications, deletions, and aneuploidy. **(a)** BAC clones representing the human genome are spotted in order onto a microarray. **(b)** The genomic sample to be tested is labelled with one colour dye (here, *red*), and the control genome sample is labelled with a second colour dye (*yellow*). **(c)** The two samples are mixed together, denatured, and then incubated on the microarray. **(d)** Automated analysis of each spot on the microarray detects the ratio of the two dyed probes that hybridize. *Orange* indicates a 1:1 ratio; other colours indicate deletion (0.5:1 ratio; *yellow*) or duplication (1.5:1 ratio; *red*) of BAC clone sequences in the test sample.

analysis. They can also use CGH to screen tissue biopsies for cancerous cells that have deleted or duplicated regions containing oncogenes or tumour-suppressor genes. The technique thus holds great promise for the detection of new genes that contribute to the genesis of cancer.

> Using microarray technology, it is now possible to detect copy number variations at a resolution not possible using traditional methods. This new genomic methodology is referred to as comparative genome hybridization (CGH).

20.4 Genome-wide Association Studies

Before the advent of microarray technology, the guiding principle of traditional disease-gene discovery relied upon identifying one or several large families in which a disease of interest was inherited by some offspring, but not others. Then, with the use of DNA markers spread across the genome, one could determine which copies of each chromosomal region had been inherited by each family member. Finally, correlations between disease transmission and a particular marked region of a chromosome would be identified. If a genetic correlation was uncovered, then one could use the tools of molecular biology to zero in on the guilty gene (discussed in detail in Chapter 15).

While useful in the study of monogenic traits, the vast majority of genetically influenced attributes that distinguish one person from another result from complex interactions among variant forms of multiple genes. The expression of most complex genetic traits is also influenced by nongenetic factors such as diet, other aspects of a person's lifestyle, and the "noise" inherent in biological systems. Because human environments are variable and not subject to control by scientists, nongenetic influences on the translation of genotype into phenotype can be substantial. In this section, we will discuss how microarray technology has been applied to deal with these confounding factors in order to better understand the genetics of complex traits.

Family-based pedigree studies are inadequate to reveal complex-trait genes

Although DNA-based pedigree analysis has been used successfully for the identification of genes involved in a variety of monogenic traits, its power to identify more complex genetic associations is severely limited for several reasons. First, the study of each new trait requires scientists to perform a new round of tedious and costly tests of individual DNA markers in each subject. Second, by definition, inheritance of a complex trait is only partially correlated with the inheritance of any single gene. Evidence for a partial correlation requires a larger sample size than is needed when a correlation is absolute. The most useful data for genetic mapping come from the analysis of multiple siblings, but in modern societies, human parents typically have only a few children. Pedigree mapping and identification of genes involved in a complex trait would require analysis of families with hundreds or thousands of members.

In the United States, where a majority of disease-gene discovery projects have been conducted, most people cannot trace their ancestors back more than a few generations, and the largest families, extending to second or third cousins, consist of a few hundred living subjects at most. Subject panels of this size were fine for identifying genes associated with simply inherited, all-or-none diseases, but they fail to provide sufficient data to identify weaker, multiple-gene correlations typical for common diseases.

The Icelandic geneticist Kari Stefansson decided to solve this problem by taking aim at the largest well-documented extended family that he knew—which was, actually, his own. Nearly all of the 300 000 citizens of Iceland, like Stefansson, can trace their ancestors back, through detailed public genealogical records, to the Vikings who settled this desolate European island over a thousand years ago. Stefansson convinced the Icelandic government to provide DeCODE Genetics with exclusive access to the health records of its citizenry in return for bringing investment capital and hundreds of high-tech jobs to the capital of Reykjavik.

DeCODE made rapid progress in identifying genes associated with 28 common diseases, including glaucoma, schizophrenia, diabetes, heart disease, prostate cancer, hypertension, and stroke, among others. In some cases, such as glaucoma and prostate cancer, DeCODE's findings could lead to diagnostic tests for identifying people at risk of developing disease. In other instances, such as schizophrenia, gene identifications have led to immediate insight about the cause of disease, which could lead to future therapies.

Buoyed by Stefansson's success, other geneticists were eager to perform large-scale family studies, yet few had similar access to ancient genealogical records. Gene "mapping" is based on the fact that long segments of chromosomes are transmitted in blocks from parents to children. Conventional wisdom held that unrelated families carried unrelated gene variants that were separately responsible for the expression of heritable characteristics. If this were true, geneticists outside Iceland might never have had the power to identify causative genes and alleles.

Haplotyping allows the world population to be seen as one giant pedigree

And then serendipity struck with an amazing discovery. With DNA microarray analysis of SNPs (see Chapter 15) in the genomes of many people, geneticists were able to detect and decipher the remnants of distant family relationships among individuals in all human populations. What primarily distinguishes so-called "unrelated" people from each other is not unique variants or alleles, but, rather, unique combinations of common SNP alleles along extended regions of genome. Fragments of genomes carried by our distant ancestors can be observed as blocks of DNA called **haplotypes** that are shared by many "unrelated" people, who are actually distant relatives.

A haplotype is created over evolutionary time by the accumulation of SNP variants, one by one, in a region of DNA that has been inherited intact over many generations, extending back thousands or tens of thousands of years. Indeed, for the purposes of genetic analysis and prediction, all 6 billion people can be treated as members of a single extended family with major branches located on each of the inhabited continents. In essence, we all carry ancestral genetic whispers of humanity's evolutionary past.

Haplotypes can be viewed as extended versions of alleles. Previously we learned that most of the genetic variation in the global human population is confined to a limited number of SNP loci, each of which can appear as one of two possible DNA bases. Allelic haplotypes are distinguished from each other at every SNP in a chromosomal region that may cover multiple genes. As a result of the common human heritage, haplotypes can be shared in their entirety by many "unrelated" people throughout the world. In many regions of the genome, 95 percent or more of the existing diversity is defined by just five to ten alternative haplotypes.

Because of the haplotype structure of the human genome, scientists can use information from a limited number of SNP loci to profile an entire human genome. A haplotype, containing dozens or hundreds of SNP alleles, can be tagged with just a few well-chosen "tag SNPs." A nearly complete whole-genome profile of any individual person can be obtained with the use of a DNA microarray that distinguishes genotypes at just 500 000 tag SNPs.

This conceptual breakthrough alone would not have been enough to transform human genetics if the process of DNA marker typing remained as tedious as it had been just a few years ago. But DNA microarrays have allowed the development of a technology for simultaneously screening large numbers of SNPs. Over the last decade, chip capacity has increased dramatically, crossing a threshold of 4.3 million SNPs in 2013. Automated data analysis of large test populations screened with these high-capacity chips has led to a flood of SNP-trait associations, and a new era of "genome-wide" genetics.

The massive data from genome-wide studies can be sifted for gene-disease correlations

The application of SNP chip genotyping to large populations of people for the purpose of discovering genetic associations between particular SNPs and traits is referred to as a **genome-wide association study**, or GWAS. Since the first GWAS publication in 2007, SNPs have been identified that are tightly linked to, and sometimes play causative roles, in a broad range of common diseases, including type 1 and type 2 diabetes; schizophrenia; bipolar disorder; glaucoma; inflammatory bowel disease; rheumatoid arthritis; hypertension; restless legs syndrome; susceptibility to gallstone formation; lupus; multiple sclerosis; coronary heart disease; colorectal, prostate, and breast cancers; and the pace at which HIV infection causes full-blown AIDS. With more and more biomedical research groups taking advantage of commercially available DNA chips, equipment, and software, every imaginable human attribute is being investigated for SNP associations (see **Figure 20.8** for an example related to body mass index). New SNP-trait associations are being published on a daily basis, and earlier associations are being refined and expanded by correlating larger sets of SNPs to more narrowly defined subtraits (**Figure 20.9**).

The open source genome

Two aspects of the genome-wide profiling approach to gene discovery are critical to its enormous success. First, essentially every published research finding is based on SNPs defined by a standardized open source system of nomenclature. Second, all newly obtained information about human genes and associated traits, along with numerous software tools used for analysis, is deposited in freely available, public databases maintained by the National Center for Biotechnology Information (NCBI) and other institutes at the National Institutes of Health (NIH). Consequently, bioinformaticians can compare any newly obtained, individual genome-wide profile against the compendium of banked data to determine a broad range of heritable characteristics. In a metaphorical sense, it is possible to "google" the genome.

Currently, NCBI hosts 29 interlinked databases. Most relevant to a GWAS program are dbSNP, a comprehensive database of all identified human SNPs (53 567 890 identified with 38 072 522 validated, as of June 2012); dbGap, a relational database of the results obtained in genome-wide association studies with the use of DNA microarrays to identify relationships between specific SNPs and specific diseases or nondisease traits; and OMIM, an online compendium of annotated records with detailed

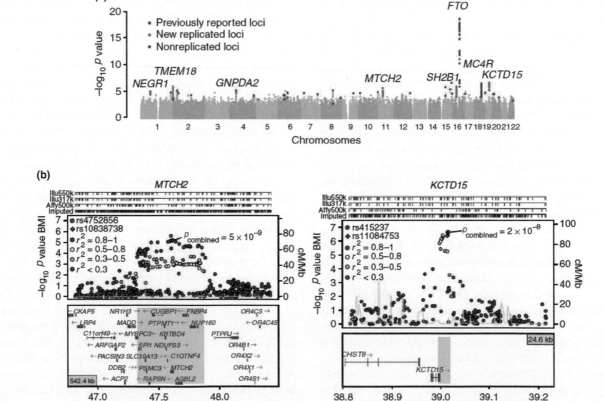

Figure 20.8 Whole-genome association study of body mass index (BMI). (a) *P* values for all SNPs tested for association with BMI across all chromosomes. Each dot represents a single SNP test. The higher the dot, the lower the *p* value. **(b)** Fine-scale mapping resolution of two BMI-associated regions.

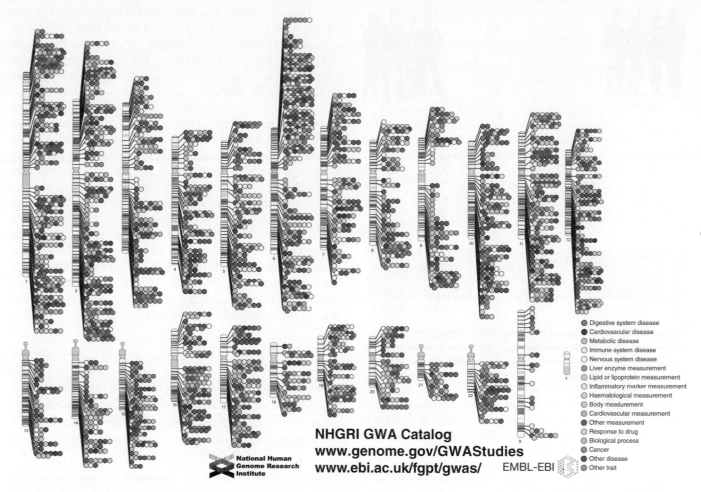

Figure 20.9 Genes associated with common complex traits in GWAS as of December 2012.

descriptions of each heritable human trait and gene that has been characterized (21 565 records, to date). NCBI has created numerous software tools that researchers can build upon to query and retrieve online genetic data automatically for specialized uses. NCBI datasets provide the foundation for other important open-access NIH-based and offsite data consolidation efforts focused on particular research problems or efforts. Three that focus on established associations between genes and traits are the Genetic Association Database (GAD) maintained by the National Institute on Aging (geneticassociationdb.nih.gov), the Human Genome Epidemiology navigator maintained by the Center for Disease Control (hugenavigator.net), and the Catalog of Published Genome-Wide Association Studies (www.genome.gov/gwastudies) maintained by the National Human Genome Research Institute. Among the most important consolidators of genetic and genomic data is the UCSC Genome Browser, which provides a graphical interface to view specific associations between traits and SNPs.

The GWAS methodology

The genome-wide association study (GWAS) approach to gene mapping is computationally intensive but conceptually simple. As a first step, a SNP DNA microarray is used to obtain whole-genome SNP profiles for each member of a test population consisting of thousands or tens of thousands of people. Each individual is also observed or tested for expression of one or more traits of interest to investigators (e.g., a disease phenotype). The vast majority of the SNPs will not be associated with differences in trait expression. But if the experiment is successful, a small number will show an association. In this way, regions of the genome that influence disease risk can be identified.

GWAS analyses are more broadly applicable and provide greater power and resolution than traditional pedigree analyses. Unlike all previous methods of gene mapping, genome-wide approaches do not depend on the analysis of closely related family members. There is no limit to the number of human subjects that can be included in a GWAS test population. In addition, direct comparative studies between affected and non-affected individuals can be performed (**Figure 20.10**). Moreover, the GWAS approach can be used to map and identify trait associated genes that follow any pattern of inheritance, simple or complex.

The first broad-based GWAS study, published in June 2007, included 17 000 British test subjects and resulted in the identification of 24 independent genetic loci associated with seven common diseases, including type 2 diabetes.

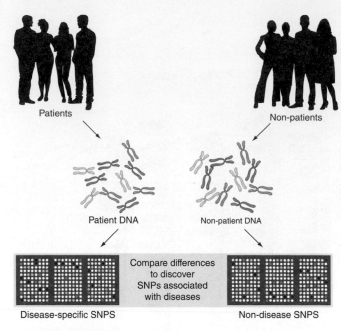

Patients

Non-patients

Patient DNA

Non-patient DNA

Compare differences to discover SNPs associated with diseases

Disease-specific SNPS

Non-disease SNPS

Figure 20.10 By analyzing the genomes of large groups of individuals with SNP DNA microarrays, genome-wide association studies can identify genomic regions associated with increased risk for various diseases.

Between 2007 and 2008, dozens of GWAS analyses have uncovered genetic variants at over 150 genes with roles in 50 common diseases and traits. Additional associations are being reported every month.

Type 2 diabetes is a prototype for exploring the power of the GWAS approach. From 2007 to 2008, dozens of additional GWAS studies of type 2 diabetes in European, Asian, or African populations led to the reproducible identification of 20 separate tagged haplotypes with an influence on the disease.

Once a tagged haplotype is found to be associated with diabetes, a more detailed study of SNPs and other variants in the region leads to the identification of the precise DNA sequence that is responsible. The specific role played by this sequence in the function of a particular gene can then be deciphered. This knowledge provides better insight into the disease and potential therapies to overcome it. Each individual diabetes-related gene influences disease risk by a small amount, but when multiple risk genes appear together in the same genome, risk increases substantially. One remarkable finding of GWAS results obtained for diverse populations is that the same disease-causing variants are present universally, although at different frequencies in different parts of the world.

The discovery of GWAS as a way of viewing the entire human population as an "extended" family pedigree has revolutionized the evaluation of disease genotypes. Identification of contributing genes may allow refinements of treatment options.

20.5 Epigenomics

So far in our analysis of genomics technologies, we have focused on phenomenon controlled at the genetic level; that is, by variation in the underlying DNA sequence. However, this may not be sufficient for a full understanding of organismal phenotype. This is because, in addition to the genetic layer of control, there also exists an "epigenetic" layer. Epigenetic modifications do not involve an alteration to the underlying DNA sequence, but instead involve reversible modifications to DNA bases or to histones. These modifications can be stably inherited and are able to modulate gene expression. In addition, they can be modulated by environmental factors. Thus, epigenetic modifications have a profound effect on the transcriptome and therefore on organismal phenotype.

Two of the most common types of epigenetic modifications are (1) the methylation of cytosines (to form 5-methylcytosine), and (2) the addition of modifying groups such as methyl and acetyl moieties to histone proteins (**Figure 20.11**). As we saw in Chapter 11, the addition of methyl groups to cytosine residues (that are part of strings of CpG dinucleotides) plays an important role in gene expression. Similarly, histone modifications strongly affect the affinity of DNA for histones, modulating gene expression as part of a process called chromatin remodelling (i.e., altering nucleosomal structure to allow or inhibit interaction

with cellular transcription factors). Furthermore, changes in the pattern of epigenetic modifications have proven to play an important role in a variety of diseases, most notably cancer. For these reasons, an examination of epigenetic modifications on a genome-wide scale has become a priority for researchers. For example, the NIH Roadmap Epigenomics Consortium was recently established to help catalogue epigenetic modifications and to determine how they relate to disease and other biological phenomena (roadmapepigenomics.org). In this section, we will describe (1) the epigenomic technologies currently in use and (2) how the derived data are utilized in a clinical setting.

DNA methylome analysis

The existence of aberrant DNA methylation patterns in cancer cells (and other diseased tissue) has made the characterization of this epigenetic modification a high priority among researchers. Not only can these differences be used as biomarkers for diagnosis and prognosis, but they can also be utilized to infer the optimal choice of drugs to be used for treatment. Two methods are commonly used to determine DNA methylation patterns on a genome-wide scale: a microarray-based method called methyl-DNA immunoprecipitation (MeDIP), and a method (based on next-generation sequencing) referred to as BS-seq.

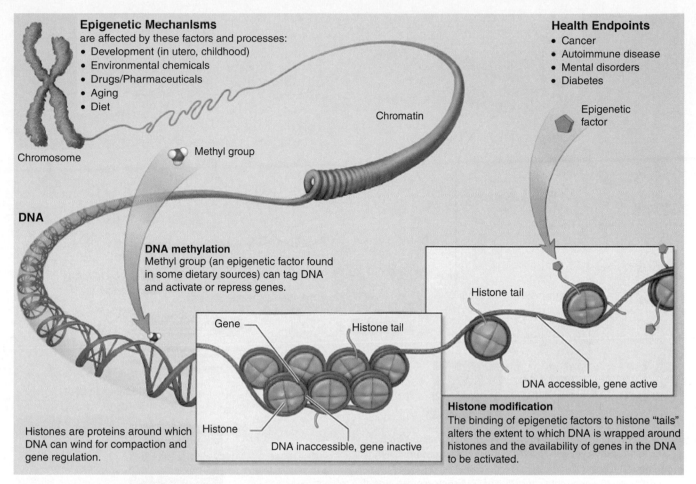

Epigenetic Mechanisms
are affected by these factors and processes:
- Development (in utero, childhood)
- Environmental chemicals
- Drugs/Pharmaceuticals
- Aging
- Diet

Chromosome

Chromatin

Methyl group

Health Endpoints
- Cancer
- Autoimmune disease
- Mental disorders
- Diabetes

Epigenetic factor

DNA

DNA methylation
Methyl group (an epigenetic factor found in some dietary sources) can tag DNA and activate or repress genes.

Histone tail

Gene

Histone tail

DNA accessible, gene active

Histones are proteins around which DNA can wind for compaction and gene regulation.

Histone

DNA inaccessible, gene inactive

Histone modification
The binding of epigenetic factors to histone "tails" alters the extent to which DNA is wrapped around histones and the availability of genes in the DNA to be activated.

Figure 20.11 The addition of methyl groups to cytosine residues as well as the modification of histone molecules represent two common epigenetic modifications. These changes have the ability to remodel chromatin and thus have an enormous impact on transcriptional regulation. These changes can be stably inherited and are often affected by environmental factors. In addition, diseased tissue (e.g., cancerous tumours) often display aberrant patterns of epigenetic modifications.

The MeDIP method is similar in principle to the ChIP–chip technique described earlier. In this case, however, an antibody with high and specific affinity for 5-methylcytosine is used. As before, genomic DNA isolated from the tissue of interest is fragmented. The fragments are then immunoprecipitated with a methylation-specific antibody, and then amplified using ligated linkers in the presence of Cy5- or Cy3-conjugated nucleotides. Next, the amplified fragments are mixed with fragments derived from a control sample (that was not treated with the antibody). The combined sample can then be hybridized to a gene chip representing all genomic sequences to identify regions exhibiting high levels of methylation (**Figure 20.12**).

An alternate approach, BS-seq, relies on the fact that treatment of DNA with sodium bisulfite converts cytosine residues to uracil. In contrast, sodium bisulfite treatment has no effect on methylcytosine residues. Since DNA polymerase recognizes the uracil as a thymine (i.e., it will insert an A as its complement), one can infer the methylation state of a given DNA fragment after a complementary strand is synthesized by examining the sequence of unmodified (i.e., the original untreated genomic sample) and modified (sodium bisulfite treated) fragments (**Figure 20.13**). As you can see, this strategy is a perfect fit for next-generation sequencing (NGS) technology. In practice, treated and untreated genomic DNA samples are fragmented and then sequenced using an NGS method like pyrosequencing. Using computer algorithms that compare the sequences, it is possible to determine the methylation state of all cytosines in the genome.

Genome-wide analysis of histone modification

The modification of histones has profound effects on the structure of chromatin (and thus on gene expression). For example, the acetylation of lysine residues on histone tails removes a positive charge, thereby reducing the affinity of the histone to DNA. This remodels the chromatin into a more relaxed state that promotes gene transcription. As you have probably surmised, a ChIP–chip

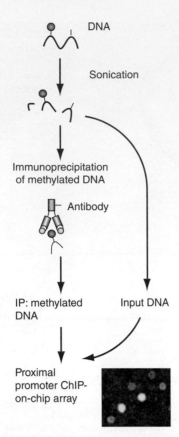

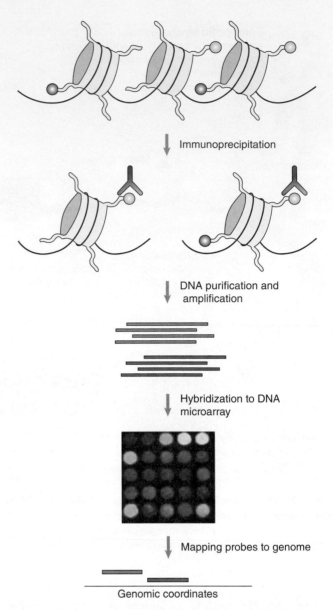

Figure 20.12 Methyl DNA immunoprecipitation uses antibodies specific to 5-methylcytosine to isolate methylated genomic DNA fragments. The identity of these stretches of DNA can be determined by hybridization to a gene chip consisting of DNA spots representing all genomic sequences (or, if interested only in promoter regions, a gene chip made up of spots representing promoter-proximal elements).

or Chip-seq strategy can easily be used to determine the genomic location of a specific modification as long as a highly specific antibody against the modification is available (**Figure 20.14**).

Clinical applications

As mentioned earlier, the detection of aberrant methylation patterns has been shown to be effective in both the

Figure 20.14 Schematic representation of the genome-wide analysis of histone modifications. Antibodies specific to the modification of interest are used to immunoprecipitate the modified DNA fragments. The genomic location of these DNA fragments can be determined by hybridization to a gene chip consisting of DNA spots representing all genomic sequences.

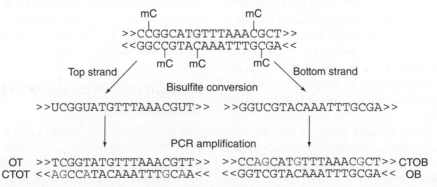

Figure 20.13 Sodium bisulfite treatment of DNA. The treatment of DNA with sodium bisulfite converts cytosine residues to uracil (in *red*) in both the original top strand (OT) and the original bottom strand (OB). Methylated cytosines (mC) are not affected by this treatment in the same way. Upon denaturation of the bisulfite-treated DNA, and subsequent DNA synthesis of new strands complementary to the original top strand (CTOT) and the original bottom strand (CTOB), it is possible to infer the methylation status of the original fragment.

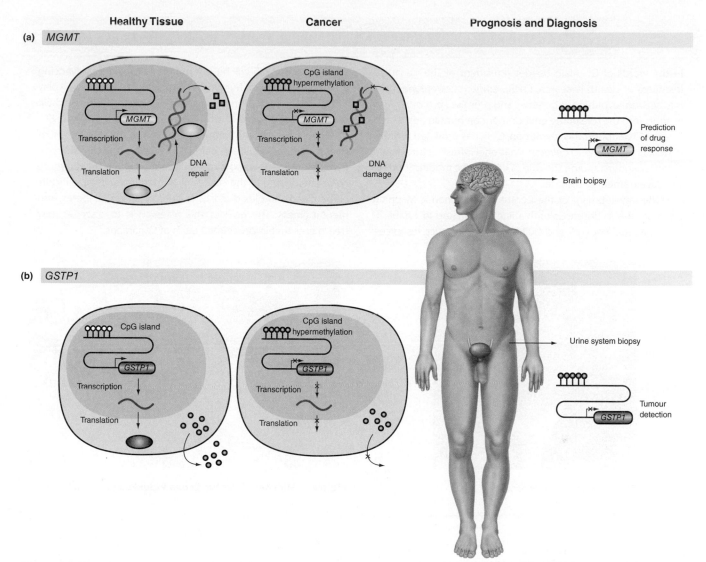

Figure 20.15 Using epigenetic modifications for diagnosis, prognosis, and the rational choice of therapeutic regimes. (a) The *MGMT* gene product removes damaging alkyl groups (*red squares*) from guanine bases. Hypermethylation of the *MGMT* promoter blocks expression, leading to the accumulation of DNA damage. Interestingly, the hypermethylated form of *MGMT* in glioblastomas is an excellent predictor of a good response to the drugs carmustine and temozolomide. **(b)** The *GSTP1* gene is required for removing carcinogens (*red circles*) from the cell. When hypermethylated, the expression of *GSTP1* is repressed, allowing the accumulation of damaging agents. Abnormal methylation of the *GSTP1* promoter is thus an early event in the development of prostate cancer and can be used as a diagnostic biomarker.

diagnosis and prognosis of disease. More recent work has shown that the same patterns can be used to predict the response of cancerous tumours to different drug regimes. For example, methylation of the promoter controlling the *MGMT* gene (O^6-methylguanine-DNA-methyltransferase) is clearly related to the progression of glioblastoma tumours and to overall patient survival rates. In fact, the methylation state of the MGMT promoter is an excellent indicator of the effectiveness of various anticancer drugs (**Figure 20.15a**). In addition to affecting treatment choices, the analysis of methylation states can also be used to aid in the early detection of cancers. For instance, the methylation state of the *GSTP1* (glutathione-S-transferase pi 1) promoter can be used as a sensitive test for the early diagnosis of prostate cancer (**Figure 20.15b**). Findings such as these underscore the potential for epigenomic data in aiding medical decision-making. However, in order for

this potential to be realized, the further characterization of reference genomes (both normal and diseased) is required. Government of Canada initiatives like the creation of the Canadian Epigenetics, Environment and Health Research Consortium (as well as other initiatives like it from around the world) will play an important role in attaining these goals (see the **Focus on Inquiry** box **"The Canadian Epigenetics, Environment and Health Research Consortium."**)

The newly emerging field of epigenomics seeks to characterize the epigenetic modifications that modulate cellular processes on a genome-wide scale. An understanding of these changes has important clinical applications with respect to diagnosis, prognosis, and treatment.

In the words of Dr. Alain Beadet (President of the Canadian Institutes of Health Research, CIHR), epigenetics/epigenomics is a "second revolution in genetics and promises profound new insights into the role of the environment on human health and disease." To support this revolution, CIHR has provided millions of dollars in funding to develop a national infrastructure aimed at better understanding the role of epigenetic modifications in the development of disease.

"The development of the Centre for Epigenomic Mapping Technologies in British Columbia is a critical step in establishing a national network and building bioinformatics resources.

Epigenetic research will have a profound impact on advancing our understanding of human genomics and this Centre will play a critical role in this innovative area of human health" said Dr. Brad Popovich (**Figure A**), Chief Scientific Officer at Genome BC.

Echoing these sentiments was the Canadian Minister of Health, Leona Aglukkaq (centre of **Figure B**). "Our Government is proud to support research that will help build a more complete picture of the causes of human illnesses, specifically chronic and complex diseases, including cancer, diabetes, and mental illness. The goal of this research is to discover new treatments that improve the health of Canadians."

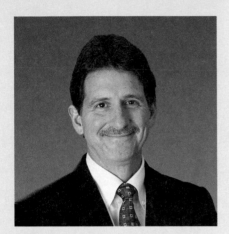

Figure A Dr. Brad Popovich.

Figure B Minister of Health, Leona Aglukkaq (centre).

Connections

Genetic analysis has been a key foundation of biological research for well over a century. During this time, such analysis has focused on the inheritance of specific traits, as well as on the function and regulation of specific genes (Chapters 2–4, 8, 10–11). With the sequencing of the human genome, and the development of new high-throughput technologies, this analysis has grown to encompass the parallel (i.e., simultaneous) study of all genes in a given genome. The sheer quantity and complexity of these data have necessitated the development of new ways to think about biology (i.e., how the interactions of individual genes/gene-products give rise to higher-order biological phenomenon). This new way of thinking defines the relatively new field of systems biology. We will learn more about this discipline and how it has changed the way geneticists go about answering biological questions in Chapter 23 (Systems Biology).

Essential Concepts

1. The accurate annotation of whole-genome sequencing data, together with the advent of DNA microarray technology, has allowed for the routine quantitative examination gene expression levels on a global scale. **[LO1–2]**

2. By hybridizing DNA samples derived from chromatin immunoprecipitations (using antibodies against specific DNA-binding proteins) to intergenic gene chips, one can identify the genomic-binding sites of any given DNA-binding protein. **[LO1, LO3]**

3. The technique of comparative genome hybridization allows the detection of copy number variations on a genome-wide scale. This method is a vast improvement over traditional karyotype analysis because of its enormously improved resolution. **[LO1, LO4]**

4. The analysis of genetic variants using SNP DNA microarrays can be used to associate specific regions of the genome with complex genetic traits such as susceptibility to disease. **[LO1, LO5]**

5. With the help of DNA microarray technology and next-generation sequencing methods, it is possible to characterize the epigenetic modifications responsible for modulating gene expression across an entire genome. **[LO1, LO6]**

I. Breast cancer has been histologically classified as invasive ductal cancer. You hypothesize that breast cancer can be further subclassified based on molecular signatures. If you can create such subcategorizations, you may be able to stratify cancers and develop specific therapies for one or more of the subcategories. You and your research colleagues have biopsy tissues from 63 invasive ductal breast cancer patients and clinical data on the course of the disease in each of these patients. You plan to use microarray technology to measure gene expression in each of these samples, and then use the results to classify the tissues. When you analyze these samples with microarrays, what other samples should you analyze and why?

Answer

Controls are an important part of any scientific experiment. You need to choose control samples to provide confidence that the subcategories you identify truly reflect a stratification of invasive ductal breast cancer. In particular, you must ensure that the biopsies you receive are neither misdiagnosed nor improperly collected. For example, if some normal breast tissue is mistakenly included as a tumour in your analysis, you may identify normal tissue as a subcategory of invasive cancer. Thus, you will want to include normal breast tissue as well as other types of breast cancer in your arrays. Your analysis will also require comparison of gene expression in the cancer samples to a standard reference; the set of normal breast cancer tissue will serve this important function. Finally, analyzing a comprehensive set of control samples will also help to develop a statistical model for the variability of your measurements and therefore increase confidence in your results. You may also wish to examine other hypotheses with your data, such as identifying markers specific to breast cancer or breast tissue. Therefore, you may wish to include some samples of other normal tissues as well as cancers of these tissues.

Problems

Vocabulary

1. For each of the terms in the left column, choose the best matching phrase in the right column.

    i. gene chip

    ii. comparative genome hybridization

    iii. ChIP-on-chip analysis

    iv. haplotypes

    v. ChIP-seq analysis

    1. a gene regulation analysis technique to determine the gene targets of any transcription factor using direct sequencing of DNA fragments

    2. a DNA microarray of DNA fragments attached to a surface

    3. a microarray-based hybridization protocol for detecting deletions, duplications, and aneuploidy

    4. a gene-regulation analysis technique to determine the gene targets of any transcription factor using next-generation sequencing technology

    5. blocks of DNA that are shared by many people

Section 20.1

2. When interpreting the results of expression profiling microarrays, what is used to measure the degree of hybridization between a given probe and its corresponding cDNA?

3. Consider a microarray experiment similar to the one shown, in which each square represents a PCR-amplified fragment of a different human gene, and the red-labelled probe is cDNA from a human lung tumour while the green-labelled probe is cDNA from normal lung tissue.

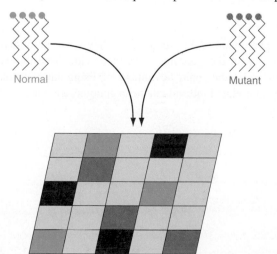

Normal   Mutant

    a. How would you interpret results in which the fluorescence signal was black? green? red? yellow?

    b. If you were searching for an anticancer drug that would inactivate a protein whose activity contributes to cancer, which of the genes represented on the microarray encode proteins you would most likely choose as a potential target for such a drug?

4. A region of the genome from two individuals is amplified by PCR so that the PCR products from one individual are labelled with rhodamine (which fluoresces in red), while the PCR products from the other person are labelled with fluorescein (which fluoresces in green). These PCR products are mixed and hybridized to an oligonucleotide microarray with the results as shown.

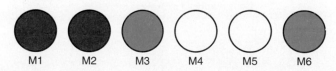

M1   M2   M3   M4   M5   M6

The oligonucleotides on the array are as follows:

M1: 5′ ACTTACCGAGAGAACCTGCG 3′

M2: 5′ ACTGACCGAGAGAGCCTGCG 3′

M3: 5′ ACTTACCGAGAGAGCCTGCG 3′

M4: 5′ ACTCACCGAGAGACCCTGCG 3′

M5: 5′ ACTCACCGAGAGATCCTGCG 3′

M6: 5′ ACTGACCGAGAGAACCTGCG 3′

a. As accurately as possible, describe the genotypes of the two individuals.

b. Why would you encounter ambiguity in assigning genotypes to these two particular individuals if you sequenced the PCR products directly, rather than by hybridizing them to an oligonucleotide microarray as above?

c. In what way would the oligonucleotide microarray approach be valuable as a diagnostic tool for human genetic diseases?

5. A *Schizosaccharomyces pombe* gene expression microarray is hybridized to Cy5-labelled cDNAs isolated from *S. pombe* cells. The spot corresponding to gene "*X*" shows greater fluorescence intensity than the spot corresponding to gene "*Y*." Based on this experimental design, is it correct to conclude that gene "*X*" is expressed at higher levels than gene *Y*? Explain why or why not. (*Hint:* The gene *X* mRNA is not necessarily similar in length to the gene *Y* mRNA.)

6. As a developmental geneticist, you are examining an important developmental gene that encodes three different splice variants in the roundworm, *C. elegans*. The gene consists of five exons. Splice variant A lacks exons 3 and 4. Splice variant B lacks exons 2 and 5. Splice variant C lacks exon 5. You design a microarray with separate spots for exon 2, exon 4, and exon 5. Then, mRNA isolated from the worm is used to create Cy3-labelled cDNAs, and mRNA isolated from larvae is used to create Cy5-labelled cDNAs. The cDNAs are mixed together and hybridized to the microarray. The exon 2 spot fluoresces bright red, the exon 4 spot fluoresces yellow, and the exon 5 spot fluoresces green. Based on these data, indicate how the expression of each splice variant changes during the development of the worm.

Section 20.2

7. You are studying the effects of an anticancer drug on pancreatic cancers. During your experiments, you obtain information from a colleague that suggests that the drug acts by inhibiting the activity of a transcription factor

called Drz1. To determine the transcriptional targets of Drz1, you perform ChIP–chip analysis with an antibody specific to Drz1. DNA retrieved from the antibody-treated sample is labelled with a red fluor, while DNA retrieved from the non-antibody-treated sample is labelled with a green fluor. The labelled DNA is then hybridized to an intergenic array. You examine 100 spots on the array and determine that 12 spots are bright red, 3 are yellow, and the remainder are bright green. Assuming that each spot on the array represents a unique probe, what do these results indicate?

8. In the ChIP-on-chip technique, the DNA to which the transcription factors are bound is fragmented into 300–500 bp lengths. The typical transcription-factor-binding site is 6–15 bp in length. How could you increase the resolution of the ChIP-on-chip technique with regard to the identification of the actual binding site?

Section 20.3

9. The accompanying figure at the bottom of this page shows a virtual karyotype obtained from a line of tumour cells derived from a human leukaemia. The left-to-right direction for each chromosome corresponds to the orientation of that chromosome from the telomere of the small arm to the telomere of the long arm. Every coloured dot corresponds to a different short region of the chromosome analyzed by a microarray technique similar to that shown in Figure 20.8.

a. Do the data indicate the existence of aneuploidy or any chromosomal rearrangements within the genome of the tumour cell?

b. What do you think you would see if you did a virtual karyotype of a cell line derived from normal, non-leukaemic cells from the same person?

c. Are there any kinds of chromosomal rearrangements that could not be detected by this virtual karyotyping method?

d. What do these data say about genes that might be responsible for the leukaemia?

e. Do these data tell us anything about the dosage of genes needed for the viability of individual cells?

Section 20.4

10. Explain what SNPs are and how they are used to identify regions in the genome associated with diseases? What would be the main advantage of using families rather than unrelated individuals for a genome-wide association study?

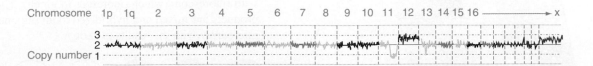

11. A group of researchers interested in autism perform a genome-wide association (GWA) study to find SNPs associated with the condition. They have SNP data for 500 000 SNPs for 2000 individuals. As part of their experimental design, they randomly split the individuals into two groups and do the GWA study separately on each group. Explain why?

12. You identify a new protein, X, that was recently discovered from annotating the newly sequenced genome of the bacterial pathogen *Leptospira interrogans*. This protein is required for the pathogenicity of the bacteria. Nothing is known about the protein, except that it carries a DNA-binding domain. What genomic techniques could you use to determine the biological function of this gene? What information would these experiments provide?

Section 20.5

13. The results of a BS-seq experiment are shown below:

Untreated DNA:	>>CCTCGGAAGGACTCAGGCCG>>
	<<GGAGCCTTCCTGAGTCCGGC<<
OT:	>>TTTTGGAAGGACTTAGGTCG<<
CTOT:	>>AAAACCTTCCTGAATCCAGC>>
CTOB:	>>CCTCAAAAGGACTCAGACCG>>
OB:	>>GGAGTTTTCCTGAGTCTGGC>>

Determine the residues that were methylated in the original sample.

14. Explain how the analysis of epigenomic data might influence medical decision-making in the future. What needs to occur in the intervening years for such a strategy to be realized?

For more information on the resources available from McGraw-Hill Ryerson, go to www.mcgrawhill.ca/he/solutions.

Problems **669**

An oil painting by artist Julie Newdoll, titled "The Worlds of Proteomics."

little islands have their own colour scheme. A protein meant to be secreted to the outside of the cell follows an elaborate path of production, first forming inside the endoplasmic reticulum, later packaged into a membrane bubble which melds with the Golgi, and finally repackaged and released to the outside of the cell. From the artist's perspective, a protein meant for the outside of the cell is rendered as a flying creature in this painting. It never ends up in the deep sea environment of the cytosol, or it would "drown." Likewise, proteins destined for the deep sea of the cytosol could not breathe outside the cell in the open air. And then there are the amphibians . . .

While a wonderful and moving piece of art, this painting also captures the complex and dynamic nature of the interacting proteins that populate our cells. As we shall see in this chapter, the analysis and characterization of protein networks has become increasingly important to biologists. This is due, in part, to research demonstrating the organization of proteins into complex structural and/or functional interaction networks. Together, these networks play a critical role in the emergence of phenotype at the molecular, cellular, and organismal levels.

In the painting that opens this chapter, the artist expresses her vision of the internal and external worlds of a typical eukaryotic cell. Her personal description of the piece follows.

When proteins are made in the cell in response to some stimuli or event, they are targeted via an address system for a specific location or locations. In this painting, the various areas in a cell are represented by various worlds—there is the world of the sea in the cytosol, that of the air outside the cell and land or earth inside the nucleus. Inside the mitochondria and the endoplasmic reticulum,

Chapter Outline
21.1 What Is Proteomics and How Does It Relate to Genetics, Medicine, and Network Biology?
21.2 Genome-Level Proteomic Interaction Analysis: The Yeast Two-Hybrid System
21.3 Genome-Level Proteomic Interaction Analysis: Affinity Purification Coupled to Mass Spectrometry
21.4 Analyzing Dynamic Changes in Protein Expression
21.5 Using Protein Arrays for Global Biochemical Analysis

Learning Objectives

1. Define the term proteome and identify the goals of proteomics research.

2. Relate the yeast two-hybrid system to the generation of global protein interaction maps.

3. Evaluate the roll of robotics in conducting high-throughput genetics research.

4. Relate the characterization of protein interaction networks (through mass spectrometry) to the emergence of cellular/organismal phenotypes.

5. Explain how isotope-coded affinity tags are employed to analyze dynamic changes in protein expression.

6. Illustrate how protein arrays are exploited to perform analytical studies.

21.1 What Is Proteomics and How Does It Relate to Genetics, Medicine, and Network Biology?

Current estimates place the number of protein-coding genes present in the human genome at approximately 20 500. Using this estimate, one might superficially presume the existence of an equal number of unique (or nonredundant) proteins. However, in reality, this number represents only the starting point in the analysis of an organism's protein complement. Why is this so? First, after being transcribed, a single primary transcript may be subject to alternative splicing (see Figure 7.10a). This process allows cells to produce different mRNAs (and thus different protein products) from a single gene. Alternative splicing is common in eukaryotes and—depending on the changes introduced to the structure of the resulting protein— may affect protein targeting (the regulated localization of proteins to distinct subcellular compartments), protein stability, and enzymatic activity, as well as the ability of the protein in question to interact with other proteins present in the cell.

Second, proteins can also be chemically modified after translation. For example, a given protein might exist in a phosphorylated or a nonphosphorylated state (other examples of posttranslational modifications include glycosylation, ubiquitination, and acetylation). Again, these modifications have the capacity to alter the structural and/or biochemical properties of the protein. Taking these factors into consideration, the estimated number of unique proteins that exist within a human cell is between 100 000 and 500 000. The complete set of unique proteins encoded by a given genome is referred to as the **proteome** of the organism.

In addition to referring to the "complete" proteome of an organism, it is also common to refer to cellular proteomes. A cellular proteome comprises the unique set of proteins encoded by a particular cell type. The cellular proteome of an epithelial cell, for example, is quite different from the cellular proteome of a fibroblast. This is due to regulated changes in gene expression that take place during development (and which allow these cells to carry out their unique roles within epithelial and connective tissues, respectively). In fact, these differences in protein complement are crucial in defining their cellular identity.

Extracellular or environmental factors can also affect the proteome. For example, the proteome of a unicellular yeast cell grown anaerobically is drastically different from the proteome of a yeast cell grown under aerobic conditions. Lastly, the analysis of proteomes from a clinical perspective has also clearly demonstrated differences in the proteomes of cells derived from normal, healthy tissue, and cells derived from diseased tissue of the same type (e.g., cancerous tumours). Taking these factors into consideration, **proteomics** can be defined as the global analysis of the proteins present in a particular cell type or organism under a defined set of environmental conditions. As defined by the Human Proteome Organization (HUPO, hupo.org), the goals of proteomics are as follows:

1. Analyze protein expression in the different cell types present in a given organism.

2. Analyze the posttranslational modifications that take place within these cell types.
3. Define protein interaction networks (i.e., determine interactions between unique proteins within cells).
4. Better understand the relationship between the structure of protein interaction networks and their biological functions.

The characterization of proteomes is of interest to geneticists as well as biochemists because of the intimate relationship between protein interaction networks and cellular function. In previous chapters, we have already seen firsthand the importance of protein interactions (e.g., in the initiation of transcription, the action of repressor proteins through quenching, and the dimerization of the Myc and Max proteins). To delve more deeply into these issues let us consider, for example, the nuclear pore complex (**Figure 21.1**).

The nuclear pore complex can be viewed as a molecular machine. It is composed of a network of over 60 unique proteins that assemble into a functional complex capable of translocating both mRNAs and proteins to and from the cytoplasm in a precise and controlled way. Just like an automobile missing its front left tire, a nuclear pore complex missing one of its protein components would likely function in an abnormal way. This, in turn, could lead to abnormal cellular physiology and ultimately an observable phenotype. This logic is supported by the observed relationship between the dysregulation of nuclear import proteins and a variety of pathological conditions. For

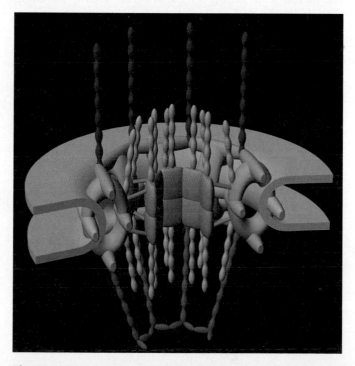

Figure 21.1 Drawing of a nuclear pore in yeast. This complex molecular machine contains about 60 proteins. One function of this machine is to translocate particular protein molecules into the nucleus.

In the same way that a house is made of bricks, a cell is made (in part) of proteins. However—similar to Poincaré's thoughts with respect to science in general—the intracellular environment of a cell is not just a collection (or "pile") of proteins. Indeed, complex and dynamic networks of interacting proteins function together to ensure proper cellular physiology. How can scientists define these networks? What are their functions? Do these networks display any characteristics (modularity, hierarchy) that can be used to provide a usable conceptual framework? Is understanding network structure important to the emergence of higher-order biological phenomena such as disease and other complex phenotypes? While the research needed to develop answers to these questions is still in its infancy, the material presented in this chapter and in Chapter 23 (Systems Biology) will provide a framework for understanding how researchers are beginning to tackle these important questions.

"Science is facts; just as houses are made of bricks, so is science made of facts; but a pile of bricks is not a house and a collection of facts is not necessarily science."

Henri Poincaré (1854–1912).

example, aberrant expression of a variety of nuclear pore complex proteins (e.g., importins, exportins, nucleoporins) are linked to a variety of pathological states, including cancer, atherosclerosis, and diabetes.

Since a cell contains many such molecular machines (e.g., the spliceosome, the proteasome, complexes involved in chromatin modification, DNA repair complexes), the characterization of these protein networks has become an integral part of understanding biology (see **Figure 21.2** and the **Focus on Inquiry** box **"The French Mathematician and Philosopher, Henri Poincaré"**). For example, the Clinical Proteomics Centre located at McGill University in Montreal (clinprot.org) is dedicated to "research aimed at elucidating the pathophysiology of disease using proteomics approaches." In the next two sections, we will discuss methods used by researchers to characterize and compare protein interaction networks.

> The structure of protein networks is intimately related to the emergence of phenotype at the cellular and organismal levels. Thus, an understanding of protein network structure is crucial in deciphering the pathophysiology of disease and other complex phenotypes.

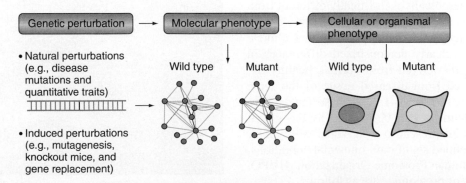

Figure 21.2 Integrating proteomic data with other biological research. By integrating data obtained from proteomics research with genetic and phenotypic data, it may be possible to elucidate how mutations—as well as normal genetic variation—are related to the structure and function of protein networks (and how, in turn, these changes are related to organismal phenotypes).

The premise of the two-hybrid system

The yeast two-hybrid system was the first method to be used in the systematic identification of protein-protein interactions on a genome-wide scale. This system uses budding yeast cells (*Saccharomyces cerevisiae*) as living test tubes to assess the ability of two proteins to physically interact. The basis of the system is the capacity to detect transcriptional activation of a "reporter" gene. One of the reporters used in this system is the *HIS3* gene. This gene encodes imidazoleglycerol-phosphate dehydratase, an enzyme that is part of a metabolic pathway that produces the amino acid histidine. Budding yeast cells grown in environments lacking histidine are dependent on the expression of the *HIS3* gene to synthesize the amino acid. On the other hand, if grown in environments that contain exogenous histidine, budding yeast cells are able to import the amino acid across the plasma membrane and into the cytosol, making the expression of the *HIS3* gene superfluous for growth. Using recombinant DNA technology, together with knowledge of the mechanisms of transcription, it is possible to make the transcriptional activation of the *HIS3* reporter gene dependent on the physical interaction of any two given proteins of interest. The system is based on the separable biochemical activities of the yeast Gal4 protein.

The Gal4 protein is a transcriptional activator that is composed of two distinct domains: a DNA-binding domain (required to bind to specific DNA sequences in the promoter region of its target genes) and an activation domain (required to interact with the basal transcriptional machinery and thus to promote the initiation of transcription). Interestingly, researchers have shown that these two domains can be physically (and functionally) separated. This fact can be exploited to create the foundation of what is known as the yeast two-hybrid assay. In this assay, two recombinant DNA constructs are created (see Chapter 14 for an explanation of recombinant DNA technology). The recombinant "bait" gene is composed of the coding sequence of the Gal4 DNA-binding domain (BD) fused to the coding sequence of a given protein of interest (for the sake of simplicity, let us refer to the first protein of interest as X). The recombinant "prey" gene, on the other hand, is composed of the coding sequence of the Gal4 activation domain (AD) fused to the coding sequence of a second protein of interest (we will call this protein, Y).

To carry out the assay, both recombinant genes are expressed within budding yeast cells possessing a copy of the *HIS3* reporter gene. The cells are then grown in media lacking histidine. Two scenarios are possible. In the first scenario, the two fusion proteins *do not* physically interact. In this case, the bait fusion protein binds its cognate binding sequence (that is part of the promoter of the *HIS3* gene). While bound to the promoter, the fusion protein is *unable* to activate transcription since the Gal4 AD is not present in the immediate vicinity. The prey fusion proteins, on the other hand, localize throughout the nucleus, as they have no DNA-binding sequences to target them to the *HIS3* promoter. Thus, under these circumstances, *HIS3* gene expression is not activated and the yeast cells are unable to proliferate.

In the second scenario, just as in the first, the bait protein binds its cognate binding sequence in the promoter of the *HIS3* gene. However, in contrast to the first scenario, let us assume that proteins X and Y are indeed physical interactors. Since proteins X and Y are now part of the same complex, the binding and activation domains are in close physical proximity, creating an entity with both DNA binding and transcriptional activation activities. In this scenario, the *HIS3* reporter gene is active, allowing the budding yeast cells to grow in media lacking histidine (**Figure 21.3**). In the final analysis, the molecular interaction of any two proteins can be inferred from the simple growth, or lack of growth, of budding yeast cells in environments lacking histidine. **Figure 21.4** shows the growth of patches of yeast cells on solid growth media lacking histidine.

Two-hybrid analysis at the genome-wide level

While we have discussed the use of two-hybrid technology to assay the interaction of any two proteins of interest, it is also possible to utilize the assay on a genome-wide

(a)

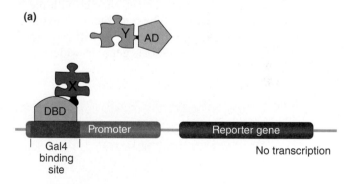

(b)

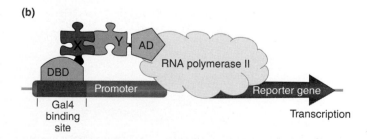

Figure 21.3 Reporter gene activation in the yeast two-hybrid system. **(a)** In the absence of interaction, the binding and activation activities of the Gal4 protein are not in close enough proximity to stimulate transcription of the reporter. **(b)** Physical interaction between the bait and prey fusions, on the other hand, will result in activation of *HIS3* transcription.

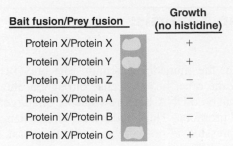

Bait fusion/Prey fusion	Growth (no histidine)
Protein X/Protein X	+
Protein X/Protein Y	+
Protein X/Protein Z	−
Protein X/Protein A	−
Protein X/Protein B	−
Protein X/Protein C	+

Figure 21.4 In the yeast two-hybrid system, the growth of yeast strains in media lacking histidine can be used to infer protein-protein interaction. Physical interaction between the bait and prey fusions will result in activation of *HIS3* transcription. This, in turn, allows the respective budding yeast strains to grow into a patch of cells visible to the naked eye. In the absence of interaction, the binding and activation activities of the Gal4 protein are not in close enough proximity to stimulate transcription of the reporter, and no growth is possible in media lacking histidine. In the example above, we can infer that protein X interacts with itself, protein Y, and protein C.

scale to make inferences regarding the underlying network structure of an organism at the protein level. These methods rely on the use of "macroarrays" and modern robotic technology.

To begin, imagine a simplistic hypothetical organism whose genome contains only 96 protein-coding genes. (As we shall see in the chapter on synthetic biology, a minimum of approximately 400–500 genes are required for life, but our abstract hypothetical organism will do to explain the methodology.) Let one of the protein-coding genes from this set be called gene "Z." Your goal is to determine which proteins from the set interact with the Z gene-product. (Remember it will be possible for Z to interact with itself, making a homodimer.) We will refer to Z as the query protein and the global, genome-wide set as the panel.

To conduct the assay one would begin by creating 97 unique recombinant DNA constructs. The first would be the bait construct, composed of the coding sequences of the Gal4 BD fused to the coding sequences of Z. The remaining 96 constructs would be fusions of the coding sequence of the Gal4 AD to *each* of the 96 coding sequences in the panel. Each of the 96 panel constructs, and the bait construct, would then be individually introduced into haploid budding yeast cells (as part of plasmids), resulting in 97 unique strains of budding yeast. One strain would express the bait fusion, and the other 96 would express the panel fusions.

Using modern robotics (e.g., the Singer ROTOR system; (www.singerinstruments.com/index.php?option=com_content&task=view&id=16&Itemid=383), it is possible to simplify and automate the process of conducting the assay. First, each of the 96 yeast strains carrying the panel constructs are grown in one well of a 96-well microtitre plate containing liquid yeast growth media (**Figure 21.5**). Next, using a robotically controlled replica pinning tool (**Figure 21.6**), cells from each of the wells of the microtitre plate are transferred to a Petri plate containing solid yeast growth media (**Figure 21.7**). The position of each unique strain in this macroarray of yeast colonies is carefully noted. Using this same method, a macroarray of yeast strains expressing the bait strain can also be made. However, for this array, each colony in the set is not unique and instead expresses the same bait fusion. Figure 21.7 shows an example of a Petri plate containing such a macroarray. (Only in this case, the system is higher throughput than our hypothetical example; this array can support up to 1536 yeast colonies!)

Once the haploid cells of the macroarray have been given the opportunity to grow and form colonies, it is time to create a diploid cell that expresses both the bait and prey construct simultaneously. To do this, our biological

(a)

(b)

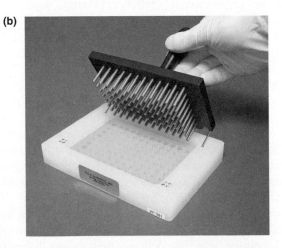

Figure 21.5 A 96-well microtitre plate. (a) Each well in this microtitre plate can be filled with liquid growth media capable of supporting budding yeast cells. Thus, one can easily culture large numbers of unique strains in an economical manner and in a small amount of space. The plate is approximately 13 cm wide and 9 cm long. Each well can hold approximately 400 microlitres of growth media. **(b)** Cells grown in this manner can be transferred to solid growth media or a fresh microtitre plate either by hand, using a replica pinning tool, or robotically (see Figure 21.6). The replica pinning tool is essentially a steel block into which carefully spaced metal pins have been embedded. The pins of the tool can be dipped into a microtitre plate containing yeast cells or stamped onto a master agar plate containing individual yeast colonies. The pins, which line up precisely with the wells of the microtitre plate, pick up a handful of cells that are then transferred to a fresh microtitre plate or to solid growth media.

Figure 21.6 Replica pinning robot. The replica pinning robot fully automates the handling of yeast, bacteria, and other biological samples.

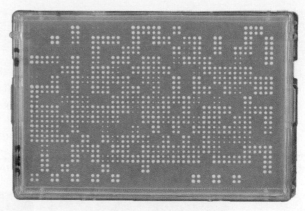

Figure 21.7 A yeast macroarray. This array, created using the robot shown in Figure 21.6, is capable of supporting the growth of up to 1536 unique budding yeast strains. Each circle represents a single budding yeast colony made up of millions of individual cells. Each colony originated from a handful of cells transferred on the tip of a pin of the replica pinning tool.

knowledge of the yeast life cycle is exploited. Haploid yeast can exist as either one of two mating types, α or *a*. Given the right environmental conditions, cells of opposite mating type can fuse to create *a*/α diploids (**Figure 21.8**).

In our example, assume that the bait macroarray strains are of the α mating type, and the panel macroarray strains are of the *a* mating type. To create diploids expressing both constructs, one can simply mix cells from each colony on the panel macroarray plate with

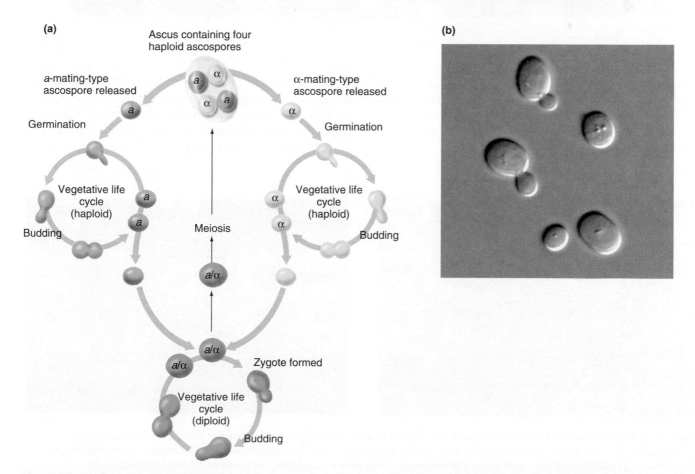

Figure 21.8 The budding yeast life cycle. **(a)** Budding yeast cells can live and grow in either a haploid or a diploid state. Diploids are formed when two cells of opposite mating types (*a* or α) fuse in a process called conjugation. Diploids have the option of entering meiosis to produce four haploid spores. **(b)** A micrograph of actively growing haploid budding yeast cells. Two cells in the process of budding can be seen in the *upper left* portion of the micrograph.

No, not the kind of robot shown in the photograph on the *left*! Instead, geneticists doing research on a global (i.e., genome/proteome-wide) scale need robotic technology to aid in the handling of yeast, bacterial, or other biological, samples. Genome-scale, two-hybrid analysis is one example of an application in which automation can save time and resources. Another application, synthetic genetic array

(SGA) analysis, will be discussed in Chapter 22 (Functional Genomics). The robots are able to automate the process of growing, arraying, and replica-pinning the samples under study. Visit the Singer instruments website to see a video of the replica pinning robot in action (www.singerinstruments. com/index.php?option=com_content&task=view&id=16& Itemid=383).

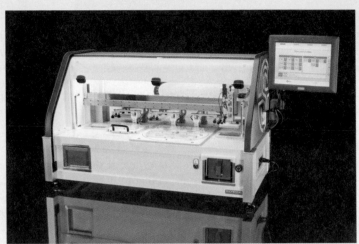

cells from each colony on the bait macroarray plate. One can then grow the mixture on a growth medium that promotes cellular fusion. In practice, this mixing of the cells can be achieved using the robotically controlled replica pinning tool.

Once the diploids have formed, it is a simple procedure to replica-pin the diploids to two fresh plates: one containing yeast growth media containing histidine, and the other containing yeast growth media lacking histidine (**Figure 21.9**). Thus, by analyzing the growth of the yeast colonies on these

(a)

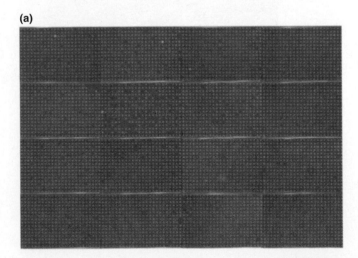

(b)

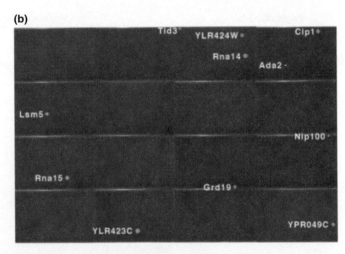

Figure 21.9 Reporter gene activation in the yeast two-hybrid system. Physical interaction between the bait and prey activates transcription of the *HIS3* reporter gene, allowing growth on media lacking histidine. The photographs show the results of a two-hybrid assay using a total of 16 macroarray plates, each bearing 384 diploid budding yeast strains. **(a)** Strains grown on media containing histidine. As expected, all the strains are able to grow since activation of the *HIS3* reporter is not required under these conditions. **(b)** The same strains replica-pinned to media lacking histidine. Only a subset of 11 strains expressing the bait protein (Pcf11, a component of the pre-mRNA cleavage and polyadenylation factor IA) and the indicated prey proteins are able to grow. Thus, intracellular interactions between proteins can be easily determined from the simple analysis of budding yeast growth.

two plates, it is possible to determine the protein interaction network for Z. Since the procedure is automated, one can then easily repeat the assay using bait constructs composed of fusions to the coding sequences of gene *A*, then *B*, etc . . .) In this way, it is possible to create a global and systematic protein interaction network for our simple organism.

21.3 Genome-Level Proteomic Interaction Analysis: Affinity Purification Coupled to Mass Spectrometry

While still commonly used, the yeast two-hybrid system does suffer from two major disadvantages; the first disadvantage being that the assay is performed in yeast cells. The intracellular environment of a budding yeast cell is quite different from that of a typical mammalian or plant cell. When testing the interactions of human proteins (or other plant or animal proteins), the effects of this unnatural environment on protein form and/or function must be considered. Perhaps, interaction can occur only "within the cozy confines of home" so to speak, or at the other extreme, perhaps the conditions present in budding yeast cells might allow two proteins to come together when they are normally unable to do so.

The second disadvantage is that the assay depends on the creation of protein fusions with the Gal4 AD and Gal4 BD. It should be noted that these domains each have a mass of approximately 20 kDa. It is thus possible that the presence of the exogenous protein tags will interfere with the form (folding) and/or function of the proteins of interest. If so, these changes might inhibit interaction, or unduly promote interaction (i.e., create erroneous false-negative or false-positive results). A second-generation technology

has now emerged that addresses these issues and which has also superseded the two-hybrid strategy for examining protein interaction networks. This technology relies on the methods of affinity purification and mass spectrometry. We will discuss each of these technologies in the following paragraphs.

During affinity purification, an antibody (or other interacting molecule with affinity for the protein of interest) is used to capture or "pull down" one protein from a complex mixture (**Figure 21.10**). The antibody in this case is referred to as a *capture molecule*. For example, consider a researcher interested in understanding the interaction network of a protein such as the one encoded by the *myc* oncogene. After creating a protein extract from the tissue under study, the researcher could employ an antibody (with specificity towards Myc) to pull down the Myc protein. Any protein physically associated with Myc will also be captured (these proteins essentially "piggyback" on the Myc protein and "come along for the ride"). Using this method, the researcher could isolate all of the proteins that associate with Myc in healthy tissue. Furthermore,

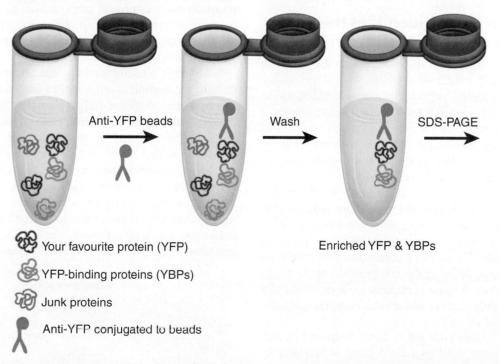

Anti-YFP beads → Wash → SDS-PAGE

Enriched YFP & YBPs

Your favourite protein (YFP)
YFP-binding proteins (YBPs)
Junk proteins
Anti-YFP conjugated to beads

Figure 21.10 Affinity purification of a protein of interest. A protein extract containing "your favourite protein" (YFP) is incubated with an antibody specific to YFP. Attached to the antibody is a bead made of Sepharose (the Sepharose beads simply help to precipitate the antibody–YFP complex). Binding of the antibody to YFP will force the complex to precipitate out of solution, allowing non-YFP-binding proteins to be washed away. The sample is now enriched in YFP and its interactors and can be further analyzed by mass spectrometry or other techniques.

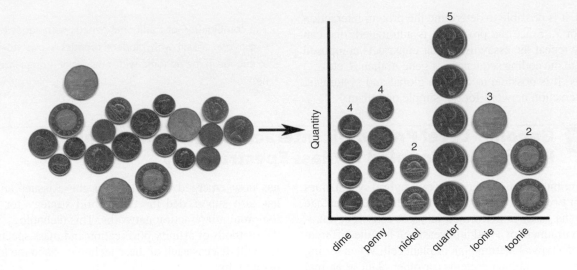

Figure 21.11 Sorting and counting coins in a piggy bank. In the same way that you might sort and count coins from a piggy bank, mass spectrometry sorts and counts peptides from a complex mixture.

the researcher could also examine how these associations are affected in abnormal or diseased (e.g., cancerous) tissue. Given the role of Myc in cancer development, these experiments might provide important clues as to why proliferation in cancer cells is uncontrolled.

In this case, the researcher, subsequent to the affinity purification, would possess two protein samples, each containing Myc and its interactors; one originating from healthy tissue and the other originating from the cancerous tissue. The goal would now shift to determining the identity of the interacting proteins in the two samples. The most commonly used method to make this determination is mass spectrometry.

A mass spectrometer identifies the components of a complex mixture

The mass spectrometer is an instrument with the ability to analyze and identify a wide variety of biological molecules, including small proteins, peptides, oligonucleotides, lipids, carbohydrates, and small RNAs. The mass spectrometer is able to do this through the sorting (according to mass) and counting of the molecules present in a complex mixture. How will this help us identify the proteins in the sample? Perhaps an analogy will help make this clear.

Consider a complex mixture of coins taken from a piggy bank. Imagine that you sort these coins into six individual piles (i.e., pennies, nickels, dimes, quarters, loonies, and toonies). You then count how many of each type of coin is in each pile. Using these data you could produce a graph (or spectrum) defining the composition of the piggy bank (**Figure 21.11**). A mass spectrometer does essentially the same thing, only it sorts and counts complex mixtures of proteins instead of coins.

Next, imagine that your piggy bank is stolen in a daring heist by the notorious Piggy Bank gang. While initially heartbroken, you rejoice upon receiving news from the authorities that they have apprehended the perpetrators. However, there is a problem. In addition to your piggy bank there are 99 other piggy banks (indistinguishable

based on their outward appearance) that have been found in the stash of the thieves. How will the police return the proper amount of money to each of the victims? Provided each victim noted the number and type of coins deposited while the piggy bank was in their possession, each victim could submit their data to the police in the form of a spectrum. Each submitted spectrum, being unique, could be used to compare with the police records (the police having sorted and counted the coins from each of the piggy banks after recovering the banks from the thieves). It would then be a simple matter of comparing the spectrums, finding the appropriate match, and determining how much money to return to each victim. In other words, by comparing your query spectrum with the complete police database of possible spectrums, you could definitively identify which piggy bank was yours.

Although a useful analogy, the above example fails to capture the complex and sophisticated nature of the mass spectrometer. While a full explanation would require an extensive discussion of chemistry and physics that are beyond the scope of this text, the description below will suffice to provide us with a "working knowledge" of the device.

Mass spectrometers consist of three components: (1) a source of ionization, which turns polypeptides into ions; (2) a mass analyzer, which separates the ion fragments according to their mass-to-charge ratio (m/z); and (3) a detector, which measures separation times, and produces a graphical representation of the results (**Figure 21.12**).

Mass Spectrometer

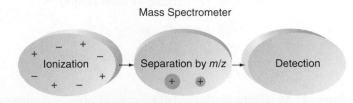

Figure 21.12 Mass spectrometer. The mass spectrometer consists of three components: (1) an ionization source, (2) a mass analyzer, and (3) a detector.

Protein Mass Mapping

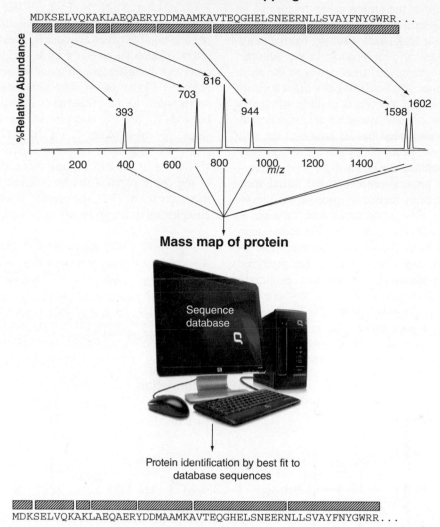

MDKSELVQKAKLAEQAERYDDMAAMKAVTEQGHELSNEERNLLSVAYFNYGWRR . . .

Mass map of protein

Sequence database

Protein identification by best fit to database sequences

MDKSELVQKAKLAEQAERYDDMAAMKAVTEQGHELSNEERNLLSVAYFNYGWRR . . .

Figure 21.13 An idealized example of a mass spectrum produced by a mass spectrometer. Once the spectrum is generated, it can be compared against a database to identify the best match.

The process of identifying the proteins in the complex mixture occurs as follows. First, cells of a particular type are isolated. In the next step, the protein of interest is affinity-purified (along with its interactors) and digested with trypsin, a protease that cleaves polypeptide chains at arginine and lysine residues. This is done because mass spectrometry can measure the masses of small molecules more accurately than large ones. Alternatively, the entire protein extract can be subfractionated according to user-defined parameters through the use of an appropriate biochemical procedure (i.e., one researcher with a particular interest in the protein interaction networks of the nucleus might employ a protocol that specifically purifies nuclear proteins). The peptides isolated from the affinity purification, or subfractionation, can then be introduced into the mass spectrometer to obtain mass measurements of tryptic peptides.

The mass spectrometer requires that molecules be ionized and transferred to a vacuum. This is achieved through treating the sample with high-intensity laser light that results in the ionized peptides acquiring a single positive charge. In this form, the spectrometer can determine the masses of the peptides by measuring their migration rates through an apparatus called a drift tube in the presence of an electric field. Since the charge of the fragments are the same, the peptides will separate according to their mass, with smaller fragments migrating more quickly, and larger fragments migrating more slowly.

At the conclusion of the analysis, the mass spectrometer will output a spectrum (such as the one shown in **Figure 21.13**) denoting the size and abundance of the peptides in the sample. This spectrum is analogous to the one generated in Figure 21.11 in our coin example. Each peak in the graph corresponds to a different peptide fragment. The height of the peak corresponds to the abundance of the fragment in the mixture—the taller the peak, the more abundant the species of polypeptide. The position of the peak along the *x*-axis denotes the size. Peaks to the right correspond to larger fragments. Peaks to the left side correspond to smaller fragments. Remarkably, the system is accurate enough to measure differences in mass on the order of a few daltons!

In the post-genomic era, where all genes have been sequenced, one can predict the amino acid structure of all encoded proteins in a genome and thus calculate the theoretical size of the peptide fragments resulting from a simulated tryptic digest of every gene product in the genome. Computer software programs can then compare the mass of each peptide experimentally analyzed in a mass spectrometer against the database of theoretical peptide masses for the proteins encoded by the genome. (This database is analogous to the database generated by the police in the coin example.) A single mass spectrometry run can assign hundreds of peptide fragments to the genes that encode them.

Remarkably, each peptide present in the initial spectrum can be analyzed even further to provide its precise amino acid sequence. This is accomplished through an apparatus known as a collision cell. After the initial separation, any fragment from the initial spectrum can be moved to the collision cell, where it is further analyzed. Although an oversimplification, this process essentially cleaves a single peptide bond in almost every polypeptide fragment in the population, generating a spectrum like the one shown in **Figure 21.14**. The peak to the far right of the spectrum ($m/z = 1598$) corresponds to the full-length polypeptide that has not been cleaved ($A_1V_2T_3E_4Q_5G_6H_7E_8L_9S_{10}N_{11}E_{12}E_{13}R_{14}$). The peak to its immediate left ($m/z = 1424$) corresponds to the fragments in which the peptide bond between E_{13} and R_{14} was cleaved. This can be inferred since the mass of an R residue (174 Da) matches precisely the difference in mass between the two fragments ($1598 - 174 = 1424$). The next peak over ($m/z = 1295$) corresponds to the fragments in which the peptide bond between E_{12} and E_{13} was broken. Again this can be inferred since the mass of an E residue (129 Da) matches precisely the difference in mass between the two fragments ($1424 - 129 = 1295$). Using this same logic, the sequence of the entire peptide can be inferred by an analysis of the mass spectrum. Thus, the precise identity of any protein in a complex mixture can be unequivocally identified.

> Mass spectrometry allows for the identification of peptide samples that have been separated according to mass and ionic charge. These data may then be stored in a database. The use of a collision cell refines the technique by isolating a fragment from the first analysis, fragmenting it further, and then analyzing it again.

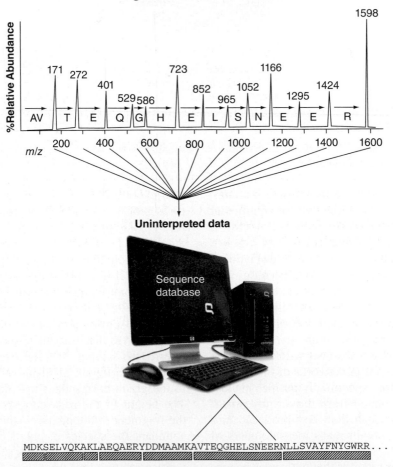

Tandem Mass Spectral Fragmentation Information

Figure 21.14 An idealized example of a mass spectrum produced after analysis of a single peptide fragment using a collision cell. Using this technology, any peptide can be sequenced and thus be unequivocally identified.

In addition to simply identifying proteins in a complex mixture, mass spectrometry can be exploited to provide quantitative data with regard to dynamic changes in protein expression. The technique called *isotope analysis* employs prepared reagents known as *isotope-coded affinity tags* (*ICATs*). The ICAT reagent has three components:

1. A biotin tag (biotin is a molecule that binds tightly to a substance called avidin); this binding provides a means for the affinity purification of proteins or peptides of interest
2. A linker to which eight hydrogens or eight deuteriums can be attached to create light (hydrogen) or heavy (deuterium) chemical isotope forms (differing by 8 Da)
3. A chemical group that reacts with the thiol (–SH) group of cysteine amino acids and thus attaches the ICAT reagent to all cysteines in a protein or peptide (**Figure 21.15**)

Let us now consider an example in which a researcher wishes to compare the proteomes of a normal and a cancerous cell type. To this end, the normal cells are labelled with the light isotope reagent and the cancerous counterparts with the heavy reagent (**Figure 21.15b**). After equal quantities of the normal and cancerous cells are mixed together, their proteins can be purified and digested with trypsin. Next, the cysteine–ICAT-labelled peptides are purified by an avidin-based affinity purification technique that captures all biotin-labelled peptides. The biotin-labelled peptides are then fractionated and successive fractions analyzed by the mass spectrometer. Peptides differing by exactly 8 Da represent two peptides derived from the expression of the same gene; the lighter one having been expressed in the normal cells, and the heavier version from the cancerous cells. The heights of the peaks provide a measure of the abundance of the peptides. Thus, the abundance of each peptide in normal versus cancerous tissue can be quantified.

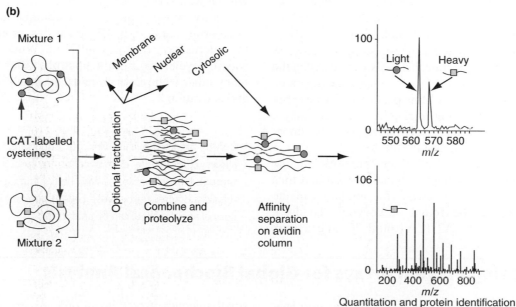

Figure 21.15 The isotope-coded affinity tag (ICAT) approach to quantifying complex protein mixtures from two different states.
(a) The isotope-coded affinity tag reagent. **(b)** The strategy for labelling the proteins of two cell types with the light and heavy reagents. The *yellow squares* indicate covalent linkage of the heavy ICAT reagent (with deuterium) from part (a) to cysteines in proteins from cancer cells. The *green circles* indicate covalent linkage of the light ICAT reagent (with hydrogen) from part (a) to cysteines in proteins from normal cells.

A key observation to emerge from the analysis of protein-protein interaction (PPI) networks is the existence of hub proteins—proteins that interact with a higher than normal number of other proteins. Interestingly, many of these hub proteins have been found to be essential for cellular growth. Why is this so? Some have suggested that hub proteins are analogous to the hub found on a wagon wheel (**Figure A**). If you were to remove the hub, the wheel would be useless. However, the wheel would still function if one, or even a few, of the spokes were removed. The tendency for hub proteins to be essential

is called the centrality–lethality rule. Protein interaction networks (**Figure B**) generated and analyzed by people like Jack Greenblatt (**Figure C**) from the Terrence Donnelly Centre for Cellular & Biomolecular Research, University of Toronto, have been crucial in developing the conceptual framework needed to understand fundamental biological principles such as the centrality–lethality rule. More of these principles will be discussed in Chapter 23 (Systems Biology).

Figure A A wagon wheel.

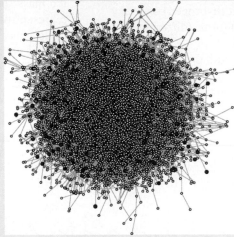

Figure B A computer-generated plot of a protein interaction network.

Figure C Jack Greenblatt.

In the example shown in Figure 21.15b, two peaks differ by exactly 8 Da. It can be inferred then that the left-most peak represents a given peptide that is expressed in the normal cells, and that the right-most peak represents the same species of polypeptide that is also expressed in the cancerous cells. Based on the height of the peak, it can be inferred that the abundance of this peptide is decreased in the cancer cells relative to normal cells. Does this play a role in the expression of the cancer phenotype? Only more direct experimentation will provide the answer.

Further fragmentation of these peptides in the collision cell will allow them to be sequenced and thus unequivocally identified. A wonderful animation of the ICAT technique can be found at www.bio.davidson.edu/courses/genomics/ICAT/ICAT.html. Taking a look

will be well worth the effort. It is interesting to note the similarity between the ICAT technique and two-colour microarray studies: ICAT quantifies proteins, whereas the microarrays quantify mRNAs. Work is now under way to develop high-throughput platforms for measuring many other features of proteins, such as phosphorylation and activation.

> Isotope analysis utilizing isotope-coded affinity tags (ICATs) allows comparative quantification of protein abundance through labelling proteins with heavy and light isotopes. In this way, differences between cells in different states, such as normal and diseased, can be analyzed.

21.5 Using Protein Arrays for Global Biochemical Analysis

In addition to using the methods described above to analyze protein network structure and protein abundance, biologists can also analyze biochemical function on a near-global scale. For example, it is possible to place

small amounts of proteins in ordered arrays on silicon or glass surfaces similar to the way in which DNA arrays are constructed. These arrays can be used to investigate the potential for each protein to be chemically modified. In the

application described in the next paragraph, a set of proteins are arrayed to determine whether they are substrates for phosphorylation. Many other biochemical parameters can be measured using protein arrays (search the Internet for many other examples). The only limitation is the imagination (and technical savvy) of the researcher.

In the experiment shown in **Figure 21.16**, budding yeast cells expressing the coding sequence (i.e., open reading frame) of a given gene from the budding yeast genome are grown in a well of a microtitre plate. Each open reading frame is under the control of a very strong promoter to allow high levels of the protein of interest to be produced. In total, 4400 individual strains (each expressing a different protein from the genome) are grown in a collection of microtitre plates. Using affinity purification, the protein of interest is extracted from each well and spotted onto the protein array chip. The chip is then incubated with one of the approximately 100 budding yeast kinases, together with radioactive phosphate. Only the proteins on the chip that are substrates of the given kinase will incorporate radioactivity. In this way, the intracellular targets of all protein kinases in a genome can be determined.

> Just like DNA, proteins can be arrayed on glass slides to allow for the global analysis of many of their enzymatic or biochemical properties.

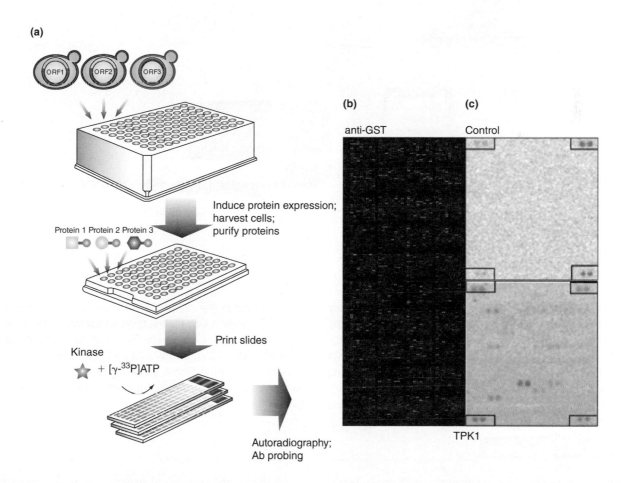

Figure 21.16 Kinase substrate protein array. (a) A schematic of the experimental procedure. **(b)** A spectral image of an entire protein chip. The chip contains an array of 40 blocks (4 blocks wide, 10 blocks tall), with each block containing 256 protein spots (each protein is spotted in duplicate to increase accuracy). While difficult to see at this magnification, the intensity of the whiteness of a spot indicates the amount of protein present. **(c)** Picture of a photographic film of two individual blocks from two separate arrays. The *top block* is from a control array performed in the absence of kinase. Spots incorporated with radioactivity will appear black. The spots in the corners are control proteins that auto-phosphorylate. The *bottom block* is an example of an experiment in which the array was incubated with the protein kinase TPK1.

Believe it or not, the analysis of your tears might play a crucial role in medical diagnosis in the near future. Tears are an easily accessible source of body fluids. The tear proteome contains approximately 500 different proteins (mucin, lysozyme, lacto-ferrin, secretory immunoglobulin A, lipocalin, and lipophilin, to name just a few). Interestingly, the expression of many of these proteins (referred to as biomarkers) varies in response to a variety of systemic medical conditions (i.e., a disease affecting the whole body). For example, the analysis of tear biomarkers has been useful in the diagnosis of various infections, diabetes, as

well as many forms of cancer. The flow chart *below* describes how this technology is used in practice. Once a tear sample is isolated from the patient, the relevant proteins are bound to a protein chip via antibodies specific to the relevant biomarkers. The bound proteins are then analyzed via mass spectrometry, providing a "fingerprint" or "barcode" that can be used to supply information regarding the presence or progression of a certain pathophysiological condition. As more and more samples are analyzed, computer algorithms will be able to make more accurate predictions based on the observed patterns of expression.

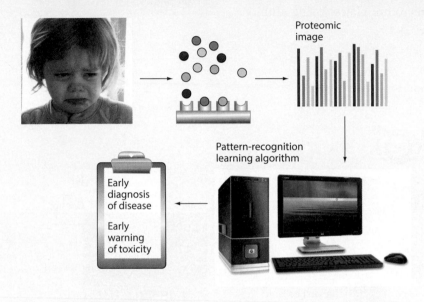

Connections

In Chapter 7, we learned about the flow of information from DNA to RNA to protein. Indeed, the deciphering of the genetic code remains one of the greatest accomplishments of the twentieth century. Nevertheless, the translation of a mRNA molecule into a polypeptide is in one sense just the beginning of the story. As we learned in this chapter on proteomics, a cell is not just a simple collection of proteins. In contrast, a

cell's protein complement is both complex and dynamic (i.e., countless protein complexes must assemble and disassemble at precise locations and at specific times to ensure proper cellular physiology). In Chapter 23 on Systems Biology, we will continue to explore these ideas by examining how changes in the architecture of protein networks affect the "emergent" properties of a given system.

Essential Concepts

1. The term proteome refers collectively to all products of translation of the protein-coding genome. High-through-put platforms have been adapted to proteomic studies in order to study protein expression, interaction, and modi-fication on a global scale. Modern robotics together with traditional genetic techniques have combined to provide powerful tools for the genome-wide analysis of protein networks. **[LO1–4]**

2. The mass spectrometer, in conjunction with fractionation methods, allows researchers to determine the abundance,

identity, and amino acid sequence of all peptides within a complex mixture. **[LO2, LO5]**

3. Isotope analysis utilizing isotope-coded affinity tags (ICATs) enables the determination of protein con-centration changes in two different cellular or tissue states. **[LO2, LO5–6]**

4. Protein arrays, constructed similarly to DNA arrays, can be probed to assay a variety of the biochemical param-eters associated with the spotted samples. **[LO2, LO7]**

I. You hypothesize that the addition of a phosphate group to a particular serine amino acid in a kinase enzyme is important for the ability of this kinase to play a role during mitosis. How could you use affinity purification coupled to mass spectrometry to determine whether or not the kinase was indeed phosphorylated at this serine during mitosis?

Answer

According to your, model the serine residue of the kinase will be phosphorylated in mitosis, but unphosphorylated in interphase. Therefore, you could first isolate protein extracts from (i) cells undergoing mitosis, and (ii) cells in interphase. Using an antibody against the kinase in question, you could immunoprecipitate the kinase from each sample, digest the product with trypsin, and then use the mass spectrometer to determine if the mass of the fragment containing the serine residue shifts in size by 80 Da (the mass of a single phosphate group).

Problems

Vocabulary

1. For each of the terms in the left column, choose the best matching phrase in the right column.

i. proteome

ii. proteomics

iii. affinity purification

iv. mass spectrometer

1. a procedure in which an antibody is used to capture a protein from a complex mixture

2. the global analysis of the proteins present in a particular cell type or organism under a defined set of conditions

3. a device that ionizes polypeptides to identify different proteins by their mass-to-charge ratio (*m/z*)

4. the complete set of unique proteins encoded by a given genome

Section 21.1

2. a. Define the terms genome, transcriptome, and proteome, and describe the relationship among them.

b. Evaluate the limitations of these terms in describing gene expression in a complex multicellular eukaryotic organism.

Section 21.2

3. Which of the statements below are true of two-hybrid analysis? In each case, provide reasons for your answer.

a. Transcription factors are covalently linked to DNA.

b. Interaction between proteins is indicated by decreased expression of a reporter gene.

c. A DNA sequence containing a gene of interest is fused to the DNA sequence of the dimerization domain of GAL4.

4. Consider the following statement:

"To activate transcription of the reporter gene in the yeast two-hybrid assay, both the bait and prey fusion proteins must enter the yeast nucleus."

a. Do you agree or disagree with this statement? Explain your answer.

b. If you agree, explain why this characteristic might pose a problem with regard to obtaining accurate results.

5. While discussing the yeast two-hybrid system with your classmate, Jake, you are surprised to hear him say that the two-hybrid assay is not a particularly useful tool since you can only analyze yeast proteins. Your other classmate, Sarah, says that this is not true and that you can study interactions between proteins from any organism of interest. Who is correct and why?

6. With respect to the two-hybrid assay,

a. discriminate between what is meant by the terms "false-negative" and "false-positive."

b. provide an example of a set of circumstances that might result in a false-negative result.

c. provide an example of a set of circumstances that might result in a false-positive result.

d. develop a series of "control" experiments that you could use to reduce the possibility of being fooled by false-positive or -negative results.

Section 21.3

7. How many of the following are typical applications of mass spectrometry? In each case, explain the reasoning behind your answer.

a. determining protein charge

b. determining the sequence of DNA

c. determining the sequence of proteins

d. quantifying mRNA levels

e. quantifying protein levels

8. Imagine that you are a researcher studying cell proliferation in pancreatic cells. In one experiment, you use affinity purification of the Rb protein from a normal pancreatic cell line, and then from a cancerous derivative cell line. The sequence of some of the peptide fragments obtained after mass spectrometry of each of the samples is shown in the chart below. Using you knowledge of the Rb protein (see Chapter 16), and your bioinformatics skills (see Chapter 20), formulate a testable hypothesis regarding the development of uncontrolled cellular proliferation in the cancerous cell line. What experiments could you perform to test the veracity of your hypothesis?

Normal pancreatic cell line	Cancerous derivative cell line
LLNDNIFHMSLLACALEVVMATYSR	LLNDNIFHMSLLACALEVVMATYSR
STSQNLDSGTDLSFPWILNVLNLK	STSQNLDSGTDLSFPWILNVLNLK
IMESLAWLSDSPLFDLIK	IMESLAWLSDSPLFDLIK
ILVSIGESFGTSEK	ILVSIGESFGTSEK
TVMNTIQQLMMILNSASDQPSENLISYFNNCTVNPK	TVMNTIQQLMMILNSASDQPSENLISYFNNCTVNPK
DDAFTDSLQLDVVGDSAVDEFEK	APEVLLQSTYATPVDMWSVGCIFAEMFR
WIGPVDFSSSDEELVDVSASVLPELK	GPRPVQSVVPEMEESGAQLLLEMLTFNPHK
SSTLPSADPQLQSQPSLNLSPVMSR	APPPGLPAETIK
APEVLLQSTYATPVDMWSVGCIFAEMFR	DLKPENILVTSGGTVK
GPRPVQSVVPEMEESGAQLLLEMLTFNPHK	FISNPPSMVAAGSVVAAVQGLNLR
LCIYTDGSIRPEELLQMELLLVNK	ACQEQIEALLESSLR
FISNPPSMVAAGSVVAAVQGLNLR	MEHQLLCCEVETIR

Section 21.4

9. Which of the following statements are true of the technique referred to by the acronym ICAT?

   a. The ICAT reagent contains a hydrophobic linker region.

   b. ICAT makes use of affinity capture coupled to mass spectrometry.

   c. ICAT quantifies changes in mRNA expression from two different cellular states.

10. The comparative quantitation of protein abundance using ICAT depends on identifying pairs of peptide fragments differing by exactly 8 Da. Think of a set of circumstances that would result in your inability to provide a comparative estimate of protein levels. Is there a simple way for this problem to be resolved? Explain.

11. Suppose you are conducting an analysis of the entire yeast transcriptome using gene expression microarray chips. You examine yeast in two different cellular states, and you determine that about 400 genes increase or decrease their levels of expression in state 1 as opposed to state 2. You now analyze 600 proteins from these two different yeast states using ICAT proteomics technology. To your amazement, you find that 50 protein levels change when their corresponding mRNA levels do not. Moreover, 60 protein levels do not change when their corresponding mRNA transcript levels remain the same. Finally, you notice that 30 protein levels change in opposite directions from that of their mRNA transcript counterparts. What simple explanation can you offer for these observations (apart from measurement error)?

12. Below is the amino acid sequence of a portion of the c-Myc protein.

```
        10          20          30          40          50          60        66
MDFFRVVENQ   QPPATMPLNV   SFTNRNYDLD   YDSVQPYFYC   DEEENFYQQQ   QQSELQPPAP   SEDIWK
```

a. Determine the mass of each of the tryptic digest products of the peptide above using the table of amino acid masses.

Amino acid	Weight	Amino acid	Weight
Glycine (G)	75	Aspartate (D)	133
Alanine (A)	89	Glutamate (E)	147
Valine (V)	117	Glutamine (Q)	146
Leucine (L)	131	Asparagine (N)	132
Isoleucine (I)	131	Lysine (K)	146
Serine (S)	105	Arginine (R)	174
Threonine (T)	119	Histidine (H)	155
Tyrosine (Y)	181	Cysteine (C)	121
Phenylalanine (F)	165	Methionine (M)	149
Tryptophan (W)	204	Proline (P)	115

b. Using your work from part *a*, determine whether or not the Myc peptide is present in the following spectrums.

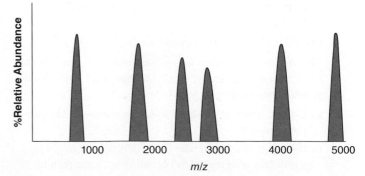

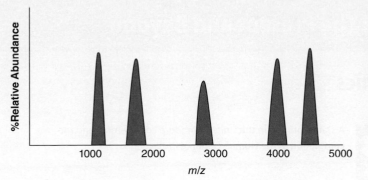

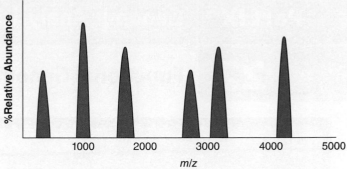

Genome **The End of the Beginning**

A stamp printed in the United Kingdom commemorating the completion of the Human Genome Project.

The publication of the human genome sequence in 2001 marked the beginning of a new era in biology, an era dominated by two distinct, yet inseparable fields—functional genomics and systems biology. The discipline of functional genomics seeks to provide an understanding of the function, regulation, and interaction of all gene-products in a genome. By its very nature it relies on high-throughput techniques that produce vast data sets describing the characteristics of the genes/gene-products governing cellular physiology. The establishment of these data sets has in turn necessitated the development of philosophies and techniques—collectively defining the discipline of systems biology—that are aimed at understanding how the interaction of cellular components give rise to higher-order biological phenomena. In this chapter, we will discuss how genome sequence data can be exploited to analyze both gene function and genome architecture. In the following chapter, we will take a closer look at how systems biologists draw these data together to better understand how living organisms organize and control biological processes.

Chapter Outline
22.1 What Is Functional Genomics?
22.2 The Genome-wide Budding Yeast Gene Deletion Set
22.3 Using RNA Interference to Probe Gene Function
22.4 Synthetic Genetic Arrays: Understanding the Principles of Genome Architecture

Learning Objectives

1. Define, compare, and contrast the disciplines of functional genomics and systems biology.

2. Relate genetic barcoding to the elucidation of gene function on a genome-wide scale.

3. Explain the role of RNAi in carrying out functional genomics studies in developmentally complex eukaryotes.

4. Distinguish between hub and nonhub genes and illustrate the importance of each class of genes with respect to determining phenotype in outbred populations.

22.1 What Is Functional Genomics?

As you have gathered from reading Chapters 19–21, whole-genome sequencing—together with the development of global genetic analysis schemes—has defined not an endpoint in the history of genetics research, but rather the beginning of a new scientific era (often referred to as the post-genomic era). In the broadest sense, research of

the post-genomic era revolves around better understanding genome function; that is, how simple strings of A's, G's, C's, and T's can direct the complex organization, regulation, and decision-making ability of all living things (from bacteria to plants to human beings). To a large extent, this involves the detailed understanding of the function and regulation of individual gene-products (i.e., their biochemical activity, the control of their expression, their localization). However, as we shall see in the next few chapters, this is just the tip of the iceberg. For instance, it is clear that the organization of genes/gene-products into networks (both physical networks and more abstract genetic interaction networks) has clear and profound consequences with regard to organismal phenotype.

Thus, while functional genomics seeks to understand gene function on a global scale, it might be more accurate to say that the discipline is interested in understanding how complex biological phenomena (e.g., developmental processes, cellular memory, morphogenesis, to name a few) emerge from the intricate, dynamic, and complex web of interacting DNA, proteins, and lipids found within our cells (**Figure 22.1**). As we shall see later in this chapter, the detailed analysis of functional genomic data has provided great insight with respect to our understanding of life, and, moreover, how this understanding can be exploited to benefit society in ways not before thought possible.

> Functional genomics is the branch of genetics that seeks to define the function, regulation, and interaction of all gene-products in a genome.

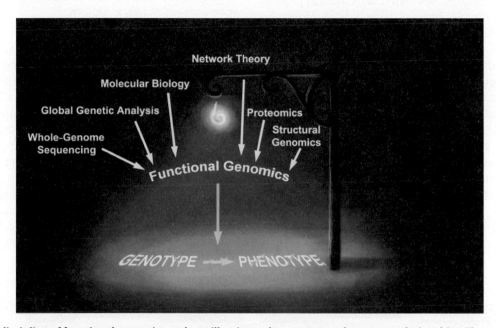

Figure 22.1 The discipline of functional genomics seeks to illuminate the genotype–phenotype relationship. Through the use of methodologies derived from both the pre-genomic and post-genomic eras, functional genomicists attempt to understand how complex biological phenomena emerge from the simple strings of A's, G's, C's, and T's that comprise the genome.

22.2 The Genome-wide Budding Yeast Gene Deletion Set

The budding yeast, *Saccharomyces cerevisiae*

As we have seen in previous chapters, model organisms have been crucial to the development of our understanding of many fundamental genetic processes. For example, the study of pea plants allowed Mendel to infer the laws of inheritance (see Chapter 2). In contrast, Jacob and Monod's detailed study of lactose metabolism in *Escherichia coli* allowed them to elucidate the mechanisms of gene regulation (see Chapter 10). For developmental genetics, *Drosophila melanogaster* has proven

to be an excellent subject with which to define the cellular networks controlling developmental transitions (see Chapter 17). Continuing in this tradition with respect to the field of functional genomics is the budding yeast, *Saccharomyces cerevisiae*.

As we saw in Chapter 16, *S. cerevisiae* is a unicellular, eukaryotic yeast that has been used extensively as a model organism. Its genome of approximately 12 Mb has been fully sequenced and encodes approximately 6000 genes. Its utility as a genetic model derives, in part, from the fact that budding yeast can exist either as a haploid, or as a diploid. This is to say, both haploid and diploid cells can

grow vegetatively and divide through mitosis in a stable fashion over many generations. Haploids can be of the *a* or the α mating type. Under certain environmental conditions, *a* and α cells can fuse to form *a*/α diploids. These diploids can grow mitotically or, if deprived of nitrogen, be induced to undergo meiosis and produce four haploid spores. These spores, if provided with favourable growth conditions, can germinate and grow vegetatively as haploids (**Figure 22.2**).

Gene knockouts

Another key property of budding yeast (with respect to its utility as a model system) relates to the frequency of homologous recombination. Homologous recombination refers to the process by which DNA fragments are integrated into a chromosome through the alignment, pairing, and recombination of the molecule with homologous genomic DNA sequences. Normally, cells use this process to repair DNA in the G$_2$ phase of the cell cycle. However, researchers can also exploit this phenomenon to genetically manipulate budding yeast cells to create strains bearing gene "disruptions" or more commonly gene "knockouts" (**Figure 22.3**). Interestingly, this technique cannot be used in more developmentally complex organisms (e.g., humans) because the frequency of homologous recombination is too low. For example, 30–85 percent of introduced DNA fragments will integrate at the homologous genomic location in the budding yeast, while the ratio of homologous integration to nonhomologous recombination (i.e., random integration) in humans is orders of magnitude lower.

Gene knockouts are created by first constructing a gene deletion cassette—a linear dsDNA fragment

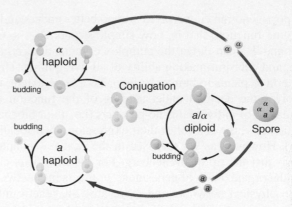

Figure 22.2 The budding yeast life cycle. Haploid budding yeast cells of the *a* or α mating type can grow vegetatively by mitosis, or be induced to fuse during the process of conjugation. *a*/α diploids can also grow vegetatively by mitosis or, if deprived of nitrogen, enter the meiotic cycle to produce four haploid spores. Once favourable nutritional conditions are sensed, these spores can germinate and again grow vegetatively as haploids.

containing a gene encoding a selectable marker. The selectable marker is typically a gene encoding resistance to an antibiotic. It is flanked on one end by 80–100 bp of sequence homologous to the region directly upstream of the start codon of the gene of interest. A further 80–100 bp of sequence homologous to the region directly downstream of the stop codon of the gene of interest is placed at the other end. The deletion cassette is then added to a suspension of yeast cells that have been treated so that they are prone to taking up foreign DNA. Cells in the suspension that homologously integrate the deletion cassette into their genomes are then selected by culturing the cells in media containing the appropriate

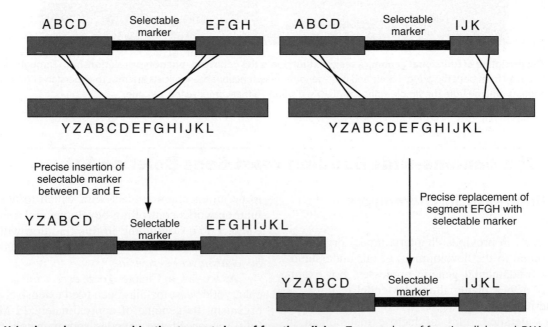

Figure 22.3 Using homologous recombination to create loss-of-function alleles. To create loss-of-function alleles, a dsDNA cassette—composed of a selectable marker flanked by DNA sequences homologous to a genomic region of interest—is introduced into a cell. Homologous recombination between the common sequences will result in the incorporation of the selectable marker at the desired locus. In the case of a disruption, no genomic sequences are lost. In the case of a knockout, the entire open reading frame is replaced by the selectable marker.

antibiotic (i.e., the antibiotic to which the selectable marker confers resistance). One commonly used selectable marker is the *kanR* gene. In yeast, this gene confers resistance to the antibiotic geneticin.

Typically, gene knockouts are created in diploid budding yeast strains. Can you think why? To answer this question, consider a scenario in which the gene of interest is essential for life (e.g., DNA polymerase or a gene encoding an enzyme critical for metabolism). If one knocked out this gene in a haploid, then the cells that had integrated the construct would fail to proliferate (since there is no alternate wild-type copy of the gene on a homologous chromosome). On the other hand, if knocked out in a diploid, a second wild-type copy would be present, allowing the cells to survive. In this case, the phenotype of the gene deletion mutant could be observed by simply inducing the heterozygous diploid to undergo meiosis to generate four haploid spores (two of which would carry the wild-type allele of the gene of interest, and the other two the knockout allele). If one observed that two of the four spores were viable, and furthermore that these viable progeny were sensitive to the antibiotic, one could conclude that the gene of interest was required for life. Such genes are referred to as essential genes.

In the budding yeast—where gene knockout strains corresponding to each gene in the genome have been constructed—approximately 19 percent of genes are essential. The remaining 81 percent of budding yeast genes are nonessential and therefore can be cultured and assayed for any observable phenotype. For example, a gene deletion mutant exhibiting hypersensitivity to a DNA-damaging agent would indicate that the gene in question played a role in DNA repair. On the other hand, a gene deletion mutant displaying morphological defects, might indicate a role for the gene in controlling either the actin and/or microtubular cytoskeletons. Thus, the genome-wide set of viable gene deletion mutants can be used as a general tool to determine gene function in the budding yeast with respect to any phenotype and under any growth/environmental condition (see **Figure 22.4** and the **Tools of Genetics** box "**Fully Automated Genome-wide Analysis of Spindle Morphogenesis**").

In early applications of this strategy, mutants in the set were typically analyzed individually; that is, a researcher would examine one or a small handful of strains at a time, until all of the viable haploid gene deletion mutants in the set were analyzed. However, subsequent researchers have made use of microarray technology to increase the analytical power, as well as the speed and efficiency of the technique. In the next section, we will see how microarray technology, together with yeast barcoding, allows for whole-genome parallel analysis of the budding yeast gene deletion set.

Barcoding

Whole-genome parallel analysis of the budding yeast gene deletion set is made possible by a technique known as genetic barcoding. The barcodes in question are unique DNA sequences that have been incorporated into each gene deletion cassette; they act as identifiers of each individual

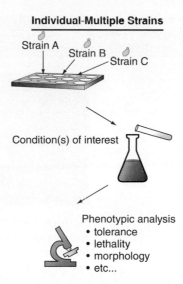

Figure 22.4 Phenotypic analysis of the budding yeast gene deletion set. The entire set of viable haploid budding yeast gene deletion mutants can be purchased commercially for only $3000. In such a set, each of the approximately 4800 viable gene deletion mutants is stored within a well of a microtitre plate. The cells in the wells are mixed with a glycerol solution that allows them to be stored at −80°C without any ill effects. Therefore, at the researcher's leisure, any or all strains from the set can be retrieved from cryogenic storage, cultured, and then examined under any growth/environmental condition for any phenotype of interest.

gene deletion strain. As shown in **Figure 22.5** the barcode sequences are added just upstream and just downstream of the *kanR* selectable marker. As usual, regions of homology to the sequences surrounding the gene of interest are still present at each end of the deletion cassette. The upstream barcode (approximately 20 nucleotides long) is referred to as the uptag and the downstream barcode (also approximately 20 nucleotides long) is referred to as the down- or dntag. These tags, just like the barcodes that identify products at the supermarket, are unique. This is to say, each gene deletion is marked by tags that do not match any sequences present in the yeast genome, or in any of the other up- or dntags.

In addition to the tags, PCR-primer-binding sequences are also added to the cassette. These flank both the uptag and the dntag and are *not* unique. In fact, each gene deletion cassette contains the identical PCR-primer-binding sequences (we shall see why later). As described previously, these barcoded gene deletion cassettes can be homologously integrated into the budding yeast to create a set of barcoded haploid gene deletion mutants. Let us now see how this set of barcoded mutants—used in conjunction with microarray technology—can be used for whole-genome parallel analysis.

Whole-genome parallel analysis

To understand the concept of whole-genome parallel analysis, first consider a specific scenario related to the repair of DNA damage. Imagine that a researcher is interested in

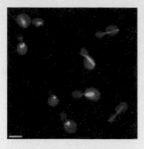

Figure A Gene deletion mutants expressing a fluorescent marker of the mitotic spindle.

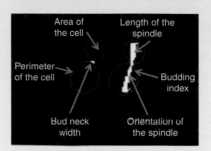

Figure B Computerized measurement of parameters related to spindle function.

One particularly ingenious use of the genome-wide deletion set—spear-headed by University of Toronto researcher Brenda Andrews and colleagues—has incorporated fluorescence microscopy and sophisticated computer software to uncover and characterize the budding yeast genes required for proper functioning of the mitotic spindle. In this study, each of the viable haploid deletion mutants (expressing a fluorescent marker of the mitotic spindle) was screened using an entirely automated pipeline. First, a robotic fluorescence microscope system was used to capture a series of images of each of the gene deletion strains (**Figure A**). Images were then passed on to a specialized software program that used sophisticated algorithms to define and measure a variety of parameters related to spindle function (e.g., length, breadth, orientation, morphology, to name a few; **Figure B**). In this way, the researchers were able to define 92 genes with roles in various aspects of spindle morphology. This example illustrates how the budding yeast gene deletion set—in the hands of imaginative and creative researchers—can be used to provide comprehensive information regarding any biological process of interest.

identifying the budding yeast genes required for repairing DNA damage induced by the drug camptothecin. To this end, equal amounts of each deletion strain are inoculated by the researcher into liquid growth media and the cells allowed to proliferate. At a given time after inoculation, the culture is split in two with one-half being treated with camptothecin (the experimental sample), and the other half remaining untreated (the control sample). Again, the cells in both cultures are given the opportunity to proliferate for a certain length of time. At the end of the growth period, the cells are isolated, lysed, and the genomic DNAs extracted. If gene "X" is required to repair camptothecin-induced DNA damage, then one would expect the gene X deletion strain to be less apt to grow upon camptothecin

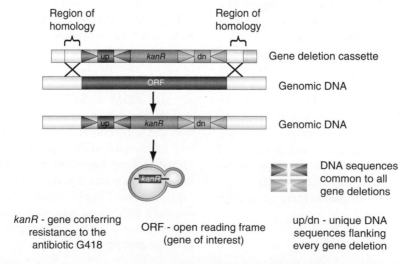

kanR - gene conferring resistance to the antibiotic G418

ORF - open reading frame (gene of interest)

up/dn - unique DNA sequences flanking every gene deletion

Figure 22.5 Genetic barcoding. Barcodes are composed of unique DNA sequences (approximately 20 bp long) that flank the selectable marker (*grey*) of the deletion cassette. Typically two such barcodes, an uptag (*red*) and a dntag (*green*), are included. Common PCR-primer-binding sites that flank each tag are also included (*triangles*). Through homologous recombination, the cassettes can be used to knock out a gene of interest (*blue*) to create a loss-of-function allele. Each knockout strain made in this manner can be uniquely identified by PCR-amplifying and analyzing the barcode sequences.

treatment relative to X^+ cells. At the same time, however, it is expected that X^- cells would grow at a rate similar to X^+ cells in the control sample (since no camptothecin is present and the gene is not required). Thus, at the conclusion of the growth period, the gene X deletion strain will be present in the control culture, but absent (or present at very low numbers) in the camptothecin-treated sample.

In contrast, cells bearing a deletion in a gene "Y," a gene that has no role in repairing DNA damage, would be expected to grow normally in the control sample, and to outgrow the gene X deletion cells in the camptothecin-treated sample (since it has the capacity to repair camptothecin-induced damage). The gene Y deletion strain would thus be represented in greater numbers in the camptothecin-treated sample relative to the gene X deletion strain. Thus, in the final analysis, there will be a deficit of genomic DNA in the experimental sample derived from those cells bearing gene deletions conferring camptothecin sensitivity, and an overabundance of genomic DNA derived from cells bearing gene deletions with no role in repairing camptothecin-induced damage. It is at this point that the barcode sequences, together with microarray technology, can be exploited to identify the camptothecin-sensitive gene deletion strains.

In this scenario, the DNA spots on each microarray consist of oligos corresponding to each of the unique up- and dntag barcode sequences. To make use of the barcodes, genomic DNA isolated from the control and experimental samples is used as a template in a PCR that amplifies the up- and dntags from each cell in the respective cultures. As part of the PCR, the Cy3 fluorophore (see Chapter 20) is incorporated into the amplicons of the control sample, and the Cy5 fluorophore into the amplicons of the experimental sample (or vice versa). The labelled PCR amplicons are then combined and hybridized to the barcode array. At this point, based on the relative amount of Cy3 and Cy5 signal emanating from the chip, the camptothecin-sensitive gene deletion mutants can be identified. For example, spots on the array displaying predominantly green fluorescence (i.e., a high Cy3/Cy5 ratio) will correspond to the barcoded genes that function to repair camptothecin-induced DNA damage (since these cells were eliminated from the experimental Cy5-labelled sample). In contrast, spots on the array displaying a much lower Cy3/Cy5 ratio would correspond to barcoded genes with no role in the DNA damage response (**Figure 22.6**). (These cells could repair their DNA and thus were not eliminated from the experimental population.)

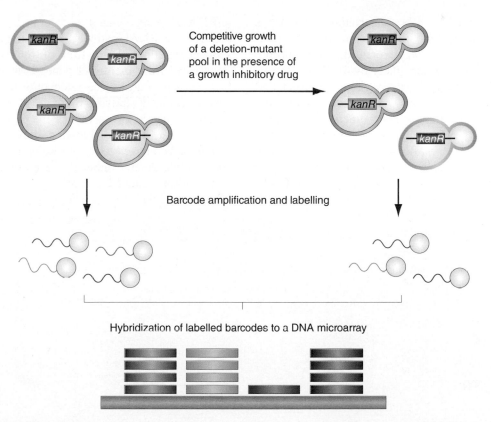

Figure 22.6 Whole-genome parallel analysis. The barcoded yeast deletion set, together with microarray technology, is a powerful tool that can be used to define gene function on a global scale. In the example, the complete set of gene deletion mutants is screened by growth in the presence of a drug/environmental condition of interest. Growth of each of the approximately 4800 deletion mutants in the population is assayed by PCR-amplifying/labelling the barcode sequences and hybridizing the amplicons to the barcode array. For the sake of simplicity, only four gene deletions strains are shown (defined by the *red, blue, green,* and *purple* gene deletion cassettes). In the given example, a growth inhibitory drug such as camptothecin is tested. Cells deleted for the blue gene drop out of the population. Thus, few barcodes from this strain are PCR-amplified and little barcode DNA hybridizes to the chip. This indicates a role for the gene in responding to the effects of the drug.

Thus, all genes with a role in repairing camptothecin-induced DNA damage could be defined using one simple and inexpensive assay (in only a day or two). Furthermore, researchers could then go on to examine distinct forms of DNA damage simply by using different damage-inducing drugs. In this way, researchers could quickly and easily identify the genes required for the repair of any particular class of DNA damage.

Importantly, any drug, growth condition, or stress could be tested in this way. For example, other researchers might seek to identify the genes required for the cellular heat-shock response. In this case, researchers would perform an experiment in which the barcoded strains are grown at room temperature (control sample) and at 42°C (experimental sample). In the final analysis, it is both possible (and practical) to use such assays to define gene function in the budding yeast on a global scale. This type of approach is often referred to as going from genome to phenome (i.e., having the ability to map all genes to many different phenotypes).

While a powerful technique, it should be noted that the approach described above is restricted to budding yeast and other organisms where the frequency of homologous recombination is high. Unfortunately, in more developmentally complex systems, like the mouse or human, the frequency of homologous recombination is too low to make the creation of gene knockouts practical on a genome-wide level. Luckily, however, our molecular understanding of RNA interference can be used in these cases to create an analogous system based on gene silencing. These techniques are discussed in the next section.

The budding yeast, *Saccharomyces cerevisiae,* has emerged as the premier model organism in functional genomics research. This is due in large part to the ease with which one can create gene deletion mutants through homologous recombination. In fact, genome-wide sets of gene deletion mutants are commercially available. The barcoding of gene deletion mutants, together with microarray technology, allows researchers to engage in whole-genome parallel analysis and thus to map the function of all budding yeast genes to any observable phenotype. This strategy is often referred to as proceeding from genome to phenome.

22.3 Using RNA Interference to Probe Gene Function

As we saw in Chapter 11, RNA interference (RNAi) is a natural biological process in which small RNA molecules silence or knock down expression of a specific target gene. While this system likely evolved as a means to protect cells from attack by RNA viruses, our detailed molecular understanding of this process can nevertheless be exploited as a tool to study gene function. Typically, a double-stranded RNA molecule—composed of both the sense and antisense strands of a portion of the target gene mRNA—is synthesized and introduced into the cell type under study. These introduced short interfering RNAs (siRNAs) are recognized as foreign by the cell and thus activate the RNAi pathway. This in turn results in the sequence-specific destruction of the mRNA molecules transcribed from the target gene. Thus, in the same way that yeast geneticists use knockouts to study gene function, researchers studying more developmentally complex organisms use RNAi to reduce the expression of a target gene to create a loss-of-function allele (**Figure 22.7**). This technique is referred to as knockdown instead of knockout since, while greatly reduced, the expression of the target gene is rarely reduced to zero.

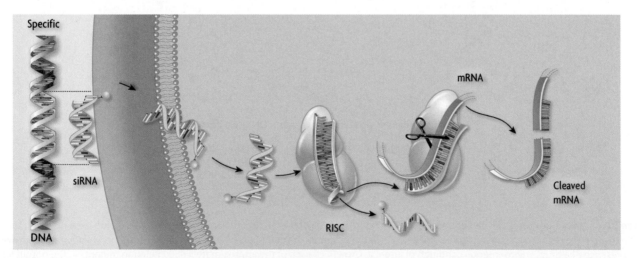

Figure 22.7 Using RNA interference to knock down gene expression. To knock down expression of a given gene, short interfering RNAs (siRNAs) targeting the gene of interest are synthesized and introduced into the cell type of interest. The introduced RNA molecules are recognized as foreign and incorporated into RISC (see Chapter 11). The double-stranded RNA is then unwound, leaving the antisense strand. The complex then uses the antisense strand to guide itself to complementary mRNAs that are subsequently destroyed by endolytic cleavage.

Two types of small RNAs are typically used to perform knockdowns: short interfering RNAs (siRNAs) and short hairpin RNAs (shRNA) (**Figure 22.8**). siRNAs are dsRNA fragments approximately 19–22 bp in length that are chemically synthesized. These molecules are directly transfected into a target cell, resulting in the transient (i.e., temporary) silencing of the target gene. shRNAs, on the other hand, are used to achieve stable (i.e., permanent) silencing. shRNAs are formed by engineering a recombinant DNA construct to express an approximately 50-nucleotide-long ssRNA molecule. The ssRNA is designed so that it contains complementary sequences that anneal to form a stem-loop structure. The loop portion is typically 5–10 nucleotides long, whereas the stem portion is 10–21 bp long. To silence the target gene, the plasmid construct is transfected into a target cell, or even packaged into a virus with the ability to infect the cell type being used. Once the shRNA is expressed and transported into the cytoplasm, the loop of the shRNA is cleaved. This creates a siRNA that enters the canonical RNAi pathway, which, in conjunction with RISC, knocks down expression of the target gene.

Remarkably, in addition to silencing genes in cells grown in culture, it is also possible to silence genes in the tissues of intact, living organisms (**Figure 22.9**). For the round worm, *Caenorhabditis elegans*, three methods have been developed to silence gene expression at the organismal level. The most straightforward method involves soaking the worms in a solution containing a siRNA for 24 hours. The worms are then transferred to Petri plates where they are cultured on agar growth medium and assayed for the phenotype of interest. An alternative method involves culturing the worms on Petri plates where *E. coli* cells have been spread. Remarkably, if the *E. coli* upon which the worms feed are themselves transformed with a plasmid expressing a given shRNA, this results in the silencing of the *C. elegans* gene of interest. Lastly, gene silencing can be achieved by directly injecting dsRNAs into the gonads of adult worms. This method leads to the silencing of the target gene in the progeny worms.

Any given phenotype can be assayed on a genome-wide scale, by simply creating RNAi constructs targeting each gene in the genome. In *C. elegans*, RNAi libraries containing approximately 20 000 individual constructs have been created and used to screen for phenotypes ranging from growth to sterility, abnormal development, longevity, and neurological defects (**Figure 22.10**).

RNAi methodologies like the ones described above have been used to study a wide variety of biological processes in a diverse assortment of model systems. One particular ingenious example (described below) has identified an interesting group of genes that are required for the survival and proliferation of cancer cells in humans (**Figure 22.11**). The identification of such genes—that is, genes affecting cancer cell survival, but not the viability of normal noncancerous cells—allows for the targeting and destruction of cancerous tissues (while at the same time leaving normal tissue unharmed).

In this study, shRNA constructs targeting 2924 genes central to the regulation of proliferation were created and cloned into plasmid vectors. These vectors were then packaged into a retrovirus capable of integrating the shRNA construct into the genomic DNA of the cell lines of interest (a breast cancer line and a normal human mammary epithelial cell line). To begin the experiment, cells were infected with the retrovirus and, after 48 hours of growth, genomic DNA was isolated from a portion of the infected cells. Next, a unique portion of each integrated construct (a barcode) was PCR-amplified and the amplicons were labelled with Cy5 (red fluorophore). The remaining cells were then cultured for a further three weeks before a second sample of genomic DNA was isolated. Again, the barcode sequences were amplified, but this time the amplicons were labelled with the green fluorophore, Cy3. The labelled amplicons were then combined and hybridized to a corresponding barcode array.

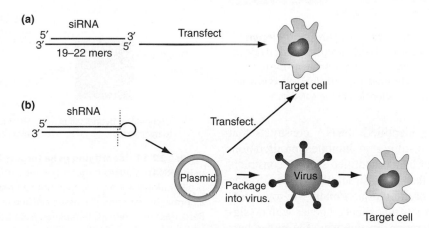

Figure 22.8 Alternative approaches to RNAi-mediated knockdown of gene expression. **(a)** siRNA molecules 19–22 bp in length are chemically synthesized and directly transfected into the target cell to achieve transient knockdown. **(b)** shRNA-expressing constructs are incorporated into plasmids and either directly transfected into the target cell, or packaged into viruses that infect the target cell. In either scenario, stable integration of the construct is achieved, resulting in permanent silencing.

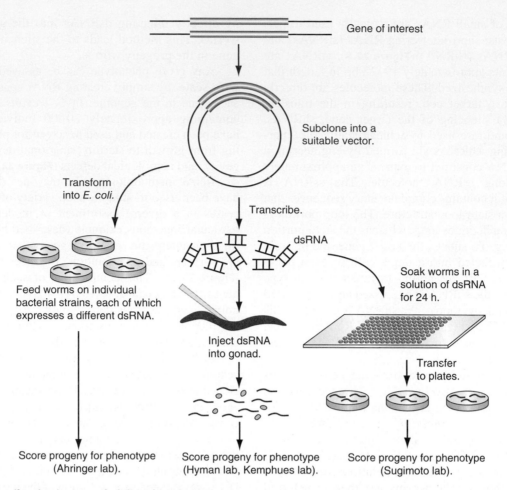

Figure 22.9 Gene silencing in *Caenorhabditis elegans*. In *C. elegans* three methods have been developed to silence genes. Worms can be (1) fed on *E. coli* cells expressing a shRNA targeting the gene of interest, (2) soaked in a solution containing a given siRNA, or (3) injected with a siRNA.

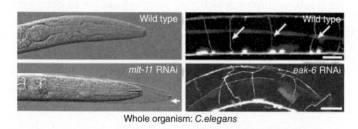

Whole organism: *C.elegans*

Figure 22.10 Examples of whole-organism RNAi in *C. elegans*.
(*Left*) A defect in molting induced by the silencing of the *mlt-11* gene. Worms in which *mlt-11* is silenced are unable to separate the old exoskeleton (*white arrow*) from the epidermis. (*Right*) Defects in axon (*white arrows*) guidance caused by silencing of the *eak-6* gene.

Red spots on the array identified shRNA constructs that silenced a given gene resulting in inhibited proliferation. Green spots, on the other hand, identified shRNA constructs with the opposite effect; that is, constructs that silenced a gene resulting in increased proliferation. Finally, yellow spots identified shRNA constructs that silenced genes with no significant affect on proliferation. In this way, the researchers were able to identify genes that (1) when silenced resulted in a decrease in proliferation of the cancer line, and (2) had

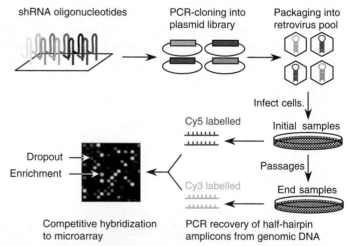

Figure 22.11 Identifying gene targets for anticancer therapy using RNAi. shRNAs targeting genes with known roles in cellular proliferation were expressed in both a breast cancer cell line and a normal breast epithelial tissue cell line using retroviral vectors. Barcodes were then PCR-amplified from both cells lines and labelled just after infection and three weeks post-infection. In this way, genes required for proliferation of the cancer cell line were identified.

no effect on the healthy HMEC control cell line. Such genes are of particular interest since they define proteins that would represent ideal targets for inhibitory drugs.

As demonstrated by this example, in the hands of skilled researchers with fertile imaginations, RNAi techniques can be exploited to provide rich functional information with respect to a wide variety of biological processes. Many other fascinating examples can be found by browsing the primary scientific literature.

In developmentally complex organisms where the frequency of homologous recombination is low, RNA interference can be used to silence or knock down the expression of target genes to create loss-of-function alleles. Such strategies have been successfully used in high-throughput, genome-wide screens to define gene function with respect to a variety of different phenotypes in many different experimental models.

22.4 Synthetic Genetic Arrays: Understanding the Principles of Genome Architecture

In addition to simply defining gene function, the global analysis of loss-of-function mutations has also provided insight into the underlying genetic architecture of the genome. By architecture, we are referring not to any physical or organizational attributes of DNA or chromatin, but rather a far more abstract concept related to the idea of genetic buffering. As we shall see in this section, the merging of this idea with a technique referred to as synthetic genetic array (SGA) analysis has revealed a previously unimagined level of genome organization. While this level of organization is entirely abstract, it nonetheless has very practical and profound implications with respect to our understanding of the genotype–phenotype relationship.

Synthetic lethality and genetic buffering

To understand these concepts we must start by first describing the phenomenon of synthetic lethality. Synthetic lethality arises when the combination of two distinct mutations—that when present in isolation have no lethal effect—results in inviability of the organism. For example, imagine two yeast genes (let us call them "A" and "B") in which loss-of-function mutations have been isolated. Haploid wild-type cells (i.e., cells of the genotype $A^+ B^+$), cells bearing only the A^- allele, and cells bearing only the B^- allele are viable. However, cells bearing both the A^- and B^- alleles are inviable. In this case, the A^- and B^- alleles are said to display a synthetic lethal interaction (**Figure 22.12a**).

The critical point with respect to our current discussion is that synthetic lethality identifies "buffering" relationships. Such a relationship means that an organism with full function of one of the two genes, and partial function of the other, remains viable. Thus, each gene buffers variation in the other. In other words, if A were fully functional, then the activity of B would be nonessential for life. Therefore, variation in the B gene (resulting in reduced activity of the encoded gene-product) could be tolerated by the organism in question (**Figure 22.12b**).

While the above ideas appear obvious and simplistic, they have profound implications for our understanding of the genotype–phenotype relationship. For instance, imagine a gene B that encodes an enzyme whose activity varies from low to high across a number of distinct alleles in a population. If these alleles were present in an individual where A was fully functional, then the phenotypic effects of the changes in activity of B would be buffered by the presence of A, and would not result in an overt phenotype. If A was also compromised, then no buffering would be possible and changes in the activity of B would be far more likely to result in an overt phenotype (**Figure 22.13**). In the final analysis, the phenotype observed in any one individual results from the complex

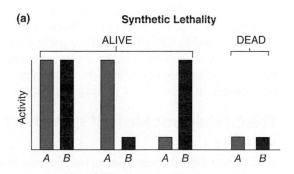

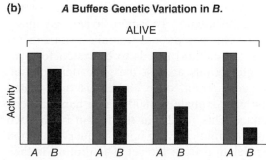

Figure 22.12 Synthetic lethality. **(a)** The presence of both the A and B loss-of-function alleles (in the same individual) results in synthetic lethality. **(b)** Such an interaction identifies a buffering relationship between A and B. This is to say, in the presence of the wild-type A gene, variation in B can be tolerated.

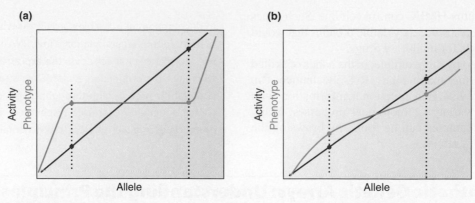

Figure 22.13 Genetic buffering. In both cases shown, the activity of the hypothetical *B* gene product (*red*) varies across a wide number of distinct alleles present in the population. **(a)** The *A* gene product is fully functional and thus effectively buffers the variation in *B* and no overt phenotype (*blue*) is realized. **(b)** The *A* gene product is nonfunctional. It is no longer capable of effectively buffering the variation in *B* and phenotypic changes are observed as a function of the activity of the *B* gene product.

interaction of two genes and, therefore, cannot be determined (or predicted) by the analysis of a single locus. It is important to note that such a phenomenon would be particularly important in outbred populations where individuals exhibit high degrees of genetic variation at many different loci.

How truly important are these hypotheses with respect to understanding the genotype–phenotype relationship? While an interesting idea, experimental evidence supporting the paradigm was lacking until recent technical advances were made in the field of functional genomics. In the next section, we will discuss the advent of synthetic genetic array technology and see how it can be used to expand upon, and to validate, the ideas presented above.

Synthetic genetic arrays

Synthetic genetic arrays (first developed and used by University of Toronto researcher Charlie Boone) represent an attempt to directly test the importance of genetic buffering relationships on phenotype. Due to the availability of a genome-wide set of haploid gene deletion mutants, initial experiments were performed using the budding yeast, *S. cerevisiae*. While the genetic selection schemes employed by SGA are quite sophisticated and complex (**Figure 22.14**), the overall strategy of the method is straightforward. To begin, a haploid query strain is chosen. This haploid query strain bears a mutation (e.g., a gene deletion) that is to be tested for synthetic lethal interactions against the gene deletion set. The haploid query strain, once chosen, is then mated to each of the approximately 4800 viable haploid gene deletion mutants. This results in the fusion of the two cells, creating a diploid carrying both the query mutation and one gene deletion from the set. The diploid is then sporulated and selection applied to specifically isolate only haploid progeny that bear both mutations. Growth (if any) of the progeny is then measured and the synthetic genetic interactions are identified. Performing such a large-scale experiment is not a trivial exercise. Such

studies are typically performed with the help of the same robotic technology discussed in Chapter 21. This technology automates and greatly simplifies the manipulation and handling of the large number of strains that are employed in these experiments.

At the end of this process, one is left with an agar plate upon which the haploid double mutants have been given the opportunity to grow. Typically, all double-mutant combinations are tested in duplicate. In **Figure 22.15** you can see an example of just such a plate. In most cases, two large, circular yeast colonies form for each genotype tested. These represent double-mutant combinations that do not result in a synthetic growth defect. In a small number of other cases, no growth is observed. This is because these particular mutant combinations are synthetically lethal, preventing the cells from growing and forming colonies. In other cases, colonies form, but they are much smaller than normal; these represent mutant combinations that, while not lethal, result in significantly hampered growth. These strains are often referred to as being synthetic sick as opposed to synthetic lethal. In any event, all synthetic interactions between the query mutation and each of the deletion mutants can be uncovered and characterized.

In the following section, we will examine what these interactions can tell us about gene function at the molecular level. Finally, we will assimilate these ideas to reveal a previously unrealized level of abstract genetic architecture that fundamentally affects the genotype–phenotype relationship in populations of individuals.

The mechanistic basis of synthetic lethality

Why would a given pair of gene deletions confer a lethal phenotype when present together in the same individual? While difficult to know for certain in the absence of detailed molecular analysis, one likely scenario is that the two gene-products function in parallel pathways that both impinge upon the same essential function. This is

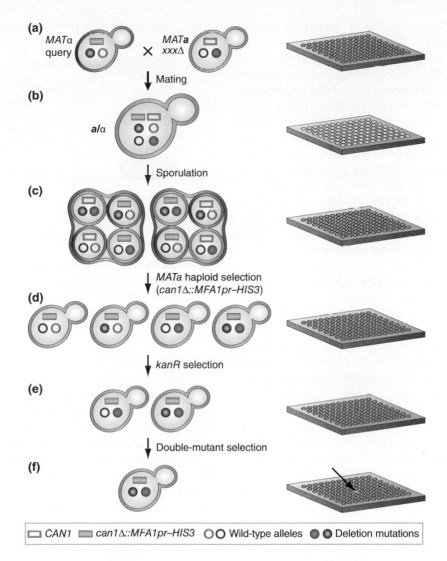

Figure 22.14 Synthetic genetic array methodology. (a) An α-mating-type strain carrying the query mutation (*filled black circle*) is crossed to an array of *a*-mating-type deletion mutants (*xxx*Δ). In each of these deletion strains, a single gene is deleted using the *kanR* selectable marker (*filled blue circle*). **(b)** The resultant heterozygous diploids are sporulated, resulting in the formation of haploid progeny. **(c)** Spores are then transferred to a growth medium that selects for *a* haploids. **(d)** The *a* meiotic progeny are transferred to a medium that contains geneticin, which selects for the query mutation. **(e, f)** The remaining cells are then transferred to the final growth medium that selects for the double-mutant cells.

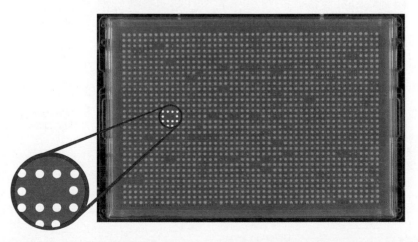

Figure 22.15 Growth assay of double mutants. All double-mutant combinations (in duplicate) are plated on agar growth media and incubated for three days (to allow the cells to grow and form colonies). Most mutant combinations are viable, resulting in the formation of two large, circular yeast colonies for each genotype tested. Other mutant combinations display a synthetic lethal interaction (*red circle*) and are not able to form colonies. In this way, all synthetic lethal interactions between the query and the genome-wide set of deletion mutants can be identified.

referred to as the between-pathway model of genetic interaction (**Figure 22.16**).

For example, the two pathways shown in Figure 22.16 are each made up of three gene-products that are required for an essential biological function (e.g., DNA replication). Fortunately for the cell, if one of these pathways is rendered incapacitated, there will still be enough activity present from the alternate pathway to ensure that the essential biological function takes place. However, if both pathways are incapacitated, then the essential biological function is abolished and the cells will be unable to proliferate. Gene deletion of any of the components of the A pathway, together with deletion of any of the components of the B pathway (indicated by the lines with no arrowheads), would confer a lethal phenotype.

Importantly, the between-pathway model of genetic interaction—together with the use of hierarchical clustering techniques—can be applied to identify the components of functional pathways. For example, **Figure 22.17** shows a small subset of the synthetic lethal interactions identified in a budding yeast study. To the right of the chart are listed the query genes that were screened. Below the chart are listed the genes from the deletion array that, when combined with the indicated query mutant, conferred a synthetic lethal interaction (*red squares*).

Essentially, the clustering algorithm has taken the data and organized it so that genes with similar patterns of genetic interactions are grouped together; for example, members of the dynein–dynactin pathway group together with one another based on their interaction with genes involved in microtubule dynamics. In this way, one can use synthetic lethal profiles to group genes into functional groups that are involved in similar biological processes. Significantly, such data can be used to assign function to uncharacterized genes. For example, the function of the *DYN3* gene (then only known as *YMR299c*) was uncharacterized at the time of the study. However, since its profile placed it together with other members of the dynein–dynactin pathway, researchers naturally hypothesized that

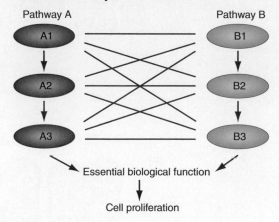

Between-Pathway Genetic Interactions

Figure 22.16 The between-pathway model of genetic interaction. Individuals bearing severe loss-of-function mutations in both pathway A and pathway B would be inviable since at least one of the two pathways is required for an essential biological function. Such a model may represent the mechanistic basis for many of the synthetic lethal interactions observed in the budding yeast.

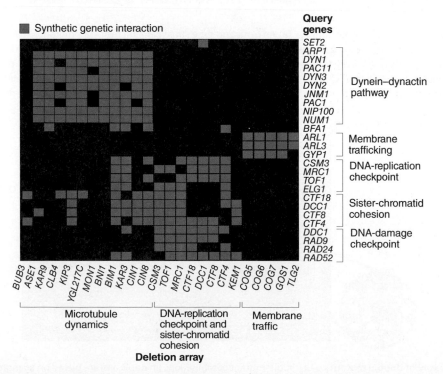

Figure 22.17 Hierarchical clustering of synthetic lethal genetic interactions. A small subset of the genetic interactions identified from a budding yeast study are shown. The clustering algorithm organizes the query and array genes into groups (displaying similar profiles) that define components of specific functional pathways.

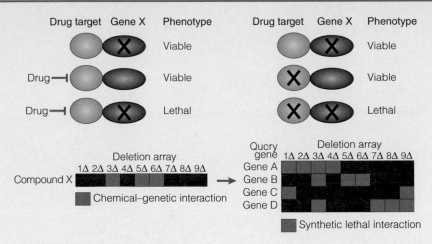

Figure A Identifying drug targets.

Can synthetic genetic array technology be applied to identify drug targets? SGA inventor Charles Boone and colleagues at the University of Toronto think so. Their methodology, referred to as chemical SGA, rests on the principle that cells treated with a drug that inhibits the function of a specific gene will display the same (or highly similar) SGA profile as cells deleted for this gene. Consider the following experiment in which each budding yeast gene deletion mutant is treated with compound X. In this hypothetical scenario (**Figure A**), treatment with the compound is lethal to deletions #3, #5, and #6 (*bottom left*). In addition to this observation, assume that the researchers have also determined that query gene B from the deletion set displays the identical synthetic lethal profile (i.e., it is lethal in combination with gene deletions #3, #5, and #6). Based on this similarity, it can be inferred that gene B is the likely target of compound X. This strategy has shown great potential in uncovering the molecular targets of a wide variety of drugs (both therapeutic and nontherapeutic). Why is this knowledge so valuable? Essentially, knowing the targets of a drug makes the process of drug discovery much more efficient. By identifying the targets associated with a particular small molecule, one can better understand its biological role within the cell. This in turn can lead to better functional assays (faster optimization of the drug for clinical use) and a better understanding of the drug's side effects.

it must also function together with these genes. Remarkably, further molecular and genetic analyses clearly showed that this was indeed the case (see the **Focus on Inquiry** box **"Chemical SGA: Identifying Drug Targets"**).

Genetic interaction maps

In addition to hierarchical clustering, SGA data can also be organized into genetic interaction maps (**Figure 22.18**). In such a map, genes are represented as circles (referred to as the nodes of the graph), and synthetic lethal interactions are defined by lines (referred to as the edges or links of the graph). A line connecting two nodes indicates that this pair of genes exhibits a synthetic lethal interaction. Interestingly, a somewhat unexpected pattern reveals itself when the data are examined in this way. Instead of a random set of interactions, where the majority of nodes have an intermediate number of links (**Figure 22.19a**), such maps display what is referred to as a scale-free architecture. This type of architecture is characterized by the majority of nodes having only a few links, and by only a few nodes having a large number of links (**Figure 22.19b**). This latter group of nodes are referred to as being the hubs of the

genetic interaction network (not to be confused with the protein interaction hubs described in Chapter 21).

Now let us think back to the beginning of the section where the concept of genetic buffering was introduced. Does the abstract organization of genes into a scale-free network have any consequences with regard to genetic buffering—and thus the genotype–phenotype relationship? First, consider a hub gene that has been mutated so that its gene-product has little or no activity. If there is a loss of function in *just one* of its many different interactors, this will result in a phenotype. Thus, the loss of a hub gene makes an individual particularly sensitive to the effects of variation in a wide subset of interacting genes. This is because, with the loss of the hub, the individual has lost the capacity to buffer the variation. In contrast, variation in a nonhub will result in a phenotype *only* if the hub is also abrogated. In the final analysis, these data suggest that with respect to the genotype–phenotype relationship, not all genes are created equal. Hub genes, due to their central role in buffering variation, may play a much more important part in determining phenotype than nonhub genes. In the next section, we will explore these ideas further by

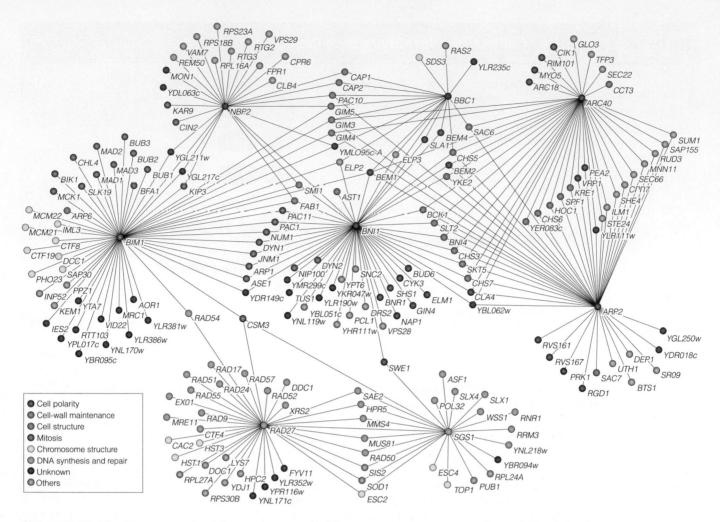

Figure 22.18 A budding yeast genetic interaction map. Budding yeast genes are represented as *circles* (nodes) and interactions as *lines* (edges or links) that connect the nodes. Shown are 291 interactions between 204 genes. Interestingly, the map displays a scale-free architecture.

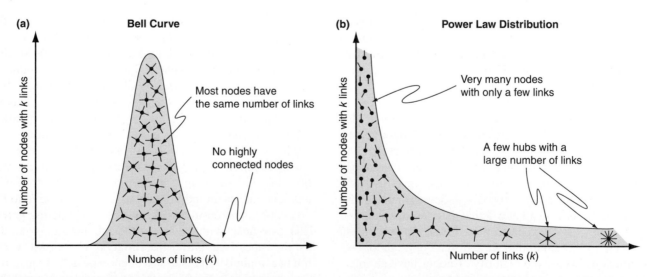

Figure 22.19 Random networks versus scale-free networks. (a) Randomized networks display a normal distribution of links (i.e., most nodes have an intermediate number of links and few nodes have either a large or a small number of links). **(b)** Genetic interaction networks, on the other hand, display a scale-free architecture (i.e., most nodes have only a small or intermediate number of links, while a few nodes, called hubs, possess a large number of links).

taking a look at SGA analysis in a more developmentally complex model system, the nematode *C. elegans*.

Synthetic genetic arrays in *C. elegans*

C. elegans is a free-living, diploid roundworm with a genome comprising 100 million bp and encoding approximately 20 000 genes (**Figure 22.20**). Unlike for budding yeast, a comprehensive collection of deletion mutants is not available. Thus, the method of choice for SGA analysis in this organism involves RNAi. In one particularly interesting example of SGA in this organism, 37 different mutant *C. elegans* lines were chosen as queries. These mutations affected a wide variety of processes ranging from development to morphology. Each mutant line was screened against a panel of 1860 RNAi constructs. A total of 68 820 (37 × 1860) crosses were performed. From these crosses, 349 synthetic lethal interactions were uncovered and characterized. Similar to budding yeast, the gene interaction network map from this experiment exhibited a scale-free architecture. Furthermore, the researchers were able to identify six hub genes. Thus, the same principles of genome architecture uncovered

Figure 22.20 The roundworm, *Caenorhabditis elegans*.

in the simpler budding yeast applies to the more developmentally complex *C. elegans*.

Importantly, in addition to the basic analysis described above, the researchers also directly tested the hypothesis that hub genes play a critical role in determining phenotype (in a general sense). To this end, they knocked down expression of each of the identified hub genes in (1) a *vab-1* mutant background, or (2) a *dpy-20* mutant background. *vab-1* encodes a kinase involved in neural morphogenesis. Loss-of-function mutations in this gene result in an abnormal notched-head phenotype that is approximately 30 percent penetrant (**Figure 22.21**). *dpy*, on the other hand,

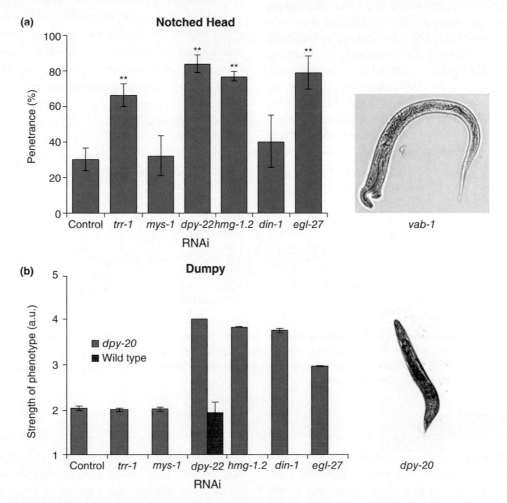

Figure 22.21 Hub genes buffer diverse phenotypes. (a) Knocking down the expression of hub genes strongly enhances the penetrance (percentage of animals displaying the phenotype) of the notched-head phenotype of *vab-1* mutants. (b) Knocking down the expression of the same group of hub genes also strongly enhances the dumpy phenotype of *dpy-20* mutants.

encodes a transcription factor with a role in the determination of body morphology. Again, similar to the *vab-1* mutant, the *dpy-20* mutation used in this study is not fully penetrant. Remarkably, the majority of the identified hub genes (four out of six in both cases) greatly exacerbated the phenotype associated with either the *vab-1* or *dpy-20* mutation (i.e., increased the phenotype's penetrance/strength). As predicted by our earlier model, these data demonstrate that hub genes do indeed have the capacity to buffer variation. Furthermore, this buffering can effect multiple, completely disparate phenotypes. As illustrated in **Figure 22.22**, these data imply that hub genes can indeed act as modifiers in a wide variety of unrelated phenotypes.

Our examination of these studies reveals two important and fundamental observations regarding our underlying genetic architecture. One of these observations could be classified as good news and the other as bad news. The bad news is that any given phenotype is far more likely to be caused by the interaction of gene variants than from mutation in a single gene. For instance, the synthetic lethal interaction rate in budding yeast is approximately 0.6 percent. Thus, there are $6000 \times 6000 \times 0.006$ or approximately 216 000 possible mutant combinations that result in a lethal phenotype. In contrast—based on analysis of the deletion set—only about 1000 genes confer lethality when deleted in isolation. Therefore, when considering the phenotype of viability, there are 216 000 paths leading to a lethal phenotype through digenic interactions, but only 1000 paths when considering a single loci. These simple calculations imply that complex phenotypes (like those associated with disease) are more often caused by combinations of mutations in many different genes (as opposed to a single gene). As you have probably realized, this greatly complicates the pursuit of genetic analysis aimed at identifying and studying disease genes.

The good news is that according to this model, there is only a small set of hub genes that can act as modifiers. In contrast to the bad news, this observation simplifies the situation. This is because our idea implies that only a relatively small number of genes (the hubs) will be key players in determining the strength of complex phenotypes. Therefore, researchers can focus their attention on this particularly important class of genes.

In the context of human disease, hubs are referred to as modifiers, and nonhubs as specifiers (since they confer the disease-specific phenotype that is modulated by the hub). As these ideas are still in their infancy, it will be interesting to see if they can indeed be applied to both the understanding and treatment of human disease. In any event, it is clear that the detailed analysis of functional genomics data will continue to strengthen our understanding of the relationship between genotype and phenotype.

Synthetic genetic array analysis is a technique used to identify synthetic lethal genetic interactions on a genome-wide scale. This analysis has revealed that gene interaction networks display a scale-free architecture in which only a few genes (hubs) possess a large number of interactions. This architecture implies that hub genes are particularly important in determining complex phenotypes since they "buffer" the genetic variation in many other genes.

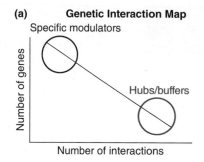 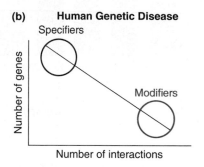

Figure 22.22 Specifiers and modifiers in human genetic disease. **(a)** *C. elegans* genetic interaction maps contain many genes that interact with only a few other genes (specific modulators) as well as highly connected genes (hubs). **(b)** Human genetic interaction maps are likely to display a similar architecture. While mutations in most genes will act as disease-specific alleles (specifiers), mutations in a small number of hubs might act as modifier genes in multiple, unrelated diseases.

Connections

In Chapter 2, we learned that genes determine traits. For example, in the 1860s Mendel was able to infer the existence of a gene that affected pea shape. However, it was not until more than a century later that the function of this gene's protein product in converting unbranched to branched starch was elucidated (see the Fast Forward box on "Genes Encode Proteins" in Chapter 2). As we saw in this chapter, it is now possible (at least in model organisms) to define gene function on a genome-wide scale through "knocking out" or "knocking down" gene activity. As with many recent advances in genetics, this capacity to systematically analyze gene function was made possible by genome sequencing projects (Chapter 19). If Mendel were alive today, he would without a doubt be awed and amazed at how his ingenious experiments in pea plants laid the foundation for these fantastic accomplishments.

1. The discipline of functional genomics emerged in the post-genomic era. It aims to define the function, regulation, and interaction of all genes in a genome. **[LO1–4]**

2. The creation of a genome-wide set of gene deletion mutants in budding yeast, together with the advent of genetic barcoding, has allowed researchers to map the function of all nonessential budding yeast genes to any phenotype of interest in a process called whole-genome parallel analysis. **[LO1–2]**

3. RNA interference is a useful genetic tool that can be used to knock down the expression of any given gene to create a loss-of-function allele. By scaling up the process to the genome-wide level, it is now possible to determine the function of genes with respect to any observable phenotype in a wide variety of complex, multicellular, experimental organisms. **[LO1, LO3–4]**

4. Synthetic genetic array analysis identifies genetic buffering relationships through the characterization of synthetic lethal interactions. This analysis has demonstrated that genetic interaction maps exhibit a scale-free architecture. This suggests that a small number of hub genes may be critical in determining the genotype–phenotype relationship with respect to a variety of seemingly unrelated biological processes. **[LO1, LO3–4]**

Solved Problems

I. You are a researcher studying how yeasts respond to high concentrations of salt. You begin by comparing gene expression levels in a haploid wild-type yeast strain grown under normal conditions and in media containing high salt concentration. You identify a group of 15 genes that are significantly up-regulated in response to high salt, and another group of 12 genes that are down-regulated in response to high salt. How could you follow up on these data to determine if any of these genes are indeed required to respond to high salt stress, or if changes in their expression are only correlated to salt exposure?

Answer

The simplest way to proceed would be to create *kanR*-marked gene deletion mutants (via homologous recombination in a wild-type diploid strain) for each of the 27 genes. Alternatively, you could also purchase ready-made strains from a commercial company. Each of the 27 heterozygous diploids could then be sporulated. Strains showing 2:2 segregation of viable geneticin-sensitive spores to inviable spores would indicate that the knocked-out gene was essential (making analysis of the knockout strain impossible). Strains showing 2:2 segregation of viable geneticin-sensitive spores to viable geneticin-resistant spores would indicate that the knocked-out gene was nonessential and could then be studied further. Once geneticin-resistant haploid segregants are isolated from each of the remaining strains, one could analyze the growth rate of each individual gene deletion mutant in comparison to a wild-type control as a function of salt concentration (from very low concentrations that are tolerable to wild type, all the way to very high concentrations that prevent the growth of wild-type strains). Haploid gene deletion mutants that exhibit hypersensitivity (or resistance) to salt would thus identify the genes from the set that did indeed have a functional role in responding to salt stress.

Problems

Vocabulary

1. For each of the terms in the left column, choose the best matching phrase in the right column.

   i. functional genomics   1. a unique 20-bp-long DNA sequence identifier that flanks the selectable marker of a deletion cassette

   ii. gene knockout   2. the study and understanding of the function, regulation, and interaction of all gene products in a genome

   iii. barcode   3. a technique by which an organism's gene is made nonfunctional by having an entire open reading frame replaced by a selectable marker

Section 22.1

2. Compare and contrast functional genomics to the traditional genetic analysis described in earlier chapters.

3. What is meant by describing a gene as essential? Describe an experimental methodology that could be used to define all the essential genes present in the genome of a model organism.

Section 22.2

4. Genes encoding heat-shock proteins are important in protecting cells from excessively high temperatures as well as a variety of other stresses. Imagine you are a geneticist studying the heat-shock response in the budding yeast, *Saccharomyces cerevisiae*. Describe an experimental method by which you could quickly and easily identify

all nonessential budding yeast genes required for survival under conditions of high temperature stress. (*Note:* Wild-type budding yeast grow normally at 25°C, but exhibit a strong heat-shock response at 42°C.)

5. You are the leader of a research team interested in elucidating how flower shape is determined in *Arabidopsis*. To begin, the team decides to determine what genes may be involved. One team member suggests conducting a forward genetic screen, while another suggests a reverse genetic approach. Discuss the benefits and drawbacks of each approach. Explain the reasoning behind your final choice.

6. Listed below are three types of commonly used micro-arrays. For each type of array listed, provide (i) a definition, and (ii) an example of an application in which each is commonly used.
   a. 3′ exon expression array
   b. intergenic gene chip
   c. barcode array

Section 22.3

7. You are a researcher interested in understanding the genes required for RNA interference in *C. elegans*. However, creating a gene deletion mutant library for this organism would be too expensive and time-consuming. Would it be possible to use a genetic approach based on gene silencing through RNAi to identify silencing genes? If so, explain your experimental design.

8. Upon the sequencing and annotation of the genome of the fission yeast *Schizosaccharomyces pombe*, a significant number of genes were found that had no known function (and did not display any sequence similarity to any genes with known function). Using the tools of global genetic analysis and functional genomics, describe how you could go about elucidating the biological role of these genes.

9. Genetic interaction networks in all eukaryotes studied to date can be classified as scale-free. Explain what is meant by the term scale-free network in a biological context.

10. According to current models, mutations affecting the activity of hub genes are thought to play an important role in a wide variety of unrelated phenotypes. What is a hub gene? Explain why a hub gene is able to buffer an organism from the phenotypic consequences of mutation in many other loci whereas a nonhub gene cannot? Comment on the role of hubs in human disease.

Section 22.4

11. Compare and contrast the role of specifier and modifier genes with respect to current models of the development of human disease.

12. Design a methodological variant of SGA analysis that makes use of barcodes to quantitate growth.

CHAPTER **23**   Systems Biology

The complexity and dynamic nature of living things has led to the development of the "systems" approach to biology. As we shall see in this chapter, this approach has become an important tool in the analysis of biological processes.

Practitioners of the newly emerging field of systems biology attempt to define all the components of a biological system and understand how they function in conjunction with one another. This pursuit relies on the tools and strategies of genomics and proteomics, as well as sophisticated computational and mathematical tools. One main theme stands out in our overview of systems biology; by studying how the gene and protein components of a system function together, we can begin to understand how the system's interacting elements give rise to its emergent properties.

Learning Objectives

1. Define the discipline of systems biology. Contrast the systems approach to those used prior to the Human Genome Project.

2. Define the concept of emergence. Explain how this concept can be applied to systems already described in this text (e.g., the *lac* operon, the *trp* operon).

3. Illustrate how hysteresis emerges through positive feedback.

4. Describe the transcriptional activation (or repression) of a gene using the Hill equations.

5. Define a network motif. Discuss the functional significance of network motifs (e.g., the feed forward loop) with respect to their role in controlling transcriptional regulatory networks.

23.1   What Is Systems Biology?

A **biological system** is a collection of interacting elements that carry out a specific biological task. These elements might include molecules such as proteins, mRNAs, or metabolites, the control regions of genes, or perhaps the cells of a particular tissue. The elements rarely act independently; rather they often function in association with one another.

Systems biology seeks to describe the multiple components of a biological system and analyze the complex interactions of these components, both within the system and in relation to the components of other systems. Systems biology utilizes both bottom-up strategies, starting with large molecular data sets, and top-down approaches employing computational modelling

and simulations. The desired outcome is to trace complex observations of phenotype back to the digital core encoded in the genome. This approach is possible because the Human Genome Project (described in Chapter 19) has provided global data-gathering tools and genomic information on a scale never before available. These tools and data are central to the current practice of systems biology.

The following outcomes of the Human Genome Project serve as the foundation for the current practice of systems biology.

- High-throughput platforms for genomics and proteomics (Chapters 19–22) enable the acquisition of global, or comprehensive data sets of differing types of biological information (all genes, all mRNAs, all proteins, and so forth).
- Powerful computational tools make it possible to acquire, store, analyze, integrate, display, and model biological information.
- Studies of simple model organisms such as *E. coli* and yeast allow scientists to compile global data sets from experimental manipulations of less complex biological systems. From their analyses of these data, they can learn how to study systems biology in more complex organisms.
- Finally, comparative genomics allows scientists to begin to determine the logic of life for individual organisms and to discover how that logic has changed in different evolutionary lineages.

Four questions help guide thinking about biological systems

Four fundamental questions can be asked about a biological system:

1. What are the *elements* of the system?
2. What are the *associations* among the elements?
3. How do *perturbations* affect the system and other systems connected to it?
4. How do a system's elements, associations, and relation to changes in the biological context explain its *emergent properties*?

We discuss each of these briefly.

What are the elements of a system?

Defining the elements or components of a system lays the fundamental groundwork for further analyses. The objective is to identify the proteins, genes, metabolites, cells, tissues, and organs that are involved. Modern systems biology uses the data sets generated by genomic and proteomic tools described in Chapters 19–22 to identify genes, mRNAs, and proteins. Our ability to interrogate large basic data sets to find the components involved makes the systems approach possible.

What associations occur among elements?

What are the protein-protein interactions, protein-DNA interactions, and interactions between molecules in the system? The isotope-coded affinity tag (ICAT) and yeast two-hybrid analysis described in Chapter 21 are two of the tools that generate data on physical interactions. It is possible to depict the interactions of a system's elements in a graphical representation of a network where the nodes, or points, represent individual proteins and the connections represent physical interactions between the proteins (**Figure 23.1**). A series of such network graphs can reveal how systems change throughout the development of an organism or during physiological responses to changing environments.

What happens when the system undergoes perturbation?

Delineating the dynamic behaviour of systems is one of the central challenges of systems biology. One way to accomplish this is to ask how the relationships of the elements change when the system is subject to specific genetic or environmental perturbations.

Systems are most often studied in one of two biological contexts: within individual cell types or within an entire organism. The comprehensive knowledge of all the genes (and, hence, all the mRNAs and their predicted proteins) in a cell or organism permits the study of systems in their biological contexts. This broad and yet detailed study may allow predictions about the effects of a disease state or other abnormal change in one system, termed a perturbation, upon other systems.

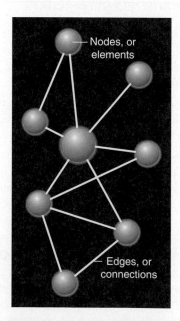

Figure 23.1 Representation of a biological network. The nodes (*blue circles*) of the graph may represent molecules such as proteins and metabolites or cells (like those of the immune or nervous systems). The *lines* connecting some nodes represent relationships between the elements.

What gives rise to a system's emergent properties?

An **emergent property** is one that arises from the operation of the system as a whole. Examples are the ability of the immune system to generate immune responses, the ability of the heart to pump blood, and the ability of a metabolic pathway to convert galactose into glucose. In some cases, an emergent property can be greater than the sum of individual properties of system components. An example is the action of the digestive system, in which many organs, tissues, and cellular components accomplish ingestion of food, absorption of nutrients, and elimination of wastes in a concerted, interdependent, and stepwise fashion.

Systems biology requires a cross-disciplinary approach

Systems biology is an immature science whose practitioners are still developing the tools and strategies of the discipline. To succeed in answering the basic questions just outlined, they must practise *cross-disciplinary biology*, in which teams of biologists, computer scientists, chemists, engineers, mathematicians, and physicists work together on common problems. Driven by the needs of systems biology, these teams must develop new high-throughput measuring instruments; use the new instruments to generate global data sets; and develop new computational tools to organize, annotate, analyze, integrate, and model the accumulated data.

New instruments and tools include more effective platforms for DNA sequencing and chip array technology that offer higher throughput, better quality data, greater sensitivity, and lower cost. Many of the techniques require the acquisition of large amounts of data and are best suited to nanotechnology. High-resolution techniques must be further developed to analyze single molecules and cells using advanced imaging techniques. More powerful computational and mathematical tools for storing, integrating, graphically displaying, analyzing, and mathematically modelling biological systems are needed. Scientists must work together to gather and annotate their data in compatible formats. The development and implementation of these new technologies demands that scientists of varying backgrounds learn to speak the language of biological science, and that biologists learn to speak the languages of physical and mathematical sciences.

Systems biology benefits from an algorithmic methodology

The modern systems approach to biology incorporates the following interconnected steps:

Step 1. *Scan the biological literature and databases* for all that is known about the system of interest. A key aspect of this search is access to the entire genome sequence of the organism under consideration because the discovery of all of the genes, RNAs, and proteins in a cell or organism, or as many as possible, is the foundation of systems analysis. Use the knowledge gained from discovery science and the literature to define the system of interest as best you can, identifying its elements, their relationships, and their contextual changes.

Step 2. *Develop a preliminary model* about how the system functions. This model may be descriptive (words), graphical (network diagrams), or mathematical, depending on how much information is available.

Step 3. *Formulate a hypothesis-driven query* about the model, and answer this query through genetic or environmental perturbations of the system. These perturbations may include alterations ranging from gene knockouts to changes to the environment such as the addition of nutrients, or a combination. In conjunction with these perturbations, collect comprehensive data sets from different levels of biological information (DNA sequence, mRNA levels, protein levels, protein-protein or protein-DNA interactions, and so forth).

Step 4. *Integrate different types of data* either graphically or mathematically and compare the results against the initially formulated model. Disparities will likely arise between the new experimental data and predictions based on the original model. Formulate new hypotheses that seek to resolve these discrepancies.

Step 5. *Perform iterative perturbations.* To test the new hypotheses, design a second round of genetic and environmental perturbations that will generate new global data sets whose integration will make it possible to resolve the discrepancies. Repeat Steps 3–5 until model and experimental data are in accord.

Step 6. *Evaluate whether the refined final model enables biologists to predict the behaviour of the system*, even with perturbations that have never before been tested. An accurate model should explain the emergent properties of the system and allow prediction of new emergent properties.

Having provided an overview of the systems biology approach, and a summary of our algorithmic methodology, we will now take a look at a biological system that you already know very well—the *lac* operon—and examine it through the eyes of a systems biologist.

- Systems biology is guided by questions regarding (1) system components, (2) their associations, (3) the effects of changes to the system, and (4) the emergent properties of a system.

- Because complex biological systems encompass so many aspects, a cross-disciplinary approach is needed.

- Systems biology employs both discovery science and hypothesis-driven science. It requires both the *acquisition of global data sets* from different levels of biological information and the graphical or mathematical *integration of different types of data*.

23.2 A Familiar Example: The *lac* Operon as a Logical Circuit

As you may remember from Chapter 10, the *lac* operon is a transcriptional unit in *Escherichia coli* that enables the coordinated regulation of the *lacZ*, *lacY*, and *lacA* genes in response to the availability of lactose and glucose. The operon was first studied by François Jacob and Jacques Monod in the 1950s as part of their Nobel Prize winning research on transcriptional regulation. Since that time, the *lac* operon has remained the subject of much curiosity and has been extensively examined by scientists interested in molecular biology and genetics. The *lac* operon represents one of the most well-studied and best understood biological systems. To familiarize you with the systems approach, we will now take another look at the *lac* operon. This time, however, we will examine it through the eyes of a systems biologist.

Step 1. Scan the biological literature and databases.

The study of the *lac* operon provided the cornerstone of our understanding of the process of gene regulation. For this reason, the *lac* operon represents one of the most thoroughly studied biological systems of the last 70 years. With the possible exception of the *lacA* gene product, the components of this system, and their biological roles, have been meticulously documented in the primary research literature. As shown in **Table 23.1**, it is a trivial exercise to summarize what is known about the *lac* operon by constructing a table listing its relevant components and their functions. In a similar manner, the interaction of all these components—protein-DNA and allosteric interactions—can also be easily summarized (**Table 23.2**).

Step 2. Develop a preliminary model.

Based on Tables 23.1 and 23.2, we can now construct a graphical representation of the system by placing the components into a pathway diagram (**Figure 23.2**). The system can be clearly viewed as being composed of two independent regulatory layers: one positive and the other negative. The negative layer functions through allolactose and its interaction with the *lacI* repressor (i.e., when allolactose is present, it binds to the repressor, inhibiting its ability to bind the *lac* operator sequence). The positive layer functions through glucose and its effects on cAMP levels (i.e., in the absence of glucose, high levels of CRP–cAMP complexes accumulate and then bind to *cis*-regulatory sequences in the *lac* operon promoter).

With the aid of this graphical representation, making predictions regarding the behaviour of the system becomes trivial. For instance, we know that high levels of *lacZ* transcription will be possible only if (1) allolactose is present (relieving inhibition of the *lacI* repressor), and (2) if high levels of cAMP are available (to promote CRP-dependent activation of transcription). We could represent this graphically as shown in **Figure 23.3**, where transcription is represented as a function of both allolactose and cAMP concentration. Furthermore, the level of transcription is represented by colour (*red* indicating high levels of transcription and *blue* representing low levels of transcription). The output of the system (*lacZ* transcription) is dependent on two inputs: allolactose and cAMP. Furthermore, output from the system is only realized when both inputs are present. Such a system formally defines a binary logical circuit called an AND gate.

Table 23.2	Interactions Between the Components of the *lac* Operon.
Protein-DNA Interactions	**Allosteric Interactions**
Repressor–Operator	Allolactose–Repressor
CRP-cAMP–Promoter	cAMP–CRP

Table 23.1	Components of the *lac* Operon and Their Biological Roles.
Part	**Functional Role**
β-Galactosidase	Lactose metabolism
Lactose permease	Import of lactose
lacI repressor protein	Transcriptional repressor
Operator	Binds *lacI* repressor protein
Allolactose	Allosteric regulation of *lacI* repressor protein
CRP	Transcriptional activator
cAMP	Allosteric regulator of CRP
CRP-binding site	Binds CRP–cAMP complexes
Glucose	Reduces cAMP levels

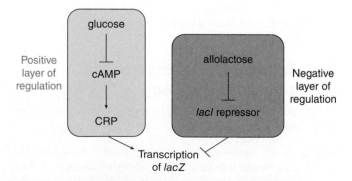

Figure 23.2 Pathway diagram describing the *lac* operon. The *lac* operon can be represented as a simple pathway diagram with two branches: a positive branch that stimulates transcription when glucose is absent, and a negative branch that induces transcription in the presence of lactose.

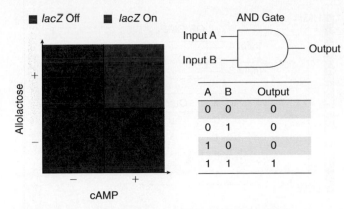

Figure 23.3 The *lac* operon can be thought of as a binary logical circuit. Transcription of the *lac* operon is stimulated only in the presence of both cAMP and allolactose. This system defines an AND gate.

Step 3. Formulate a hypothesis-driven query.

Based on our graphical analyses, we can now formally state our hypothesis regarding the emergent properties of the system. Remember, an emergent property is one that arises from the function of the system as a whole. While perhaps somewhat obvious in this simple example, we can nonetheless theorize that the system functions as a binary logical circuit (i.e., that an AND gate emerges from the set of interacting parts).

Step 4. Integrate different types of data.

To test the hypothesis, we will need detailed data documenting the level of *lacZ* transcription as a function of both allolactose and cAMP concentrations. Luckily, just such a detailed analysis has been performed and the data set published. **Figure 23.4** shows the empirical data from this study displayed in a three-dimensional graph. The concentration of cAMP is graphed on the *x*-axis and the concentration of allolactose on the *y*-axis. The third dimension of the graph (i.e., the level of *lacZ* transcription) is represented by colour (according to the scale that appears to the right

of the graph). A deep red colour indicates high levels of transcription, yellow indicates an intermediate level of transcription, and blue indicates low levels of transcription. As you can see, the empirical data matches the prediction well. It is clear that high levels of allolactose or cAMP alone are not enough to realize high levels of transcription. However, high concentrations of both together are indeed sufficient to significantly stimulate the production of *lacZ* transcription.

Step 5. Perform iterative perturbations.

Currently, our model matches the data quite reasonably. However, like any good model, its extension can lead to further testable predictions. For example, if it is possible for an AND gate to emerge from interaction of the parts of this system, would it be possible—by modifying these same interactions—to change the logic of the circuit? To test this hypothesis, the researchers decided to take a genetic approach. They randomly mutated residues in the *cis*-control regions of the *lac* operon (**Figure 23.5**) and then analyzed the effects of these mutations on *lacZ* transcription. One particularly intriguing mutant—where the behaviour of the system was dramatically altered—is shown in **Figure 23.6**. In this incarnation of the system, *either* high allolactose *or* high cAMP concentrations resulted in high levels of *lacZ* transcription. In other words, the emergent property of the system was indeed altered through modulating the nature of the interactions. In this case, the AND gate was converted to an OR gate (i.e., output was realized only if one, or the other, input was present).

Step 6. Evaluate whether the refined final model enables biologists to predict the behaviour of the system.

Although it was possible for the researchers to convert the logical behaviour of the *lac* system from an AND gate to an OR gate, it was not possible to rationally design the mutant. While the researchers were able to screen a large number of randomly mutated strains to isolate a given circuit exhibiting the desired behaviour, they could not (beforehand) design a promoter sequence to produce the desired levels of *lacZ*

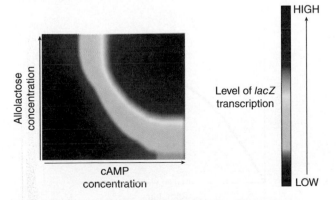

Figure 23.4 Levels of *lacZ* transcription in wild-type *E. coli.* The level of *lacZ* transcription was empirically determined as a function of both allolactose and cAMP concentrations. Transcript levels are displayed by colour according to the scale to the right of the graph.

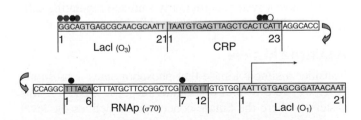

Figure 23.5 The *lac* operon promoter region. The location of point mutations (*dots*) in the *lac* operon promoter region. *Red dots* indicate mutations used in all variants. *Black dots* indicate positions that were change to A or T with equal probability. *White* indicates positions that were changed to A, C, or T with equal probability. Also shown are the positions of the two LacI-binding sites, O_1 and O_3, the CRP-binding site, and the RNAp-binding site (−10 and −35 regions). The *black arrow* indicates the transcription start site of the *lacZ* gene.

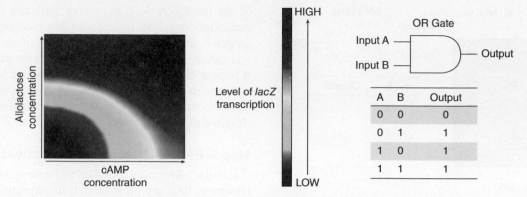

Figure 23.6 Levels of *lacZ* transcription in a mutant strain of *E. coli* plotted as a function of allolactose and cAMP concentrations. The mutations in this strain have converted the logic of the system from an AND gate to an OR gate.

transcript. Conversely, given a specific mutated promoter sequence, they could not accurately predict the transcriptional outcome. These failures highlight the great difficulty system biologists face when dealing with complex networks of interacting components. In the next section, we will take a look at another complex system in which the emergent properties are even more unexpected and difficult to predict.

> By applying the systems approach, it is possible to uncover the emergent properties of a system. In the case of the *lac* operon, the dynamic interactions of the constituents lead to the emergence of a binary logical circuit called an AND gate. Altering the nature of the interactions can change the emergent properties of the system.

23.3 A Not So Familiar Example: Hysteresis

We have learned that systems biologists choose to look at systems as a whole (in contrast to focusing on their individual parts). Part of the reason for this is that unexpected behaviours/properties can emerge from the set of interacting parts. In the case of the *lac* operon, the binary logic circuit that emerged from the system may have been obvious to you even before you began considering the example. However, in other instances, the behaviours that emerge are far more unexpected and far more difficult to predict. Our next example illustrates these points and shows that often times biological systems exhibit behaviours so complex that they are beyond human intuition. We will begin by examining a simple dynamic system and then see how its behaviour changes as new parts are added. In the end, we will have a system particularly suited to regulating cell-cycle transitions.

A simple kinase

Consider a simple kinase that phosphorylates a substrate, A. If we were to plot a stimulus response curve for this system—where the kinase concentration defines the level of stimulus, and the level of phosphorylated A defines the response—a hyperbolic curve would be generated like the one shown in **Figure 23.7**. (An increase in the stimulus results in a relatively steady increase in the phosphorylation of A to an upper limit.)

Next, consider the consequences of complicating the system by adding a protein, I, that is able to bind the kinase and inhibit its function. Now the free (active) kinase

exists in equilibrium with the inhibitor-bound form. In this scenario, the stimulus response curve is ultrasensitive (**Figure 23.8**). At low (or high) stimulus levels, increases in the stimulus result in only small effects on the level of response. However, at intermediate stimulus levels, changes in the stimulus result in much more drastic effects on the level of response. Thus, the system is acting more or less like a switch.

At this point in the development of our system, it is still possible to use intuition to see how this behaviour emerges. Early on in the curve (when the concentration of the kinase is low), there is sufficient inhibitor in the system

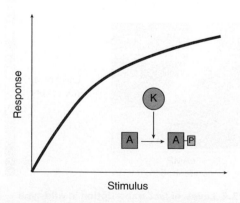

Figure 23.7 A stimulus response curve describing the phosphorylation of a substrate, A, by a kinase, K. The concentration of kinase defines the level of stimulus. The concentration of phosphorylated substrate defines the response.

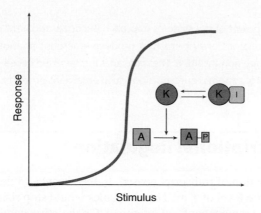

Figure 23.8 A stimulus response curve describing the phosphorylation of a substrate, A, by a kinase, K, in the presence of a kinase inhibitor, I. The concentration of kinase defines the level of stimulus. The concentration of phosphorylated substrate defines the response.

to bind the kinase and prevent it from phosphorylating A. As the midpoint of the curve is approached, however, kinase levels begin to approach, and then exceed, the level of the inhibitor (resulting in more and more free kinase and a rapid phosphorylation of A). Lastly, towards the end of the curve, there is more than enough free kinase to phosphorylate all of the A in the system. Thus, increases in the concentration of the kinase have little effect.

In our next scenario, we will complicate the system further by adding a layer of positive feedback. In this case, we will assume that the phosphorylated substrate, A, promotes the conversion of the inhibitor-bound kinase to free (unbound) kinase. The action of the kinase creates more phosphorylated substrate, which in turn creates more free kinase, which in turn creates more phosphorylated substrate, etc. At this stage of the development of our system, it exhibits an intriguing behaviour known as bistability (**Figure 23.9**). The system—due to the presence of the positive feedback loop—goes from little, or no, response to a strong response so quickly that no stable intermediate state

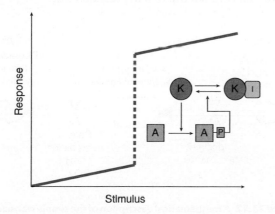

Figure 23.9 A stimulus response curve describing the phosphorylation of a substrate, A, by a kinase, K, in the presence of a kinase inhibitor, I, and positive feedback from phosphorylated A. The concentration of kinase defines the level of stimulus. The concentration of phosphorylated substrate defines the response.

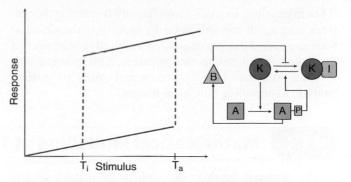

Figure 23.10 A stimulus response curve describing the phosphorylation of a substrate, A, by a kinase, K, in the presence of a kinase inhibitor, I, and both direct and indirect positive feedback from phosphorylated A. The concentration of kinase defines the level of stimulus. The concentration of phosphorylated substrate defines the response.

exists. In other words, the system will always either be in an ON state or in an OFF state.

Our last scenario complicates the network further by adding a second layer of positive feedback. Now, not only does the phosphorylated substrate promote dissociation of the kinase from the inhibitor, but it also prevents kinase reassociation to the inhibitor through an intermediate, B (**Figure 23.10**). In this instance, a difficult to predict and non-intuitive behaviour referred to as **hysteresis** is exhibited by the system. The system responds to the stimulus in a manner that is dependent on its past history. For example, at a given level of stimulus it is possible for the system to be in either of two unique states; the choice of which is determined by what state the system was in previously. If the system was initially in an inactive state (i.e., low stimulus/low response), the stimulus concentration would need to rise to the activation threshold (T_a) before the system could turn itself ON. If the system was in an active state (i.e., high stimulus/high response), the stimulus levels would need to decrease to a level *below* the T_a for the system to turn itself OFF. This level is called the inactivation threshold (T_i). Unlike the bistable system shown in Figure 23.9—where the activation and inactivation thresholds are equal—in a hysteretic system, the inactivation threshold is below that of the activation threshold. The consequence of such an architecture is that the system requires more stimulus to turn itself ON than it does to *keep* itself ON. For this reason, such systems are referred to as molecular ratchets (since they allow for movement in one direction, but not back).

Interestingly, such architectures are commonly observed in regulatory networks controlling cell-cycle transitions; a situation in which reversal would have disastrous consequences for a cell. (Imagine the chaos that would ensue if chromosomes decondensed in the midst of anaphase.) In such cases, once an event is initiated, the cell benefits from committing itself to complete the process.

In the final analysis, just like the binary logic circuits that emerged from the *lac* operon, a molecular ratchet emerges from the system diagrammed in Figure 23.10. Unlike in the case of the *lac* operon, the behaviour that emerged was not so obvious and would have been difficult,

if not impossible, to predict from simply inspecting the network diagram. It was only upon the analysis of the stimulus/response curves that the unusual, non-intuitive behaviour of the molecular ratchet became apparent. This example thus serves to illustrate the importance and utility of systems analysis in providing biological insight.

> The power of the systems approach becomes apparent upon the analysis of systems that produce emergent phenotypes that are non-intuitive. The molecular ratchet used in cell-cycle control systems represents one such emergent property.

23.4 Mathematical Modelling of Transcriptional Regulation

In the previous section, our ability to mathematically model the system with stimulus response curves proved crucial to uncovering the molecular ratchet that emerged. The art and science of mathematical modelling is both a broad and complex topic, so a complete and in-depth discussion of this discipline would be far beyond the scope of this text. However, using our knowledge of transcriptional control and high-school level calculus, we can illustrate some of the basic mathematical principles involved. This will be done by creating models of simple transcriptional networks. In the remainder of this section, we will thus strive to take a few "baby steps" along the path to creating detailed dynamic models of complex biological systems.

Modelling basal transcription of a gene

Before we can begin to model the regulation of a gene, we must be able to model simple unregulated (i.e., basal) transcription. Let us consider a hypothetical gene, "Y." The gene contains a binding site for RNA polymerase which, when bound, supports basal levels of transcription (**Figure 23.11**). Now consider how one would calculate the number of mRNA molecules transcribed from this gene. Essentially, two factors will determine how many mRNA molecules accumulate within the cell: (1) the mRNA production rate, and (2) the mRNA degradation rate. In **Figure 23.12**, we incorporate these rates to illustrate how they would affect the accumulation of Y mRNA molecules.

In Figure 23.12, imagine that the box represents a cell within which Y mRNA molecules accumulate. In this scenario, the basal transcription of gene Y is considered the source of the Y mRNA. For example, if gene Y continuously produced transcripts, but these transcripts were not degraded, the amount of Y mRNA molecules within the box would accumulate to ever higher levels *ad infinitum*. Now imagine that after a certain period of time, the

production of transcript ceased. In the absence of degradation, the level of Y mRNA molecules would stop rising and instead remain constant over time. On the other hand, if the message was degraded (e.g., through the action of nucleases), then the level of Y mRNA molecules would begin to drop and eventually reach zero. Thus, degradation through nucleases can be considered to be a sink for Y mRNA.

Now let us make our model more quantitative by providing units for our production and degradation rates. For production, we will use the units of molecules/min (i.e., the number of molecules of mRNA that are synthesized every minute). For degradation, the units will be expressed as a percentage (e.g., 10 percent of the pool of mRNA molecules might be said to degrade every minute). Why is the rate of degradation not expressed in molecules/min? This is because the rate of degradation will depend on how many molecules of Y mRNA are present at any given instant in time. For instance, if the degradation rate is 10 percent (or 0.1/min), then this would mean that, if there were 1000 molecules of Y mRNA present at a given instant in time, the degradation rate would be 100 molecules/min at that instant. If at another instant of time, there were 100 molecules of Y mRNA, then the degradation rate would be only 10 molecules/min at that instant. Thus, we can say that the degradation rate is 10 molecules per minute for every 100 molecules present or

$$\frac{10\ \text{molecules}}{\text{minute} \cdot 100\ \text{molecules}}$$

As you can see, the units of molecules cancel out, leaving us with our original 0.1/min degradation rate.

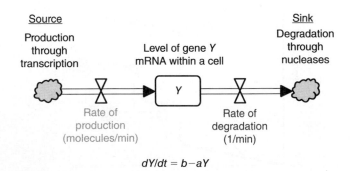

$$dY/dt = b - aY$$

Figure 23.12 A mathematical description of the simple transcription of gene Y. Y mRNA transcripts are produced through transcription (the source) at a rate denoted by the variable b. Y mRNA transcripts are degraded through the action of nucleases (the sink) at a rate denoted by the variable a. The accumulation of Y mRNA within the cell can be described by the differential equation, $dY/dt = b - aY$.

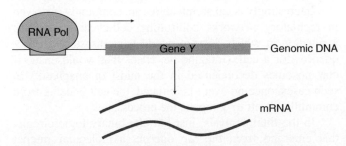

Figure 23.11 A schematic illustrating the simple transcription of a gene, Y. The binding of RNA polymerase to the promoter of gene Y stimulates transcription of Y mRNA.

In Figure 23.12, production and degradation are symbolized by the double triangle symbol. This symbol is intended to represent a valve that controls the flow of mRNA into (or out of) the box. When there are high rates of production, the valve would be more fully open (allowing more Y mRNA to accumulate within the box). If there were low production rates, then the valve would be more fully closed (restricting the flow of Y mRNA into the box). Similarly, the valve controlling the flow of Y mRNA into our sink represents the rate of degradation (with higher rates indicating a more fully opened valve and higher flows of Y mRNA out of the box).

We can now define a differential equation representing the transcription of gene Y:

$$\frac{dY}{dt} = b - aY$$

where Y is equal to the number of Y mRNA molecules, b is equal to the production rate (in molecules/min), and a is equal to the degradation rate of Y mRNA molecules (in units of 1/min). dY/dt, of course, represents the rate of change of Y mRNA molecules with respect to time.

With respect to the differential equation, the most crucial piece of information that we can extract is the steady-state level of Y mRNA molecules within the cell. (That is, once the system reaches equilibrium, how many molecules are present?). Using this equation, such a determination becomes trivial. If the system is at steady state, one can conclude that $dY/dt = 0$. Therefore,

$$0 = b - aY$$

and, through algebraic manipulation, we can show that Y at steady state is equal to b/a (i.e., $Y_{st} = b/a$). Therefore, the steady-state level of message is equal to the production rate divided by the degradation rate. In the next section, we will use computer simulations to further examine the consequences of our mathematical model.

Simulating our model

Now that we have an equation describing simple transcription, it is possible to create computer simulations in which we are able to modulate the parameters and observe the consequences to the system. In **Figure 23.13**, three genes, all of which begin transcribing message at $t = 0$, are represented. The genes differ only in their rate of production and/or rate of degradation. As you can see, the level of mRNA eventually reaches a steady state determined solely by the production and degradation rates of the respective genes. Furthermore, if we run a second simulation (**Figure 23.14**), but this time alter the initial levels of mRNA present, the same steady-state levels are attained regardless of the initial levels of the system. No matter what the starting amount of Y mRNA, our system will reach a steady-state level (with respect to the level of mRNA) that can by calculated by $Y_{st} = b/a$.

Modelling inducible and repressible transcription

Now that we have a mathematical framework for basal transcription, we can consider the effects of *trans*-acting factors. In **Figure 23.15a**, a schematic of simple transcriptional activation is shown. In this scenario, our gene of interest, Y, is regulated by an activator protein, X, that is able to bind a *cis*-acting control element in the promoter of gene Y. In response to a signal, which we will call S_x, X is converted into an active form, X^*, and binds to the control element, thus increasing the rate of transcription. In **Figure 23.15b**, transcriptional repression is shown. In this case, our signal, S_x, converts a repressor, X, to an active form denoted by X^*. The active repressor then binds a negative *cis*-acting control element, thus inhibiting transcription.

To build our mathematical construct, we will begin by examining the system from a qualitative perspective. In the case of activation, we know that the rate of production of Y mRNA molecules (or Y promoter activity) is dependent on, or is a function of, X^*. This is to say, Y promoter activity is proportional to the amount of X^* in the system. In the case of repression, Y promoter activity is proportional to $1/X^*$. Thus, we know a mathematical relationship between the concentration of X^* and the promoter activity of the Y gene exists. The objective now is to determine what the function is.

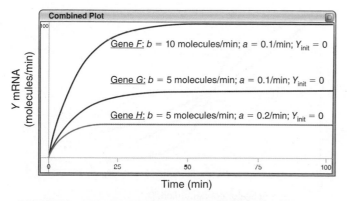

Figure 23.13: b = 10 molecules/min; a = 0.1/min; Y_init = 0 (Gene F), b = 5 molecules/min; a = 0.1/min; Y_init = 0 (Gene G), b = 5 molecules/min; a = 0.2/min; Y_init = 0 (Gene H)

Time (min)

Figure 23.13 The numerical results of a simulation of the simple transcription of three genes that differ with respect to production and/or degradation rates. In each case, a steady-state level is reached that can be calculated by the equation, $Y_{st} = b/a$.

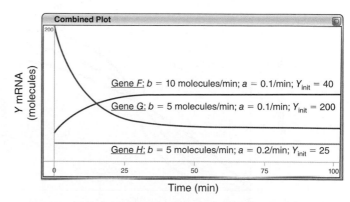

Figure 23.14: Gene F: b = 10 molecules/min; a = 0.1/min; Y_init = 40; Gene G: b = 5 molecules/min; a = 0.1/min; Y_init = 200; Gene H: b = 5 molecules/min; a = 0.2/min; Y_init = 25

Time (min)

Figure 23.14 The numerical results of a simulation of the simple transcription of three genes that differ with respect to the initial levels of mRNA, and production and/or degradation rates. In each case, a steady-state level is reached that can be calculated by the equation, $Y_{st} = b/a$.

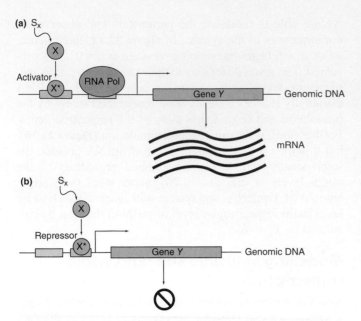

(a) S_x

Activator

RNA Pol

Gene Y — Genomic DNA

mRNA

(b) S_x

Repressor

Gene Y — Genomic DNA

Figure 23.15 Schematics illustrating the regulated transcription of gene Y. (a) The binding of the activator, X^*, to the promoter of gene Y stimulates the action of RNA polymerase, resulting in increased levels of transcription of Y mRNA. The signal, S_x, converts inactive X to its active form X^*. **(b)** The binding of the repressor, X^*, to the promoter of gene Y prevents RNA polymerase from transcribing gene Y. The signal, S_x, converts inactive X to its active form, X^*.

While every relationship between a *trans*-acting factor and its target gene is likely unique, a function that does a good overall job of approximating the effects of activators and repressors is the Hill function (**Figure 23.16**). The Hill function for activation is

$$\text{Promoter activity of } Y = \frac{\beta(X^*)^n}{K^n + (X^*)^n}$$

The Hill function for repression is

$$\text{Promoter activity of } Y = \frac{\beta}{1 + \left(\frac{X^*}{K}\right)^n}$$

These equations are composed of only four parameters: X^*, β, K, and n. X^* represents the concentration of activator, while β is a parameter that represents the maximum promoter activity of gene Y. Why is there such a limit? This is because, even under optimal growth conditions, some cellular constituent (nucleotides, or a critical enzyme) will eventually become limiting for transcription. K represents the concentration of X^* at which the promoter activity of Y is at half-maximal levels. Lastly, the parameter n represents the order of the equation; this can be thought of as a measure of how switchlike the gene is. For example, for $n = 1$, you can see a relatively slow and steady increase in promoter activity rates as X^* concentrations increase. However, at higher orders of n the shape of the curve changes. Now as X^* increases, there is little response at first, but then a very strong response around K. Thus, the higher the order, the more switchlike the changes in promoter activity.

The Hill function for activation and the Hill function for repression allow us to move from simple qualitative descriptions of transcription to exact quantitative models. While more sophisticated dynamic models of complex systems are beyond the scope of this text, the fundamental mathematical reasoning behind such endeavours makes use of many of the same principles that we have just discussed.

In the next section, we will discuss how the quantitative analysis of relatively simple transcriptional networks in *E. coli* led to the discovery of another central concept of the discipline of systems biology, network motifs.

> The basal transcription of a gene can by modelled using the differential equation, $dY/dt = b - aY$. At steady state, the amount of mRNA molecules within a cell is equal to the mRNA production rate divided by the mRNA degradation rate. The action of transcriptional activators (and repressors) can be modelled using the Hill function.

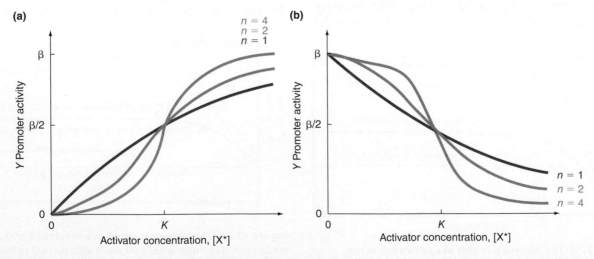

(a)

$n = 4$
$n = 2$
$n = 1$

Y Promoter activity

β

$\beta/2$

0

0   K

Activator concentration, [X*]

(b)

Y Promoter activity

β

$\beta/2$

0

$n = 1$
$n = 2$
$n = 4$

0   K

Activator concentration, [X*]

Figure 23.16 The regulation of gene Y promoter activity as described by the Hill equations. (a) The promoter activity of gene Y, as a function of increasing concentration of the activator protein, X^*. **(b)** The promoter activity of gene Y, as a function of increasing concentration of the repressor protein, X^*. n refers to the order of the Hill equation.

23.5 Network Motifs

Network motifs were first discovered through the quantitative analysis of transcription networks in *E. coli*. In these studies, researchers wished to determine if there was any meaningful, biologically relevant organization to the transcriptional networks they were examining. Remarkably, they discovered that certain network architectures appeared over and over again. In fact, they appeared much more often than one would expect by chance alone. These commonly appearing architectures are referred to as network motifs.

A network motif is conceptually analogous to structural motifs characterized in proteins. For example, the leucine zipper motif found in the Myc protein can be defined by the appearance of the following sequence: $K-X_6-L-X_6-L-X_6-L-X_6-L-X_6-L$ (where K = lysine, L = leucine, and X = any amino acid). This motif is related to the function of the Myc protein in that it allows for its dimerization with either another Myc molecule or with Max (see Chapter 11). This motif is found not only in the Myc protein, but also in many different proteins across a wide range of organisms. In each case, it is involved in mediating protein-protein interactions. By examining the amino acid composition of a given protein for residues that appear together more often than one would predict by chance, it is possible to infer functional information about the protein.

A network motif uses the same logic, except that the structure of the network, as opposed to the structure of the protein, is examined. For example, consider hypothetical networks consisting of three nodes. There are many different ways in which a three-node network could be organized. In **Figure 23.17** we see four possibilities—a bifurcating network, a linear network, a cyclic network, and a structure known as a feed forward loop. In each case, imagine that each node represents a gene-product that regulates the transcription of one of the other two genes (e.g., in the linear example, the *A* gene-product promotes transcription of *B*, and the *B* gene-product promotes transcription of *C*).

Remarkably, when one examines transcription networks in living organisms like *E. coli*, certain unexpected patterns reveal themselves. For example, the feed forward loop appears over and over again in the transcription networks of *E. coli* (42 instances to be exact). On the other hand, examples of the other three architectures are rarely or never observed. Does this mean that the architecture of the feed forward loop is related to the function of the

transcription network? Does it confer specific properties that make it particularly suited to dealing with a certain biological problem? An examination of a typical feed forward loop in *E. coli* will help us answer these questions.

The *araBAD* operon is controlled by a feed forward loop

As we saw in Chapter 10, the *araBAD* operon encodes enzymes involved in metabolizing the carbon source, arabinose. Transcription of the operon is positively regulated by the *araC* gene, which when bound to arabinose, stimulates expression of the operon. In addition, in the absence of glucose, CRP (in complex with cAMP) also acts as a positive regulator of *araBAD* transcription. Interestingly, CRP–cAMP complexes also positively regulate the transcription of the *araC* gene. This transcription network matches the architecture of the feed forward loop. CRP is at the top of the network and positively regulates transcription of *araC*, which in turn stimulates transcription of *araBAD*. Furthermore, CRP can also directly promote transcription of *araBAD* (in addition to its indirect positive effect through its action on *araC*). Generally, the top component of a feed forward loop is referred to as the master regulator, the middle component the secondary regulator, and the bottom component the target gene (**Figure 23.18**).

Now let us see what properties this network structure confers by examining the results of computer simulations. In the graphs shown in **Figure 23.19**, the levels of CRP–cAMP are plotted against the levels of *araC* and *araBAD* transcription. At various intervals, the activating signals (cAMP and arabinose) are added, either for a brief pulse or for an extended period. The brief pulse activates CRP, stimulating *Y* transcription, but does not affect *araBAD* transcription. In contrast, a persistent signal not only activates *Y* transcription, but also—after a delay—activates *Z* transcription.

Thus, out of this particular set of interacting parts emerges a system that activates itself, but only after a time

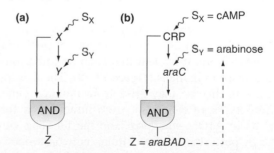

Figure 23.18 The regulation of the *araBAD* operon is structured as a feed forward loop. (a) A generic feed forward loop. In response to the appropriate signals (S_x and S_y), *X* and *Y* both directly activate the transcription of *Z*. In addition, *X* is able directly activate the transcription of *Y*. **(b)** The *araBAD* feed forward loop. In response to cAMP, CRP is able to actively stimulate transcription of both *araC* and *araBAD*. In response to arabinose, *araC* positively regulates the transcription of *araBAD*.

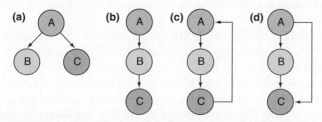

Figure 23.17 Four possible configurations of a three-node network. (a) A bifurcating configuration. **(b)** A linear configuration. **(c)** A cyclic configuration. **(d)** A feed forward loop.

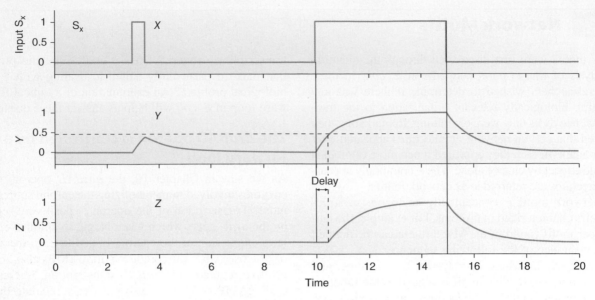

Figure 23.19 **Simulating the response of a feed forward loop to both a transient and persistent signal, S_x.** The architecture of the feed forward loop ensures a delay in Z transcription after the addition of the activating signal. Thus, the feed forward configuration ensures that the system responds only to persistent stimuli.

delay. Why would such a design be advantageous to *E. coli*? This architecture serves to prevent the spurious activation of the pathway. (The system ensures that transcription is turned on if, and only if, a cell is sure that its source of arabinose has proven to be reliable.) In the same way that a mining company would only invest in building an oil well if there was a large supply of oil at the site (as opposed to a few small pockets of oil), *E. coli* cells only invest in transcribing the *araBAD* operon when arabinose is consistently present.

In addition to feed forward loops, a variety of other network motifs have been characterized. These include motifs referred to as oscillators, buzzers, sniffers, and toggle switches. A detailed examination of all of these is beyond the scope of this text, but you are encouraged to search the Internet to learn more on your own. Nonetheless,

it is clear that an understanding of these network motifs will be crucial in understanding the function of biological organisms, as well as the genotype–phenotype relationship (see the **Focus on Inquiry** box **"Network Motifs and Cancer Biology"**).

> Transcriptional networks in *E. coli* can be categorized into one of a small set of commonly recurring architectures called network motifs. Each motif has evolved to perform a specific function. For example, one of the most common network motifs, the feed forward loop, is able to activate itself in response to a signal, but only after a time delay. This architecture thus ensures that spurious activation of the network cannot occur.

23.6 A Systems Approach to Disease

Genetic and environmental perturbations can cause cellular networks to alter their patterns of gene expression. Furthermore, it is clear that such perturbations can lead to disease in humans (**Figure 23.20**). The disruptions that result in disease may arise from mutated genes, as in various types of cancer, or from infection by foreign agents, as in AIDS, smallpox, and the flu. The view of disease as perturbations in cellular networks opens the door to new approaches to diagnostics, therapeutics, and ultimately prevention.

Identification of biomarkers is a first step

Molecules that are present under specific conditions or when a disease is present are referred to as **biomarkers**. In recent years, the search for biomarkers associated with early stages of disease has intensified because many diseases,

including cancer, are highly treatable when detected at early stages. Patterns of mRNAs and proteins present in the body can be referred to as molecular fingerprints. Proteins are more stable and live longer than RNA molecules and are therefore the focus of most biomarker research, although advances have been made analyzing both mRNAs (the transcriptome) and siRNAs.

Many different bodily fluids, such as blood, urine, tears, and saliva, can be sampled to identify altered molecular fingerprints. Of these, blood is probably the most information rich in that it bathes all tissues in the body. Through gene expression and proteomic approaches, researchers are discovering organ-specific products that are secreted into the blood. When an organ malfunctions, protein levels may change, indicating the start of disease or the progression of an organ-specific disease.

Figure A Edward Wang.

Edwin Wang (**Figure A**) is a Professor at the McGill University Centre for Bioinformatics. He is also a senior investigator at the National Research Council of Canada. Over the last decade,

Dr. Wang has explored how systems biology concepts such as network motifs can be applied to improve our understanding of cancer biology. The study of cancer is particularly suited to the systems approach because of the great number of cellular components involved and their incredibly intricate and complex interactions. **Figure B** shows how the systems approach evolved from a traditional pathway-specific view, to a network view (in which data from unbiased, high-throughput functional genomic/proteomic data have been incorporated), to a systems view that utilizes network analysis to identify network motifs, and mathematical analysis to run simulations and formulate testable hypotheses. As demonstrated by Dr. Wang, an understanding of these concepts with respect to cancer biology is crucial to the development of novel therapeutic approaches.

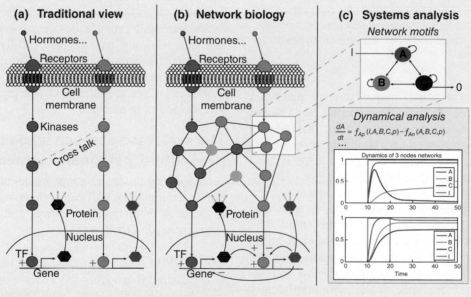

Figure B The development of the systems approach.

As one example, the altered patterns of gene expression in the disease-perturbed networks of a cancerous prostate gland cause changes in the levels and types of proteins expressed by various prostate cells. Normal prostate cells secrete a protein called prostate-specific antigen (PSA) into the blood. PSA is one component of the blood protein fingerprint for the prostate. In cancer, the levels of PSA increase, and hence, blood measurements of this protein are routinely used to detect prostate cancer.

A more comprehensive systems approach is identification of many organ-specific proteins and monitoring levels of all these proteins to determine if the organ is diseased. A significant fraction (roughly 10 percent) of the proteins with disease-perturbed expression levels are secreted into the blood. There are probably 50 additional

proteins in the prostate organ fingerprint beyond PSA, and measuring each of these in the future will give much more accurate diagnoses of prostate cancer. In addition, studies show that changes in blood concentration of several prostate-specific blood proteins reflect various stages of prostate cancer.

Each major organ or cell type in the body undoubtedly produces a molecular fingerprint. In this way, the blood becomes a window into health and disease.

Disease stratification may be identified

Many major human diseases, such as cancer or heart disease, can be divided into different disease subtypes that result in the same general phenotype. Cancer is typified by uncontrolled cellular growth, but we know that many

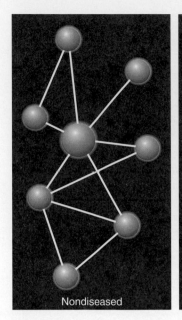
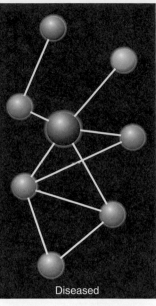

Nondiseased   Diseased

Figure 23.20 A normal and a diseased cell. The *blue circles* represent nodes (proteins), and the *lines* depict the connections (interactions) between nodes in this hypothetical network. Normal nodes are in *blue*; nodes perturbed by disease to differing extents are in *red*, *orange*, and *green*.

different genetic modifications may have taken place in a particular cell to cause it to become cancerous (see Chapter 16). Profiling the disease through molecular markers and analysis of the transcriptome and proteome gives important information that can be used to classify the type of cancer, as well as how far the disease has progressed.

Stratification of disease is seen in breast cancer. The diagnosis of breast cancer indicates the presence of a tumour—uncontrolled cell division—in the breast. Not all tumours are alike, but many can be grouped together based on the origin of the disease (what genetic changes occurred, what systems are perturbed). This knowledge allows us to more specifically and intelligently treat the patient. For example, in about 20 percent of breast cancer cases, there is an increased expression of a protein called HER2, which is found on the surface of cancer cells. The HER2-positive tumours grow faster than other types and are more likely to recur. Knowing if the tumour is in the HER2-positive subclass directs the physician to treat the cancer differently. The drug trastuzumab (brand name Herceptin®) targets the HER2 protein and decreases the recurrence of the disease. Thus, by understanding system networks, both diagnostics and treatments may improve.

Knowledge of protein interactions can identify drug targets

Some of the proteins in a molecular fingerprint can point to the protein networks that have been perturbed by disease. An understanding of the protein interactions in these networks can lead to new candidates for *drug targets*: proteins whose

interactions with specific drugs will either kill the cell (in the case of cancer) or alter the function of the network back toward normal. One can even imagine the future creation of drugs able to prevent disease by keeping networks from becoming perturbed in the first place.

This systems approach will undoubtedly lead to the integration of diagnosis and therapy. It will also lead to a revolution in medicine in which predictive, preventive, and personalized modes will replace the largely reactive current model in which physicians begin treatment only when a person is sick.

Advances in technology are needed for new medicine

The new medicine will require new technologies. Over the next ten years or so, nanotechnology will revolutionize DNA sequencing, making it possible to rapidly sequence individual human genomes for well under $1000. More sensitive *molecular imaging* techniques will permit the non-invasive visualization of drug activity and function in model organisms and humans. *Microfluidics* and *nanotechnology* will produce devices that measure, identify, and inexpensively quantify (in a fully automated high-throughput platform) thousands of proteins from a small drop of blood or assess the information content of individual cells. We consider next one possible scenario for this revolution in medicine.

The systems approach leads to predictive, preventive, personalized medicine

As the ability to integrate vast amounts of biological data improves, medicine will continue to undergo a revolution in terms of prediction, prevention, and personalization of intervention and treatment.

Prediction

It is possible that within ten years, physicians will have in place two major approaches to medical prediction. First, all patients will have their genome sequence determined by a nanotechnology device. It is now possible to sequence an individual human genome in days at a small fraction of the initial cost of the Human Genome Project. The time and cost of this endeavour will continue to decrease in coming years. From genome information and genetic analyses using SNPs and other molecular markers, we continue to identify more about single alleles or combinations of alleles that determine susceptibility to diseases or conditions.

From the individual's genomic sequence, it will then be possible to glean the information for predicting the individual's future health. For instance, a woman might learn that she has a 30 percent chance of developing cardiovascular disease by age 50; a 40 percent chance of getting ovarian cancer by age 60; and a 40 percent chance of developing rheumatoid arthritis by age 65. This knowledge may allow her and her doctor to plan

strategies for her lifestyle and for medical interventions throughout her life to maximize her health and lower her risk of disease.

Second, quantitative measurements of the 1000–2000 proteins in a droplet of blood could be sent by wireless transmission to a server that will process the data in this molecular fingerprint and send the client/patient and the physician an email stating, for example, "You are fine—do this again in six months" or "Consider additional tests for clarification of results." These blood fingerprints will allow the very early detection of disease as well as the stratification of particular disease types. The fingerprints will also make it possible to follow a patient's response to therapy and to detect adverse drug reactions in early stages. Because predictive medicine without the ability to treat or prevent health problems is unsatisfactory to most patients, better therapies and prevention will have to emerge along with the predictive tools.

Prevention

The development of efficient nanolaboratories that are able to measure the protein and mRNA levels as well as protein-protein and protein-DNA interactions in individual cells will give rise to a new kind of preventive medicine that will work hand in hand with lifestyle measures used today. The new prevention strategies will rely on the delineation of the networks in normal and diseased cells; analyses that clarify their differences; and the identification of key proteins (central nodes in the networks) as potential drug targets.

For instance, neuropsychiatrists had known for more than a decade that the protein serotonin (a neurotransmitter) plays a crucial role in the network whose perturbation contributes to clinical depression in humans. However, they did not know for sure how the serotonin functioned or what went wrong. Studies show that a protein designated p11 is another key player in that same network. It appears to modulate serotonin activity by influencing the number of serotonin receptors in the membranes of brain neurons that modulate mood. Although current drugs, such as Prozac and other selective serotonin reuptake inhibitors, slow the resorption of serotonin from the synapses, future antidepressive drugs that target the p11 node of the network may help the serotonin present in the synapses do its job more effectively.

Over the next two decades, the systems approach will not only produce more effective therapeutic agents for treating existing diseases, it will also lead to the development of drugs that can prevent disease by intervening to keep networks from becoming perturbed. Physicians will then be able to counsel patients with greater insight and sophistication. As a hypothetical example, a doctor might explain that although a woman has a 40 percent chance of developing ovarian cancer by the age of 60, by taking a specific medication beginning when she is 40, she can essentially prevent the disease by reducing that probability to 2 percent.

Personalization

Because the genome of one person differs from that of another by about 6 million base pairs, we are each susceptible to differing combinations of diseases. Increasingly, medical practitioners will be able to practise personalized medicine by applying the power of predictive and preventive medicine to our individual needs.

For treatment of breast cancer, we have already seen the power of the systems/biomarker approach. A set of 70–75 markers has been identified that can predict risk of metastasis of breast cancer. If the risk of metastasis is very low, surgery and tamoxifen treatment may be sufficient, making the systemic treatments unnecessary. The patient and doctor must then decide, based on the risk assessment, whether the systemic treatments (which have significant side effects) will be part of the patient's treatment.

> A systems approach to diseases involves identification of biomarkers that indicate the presence of disease at an early stage; classification of disease subtypes, which allows specific interventions; and clarification of protein interactions and pathways to help identify potential drug targets. Predictive, preventive, and personalized medicine will transform the health care industry and the practice of medicine. Striking changes based on the systems approach to biology will have a profound effect on the use of drugs and treatment options, as well as in the prevention of disease.

Connections

It is often said that necessity is the mother of invention. This is particularly true of the discipline of systems biology. With the vast amounts of data being generated as a result of genome sequencing projects (Chapter 19), as well as the continued development of functional genomic and proteomic technologies (Chapters 20–22), it has become apparent to many that new philosophies and approaches are needed to derive fruitful biological insight from these genome-level studies. By combining the power of mathematical analysis and modern computing, it is now possible for systems biologists to define emergent, systems-level properties (e.g., hysteresis) that play a crucial role in the genotype–phenotype relationship and which previously could not even have been imagined. Such approaches will undoubtedly be a crucial determinant in the success of future research in biology and genetics.

1. The practice of systems biology requires one to identify the elements of a biological system; measure their changing relationships; measure their relationships to the other systems functioning in the same context (organism or cell); and with this information, attempt to explain the system's emergent properties. The key point is that biological systems are dynamic entities that reflect changes that range across evolutionary, developmental, and physiological responses. **[LO1–2]**

2. Biological information consists of the digital information of the genome and environmental signals from outside the genome, which modify the genome's output. Gene regulatory networks integrate the inputs of information from signal transduction pathways and transmit information to the batteries of genes that encode protein networks. These networks carry out metabolism, development, and physiology. Ability to integrate biological information from many hierarchical levels is critical for understanding the system. **[LO1–2, LO5]**

3. Researchers use genetic and environmental perturbations to study biological systems. A preliminary model of the system is created from pre-existing knowledge; the system is then perturbed in a known way. Genomic, proteomic, genetic, and biological assay data sets are collected and integrated, and comparison is made between a visualization of the integrated data and the model. Where discrepancies arise, new hypotheses are formulated that can be tested by another round of perturbations. This process is repeated until the experimental data and the model are in accord. **[LO1–2, LO5]**

4. Two of the most critical goals of the systems approach are (1) to uncover the emergent properties of a system and (2) to understand how alterations in the system affect these emergent properties. The power of the systems approach becomes apparent upon the analysis of systems that produce emergent phenotypes that are non-intuitive. **[LO2–3]**

5. Mathematical modelling of dynamic systems has become an increasingly important tool for the systems biologist. For example, simple transcriptional networks can be easily modelled using differential equations and the Hill function. The Hill function for an activator describes the effects of increasing activator concentration on the promoter activity of a gene. The Hill function for a repressor describes the effects of increasing repressor concentration on the promoter activity of a gene. **[LO1, LO4]**

6. Network motifs can be described as commonly occurring network architectures. In a biological context, each motif has evolved to perform a specific function. For example, the feed forward loop design ensures that spurious activation of the network does not occur. **[LO1, LO5]**

7. A systems approach to disease encompasses the idea that disease arises from perturbed networks. From this simple idea come powerful new approaches to diagnosis, therapy, and prevention. The systems approach to disease is catalyzing a change from the current reactive mode of medicine to a future of predictive, preventive, and personalized medicine. **[LO1–2, LO5]**

Solved Problems

I. You hypothesize that breast cancer that has been histologically classified as invasive ductal cancer can be further subclassified based on molecular signatures. If you can create such subcategorizations, you may be able to stratify the cancers and develop specific therapies for one or more of the subcategories. You and your fellow researchers have biopsy tissue from 63 invasive ductal breast cancer patients and clinical data on the course of disease in each of these patients. You plan to use microarray technology to measure gene expression in each of these samples, and then use the results to classify the tissues. When you analyze these samples with microarrays, what other samples should you analyze and why?

Answer

Controls are an important part of any scientific experiment. You need to choose control samples to provide confidence that the subcategories you identify truly reflect a stratification of invasive ductal breast cancer. In particular, you must ensure that the biopsies you receive are neither misdiagnosed nor improperly collected. For example, if some normal breast tissue is mistakenly included as a tumour in your analysis, you may identify normal tissue as a subcategory of invasive cancer. Thus, you will want to include normal breast tissue as well as other types of breast cancer in your arrays. In addition, your analysis will require comparing gene expression in the cancer samples to a standard reference, so your set of normal controls will serve this important function. Finally, analyzing a comprehensive set of control samples will also help to develop a statistical model for the variability of your measurements and thus increase confidence in your results.

You may also wish to approach other hypotheses with your data. For example, you may want to identify markers specific to breast cancer or breast tissue. Therefore, you may wish to include some samples of other normal tissues as well as cancers of these tissues.

II. The diagram that follows is a small portion of the much larger computerized model of an interaction network in the yeast *Saccharomyces cerevisiae*. The nodes in the diagram represent either proteins or the genes encoding those proteins. *Blue lines* connecting nodes indicate protein-protein interactions. *Red arrows* connecting

nodes show protein-DNA interactions, with the protein at the base of the arrow and the DNA sequence at the arrowhead.

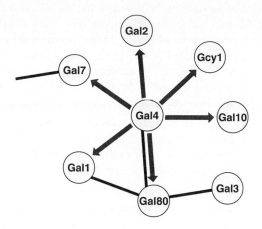

a. What kinds of techniques discussed in this chapter might have been involved in the collection of data for this interaction network?

b. What proteins in the diagram are likely to be in a complex with each other?

c. What proteins are likely to act as a transcription factor, and how would this transcription factor operate?

d. The proteins and genes indicated in the diagram enable yeast cells to utilize the sugar galactose as a carbon source. You hypothesize that these genes are regulated at the transcriptional level by the type of carbon source: For example, expression of many of these genes might be increased if the medium contained galactose, but expression might be repressed if the medium contained glucose. What kinds of genome-scale experiments might you do to test such hypotheses?

Answer

a. The data describing protein-protein interactions could have been derived in either of two ways. First, researchers might have used the affinity capture/mass spectrometry technique (see Chapter 21). A second method to uncover protein-protein interactions is the yeast two-hybrid approach. Databases of the yeast interactome already exist that report the results of systematic tests of each yeast gene fused either to a DNA-binding domain or to a DNA activation domain. Protein-DNA interactions could be identified by the ChIP-on-chip technique (also described in Chapter 21) if antibodies that recognize the transcription factor(s) among the proteins in the diagram were available.

b. The proteins that could be in a complex must be linked by protein-protein interactions. These include Gal4, Gal80, Gal1, and Gal3.

c. The diagram indicates that the Gal4 protein can bind to DNA sequences in the vicinity of several genes, including the genes encoding Gal1, Gcy1, Gal2, Gal7, Gal10, and Gal80; the DNA sequences are presumably near the promoters of these genes. The binding of Gal4 to these DNA sequences in theory would regulate the transcription of these other genes.

d. One type of experiment to test your hypothesis is a microarray analysis. You would grow yeast cells in the presence of either galactose or glucose, and then compare the levels of the mRNAs for all yeast genes under these two conditions, with particular attention to those genes in the diagram that are potential targets for regulation by a protein complex containing Gal4. Even more interesting would be the microarray analysis of transcription in yeast that had some elements of the galactose utilization system knocked out by mutation.

Problems

Vocabulary

1. For each of the terms in the left column, choose the best matching phrase in the right column.

   i. biological system

   ii. systems biology

   iii. emergent property

   iv. biomarker

   1. a system trait that arises from the operation of the system as a whole
   2. a molecule that is present under specific conditions or when a disease is present
   3. a discipline that seeks to describe the multiple components of a biological system and to analyze the complex interactions of these components within the system and in relation to the components of other systems
   4. a collection of interacting elements that carry out a specific biological task

Section 23.1

2. What is an emergent property? Describe an example of an emergent property from a system that you are familiar with from your everyday life.

3. What are the four fundamental concepts related to defining a biological system?

4. Systems biology is a cross-disciplinary field.

   a. What role do engineers have in this field?

   b. What role do mathematicians have in this field?

5. If you were to catalogue *cis*-control elements, would you be studying protein networks or gene regulatory networks? What other information would be required to gain an understanding of the kind of network you are studying?

6. How has the Human Genome Project enabled systems biology in the twenty-first century?

7. Answer the following as true or false:

   a. Systems biology employs both discovery science and hypothesis-driven science.

b. Gene regulatory networks integrate information they receive from signal transduction networks and transmit it to protein networks.

c. A transcription factor binds only a single *cis*-control region in the genome.

d. A DNA sequence may be modified by environmental information.

e. The yeast cell contains only about 6000 proteins.

f. The integration of different types of global sets may be carried out with graphical networks.

g. The proteome of the organism is the sum of the proteomes of all cells in all developmental or physiological states.

h. The mass spectrometer currently has the capacity to quantify globally all of the proteins in a given cell type.

i. Protein chips are as global in their measurement capacity as DNA chips.

j. The global localization procedure for identifying transcription-factor-binding sites also works for other proteins or protein complexes directly bound to DNA or indirectly bound to DNA through other DNA-binding proteins.

k. The galactose utilization system is interconnected to many other cellular systems in yeast.

Section 23.2

8. Using a similar approach to the one described with respect to the *lac* operon, analyze the *trp* operon through the eyes of a systems biologist. What properties emerge from this set of interacting parts? How would increasing the number of stop codons in the leader sequence affect these emergent properties? How would changing the affinity of complementary binding between sequences 3 and 4 affect these emergent properties?

Section 23.3

9. Your friend states that in his immunology research, he is measuring the levels of 100 cytokines in response to knockout perturbations of interesting genes in his system. He claims that he is doing systems biology. Is he correct? Explain.

10. Define hysteresis and describe one non-biological example of the phenomenon from the disciplines of physics or engineering. In a biological context, explain what properties of hysteretic systems make them particularly suited to controlling cell-cycle or developmental transitions.

Section 23.4

11. Consider a cell in which a species of mRNA is being produced at a rate of 5 molecules/min (i.e., five individual mRNA molecules are transcribed from the gene every minute). This same species of message is translated into a polypeptide at a rate of 60 molecules/h (i.e., 60 polypeptides can be made from a single message every hour). The degradation rates of the message and the encoded protein are 0.2/min and 0.25/min, respectively. Interestingly, the polypeptide produced acts as a transcriptional activator of a target gene called *SLA3*. The transcriptional activation of the *SLA3* gene can be described by the Hill function, using the parameters $K = 50$, $\beta = 300$, $n = 1$.

(*Note:* $Y_{m(steady\ state)}$ = steady-state level of the message; $Y_{p(steady\ state)}$ = steady-state level of the protein; b_m = rate of production of the message; a_m = rate of degradation of the message; b_p = production rate of the protein; a_p = degradation rate of the protein).

a. Which Hill equation would most accurately describe the effects of the activator protein on the promoter activity of the *SLA3* gene?

b. In the Hill equation that describes the effects of the activator protein on the promoter activity of the *SLA3* gene, explain what the parameter K represents.

c. Calculate the steady-state level of the message encoding the transcriptional activator.

d. Assuming that mRNA levels have reached steady state, develop an equation that could be used to determine the steady-state level of the protein.

e. Assuming the system reaches steady state, how would the expression level of the *SLA3* gene be affected by a mutation that increased the transcriptional activator protein's degradation rate by a factor of 2?

Section 23.5

12. Consider a feed forward loop that, instead of the AND logic of the arabinose operon, uses OR logic. How would this change affect the emergent properties of the system? Think of a biological situation where such a regulatory configuration would be advantageous.

Section 23.6

13. Answer the following as true or false:

a. A systems approach to disease embodies the concept that diseased cells have some abnormal networks.

b. A protein molecular fingerprint in the blood has the capacity to assess the state (e.g., health or disease) of the cell type from which it was secreted.

c. Systems approaches to disease provide new approaches for the discovery of drug targets.

d. Predictive medicine without the ability to treat the predicted disease raises ethical concerns about whether insurance companies could use this information to modify insurance rates.

e. Predictive, preventive, and personalized medicine will require medical education to be greatly modified.

 For more information on the resources available from McGraw-Hill Ryerson, go to www.mcgrawhill.ca/he/solutions.

CHAPTER **24** | **Synthetic Biology**

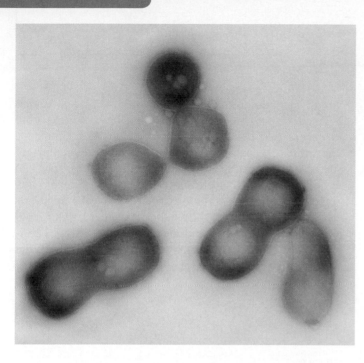

In May of 2010, J. Craig Venter and colleagues unveiled their newest creation, a strain of *Mycoplasma mycoides* controlled by a genome that was chemically synthesized and assembled in the lab. This new strain is referred to as *M. mycoides* JCVI-syn1.0 or Synthia. These bacterial cells represent the first to be born, not of another cell, but of the combined efforts of the research team. This and similar work have defined a new branch of genetics called synthetic biology. The significance and ramifications of Venter's remarkable achievement, along with other examples of synthetic biology, will be discussed in detail later in this chapter.

On May 21, 2010, scientists from the J. Craig Venter Institute took a giant leap forward in genetics research. On this day, they published a landmark article describing the creation of the world's first synthetic organism (dubbed Synthia or JCVI-1.0syn). Amazingly, these researchers were able to synthesize a complete genome (approximately 1 million base pairs in size) and then transfer it to a host cell; in effect rebooting the host so that it was under the control of the introduced genome. JCVI-1.0syn cells represent the first to be born, not of another cell, but from the combined efforts of the research team.

In contrast to simply sequencing (i.e., reading) a genome, these researchers were able to take the next critical step in the evolution of genetics; this being the synthesis (i.e., writing) of a complete genome. As proof-of-principle, this work has opened the door to a brand new world of biology; a world where researchers might one day sit down at their computers to design artificial genomes (coding for user-designed characteristics) in order to create completely novel organisms with properties of medical or industrial importance. For example, J. Craig Venter's newly formed company, Synthetic Genomics Incorporated (SGI), is working closely with industry to create synthetic organisms capable of producing an entirely new generation of biofuels (**Figure 24.1**). In the remainder of this chapter, we will examine the field of synthetic biology from its beginnings over a decade ago to the landmark creation of Synthia.

Chapter Outline
24.1 What Is Synthetic Biology?
24.2 Synthetic Biology: The First Synthetic Gene Circuits
24.3 Applications of Synthetic Biology
24.4 Multiplex Automated Genome Engineering
24.5 Synthetic Genomics

Figure 24.1 J. Craig Venter taking questions from reporters regarding his latest research.

Learning Objectives

1. Define synthetic biology and relate this new discipline to the Human Genome Project.
2. Evaluate the potential benefit (and limitations) of synthetic strategies aimed at creating novel genetic circuits with user-defined functions.
3. Compare and contrast genome engineering to traditional recombinant DNA technology.
4. Describe the process used to synthesize and assemble entire genomes from "scratch." Discuss the ethical ramifications of such a feat.

24.1 What Is Synthetic Biology?

Synthetic biology defines a new and exciting discipline that combines knowledge derived from the fields of molecular biology and genetics with the principles of engineering. In the same way that a civil engineer might construct a tower or a bridge out of steel and bricks, a synthetic biologist might construct a biological system (e.g., a network or pathway) using characterized genes and regulatory DNA sequences. In other words, the genes characterized by researchers in the past have formed a useful toolkit that can be employed by today's geneticists to create novel regulatory modules with specific functions.

In fact, the BioBricks Foundation, an organization devoted to developing synthetic biology practices (and to ensure that these practices are ethical), maintains a repository of characterized biological components that any researcher can browse online (**Figure 24.2**). Just like an individual involved in the construction of a complex machine might go to a toolbox and pick out the appropriate tool or component for a job, or a child might go to his or her LEGO® box and pick out just the right building block, a geneticist can go to the BioBricks website (biobricks.org) to pick out the promoter, membrane

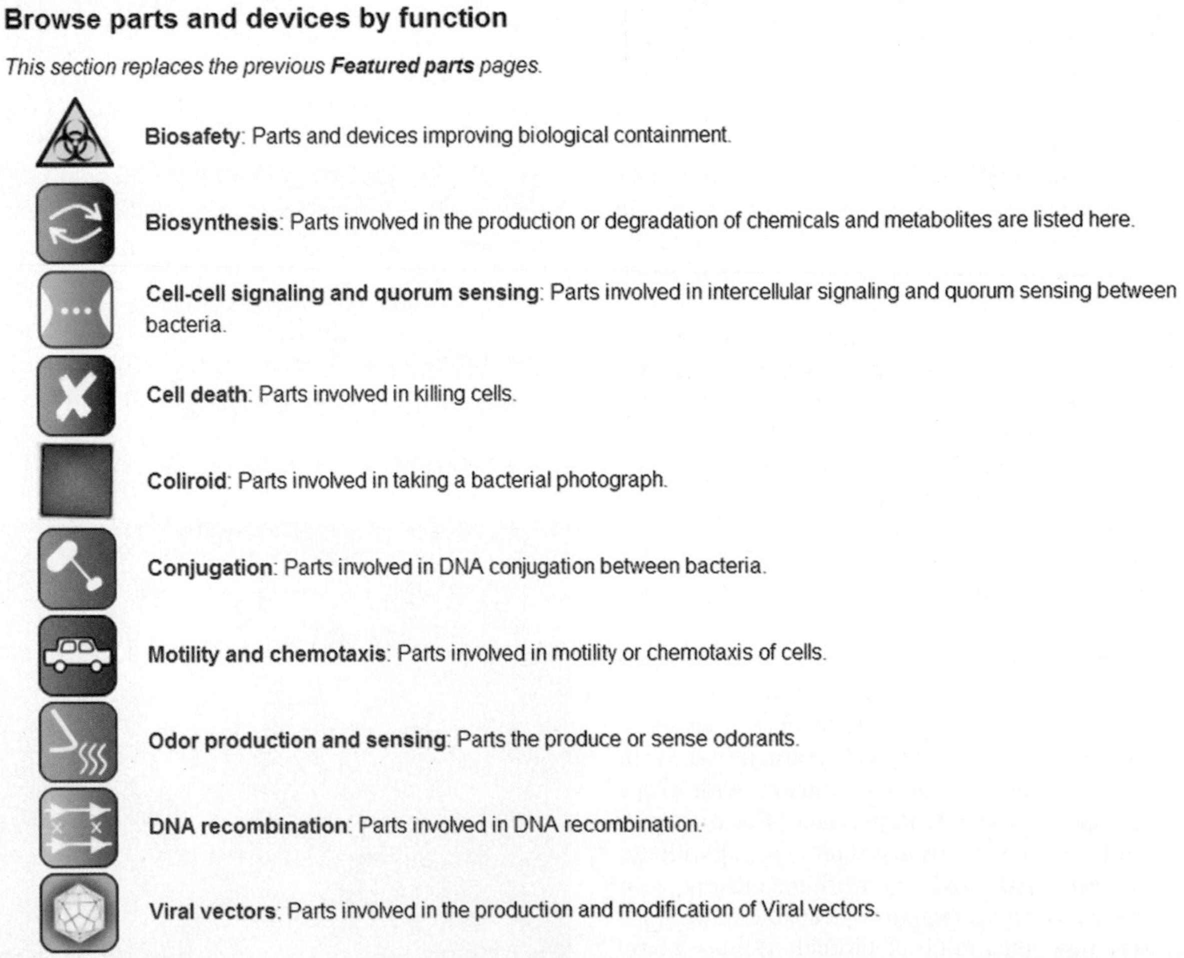

Browse parts and devices by function
This section replaces the previous Featured parts pages.
Biosafety: Parts and devices improving biological containment.
Biosynthesis: Parts involved in the production or degradation of chemicals and metabolites are listed here.
Cell-cell signaling and quorum sensing: Parts involved in intercellular signaling and quorum sensing between bacteria.
Cell death: Parts involved in killing cells.
Coliroid: Parts involved in taking a bacterial photograph.
Conjugation: Parts involved in DNA conjugation between bacteria.
Motility and chemotaxis: Parts involved in motility or chemotaxis of cells.
Odor production and sensing: Parts the produce or sense odorants.
DNA recombination: Parts involved in DNA recombination.
Viral vectors: Parts involved in the production and modification of Viral vectors.

Figure 24.2 A screen capture from the BioBricks website. By visiting the BioBricks website, researchers can browse through a catalogue of biological parts that can be employed in the creation of novel biological networks. At present, the repository houses thousands of parts or "biobricks" organized according to their biological role. Repositories such as these represent the toolkit being used by the current generation of geneticists.

Figure A **The iGEM logo.**

The International Genetically Engineered Machine (iGEM) competition brings together undergraduate university students from all over the world and challenges them to create a novel biological system using the BioBrick repository of interchangeable parts (**Figure A**). Started in 2003 with only a handful of individuals, the competition has expanded to three regional competitions (Europe, Asia, and the Americas), together representing over 130 teams. The finalists gather at the Massachusetts Institute of Technology every year in November to determine the world champions. The 2012 University of Calgary team (**Figure B**) made it all the way to the finals. Their entry was a project in which they developed organisms to detect and destroy toxins generated from the oil extraction process used in northern Alberta's oil sands. You can learn more about their discoveries at http://2012.igem.org/Team:Calgary.

Figure B **iGEM competition team from the University of Calgary (2012).**

receptor, or transcription factor best suited for the biological system under construction (see the **Focus on Inquiry** box **"The iGEM Competition"**).

At one extreme of the spectrum, a synthetic biologist might use his or her knowledge to make only a handful of modifications to a cell (see Sections 24.2 and 24.3). Further along the spectrum, a synthetic biologist might also choose to make more comprehensive modifications through the use of a procedure like multiplex automated genome engineering (see Section 24.4). Lastly, at the very end of the spectrum, a researcher might decide to simply synthesize an entire genome from "scratch" (see Section 24.5). For all these methodologies, however, the ultimate goal is the same: to create a biological device, with user-defined properties, that can serve a useful purpose to society.

Generally speaking, the process of designing a biological system begins by combining the basic design units of systems that have already been characterized. By using the appropriate biobricks, it is possible to coordinate the biochemical reactions that form the pathways, that make up the networks that define the cell types, which in turn become tissues. This approach is not fundamentally different from the approach taken to build a computer; where transistors, capacitors, and resistors are used to form the circuits, that make up the circuit boards, that define the computer, that is part of the world wide web (**Figure 24.3**). While still in its infancy, the discipline of synthetic biology has the potential to revolutionize how we think about life. Furthermore, as we shall see later in this chapter, it is also likely to influence how society chooses to solve some of its most pressing problems (see Section 24.5).

> The discipline of synthetic biology is a fusion of both biological and engineering principles. Using a hierarchical approach, synthetic biologists make use of repositories of standard parts (or biobricks) to create novel organisms with user-defined properties to benefit society.

24.2 Synthetic Biology: The First Synthetic Gene Circuits

The discipline of synthetic biology is rooted in (1) the work of Jacob and Monod, who demonstrated the circuit-like control of gene regulation, and (2) the field of recombinant DNA technology, which discovered and described methods for the experimental manipulation of DNA and the genetic engineering of organisms. Using this work as a foundation, synthetic biologists have continued to extend these ideas in order to manipulate organisms on a much grander scale (i.e., to engineer multiple biological components into complex artificial systems with specific, user-defined functions). By attempting to construct such systems, researchers are not only able to create practical products, but also (based on their success or failure) to test their knowledge of genetics. In fact, the construction of artificial systems can be thought of as one of the best ways to test one's knowledge. For instance, the ability of aeronautical engineers to build complex aircraft like the Boeing 747 is excellent evidence of their understanding of

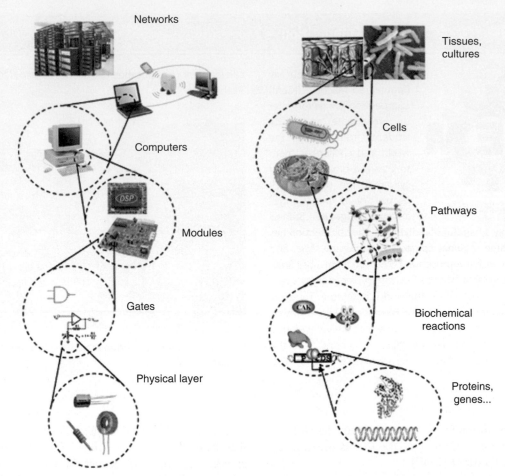

Figure 24.3 Building biological systems with biobricks uses the same hierarchical strategy employed in the construction of computers.

the principles of flight. In the remainder of this section, we will examine some of the first successful attempts at the creation of novel, rationally designed, genetic circuits.

Toggle switches

One of the first artificially created biological devices was the toggle switch. Such switches are important to biology because they represent one of the most basic elements for memory storage. Memory storage is possible because toggle switches demonstrate bistable behaviour; that is, they can be in one of two stable states. Furthermore, they can be induced to switch between states based upon the inputs to the system (**Figure 24.4a**). Interestingly, such systems have proven to be crucial for a wide variety of developmental processes in biology (where cells irreversibly change from an initial undifferentiated state to a differentiated one).

From a theoretical perspective, one could create a toggle switch with just two genes encoding repressors (we will call them repressor A and repressor B). If the repressor A protein bound the promoter of the repressor *B* gene, and the repressor B protein bound the promoter of the repressor *A* gene, then a bistable system could be established if one had inputs (e.g., allosteric small molecule inhibitors) capable of inhibiting repressor A and repressor B, respectively (we will call them Input 1 and Input 2). In this scenario, exposure to Input 1 leads to the expression of repressor A, thus turning the expression of the repressor *B* gene off. In fact, even if the input were removed, the system would remain in this state. If the system was exposed to Input 2, then the expression of repressor B would be turned on, turning off repressor gene *A* expression. Again, even if Input 2 were removed, the system would remain in this state.

Now imagine the same system, only this time, assume that the repressor B coding sequence is fused to the coding sequence of a GFP reporter protein. When repressor B is expressed, the cell fluoresces a bright green colour. In this case, upon transient exposure to Input 1, the system can be thought of as being toggled to the off state (no fluorescence). Likewise, upon transient exposure to Input 2, the system can be thought of as being toggled to the on state (green fluorescence). Remarkably, using parts from prokaryotic regulatory systems, just such a toggle switch has been created (**Figure 24.4b**).

The first repressor of the system is one you know well, the *lacI* repressor. The second is the *tetR* repressor. The *tetR* repressor is part of an operon involved in conferring resistance to the antibiotic tetracycline. When *E. coli* cells are grown in the absence of tetracycline, the *tetR* repressor is bound to the *tetO-1* promoter keeping expression of the *tetA* gene off. (The *tetA* gene encodes an efflux pump

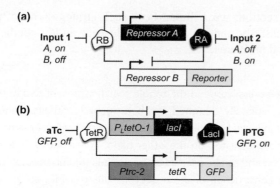

Figure 24.4 A genetic toggle switch. (a) A theoretical toggle switch composed of two repressor proteins, A and B. The expression of the repressor *A* gene is controlled by the repressor B protein (RB), and the expression of the repressor *B* gene is controlled by the repressor A protein (RA). RA activity is controlled by Input 2, while RB activity is controlled by Input 1. **(b)** An actual toggle switch composed of the *lacI* and *tetR* repressors. *lacI* expression is controlled by the *tetR* repressor, and *tetR* expression is controlled by the *lacI* repressor. LacI is allosterically inhibited by the presence of the inducer IPTG (isopropylthio-β-galactoside, a structural analogue of allolactose), while TetR is allosterically inhibited by aTc (anhydrotetracycline).

that is able to transport tetracycline out of the cell, therefore conferring resistance.) In the presence of the drug, however, imported tetracycline binds to the *tetR* repressor inhibiting its ability to bind to the *tetA* operator sequence, thus allowing transcription of *tetA*.

In the artificially constructed system shown in Figure 24.4b, the *lacI* repressor gene is put under the control of a promoter ($P_L tetO-1$) that is regulated by the *tetR* operator sequence. Similarly, the *tetR* repressor gene is placed under

the control of a promoter (*Ptrc-2*) that is regulated by the *lacI* operator sequence. The *tetR* coding sequence is fused to the coding sequence of the *GFP* gene. In this way, researchers can control the on/off state of the system through the addition of the two inputs: aTc (a derivative of tetracycline) or IPTG (a molecule structurally similar to allolactose). This is to say, transient exposure to aTc turns the system off (no fluorescence) and transient exposure to IPTG turns the system on (green fluorescence). While this system serves no practical purpose, this ingenious biological engineering experiment helped prove a principle: Cells could indeed be rationally altered to carry out user-defined functions.

The repressilator

Another early example of a synthetically designed biological system is the repressilator (**Figure 24.5**). In this case, the system acts as an oscillating timing mechanism. Such systems are crucial in biology for processes that cycle with a predetermined rhythm (e.g., the cell cycle or circadian cycles). In order to construct an oscillating gene circuit, researchers used three repressors: the *lacI* repressor, the *tetR* repressor, and the *cI* repressor (a repressor normally involved in controlling aspects of the lambda phage life cycle). The *tetR* repressor binds the *tetR* operator that is controlling the expression of the *lacI* repressor, and the *lacI* repressor binds the operator sequence that is controlling the expression of the *cI* repressor. The circuit is completed through the binding of the *cI* repressor to the *cI* operator that is controlling the expression of the *tetR* repressor. In addition, the *tetR* repressor was made to inhibit the expression of a *GFP* reporter gene through the incorporation of a *tetR* operator sequence in the *GFP* gene's promoter. The first promoter in the cascade drives

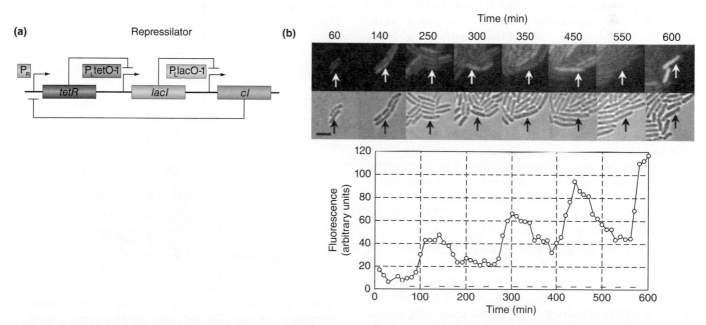

Figure 24.5 The repressilator. (a) The oscillating repressilator is composed of three repressor genes, each controlled by the repressor encoded by another gene in the system. **(b)** The repressilator in action. The GFP expression levels of an *E. coli* cell (indicated by the *white arrow*) is monitored over the course of the experiment. The *middle* panel shows the same cell (*black arrow*) using light microscopy. Quantitative measurements of GFP fluorescence are shown in the *bottom* panel. GFP expression cycles with a period of approximately 150 minutes.

expression of the second promoter's repressor, the second promoter drives expression of the third promoter's repressor, and the third promoter drives expression of the first promoter's repressor. Therefore, the expression of each of the encoded proteins is forced to oscillate in a regular cycle with a definable period. Furthermore, through the incorporation of the GFP reporter, this oscillation can be observed experimentally by simply monitoring GFP fluorescence.

While another interesting proof-of-principle, the repressilator again served no practical purpose. For synthetic biology approaches to be appreciated and accepted by a wider audience, it was clear that they needed to be put to use by researchers in order to benefit society. In the next section, we will see how such strategies have been successfully applied to solve industrial and/or medically related problems.

> The repressilator and the toggle switch represent two of the first rationally designed synthetic genetic circuits. Toggle switches can be exploited to act as a form of cellular memory. The repressilator, on the other hand, cycles with a defined period and can be used as an intracellular timing mechanism. The construction of these gene circuits provides an underlying proof-of-principle for the discipline of synthetic biology.

24.3 Applications of Synthetic Biology

Altering metabolism

The production of metabolites of industrial (e.g., biofuels) or medical (e.g., pharmaceuticals) importance is one of the most common applications of synthetic biology. For example, it is now possible to use synthetically engineered budding yeast strains to produce artemisinic acid, a precursor to artemisinin (a critically important antimalarial drug). Since previous methods of extracting artemisinic acid from its natural source, *Artemissia annua* (also known as sweet wormwood), are both complex and expensive, researchers have engineered yeast strains to make the process more efficient and thus more economically viable. Three classes of modifications were introduced into yeast (**Figure 24.6**). First, the mutant *ucp2-1* allele was incorporated into the strain, resulting in increased production of farnesyl pyrophosphate (FPP, a precursor of artemisinins). Second, the squalene synthetase gene (under control of a methionine repressible promoter) was added. In the presence of methionine, this addition results in the downregulation of the pathway responsible for converting FPP to ergosterol (therefore increasing the flux to artemisinic acid production). Third, *Artemissia annua* genes encoding amorphadiene synthase (ADS), cytochrome p450 (cyp71), and cytochrome p450 oxidoreductase were also added. By incorporating these modifications, the researchers were able to produce artemisinic acid with relative ease and at a fraction of the cost of previous methods. Fortunately, this has made the drug much more readily available for the treatment of malaria in developing countries.

Drug discovery

In addition to the production of important pharmaceuticals, synthetic strategies can also be used in the drug discovery process. Generally speaking, given a detailed understanding of the molecular details of a process, it is possible to design artificial circuits that can be used for drug screening. For example, the antibiotic ethionamide is used as a treatment for tuberculosis. For the drug to be effective, however, it must first be converted to a cytotoxic form by an enzyme coded for by the *Mycobacterium tuberculosis*

ethA gene. Unfortunately, expression of the *ethA* gene is repressed by the product of the *ethR* gene. One common route of ethionamide resistance in *M. tuberculosis* is through the selection of variants that have high *ethR* expression levels. For these reasons, drugs capable of inhibiting EthR are highly sought after since they greatly enhance the effectiveness of ethionamide treatment.

The circuit described in **Figure 24.7** provides a simple platform to identify *ethR* inhibitors. The first component of

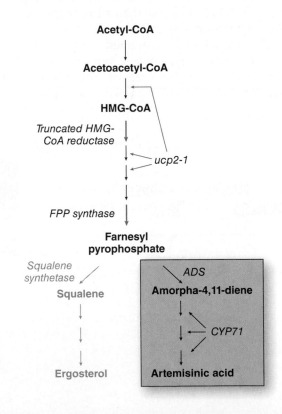

Figure 24.6 Synthetic metabolic pathway for the production of artemisinic acid in yeast. The *ucp2-1* allele results in increased synthesis of farnesyl pyrophosphate (*blue arrows*). The addition of the squalene synthetase gene reduces the flux of farnesyl pyrophosphate to ergosterol (*light grey*). The addition of the amorphadiene synthase pathway (*mauve box*) converts farnesyl pyrophosphate to artemisinic acid.

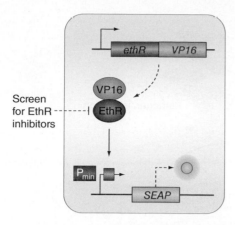

Figure 24.7 A synthetic circuit for the identification of EthR inhibitors. In the presence of drugs that inactivate EthR, expression of the SEAP (human placental secreted alkaline phosphatase) reporter is inhibited.

the circuit is a recombinant gene encoding a fusion of the *ethR* coding sequence to the mammalian *trans*-acting transcription factor, VP16. (This fusion protein binds the *ethR* operator sequence, but—due to the presence of the VP16 moiety—now acts as an activator of transcription.) The second component of the system is a reporter gene, *SEAP* (human placental secreted alkaline phosphatase), which is placed under the control of a minimal promoter, P_{min} (this promoter possesses a binding site for the EthR–VP16 fusion protein). In the presence of drugs that inhibit EthR activity, the reporter gene is turned off. Using such a system, it is possible for researchers to quickly and easily screen a variety of compounds for the desired effect on EthR activity. In fact, this method has already been used to identify the drug phenyl ethyl butyrate, which now acts as a new line of defence against antibiotic-resistant *M. tuberculosis* strains.

Interfacing synthetic circuits

Perhaps the most exciting recent development in the field of synthetic biology has been the creation of synthetic circuits with the ability to interface with natural cellular processes. Such systems are able to (1) sense changes in the intercellular environment, and (2) coordinate expression of a reporter (or even a therapeutic drug) with the intracellular changes that are associated with the disease or infection. For example, in **Figure 24.8**, a synthetic circuit is shown that senses an endogenous cellular signal—anything from a transcription factor to a metabolite—that is associated with the disease phenotype of interest. In response to this signal, the circuit can be made to activate (1) a reporter (calling attention to the deleterious intracellular changes) and/or (2) a gene whose expression counteracts the deleterious alteration.

One particularly ingenious example of such an interfacing system is the creation of synthetic killer microbes that are capable of eradicating the human pathogen *Pseudomonas aeruginosa*. *P. aeruginosa* is capable of infecting both the human respiratory and gastrointestinal tracts. Such infections can be quite serious, especially in immunocompromised patients. In fact, *P. aeruginosa* is responsible for

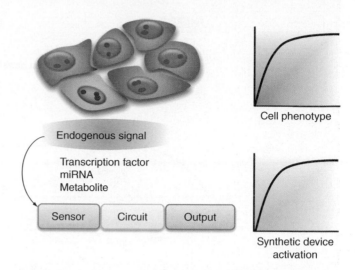

Figure 24.8 A generic example of an interfacing synthetic circuit. The synthetic genetic circuit—in the presence of an endogenous signal that is associated with the disease—activates the expression of a reporter gene (to call attention to the change) and/or activates the expression of a gene able to counteract the effects of the disease.

approximately 10 percent of all hospital acquired infections. To make matters worse, it is very common for *P. aeruginosa* isolates to rapidly acquire antibiotic resistance.

To combat *P. aeruginosa*, researchers in Singapore developed a synthetic *E. coli* strain capable of sensing and killing pathogenic *P. aeruginosa* strains (**Figure 24.9**). Since *P. aeruginosa* infects the gastrointestinal tract, and *E. coli* is a natural inhabitant of this compartment, they hypothesized that it would be possible to prevent and/or treat infections by colonizing the human gut with engineered *E. coli*. The system works through the incorporation of a synthetic circuit composed of three genes: the *lasR* gene (the sensor), the pyocin S5 gene (the killer), and the E7 lysis protein gene (the release mechanism). The *lasR* gene is expressed constitutively under the control of the *tetR* promoter. Furthermore, the LasR protein is able to bind a compound, $3OC_{12}HSL$, that is released by *P. aeruginosa* cells. When levels of this compound reach a certain threshold level, LasR–$3OC_{12}HSL$ complexes bind *luxR* promoters, activating the expression of the genes encoding pyocin S5 and the E7 lysis protein. The E7 protein then lyses the *E. coli* cells, releasing the pyocin S5 into the gut. Importantly, this compound exhibits strong bactericidal activity against *P. aeruginosa*, but does not affect *E. coli* cells. This ingenious interfacing system (that incorporates itself into the normal human gut flora) is able to both sense the presence of *P. aeruginosa* and eradicate the infection.

> Synthetic strategies have been used to increase the efficiency of drug production, for drug discovery, and for the creation of systems capable of interfacing with normal cellular processes. These latter systems are particularly useful since they can coordinate the expression of reporters (or therapeutic drugs) with intracellular changes associated with disease.

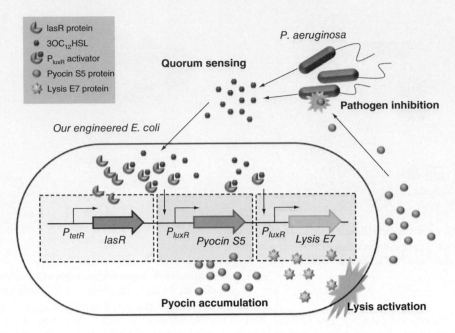

Figure 24.9 The engineering of synthetic killer microbes. In response to the secretion of the *P. aeruginosa* metabolite 3OC$_{12}$HSL, engineered *E. coli* cells activate expression of the E7 lysis protein and the antibiotic pyocin S5. This lyses the *E. coli* cells, releasing the antibiotic to combat the *P. aeruginosa* infection.

24.4   Multiplex Automated Genome Engineering

In addition to creating small synthetic gene circuits like the ones described in Section 24.3, it is also possible to incorporate genetic modifications on a much grander scale through a novel synthetic strategy referred to as multiplex automated genome engineering (MAGE). Conceptually, the methodology is simple. Based on complementarity, small ssDNA oligonucleotides are targeted to specific genomic locations (where they act as Okazaki fragments during the synthesis of the lagging strand). By incorporating small mismatches, insertions, and/or deletions in the oligonucleotides (but making sure there is enough homology remaining to efficiently target the oligo to the correct genomic location), it is possible for the added oligos to be incorporated into the newly replicated DNA, creating the desired genetic alteration. This is achieved with the help of bacteriophage λ-Red ssDNA-binding protein β, a protein that promotes strand annealing. Remarkably, this technique can incorporate up to 50 different genomic alterations at a given time.

The power of the technique is derived from the use of degenerate oligo pools that incorporate a set of defined and targeted genetic changes. By continually adding synthetic oligos to a population of continually growing *E. coli* cells, it is possible to rapidly generate a great amount of diversity at a variety of chromosomal locations. By selecting for the desired phenotype, researchers employing the technique can isolate variants that display user-defined characteristics. Essentially, the technique takes engineering principles and places them within the context of evolution to create an artificial evolution machine. The following example may clarify what is meant by the use of this term.

As a proof-of-principle, the researchers chose to create an *E. coli* strain capable of producing large amounts of an isoprenoid, lycopene (a commercially and pharmaceutically important antioxidant). To increase lycopene production, the researchers targeted 24 genes that were known to be involved in various aspects of lycopene synthesis. Then, using a MAGE machine (**Figure 24.10**)—an instrument that automates the process—the researchers were able to isolate strains expressing five times the normal amount of lycopene. This was achieved in only three days and with only $1000 worth of reagents! As demonstrated by this example, MAGE represents a revolutionary new methodology with the potential to transform how both medical and industrial researchers produce a wide variety of compounds.

> Multiplex automated genome engineering is a revolutionary new technique that accelerates evolution by repeatedly introducing synthetic ssDNA oligonucleotides at many targeted locations in a large population of cells. This generates a high degree of genetic diversity and a richly heterogeneous population. Cells with the desired user-defined properties can then be selected from this diverse population.

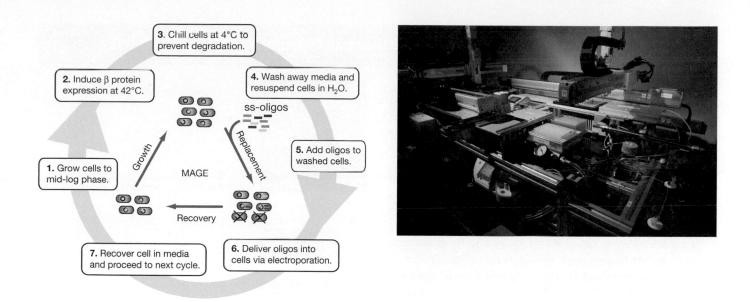

Figure 24.10 Multiplex automated genome engineering. (a) *E. coli* cells are grown to mid-log phase and then induced to express bacteriophage λ-Red ssDNA-binding protein β. Cells are then chilled and suspended in water. Added oligos are then introduced by electroporation. The process can be repeated for as many cycles as required. **(b)** A photograph of an actual MAGE machine constructed by the researchers. This instrument automates the growth and handling of the *E. coli* cells.

24.5 Synthetic Genomics

Perhaps the most ambitious synthetic strategy is the one championed by J. Craig Venter and colleagues. Referred to as synthetic genomics, this methodology involves synthesizing entire genomes from scratch in the laboratory. This strategy represents a huge landmark, not only for genetics and science in general, but for society as a whole. This methodology, in contrast to the previous 140 years of genetics research (which has concerned itself with the reading/understanding of genomes), focuses on the writing and construction of genomes. In the same way that a child learns to read and then eventually progresses to the point where he or she can write his or her own compositions, the field of genetics has itself matured and moved from genome sequencing to genome construction. How has this become possible? In this section, we will see how synthetic genomics, which only a few short years ago would have seemed like science fiction, has become a reality.

Genome transplantation

In order to create a synthetic organism three feats must be accomplished: (1) The researcher must have sufficient biological knowledge to write a complete functional genome, (2) the written genome must be synthesized and assembled, and (3) the genome must be transplanted into a host cell. While the latter two feats are technically difficult (but conceptually simple), the first point, in contrast, is technically simple (i.e., as simple as typing the sequence of A's, G's, C's, and T's into a computer), but conceptually very complex. In fact, this first critical step is the greatest limitation of the technique as it stands today. To begin our

exploration of Venter's work, we will begin by examining the simplest of these three feats, genome transplantation.

For a challenge of this magnitude, Venter and colleagues decided to start with the simplest organisms available, bacteria of the genus *Mycoplasma*. The *Mycoplasma* include some of the simplest free living organisms found on Earth (i.e., they are the simplest cells that satisfy all of the criteria of life). This conclusion is supported by the fact that bacteria of this genus possess incredibly small genomes (between 500 and 1000 genes and 0.6–1.0 Mb in size).

To prove that genome transplantation was possible, the researchers attempted to transplant the genome of one *Mycoplasma* species, *Mycoplasma mycoides* (the donor), to another *Mycoplasma* species, *Mycoplasma capricolum* (the recipient or host). In effect, the researchers were attempting to convert one species into another by substituting the recipient's biological instructions. Conceptually, this procedure is similar to rebooting a computer; with the genomic sequence being thought of as the software, the recipient cell being thought of as the hardware (i.e., the computer), and the process of transplantation being thought of as a reboot.

In their first landmark study, Venter and colleagues were able to develop procedures that allowed them to isolate the donor genome (which was modified to contain a tetracycline resistance gene) and introduce it into the recipient cell. Cells bearing just the donor genome were then selected for growth on media containing the antibiotic tetracycline. After many weeks, the researchers were able to show, through a variety of phenotypic tests, that the *M. capricolum* cells had indeed been converted to *M. mycoides*. Thus,

In addition to the genomic sequences defining *M. mycoides*, Synthia also contained four watermark sequences. Using the word in its traditional sense, a watermark can be defined as a translucent design imprinted on a piece of paper. By holding the paper up to the light, one can see the design and unambiguously identify who manufactured the paper. In a genetics context, a watermark is a stretch of DNA that serves to identify the organism as synthetic (i.e., it allows one to distinguish Synthia cells from normal *M. mycoides* cells). Instead of incorporating a set of random barcode sequences, Venter chose to create a code for translating DNA into the English language (i.e., a cipher for incorporating text messages into DNA). To this end, a triplet code—in which each letter of the English language, as well as a series of punctuation marks—was assigned a unique three-letter DNA triplet. Three of the four watermarks contained the following quotations:

"SEE THINGS NOT AS THEY ARE, BUT AS THEY MIGHT BE." – Robert Oppenheimer

"WHAT I CANNOT BUILD, I CANNOT UNDERSTAND." – Richard Feynman

"TO LIVE, TO ERR, TO FALL, TO TRIUMPH, TO RECREATE LIFE OUT OF LIFE." – James Joyce

The fourth watermark identified the J. Craig Venter Institute (JCVI) as the synthesizer of the genome and provided both an html code that pointed to the JCVI website and an email address. Thus, Synthia did indeed represent the first example of a nonhuman organism with its own website and email address!

researchers for the first time in human history were able to convert one species into another (**Figure 24.11**).

In the experiment described above, the researchers essentially borrowed a genome that already existed. This was done as proof-of-principle (i.e., to determine whether genome transplantation was possible). Imagine, however, if one had the biological knowledge to design an original, never-before-seen, donor genome—a genome that incorporated the researcher's choice of genes and regulatory sequences, and which was designed to confer the characteristics required for the application at hand. Such an accomplishment would represent the first truly synthetically constructed life-form. Our attention now turns to this next critical question: How can we create (i.e., synthesize and assemble) a user-defined genome from scratch?

Genome synthesis and assembly

As proof-of-principle, J. Craig Venter and colleagues first devised a strategy to synthesize and assemble a synthetic copy of the *M. mycoides* genome. This synthetic genome, in addition to carrying the normal *M. mycoides* complement of genes and regulatory sequences, was designed to contain four watermark sequences that unambiguously defined the genome as synthetically constructed. (See the **Focus on Critical Thinking** box **"The Only Nonhuman Organism with Its Own Website?!"**) To achieve this goal, they first

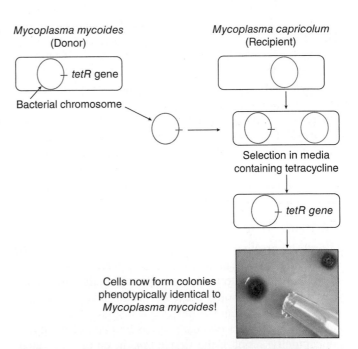

Figure 24.11 Genome transplantation. In Venter's landmark research, genomic DNA from *M. mycoides* cells was isolated and then introduced into *M. capricolum* cells. After selection in media containing tetracycline, cells containing only the *M. mycoides* genome were isolated. Remarkably, these cells were genetically and phenotypically identical to the original *M. mycoides* cells.

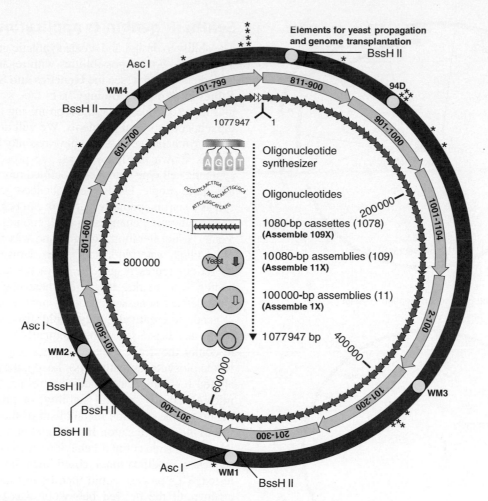

Figure 24.12 Synthesis and assembly of a genome. An oligo synthesizer machine was used to create a series of small oligos representing the entire synthetic genome. The oligos were assembled into approximately 1-kb-long cassettes (*orange arrows*) through polymerase cycling assembly. These 1-kb cassettes were assembled into approximately 10-kb-long cassettes (*blue arrows*) through homologous recombination in yeast. Then these 10-kb cassettes were then assembled into approximately 100-kb-long cassettes (*green arrows*), which were in turn assembled into a complete genome (*red circle*) via homologous recombination. The locations of watermarks are indicated by *yellow circles*.

created 1078 individual dsDNA cassettes (each 1080 bp long) that together represented the entire *M. mycoides* genome. Each individual cassette possessed 80 bp of overlap between itself and its adjacent cassettes (**Figure 24.12**). (In Figure 24.12, each *orange arrow* represents one such 1080-bp cassette.) Each of these cassettes was constructed with the use of ssDNA oligos (synthesized via an oligo synthesizer) that were assembled in a process called polymerase cycling assembly (**Figure 24.13**).

The next step of the hierarchical assembly process was to combine these cassettes, ten at a time, to produce 109 approximately 10-kb-long assemblies (*blue arrows*). These 10-kb cassettes were created with the help of the budding yeast, *Saccharomyces cerevisiae*, which recognizing the homology, assembled the fragments into one contiguous 10-kb-long cassette through homologous recombination (**Figure 24.14**). The 10-kb cassettes were then assembled (again through homologous recombination in yeast) into 11 approximately 100-kb-long assemblies (*green arrows*). Finally, the 100-kb cassettes were homologously recombined and propagated in yeast to reconstitute the complete 1.0-Mb-long *M. mycoides* genome.

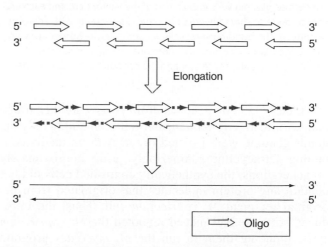

Figure 24.13 Polymerase cycling assembly. ssDNA oligos representing a given stretch of DNA are designed so that they share complementarity and overlap. The overlapping oligos prime DNA synthesis in a PCR. After many cycles of denaturation, annealing, and extension, a contiguous DNA fragment results.

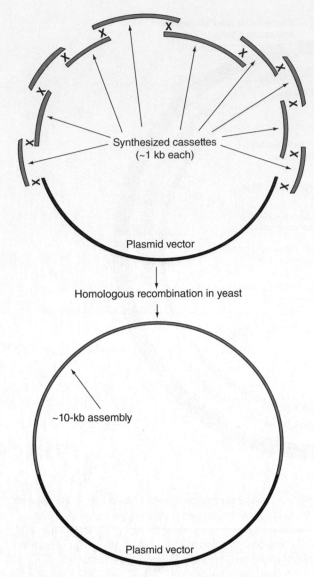

Figure 24.14 The use of homologous recombination to assemble DNA fragments in yeast. dsDNA cassettes approximately 1 kb long (*blue*) were introduced into budding yeast cells along with a linearized plasmid vector. Each end of the vector (*red*) and each end of each cassette shared overlapping complementarity with adjacent cassette(s). The budding yeast cells, recognizing the homology, were able to recombine the fragments into one contiguous circular DNA molecule containing the vector and the contiguous 10-kb cassette.

Intact synthetic genomes from yeast were then transplanted into *M. capricolum*, and cells containing only the donor genome were isolated (by growth on media containing tetracycline). Amazingly, after approximately 30 generations, the synthetically controlled cells did not contain any protein molecules that originated from the original recipient *M. capricolum* cell. Thus, the introduced software had indeed rebooted the *M. capricolum* cells, inducing them to run the *M. mycoides* program (i.e., the software built its own hardware). The first synthetically controlled organism was thus created and dubbed Synthia or JCVI-1.0syn.

Synthetic genomics applications

The ability to design and create synthetic genomes opens up a Pandora's box of possibilities with regard to their use in biological research (see the **Genetics and Society** box "**The Ethics of Synthetic Biology**"). In fact, synthetic technologies represent one of the most promising avenues for solving a host of societal problems. We will describe just a few of the synthetic projects that have recently been initiated.

As shown in **Figure 24.15**, the process of creating a synthetic cell type with a user-defined function is conceptually quite simple. Using the sequenced genome of a given cell as a starting point, researchers can begin the process by making iterative changes to the genome. For example, in Venter's first iteration, the genome was modified by simply adding watermark sequences. However, such minor alterations represent just the tip of the iceberg. Instead of adding watermarks, the researchers could just as easily have deleted nonessential genes unnecessary for the task-at-hand. Alternatively, they could also have added genes (from any organism with a sequenced genome) in order to confer the desired characteristics (e.g., genes encoding whole biosynthetic pathways). Lastly, they could have also altered regulatory sequences controlling gene expression, mRNA stability, protein stability, or protein localization (again to fine-tune the metabolism of the synthetic organism to the desired state). If the initial iteration failed to produce the desired cellular behaviour, a second iteration (with sequence modifications) could then be constructed and tested. This process could then be repeated over and over again until the desired behaviour was produced. Amazingly, such approaches are currently being employed in an attempt to solve some of society's most pressing problems.

For example, with dwindling oil supplies, one pressing societal need is the production of a new generation of fuels. Several synthetic approaches to this problem have currently been initiated. In one such approach, synthetic organisms are being created to convert plant materials (i.e., lignocellulosic biomass) to advanced fuels (**Figure 24.16**). In an alternate approach, Venter and his team at Synthetic Genomics Incorporated (SGI) are in the process of designing organisms that are able to convert raw coal to fuel and other petrochemicals. Perhaps the most promising avenue is the direct conversion of atmospheric carbon dioxide to fuel using modified algal cells. In fact, at the time of writing, SGI in collaboration with Exxon Mobil has developed algae capable of secreting oils, thus improving the recovery process. It is envisioned that these oils (a biocrude) could then be used as a feedstock in the refinery process.

The use of algae in this way has many advantages. First, algae, with only sunlight as an energy source, will be able to remove the greenhouse gas carbon dioxide from the atmosphere. Second, algae produce significantly more biomass and oil compared with terrestrial crops, setting the stage for the efficient production of both food and fuel. Third, algae can thrive in waste water (or even sewage), freeing up land that can be used for agriculture.

Genetics and Society The Ethics of Synthetic Biology

The ability to create synthetic life-forms raises some important ethical considerations. In fact, upon publication of this work, J. Craig Venter and colleagues received phone calls from both the President of the United States and the Pope to discuss these issues! Fortunately, in the run-up to the publication of this work, an ethical review had been initiated by the Alfred P. Sloan Foundation, a nonprofit organization with the goal of supporting basic research and science education. The foundation's ethical review addressed two main concerns: (1) the potential for illegitimate use of such technologies for bioterrorism, and (2) the environmental impact that might result from the escape of a synthetic organism into the wild. In their report, the foundation carefully assessed both the risks and benefits of the technology, and in addition proposed guidelines for the governance of synthetic biology research. In addition, upon publication of Venter's work in May of 2010, United States

President Barack Obama directed the Presidential Commission for the Study of Bioethical Issues (chaired by Dr. Amy Gutmann, **Figure A**) to provide a report on the potential benefits and risks of the technology within six months. In December of the same year, the Commission provided the report and concluded that "synthetic biology offers extraordinary promise to create new products for clean energy, pollution control, and medicine, to revolutionize chemical production and manufacturing, and to create new economic opportunities" (**Figure B**). Furthermore, the Commission cited 18 recommendations to provide regulatory guidelines, enhance biosecurity protocols, and protect the environment. As with all technologies, synthetic biology holds great promise, but, as pointed out by the Commission, "With this promise comes a duty to attend carefully to potential risks, be responsible stewards, and consider thoughtfully the implications for humans, other species, nature, and the environment."

Figure A Amy Gutmann.

Figure B Report by the Presidential Commission for the Study of Bioethical Issues on Venter's work.

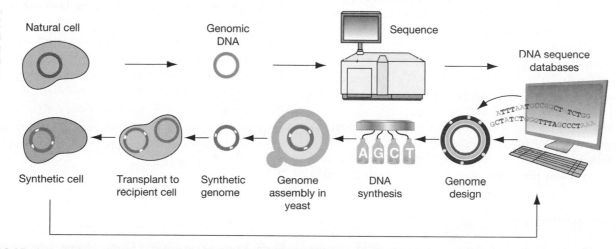

Figure 24.15 An overview of the iterative process used to create synthetic organisms.

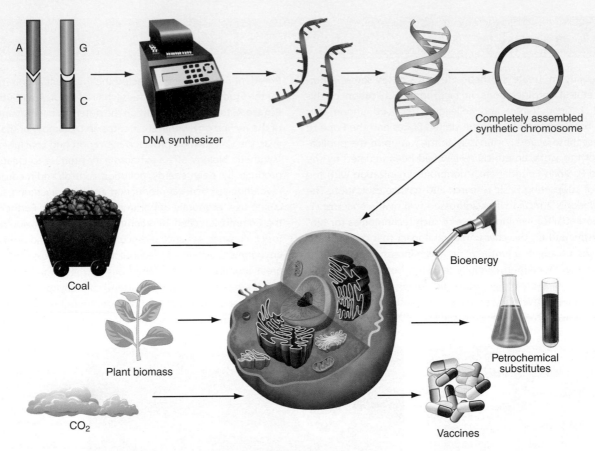

Figure 24.16 Possible applications for synthetic genomics technologies.

In addition to this collaboration with Exxon Mobil, Venter is also working in collaboration with the pharmaceutical company Novartis—through his company, Synthetic Genomics Vaccines Incorporated (or SGVI)—to improve the speed with which the flu vaccine is distributed. Normally, the World Health Organization identifies and distributes a reference virus that is used by companies like Novartis to make the annual seasonal vaccine (or, when necessary, a pandemic vaccine). Instead of waiting for the reference strain, Novartis and SGVI have created a bank of synthetically constructed seed viruses composed of a multitude of flu virus variants. These viruses are ready to go into production as soon as the seasonal or pandemic flu strain is identified (saving up to two months in production time). In fact, in April of 2013, the SVGI was able to quickly provide an H7N9 avian flu seed virus to the American Centers for Disease Control, as soon as it was identified as being responsible for an outbreak of the avian flu in China, Taiwan, and Australia.

While it is unclear what the future holds with respect to the applications of synthetic research, the creation of Synthia makes it clear that technical barriers related to the creation of synthetic life-forms no longer exist. As discussed in the Genetics and Society Box, Venter's methodology, like any new invention, has the potential for great societal benefits. Unfortunately, it could also be used for more nefarious purposes. However, by following the guidelines set out by the Presidential Commission of the United States, together with effective governance, it is expected that the benefits of these new technologies will far outweigh the risks.

Synthetic genomics is a rapidly evolving discipline in which entire genome sequences are designed, synthesized, assembled, and then transplanted into a host cell. At this point, the newly introduced genome can begin to direct the metabolism of the cell according to the incorporated instructions. The ultimate goal is to rationally design synthetic organisms so that they function to benefit society. For example, synthetic genomic technologies are currently being used in an attempt to create organisms capable of removing the greenhouse gas carbon dioxide from the atmosphere, and converting it into a novel generation of biofuels.

Connections

For most of its history, genetics has been concerned with the analysis of existing biological information. That is, the study of the genes and genomes of organisms that naturally inhabit Earth. However, as we saw in this chapter on synthetic biology, this notion has now been turned on its head. Instead of simply "reading" DNA sequences, it is now possible to

"write" entire genomes; thereby creating entirely synthetic, never-before-seen, organisms. These technologies, if applied conscientiously, have the potential to solve some of the most pressing needs of society by revolutionizing the concepts of chemical production and manufacturing. Thus, in a mere century-and-a-half, society has progressed from Mendel's laws and the abstract concept of a gene (Chapter 2), to the discovery of the structure of DNA (Chapter 5), to the analysis of entire genomes (Chapters 19–23), to the synthesis of artificial genomes encoding novel organisms with user-defined characteristics. One can only imagine the biological insight that will emerge from the next 150 years of genetics research.

Essential Concepts

1. The discipline of synthetic biology aims to create novel organisms with rationally designed characteristics that can be put to use to benefit society. **[LO1–4]**

2. One approach to creating synthetic organisms involves the use of a repository of standard parts (i.e., genes, promoter sequences) that have been characterized through previous research. These parts are referred to as biobricks and can be used to create novel genetic circuits with user-defined properties. **[LO2]**

3. Synthetic biology strategies have already been used successfully to create organisms for a wide variety of applications, including drug production and drug discovery. Synthetic circuits capable of interfacing with natural cellular processes hold particular promise. **[LO2–3]**

4. Multiplex automated genome engineering is a revolutionary new methodology that allows for the accelerated evolution of target cells. The system works by introducing synthetic ssDNA oligonucleotides at many targeted locations in the genome of a large population of cells. This generates a high degree of genetic diversity and permits the researcher to select for cells with the desired user-defined properties. **[LO3]**

5. Synthetic biology technologies have evolved to the point where it is now possible to design, synthesize, and assemble entire genomes. This allows for the rational design of synthetic organisms that can be used to benefit society (e.g., to create new biofuels). Continued openness and discussion among researchers and the public is needed to ensure that these technologies are used in an ethical manner. **[LO1, LO4]**

Solved Problems

I. Consider the two synthetic circuits described below. Compare and contrast the circuits in terms of how switch-like the two systems would be. How could you create a system that was even more switchlike? Are these systems related to the Hill equations described in Chapter 23? If so, how?

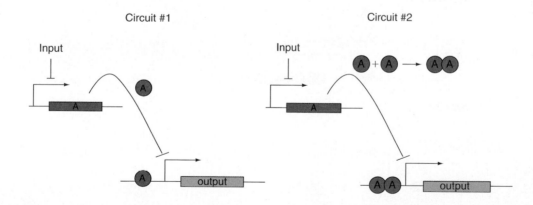

Answer

Circuit #1 consists of a regulatory gene that encodes a repressor; the repressor is able to inhibit the expression of the output gene. The input signal inhibits the expression of the repressor, thereby relieving inhibition of the output gene (turning on its expression). Thus, one would expect to see an increase in expression roughly proportional to the level of input and the circuit would not be very switchlike. Circuit #2 is similar, except that the repressor must form a homodimer in order to function. At low input levels, there will be high levels of expression. However, as the input signal rises, and expression of the repressor gene is tuned down, the likelihood of formation of homodimers will decrease as a *square* of the input signal. Relative to Circuit #1, Circuit #2 would be more switchlike. One could design a system that was even more switchlike by using a repressor that forms a homotetramer. Interestingly, Circuit #1 would correspond to a Hill function with $n = 1$. Circuit #2 would correspond to a Hill function with $n = 2$.

Vocabulary

1. For each of the terms in the left column, choose the best matching phrase in the right column.

i. synthetic biology	**1.** the discipline in which entire genome sequences are designed, synthesized, assembled, and then transplanted into a host cell
ii. multiplex automated genome engineering	**2.** a process whereby single-stranded DNA oligos are used to produce contiguous DNA fragments
iii. polymerase cycling assembly	**3.** a method whereby small single-stranded DNA oligonucleotides are targeted to specific genomic locations
iv. synthetic genomics	**4.** the discipline of constructing a biological system using characterized genes and regulatory DNA sequences

Section 24.1

2. Using your knowledge of synthetic biology, together with your understanding of network motifs, design a biological circuit that would output green fluorescence in *E. coli* cells in response to persistent (but not transient) exposure to arabinose.

Sections 24.2 and 24.3

3. Consider the two synthetic circuits shown below. Compare and contrast the circuits in terms of (i) how switchlike they would be, (ii) how long it would take for an input signal to be propagated into an output signal, and (iii) how each system would respond to small random fluctuations in input signal and/or repressor concentrations.

4. Using the biobricks described in Figures 24.4 and 24.5, design synthetic regulatory systems that show the following behaviour. You may use (or not use) any of the available biobricks an unlimited number of times.

a. A system that is always ON.

b. A system that is ON only if aTc is present.

c. A system that is ON only if both IPTG and aTc are present.

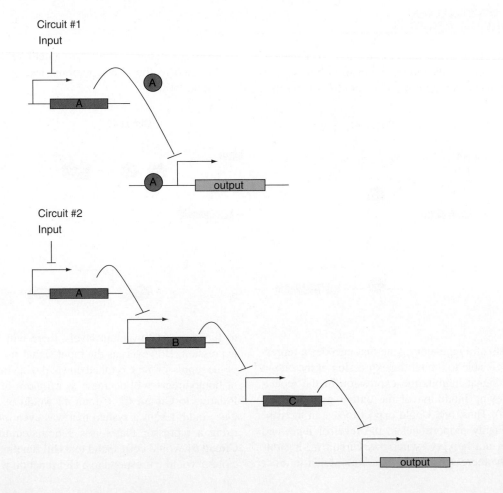

5. In the MAGE experiments that resulted in increased lycopene production, the researchers were fortunate in that lycopene levels could be assayed simply by monitoring the level of red pigmentation of the *E. coli* colonies. Comment on the utility of MAGE in situations where simple phenotypic assays (such as the appearance of red pigments) are not available for the desired characteristics.

Section 24.5

6. As an expert in the field of synthetic biology, the Minister of Science and Technology invites you to lunch to discuss the introduction of legislation to regulate synthetic biology research. What are your recommendations to the Minister? Explain your reasoning.

7. Describe the rationale behind the choice of *Mycoplasma* spp. as subjects in the synthetic genomic experiments conducted by J. Craig Venter and colleagues.

8. You are the chief executive and scientific officer of a new start-up company that seeks to develop a synthetic system to convert plant biomass into fuel that could be used to operate vehicles. To generate capital, you agree to appear on the television show *Dragon's Den*. Describe in general terms how you would market your idea. (*Note:* you only have five minutes to make your presentation, so you will not have time to go into the technical details.)

9. What is polymerase cycling assembly? Explain how it is used with respect to the creation of synthetic genomes.

10. As a newly hired research scientist at the J. Craig Venter Institute, you are given the assignment of creating a synthetic version of a 24-kb genome of a newly discovered coronavirus. It is hoped that subsequent combinatorial modifications of the synthetic viral genome will shed light on its ability to replicate within lung epithelial cells. Jake, the lab technician across the hall, sequenced the genome yesterday using pyrosequencing and has emailed you his data. Describe how you would go about creating a synthetic version of this genome using the raw sequence. Once created, how would you exploit the system to elucidate the genetic basis of pathogenicity of the virus?

11. A journalist from *The Globe and Mail* recently saw an episode of the television news program *60 Minutes* entitled "Designing Life." In this documentary, synthetic genomic technologies are described as potentially dangerous to the entire planet. As an expert in the field of synthetic genomics, the journalist wants your opinion for an upcoming feature story. What do you tell the journalist?

connect® For more information on the resources available from McGraw-Hill Ryerson, go to www.mcgrawhill.ca/he/solutions.

Problems   741

Guidelines for Gene Nomenclature

There are inconsistencies within the various branches of genetics on some nomenclature—because it is a relatively new area of scientific investigation, the consistency present in more basic sciences has not been established. The authors debated whether they should try to impose a consistency on the entire topic area and decided against that path. As the study of genetics matures, the process itself will create a more consistent nomenclature. The following guidelines can be applied to all chapters in this book.

General Rules

- Names of genes are in italics (*lacZ*, *CDC28*)
- Names of proteins are in regular (Roman) type with an initial cap (LacZ, Cdc28)
- Chromosomes: sex chromosomes are represented by a capital letter in Roman type (X, Y); autosomes are designated by a cardinal number (1, 2, 21, 22)
- Names of transposons are in Roman type (Tn10)

Specific Rules for Different Organisms

Human gene symbols are designated by uppercase Latin letters or by a combination of uppercase letters and Arabic numerals. The initial character should always be a letter and the whole symbol should have six or fewer characters. Greek letters in older gene symbols should be changed to the Latin equivalent. Thus, the haemoglobin-alpha gene was originally assigned the symbol HBα. The revised symbol is HBA. Alleles are limited to three characters using only capital letters or Arabic numerals. The allele designation is written on the same line as the gene symbol separated by an asterisk (e.g., *PGM1**1); the allele is printed as *1. More detailed nomenclature information is available at HGNC Guidelines (www.genenames.org/guidelines.html).

Brief Answer Section

Chapter 2

1. a. i. 4; ii. 3; iii. 6; iv. 7; v. 11; vi. 13; vii. 10; viii. 2; ix. 14; x. 9; xi. 12; xii. 8; xiii. 5; xiv. 1.

    b. i. 2; ii. 5; iii. 10; iv. 7; v. 6; vi. 8; vii. 11; viii. 3; ix. 4; x. 1; xi. 9.

3. For peas: (1) rapid generation time; (2) can either self-fertilize or be artificially crossed; (3) large numbers of offspring; (4) can be maintained as pure-breeding lines; (5) maintained as inbred stocks and two discrete forms of many phenotypic traits are known; (6) easy and inexpensive to grow. In contrast, for humans: (1) generation time is long; (2) no self-fertilization, it is not ethical to manipulate crosses; (3) produce only a small number of offspring per mating; (4) although people that are homozygous for a trait exist, homozygosity cannot be maintained; (5) populations are not inbred so most traits show a continuum of phenotypes; (6) require a lot of expensive care to "grow." One advantage to the study of genetics in humans is that a very large number of individuals with variant phenotypes can be recognized. Thus, the number of genes identified in this way is rapidly increasing.

5. Short hair is dominant to long hair.

7. The genotype can be determined by performing a testcross; that is, crossing your fly with the dominant phenotype (but unknown genotype) to a fly with the recessive (short wing) phenotype. If your fly has the homozygous dominant genotype, the progeny in this case would be Ww and would have the dominant phenotype. If your fly had a heterozygous genotype, 1/2 of the progeny would be normal (Ww) and 1/2 of the progeny would be short (ww).

9. The dominant trait (short tail) is easier to eliminate from the population by selective breeding. You can recognize every animal that has inherited the allele, because only one dominant allele is needed to see the phenotype. Those mice that have inherited the dominant allele can be prevented from mating.

11. a. Dry is recessive while sticky is dominant. b. The 3:1 and 1:1 ratios are obscured because the offspring are the combined results of different crosses.

13. a. 1/6. b. 1/2. c. 1/3. d. 1/36. e. 1/2. f. 1/6. g. 1/9.

15. a. 2. b. 4. c. 8. d. 16.

17. a. $aa\,Bb\,Cc\,DD\,Ee$. b. $a\,B\,C\,D\,E$ or $a\,B\,c\,D\,E$ or $a\,B\,C\,D\,e$ or $a\,B\,c\,D\,e$ or $a\,b\,C\,D\,E$ or $a\,b\,C\,D\,e$ or $a\,b\,c\,D\,E$ or $a\,b\,c\,D\,e$.

19. They must both be carriers (Pp); the probability that their next child will have the pp genotype is 1/4.

21. a. Rough and black are the dominant alleles (R = rough, r = smooth; B = black, b = white). b. 1/4 rough black : 1/4 rough white : 1/4 smooth black : 1/4 smooth white.

23. a. 3/16. b. 1/16.

25. P = purple, p = white; S = spiny, s = smooth. a. $Pp\,Ss \times Pp\,Ss$. b. $PP\,Ss \times P-\,ss$ or $P-\,Ss \times PP\,ss$. c. $Pp\,S- \times pp\,SS$ or $Pp\,SS \times pp\,S-$. d. $Pp\,Ss \times pp\,Ss$. e. $Pp\,ss \times Pp\,ss$. f. $pp\,Ss \times pp\,Ss$.

27. Cross 1: male = $tt\,Nn$, female = $tt\,Nn$; Cross 2: male = $Tt\,nn$, female = $tt\,Nn$; Cross 3: male = $Tt\,nn$, female = $Tt\,Nn$; Cross 4: male = $Tt\,nn$, female = $Tt\,NN$.

29. a. Recessive; two unaffected individuals have an affected child. It was a consanguineous marriage that produced the affected child. II-1

and V-2 are affected (aa); all unaffected individuals except II-2, II-4, III-4, III-5, and possibly V-1 are carriers (Aa). b. Dominant. The trait is seen in each generation and each affected child has an affected parent; if the trait were recessive it would not be possible for III-3 to be unaffected even though both his parents are affected. All affected individuals are Aa, though III-4, III-5, and III-6 could be AA; carrier, is not applicable when the mutation is dominant. c. Recessive. Unaffected parents have an affected child. I-2 and III-4 are affected (aa); II-4 and II-5 are carriers (Aa); all others could be AA or Aa, but I-1 is almost certainly AA if the disease is rare.

31. a. 2/3. b. 1/9. c. 4/9.

33. Recessive; common.

35. a. 1/16 = 0.0625. b. 0.067.

37. In about 40 percent of the families, both parents were Mm heterozygotes. In the remaining 60 percent of the families, at least one parent was MM.

39. One gene, two alleles, incomplete dominance; 1/2 $c^r c^w$ (yellow) : 1/4 $c^r c^r$ (red) : 1/4 $c^w c^w$ (white).

41. Long is completely dominant to short. Flower colour trait shows incomplete dominance of two alleles.

43. a. Single-gene inheritance with incomplete dominance; heterozygotes have intermediate serum cholesterol levels; homozygotes have elevated levels. The following people must have the mutant allele but do not express it (incomplete penetrance): Family 2 I-3 or I-4; Family 4 I-1 or I-2. b. Other factors are involved, including environment (particularly diet) and other genes.

45. a. ii (phenotype O) or $I^A i$ (phenotype A) or $I^B i$ (phenotype B). b. $I^B I^B$, $I^B i$, or $I^B I^A$. c. ii (phenotype O).

47. a. Marbled and dotted. b. 1/4 spotted dotted : 1/2 marbled : 1/4 spotted.

49. a. Coat colour is determined by three alleles of a single gene arranged in a dominance series with C (for chinchilla), which is dominant to c^h (for himalaya), which is dominant to c^a (for albino). b. Cross 1: $c^h c^a \times c^h c^a$; Cross 2: $c^h c^a \times c^a c^a$; Cross 3: $Cc^h \times C(c^h$ or $c^a)$; Cross 4: $CC \times c^h(c^h$ or $c^a)$; Cross 5: $Cc^a \times Cc^a$; Cross 6: $c^h c^h \times c^a c^a$; Cross 7: $Cc^a \times c^a c^a$; Cross 8: $c^a c^a \times c^a c^a$; Cross 9: $Cc^h \times c^h(c^h$ or $c^a)$ or $Cc^a \times c^h c^h$; Cross 10: $Cc^a \times c^h c^a$. c. 3/4 chinchilla (CC, Cc^h, and Cc^a) and 1/4 himalaya ($c^h c^a$) or 3/4 chinchilla (CC and Cc^a) and 1/4 albino ($c^a c^a$).

51. a. 2/3 curly : 1/3 normal. b. Cy/Cy is lethal. c. 90 curly-winged and 90 normal-winged flies.

53. a. The 2:1 phenotypic ratio shows that the montezuma parents were heterozygous, Mm, and homozygosity for M is lethal. b. 1/2 montezuma, normal fin : 1/2 green, normal fin. c. 6/12 montezuma normal fin : 2/12 montezuma ruffled fin : 3/12 green normal fin : 1/12 green ruffled fin.

55. Two genes are involved. The black mare was $A-bb$ and the chestnut stallion was $aaB-$, the F_1 bay horses were $AaBb$, and the F_2 liver horses were $aabb$.

57. a. There are two genes involved; homozygosity for the recessive allele of either or both genes causes yellow colour. Green parent is $AABB$; yellow parent is $aabb$. b. $AaBb$, $aaBb$, $Aabb$, and $aabb$ in equal proportions : 1/4 green : 3/4 yellow fruit.

59. Dominance relationships are between alleles of the same gene. Only one gene is involved. Epistasis involves two genes. The alleles at one gene affect the expression of a second gene.

61. 1/4 would appear to have O type blood, 3/8 have A, 3/8 have AB.

63.

	I-1	I-2	I-3	I-4	II-1	II-2	II-3	III-1	III-2
Phenotypes	AB	A	B	AB	O	O	AB	A	O
Genotypes	I^AI^B	I^Ai or I^AI^A	I^Bi or I^BI^B	I^AI^B	ii	I^AI^A or I^Ai or I^AI^B	I^AI^B	I^Ai	I^AI^A or I^AI^B or I^Ai or I^Bi or I^BI^B
	Hh	*Hh*	*H−*	*H−*	*H−*	*hh*	*Hh*	*Hh*	*hh*

One or both of I-3 and I-4 must carry *h*.

65. 2/6 yellow : 3/6 albino : 1/6 agouti.

67. a. $A^yaBbCc \times AabbCc$. b. six phenotypes: albino, yellow, brown agouti, black agouti, brown, black.

69. a. 27/64 wild type, 37/64 mutant. b. *AA Bb Cc*.

71. a. Two genes are involved. $A-B-$ and $aa\ B-$ are WR, $A-bb$ is DR, and *aa bb* is LR. b. WR-1 is *AA BB*; WR-2 is *AA bb*; DR is *aa BB*, and LR is *aa bb*. c. *Aa Bb* (WR) × *aa bb* (LR).

73. 44/56.

Chapter 3

1. i. 13; ii. 7; iii. 11; iv. 10; v. 12; vi. 8; vii. 9; viii. 1; ix. 6; x. 15; xi. 3; xii. 2; xiii. 16; xiv. 4; xv. 14; xvi. 5.

3. a. 7 centromeres. b. 7 chromosomes. c. 14 chromatids. d. 3 pairs. e. 4 metacentric and 3 acrocentric. f. females are XX.

5. a. iii. b. i. c. iv. d. ii. e. v.

7. a. 1, 1 → 2, 2, 2, 2, 1, 1. b. Yes, yes, yes, yes → no, no, no, no → yes. c. No, no, no, no → yes, yes, yes, yes → no. d. Yes, yes, yes, yes → no, no, no, no → yes.

9. Meiosis produces four cells (*n*, haploid), each with seven chromosomes.

11. a. Mitosis, meiosis I, II. b. Mitosis, meiosis I. c. Mitosis. d. Meiosis II and meiosis I. e. Meiosis I. f. None. g. Meiosis I. h. Meiosis II, mitosis. i. Mitosis, meiosis I.

13. a. Metaphase or early anaphase of meiosis I in a male (assuming X−Y sex determination in *Tenebrio molitor*). b. Sister chromatids, centromeres, and telomeres (among others). c. Five.

15. It is very realistic to assume that homologous chromosomes carry different alleles of some genes. In contrast, recombination almost always occurs between homologous chromosomes in any meiosis; thus the second assumption is much less realistic. The couple could potentially produce $2^{23} \times 2^{23} = 2^{46}$ or 70 368 744 177 664 different zygotic combinations.

17. Yes. Meiosis requires the pairing of homologous chromosomes during meiosis I.

19. a. 400 sperm. b. 200. c. 100. d. 100. e. 100. f. None.

21. a. Only females. b. Males. c. Males. d. 1/5 ZZ males and 4/5 ZW females.

23. a. Brown females and ivory-eyed males. b. Females with brown eyes and males with ivory or brown eyes in ratio of 1:1.

25. a. Nonbarred females and barred males. b. Barred and nonbarred females and barred and nonbarred males.

27. The bag-winged females have one mutation on the X chromosome that has a dominant effect on wing structure and that also causes lethality in homozygous females or hemizygous males.

29.

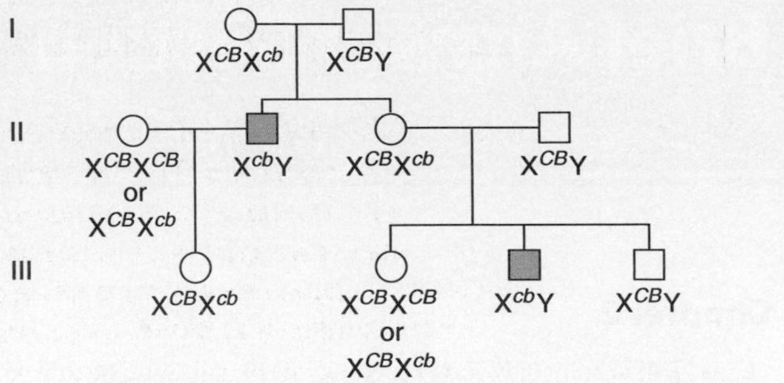

31. a. Recessive. b. Autosomal recessive. c. *aa*. d. *Aa*. e. *Aa*. f. *Aa*. g. *Aa*. h. *Aa*.

33. Vestigial wings is autosomal; yellow body colour is X-linked recessive.

35. a. X-linked dominant inheritance. b. Can exclude sex-linked recessive inheritance because affected females have unaffected sons. Can exclude autosomal recessive inheritance because the trait is rare and affected females have affected children with multiple husbands. Can exclude autosomal dominant inheritance because all the daughters but none of the sons of an affected male are affected. c. III-2 had four husbands and III-9 had six husbands.

37. a. 3. b. 1 or 3.

39. a. Purple is caused by homozygosity for a recessive allele of an autosomal gene (*p*), but the X-linked recessive white mutation is epistatic to *p* and to p^+. b. F_1 progeny: 1/2 white-eyed males and 1/2 wild-type (red) females; F_2 progeny: 1/4 white males, 1/4 white females, 3/16 red males, 3/16 red females, 1/16 purple males, 1/16 purple females.

41. a. Individual III-5. b. The *BRCA2* mutation has a dominant effect on causing cancer. c. The data do not clearly distinguish between X-linked and autosomal inheritance; *BRCA2* is actually on autosome 13. d. The penetrance of the cancer phenotype is incomplete. e. The expressivity is variable. f. Ovarian cancer is sex-limited, the penetrance of breast cancer may be sex-influenced. g. Low penetrance of the cancer phenotype, particularly among men.

Chapter 4

1. i. 7; ii. 3; iii. 1; iv. 8; v. 4; vi. 5; vii. 2; viii. 6.

3. a. Parental gametes are $B_1\ D_1$ and $B_3\ D_3$. b. Recombinant gametes will be $B_1\ D_3$ and $B_3\ D_1$. c. The *B* and *D* DNA loci are linked.

5. a. *Oo Bb*. b. 9:3:3:1. c. not significant. d. between 0.5 and 0.1.

7. a. Notice that the null hypothesis is the same in both cases: that the genes are assorting independently. b. Using two classes is a more sensitive test for linkage than using four classes. c. In a situation in which certain classes are subviable, you might see linkage with the two-class test, but you would miss the even more important point that one allele causes reduced viability. This ability to see the relative viability of the alleles is an advantage to the four-class method.

9. a. $Gs\ Bhd^+ / Gs^+\ Bhd$ ♀ × $Gs^+\ Bhd^+$ / Y ♂ → 49 $Gs\ Bhd^+$ ♂ : 48 $Gs^+\ Bhd$ ♂ : 2 $Gs\ Bhd$ ♂ : 1 $Gs^+\ Bhd^+$ ♂. The RF is 3 m.u. b. Genotypes, phenotypes, and frequencies of the female progeny would be the same as their brothers.

11. 10 percent.

13. a. A = normal pigmentation, a = albino allele, $Hb\beta^A$ = normal globin, $Hb\beta^S$ = sickle cell allele; 49.5% $a\ Hb\beta^A$, 49.5% $A\ Hb\beta^A$, 0.5% $a\ Hb\beta^A$, 0.5% $A\ Hb\beta^A$. b. 49.5% $a\ Hb\beta^A$, 49.5% $A\ Hb\beta^A$, 0.5% $a\ Hb\beta^A$, 0.5% $A\ Hb\beta^A$. c. 0.0025.

15. a. 1/4 black, 1/2 albino, 1/4 brown. b. 34 m.u. apart.

17. a. Gametes: 20% *A b* and *a B*, 30% *A B* and *a b*. F_2 generation: 59% $A-\ B-$, 16% $A-\ bb$ and $aa\ B-$, 9% *aa bb*. b. Gametes: 30% *A b* and *a B*, 20% *A B* and *a b*. F_2 generation: 54% $A-\ B-$, 21% $A-\ bb$ and $aa\ B-$, 4% *aa bb*.

19. a. The two genes are assorting independently. b. The two genes are on the same chromosome; yes. c. Recombination occurs at the four-strand stage of meiosis, and so many crossovers occur between genes

when they are far apart on the same chromosome that the linkage between alleles of these genes will be randomized. d. By summing up the values obtained for smaller distances separating other genes in between those at the ends.

21. The order of the genes is *HIS4 − LEU2 − MAT − THR4*.

23. The best map of these genes is as follows:

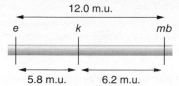

25. a. 360 $a^+ b^+ c^+$; 360 $a b c$; 90 $a^+ b c$; 90 $a b^+ c^+$; 40 $a^+ b^+ c$; 40 $a b c^+$; 10 $a^+ b c^+$; 10 $a b^+ c$. b. 500 $a^+ b^+ c^+$; 500 $a b c$.

27. a. *sc ec cv / + + +* and *b / +*. b. *sc ec cv / + + +*, $sc−ec = 9$ m.u., $ec−cv = 10.5$ m.u. c. Predicted DCO = 0.009, observed DCO = 0.001, interference = 0.89.

29. a. 39%. b. 39%. c. 0.5%. d. 8%.

31. $\dfrac{dwp / pld^+ rv^+ rmp}{dwp^+ / pld\, rv\, rmp^+}$; $rv−rmp = 10$ m.u.; $pld / dwp−rv = 5$ m.u.; interference = 0.

33. a. The sectors consist of $ade2^- / ade2^-$ cells generated by mitotic recombination. b. The sector size depends on when the mitotic recombination occurred during the growth of the colony. There should be many more small sectors because the mitotic recombinations creating them occur later in colony growth when there are many more cells.

35. a. Two mitotic crossovers occurred in succession in the same cell lineage. The first was between the *sn* and *y* genes, creating a patch of yellow tissue. The second was between the centromere and *sn*, creating a "clone within a clone" of yellow, singed cells. b. Yes.

Chapter 5

1. i. 6; ii. 10; iii. 8; iv. 2; v. 4; vi. 7; vii. 9; viii. 11; ix. 3; x. 5; xi. 1.

3. *c*.

5. Tube 1, nucleotides; tube 2, base pairs (without the sugar and phosphate) and sugar phosphate chains without the bases; tube 3, single strands of DNA.

7. a. 20% C. b. 30% T. c. 20% G.

9. Single-stranded.

11. 5′ CAGAATGGT~~GCTCTG~~CTAT 3′.

13. 3′ GGGAACCTTGATGTTTCGGCTCTAATT 5′.

15. a. 1/4 of DNA in a band at the bottom of gradient, 3/4 of DNA in a band near the top. b. 1/4 of DNA in a band at the middle of the gradient, 3/4 of the DNA in a band near the top. c. A broad smear of DNA. d. 1/8 of DNA in a band at the bottom of the gradient, 7/8 of DNA in a band near the top.

17. Round 1: each chromatid contains label on one of its two strands; Round 2 (if unlabelled strand used as template): double-stranded chromatid unlabelled; Round 2 (if labelled strand used as template): one strand of chromatid labelled, the other unlabelled, thus option c is correct.

19. a. Interaction with phosphate groups at the 5′ end of incoming nucleotide triphosphate and the 3′ end of previously incorporated nucleotide automatically dictates the 5′-to-3′ direction growth of the DNA chain. b. Phosphodiester bond using the energy of ATP hydrolysis. DNA polymerase would have no energy source to catalyze the formation of the phosphodiester bond to a nucleotide without a high-energy bond.

21. a. 3. b. 6. c. The one in the middle.

23. There is no way to synthesize complementary DNA for the 5′-most end of a newly synthesized linear chromosome since there is no preceding Okazaki fragment to be extended.

25. a. No new DNAs will be formed. b. No new DNAs will be formed. c. 5′ CTACTACGGATCGGGACTGG 3′. d. 5′ CCAGTCCCGATCCGTAGTAG 3′.

Chapter 6

1. i. 4; ii. 7; iii. 8; iv. 2; v. 3; vi. 5; vii. 1; viii. 6.

3. Interphase: 40-fold compaction; metaphase: 10 000-fold compaction.

5. a. 1.2×10^8 molecules of H2A protein. b. During or just after S phase. c. More templates that the cells can transcribe simultaneously, allowing the more rapid production of histone proteins.

7. H1 is on the outside of the complex and locks the DNA to the core. It interacts with H1 proteins from other nucleosomes forming the centre of the coil that is thought to form the 300 Å fibre. The other histone proteins are coated with DNA and cannot form the 300 Å fibre.

9. Mutate the DNA sequence so that the twelfth amino acid encoded is not lysine but another similar amino acid.

11. a. 1. b. 0. c. 1. d. 1. e. 3. f. 0.

13. Enough cells will have the *CB* allele active to provide sufficient colour vision.

15. a. All progeny have mutant coat colour. b. All progeny have wild-type coat colour. c. It is impossible to determine dominance or recessiveness since in this scenario you cannot examine the phenotype of heterozygotes under conditions where both alleles are expressed. d. In marsupials, the paternal X chromosome is always inactivated.

17. a. Barr body. b. Packaging of newly replicated DNA.

19. 5500 origins of replication.

21. a. 69 kb/s. b. 7.3 bp/s.

23. a. The CENP-A mutant dies while the CENP-B mutant is viable. Chromosome loss at elevated temperature cannot be measured in CENP-A because the cell dies. The CENP-B mutant, on the other hand, shows increased chromosome loss. b. Cells with a marker that is on a chromosome, or on an artificial linear chromosome (YAC), or on a circular plasmid containing a centromere.

25. In rare cases, the heteroduplex is not corrected and a single bacteriophage particle can be generated with a chromosome that contains the mismatch. One strand of DNA would be c^+ and the other strand would be c.

Chapter 7

1. i. 5; ii. 10; iii. 8; iv. 12; v. 6; vi. 2; vii. 9; viii. 14; ix. 3; x. 13; xi. 1; xii. 7; xiii. 15; xiv. 11; xv. 4; xvi. 16.

3. a. GU GU GU GU GU or UG UG UG UG.

   b. GU UG GU UG GU UG GU UG GU.

   c. GUG UGU GUG U etc.

   d. GUG UGU GUG UGU GUG UGU GUG UGU GU (depends on where you start).

   e. GUG UGU GU or UGU GUG UG (depends on where you start).

5. Gene *F*: bottom strand; Gene *G*: top strand.

7. Base-pairing between the codon in the mRNA and the anticodon in the tRNA is responsible for aligning the tRNA that carries the appropriate amino acid to be added to the polypeptide chain.

9. a. Translation. b. Tyrosine (Tyr) is the next amino acid to be added to the C terminus of the growing polypeptide, which will be nine amino acids long when completed. c. The carboxy-terminus of the growing polypeptide chain is tryptophan. d. The first amino acid at the N terminus would be f-met in a prokaryotic cell and met in a eukaryotic cell. The mRNA would have a cap at its 5′ end and a poly-A tail at its 3′ end in a eukaryotic cell but not in a prokaryotic cell. If the mRNA were sufficiently long, it might encode several proteins in a prokaryote but not in a eukaryote.

11. a. 1431 base pairs.

    b. 5′ ACCCUGGACUAGUGGAAAGUUAACUUAC 3′.

    c. N Pro Trp Thr Ser Gly Lys Leu Thr Tyr C.

13. Mitochondria do not use the same genetic code; mutate the 5′ CUA 3′ codons in the mitochondrial gene to 5′ ACN 3′.

15. Order: c e i f a k h d b j g.

17. a. Very severe. b. Mild. c. Very severe. d. Mild. e. No effect. f. Mild to no effect. g. Severe. h. Severe or mild.

19. Mutations possibly causing a detectable change in protein size: d, e, g, and i. In protein amount (assumes all mutant proteins are equally stable): e, f, j, and k. In mRNA size: i and j. In mRNA amount (assumes all mutant mRNAs with poly-A tails are equally stable): f and j.

21. If the met$^+$ phenotype is due to a true reversion: $met^- \times met^+ \rightarrow met^+ / met^- \rightarrow 2\ met^+$; $2\ met^-$. If there is an unlinked suppressor mutation: $met^-\ su^-$ (phenotypically met$^+$) $\times$ $met^+\ su^+$ (wild type) $\rightarrow$ met^- / met^+; $su^- / su^+ \rightarrow 3/4\ met^+$; $1/4\ met^-$.

23. a. 3′ AUC 5′. b. 5′ CAG 3′. c. Minimum two genes.

25. a. Missense mutations change identity of a particular amino acid inserted many times in many normal proteins but nonsense suppressors only make proteins longer. b. (i) A mutation in a tRNA gene in a region other than that encoding the anticodon itself, so that the wrong aminoacyl-tRNA synthetase would sometimes recognize the tRNA and charge it with the wrong amino acid; (ii) a mutation in an aminoacyl-tRNA synthetase gene, making an enzyme that would sometimes put the wrong amino acid on a tRNA; (iii) a mutation in a gene encoding either a ribosomal protein, a ribosomal RNA, or a translation factor that would make the ribosome more error-prone, inserting the wrong amino acid in the polypeptide; (iv) a mutation in a gene encoding a subunit of RNA polymerase that would sometimes cause the enzyme to transcribe the sequence incorrectly.

27. a. 5′ UUA 3′. b. 5′ UAG 3′ and 5′ UAA 3′ (due to wobble at the codon's 3′-most nucleotide). c. Gln, Lys, Glu, Ser, Leu, and Tyr.

29. Hb^C therefore precedes Hb^S in the map of the β-*globin* gene.

31. 5′ GGN GCA CCA AGG AAA 3′

33.

Stop Codon \ Change	UAA		UAG		UGA	
1st position	A̲AA	Lys	A̲AG	Lys	A̲GA	Arg
	C̲AA	Gln	C̲AG	Gln	C̲GA	Arg
	G̲AA	Glu	G̲AG	Glu	G̲GA	Gly
2nd position	UU̲A	Leu	UU̲G	Leu	UU̲A	Leu
	UC̲A	Ser	UC̲G	Ser	UC̲A	Ser
	UG̲A	STP	UG̲G	Trp	UA̲A	STP
3rd position	UA̲U	Tyr	UA̲A	STP	UG̲U	Cys
	UA̲C	Tyr	UA̲C	Tyr	UG̲C	Cys
	UA̲G	STP	UA̲U	Tyr	UG̲G	Trp

35. a. UGG changed to UGA or UAG so the DNA change was G to A. b. If the second base of the Trp codon UGG changes to A, a UAG stop codon will result. If the third base of the Trp codon UGG changes to A, a UGA stop codon will result. Mutation of A to T in the first base of the Lys codon leads to UAA. If the Gly codon is GGA, mutation of the first G to T creates a UGA stop codon.

37. three

39. a. Mutant 1: transversion changes Arg to Pro; mutant 2: single-base-pair deletion changes Val to Trp and then stop; mutant 3: transition Thr (silent); mutant 4: single-base-pair insertion changes several amino acids then stop; mutant 5: transition changes Arg to stop; mutant 6: inversion changes identity of six amino acids. b. EMS: 1, 3, 5; Proflavin: 2, 4.

Chapter 8

1. 1. i, ii, viii; 2. ii, iii; 3. vi, x; 4. vii, x; 5. i, ii, viii; 6. vi, ix, xi; 7. vii, xi; 8. iv, ix; 9. iv, v, vi, vii, ix.

3. The wild-type sequence is
5′ ACCGTA̲GTCGACT̲GGTA̲AACTTTGCGCG.

5. 9.5×10^{-5}; higher than normal rate.

7. If phages induce resistance, several resistant colonies appear in random positions on each of the replica plates. If the mutations pre-exist, the resistant colonies would appear at the same locations on each of the three replica plates.

9. Female A has a white-eyed mutation on the X. Female B has a recessive lethal mutation on the X. Female C is mosaic with a lethal mutation on one strand and wild-type sequence on the other strand of one X chromosome, or she is heterozygous for an incompletely penetrant lethal mutation.

11. a. 857 essential X-linked genes. b. 37.5 percent of the genes on the X chromosome are essential. c. The X-ray-induced mutation rate = 1.4×10^{-4}, a 40-fold increase.

13. a. Two-way mutagen. b. One-way mutagen. c. Two-way mutagen. d. Two-way mutagen. e. Two-way mutagen.

15. a. Nucleotide excision repair and the SOS-type error-prone repair. b. AP endonuclease and other enzymes in the base excision repair system could remove the damage and the SOS repair systems can work at AP sites, adding any of the four bases at random.

17. Yes; liver converts substance X into a mutagen.

19. a. Complementation test; − is a lack of complementation; + means that the two mutations complemented each other; boxes filled in green represent crosses already done, but the parents were switched. b. $1 \times 4 = -$, $1 \times 6 = +, 2 \times 3 = +, 2 \times 4 = +, 2 \times 5 = +, 3 \times 5 = +, 3 \times 6 = +, 4 \times 5 = +, 4 \times 6 = +$. c. Three genes: (1, 3, and 4), (2 and 6), (5).

21. a. Two. b. (1, 4), (2, 3, 5).

23. 45 purple: 16 green: 3 blue.

25. a. In all four crosses, there are two unlinked genes involved with complete dominance at both loci. b. (Each arrow represents a biochemical reaction catalyzed by one of the two gene products.) Cross 1: colourless → blue → purple; Cross 2: colourless1 → colourless2 → purple; Cross 3: colourless1 → red and colourless2 → blue, with red + blue = purple; Cross 4: colourless1 → purple and colourless2 → purple. c. Cross 2. d. F$_2$ only. (Cross 1) 2 purple : 1 blue : 1 white; (Cross 2) 1 purple : 1 white; (Cross 3) 2 purple : 1 red : 1 blue; (Cross 4) all purple.

27. a.   18   14   9   10   21
      X → D → B → A → C → thymine

    b. 9 and 10 accumulates B; 10 and 14 accumulates D.

29. a. Successful, immediate, prolonged. b. Unsuccessful. c. Successful, delayed, prolonged. d. Successful, immediate, prolonged. e. Unsuccessful. f. Unsuccessful. g. Successful, delayed, prolonged. h. Unsuccessful. i. Successful, immediate, short term. j. Successful, immediate, prolonged.

31. a. Two. b. 1/16 α1α1 β1β1 : 1/8 α1α2 β1β1 : 1/16 α2α2 β1β1 : 1/8 α1α1 β1β2 : 1/4 α1α2 β1β2 : 1/8 α2α2 β1β2 : 1/16 α1α1 β2β2 : 1/8 α1α2 β2β2 : 1/16 α2α2 β2 β2.

33. One chromosome with β β/δ δ; another with β/δ only (where / signifies a protein part of which, for example, the N-terminal part is one type of globin and the other part is the other type of globin).

Chapter 9

1. i. 4; ii. 8; iii. 6; iv. 5; v. 7; vi. 3; vii. 2; viii. 1.

3. In a duplication, there would be a repeated set of bands; in a deletion, bands normally found would be missing.

5.

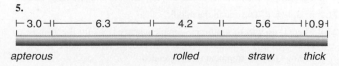

| ⊢ 3.0 ⊣|⊢ 6.3 ─────|⊢ 4.2 ───|⊢ 5.6 ────|⊢0.9⊣ |

apterous   rolled   straw   thick

7.

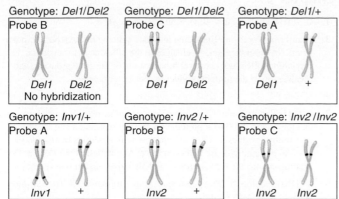

Genotype: *Del1/Del2*
Probe B
Del1 Del2
No hybridization

Genotype: *Del1/Del2*
Probe C
Del1 Del2

Genotype: *Del1/+*
Probe A
Del1 +

Genotype: *Inv1/+*
Probe A
Inv1 +

Genotype: *Inv2/+*
Probe B
Inv2 +

Genotype: *Inv2/Inv2*
Probe C
Inv2 Inv2

9. a. The parental types $y^+ z^1 w^{+R} spl^+$/ Y (zeste) and $y z^1 w^{+R} spl$ / Y (yellow zeste split). b. Crossing-over anywhere between the *y* and *spl* genes. c. Mispairing and unequal crossing-over between the two copies of the w^+ gene. d. 5.9 m.u.

11. a. 2, 4. b. 2, 4. c. 2. d. 1, 3.

13. a. 1, 3, 5, and 6. b. 2 and 4. c. 1 and 3. d. 5 and 6; 2 and 4.

15. a. 1/4 fertile green, 1/4 fertile yellow-green, 1/4 semisterile green, 1/4 semisterile yellow-green. b. 1/2 fertile yellow-green, 1/2 semisterile green. c. From crossing-over events between the translocation chromosome and homologous region on the normal chromosome.

17. 1/2 *Lyra* males : 1/2 *Lyra$^+$* (wild-type) females.

19. The 11-base-long primers must be 5′ GTTCGCATACG 3′ and 5′ GTGTACGCACG 3′.

21. *Ds* is a defective transposable element and *Ac* is a complete, autonomous copy.

23. Use a probe made of DNA from the sequence preceding the 200 A residues to hybridize to genomic DNA on Southern blots or to chromosomes by *in situ* hybridization.

25. a. The dark purple *K. waltii* genes are duplicated in *S. cerevisiae*. b. At some time after the evolutionary lines for these two species separated, a portion of the *S. cerevisiae* genome was duplicated in a progenitor of *S. cerevisiae*. Over time, one copy was lost of many of the duplicated genes. Occasionally both copies of a gene were retained.

27. a. 7. b. Sand oats : diploid, slender wild oats : tetraploid, cultivated oats : hexaploid. c. Sand oats : 7, slender wild oats : 14, cultivated oats : 21. d. Same answer as c.

29. a. (i) aneuploid, (ii) monosomic for chromosome 5, (iii) embryonic lethal. b. (i) aneuploid, (ii) trisomic for chromosomes 1 and 5, (iii) embryonic lethal. c. (i) euploid, (ii) autotriploid, (iii) viable but infertile. d. (i) euploid, (ii) autotetraploid, (iii) viable and fertile.

31. A: meiosis II in father. B: meiosis I in mother. C: meiosis I in father. D: meiosis II in mother.

33. You would actually expect more monosomies than trisomies, because meiotic nondisjunction would produce equal frequencies of monosomies and trisomies, but chromosome loss would produce only monosomies. The low frequency of monosomies observed is because monosomic zygotes usually arrest development so early that a pregnancy is not recognized. This may be due to a lower tolerance for imbalances involving only a single copy of a chromosome than for those involving three copies, or because recessive lethal mutations are carried on the remaining copy.

35. a. Mate putative mutants that are $ey ci^+$ / $ey^+ ci$ with flies that are *ey ci / ey ci*. b. Nondisjunction during meiosis I will produce wild-type progeny. nondisjunction during meiosis II cannot be recognized. c. The expected progeny is 2 eyeless : 2 cubitus interruptus : 1 eyeless, cubitus interruptus : 1 wild type. d. Mate putative mutants that are $ey ci^1$/ $ey^1 ci$ with flies that have an unmarked attached chromosome 4. Nondisjunction during meiosis II would yield eyeless or cubitus

interruptus progeny, but you could not recognize progeny resulting from nondisjunction during meiosis I. If the attached chromosome 4 carried two copies of *ey* and two copies of *ci*, you could recognize and discriminate some of the products of nondisjunction during the two meiotic divisions.

37. Treat with colchicine.

39. In autopolyploids, the banding patterns of homologues should be the same; in allopolyploids, different banding patterns will be seen for chromosomes from different species.

Chapter 10

1. i. 4; ii. 8; iii. 5; iv. 2; v. 7; vi. 1; vii. 3; viii. 6.

3. a. i, ii, iii. b. iv, v, vi.

5. Mutations in the promoter region can only act in *cis* to the structural genes immediately adjacent to this regulatory sequence. This promoter mutation will not affect the expression of a second, normal operon.

7. b.

9. The non-lysogenic recipient cell did not have the cI (repressor) protein, so the incoming infecting phage could go into the lytic cycle.

11.

	β-galactosidase	Permease
a.	constitutive	constitutive
b.	constitutive	inducible
c.	inducible	inducible
d.	no expression	constitutive
e.	no expression	no expression

13.

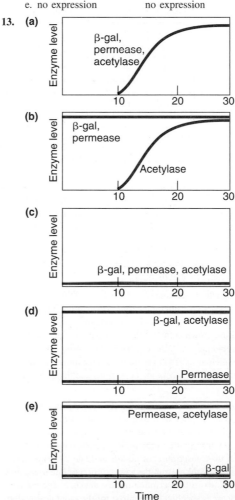

(a) Enzyme level — β-gal, permease, acetylase; Time 10, 20, 30

(b) Enzyme level — β-gal, permease; Acetylase; Time 10, 20, 30

(c) Enzyme level — β-gal, permease, acetylase; Time 10, 20, 30

(d) Enzyme level — β-gal, acetylase; Permease; Time 10, 20, 30

(e) Enzyme level — Permease, acetylase; β-gal; Time 10, 20, 30

15. If the three genes make up an operon, they are co-transcribed as one mRNA and only one band should appear on a hybridization analysis using any of the three genes as a probe versus mRNA. If the genes are not part of an operon, there would be three differently sized hybridizing bands.

17. a. 4. b. 6. c. 7. d. 2. e. 3. f. 5. g. 1.

19. a. i, iii, v, and vi. b. Mutations ii, iii, and iv. c. Mutation l is i, mutation 6 is ii, mutation 2 is iii, mutation 4 is iv, mutation 5 is v, and mutation 3 is vi.

21. a. Mutations in O_2 or O_3 alone have only small effects on synthesis levels. b. Small DNA insertions between O_1 and O_2 may change the face and either change the ability of the repressor to bind one of the sites or change the ability of the bound repressor to bend the DNA leading to an O^c mutant phenotype. c. It would be insensitive to a I^s repressor protein.

23. The protein-coding region of your gene must be in the same frame as the *lacZ* gene.

25. Seven His codons (CAC or CAU), in a row.

27. a. This seems to be a biosynthetic operon; the operon is repressible.

b.

Condition	Gene *A*	Gene *B*	Gene *C*	Gene *D*
Wild type	completely repressible	constitutive	completely repressible	completely repressible
Nonsense in *A*	not expressed	constitutive	completely repressible	not expressed
Nonsense in *B*	partially repressible	not expressed	partially repressible	partially repressible
Nonsense in *C*	not expressed	constitutive	not expressed	not expressed
Nonsense in *D*	completely repressible	constitutive	completely repressible	not expressed
Deletion of region incl. *E*	partially repressible	constitutive	partially repressible	partially repressible
Deletion of *F*	partially repressible	constitutive	partially repressible	partially repressible
Deletion of *G*	not expressed	constitutive	not expressed	not expressed

c.

29. Negative regulator; since the loss of LexA function leads to the new expression of many genes, the wild-type LexA protein must bind to the operators of these genes to shut them off.

31. a. Use two probes, where one consists of labelled cDNA corresponding to the mRNA extracted from the culture grown at the higher osmolarity, and the other consists of cDNA corresponding to the mRNA in the culture grown at the lower osmolarity. b. Each spot on the microarray would have a DNA sequence representing a single *E. coli* gene. c. Use microarrays to compare the gene expression changes in cells grown under different osmotic conditions and those that are heat-shocked.

33. a. All of these turn out to be early genes. b. The early genes have the promoters. c. Transcription of the large majority of *E. coli* genes would be drastically decreased. d. Loss-of-function mutations in the *motA* gene prevent transcription of the middle genes, those in the *asiA* gene should lower the transcription of middle and late *T4* genes, and those in the *55* gene should prevent the transcription of late transcripts but have little effect on the transcription of host genes. e. The *regA*-encoded ribonuclease is specifically required for the rapid destruction of *T4* early mRNAs.

Chapter 11

1. i. 7; ii. 4; iii. 6; iv. 2; v. 9; vi. 8; vii. 5; viii. 3; ix. 1.

3. Transcript processing (including alternate splicing of the RNA), export of mRNA from the nucleus, changes in the efficiency of translation (including miRNAs), chemical modification of the gene products, and localization of the protein product in specific organelles.

5. a. i. b. ii.

7. A *GAL80* mutation in which the protein is not made or is made but cannot bind to the GAL4 protein will prevent repression and lead to constitutive synthesis. A *GAL4* mutation, which inhibits binding to the GAL80 protein, will also be constitutive. A mutation of the DNA at the binding site for the GAL4 protein will also give constitutive synthesis.

9. a. DNA binding. b. DNA binding. c. Dimer formation. d. Transcription activation. e. DNA binding.

11. If *Id* acts by quenching it interacts with *MyoD*, whereas if it blocks access to an enhancer it binds to DNA. Experimentally, look for binding to the regulatory region of *MyoD*.

13. a. Have a *Drosophila* promoter sequence with the promoter added somewhere upstream of the DNA encoding the initiating AUG. Other helpful elements are a *Drosophila* poly-A addition sequence and a transcription termination signal downstream of the *lacZ* coding sequence. b. The type of construct you made is called an *enhancer trap*; these different insertions signal a position in the genome adjacent to a tissue-specific enhancer. In strains in which *lacZ* is expressed in the head, your construct must have integrated into the genome very near to an enhancer that helps activate transcription in the head. In other strains, your construct integrated into the genome near enhancers that are specific for other tissues like the thorax. Since the density of enhancer elements in the genome is low, most of the time new integrations of your construct would be located too far from an enhancer, so there would be no *lacZ* expression and no blue colour. This technique allows you to find genes that are normally expressed in a tissue-specific manner.

15. Differing levels of gene expression depending on their association with highly compacted, heterochromatic DNA versus euchromatic are the evidence. One example is position-effect variegation in *Drosophila*, another is Barr body formation in human females. Decompaction affects the location of the nucleosomes, and gives rise to DNase I hypersensitive sites where nucleosomes have been removed and the DNA is available for binding by RNA polymerase or regulatory proteins. Transcriptional silencing, on the other hand, involves methylation of the DNA.

17. Liver cell DNA has a DNaseI hypersensitive (DH) site 4 kb from one end of the *Eag*I fragment. This site is probably the promoter region for your gene.

19. a. 1, 2, and 4. b. 1 and 4. c. 1 and 4. d. 3.

21. a. True. b. True. c. False. d. False.

23. { } represents an allele that is transcriptionally inactivated (imprinted). a. Bill Sr.'s genotype is *50K*/{*60K*} and Joan's genotype (*60K*/{?}). b. Joan's genotype is *60K*/{*50K*} and Bill Sr.'s genotype is *50K*/{*60K*}.

25. a. The allele of the gene is not expressed in the germ cells of male I-2. b. The allele of the gene from male I-2 will not be expressed in the somatic cells of II-2. c. The allele of the gene from male I-2 will be expressed in the germ cells of II-2. d. The allele of the gene from male I-2 will not be expressed in the somatic cells of II-3. e. The allele of the gene from male I-2 will not be expressed in the germ cells of II-3. f. The allele of the gene from male I-2 will be expressed in the somatic cells of III-1. g. The allele of the gene from male I-2 will not be expressed in the germ cells of III-1.

27. Introns are spliced out, ribonuclease cleaves the primary transcript near the 3′ end and a poly-A tail is added, 5′ methyl CAP is added.

29. The 5′ and 3′ untranslated regions could be cloned at the 5′ or 3′ ends of a reporter gene that is transformed back into *Drosophila* early embryos to see if either of the sequences affects the translatability of the reporter protein.

31. The protein in the fat cells may be posttranslationally modified (e.g., phosphorylated or dephosphorylated) so that it is only active in fat cells. Alternatively, the protein may need a cofactor to be activated, and this cofactor is only transcribed in fat cells.

33. a. The difference in first detection of the mRNA and protein probably results from the different sensitivity in detecting mRNA versus protein. The difference in duration of the mRNAs versus proteins is the proteins are more stable than the mRNAs so they remain in the cells for several days longer. If the normal protein disappears at day 10.5, then the *lacZ* mRNA is more stable; or the β-galactosidase protein is more stable; or the transgene is transcribed until day 12. b. I would expect onset. The *cis*-controlling DNA elements controlling *lacZ* reporter gene transcription should allow transcription of the reporter gene to be turned "on" and "off" with the same schedule as transcription of the interesting gene. If mRNA or protein stability for the two genes differs, however, the activity of β-galactosidase may cease sooner or continue longer than the activity of the protein product of the gene of interest.

Chapter 12

1. i. 3; ii. 5; iii. 8; iv. 7; v. 6; vi. 1; vii. 9; viii. 2; ix. 4.

3. a, e.

5. a. Initial population not in equilibrium. b. Genotype frequencies in the F_1 will be $0.36\ MM + 0.48\ MN + 0.16\ NN = 1$, allele frequencies in the F_1 generation $M = 0.6$ and $N = 0.4$. c. The same as in part *b*.

7. Each allele frequency has a different set of genotype frequencies at equilibrium.

9. a. $N = 0.1$. b. 478 MN children on the island. c. $N = 0.525$.

11. a. $p^2 + 2pq + q^2 + 2pr + r^2 + 2qr = 1$. b. 0.516 A, 0.122 B, 0.075 AB, and 0.287 O.

13. a. $C = 8324/9049 = 0.92$, $c = 725/9049 = 0.08$. b. Sample does not demonstrate Hardy–Weinberg equilibrium. c. Frequency of $c^P = 0.018$, frequency of $c^d = 0.064$, frequency of the C allele $= 0.918$. d. In boys, $C = 0.918$ (normal vision), $c^d = 0.064$ (colour blind) and $c^P = 0.018$ (colour blind). In girls, the genotype frequencies are $CC = 0.843$ (normal vision), $Cc^d = 0.118$ (normal vision), $Cc^P = 0.033$ (normal vision), $c^P c^P = 3.3 \times 10^{-4}$ (colour blind), $c^d c^d = 0.004$ (colour blind), and $c^d c^P = 0.002$ (normal vision). e. The population is in equilibrium. As seen in part *c*, the allele frequency of C is the same in boys and girls and the allele frequency of c in the boys is the same as the total frequencies of $c^d + c^P$ in girls.

15. a. The genotype frequencies in the F_2 are $0.33\ vg^+ vg^+$ and $0.67\ vg^+ vg$. The allele frequencies in the F_2 for $vg^+ = 0.33 + 1/2\ (0.67) = 0.67$ and for $vg = 1/2\ (0.67) = 0.33$. b. The genotype frequencies in the F_3 progeny are $0.449\ vg^+ vg^+ + 0.442\ vg^+ vg + 0.109\ vg\ vg = 1$, or 0.891 wild type and 0.109 vestigial. c. F_4 allele frequencies are $vg^+ = 0.753$ and $vg = 0.249$. d. If all of the F_4 flies are allowed to mate at random then there is no selection and the population will be in Hardy–Weinberg equilibrium: $0.566\ vg^+ vg^+ + 0.373\ vg^+ vg + 0.062\ vg\ vg = 1$; $vg^+ = 0.753$ and $vg = 0.247$.

17. Selection against the homozygous recessive genotype will decrease the frequency of the recessive allele in the population, but it will never totally remove it, as the recessive allele is hidden in the heterozygote; a recessive allele sometimes confers an advantage when present in the heterozygote; mutation can produce new recessive alleles in the population.

19. a. $b = \sqrt{0.25} = 0.5$, $B = 0.5$. b. Δq for tank 1 $= -0.1$, q for all tanks $= 0.5$; Δq for tank 2 $= 0$; Δq for tank 3 $= 0.05$.

	Tank 1	Tank 2	Tank 3
b. Δq	−0.1	0.0	0.05
c. w_{Bb}	1.0	1.0	1.0
d. w_{bb}	<1.0	1.0	>1.0

21. a. Fitness value (w) $= 0$ and the selection coefficient (s) $= 1$ for the affected genotype. There is no selection pressure against the carrier or the homozygous unaffected genotypes, so for both of these $w = 1$ and $s = 0$. b. $\Delta q = -1.54 \times 10^{-3}$. c. 1.02×10^{-3}, which is smaller than the observed q of 0.04. d. CF^+/CF^- heterozygotes may be better able to survive outbreaks of cholera.

23. b.

25. a. Height has the highest broad-sense heritability and weight has the lowest broad-sense heritability. b. The data from the CDC are roughly in line with the conclusions from part *a*.

27. a. Founder effect. b. Advantages: genetic homogeneity and fewer genes that may affect a quantitative trait; disadvantages: some mutations are not found in the population that are in the general population.

29. a. $2n + 1$, where $n =$ number of genes. b. $(1/4)^n = 1/256$, so $n = 4$.

Chapter 13

1. i. 4; ii. 6; iii. 5; iv. 2; v. 1; vi. 7; vii. 3.

3. a. Different constraints on the functions of each of the proteins. b. Rates are more constant because these base changes do not affect function of the gene product.

5. Suggests there is some benefit to the *CF* allele in the heterozygous state.

7. Duplication followed by evolutionary divergence.

9. a. 240 million years. b. Two; *C* allele arose 30 million years ago; *B* allele arose 1 million years ago. c. Duplication of *B*: transposition; duplication of *C*: misalignment and crossing-over.

11. This gene was introduced from a different species.

13. a. Exons. b. Genes.

15. They mediate genome rearrangements or contribute regulatory elements adjacent to a gene.

17. a. SINEs or LINEs. b. Centromere satellite DNA.

Chapter 14

1. i. 10; ii. 1; iii. 9; iv. 7; v. 6; vi. 2; vii. 8; viii. 3; ix. 5; x. 4.

3. Shorter molecules slip through pores more easily, larger molecules get caught.

5. a. A. b. 10 kb. c. 10 kb.

d.

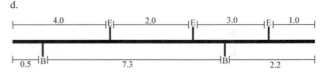

7. Selectable markers are genes that allow a vector to impart protection from an antibiotic on a host cell. When cells are transformed by a vector with a selectable marker, and then exposed to the appropriate antibiotic, only cells that have the vector will survive.

9. a. All. b. 1/4. c. None. d. 1/2 chance. e. 3/4.

11. a. Five. b. Divide the number of base pairs in the genome by the average insert size, then multiply by 5.

13. After cloning: *Eco*RI: 42 and 2400 bp fragments; *Mbo*I: 705, 944, 500, and 300 bp or 905, 744, 500, and 300 bp fragments.

15. a. Alkaline phosphatase removes the 5′-phosphate groups so ligase cannot join a hydroxyl group to the dephosphorylated 5′ ends. The ligation with the nonphosphorylated vector reanneals to itself at a high frequency, leading to 99/100 blue colonies. The phosphorylated vector formed 99/100 white colonies, showing that almost all of the vectors had an insert. b. The dephosphorylation of the vector increased the number of clones (vector + insert) 100-fold. c. If the insert were dephosphorylated, it will not self-ligate, but the vector **will** self-ligate. The vector has the antibiotic resistance gene and ORI, so the "empty" vector will be propagated in *E. coli*, generating a high level of "background."

17. Probes should be between about 15 and 18 nucleotides long. (i) If you knew the sequence of the protein from several bacterial species, you could choose a very highly conserved region on which to base a probe. (ii) Find a region of five or six contiguous amino acids with low degeneracy.

19. a. Whether the PKU syndrome in this patient is caused by a mutation in the phenylalanine hydroxylase gene. b. 300 template molecules in 1 ng of DNA. c. 110 ng of a 1-kb section of the genome after the PCR.

21. a. The chance that one of the two primers will anneal to a random region of DNA that is not the targeted CFTR exon would be $(1/4)^{18}$, or about 1 chance in 7×10^{10} so an 18-base sequence will be present once in every 70 billion nucleotides. b. (i) The chance probability of a 16-base sequence in random DNA is $(1/4)^{16}$, or 1 chance out of 4×10^9. (ii) The longer the primers the more expensive they are to synthesize, the longer the primers the more likely they are to anneal with each other. If the primer is too long it can hybridize with DNA with which it is not perfectly matched. Thus, longer primers might anneal to other regions of the genome than the region you actually want to amplify. c. The 5′ end mismatches at the 3′ end and would prevent DNA polymerase from adding any new nucleotides to the chain. d. Add 5′ GAATTC 3′ to the 5′ end of each primer sequence during synthesis of the oligos.

23. The entire DNA sequence of the genomes is now available in well-studied organisms such as *C. elegans*, *D. melanogaster*, yeast, and mice. To study any region in these genomes, design PCR primers based on the genomic sequence. Having the genome sequence of an organism increases the importance of PCR. Restriction digestions remain the basis for many important applications of DNA cloning.

25. a. Newly synthesized strand: 5′ TAGCTAGGCTAGCCCTTTATCG 3′, template strand: 3′ ATCGATCCGATCGGGAAATAGC 5′. b. 5′ CGAUAAAGGGCUAGCCUAGCUA 3′. c. There are stop codons in each frame so it is unlikely that this is an exon sequence of a coding region.

27. a. This terminal ddA, which is linked to a green fluorescent label, therefore becomes the 3′ end of this molecule. b. The sequence is 5′. . . ACCTATTTTACAGGAATT . . . 3′. c. "Residue Position" indicates a peak at a specific location in the scan, so the size of the single-stranded DNA fragment is represented by the residue position. d. The double peak at position 370 is most likely caused by the fact that the original DNA actually had two different DNA sequences. One chromosome carries a T–A base pair at this location while the homologue had a G–C base pair.

Chapter 15

1. i. 4; ii. 7; iii. 5; iv. 2; v. 6; vi. 1; vii. 3

3. Anonymous DNA markers are the DNA sequence of an individual. The terms dominant and recessive can only be used when discussing the phenotype of an organism, so in one sense this question is meaningless. Geneticists often say that DNA markers are inherited in a co-dominant fashion to denote that both the alleles can be seen in the DNA sequence.

5. a. Different numbers of simple sequence repeats. b. Slippage of DNA polymerase during replication. c. A different mechanism: unequal crossing-over.

7. a. The polymorphism is within the short DNA sequence that is used as a probe. b. The polymorphism is in the nucleotide adjacent to the sequence used as a primer. c. The SNP polymorphism can be kilobases away from the probe sequence in a restriction site recognized by the restriction enzyme used to digest the genomic DNA.

9. The sequences of the ASOs would be 3′ GATATTTACCCGATCC-GCA and 3′ GATATTTACGCGATCCGCA.

11. Sperm collected from man, eggs collected from woman. After *in vitro* fertilization, embryos are allowed to develop to the eight-cell stage. A single cell from each eight-cell embryo is removed. DNA is prepared and genotype is analyzed using PCR and *Mst*II digestion. Embryos with the desired genotype are implanted into the woman's uterus.

13. Co-worker 3 has the same DNA fingerprint as the crime sample and must be the perpetrator of the crime. The probability is essentially 100 percent.

15. a. Individuals A, B, C, and E. b. Individuals D and F. c. 48 bp.

17. a. 10 kb. b. 10 kb. c. 0 percent. d. 50 percent.

19. Members of the disease family must be segregating two or more alleles at each DNA marker that is chosen.

21. a. 0 percent chance. b. 0.0075 probability of an affected child.

23. Mating W is not informative; mating X is informative—both parents are doubly heterozygous; mating Y is non-informative; mating Z is non-informative.

25. Identify sequences that are transcribed into RNA. Use computational analysis to identify sequences that are conserved between distantly related species. Use computational analysis to identify sequences that are open reading frames with appropriate codon usage and splice sites.

27. a. A, C, and E. b. Three different genes. c. Yes. d. Fragments C and E. e. Gene recognized by fragment E. f. If there is a mouse model of this disease you would transform the mice with the cDNA clone of the candidate gene and look for the normal human gene to rescue the mutant phenotype in the mice.

29. a. Autosomal dominant. b. Yes, II-2, II-3, and III-1.

31. a. 12 500 different haplotypes. b. 156 250 000 possible diplotypes. c. Father's genotype: A25 C4 B7 / A23 C2 B35, mother's genotype: A24 C5 B8 / A3 C9 B44. d. 1/4.

Chapter 16

1. i. 7; ii. 6; iii. 8; iv. 2; v. 1; vi. 9; vii. 3; viii. 5; ix. 4.

3. a. A *RAS* mutant that stays in the GTP-bound state is permanently activated and will cause the cell to continue dividing. b. Under the restrictive conditions the cells will not divide.

5. a. Effect of the T antigen is minimized. b. Decrease the ability of *p53* to function in cell cycle control. c. In a functional domain other than those that bind the T antigen and the transcription factor.

7. a. Hybridization to DNA from tumour and normal cells. Use two different probes with one representing the specific sequence that you are analyzing and the other representing an unamplified control sequence. b. Look for alterations in the chromosomal banding patterns in a karyotype analysis.

9. Studies can be set up within the recent Indian immigrant population versus the Canada-based Indian native population examining the effect of a Westernized diet versus a diet resembling that of the ethnic group. Also, the effects of the same diets on non-Indian Canadians should be examined. To assess the role of genetic differences, you need to keep other factors, for example, diet, as constant as possible. You could look at the incidence in Indians and Canadians who have similar diets.

11. Order: d, a, b, c.

13. a. Examination of the pedigree suggests that predisposition to colorectal cancer in this family could be an autosomal dominant trait. If this is true, then individuals II-2 and either I-1 or I-2 must have the mutation, but not express it. b. Individuals I-1, I-2, and II-2 are not among the high coffee consumers. Perhaps the predisposition to colorectal cancer is a combination of a particular genotype and the environmental factor of consumption of the special coffee.

15. Technique d.

17. If one PCR primer binds to one of the chromosomes at one side of the translocation while the other primer binds to the other chromosome on the other side of the breakpoint, then your PCR primers will span the translocation. This would amplify a PCR fragment only if there were still cells in the blood that had the translocated chromosomes.

19. Both. These instabilities can be caused by somatic or germline mutations in genes such as *p53* or the genes for DNA repair enzymes; genome and karyotype instabilities can then result in additional problems that can contribute to cancer progression. Mutations in DNA repair enzymes lead to a high rate of mutation and such mutations might inactivate a tumour-suppressor gene or activate a proto-oncogene.

21. a. M (mitosis). b. M. c. S phase. d. G_1 phase.

23. Three complementation groups.

25. Cyclical regulation of transcription, cyclical regulation of translation, and cyclical control of posttranslational modifications.

27. a. 2. b. 3. c. 1.

29. In the M phase, checkpoint molecules made by unattached kineto-chores prevent the anaphase-promoting complex (APC) from being activated. The APC must become activated at the beginning of ana-phase to destroy M-phase cyclin, allowing cells to leave M phase. The activated APC adds ubiquitin to protein substrates. When this happens, the ubiquinylated proteins are rapidly destroyed by the proteosome. One simple hypothesis is that cohesin is also targeted by the APC since it must be destroyed at the beginning of anaphase.

Chapter 17

1. i. 11; ii. 4; iii. 8; iv. 7; v. 12; vi. 9; vii. 13; viii. 3; ix. l; x. 5; xi. 6; xii. 10; xiii. 2.

3. a. In *C. elegans*, laser ablation at this early stage of development would almost certainly be lethal, while in mice the loss of one out of four early embryonic cells would have no effect. b. It would be lethal to *C. elegans*, and it is possible that the separated cells could develop into a mouse. c. In *C. elegans*, it would likely be lethal. In mice, such a fusion would be tolerated, giving rise to a chimeric animal if the two embryos had different genotypes.

5. Make mRNA preparations from homozygotes for the new null allele and then analyze these preparations on Northern blots, or use RT-PCR—if you cannot detect *rugose* mRNA the mutation is null because the gene is not transcribed. Alternatively, analyze protein extracts from the homozygous mutant animals by Western blot using the antibody against the Rugose protein as a probe.

7. a.

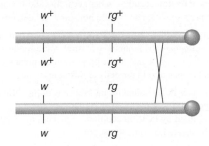

b. As a result of the mitotic crossover, developing ommatidia in the eye would be simultaneously homozygous for the mutations in *rugose* and *white*, while adjacent ommatidia would be heterozygous for the wild-type and mutant alleles of both genes. If the red ommatidia are abnor-mal even though their genotype predicts a normal structure, then the lack of Rugose in the adjacent white ommatidia affects the red omma-tidia. c. If these white patches were normal in appearance, then Rugose does not have an important role in eye development. If the white patch is abnormal, then Rugose is important for eye development.

9. Mutate possible regulatory DNA elements, for example, by deleting various DNA sequences near the 5′ end of the *myo-2* gene. Transform these deleted *myo-2::GFP* transgenes back into the worm. Look to see if and how the normal pattern of GFP expression in the pharynx is affected in these worms.

11. Make DNA constructs that place a wild-type genomic copy of gene *X* adjacent to *myo-2::GFP*. You then transform these constructs into worms that are homozygous for a null allele of gene *X* (and that did not contain any *GFP* source). The constructs form extrachromosomal arrays as described. Pharyngeal cells containing the arrays would be wild type for gene *X* and express GFP. Pharygeal cells that had lost the arrays would be homozygous mutant for gene *X* and would not express GFP.

13. a. If the mutation was due to an insertion of the transgene, the MMTV/*c-myc* transgene should segregate with the phenotype. b. Clones containing the *c-myc* fusion could be identified by hybridiza-tion of MMTV sequences to a library of genomic clones produced from the cells of the mutant mouse. c. The sequence of the gene into which the MMTV/*c-myc* fusion inserted could be analyzed in the *ld* mutant to determine if there were mutations in the gene.

15. a. Promoter-binding sites for transcription factors such as Bicoid and binding sites for other transcription factors that ensure the *hb* gene is transcribed in the proper cells in the mother. b. The amino acids in Hunchback that comprise DNA-binding domains and domains involved in the transcriptional regulation of gap and pair-rule genes. c. Translational repression carried out by Nanos protein.

17. The cytoplasm from the anterior of a wild-type embryo could be injected into the anterior end of a *bicoid* mutant embryo to see if there was rescue of the mutant phenotype. Alternately, purified *bicoid* mRNA injected into the anterior end of a *bicoid* mutant embryo would be a more definitive experiment. Finally, purified *bicoid* mRNA could be injected into the posterior end of a wild-type embryo.

19. a. Wild-type Knirps protein is needed to restrict the posterior limit of the zone of Krüppel expression. b. Yes, Hunchback protein would be seen throughout the embryo.

21. A mutation in the gene encoding a maternally supplied component that affects early development must be in the mother's genome. If the mutation affecting early development is in a gene whose transcription begins after fertilization, then the mutation must be in the genome of the zygote (these are thus sometimes called zygotic genes). You would need two different kinds of genetic screens to generate mutations either in the mother's genome or the zygote's genome.

23. a. The presence of PAR-3 and absence of PAR-2 from these cells indirectly dictates their ability to translate *glp-1* mRNA into GLP-1 protein. b. It appears that the stable localization of APX-1 at the membrane requires an interaction with GLP-1. Such an interaction could occur through the extracellular domains of both proteins. c. The receptor is the GLP-1 protein. Thus APX-1 would be the ligand. d. (i) The ablation of P_2 would make ABp and its descendants have the same fate shown by ABa and its descendants. (ii) A null mutation of *apx-1* would have the same effect. (iii) A null mutation of *glp-1* would have the same effect. (iv) A null mutation of that gene would have the same effect.

Chapter 18

1. i. 4; ii. 5; iii. 2; iv. 7; v. 6; vi. 3; vii. 1.

3. 200 colonies on the first plate and 20 colonies on the second plate.

5. a. iv. b. iii. c. ii.

7. The *purE* and *pepN* genes will be co-transformed at a lower frequency if the *H. influenzae b* pathogenic strain was used as a host donor strain.

9. Plasmid transformation into *Shigella dysenteriae*, bacteriophage infec-tion of *Staphylococcus*, *Streptococcus*, or *E. coli* species, and transpo-sition of DNA (pathogenicity island) into *Vibrio cholerae*.

11. Do a mating between the mutant cell with three or four copies of F and a wild-type F⁻ recipient. A mutation in the F plasmid means the exconjugant will have the higher copy number. If the mutation is in a chromosomal gene, the higher copy number phenotype would not be transferred into the recipient. You could isolate the F-plasmid DNA from the mutant cell and then transform this plasmid into new recipi-ent cells. By examining the number of copies of the F factor in the transformed cells, you could tell whether the trait was carried by the plasmid.

13. a. The mechanisms are (i) transformation, (ii) conjugation, and (iii) transduction. b. If the donor DNA used in the transformation includes plasmids, recipient cells may take up the entire plasmid and acquire the characteristics conferred by the plasmid genes. Conjugation requires the presence of a conjugative plasmid in the donor cell. c. The mechanism requiring a bacteriophage is transduction. d. The mechanisms are natural transformation with DNA fragments, conjugation with an Hfr, and generalized transduction.

15. Transform the plasmid into a nontoxin-producing recipient strain and assay for toxin production.

17. Order: *ilv bgl mtl.*

19. a.

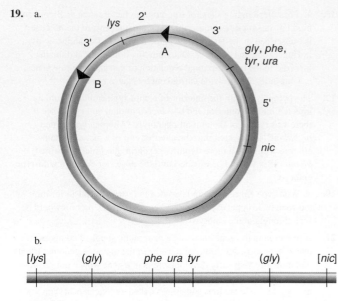

b.

[*lys*]	(*gly*)	phe ura tyr	(*gly*)	[*nic*]

c. To map the *gly* gene with respect to other markers, select for Gly⁺ transductants on min + lys + phe + tyr + ura + nic. Then score the other markers to determine which genes are co-transduced with Gly⁺ at the highest frequency.

21. This assay detects *recA⁻* mutants in the F⁻ cell based on the inability to form stable exconjugants.

23. a. *In vitro* portion. b. *In vivo*. c. *In vivo*. d. *In vitro*.

25. a. Both. b. Both. c. Neither. d. Both.

27. a.

		Trp	His	Ile	Met
mRNA	5′	UGG	CAU/C	AUU/C/A	AUG
nDNA mRNA-like	5′	TGG	CAT/C	ATT/C/A	ATG
nDNA template	3′	ACC	GTA/G	TAA/G/T	TAC

b.

		Trp	His	Ile	Met
mRNA	5′	UGA/G	CAU/C	AUC/U	AUG/A
mtDNA mRNA-like	5′	TGA/G	CAT/C	ATC/T	ATG/A
mtDNA template	3′	ACT/C	GTA/G	TAG/A	TAC/T

29. a and d.

31. a. Introns would have to be removed, some of the codons in the nuclear gene would have to be changed since the genetic code in the nucleus and in mitochondria is not identical. A mitochondrial translational start site and a transcriptional termination site must be added. The open reading frame of the altered nuclear gene would have to be placed under the control of a mitochondrial promoter, and the cloned gene must be introduced into the yeast mitochondria. b. Such a strain can then be used to select for function of the mitochondrial genetic system in mutants that are unable to respire. Try to find arginine auxotrophs that could no longer make arginine because there was a DNA change that obliterated the function of the promoter. Such mutations allow analysis of the function of regulatory elements in the mitochondrial genome.

33. The small size of the sperm can mean that organelles are excluded. Cells can degrade organelles or organellar DNA from the male parent. Early zygotic mitoses distribute the male organelles to cells that will not become part of the embryo. The details of the fertilization process may prevent the paternal cell from contributing any organelles (only the sperm nucleus is allowed into the egg), and in some species that zygote destroys the paternal organelle after fertilization.

35. If the mutation is very debilitating to the cell, either because of the loss of energy metabolism (in the case of mitochondria) or of photosynthetic capability (in the case of chloroplasts), a cell that is homoplasmic for the mutant genome will die.

37. a. The zygote that formed this plant was heteroplasmic, containing both wild-type chloroplast genomes and mutant chloroplast genomes. These two types of genomes can segregate as tissue is propagated

mitotically. b. Many of the ovules generated on this branch are heteroplasmic and give rise to variegated plants. Some of the cells on the variegated branch gave rise to ovules that segregated one or the other chloroplast genomes. The phenotype of the progeny will reflect the type(s) of chloroplast genome(s) in the ovules. White-leaved plants cannot make chlorophyll and since they cannot conduct photosynthesis, they will die. c. Part of the plant can conduct photosynthesis and then these carbohydrates can be made and transported to tissues that are unable to conduct photosynthesis.

39. a. If one of the parental inbred lines is male sterile, then this line cannot self-fertilize and the seed companies would not have to do anything more to prevent self-fertilization. b. The sterile inbred line in part *a* must also be homozygous for the recessive *rf* allele of *Restorer*. The other inbred line, the male parent supplying the pollen for the cross, would have to have at least one (and preferably two) dominant *Rf* alleles of *Restorer*. c. Make a fertile "Maintainer" line that has mitochondria with a normal (non-CMS) genome but whose nuclear genomes are the same as the CMS plants and also *rf/rf*. The *rf/rf* CMS plants are used as the female parent (they cannot produce pollen). When pollen from Maintainer plants fertilizes the CMS plants, the progeny will have CMS mitochondria and will also be *rf/rf*. In other words, these progeny will be identical to their maternal parents. d. Farmers are now dependent upon the seed companies to provide their seed. The genetic variation of the corn crop overall is reduced.

41. a. The mother (I-1) may have had very low levels of mutant mitochondrial chromosomes or there may have been a spontaneous mutation either in the mitochondrial genome of the egg that gave rise to individual II-2 or in the early zygote of individual II-2. b. You could look at the mitochondrial DNA from somatic cells from various tissues in the mother. If the mutation occurred in her germ line and was inherited by II-2, the mother's somatic cells would not show any defective DNA.

43. The variation in affected tissues is due to differences in where and when during development the mutation occurred. Variation in the severity of the disease can be due to the proportion of mutant genomes (the degree of heteroplasmy) in the cells of different tissues.

45. Gel electrophoresis is better suited for an overview of the differences between mitochondrial genomes. Deletions can be very large and might not be amplified by PCR. In addition, sequences to which primers bind might be deleted in some mutations

Chapter 19

1. i. 5; ii. 3; iii. 4; iv. 1; v. 2 ; vi. 6.

3.

5. Two.

```
Contig #1:  5′ CAAATAGCAGCAAATTACAGCAATATGAAGAGAT
               CATACAGTCCACTGAA 3′
            3′ GTTTATCGTCGTTTAATGTCGTTATACTTCTCTA
               GTATGTCAGGTGACTT 5′

Contig #2:  5′ GTAGTATCTCCTTTTAAAAATCTCATTTCCTTTA
               GGGCATTTTCAAATTC 3′
            3′ CATCATAGAGGAAAATTTTTAGAGTAAAGGAAAT
               CCCGTAAAAGTTTAAG 5′
```

7. Some genomic sequences cannot be cloned (e.g., heterochromatin) and some sequences rearrange or delete when cloned (e.g., some tandemly arrayed repeats).

9. a. The predicted amino acid sequence can be examined for protein domains or functional units. Using comparisons between sequenced genomes of different species, you can find orthologous genes. b. Use a Northern blot analysis to discover in which tissues the gene is transcribed. Or knock out a gene and look for a phenotype caused by the knockout.

11. a. Protein is a transcription factor. b. Genes arose by duplication.

13. Next-generation sequencing is an umbrella term that has come to mean post-Sanger sequencing methods. These methods are characterized by their high-throughput and massively parallel design leading to huge gains in speed of sequencing and reductions in cost.

15. Incorporation of a nucleotide generates pyrophosphate, which is converted to ATP (by ATP sulfurylase). The ATP is then used to convert luciferin to oxyluciferin (by luciferase). Light is a by-product of this last reaction.

Chapter 20

1. i. 2; ii. 3; iii. 4; iv. 5; v. 1.

3. a. For black, this gene is either. For green, the mRNA for this gene accumulates to higher levels in normal tissue than in the tissue from the tumour. For red, the mRNA for this gene accumulates to higher levels in cancerous tissue than in normal tissue. For yellow, the mRNA for these genes accumulates to the same level in both kinds of tissue. b. I would choose a red signal.

5. No. This could only be concluded if the mRNAs happened to be of equivalent length.

7. The identity and location of 15 Drz1-binding sites.

9. a. Chromosomes 12 and X are present in three copies, portions of chromosomes 11 and 13 are found in only one copy. b. Two copies of all regions in the genome. c. Balanced inversions or translocations. d. Any region of the genome with a change in dosage may contain a gene that contributes to leukaemia. e. Changes in gene dosage discussed in part *d* cannot affect the viability of cells since tumour cells with these changes grow and divide perfectly well.

11. To ensure the internal validity of the experiment by showing that random differences in subsample composition do not affect the identification of important loci.

13. Untreated DNA: >>CCTCGGAAGGA<u>C</u>TCAGGC<u>C</u>G>>
    <<GGAGCCTT<u>CC</u>TGAGT<u>C</u>CGG<u>C</u><<

Chapter 21

1. i. 4; ii. 2; iii. 1; iv. 3.

3. a. False. b. False. c. False.

5. Sarah. While the assay is performed in yeast, proteins from any organism can be expressed.

7. c and e.

9. b.

11. For these 80 proteins, the regulation of protein levels must occur at the posttranscriptional level. This can involve differences in the rates of translation of different mRNAs or differences in the rates at which different proteins are degraded.

Chapter 22

1. i. 2; ii. 3; iii. 1.

3. A gene whose function is required for viability under all growth conditions.

5. In forward genetic screens, random mutations are generated (e.g., by treatment with a mutagen) in a population of cells or individuals. Those cells or individuals displaying a specific phenotype are then isolated for further study. In this method, many different alleles (both gain- and loss-of-function) of a single gene can be isolated. However, the gene that has been mutated is not immediately known. In fact, identifying

the affected gene may be difficult and cumbersome. In reverse genetics, the phenotypic effects of disrupting the function of a known gene (i.e., a sequenced and annotated gene) is determined. In this method, one knows from the outset the identity of the gene. However, most methods (e.g., RNAi, knockouts) create only loss-of-function mutations.

7. Yes. Create a strain that expresses a GFP reporter gene and a dsRNA targetting the GFP mRNA. This strain should not display any green fluorescence. Use RNAi to systematically knock down the expression of all *C. elegans* genes one by one. Genes involved in RNAi will restore GFP expression when knocked down.

9. A genetic interaction network in which most genes have only a few interactions and a small number of genes (the hubs) have many interactions.

11. Based upon the results of SGA analysis in model systems, it is thought that most genes will act as disease-specific alleles (specifiers), whereas a small number of hub genes will act as modifiers for multiple, unrelated diseases.

Chapter 23

1. i. 4; ii. 3; iii. 1; iv. 2.

3. The elements of the system; the physical associations among the elements; the biological context of the system; how the association of the system's elements and their relation to changes in the biological context explain its emergent properties.

5. A catalogue of *cis*-control elements would be part of a gene regulatory network. To gain a further understanding of the network, you need to understand what environmental information affects the network via signal transduction pathways and other inputs, as well as how the network integrates and modifies these inputs and the output of the transformed information to various protein networks.

7. a. True. b. True. c. False. d. False. e. True. f. True. g. True. h. False. i. False. j. True. k. True.

9. What exactly is the biological system that he is perturbing? Does he know all of the elements in this system? Some of the cytokines may be the output of the system, but does he know all of the outputs? Or all of the inputs? If he is measuring protein levels, he has only one type of biological information. Also, your friend is not looking at the other systems whose behaviours will be altered by the knockout perturbations. Your friend's research is not systems biology.

11. a. Promoter activity $= B(X^*)^n / K^n + (X^*)^n$. b. The concentration of activator at which the promoter activity of *SLA3* is at half maximal value. c. 25 molecules. d. $Y_{p(\text{steady state})} = (b_m/a_m)/a_p$. e. The promoter activity of the *SLA3* gene would decrease ~25%.

13. a. True. b. True. c. True. d. True. e. True.

Chapter 24

1. i. 4; ii. 3; iii. 2; iv. 1.

3. i. Circuit #2 would be more switchlike. ii. It would take longer for circuit #2. iii. Circuit #2 would be more resistant to small random fluctuations.

5. The strategy would be much slower and cumbersome since it would not be a trivial exercise to select for the appropriate variants.

7. *Mycoplasma* spp. are among the simplest free-living organisms on the planet. Genome sizes for this group are very small and range from 0.6–1.0 Mbp containing about 500–1000 genes. Thus, these bacteria represent a logical first step in the creation of synthetic genomes.

9. A variation of PCR in which a series of short, overlapping oligos prime DNA synthesis, ultimately producing a contiguous piece of DNA. PCA is used in synthetic genomics to create DNA cassettes bearing user-defined sequences.

11. While any technology can be used negatively, ethical reviews have concluded that these technologies show great promise with respect to both technological and medical applications. Furthermore, these same ethical reviews have led to the adoption of regulatory guidelines for the safe use of the approach.

Glossary

Note: An italicized word in a definition indicates that word is defined elsewhere in the glossary.

A

α-*globin* locus a chromosomal region carrying all of the α-globin-like genes.

***Ac* (*Activator*)** a transposable element in corn that promotes chromosomal breaks or mutations.

acentric fragment a *chromatid* fragment lacking a *centromere*; usually the result of a crossover in an *inversion loop*.

acrocentric a *chromosome* in which the *centromere* is close to one end.

activator a type of *transcription factor* that can bind to specific *cis*-acting enhancer elements and increase the level of transcription from a nearby gene.

adaptation the ability to stop responding when the stimulus is present.

adenine (A) a nitrogenous base; one member of the *base pair* A–T (adenine–thymine).

adjacent-1 segregation pattern one of two patterns of *segregation* resulting from the normal disjunction of homologues during *meiosis I*. Homologous *centromeres* disjoin so that one *translocation* chromosome and one normal chromosome go to each pole. Contrast with *alternate segregation pattern*.

adjacent-2 segregation pattern a pattern of *segregation* resulting from a *nondisjunction* in which homologous *centromeres* go to the same pole during *meiosis I*.

agouti a signalling *protein* that functions as a regulator of pigmentation in the hair follicle melanocytes.

allele frequency the proportion of all copies of a *gene* in a *population* that are of a given allele type.

allele-specific oligonucleotides (ASOs) short *oligonucleotides* of 30–40 bases that hybridize to only

one of the two alleles at a *SNP* locus under appropriate conditions.

alleles alternative forms of a single *gene*.

allopolyploids *polyploid* hybrids in which the chromosome sets come from two or more distinct, though related, species.

allosteric proteins proteins that undergo reversible changes in conformation when bound to another molecule.

alternate segregation pattern one of two patterns of *segregation* resulting from the normal disjunction of homologues during *meiosis I*. Two *translocation* chromosomes go to one pole, while two normal chromosomes move to the opposite pole, resulting in *gametes* with the correct haploid number of genes.

alternative hypothesis a statistical hypothesis to be tested and to be compared with the *null hypothesis*.

alternative splicing the production of different mature *mRNAs* from the same *primary transcript* by joining different combinations of *exons*.

Ames test screens for chemicals that cause mutations in bacterial cells.

amino acids the building blocks of proteins.

aminoacyl (A) site the site on a *ribosome* to which a *charged tRNA* first binds.

aminoacyl-tRNA synthetases enzymes that catalyze the attachment of tRNAs to their corresponding amino acids, forming *charged tRNAs*.

amniocentesis a medical procedure in which a sample of amniotic fluid is taken from a pregnant woman to determine the condition of an unborn baby. A hollow needle is inserted through the woman's abdomen and uterine wall, and the fluid, which contains cells shed by the embryo, is drawn off.

amorphic mutations changes that completely block the function of a *gene* (synonym for *null mutations*).

amphidiploids organisms produced by two *diploid* parental species; they contain two diploid genomes, each one derived from a different parent.

amplification the spread of a molecular variant throughout a population over many generations.

anaphase I the phase of *meiosis I* during which the *chiasmata* joining *homologous chromosomes* dissolve, allowing maternal and paternal homologues to move toward opposite spindle poles; the *centromeres* do not divide so that the *chromosomes* moving toward the poles each consist of two *chromatids*.

anaphase II the phase of *meiosis II* when the dismantling of *cohesin* complexes allows *sister chromatids* to move to opposite spindle poles.

anaphase the stage of *mitosis* in which the connection of *sister chromatids* is severed, allowing the chromatids to be pulled to opposite spindle poles.

aneuploid an individual whose chromosome number is not an exact multiple of the *haploid* number for the species.

angiosperm a plant descriptive meaning the seeds of the plant are enclosed in an ovary within the flower.

angstrom (Å) a unit of length equal to one ten-billionth of a metre.

anonymous locus a designated position on a chromosome with no known function. See *locus*.

Antennapedia complex (ANT-C) in *Drosophila*, a region containing several *homeotic genes* specifying the identity of segments in the head and anterior thorax.

anticodons groups of three *nucleotides* on *transfer RNA* (*tRNA*) molecules that recognize codons on the mRNA by *complementary base-pairing* and *wobble*.

antigen a foreign substance (or a particular part of a foreign substance) that induces an *immune response* when introduced into the body.

antisense the sequence of a single-stranded RNA or DNA molecule that

is complementary to—and can base-pair with—a portion of a transcribed RNA molecule.

apoptosis programmed cell death; a process in which DNA is degraded and the nucleus condenses; the cell may then be devoured by neighbouring cells or phagocytes.

Archaea one of the three major evolutionary lineages (domains) of living organisms.

artificial selection the purposeful control of mating by choice of parents for the next generation. Contrast with *natural selection.*

artificial transformation a process to transfer genes from one bacterial strain to another, using laboratory procedures to weaken cell walls and make membranes permeable to DNA. Contrast with *natural transformation.*

astral microtubules short, unstable microtubules that extend out from a *centrosome* toward the cell's periphery to stabilize the *mitotic spindle.*

attenuation a type of gene regulation in which *transcription* of a gene terminates in the regulatory region before a complete mRNA *transcript* is made.

autocrine stimulation a process by which many tumour cells make their own signals to divide, rather than waiting for signals from neighbouring cells.

autonomous elements intact transposable elements that can move from place to place in the genome by themselves. Contrast with *nonautonomous elements.*

autopolyploids a kind of *polypoid* that derives all of its chromosome sets from the same species.

autosome a *chromosome* not involved in sex determination. The *diploid* human genome consists of 46 chromosomes, 22 pairs of autosomes, and 1 pair of sex chromosomes (the X and Y chromosomes).

autotroph a plant that is nutritionally self-sufficient; it produces its own food via photosynthesis.

auxotroph a mutant microorganism that can grow on minimal medium only if it has been supplemented with one or more growth factors not required by wild-type strains.

B

β-*globin* locus a chromosomal region carrying all the β-globin-like genes.

bacterial chemotaxis bacterial movement up and down gradients of chemical attractants or repellents.

bacterial chromosome an essential component of a typical bacterial *genome*; usually a single circular molecule of double-helical DNA.

bacteriophage See *phage.*

balancer chromosome a special *chromosome* created for use in genetic manipulations; helps maintain *recessive lethal* mutations in stocks.

balancing selection a situation in which heterozygotes have a higher selective advantage, or fitness, than either homozygote.

Barr bodies inactive X chromosomes observable in interphase cells as darkly stained *heterochromatin* masses.

barriers sites that block the spread of heterochromatin by various methods such as binding of specific proteins, attachment to a nuclear pore, or transcription of small RNA molecules.

basal factor a type of *transcription factor* that can associate nonspecifically with all *promoters* in a *genome*. A complex of basal factors associated with a promoter is required for the initiation of *transcription* by *RNA polymerase*. See *basal transcription apparatus.*

basal transcription apparatus a complex protein machine that interacts with complexes of *transcription factors* and *cis-control elements* to mediate the synthesis of an RNA *transcript*. This RNA transcript is spliced and edited to produce a *messenger RNA (mRNA).*

base analogues *mutagens* that are so similar in chemical structure to the normal nitrogenous bases in DNA that the replication machinery can incorporate them into DNA in place of the normal bases. This can cause base *substitutions* on the complementary strand synthesized in the next round of DNA replication.

base excision repair an homology-dependent mechanism in which specific enzymes cleave an altered base from the sugar of its *nucleotide* to create an apurinic or apyrimidinic (AP) site in the DNA chain; nick the DNA backbone at the AP site; and remove nucleotides from the vicinity of the nick. DNA polymerase fills in this gap by copying the undamaged strand, restoring the original DNA sequence. See *nucleotide excision repair.*

base pair (bp) two nitrogenous bases on complementary DNA strands held together by *hydrogen bonds. Adenine* pairs with *thymine*, and *guanine* pairs with *cytosine.*

base sequence the order of nucleotide bases in a DNA or RNA strand.

base sequence analysis a method for determining a *base sequence.*

basic chromosome number (x) the number of different chromosomes that make up a single complete set. See *x.*

B-form DNA the most common form of DNA in which molecular configuration spirals to the right. Compare with *Z-form DNA.*

bidirectional describes a mechanism of *DNA replication* in which two *replication forks* move in opposite directions away from the same *origin of replication.*

biochemical pathway an orderly series of reactions that allows an organism to obtain simple molecules from the environment and convert them step by step into successively more complicated molecules.

bioinformatics the science of using computational methods to decipher the biological meaning of information contained within organismal systems.

biological context the context in which a *biological system* operates within a cell or organism.

biological information consists of the digital information of the *genome* and environmental signals that activate or modulate the output of genomic information. These two types of information interact to mediate biological activity across the three time scales of *evolution*, development, and physiological responses.

biological system any complex network of interacting molecules or groups of cells that function in a coordinated manner through dynamic signalling.

biomarkers molecules that are present under specific conditions or when a disease is present.

biosynthetic See *biochemical pathway*.

biotechnology a set of biological techniques developed through basic research and now applied to research and product development. In particular, the use by industry of recombinant DNA, cell fusion, and new bioprocessing techniques.

bithorax complex (BX-C) in *Drosophila*, a cluster of *homeotic genes* that control the identity of segments in the abdomen and posterior thorax.

bivalent a pair of synapsed homologous chromosomes during *meiosis I*.

blastocysts describes the embryo at the 16-cell stage of development through the 64-cell stage, when the embryo implants.

blastomeres early embryonic cells.

blunt end the 5′ or 3′ end of a double-stranded DNA molecule without *sticky ends*.

branched-line diagram a method for systematically listing the expected results of multigene crosses.

branch sites a special sequence of RNA nucleotides within an *intron* that helps form the "lariat" intermediate required for *RNA splicing*.

broad-sense heritability (H^2) the proportion of *total phenotypic variance* within a single generation that is ascribable to *genetic variance*. Compare with *narrow-sense heritability*.

C

cancer genes *genes* having mutant *alleles* that lead to cancer.

carriers *heterozygous* individuals of normal *phenotype* that have a *recessive allele* for a trait.

catabolite repression repression of expression in sugar-metabolizing *operons* like the *lac* operon when glucose or another preferred catabolite is present.

CDK inhibitors inhibit the activity of *cyclin-dependent kinases*.

cDNA complementary DNA has a base sequence that is complementary to that of the mRNA template and contains no *introns*.

cDNA library a large collection of cDNA clones that are representative of the *mRNAs* expressed by a particular cell type, tissue, organ, or organism.

cell autonomous trait a trait for which the *phenotype* expressed by a cell depends solely on the *genotype* of that cell and not that of any other nearby cell. Compare with *nonautonomous trait*.

cell-bound signals signals requiring direct contact between cells for transmission.

cell cycle the repeating pattern of cell growth, replication of genetic material, and *mitosis*.

cellular blastoderm in *Drosophila* embryos, the one-cell-deep epithelial layer resulting from cellularization of the *syncytial blastoderm*.

cellular clone an isolated colony of cells that are all descendants from a single progenitor cell.

centimorgan (cM) a unit of measure of *recombination frequency*. One cM is equal to a 1 percent chance that a marker at one genetic *locus* will be separated from a marker at a second locus due to *crossing-over* in a single generation.

central dogma the flow of information from *DNA* to *RNA* to *protein*.

centrioles short cylindrical structures that help organize microtubules. Two centrioles at right angles to each other form the core of a *centrosome*. Each centrosome serves as a pole of the *mitotic spindle*.

centromere a specialized chromosome region at which *sister chromatids* are connected and to which spindle fibres attach during cell division.

centrosomes microtubule organizing centres at the poles of the *spindle apparatus*.

charged tRNA a tRNA molecule to which the corresponding *amino acid* has been attached by an *aminoacyl-tRNA synthetase*.

checkpoints mechanisms that prevent cells from continuing to the next phase of the *cell cycle* until a previous stop has been successfully completed, thus safeguarding genomic integrity.

chiasmata observable regions in which nonsister chromatids of homologous chromosomes cross over.

chimera genetic *mosaics* in which cells of different genotype originate from two different individuals.

chi-square (χ^2) test a statistical test to determine the probability that an observed deviation from an expected outcome occurs solely by chance.

chorionic villus sampling (CVS) a medical procedure in which a sample of chorionic villi cells from the placenta at the point where it attaches to the uterine wall is taken from a pregnant woman to determine the condition of an unborn baby. Samples can be collected transcervically or transabdominally. Relative to amniocentesis, this procedure can be performed at an earlier stage during pregnancy and larger fluid samples can be obtained.

chromatid one of two copies of a *chromosome* that exist immediately after *DNA replication*. See *sister chromatids*.

chromatin the generic term for any complex of *DNA* and *protein* found in a cell's nucleus.

chromatin immunoprecipitation (ChIP) a procedure used by scientists to identify specific DNA sequences that are bound, *in vivo*, to proteins of interest.

chromocentre the dense heterochromatic mass formed by the fusing of the *centromeres* of *polytene chromosomes*.

chromosomal interference the phenomenon of crossovers not occurring independently. Refer to *crossing-over*.

chromosomal puff the region of a *polytene chromosome* that swells to form a large, diffuse structure when high rates of gene *transcription* cause bands to decondense.

chromosomal rearrangement a change in the order of *DNA sequences* along one or more *chromosomes*.

chromosome loss a mechanism causing *aneuploidy* in which a particular *chromatid* or *chromosome* fails to become incorporated into either daughter cell during cell division.

chromosome theory of inheritance the idea that *chromosomes* are the carriers of *genes*.

chromosomes organelles that package and manage the storage, duplication, expression, and *evolution* of DNA.

***cis*-acting elements** short DNA sequences (6–15 base pairs long) that constitute the control elements adjacent to *genes*. Through their binding to *transcription factors*, *cis*-acting elements control or modulate transcription initiation at one or more nearby *genes*. *Promoters*, *enhancers*, and *locus control regions* are three types of *cis*-acting elements. Also called *cis*-controlling elements.

cistron a term sometimes used as a synonym for *complementation group* or *gene*.

cleavage stage in early embryonic development, the stage of the first four equal cell divisions.

clone a group of biological entities—cells or DNA molecules—that are genetically identical to each other.

cloning the process by which cellular clones or DNA clones are formed.

cloning vector a DNA molecule into which another DNA fragment of appropriate size can be integrated without loss of the vector's capacity for replication. Vectors introduce foreign DNA into host cells, where it can be reproduced in large quantities. See *vector*.

coactivator a type of *transcription factor* or other molecule that binds to *transcription factors* rather than to *DNA* and plays a role in increasing levels of *transcription*.

codominant the expression of a *heterozygous genotype* resulting in hybrid offspring that resemble both parents equally for a particular trait.

codon a nucleotide triplet that represents a particular *amino acid* to be inserted in a specific position in the growing amino acid chain during *translation*. Codons can be either in the *mRNA* or in the *DNA* from which the RNA is transcribed.

coefficient of coincidence the ratio between the actual frequency of double crossovers observed in an experiment and the number of double crossovers expected on the basis of independent probabilities.

cohesin a multisubunit protein complex that associates with *sister chromatids* in eukaryotic cells and holds the chromatids together until *anaphase*; can be found at both the *centromere* and along the *chromosome* arms.

cohesive ends short, single-stranded unpaired flaps protruding from the ends of a cut DNA molecule. Each flap can reform *hydrogen bonds* with a complementary sequence protruding from the end of another piece of DNA.

colinearity the parallel between the sequence of *nucleotides* in a *gene* and the order of *amino acids* in a *polypeptide*.

colony a mound of genetically identical cells.

competent the description of a state of cells able to take up DNA from the medium.

complement (used as a verb) when a *heterozygote* for two *recessive mutations* displays a normal *phenotype* because the dominant *wild-type alleles* on each of the two homologues make up for a defect in the other *homologous chromosome*. Mutations that complement are usually in different genes, whereas mutations that fail to complement are usually in the same gene.

complementary base-pairing during *DNA replication*, base-pairing in which a complementary strand aligns opposite the exposed bases on the parent strand to create the nucleotide sequence of the new strand of DNA. Refer to *base pair, complementary sequences*.

complementary gene action genes working in tandem to produce a particular trait.

complementary sequences nucleic acid base sequences that can form a double-stranded structure by matching base pairs; the complementary sequence to 5′ GTAC 3′ is 3′ CATG 5′.

complementation the process in which heterozygosity for *chromosomes* bearing mutant *recessive alleles* for two different *genes* produces a normal *phenotype*.

complementation group a collection of *mutations* that do not complement each other. Often used synonymously for *gene*.

complementation table a method of collating data that helps visualize the relationship among a large group of mutants.

complementation testing a technique based on the idea of a *gene* as a unit of function; it determines whether or not two *mutations* are *alleles* of the same gene.

complete coverage in mapping of DNA, when the number of *markers* on a *linkage map* is sufficient so that the *locus* controlling any *phenotype* can be linked to at least one of those markers.

complete digest digesting a sample of DNA molecules with a *restriction enzyme* such that cleavage has occurred at every DNA site recognizable by the enzyme.

complete genomic library a hypothetical collection of *DNA clones* that includes one copy of every sequence in the entire *genome*.

complex refers to the multiple types of variation that can exist at alternative *alleles*, including more than one nucleotide *substitution*, a substitution in combination with a small *deletion*, a *duplication*, or another *insertion*.

complex haplotype a set of linked DNA variations along a *chromosome*, with the possibility of many differences in alternative *alleles*. See *haplotype*.

complex trait See *continuous trait*.

concerted evolution a process that allows changes in single genetic elements to spread across a complete set of *genes* in a particular *gene family*.

condensation a cellular process of *chromatin* compaction that results in the visible emergence of individual *chromosomes*.

condensin a multisubunit complex of proteins in eukaryotic cells that compacts *chromosomes* during *mitosis*.

conditional lethal an *allele* that is lethal only under certain conditions.

conjugation one of the mechanisms by which bacteria transfer genes from one strain to another; in this case, the *donor* carries a special type of *plasmid* that allows it to transfer DNA directly when it comes in contact with the recipient. The recipient is known as an exconjugant. Contrast with *transformation*.

conjugative plasmids *plasmids* that initiate *conjugation* because they carry the genes that allow the *donor* to transfer genes to the *recipient*.

consanguine related by a common ancestor.

consanguineous mating mating between blood relatives.

conservative substitutions mutational changes that substitute an amino acid in a protein with a different amino acid with similar chemical properties. Compare with *nonconservative substitutions*.

conservative transposition *transposition* in which the transposable elements actually leave their original position when they move. See *replicative transposition*.

conserved synteny the state in which the same two or more *loci* are found to be linked in several species. Compare with *syntenic*, *syntenic segments*.

constitutive expression refers to a state of gene activation that remains at a constant high level and is not subject to modulation.

constitutive heterochromatin chromosomal regions that remain condensed in *heterochromatin* at most times in all cells.

constitutive mutants synthesize certain enzymes all the time, irrespective of environmental conditions.

constitutive mutation a mutation in a *cis*-acting or *trans*-acting element that causes an associated gene to remain in a state of activation irrespective of environmental or cellular conditions that modulate gene activity in nonmutant cells.

contig a set of two or more partially overlapping cloned DNA fragments that together cover an uninterrupted stretch of the *genome*.

continuous trait an inherited trait that exhibits many intermediate forms; determined by segregating alleles of many different *genes* whose interaction with each other and the environment produces the *phenotype*. Also called a *quantitative*, *multifactorial*, or *complex* trait. Compare with *discontinuous trait*.

contractile ring a transitory organelle composed of actin microfilaments aligned around the circumference of a dividing animal cell's equator; contraction of the filaments pinches the cell in two.

convergent evolution the process in which structurally unrelated but functionally analogous organs emerge in different species as a result of *natural selection*.

copy number polymorphism (CNP) a category of genetic variation arising from large regions of duplication or deletion, occurring with an overall frequency of greater than 1 percent in a population. Compare with *copy number variant*.

copy number variant (CNV) a category of genetic variation arising from large regions of duplication or deletion, occurring with an overall frequency of less than 1 percent in a population. Compare with *copy number polymorphism*.

core histones proteins that form the core of the *nucleosome*: H2A, H2B, H3, and H4.

corepressor a type of *transcription factor* or other molecule that binds to transcription factors rather than DNA and prevents transcription above basal levels.

cosmids hybrid *plasmid-phage vectors* that make use of a virus capsule to infect bacteria; constructed with plasmid-derived selectable markers and two specialized segments of phage l DNA known as *cos* (for *co*hesive end) sites.

co-transformation the simultaneous transformation of two or more genes. See *transformation*.

crisscross inheritance an inheritance pattern in which males inherit a trait from their mothers, while daughters inherit the trait from their fathers.

cross the deliberate mating of selected parents based on particular genetic traits desired in the offspring.

cross-disciplinary biology a type of biology in which teams of biologists, computer scientists, chemists, engineers, mathematicians, physicists, and others work together on the problems of *systems biology*.

cross-fertilization brushing pollen from one plant onto a female organ of another plant.

cross-fostering the random relocation of offspring to the care of other parents, typically done with animal studies to randomize the effects of environment on outcome.

crossing-over during *meiosis*, the breaking of one maternal and one paternal *chromosome*, resulting in the exchange of corresponding sections of DNA and the rejoining of the chromosome. This process can result in the exchange of *alleles* between chromosomes. Compare with *recombination*.

crossover suppression the result of heterozygosity for *inversions*, in which no viable recombinant progeny are possible.

crossover suppressors chromosomal inversions that act to inhibit *crossovers*.

C terminus the end of the *polypeptide* chain that contains a free carboxylic acid group.

cyclin-dependent kinases (CDKs) *protein kinases* that are dependent on another protein known as a *cyclin* for the targeting of their activity to a specific substrate. CDKs regulate the transition from G_1 to S and from G_2 to M through phosphorylations that activate or inactivate target proteins.

cyclins a family of proteins that combine with *cyclin-dependent kinases* and thereby determine the substrate specificity of the kinases. By directing kinases to the right substrates, the cyclins help regulate passage of the cell through the *cell cycle*. Concentrations of the various cyclins rise and fall throughout the cell cycle.

cytokinesis the final stage of cell division, which begins during *anaphase* but is not completed until after *telophase*. In this stage, the daughter nuclei emerging at the end of telophase are packaged into two separate daughter cells.

cytosine (C) a nitrogenous base; one member of the base pair G–C (guanine–cytosine).

D

dauer larva in nematodes, an alternate L3 form that does not feed and has a specialized cuticle that resists desiccation; dauer larvae can survive more than six months when food is scarce.

deamination the removal of an amino ($-NH_2$) group from normal DNA.

degenerate genetic code a *genetic code* in which a single amino acid may be coded for by more than one codon

degrees of freedom (df) the measure of the number of independently varying parameters in an experiment.

deletion occurs when a block of one or more nucleotide pairs is lost from a DNA molecule. Compare with *insertion*.

deletion loop an unpaired bulge of the normal *chromosome* that corresponds to an area deleted from a paired homologue. Contrast with *duplication loop*.

denaturation (denature, denatured) the disruption of hydrogen bonds within a macromolecule that normally uses hydrogen bonds to maintain its structure and function. Hydrogen bonds can be disrupted by heat, extreme conditions of pH, or exposure to chemicals such as urea. When normally soluble proteins are denatured, they unfold and expose their nonpolar amino acids, which can cause them to become insoluble. When DNA is denatured, double-stranded molecules break apart into two separate strands.

deoxyribonucleic acid (DNA) See *DNA*.

deoxyribonucleotide See *nucleotide*.

depurination DNA alteration in which the hydrolysis of a purine base, either A or G, from the deoxyribose–phosphate backbone occurs.

developmental geneticists biologists who use *genetics* to study how the fertilized egg of a multicellular organism becomes an adult.

developmental genetics the use of *genetics* to study how the fertilized egg of a multicellular organism becomes an adult.

developmental hierarchy a *developmental pathway* in which the product of one *gene* regulates the expression of another gene.

developmental pathway a detailed description of how many *genes* interact biochemically to produce a particular outcome in development.

diakinesis a substage of *prophase I* during which *chromosomes* condense to the point where each *tetrad* consists of four separate *chromatids*; at the end of this substage, the nuclear envelope breaks down and the microtubules of the *spindle apparatus* begin to form.

dicentric chromatid a *chromatid* with two *centromeres*.

dideoxynucleotide a nucleotide analogue lacking the 3′-hydroxyl group that is critical for the formation of *phosphodiester bonds*. The four types of dideoxynucleotides are abbreviated as ddTTP, ddATP, ddGTP, and ddCTP. Dideoxynucleotides are key components of the most common methods of DNA sequencing.

digestion the enzymatic process by which a complex biological molecule (DNA, RNA, protein, or complex carbohydrate) is broken down into smaller components.

dihybrid an individual that is *heterozygous* for two *genes* at the same time.

dimerization domain the region of a *polypeptide* that facilitates interactions with other molecules of the same polypeptide or with other polypeptides. Certain motifs such as the leucine zipper often serve as dimerization domains.

dipeptide two *amino acids* connected by a *peptide bond*.

diploid *zygotes* and other cells carrying two matching sets of chromosomes are described as diploid. Sometimes denoted as 2*n*. Compare with *haploid*.

diplotene a substage of *prophase I* during which there is a slight separation of regions of *homologous chromosomes* but the aligned homologous chromosomes of each *bivalent* remain tightly merged at *chiasmata*.

discontinuous traits *phenotypes* that are expressed in clear-cut variations. Compare with *continuous trait*.

discrete trait an inherited trait that clearly exhibits an either/or status (i.e., purple versus white flowers).

diversification the evolution of many variants from one progenitor molecule.

DNA deoxyribonucleic acid; the molecule of heredity that encodes genetic information.

DNA clone a purified sample containing a large number of identical *DNA* molecules.

DNA fingerprint the pattern produced by the simultaneous detection of *genotype* at a group of unlinked, highly polymorphic loci.

DNA marker See *marker*.

DNA polymerase III a complex enzyme that forms a new DNA strand during *replication* by adding *nucleotides*, one after the other, to the 3′ end of a growing strand.

DNA polymorphisms two or more *alleles* at a *locus* detected with any method that directly distinguishes differences in DNA sequence. The sequence variations of a DNA polymorphism can occur at any position on a chromosome and may, or may not, have an effect on *phenotype*. See *polymorphisms*.

DNA probe a purified fragment of DNA labelled with a radioactive isotope or a fluorescent dye and used to identify complementary sequences by means of *hybridization*.

DNA replication the process by which a double-helical DNA molecule is duplicated into two identical double-helical DNA molecules.

DNase hypersensitive (DH) sites sites on DNA that contain few, if any, *nucleosomes*; these sites are susceptible to cleavage by DNase enzymes.

DNA sequence the relative order of *base pairs*, whether in a fragment of DNA, a *gene*, a *chromosome*, or an entire *genome*.

DNA topoisomerases a group of enzymes that help relax *supercoiling* of the DNA helix by nicking one or both strands to allow the strands to rotate relative to each other.

DNA transposons units of DNA that move from place to place within the *genome* without an RNA intermediate, sometimes causing a change in gene function when they insert themselves in a new chromosomal location.

domain a discrete region of a *protein* with its own function. The combination of domains in a single protein determines its overall function.

domain architecture the number and order of a protein's functional domains.

dominance series dominance relations of all possible pairs of *alleles* are arranged in order from most dominant to most recessive.

dominant allele an *allele* whose *phenotype* is expressed in a *heterozygote*. See *recessive allele*.

dominant epistasis the effects of a *dominant allele* at one *gene* hide the effects of alleles at another gene. Compare with *recessive epistasis*.

dominant negative a mechanism of dominance in which some alleles of genes encode subunits of *multimeric proteins* that block the activity of the subunits produced by *wild-type alleles*.

dominant-negative alleles *alleles* that block the activity of *wild-type alleles* of the same gene, causing a loss of function even in *heterozygotes*. Also known as antimorphic mutations.

dominant-negative mutants loss-of-function mutations that have a dominant effect by counteracting the functions of proteins encoded by wild-type alleles. See *dominant negative*.

dominant suppression an *epistatic* interaction whereby a dominant allele on one gene hides the effects of a dominant allele of another gene. Compare with *dominant epistatis*.

dominant trait the trait that appears in the F_1 hybrids (*heterozygotes*) resulting from a mating between pure-breeding parental strains showing antagonistic *phenotypes*.

donor in gene transfer in bacteria, the cell that provides the genetic material. See *recipient, transformation, conjugation,* and *transduction*.

dosage compensation a mechanism that equalizes levels of *X-linked* gene expression independent of the number of copies of the X chromosome; in mammals, the dosage compensation mechanism is *X-chromosome inactivation*.

double helix the shape that two linear strands of DNA assume when bonded together.

double minutes small chromosome-like bodies lacking centromeres and telomeres.

downstream the direction travelled by RNA polymerase as it moves from the promoter to the terminator. Compare with *upstream*.

Drosha a nuclear enzyme component of the *RNAi* machinery present in all eukaryotic cells. Drosha recognizes and processes stem-loop structures associated with primary *miRNA*-containing transcripts. Drosha products are transported into the cytoplasm where they are further processed by *Dicer* into mature miRNAs.

Ds (Dissociation) a transposable element in corn that causes chromosomal breaks or mutations in the presence of an *Ac (Activator)* element.

duplication events that result in an increase in the number of copies of a particular chromosomal region. See *tandem duplications, nontandem duplications*.

duplication loop a bulge in the *duplication*-bearing chromosome that has no similar region with which to pair in the unduplicated normal *homologous chromosome*. Contrast with *deletion loop*.

E

ectopic gene expression gene expression that occurs outside the cell or tissue where the gene is normally expressed.

electrophoresis a method of separating large molecules (such as DNA fragments or proteins) from a mixture of similar molecules. An electric current is passed through a medium containing the mixture, and each kind of molecule travels through the medium at a different rate, depending on its size and electrical charge. See *gel electrophoresis*.

elements components of a *biological system*.

elongation the phase of *DNA replication, transcription,* or *translation* that successively adds nucleotides or amino acids to a growing macromolecule. Compare with *initiation*.

elongation factors proteins that aid in the *elongation* phase of *translation*.

embryonic stem (ES) cells undifferentiated cells, originally derived from early embryos that can grow in cell culture and remain undifferentiated.

emergent properties traits and behaviours that arise from the operation of a *biological system* as a whole; *immunity* and *tolerance* are two emergent properties of the immune system.

endosymbiont theory proposes that chloroplasts and mitochondria originated when free-living bacteria were engulfed by primitive nucleated cells. Host and guest formed cellular communities in which each member adapted to the group arrangement and derived benefit from it.

enhanceosome a completely assembled set of *transcription factors* (which may include *activators, repressors, coactivators,* and *corepressors*) associated with an *enhancer* or *locus control region* in a structure that is able to modulate transcription activity.

enhancer mutations mutations in a *modifier gene* that worsen the phenotypic effects of a mutation in another gene.

enhancer trapping in *Drosophila*, the identification of P-element insertion lines with particular β-galactosidase expression patterns.

enhancers *cis*-acting elements that can regulate *transcription* from nearby genes. In yeast, enhancers are called upstream activation sites, or UASs. Enhancers function by acting as binding sites for *transcription factors*.

environmental variance (V_E) deviation from the mean attributed to the influence of external, non-inheritable factors. Compare with *genetic variance*.

epigenetic a state of gene functionality that is not encoded within the DNA sequence but that is still heritable from one generation to the next. It can be accomplished and maintained through a chemical modification of DNA such as methylation.

episomes *plasmids*, like the *F plasmid*, that can integrate into the host genome.

epistasis a gene interaction in which the effects of an *allele* at one gene hide the effects of alleles at another gene. See *dominant epistasis* and *recessive epistasis*.

epistatic describes an *allele* of one gene that masks the effects of one or more alleles of another gene.

equational division cell division that does *not* reduce the number of *chromosomes*, but instead distributes *sister chromatids* to the two daughter cells. *Mitosis* and *meiosis II* are both equational divisions.

equilibrium frequency the q value at which $\Delta q = 0$; the *allele* frequency required to maintain an allele in the *population*.

euchromatin a chromosomal region of cells that appears much lighter and less condensed when viewed under a light microscope. Contrast with *heterochromatin*.

eukaryotes one of the three major evolutionary lineages of living organisms known as domains; organisms whose cells have a membrane-bounded nucleus. Contrast with *prokaryotes*.

eukaryotic describes cells that have a membrane-bounded nucleus.

eukaryotic gene regulation the control of gene expression in the cells of *eukaryotes*.

euploid describes cells containing only complete sets of *chromosomes*.

evolution the change in the genetic composition of a population resulting from the *selection* of individuals possessing DNA variants that make them particularly suited to survive and reproduce in a given environment.

excision repair a DNA repair mechanism in which specialized proteins recognize damaged base-pairing and remove the damaged section so that *DNA polymerase* can fill in the gap by complementary base-pairing with the information from the undamaged strand of DNA. See *base excision repair* and *nucleotide excision repair*.

exconjugants *recipient* cells resulting from gene transfer in which *donor* cells carrying specialized *plasmids* establish contact with and transfer DNA to the recipients.

exit (E) site one of three *transfer RNA* binding sites in ribosomes. The E site is occupied by tRNAs during the period just after their disconnection from *amino acids* by the action of *peptidyl transferase* and just before the release of the tRNAs from the *ribosome*.

exons sequences that are found both in a gene's DNA and in the corresponding mature *messenger RNA (mRNA)*. See *introns*.

exon shuffling the exchange of *exons* among different genes during evolution, producing mosaic proteins with different combinations of functions.

expressed sequence tag (EST) a 600–1000-bp-long single-sequence run on cDNA inserts.

expressivity the degree or intensity with which a particular *genotype* is expressed in a *phenotype*.

extracellular signals signals that are able to diffuse from cell to cell or by cell-to-cell contact; can occur in the form of steroids, peptides, and proteins and are collectively known as hormones.

F

5′ and 3′ untranslated regions (5′ and 3′ UTRs) sequences at the 5′ and 3′ ends of mRNAs that are not translated, but play important roles in regulating the efficiency of *translation*.

5′ untranslated leader in *eukaryotes*, the mRNA region between the 5′ *methylated cap* and the *initiation codon*.

filial generations the successive offspring in a controlled sequence of breeding, starting with two parents (the P generation) and selfing or intercrossing the offspring of each subsequent generation. An offspring generation is denoted by $F_\#$.

fine structure mapping recombination mapping of *mutations* in the same gene; in some *bacteriophage* experiments, fine structure mapping can resolve mutations in adjacent nucleotide pairs.

first filial (F_1) generation progeny of the *parental generation* in a controlled series of crosses.

FISH (fluorescence *in situ* hybridization) a *physical mapping* approach that uses fluorescent tags to detect *hybridization* of nucleic acid probes with *chromosomes*.

fitness the relative advantage or disadvantage in reproduction that a particular *genotype* provides to members of a *population* in comparison to alternative genotypes at the same *locus*.

fitness cost negative impact of the development of a homozygous resistance *genotype*.

fixed describes a population in which no further changes in *allele frequency* can occur (in the absence of migration or mutation).

fluctuation test the Luria–Delbrück experiment to determine the origin of bacterial resistance. Fluctuations in the numbers of resistant colonies growing in different Petri plates showed that resistance is not caused by exposure to bactericides.

forward mutation a mutation that changes a *wild-type allele* of a gene to a different *allele*.

founder cells in nematode development, progenitors of the major embryonic lineages.

founder effect a variation of *genetic drift*, occurring when a few individuals separate from a larger *population* and establish a new one that is isolated from the original population, resulting in altered *allele frequencies* in the new population.

F plasmid a *conjugative plasmid* that carries many genes required for the transfer of DNA. Cells carrying the F plasmid are called F^+ cells. Cells without the plasmid are called F^- cells.

F′ plasmids *F-plasmid* variants that carry most F-plasmid genes plus some bacterial DNA; particularly useful in genetic complementation studies.

frameshift mutations *insertions* or *deletions* of base pairs that alter the grouping of *nucleotides* into *codons*. Refer to *reading frame*.

free duplications small DNA fragments maintained extrachromosomally in a genetic stock.

fusion proteins *proteins* encoded by parts of more than one gene.

G

G_0 in the cell cycle, a resting form of G_1. Cells in G_0 normally do not divide.

G_1 the stage of the cell cycle from the birth of a new cell until the onset of chromosome replication at *S phase*.

G_2 the stage of the cell cycle from the completion of *chromosome replication* until the onset of cell division.

gain-of-function mutations rare *mutations* that enhance a gene's function or confer a new activity on the gene's product.

gametes specialized cells (eggs and sperm) that carry genes between generations.

gametogenesis the formation of *gametes*.

gastrulation the folding of the cell sheet early in embryo formation; usually occurs immediately after the blastula stage of development.

gel electrophoresis a process used to separate DNA fragments, RNA molecules, or polypeptides according to their size. Electrophoresis is accomplished by passing an electric current through agarose or polyacrylamide gels. The electric current forces molecules to migrate through the gel at different rates dependent on their sizes.

gene amplification an increase from the normal two copies to hundreds of copies of a *gene*; often due to *mutations* in *p53*, which disrupt the G_1-to-S *checkpoint*.

gene conversion any deviation from the expected 2:2 *segregation* of parental *alleles*.

gene dosage the number of times a given *gene* is present in the cell nucleus.

gene expression the process by which a gene's information is converted into *RNA* and then (for protein-coding genes) into a *polypeptide*.

gene family set of closely related genes with slightly different functions that most likely arose from a succession of gene *duplication* events.

gene function generally, to govern the synthesis of a *polypeptide*; in Mendelian terms, a gene's specific contribution to *phenotype*.

gene pool the sum total of all *alleles* carried in all members of a *population*.

generalized transduction a type of *transduction* (gene transfer mediated by bacteriophages) that can result in the transfer of any bacterial gene between related strains of bacteria.

gene regulatory network a set of interacting *transcription factors* and their cognate *cis-control elements* that receive diverse inputs of biological information; integrate and modify those inputs; and transmit the transformed information to various *protein networks*. The fundamental link in a gene regulatory network is the interaction of a transcription factor with its cognate *cis*-control element.

genes basic units of biological information; specific segments of DNA in a discrete region of a *chromosome* that serve as a unit of function by encoding a particular *RNA* or *protein*.

gene superfamily a large set of related genes that is divisible into smaller sets, or families, with the genes in each family being more closely related to each other than to other members of the larger superfamily. The single-gene or *multigene families* that compose a superfamily reside at different chromosomal locations. The families of genes encoding the globins and the *Hox* transcription factors are examples of gene superfamilies.

genetic code the sequence of *nucleotides*, coded in triplets (*codons*) along the *mRNA*, that determines the sequence of *amino acids* in *protein* synthesis.

genetic drift a change in allele frequencies as a consequence of the randomness of inheritance from one generation to the next.

genetic imbalance a situation when the *genome* of a cell or organism has more copies of some genes than other genes due to *chromosomal rearrangements* or *aneuploidy*.

genetic linkage See *linkage*.

genetic markers *genes* identifiable through phenotypic variants that can serve as points of reference in determining whether particular progeny are the result of *recombination*. Compare with *physical markers*.

genetic mosaic organisms in which some cells have one genotype, whereas other cells have a different genotype.

genetic relatedness the average fraction of common *alleles* at all gene loci that individuals share because they inherited them from a common ancestor.

genetics the science of heredity.

genetic screen an examination of each colony in a *population* for its *phenotype*.

genetic variance (V_G) deviation from the mean attributable to inheritable factors. Compare with *environmental variance*.

genetic variant describes *alleles* of a *polymorphic locus*.

genome the entire collection of *chromosomes* in each cell of an organism.

genome-wide association study (GWAS) the application of *SNP* chip genotyping to large populations of people for the purpose of discovering genetic associations between particular SNPs and traits.

genomic equivalent the number of clones—with inserts of a particular size—that would be required to carry a single copy of every sequence in a particular *genome*.

genomic imprinting the phenomenon in which the expression pattern of a *gene* depends on the parent that transmits it.

genomic islands large segments of *DNA* (10–200 kb in size), showing properties that suggest that they originated from the transfer of foreign DNA into a bacterial cell.

genomic library a long-lived collection of *cellular clones*, containing copies of every DNA sequence in the whole *genome* inserted into a suitable *vector*.

genomics a branch of biology dedicated to the study of whole *genomes*.

genotype the actual *alleles* present in an individual.

genotype frequency the proportion of total individuals in a *population* that are of a particular *genotype*.

genotypic class a grouping defined by a set of related *genotypes* that will produce a particular *phenotype*. The term is most useful in describing progeny of *dihybrid* or *multihybrid* crosses involving complete dominance; for example, in a cross between *Aa Bb* individuals, the genotypic classes are *A– B–*, *A– bb*, *aa B–*, and *aa bb*.

germ cells specialized cells that incorporate into the reproductive organs, where they ultimately undergo *meiosis*, thereby producing *haploid gametes* that transmit genes to the next generation. Compare with *somatic cells*.

germ line all the *germ cells* in a sexually reproducing organism. In animals, it is set aside from the *somatic cells* during embryonic development. The germ cells in it divide by *mitosis* to produce a collection of specialized *diploid* cells that then divide by *meiosis* to produce *haploid* cells, or *gametes*. It thus includes the precursors of the gametes such as *oogonia, spermatagonia, primary*

and *secondary oocytes*, and *primary* and *secondary spermatocytes* as well as the *gametes*.

germline gene therapy a genetic engineering technique that modifies the DNA of *germ cells* that are passed on to progeny.

germline mutations genetic alterations that occur in cells that give rise to *gametes* (sperm and oocytes); if a mutant gamete contributes to the formation of a *zygote*, all of the cells of the zygote will carry the mutation.

growth factors extracellular hormones and cell-bound signals that stimulate or inhibit cell proliferation.

guanine (G) a nitrogenous base; one member of the base pair G–C (guanine–cytosine).

gynandromorph a rare genetic *mosaic* with some male tissue and some female tissue, usually in equal amounts.

H

hairpin loops structures formed when a single strand of DNA or RNA can fold back on itself because of *complementary base-pairing* between different regions in the same molecule.

haploid a single set of *chromosomes* present in the egg and sperm cells of animals and in the egg and pollen cells of plants. Sometimes denoted as *n*. Compare with *diploid*.

haploinsufficiency a rare form of dominance in which an individual *heterozygous* for a *wild-type allele* and a *null* allele shows an abnormal *phenotype* because the level of gene activity is not enough to produce a normal phenotype.

haplotype a specific combination of *linked alleles* in a cluster of related *genes*. A contraction of the phrase "haploid genotype."

Hardy–Weinberg law defines the relationships between *genotype* and *allele frequencies* within a generation and from one generation to the next.

hemizygous describes the *genotype* for genes present in only one copy in an otherwise *diploid* organism, such as *X-linked* genes in a male.

heredity the way *genes* transmit physiological, physical, and behavioural traits from parents to offspring.

hermaphrodite an organism that has both male and female organs and produces both male *gametes* (sperm) and female gametes (eggs). The organism can have both types of organs at the same time (simultaneous hermaphrodite) or have one type early in life and the other type later in life (sequential hermaphrodite).

heterochromatic DNA genomic DNA from heterochromatic regions; this DNA is often difficult to clone.

heterochromatin highly condensed chromosomal regions within which genes are usually transcriptionally inactive.

heterochronic mutations mutations resulting in the inappropriate timing of cell division and cell-fate decisions during development.

heteroduplex region a region of double-stranded DNA in which the two strands have nonidentical (though similar) sequences. Heteroduplex regions are often formed as intermediates during *crossing-over*.

heterogametic sex the gender of a species in which the two *sex chromosomes* are dissimilar; for example, males are the heterogametic sex in humans because they have an X and a Y chromosome. Compare with *homogametic sex*.

heterogeneous trait occurs when a *mutation* at any one of a number of genes can give rise to the same *phenotype*.

heteromers multimeric proteins composed of nonidentical *subunits*. Compare with *homomers*.

heterozygote an individual with two different *alleles* for a given *gene* or *locus*.

heterozygous a *genotype* in which the two copies of the gene that determine a particular trait are different *alleles*. See *hybrid*.

heterozygous carrier unaffected parents who bear a *dominant* normal *allele* that masks the effects of an abnormal *recessive* one.

Hfr bacteria that produce a *high frequency* of *recombinants* for chromosomal genes in mating experiments because their *chromosomes* contain an integrated *F plasmid*.

high-density linkage map a *linkage map* that shows one *gene* or *marker* for each centimorgan of a *genome*.

histocompatibility antigens cell surface molecules that play a critical role in stimulating a proper *immune response*.

histones small DNA-binding proteins with a preponderance of the basic, positively charged amino acids lysine and arginine. Histones are the fundamental protein components of *nucleosomes*.

homeobox in *homeotic genes*, the region of homology, usually 180 bp in length, that encodes the *homeodomain*.

homeodomain a conserved DNA-binding region of *transcription factors* encoded by the *homeobox* of *homeotic genes*.

homeotic gene a gene that plays a role in determining a tissue's identity during development.

homeotic mutation a mutation that causes cells to misinterpret their position in the blueprint and become normal organs in inappropriate positions; such a mutation can alter the overall body plan.

homeotic selector genes genes that control the identity of body segments.

homogametic sex the gender of a species in which the two *sex chromosomes* are identical; in humans, females are the homogametic sex because they have two X chromosomes. Compare with *heterogametic sex*.

homologous chromosomes (homologues) *chromosomes* that match in size, shape, and banding. A pair of chromosomes containing the same linear gene sequence, each derived from one parent.

homologous genes genes in different species with enough sequence similarity to be evolutionarily related.

homologues genes or regulatory DNA sequences that are similar in different species because of descent from a common ancestral sequence.

homomers *multimeric proteins* composed of identical *subunits*. Compare with *heteromers*.

homozygote an individual with identical *alleles* for a given *gene* or *locus*.

homozygous a *genotype* in which the two copies of the gene that determine a particular trait are the same.

horizontal gene transfer See *lateral gene transfer*.

hormone a small molecule or polypeptide that is made and released by certain secretory cells in multicellular organisms. Secreted hormones diffuse or move via bodily fluids to other cell types that contain specific receptors. Hormone-receptor binding can elicit changes in gene function and differentiation in the target cells.

hot spots sites within a gene that mutate more frequently than others, either spontaneously or after treatment with a particular *mutagen*.

Human Genome Project an initiative to determine the complete sequence of the human *genome*.

hybrid dysgenesis the phenomenon in which high *transposon* mobility causes reduced fertility in hybrid progeny; the result of crossing *Drosophila* males carrying the *P* element with females that lack the *P* element.

hybridization the natural propensity of complementary single-stranded molecules of *DNA* or *RNA* to base-pair and form stable double helixes.

hybrids offspring of genetically dissimilar parents; often used as a synonym for *heterozygotes*.

hydrogen bonds weak electrostatic bonds that result in a partial sharing of hydrogen atoms between reacting groups.

hypermorphic mutation a *gain-of-function mutation* that generates either more protein than the *wild-type allele* or the same amount of a more efficient protein. If excess protein activity alters *phenotype*, the hypermorphic allele is dominant. Compare with *hypomorphic mutation*.

hypomorphic mutation a *loss-of-function mutation* that produces either less of a protein or a protein with very weak but detectable function. Compare with *hypermorphic mutation*.

hysteresis a system that has a different threshold for the transition from the OFF to the ON state, compared with the transition from the ON to the OFF state.

I

idiogram a black-and-white diagram of the chromosomes converted from the light and dark bands observed under the microscope.

imaginal discs flattened epithelial sacs that develop from small groups of cells set aside in the early embryo from which adult-specific structures like wings, legs, eyes, and genitalia develop; in *Drosophila*, they undergo extensive growth and development during the larval and pupal stages.

immune response a physiological response to the immune system's activation by *antigen*. Immune responses include the production of antibodies by B lymphocytes and the ability of killer T cells to destroy foreign or cancerous cells by direct cell-to-cell contact.

immunity the ability to generate immune responses to infectious agents or vaccines. Such a response protects against serious disease.

incomplete dominance the expression of a *heterozygous phenotype* resulting in offspring whose phenotype is intermediate between those of the parents.

indels *deletions*, *duplications*, and *insertions* at nonrepeat loci.

inDels or DIPs genetic variants with short deletions and insertions; type of SNPs.

independent assortment the random distribution of different genes during gamete formation. See *law of independent assortment*.

induced pluripotent stem (iPS) cells cells created by the introduction of *transcription factors* to adult somatic cells to make them behave like pluripotent ES cells.

inducer a specific molecule responsible for stimulating production of a given *protein*.

induction the process by which a specific molecule stimulates synthesis of a given *protein*.

inflorescence in a plant, the flower or group of flowers at the tip of a branch.

initiation the first phase of *DNA replication*, *transcription*, or *translation* needed to set the stage for the addition of nucleotide or amino acid building blocks during *elongation*.

initiation codon a nucleotide triplet that marks the precise spot in the nucleotide sequence of an *mRNA* where the code for a particular *polypeptide* begins. Compare with *nonsense codon*.

initiation factors a term usually applied to proteins that help promote the association of *ribosomes*, *mRNA*, and initiating *tRNA* during the first phase of *translation*.

insertion the addition to a DNA molecule of one or more nucleotide pairs.

insertional mutagenesis a method of mutagenesis in which a foreign DNA sequence (viral, *plasmid*, *transposon*, or cloned fragment) is used as the mutagenic agent. Mutations result when the foreign sequence integrates into a gene. The disrupted gene can be easily identified and cloned based on its association with the foreign DNA.

insertion sequences (ISs) small *transposable elements* that dot the chromosomes of many types of bacteria; they are *transposons* that do not contain selectable markers.

insulator a transcriptional regulation element in *eukaryotes* that stops communication between *enhancers* on one side of it with *promoters* on the other side. Insulators play an important role in limiting the *chromatin* region over which an enhancer can operate.

integrative and conjugative element (ICE) a type of *pathogenicity island* that contains features of conjugative plasmids (like the F factor); it encodes an integrase that allows the DNA to integrate or excise from a chromosome.

intercalators a class of chemical *mutagens* composed of flat, planar molecules that can sandwich themselves between successive *base pairs* and disrupt the machinery of *replication*, *recombination*, or repair.

intergenic gene conversion the information flow between related DNA sequences that occurs through an alternative outcome of the process responsible for *unequal crossing-over*.

interkinesis the brief *interphase* between *meiosis I* and *meiosis II*.

interphase the period in the cell cycle between divisions.

intragenic suppression the restoration of gene function by one *mutation* cancelling the effects of another mutation in the same *gene*.

introns the DNA base sequences of a *gene* that are spliced out of the primary *transcript* and are therefore not found in the mature mRNA. See *exons*.

inversion a 180-degree rotation of a segment of a *chromosome* relative to the remainder of the chromosome.

inversion heterozygotes cells or organisms in which one *chromosome* is of a normal gene order, while the *homologous chromosome* carries an *inversion*.

inversion loop formed in the cells of an *inversion heterozygote* when the inverted region rotates to pair with the similar region in the normal homologue.

K

karyotype the visual description of the complete set of *chromosomes* in one cell of an organism; usually presented as a photomicrograph with the chromosomes arranged in a standard format showing the number, size, and shape of each chromosome type.

kinetochore a specialized chromosomal structure composed of DNA and proteins that is the site at which *chromosomes* attach to the spindle fibres. See *kinetochore microtubules*.

kinetochore microtubules microtubules of the *mitotic spindle* that extend between a *centrosome* and the *kinetochore* of a *chromatid*. *Chromosomes* move along the kinetochore microtubules during cell division.

knockout construct a DNA construct that is introduced into a cell type of interest to replace (via homologous recombination) a specific open reading frame from its start to stop codons. The construct generally contains a gene encoding a selectable marker.

knockout mice mice *homozygous* for an induced mutation in a targeted gene; the mutation destroys (knocks out) the function of the gene.

L

***lac* operon** a single DNA unit in *E. coli*, composed of the *lacZ*, *lacY*, and *lacA* genes together with the *promoter* (*p*) and the operator (*O*), that enables the simultaneous regulation of the three structural genes in response to environmental changes.

lagging strand during *replication*, the DNA strand whose polarity is opposite to that of the *leading strand*. The lagging strand must be synthesized discontinuously as small *Okazaki fragments* that are ultimately joined into a continuous strand.

lambda (λ) a naturally occurring double-stranded DNA virus that infects *E. coli*. A common *plasmid vector* used to clone DNA from other organisms. Also called a *phage lambda*.

lambda (λ) repressor protein a protein that binds to *operator sites* on phage λ DNA, preventing the transcription of genes needed for the lytic cycle; this protein makes λ lysogens immune to infection with incoming λ phage.

late-onset a genetic condition in which symptoms are not present at birth, but manifest themselves later in life.

lateral gene transfer the introduction and incorporation of DNA from an unrelated individual or from a different species.

law of independent assortment during *gamete* formation, different pairs of *alleles* segregate independently of each other.

law of segregation the two *alleles* for each trait separate (segregate) during *gamete* formation, and then unite at random, one from each parent, at fertilization.

lawn bacteria immobilized in a nutrient agar, used as a field on which to test for the presence of viral particles.

leader sequence a DNA sequence that precedes the coding sequence and contains signals that regulate *transcription termination* in an *operon* controlled by *attenuation*.

leading strand during *replication*, a DNA strand synthesized continuously 5′ to 3′ toward the unwinding Y-shaped *replication fork*. Compare with *lagging strand*.

leptotene first definable substage of *prophase I* during which the long, thin, already duplicated *chromosomes* begin to thicken.

library See *genomic library* or *cDNA library*.

ligands molecules such as *growth factors* or *hormones* that are produced by one cell and bind to receptors on a different cell, initiating a *signal transduction pathway*.

LINE long *inter*spersed *e*lements; one of the two major classes of *transposable elements* in mammals. Contrast with *SINE*.

lineage compartments regions of an organism in which cells have a restricted developmental potential.

linkage the proximity of two or more *markers* on a *chromosome*; the closer together the markers are, the lower the probability that they will be separated by *recombination*. Genes are *linked* when the frequency of parental-type progeny exceeds that of recombinant progeny (i.e., genes that travel together more often than expected).

linkage disequilibrium when *alleles* at separate *loci* (such as *marker* alleles and disease alleles) are associated with each other at a significantly higher frequency than would be expected by chance. Linkage disequilibrium at particular loci in a *population* can be evidence of common ancestry.

linkage group a group of genes chained together by linkage relationships. See *linkage*.

linkage map depicts the distances between *loci* as well as the order in which they occur on a *chromosome*.

linked describes *genes* whose *alleles* are inherited together more often than not; linked genes are usually located close together on the same *chromosome*.

linker DNA a stretch of less than 40 base pairs of DNA that connect one *nucleosome* with the next.

locus a designated location on a *chromosome*. See α-*globin locus* and β-*globin locus*.

locus control region (LCR) a *cis*-acting regulatory element that operates to enhance *transcription* from individual genes within a gene complex such as the β-globin complex.

loss-of-function mutation a DNA *mutation* that reduces or abolishes the activity of a gene; most (but not all) loss-of-function alleles are *recessive*. Compare with *gain-of-function mutation*.

lysate a population of *phage* particles released from the host bacteria at the end of the *lytic cycle*.

lysogenic bacterium a bacterial cell that carries a *prophage*; lysogen.

lysogenic cycle the cycle during which *bacteriophage* DNA integrates into the host *genome* such that it multiplies along with that genome, but does little or no harm to the host.

lysogeny the integration of *bacteriophage* DNA into the host *chromosome*.

lytic cycle a bacterial cycle of *phage*-infected cells resulting in cell lysis and release of progeny phage. Compare with *lysogenic cycle*.

M

major groove in a space-filling representation of the DNA *double helix* model, the wider of the two grooves resulting from the vertical displacement of the two backbone threads. See *minor groove*.

MAP kinase cascade a trio of three protein kinases that activates a transcription factor.

mapping function a mathematical equation that compensates for the inaccuracies inherent in relating *recombination frequencies* to physical distance.

mapping panels a set of DNA samples used in multiple *linkage* analysis tests.

map unit (m.u.) synonymous with *centimorgan*.

marker an identifiable physical location on a *chromosome*, whose inheritance can be monitored. Markers can be expressed as regions of DNA (genes) or any segment of DNA with variant forms that can be followed.

maternal-effect mutations recessive mutations in maternal genes that influence embryonic development.

maternally supplied components molecules synthesized by the mother that are supplied to the egg and that are needed for early development of the progeny.

mean a statistical average; found by summing all values and dividing by the number of values.

meiosis the process of two consecutive cell divisions starting in the *diploid* progenitors of sex cells. Meiosis results in four daughter cells, each with a *haploid* set of *chromosomes*.

meiosis I (or division I of meiosis) the parent nucleus divides to form two daughter nuclei; during meiosis I, the previously replicated *homologous chromosomes* segregate to different daughter cells.

meiosis II (or division II of meiosis) each of the two daughter nuclei resulting from *meiosis I* divide to produce four nuclei; because the *chromosomes* do not duplicate at the start of meiosis II, these four daughter nuclei are *haploid*.

Mendel's first law the law of segregation states that the two alleles for each trait separate (segregate) during gamete formation, and then unite at random, one from each parent, at fertilization.

Mendel's second law the law of independent assortment states that during gamete formation, different pairs of *alleles* (genes) segregate independently of each other.

messenger RNA (mRNA) RNA that serves as a template for protein synthesis. See *genetic code*.

metabolism the chemical and physical reactions that convert sources of energy and matter to fuel growth and repair within a cell.

metacentric a *chromosome* in which the *centromere* is at or near the middle.

metamorphosis a dramatic reorganization of an organism's body plan; for example, transition from larval stage to adult insect stage.

metaphase a stage in *mitosis* or *meiosis* during which the *chromosomes* are aligned along the equatorial plane of the cell.

metaphase I the phase of *meiosis I* when the *kinetochores* of *homologous chromosomes* attach to microtubules from opposite spindle poles, positioning the *bivalents* at the equator of the *spindle apparatus*.

metaphase II the second phase of *meiosis II* during which *kinetochores* of *sister chromatids* attach to microtubule fibres emanating from opposite poles of the *spindle apparatus*. Two characteristics distinguish metaphase II from its counterpart in *mitosis*: (1) the number of chromosomes is one-half that in mitotic metaphase of the same species, and (2) in most chromosomes, the two sister chromatids are no longer identical because of the recombination through crossing-over that occurred in *meiosis I*.

metaphase plate the imaginary equator of the cell toward which *chromosomes* move during *metaphase*.

metastable epiallele an allele that can be modified in a variable and reversible epigenetic manner, resulting in phenotypic variation even when cells are genetically identical.

metazoans multicellular animals that first appeared about 570 million years ago.

methylated cap formed by the action of capping enzyme and methyl transferases at the 5′ end of eukaryotic *mRNA*, critical for efficient *translation* of the mRNA into protein.

methyl-directed mismatch repair a DNA repair mechanism that corrects mistakes in *replication*, discriminating between a newly synthesized strand and its parental DNA strand by the methyl groups on the parental strand.

microevolution alterations of a population's *gene pool* due to the changes in the frequency of alleles within a population.

micro-RNA (miRNA) an *RNA* molecule 21–24 bases in length that is encoded in the genome of an organism and used by a cell to modulate gene expression through the process of *RNA interference*.

microsatellite a DNA element composed of 15–100 tandem repeats of one-, two-, or three-base-pair sequences.

minimal tiling path the result of the final step in the *shotgun* sequencing strategy, a minimally overlapping set of BAC clones.

minisatellite a DNA element composed of 10–40-bp tandem repeating units of identical sequence.

minor groove in a space-filling representation of the DNA *double helix* model, the narrower of the two grooves resulting from the vertical displacement of the two backbone threads. See *major groove*.

miRISC a functional *RNA-induced silencing complex* (*RISC*) loaded with a *micro-RNA*.

missense mutations changes in the *nucleotide* sequence of a *gene* that change the identity of an *amino acid* in the *polypeptide* encoded by that gene.

mitosis the process of division that produces daughter cells that are genetically identical to each other and to the parent cell.

mitotic nondisjunction the failure of two *sister chromatids* to separate during mitotic *anaphase*; generates reciprocal *trisomic* and *monosomic* daughter cells.

mitotic spindle a structure composed of three types of microtubules (*kinetochore microtubules*, *polar microtubules*, and *astral microtubules*). The mitotic spindle provides a framework for the movement of *chromosomes* during cell division.

model organisms organisms whose characteristics (e.g., short generation time) facilitate experimental laboratory research.

modification the phenomenon in which growth on a restricting host modifies a virus so that succeeding generations grow more efficiently on that same host.

modification enzymes enzymes that add methyl groups to specific DNA sequences, preventing the action of specific *restriction enzymes* on that DNA.

modifier genes genes that produce a subtle, secondary effect on *phenotype*. There is no formal distinction between major and modifier genes, rather it is a continuum of degrees of influence.

molecular marker a segment of DNA found at a specific site in a *genome* that has variants which can be recognized and followed. See *physical marker*, *genetic marker*.

monohybrid crosses crosses between parents that differ in only one trait.

monohybrids individuals having two different alleles for a single trait.

monomorphic a gene with only one *wild-type allele*.

monoploid describes cells, nuclei, or organisms that have a single set of unpaired *chromosomes*. For *diploid* organisms, monoploid and *haploid* are synonymous.

monosomic an individual lacking one *chromosome* from the *diploid* number for the species.

Monte Carlo simulation a computer simulation that uses a random number generator to choose an outcome for probabilistic events in a dynamic system defined by predetermined rules of probability.

morphogens substances that define different cell fates in a concentration-dependent manner.

mosaic an organism containing tissues of different *genotypes*.

mosaic determination where the embryo is a collection of self-differentiating cells that at their formation receive a specific set of molecular instructions governing their unique fates. Contrast with *regulative determination*.

M phase the part of a *cell cycle* in which one cell divides into two.

mRNA See *messenger RNA*.

multifactorial trait See *continuous trait*.

multigene family a set of *genes* descended by *duplication* and *diversification* from one ancestral gene. The members of a multigene family may be either clustered on the same chromosome or distributed on different chromosomes.

multi-hit model a model about how cancer forms based on the assumption that multiple mutations in an array of genes must accumulate over time in the clonal descendants of a single cell before the cancer phenotype appears.

multihybrid crosses crosses between parents that differ in three or more traits.

multimeric protein a protein made from more than one *polypeptide*; each polypeptide in the multimeric protein is called a *subunit*.

multiple alleles a set of *alleles* of a gene with more than two variant forms.

mutable site the location on a chromosome where a chromosomal break or mutation occurs when a transposable element is inserted at that particular site.

mutagen any physical or chemical agent that raises the frequency of *mutations* above the spontaneous rate.

mutant allele (1) an *allele*, or DNA variant, whose frequency is less than 1 percent in a *population*; (2) an allele that dictates a phenotype seen only rarely in a population. See *allele frequency*.

mutant tumour-suppressor genes recessive *mutant alleles* that contribute to the formation of cancer.

mutations heritable changes in *DNA sequence*.

N

n the number of *chromosomes* in a normal *gamete*; for organisms that are not *polyploids*, n is the number of chromosomes in any *haploid* cell and is also equal to x (the number of chromosomes in a single complete set of nonhomologous chromosomes). n and x are *not* identical in polyploid organisms.

$2n$ the number of *chromosomes* in a normal *diploid* cell.

narrow-sense heritability (h^2) the proportion of *total phenotypic variation* that is attributable to additive *genetic variation*. Compare with *broad-sense heritability*.

natural selection in nature, the process that progressively eliminates individuals whose fitness is low and chooses individuals of high fitness to survive and become the parents of the next generation. Contrast with *artificial selection*.

natural transformation a process by which a few species of bacteria transfer genes from one strain to another by spontaneously taking up DNA fragments from their surroundings. Contrast with *artificial transformation*.

neomorphic mutations rare mutations that generate a novel *phenotype* due to production of a protein with a new function or due to *ectopic expression* of the protein.

N-formylmethionine (fMet) a modified methionine whose amino end is blocked by a formyl group; fMet is carried by a specialized *tRNA* that functions only at a *ribosome*'s *translation* initiation site.

nodes (of a *phylogenetic tree*) representations of taxonomic units such as species, *populations*, individuals, or *genes*.

non-allelic noncomplementation (also called second-site noncomplementation) the failure of *mutations* in two different genes to complement each other, causing a mutant *phenotype* in *heterozygotes*. Usually indicates that the two polypeptide products cooperate with each other, for example as *subunits* of the same *multimeric protein*.

nonautonomous elements defective *transposable elements* that cannot move unless the genome contains

nondefective *autonomous elements* that can supply necessary functions like *tranposase* enzymes.

nonautonomous trait when the *phenotype* expressed by a particular cell depends upon the genotypes of neighbouring cells. Compare with *cell autonomous trait*.

nonconservative substitutions mutational changes that substitute an *amino acid* in a *protein* with a different amino acid that has different chemical properties. Compare with *conservative substitutions*.

nondisjunction a failure in *chromosome segregation* during *meiosis*; responsible for defects such as trisomy. See *trisomic*.

nonhomologous end-joining a mechanism for stitching back together ends formed by double-strand breaks. It relies on proteins that bind to the ends of the broken DNA strands and bring them close together. Overhanging ends are often "trimmed" during nonhomologous end-joining, resulting in DNA loss.

nonhomologous unequal crossing-over an exchange between nonequivalent chromosomal segments with little sequence homology. A nonhomologous crossover may be mediated by at least a short stretch of sequence homology, coding or noncoding, at the crossover's two sites of initiation. For example, related repeat sequences in the *genome* may mediate nonhomologous unequal crossing-over.

non-Mendelian inheritance a pattern of inheritance that does not follow Mendel's laws and does not produce Mendelian ratios among the progeny of various crosses.

nonreciprocal interchromosomal translocation a large-scale mutation in which part of one chromosome becomes attached to another region of the same chromosome.

nonreciprocal intrachromosomal translocation a large-scale mutation in which part of one chromosome becomes attached to a nonhomologous chromosome.

nonsense codons the three stop codons that terminate *translation*. Compare with *initiation codon*.

nonsense mutations mutational changes in which a codon for an amino acid is altered to a premature *stop codon*, resulting in the formation of a *truncated protein*.

nonsense suppressor tRNAs *tRNAs* encoded by mutant tRNA genes; these tRNAs contain *anticodons* that can recognize *stop codons*, thus suppressing the effects of *nonsense mutations* by inserting an *amino acid* into a *polypeptide* in spite of the stop codon.

nontandem duplications two or more copies of a region that are not adjacent to each other and may lie far apart on the same chromosome or on different chromosomes. Contrast with *tandem duplications*.

Northern blot analysis a protocol for determining whether a fragment of DNA is transcribed in a particular tissue. In this protocol, RNA *transcripts* in the cells of a particular tissue are separated by *gel electrophoresis*. Compare with *Southern blot*.

N terminus the end of a *polypeptide* chain that contains a free amino group that is not connected to any other *amino acid*.

nuclear envelope an envelope composed of two membranes that surrounds the nucleus of a eukaryotic cell.

nuclear lamins a group of proteins that underlie the inner surface of the nuclear membrane; provide structural support for the nucleus and may also provide sites for the assembly of proteins that function in *DNA replication*, *transcription*, RNA transport, and chromosome structure.

nucleoid body a folded bacterial *chromosome*.

nucleolar organizer clusters of *rRNA* genes on long loops of DNA within a *nucleolus*.

nucleolus a large sphere-shaped organelle visible in the nucleus of *interphase* eukaryotic cells under a light microscope; formed by the *nucleolar organizer*.

nucleosome a rudimentary DNA packaging unit; composed of DNA wrapped around a *histone protein* core.

nucleotide a subunit of DNA or RNA consisting of a nitrogenous base (adenine, guanine, thymine, or cytosine in DNA; adenine, guanine, uracil, or cytosine in RNA), a phosphate group, and a sugar (deoxyribose in DNA; ribose in RNA).

nucleotide excision repair a homology-dependent mechanism that removes DNA alterations or damage, such as thymine-thymine dimers that *base excision repair* cannot take care of. It depends on process-specific enzyme complexes that patrol the DNA for irregularities and cut the damaged strand in two places that flank the damage, releasing a short region containing the damaged strand. DNA polymerase fills in the resultant gap.

null hypothesis a statistical hypothesis to be tested and either accepted or rejected in favour of an alternative.

null mutations *loss-of-function mutations* that block the function of a protein encoded by the *wild-type allele*. Such mutations either prevent synthesis of the protein or promote synthesis of a protein incapable of carrying out any function. See *amorphic mutations*.

O

octant stage embryo the early stage of embryo formation during cell divisions of a plant.

Okazaki fragments during *DNA replication*, small fragments of about 1000 bases that are joined after synthesis to form the *lagging strand*.

oligonucleotide a short single-stranded DNA molecule (containing less than 50 bases); can be synthesized by an automated DNA synthesizer. Oligonucleotides are used as *DNA probes* and as *primers* for DNA sequencing or PCR.

oligopeptide several *amino acids* linked by *peptide bonds*.

oncogene a gene, one or more forms of which is associated with cancer. Many oncogenes are involved, directly or indirectly, in controlling the rate of cell growth.

oogenesis the formation of the female gametes (eggs).

oogonia *diploid germ cells* in the ovary.

open reading frames (ORFs) DNA sequences with long stretches of codons in the same reading frame uninterrupted by *stop codons*; suggest the presence of genes.

operator site a DNA sequence near a *promoter* that can be recognized by a *repressor* protein; binding of repressor to the operator site blocks *transcription* of the gene.

operon a unit of DNA composed of specific genes, plus a *promoter* and/or *operator*, that acts in unison to regulate the response of the structural genes to environmental changes.

operon theory a model of *gene* regulation, which suggests that a single signal can simultaneously regulate the expression of several genes that are clustered together on a *chromosome* and involved in the same process.

origin of replication a short sequence of *nucleotides* at which the *initiation* of DNA *replication* begins.

orthologous genes with sequence similarities in two different species that arose from the same gene in the two species' common ancestor.

ovum a *haploid* female *germ cell* (the egg).

oxidative phosphorylation a set of reactions requiring oxygen that creates portable packets of energy in the form of ATP.

P

pachytene a substage of *prophase I* that begins at the completion of *synapsis* and includes the *crossing-over* of genetic material that results in *recombination.*

paracentric inversions *inversions* that exclude the *centromere.* Compare with *pericentric inversions.*

paralogous genes that arise by *duplication* within the same species, often within the same *chromosome*; paralogous genes often constitute a *multigene family.*

parental classes combinations of *alleles* present in the original *parental generation.*

parental (P) generation pure-breeding individuals whose progeny in subsequent generations will be studied for specific traits. Refer to *filial generations.*

parental types *phenotypes* that reflect a previously existing parental combination of genes that is retained during gamete formation.

parthenogenesis reproduction in which offspring are produced by an unfertilized female. Parthenogenesis is common in ants, bees, wasps, and certain species of fish and lizards.

pathogenicity islands segments of DNA in disease-causing bacteria that encode several genes involved in pathogenesis. Pathogenicity islands appear to have been transferred into the bacteria by lateral gene transfer from a different species.

pedigree an orderly diagram of a family's relevant genetic features, extending through as many generations as possible.

penetrance indicates how many members of a *population* with a particular *genotype* show the expected *phenotype.*

peptide bond a covalent bond that joins *amino acids* during protein synthesis.

peptide motifs stretches of *amino acids* conserved in many otherwise unrelated *polypeptides.*

peptidyl (P) site the site on a *ribosome* to which the initiating *tRNA* first binds and at which the tRNAs carrying the growing *polypeptide* are located during *elongation.*

peptidyl transferase the enzymatic activity of the *ribosome* responsible for forming *peptide bonds* between successive *amino acids.*

pericentric inversions *inversions* that include the *centromere.* Compare with *paracentric inversions.*

permissive condition an environmental condition that allows the survival of an individual with a *conditional lethal* allele. Contrast with *restrictive condition.*

phage short for *bacteriophage*; a *virus* for which the natural host is a bacterial cell; literally "bacteria eaters."

phage induction the process of phage DNA excision from the bacterial chromosome in a *lysogen* and entry into the *lytic cycle.*

phenocopy a change in *phenotype* arising from environmental agents that mimic the effects of a mutation in a gene. Phenocopies are not heritable because they do not result from a change in a gene.

phenotype an observable characteristic.

phenotype frequency the proportion of individuals in a population that express a particular phenotype.

phenotype variance (V_p) See *total phenotype variance.*

pheromones molecules produced by one sex that serve as agents to elicit mating behaviours in individuals of the opposite sex.

phosphodiester bonds covalent bonds joining one *nucleotide* to another. Phosphodiester bonds between nucleotides form the backbone of DNA.

phylogenetic tree a diagram composed of *nodes*, which represent the taxonomic units, and *branches*, which define the relationship of these units.

physical associations protein or regulatory interactions in a *biological system.*

physical map a map of locations of identifiable landmarks on DNA (e.g., restriction enzyme cutting sites, genes). For the human genome, the lowest-resolution physical map is the banding patterns on the 24 different chromosomes; the highest-resolution map is the complete *nucleotide* sequence of the *chromosomes.* See *karyotype.*

physical markers cytologically visible abnormalities that make it possible to keep track of specific *chromosome* parts from one generation to the next. Compare with *genetic markers.*

pilus a stiff, thin strand of protein that protrudes from an F^+ bacterial cell and binds to the cell wall of an F^- cell. Retraction of the pilus into the F^- cell draws the two cells close together in preparation for gene transfer.

plaque a clear area on a bacterial *lawn*, devoid of living bacterial cells, containing the genetically identical descendants of a single *bacteriophage.*

plasmids minute circles of double-stranded DNA that can gain admission to and replicate in the cytoplasm of many kinds of bacterial cells, independently of the *bacterial chromosome.*

pleiotropy a phenomenon in which a single gene determines a number of distinct and seemingly unrelated characteristics.

point mutation a mutation of only one nucleotide pair.

polar body a cell produced by *meiosis I* or *meiosis II* during oogenesis that does not become the *primary* or *secondary oocyte*.

polarity an overall direction.

polar microtubules microtubules that originate in *centrosomes* and are directed toward the middle of the cell; polar microtubules that arise from opposite centrosomes interdigitate near the cell's equator and push the spindle poles apart during *anaphase*.

poly-A tail the 3′ end of eukaryotic *mRNA* consisting of 100–200 angstroms, believed to stabilize the mRNA and increase the efficiency of the initial steps of *translation*.

polycistronic mRNA an *mRNA* that contains more than one protein-coding region; often the transcriptional product of an *operon* in bacterial cells.

polygenic trait a trait controlled by multiple genes.

polymer a linked chain of repeating subunits that form a larger molecule; DNA is a type of polymer.

polymerase chain reaction (PCR) a fast inexpensive method of replicating a DNA sequence once the sequence has been identified; based on a reiterative loop that amplifies the products of each previous round of *replication*.

polymerases, DNA or RNA enzymes that catalyze the synthesis of nucleic acids on pre-existing nucleic acid templates, assembling *RNA* from ribonucleotides or *DNA* from deoxyribonucleotides.

polymerization the linkage of subunits to form a multi-unit chain. In DNA *replication*, the polymerization of nucleotides occurs through the formation of *phosphodiester bonds* by DNA polymerase III.

polymorphic locus a *locus* with two or more distinct *alleles* in a *population*.

polymorphism a variant of a gene or noncoding region that has two or more *alleles*. Molecular geneticists use this term to describe a variant of a *locus* within a *population* of organisms that has two or more alleles. Population geneticists reserve the term for variants at a locus where two or more alleles are present at a frequency of 1 percent or greater; for example, to describe the alternative forms of a gene that has more than one *wild-type allele*.

polypeptides amino acid chains containing hundreds to thousands of *amino acids* joined by *peptide bonds*.

polyploids *euploid* species that carry three or more complete sets of *chromosomes*.

polyproteins *polypeptides* produced by *translation* that can subsequently be cleaved by protease enzymes into two or more separate proteins.

polyribosomes structures formed by the simultaneous *translation* of a single *mRNA* molecule by multiple *ribosomes*.

polytene chromosome a giant *chromosome* consisting of many identical *chromatids* lying in parallel register.

population a group of interbreeding individuals of the same species that inhabit the same space at the same time.

population bottleneck occurs when a large proportion of individuals in a population perish, as a consequence of environmental disturbances, resulting in surviving individuals who are essentially equivalent to a founder population.

population genetics a scientific discipline that studies what happens in whole *populations* at the genetic level.

positional cloning the process that enables researchers to obtain the clone of a gene without any prior knowledge of its protein product or function. It uses genetic and physical maps to locate *mutations* responsible for particular *phenotypes*.

position-effect variegation the variable expression of a gene in a population of cells, caused by the gene's location near highly compacted *heterochromatin*.

posttranslational modifications changes, such as cleavages that remove *amino acids* and additions of chemical constituents to specific amino acids, that occur to a *polypeptide* after *translation* has been completed.

primary nondisjunction *nondisjunction* that results in aneuploid gametes.

primary oocytes/spermatocytes *germline* cells in which *meiosis I* occurs.

primary structure the linear sequence of *amino acids* within a *polypeptide*.

primary transcript the single strand of *RNA* resulting from *transcription*.

primer a short, pre-existing *oligonucleotide* chain to which new DNA can be added by DNA polymerase.

primer walking a common approach to directed sequencing of DNA in which *primers* are synthesized using the information from each previous round of DNA sequencing.

product rule states that the probability of two or more independent events occurring together is the product of the probabilities that each event will occur by itself.

programmed cell death (PCD) See *apoptosis*.

prokaryotes one of the three major evolutionary lineages of living organisms known as domains; characterized by the lack of a nuclear membrane. Contrast with *eukaryotes*.

prokaryotic describes cells that do not have a membrane-bounded nucleus.

prokaryotic gene regulation the control of gene expression in a prokaryotic cell by activating, increasing, diminishing, or preventing the *transcription* and *translation* of specific *genes* or groups of genes.

prometaphase the stage of *mitosis* or *meiosis* just after the breakdown of the *nuclear envelope*, when the chromosomes connect to the *spindle apparatus* and begin to move toward the *metaphase plate*.

promoters DNA sequences near the beginning of *genes* that signal *RNA polymerase* where to begin transcription. Compare with *terminators*.

prophage the integrated copy of the temperate bacteriophage.

prophase the phase of the *cell cycle* marked by the gradual emergence of the individual *chromosomes* from the undifferentiated mass of *chromatin*, indicating the beginning of *mitosis*.

prophase I the longest, most complex phase of *meiosis* consisting of several substages.

prophase II the first phase of *meiosis II*; if the *chromosomes* decondensed during interkinesis, they recondense. At the end of prophase II, the *nuclear envelope* breaks down and the *spindle apparatus* re-forms.

protein domains discrete functional units of a protein, encoded by discrete regions of DNA.

protein kinases enzymes that add phosphate groups to their protein substrates.

protein network a set of proteins (and often other molecules such as metabolites) that interact in executing a particular biological function.

proteins large polymers composed of hundreds to thousands of *amino acid* subunits strung together in long chains. Proteins are required for the structure, function, and regulation of the body's cells, tissues, and organs.

proteome the complete set of unique *proteins* encoded by a given *genome*.

proteomics the global analysis of the *proteins* present in a particular cell type or organism under a defined set of environmental conditions.

proteasome a large multiprotein complex in the cytoplasm of eukaryotic cells that contains proteolytic enzymes that degrade proteins tagged with *ubiquitin*.

proto-oncogene a *gene* that can mutate into an *oncogene*—an *allele* that causes a cell to become cancerous.

prototroph a microorganism (usually wild type) that can grow on minimal medium in the absence of one or more growth factors. Compare with *auxotroph*.

pseudodominance the expression of a *phenotype* caused by a *recessive allele* in a deletion heterozygote because the other homologue has no copy of that gene.

pseudogene a nonfunctioning gene. The result of *duplication* and divergence events in which one copy of an originally functioning *gene* has undergone *mutations* such that it no longer has an intact *polypeptide* coding sequence.

pseudolinkage the characteristic of a heterozygote for a *reciprocal translocation*, in which genes located near the translocation breakpoint behave as if they are linked even though they originated on nonhomologous chromosomes.

pulse-field gel electrophoresis a special type of electrophoretic protocol that allows an extreme extension of the normal separating capacity of the gels. In this protocol, the DNA sample is subjected to pulses of electric current that alternate between two directions. The range of sizes separated in this manner is a function of pulse length. Compare with *electrophoresis*.

punctuated equilibrium the tendency of evolution to proceed through long periods of stasis (lack of change) followed by short periods of explosive change.

pure-breeding lines families of organisms that produce offspring with specific parental traits that remain constant from generation to generation.

p value the numerical probability that a particular set of observed experimental results represents a chance deviation from the values predicted by a particular hypothesis.

Q

quantitative trait See *continuous trait*.

quantitative trait loci (QTLs) loci that influence the expression of *continuous traits*.

quaternary structure a structure made up of the three-dimensional configuration of subunits in a *multimeric protein*.

R

radial loop-scaffold model the model of looping and gathering of DNA by nonhistone proteins that results in high compaction of *chromosomes* at *mitosis*.

random amplification of polymorphic DNA (RAPD) a protocol designed to detect single-base changes at *polymorphic* loci throughout a *genome*.

random walk the description of movement of bacteria to achieve chemotaxis reflecting nonpredictable changes in the direction of movement. Contrast with *biased random walk*.

reading frame the partitioning of groups of three *nucleotides* from a fixed starting point such that the sequential interpretation of each triplet generates the correct order of *amino acids* in the resulting *polypeptide* chain.

rearrangements events in which the *genome* is reshaped by the reorganization of *DNA sequences* within one or more *chromosomes*.

receptors proteins embedded in the membrane of a cell that bind to *growth factors* and other *ligands*, initiating a *signal transduction pathway* in the cell.

recessive allele an *allele* whose *phenotype* is not expressed in a heterozygote. See *dominant allele*.

recessive epistasis a special case of *epistasis*, in which the *allele* causing the epistasis is *recessive*. Compare with *dominant epistasis*.

recessive lethal allele an *allele* that prevents the birth or survival of *homozygotes*, though *heterozygotes* carrying the allele survive.

recessive trait the trait that remains hidden in the F_1 hybrids (*heterozygotes*) resulting from a mating between pure-breeding parental strains showing antagonistic *phenotypes*; the recessive trait usually reappears in the *second filial* (F_2) *generation*.

recipient during gene transfer in bacteria, the cell that receives the genetic material. See *donor, transformation, conjugation,* and *transduction*.

reciprocal crosses *crosses* in which the traits in the males and females are reversed, thereby controlling whether a particular trait is transmitted by the egg or the pollen.

reciprocal translocation results when two breaks, one in each of two *chromosomes*, yield DNA fragments that do not religate to their chromosome of origin; rather, they switch places and become attached to the other chromosome. Compare with *translocation*.

recombinant classes reshuffled combinations of *alleles* that were not present in the *parental generation*.

recombinant DNA molecules a combination of DNA molecules of different origin that are joined using *recombinant DNA technologies*.

recombinant DNA technology methodologies used by geneticists to cut, copy, and paste together DNA fragments from multiple sources; these methodologies allow researchers to bring together genetic material from different sources to create novel *sequences* not before seen in nature.

recombinant types *phenotypes* reflecting a new combination of genes that occurs during gamete formation.

recombinants *chromosomes* that carry a mix of *alleles* derived from different *homologous chromosomes.*

recombination the process by which offspring derive a combination of genes different from that of either parent; the generation of new allelic combinations. In higher organisms, this can occur by *crossing-over.*

recombination frequency (RF) the percentage of *recombinant* progeny; can be used as a gauge of the physical distance separating any two genes on a chromosome. See *centimorgan.*

recombination nodules structures that appear during *pachytene* of *prophase I.* An exchange of parts between non-sister *chromatids* occurs at recombination nodules.

recombineering an approach to determining the function of bacterial genes by making a knockout mutation in the chromosomal gene using recombinant DNA techniques and the homologous recombination machinery of bacteriophages.

reduction(al) division cell division that reduces the number of *chromosomes*, usually by segregating *homologous chromosomes* to two daughter cells. *Meiosis I* is a reductional division.

regulative determination where the embryo is a collection of cells that can alter, or "regulate," their fates according to the environment, for example, to make up for missing cells. Contrast with *mosaic determination.*

regulon a group of *genes* whose expression is regulated by the same regulatory proteins.

release factors *proteins* that recognize *stop codons* and help end translation.

remodelling complex a complex of proteins that remove promoter-blocking nucleosomes or reposition them in relation to the gene, helping to prepare a gene for transcriptional activation.

replication See *DNA replication.*

replication bubble an unwound area of the original DNA *double helix* during *replication.*

replication fork a Y-shaped area consisting of the two unwound DNA strands branching out into unpaired (but complementary) single strands during *replication.*

replicative transposition *transposition* in which the *transposable element* moves by making a copy that becomes inserted into a different chromosomal site while the initial transposable element stays in its original position. Contrast with *conservative transposition.*

replicon (replication unit) the DNA running both ways from one *origin of replication* to the endpoints where it merges with DNA from adjoining *replication forks.*

reporter gene a protein-encoding gene whose expression in the cell is quantifiable by sensitive and reliable techniques of protein detection.

repressor a type of *transcription factor* that can bind to specific *cis*-acting elements such as *operator sites* and thereby diminish or prevent *transcription.*

reproductive cloning the creation of a cloned embryo by insertion of the nucleus of a *somatic cell* from one individual into an egg cell whose nucleus has been removed. The hybrid egg is stimulated to begin embryonic cell divisions, and the resulting cloned embryo is transplanted into the uterus of a foster mother and allowed to develop to term. Contrast with *therapeutic cloning.*

reproductive development the period in plant development that produces flower and seed.

response to selection (R) the amount of evolution, or change in mean trait value, resulting from *selection.* Compare with *selection differential.*

restriction the ability of a bacterial strain to prevent the replication of an infecting virus.

restriction enzymes proteins made by bacteria that recognize specific, short *nucleotide* sequences and cut DNA at those sites.

restriction fragment length polymorphism (RFLP) the variation between individuals in DNA fragment size cut by specific *restriction enzymes*; polymorphic sequences that result in RFLPs are used as markers on both *physical maps* and *genetic linkage* maps.

restriction fragments DNA fragments generated by the action of *restriction enzymes.*

restriction map a linear diagram that illustrates the positions of *restriction enzyme* recognition sites along a DNA molecule.

restrictive condition an environmental condition that prevents the survival of an individual with a *conditional lethal* allele. Contrast with *permissive condition.*

retrotransposons genetic elements that transpose via *reverse transcription* of an RNA intermediate. One class of *transposable elements.* Contrast with *DNA transposons.*

retroviruses viruses that hold their genetic information in a single strand of *RNA* and carry the enzyme *reverse transcriptase* to convert that RNA into DNA within a host cell.

reverse mutation a mutation that causes a novel *mutant allele* to revert back to wild type.

reverse transcriptase an RNA-dependent DNA *polymerase* that synthesizes DNA strands complementary to an RNA *template.* The product of reverse transcriptase is a *cDNA* molecule.

reverse transcription the process by which *reverse transcriptase* synthesizes DNA strands complementary to an RNA *template.* The product of reverse transcription is a cDNA molecule.

reversion See *reverse mutation.*

rhodopsin the *protein* that receives photons and triggers the processing of information in rod cells.

ribonucleic acid (RNA) a *polymer* of ribonucleotides found in the nucleus and cytoplasm of cells; it plays an important role in protein synthesis. There are several classes of RNA molecules, including *messenger RNA (mRNA)*, *transfer RNA (tRNA)*, ribosomal RNA (rRNA), and other small RNAs, each serving a different purpose.

ribosomal RNAs (rRNAs) *RNA* components of *ribosomes*, which are composed of both rRNAs and proteins.

ribosome-binding sites regions on *prokaryotic mRNAs* containing both an *initiation codon* and a *Shine–Dalgarno box*; *ribosomes* bind to these sites to start *translation.*

ribosomes cytoplasmic structures composed of *ribosomal RNA* (rRNA) and protein; the sites of protein synthesis.

ribozymes RNA molecules that can act as enzymes to catalyze specific chemical reactions.

RISC See *RNA-induced silencing complex*.

RNA See *ribonucleic acid*.

RNA-dependent DNA polymerase See *reverse transcriptase*.

RNA editing specific alteration of the genetic sequence carried within an *RNA* molecule after *transcription* is completed.

RNA-induced silencing complex (RISC) a large enzymatic complex in the cytoplasm of all eukaryotic cells that binds to an *miRNA* and performs sequence-specific *RNA interference*.

RNA interference (RNAi) the sequence-specific modulation of eukaryotic gene expression by a 21–24-nucleotide-long RNA molecule referred to as *micro-RNA* if it is encoded within the genome and *short interfering RNA* if it is introduced into the cell by scientists or infectious agents. These small, specialized RNAs prevent the expression of specific genes through complementary base-pairing. In the most common natural form of RNAi, primary miRNA-containing transcripts are processed sequentially with the enzymes *Drosha* and *Dicer* to produce a mature miRNA that is loaded onto an *RNA-induced silencing complex* (*RISC*). RISC binds to complementary mRNA targets, causing mRNA degradation or reduced translational activity.

RNA-like strand a strand of a double-helical DNA molecule that has the same nucleotide sequence as an *mRNA* (except for the substitution of T for U) and that is complementary to the *template strand*.

RNA polymerases enzymes that transcribe a DNA sequence into an RNA transcript. *Eukaryotes* have three types of RNA polymerases called pol I, pol II, and pol III that are responsible for transcribing different classes of genes.

RNA splicing a process that deletes *introns* and joins together successive *exons* to form a mature *mRNA* consisting of only exons.

RNA world a hypothetical primordial world in which RNA became the first replicator.

Robertsonian translocation *translocation* arising from breaks at or near the *centromeres* of two *acrocentric chromosomes*. The reciprocal exchange of broken parts generates one large *metacentric* chromosome and one very small chromosome.

S

satellite DNAs blocks of repetitive, simple noncoding sequences, usually around *centromeres*; these blocks have a different *chromatin* structure and different higher-order packaging than other chromosomal regions.

saturated gene map a high-density linkage map.

saturation mutagenesis an attempt to isolate *mutations* in all of the *genes* that direct a particular biological process.

scaffold the structural backbone of a chromosome that is made of nonhistone proteins.

scaffold-associated regions (SARs) special, irregularly spaced repetitive base sequences of DNA that associate with nonhistone proteins to define chromatin loops. SARs are most likely the sites at which DNA is anchored to the condensation scaffold.

secondary nondisjunction a second occurrence of failures in *chromosome segregation* during *meiosis*. See *nondisjunction*.

secondary oocytes/spermatocytes *germline* cells in which meiosis II occurs.

secondary structure localized region of a *polypeptide* chain with a characteristic geometry, such as an α helix or β-pleated sheets.

second filial (F₂) generation progeny resulting from self-crosses or inter-crosses between individuals of the F₁ generation in a series of controlled matings.

sectors portions of a growing *colony* of microorganisms that have a different *genotype* than the remainder of the colony.

segmentation gene in *Drosophila*, a large group of genes responsible for subdividing the body into an array of essentially identical body segments.

segregation the equal separation of alleles for each trait during *gamete* formation, in which one *allele*, and only one allele, of each *gene* goes to each gamete.

selectable markers antibiotic resistance genes and other *vector* genes that make it possible to pick out cells harbouring a particular DNA molecule.

selection a process that establishes conditions in which only the desired mutant will grow.

selection differential (S) a measure of the strength of selection on a trait. Compare with *response to selection*.

self-fertilization (selfing) fertilization in which both egg and pollen come from the same plant.

semiconservative replication a pattern of double helix duplication in which *complementary base-pairing* followed by the linkage of successive nucleotides yields two daughter double helixes that each contain one of the original DNA strands intact (conserved) and one completely new strand.

semisterility a condition in which the capacity of generating viable offspring is diminished by at least 50 percent.

sense describes a laboratory-designed single-strand RNA molecule, or portion of an RNA molecule, with a sequence equivalent to that present in a cellular mRNA. A laboratory-designed *short interfering RNA* (*siRNA*) mediator of RNA interference will contain both sense and *antisense* components.

sequence tagged sites (STSs) one-of-a-kind markers that tag positions along the DNA molecule.

sequencing determining the order of *nucleotides* (base sequences) in a *DNA* or *RNA* molecule or the order of *amino acids* in a *protein*.

serial analysis of gene expression (SAGE) a sequencing technique for determining the quantities of different RNAs in a mixture.

sex chromosomes the X and Y *chromosomes* in human beings, which determine the sex of an individual. Compare with *autosome*.

sex-influenced traits traits that can show up in both sexes but are expressed differently in each sex due to hormonal differences.

sex-limited traits traits that affect a structure or process that is found in one sex but not the other.

Shine–Dalgarno box a sequence of six *nucleotides* in *mRNA* that is one of two elements comprising a *ribosome-binding site* (the other element is the *initiation codon*).

shotgun a sequencing approach in which the overlapping fragments to be sequenced have been randomly generated by shearing (via sonication) or by partial digestion with *restriction enzymes*.

signal transducers cytoplasmic proteins that relay signals inside the cell.

signal transduction the activation and inhibition of intracellular targets after binding of *growth factors* to *receptors*.

signal transduction pathway a form of molecular communication in which the binding of proteins to receptors on cell surfaces constitutes a signal that is converted through a series of intermediate steps to a final intracellular regulatory response, usually the activation or repression of particular target genes.

silent mutations *mutations* without effects on *phenotype*; usually denotes *point mutations* that change one of the three bases in a *codon* but that do not change the identity of the specified *amino acid* because of the *degeneracy* of the *genetic code*.

simple sequence repeat (SSR) a two- or three-base-long *SNP*.

SINE short *in*terspersed *e*lements; one of the two major classes of *transposable elements* in mammals. Contrast with *LINE*.

single nucleotide polymorphism (SNP) a single *nucleotide* locus with two naturally existing *alleles* defined by a single base pair *substitution*. SNP loci are useful as DNA-based *markers* for formal genetic analysis.

site-directed mutagenesis a technique that allows researchers to alter the sequence of a *gene* at a specific nucleotide position and then to study the effects of this change on protein function.

sister chromatids the two identical copies of a chromosome that exist immediately after DNA replication. Sister chromatids are held together by protein complexes called *cohesins*.

small interfering RNA (siRNA) an RNA molecule 21–24 bases in length that originates outside an organism and can co-opt the natural *RNAi* machinery to effect *RNA interference*. Also called short interfering RNA.

SNP See *single nucleotide polymorphism*.

somatic cells any cell in an organism except *gametes* and their precursors. Compare with *germ cells*.

somatic gene therapy remedial measures in which a replacement *gene* is inserted into the affected tissue where the gene is expressed to compensate for a faulty gene.

somatic mutations *genetic* alterations that occur after conception in non-germline cells and are passed to the progeny of the mutated cells through *mitosis*. These mutations affect all cells descended from the mutated cell, but cannot be passed on to the next generation.

SOS system an emergency repair system in bacteria that relies on error-prone ("sloppy") DNA *polymerases*; these special SOS polymerases allow cells with damaged DNA to divide, but the daughter cells carry many new mutations.

Southern blot a protocol for transferring DNA sequences separated by *gel electrophoresis* onto a nitrocellulose filter paper for analysis by *hybridization* with a *DNA probe*. Compare with *Northern blot analysis*.

specialized transducing phage *bacteriophage* carrying mainly phage DNA but also one or a few of the bacterial genes that lie near the site of *prophage* insertion. They can transfer these genes to another bacterium in the process known as *specialized transduction*.

specialized transduction *bacteriophage*-mediated transfer of a few bacterial genes located next to the bacteriophage DNA in the bacterial chromosome.

sperm a *haploid* male *gamete* produced by *meiosis*.

spermatids *haploid* cells produced at the end of *meiosis* that will mature into *sperm*.

spermatogenesis the production of *sperm*.

spermatogonia *diploid* germ cells in the testes.

S phase the stage of the *cell cycle* during which *chromosome replication* occurs.

spindle apparatus a microtubule-based structure responsible for *chromosome* movements and *segregation* during cell division.

splice acceptors *nucleotide* sequences in a *primary transcript* at the border between an *intron* and the downstream *exon* that follows it; required for proper *RNA splicing*.

splice donors *nucleotide* sequences in a *primary transcript* at the border between an *intron* and the upstream *exon* that precedes it; required for proper *RNA splicing*.

spliceosome a complicated intranuclear machine that ensures that all of the splicing reactions take place in concert.

stem cells relatively undifferentiated cells that undergo asymmetric mitotic divisions. One of the daughter cells produced by such a division is another stem cell, while the other daughter cell can differentiate. In this way, stem cells are self-renewing but can also give rise to differentiated cells.

sticky end the result achieved after digestion by many *restriction enzymes* that break the *phosphodiester bonds* on the two strands of a double helix DNA molecule at slightly different locations. The resulting double-stranded DNA molecule has a single protruding strand at each end that is usually one to four bases in length.

stop codons See *nonsense codons*.

substitution occurs when a base at a certain position in one strand of the DNA molecule is replaced by one of the other three bases.

substrate the target of an enzyme, usually cellular proteins.

subunit a single *polypeptide* that is a constituent of a *multimeric protein*.

sum rule the probability that either of two mutually exclusive events will occur is the sum of their individual probabilities.

supercoiling additional twisting of the DNA molecule caused by movement of the *replication fork* during unwinding. Refer to *DNA topoisomerases*.

suppressor mutations *mutations* that alleviate the phenotypic abnormality caused by another mutation. The two mutations can be in the same gene (see *intragenic suppression*), or the suppressor mutation can be in a second, *modifier gene.*

suspensor in plants, a structure analogous to the umbilical cord in mammals.

synapsis the process during which *homologous chromosomes* become aligned and zipped together; occurs during *zygotene* of *prophase I.*

synaptonemal complex a structure that helps align *homologous chromosomes* during *prophase* of *meiosis I.*

syncytial blastoderm in *Drosophila* embryos, formed when most of the nuclei migrate out to the cortex just under the surface of the embryo.

syncytium an animal cell with two or more nuclei.

syntenic the relationship of two or more *loci* found to be located on the same chromosome. Compare with *conserved synteny* and *syntenic segments.*

syntenic blocks blocks of *linked loci* within a *genome.*

syntenic segments in the comparison of two *genomes*, large blocks of DNA sequences in which the identity, order, and transcriptional direction of the *genes* are almost exactly the same.

synthetic lethality when an individual carrying mutations in two different genes dies, even though individuals carrying either mutation alone survive.

systems biology a discipline that seeks to describe the multiple components of a *biological system* and analyze the complex interactions of these components both within the system and in relation to the components of other systems.

T

TAFs *T*BP-*a*ssociated *f*actors; one type of basal factor.

tandem duplications repeats of a chromosomal region that lie adjacent to each other, either in the same order or in reverse order. Contrast with *nontandem duplications.*

target-site duplication identical nucleotide sequences that are oriented in the same direction and repeated. Each type of *transposable element* differs in the length of its target-site duplication. Also called direct repeats.

TBP *TATA* box-binding *p*rotein; a key basal factor that assists the binding of RNA polymerase II to the promoter and the initiation of basal levels of *transcription.*

telomerase an enzyme critical to the successful replication of *telomeres* at *chromosome* ends.

telomeres specialized terminal structures on eukaryotic *chromosomes* that ensure the maintenance and accurate replication of the two ends of each linear chromosome.

telophase the final stage of *mitosis* in which the daughter *chromosomes* reach the opposite poles of the cell and re-form nuclei.

telophase I the phase of *meiosis I* when nuclear membranes form around the *chromosomes* that have moved to the poles; each incipient daughter nucleus contains one-half the number of chromosomes in the original parent cell nucleus, but each of these chromosomes consists of two *sister chromatids* held together by *cohesin* protein complexes.

telophase II the final phase of *meiosis II* during which membranes form around each of the four daughter nuclei, and *cytokinesis* places each nucleus in a separate cell.

temperate bacteriophages after infecting the host, these *phages* can enter either the *lytic cycle* or the alternative *lysogenic cycle*, during which their DNA integrates into the host *genome.* Compare with *virulent bacteriophages.*

template a strand of *DNA* or *RNA* that is used as a model by DNA or RNA *polymerase* or by *reverse transcriptase* for the creation of a new complementary strand of DNA or RNA.

template strand the strand of the double helix that is complementary to both the RNA-like *DNA* strand and the *mRNA.*

terminalization the shifting of the *chiasmata* from their original position at the *centromere* toward the *chromosome* end or *telomere.*

termination the phase of *translation* that brings *polypeptide* synthesis to a halt.

terminators sequences in the *RNA* products that tell RNA *polymerase* where to stop *transcription.* Compare with *promoters.*

tertiary structure the complete three-dimensional arrangement of a *polypeptide.*

testcross a *cross* used to determine the *genotype* of an individual showing a *dominant phenotype* by mating with an individual showing the *recessive phenotype.*

tetraploid describes cells or organisms with four complete sets of chromosomes. See *haploid, diploid,* and *triploid.*

tetrasomic an otherwise *diploid* organism with four copies of a particular *chromosome.*

therapeutic cloning the creation of an embryo by the method described for *reproductive cloning*, with the exception that the embryo is not allowed to develop to term. Instead it is cultured in a Petri plate to the *blastocyst* stage, at which point the *embryonic stem* (*ES*) *cells* are collected and placed in culture. The cultured ES cells can be induced to differentiate into many kinds of cells that might be of therapeutic value.

thymine a nitrogenous base; one member of the base pair A–T (adenine–thymine).

Tn element a complex type of transposable element in bacteria that carries a gene for *transposase* and which may contain genes conferring resistance to antibiotics or toxic metals such as mercury.

tolerance the ability to prevent the body from making *immune responses* to its own proteins.

topoisomerase See *DNA topoisomerases.*

total phenotype variance (V_P) population deviation calculated as the sum of the *genetic variance* and the *environmental variance.*

totipotent the description of cell state during early embryonic development in which the cells have not yet differentiated and retain the ability to produce every type of cell found in the developing embryo and adult animal.

trans-acting element a gene that codes for a *transcription factor*.

transcript the product of *transcription*.

transcription the conversion of *DNA*-encoded information to its *RNA*-encoded equivalent.

transcriptional silencing hypercondensation of *chromatin* domains makes it impossible to activate genes within those domains, no matter what *transcription factors* are active in the cell.

transcription bubble the region of DNA unwound by RNA *polymerase*.

transcription factor a protein (or RNA) whose binding to or indirect association with a *cis-control element* helps regulate the timing, location, and level of a particular gene's *transcription*. Functional categories include *activators*, *repressors*, *coactivators*, and *corepressors*.

transcriptome the population of *mRNAs* expressed in a single cell or cell type.

transductants cells resulting from gene transfer mediated by *bacteriophages*.

transduction one of the mechanisms by which bacteria transfer genes from one strain to another; *donor* DNA is packaged within the protein coat of a *bacteriophage* and transferred to the recipient when the phage particle infects it. Recipient cells are known as *transductants*.

transfection the transformation of mammalian cells via the uptake of DNA from the medium.

transfer RNA (tRNA) a small RNA adaptor molecule that places specific *amino acids* at the correct position in a growing *polypeptide* chain.

transformants cells that have received naked *donor* DNA.

transformation one of the mechanisms by which bacteria transfer genes from one strain to another; occurs when DNA from a *donor* is added to the bacterial growth medium and is then taken up from the medium by the recipient. The recipient cell is known as a *transformant*. Contrast with *conjugation*.

transgene any piece of foreign *DNA* that researchers have inserted into the *genome* of an organism.

transgenic any individual carrying a *transgene*.

transgenic technology the tools for inducing a specific change in a gene and confirming that this change causes the predicted *phenotype*. Transgenic technology includes experimental methods that allow scientists to add laboratory-constructed *DNA sequences* to the *genomes* of animals or plants.

transition a type of *substitution* mutation that occurs when one purine (A or G) replaces the other purine, or one pyrimidine (C or T) replaces the other pyrimidine. Contrast with *transversions*.

translation the process in which the *codons* carried by *mRNA* direct the synthesis of *polypeptides* from *amino acids* according to the *genetic code*. Compare with *transcription*.

translocation a rearrangement that occurs when parts of two nonhomologous *chromosomes* change places. Compare with *reciprocal translocation*.

translocation Down syndrome occurs in individuals affected by Down syndrome who have inherited three copies of a part (rather than all) of chromosome 21 because one of their parents was *heterozygous* for a *translocation* involving chromosome 21.

transposable elements (TEs) DNA segments several hundred to several thousand base pairs long that move about in the *genome*, regardless of mechanism.

transposase a protein that catalyzes *transposition* through its recognition of the inverted repeats of *DNA transposons*.

transposition the movement of *transposable elements* from one position in the *genome* to another.

trans-splicing a rare type of *RNA splicing* that joins together *exons* of the *primary transcripts* of two different genes.

transversions a type of *substitution* mutation that occurs when a purine (A or G) replaces a pyrimidine (C or T) or when a pyrimidine replaces a purine.

triploid describes cells or organisms with three complete sets of *chromosomes*. See *haploid*, *diploid*, and *tetraploid*.

trisomic an individual having one extra *chromosome* in addition to the normal *diploid* set of the species.

truncated proteins *polypeptides* with fewer amino acids than normal encoded by genes containing *nonsense mutations*.

tumour suppressor See *mutant tumour-suppressor gene*.

twin spots adjacent islands of tissue that are phenotypically distinct from each other and from the surrounding tissue; can be produced as a result of mitotic recombination.

two-point testcross a cross in which a dihybrid F_1 individual is crossed with a true-breeding individual showing recessive phenotypes.

U

ubiquitin a highly conserved protein whose covalent attachment to other proteins marks them for degradation by the *proteasome*.

unequal crossing-over a change in DNA caused by out-of-register *recombination* during meosis in which one *homologous chromosome* ends up with a *duplication*, while the other homologue sustains a *deletion*.

upstream movement opposite the direction *RNA* follows when moving along a gene. Compare with *downstream*.

uracil a nitrogenous base normally found in *RNA* but not in *DNA*; uracil is capable of forming a *base pair* with adenine.

V

variance statistical measurement of deviation from the mean (middle); typically expressed as plus or minus terms in relation to the mean.

vector a specialized DNA sequence that can enter a living cell, signal its presence to an investigator by conferring a detectable property on the host cell, and provide a means of replication for itself and the foreign DNA inserted into it. A vector must also possess distinguishing physical traits by which it can be purified away from the host cell's *genome*. See *cloning vector*.

virulent bacteriophages after infecting the host, these phages always enter the *lytic cycle*, multiply rapidly, and kill the host. Compare with *temperate bacteriophages*.

virus a noncellular biological entity that can reproduce only within a host cell. Viruses consist of nucleic acid covered by protein. Inside an infected cell, the virus uses the synthetic capability of the host to produce progeny virus.

W

wild-type allele (1) an *allele*, or DNA variant, whose frequency is more than 1 percent in a *population*; (2) an allele that dictates the most frequently observed phenotype in a population. Wild-type alleles are often designated by a superscript "plus" sign ($^+$).

wobble the ability of the 5′ nucleotide of an anticodon to interact with more than one nucleotide at the 3′ end of codons; helps explain the *degeneracy* of the *genetic code*.

X

x indicates the number of chromosomes in a complete set of nonhomologous chromosomes.

X-chromosome inactivation in mammals, a mechanism of *dosage compensation* in which all X *chromosomes* in a cellular genome but one are inactivated at an early stage of development through the formation of heterochromatic *Barr bodies*.

X-chromosome reactivation in mammals, a mechanism by which X *chromosomes* that were inactivated become reactivated in *oogonia* so that the *haploid* cells in the *germ line* all have an active X chromosome.

X-linked carried by the X *chromosome*.

Y

YAC See *yeast artificial chromosome*.

yeast artificial chromosome (YAC) a vector used to clone DNA fragments up to 400 kb in length; it is constructed from *telomeric, centromeric,* and *origin-of-replication* sequences needed for replication in yeast cells. Compare with *cloning vector, cosmid*.

Z

Z-form DNA DNA in which the *nucleotide sequences* cause the structure to assume a zigzag shape due to the helixes spiralling to the left. Compare with *B-form DNA*. The significance of this variation on DNA structure is unknown at this time.

zygote the *diploid* cell formed by the fertilization of the egg by the *sperm* during sexual reproduction.

zygotene a substage of *prophase I* when *homologous chromosomes* become zipped together in *synapsis*.

Credits

Chapter 1

Opener: Russ London/Wikipedia; **1.1(a):** © David M. Phillips/Visuals Unlimited; **1.1(b):** © Vol. 44/Photo Disc/Getty Images; **1.1(c):** © CORBIS RF; **1.1(d):** GenomeSystems Inc./Photo provided by Kearns Communication Group; **1.1(e):** © Vol. 24/PhotoDisc RF; **Page 3(a):** Courtesy of the J. Craig Venter Institute; **3(b):** Electron micrographs were provided by Tom Deerinck and Mark Ellisman of the National Center for Microscopy and Imaging Research at the University of California at San Diego. Courtesy of the J. Craig Venter Institute; **3(c):** Courtesy of the J. Craig Venter Institute; **3(d):** Electron micrographs were provided by Tom Deerinck and Mark Ellisman of the National Center for Microscopy and Imaging Research at the University of California at San Diego. Courtesy of the J. Craig Venter Institute; **3(e):** Courtesy of the J. Crain Venter Institute; **1.3:** Jim Karagiannis; **1.4:** © Biophoto Associates/Photo Researchers, Inc.; **1.9(a):** © Carolina Biological/Photo Researchers, Inc.; **1.9(b):** © Vol. OS02/PhotoDisc RF; **1.9(c):** Kristina Yu, © Exploratorium, www.exploratorium.edu; **1.9(d):** Kristina Yu, © Exploratorium, www.exploratorium.edu; **Table 1.1(a):** © J. William Schopf, UCLA. Reprinted with permission from *Science* 260: 640–646, 1993. "Microfossils of the Early Archean Apex Chart: New Evidence of the Antiquity of Life." © 1993 American Association for the Advancement of Science; **Table 1.1(c–d):** Prof. Andrew Knoll; **Table 1.1(e):** © Brand X Pictures/PunchStock; **Table 1.1(f):** © Alan Sirulnikoff/Photo Researchers, Inc.; **1.11:** © Edward Lewis, California Institute of Technology; **1.12(a):** © David M. Phillips/Visuals Unlimited; **1.12(b):** Lee Hartwell; **1.12(c):** © Sinclair Stammers/SPL/Photo Researchers, Inc.; **1.12(d):** Courtesy Debra Nero/Cornell University; **1.12(e):** © Myung Shin/Bergman Collection; **1.14(a):** iStockphoto/Thinkstock; **1.14(b):** Jason Merritt/Getty Images; **1.15:** Courtesy of the Huntington Society of Canada and the Canadian Coalition for Genetic Fairness.

Chapter 2

Opener(a): Rasbak/Wikipedia Commons; Opener(b): Rasbak/Wikipedia Commons; Opener(c): Robert Calentine/Visuals Unlimited, Inc.; **2.1:** © Bruce Aryes/Stone/Getty Images; **2.2:** © Science Photo Library/Photo Researchers, Inc.; **2.3:** © Saudjie Cross Siino/Weathertop Labradors; **2.4(a):** © Malcolm Gutter/Visuals Unlimited; **2.4(b):** James King-Holmes/Photo Researchers, Inc.; **2.5:** © Klaus Gulbrandsen/SPL/Photo Researchers, Inc.; **2.6:** © Dwight Kuhn Photography; **Page 22(a):** Reprinted from Bhattacharyya MK, et al. *Cell.* 1990 Jan 12: 60(1):115–22, with permission from Elsevier; **2.18(a):** © Science Photo Library/Photo Researchers, Inc.; **2.18(b-d):** © Mendelianum Institute, Moravian Museum; **2.22:** © Jerry Marshall; **2.24:** © John D. Cunningham/Visuals Unlimited; **2.28(a) top-left:** © McGraw-Hill Higher Education Group, Inc./Jill Birschbach, photographer, arranged by Aleandra Dove, McArdle Laboratory; **2.28(a) top-right:** © Charles River Laboratories; **2.28(a) bottom:** © Charles River Laboratories; **2.30(a):** © Stanley Flegler/Visuals Unlimited; **2.32(a):** © William H. Allen, Jr./Allen Stock Photography; **2.39(a):** © Radu Sigheti/

Reuters/Landov; **2.40(a):** © Renee Lynn/Photo Researchers, Inc.; **Page 53:** TODD KOROL/Reuters/Landov.

Chapter 3

Opener: © 2007 Rockefeller University Press. Originally published in *The Journal of Cell Biology,* Vol. 177, No. 2, pp. 231–242 (Figure 2); **3.1:** © Richard Hutchings/Photo Researchers, Inc.; **3.4:** © Scott Camazine/Photo Researchers, Inc.; **Page 76(b):** Karmes StayWell, 780 Township Line Road, Yardley, PA 19067 (267) 685-2500; **3.5(a):** © Biophoto Association/Photo Researchers, Inc.; **3.7:** Photographs by Dr. Conly L. Rieder, Division of Molecular Medicine, Wadsworth Center, NYS Dept. of Health, Albany, NY; **3.8(a):** © David M. Phillips/Visuals Unlimited; **3.8(b):** © R. Calentine/Visuals Unlimited; **3.9:** © Dr. Byron Williams/Cornell University; **Page 84(a–b):** © Dr. Michael Goldberg/Cornell University; **3.14:** Republished with permission of *Annual Review of Genetics,* from *The Synaptinemal Complex,* M. Westergaard, and D.V. Wettstein, Vol. 6: 71–110 (December 1972), permission conveyed through Copyright Clearance Center, Inc.; **3.15:** © Dr. Leona Chemnick, Dr. Oliver Ryder/San Diego Zoo, Center for Reproduction of Endangered Species; **3.21:** Colour deficit simulation, courtesy of Vischeck (www.vischeck.com). Source image courtesy of NASA; **3.23(a):** Tim Rooke/Rex Features/CP Images; **3.23(b):** WPA Pool/Getty Images.

Chapter 4

Opener: © Eye-Stock/Alamy; **4.7:** "Meiosis" by Prof. Bernard John, Cambridge University Press, Fig. 2.3(a), p. 33. Reprinted with the permission of Cambridge University Press; **4.16:** Geng J, Javed N, McVetty PBE, Li G, Tahir M (2012) *An Integrated Genetic Map for Brassica napus.* Derived from Double Haploid and Recombinant Inbred Populations. Hereditary Genetics 1:103. doi: 10.4172/2161-1041.1000103; **Page 129(a):** Permission per Lap-Chee Tsui; **4.19:** Image courtesy of B.A. Montelone, Ph.D. and T.R. Manney, Ph.D.

Chapter 5

Opener(a): LoKiLeCh/Wikipedia; **Opener(b):** SCIENCE PHOTO LIBRARY; **Opener(c):** Ploclaim/Wikipedia; **Opener(d):** TORUNN BERGE/SCIENCE PHOTO LIBRARY; **Opener(e):** Bruce Rolff/Shutterstock; **5.1(a):** © George Bernard/Animals Animals; **5.1(b):** © William Hauswirth; **5.1(c):** © Archivo Iconografico, S.A./CORBIS; **5.2:** Johannes-Maria Schlorke Fotografie; **5.4:** Reprinted from *Mutation Research/Genetic Toxicology and Environmental Mutagenesis,* Vol. 743, Issues 1–2, 18 March 2012, pp. 20–24. O. Herrero, J.M. Pérez Martin, P. Fernández Freire, L. Carvajal López, A. Peropadre, M.J. Hazen, "Toxicological evaluation of three contaminants of emerging concern by use of the Allium cepatest." © 2012, with permission from Elsevier; **5.5:** Used with permission from Arnold et al.; New Associations with Pseudomonas Luteola Bacteremia: A Veteran with a

History of Tick Bites and a Trauma Patient with Pneumonia. *The Internet Journal of Infectious Diseases* 2005: Volume 4, Number 2; **5.7(b):** © PHANIE/Photo Researchers, Inc.; **Page 148(a):** National Institutes of Health (NIH); **Page 148(b):** NATIONAL LIBRARY OF MEDICINE/SCIENCE PHOTO LIBRARY; **Page 148(c):** © 1994 Rockefeller University Press. Originally published in *The Journal of Experimental Medicine,* Vol. 179, pp. 385–394; **Page 148(d):** NATIONAL LIBRARY OF MEDICINE/SCIENCE PHOTO LIBRARY; **5.10:** © Science Source/Photo Researchers, Inc.; **Page 152:** PHOTO © ESTATE OF FRANCIS BELLO/SCIENCE PHOTO LIBRARY; **5.13(a):** © A. Barrington Brown/Science Source/Photo Researchers, Inc.; **5.15(a):** © Biophoto Associates/Science Source/Photo Researchers, Inc.; **5.15(b):** © Microworks/Dan/Phototake.com; **5.15(c):** © Ross Inman & Maria Schnos, University of Wisconsin, Madison, WI; **5.15(d):** © Jack D. Griffith/University of North Carolina Lineberger Comprehensive Cancer Center.

Chapter 6

Opener: POWER AND SYRED/SCIENCE PHOTO LIBRARY; **Page 170:** D. Leja, NHGRI, NIH; **6.1:** © Dr. Don Fawcett/J.R. Paulson, U.K. Laemmli/Photo Researchers, Inc.; **6.2:** © Daniel A. Starr/University of Colorado; **6.3:** © Ada L. Olins/Biological Photo Services; **6.5:** © Dr. Gerard J. Bunick, Oak Ridge National Laboratory; **6.6(a) left:** © Dr. Barbara Hamkalo/University of California—Irvine, Department of Biochemistry; **6.6(b) right:** © Dr. Don Fawcett and A. Olins/Photo Researchers, Inc.; **6.7:** © Dr. Don Fawcett/J.R. Paulson, U.K. Laemmli/Photo Researchers, Inc.; **6.8:** © Doug Chapman, University of Washington Medical Center Cytogentics Laboratory; **Page 175(a) :** The Canadian Medical Hall of Fame and Irma Council; **Page 175(b):** CHRIS BJORNBERG/SCIENCE SOURCE/SCIENCE PHOTO LIBRARY; **6.10:** © H. Kreigstein and D.S. Hogness, "Mechanism of DNA Replication in Drosophilia Chromosomes: Structure of Replication Forks and Evidence of Bidirectionality," *Preceedings of the National Academy of Sciences, USA* 71(1974): 135–139; **6.12:** © Dr. Robert Moyzis/University of California—Irvine, Department of Biochemistry.

Chapter 7

7.3: © Professor Oscar Miller/SPL/Photo Researchers, Inc.; **Page 201:** THE CANADIAN PRESS/Geoff Robins; **7.9(b):** © Dr. Thomas Maniatis, Thomas H. Lee Professor of Molecular and Cellular Biology, Harvard University; **7.16:** Cech TR. *Science* 11, August 2000: Vol. 289, No. 5481, pp. 878–879; **7.19:** Kbradnam/Wikipedia Commons; **7.24(a):** Courtesy of Dr. Karen Artzt/Artzt Lab/The University of Texas at Austin; **7.24(c):** © Tom Vasicek; **7.24(d) left:** © Science VU/Dr. F.R. Turner/Visuals Unlimited; **7.24(d) right:** © Eye of Science/Photo Researchers, Inc.

Chapter 8

Opener: © Daniel Karmann/dpa/CORBIS; **8.1:** © Dr. Don Fawcett/J.R. Paulson & U.K. Laemmli/Photo Researchers, Inc.; **8.3:** © Charles River Laboratories; **Page 245(a):** © Science VU/Visuals Unlimited; **8.15:** © Dr. Ken Greer/Visuals Unlimited; **8.17:** © Carolina Biological/Photo Researchers, Inc.; **Page 265(a):** CP PHOTO/Chuck Stoody; **Page 265(b):** Barry Shell; **8.27:** Colour

deficit simulation, courtesy of Vischeck (www.vischeck.com). Source image courtesy of NASA.

Chapter 9

Opener: Multimed Inc.; **Page 281(c):** © Courtesy of The Centers for Disease Control; **9.7:** © Dr. Ross MacIntyre, Cornell University; **9.9(a):** © Dr. Michael Goldberg/Cornell University; **9.10:** Courtesy of Dr. Philip Cotter; **9.13 top:** © Cabisco/Visuals Unlimited; **9.13 middle:** Courtesy of Dr. Brian R. Calvi; **9.13 bottom:** © Cabisco/Visuals Unlimited; **9.15:** © Provided by M.J. Moses, Duke University, from Poorman, Moses, Davisson, and Roderick: Chromosoma (Berl. 83:419–429 (1981)); **9.18(b):** © Lisa G Shaffer, Ph.D./Baylor College of Medicine; **9.20(a) left:** © J. Carillo-Farg/Photo Researchers, Inc.; **9.20(a) right:** © Dr. E. Walker/SPL/Photo Researchers, Inc.; **9.21(d):** © M.G. Neuffer, University of Missouri; **9.23:** © Dr. Michael Goldberg/Cornell University; **Page 295(a):** NATIONAL LIBRARY OF MEDICINE/SCIENCE PHOTO LIBRARY; **9.24:** © Dr. Nina Fedoroff; **9.29:** Britton-Davidian et al., "Environmental Genetics: Rapid Chromosomal Evolution in Island Mice," *Nature* 403, 158 (13 January 2000); **9.36:** © Leonard Lessin/Peter Arnold, Inc./PhotoLibrary.com; **9.37:** © Davis Barber PhotoEdit.

Chapter 10

Opener: With kind permission from Springer Science + Business Media: "The lactose repressor system: Paradigms for regulation, allosteric behavior and protein folding," Wilson C.J., Zhan H., Swint-Kruse L., Matthews K.S. Cell Mol Life Sci. 2007 Jan; 64(1): 3–16. © 2007; **10.1:** © London School of Hygiene & Topical Medicine/SPL/Photo Researchers, Inc.; **10.4:** © Bettman/UPI/CORBIS; **Page 329(a):** Courtesy of Dr. Jack Greenblatt; **Page 329(b):** Figure 9b from "Elongation factor NusG interacts with termination factor rho to regulate termination and antitermination of transcription," J. Li, S.W. Mason, and J. Greenblatt. *Genes and Development* © 1993, pp. 161–172; **Page 345(a):** © Dr. Ann Hirsch, UCLA; **Page 354:** From Luke K. Kogan Y. et al., Microarray analysis of gene expression during baceriophage T4 infection. *Virology.* 2002 August 1; 299(2) 181–191. Copyright © 2002, with permission from Elsevier Science.

Chapter 11

Opener: Courtesy of Dr. Randy Jirtle; **Page 367(a):** Reprinted by permission from Macmillan Publishers Ltd. "Environmental epigenomics and disease susceptibility," Randy L. Jirtle and Michael K. Skinner. *Nature Reviews Genetics* 8, 253–262 (April 2007), doi:10.1038/nrg2045; **Page 367(b):** Mol. Cell. Biol. August 2003, Vol. 23, No. 15 5293–5300, doi: 10.1128/MCB.23.15.5293-5300.2003. Reproduced with permission from American Society for Microbiology; **Page 368(c):** Mol. Cell. Biol. August 2003, Vol. 23, No. 15 5293-5300, doi: 10.1128/MCB.23.15.5293-5300.2003. Reproduced with permission from American Society for Microbiology; **Page 374(a):** Fred Lum/The Globe and Mail/CP Images; **Page 374(b):** Reprinted by permission from Macmillan Publishers Ltd. "Epigenetics as a unifying principle in the aetiology of complex traits and diseases," Arturas Petronis. *Nature* 465, 721–727 (10 June 2010) doi: 10.1038/nature09230.

reprinted with permission from Company of Biologists Ltd.; **17.15(a–b):** Reprinted from *CELL,* Vol. 87, No. 5, Basler, Zecca, Struhl, "Direct and Long-Range Action of a Wingless Morphogen Gradient," pp. 833–844, © 1996 with permission from Elsevier Science; **17.17(b):** © Bill Sullivan, University of California—Santa Cruz; **17.18(a–c):** © Dr. Rudi Turner and Dr. Tom Kaufman, Indiana University; **17.19(a):** © Steve Small, New York University; **17.19(b):** © David Kosman, John Reinitz; **17.21(a):** © David Kosman, John Reinitz; **17.21(b):** © Dr. Eric Wieschaus, Princeton University. Reprinted with permission from *Nature* 287: 795–801, "Mutations affecting Segment Number and Polarity in Drosophilia," Nusslein-Volhard, C., and E. Wieschaus, 1980. © 1980 Nature Publishing Group; **17.22(a):** © David Kosman, John Reinitz; **17.23(a):** © Steve Small, New York University; **17.25(b):** © Edward Lewis, California Institute of Technology; **17.29:** © St. Bartholomew Hospital/Photo Researchers, Inc.; **17.30:** © Dennis Kunkel Microscopy, Inc.; **17.31:** © Drs. Scott Weatherbee and Sean B. Carroll; **17.33:** © AP/Wide World Photo; **Page 585:** From Hoda K. Shamloula, Mkajuma P. Mbogho, Angel C. Pimentela, Zosia M. A. Chrzanowska-Lightowlersb, Vanneta Hyatta, Hideyuki Okanoc, and Tadmiri R. Venkatesh. Rugose (rg) a Drosophilia A Kinase Anchor Protein, Is Required for Retinal Pattern Formation and Interacts Genetically with Multiple Signalling Pathways. *Genetics,* Vol. 161, 693–710, June 2002. © Genetics Society of America; **Page 586:** Image courtesy of John M. Kemner; **Page 588:** © Dr. Ken Kemphues and Bijan Etemad-Moghadam, Cornell University.

Chapter 18

Opener: Mattosaurus/Wikipedia Commons; **18.1:** © Stephen Frish Photography; **18.2:** © Dr. Jeremy Burgess/SPL/Photo Researchers, Inc.; **18.3:** © David M. Phillips/Visuals Unlimited; **18.6:** © Dr. Gopal Murti/SPL/Photo Researchers, Inc.; **18.10:** © U.S. Landmarks and Travel 2/V74 Robert Glusic/Getty Images; **18.11:** © SPL/Science Source/Photo Researchers, Inc.; **18.25:** © Jack D. Griffith/University of North Carolina Lineberger Comprehensive Cancer Care Center; **Page 610:** Proc Natl Acad Sci U.S.A. 2006 January 10; 103 (2): 425–430, Glass et al., "Essential genes of a minimal bacterium." © 2006 National Academy of Sciences, U.S.A.; **18.29:** © Newcomb & Wergin/Biological Photo Service; **18.30:** Electron micrograph by Dr. Stephen Hajduk/University of Alabama at Birmingham; **18.33:** © Eric L. Heyer/Grant Heilman Photography; **18.36:** © Jim Strawser/Grant Heilman Photography; **18.38:** Reprinted from *Trends in Genetics,* January 1989, Vol. 5, No. 1, Dr. Douglas C. Wallace, p. 11. © 1989 with permission from Elsevier Science; **Page 622:** Hospital Archives, The Hospital for Sick Children, Toronto. Robert Teteruck, photographer.

Chapter 19

Opener(a): © RGB Ventures LLC dba SuperStock/Alamy; **Opener(b):** © Trinity Mirror/Mirrorpix/Alamy; **19.1:** Wetterstrand KA. DNA Sequencing Costs: Data from the NHGRI Genome Sequencing Program (GSP). Available at www.genome.gov/sequencingcosts. Courtesy: National Human Genome Research Institute; **Page 641:** Alex Wong/Getty Images; **19.11(a–b):** Reprinted by permission from Macmillan Publishers Ltd. "Sequencing technologies—the next generation," Michael L. Metzker. *Nature Reviews Genetics* 11, pp. 31–46 (January 2010) doi: 10.1038nrg2626; **19.12:** Jerod Harris/WireImage/Getty Images.

Chapter 20

Opener(a): Mopic/Shutterstock; **Opener(b):** McGraw-Hill Companies; **Opener(c):** Juan Gaertner/Shutterstock; **Opener(d):** Image Source/Getty Images; **Opener(e):** Monkey Business Images/Shutterstock; **Opener(f):** Gaeddal/Wikipedia Commons; **Opener(g):** Courtesy: National Institute of Health; **Opener(h):** Bill Branson/National Cancer Institute; **20.2(a):** PATRICK DUMAS/EURELIOS/SCIENCE PHOTO LIBRARY; **20.2(b):** Andre Nantel/Shutterstock; **20.4:** Roche Diagnostics GmbH; **20.5:** © ZUMA Press, Inc./Alamy; **Page 655:** From "Molecular Classification of Cancer: Class Discovery and Class Prediction by Gene Expression Monitoring," T.R. Golub, D.K. Slonim, P. Tamayo, C. Huard, M. Gaasenbeek, J.P. Mesirov, H. Coller, M.L. Loh, J.R. Downing, M.A. Caligiuri, C.D. Bloomfield, and E.S. Lander. *Science* 15 October 1999: 286 (5439), 531–537. DOI: 10.1126science.286.5439.531. Reprinted with permission from AAAS; **20.6:** Christine A. White and Lois A. Salamonsen, "A guide to issues in microarray analysis: Application to endometrial biology." *Reproduction* 130 (1) 1–13, doi: 10.1530/rep.1.00685. Copyright © 2005 Society for Reproduction and Fertility; **20.8:** Reprinted with permission from *Nature Genetics* (2009) Vol. 41 (1) pp. 25–34. © 2009 Nature Publishing Group; **20.9:** National Human Genome Research Institute, www.genome.gov/GWAStudies, www.ebi.ac.uk/fgpt/gwas; **20.12:** Reprinted by permission from Macmillan Publishers Ltd. "Cancer epigenomics: DNA methylomes and histone-modification maps," Manel Esteller. *Nature Reviews Genetics* 8, pp. 286–298 (April 2007) doi: 10.1038/nrg2005; **20.13:** Reprinted by permission from Macmillan Publishers Ltd. "DNA methylome analysis using short bisulfite sequencing data," Felix Krueger, Benjamin Kreck, Andre Franke, and Simon R. Andrews. *Nature Methods* 9, pp. 145–151 (2012) doi: 10.1038/nmeth.1828; **20.14:** Reprinted by permission from Macmillan Publishers Ltd. "Genome-wide approaches to studying chromatin modifactions," Dustin E. Schones and Keji Zhao. *Nature Reviews Genetics* 9, pp. 179–191 (March 2008) doi: 10.1038/nrg2270; **20.15:** Reprinted by permission from Macmillan Publishers Ltd. "DNA methylation profilinig in the clinic: Applications and challenge," Holger Heyn and Manel Esteller. *Nature Reviews Genetics* 13, pp. 679–692 (October 2012) doi: 10.1038/nrg3270; **Page 666(a):** Courtesy of Dr. Brad Popovich; **Page 666(b):** Mario Beauregard/CP Images.

Chapter 21

Opener: Artwork by Julie Newdoll, www.brushwithscience.com originally commissioned for the April 2011 issue of *Molecular and Cellular Proteomics* (Vol. 10, Issue 4); **21.1:** John Aitchison; **Page 672:** SCIENCE PHOTO LIBRARY; **21.3:** "Yeast two-hybrid, a powerful tool for systems biology," Brückner A., Polge C., Lentze N., Auerbach D., Schlattner U. *Int J Mol Sci.* 2009 June 18, 10 (6): pp. 2763–2788, doi: 10.3390/ijms10062763; **21.5 right:** V&P Scientific, Inc.; **21.6:** Masur/Wikipedia Commons; **21.7:** Singer Instrument Co. Ltd.; **21.8(b):** Masur/Wikipedia Commons; **Page 676(a):** © AF archive/Alamy; **Page 676(b):** Singer Instrument Co. Ltd.; **21.9:** Reprinted by permission from Macmillan Publishers Ltd. "A comprehensive analysis of protein–protein interactions in Saccharomyces cerevisiae," Peter Uetz, Loic Giot, Gerard Cagney, Traci A. Mansfield, Richard S. Judson, James R. Knight, Daniel Lockshon, Vaibhav Narayan, Maithreyan Srinivasan, Pascale Pochart, Alia Qureshi-Emili, Ying Li, Brian Godwin, Diana Conover, Theodore Kalbfleisch, Govindan Vijayadamodar, Meijia Yang, Mark Johnston, Stanley Fields, and Jonathan M. Rothberg. *Nature* 403,

pp. 623–627 (10 February 2000) doi: 10.1038/35001009; **21.10:** "Schwartz's Principles of Surgery," Ninth Edition. F. Brunicardi, Dana Andersen, Timothy Billiar, David Dunn, John Hunter, Jeffrey Matthews, Raphael E. Pollock. 2009. © McGraw-Hill Education; **21.13:** Yates J.R. (1998), Mass spectrometry and the age of the proteome. *J. Mass Spectrum,* 33: pp. 1–19, doi: 10.1002/(SICI)1096-9888(199801)33:1<1:AID-JMS624>3.0.CO;2-9; **21.14:** Yates J.R. (1998), Mass spectrometry and the age of the proteome. *J. Mass Spectrum,* 33: pp. 1–19, doi: 10.1002/(SICI)1096-9888(199801)33:1<1:AID-JMS624>3.0.CO;2-9; **Page 682(a):** John Vetterli/Wikipedia Commons; **Page 682(b):** Fig. 1 from "Positive selection at the protein network periphery: Evaluation in terms of structural constraints and cellular context," Philip M. Kim. *PNAS* December 18, 2007, vol. 104, no. 51, 20274–20279. © 2007 National Academy of Sciences, U.S.A.; **21.16:** Reprinted by permission from Macmillan Publishers Ltd. "Global identification of protein kinase substrates by protein microarray analysis," Janie Mok, Hogune Im, and Michael Snyder. *Nature Protocols* 4, pp. 1820–1827 (2009), doi: 10.1038/nprot.2009.194; **Page 684(a):** Crimfants/Wikipedia Commons.

Chapter 22

Opener: Neftali/Shutterstock; **22.3:** "Using Homologous Recombination to Manipulate the Genome of Human Somatic Cells," Matthew Porteus. Biotechnology and Genetic Engineering Reviews, Vol. 24, Issue 1, 2007, pp. 195–212, doi: 10.1080/02648725.2007.10648100; **Page 692:** © Rockefeller University Press. Originally published In the *Journal of Cell Biology,* Vol. 188, No. 1, pp. 69–81 (Figure 1); **22.7:** Illustration by Fairman Studios, courtesy of Alnylam Pharmaceuticals; **22.8:** Reprinted by permission from Macmillan Publishers Ltd. "Building mammalian signalling pathways with RNAi screens," Jason Moffat and David M. Sabatini. *Nature Reviews Molecular Cell Biology* 7, pp. 177–187 (March 2006), doi: 10.1038/nrm1860; **22.9:** Reprinted by permission from Macmillan Publishers Ltd. "http://C.Elegans: Mining the functional genomic landscape," Stuart K. Kim. *Nature Reviews Genetics* 2, pp. 681–689 (September 2001), doi: 10.1038/35088523; **22.10:** Reprinted by permission from Macmillan Publishers Ltd. "The art and design of genetic screens: RNA interference," Michael Boutros & Julie Ahringer. *Nature Reviews Genetics* 9, pp. 554–566 (July 2008), doi: 10.1038/nrg2364; **22.11:** From "Cancer Proliferation Gene Discovery Through Functional Genomics," Schlabach et al., *Science,* 1 February 2008: 319 (5863), pp. 620–624, doi: 10.1126/science.1149200. Reprinted with permission from AAAS; **22.15:** Masur/Wikipedia Commons; **22.17:** Reprinted by permission from Macmillan Publishers Ltd. "Exploring genetic interactions and networks with yeast," Charles Boone, Howard Bussey & Brenda J. Andrews. *Nature Reviews Genetics* 8, pp. 437–449 (June 2007), doi: 10.1038/nrg2085; **Page 701:** Reprinted by permission from Macmillan Publishers Ltd. "Exploring genetic interactions and networks with yeast," Charles Boone, Howard Bussey & Brenda J. Andrews. *Nature Reviews Genetics* 8, pp. 437–449 (June 2007), doi: 10.1038/nrg2085; **22.18:** Reprinted by permission from Macmillan Publishers Ltd. "Exploring genetic interactions and networks with yeast," Charles Boone, Howard Bussey & Brenda J. Andrews. *Nature Reviews Genetics* 8, pp. 437–449 (June 2007), doi: 10.1038/nrg2085; **22.19:** © 2007 IEEE. Reprinted, with permission, from "The Architecture of Complexity," Albert-Laszlo Barabasi. Control Systems, IEEE (Volume 27, Issue 4), pp. 33–42; **22.20:** Kbradnam/Wikipedia Commons; **22.21:** Reprinted by permission from

Macmillan Publishers Ltd. "Systematic mapping of genetic interactions in Caenorhabditis elegans identifies common modifiers of diverse signalling pathways," Ben Lehner, Catriona Crombie, Julia Tischler, Angelo Fortunato, and Andrew G. Fraser. *Nature Genetics* 38, pp. 896–903 (2006), published online: 16 July 2006, doi: 10.1038/nrg1844; **22.22:** Reprinted by permission from Macmillan Publishers Ltd. "Systematic mapping of genetic interactions in Caenorhabditis elegans identifies common modifiers of diverse signalling pathways," Ben Lehner, Catriona Crombie, Julia Tischler, Angelo Fortunato, and Andrew G. Fraser. *Nature Genetics* 38, pp. 896–903 (2006), published online: 16 July 2006, doi: 10.1038/nrg1844.

Chapter 23

Opener: Andrey Burmakin/Shutterstock; **23.4:** Adapted from "Plasticity of the cis-Regulatory Input Function of a Gene," Avraham E. Mayo, Yaakov Setty, Seagull Shavit, Alon Zaslaver, Uir Alon. Plos *Biology,* April 2006, Vol. 4, Issue 4; **23.5:** "Plasticity of the cis-Regulatory Input Function of a Gene," Avraham E. Mayo, Yaakov Setty, Seagull Shavit, Alon Zaslaver, Uir Alon. Plos *Biology,* April 2006, Vol. 4, Issue 4; **23.6:** Adapted from "Plasticity of the cis-Regulatory Input Function of a Gene," Avraham E. Mayo, Yaakov Setty, Seagull Shavit, Alon Zaslaver, Uir Alon. Plos *Biology,* April 2006, Vol. 4, Issue 4; **23.18 & 23.19:** Reprinted by permission from Macmillan Publishers Ltd. "Network motifs: Theory and experimental approaches," Uri Alon. Nature Reviews Genetics 8, pp. 450–461 (June 2007), doi: 10.1038/nrg2102; **Page 719(a):** Courtesy of Edwin Wang; **Page 719(b):** Reproduced from *Integr. Biol.* 2011, 3, pp. 724–732, doi: 10.1039/C0IB00145G with permission of The Royal Society of Chemistry.

Chapter 24

Opener: Electron micrographs were provided by Tom Deerinck and Mark Ellisman of the National Center for Microscopy and Imaging Research at the University of California at San Diego. Courtesy of the J. Craig Venter Institute; **24.1:** AP Photo/Michel Euler/CP Images; **24.2:** The BioBricks Foundation; **Page 727(a):** iGEM; **Page 727(b):** iGEM; **24.3:** "Synthetic biology: New engineering rules for an emerging discipline," Ernesto Andrianantoandro, Subhayu Basu, David K. Karig & Ron Weiss. *Molecular Systems Biology* 2 Article number: 2006.0028 doi: 10.1038/msb410073. Published online: 16 May 2008; **24.4:** © 2009 Rockefeller University Press. Originally published in *The Journal of Cell Biology,* Vol. 187, pp. 589–596; **24.5(a):** Reprinted by permission from Macmillan Publishers Ltd. "Synthetic biology: Applications come of age," Ahmad S. Khalil & James J. Collins. *Nature Reviews Genetics* 11, pp. 367–379 (1 May 2010), doi: 10.1038/nrg2775; **24.5(b):** Reprinted by permission from Macmillan Publishers Ltd. "A synthetic oscillatory network of transcriptional regulators," Michael B. Elowtiz & Stanislas Leibler. *Nature* 403, pp. 335–338 (20 January 2000), doi: 10.1038/35002125; **24.6:** Figure 4 from "Designing biological systems," Drubin, Way, and Silver. *Genes and Development* © 2007. pp. 242–254; **24.7:** Reprinted by permissioned from Macmillan Publishers Ltd. "Synthetic biology: Applications come of age," Ahmad S. Khalil & James J. Collins. *Nature Reviews Genetics* 11, pp. 367–379 (1 May 2010), doi: 10.1038/nrg2775; **24.8:** © 2009 Rockefeller University Press. Originally published in *The Journal of Cell Biology,* Vol. 187, pp. 589–596; **24.9:** "Engineering microbes to sense and eradicate Pseudomonas

aeruginosa, a human pathogen," Chang et al., *Molecular Systems Biology* 7, Article number 521, doi: 10.1038/msb.2011.55. Published online: 16 August 2011; **24.10(a):** Reprinted by permission from Macmillan Publishers Ltd. "Programming cells by multiplex genome engineering and accelerated evolution," Harris H. Wang, Farren J. Isaacs, Peter A. Carr, Zachary Z. Sun, George Xu, Craig R. Forest & George M. Church. *Nature* 460, pp. 894–898 (13 August 2009), doi: 10.1038/nature08187; **24.10(b):** Wyss Institute for Biologically Inspired Engineering at Harvard University; **Page 734(a):** © GL Archive/Alamy; **Page 734(b):** © Atomic/Alamy; **Page 734(c):** JOE MUNROE/Landov; **24.12:** From "Creation of a Bacterial Cell by a Chemically Synthesized Genome," Daniel G. Gibson, John I. Glass, Carole Lartigue, Vladimir N. Noskov, Ray-Yuan Chuang, Mikkel A. Algire, Gwynedd A. Benders, Michael G. Montague, Li Ma, Monzia M. Moodie, Sanjay Vashee, Radha Krishnakumar, Nacyra Assad-Garcia, Cynthia Andrews-Pfannkoch, Evgeniya A. Denisova, Lei Young, Zhi-Qing Qi, Thomas H. Seagall-Shapiro, Christopher H. Calvey, Prashanth P. Parmar, Clyde A. Hutchison III, Hamilton O. Smith, and J. Craig Venter. *Science* 2, July 2010: 329 (5987), pp. 52–56. Published online 20 May 2010 (doi: 10.1126/science.1190719). Reprinted with permission from AAAS; **Page 737(a):** University of Pennsylvania; **24.15:** Courtesy of the J. Craig Venter Institute.

Index

C

C. *See* Cytosine (C)
C-abl protein, 290–291
Caenorhabditis elegans, 376
    collagen gene of, 214*f*–215
    double mutants, 567–568
    gene expression in, 213–215
    genome sequence of, 437, 634*t*
    as model organism, 213, 554–555, 634
    nucleotide base composition of DNA
        of, 150*t*
    polycistronic transcripts, 215
    RNA interference (RNAi) in, 561–562,
        695–697
    sex determination in, 78*f*
    synthetic genetic arrays in, 703–704
    temperature-shift analysis, 560
    trans-splicing, 203–204*f,* 215
CAMP receptor protein (CRP), 333
Canadian Coalition for Genetic Fairness
    (CCGF), 12–13*f*
Canadian Epigenetics, Environment and
    Health Research Consortium, 665–666
Cancer. *See also* Mutagens
    age, and incidence of, 530*f*
    breast cancer, 11–12, 509–510, 544
    colorectal cancer, 252, 544
    detection of, 655, 664–665
    development over time, 529
    and DNA chips, 10–11
    genetic testing, in prediction and treat-
        ment of, 544
    incidence of, in various countries, 529*t*
    inherited predisposition, 529
    lung cancer rates, 530*f*
    and microarray scanners, 655
    mistakes in gene rearrangements, 281
    and mitotic recombination, 132
    mutations, in cancer phenotype,
        524–530
    and network motifs, 719
    percentage of new cancers in Canada,
        522*f*
    proteomes, normal, compared to,
        681–682
    RNAi, anticancer therapy, 695–696
    skin cancer, 252
    translocations in somatic cells,
        290–291
    Xeroderma pigmentosum, 252
Cancer drugs, 10, 11*f*
Cancer genes, 530
Capping enzyme, 199
Capture molecule, 677
Carcinogens, Ames test for, 252–253
Cardiovascular disease, penetrance and
    expressivity of, 53
Caron, Jean-Bernard, 428
Carriers, of cystic fibrosis, 32
Catalog of Published Genome-Wide
    Association Studies, 661
Cats
    cloned kitten, 583
    Siamese cat, coat colour, 52
CDNA (complementary DNA), 460

CDNA libraries, 460–461
    vs. genomic libraries, 461–462*f*
C. elegans. *See Caenorhabditis elegans*
Cell-bound signals, 522–523
Cell cycle
    control of, 536–544
    described, 536
    and DNA repair genes, 539*t*
    in *Drosophila melanogaster,* 545*f*
    G_1-to-S transition, 540
    G_2-to-M transition, 540 need symbol
    mitosis, 79–80
    mutations, analysis of, 537–538
    mutations, isolation of, 537–538
    phases of, 536–537
Cell-cycle checkpoints
    chromosome separation correction,
        83–85
    G_1-to-S checkpoint, 541, 542*f*
    G_2-to-M checkpoint, 541–542, 543*f*
        need symbol
    genomic stability, 541–544
    M checkpoint, 542
    necessity of, 542–544
Cell division
    and cancer, 524–536
    cell-bound signals, 522–523
    extracellular signals, 522–523
    kinases in yeasts, 537
    and mitosis, 79–85
    molecular interactions, 523–524
    normal control of, 536–545
    overview, 522–524
    tumour suppressor genes, 536
Cell plates, 82*f,* 83
Cell proliferation
    enhanced mutation potential, 533
    increased, from mutant tumour-
        suppressor alleles, 533,
        535–536
    from oncogenes, 530–533
    telomerase, 179
Cells
    abnormalities, 524
    donor cell, 599
    earliest cells, 427
    germ cells, 85
    normal *vs.* diseased cells, 718–720
    protein makeup, 672
    recipient cell, 599
    tumour cells, *vs.* normal cells, 524,
        525*f*–527*f*
Cellular blastoderm, 569
Cellular clones, 458
Center for Disease Control, 661
Centimorgan (cM), 120
Central Dogma, 193–194
Central dogma, 6
Centrality-lethality rule, 682
Centre for Epigenomic Mapping
    Technologies, 666*f*
Centrioles, 80, 81*f,* 86*f*
Centromeres, 74–75*f,* 89, 171*f*
    characteristics of, 180
    histone variants at, 182
    mechanics of segregation, 180–181

repeat sequences, 443
    structure, 181–182
Centromeric fibres, 82
Centrosome, 80
CF allele, 32–33
CGH. *See* Comparative genomic hybridiza-
    tion (CGH)
Chargaff, Erwin, 150
Charged tRNA, 205
Chase, Martha, 145, 148
Chiasmata
    chromosomal recombination, 183
    homologues, held together by, 89, 118
    multiple in grasshopper, 118*f*
    recombination markers, 119–120
    terminalization, 120
Chicken, feather colour, 47, 48*t*
Chimeras, 566
Chimpanzees, 429–431
Chip, DNA, 10–11, 652–654
ChIP-on-Chip analysis, 656–657, 663–664
ChIP-Seq analysis, 656–657, 664
ChIP technology, 381, 656–657
Chi-square test
    alternative hypothesis, 115
    application to linkage, example, 117–118
    degrees of freedom (df) in, 116, 117
    and linkage analysis, 115–118
    null hypothesis, 115
Chlamydomonas reinhardtii, 615
Chloroplast genome
    biparental inheritance, 620
    of liverwort, 615*t*
    maternal inheritance of, 618–619
    neurodegenerative disease in humans,
        619
    non-Mendelian inheritance, 617–620
    and nuclear genome, 615–616
    of rice, 615*t*
    size of, 615
    techniques for introducing genes and
        DNA fragments, 615
    of tobacco, 615*t,* 618–619
    transformation of, 615
    uniparental inheritance in, 620
Chloroplasts
    DNA of, 612
    DNA sizes of, 615*t*
    genetics of, 612–617
    gene transfer between nucleus, 616
    illustrated, 612*f*
Chorionic villus sampling (CVS), 76–77
    Down syndrome, 75
Chromatid, 79, 80
    sister, 74, 75*f,* 79*f,* 80–83, 86*f,* 90, 92*f*
Chromatin, 79
    defined, 169
    hypercondensation of, 369–370
    loops of chromosomes, 174*f*
    remodelling, activates transcription,
        368–370
    reproduction, and chromosome
        duplication, 179
    structure and gene expression, 176
    SWI-SNF remodeller, 369
    transcription reduction, 366

Homogeneously staining region (HSR), 541

Homologous chromosomes. *See* Homologues

Homologues, 75
in anaphase I, 89
in metaphase I, 89
and recombination, 118–119

Homology mapping of genomes, 639–640

Homoplasmic cells, 619

Homo sapiens, 429–431

Homozygosity, lethal effects for deletions, 279

Homozygotes, 25

Homozygous, 25

Honeybee, 150*t*

Horizontal gene transfer, 598

Horizontal pattern of inheritance, 32–33, 51

Horowitz, Norman, 257

Host cells, transformation of, 458

Hot spots of mutation, 255

Housekeeping genes, 10

Hox genes
of *Drosophila melanogaster*, 441, 578–580
of mouse, 441

HTH (Helix-turn-helix) motif, 334–335

Human (liver), nucleotide base composition of DNA of, 150*t*

Human brain, and biological information, 6

Human cloning, and stem cells, 558–560

Human cystic fibrosis transmembrane receptor (CFTR) locus, 394

Human disease-related genes, identification by linkage mapping, 127*t*

Human dystrophin gene, and RNA splicing, 202*f*

Human Gene Mutation Database, 643

Human Genome Epidemiology, 661

Human Genome Project, 9
cost of, 633–634
described, 632
number of genes, 639
outcomes, and foundation for systems biology, 708
reference sequence, is "composite" of individuals, 633, 643–644
repercussions of, 644–646

Human genomes
homology mapping of, 639–640
large-scale deletions and duplications differentiate, 502–505
vs. mouse genome, 276–277

Human immunodeficiency virus (HIV). *See* HIV/AIDS

Human lymphocyte antigens (HLAs), 621

Humans
aneuploidy in, 312–313*t*
Aniridia gene, 564
chimpanzees, differences, 429–431
chromosomes, 4, 72, 236–237
codominance of red blood cells, 35
colour blindness, 100, 110, 111*f*, 133
development, 552–553
evolution of, 429–431
fertilized human egg, 79

genetic code variations, in human mitochondria, 614*t*
genetic heterogeneity in, 48*f*
genome length, 433
genome sequence of, 634*f*, 639
heterochromatic chromosomes, 174
inheritance, 29–33
Mendelian pattern of inheritance in, 29–33
mitochondrial DNA size, 613*t*
oogenesis in, 93–94
pedigree analysis, 50–51
sex determination of, 77–78
single-gene traits, 30*t*
single nucleotide polymorphisms (SNPs) in, 492–494
spermatogenesis in, 94
T-cell receptor genes, 276
vs. mouse genome, 276–277
X chromosomes of, 72, 77, 111, 133

Hunchback (hb) gene, of *Drosophila melanogaster*, 572

Hunt, Tim, 534

Hunter syndrome, 133*f*

Huntington, George, 31

Huntington disease (HD)
gene mapping, 507–508*f*
HD gene, 30*t*, 31–32
pedigree analysis, 31–32
and simple sequence repeats (SSRs), 499–500

Hutchison, Clyde A., 3

Hybrid dysgenesis, 300

Hybridization, 462–468
comparative genomic hybridization (CGH), 311–312*f*, 657–658
defined, 10
polymerase chain reaction (PCR), 468

Hybrids
defined, 19
parents with antagonistic traits, 20*f*
of peas, 24*f*

Hybrid sterility, 91*f*

Hydrogen bonds, in DNA, 152

Hydroxyurea, 43

Hypercholesterolaemia, 30*t*

Hypermorphic mutations, 217

Hypomorphic mutations, 216, 556

Hypophosphataemia, 100

Hysteresis, 712–714

I

ICATs (isotope-coded affinity tags), 681–682

Illegitimate recombination, 433–434

Immune system, and programmed DNA rearrangements, 280–281

Imprinting, genomic, 370–373

Incomplete dominance, 34–36, 35*f*, 49–50, 49*f*
and mutations, 217

Incomplete penetrance, 51, 53, 509–510

InDels (short deletions and insertions), 492, 496–497*f*

Independent assortment, of gene pairs, 26

Induced pluripotent stem (iPS) cells, 556

Inducer, 326

Induction
of gene expression by lactose, 326, 330
of *lac* genes, 328*f*

Information. *See* Biological information

Information flow, of transcription, 196*f*

Information theory, 152

Infra-abdominal (iab) genes, of *Drosophila melanogaster*, 578

Ingram, Vernon, 258

Inheritance. *See also* Heredity
blended inheritance, 17, 29
and genes, 21
historical puzzle of, 16–19
horizontal pattern, 32–33, 51
in humans, 29–33
single-gene, extensions to Mendel for, 33–42
vertical pattern, rare dominant trait, 31–32

Initiation factors, in translation, 209*f*

Initiation stage
of DNA replication, 160, 161*f*
of transcription, 196*f*–197*f*, 198
of translation, 208, 209*f*

Insecticide resistance, 408–409

Insertion
defined, 238
screening for, in DNA molecules, 458–459

Insertion sequence (IS), in bacterial genomes, 595–596

In silico proteins, 381

In situ hybridization, 284–285, 565, 581
DiGeorge syndrome, 285

Insulators, 370–371

Insulin, synthesis of, with recombinant DNA technology, 451

Integrative and conjugative element (ICE), 608

Intelligent design *vs.* evolution, 435

Intercalators, 246–248

Interference. *See also* RNA interference (RNAi)
and gene mapping, 124–125
of phylogenetic relationships, 436

Intergenic gene conversion, 438–439

Intergenic mutation, 255

Intergenic nucleotide, 255

Interkinesis, 87*f*, 89

International Genetically Engineered Machine (iGEM), 727

Interphase
cell cycle, 79–80
gap 1 (G_1), 79–80
gap 2 (G_2), 79–80
synthesis (S), 79–80

Interrupted-mating experiments, 605*f*

Intersex (ix) gene, 384

Intragenic mutation, 255

Intragenic suppression, 223

Introns
defined, 8, 170, 201
in transcription, 200–202

Phages. *See* Bacteriophages

Pharmaceuticals. *See* Medicine

Phenocopy, 509*t,* 510, 562

Phenotype frequency, 393, 398

Phenotypes

    abnormal phenotype, 72

    complementation test, 48

    defined, 25

    in developmental pathways, 565–567

    in *Drosophila melanogaster,* 382*t*

    environmental effects on, 52–53

    gene interactions, and novel
        phenotypes, 43–44

    inversions, 287

    and modifier genes, 51–52

    production, and genotype, 51–53

    of sexual identity, 94

    vs. genotype, 25

Phenotypic variation, 391, 411–412

Phenylalanine hydroxylase, 52

Phenylketonuria (PKU), 30*t*, 52

Phosphodiester bonds, 143

Phosphorylation, 380–381

Photolyase enzymes, 249

Photorepair, 249

Phylogenetic footprinting, 381

Phylogenetic trees, 436

Physical markers, 118

Pie diagrams, 397, 405*f*

Plant cell, 82*f*

Plasmids, 457–458, 597–598

Plasmodium falciparum, mitochondrial
    DNA size, 613*t*

Platforms, 651

Pleiotropy, 39, 41

Poincaré, Henri, 672

Point mutations, 222

Polar body, 93

    first polar body, 93*f*

    second polar body, 93*f*

Polarity

    of DNA, 162*f*

    in mRNA corresponds to N-to-C
        polypeptide, 226

    segment polarity genes, 573, 575*f*, 576

Pole bean, height of, 55

Poly-A binding protein, 199

Poly-A tail of mRNA, 199

Polycistronic, 213

Polycistronic transcripts, in *Caenorhabditis
    elegans,* 215

Polygenic inheritance, 511–512

Polygenic traits, 55

Polymer, 143

Polymerase, of DNA, 160, 161*f*–162*f*, 163

Polymerase chain reaction (PCR), 457,
    469*f*–470*f*

    analysis of restriction-site-altering
        SNPs, 495

    defined, 468

    DNA hybridization, 468

    for HIV detection, 471

    products, like cloned restriction
        fragments, 468, 471

    uses of, 471

Polymerase cycling assembly, 735*f*

Polymorphic gene, 38, 237, 491

Polypeptides

    amino acid chains, 258–259

    mRNAs translated into, 208

    polarities, correlation to DNA and
        mRNA, 226

    polarity in mRNA corresponds to
        N-to-C polypeptide, 226

    structure after translation, 208

    subunits of, 261

Polyploids, 306, 310–311

Polyprotein, 208

Polyribosome, 210*f*

Polytene chromosomes, 283–284

Population

    allele frequency, in finite populations,
        399–401

    defined, 392

    deviations, and Hardy-Weinberg law,
        398–399

    fixed locus, 400

    founder effect, 400–401

    and haplotypes, 659

    Hardy-Weinberg law description, 398

    size and time to fixation, 400

Population bottlenecks, 400–401

Population genetics, 33, 391

Positional cloning, 505–508. *See also*
    Cloning

    and DNA markers, identify disease
        genes, 506–508

    Huntington disease (HD) gene,
        507–508*f*

Positive autoregulation, 433

Postbithorax (pbx) gene, of *Drosophila
    melanogaster,* 577

Posttranscription regulators, 380

Posttranslational modifications, 208, 211*f*

PPI (protein-protein interaction), 682

Predicted values, and chi-square test, 115

Predictive/preventive medicine

    molecular studies in, 11–12

    and systems biology, 720–721

Preimplantation genetic diagnosis (PGD),
    511

Premutation alleles, 245

Prenatal diagnosis, 76–77

Presidential Commission for the Study of
    Bioethical Issues, 737

Primary nondisjunction, 98–99

Primary oocyte, 93–94

Primary spermatocyte, 94

Primary structure

    of proteins, 260–261*f*

    tRNA, 205

Probability

    basic rules of, 23–24

    and Mendel's laws, 28

Proboscipedia (pb) genes, of *Drosophila
    melanogaster,* 578

Product rule, 23–24

Proflavin mutagen, 222

Progeny, recombinant, 110–111

Programmed cell death (PCD), 541

Prokaryotes

    gene expression in, 212–213

    gene regulation of, 323–324, 324–325

    unicellular, 323

Prokaryotic cells, 8

Prometaphase, 80–82, 81*f*

Promoters

    cis-acting regulatory region, 358–359

    defined, 197*f*, 198

    of *Escherichia coli* genes, 198*f*

    identifying, 359

Prophage, 606

Prophase, 80–81*f*

Prophase I, of meiosis I, 85–88

Prophase II, and meiosis II, 86*f,* 90

Protein expression

    analyzing dynamic changes in,
        681–682

    and mass spectrometry, 681

Protein kinases, 538

Protein mass mapping, 679*f*

Protein-protein interaction (PPI), 682

Proteins. *See also* Proteomics

    amino acids in, 5

    arrays, used for global biochemical
        analysis, 682–684

    cellular components of, 219*t*

    centromere proteins, 171*f*

    characteristics of life, 5–6*f*

    and chromatin, 79

    and chromosomal DNA, 169

    cohesin, 180

    colinearity of, 221

    condensins, 172

    defined, 5

    genes encode, 22–23

    histones, 168, 169

    light-sensitive proteins in retina,
        264–267

    linear polymers of amino acids linked
        by peptide bonds, 258–259

    mass map of, 679*f*

    molecules, 5–6*f*

    native configuration of proteins, 260

    nonhistone proteins, 169, 173*f*

    primary structure of, 260–261*f*

    quaternary structure of, 261–262

    regulatory. *See* Regulatory proteins

    remodelling complex, 176

    and RNA, 6–7

    secondary structure of, 260–261*f*

    Shugoshin protein, 181

    tertiary structure of, 260–261*f*

    and translation, 380–381, 671

    transposase, 298–300

Protein tyrosine kinase, 290

Proteomes, 671, 681–682

Proteomics. *See also* Proteins

    affinity purification coupled to mass
        spectrometry, 677–680

    data and biological research, 672

    defined, 671–672

    described, 670

    nuclear pore complex, 671–672

    and tears, 684

    yeast two-hybrid system, 673–677

Proteosomes, 380

Proto-oncogenes, 290, 531

Shugoshin protein, 181
Siamese cat, coat colour, 52
Sickle-cell anaemia, 15, 30*t*, 447
   and balancing selection, 404–406
   dominance relations, 42
   equilibrium frequency of alleles, 405–406
   and malaria, 406, 407*f*
   multiple alleles, 41
   mutation, and DNA sequencing, 478–480
   mutation, and molecular basis of, 258, 260
   pleiotropy of, 41
   recessive lethality, 41–42
   resistance to malaria, 42
   SNP detection with polymerase chain reaction (PCR), 495
Sigma (σ) factors, 198
   Escherichia coli and heat shock, 342–343
   global regulation by alternative, 343
   RpoS sigma factor, 343
Signal transducers, 522
Signal transduction, 523, 533*t*
Signal transduction pathways, 576
Silent mutations, 215
Simple kinase, 712–714
Simple sequence repeats (SSRs)
   in genomes, 498
   and Huntington disease (HD), 499–500
   SNP category, 492
   tandem repeats, 443
SINEs (short interspersed elements)
   creation of, 442
   defined, 296–297*f*, 297
   potential selective advantage of, 443
SingerROTOR system, 674
Single-gene, extensions to Mendel, 33–42
Single-gene traits, 509*t*
Single nucleotide polymorphisms (SNPs)
   defined, 492–493*f*
   detection, with allele-specific oligonucleotide hybridization, 495–496
   detection, with DNA microarrays, 496
   distributions, 493
   DNA polymorphism, 492
   human SNPs, origin of, 492–494
   polymerase chain reaction (PCR) of restriction-site-altering, 495
   and small-scale-length variations, 492–500
   Southern blot analysis of restriction-site-altering, 494–495
Single-stranded primary transcript, 197*f*
Sister chromatids, 74, 75*f*, 79*f*, 80–83, 86*f*, 90, 92*f*
Site-directed mutagenesis, 265, 478
Skin cancer, 252
Small interfering RNAs (siRNAs), 376, 379, 695–697
Small nuclear ribonucleoproteins (SnRNPs), 202–203, 204*f*
Smith, Hamilton O., 3, 610
Smith, Michael, 262, 265, 478
Smooth bacteria, 144, 146*f*

Snapdragons, 34, 35*f*
SNP. *See* Single nucleotide polymorphisms (SNPs)
SnRNPs (small nuclear ribonucleoproteins), 202–203, 204*f*
Social issues
   and genetics, 12–13
   germline gene therapy, 646
   patentability of DNA, 645–646
   preimplantation genetic diagnosis (PGD), 511
   somatic gene therapy, 646
Somatic cells, 85, 521
Somatic gene therapy, 646
Somatic mutations, 252
   and cancer, 521–545
Sonic hedgehog gene, of chicken, 576
SOS system, 251
Southern blots, 465–468
   of restriction-site-altering SNPs, 494–495
Specialized transducing phage, 606–607*f*
Specialized transduction, 606
Species reference sequence (RefSeq), 637–638
Species variation
   in chromosomes, 74–75
   sex determination and, 78–79
Spectinomycin, 590
Spermatids, 94, 95*f*
Spermatocytes, 94
Spermatogenesis in humans, 94
Spermatogonia, 94, 95*f*
Sperm nuclei, 73–74
Sperm production, 94, 95*f*
S phase of cell cycle, 536–537
Spindle morphogenesis, 692
Spindle poles, 81*f*, 82, 89
Splice acceptors, 202
Splice donors, 202
Spliceosomes, 202–203, 204*f*
Spontaneous mutations, 239–244
SSRs (simple sequence repeats)
   in genomes, 498
   and Huntington disease (HD), 499–500
   SNP category, 492
   tandem repeats, 443
Stahl, Franklin, 158
Staphylococcus afermentans, nucleotide base composition of DNA of, 150*t*
Stefansson, Kari, 659
Stem cells
   and embryonic development, 559–560
   and human cloning, 558–560
Stern, Curt, 118
Steroid hormone receptors, 361*f*
Sticky ends
   and base-pairing, 457
   restriction fragments, 449
Stimulus response curves, 712–714
Streptococcus pneumoniae bacteria, transformation in, 144, 146*f*
Studies, breeding, 49–50
Sturtevant, Alfred H., 120, 122, 127
Substitution
   defined, 237

nonsynonymous substitutions, 431
synonymous substitutions, 431
Subunits
   of polypeptides, 261
   of snRNPs (small nuclear ribonucleoproteins), 204*f*
Sum rule, 23–24
Supercoiling, 172
Super-repressor mutant, 330–331
Sutton, Walter S., 77
Suzuki, David, 52, 53
Sweet peas. *See also* Garden peas
   autosomal traits, 113–114
   colour, 44, 45*f*
   complementary gene action in, 45*f*
   and Punnett square, 113
SWI-SNF remodeller, 369
Sxl (Sex-lethal) gene, in *Drosophila melanogaster* melangaster, 375, 381–383
Synapsis, 88
Syncytial blastoderm, 569
Syncytium, 83
Synonymous substitutions, 431
Syntenic genes, 111–112
Syntenic segments, 276
Synthesis (S), 79–80
Synthetic biology
   applications of, 730–732
   defined, 725–727
   and ethics, 737
   first synthetic gene circuits, 727–730
   interfacing synthetic circuits, 731
   metabolites, 730
   multiplex automated genome engineering (MAGE), 732–733*f*
   regulatory guidelines, 737
   repressilator, 729–730
   toggle switches, 728–729
Synthetic genetic array (SGA) analysis, 676
   in Caenorhabditis elegans, 703–704
   chemical SGA, used to identify drug targets, 701
   genetic interaction maps, 701–703
   and genome architecture, 697–704
   hierarchical clustering, 700*f*
   methodology, 699*f*
   in yeast, 698
Synthetic genomics
   applications of, 736–738
   defined, 3, 733
   genome synthesis and assembly, 734–736
   genome transplantation, 733–734
   iterative process, 737*f*
   regulatory guidelines, 737
Synthetic Genomics Incorporated (SGI), 725
Synthetic Genomics Vaccines Incorporated (SGVI), 738
Synthetic lethality
   and genetic buffering, 697–698
   mechanistic basis of, 698–701
Synthia (JCVI-1.0syn), 3, 725
   watermarks, 734

Xeroderma pigmentosum, 252
X-Gal, 458–459
X-linked traits
    in humans, 99–100, 110, 111*f*
    recessive inheritance, 100–101
X-ray diffraction pattern, of DNA,
    149–150*f*
X-rays
    and DNA damage, 184, 188
    mutagenicity of, 244, 246
X reactivation, 304
X symbol, for basic chromosome number,
    306
XX females, 99*f*
XXY females, 99*f*

Y

Yanofsky, Charles, 221, 340
Y chromosomes, of humans, 72, 77
Yeast. *See also* Yeast two-hybrid system

ADE2 gene of, 131
centromeres, 181
colonies, 131–132, 617
DNA fragments in, 735–736
and DNA microarray, 652–653
gene knockouts, 690–691, 694
genetic barcoding, 691–692*f*
genetic interaction map, 702*f*
genome sequence of, 9–10*f*, 634*t*
homologous recombination in,
    690–691
kinases in yeasts, 537
kinase substrate protein array, 683
life cycle of, 675, 690*f*
macroarray of, 674–676
mitochondrial DNA size, 613*t*
mitotic recombination, and sectored
    yeast colonies, 131–132
as model organism, 554–555, 689
non-Mendelian inheritance in, 617–618
nucleotide base composition of DNA
    of, 150*t*

replication, 177
synthetic genetic array (SGA) analysis
    in, 698
Yeast two-hybrid system
    analysis at genome-wide level, 673–677
    disadvantages of, 677
    premise of, 673
    reporter gene, 673*f*, 676*f*
Y-linked traits, in humans, 99–101

Z

Z form DNA, 153, 156*f*
Zw10 gene, 84
Zyg-9 gene, 560
Zygote
    defined, 21, 72
    diploid zygotes, 74
Zygotene, 86*f*, 88
Zygotic genes, of *Drosophila melanogaster,* 573–576, 582